ENCYCLOPEDIA OF
TIDEPOOLS AND ROCKY SHORES

ENCYCLOPEDIA OF TIDEPOOLS AND ROCKY SHORES

EDITED BY

MARK W. DENNY

Stanford University

STEVEN D. GAINES

University of California, Santa Barbara

UNIVERSITY OF CALIFORNIA PRESS

Berkeley   Los Angeles   London

University of California Press, one of the most distinguished university presses in the United States, enriches lives around the world by advancing scholarship in the humanities, social sciences, and natural sciences. Its activities are supported by the UC Press Foundation and by philanthropic contributions from individuals and institutions. For more information, visit www.ucpress.edu.

Encyclopedias of the Natural World, No. 1

University of California Press
Berkeley and Los Angeles, California

University of California Press, Ltd.
London, England

Library of Congress Cataloging-in-Publication Data

Encyclopedia of tidepools and rocky shores / Mark W. Denny and Steven D. Gaines, editors.
        p.   cm.
    Includes bibliographical references.
    ISBN 978-0-520-25118-2 (cloth: alk. paper)
    1. Tide pool ecology—Encyclopedias.   2. Tide pools—Encyclopedias.
3. Seashore ecology—Encyclopedias.   4. Seashore—Encyclopedias.   I. Denny,
Mark W., 1951- II. Gaines, Steven D. (Steven Dean), 1955-

QH541.5.S35E53 2007
577.69'9--dc22   2006035368

Manufactured in Singapore
    10   09   08   07
    10   9   8   7   6   5   4   3   2   1

The paper used in this publication meets the minimum requirements of ANSI/NISO Z39.48-1992 (R 1997) (*Permanence of Paper*).

Cover photographs: At bottom: J. Rotman © CORBIS. In top row, from left: tidepools at Point Piños, Pacific Grove, California, photograph by Adam Schneider, http://adamschneider.net; mussel extending byssal threads, photograph courtesy of Laura Coutts; Point Piños, photograph by Adam Schneider; red and purple urchins, photograph courtesy Richard B. Emlet.

CONTENTS

CONTENTS BY SUBJECT AREA

LARRY G. ALLEN
Department of Biology
California State University, Northridge
Fish

RICHARD F. AMBROSE
Department of Environmental
    Health Sciences
University of California, Los Angeles
Habitat Restoration

ROLAND C. ANDERSON
The Seattle Aquarium
Seattle, Washington
Octopuses

WILLIAM C. AUSTIN
Khoyatan Marine Laboratory
Sidney, British Columbia, Canada
Sponges

BONNIE A. BAIN
Biology Department
Southern Utah University
Cedar City, Utah
Pycnogonids

RICHARD T. BARBER
NSOE Marine Laboratory
Duke University
Beaufort, North Carolina
Seawater

JOHN A. BARTH
College of Oceanic and
    Atmospheric Sciences
Oregon State University, Corvallis
Upwelling

RAYMOND T. BAUER
Department of Biology
University of Louisiana, Lafayette
Shrimps

KEVIN S. BEACH
Department of Biology
University of Tampa, Florida
Algal Color

MARK D. BERTNESS
Department of Ecology and Evolutionary Biology
Brown University
Providence, Rhode Island
Facilitation

CAROL A. BLANCHETTE
Marine Science Institute
University of California, Santa Barbara
Seagrasses

MICHAEL L. BOLLER
Hopkins Marine Station
Stanford University
Pacific Grove, California
Wave Forces, Measurement of

CAREN E. BRABY
Monterey Bay Aquarium Research Institute
Moss Landing, California
Cold Stress

MATTHEW E. S. BRACKEN
Bodega Marine Laboratory
University of California, Davis
Bodega Bay, California
Excretion

GEORGE M. BRANCH
Department of Zoology
University of Cape Town, South Africa
Limpets

MARK A. BRZEZINSKI
Department of Ecology, Evolution, and Marine Biology
University of California, Santa Barbara
Phytoplankton

RODRIGO H. BUSTAMANTE
CSIRO Marine Research
Cleveland, Queensland, Australia
Iguanas, Marine

ROY L. CALDWELL
Department of Integrative Biology
University of California, Berkeley
Mantis Shrimps

GERARD M. CAPRIULO
Department of Biology
St. Mary's College of California, Moraga
Protists

JAMES T. CARLTON
Williams-Mystic Program
Williams College, Mystic, Connecticut
Introduced Species

ROBERT C. CARPENTER
Department of Biology
California State University, Northridge
Algal Turfs

JUAN CARLOS CASTILLA
Departamento de Ecologia
Pontificia Universidad Católica de Chile, Santiago
Economics, Coastal

FRANCIS CHAN
Department of Zoology
Oregon State University, Corvallis
Nutrients

FRANCISCO CHAVEZ
Monterey Bay Aquarium Research Institute
Moss Landing, California
Variability, Multidecadal

ROSS A. COLEMAN
Centre for Research on Ecological Impacts of
    Coastal Cities
University of Sydney, Australia
Homing

JOSEPH H. CONNELL
Department of Ecology, Evolution, and Marine Biology
University of California, Santa Barbara
Competition

DANIEL COSTA
Institute of Marine Sciences
University of California, Santa Cruz
Seals and Sea Lions

ISABELLE M. CÔTÉ
Department of Biological Sciences
Simon Fraser University
Burnaby, British Columbia, Canada
Blennies

JAMES A. COYER
Department of Marine Biology
University of Groningen, The Netherlands
Algal Biogeography

MOLLY CUMMINGS
Section of Integrative Biology
University of Texas, Austin
Light, Effects of

EDWARD B. CUTLER
Department of Invertebrate Zoology
Harvard University
Cambridge, Massachusetts
Sipunculans

MARYMEGAN DALY
Department of Evolution, Ecology,
    and Organismal Biology
Ohio State University, Columbus
Cnidaria, Overview

WILLEM H. DE SMET
Department of Biology
University of Antwerp, Belgium
Rotifers

MARK W. DENNY
Hopkins Marine Station
Stanford University
Pacific Grove, California
Evaporation and Condensation
Surface Tension
Tides

MEGAN N. DETHIER
Friday Harbor Laboratories
University of Washington, Friday Harbor
Algal Crusts and Lichens
Monitoring: Techniques

ANDREW P. DEVOGELAERE
Monterey Bay National Marine Sanctuary
Monterey, California
Marine Sanctuaries and Parks

WILLIAM J. DOUROS
Monterey Bay National Marine Sanctuary
Monterey, California
Marine Sanctuaries and Parks

ROBERT DUDLEY
Department of Integrative Biology
University of California, Berkeley
Air

SARAH E. A. LE V. DIT DURELL
Centre for Ecology and Hydrology, United Kingdom
Dorchester, Dorset, United Kingdom
Oystercatchers

THOMAS A. EBERT
Department of Zoology
Oregon State University, Corvallis
Sea Urchins

KEVIN J. ECKELBARGER
Darling Marine Center
University of Maine, Orono
Marine Stations

GINNY L. ECKERT
Department of Biology
University of Alaska, Juneau
Sea Cucumbers

DOUGLAS J. EERNISSE
Department of Biological Science
California State University, Fullerton
Chitons

RICHARD B. EMLET
Oregon Institute of Marine Biology
University of Oregon, Charleston
Echinoderms, Overview

JAMES A. ESTES
United States Geological Survey
Santa Cruz, California
Sea Otters

RON ETTER
Department of Biology
University of Massachusetts, Boston
Snails

PATRICK J. EWANCHUK
Department of Biology
Providence College
Providence, Rhode Island
Predator Avoidance

MIRIAM FERNÁNDEZ
Department of Ecology
Pontificia Universidad Católica de Chile, Santiago
Biodiversity, Global Patterns of

PIETER AREND FOLKENS
Alaska Whale Foundation
Seattle, Washington
Symbolic and Cultural Uses

BENJAMIN A. FOOTE
Department of Biological Sciences
Kent State University
Kent, Ohio
Flies

DOUGLAS FUDGE
Department of Integrative Biology
University of Guelph
Ontario, Canada
Materials, Biological

STEVEN D. GAINES
Marine Science Institute
University of California, Santa Barbara
Dispersal
El Niño
Marine Reserves
Recruitment

BRIAN GAYLORD
Bodega Marine Laboratory
University of California, Davis
Bodega Bay, California
Hydrodynamic Forces

JONATHAN B. GELLER
Moss Landing Marine Laboratories
Moss Landing, California
Bivalves
Symbiosis

WILLIAM GILLY
Hopkins Marine Station
Stanford University
Pacific Grove, California
Ricketts, Steinbeck, and Intertidal Ecology

MICHAEL A. GLASSOW
Department of Anthropology
University of California, Santa Barbara
Food Uses, Ancestral

PETER W. GLYNN
Rosenstiel School of Marine and Atmospheric Science
University of Miami, Florida
Corals

JOHN M. GOSLINE
University of British Columbia
Vancouver, Canada
Materials, Biological

MICHAEL GRAHAM
Moss Landing Marine Laboratories
Moss Landing, California
Sea Level Change, Effects on Coastlines

BRIAN A. GRANTHAM
Coastal and Estuarine Assessment Unit
Washington State Department of Ecology, Olympia
Water Properties, Measurement of

GARY GRIGGS
Institute of Marine Sciences
University of California, Santa Cruz
Beach Morphology
Coastal Geology
Tidepools, Formation and Rock Types

RICK GROSBERG
Section of Evolution and Ecology
Division of Biological Sciences
University of California, Davis
Tunicates

E. C. HADERLIE
Department of Oceanography
Naval Postgraduate School
Monterey, California
Stone Borers

BENJAMIN S. HALPERN
National Center for Ecological Analysis and Synthesis
University of California, Santa Barbara
Food Webs

PATRICIA M. HALPIN
Materials Research Laboratory
University of California, Santa Barbara
Herbivory

STEVEN C. HAND
Department of Biological Sciences
Louisiana State University, Baton Rouge
Desiccation Stress

CHRISTOPHER HARLEY
Department of Zoology
University of British Columbia
Vancouver, Canada
Zonation

PATRICK A. HARR
Department of Meteorology
Naval Postgraduate School
Monterey, California
Storms and Climate Change

JEAN M. HARRIS
University of Cape Town, South Africa
Food Uses, Modern

CHRISTOPHER HARROLD
Monterey Bay Aquarium
Monterey, California
Aquaria, Public

STEVE HAWKINS
Marine Biological Association of the
    United Kingdom
Plymouth, Devon, United Kingdom
Monitoring: Long-Term Studies

BRIAN HELMUTH
Department of Biological Sciences
University of South Carolina, Columbia
Heat and Temperature, Patterns of
Temperature Change
Temperature, Measurement of

LEA-ANNE HENRY
Scottish Association for Marine Science,
    United Kingdom
Oban, Argyll, Scotland, United Kingdom
Hydroids

YAYOI M. HIRANO
Chiba University
Chiba, Japan
Hydromedusae
Stauromedusae

KEITH HISCOCK
Marine Biological Association of the
    United Kingdom
Plymouth, Devon, United Kingdom
Monitoring: Long-Term Studies

F. G. HOCHBERG
Santa Barbara Museum of Natural History
Santa Barbara, California
Octopuses

JENS T. HØEG
Department of Cell Biology and
  Comparative Zoology
University of Copenhagen, Denmark
Rhizocephalans

PATRICIA A. HOLDEN
Donald Bren School of Environmental
  Science and Management
University of California, Santa Barbara
Microbes

JUSTIN HOLL
Pontificia Universidad Católica de Chile, Santiago
Biodiversity, Global Patterns of

MICHAEL H. HORN
Department of Biology
California State University, Fullerton
Fish

ALLISON M. HORST
University of California, Santa Barbara
Microbes

LUKE J. H. HUNT
Hopkins Marine Station
Stanford University
Pacific Grove, California
Surveying

TOHRU ISETO
The Kyoto University Museum
Kyoto University, Japan
Entoprocts

GEORGE JACKSON
Department of Oceanography
Texas A&M University, College Station
Diffusion

GLENN JAECKS
Slovenian Academy of Science and Arts
Ljubljana, Slovenia
Brachiopods

SÖNKE JOHNSEN
Department of Biology
Duke University
Durham, North Carolina
Light, Effects of

LADD E. JOHNSON
Department of Biology
Université Laval, Quebec, Canada
Ice Scour

JENNIFER JOST
Department of Biological Sciences
University of South Carolina, Columbia
Temperature, Measurement of

DAVID JULIAN
Department of Zoology
University of Florida, Gainesville
Echiurans

SARA KIMBERLIN
Pontificia Universidad Católica de Chile, Santiago
Biodiversity, Global Patterns of

PAUL D. KOMAR
College of Oceanic and Atmospheric Sciences
Oregon State University, Corvallis
Ocean Waves

ARMAND M. KURIS
Department of Ecology, Evolution and Marine Biology
University of California, Santa Barbara
Parasitism

KEVIN D. LAFFERTY
United States Geological Survey
Santa Barbara, California
Diseases of Marine Animals

JOHN J. LEE
Biology Department
City College, City University of New York
Protists

JAMES J. LEICHTER
Integrative Oceanography Division
Scripps Institution of Oceanography
La Jolla, California
Waves, Internal

SALLY P. LEYS
Department of Biological Sciences
University of Alberta
Edmonton, Canada
Sponges

DAVID R. LINDBERG
Department of Integrative Biology
University of California, Berkeley
Fossil Tidepools
Molluscs, Overview

MATS LINDEGARTH
Tjärnö Marine Biological Laboratory
Göteborg University, Sweden
Wave Exposure

COLIN LITTLE
School of Biological Sciences
University of Bristol, United Kingdom
Foraging Behavior

DAVID P. LOHSE
Department of Ecology and Evolutionary Biology
University of California, Santa Cruz
Barnacles

MARTHA L. MANSON
Monterey Bay Aquarium
Monterey, California
Aquaria, Public

KAREN L. MARTIN
Natural Science Division, Seaver College
Pepperdine University
Malibu, California
Rhythms, Non-Tidal

MARLENE M. MARTINEZ
Department of Biology
American River College
Sacramento, California
Locomotion: Intertidal Challenges

PATRICK T. MARTONE
Stanford University
Stanford, California
Algae, Calcified

PATRICIA MASTERSON
Marine Biological Association of the United Kingdom
Plymouth, Devon, United Kingdom
Monitoring: Long-Term Studies

LAURIE A. MCCONNICO
Department of Biology
San Diego City College, California
Algal Economics

CATHERINE S. MCFADDEN
Department of Biology
Harvey Mudd College
Claremont, California
Octocorals

IAIN MCGAW
School of Life Sciences
University of Nevada, Las Vegas
Circulation

REBECCA J. MCLEOD
University of Otago
Dunedin, New Zealand
Salinity Stress

MARGARET ANNE MCMANUS
Department of Oceanography
University of Hawaii, Honolulu
Nearshore Physical Processes, Effects of

KRISTINA MEAD
Department of Biology
Denison University
Granville, Ohio
Fertilization, Mechanics of

MICHAEL A. MENZE
Department of Biological Sciences
Louisiana State University, Baton Rouge
Desiccation Stress

CLAUDIA E. MILLS
Friday Harbor Laboratories
University of Washington, Friday Harbor
Hydromedusae
Stauromedusae

STEPHEN MONISMITH
Department of Civil and Environmental
    Engineering
Stanford University
Stanford, California
Turbulence

RICH MOOI
Department of Invertebrate Zoology
    and Geology
California Academy of Sciences, San Francisco
Museums and Collections

PIPPA MOORE
Marine Biological Association of the United Kingdom
Plymouth, Devon, United Kingdom
Monitoring: Long-Term Studies

AMY L. MORAN
Clemson University
Clemson, South Carolina
Size and Scaling

STEVEN G. MORGAN
Bodega Marine Laboratory
University of California, Davis
Bodega Bay, California
Rhythms, Tidal

JAMES G. MORIN
Department of Ecology and Evolutionary Biology
Cornell University
Ithaca, New York
Bioluminescence

STEVEN N. MURRAY
Department of Biological Science
California State University, Fullerton
Habitat Alteration
Succession

SERGIO A. NAVARRETE
Department of Ecology
Pontificia Universidad Católica de Chile, Santiago
Biodiversity, Maintenance of

TODD NEWBERRY
Long Marine Laboratory
University of California, Santa Cruz
Tunicates

KARINA J. NIELSEN
Department of Biology
Sonoma State University
Rohnert Park, California
Algae, Overview

WENDELL A. NUSS
Department of Meteorology
Naval Postgraduate School
Monterey, California
Fog
Wind

JAMES NYBAKKEN
Moss Landing Marine Laboratories
Moss Landing, California
Nudibranchs and Related Species

TODD H. OAKLEY
Department of Ecology, Evolution,
   and Marine Biology
University of California, Santa Barbara
Vision

F. PATRICIO OJEDA
Department of Ecology
Pontificia Universidad Católica de Chile, Santiago
Territoriality

JÖRG OTT
Department of Marine Biology
University of Vienna, Austria
Micrometazoans

ROBERT T. PAINE
Department of Biology
University of Washington, Seattle
Predation

STEPHEN R. PALUMBI
Hopkins Marine Station
Stanford University
Pacific Grove, California
Genetic Variation and Population Structure

JOHN S. PEARSE
Institute of Marine Sciences
University of California, Santa Cruz
Reproduction, Overview

VICKI BUCHSBAUM PEARSE
Institute of Marine Sciences
University of California, Santa Cruz
Sea Anemones

D. H. PEREGRINE
School of Mathematics
University of Bristol, United Kingdom
Rogue Waves

THOMAS P. PESCHAK
Marine Biology Research Institute
University of Cape Town, South Africa
Vertebrates, Terrestrial

ANNA PFEIFFER-HOYT
Coastal Physical Oceanography and Marine Ecosystems
University of Hawaii, Honolulu
Nearshore Physical Processes, Effects of

CATHERINE A. PFISTER
Department of Ecology and Evolution
University of Chicago
Sculpins

NICOLE PHILLIPS
School of Biological Sciences
Victoria University, Wellington, New Zealand
Metamorphosis and Larval History

BARBARA B. PRÉZELIN
Department of Ecology, Evolution, and Marine Biology
University of California, Santa Barbara
Algal Blooms

PETER T. RAIMONDI
Ecology and Evolutionary Biology
University of California, Santa Cruz
Barnacles

DANIEL RITTSCHOF
Marine Laboratory, NSOE
Duke University
Beaufort, North Carolina
Hermit Crabs

LORETTA ROBERSON
Institute of Neurobiology
University of Puerto Rico, San Juan
Materials: Strength

CARLOS ROBLES
CEA-CREST
California State University, Los Angeles
Lobsters

JOSÉ M. ROJAS
Department of Ecology
Pontificia Universidad Católica de Chile, Santiago
Territoriality

ANNE WERTHEIM ROSENFELD
Kentfield, California
Photography, Intertidal

KAUSTUV ROY
Division of Biological Science
University of California, San Diego
Ecosystem Changes, Natural vs. Anthropogenic

STEVEN S. RUMRILL
South Slough National Estuarine Research Reserve
Charleston, Oregon
Salinity, Measurement of

RAPHAEL SAGARIN
Nicholas Institute for Environmental Policy Solutions
Duke University
Durham, North Carolina
Climate Change, Overview

ERIC SANFORD
Bodega Marine Laboratory
University of California, Davis
Bodega Bay, California
Sea Stars

DAVID R. SCHIEL
School of Biological Sciences
University of Canterbury
Christchurch, New Zealand
Abalones
Kelps

ANDREAS SCHMIDT-RHAESA
Zoological Museum
Hamburg, Germany
Worms

KIMBERLY R. SCHNEIDER
Department of Biological Sciences
University of South Carolina, Columbia
Heat and Temperature, Patterns of

STEPHEN C. SCHROETER
Marine Science Institute
University of California, Santa Barbara
Competition

PATRICIA M. SCHULTE
Department of Zoology
University of British Columbia
Vancouver, Canada
Water Chemistry

ANJA SCHULZE
Smithsonian Marine Station
Fort Pierce, Florida
Sipunculans

ALAN SHANKS
Oregon Institute of Marine Biology
University of Oregon, Charleston
Projectiles, Effects of

J. MALCOLM SHICK
School of Marine Sciences
University of Maine, Orono
Ultraviolet Stress

SUSAN SHILLINGLAW
Department of English
San Jose State University, California
Ricketts, Steinbeck, and Intertidal Ecology

STEPHEN M. SHUSTER
Department of Biological Sciences
Northern Arizona University, Flagstaff
Sex Allocation and Sexual Selection

KERRY SINK
University of Cape Town, South Africa
Food Uses, Modern

CHRISTINA J. SLAGER
Monterey Bay Aquarium
Monterey, California
Aquaria, Public

ANDREW SMITH
Department of Biology
Ithaca College, New York
Adhesion

CELIA M. SMITH
Department of Botany
University of Hawaii, Honolulu
Algal Color

GEORGE SOMERO
Hopkins Marine Station
Stanford University
Pacific Grove, California
Heat Stress

WAYNE P. SOUSA
Department of Integrative Biology
University of California, Berkeley
Disturbance

ALAN SOUTHWARD
Marine Biological Association of
   the United Kingdom
Plymouth, Devon, United Kingdom
Monitoring: Long-Term Studies

JOHN J. STACHOWICZ
Section of Evolution and Ecology and Center for
   Population Biology
University of California, Davis
Biodiversity, Significance of Camouflage Mutualism

ROBERT S. STENECK
Darling Marine Center
University of Maine
Walpole, Maine
Algae, Calcified

KATJA STERFLINGER
Institute for Applied Microbiology (ACBR)
University of Natural Resources
   and Applied Life Sciences
Vienna, Austria
Boring Fungi

JONATHON STILLMAN
Romberg Tiburon Center
San Francisco State University, California
Crabs

SUSAN L. SWARBRICK
Marine Science Institute
University of California, Santa Barbara
Competition

LAUREN SZATHMARY
Department of Biological Sciences
University of South Carolina, Columbia
Temperature Change

MICHAEL TEMKIN
Department of Biology
St. Lawrence University
Canton, New York
Bryozoans

NORA B. TERWILLIGER
Department of Biology
University of Oregon, Charleston
Arthropods, Overview

RICHARD C. THOMPSON
School of Biological Sciences
University of Plymouth, Devon, United Kingdom
Biofilms

CAROL S. THORNBER
Department of Biological Sciences
University of Rhode Island, Kingston
Algal Life Cycles

EDWARD B. THORNTON
Department of Oceanography
Naval Postgraduate School
Monterey, California
Surf-Zone Currents

HENDRIK L. TOLMAN
NOAA/National Centers for Environmental Prediction
Camp Springs, Maryland
Storm Intensity and Wave Height

MARK E. TORCHIN
Smithsonian Tropical Research Institute
Republic of Panama
Rhizocephalans

GEOFFREY C. TRUSSELL
Department of Biology
Northeastern University
Nahant, Massachusetts
Predator Avoidance

JOHN UGORETZ
California Department of Fish and Game
Monterey, California
Management and Regulation

A. J. UNDERWOOD
Centre for Research on Ecological Impacts
   of Coastal Cities
University of Sydney, Australia
Monitoring, Overview Monitoring: Statistics

LUCA A. VAN DUREN
Department of Ecosystem Studies
Netherlands Institute of Ecology, Yerseke
Boundary Layers

LUIS R. VINUEZA
Department of Zoology
Oregon State University, Corvallis
Iguanas, Marine

PIETER T. VISSCHER
Department of Marine Sciences
University of Connecticut, Groton
Stromatolites

STEVEN VOGEL
Department of Biology
Duke University
Durham, North Carolina
Body Shape

JOHN WARES
Department of Genetics
University of Georgia, Athens
Genetic Variation, Measurement of

TONY WARRINGTON
California Department of Fish and Game
Sacramento, California
Management and Regulation

LIBE WASHBURN
Department of Geography
University of California, Santa Barbara
Currents, Coastal

JAMES M. WATANABE
Hopkins Marine Station
Stanford University
Pacific Grove, California
Invertebrates, Overview

LES WATLING
Darling Marine Center
University of Maine
Walpole, Maine
Amphipods, Isopods, and Other Small Crustaceans

PAUL WEBB
School of Natural Resources and Environment
University of Michigan, Ann Arbor
Buoyancy

ALI WHITMER
Marine Science Institute
University of California, Santa Barbara
Education and Outreach

GARY C. WILLIAMS
California Academy of Sciences, San Francisco
Collection and Preservation of Tidepool Organisms

GRAY A. WILLIAMS
The Swire Institute of Marine Science
University of Hong Kong, China
Foraging Behavior

RORY P. WILSON
School of Biological Sciences
University of Wales
Swansea, United Kingdom
Penguins

STEPHEN R. WING
Department of Marine Science
University of Otago
Dunedin, New Zealand
Salinity Stress

JON D. WITMAN
Department of Ecology and Evolutionary Biology
Brown University
Providence, Rhode Island
Benthic-Pelagic Coupling

H. ARTHUR WOODS
Division of Biological Sciences
University of Montana, Missoula
Size and Scaling

J. TIMOTHY WOOTTON
Department of Ecology and Evolution
University of Chicago
Birds
Gulls

DANIELLE C. ZACHERL
Department of Biological Science
California State University, Fullerton
Dispersal, Measurement of

CHERYL ANN ZIMMER
Department of Ecology and Evolutionary Biology
University of California, Los Angeles
Larval Settlement, Mechanics of

RICHARD K. ZIMMER
Department of Biological Sciences
University of California, Los Angeles
Chemosensation

RUSSEL L. ZIMMER
Department of Biological Sciences
University of Southern California, Los Angeles
Phoronids

RICHARD C. ZIMMERMAN
Department of Ocean, Earth, and
    Atmospheric Sciences
Old Dominion University
Norfolk, Virginia
Photosynthesis

The *Encyclopedia of Tidepools and Rocky Shores* is a comprehensive, complete, and authoritative reference dealing with all of the physical and biological aspects of tidepool and rocky shore habitats. Articles are written by researchers and scientific experts and provide a broad overview of the current state of knowledge on these fascinating places. Chemists, biologists, ecologists, geologists, and zoologists have contributed reviews intended for students as well as the interested general public.

In order for the reader to easily use this reference, the following summary describes the features, reviews the organization and format of the articles, and is a guide to the many ways to maximize the utility of this *Encyclopedia*.

SUBJECT AREAS

The *Encyclopedia of Tidepools and Rocky Shores* includes 186 topics that review the various ways scholars have studied rocky coastal habitats. The *Encyclopedia* comprises the following subject areas:

- Geology
- Oceanography
- Weather and Climatology
- Plants, Algae, Fungi, and Microbes
- Invertebrate Animals
- Vertebrate Animals
- Ecology and Behavior
- Physiology and Ecophysiology
- Human Uses and Interactions
- Research and Methodology

ORGANIZATION

Articles are arranged alphabetically by title. An alphabetical table of contents begins on page v, and another table of contents with articles arranged by subject area begins on page xi.

Article titles have been selected to make it easy to locate information about a particular topic. Each title begins with a key word or phrase, sometimes followed by a descriptive term. For example, "Rhythms, Tidal" is the title assigned rather than "Tidal Rhythms," because *rhythms* is the key term and, therefore, more likely to be sought by readers. Articles that might reasonably appear in different places in the *Encyclopedia* are listed under alternative titles, of which one title appears as the full entry; the alternative title directs the reader to the full entry. For example, the alternative title "Biological Materials" refers readers to the entry entitled "Materials, Biological."

ARTICLE FORMAT

The articles in the *Encyclopedia* are all intended for the interested general public. Therefore, each article begins with an introduction that gives the reader a short definition of the topic and its significance. Here is an example of an introduction from the article "Camouflage":

> Camouflage is a means by which animals avoid detection by other animals by blending in with environment. As a result, the animal may either not be perceived at all or be perceived to be something it is not, and thus ignored or avoided. Although humans most often think of camouflage as visual, many marine organisms have poorly developed visual systems, so chemically mediated camouflage is common as well. Both chemical and visual camouflage protect animals from predators as well as facilitate predation by "sit-and-wait" predators. Camouflage can be a major reason why some organisms are able to live in areas with very high densities of predators.

Within most articles and especially the longer articles, major headings help the reader identify important subtopics within each article. The article "Corals" includes the following headings: "The Coral Animal," "Biogeographic Patterns," "Coral Communities," "Coral Reef Morphology," and "Outlook."

CROSS-REFERENCES

Many of the articles in this *Encyclopedia* concern topics for which articles on related topics are also included. In order to alert readers to these articles of potential interest, cross-references are provided at the conclusion of each article. At the end of "Chitons," the following text directs readers to other articles that may be of special interest:

SEE ALSO THE FOLLOWING ARTICLES

Fossil Tidepools / Homing / Molluscs / Rhythms, Nontidal

Readers will find additional information relating to chitons in the articles listed.

BIBLIOGRAPHY

Every article ends with a short list of suggestions for "Further Reading." The sources offer reliable in-depth information and are recommended by the author as the best available publications for more lengthy, detailed, or comprehensive coverage of a topic than can be feasibly presented within the *Encyclopedia*. The citations do not represent all of the sources employed by the contributor in preparing the article. Most of the listed citations are to review articles, recent books, or specialized textbooks, except in rare cases of a classic ground-breaking scientific article or an article dealing with subject matter that is especially new and newsworthy. Thus, the reader interested in delving more deeply into any particular topic may elect to consult these secondary sources. The *Encyclopedia* functions as ingress into a body of research only summarized herein.

GLOSSARY

Almost every topic in the *Encyclopedia* deals with a subject that has specialized scientific vocabulary. An effort was made to avoid the use of scientific jargon, but introducing a topic can be very difficult without using some unfamiliar terminology. Therefore, each contributor was asked to define a selection of terms used commonly in discussion of their topic. All these terms have been collated into a glossary at the back of the volume after the last article. The glossary in this work includes over 900 terms.

INDEX

The last section of the *Encyclopedia of Tidepools and Rocky Shores* is a subject index consisting of more than 3,200 entries. This index includes subjects dealt with in each article, scientific names, topics mentioned within individual articles, and subjects that might not have warranted a separate stand-alone article.

ENCYCLOPEDIA WEBSITE

To access a website for the *Encyclopedia of Tidepools and Rocky Shores*, please visit

http://www.ucpress.edu/books/pages/10341.html

This site provides a list of the articles, the contributors, several sample articles, other materials, and links to a secure website for ordering copies of the *Encyclopedia*. The content of this site will evolve with the addition of new information.

Rocky shores lie at the interface between the land and the sea, experiencing a Jekyll-and-Hyde existence of alternating terrestrial and marine habitats. When the tide is in, coastal plants and animals are bathed by seawater that exposes them to predators, moderates temperatres, delivers food, transports propagules, and imposes large hydrodynamic forces. When the tide is out, the same rocky-shore species are subjected to terrestrial predators, desiccation, temperature extremes, intense solar radiation, and occasional dousing by freshwater. Intertidal organisms must survive in both terrestrial and marine worlds, transitioning as often as twice each day as the tides ebb and flow.

One might easily suppose that the physical and biological rigors of the shore would be sufficient to exclude all but a few plants and animals. What terrestrial animal could withstand both the pounding of waves and the cyclical change in temperature characteristic of the shore? What delicate sea creature could survive being baked for hours in the sun? Yet, despite such adversity, rocky shores are home to a striking diversity of species. As John Steinbeck famously noted upon approaching a wave-swept shore in Mexico: "The exposed rocks had looked rich with life under the lowering tide, but they were more than that: they were ferocious with life." Mussels, sculpins, kelps, and urchins; anemones and sea stars of all sizes and colors; barnacles, worms, limpets, and abalone; algae that look like corals; others that look like tar—the diversity of rocky shores rivals that of tropical rainforests.

This diversity is a rich source of wonder and the cause for curiosity. For many people, rocky shores at low tide provide the only opportunity to see and interact with marine species firsthand. Go down to the shore and poke around. Kneel down and stare into a tidepool. Comb through the seaweeds to see what lies beneath. Undoubtedly, questions will rapidly spring to mind. At first, they may be specific and small-scale. What species is that snail? Do seagrasses really have flowers? What do sea anemones eat? But small questions lead to larger ones: How do these marine organisms survive exposure to air? How did these intertidal animals and algae, seemingly immobile, get here? How do interactions between species affect where they occur and how abundant they are? In this encyclopedia, we strive to provide answers to these types of questions, both small and large.

The extent to which we can provide these answers—as indicated by the heft of this tome—is testament to the intellectual attraction of wave-swept shores. Rocky coasts are not merely a curious oddity among Earth's habitats; they have proven scientific value. To cite a prime example, thanks to their steep environmental gradient, their two-dimensional structure, and the rapid turnover of their abundant sessile or slow-moving organisms, wave-swept rocky shores make it practical for ecologists to conduct experiments that would be difficult or impossible elsewhere. Much of what we know about the ecological importance of processes such as competition, recruitment, predation, and patch dynamics has been tested on rocky shores. Although rocky shores are a comparatively minor habitat on Earth in terms of area, they have played a disproportionately large role in our understanding of ecological systems. As we move into an era of accelerated global climate change and expanded human domination of ecosystems, the extensive past work in the intertidal zone may serve as a valuable baseline against which to measure the effects of environmental shifts.

Our goal here of providing answers is constrained by several factors. First, many questions in intertidal science remain unresolved. As you delve into this book, you will find many articles noting the limitations of present knowledge. But that is how science works. Noting the things we do not yet understand is as important as cataloging the things we do know. If you find yourself frustrated by the lack of an answer to a particular question, jump right in and explore the topic yourself. We hope that the information provided here will at least provide a basis from which to proceed.

Second, even for a volume that professes to be encyclopedic, it was impossible to cover everything. For example, with few exceptions, we have limited ourselves to discussion of intertidal and shallow subtidal zones of rocky shores. As a consequence, there is little here that deals with sandy beaches and coral reefs. Even within this restricted scope, choices had to be made. A few topics were left out because we could not find authors to write them. A few topics were included that might appear whimsical. But most topics were chosen in an attempt to provide coverage that is both broad and deep of the multitudinous aspects of near-shore science. We hope that we have chosen wisely.

The *Encyclopedia of Tidepools and Rocky Shores* is intended for a general audience. In fact, the topic is so broad that it would be impossible to write a volume aimed at specialists. An article written for a specialist on the taxonomy of seaweeds, for instance, would not be accessible to a specialist on the hydrodynamics of breaking waves, and *vice versa*. To aid in the flow of information across topics, we have included an extensive glossary.

We thank the staff at the University of California Press for their invaluable assistance in the preparation of this volume. The idea for a one-volume encyclopedia of near-shore science originated with Chuck Crumly, who then managed to convince two overextended academics that editing an encyclopedia would be fun. Scott Norton and Ted Young deftly shepherded the manuscript through production. Without Gail Rice's experience, organizational skills, and persistence in shepherding articles from inception to production, this book would not have been possible. And we thank our families—Sue, Katie, and Jim; Peggy, Erin, and Andrea—for their forbearance of the endless evenings we spent with our attention diverted by wave-swept shores.

**Mark W. Denny and
Steven D. Gaines**

ABALONES

DAVID R. SCHIEL

University of Canterbury, Christchurch, New Zealand

Abalones are marine snails that play only a minor role in the functioning of marine communities yet are culturally and commercially important and are iconic species of kelp-dominated habitats. They were once tremendously abundant along the shores of much of the temperate zone, with large aggregations piled two or more layers deep along large stretches of rocky sea floor (Fig. 1). Abalones were prized for their meat and shells by indigenous communities worldwide, particularly along the coasts of California, Japan, and New Zealand. Their shells are well represented in middens, such as on the Channel Islands off southern California, that date as far back as 10,500 years ago. The tough shells, lined with iridescent mother of pearl, were used as ornaments and to fashion practical implements such as bowls, buttons, and fish hooks. In New Zealand Maori culture, abalone shell is extensively incorporated into carvings and ornaments. These historical uses have translated into high demand in modern times to such an extent that most abalone populations are greatly depleted from their historical abundances, with some species even considered to be in danger of extinction. There are now management practices in place worldwide to sustain and extend populations, based on knowledge about their biology and ecology.

DISTRIBUTION AND BIOLOGY

Abalone is the generic name that refers to the group of single-shelled molluscs of the class Gastropoda, family Haliotidae, and the single genus *Haliotis*. There are many

FIGURE 1 Once-abundant coastal populations of most abalone species are now severely depleted, and dense aggregations are usually confined to small patches in remote places. Shown here, New Zealand abalone (*Haliotis iris*, locally called "paua") of about 16 cm in shell length at a remote offshore island in a 5-m depth of water. Photograph by Reyn Naylor, National Institute of Water & Atmospheric Research Ltd, New Zealand.

regional names for *Haliotis* species, including paua from New Zealand, perlemoen in South Africa, and ormers in Europe. There is no consensus on the number of species of abalone, but estimates range from 70 to around 130. There are many subspecies, and some co-occurring species form hybrids with gradations of characteristics. The taxonomy of many of these variants is unresolved.

Abalones occur from the warm waters of tropical coral reefs through to the cool temperate zone. Their major abundances are along coastlines where kelp beds occur, particularly New Zealand, Australia, South Africa, Korea, Japan, China, the west coast of North America, and the Atlantic coast from Senegal to Spain, France, and the British Isles. An anomalous occurrence is the commercially fished species *H. mariae*, which is restricted to two areas of the tropical waters of Oman where cool, nutrient-rich, upwelled water sustains a kelp forest. *Haliotis* is an ancient genus, 60 to 100 million years old. One current view, based on DNA analyses of species worldwide, is that it originated in the ancient land mass of the Southern Hemisphere and speciated as the continents drifted into their present positions. There is disparity in sizes of tropical and cold-water species. Tropical species rarely exceed 50 mm in length; large-bodied species occur only in cold-water areas dominated by kelp and other macroalgae, which provide energy-rich food. Many of the commercially important species reach large sizes. For example, *H. iris* from New Zealand, *H. ruber* from Australia, *H. discus hannai* from Japan, and *H. midae* from South Africa can reach lengths of around 180 mm, and the giant of them all, *H. rufescens* of California, can grow to more than 300 mm. Different species of abalone occur from the intertidal zone to depths of hundreds of meters. However, their major distribution is in near-shore waters generally less than 30 m in depth. Their sleek hydrodynamic shells make abalones well suited for survival in turbulent waters.

Abalones have a relatively simple biology. A muscular foot, which comprises most of the weight of the animal, holds them tenaciously to rock surfaces by means of a contact mucilage and powerful contractions. The foot can raise the shell several centimeters above the substratum, which allows the abalone to trap drift algae as it tumbles along the sea floor. As in other gastropods, food is rasped by a radula, a double-rowed band of spiked chitinous material inside the mouth. Respiration occurs through the gills, which sit just below a series of holes that radiate in an arc along the top of the shell. These holes are also avenues for expelling waste products and releasing gametes during reproduction.

Growth occurs by adding to the shell margin to accommodate the expanding body size. The shell itself is a marvel of composite construction. It is secreted by epithelial cells of the mantle, a thin tissue that lines the inner surface of the shell, and is composed of layers of two crystalline forms of calcium carbonate embedded in a protein called conchiolin. These are laid down in a lattice construction that makes the shell very tough to crack or shatter. The outer layers are mainly calcite and the inner layer is nacreous aragonite, which can thicken throughout the life of the animal, enabling not only growth but also repairs to cracks and eroded areas of shell. This nacre gives the inner shell its characteristic luster. Species with beautiful iridescent coloring, such as *H. iris* with its swirling blends of green, blue, purple, and pink, are used to grow mabé or blister pearls, which are fashioned into jewelry.

Growth rates are influenced by a wide range of environmental conditions, especially temperature, food availability, and water circulation. In good conditions, larger species may increase an average of 25 to 30 mm in length annually for 5 or 6 years. However, in more quiescent areas, where algal drift is not plentiful and sediments may accumulate on the sea floor, growth can be only a few millimeters per year. Linear growth, however, may be deceptive because abalones have a geometric increase in body size with linear growth of the shell. A California red abalone *(H. rufescens)* of 200-mm shell length, for example, may weigh five to ten times more than one of 100 mm. Large individuals of many species can be 30 or more years old, although these are now rare.

Reproduction in abalones tends to be highly seasonal and related to water temperature, although there is considerable variation among species. For example, three species of California abalone have different spawning periods. White abalone, *H. sorenseni*, are highly synchronized in an annual spawning event that occurs during a few weeks of winter. Green abalone, *H. fulgens*, spawn through summer and early autumn, and pink abalone, *H. corrugata*, are the least synchronized and have spawning episodes throughout the year. These differences may relate to aggregation behavior and the likelihood of encountering other abalone, which is essential for reproductive success.

Abalones are dioecious, and populations generally have an equal proportion of males and females. Gonads usually form when animals are 2 to 3 years old. The sexes are readily determined by pushing the foot aside and looking at the color of the gonad, which surrounds the liver in a horn-shaped structure between the foot and the shell. Abalones are broadcast spawners, and reproduction begins with a few individuals releasing eggs and sperm directly into the water through the respiratory holes. This triggers others to move closer and release their gametes. In large gatherings, there can be a cloud of sperm wafting in the water current above the abalones, which enhances the chances of successful fertilization. Small females may release only a few thousand eggs, but the largest individuals may release 7 to 10 million eggs. Because these large abalone have traditionally been targeted in fisheries, there has been a tremendous loss of reproductive potential in abalone species worldwide.

ECOLOGY AND FISHERIES

There are four major bottlenecks in the life cycle of abalone: fertilization, larval development, the settlement process, and movement from juvenile to adult habitats. Each of these has critical features essential for continued survival. Once fertilization occurs, eggs of about 0.2 mm in diameter begin a perilous journey, but they do have resources

FIGURE 2 (A) One-week-old red abalone *(Haliotis rufescens)*, about 1 mm long, on a thin crust of calcareous red algae; chemicals produced by such algae help trigger settlement of late-stage larvae from the plankton. (B) Juvenile red abalone, about 1 cm long, cryptically colored against a background of encrusting red and filamentous algae, note tentacles protruding from the head region. Photographs by Jessie Alstatt, Science Director, Santa Barbara Channelkeeper.

in the yolk that will sustain them for the larval period. The developing larvae do not feed, but they can absorb nutrients from the water. They go through a series of stages, forming the necessary structures for life on the sea floor. Later veliger stages have a foot, larval shell, and feeding apparatus. These larvae drift in coastal currents but have the ability to swim up and down in the water column using the cilia of their velum. After 1 to 2 weeks, depending on the seawater temperature, larvae are ready to settle. The consensus is that most larvae do not travel more than a few kilometers from spawning sites because of their quick development time and ability to stay in the near-shore zone.

Finding a suitable settlement site is no easy matter. Larvae sink to the sea floor and begin testing surfaces by touching them, hovering over them, and moving along to another place. Many species of abalone preferentially settle on thin encrusting red algae, which contain a chemical trigger that induces settlement (Fig. 2). Once this occurs,

the larva drops its velum and begins life as a juvenile, initially feeding on algal films and diatoms. Many species of abalone migrate to the undersides of large rocks in 1 to 10 meters' depth, where they will live and grow for up to 3 years. They are very vulnerable to a wide range of predators at this stage, including the numerous predacious reef fish, such as the large sheephead wrasse *Semicossyphus pulcher,* that patrol kelp forests, and the lobsters, crabs, and carnivorous whelks that share their secretive boulder habitats. As they grow, abalones feed mostly on drift seaweed that becomes trapped in cracks and crevices.

When abalones reach around 70 mm in length, they begin to emerge from their juvenile habitats to spend their adult period on more open areas of reef. Here they lead a mostly sedentary life, feeding passively on drift seaweed. As they get beyond about 120 mm, abalones worldwide have few natural predators. A major exception is along the west coast of North America, where sea otters *(Enhydra lutris)* occur. Otters have a high energy demand, consuming up to one-fourth of their body weight in food daily, and their favorite food is abalone, which they remove by repeatedly smashing the shell with a rock. Where otters are abundant, species such as the California red abalone, *H. rufescens,* tend to recede into cracks and crevices at least an otter's arm length deep.

PROBLEMS AND SOLUTIONS

No story about abalone is complete without mention of the history of extraction. Put simply, abalones have been overfished worldwide. The California fishery (Fig. 3) shows a pattern common elsewhere. A species is heavily exploited, then the fishery moves on to other species

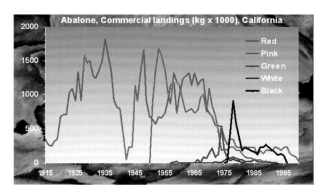

FIGURE 3 Commercial landings of California abalone over the twentieth century, showing serial depletion of five species. White abalone *(Haliotis sorenseni)* are now so rare they are listed as an endangered species. Background shows the iridescent nacre on the inside of an abalone shell. Graphic by Kim Seaward, Marine Ecology Research Group, Christchurch, New Zealand. Data derived from California Department of Fish and Game reports.

sequentially. Abalones have been fished commercially since at least 1850 in California. Initially this involved intertidal populations of green and black abalone *(H. fulgens* and *H. cracherodii)*, but then was extended to subtidal populations of these and other species, especially the red abalone *H. rufescens*. More than a million pounds of several abalone species were extracted annually from the early 1900s, with peaks of more than five million pounds in the 1950s and 1960s. Catches declined to less than a million pounds again by the early 1980s. Finally, in 1997 commercial and sport fishing for abalone was stopped in southern California. It is still allowed north of San Francisco but only by free diving. These sorts of stock depletions and catch restrictions apply to most abalone fisheries now.

Why does it take so long for restrictions to come in? Part of the reason is that fisheries for sedentary species such as abalone tend to be serially depleted. Catch rates tend to stay high until populations are near collapse because people become very efficient at minimizing the effort involved in spotting and catching abalone. Other reasons are the magnitude and variety of modern pressures that interact to affect populations, such as market demand, commercial harvest, increasing sport harvest, illegal fishing, expanding sea otter populations, habitat degradation (especially from pollution and sedimentation of juvenile habitats), disease, variable recruitment, illegal fishing, and often weak management and enforcement. These interactive effects have positive feedbacks on populations. Although there are minimum size limits in commercial fisheries, usually set at around 120 to 130 mm when animals have had several years to reproduce, these alone have not saved populations. Furthermore, the removal of the largest individuals takes away most of the egg production of a population. Thinning of aggregations to isolated individuals reduces the chances of fertilization. Degradation of settling and juvenile habitats results in diminished recruitment. As seawater warms, there may also be a higher incidence of disease. The incidence and spread of "withering foot syndrome," caused by a bacterium that is lethal in temperatures over 18 °C, has severely depleted populations of the black abalone *H. cracherodii* throughout southern California.

Strong management of depleted stocks worldwide, an expanding aquaculture of abalone (particularly in Asia), and reseeding of depleted coastal areas may lead to rejuvenation of natural populations while filling market demand. Strict TACs (total allowable catches) and regional catch restrictions have greatly helped in the sustainability of some fisheries, such as in New Zealand and Australia, whereas closures and marine protected areas are seen as avenues to recovery in severely depleted areas. An optimistic projection would see abalones returned to at least being common again in coastal waters. For the present, however, it is clear that much of the adaptability of populations and buffering from climate and other environmental vagaries is compromised for many of the world's abalone.

SEE ALSO THE FOLLOWING ARTICLES

Adhesion / Food Uses, Ancestral / Management and Regulation / Snails / Symbolic and Cultural Uses

FURTHER READING

Altstatt, J. M., R. F. Ambrose, J. M. Engle, P. L. Haaker, K. D. Lafferty, and P. T. Raimondi. 1996. Recent declines of black abalone *Haliotis cracherodii* on the mainland coast of central California. *Marine Ecology Progress Series* 142: 185–192.

Erlandson, J. M., T. C. Rick, J. A. Estes, M. H. Graham, T. J. Braje, and R. L. Vellanoweth. 2005. Sea otters, shellfish, and humans: 10,000 years of ecological interaction on San Miguel Island, California, in *Proceedings of the Sixth California Islands Symposium, Ventura, California, December 1–3, 2003.* D. K. Garcelon and C. A. Schwemm, eds. Arcata, CA: Institute for Wildlife Studies, 9–21.

Multi-Agency Rocky Intertidal Network Scientific Publications, abalone. http://www.marine.gov/publications.htm.

Shepherd, S. A., M. J. Tegner, and S. A. Guzmán del Próo, eds. 1992. *Abalone of the world: biology, fisheries and culture. Proceedings of 1st International Symposium on Abalone.* Oxford: Fishing News Books.

Vilchis, L. I., M. J. Tegner, C. S. Man, K. L. Riser, T. T. Robbins, and P. K. Dayton. 2005. Ocean warming effects on growth, reproduction, and survivorship of southern California abalone. *Ecological Applications* 15: 469–480.

ADHESION

ANDREW SMITH

Ithaca College

To survive on intertidal rocks, organisms generally must have a way of holding their positions. Because they are often specialized to live on the limited substrate and unique conditions of the intertidal zone, being swept away would be tantamount to death. Thus, many of the best-known biological adhesives are produced by intertidal animals. These adhesives can be solids or gels, and they form strong attachments to irregular surfaces despite the presence of water.

MARINE ADHESIVE MECHANISMS

There are a wide variety of organisms that attach strongly to intertidal surfaces (Fig. 1). Animals such as barnacles permanently cement themselves to hard surfaces. Mussels

FIGURE 1 A wide variety of organisms adhere strongly to rocks in the intertidal environment. Barnacles attach with a solid cement under their base (A). Mussels use byssal threads that adhere to the surface via small plaques (B). Sea urchins secrete an adhesive material underneath their tube feet (C). Limpets can secrete an adhesive gel underneath their foot or use suction (D). Photographs by K. Kamino (A), J.H. Waite (B), Romana Santos (C), and L. Miller (D).

are also well known for using a permanent, solid adhesive. This adhesive is macroscopically more complex in that it involves hardened plaques that attach to the surface and connect to the organism through tough threads. This structure is called the byssus. Many other organisms use glues that are less solid. A large number of molluscs use glues that are gels. These gels consist of networks of large polymers that trap water. The tube feet of echinoderms also use gels for attachment. The stiffness of these gels can vary considerably, with some being remarkably firm and elastic. Finally, many algae can attach to rocks using tough holdfasts.

Whether the adhesives are solid cements or relatively soft, wet, gel-based structures, they face similar issues. They must be able to adhere to wet, irregular surfaces, and they must maintain their integrity despite the presence of water.

BASIC PRINCIPLES OF GLUING

Two primary processes that govern the effectiveness of a glue are adhesion and cohesion. Adhesion is the ability of the glue to stick to a surface. It involves spreading across a surface and bonding to it. Cohesion provides the mechanical strength for the glue to resist deformation. It depends on interactions within the glue, specifically between the molecules that make up the bulk of the glue. Many artificial glues start in liquid form so that they can

spread across a surface easily. Once the glue has bonded adhesively to the surface, the glue then cures; that is, it hardens to provide cohesive strength. Glues can fail at the adhesive interface—for instance, by peeling off a waxy surface. They can also fail cohesively, in which case the glue itself breaks or deforms, leaving a patch of glue still firmly adhering to the surfaces.

Adhesion

Glues often have a relatively low viscosity (resistance to flow) at first, so that they can spread across a surface. The ability of a glue to flow during the initial stages of adhesion is essential. This is because it is extraordinarily difficult for two solid objects to come into full contact, as a result of their surface roughness (Fig. 2A). Even surfaces that look and feel perfectly smooth at the macroscopic level typically have a great deal of microscopic roughness. Nanometer-scale roughness, which is nearly unavoidable, is sufficient to prevent most of the area on the surfaces from coming into direct contact. The surfaces may touch and adhere at some points, but those points cover only a small fraction of the available area. A viscous material, however, can flow to fill all the gaps and indentations. Thus, it can achieve far more intimate contact with the surface (Fig. 2B). Pressure-sensitive adhesives such as tapes and tacky glues also depend on flow to achieve this contact. These adhesives have a high viscosity, but they are still

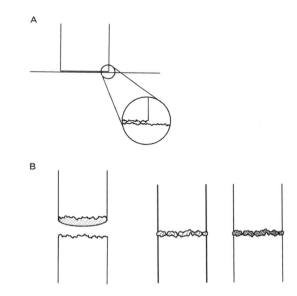

FIGURE 2 Most solids, regardless of how smooth, cannot achieve intimate contact with other solids because of nanoscale surface roughness (A). Adhesives help by flowing to fill the gaps and achieve full contact (B). After wetting the surfaces, adhesives often harden to provide cohesive strength. Illustration by C. Chew.

sufficiently soft and deformable that a reasonable pressure will force them to flow into the gaps and indentations on the surface. Without such pressure, the area of contact is insufficient to provide a strong adhesive force.

Some biological adhesives, such as those from barnacles and mussels, appear to be secreted in a viscous form that hardens over time. Other animals secrete highly viscous gels that may or may not stiffen after secretion. These gels may also be forced into contact with the surface in the same way as a pressure-sensitive adhesive. Underwater, spreading may be assisted by the fact that the surface is already wet. Instead of relying on bulk flow of the material, the polymers can diffuse through water to the wetted surface. Of course, this merely replaces one problem with another, because the adhesive must then displace the water to contact the surface directly.

In addition to being able to flow, a glue must be able to wet the surface. Wetting involves the interactions between a fluid and a surface. If the fluid and the surface are incompatible, wetting will not occur. For example, water cannot interact with Teflon or wax, so it beads up on those surfaces rather than spreading. If a glue does not wet a surface, adhesion will be weak.

Whether or not wetting occurs is determined by the potential energy of the exposed surfaces relative to the interfacial energy between the adhesive and substrate (Fig. 3). The potential energy of an exposed surface results from the fact that it often takes energy to create new surfaces. A high-energy surface results when molecules that had been interacting with similar molecules, as in a bulk solid, are exposed to a medium that they cannot interact with, such as air. Energy is required to create such a surface. A more stable, low-energy surface results when the solid interacts well with the medium, as polar solids do with water.

When an organism places an adhesive on a surface, the adhesive will wet or fail to wet the surface, depending on which minimizes the total potential energy of all the surfaces. Thus, wetting occurs when the solid–adhesive interface has a lower energy than the other surfaces exposed to the medium.

Good wetting usually occurs when there is chemical interaction between molecules in the adhesive and the surface. Such chemical bonding can occur in a variety of ways. Hydrophobic regions may come together to minimize contact with water, oppositely charged regions may interact, and more specialized forms of bonding may occur, possibly involving interactions with metals or oxides on the surface. Additionally, most molecules can interact through van der Waals forces, which occur when two molecules come into close proximity and the charge on one causes a corresponding redistribution of charge on the other. This creates locally charged regions that can interact. Van der Waals forces can occur between most surfaces; the key is to make sure the surfaces get into close enough contact over most of their area.

Adhesive interactions are complicated greatly by the presence of water. Even if a glue can spread and bond to a dry surface, it may not work underwater. Unless the water bound to a surface is displaced, the glue will adhere to a thin layer of water rather than directly to the surface. This forms a substantial weak link in the adhesion, because water can flow, causing adhesive failure. The basic problem is that water interacts strongly with polar and charged surfaces. For an adhesive to work well, it must displace this bound water by having an even stronger affinity for the surface. Furthermore, many adhesive interactions involve attraction between charged molecules. Because water is polar, it will also be attracted to charged regions of the glue polymer. Thus, it will cover and effectively mask the binding regions of the adhesive molecule as well as masking the surface to which the glue is trying to adhere. In energetic terms, the surrounding water may make the separate solid and glue surfaces lower in energy and thus more stable than the solid–glue interface (Fig. 3).

In addition to the presence of water, any glue must contend with other impurities. There will always be a layer of organic material fouling surfaces underwater, and this may need to be displaced. Glues may also adhere to dirt and small bits of organic material that are loosely bound to the surface instead of to the firm surface underneath. Given these considerations, the ability of marine organisms to adhere strongly in the intertidal environment is impressive.

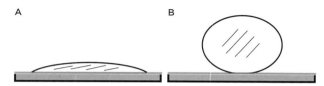

A   B

FIGURE 3 The ability of an adhesive to wet a solid depends on the surface energies of the materials involved. If the solid–adhesive interface (red) is stable due to strong interactions, it will have a low energy. If this energy is lower than the energy of the solid in contact with the surrounding water (green), wetting will occur (A). In essence, a stable, low-energy interface replaces the higher-energy surface. If the energy of the solid–adhesive interface is higher than the solid's surface energy in water, wetting will not occur (B). Note that a more detailed analysis would also factor in the energy of the adhesive–liquid interface.

While the nature of the chemical bonds allowing this adhesion is often not fully clear, a good deal of work has identified several possibilities. Mussels use a number of small proteins that are rich in the rare amino acid 3,4-dihydroxyphenylalanine (DOPA). This amino acid may form strong interactions through a variety of mechanisms. With its two hydroxyl groups it is strongly polar and readily forms hydrogen bonds to polar surfaces. It can also form complexes with metals such as iron that may be present on the surface. Like mussels, barnacles use a number of relatively small proteins at their adhesive interface, and some appear specialized for adhering to different surfaces. Many adhesive proteins have a notably high percentage of polar regions, particularly hydroxyl groups, and charged regions. These likely assist in displacing water from the surface and binding to polar surfaces such as the rocks of a tidepool.

Cohesion

In addition to adhering to a surface, a glue must have sufficient mechanical strength to resist deformation. The proteins that make up the bulk of the secretion are likely to be fibrous or take on an extended configuration. Typically, these fibers or sheets are linked together to create a highly viscous or solid material. These may be linked into a random network as shown in Fig. 4, or they may be in a highly ordered array. Such linkages may result from

FIGURE 4 The glues of marine invertebrates typically consist of crosslinked networks of proteins. If one were able to view the adhesive gel of a snail at the molecular level, it might appear as shown in the expanded view. Crosslinks between the proteins (blue) provide cohesive strength, and bonding between the proteins and the surfaces (red) provides adhesive strength. In other animal glues, the proteins may be crosslinked into a more orderly array. Illustration by C. Chew.

chemical crosslinks between polymers or from physical tangling between long polymers that are packed into a high concentration. As with adhesion, the presence of water is a serious detriment to good cohesive strength. Water within the glue may mask crosslinking sites just as it does in adhesion. Thus, the crosslinks within the glue must be able to form and avoid degradation despite the presence of water.

The physical network of polymers can take a variety of forms. In the case of mussels, the byssal threads are composed of collagenous fibers with either silk-like or elastin-like regions, thus conferring stiffness or elasticity. These fibers are crosslinked into tough threads. The plaques where the byssal threads meet the substrate include smaller proteins that are rich in the amino acid DOPA, which may be involved in crosslinking. DOPA residues can link together covalently in a process known as quinone tanning. Some algae may depend on a similar type of crosslinking to join polysaccharide chains. Barnacle adhesives are also made of crosslinked proteins, with large (~100 kD) proteins linked by hydrophobic interactions and other possible interactions. Many snails adhere using glues that are gels. Such gels are often based on giant proteins or protein–polysaccharide complexes (>1000 kD). These giant molecules would normally form a loose network by tangling interactions, but they also appear to be crosslinked by other proteins that are present in the glue. In some molluscs, such as limpets, the bulk of the glue consists of proteins that are closer to 100 kD in size, and these also appear crosslinked. The mechanism of crosslinking in these gels is unknown, but it appears to depend on charged molecules. Proteins seem to make up the key constituents of a number of other well-studied algal and echinoderm adhesives. Overall, for many of the biological adhesives that are used in the marine environment, crosslinked proteins seem to be essential for providing cohesive strength.

In addition to cohesive interactions between molecules, there are a wide variety of engineering factors that affect the ability of the glue to resist deformation. Although great stiffness would seem desirable in an adhesive, it doesn't always provide optimum performance. Many glues deform markedly during detachment. This deformation involves breaking crosslinks, rearranging large molecules, and dragging them past each other. This process can dissipate a great deal of energy before dislodgement occurs. This may be one reason why flexible, gel-based glues are so effective. Even relatively stiff glues often have mechanisms for energy dissipation. Failure is rarely through simple crack formation or uniform flow of a viscous glue.

Finally, it is worth noting that biological adhesives are often complex mixtures of polymers. Some polymers may be involved in adhesion, some in cohesion, and others may perform other tasks. For example, echinoderms such as sea stars appear to use a dual-gland system in which one gland secretes an adhesive material and a neighboring gland secretes a separate, de-adhesive material. The deadhesive appears to cause release of the attachment.

SUCTION ADHESION

A number of marine animals use suction to adhere, most notably octopuses and squid, but also a number of gastropods and possibly a wide variety of other animals as well. Suction in the marine environment is quite different from suction in dry environments. In dry environments, suction cups are filled with air and work by creating a partial vacuum. Underwater, the suction cups are filled with water. The cup exerts an expanding force on the water, but water has cohesive strength and is essentially inexpansible at physiological pressures. Thus, when the suction cup pulls on the water, it reduces the pressure in the trapped water with no detectable expansion.

Suction works well underwater because water adheres well to the underwater surfaces. Furthermore, because the water is trapped underneath the sucker, it cannot flow. The cohesive strength of the bonds between water molecules is quite large. Thus, water resists expansion remarkably well, as long as it adheres to all the surfaces. There are two major causes of failure of suction underwater: (1) failure to maintain a seal, so that the water can flow in, or (2) adhesive failure, in which the water does not fully wet the surface and thus an air pocket in the unwetted area expands rapidly as the pressure in the surrounding water drops. This often results from the presence of microscopic air bubbles trapped on particles. These microscopic air pockets form weak spots in the adhesion of water to the surface. Despite the presence of such weak spots, suction underwater appears to be considerably more effective than suction in air.

SEE ALSO THE FOLLOWING ARTICLES

Barnacles / Bivalves / Locomotion: Intertidal Challenges / Materials, Biological / Materials: Strength

FURTHER READING

Flammang, P. 1996. Adhesion in echinoderms. *Echinoderm Studies* 5: 1–60.
Gay, C. 2002. Stickiness: some fundamentals of adhesion. *Integrative and Comparative Biology* 42: 1123–1126.
Smith, A. M. 2002. The structure and function of adhesive gels from invertebrates. *Integrative and Comparative Biology* 42: 1164–1171.
Smith, A. M., and J. A. Callow. 2006. *Biological adhesives*. Berlin: Springer.
Waite, J. H., N. H. Andersen, S. Jewhurst, and C. Sun. 2005. Mussel adhesion: finding the tricks worth mimicking. *Journal of Adhesion* 81: 1–21.

AIR

ROBERT DUDLEY

University of California, Berkeley

The intertidal zone is characterized by regularly alternating exposure to air and water. Intertidal plants and animals can also be partially submerged for extended periods and thus will be simultaneously exposed to both media. Air differs physically from seawater in diverse and important ways that influence the forces experienced by organisms, their ability to effect gas exchange, and their overall thermal balance with respect to the surrounding environment. Unlike the chemical composition of seawater, the gaseous constituents of the atmosphere have changed dramatically through geological time. Such variation has had important biophysical and physiological consequences for intertidal organisms.

THE PHYSICAL PROPERTIES OF AIR

The physical properties of air derive from its constituent molecules, predominantly nitrogen (~78%) and oxygen (~21%) in today's atmosphere. Additional constituents such as argon (~1%) and carbon dioxide (~0.03%) are much less important from a physical perspective, although levels of the latter molecule have critical consequences for photosynthetic capacity by plants.

For humans, the most obvious difference between water and air is their relative density (the mass per unit volume), with water being some 800 times denser than air. Interestingly, both air and water behave as fluids rather than solids in that they resist the rate of deformation in response to an applied force, rather then resisting deformation per se. Such resistance is a function of the fluid's dynamic viscosity, the value of which for water is approximately 55 times greater than that of air. Also, the dynamic viscosity of air is relatively insensitive to temperature, whereas that of water is strongly (and inversely) thermally dependent. Comparable differences between the two fluid media characterize a number of physical parameters (Table 1). Of particular importance for heat exchange between organism and environment in the intertidal zone is the heightened thermal conductivity of water. Finally, diffusion of oxygen is dramatically slower in water than in air because of the much higher density of the former medium. Overall, air and water represent two dramatically different physical environments from the perspectives of energy, gas, and momentum exchange between organism and surrounding medium.

TABLE 1

Ratios for Key Physical Parameters of Their Values in Water to Those in Air

Physical Parameter	Ratio of Water to Air
density	800
dynamic viscosity	55
thermal conductivity	23
oxygen diffusivity	0.0001
oxygen concentration	0.0265
carbon dioxide concentration	0.708

NOTE: Values refer to a temperature of 20 °C.

EFFECTS OF INTERMITTENT EXPOSURE TO AIR

If we consider a sea anemone, first suspended within the depths of a tidepool, and then exposed to air as the tide recedes, the physical consequences of these two contrasting environments become clear. The much greater density of water renders the anemone effectively buoyant within water (Fig. 1A), whereas gravity dominates in air and induces retraction of tentacles and a flaccid body posture (Fig. 1B). Given the linear dependences of hydrodynamic forces on fluid density, water currents within the tidepool often exert substantial forces on the sea anemone, whereas wind motions typically exert but minimal aerodynamic drag because of the much lower density of air. Issues of structural support against gravity thus become paramount when intertidal taxa are exposed to air, and attachment mechanisms to substrate together with postural changes become evident, as in the case of the drooping sea anemone. As a further, botanical example, stands of marine algae collapse upon emergence from water, precluding photosynthesis by all but the uppermost fronds.

In water, plants and animals of the intertidal are isothermal (i.e., at the same temperature) relative to the surrounding water because of its high rate of heat conduction.

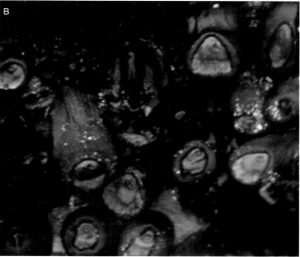

FIGURE 1 Sea anemones submerged within a tidepool (A) and exposed to air (B). Photographs courtesy of Jonathon Stillman.

In air, the situation is more complicated. Direct absorption of solar radiation, now unattenuated by water, increases body temperature. Heat loss, by contrast, is typically much lower in air than in water. Intertidal plants and animals thus typically heat up when exposed to the air as the tide recedes, although convective cooling by the moving wind may partially mitigate this effect, particularly on overcast days. Nonetheless, vaporization of endogenous water reserves is one major strategy used by intertidal taxa to overcome heating when exposed to air, a phenomonen clearly irrelevant when an organism is immersed within a tidepool.

HISTORICAL VARIATION IN ATMOSPHERIC COMPOSITION

One of the most interesting features of the atmosphere has been the variation over geological timescales in its gaseous composition. Nitrogen content of the atmosphere is believed to have been fairly constant since the formation of the earth, but both oxygen and carbon dioxide content have changed dramatically through time. Oxygen concentration is thought to have been only about 15% throughout the early Phanerozoic but, starting in the mid-Devonian (~380 million years ago), exhibited a substantial rise associated with terrestrialization by plants, reaching values potentially as high as 35% by the end of the Carboniferous (~290 million years ago) in what is known as the late Paleozoic oxygen pulse (Fig. 2). This increase in oxygen content, coupled with constant nitrogen content, also yielded a more dense atmosphere at any given elevation. Viscosity of the atmosphere, by contrast, would have remained fairly constant. Overall, heightened oxygen levels would have increased the amounts of this gas available to terrestrial plants and animals via diffusion from the atmosphere, although increased atmospheric density would partly mitigate this effect via a reduced diffusion coefficient. Because the amount of oxygen dissolved in seawater is an approximately linear function of that in the surrounding air under equilibrium conditions, elevated atmospheric oxygen would also have increased availability of this gas to plants and animals in water.

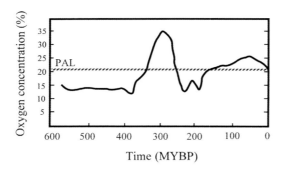

FIGURE 2 Variation in oxygen content of the atmosphere through Phanerozoic time, relative to the present atmospheric level (PAL). MYBP = million years before present.

Following the end-Carboniferous peak in atmospheric oxygen, levels declined as a result of complicated, interacting geological and biological causes, and by the end-Permian (~250 million years ago) atmospheric oxygen reached levels as low as 15% (see Fig. 2). Thereafter, oxygen concentration rose slowly to reach a secondary peak of 25–30% by the end of the Cretaceous (~65 million years ago) and then declined to present-day values of about 21%. At sea level, therefore, the density of air has fluctuated considerably both upward and downward at different stages of the Phanerozoic.

The picture for carbon dioxide, by contrast, is one of approximately continuous drawdown in the Phanerozoic from highs of about 0.5% to the present-day level of 0.03%. Such changes in the availability of carbon dioxide, the raw material of photosynthesis, have had important consequences for the physiology of marine algae and other plants, as evidenced by paleontological analyses of stomatal density and other proxies of photosynthetic activity.

The concept that air has varied historically in oxygen content is, in scientific terms, one that is fairly recent, and the full implications for organismal physiology and biophysics are only now being realized. The effects of variable air density on heat exchange, together with variable oxygenation of tidepools, are the most likely to have been relevant to intertidal organisms.

SEE ALSO THE FOLLOWING ARTICLES

Desiccation Stress / Diffusion / Heat Stress / Seawater

FURTHER READING

Berner, R. A. 2004. *The Phanerozoic carbon cycle.* Oxford: Oxford University Press.
Denny, M. W. 1993. *Air and water.* Princeton, NJ: Princeton University Press.
Ehleringer, J. R., Cerling, T. E., and Dearing, M. D., eds. 2005. *A history of atmospheric CO_2 and its effects on plants, animals, and ecosystems.* New York: Springer.
Gates, D. M. 1980. *Biophysical ecology.* New York: Springer.
Graham, J. B., R. Dudley, N. Aguilar, and C. Gans. 1995. Implications of the late Palaeozoic oxygen pulse for physiology and evolution. *Nature* 375: 117–120.
Lane, N. 2003. *Oxygen: the molecule that made the world.* Oxford: Oxford University Press.
Vogel, S. 1994. *Life in moving fluids,* 2nd ed. Princeton, NJ: Princeton University Press.

ALGAE, OVERVIEW

KARINA J. NIELSEN

Sonoma State University

Marine algae are photosynthetic organisms that fuel the base of the food chain in marine ecosystems and provide habitat for a huge diversity of intertidal and subtidal organisms. They include the microscopic phytoplankton that drift in the pelagic zone and form thin films on rocks, as well as the large seaweeds and kelp forests that line many rocky shores. Algae are an evolutionarily, ecologically, and functionally diverse group of organisms that play critical roles in the structure and functioning of rocky-shore ecosystems.

ALGAL DIVERSITY

Intertidal algae encompass an extraordinarily diverse array of photosynthetic organisms, found growing on rocky shores, that are not true plants, technically speaking (Fig. 1). They include the macroscopic and multicellular seaweeds as well as the microscopic single-celled phytoplankton (although the Latin word *algae* literally means "seaweed," and these will be the focus of this chapter). The algae are an evolutionary hodgepodge (not a monophyletic group) that includes all eukaryotic, photosynthetic organisms that lack leaves, roots, flowers, and other organ structures that define true plants and the prokaryotic cyanobacteria. With the single exception of the surfgrasses (found only along the west coast of North America, where they thrive and compete with seaweeds for space on the shore), true plants are not found growing on rocky shores. The phyletic (or deep) diversity of algae that live along wave-swept shores (and of all photosynthetic organisms in the ocean) is far greater than that found among all the photosynthetic organisms that inhabit land. This is the same pattern found when contrasting animal diversity on land and in the sea, and it reflects life's early origin and diversification in the ocean. The genetic differences among the various taxa grouped together as the algae are far greater than the genetic differences among the admittedly more species-rich group of true plants found on land. It is the combination of deep evolutionary diversity, a crowded coexistence on rocky

FIGURE 1 Diverse assemblage of algae uncovered on a rocky shoreline during an extreme low tide. Photograph courtesy of Jacqueline L. Sones.

shores, and a disturbance-prone environment that makes the study of tidepool algae especially fascinating.

ALGAL ORIGINS

All photosynthetic organisms, including seaweeds, are ultimately derived from early anaerobic, photosynthesizing bacteria (prokaryotes) that appeared on Earth approximately 3.8 billion years ago. Oxygen-producing, photosynthetic organisms (very similar to the cyanobacteria found living on Earth today) appeared in the ocean about 100–200 million years later and completely transformed Earth's atmosphere by filling it with oxygen (an event often referred to as the oxygen revolution), and as a result had a pervasive influence on the evolution of all life forms on Earth. By about 1.7 billion years ago, multicellular photosynthetic organisms had evolved and can be found in the fossil record. The algae evolved into about 12 phyla (or divisions) of uni- and multicellular photosynthetic organisms, with all but one being eukaryotic. Several of the major lineages of algae arose as the result of endosymbiotic events whereby a single eukaryotic cell engulfed and co-opted either a cyanobacterium or a eukaryotic alga that ultimately became the plastid (a subcellular structure or organelle) used to effect photosynthesis. Evidence of these historic endosymbiotic events can be found in analyses of genetic relationships, subcellular structures, photosynthetic pigments, and biochemistry of the algae. The seaweeds that we find on rocky shores today belong to three phyla or lineages of algae: the Rhodophyta, the Chlorophyta and the Heterokonta.

REDS, GREENS, AND BROWNS

Seaweeds are commonly described as coming from one of three color groups: reds, greens, and browns. These colorful names (taxonomically Rhodophyta, Chlorophyta, and Heterokonta, respectively) reflect their evolutionary relationships and the characteristic suite of pigments each group uses to collect and dissipate light energy. However, these colorful groupings can often be misleading to the novice trying to identify seaweeds on the shore; some "red" seaweeds can look brown or even black and some can look green, for example. The various colors that we see among the seaweeds result from the specific complement of pigments they possess and the relative abundances of each that are characteristic for a species. The relative abundance of the different pigments can also vary substantially within a species, or even an individual alga, in response to environmental conditions.

All the seaweeds (as well as land plants and cyanobacteria) have chlorophyll *a,* the primary pigment responsible for photosynthesis. There are two other common chlorophyll molecules: chlorophyll *b*, found in land plants and the green algae, and chlorophyll *c*, found in the brown algae. In addition to the chlorophylls, all algae have accessory pigments called carotenoids that can either transfer light energy to chlorophyll *a*, to enter the photosynthetic pathway, or direct it away from chlorophyll *a* when excess light is absorbed, protecting the molecules of the photosystem and dissipating the excess energy as heat. Fucoxanthin, for example, is the caratenoid that imparts a brown color to the kelps and other brown algae. The red algae (and cyanobacteria) also have another group of accessory pigments called phycobilins that give them their distinctive reddish hue. The relative abundance and presence or absence of these accessory pigments in combination with the chlorophylls determines the characteristic colors of the seaweeds we see on the shore.

STORING ENERGY AND CELLULAR WALLS

In addition to the differences we see in the pigments among the three major lineages of seaweeds, there are characteristic differences in the biological molecules these organisms use to store the products of photosynthesis and for structural support. Seaweeds (and sea- and surfgrasses) produce a variety of polysaccharides that form part of the matrix of their cell walls (unlike freshwater algae and land plants). Some are gel-forming compounds that confer both structural support and elasticity—clearly a useful characteristic for algae living on wave-swept rocky shores. The red algae store starch as granules within the cell's cytoplasm in a unique form called floridean starch. They form their cell walls from cellulose and a matrix of polysaccharide compounds, including the agars that are used in biotechnology applications (e.g., gel electrophoresis and culturing bacteria) and the highly sulfated carrageenans used as food thickeners. Some red seaweeds also impregnate their cell walls with calcium carbonate; these typically appear pinkish-red in color, are tougher than their noncalcified relatives, and are collectively called the corallines (Fig. 2). Green and brown seaweeds also use cellulose to construct their cell walls, and a few species incorporate calcium carbonate. Most brown seaweeds produce alginates as part of their cell wall matrix; alginates are extracted for use in textile production, to make medical dressings, and as food stabilizers and thickeners. For energy reserves brown seaweeds use lipid droplets or soluble carbohydrates called laminarans within the cytoplasm, rather than the starches red and green seaweeds use. The green seaweeds are further distinguished

FIGURE 2 Pink coralline algae surrounding the herbivorous chiton *Katharina tunicata*. There are both crustose and upright forms of coralline algae, as well as fleshy red algal crusts. Also visible are the grooved fronds of two young sporlings of the sea palm, *Postelsia palmaeformis*, blades of the red alga *Mazzaella flaccida*, and the branched red alga *Microcladia borealis*. Photograph by the author.

by storing starch within their plastids, a characteristic they share with land plants.

SEAWEED ARCHITECTURE, GROWTH, AND FUNCTIONAL FORMS

Seaweeds have an amazing diversity of forms given their anatomical simplicity. In contrast to true plants, they lack true tissues and organs, and have very few specialized cell types, typically just vegetative and reproductive cells, although there are some interesting exceptions. Specialized transport cells called trumpet hyphae or sieve elements (which are analogous to the sieve elements found in the vascular system of true plants and are shaped like a trumpet) are found in some kelps (members of the brown order Laminariales), including the giant kelp-forest-forming seaweeds *Macrocystis* and *Nereocystis*.

While many people are familiar with the spectacular beauty of kelp forests, uninitiated visitors to the seashore may simply view seaweeds as a slippery and sometimes smelly nuisance, especially when tangled masses have been washed ashore as wrack and begin to decompose after being dislodged by storms or as a result of an unusual bloom (this latter phenomenon often occurs as a result of nutrient pollution and may be especially problematic in systems where herbivores have also been overexploited by humans). Many delicate and intricately branched forms become plastered to the rocks during emersion at low tide, forming an amorphous and relatively unappealing-looking mat. However, seashore enthusiasts often come to appreciate the architectural and anatomical diversity of seaweeds after observing them more closely, perhaps with the aid of a magnifying glass or a microscope, or by observing seaweeds suspended in the water of a tidal pool.

The thallus (or "body") of an alga can be as simple as a chain of single cells arranged in a linear filament, or one or two layers of cells arranged in a sheetlike blade, where all the cells are essentially identical. More complex forms can have several layers of cells, differentiated into an outer cortex of highly pigmented cells and an inner medulla of larger, nonpigmented cells. In addition to sheetlike forms there are branched, tubular, and lobed forms, as well as saclike forms that can hold water (Figs. 3, 4). Most forms attach to the rock at a single point by a holdfast, which can be a simple disclike structure or a rootlike structure made up of

FIGURE 5 The sea palm, *Postelsia palmaeformis*, shares major anatomical features with other seaweeds including a holdfast at its base composed of haptera that attaches it firmly to the rock (or, in this case, a mussel), a tall, stalklike stipe, and a pom-pom-like crown of fronds. Photograph courtesy of Sarah Ann Thompson.

FIGURE 3 Diversity of functional forms of red, green, and brown algae. Photograph by the author.

FIGURE 4 A bed of the saccate red alga *Hallosaccion glandiforme*. Photograph courtesy of Sarah Ann Thompson.

haptera (Fig. 5). Some are entirely attached to the rock with encrusting thalli several cell layers thick; these can be soft and fleshy, or hard and calcareous, and either smooth or rugose in form (see Fig. 2). Filaments can also form intricately branched thalli with exquisite branching patterns (Fig. 6). Some branched forms are not just simple filaments but have intricate banding patterns caused by additional layers of outer (cortical) cells organized around nodes, or the cells can be arranged to look like stacked wagon wheels in a form described as polysiphonous, or they may be thicker and differentiated into an outer cortex and an inner medulla, several layers of cells thick. Some sheetlike seaweeds grow from the fusion of many chainlike filaments, while some large, thickly branched seaweeds are formed from a few multinucleated siphons (essentially one to a few giant cells wrapped, folded, and intertwined together) (Fig. 7). Seaweeds grow in different ways; some, such as the sea lettuce (*Ulva*), grow by dividing cells found throughout the thallus, while some have cells that divide only at the

FIGURE 6 The red alga *Platythamnion villosum* is a delicately branched filament made up of chains of single cells. This specimen is a tetrasporophyte, one of the diploid, free-living phases of the typical red algal life history. The dark circular structures attached to the branchlets are the tetraspores (groups of four haploid spores that are produced by meiosis), which will eventually be liberated and give rise to male and female gametophytes of identical morphology in this species. Small, clear gland cells are also visible at the ends of some of the branchlets. Photograph by the author.

FIGURE 7 The green alga *Codium fragile*, which is composed of just a few large multinucleated cells. Photograph courtesy of Sarah Ann Thompson.

edges of the blades (the apices of the joined filaments in some red seaweeds). Many of the large kelps have regions of actively dividing cells between the stemlike stipe and the blade (see Fig. 5), while the bladderwracks (members of the brown-order Fucales) (Fig. 8) have a single cell that divides at the tip of each branch.

FIGURE 8 The bladderwrack, *Fucus garderii* (broader blades), and another member of the order Fucales, *Pelvetiopsis limitata* (thinner) growing in distinct intertidal zones. Photograph by the author.

Large seaweeds such as the bladderwracks and kelps often bear one or more gas-filled bladders called pneumatocysts (Fig. 9), or, like the Southern Hemisphere species *Durvillaea antarctica* (Fig. 10), grow with many sponge-like pockets (similar to the closed-celled neoprene used to make wet suits and mouse pads) that provide added buoyancy to keep the alga suspended at the water's surface, where light for photosynthesis is abundant (Fig. 11). Those without such buoyancy aids often have stiffer, yet very flexible, stipes, such as those of *Postelsia palmaeformis* (an alga resembling a miniature palm tree that could have been drawn by Dr. Seuss; see Fig. 5) and *Lessonia nigrescens* (an alga resembling an intertidal bush), that help keep their photosynthesizing fronds suspended and intact while being battered by waves (Fig. 12).

Although there is stunning diversity in seaweed forms, there are many recurrent themes in their architecture that are the result of convergent solutions to similar environmental challenges and phylogentic (evolutionary) constraints. Seaweeds growing on rocky shores must acquire light and nutrients to grow while also staying attached to the rock, often in the face of the enormous forces of lift, drag, and acceleration imposed upon them by breaking waves. They simultaneously must overcome competitors

FIGURE 9 Balloon-like pnuematocysts along the sides of the strap-like stipe of the brown alga *Egregia menziesii*. Photograph courtesy of Sarah Ann Thompson.

FIGURE 10 The brown alga *Durvillaea antarctica* floating at the surface of the water, where light is plentiful. Photograph by the author.

FIGURE 11 Pnuematocysts keep the alga *Egregia menziesii* floating at the water's surface. The stiif stipe of *Pterygophora* keeps it elevated in the water. In the lower left, *Cystoseira* uses chains of smaller floats to keep its fronds aloft. Photograph by the author.

for limited resources (primarily space, light, and nutrients), and they must escape from being eaten by hungry herbivores. Scientists often aggregate seaweeds with similar architecture or body forms into groups that reflect functional forms (e.g., blades, tubes, filaments, or crusts) rather than evolutionary relationships, especially when studying seaweed physiology or susceptibility to herbivores.

FIGURE 12 The large, shrublike brown alga *Lessonia nigrescens* growing on rocks covered with pink, encrusting coralline algae. Photograph by the author.

LIFE CYCLES

Describing the "typical" life cycle for seaweeds is no easy task. Reproduction of new individuals can occur via sexual or asexual reproduction. Asexual reproduction can occur in some seaweeds by fragmentation, whereby a portion of an intact seaweed breaks off and grows into a fully functional mature seaweed from the fragment, or by parthenogenesis, whereby reproductive cells that are not successfully fertilized germinate and grow nonetheless.

Many species have a life cycle that alternates between a diploid phase (with paired chromosomes) and a haploid phase (with half the number of unpaired chromosomes, the complement of chromosomes typically found in sperm and egg cells). These two phases, with different numbers of chromosomes, may both look exactly the same (isomorphic), or they may be very different (e.g., a crustose and a branched form, or a tall, bush-like form and a microscopic filament), in which case the species is called heteromorphic. The haploid phases (gametophytes) exist as separate male and female individuals, producing male and female reproductive cells (gametes). The male and female gametes generally unite and grow into the diploid form (sporophyte), although sometimes—and this is surprisingly common among the algae—they can grow into another free-living haploid form again if fertilization does not occur. It is in the diploid form that meiosis occurs (a form of cell division in which the resulting cells have only half the number of chromosomes). These cells may be flagellated (zoospores) and swim around in the water before settling down and attaching to the substratum. Zoospores, however, are only found among the brown and green seaweeds.

In contrast, the red seaweeds have no flagellated or motile cells during any part of their life cycle, a distinctive product of their unique evolutionary history. They also alternate between phases with different numbers of chromosomes, but instead of just two phases they typically alternate among three phases. Two are diploid: the tetrasporophyte, which produces tetraspores via meiosis (see Fig. 6) and the carposporophyte. One is haploid: the male and female gametophytes. Interestingly, one of the diploid phases, the carposporophyte, is considered a parasite of the female gametophyte, because it is not free-living and instead grows inside the female gametophyte subsequent to fertilization of an egg, essentially cloning many spores from the single fertilized egg that eventually are released, settle, and grow up to be free-living tetrasporophytes (whose spores germinate into the gametophytes). This extra phase is thought to be an evolutionary solution to the problem of limited fertilization success among the red algae because they lack flagellated sperm cells. Presumably the nonswimming male gametes (spermatia) have a harder time getting themselves over to a receptive egg. The carposporophyte provides a means to increase the number of individuals produced from a single fertilization event, theoretically compensating for the lowered frequency of successful fertilization among the red algae. An interesting consequence of this unique life history strategy is that the high frequency of parasitic red algal species (though these are generally colorless) is thought to have been evolutionarily facilitated by the presence of this third "parasitic" phase.

Some brown seaweeds have a more familiar life cycle, one very similar to that found in animals. The mature diploid sporophytes simply produce eggs and sperm by meiosis, these are released to the environment, and the flagellated sperm fertilizes the egg cell, completing the cycle by growing into the sporophyte again. There is no free-living gametophyte phase. Among many species of brown algae, egg cells have been found to produce sexual pheromones that induce flagellated sperm cells to swim toward the egg cells and swarm around them until one has succeeded in fusing with the egg.

Seaweeds can be annuals, completing their entire life cycle every year, or perennials, persisting and holding onto precious real estate on the shore for several years at a time. Reproduction is often stimulated by predictable environmental cues such as changes in day length and temperature as the seasons change.

COPING WITH THE ENVIRONMENT AND OTHER ORGANISMS: PATTERNS ON THE SEASHORE

One of the most ubiquitous patterns found along rocky shorelines worldwide is the zonation of organisms, including seaweeds, into characteristic bands from low on the shore to high on the shore (Fig. 13). These zones are created by the interplay of stresses created by the regular pattern of advancing and retreating tides each day and the physiological tolerances of each species, in concert with the biological interactions among the different species of seaweeds and other organisms. Species can engage in positive or negative interactions with each other: some species can facilitate the success of seaweeds (a positive interaction), while others compete with them for space on

FIGURE 13 Intertidal zonation of seaweeds, surfgrasses, and sessile invertebrates along wave-swept rocky shores. Photograph by the author.

the shore or graze them off the rocks (negative interactions). The outcome of these interactions, such as who wins the battle for space, may be altered by such factors as how hot or dry the environment is, how often it is disturbed by waves, or by how many herbivores or even predators are found in a place.

Tides, Waves, and Environmental Gradients

Seaweeds living low on the shore do not dry out as often as seaweeds high on the shore do, because the former spend a greater proportion of each day covered by water. As a result they may grow faster and do better in the face of competition from another species trying to encroach on the same patch of rock, or they may be better able to outgrow the limpets or snails that are slowly nibbling away at their fronds. Because seaweeds low on the shore spend more time in the water, they also have more time to acquire essential nutrients from the water, such as nitrogen and phosphorus. They are also subjected to fewer extremes of temperature than seaweeds higher on the shore, which may be exposed to air for many hours, and quite possibly to the heat of the noonday sun or the bitterest cold at midnight.

Sunlight can be a mixed blessing for seaweeds. Too little sun, and seaweeds cannot photosynthesize enough to grow and reproduce, limiting how deep in the water they can live, but too much light can outstrip a seaweed's capacity to use or dissipate all the light energy it captures with its pigments. Excess light energy that is not used or dissipated causes damage to subcellular structures and molecules involved in photosynthesis, ultimately limiting how high on the shore a seaweed can live. Algae can avoid absorbing too much light by altering the complement or amount of pigments they produce, or by physically rearranging the pigment-containing organelles within their cells so that they are not completely facing the sun. A seaweed's architecture or functional form can also influence the proportion of incident sunlight absorbed. Thin blades expose virtually all their cells to light, while species composed of many cell layers expose only a fraction of their cells to full sunlight, because much of it is captured by the first few layers of cells. Branched forms allow some light to penetrate to branches at lower levels rather than being completely intercepted at the surface.

When more light is absorbed than can be used in photosynthesis, free radicals are often produced, which can damage molecules involved in photosynthesis and other essential cellular functions. Seaweeds have intricate biochemical mechanisms to scavenge these free radicals and deactivate them, protecting themselves from the damaging effects that may ensue to some degree, but these biochemical mechanisms can be compromised by other environmental stresses, including limited nitrogen, desiccation, and extremes in temperature—all of which tend to be exacerbated in concert with increasing exposure to sunlight higher on the shore. When these stresses, alone or in combination, pass a critical threshold, the seaweed will ultimately succumb and die, leaving behind its withered, bleached, and ghostly-looking thallus for a time before it is inevitably washed off the rock. Before that critical life-or-death threshold is passed, growth and reproduction may be reduced as the seaweed devotes energy to repairing stress-related damage, compromising its ability to compete successfully with space-hogging, sunlight-stealing neighbors or to outgrow munching herbivores.

Crashing waves can be both destructive and necessary for seaweeds. Some rely on waves to rip other organisms off the rock (such as the mussel beds that dominate space on many shores), creating a bare space they can colonize before becoming encroached upon again by a superior competitor. Seaweeds such as the annual sea palm (*P. palmaeformis;* see Fig. 5) use the predictable disturbance of winter storms to great effect. Their microscopic, filamentous reproductive phases (gametophytes) germinate from spores that settle onto the rock below the mussel bed and derive shelter there from heat and desiccation. But then, in a marvelous twist of fate, of all the juvenile sporophytes conceived beneath the mussel bed, the ones most likely to survive and reproduce are the ones that settle where winter storms will eventually rip the cover of the mussel bed away, allowing light to reach the growing sporophyte and creating space for them to occupy as they grow.

Waves can also rip and tatter seaweeds whose mechanical properties or growth form are ill-suited to the incredible forces of drag, lift, and acceleration imposed by the larger, breaking waves that impinge on more wave-exposed portions of the shore. Thus seaweeds whose forms are less robust to the full force of breaking waves are found on wave-protected shores or coves, whereas others that have evolved more hydrodynamic forms, or have material properties that allow them to both sustain their iron grip on the rock and simultaneously be flexible and elastic in the face of crashing waves, occupy the most wave-exposed outer rocks and headlands.

Seaweeds living in the more wave-exposed locations do not have it all bad, however. They can live higher on the shore because the sea spray and splash from the crashing waves keep things moist. Also, for some seaweeds the motion of waves tossing their fronds around can help to maximize the amount of sunlight they can absorb and use, thereby increasing their photosynthetic rate. They also benefit from the plentiful supply of nutrients delivered by the high water flow in these locations. Seaweeds in very quiet coves or areas of the shore where water flow is much lower may suffer from nutrient limitation. Morphological features can overcome some of these problems of living in low-flow habitats. Bumps or projections help create turbulent mixing of water near the nutrient-absorbing surfaces of the seaweed. Additional surface area for absorbing nutrients can also be provided in the form of specialized hairs or projections.

Sand, cobbles, and boulders are all moved around by waves and ocean currents. Sand in fast-flowing water can scour delicate algae and vulnerable young forms from the rock. It can also periodically bury rocks where seaweeds grow in the course of the seasonal movement of sand on and off of beaches every year. Seaweeds that can tolerate the scouring and anoxic conditions that develop on these sand-influenced rocky shores are the only ones that persist here (these are often referred to as psammophilic, or sand-loving, species). Cobbles can be set to churn and tumble in tidal pools, eliminating all but the toughest calcified crusts. Boulder fields with rocks of different sizes are interesting habitats; larger, heavier boulders are moved from time to time by waves, but less frequently than smaller boulders and cobbles. Interestingly, these different frequencies of disturbance for boulders of different sizes result in a mosaic of higher algal diversity than if the boulders were never moved by waves at all or if they were all moved around with equal frequency. This results from the combined effect of disturbance and the predictable sequence of colonization by seaweeds as a community develops on a patch of bare rock (this process is called ecological succession). A similar phenomenon occurs on rocky benches, where different-sized patches of mussel beds are removed periodically by large waves or by mussel predators (typically sea stars), opening up new space for colonization by seaweeds and other organisms. A high-diversity community forms along the shore from the mosaic of patches of different ages and sizes.

Interacting with Other Seashore Denizens

A space of one's own is often the hardest thing to acquire on a rocky shore. Imagine being an algal spore, trying to find a suitable place to get attached to and grow up in. There are many animals already living attached to the rocks, including some, such as mussels and barnacles, that filter water to capture particles just your size and flavor for food; a spore must escape these filter feeders before it can even settle down out of the water. Some

FIGURE 14 *Endocladia muricata,* a red alga, growing atop suspension-feeding mussels. Photograph by the author.

seaweeds pull off this escape act by growing atop mussels that might otherwise consume them (Fig. 14). A few have even devised ways to survive transiting through the digestive tract of molluscs or other grazers, and in some cases they germinate more readily after doing so. Apparently the spores benefit from the close association with the nutritious waste products that now surround them. Eventually, though, to survive a spore must settle down in a place that is not too hot, dry, and sunny. A nice, moist nook or cranny in a rock, in between the crowds of barnacles, or among the stipes or holdfasts of more mature seaweeds might do, but only if these neighbors do not overgrow the young sporeling or emit some kind of toxic chemical to deter it from encroaching on their space. Once settled, the germinating spore and tiny juvenile that emerges has to be lucky enough to keep from being scraped off the rock by the rough, rasping tongues of snails, chitons, and limpets or the gnawing jaws of sea urchins or fishes. Once established, the young alga must maintain its space on the shore by avoiding or outgrowing consumers and continuing to outcompete its competitors.

Some seaweeds unwittingly facilitate the recruitment of their best competitors; their very form condemns them to the fate of encouraging an ecological succession that spells their ultimate doom. For example, juvenile mussels prefer to settle onto finely branched, turf-forming algae and ultimately overgrow them. The seeds of surfgrasses have specialized projections that allow them to catch on finely branched red algal turfs or coralline algae, and they too ultimately overgrow the seaweeds that snagged them from the water. Sometimes, though, seaweeds win the battle; for example, the holdfast of a fast-growing kelp can grow right over smaller barnacles, mussels, and other seaweeds. Later, the shady, moist environment below the canopy formed by larger species, such as kelps or bladderwracks, can also become the perfect habitat for understory species that could not survive in that location otherwise. Encrusting forms frequently conquer space by overgrowing other organisms; along zones of intense competition between different encrusting species (including seaweeds, sponges, and other colonial invertebrates), one can typically discern who is currently winning the battle by seeing who is growing on top of whom. On wave-exposed rocky shores, the predictable sequence of algal succession starts with so-called ephemeral or early successional species, including colonial diatoms and anatomically simple seaweeds that grow as thin sheets, tubes, and filaments. These seaweeds tend to reproduce often; thus their reproductive propagules are readily available to colonize newly opened spaces on the shore. These species tend to be highly palatable to a wide range of consumers, including snails, limpets, chitons, fish, and crustaceans, as well as being weak competitors for space. These characteristics predispose them to being ephemeral in nature. Eventually more complex algal forms or sessile invertebrates, more resistant to being eaten or outcompeted for resources, come to dominate the shore.

Strategies to Avoid Consumption

Seaweeds use a wide array of strategies to escape from being eaten by snails, limpets, chitons, sea urchins, fish, crustaceans, and other rocky shore grazers. Some of these strategies may be simply the result of evolutionary luck, while others have evolved in response to strong selective pressure. Calcification in algae, for example, is unlikely to have evolved in direct response to consumer pressure, because it appears to be a by-product of photosynthesis, but it can confer a distinct advantage to seaweeds facing the rasping bites of snails or other consumers by making the alga both tougher and less nutritious. Coralline algal "barrens" are often all that are left behind after grazers ravage a tidepool or a kelp bed. Anatomically simple forms such as filaments and thin blades tend to be most vulnerable to consumers. Crusts, blades with toughened outer cuticles, and more leathery forms are typically less favored because of their inherent toughness.

Chemical defenses are the weapon of choice for some seaweeds to deter would-be consumers, enabling surprisingly delicate and otherwise potentially very palatable species to survive even in the midst of their would-be consumers. There are several classes of chemicals, usually produced as secondary metabolites, that have antiherbivore properties, including terpenoids, phlorotannins, DMS

(dimethyl sulfide), and even sulfuric acid. Brown seaweeds in the genus *Desmerestia* avoid being eaten by sea urchins by virtue of the sulfuric acid they contain; the five-part calcium carbonate jaw that urchins use to mow down kelp beds is readily dissolved by these acid "brooms." Toxic chemicals may be produced constitutively (present all the time), although in some the chemicals may need to be activated: nonactive forms are stored and are activated only via a chemical reaction that occurs when cells are damaged. Some delicate and seemingly palatable species of green and red algae in the genera *Ulva, Enteromorpha,* and *Polysiphonia* produce DMS as an activated defense and are actively avoided by urchins. Other seaweeds produce their defensive compounds only in response to stimuli related to grazing damage; this is called an inducible defense. The advantage is that the alga need to expend energy to produce the noxious chemical only after being subjected to a real and imminent threat. The bladderwrack *Fucus garderi* is known to respond in this way when grazed upon by small snails. Other organisms often take advantage of the "free" noxious chemicals found in seaweeds to defend themselves from predators. For example, decorator crabs found along the Gulf and Southern Atlantic coasts of North America preferentially dress themselves in the noxious seaweed *Dictoyota mensualis* to avoid being eaten by omnivorous fish.

Having two distinct anatomical forms within a life cycle is another way that seaweeds obtain refuge from their consumers. For example, the brown algae *Petalonia fascia* and *Scytosiphon lomentaria* alternate between a crustose form and a more delicate form consisting of a thin blade or tubular upright. In the presence of grazers the crustrose phase persists while the upright phase is absent or scarce. Since grazers of these species tend to be most abundant and active in the summer months, the upright phase of the algal life history was thought to be responding to seasonal cues and would never appear in the summer. However, removing grazers during the summer months induces the uprights to appear and thrive. The crusts are much more resistant to grazing but grow more slowly than the upright forms that are preferred by most grazers. Being a shape-shifter may confer a distinct evolutionary advantage on an alga, especially when one shape is more likely to survive a regularly occurring onslaught of hungry herbivores.

Biodiversity

The diversity of seaweeds found along rocky shores is tremendous. However, the distribution of seaweed diversity among the various rocky shorelines of the world's oceans is not uniform, nor is it entirely random either. Diversity patterns emerge at different spatial scales and have different underlying causes. Seaweed diversity tends to be greatest lower on the shore and in more wave-exposed locations. Physical disturbances by waves tend to increase diversity by removing competitively dominant species and making room for subordinate species to gain a foothold, albeit a temporary one. Lower on the shore the abundance of herbivores may have similar effects, especially if they prefer to eat species that would otherwise dominate the shoreline. The interplay of physical and biological factors along environmental gradients of tidal emersion and wave exposure along a stretch of seashore are complex, but this interplay plays a major role in determining the number and kinds of seaweeds encountered in tidepools.

Less well understood is the cause of the variation seen in the numbers of seaweed species found in tropical vs. temperate vs. polar shorelines. In terrestrial ecosystems and for many marine species, a greater number of species can be found in the tropics than at higher latitudes. Puzzlingly, seaweeds are a major exception to this pattern. Along the Atlantic coasts of Europe and North America, seaweed diversity increases as expected as one travels from the poles to the tropics, but along the Pacific coasts of North and South America seaweed diversity declines in tropical latitudes. Understanding why seaweeds exhibit these seemingly anomalous latitudinal diversity patterns along some shorelines may help us to understand the underlying causes of large-scale diversity patterns in nature more generally.

CONSERVATION

Seaweeds, despite their "weedy" name, are not invulnerable to human impacts. We collect algae from wild populations to extract their specialized biological molecules to use in a variety of industrial processes, from making textiles to thickening milkshakes to herbal remedies. Many coastal cultures have culinary traditions that include seaweeds, and modern interest in seaweeds as a healthy and tasty part of our human diet is increasing. Asian cultures use seaweed regularly in a wide range of dishes, including the increasingly ubiquitous miso soup, sushi, and seaweed salad that we see on restaurant menus. We wrap our seafood and rice in nori (made from the red algae in the genus *Porphyra*) to create sushi rolls. In Chile, the honeycombed *Durvilleae antarctica* is collected on the coast and brought inland to be sold in city markets and made into an *ensalada de ulte*. A few species are cultivated for harvesting, including *Porphyra, Gelidium, Gracilaria, Laminaria,* and *Undaria*, but a large proportion of what we use is collected from wild populations. Careful planning is needed to maintain sustainable levels of exploitation of these natural populations as interest in seaweed products increases.

Seaweeds can create problems as invasive species when they are inadvertently introduced by people to locations outside their natural range. Perhaps the most infamous example is the case of the green alga *Caulerpa taxifolia* in the Mediterranean. Where it was introduced, it carpeted vast expanses of sea floor, excluded native seaweeds, and reduced the availability of suitable habitat and forage for native animals. Recent introductions have occurred in other locations, including a harbor in San Diego County, California, most likely via the aquarium trade, as this alga and its close relatives are commonly used in saltwater aquaria. *Sargassum muticum,* a brown alga native to northwestern Pacific shores, can now be found along the shores of the northeastern Pacific, the Atlantic, and the Baltic Sea. It probably arrived to these shores as packaging used in transporting oysters for aquaculture. The kelp *Undaria pinnatifida,* a native of Asia, where it is extensively cultivated for food and sold as *wakame,* is another species that is now invasive in Europe, New Zealand, Argentina, and California. A subspecies of the green alga *Codium fragile,* native to Asia, is possibly the most invasive alga known. It was transported outside its range, most likely via the aquaculture trade or on the hulls or in the ballast water of ships. This species is now found along the shores of Africa, Australia, Europe, and North and South America and has become an economic problem, because it fouls shellfish beds, especially in the northwest Atlantic. Eradication of invasive species is generally fraught with enormous and often insurmountable challenges and costs; preventing inadvertent introductions through education and regulation is clearly the best hope to prevent future invasions.

Development and nutrient pollution can strongly alter the abundance and distribution of seaweeds. For example, in the Baltic, eutrophication has promoted the growth of phytoplankton and other smaller, more ephemeral seaweeds. Because these smaller photosynthetic organisms live suspended in the water (phytoplankton) or often grow as epiphytes (on top of other organisms), they can intercept the light before it reaches the larger seaweeds growing on the rocks below. The result is that depth distribution and abundance of *Fucus,* an important habitat-forming species in the Baltic, is now sharply reduced.

Global changes that impact the ocean environment such as global warming have the potential to impact seaweed communities as well. Ocean productivity is linked to atmospheric phenomena such as El Niños; these and other climatic fluctuations influence the amount of nutrients that are brought up from the ocean's depth and fuel the growth of all photosynthetic marine organisms. Changes in the abundance of phytoplankton or nutrients in the ocean water that overlay seaweed-covered shores are likely to significantly alter the ecological character of these shorelines.

The slippery and slimy seaweeds that can sometimes make walking along the shore at low tide a challenging affair are a fascinating yet often overlooked component of healthy, functioning marine ecosystems. We know that many species depend upon seaweeds for food and habitat. We know their moist cover provides a desirable refuge for many intertidal inhabitants during low tide. Despite the enormous productivity and diversity of seaweeds, we still know surprisingly little about the contribution that seaweeds make to coastal ecosystem production and functioning. The enormous diversity in form, life history, ecology, and evolutionary history of seaweeds make them both challenging and very rewarding to study.

SEE ALSO THE FOLLOWING ARTICLES

Biodiversity, Significance of / Food Uses, Modern / Introduced Species / Zonation

FURTHER READING

Falkowski, P.G., and J.A. Raven. 1997. *Aquatic photosynthesis.* Malden, MA: Blackwell Scientific.

Graham, L.E., and L.W. Wilcox. 2000. *Algae.* Upper Saddle River, NJ: Prentice Hall.

Lobban, C.S., and P.J. Harrison. 1997. *Seaweed physiology and ecology.* Cambridge, UK: Cambridge University Press.

Thomas, D.M. 2002. *Seaweeds.* Washington, DC: Smithsonian Institution Press.

Van Den Hoek, C., D.G. Mann, and H.M. Jahns. 1995. *Algae: an introduction to phycology.* Cambridge, UK: Cambridge University Press.

ALGAE, CALCIFIED

ROBERT S. STENECK

University of Maine

PATRICK T. MARTONE

Stanford University

Calcified algae are a unique subset of marine seaweeds that incorporate calcium carbonate—essentially, limestone—into their thalli. As a group, they are quite diverse, because calcification has evolved independently in the three major divisions of macroalgae: Rhodophyta, Chlorophyta, and Ochrophyta (red, green, and brown algae, respectively). Today, calcified algae dominate biotic communities in many subtidal, intertidal, and tidepool environments worldwide. They build reefs, contribute to sediments, and are home to numerous plants and animals. In sum, their unique attributes enable them to play key ecological and geological roles in marine ecosystems.

THE DIVERSITY AND IMPORTANCE OF CALCIFIED ALGAE

Among the different groups of calcified algae, the mode and extent of calcification varies widely. For example, the brown alga *Padina* develops a thin white calcified coating, whereas the green alga *Acetabularia* and the red alga *Liagora* incorporate low concentrations of calcium carbonate directly into their flexible thalli. Other, more rigid but still flexible, calcified algae include the red alga *Galaxaura* and the green algae *Udotea* and *Penicillus*. The most heavily calcified algae include the green alga *Halimeda* and the so-called "coralline" red algae, which impregnate every cell wall with calcium carbonate and can even resemble stony corals.

These heavily calcified algae are most abundant and, arguably, most important. They exist in two fundamentally different forms. One has calcified segments separated by flexible joints called genicula. These "articulated" calcified algae include the green alga *Halimeda* (Fig. 1A) and red algal genera such as *Amphiroa, Corallina, Calliarthron,* and *Bossiella* (Fig. 1B). The other growth form lacks genicula and typically grows as an encrusting pink patch on hard substrata (Figs. 1C–E) but can also be found unattached in sediment habitats (Fig. 1F). Algae with this nongeniculate morphology, or "crustose" coralline red algae, include common genera such as *Lithothamnion, Clathromorphum, Lithophyllum,* and *Phymatolithon.* These two heavily calcified growth forms are ubiquitous, growing throughout the euphotic zone from the Arctic to the Antarctic, from temperate regions to the tropics. Most calcified algae grow on hard substrata, but some live on other plants or anchor in shallow marine sediments.

Among calcified algae, crustose coralline red algae and the articulated green *Halimeda* stand out as ecologically and geologically important. *Halimeda* is abundant in coral reef environments and, by some estimates, generates most of the total calcium carbonate there. Accumulated *Halimeda* segments produce the sand on most of the world's coral reefs, lagoons, and beaches. Crustose coralline red algae are perhaps the most abundant organism (plant or animal) to occupy hard substrata within the world's

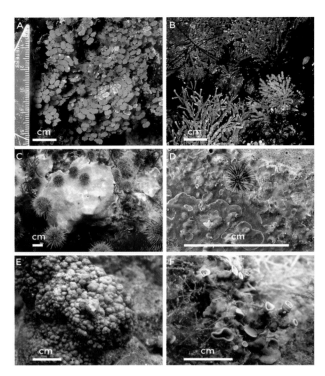

FIGURE 1 Morphological and ecological variety of calcified algae. Bars = 1 cm. (A) Articulated green alga (*Halimeda* from Honduras in the Caribbean) growing on a coral reef. (B) Articulated red algae (*Calliarthron* and *Bossiella* from Monterey Bay, California) growing in low intertidal zone tidepools. (C) A massive crustose coralline (*Clathromorphum* from Alaska's Aleutians islands) that can grow to be a meter thick despite grazing sea urchins. (D) A thin crustose coralline (*Titanoderma* from Bonaire, Netherlands Antilles in the Caribbean) with a newly settled reef-building coral. (E) A slightly branched crustose coralline (*Hydrolithon* from Honduras in the Caribbean). (F) A branched or rolled maerl morphology (*Lithothamnion* from Guatemala in the Caribbean) growing unattached on sediment. All photographs by R. Steneck.

marine photic zone. They are abundant on most rocky subtidal areas, intertidal shores, and tide pools, lending a pink hue to these environments. They have been collected in the Bahamas by a submarine at more than 260 m water depth, making them the deepest attached benthic algae in the world. Growth bands from living corallines in Alaska's Aleutian Islands reveal they can live to be at least 700 years and probably well over 1000 years, making them the longest-lived algae and one of the longest-lived marine organisms known. Vast regions of Japan; the North Pacific, North Atlantic, Tropical Indopacific, Mediterranean, and Caribbean; and Antarctica have 50–100% of shallow hard substrata covered by crustose coralline algae. In tropical wave-exposed areas, encrusting coralline algae create one of the most ecologically important noncoral constructed reefs, called "algal ridges." These specialized reefs have been constructed by coralline algae accumulating over thousands of years, resulting in a calcium carbonate reef over 10 meters thick. Algal ridges create their own rocky intertidal zone by projecting as much as one to two meters above mean low water.

ORIGINS: PHYLETIC AND MORPHOLOGICAL EVOLUTION

Calcareous red algae have left behind a fossil record that extends back to Precambrian times (over 600 million years ago). Thus, their evolutionary history exceeds that of most extant organisms. During the Paleozoic era (570–245 million years ago), a variety of calcified articulated and crustose taxa evolved and went extinct. About 360 million years ago, calcified crusts with modern anatomical characters, similar to those of present-day coralline red algae, evolved. They formed moundlike reefs during the Carboniferous period, well before dinosaurs first evolved. These early nongeniculate calcareous algae were morphologically simple, resembling a potato chip.

Nongeniculate corallines today exhibit considerable variation in form. Some species encrust hard substrata as a thin or meshlike crust only 20 μm thick (Fig. 1D), whereas other species can grow to nearly a meter in thickness (Fig. 1C). These corallines can grow over hard substrata as entirely adherent or as leafy crusts resembling their ancestral potato-chip-like cousins. Some develop protuberances or nonflexible branches that give the group further morphological variety. Nongeniculate morphologies range from subtle, low-profile bumps (e.g., Fig. 1E) to conspicuous spindly shapes, to an elaborate matrix of interconnected branches forming hemispherical a heads half-meter in diameter. Branches themselves can be simple pinnacles, ornamented with secondary protuberances, bladelike or even rolled leafy forms creating tubular branches (Fig. 1F). However, the biomechanical constraint of being heavily calcified and inflexible prevents branches from extending too far into fast-moving currents, and most are relatively diminutive—well less than 1 cm in height.

Erect fronds of articulated algae overcome the biomechanical limitations of calcification by producing flexible genicula between calcified segments. This jointed architecture evolved convergently among the green algae, such as *Halimeda* (Fig. 1A) and the coralline red algae (Fig. 1B). Furthermore, paleontological, developmental, and phylogenetic analyses suggest that, even among coralline red algae, articulated fronds evolved from crusts at least three separate times in evolutionary history. Such a striking example of convergent evolution suggests that the development of flexible joints is an adaptive solution for attaining vertical height under the constraints of calcification. Articulated fronds can be diminutive, such as those produced by the coralline *Yamadaea*, which consist of only a couple segments that extend a few millimeters above the basal crust, or rather large, as in the green *Halimeda* (Fig. 1A) or the coralline *Calliarthron* (Fig. 1B) whose fronds can grow more than 20 cm long. Segment morphologies range from cylindrical to flattened to highly ornate, with a single frond often spanning the entire morphological range from base to apex.

Most calcified algae are firmly attached to hard substrata, but some corallines grow unattached as large balls, called rhodoliths, or as smaller branched forms, called maerl, which look like (and are the size of) a child's "jacks." Often these growth forms develop by breaking free from the substratum and growing unattached on the sea floor while rolling periodically from water motion. Rhodoliths can range from golf ball to basketball size, but the majority are baseball sized. The more diminutive maerls produce biogenic sediments resembling calcified tumbleweeds. Both rhodolith and maerl deposits are scattered globally. A so-called "coral" beach on the northwest coast of Scotland is actually composed of maerl fragments of a free-branching coralline alga.

ECOLOGY: DOMINANCE, HABITATS, AND INTERACTIONS

Calcified algae are ubiquitous biogeographically and span the depth gradient from the intertidal zone to the deepest reaches of the benthic euphotic zone. It is their remarkable abundance and absence under certain conditions that tells us much about the ecology of this group.

Although calcified algae can dominate tidepools and shallow subtidal habitats, they are less common or absent from middle to upper intertidal regions because they are susceptible to drying out (desiccation). Unlike noncalcified "fleshy" seaweeds, whose thalli may be as much as 80–90% water, some articulated corallines, such as *Calliarthron,* are less than 30% water and dry out very quickly. Densely branched calcified turfs, such as some *Corallina* species, resist desiccation by retaining water within their fronds during low tide, like paint between the bristles of a paintbrush. As a result, this growth form can live much higher in the intertidal zones than other coralline algae can. The coincidence of low tides and high temperatures can cause emergent corallines to bleach, often killing part, but not necessarily all, of their thalli.

The abundance and ecological success of crustose coralline algae is at first glance enigmatic. As a group, they are among the slowest-growing algae in benthic marine photic zone and are frequently overgrown and outcompeted for space by fleshy algae. Yet they thrive under conditions of frequent and intense physical and biological disturbance. Calcareous algae are the only forms found where sand and small rocks scour the sea floor, and they thrive where herbivory is most intense. Coralline algae often dominate wave-exposed habitats, such as the shallow seaward face of algal ridges, where water velocities dislodge other organisms or prevent them from persisting.

Shallow-water crustose corallines also appear to have a symbiotic dependence on intense and frequent grazing by herbivores, such as limpets, sea urchins, and parrotfish. For example, the long spined sea urchin *Diadema antillarum* was extremely abundant and the dominant herbivore throughout the Caribbean until 1983 and 1984, when it suffered a mass mortality throughout the Caribbean, during which over 90% of the population died. As a result, fleshy algae rapidly increased in abundance, and the entire coralline community declined 80–100% at monitored sites on the coral reefs of St. Croix and Jamaica.

The relationship between scraping herbivores and corallines is a long-standing one. Paleontological studies have found that as sea urchins and grazing parrotfish evolved and became abundant in shallow seas, so too did the crustose corallines diversify and come to dominate many coastal zones. In the western North Atlantic, a particularly tight algal–herbivore association evolved. The species *Clathromorphum circumscriptum* (closely related to the species depicted in Fig. 1C) is commonly associated with the limpet *Tectura testudinalis*. Limpet grazing benefits coralline thalli by removing epiphytes that would otherwise shade or smother the calcified thalli, while the regions where the alga grows (its meristem) and reproduces (its conceptacles) remain safely beneath the heavily grazed thallus surface. This coralline is also a nursery habitat for limpets, and if they are removed, the *Clathromorphum* dies. There are many examples of similarly tight associations. For example, the chiton *Choneplax lata* bores into and eats the tropical coralline *Porolithon pachydermum*, keeping the alga free of epiphytes; the tropical crab *Mithrax sculptus* lives between and is protected by the calcified branches of the crustose coralline *Neogoniolithon strictum* and performs a similar cleaning duty. Thus, many plant–herbivore interactions between crustose corallines and their grazers are more of a positive facilitation than the negative interaction most commonly seen between fleshy algae and their herbivores.

Unlike their crustose counterparts, erect calcified fronds are more often fodder for hungry herbivores, although their calcium carbonate makes them far less preferable than fleshy seaweeds. Being more susceptible to herbivores, articulated calcified algae use a wider variety of herbivore deterrents. For example, the articulated green alga *Halimeda* fortifies its thallus with chemical herbivore deterrents as it produces new (uncalcified and relatively vulnerable) segments at night, when herbivory is low or absent. By the next day the segments have hardened, and the combination of calcium carbonate and chemical deterrents is sufficient to minimize subsequent herbivore damage. Besides being relatively inedible, the calcium carbonate in algae can deter grazing fish that use acid to digest their algal prey. Thus, even the lightly calcified algae, such as the brown alga *Padina,* may receive some protection from herbivores.

Many organisms have evolved to live in or on calcified algae as an alternative hard substratum. For instance, certain species of bryozoans, hydroids, fleshy seaweeds, and calcified crusts grow directly on articulated coralline fronds in tidepools. Amphipods and polychaetes wrap themselves in calcified articulated fronds, and worms burrow into calcified crusts. Abalone, sea stars, limpets, chitons, and reef corals often recruit to coralline algae. Reef corals, in particular, chemically detect, metamorphose, and settle on (Fig. 1D) or near coralline algae, which presumably indicate favorable coral habitat. Similarly, many corallines grow as epiphytes on sea grasses and other algae and on the shells of snails, mussels, and barnacles. Occasionally the thickness of coralline accumulations far exceeds that of the shell of the organism on which it is growing.

Several species of calcareous green algae including *Halimeda* and less heavily calcified forms of *Udotea* and *Penicillus* are uniquely capable of colonizing sandy substrates in tropical lagoons. These "rhizophytic" algae anchor themselves in the sediment with hairlike cells called rhizoids. Rhizoids can extract nutrients from substrates as do the roots of higher plants. These rhizophytic algae add organic matter and stabilize sediments, thereby facilitating the colonization and succession of sea grasses. Similarly, articulated coralline algae facilitate the succession of California's intertidal seagrass *Phyllospadix* by literally snagging its seeds in their fronds. This allows the angiosperm to take root and come to dominate patches in the intertidal zone.

By incorporating calcium carbonate into their thalli, this phyletically and morphologically diverse group of algae became unique and ecologically important. As a group, calcified algae occupy more biogeographic zones and live in a wider range of habitats than most other algae or other primary producers in the sea. They coexist with deep grazing herbivores, live on many types of substrata, and provide critical habitat for numerous other organisms.

Increasing metabolic costs associated with global climate change may offset the advantages of calcification. Carbon dioxide in our atmosphere combines with water to form carbonic acid, which rains back to Earth. As this greenhouse gas builds up in our atmosphere, the world's oceans are becoming increasingly acidic, which in turn increases the energy required to calcify. Recent increases in disease may indicate that oceans are becoming a more stressful environment for calcified algae.

SEE ALSO THE FOLLOWING ARTICLES

Algal Biogeography / Algal Crusts / Corals / Fossil Tidepools

FURTHER READING

Abbott, I. A., and G. J. Hollenberg. 1976. *Marine algae of California.* Stanford, CA: Stanford University Press.

Borowitzka, M. A. 1982. Morphological and cytological aspects of algal calcification. *International Review of Cytology* 74: 127–162.

Hillis-Colinvaux, L. 1980. Ecology and taxonomy of *Halimeda:* primary producer of coral reefs. *Advances in Marine Biology* 17: 1–327.

Johansen, H. W. 1981. *Coralline algae, a first synthesis.* Boca Raton, FL: CRC Press.

Kozloff, E. 1993. *Seashore life of the northern Pacific coast.* Seattle: University of Washington Press.

Steneck, R. S. 1986. The ecology of coralline algal crusts: convergent patterns and adaptive strategies. Annual Review of Ecological Systems 7: 273–303.

Steneck, R. S., and V. Testa. 1997. Are calcareous algae important to reefs today or in the past? Symposium summary. *Proceedings of the 8th International Coral Reef Symposium* 1: 685–688.

Taxonomic information for genera: www.algaebase.org/

van den Hoek, C., D. G. Mann, and H. M. Jahns. 1995. *Algae: an introduction to phycology.* Cambridge, UK: Cambridge University Press.

Woelkerling, W. J. 1988. *The coralline red algae: an analysis of genera and subfamilies of nongeniculate Corallinaceae.* New York: Oxford University Press.

ALGAL BIOGEOGRAPHY

JAMES A. COYER

University of Groningen, The Netherlands

Algal or seaweed biogeography is a discipline that addresses two essential questions: What are the patterns of species distributions, and how are these distributions influenced by the history of the earth? The former question addresses species ranges as a function of contemporary interactions with their environment (ecological), whereas the latter attempts to reconstruct the origin, dispersal, and extinction of species (historical). In recent decades, a third question has become increasingly important: How are biogeographic patterns influenced by human activities?

WHAT IS BIOGEOGRAPHY?

Biogeography is a hypothesis-driven observational science, rather than an experimental science, because its spatial and temporal scales are far too large for experimental manipulation. Sometimes, however, nature provides an experiment in the form of large-scale natural disturbances such as the emergence of new volcanic islands, widespread destruction by a volcanic eruption, or ice ages. Experiments of a different sort are presented when human activities lead to the elimination or introduction of species and the connection or disruption of previously separated or connected populations. Both natural and human-induced disturbances offer unique opportunities for biogeographers to understand the whys, hows, and why nots of species distributions.

Three stimuli have led to the emergence of biogeography as a widely respected science since the 1950s and 1960s. First, biogeography evolved from a descriptive science coupled with traditional taxonomy, to a concept-orientated discipline concerned with developing and testing biogeographic hypotheses using mathematical, statistical, and modeling analyses. Second, new technologies offered biogeographers powerful tools for data collection. For example, computers permit rapid compilation, manipulation, and analysis of enormous quantities of many different data sets, including species distributions and climate, oceanographic, and geological data. Orbiting satellites, surface and submerged buoys and vessels, and ground-based systems automatically collect real-time measurements of the earth's environment. Finally, numerous molecular methods and corresponding analyses have been developed over the past decade that enable biogeographers to develop increasingly accurate spatial and temporal reconstructions of a species' history and migrations. As a result of molecular methodology, a new specialization of biogeography has emerged, termed phylogeography, which tracks, in space and time, "lineages" within a species based on the similarity of various DNA sequences. The molecular methods also have been important for studying those species (seaweeds and many invertebrates) that do not leave a fossil record because of their body structures (e.g., soft) or habitats (e.g., wave-swept rocky substrates).

The third factor contributing to the growth of biogeography as a discipline is that human societies have recognized the critical importance of understanding and managing their impact on the earth. Interest in the late 1960s and early 1970s focused on relatively small-scale problems, but beginning in the 1980s, it became apparent that humans were radically transforming the earth on a regional and global scale. Consequently, there is a critical need for biogeographers to measure and predict these changes as well as to find ways to manage or minimize the effects.

BIOGEOGRAPHIC DATA

The Temperature Contribution

What kinds of data are collected by marine biogeographers, and how are the data analyzed? For seaweeds inhabiting wave-swept rocky intertidal shores, biogeographers compile maps that correlate species presence/absence (preferably biomass or areal cover) with key variables such as air and water temperature and degree of physical isolation. Both air and water temperatures are important factors influencing the distribution of intertidal organisms, because they profoundly affect survivorship, growth, and reproduction.

Long-term temperature data are used to generate surface seawater isotherms (boundaries of the same average water temperature averaged over many years), which are combined with both experimentally determined temperature limitations and distributional maps of the species to identify species ranges. By examining maps for several species, biogeographers can identify areas that are characterized by drastic changes in species composition over a relatively small distance.

The Geological Contribution

The notion that continents moved to their present position from a different configuration in the past was presented to a very skeptical scientific community in 1912 as the theory of continental drift. The theory was not widely accepted until the 1960s, however, when the theory of plate tectonics provided a realistic mechanism for the movements of continents. Briefly, the theory of continental drift states that, about 230 million years ago, the land masses observed today formed a single land mass called Pangaea (Fig. 1A), which, about 225–200 million years ago, split into Laurasia (the modern Northern Hemisphere continents) and Gondwanaland (modern Antarctica, South America, Africa, Australia, and India) (Fig. 1B). Gondwanaland began to separate

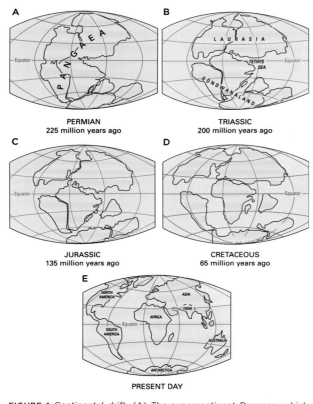

FIGURE 1 Continental drift. (A) The supercontinent Pangaea, which began to fragment approximately 225 million years ago during the Permian Period of geological history. (B–D) Configurations of land masses during subsequent periods; (E) present land masses. Because continental drift is a dynamic process, the configuration of continents will change continuously, albeit in a predictable manner, as subsequent millennia accumulate. Map from Kious and Tilling (1996).

into its component continents about 200 million years ago (Fig. 1C), whereas North America separated from the European/Asian land mass about 100 million years ago (Fig. 1D). The present distribution of land masses (Fig. 1E) was apparent about 10 million years ago, but the process of land mass expansion and contraction is a dynamic process. For example, the Pacific Ocean is becoming smaller each year and will disappear in about 200 million years, whereas the Atlantic Ocean is getting larger.

Vicariance and dispersal are two mechanisms proposed to explain biogeographic patterns. According to the vicariance mechanism, species distributions reflect the splitting of an ancestral population by the development of environmental barriers (physical obstructions or perhaps drastic differences in air or water temperature) that prevented interbreeding between the separated subpopulations. These separated subpopulations

evolved into separate species, all of which are closely related because of their common ancestor. The dispersal mechanism maintains that ancestral populations dispersed from a center of origin to found new species and that it is not necessary to invoke physical barriers as a separating mechanism. Although the relative contribution of each mechanism has been heavily debated, both can explain the biogeographic distributions that we now observe.

GENERAL PATTERNS OF SEAWEED DISTRIBUTION

Several general patterns of seaweed distribution are of interest to marine biogeographers. One pattern is that tropical regions exhibit a high degree of similarity in species composition on a worldwide basis. The most accepted explanation is that, for the last 100–150 million years there always has been a warm-water belt encircling the globe, thereby ensuring connection between widespread populations and maintenance of gene flow. The belt has been closed by land bridges only relatively recently: once when the connection between the Mediterranean Sea and Indian Ocean (the ancient Tethys Sea) was closed about 17 million years ago, and again when the connection between North and South America (Isthmus of Panama) was formed 3–4 million years ago.

Another pattern is that cold-water species of the Northern Hemisphere differ fundamentally from those in the Southern Hemisphere, because increased cooling throughout the Tertiary Period (65 to 2 million years ago) and subsequent glaciations of the Quaternary Period (2 million years ago to present) allowed species to evolve in each hemisphere. Although some cold-water species might have moved between the hemispheres during the glacial periods, when the tropical warm-water belt became narrower and cooler, no exchange could occur during the interglacial period, when the tropical region became much broader and warmer. Consequently, the species in each hemisphere evolved separately, leading to the fundamental differences observed today.

A different pattern is that highly related or identical seaweed species occur over vast distances in the Southern Hemisphere. Because the West Wind Drift (a relatively fast west-to-east current) is not blocked by barrier-forming continents in the Southern Ocean, dispersal of marine organisms is unimpeded, thereby promoting a high degree of connectivity (and therefore gene flow) between widely separated populations. For example, in the ecologically important giant kelp *Macrocystis pyrifera,* sequence analysis of a nuclear gene region has shown that there are far fewer differences between populations in Chile, South Africa, Australia, and New Zealand than there are between populations in the much smaller region from Baja California, Mexico, to Central California, United States.

A final pattern of interest is that species of seaweeds and invertebrates in the North Pacific are vastly different and more diverse than those in the North Atlantic, because the North Pacific is much older than the North Atlantic. Nevertheless, there has been exchange of species between the two areas at various times during the last 3 million years, and the advent of molecular methodology has stimulated recent research on the North Pacific–North Atlantic exchange.

THE NORTH PACIFIC–NORTH ATLANTIC EXCHANGE

The much higher percentage of land mass in the Northern Hemisphere relative to the Southern Hemisphere, coupled with sea level changes associated with the Ice Ages, increases the complexity of oceanic circulation and severely restricts exchange between the North Pacific and North Atlantic (compared to the ease of exchange between the South Pacific and South Atlantic). Five phases over the past 265 million years are central to understanding the North Pacific–North Atlantic exchanges with respect to marine organisms, including intertidal seaweeds. During Phase 1 (some 230–65 million years ago), the Pacific Ocean was wide open to the cold-water Arctic Ocean, whereas the North Atlantic was not formed until about 165 million years ago, when North America and Europe began to drift apart. In Phase 2 (about 65–50 million years ago), the North Pacific was separated from the Arctic Ocean by the Bering Land Bridge, creating a vast cool-water area that undoubtedly facilitated evolution of cool-water species. The isolated and cold-water Arctic Ocean may have served as a refugium for species that later colonized the North Atlantic. During Phase 3 (50–6 million years ago), the North Atlantic was connected to the Arctic Ocean, whereas the North Pacific was still separated by the Bering Land Bridge. Two major cooling events occurred in this phase, one about 40 million years ago and another from 10 to 5 million years ago; the latter leading to global glaciation and worldwide fluctuations in sea level. In Phase 4 (6–3 million years ago) the Bering Strait opened, connecting the North Pacific and North Atlantic via the Arctic Ocean; additionally, the Isthmus

of Panama formed, separating the tropical Pacific and Atlantic Oceans. These two events led to a profound change in the major surface currents of all oceans and a further decline in average sea surface temperatures. Examination of over 100 fossil mollusc species has revealed that before the onset of Pleistocene glaciation (1.6 million to 11,000 years ago), at least eight times more species migrated from the North Pacific to the North Atlantic than vice versa.

QUATERNARY GLACIATIONS AND THE LAST GLACIAL MAXIMUM

For many biogeographers, Phase 5 is the most exciting in terms of the North Pacific–North Atlantic exchange. During this phase (2 million to 18,000 yrs ago), and especially in the last 130,000 years, the Bering Land Bridge repeatedly emerged and submerged as a result of numerous glacial (ice-building, sea level decline) and interglacial (ice reduction, sea level rise) periods, leading to closures and openings of the North Pacific–North Atlantic connection. The advance and retreat of vast ice sheets over much of North America and Northern Europe profoundly shaped the distributions of virtually all marine and terrestrial species currently found in both areas. During cold glacial periods, the diminution of the warm-water tropical belt allowed the dispersal of Arctic and Antarctic species across the equatorial area, which then became isolated during the following interglacial. Additionally, lower sea levels (130 m lower than present) during the last glacial maximum (LGM) (20,000–18,000 years ago) increased the potential for dispersal of intertidal species. As sea levels rose when the ice receded, broad species distributions were fractured, leading to increased isolation of populations and subsequent speciation.

How have the last glacial–interglacial periods influenced the distributions of rocky shore intertidal seaweeds in the Northern Hemisphere? A traditional analysis of the number of seaweed species (ignoring those species introduced by humans) reveals a significant gradient: highest numbers in Northern Europe, intermediate in Iceland, and fewest in the Canadian Maritime Provinces and New England. The gradient also is apparent in invertebrate species and is most easily explained by post-LGM dispersal from Europe. During the LGM the North American ice sheet extended as far south as Long Island, New York. Because no rocky substrates exist south of Long Island, essentially all species associated with rocky substrates in the northwest Atlantic became extinct. Iceland also was covered by ice, leading to a similar extinction of rocky shore species. However, ice sheets in Northern Europe covered only Scandinavia and Great Britain, leaving uncovered to the south vast areas of rocky shores, which served as glacial refugia. As the ice sheets retreated in Northern Europe from 20,000 to 7000 years ago during the current interglacial period, species from the refugia in Southern Europe colonized first Northern Europe, then Iceland, and finally, the Canadian Maritimes/New England area.

Many species of seaweeds occur both in the North Pacific and North Atlantic Oceans, but in most cases, identity of the species is based on morphological similarity. It is very possible that species common to both oceans may in fact be distinct species despite similar morphologies. With the development of molecular methods, seaweed biogeographers have been able to detect the presence of sibling or cryptic species, as well as to examine colonization and migration pathways in the North Pacific and North Atlantic.

Fucus (Heterokontopyta, Phaeopyceae, Fucales)

Rockweeds are a dominant component of the biomass associated with rocky shores throughout the North Pacific and North Atlantic Oceans and are important facilitator species, because they provide shelter and food for virtually all intertidal invertebrates. The genus *Fucus* is one group of rockweeds that has been extensively studied in the last decade. Molecular work has determined that the Fucaceae, the family containing the genus *Fucus* as well as several other genera, probably evolved and diversified in southern Australia. An ancestral *Fucus* species migrated across the equatorial region to the Northern Pacific from 35 to 7 million years ago during a period of global cooling. After the Bering Strait opened (6–3 million years ago), one or two ancestral *Fucus* species migrated to the North Atlantic via the Arctic Ocean and radiated into a wide range of habitats. Currently, far more species of *Fucus* are present in the North Atlantic, where they are found from the high intertidal to shallow subtidal of rocky open coast regions as well as in calm, brackish-water environments, than in the North Pacific.

During the LGM, the advancing ice and cold water forced most European species of *Fucus* to glacial refugia in the south (including the Mediterranean). As the ice retreated during the early portion of the current interglacial period, populations began a rapid colonization of areas to the north, but at the same time the former refugial areas became too warm for the cold-water *Fucus* species. Today, *Fucus* species are restricted to only three areas of the former southern refugium: the Canary Islands, the northwestern shores of Morocco (an area of localized upwelling and cool water), and the northern Adriatic Sea. The latter population is a glacial relict, surviving from a formerly Mediterranean-wide distribution. It clearly has been geographically isolated from all other *Fucus* species and populations for some 18,000 years, and just as clearly, it has diverged as a new species during this time. The Canary Island and Moroccan populations also are highly isolated and are likely to be nascent species.

Some areas, previously thought to be devoid of intertidal seaweeds because of ice coverage and low temperature, recently have been revealed to be glacial refugia. For example, several different types of mitochondrial DNA sequences are present in populations of *F. serratus* in the southwestern Ireland and western English Channel areas, implying persistence of populations during the LGM. As the ice receded, individuals from southwestern Ireland possessing one type of mitochondrial DNA rapidly colonized the northern half of Great Britain, the entire western coast of Norway up to the Russian border, and the Kattegat strait between Denmark and Sweden. Thus colonization was due to a single group of highly related colonizers (=colonization sweep) rather than to several groups of unrelated individuals.

Chondrus (Rhodopyta, Gigartinales)

Chondrus is another ecologically and economically important intertidal seaweed genus. Several species are present in the western North Pacific, but only one, *C. crispus,* is found in the North Atlantic, where it often forms large and nearly monospecific stands in the low intertidal zone. Recent molecular analysis has revealed that, like *Fucus,* the genus *Chondrus* evolved in the North Pacific and an ancestral species migrated into the North Atlantic via the Arctic Ocean within the past 5 million years. But unlike *Fucus,* the ancestor did not radiate into numerous species or habitats in the North Atlantic.

Palmaria palmata (Rhodophyta, Palmariales)

Molecular analyses of the red alga *Palmaria palmata,* known as dulse and commonly found on Atlantic rocky shores, also are enhancing our understanding of post-LGM biogeography. Many different types of chloroplast genes are found in *Palmaria* populations in the western English Channel area, but only one occurs in the areas further north. The pattern is essentially identical to that displayed by *F. serratus,* thereby strengthening considerably the argument for glacial refugia along the southwestern Irish coast and in the western English Channel, followed by a rapid colonization sweep to the north by a single, related group of individuals.

Acrosiphonia arcta (Chlorophyta, Acrosiphoniales)

Molecular methods have traced the biogeographic track of the intertidal green alga *Acrosiphonia arcta:* from Chile, to the eastern North Pacific Ocean from California to Alaska, to Greenland and Iceland, and finally, to northern Europe. Although future molecular analyses may well reveal the presence of several cryptic species within this wide distribution, it is clear that the genus has migrated from the Southern Hemisphere to the Northern Hemisphere and from the North Pacific to the North Atlantic. Again, as for all species of seaweeds, only molecular methods can reveal the history, because fossil records are highly unlikely.

HUMAN INFLUENCES ON BIOGEOGRAPHY

In recent years it has become increasingly clear that humans have profoundly impacted the distributions of marine species through local extinction, inadvertent species introductions, and modifications of species habitats. Shipping and the aquaculture and fishery industries are the primary means by which marine species have been introduced to new areas. Because marine species have been transported from one area to another by maritime activities for millennia, several marine biogeographers believe that the long history of species introductions almost certainly has corrupted our present worldwide views of coastal biogeography.

Although biogeographers can sometimes recognize that a marine species has been introduced within recent times, they often are unable to identify the source of the introduced species. Molecular techniques and analyses, however, have provided clues. For example, the rockweed *F. serratus* was documented as an introduction to Nova Scotia in the late 1800s, and molecular analysis has revealed Northern Ireland as the likely source. Similar molecular methodology combined with historical records has determined that the introduction of *F. serratus* into Iceland originated from the Oslo Fjord region of Norway sometime between ca. 1750 to 1900, whereas its introduction to the Faeroes originated from an Icelandic population between 1982 and 1997.

Another example of a human-mediated introduction is the green seaweed *Codium fragile* ssp. *tomen-*

tosoides. Originally a native of the cool waters off northern Japan, it has expanded its range throughout the world, first appearing on the Dutch coast around 1900 (probably associated with oysters imported for aquaculture), then rapidly spreading along European coastlines and the Mediterranean Sea. It was reported in Long Island Sound (United States) in 1957 and expanded north into the Gulf of St. Lawrence by 1996. In the Pacific Ocean it was introduced to San Francisco Bay (United States) and New Zealand in the 1970s, Australia in the mid-1990s, and Chile in 2001. Molecular analysis has revealed at least two separate introductions: the North Atlantic and Chilean populations originated from a set of one to three regions of Japan, whereas the Mediterranean populations stemmed from a different set of one to three regions of Japan.

The field of biogeography currently is experiencing a vigorous period of productivity, largely because of advances in technology and analysis. Consequently, biogeographic studies of seaweeds have increased markedly in the last decade, and the growth is expected to be exponential in the years to come. Regardless of whether or not a species distribution has been influenced by human maritime activities, source-sink relationships can be identified, and pathways of dispersal will be determined. Biogeographers also have the ability to distinguish the presence of cryptic species from species groups with near-identical morphology. Consequently, our present views of biogeographic patterns, global biodiversity, and speciation undoubtedly will expand.

SEE ALSO THE FOLLOWING ARTICLES

Biodiversity, Global Patterns of / Dispersal / Introduced Species

FURTHER READING

Adey, W. H., and R. S. Steneck. 2001. Thermogeography over time creates biogeographic regions: a temperature/space/time-integrated model and an abundance weighted test for benthic marine algae. *Journal of Phycology* 37: 677–698.

Avise, J. C. 2000. *Phylogeography: the history and formation of species.* Cambridge, MA: Harvard University Press.

Brown, J. H., and M. V. Lomolino. 1998. *Biogeography,* 2nd ed. Sunderland, MA: Sinauer.

Dunton, K. 1992. Arctic biogeography: the paradox of the marine benthic fauna. *Trends in Ecology and Evolution* 7: 183–189.

Hoek, C. V. D., and A. M. Breeman. 1990. Seaweed biogeography of the North Atlantic: where are we now? in *Evolutionary biogeography of the marine algae of the North Atlantic.* D. J. Garbary and G. R. South, eds. Berlin: Springer-Verlag, 55–86.

Ingólfsson, A. 1992. The origin of the rocky shore fauna of Iceland and the Canadian Maritimes. *Journal of Biogeography* 19: 705–712.

Kious, W. J., and R. I. Tilling. 1996. *This dynamic earth: the story of plate tectonics.* Washington, DC: U.S. Geological Survey. http://pubs.usgs.gov/publications/text/dynamic.pdf.

Lindstrom, S. C. 2001. The Bering Strait connection: dispersal and speciation in boreal macroalgae. *Journal of Biogeography* 28: 243–51.

Lüning, K. 1990. *Seaweeds: their environment, biogeography and ecophysiology.* New York: Wiley.

Quammen, D. 1996. *The song of the dodo.* New York: Scribner.

ALGAL BLOOMS

BARBARA B. PRÉZELIN

University of California, Santa Barbara

When the biomass of a group of phytoplankton increases excessively at a spot in the ocean, the microalgae are said to bloom. Algal blooms are highly colored and can be seen by the naked eye and monitored by satellites. There are many types of algal blooms, differing in how they form, the kinds of phytoplankton they contain, and the effects they have on secondary production.

GENERALITIES

Algal blooms begin with the rapid growth of phytoplankton in response to an ample supply of light for effective photosynthesis and an oversupply of plant nutrients for cell growth and division. Physical processes are very important in establishing these bloom conditions. High algal biomass (Fig. 1) accumulates only if primary production of phytoplankton particulate organic carbon

FIGURE 1 A spring bloom of phytoplankton dominated by diatoms (mostly *Skeletonema costatum*) in Tronheim Fjord, Norway. Photograph courtesy of G. Johnsen.

(POC) is faster than the sum of all the ways phytoplankton POC can be lost from the environment (respiration, excretion, grazing, sinking, advection). Most algal blooms support major increases in secondary production (biomass accumulated at higher trophic levels). As blooms end and phytoplankton die, they release large amounts of dissolved organic carbon (DOC) that support microbial communities and the regeneration of plant nutrients. The high metabolic activity of the microbes consumes oxygen gas from the surrounding water. If the oxygen is not resupplied by water motion, bloom waters approach anoxia (no oxygen) and animals are suffocated. In aging blooms, dead and dying phytoplankton tend to aggregate and sink as large particles (marine snow). Algal blooms export a large amount of particulate organic carbon to benthic communities and can be a significant food source for detritivores. At any time, blooms may be physically swept away by currents as coastal circulation changes.

SPRING DIATOM BLOOMS

The most common and most predictable algal blooms occur in the spring months at temperate latitudes and during spring–summer months at polar latitudes (Fig. 2). They last from weeks to months. These spring blooms are highly productive, accounting for more than 10% of the annual primary production by phytoplankton at some locations. Grazing of phytoplankton appears especially efficient in typical spring blooms compared to blooms associated with upwelling.

The seasonality of spring blooms help explain how they are formed (Fig. 3). As winter and spring winds subside, nutrients are abundant and increased sunlight penetrates and starts warming the upper water column. Warm water is less dense than cold water and thus remains at the surface (thermal stratification), forming an upper mixed layer (UML) in which phytoplankton and nutrients are confined. Increased warming causes shallowing of the mixed layer depth (MLD), bringing the thermocline closer to the surface, and the seasonal increase in sunlight causes deepening of the photic zone. For effective photosynthesis and plant growth, the stable density stratification of water must occur near enough to the surface so that phytoplankton are not mixed out of the photic zone.

To understand and predict the timing of the start of spring blooms, the concept of a critical depth was introduced by Harald Sverdrup in the 1950s. He defined critical depth as the depth above which total phytoplankton photosynthesis in the water column is equal to total

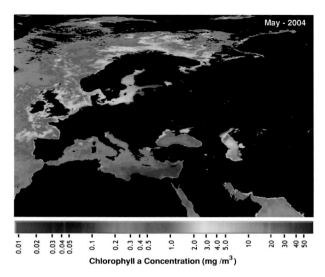

FIGURE 2 Remote image of chlorophyll *a* distribution of the marine waters of northwest Europe, averaged for May 2005. Note the many distinct phytoplankton blooms and their large-scale coverage. Image courtesy of Modis Land Response Team, NASA/GSFC.

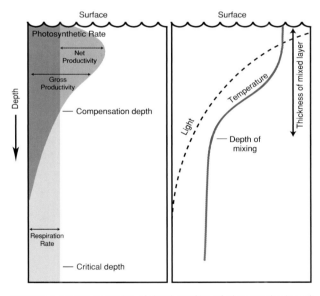

FIGURE 3 (Right) Examples of depth profiles of photosynthesis and respiration rates of phytoplankton and how calculations would be made of gross and net primary productivity. (Left) Examples of depth profiles of light availability and water temperature. See text for discussion of critical depth and upper mixed layer.

phytoplankton respiration in the water column. Subtracting rates of phytoplankton respiration from rates of photosynthesis gives the value of net primary production (NPP) (Fig. 3), a measure of the growth potential of phytoplankton. Net primary productivity occurs only above the critical depth. If the thermocline is deeper than the critical depth, phytoplankton will be mixed below the critical depth and total algal biomass will not increase in the upper mixed layer. Bloom conditions are met when the MLD becomes shallower (by further warming) or the euphotic zone becomes deeper (by increasing sunlight). Early field observations of spring blooms successfully used the concept of critical depth to explain the differences in timing of spring blooms observed onshore and offshore and at different latitudes.

Today, a modification of Sverdrup's critical depth is used to predict the timing of spring blooms. It is now recognized that when DOC is abundant, some phytoplankton take DOC into the cells and thereby supplement photosynthesis with heterotrophy. Their combined rates determine the growth rate of phytoplankton. Furthermore, the rate of loss of phytoplankton organic carbon by respiration has been expanded to include losses due to the release of dissolved organic carbon from phytoplankton by excretion. Excretion can be significant as phytoplankton transition from exponential to stationary phases of growth. Other loss processes of phytoplankton particulate inorganic carbon include grazing, sinking, and advection out of the area.

Spring blooms have recognizable stages. In the first stage, phytoplankton growth accelerates and phytoplankton particulate inorganic carbon increases. With abundant silicate in the nutrient-rich waters, diatoms grow faster than other phytoplankton groups and come to dominate the microalgae assemblage in the bloom. These diatoms are a high-quality food source for herbivores. Excretion is small, and the little dissolved organic carbon released is highly susceptible to photochemical and microbial degradation. The second stage is characterized by declining nutrient supply and algal growth rates. Grazing is sufficient to lower total phytoplankton particulate inorganic carbon. A smaller fraction of net primary productivity is going into food chains, because excretion increases greatly and the released dissolved organic carbon pools promote carbon cycling in the microbial loop. The third stage is marked by nutrient stress, senescence, sinking, and induction of survival strategies by bloom phytoplankton. Excretion rates are low unless weakened phytoplankton cells are lysed by viruses. Detritivores begin feeding on abundant aggregates of dead and dying cells (marine snow) that sink and settle on benthic communities. Spring blooms can briefly cycle back to earlier stages if short periods of wind mixing disrupt the thermocline and additional plant nutrients are mixed upward. These episodic inputs of new nutrients can stimulate new production and a brief resurgence of rapidly growing diatoms, thereby prolonging the lifetime of spring blooms from weeks to months.

OTHER TYPES OF PHYSICALLY FORCED ALGAL BLOOMS

Fall turnover occurs when air temperatures begin to cool surface water temperatures. The colder water sinks and disrupts the thermocline. Wind-driven mixing accelerates fall turnover. Nutrients from the depth are mixed upward and cause an algal bloom that promotes secondary production. Due to the loss of the UML and the seasonal decline in sunlight, turnover blooms tend to be smaller in algal biomass and do not last as long as spring blooms.

Wind-driven upwelling along coastlines brings large volumes of cold and nutrient-rich waters from depth to the surface. Large near-surface blooms of diatoms result and are evident when matching satellite images of chlorophyll (Chl) biomass with images of cold surface temperatures. Bathymetric upwelling (or upward mixing) over the continental shelf brings deep waters closer to the surface, but not necessarily all the way. Blooms of diatoms can form at depth, and their presence may be missed by satellite images. The circulation patterns associated with both types of upwelling include transfer of large amounts of phytoplankton particulate inorganic carbon to downstream benthic communities. When upwelling ceases, diatom growth continues until nutrients are depleted and then they sink in large quantities. There are major wind-driven upwelling sites along the coasts of California, Peru, and parts of Africa. The high NPP in these waters supports major fisheries. Bathymetric upwelling occurs where bottom currents or subsurface intrusions of offshore currents upwell in response to interactions with the bottom. Areas of significant bathymetric upwelling associated with high NPP are found along coasts of the southeastern United States, eastern Australia, and the west Antarctic Peninsula. There are many other locations where both types of upwelling have not been studied.

Ice-covered shores can form algal blooms in the thin layer of fresher surface waters that is generated as the sea ice melts in the spring. Ice edge blooms occur before spring blooms in polar fjords and bays. Ice edge blooms can persist between wind-mixing events, as long as there is a melting marginal ice zone (MIZ) associated with the retreating sea ice and sufficient light to promote photosynthesis. The MIZ of the Antarctic and Arctic cover enormous areas of the coastal ocean. The algal biomass is a major food supply in early spring for zooplankton and herbivorous fish and larvae. MIZ phytoplankton biomass settles to the bottom and supports rich benthic communities.

There are many other algal blooms associated with other distinct physical and chemical environments in the ocean. For instance, cryptophyte blooms form in glacial ice melts that cover ice-free environments along the west Antarctic Peninsula in spring and summer. Water circulation generates mesoscale and submesoscale eddies that bring plant nutrients into the photic zone and promote episodic increases in NPP. The outflow of rivers and streams into the ocean promote algal blooms in waters generally heavy in organic and mineral particulate matter. Rain can stimulate miniblooms (often picophytoplankton). The highly adaptable, opportunistic, and diverse nature of marine phytoplankton communities contribute to their success in creating bloom environments.

HARMFUL ALGAL BLOOMS

Harmful algal blooms (HABs) disrupt existing ecosystem balance in many different ways. They may reduce or eliminate sunlight reaching submerged plants, lead to anoxic conditions that suffocate animal and plants, or sink en masse to bury benthic communities. There are HABs that produce chemical toxins that sicken and kill animals, including humans. The frequency of toxic HABs appears to be increasing and may be linked to eutrophication in the coastal zone.

Red tides are blooms of dinoflagellates that are rusty red in coloration. They occur worldwide. Not all red tides are harmful, but there are times when anoxia develops at the end of a nontoxic red tide and results in massive fish kills. Anoxia is associated with red tides occurring in the summer, when they occur in warmer surface water that has been stratified a long time without disruption. Red tides arise at other times of year and may lead to anoxia if stratified waters become eutrophic as a result of runoff from agricultural land and sewage outfalls. These runoff events rarely promote diatom blooms, because there is usually little silicate in polluted waters.

Certain dinoflagellate species produce a group of potent neurotoxins called saxotoxins. They are among the most deadly algal toxins and are the cause of paralytic shellfish poisoning (PSP). Saxotoxins build up harmlessly in grazing herbivores but then are passed up the food chain to poison carnivores, including humans. PSP symptoms include numbness, shaky and unsteady movements, incoherence, and, in extreme cases, respiratory paralysis and death. For this reason, harvesting of local shellfish (e.g., mussels, clams, abalone) is prohibited at times when toxic dinoflagellates are present. Saxotoxins are also volatile and may become airborne in regions of high wave action. If inhaled, they irritate the respiratory systems of mammals, including humans.

Another type of toxin, domoic acid (DA), is produced by several diatom species in the genus *Pseudo-nitzschia*. DA is a neuroexcitory amino acid that causes amnesic shellfish poisoning (ASP) in humans. Symptoms include gastrointestinal disorders (vomiting, diarrhea, abdominal pain) and neurological problems (confusion, memory loss, disorientation, seizure, coma). *Pseudo-nitizschia australis* has been increasingly present in California coastal waters following its first discovery in Monterey Bay in 1991, where it caused massive death of seabirds feeding on anchovies that had consumed this toxic diatom. Abundant *P. australis* appears in upwelled waters along the central California coast, and major blooms have been associated with the spring upwelling season. The toxin builds up in several types of fish and shellfish that, when consumed, can account for the death of thousand of marine animals each year, including seals, sea lions, sea otters, whales, dolphins, and seabirds. The magnitude of present toxigenic effects of DA on marine animals and seabirds alter the ecological balance of marine ecosystems along the coast of California.

Brown tides are not known to produce a toxin, but they are harmful to nearby animals and plants in other ways. Blooms are caused by a very small pelagophyte, *Aureococcus anophagefferens*. Brown tides have been observed annually in some coastal embayments and estuaries of the northeastern United States and along the eastern coastline of the United States since 1985. The toxic pelagophyte prefers lower salinity and shallower waters where nutrient loading of sediments is high. Brown tides do not end in anoxia, because they occur where tidal flushing and constant mixing of shallow waters keep their habitat well aerated. Brown tides, however, inhibit grazing by filter feeders (perhaps because of their small size) and block sunlight from reaching submerged plant communities. Brown tides coincide with hard-clam spawning and have been linked to a reduction in larval recruitment. Brown tides have also been linked to the death of large areas of seagrass.

Cyanobacterial blooms are also widespread in the sea. Examples of toxic coastal blooms of cyanobacteria can be found in the Baltic Sea (Fig. 2) and along parts of the Australian and New Zealand coastlines during summer months. They begin with resting cells of *Nodularia spumigena* or *Aphanizomenon flos-aquae* that have over-wintered on the sediment and in the water column, respectively. The cue for *N. spumigena* resting cells to reenter their vegetative state is a drop in water salinity brought on by springtime river runoff and ice melt. *N. spumigena* reform resting cells when coastal waters become saltier in late summer. *N. spumigena* produces the toxin nodularin, which inhibits protein phosphatases, which are important regulatory enzymes. Ingestion of water containing *N. spumigena* kills livestock and dogs. Fish and crabs try to avoid these blooms. Humans are prohibited from eating shellfish during times of blooms. *A. flos-aquae* does not produce nodularin but makes other compounds that affect zooplankton and may be responsible for the death of some fish or crustaceans. Both types of cyanobacteria blooms are associated with decreased grazing of some herbivores and inhibition of some microbial activities.

SEE ALSO THE FOLLOWING ARTICLES

Diseases of Marine Animals / Microbes / Nutrients / Phytoplankton / Temperature Change / Upwelling

FURTHER READING

Anderson, D. M., P. M. Gilbert, and J. M. Burkholder. 2002. Harmful algal blooms and eutrophication: nutrient sources, composition and consequences. *Estuaries* 25: 704–726.
Livingston, R. J., and L. J. Livingston. 2002. *Trophic organization in coastal systems*. Boca Raton, FL: CRC Press.
Mann, K. H. 2000. *Ecology of coastal waters*. Oxford, UK: Blackwell.
Miller, C. B. 2003. *Biological oceanography*. Malden, MA: Blackwell.
Reynolds, C. S. 2006. *Ecology of phytoplankton*. Cambridge, UK: Cambridge University Press.

ALGAL COLOR

CELIA M. SMITH
University of Hawaii

KEVIN S. BEACH
University of Tampa

The colors of intertidal algae range across the visible spectrum from red to green to black. These colors reflect several things: the pigments for photosynthesis among three evolutionary groups of marine plants, the fundamental organization of their bodies, and their physiological ecology including how plants respond to solar radiation (energy budgets) and stress responses to high light that are characteristic of many tidal settings.

EVOLUTIONARY DIVERSITY

Among the plants that emerge with low tides along coastlines worldwide, there are three major groupings of eukaryotic algae, green algae (Chlorophyta), macroscopic brown algae (class Phaeophyceae, Ochrophyta), and red algae (Rhodophyta). All of these possess chlorophyll *a* (Chl *a*) as the pigment that initiates the first steps of photosynthesis, yet only a few are green like the Chl *a* molecule. The wide range of colors arise from different light-harvesting pigments that funnel light energy to Chl *a* in the process of photosynthesis (Table 1). These light-harvesting pigments are commonly the basis of photosynthesis under low ambient light levels. Ironically, most tidal algae run photosynthesis as shade plants, as if they were growing in the dark understory of kelp forests. This feature puts most tidal algae at risk in sunny intertidal settings.

TABLE 1

Major Photosynthetic and Photoprotective Pigments for Macroalgae by Evolutionary Division

Algal division and common names	Major pigments
Chlorophyta—green algae	chlorophyll *a, b*, siphonein, siphonoxanthin, β-carotene
Ochrophyta (Phaeophyceae)— brown algae	chlorophyll *a*, chlorophyll c_1, c_2, fucoxanthin, β-carotene
Rhodophyta—red algae	chlorophyll *a*, phycobilins (phycocyanin, phycoerythrin, allophycocyanin), α- and β-carotene

Intertidal and shallow subtidal regions in tropical coasts, for instance, are subjected to high levels of light that may even inhibit photosynthesis, leading to photobleaching and death of tissues. Life in higher tidal elevations increases stresses of UV radiation and leads to excessively high light levels. Tidal algae respond with complex and surprising physiological adjustments in these harsh habitats.

A surprising way in which many tidal algae can change color over the course of a low tide is that they dry out in air. Water loss begins instantly as the surface of a plant comes into contact with dry air, regardless of their evolutionary status. Algal tissues even shrink as water loss extends over long tidal exposures. As they dry, many become black (an optical effect that results from decreases in cell volume as tissue water is lost) and almost brittle. To tell which group you are looking at, make sure the plants are well soaked in seawater!

FUNCTIONAL ANATOMY

Besides an evolutionary grouping, algae can also be grouped by their body plan. For the larger forms, photosynthetic tissues are positioned as the outermost tier of cells interacting with the air and marine environments. Typically for the vast majority of large algae, including the phaeophyte *Sargassum*, tissues are composed of exceptionally small, densely packed pigmented cells (cortex) that grade to colorless larger cells (medulla). This tissue type is called a true parenchyma. A parenchymatous organization is the typical anatomy many other brown algae such as *Chnoospora*, a tropical wave zone alga, and temperate intertidal fucoids, for example, *Fucus* or *Pelvetiopsis*.

Several morphological oddities can be found among tidal algae, such as filaments of green algae such as *Cladophora* that, in aggregate, form turfs of stiff uniseriate filaments. A profoundly different body plan comes with siphonous green algae, for example, *Codium*. Siphonous green algae have a startling simplicity—their entire form is a single cell! Upon microscopic examination, their chloroplasts are clearly olive green. Yet in the field, many of these plants appear to be nearly black. The difference in apparent color (black) and actual pigments (green) arises from the density of chloroplasts being so great that they absorb almost all incoming light.

HOW TO INTERPRET ALGAL COLORS IN LIGHT OF ENERGY BUDGETS AND STRESS BIOLOGY

Most marine algae can be identified by their color—green, brown, or red, especially if you are in cool-water intertidal areas. The colors we see are reflected light that these plants are *not* using for photosynthesis. This may sound counterintuitive until you consider that the colors absorbed by these plants are used up as they are converted to excitation energy for photochemistry—called absorptance. Other colors of light have two major routes—those colors not absorbed can pass through the plant (transmittance) or they can be bounced back out of the plant (reflectance). Reflectance actually varies substantially among the three groups of algae as well as between forms that are calcified and not calcified. A selection of these is shown in Fig. 1 to highlight known differences. Absorptance and transmittance can vary as well.

Because these differences in color among tidal and shallow subtidal algae are so easily seen, early researchers in photosynthesis turned to marine algae to quantify how much light different algae absorb for photosynthesis. Drs. Francis Haxo and Lawrence Blinks, working at Hopkins Marine Station in California, pioneered comparative photosynthesis research with their study of absorbance and action spectra for photosynthesis of a selection of brown, green, and red algae; they were the first to quantify the link between high levels of absorbance and

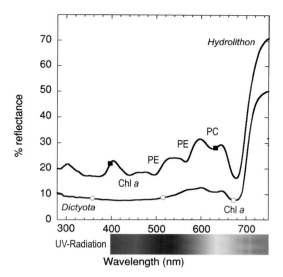

FIGURE 1 *In vivo* reflectance spectra for intertidal and shallow sublittoral species show that algae possess a wide range of reflectance properties. Spectra acquired from *Hydrolithon onkodes*, a coralline alga (■), and *Dictyota acutiloba*, a foliose brown alga (○), show differences in the color and quantity of light reflected from their surfaces. Pigments responsible for reflectance minima indicated by abbreviations. Pigments include chlorophyll *a* (Chl *a*), phycoerythrin (PE), and phycocyanin (PC).

high levels of photosynthesis for algae from kelp forest regions. Recent studies extend this work by examining the overall pattern of light energy as a budget for tidal and shallow reef algae by documenting comparative absorptance, reflectance, and transmittance among algae from high light environments.

Species of tropical tidal algae can absorb from about two-thirds to more than 90% of the light that hits their surfaces. For many algae, reflectance can range from less than 10% to about one-fourth of the light that hits their surfaces (Fig. 1). Transmittance through algal tissues ranges from zero for thick crusts to about one-fourth of the surface light for thin sheets like *Ulva*. Emerging from these studies is a new appreciation of the extent to which plant form regulates pigment densities, photosynthesis, and limits damage from high light. Many of these factors are likely to lead to ecological success.

To illustrate how variable absorbance is across the spectrum, quick inspection of Fig. 2 shows the ability of living *Ulva* to absorb some colors of white light varies from high energy blue photons to lower energy red photons by the action of Chl *a* and *b* (Table 1). This kind of spectrum is called an *in vivo* spectrum because the tissue is alive at the time of the measurement. Two areas of broad absorbance emerge—the blue (400 to 500 nm) and red (600 to 700 nm) regions. Not surprising for this green alga is the low level of absorbance for the central green (500 to 600 nm)

region, because this alga reflects green back to your eye, as was shown by Haxo and Blinks in 1950. This alga is only two cells thick, transmits roughly 27%, and reflects about 7%, with the principal proportion of reflectance in the green spectrum region. Other green algae show differing levels of absorbance across the white light spectrum (e.g., *Caulerpa racemosa;* □, Fig. 2), but they all exhibit the green-window effect—little absorbance in the 500 to 600 nm range.

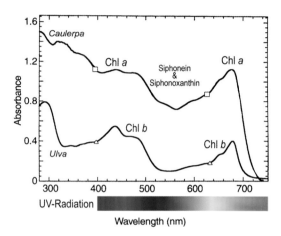

FIGURE 2 *In vivo* absorbance spectra for intertidal and shallow reef green algae *Caulerpa racemosa* (□) and *Ulva fasciata* (△) show broad absorbances in blue and red colors but substantially less absorbance in green wavelengths by these plants. Pigments responsible for absorbance maxima indicated by abbreviations. Pigments include chlorophyll *a* (Chl *a*), chlorophyll *b* (Chl *b*), siphonein, and siphonoxanthin. (Modified from Beach *et al.* 1997.)

The *in vivo* absorbance spectrum for a brown alga such as *Sargassum* or *Chnoospora* documents some of the differences that you can see by eye (Fig. 3)—the brown color of the plant body tells you that different wavelengths—brown in color—are reflected by brown algae (Fig. 2). These visual changes are the result of increased absorbance of green wavelengths by an unusual pigment, fucoxanthin, that functions in photosynthesis of brown algae.

Absorbance properties of pigments in living organisms are generally additive. As a plant makes more pigments for photosynthesis or photoprotection, the result is to add the absorbance of fucoxanthin on top of the broad blue and narrower red absorbances of the two chlorophyll pigments, Chl *a* and *c*, also found in brown algae (Table 1). Fucoxanthin absorbance adds a new spectral region to the colors absorbed by chlorophylls, extending abilities of brown algae such as *Sargassum* to harvest this abundant color in the ocean as well as coastal environments.

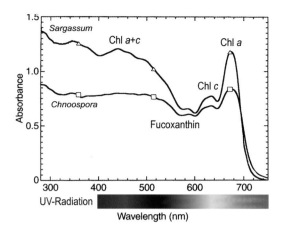

FIGURE 3 *In vivo* absorbance spectra for intertidal and shallow reef brown algae *Chnoospora minima* (□) and *Sargassum polyphyllum* (△) show broad absorbances that include some of the green spectral region. *Sargassum* can dominate tidal benches in tropical settings and absorb significant amounts of light intensity as well as color via its parenchymatous anatomy and pigments. Pigments responsible for absorbance maxima indicated by abbreviations. Pigments include chlorophyll *a* (Chl *a*), chlorophyll *c* (Chl *c*), and fucoxanthin. (Modified from Beach *et al.* 1997.)

In vivo absorbance spectra for red algae such as *Acanthophora, Hypnea*—or many deep red plants and kelp forest understory genera—appear exceptionally complex, with many minor peaks and no green window whatsoever. Overall absorbance can be high, well above 1.0 absorbance units (Fig. 4). There are a series of absorbance maxima in the range from 490 to 650 nm, absorbing

nearly completely in the green region of the spectrum. This green absorbance arises from two additional pigments not found in green or brown algae: phycoerythrin (PE) and phycocyanin (PC) (Table 1). When phycobilin absorbance is added to the absorbance of Chl *a, in vivo* measurements of absorbance are high and effective across nearly the entire visible spectrum. With PE and PC deployed, red algae are very effective at photosynthesis in low levels of ambient light, from deep kelp forest understories to exceptional depths of more than 260 m in clear tropical waters.

PUTTING ALGAL COLOR IN PERSPECTIVE

Dramatic color changes take place when algae grow for extended periods in high-light environments. In many tropical intertidal regions, algae can dominate with a rich assemblage of species and biomass despite the high light stress. The physiological challenges of too much sun, however, are clear. To the human eye, tropical tidal regions are dominated by bright green or golden brown plants. Upon closer inspection, most of these species are actually red algae. They have sunny canopies and shaded bases: such as the golden canopy and purple understory of *Ahnfeltiopsis concinna* (Figs. 5, 6). Our research tool, *in vivo* absorbance spectra, documents remarkable physiological adjustments by the canopies of red algae to life in high-light habitats.

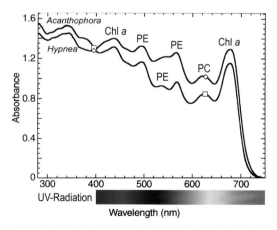

FIGURE 4 *In vivo* absorbance spectra for intertidal and shallow reef red algae *Acanthophora spicifera* (□) and *Hynpea musciformis* (◇). These two red algae can dominate tidal benches or shallow reef regions, adjusting their pigments in response to light levels. Pigments responsible for absorbance maxima indicated by abbreviations. Pigments include chlorophyll *a* (Chl *a*), phycoerythrin (PE), and phycocyanin (PC). Modified from Beach *et al.* (1997).

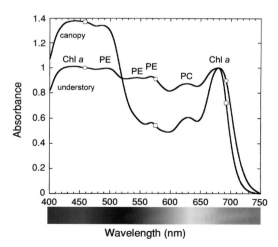

FIGURE 5 *In vivo* absorbance spectra for *Ahnfeltiopsis concinna* canopy (□) and understory (○). Among red algae, this alga is remarkable in the extent of color change over a short 10-cm distance from canopy to understory portions of the same plant. Pigments responsible for absorbance maxima indicated by abbreviations. Pigments include chlorophyll *a* (Chl *a*), phycoerythrin (PE), and phycocyanin (PC). Reproduced from Beach and Smith (1996) with permission of the *Journal of Phycology*.

FIGURE 6 *Ahnfeltiopsis concinna*, a common intertidal alga, with canopy (yellow) and understory (red) tissues exposed by the photographer who turned over the upper canopy. Tidal benches with this alga are single-species stands where the canopy shades the basal tissue so substantially that the plant synthesizes a full complement of shade pigments to live in near-darkness. Photograph by K. Beach. Reproduced with permission of the *Journal of Phycology*.

In response to high light, *A. concinna* and other red algae turn golden via significant reductions in all shade pigments (Figs. 5, 6) and synthesis of α-carotene, which functions in photoprotection. Remarkably, in the shade of these photoprotected orange canopy tissues, the bases of the same red algae synthesize the full complement of shade pigments, only a few centimeters away from the canopy (Fig. 6). This adjustment can be seen in species of temperate tidal communities from temperate *Mastocarpus papillatus* to tropical *Ahnfeltiopsis concinna* in which golden upper, sunny canopies shade bases of the same plant that lives in extreme shade. No other plant group has been shown to have this range of adjustment to gradients in light; certainly this is one reason that red algae have long dominated tidal shores worldwide and been a source of fascination for marine plant researchers over the decades.

SEE ALSO THE FOLLOWING ARTICLES

Light, Effects of / Photosynthesis

FURTHER READING

Beach, K. S., H. B. Borgeas, N. J. Nishimura, and C. M. Smith. 1997. *In vivo* absorbance spectra and UV-absorbing compounds in tropical reef macroalgae. *Coral Reefs* 16: 21–28.

Beach, K. S., H. B. Borgeas, and C. M. Smith. 2006. Ecophysiological implications of the measurement of transmittance and reflectance of tropical macroalgae. *Phycologia* 45: 450–457.

Beach, K. S., and C. M. Smith. 1996. Ecophysiology of tropical rhodophytes. I: Microscale acclimation in pigmentation. *Journal of Phycology* 32: 701–710.

Haxo, F. T., and L. R. Blinks. 1950. Photosynthetic action spectra of marine algae. *Journal of General Physiology* 33: 389–422.

Littler, M. M, D. S. Littler, S. M. Blair, and J. N. Norris. 1985. Deepest known plant life discovered on uncharted seamount. *Science* 227: 57–59.

Smith, C. M. 1992. Diversity in intertidal habitats: an assessment of the marine algae of select high islands in the Hawaiian archipelago. *Pacific Science* 46: 466–479.

Smith, C. M., and R. S. Alberte. 1994. Characterization of *in vivo* absorption features of chlorophyte, phaeophyte and rhodophyte algal species. *Marine Biology* 118: 511–521.

Vroom, P. S., and C. M. Smith. 2001. The challenge of siphonous algae. *American Scientist* 89: 524–531.

ALGAL CRUSTS AND LICHENS

MEGAN N. DETHIER

University of Washington

Algal crusts constitute a diverse group of unrelated species that have all adopted a similar encrusting growth form and share a number of ecological characteristics. They are some of the most abundant but most easily overlooked occupants of rocky shores worldwide. They often provide striking coloration to marine habitats, especially the pink and bright red "paint" seen so commonly in pools. While sharing a growth form, algal crusts nonetheless exhibit interesting variation in growth rates, susceptibility to grazers and environmental stresses, and other ecological characteristics.

RECOGNIZING ALGAL CRUSTS

Algal crusts are among the few organisms on a rocky shore that you can step on and not ever notice that you had done so—and that would not be much affected by your having done so. They are constructed like a shag or pile carpet, composed of microscopic filaments of cells, with the filaments either held together loosely (like a rich shag) or actually glued together (like a commercial carpet) (Fig. 1). The whole body, or thallus, adheres tightly to the rock, except that some species have a raised growing edge. The total thickness of this carpet is usually less than a millimeter. Patches of crust grow laterally as expanding disks; although growth is very slow, many disks may end up running together, or growing for a long time, such that rocky surfaces may end up having wall-to-wall crust carpet.

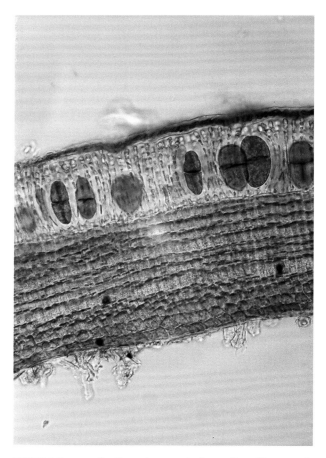

FIGURE 1 Cross section through a reproductive portion of the encrusting red alga *Peyssonnelia pacifica*. Red ovals (tetrasporangia) at the top each produce four spores that can grow into new thalli. Photograph by the author.

This growth form is seen in all the major groups of benthic algae—reds, browns, and greens—and also in cyanobacteria (formerly known as blue-green algae) and in marine lichens. Evolutionarily, this means that the crustose morphology has been converged on by unrelated groups, suggesting that there are some particular advantages

to this growth form. But of course, not all algae are crusts, suggesting that there are disadvantages, too.

FAVORABLE ENVIRONMENTS

Algal crusts can be found in marine communities worldwide, from the tropics to the poles and from the splash zone on the shore down into the subtidal zone. They are probably the most abundant algal form in the ocean, on the basis of the surface area covered (although not on the basis of mass). They are often especially common in the intertidal zone, including tidepools, and in certain other seemingly disparate marine habitats: caves, the very lowest limits of the photic zone, areas scoured by ice, cobbles and boulders that roll around in storms, rock underneath hordes of sea urchins, and surfaces underneath thick canopies of kelp. What do these environments have in common that cause them to be dominated by crusts rather than other marine organisms?

Two environmental factors appear to favor crusts over other algal growth forms, and they are linked logically to the advantages and disadvantages of this morphology. First, crusts are common in areas where algal productivity is limited by stressful environmental conditions, such as where photosynthesis is greatly reduced. These include areas of great desiccation (such as the high intertidal zone) or very low light (such as in caves, the deep photic zone, or underneath algal canopies). Other algal forms, which have higher photosynthetic rates and apparently greater minimum requirements, cannot survive in these marginal habitats. Crusts, however, have very low photosynthetic rates, and this apparently enables them to adopt a slow and steady approach to living in stressful habitats. Crusts are even known to be able to survive being overgrown by other organisms, such as mussel beds, and to persist for years in the apparent absence of light.

The other category of environment dominated by crusts is very different: areas of high disturbance, defined as processes that remove biomass. Disturbance agents can be biological, such as sea urchins or other herbivores, or physical, such as scouring ice or sand, crashing waves, or rolling boulders. The abundance of crusts in such habitats points to one of their main strengths: they are almost impossible to kill. Because they adhere to the rock with almost half their surface area, pulling them off the rock is nearly impossible—in contrast, most algae are attached only by a small holdfast and have a stipe, or stem, that is vulnerable to breakage. Organisms that eat crusts, such as sea urchins and various limpets and other snails, can remove tissue, but they usually only graze the surface, allowing the filaments beneath to regenerate. Even the occasional grazer that bites all the way through to the rock is unlikely to do so over an entire patch of crust; thus, it leaves behind sections of the thallus, which can regrow. In contrast, a group of hungry urchins can easily wipe out a kelp forest or bed of bushy red algae by biting off the stipes or crawling over and consuming all the blades, leaving little opportunity for regeneration.

The ecological tradeoff for this tolerance of stress and disturbance is that crusts grow very slowly. Only the top cells of each filament are in direct contact with light, gases, and nutrients, greatly reducing potential photosynthetic rates. Lateral growth rates have been measured to be less than a millimeter a year for some species, and even the really "speedy" crusts cannot grow faster than several centimeters a year. In addition, because they grow only flat on the surface, they are vulnerable to being shaded out or overgrown by any alga whose growth form takes it up, not out. Crusts can also be overgrown by sessile invertebrates such as sponges or bryozoans. Thus they are relegated to environments either where larger algae are removed by grazers or other forms of disturbance, or where those better competitors cannot survive because conditions are too stressful. Crusts also compete for space with each other; tidepool observers may see areas of contact between patches of crust in which each edge is growing up, trying to grow over the other (Fig. 2). Often the thicker crust wins.

FIGURE 2 A field experiment examining competition among species of crustose corallines; note the white edges growing upwards at the zone of contact. Experiment by R. T. Paine, Tatoosh Island. Photograph by the author.

VARIATION

Within the crustose growth form there is quite a lot of variation, including the tightness of the filaments mentioned in the first section. Some species have filaments running not only vertically but laterally, using

the latter to expand the thallus out over the rock. Some species with "tight" filaments actually have connections between the cells in adjacent filaments, which in some cases may allow the transfer of nutrients among filaments. One of the most striking morphological variants in crusts is that in some species, most commonly red algae in the order Corallinales, the cell walls contain calcium carbonate. This creates the beautiful pink "paint" that characterizes so many marine communities, and it also creates an algal thallus that is very hard relative to uncalcified seaweeds. This hardness, along with the low digestibility and nutrient content that presumably comes with being calcified, means that coralline crusts are often the last organisms eaten by grazers. As a result, areas with large herbivore populations, such as hordes of urchins in the shallow subtidal zone (Fig. 3) or of limpets in tidepools (Fig. 4) are often dominated by coralline crusts—in fact, these are often the only algae present in such circumstances.

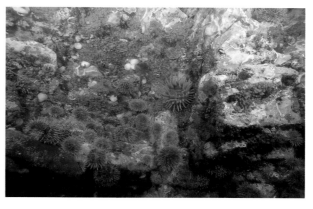

FIGURE 3 An "urchin barren" in the shallow subtidal zone of the Aleutian Islands, Alaska; the only algae present with the abundant urchins are crustose corallines. Photograph by D. Duggins.

FIGURE 4 A small tidepool on the coast of Washington state dominated by limpets and coralline algae. Photograph by the author.

Lichens are another special case, consisting of a symbiotic relationship between photosynthetic cells (of either green algae or cyanobacteria) and fungal hyphae. They may be constructed rather like algal crusts except that the fungal parts are interwoven between the algal filaments, or they may be made of more discrete layers of the different symbionts. The overall form of many intertidal lichens, however, is crustose, with one whole surface tightly adherent to the rock. From a distance, intertidal lichens look like any other crust; that is, they form a solid or patchy band of color on the rock, but the colors are often black, gray, or orange rather than the more common pink, red, or brown of true algal crusts. In addition, they tend to be most common higher on the shore, forming (along with cyanobacteria) the transition zone between truly marine and truly terrestrial habitats.

These maritime lichens, like those that live in other marginal habitats, have a tremendous ability to withstand conditions that would kill other organisms: salinity conditions that change from totally fresh to being covered with salt crystals; long dry periods; and temperatures that change from great heat to great cold. This physiological tolerance is balanced, as for other kinds of crusts, by very slow growth rates and presumably an inability to compete with larger organisms. Most lichens are found only high on the shore, although transplants of some species into areas constantly submerged by the sea show that they can survive and grow there; thus they are probably excluded from lower areas either by competition from other sessile organisms or by herbivores. Limpets, periwinkle snails, and probably other marine grazers will eat some lichens, although experiments with one species whose algal symbiont is a cyanobacterium suggested that it is relatively inedible (many cyanobacteria produce toxic chemicals).

DEFENSES AGAINST HERBIVORES

Because crusts usually grow right on the substrate, they are in some sense more available to a wide range of herbivores than are other algae; chitons, limpets, and sea urchins, for example, seldom climb up erect algae but will readily rasp away at species on the rock surface. As previously discussed, to a great extent crusts can just tolerate this grazing by regenerating from remaining tissue. Others, especially calcified crusts, resist grazing by being very hard. Still others may actually be able to avoid significant losses to herbivores by defending themselves chemically, although this phenomenon has been better studied in noncrustose species. There is evidence that some brown crusts, like many larger brown algae, produce phlorotannin compounds, which deter various grazers. Some crustose corallines avoid grazing by creating a very bumpy surface (e.g., Fig. 2), making it hard for grazers to consume tissue except off the very tips of the bumps.

Interestingly, many algal crusts appear to actually require herbivores for their survival, or at least they fare much better under moderate grazing. As previously discussed, crusts thrive under various forms of disturbance (including grazing) because disturbing forces remove larger algae that would otherwise outcompete them. Crusts are also susceptible to simply being fouled; that is, having other organisms such as microalgae or small barnacles settle on them, reducing their access to light, nutrients, and gases. For many species, having limpets or other grazers periodically clean their surface is advantageous. An extreme example of this is a limpet-coralline crust association in the North Atlantic that appears to be a mutualism; the limpet gains nutrition (and a nice smooth surface to hold onto) by grazing the surface cells of the coralline. The coralline not only gets cleaned but also may gain a reproductive advantage because the limpet grazing opens up the subsurface pits from which the crust releases its gametes. Other crusts, in contrast, appear to spontaneously slough surface cells, thus keeping themselves clean without benefit of grazers.

There are other examples of grazers, including chitons and limpets, that actually specialize on consuming algal crusts, even the seemingly un-nutritious corallines. Several herbivores, including some abalone, have larval stages that settle preferentially on their crustose food, cued in by molecules released from the surface. Thus, crusts are involved in a variety of positive interactions with other marine organisms.

LIFE HISTORIES

Although all crusts look superficially similar except in color, the foregoing discussion shows that they vary in construction, edibility, and growth rate. The method of reproduction of different crusts follows normal patterns for the taxonomic groups they belong to, although for some species (e.g., the common intertidal red crusts *Hildenbrandia* spp.) no sexual reproduction has ever been seen. All crusts release some kind of propagule, either spores or gametes (e.g., Fig. 1), which allow them to colonize new areas. We know little about dispersal ability of different species, although some seem to recruit readily to new areas whereas others have only been seen spreading vegetatively from established patches. Life span also varies hugely among species. Some crusts are annuals, passing through their entire lives in only a few months, whereas others are estimated to be decades old (based on measured growth rates and patch sizes). And when we see many meters of rock covered with crust that looks like one patch and is known to grow less than a millimeter a year, we suspect that some live for centuries!

Several crusts are involved in heteromorphic algal life histories, in which one species passes through two very different-looking forms within one life cycle. A number of red and brown algae alternate between one phase (e.g., the haploid) that is an erect alga and another that is a crust. This strategy nicely illustrates some of the tradeoffs involved in different algal forms. The erect form can grow fairly fast, compete for space, and reproduce quickly, but it is susceptible to herbivores and other forms of disturbance. The crustose form grows extremely slowly and is a poor competitor, but it can persist through many sorts of stresses and disturbances in its habitat. Having a crust as part of a complex life history thus may constitute a way for an alga to hedge its bets against conditions encountered in unpredictable marine environments.

SEE ALSO THE FOLLOWING ARTICLES

Adhesion / Algal Life Cycles / Desiccation Stress / Disturbance / Herbivory

FURTHER READING

Ahmadjian, V., and M. E. Hale, eds. 1973. *The lichens.* New York: Academic Press.
Airoldi, L. 2000. Effects of disturbance, life histories, and overgrowth on coexistence of algal crusts and turfs. *Ecology* 81: 798–814.
Dethier, M. N. 1994. The ecology of intertidal algal crusts: variation within a functional group. *Journal of Experimental Marine Biology and Ecology* 177: 37–71.
Dethier, M. N., and R. S. Steneck. 2001. Growth and persistence of diverse intertidal crusts: survival of the slow in a fast-paced world. *Marine Ecology Progress Series* 223: 89–100.
Littler, M. M., and D. S. Littler. 1980. The evolution of thallus form and survival strategies in benthic marine macroalgae: field and laboratory tests of a functional form model. *American Naturalist* 116: 25–44.
Steneck, R. S. 1982. A limpet-coralline algal association: adaptations and defenses between a selective herbivore and its prey. *Ecology* 63: 507–522.
Steneck, R. S. 1986. The ecology of coralline algal crusts: convergent patterns and adaptive strategies. *Annual Review of Ecology and Systematics* 17: 273–303.
Steneck, R. S., and M. N. Dethier. 1994. A functional group approach to the structure of algal dominated communities. *Oikos* 69: 476–498.
Taylor, R. M. 1982. *Lichens (Ascomycetes) of the intertidal region.* NOAA Technical Report, NMFS Circular 446. Washington, DC: National Marine Fisheries Center.

ALGAL ECONOMICS

LAURIE A. MCCONNICO
San Diego City College

Marine macroalgae or seaweeds are ecologically important components of intertidal and subtidal communities that are also economically valuable. Harvesting and cultivation of seaweeds for food and products from them is a multibillion-dollar-per-year industry. Although the value of algae can be estimated in the currency of market economics, their cultural significance, aesthetic appeal, and ecological function in marine environments is no doubt much more valuable but also more difficult to assess.

HARVESTING AND CULTIVATION

Seaweed harvesting from natural populations has been practiced since the fourth century, while cultivation or farming dates back to the seventeenth century. Today over 7.5 million tons of wet seaweed are harvested and cultivated annually, generating ~US$6 billion in revenue each year. The global demand for seaweeds and products from them has increased since the 1950s to the extent that sustainable harvesting of natural populations is often no longer feasible. Although natural harvesting (Fig. 1) occurs worldwide, much of the seaweed supply is now from aquaculture.

FIGURE 1 Kelp-cutting vessel harvesting giant kelp, *Macrocystis,* off the Pacific coast of California. Alginates extracted from the harvested kelp are used in many foods and cosmetics. Photograph by Michael S. Foster.

Some of the most important commercially cultivated genera are *Porphyra* (nori), *Laminaria* (kombu; Fig. 2), *Undaria* (wakame), *Gracilaria,* and *Eucheuma*. Japan, China, and Korea are the leading producers of the first three genera, *Gracilaria* is cultivated in the Indo-Pacific and Chile, and the majority of the world's *Eucheuma* supply comes from the Philippines. The red alga *Palmaria palmata* is also harvested and sold as dulse, but generally on a smaller economic scale and by hand-harvesting natural populations.

FIGURE 2 Ocean cultivation of the kelp *Laminaria* (kombu) seeded on ropes and out-planted in Japan. Photograph courtesy of Michael S. Foster.

ALGAL USES AND PRODUCTS

Macroalgae are consumed as food and are used in a wide variety of applications from fertilizer to fuel. Of the ~US$6 billion per year generated by the seaweed industry, ~US$5 billion is from edible seaweeds. *Porphyra, Laminaria,* and *Undaria* are some of the most important seaweeds in this market and are used locally and exported for use primarily to add flavor and texture to soups, sushi, and other dishes. Algal extracts, including agar and carrageenan (both from red algae) and alginate (from brown algae), account for ~US$585 million. These hydrocolloids are used as thickening and gelling agents and as stabilizers in a variety of products including ice cream, salad dressings, and shampoo. Agarose, derived from agar, is used to make gels to grow bacteria, a technique critical to the biotechnology industry. The very high-grade agar used is one of the most expensive seaweed products.

To a lesser extent, seaweeds continue to be used as fertilizers and soil amendments. Seaweed extracts are used in the cosmetic and diet industry, promising to enhance beauty and facilitate weight loss. These claims, like many of those made for seaweeds as "healthy foods," are typically not validated with rigorous studies or are not made in a cost/benefit analysis with alternatives. The use of seaweeds as a source of biofuel and in wastewater management has also been investigated, but it has yet to be done on a large scale. Their use as biofilters in integrated aquaculture has, however, been successful. When seaweed culture is combined with fish farming, effects of nutrient discharge from fish are minimized. These advances in aquaculture could have positive economic and ecological effects both in open-ocean culture and on land.

CULTURAL VALUE AND INDIGENOUS USES OF ALGAE

Indigenous people throughout the world have used seaweeds as a staple food source. This is particularly true of Asian diets. Typically seaweeds were harvested by hand from local near-shore environments. *Porphyra,* one of the historically most important commercial species, was harvested by hand as early as the fourth century, and primitive shallow-water culture, using bundles of sticks as a substrate for nori to grow on, began in the seventeenth century. The demand for nori was greater than the supply because cultivation was inhibited by the incomplete understanding of *Porphyra's* life history. Large-scale cultivation techniques were revolutionized in 1949, when Katherine Drew Baker discovered the microscopic phase of *Porphyra's* life history, which enabled controlled "seeding," leading to vastly increased production and the global increase in nori supply (Fig. 3). The discovery of the microscopic stage was so important to the nori industry that a statue of Baker was erected in Tokyo Bay to honor her contribution to cultivation.

Native Hawaiian diets also historically included seaweeds, known as limu (edible seaweed), although Western influence has reduced the frequency with which seaweed meals are currently consumed. Traditionally, women harvested drift seaweeds from the shore to support their families by collecting and preparing seaweeds for market. Prior to the introduction of Christianity, eating and selling seaweeds was essential to the survival of these women, because they were prohibited from consuming many fish species and local fruits. Today 18 species of seaweeds are still harvested from beach drift and used at home or sold in markets. Typically the limu is prepared with spices and eaten with fish or poi (ground taro root), and is thought to contribute vitamins and minerals to diets.

Indigenous people have also collected and harvested seaweeds for a variety of nonfood uses. During the sixteenth

FIGURE 3 Ocean cultivation of *Porphyra* (nori) gametophytes on nets seeded with spores and out-planted in Japan. Photograph courtesy of Michael S. Foster.

century in the British Isles, brown algae were collected from beach wrack and harvested intertidally for use as fertilizers to grow vegetable crops such as potatoes. Indigenous collections and cultivation of seaweeds for fertilizer are still practiced throughout the world. Organic farmers in Negril, Jamaica, regularly harvest seaweeds, and *Eucheuma,* a source of hydrocolloids cultivated in this region, is essential in the production of a native drink believed to be an aphrodisiac (Fig. 4). French peasants in the seventeenth century harvested intertidal rockweeds (fucoids) for use in glass production and as a pottery glaze. Romans were known to have used algal pigments as dyes for wool.

Many cultures have, and continue to consume seaweeds because of their presumed medicinal properties. The Chinese have relied upon the high iodine content in *Laminaria* to prevent goiter. Asian cultures believed the high potassium chloride associated with a seaweed diet could also prevent hay fever, and in traditional Hawaiian cultures seaweeds were applied to heal open wounds caused by scraping against coral. Finally, seaweeds have spiritual significance for Polynesians, who used them in religious and cultural ceremonies.

Although people in the United States do not cultivate or harvest seaweeds as extensively as is done in other cultures, they have incorporated algae into their diets and found many creative ways to market and display seaweeds. There is an interesting market niche for pressed algae turned into art. Typically algae are pressed to acid-free paper (techniques similar to those for terrestrial plants) as a means of cataloging and preserving specimens. However, because of their interesting shapes, styles, and colors the seaweeds can be arranged and displayed as unique artistic pieces or bookmarks. Dried seaweeds are also handcrafted into baskets, dolls, rattles, and other decorative items. Specimens used in these applications are typically from sustainable harvest of beach wrack. They produce small revenue and are mostly made for pleasure.

FIGURE 4 A Reef Ranger with Negril Marine Park (A) cultivates and (B) dries *Eucheuma* in Little Bay, Jamaica, for use in the production of traditional island beverages. Photographs by Brian E. Lapointe.

TOURISM: AESTHETIC AND ECONOMIC VALUES

No trip to the shore and rocky intertidal pools would be complete without seaweeds. For better or worse, they provide much of the smell, color, and excitement, with their funny shapes and often gooey, rubbery, or spiky textures. Although the value of simply existing and enjoying these intertidal features is difficult to assess, it can be related to tourism dollars. The cost of a local trip to the intertidal is generally minimal or free except for vehicle costs, parking, and perhaps state park entrance fees (US$3–10). Considering how low this cost is in comparison to prices for alternative activities such as going to a public aquarium (US$10–20) or the movies (US$8), seeing a sporting event (US$10–100), or visiting a theme park (US$30–50), it is probable that the millions of people visiting coastal areas every year are willing to pay more to enjoy an intertidal experience. This is also likely, considering that tourism in coastal regions around the world is a multibillion-dollar industry. Of course, unregulated coastal access comes at a price to the environment in the form of trampling by tourists, disturbance of marine organisms, and the enormous amounts of trash that must be collected during coastal cleanups each year.

Conversely, it is also possible to assess the negative value of algae and the revenue losses they generate. Macroalgal blooms, for example, are increasingly more common throughout the world and can have tremendous impacts on the environment and the economy. The blooms are often caused by eutrophication; bloom species are typically opportunists and sometimes even invasive species whose life histories enable them to exploit changes in the marine environment. In almost all cases these green (e.g., *Ulva, Codium,* and *Cladophora*) and red (e.g., *Gracilaria, Hypnea, Eucheuma,* and *Kappyphycus*) algal blooms result in massive accumulations of drift that blanket the beaches and shallow bottoms (Fig. 5). The foul smell of degrading

seaweeds, physical barriers they create on the beach, and the reduced water clarity and potential health risks associated with them all detract from the appeal of a trip to the shore and have adverse impacts on tourism as well as other near-shore commercial activities. In Hawaii alone it is estimated that over US$20 million per year are lost in tourism and property values along the Maui coast because of algal blooms. Similar impacts are occurring in the Florida Gulf Coast.

Cleanup efforts are often undertaken to eradicate algae from impacted beaches and attract people back to the shore. These efforts, although often done by volunteers, can be costly (more than US$200,000 annually) and are time-consuming. The impacts go beyond tourism losses, because poor water quality associated with the blooms can also reduce fishery and shellfish hatchery yields. Attempts to improve water quality and prevent future blooms by controlling point sources can cost millions of dollars.

ECOSYSTEM SERVICES

As the dominant organisms on rocky intertidal shores, macroalgae contribute greatly to the species diversity of these environments and perform many ecosystem functions. Together with microalgae they are the oceans' primary producers and are an important source of O_2 production. They are useful as CO_2 scrubbers and can act as natural biofilters by absorbing heavy metals from the water column. Because of these functions, as well as their responses to pollution and water clarity, algal populations can serve as indicators of water quality. Marine plants also stabilize sediments, and calcified seaweeds can be important sources of carbonate sand.

Seaweed communities are important structural components of the intertidal zone. They provide food and shelter to a variety of organisms and are favorable recruitment substrates for many invertebrates. Certain calcified red algal species are known to serve as cues for the recruitment of invertebrate larvae such as abalone, and they are recruitment sites for kelps. Many fish and invertebrates take refuge in and amongst macroalgae to escape predators and desiccation during low tides. In some cases invertebrates and fish live permanently in the interstitial spaces created by algal branches and holdfasts. The diversity of organisms living in association with algal habitats is frequently much greater than in habitats without macroalgae. For these reasons, it is arguable that perhaps the most important value of seaweeds is not at all economic, but rather their contribution to productive and diverse marine ecosystems.

FIGURE 5 Red macroalgal (primarily *Gracilaria* and *Hypnea*) drift accumulation from blooms in Lee County along the Gulf Coast of Florida. Photograph courtesy of Brian E. Lapointe.

SEE ALSO THE FOLLOWING ARTICLES

Algal Blooms / Economics, Coastal / Food Uses, Modern / Kelps / Symbolic and Cultural Uses

FURTHER READING

Abbott, I. A. 1984. *Limu: an ethnobotanical study of some Hawaiian seaweeds.* Kalaheo, HI: Pacific Tropical Botanical Garden.

Chapman, V. J., and D. J. Chapman. 1980. *Seaweeds and their uses.* London: Chapman and Hall.

Graham, L. E., and L. W. Wilcox. 2000. *Algae.* Upper Saddle River, NJ: Prentice Hall.

Lapointe, B. E., and K. Thacker. 2002. Community-based water quality and coral reef monitoring in the Negril Marine Park, Jamaica: Land based nutrient inputs and their ecological consequences, in *The Everglades, Florida Bay, and Coral Reefs of the Florida Keys. An ecosystem sourcebook.* J. W. Porter and K. G. Porter, eds. Boca Raton, FL: CRC Press, 939–964.

Lobban, C. S., and P. J. Harrison. 1994. *Seaweed ecology and physiology.* Cambridge, UK: Cambridge University Press.

McHugh, D. J. 2003. *A guide to the seaweed industry.* FAO Fisheries Technical Paper. No. 441. Rome: Food and Agricultural Organization.

Ohno, M., and A. T. Critchley, eds. 1993. *Seaweed cultivation and marine ranching.* Yokosuka, Japan: Japan International Cooperation Agency.

Hawaii Coral Reef Initiative Program. http://www.hawaii.edu/ssri/hcri/ev/kihei_coast.htm.

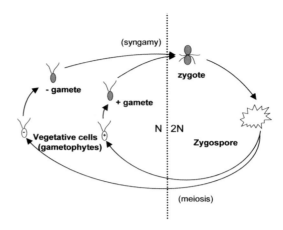

FIGURE 1 The zygotic-meiosis (haplontic) life cycle of *Chlamydomonas*, a green alga. Zygospore meiosis produces zoospores, which for this species are its vegetative cells. With the proper environmental trigger, the vegetative cells (gametophytes) turn into + or – gametes and fuse to create zygotes. The *Chlamydomonas* life cycle is haplontic because the vegetative cells can persist and self-replicate asexually for (relatively) long periods of time, while the diploid phase is much shorter lived. N represents haploid stages; 2N represents diploid stages.

ALGAL LIFE CYCLES

CAROL S. THORNBER

University of Rhode Island

Complex life cycles are found in many marine macroalgal (seaweed) species, which include green algae (Chlorophyta), red algae (Rhodophyta), and brown algae (Phaeophyceae). These algal life cycles frequently involve an alternation between two independent, free-living phases that differ in ploidy (the number of copies of each chromosome); one phase is haploid (one copy), and the other is diploid (two copies). Within this basic sequence of haploid/diploid alternation, there is great variability in the size, maximum lifespan, ecological niche, and reproductive biology of each phase. Based upon the general characteristics of the two independent phases, the life cycles of marine algae are frequently categorized into three main types.

TYPES OF LIFE CYCLES

In zygotic or haplontic life cycles the haploid phase is the dominant, frequently macroscopic phase (Fig. 1). Haploid gametophytes produce haploid gametes; these gametes may be either isogamous (the same size) or anisogamous (different sizes). Gametes fuse to create a diploid zygote, which soon thereafter produces motile zoospores via meiosis. Individual zoospores will then grow into new gametophytes. Algae with this type of life cycle include many green freshwater genera (e.g., *Chara, Coleochaete*) and green unicellular flagellates (e.g., *Chlamydomonas*). Although there are few marine macroalgal tidepool species with truly haplontic life cycles, the tropical marine green algal genus *Acetabularia* is one that has a modified haplontic cycle.

The diploid phase is the dominant, macroscopic phase in gametic (or diplontic) life cycles (Fig. 2); this is also the same basic life cycle found in humans and many

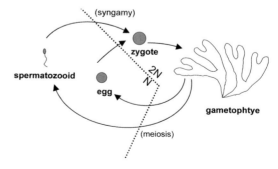

FIGURE 2 A typical gametic-meiosis (diplontic) life cycle of *Fucus*, a brown alga. The macroscopic diploid (gametophyte), when mature, produces gametes that fuse to create new diploids. In *Fucus*, these gametes are called spermatozooids (male) and eggs (female).

animals. Multicellular diploids produce haploid gametes via meiosis; these gametes are released into the water column and fuse soon thereafter to create new multicellular, longer-lived diploid organisms. Gametes are the only haploid stage in this life cycle. Species with diplontic life cycles include the brown algae *Durvillaea* and *Fucus* as well as some benthic diatoms (Bacillariophyceae).

In the sporic (or haplodiplontic) life cycle, the most commonly found in marine macroalgae, both haploid and diploid phases are macroscopic and live independently from one another (Figs. 3–5). This is the most commonly occurring life cycle in marine macroalgae. Multicellular haploid gametophytes produce haploid gametes, which shortly thereafter fuse and create diploid sporophytes.

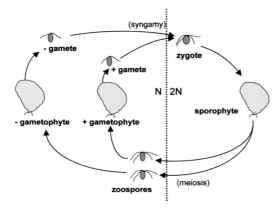

FIGURE 3 The isomorphic sporic (haplodiplontic) life cycle of *Ulva*, a green alga. Gametophytes and sporophytes are both macroscopic and are indistinguishable from one another in the field.

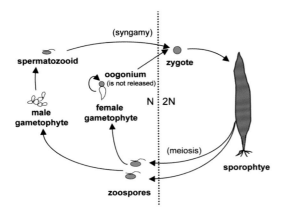

FIGURE 4 The heteromorphic sporic (haplodiplontic) life cycle of *Laminaria*, a brown alga. The sporophyte is macroscopic and much larger than the small, few-celled gametophyte. In kelps, gametes are called spermatozooids (male) and oogonia (female). The oogonium (small gray circle) remains attached to the female gametophyte, so the new sporophyte grows on top of the (then deceased) female gametophyte.

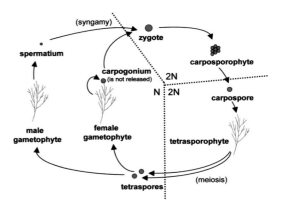

FIGURE 5 The triphasic red algal (class Florideophyceae) sporic (haplodiplontic) life cycle of *Polysiphonia*, an isomorphic red alga. The tetrasporophyte and gametophytes are macroscopic and visually identical, except when reproductive. In this class, gametes are called spermatia (male) and carpogonia (female). The carpogonium (gray circle on top of the female) remains attached to the female gametophyte, so the zygote and carposporophyte grow attached to the female blade. Note the lack of flagella on all stages, which is a red algal characteristic.

Depending upon the species, gametes may be released into the water column or may remain on the gametophyte. When mature, multicellular sporophytes produce haploid spores via meiosis; these spores are released into the water column and subsequently become new haploid gametophytes. The two phases can either be morphologically quite similar (isomorphic, Fig. 3), including *Dictyota* and *Ulva,* or they can be morphologically dissimilar (heteromorphic, Fig. 4, including *Scytosiphon* and *Laminaria*) from one another.

Most red algae (all those in the class Florideophyceae) have a haplodiplontic life cycle, with an additional short-lived phase (carposporophyte, Fig. 5) that may serve to increase reproductive output. In these species, female gametes (carpogonia) remain on the female gametophyte and are fertilized there by male gametes (spermatia). The resulting diploid zygotes quickly grow into small clusters of thousands of diploid spores (carposporophytes) while remaining attached to female blades. These spores are released into the water column and become new free-living tetrasporophytes (so called because they produce spores in packets of four). Species with a triphasic life cycle are either isomorphic *(Polysiphonia)* or heteromorphic (for example, as a haploid blade and diploid crust for *Mastocarpus,* and as a microscopic female haploid and macroscopic male haploid and tetrasporophyte for *Palmaria*) for the two independent phases.

These three life cycle types only partially segregate by taxonomic order; in brown algae, different life cycles exist in closely related species, indicating multiple evolutionary origins. Also, the distinct, multiple phases found in many algal species can provide unique opportunities for investigating scientific questions on topics ranging from intra-specific population dynamics to theories regarding the evolution of life cycles and sex.

ECOLOGICAL IMPLICATIONS

The life cycle of an algal species can substantially affect its ecological dynamics. Heteromorphic species have two phases with very distinct ecological niches, and the phases may differ in their susceptibility to herbivory, temperature fluctuations, or other physical or biological stressors. By contrast, the two phases of isomorphic species generally have more similar ecological niches, such as similar intertidal location or chemical content, although subtle but important differences may exist between phases, such as per capita fecundity and mortality.

Some algal species have the ability to regulate when switching between different phases will occur. Sporophytes of *Ectocarpus siliculosus,* a common filamentous brown alga with an isomorphic life cycle, produce diploid spores that become new sporophytes during the summertime. This cycle can be repeated several times over the course of a summer, while temperatures are warm. However, during the wintertime, when temperatures are cooler, the sporophytes will produce haploid spores that become gametophytes. This process is not unique to *Ectocarpus;* temperature regulation of life cycles has been demonstrated for other algae, such as the brown alga *Colpomenia sinuosa.*

OTHER METHODS OF REPRODUCTION IN ALGAE

Many other algal species maintain populations exclusively (or nearly so) of either sporophytes or gametophytes, but not both, indicating that they must be able to routinely skip parts of their life cycle. This ability to perform apomixis has been observed in several species and should be considered when studying algal population dynamics, because in some species a particular phase, or stage, may be frequently (or always) skipped. Why certain life stages may be dropped in some algal species and not in closely related others remains an area of active debate.

Some macroalgae can also successfully asexually reproduce via fragmentation or production of stolons or rhizoid production. These processes are more frequently observed in green algae, but they can occur in some brown and red algal species as well.

SEE ALSO THE FOLLOWING ARTICLES

Competition / Dispersal / Kelps / Reproduction

FURTHER READING

Hawkes, M. W. 1990. Reproductive strategies. in *Biology of the red algae.* K. M. Cole and R. G. Sheath, eds. New York: Cambridge University Press, 455–476.

Hughes, J. S., and S. P. Otto. 1999. Ecology and the evolution of biphasic life cycles. *American Naturalist* 154: 306–320.

Lee, R. E. 1999. *Phycology.* New York: Cambridge University Press.

Thornber, C. S. 2006. Functional properties of the isomorphic biphasic algal life cycle. *Integrative and Comparative Biology* 46(5): 605–614.

Thornber, C. S., and S. D. Gaines. 2004. Population demographics in species with biphasic life cycles. *Ecology* 85: 1661–1674.

van den Hoek, C., D. G. Mann, and H. M. Jahns. 1995. *Algae: an introduction to phycology.* New York: Cambridge University Press.

West, J. A., G. C. Zuccarello, and M. Kamiya. 2001. Reproductive patterns of *Caloglossa* species (Delesseriaceae, Rhodophyta) from Australia and New Zealand: multiple origins of asexuality in *C. leprieurii.* Literature review on apomixis, mixed-phase, bisexuality and sexual compatibility. *Phycological Research* 49: 183–200.

ALGAL TURFS

ROBERT C. CARPENTER

California State University, Northridge

Many rocky and other hard substrata on temperate and tropical shores are covered with an assemblage of algal species that form a cushion or mat. In some environments, these algal assemblages are dominated by one or a few species, although in other environments, they are composed of a diversity of species. These assemblages or communities, collectively referred to as algal turfs, are found in both the intertidal and the shallow sublittoral and can be a major space occupier in some zones. The term *algal turf* is often used to describe two different kinds of algal communities. One type of algal turf is composed of species with simple, filamentous morphologies that form a thin veneer on hard substrata. A different kind of algal turf is formed by much larger (>1 cm), macroscopic algae that grow in a densely packed mat. Although both types of algal turfs occur in a variety of marine environments, the processes that lead to their formation may vary, and both their community structure and function may be very different. Similar to other intertidal and shallow subtidal organisms, the distribution, abundance, and physiology of algal turfs are influenced strongly by both the physical environment and biological interactions with other organisms.

TEMPERATE ALGAL TURFS

On rocky shores in the temperate zone, algal turfs are found on exposed surfaces at low tide, in tidepools, and on shallow subtidal substrata. Algal turfs often develop in the low intertidal zone where they can occupy the majority of rocky surfaces. These algal turfs most often are comprised of larger, macroscopic species with more complex (at the cellular level), branching morphologies. These assemblages often consist of one or a few species with thalli that are packed densely in an interwoven mat (Fig. 1). Some of the more common taxa forming intertidal algal turfs are articulated coralline algae (e.g., *Corallina* spp.) and other red algae (e.g., *Chondracanthus* spp., *Pterocladiella* spp., and *Gelidium* spp.). Turf-forming thalli usually are shorter and more highly branched than thalli of the same taxa growing subtidally.

FIGURE 1 Algal turf assemblage exposed at low tide in a temperate low intertidal habitat (Santa Catalina Island, CA). Photograph by the author.

Intertidal algal turfs are affected most by the rigors of the physical environment in this habitat. Desiccation can be the most important factor that limits the distribution of algae in the intertidal zone. Because a dense mat of thalli traps water and helps to maintain a more humid microenvironment, turf-forming algae experience much less desiccation than single thalli growing in the same habitat. Other limiting physical factors in the intertidal can be high light and temperature. The negative effects of these factors are also mitigated within a dense assemblage of algae, because much of the photosynthetic surface of individual thalli either is self-shaded or shaded by adjacent individuals, and the trapped water reduces temperatures within the turf. Nevertheless, the surface of the turf is exposed to high levels of light and desiccation at low tide and often the apical

portions of these turf-forming thalli are bleached (appear white or yellow). Rates of photosynthesis are decreased and respiration increased by the physical conditions during tidal exposure. As a result, growth rates of species comprising intertidal algal turfs are reduced.

Algal turfs growing in tidepools or subtidally are not affected as much by the physical environment (although light still is important) and biotic interactions probably are more critical in determining their distribution and abundance. In this physiologically more benign environment, more species compete for limited space, so the abundance of algal turfs depends, in part, on the outcome of this competition. Competitive interactions often are determined by the relative rates of growth of the competitors, and for algae growth is related (inversely) to the morphological complexity of the thalli. As a result, algae that are simple filaments will have an advantage in competing with other taxa and dense assemblages of these filaments form a second type of algal turf. Another biological interaction that is more common in tidepools and subtidally is herbivory. Released from the constraints of intertidal exposure, herbivores (e.g., snails, sea urchins) can be more common and active in foraging on algae. How algae persist in the face of intense herbivory also depends on how fast they can grow and replace lost tissue. Simple filaments comprising algal turfs quickly regenerate biomass lost to herbivores. So both increased competition and herbivory in tidepool and subtidal habitats favor algal turfs that are diminutive in size, but that grow quickly.

Algal turfs provide food and habitat for a diversity of other species. The trophic importance of subtidal algal turfs in temperate environments probably has been underestimated. Recent studies suggest that rates of algal turf primary production are substantial, yet biomass remains low, indicating that much of this production is removed by herbivores. The structure of intertidal algal turfs also provides a refuge for many associated organisms. Mitigation of temperature and desiccation stress within the turf allows many small crustaceans (e.g., amphipods) and polychaetes to persist despite aerial exposure at low tide. At high tide, the turf may provide a refuge from predation for these same organisms.

TROPICAL ALGAL TURFS

Algal turfs in tropical environments have been studied much more than on temperate shores. Algal turfs in the rocky intertidal are not common in the tropics, most likely due to intense solar heating and the resulting desiccation. However, algal turfs are among the more common space occupiers in subtidal habitats, especially on coral reefs. The most common type of algal turf community in the tropics consists of many small, filamentous species that generally are <1 cm tall (Figs. 2, 3). These turfs commonly are composed of >10 different species with representatives from each of the major algal phyla (Chlorophyta, Phaeophyta, and Rhodophyta). Coral reef algal turfs are the most productive component of the coral reef and can contribute up to 80% of the total reef primary production. Even more than their temperate counterparts, reef algal turfs are low biomass, high turnover communities with up to 100% of their new biomass production removed by herbivores. As a result of the increased abundance and diversity of herbivores in the tropics, algal turf species are among the few algae able to persist in the face of intense herbivory. As a result of their abundance and high rates of productivity, algal turfs are a very important trophic link in coral reef food webs. Additionally, on some reefs where the reef flat is shallow or may be exposed at extreme low tides, algal turfs include species of cyanobacteria that are able to fix nitrogen. The highest rates of nitrogen fixation measured in any biological community have been associated with these coral reef algal turfs. So in addition to providing carbon to consumers, these algal turfs also contribute large amounts of nitrogen to a community that is nitrogen limited in the oligotrophic waters typical of the tropics.

The algal turfs that are common on coral reefs are too small to provide much habitat for other organisms. Some small crustaceans are associated with these turf assemblages, but likely are subject to frequent disturbance from herbivores.

Intense herbivory in the subtidal zone also may select for turf formation by larger algal species. While much less common than the aforementioned algal turf assemblages, macroalgal turfs also occur in reef habitats. These larger turfs can include both calcified (e.g., *Halimeda* spp.) and uncalcified (e.g., *Dictyota* spp.) taxa and provide some protection from herbivores as a result of the dense packing of branches within the turf. Conversely, herbivory on the apical portions of the algal thalli might promote turf formation by increasing production of lateral branches, much like trimming does to a hedge. Whether it is a cause or effect, it appears that herbivory is the dominant process associated with algal turf communities in the tropics.

Because of their larger size, macroalgal turfs on coral reefs can provide habitat and a refuge from predation for associated mobile organisms. This remains an understudied aspect of these turf assemblages, assemblages that might be viewed as islands of habitat distributed across large areas of reef that experience relatively high predation.

Algal turfs are a multispecific assemblage that represents a functional–form group rather than a defined group of algal species. The environmental factors associated with turf formation in algae are either temperature or desiccation stress (intertidal) or herbivory (subtidal). In each case, turf formation mitigates the negative effects of these factors on algal physiology and growth and promotes persistence in hard substratum habitats in both temperate and tropical environments where they provide food and habitat for other associated organisms.

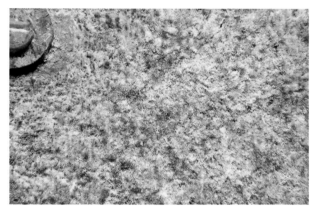

FIGURE 2 Subtidal algal turf growing on the coral substratum. Washer in photograph is 1.2 cm across. Photograph by the author.

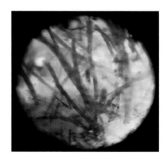

FIGURE 3 Close-up of a coral reef algal turf. Field of view is 3 mm across. Photograph by the author.

SEE ALSO THE FOLLOWING ARTICLES

Corals / Desiccation Stress / Food Webs / Herbivory

FURTHER READING

Adey, W. H., and T. Goertemiller. 1987. Coral reef algal turfs: master producers in nutrient poor seas. *Phycologia* 26: 374–386.

Airoldi, L. 2001. Distribution and morphological variation of low-shore algal turfs. *Marine Biology* 138: 1233–1239.

Copertino, M. S., A. Cheshire, and J. Watling. 2006. Photoinhibition and photoacclimation of turf algal communities on a temperate reef, after *in situ* transplantation experiments. *Journal of Phycology* 42: 580–592.

Copertino, M. S., S. D. Connell, and A. Cheshire. 2005. Prevalence and production of turf-forming algae on a temperate subtidal coast. *Phycologia* 43: 241–248.

Gorgula, S. K., and S. D. Connell. 2004. Expansive covers of turf-forming algae on human-dominated coast: the relative effects of increasing nutrient and sediment loads. *Marine Biology* 145: 613–619.

Hay, M. E. 1981. The functional morphology of turf-forming seaweeds: persistence in stressful marine habitats. *Ecology* 62: 739–750.

AMPHIPODS, ISOPODS, AND OTHER SMALL CRUSTACEANS

LES WATLING

University of Maine, Walpole

Although many groups of crustaceans can be found in tidepools and on rocky shores, among the most numerous are the amphipods, isopods, copepods, and ostracods. Most are grazers, living on small macroalgal fragments, microalgae, or detrital particles found on the bottom of pools. All members of these groups found in tidepools are very hardy, being able to tolerate a wide range of physical and chemical conditions. In some cases, the species are not very abundant outside the tidepool environment, and in other cases the species are abundant in tidepools because their preferred food is more common in pools than on the open shore.

TAXONOMY AND MORPHOLOGY

Amphipods and isopods are malacostracan crustaceans usually grouped together as orders in the superorder Peracarida because of their penchant for raising their young in a brood pouch formed by specialized ventral plates on the female. Copepods are a very diverse group of small maxillipodan crustaceans that occupy nearly all marine and freshwater habitats and have a diversity of life habits, including becoming parasitic. Ostracods are small bivalved crustaceans usually allied with the maxillopodans. They are an ancient group, known from the earliest Palaeozoic Era, and can be found in all marine and freshwater habitats. All four groups have species that are well adapted to the wide range of physical and chemical conditions often found in tidepools.

Amphipods have a very characteristic morphology. In many species the body is flattened laterally, making them higher than wide, and as a result they are often seen crawling over the rocks in tidepools, hanging on with the legs of one side of their body. Such a body design may seem to put the animal at a disadvantage, but in fact, amphipods have the most flexible of all crustacean body plans and as a result have been able to exploit a wide variety of habitats. Short, stiff rods, called uropods, are often used to help propel the amphipod away from the substrate in pools. Amphipods living on rocky shores are typically about 0.5 to 2 cm in length.

Isopods have a body design quite different from that of amphipods. They are usually thought of as having a body that is flattened dorsoventrally, but some, in fact, are more cylindrical. Isopods typically have five pairs of pleopods and one pair of uropods on the abdomen. Isopods use their pleopods as the primary source of oxygen uptake, so the pleopods are often well protected under an opercular covering made from the uropods. Isopods of rockpools can be very small, about 0.7 mm in length, but those of the open rock surfaces are often 3–4 cm long.

Copepods are much smaller than the previous two groups, generally only 1–2 mm in length. The body consists of a head with five pairs of appendages to which is fused a varying number (generally one or two) of thoracic somites. The remaining four or five free thoracic somites form a pereon bearing paired swimming legs. Copepods are easy to spot in tidepools by their herky-jerky motion over the substrate, and females always carry their eggs in paired egg strings, called ovisacs, attached to the first abdominal somite.

Ostracods are also generally less than one to a few millimeters in length. The body is encased in an enlarged carapace that is hinged like the two valves of a clam, giving them a bean shape. The thorax is reduced, and, as a result, the enlarged first and second antennae assist with locomotion. When ostracods walk across the substrate, carrying their large carapace, they often rock side to side. In contrast to the other three small crustaceans groups noted previously, ostracods living on rocky shores do not have the capability of swimming in tidepool waters.

FOOD HABITS

Amphipods and Isopods

The primary food sources for both amphipods and isopods are algae and plant detritus, although some species in both groups are capable of capturing and eating small or weakened animals, the exuviae of other crustaceans, or scavenging those freshly dead. Amphipods feed on algae by gripping pieces of the thallus with their gnathopods and holding it such that bits can be cut off with the bladelike mandible incisors. In some cases, an amphipod will first use the mandibles to scrape all the microalgae, such as diatoms, off the macroalgal thallus, and only when that high-quality food is consumed will they attack the macroalga itself. Isopods, not having their first pereopods modified as gnathopods, appress the mandible directly on the algal thallus and either scrape the surface or consume the macroalga. The large isopod *Idotea granulosa* was seen to prefer the growing tips of *Ascophyllum nodosum*. In very high intertidal pools, the major food source is either microalgae attached to rock surfaces or detrital plant fragments. Small isopods, such as species of *Jaera,* do well in these pools, as does the amphipod *Gammarus duebeni*. In both cases a very wide range of organic particles are eaten, ranging from fresh microalgae scraped from the rock surface, to pollen grains and other terrestrial plant products, to the larvae of tidepool insects.

Copepods and Ostracods

Because of their very small body size, these groups probably consume only microbes and microalgae living on the thalli of larger algal species, on the rock surfaces, or on detritus in the tidepool sediments. Some copepods have evolved specializations of their appendages that allow them to live on the macroalgae from which they are grazing microbes, without being swept away by wave forces. Such specializations include flattening of the body, enlargement of the maxilliped or modification of a walking leg to form a grasping structure, and secretion of encapsulating mucus that the copepod uses to glue itself to an algal thallus. Ostracods rarely have similar kinds of modifications, and they usually live in wave-protected microenvironments at the bases of macroalgae or in tidepool bottom sediments. Here, they feed on diatoms, animal carrion, plant fragments, and detrital products that accumulate in the pools.

ENVIRONMENTAL PHYSIOLOGY

Amphipods and Isopods

Tidepools and rocky shores are sites of extreme physical and chemical environmental conditions. Those animals that are able to live, and in some cases to thrive, in these habitats are usually those that have the ability to tolerate wide variations in temperature, salinity, and dissolved oxygen. For example, the amphipod *Gammarus duebeni* can live in water of 0.2 to 66 parts per thousand salinity. In seawater, *G. duebeni* produces urine that is isotonic with the blood, but when in freshwater the animal can

reduce the amount of excreted salt, with the result that the urine is hypotonic to the blood. In tidepools, as the water becomes saltier through evaporation, the urine becomes hypertonic relative to the blood.

Copepods

Harpacticoid copepods of the genus *Tigriopus* are very common, and highly characteristic, species of high intertidal rockpools. They live in pools that receive salt water only during high spring tides, so the salinity of the pool may vary from near fresh during rain storms to hypersaline during hot dry spells. When the salinity exceeds 90 parts per thousand, *T. fulvus* become quiescent, but resumes normal activity if the salinity drops within about 30 hours. This copepod can tolerate a salinity of 180 parts per thousand for about 3 hours. Lethal temperatures vary with salinity, being lower when the salinity is lower. At a salinity of 4 parts per thousand, death occurs at about 34 °C; but at a salinity of 90 parts per thousand death does not occur until a temperature of nearly 42 °C is reached. *Tigriopus californicus* is quite capable of living in pools lower in the intertidal zone, but because the lower pools contain many copepod predators, ranging from fish to crustaceans to sea anemones, it thrives only in the high pools where conditions are too extreme for the predators. Oxygen can be limiting in high intertidal rockpools, especially at night after the oxygen produced during the day by photosynthesizers has all been consumed. Many copepods do not have a respiratory pigment in their blood, yet at temperatures between 5 and 30 °C *T. brevicornis* is able to maintain a constant rate of oxygen consumption even as oxygen levels become quite low. Under severe hypoxia, these copepods enter a dormant state.

Ostracods

This group of small crustaceans are common inhabitants of algal covered rocky shores. Most live in the sediment at the bases of macroalgae, or on the thalli of macroalgae where there is protection from wave action. A few stray into high rockpools, but most do not have the ability to tolerate the extremes of conditions found in these pools for more than a few days. A few species, however, are well adapted for life in high rockpools, the best known being those from the Baltic shores of Sweden. The salinity in these pools varied from 2 to 8 parts per thousand. One species, *Heterocypris salinus,* could tolerate salinities as high as 32 parts per thousand, but only when temperatures were low. A few other northern European species are also euryhaline, being able to tolerate salinities from 2 to

50 parts per thousand, but most intertidal algal dwelling species are truly marine.

LIFE CYCLES

Amphipods

Typically, tidepool amphipods carry their young in a ventral brood pouch, which then hatch as miniature adults. The period of development varies with temperature. At 18 °C the incubation period is 14 days, but development slows with decreasing temperature, reaching 54 days at about 5 °C. A female may have several broods, especially as the temperature rises through the summer. On hatching, it takes about 23 to 30 weeks to reach maturity at a temperature of 15 to 20 °C. In nature an individual may live for 15 to 18 months. One curious feature of some amphipods is that the sex of the offspring changes with temperature. Under normal North Atlantic conditions, with a salinity of about 10 parts per thousand, all the young will be males if the temperature is below 5 °C a few days before the eggs are deposited in the brood pouch. On the other hand, if the temperature is above 6 °C, all the young will be females. Between 5 and 6 °C, the broods are mixed with respect to sex. Males are usually larger than the females, and some male amphipods carry individual females between their gnathopods for a few days before the eggs are deposited. This is called amplexus, and is a common sight in tidepools during the summer. The female needs to shed her exoskeleton before egg laying and so is vulnerable to predation by other amphipods. She apparently secretes some behavior-modifying substances in her urine, which triggers the guarding behavior in the male and, in particular, curtails his feeding responses. After molting, the male deposits sperm into the brood pouch where the eggs are fertilized. Soon afterward, the female exoskeleton begins to harden and the male is encouraged to leave.

Isopods

Like amphipods, female isopods are brooders, carrying their young in a ventral brood pouch. Although the larger isopods, such as those in the genus *Idotea* that are common among the algae of rocky shores, are good swimmers, at least for short distances, the smaller isopods such as the *Jaera* species are less able. Consequently, for poor swimmers, mates will most likely be other residents of the same rockpool, resulting in relatively high levels of genetic relatedness. Mating among isopods involves depositing sperm into the brood pouch of the female, usually after the female has molted the posterior half of the exoskeleton. At this time (within 2 to 12 hours after

molting) the oviducts are soft and eggs can freely pass into the brood pouch, being fertilized on the way. For many isopods there is a linear relationship between egg number and female size, with the smaller females carrying 11–15 eggs and the larger females about 60 eggs. During development there is a progressive loss of eggs, with brood mortality reaching about 50% by the time of hatching. An isopod hatches from the brood pouch as a manca, resembling the adult in all respects except that the last pair of walking legs is missing. Within three molts, an isopod is a full-fledged juvenile. It is often not known how many additional molts are passed through until sexual maturity is reached, but in many isopods only three to five more molts are required. Laboratory cultures of *Jaera albifrons* produced sexually mature females after 40 to 66 days. One female may produce a new brood every 20 days or so and perhaps four to six broods during her lifespan. Northern Hemisphere females born from April to late summer will produce broods until October and then die; those born in September will overwinter in the pool and produce their first broods the following April.

Copepods

Whether free swimmers or benthic species, copepods hatch from the egg as a nauplius and develop through four or five additional naupliar stages and usually five copepodite stages before maturing to a reproductive adult. One generation for a copepod, then, is the time from production of an egg by a female until that egg develops into another female who produces another set of eggs. As one might expect, generation times vary with temperature, which in many rockpools gradually increases with the onset of summer. For example, at 15 °C, development time of *Tigriopus californicus* is 32 days, but at 25 °C it was only 18 days. Females of *T. brevicornis* produce an average of 25 to 35 eggs during each egg production event, with the higher numbers produced as the pool warms in the summer. These copepods can survive freezing for as long as 70 hours if they are acclimated to colder waters, are moderately well fed, and the rockpool salinity is relatively high.

Ostracods

These small crustaceans either carry their eggs inside the bivalved carapace or attach the eggs to the substrate. Some rockpool parthenogenetic species lay 30 to 40 eggs, usually on small macroalgae or on the rocky substrate. The young hatch as a nauplius (that is, they have only three pairs of limbs), and the adult stage is usually reached in most intertidal species after the eighth and final molt. Many species hatch in the spring and become adults capable of producing eggs in 35–45 days. In some species, juveniles overwinter in pools, molting to the final adult stage the following spring.

SEE ALSO THE FOLLOWING ARTICLES

Body Shape / Foraging Behavior / Locomotion: Intertidal Challenges / Salinity Stress

FURTHER READING

Gunnill, F. C. 1984. Differing distributions of potentially competing amphipods, copepods, and gastropods among specimens of the intertidal alga *Pelvetia fastigiata. Marine Biology (Berlin)* 82: 277–292.

Hull, S. L. 1997. Seasonal changes in diversity and abundance of ostracods on four species of intertidal algae with differing structural complexity. *Marine Ecology Progress Series* 161: 71–82.

Lancellotti, D. A., and R. G. Trucco. 1993. Distribution patterns and coexistence of six species of the amphipod genus *Hyale. Marine Ecology Progress Series* 93: 131–141.

McAllen, R. 2001. Hanging on in there—position maintenance by the high-shore rockpool harpacticoid copepod *Tigriopus brevicornis. Journal of Natural History* 35: 1821–1829.

McAllen, R., and W. Block. 1997. Aspects of the cryobiology of the intertidal harpacticoid copepod *Tigriopus brevicornis* (O. F. Mueller). *Cryobiology* 35: 309–317.

McAllen, R., and A. Taylor. 2001. The effect of salinity change on the oxygen consumption and swimming activity of the high-shore rockpool copepod *Tigriopus brevicornis. Journal of Experimental Marine Biology and Ecology* 263: 227–240.

Whorff, J. S., L. L. Whorff, and M. H. Sweet III. 1995. Spatial variation in an algal turf community with respect to substratum slope and wave height. *Journal of the Marine Biological Association of the United Kingdom* 75: 429–444.

ANEMONES

SEE SEA ANEMONES

AQUARIA, PUBLIC

CHRISTOPHER HARROLD, MARTHA L. MANSON, AND CHRISTINA J. SLAGER

Monterey Bay Aquarium

By recreating the natural world indoors, aquarium exhibits safely introduce visitors to the plants and animals of wave-swept shores (Fig. 1). Communicating the wonder of such dynamic habitats in an exhibit challenges biologists and engineers. Complex systems support tidepool communities in nature; indoors, mechanical systems create wave motion and tide cycles and deliver nutrients. Further, although some plants and animals reproduce in exhibits, others must be collected. Protecting healthy tidepools is vital to conservation-minded institutions; collecting in a sustainable way becomes a logistical and ethical imperative.

FIGURE 1 Aquarium visitors enjoy the Kelp Forest exhibit narration at the Monterey Bay Aquarium in Monterey, California. Over 1 million people visit the Monterey Bay Aquarium every year; 142 million people per year visit aquariums and zoos accredited by the American Association of Zoos and Aquariums. © Monterey Bay Aquarium Foundation.

CREATING ROCKY-SHORE EXHIBITS

Intertidal invertebrates, aquatic plants, and small fishes are adapted to survive in harsh conditions at the edge of the sea. Recreating their habitat in a tidepool exhibit involves a raft of specialists. Fiberglass and concrete artificial rocks create the pools and substrate for plants and animals to attach themselves. Pumps and dump mechanisms generate water motion, simulating natural waves and tides (Fig. 2). Close observation of real tidepool communities guides aquarists in collecting and placing different species. In instances where unfiltered seawater circulates through the exhibit, plants and animals come in with the water and settle. Over time, an exhibit mimicking natural diversity develops.

FIGURE 2 The rocky intertidal exhibit at the Monterey Bay Aquarium. Wave surge, wave crash, and tidal level are all at work in this exhibit. © Monterey Bay Aquarium Foundation.

RECREATING THE PHYSICAL ENVIRONMENT

The aquarist's challenge is to create a physical environment that meets the needs of the plants and animals on display while also creating a compelling visitor experience. The physical parameters of greatest concern are substrate, light, temperature, water quality, and water motion.

Substrate

Artificial substrates in rocky shore exhibits are made of resin and fiberglass, epoxy resin, or reinforced concrete. Substrate texture can be recreated with amazing accuracy and realism using latex molds taken from rock surfaces in nature. With judicious use of molds, dyes and resin, or concrete, exhibit "rockwork" is virtually indistinguishable from the real thing. Benthic plants and animals readily settle, attach, and thrive.

Light

Exhibit lighting is among the most important factors influencing the aesthetic quality of rocky shore exhibits, while helping to create a healthy environment for the plants and animals on display. Exhibit lighting can be natural sunlight or artificial light. Natural lighting is surprisingly problematic, because it usually promotes heavy growth of undesirable fast-growing algae and diatoms, which cover both plants and animals and contribute to turbid water and poor viewing conditions. Artificial lighting is generally provided by fluorescent, incandescent, or metal halide lights. Sometimes theatrical gels are used to adjust the color hue. Some invertebrates, such as tropical corals, tridacnid bivalves, and some temperate-water anemones, host zooxanthellae, which require high light levels in specific wavelengths.

Water Quality

Aquarium life support systems maintain the water quality in rocky shore exhibits. There are three fundamental types of life support systems:

1. In open life support systems, water is pumped from a natural body of water, passed through filters to remove particulate material, piped into exhibits on a once-through basis, and returned to its source. Filtration may be temporarily bypassed to provide food for filter-feeding invertebrates and to encourage the establishment and growth of natural benthic communities.
2. In closed life support systems, water is recirculated through filters and the exhibits many times. In closed marine systems, the seawater composition may be artificially composed or imported from the ocean. The filtration system includes: physical filtration to remove suspended particulate material; biofiltration to detoxify the accumulated nitrogenous waste byproducts of metabolism; and aeration to replace lost oxygen and remove accumulated carbon dioxide.
3. In semiclosed life support systems, water is recirculated and processed as in a closed life support system, and in addition a small proportion of the water volume is continuously replaced with fresh water from a natural body of water.

Temperature

The water temperature of exhibits must be maintained within the tolerances of the plants and animals on display. In open systems, temperature can be maintained by simply providing adequate water flow. In closed or semiclosed systems, temperature may be actively maintained with refrigeration units, heat pumps, or heaters.

Water Motion

Water motion creates a naturalistic exhibit, adequate gas exchange, water quality, and clarity, and it encourages natural growth of organisms on exhibit surfaces.

1. Unidirectional, current-type water motion is generally provided by the entering (incurrent) and exiting (excurrent) water flows required for basic water quality.
2. Back-and-forth, surge-type water motion is created by a wide variety of mechanisms involving pumps, valves, and automated control devices. One example is shown in Fig. 3A.
3. Sudden, breaking-wave crash water motion is simulated by the simultaneous release of hundreds of gallons of seawater into the rocky shore exhibit (Fig. 3B).
4. Long-wavelength tidal cycles are required to maintain intertidal zonation patterns seen in nature. The tidal level in an exhibit can be controlled and varied using a variable-height Hartford loop (Fig. 3C).

RECREATING THE BIOLOGICAL ENVIRONMENT

Most plants and animals from rocky shores are robust and hearty and therefore easily maintained in an exhibit setting. Exhibits are generally stocked with organisms collected from the wild, though natural recruitment into exhibits can occur in open seawater systems. Natural intertidal zonation patterns may develop with suitable wave crash and tidal water motion.

Many aquariums have "live touch" experiences, in which visitors can handle exhibit plants and animals in supervised settings (Fig. 4). Care must be taken to

A

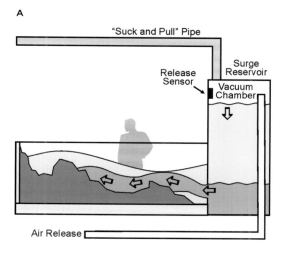

B

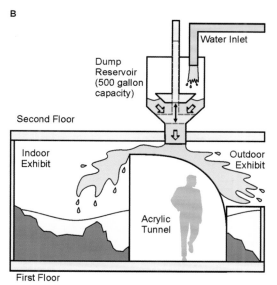

C

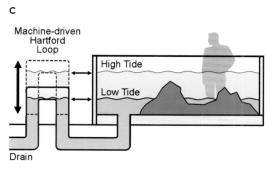

FIGURE 3 Three mechanisms that generate water motion in a rocky shore exhibit: (A) surge generator; (B) wave crash generator; (C) tide level generator. © Monterey Bay Aquarium Foundation.

FIGURE 4 At this tidepool exhibit, aquarium visitors can hold and touch hardy species of intertidal plants and animals under close supervision by aquarium staff. © Monterey Bay Aquarium Foundation.

maintain high water quality and avoid stressing animals or posing risks to visitors.

Stocking exhibits with living rocky-shore marine life raises two marine conservation issues that aquariums should proactively address. One is the impact of collecting marine life on native populations and habitats. Rocky-shore habitats are increasingly affected by human visitors. Aquariums should play a key role in minimizing or preventing significant human impacts. Collecting impacts can be reduced by minimizing demand (e.g., ensuring maximum longevity of marine life on display, captive breeding) and minimizing impacts (e.g., broadly distribute collecting activity, minimize physical impacts from collecting equipment).

The second conservation issue is the release of exotic or invasive exhibit organisms or their reproductive products into the aquatic environment. Only a few aquariums have systems designed to destroy or neutralize plant and animal reproductive products before discharge. This threat must be taken seriously by the aquarium industry, because the potential for harm is great (e.g., the release and subsequent invasion of the green alga *Caulerpa* in the Mediterranean).

INSPIRING THE VISITORS

Aquariums go to great lengths to recreate and maintain tidepool habitats for public display. They meticulously craft realistic replicas that duplicate rocky coastlines. What are they trying to accomplish with all this effort?

In the broadest sense, aquariums attempt to create a visitor experience that is enjoyable, informative, and that ultimately inspires visitors to preserve natural habitats for future generations. Aquariums want visitors to understand the complexities and adaptations of tidepool life, to experience the simple pleasure of contact, and to learn how to respectfully visit tidepools and gently handle tidepool animals. Rocky shores and tidepools, like so many other natural environments, are in peril from human impact. Unchecked, waste disposal, habitat destruction, and resource exploitation threaten to forever change or even destroy the world's tidepools and rocky shores.

Worldwide, over 142 million people visit accredited aquariums and zoos every year. People visit for many reasons—to experience the diversity of wildlife up close; to enjoy a safe, family experience; to be dazzled by the beauty of aquatic life; and to learn more about animals and nature. For many who do not live in close proximity to a coastline, an aquarium visit may be their only experience with live ocean animals. Further, typically, visitors believe what they learn during an aquarium visit. A recent Pew Charitable Trusts poll determined that the public considers aquariums and zoos to be reputable and trustworthy sources of environmental information. Consequently, aquariums have a tremendous opportunity to influence people's beliefs and behaviors. If they succeed in creating a sense of wonder about marine life and in educating visitors about the threats to aquatic ecosystems, aquariums can shape a visitor experience that ultimately compels millions of people to conserve our planet's threatened aquatic ecosystems.

SEE ALSO THE FOLLOWING ARTICLES

Education and Outreach / Light, Effects of / Museums and Collections / Water Chemistry

FURTHER READING

Meinesz, A., J. de Vaugelas, B. Hess, and X. Mari. 1993. Spread of the introduced tropical green alga *Caulerpa taxifolia* in northern Mediterranean waters. *Journal of Applied Phycology* 5: 141–147.
Taylor, L. 1993. *Aquariums: windows to nature.* Upper Saddle River, NJ: Prentice Hall.

ARTHROPODS, OVERVIEW

NORA B. TERWILLIGER

University of Oregon

The largest group of animals is the phylum Arthropoda, and it includes the crabs, shrimps, spiders, scorpions, and insects, as well as millipedes and centipedes. Arthropods are estimated to make up more than 80% of all known animals, in part because of their ability to inhabit sea, land, and air. In the rocky intertidal habitat, there are few representatives of terrestrial and aerial forms, but arthropods are still major members of the tidepool communities in both numbers and species diversity.

WHO THE ARTHROPODS ARE

The classification of the Arthropoda is complex and controversial. Current phylogenies are working to integrate morphological characteristics, including early development and functional morphology of locomotion, with contemporary molecular phylogenetics. The Chelicerata generally include the marine horseshoe crabs and pycnogonids or sea spiders as well as primarily terrestrial arachnids (spiders, scorpions, mites, and ticks). Pycnogonids and mites are the primary representatives of the Chelicerata in the rocky intertidal. The Hexapoda (six-legged insects) and Myriapoda (many-legged millipedes and centipedes) are rare in the rocky intertidal. The Crustacea (crabs, shrimps, barnacles, etc.) amply represent the Arthropoda in this challenging marine habitat.

TIDEPOOL CRUSTACEANS

Crustaceans of the rocky intertidal include a diverse group of large and small organisms. The sessile Cirripedia or barnacles, both stalked and acorn, that settle in the high intertidal are obvious inhabitants, whereas other barnacle species are found lower down in the mid- or low intertidal zones. Parasitic barnacles live on or in other rocky intertidal crustaceans, including hermit crabs and shrimps. Tiny bivalved ostracods and free-swimming or parasitic copepods are present in tidepools or under-boulder sediments, as are the leptostracans, the less obvious and most basal of the Malacostracan group of crustaceans. In their classic book *Between Pacific Tides*, first published in 1939, Ed Ricketts and Jack Calvin (1985) described a typical leptostrachan species *Nebalia* as looking like "a small beach hopper slipped inside a clam shell so small that its legs and hinder parts are left outside." Many species of isopods and amphipods swim in the tidepools, cling to the algae and rocks, and crawl about on other rocky-intertidal invertebrates. Shrimps dart back and forth across tidepools, whereas the largest representatives of the arthropods in this habitat, the crabs, are relatively reclusive. Rapidly moving snail shells are occupied by hermit crabs instead of the original shell makers. When an observer turns over a rock, numerous porcelain crabs *(Petrolisthes)* and grapsids *(Hemigrapsus, Pachygrapsus)* scurry away. The latter can be glimpsed sliding down the side of a rock and out of sight on an overcast, foggy day or on an evening tidepool foray. Bigger boulders often shield larger *Cancer* crab species. Rock holes formed by boring clams, sea urchins, or geological processes are often filled with a variety of crab species, including anomuran crabs such as *Oedignathus* that protects its soft abdomen with powerful claws to block the entrance of its hole. Beautifully camouflaged kelp crabs, decorator crabs, and umbrella crabs cling to the swaying kelp or walk slowly over the rocks, using the pointed tips, or dactyls, of their walking legs.

KEY CHARACTERISTICS OF ARTHROPODS

The key features of arthropods include paired, jointed appendages or legs (Greek *arthros* = joint, *pod* = foot), a hard, chitinous outer covering or exoskeleton that is molted, and a segmented body plan. The appendages are made up of a linear series of cylindrical sections or articles of hard exoskeleton. Each article is connected to the next by a specialized, flexible region of exoskeleton, and extensor and flexor muscles inside the articles extend across the flexible joints and allow antagonistic movement. The joints of a crab's walking leg, for example, flex in alternating planes, resulting in a limb with a great range of movement. Arthropods have modified the basic jointed appendage into a huge array of specialized structures that function in locomotion, feeding, grooming, sensory input, and reproduction. One individual may have many types of modified appendages, and the functional specialization of arthropod limbs is often referred to as the Swiss Army knife or Leatherman tool approach to evolutionary diversity.

MOLTING

The exoskeleton of arthropods provides protection from predators, abrasion, desiccation, and invading organisms (Fig. 1). It also gives structural support and allows for the antagonistic muscle movement. The rigid nature of the exoskeleton, however, and its location outside the living tissues requires that it must be shed or molted and replaced

FIGURE 1 The hard exoskeleton of the Dungeness crab *Cancer magister* covers its entire body, including claws, feeding appendages, sensory antennae, and even its eyestalks. Photograph by Margaret Ryan.

by a new, larger exoskeleton in order for the animal to grow. This process of molting, or ecdysis, is a dangerous yet repetitive part of an arthropod's life cycle. Molting is induced by a family of steroid hormones, the ecdysteroids. Recent molecular studies have indicated that hormonally regulated molting of an exoskeleton or cuticle is present in several other animal taxa. Many biologists now consider the protostome invertebrates to be classified as the Ecdysozoa (animals including arthropods, tardigrades, onychophorans, chaetognaths, nematodes, nematomorphs, priapulids, kinorhynchs, and loriciferans that produce ecdysteroid hormones and periodically molt an exoskeleton) and all other protostomes, the Lophotrochozoa.

Molting requires precise coordination of almost all major organ systems to simultaneously form a new exoskeleton and remove the old one. In many species, molting is closely integrated with the reproductive cycle as well. The sequence of events has been carefully documented and the phases divided into premolt, ecdysis, postmolt, and intermolt. A crustacean remains in intermolt due to circulating levels of neuropeptide hormones, especially a molt-inhibiting hormone (MIH), that prevent synthesis of the ecdysteroid molting hormone. It is generally thought that when levels of MIH decline, ecdysteroid synthesis increases and the animal enters premolt.

Secretion of a crustacean's new exoskeleton begins during premolt while the animal is still stuffed inside the old one, filling available space to capacity. First, proteolytic enzymes are secreted by the cells of the epidermis (sometimes referred to as hypodermis) that lies just inside the exoskeleton. These enzymes digest the inner lining of the exoskeleton to disconnect the extracellular exoskeleton from the living tissues and to make space to build a new exoskeleton. Structural proteins and chitin are then transported across the epidermis into this space to form the layers of the new exoskeleton. After a thin epicuticle is secreted into the extracellular space between epidermis and old exoskeleton, a thicker exocuticle layer appears underneath the new epicuticle. In addition to chitin and cuticular proteins that originate in the epidermal cells, the exoskeleton includes proteins such as cryptocyanin that are synthesized in the hepatopancreas and transported via the hemolymph across the epidermis to the forming exoskeleton. While the old exoskeleton is being partly broken down and the new one forming beneath it, claw muscles, especially in large-clawed species like the lobster, are selectively atrophied so that the claw can fit through the narrow wrist joint at ecdysis. Missing limbs are regenerated during premolt, and calcium from the old exoskeleton is stored in cells of the hepatopancreas in some species.

When the new exoskeleton is sufficiently formed, the crustacean needs to escape rapidly from the confines of its loosened old exoskeleton during ecdysis. The animal takes up water to increase internal pressure and split open the old exoskeleton along an ecdysial line between the posterior carapace and abdomen (Fig. 2). The old exoskeleton includes the lining of the foregut, the hindgut, and the covering over the eyes, so once ecdysis is initiated, the animal is temporarily blind with no functional mouth or anus. The new exoskeleton is soft and affords no protection against predators, and although it provides a fluid-filled hydrostatic skeleton temporarily, the animal is relatively floppy and defenseless (Fig. 3). Thus ecdysis must be completed rapidly.

During the postmolt stage, the new exoskeleton must harden—but only after all the limbs have been extracted from the old exoskeleton and after the new layer has expanded to its larger size. Immediately after ecdysis, the crab absorbs huge quantities of water to stretch the new, larger, temporarily flexible exoskeleton. Sclerotization or

FIGURE 3 Immediately after escaping from its old skeleton (ecdysis), *Homarus americanus* is the same size as before molting, and the new exoskeleton is soft. The newly molted lobster is able to move, using its fluid-filled body as a hydrostatic skeleton, but its claws are still small. The claw muscles are too flaccid and the shell is too soft to keep the claw tips rigid and uncrossed (arrows). During the next few hours, the exoskeleton will expand and begin hardening. Photograph by Margaret Ryan.

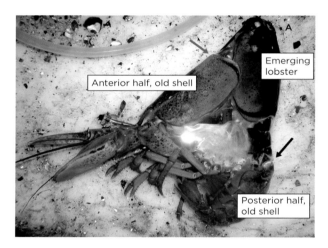

FIGURE 2 American lobster *Homarus americanus* in midmolt. Old exoskeleton has split at the suture line between posterior end of carapace (red arrow) and anterior end of abdomen (black arrow). Anterior head of lobster (A) and part of abdomen have already emerged from old exoskeleton, but claws and walking legs are still being withdrawn. Photograph by Margaret Ryan.

hardening occurs through a process of crosslinking the proteins and chitin molecules of the new exoskeleton. The enzyme phenoloxidase plays a major role in sclerotization. Synthesis of more exocuticle continues, and then secretion of a third layer, the endocuticle, begins beneath the new exocuticle. The endocuticle layer continues to increase in thickness, while calcification and sclerotization of the exoskeleton progress, until the crab enters intermolt. Replenishment and growth of the claw and other muscles occur during postmolt and especially intermolt, until the

crab enters its next molt cycle (Fig. 4). The length of time from one molt to the next, the intermolt duration, varies with age, nutritional status, and limb regeneration. Juvenile crustaceans molt much more frequently than adults.

Ecdysis and postmolt are dangerous times for crustaceans in the rocky intertidal and elsewhere, because the animals are vulnerable to predators and also subject to life-threatening

FIGURE 4 Old exoskeleton left by recently molted Dungeness crab *Cancer magister*. In addition to the external covering of body and gills, numerous connecting structural elements supported internal organs. Photograph by Margaret Ryan.

mechanical problems if the escape–expansion–hardening processes are not perfectly synchronized. The physiological challenges of rocky-intertidal life, including thermal, osmotic, and hypoxic stress on the rocks and in the tidepools, as well as the dynamics of wave action and surge channels, must be factored into the timing of ecdysis by rocky-intertidal arthropods.

REPRODUCTION

Reproduction and development in arthropods is complex and often includes multiple larval forms. Reproduction is usually closely linked to the molt cycle, with transfer of sperm from male to female usually occurring just after the female has molted, while her new exoskeleton is soft.

Barnacles are unique among the rocky intertidal crustaceans because they are hermaphrodites. Sperm are directly transferred from one barnacle via its remarkably long penis into the mantle cavity of another. Since most adult barnacles are permanently attached to the substrate, settlement in the proximity of other barnacles is obviously a requirement for reproductive success. The fertilized embryos are retained in lamellae in the mantle cavity of the second barnacle until they hatch as free-swimming naupliar larvae. The nauplius undergoes several molts, adding appendages each time, before metamorphosing into a swimming cyprid, the stage that eventually settles onto a hard substrate, glues itself down, and metamorphoses into a juvenile barnacle. The progression through these morphologically and physiologically different developmental stages is referred to as indirect development. The diversity of barnacle types, including sessile, stalked, free-living, commensal, and parasitic, is reflected in differences in the general reproductive pattern.

Crabs and lobsters, like barnacles, undergo indirect development, and reproduction is closely coordinated to the female's molt cycle. During premolt, the female releases pheromones that attract the male. He embraces her until she is ready to molt, releases her when she is ready to shed her old exoskeleton, then embraces her again to insert his sperm into two storage pockets on her soft ventral thorax. He stays with her to protect his investment for several days while her exoskeleton hardens, and then he departs (he molts on his own several months later). Oogenesis and oocyte growth then continue for several months in the female. The oocytes are finally fertilized by the stored sperm as they are extruded from the crab's ovary and pass out onto her ventral surface. She scoops them up with her abdominal appendages or pleopods. The sticky covering added to the fertilized embryos as they were extruded attaches the embryos to the fine processes of the pleopods, and the embryos form a large egg mass or berry on the ventral surface of the abdomen. In

a large *Cancer* crab, for example, the berry may include 1–5 million embryos. The female crab aerates the embryos by flexing her abdominal muscles back and forth. After several months of development, the embryo hatches as a swimming zoea. Usually there are multiple zoeal stages, depending on the species, followed by a metamorphosis into a swimming megalopa (Fig. 5). The planktonic megalopa then metamorphoses into a bottom-dwelling juvenile crab (Fig. 6). The first-instar juvenile crab molts within several weeks to a larger, second instar. With each succeeding molt, the interval between molts increases, so that an adult crab usually molts only once per year.

FIGURE 5 The megalopa stage of *Cancer magister*, the Dungeness crab, is an active swimmer that will metamorphose into a first-instar juvenile crab. Photograph by the author.

FIGURE 6 The first instar juvenile stage of *Cancer magister* is primarily a walker instead of a swimmer. Although it has the general shape and full complement of adult appendages, the physiology of the juvenile differs from that of the adult. It will grow and mature with each successive molt. Reproduced with permission of the Company of Biologists.

Amphipods and isopods undergo direct development. The male isopod transfers sperm to the female, after which she releases her oocytes. The fertilized embryos are then deposited into her special brood pouch or marsupium. The marsupium is located on her ventral surface and is a space enclosed by special flaplike extensions of her thoracic limbs that form during her previous molt. The embryo develops within its egg case, gradually acquiring the form of a miniature adult. The newly hatched juvenile, known as a manca, remains in the brood pouch for a short time before crawling away as a young isopod or amphipod.

SEE ALSO THE FOLLOWING ARTICLES

Barnacles / Crabs / Lobsters / Pycnogonids / Reproduction / Shrimps

FURTHER READING

Bliss, D. E., editor-in-chief. 1982. *The biology of crustacea*. Orlando, FL: Academic Press.

Jensen, G. C. 1995. *Pacific Coast crabs and shrimps*. Monterey, CA: Sea Challengers.

Ricketts, E. F., J. Calvin, and J. W. Hedgpeth (revised by D. W. Phillips). 1985. *Between Pacific tides*. Stanford, CA: Stanford University Press.

Ruppert, E. R., R. S. Fox, and R. B. Barnes. 2004. *Invertebrate zoology: a functional evolutionary approach,* 7th ed. Belmont, CA: Brooks/Cole–Thompson Learning.

Terwilliger, N., and M. Ryan. 2006. Functional and phylogenetic analyses of phenoloxidases from brachyuran *(Cancer magister)* and branchiopod *(Artemia franciscana, Triops longicaudatus)* crustaceans. *Biological Bulletin* 210: 38–50.

Terwilliger, N., M. Ryan, and D. Towle. 2005. Evolution of novel functions: cryptocyanin helps build new exoskeleton in *Cancer magister. Journal of Experimental Biology* 208: 2467–2474.

B

BARNACLES

DAVID P. LOHSE AND PETER T. RAIMONDI
University of California, Santa Cruz

Barnacles are crustacean arthropods, which means they are distantly related to such animals as crabs, lobsters, and shrimp. However, unlike their mobile cousins, barnacles have adopted a sessile existence. Barnacles are found on hard substrates in virtually all marine habitats and on all levels of the shore, often in vast numbers. This, coupled with the fact that they have a typical marine life cycle with an easily identifiable, planktonic larval stage, has made them a model study organism. Consequently, much is known about the biology of barnacles.

LIFE CYCLE

The life cycle of a typical barnacle includes two stages: a free-swimming larval stage and a sessile adult stage. As with all crustaceans, a barnacle's body is encased in a hard exoskeleton made of chitin. What distinguishes this group from other crustaceans, however, is that as adults they also secrete an outer calcareous "shell," called a test, which provides them additional protection against both predators and the elements. This test is made up of several plates that are attached either directly to the substrate (acorn barnacles) or to a fleshy stalk (gooseneck barnacles). Although some barnacle species are parasitic as adults, most are free-living filter feeders. Instead of being used for locomotion, their legs have been modified into a netlike structure called a cirral net. When under water, this net is extended out through an aperture in the test, called the operculum, to filter food out of the water column (Fig. 1).

Adult barnacles are generally hermaphroditic (although there are some amazing twists on this), meaning they have both male and female reproductive organs. Prior to mating, one individual, the "male," first polls the reproductive state of its immediate neighbors. It does this by using its elongated penis to "tap" nearby barnacles, a process thought to determine whether they carry eggs. If eggs are present, insemination occurs. Because fertilization is internal, the penis needs to be quite long and can be twice the length of the barnacle itself (Fig. 2).

Following fertilization, eggs are brooded within the test, where they develop into naupliar larvae, which are released into the water. This stage is the start of the mobile and dispersing larval period, which lasts between 10 and 45 days. While in the plankton, barnacles usually progress through six naupliar larval stages, all of which feed on phytoplankton and deposit lipid (fat) reserves. The final larval stage, the cyprid, is a nonfeeding stage, and it survives by utilizing the fat reserves built up

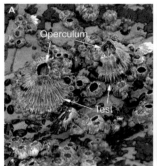

FIGURE 1 Some of the morphological features of (A) acorn and (B) stalked barnacles. Note that the cirral net pictured is only partially deployed; when fully deployed, it adopts is fanlike shape. Photograph by D. Lohse.

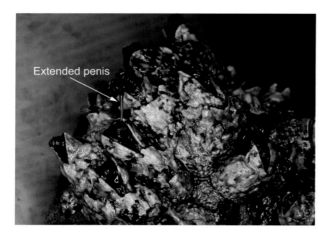

FIGURE 2 Acorn barnacles *(Catomerus polymerus)* copulating. Photograph by P. Raimondi.

during the previous stages (Fig. 3). The cyprid is the stage that ultimately settles on the shore and undergoes metamorphosis into the adult form. Because adults are sessile, finding a good spot to settle (the location it will reside in for the rest of its life) is very critical. As such, cyprids utilize both chemo- and neuro-receptors to

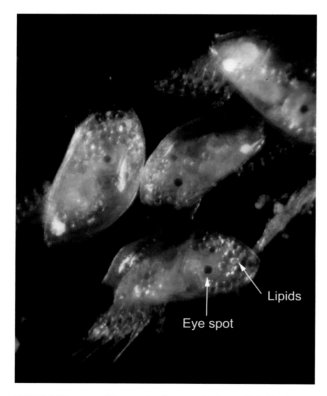

FIGURE 3 Close-up of the nonfeeding, cyprid stage. Note the deposits of lipids, which are food reserves accumulated during the previous larval stages, and the eye spot, which aids in the detection of light. Photograph by P. Raimondi.

detect environmental cues that facilitate finding a suitable settlement site.

SETTLEMENT CUES

Since relocating after settlement is impossible, selecting a good settlement site is probably the most critical decision a barnacle larva will make. To aid in this decision, barnacles use cues from the environment to distinguish good sites from bad. A good cue is one that reliably predicts locations on the shore where individuals are likely to survive and reproduce. Hence, one of the most potent settlement cues used by cyprids is the presence of other individuals of the same species; if a site already has barnacles, then this is a strong indication that survivorship and mating opportunities are good. Other cues that have been demonstrated to influence the selection of a settlement site include bacterial films, surface texture, and even the presence of predators.

INTERACTIONS

Many studies have shown that once a barnacle settles on the shore, it must deal with a suite of physical and biological factors in order to survive, grow, and reproduce. One of the earliest, and perhaps most influential, of these studies was done by Joseph Connell in the mid-1950s on the rocky shores of Scotland. In this system were two species of barnacles: *Chthamalus stellatus* and *Semibalanus* (formerly the genus *Balanus*) *balanoides*. Connell observed that as adults, *Chthamalus* were restricted to a region of the shore above *Semibalanus,* but as juveniles their distribution extended down into the region of the shore occupied by *Semibalanus*. To determine why the distribution of adult *Chthamalus* was smaller than that of the juveniles, Connell performed the first manipulative field experiment in ecology. He first transplanted rocks covered with *Chthamalus* into the region where *Semibalanus* was abundant. Then, after allowing *Semibalanus* to settle on them, he removed the *Semibalanus* from half of each rock. Thus, half of each rock contained a mixture of *Semibalanus* and *Chthamalus,* while the other half had only *Chthamalus*. This treatment tested the idea that *Semibalanus* had a negative affect on the survival of *Chthamalus*. Cages to keep out predators were erected over half of the rocks to determine whether predation could explain the distribution of *Chthamalus*. Subsequent observations revealed that, when by themselves, *Chthamalus* survived just fine, but those in the mixture treatment were either crushed or undercut by the faster-growing *Semibalanus*. The presence/absence of predatory snails had little effect on this outcome. Based on this, Connell concluded that *Chthamalus* was actually capable of living lower on the shore but was prevented from doing so by the competitively superior *Semibalanus*. Additional observations suggested that the upper limit of *Chthamalus* was determined by physical factors such as desiccation. Thus, this study showed that competition for space is important in determining the structure of intertidal communities and that lower limits are determined by biological factors, upper limits by physical factors.

Studies done since Connell's have shown that barnacle abundances are affected by a variety of other processes in addition to interspecific competition. For example, barnacles are attacked by a myriad of predators including sea stars, flatworms, nemerteans, predatory snails, and, in the case of gooseneck barnacles, even shore birds. Most predatory snails attack barnacles by drilling through the barnacle's test. The time needed to drill through the test can take many hours and increases with the size of the barnacle. Beyond a certain size the time needed to successfully attack becomes so great that the predator would have to continue drilling even while exposed during low tides. Since doing so would increase the snails' risk of desiccation, they usually do not attack barnacles beyond a certain size. Thus, one way barnacles can protect themselves from predation is by growing rapidly enough to reach this size.

One species of barnacle found in the Gulf of Mexico, *Chthamalus anisopoma,* has evolved a different method of protecting itself from attack. Instead of drilling, its principal predator, *Acanthina angelica,* sits atop the barnacle and uses a large "tooth" to pry its way through the operculum. However, if exposed to *Acanthina* when young, *C. anisopoma* can adopt a "bent" morphology. That is, they grow in such a way that the operculum shifts from the top to the side of the test (Fig. 4). Since *Acanthina* has great difficulty accessing the operculum in this location, adopting this bent morphology greatly reduces the risk of attack by *Acanthina*. However, this protection comes with a cost: to grow bent requires the loss of part of the gonad. Thus, while a bent individual may have a higher chance of survival than a normal barnacle, it also has a lower reproductive potential. This trade-off has important evolutionary consequences and prevents the entire population from adopting the bent morphology. Since the abundance of mussels and algae in this system is affected by the relative abundance of the two types of barnacles, this, in turn, also has important consequences for the structure of the community.

Other studies have shown that, high on the shore, limpets can also affect barnacle abundances. However, exactly how they do so depends upon several different direct and indirect interactions. Barnacles and limpets compete

FIGURE 4 Top and side views of the conic (left) and bent (right) morphs of *Chthamalus anisopoma* showing the different locations of the operculum. Photograph by D. Lohse.

directly for space, so a decline in abundance of one can lead to an increase in abundance of the other. However, barnacles offer juvenile limpets protection from desiccation, so limpet recruitment is enhanced by the presence of barnacles. Thus, an increase in barnacles could actually cause the number of limpets to increase. Likewise, because barnacles and algae compete for space, by grazing on algae limpets actually assist the barnacles by reducing the abundance of a competitor. Thus, whether and how much a change in limpet abundance affects the number of barnacles depends upon the relative strengths of each interaction that makes up the complex interaction web among limpets, barnacles, and algae.

Intraspecific interactions also play an important role in determining barnacle abundances on the shore. Because barnacles need to be near each other to reproduce, and living together can also reduce the risks of desiccation, it is beneficial to live near conspecifics. However, because this strategy is true for all individuals, it is not uncommon for barnacles to experience severe crowding. Studies have shown that crowding can lead to higher mortality, slower growth, and lower reproductive output. Thus, this type of intraspecific competition has profound negative effects on an individual's fitness.

In some cases, crowding manifests itself through the production of hummocks. Hummocks are areas in which the densely packed barnacle cover has developed a dome-like, or mounded, appearance (Fig. 5). Although barnacles typically grow by increasing in diameter, crowded individuals are prevented from doing so by their neighbors. Instead, they grow upward, which causes them to adopt a cylindrical shape instead of the typical conic morphology. Not all

FIGURE 5 Barnacle hummocks on the shore. The inset shows a crowded barnacle upon which several other barnacles have settled. Note the difference in their morphology: crowded (tall, cylindrical) vs noncrowded (cone shape). Photograph by D. Lohse.

crowded individuals grow at the same rate, and, as a result of differences in water flow and food availability, those that grow the fastest (and are therefore the tallest) end up surrounded by slower-growing (and therefore shorter) individuals. The end result is a hummock. Although individuals in hummocks can be quite tall, their area of attachment to the substrate tends to be relatively small. Further, because they are crowded, their tests tend to be thinner, which increases their fragility. Thus, hummocks tend to be inherently unstable. Any disturbance that kills even just a few barnacles can end up destabilizing the entire formation; the loss of a few individuals can result in the denudation of barnacles from large areas of rock.

Interestingly, in some cases it is the barnacles themselves that sow the seeds of their own destruction. Although many species compete for space with barnacles, for some their presence actually facilitates settlement. For example, along the shores of California the settlement of both mussels *(Mytilus)* and algae *(Endocladia)* is enhanced by the presence of barnacles. However, because this interaction ultimately results in the barnacles being smothered, it is usually detrimental to the barnacles. Such facilitation is often needed for succession to occur following a disturbance.

NEW DIRECTIONS IN RESEARCH

While past studies have shown that processes such as competition and predation affect barnacle abundances on the shore, recent studies have demonstrated that what happens to larvae while in the plankton is also important. For example, while in the plankton, larvae are subject to predation by planktonic predators and to being carried about by currents and other oceanographic processes. In fact, given the length of time they spend in the plankton (10–45 days), and assuming a current velocity of 10 cm/s, barnacle larvae could end up 80 to 400 km from where they were released by their parent. Thus, current thought is that the larvae that settle at a site are probably not produced by the local population. This decoupling of larval input from local production, and how it affects population abundances on the shore, is of great interest to marine ecologists.

Interestingly, recent genetic data suggests that, although they have the potential to travel great distances, barnacle larvae typically disperse much less than this. Recent advances in coupled biological–oceanographic models, which link larval behavior with realistic depictions of oceanography, suggest that directed swimming behavior (e.g., swimming to maintain position in surface water) may be responsible for the local retention of larvae. These models have implications beyond barnacles and may be fundamentally important to the management of harvested species such as rockfish, abalone, and sea urchins.

SEE ALSO THE FOLLOWING ARTICLES

Competition / Facilitation / Larval Settlement, Mechanics of / Limpets / Recruitment / Succession

FURTHER READING

Bertness, M. D., S. D. Gaines, and S. E. Yeh. 1998. Making mountains out of barnacles: the dynamics of acorn barnacle hummocking. *Ecology* 79: 1382–1394.

Caley, M. J., M. H. Carr, M. A. Hixon, T. P. Hughes, G. P. Jones, and B. A. Menge. 1996. Recruitment and the local dynamics of open marine populations. *Annual Review of Ecology and Systematics* 27: 477–500.

Pawlik, J. R. 1992. Chemical ecology of the settlement of benthic marine invertebrates. *Oceanography and Marine Biology Annual Review* 30: 273–335.

Raimondi, P. T., S. E. Forde, L. F. Delph, and C. M. Lively. 2000. Processes structuring communities: evidence for trait-mediated indirect effects through induced polymorphisms. *Oikos* 91: 353–361.

Underwood, A. J. 2000. Experimental ecology of rocky intertidal habitats: What are we learning? *Journal of Experimental Marine Biology and Ecology* 250: 51–76.

BEACH MORPHOLOGY

GARY GRIGGS

University of California, Santa Cruz

Beaches are the loose deposits of sand, gravel, or shells that cover the shoreline in many places. Beaches serve as buffer zones or shock absorbers that protect the coastline, sea cliffs, or dunes from direct wave attack. They also provide important coastal recreational areas for millions of residents and visitors around the world, and they are the most intensively used parts of the coastal environment. The landward edge of the beach may be a seacliff, sand dune, vegetation line, or seawall. In the seaward direction beaches extend to water depths of about 10 meters beyond where there is usually little seasonal sediment movement due to normal wave action. Although beaches may appear wide and stable during the summer months under periods of low, gentle waves, they can erode very quickly when attacked by hurricanes, large storms, or heavy surf. In winter they change their entire character or may even disappear altogether. Beaches can undergo significant change in response to human activity in addition to their seasonal and storm cycles. The sources and losses of beach sand, the interaction of waves, tidal action, and wind are each important in determining the composition, size, and shape of any individual beach. About 5280 km, or 30%, of the entire continental coastline of the United States consists of beaches.

FORMATION OF BEACHES AND SOURCES OF BEACH SAND

Beaches form where there is enough loose or unconsolidated material available and where there is some suitable coastal environment in which the waves and coastal landforms will allow these sediments to accumulate. The materials making up beaches can vary in size from very fine-grained sand to pebbles, cobbles, and even boulders (Fig. 1A–C). The sediment may be either terrigenous or land-derived, having been transported to the shoreline by rivers and streams, or through erosion of the coastal bluffs or cliffs. In tropical areas or where land-derived material is lacking, beach sediment may be biogenous in origin and consist of broken bits of coral or shells of nearshore organisms (Fig. 2A, B). On volcanic islands such as Hawaii, beaches may be black and consist almost entirely of broken lava and volcanic minerals (Fig. 3).

FIGURE 1 (A) A fine-grained sand beach in the Monterey Bay area of Central California. (B) A beach in southeast Alaska consisting of rounded granitic pebbles. (C) A beach at Newport, Oregon, consisting of rounded basaltic cobbles. Photographs by the author.

Fine-grained sand beaches tend to be quite flat (usually a slope of only a few degrees), whereas coarser pebble or cobble beaches are usually much steeper, often as much as 15 to 20 degrees or more. These slope differences are due to the differing permeabilities of the materials and the resulting balance between wave uprush and backwash. Where the beach consists of fine-grained sand, permeability is low, and little of the broken wave that washes up the beach face can percolate into the sand. Thus the backwash, or the water returning to the surf zone, will

FIGURE 2 (A) A beach at Doubtless Bay, New Zealand, consisting almost entirely of clam shells. (B) A beach on the island of Raratonga consisting of coral and the remains of other tropical organisms as well as small bits of black lava. Photographs by the author.

carry the fine-grained sediment in suspension back down the beach face, maintaining a low-sloping beach. With a cobble beach, in contrast, the voids between the cobbles are large, permeability is high, and much of the uprushing wave percolates into the beach face, so much less water washes back down the beach face, reducing the seaward movement of the cobbles. Thus the coarser materials can accumulate on the beach, where they are transported by larger winter waves.

THE MOVEMENT OF BEACH SAND AND BEACH SHAPE

The sediment on beaches is in constant motion, being driven by the waves breaking on the shoreline and also wave-induced near-shore currents. The lower, less energetic, and longer-period summer waves tend to pick up the fine sand that is suspended where the waves break and to carry it up the beach, where it is deposited before the next wave breaks. Through the late spring, summer, and early fall months, these lower-energy waves transport the sand onshore, building up a high berm and broadening the beach (Fig. 4A). By the end of summer many beaches

are at their widest extent, providing lots of area for beach goers. Where the beach is very wide and a persistent onshore wind blows, dunes may even begin to form on the back beach, and in some cases, where there are no topographic barriers and an abundance of sand, large dune fields may form and migrate inland.

FIGURE 3 In Tahiti, the primary source of beach material is black volcanic rock, resulting in a black sand beach. Photograph by the author.

With the arrival of the first winter storms, which bring higher, shorter-period, and more energetic waves, the sand is thrown into suspension, scoured off the beach face, and carried back into the surf zone, leading to erosion or narrowing of the summer berm. As the winter storm waves continue, the beach face is lowered, the sand moves offshore to form a series of bars and troughs, which serve to protect the beach by causing the winter waves to break further offshore. Winter beaches are then narrower, usually coarser-grained, and may be completely eroded (Fig. 4B, where the sand has not disappeared but has only been carried offshore into bars and troughs that are parallel to the

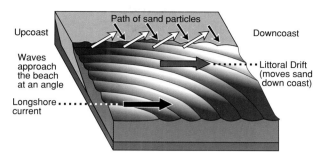

FIGURE 5 A longshore current develops along the shoreline where waves approach the beach and break at an angle. This flow of water moves sand alongshore as littoral drift. Drawing by Shannon Griggs.

FIGURE 4 Beach in Santa Cruz, California, (A) in October 1997, when it was wide and sandy at the end of summer; (B) in February 1998, with much sand removed and waves attacking the cliffs. Photographs by the author.

beach). Most beaches undergo some seasonal change each winter and then return or reform the following spring and summer. In some cases, beaches may be 2 to 4 meters lower in the winter than in the summer months, and pebbles or cobbles may be exposed as the sand is moved offshore.

In addition to the seasonal on- and offshore movement of sand, in any location where the waves approach the shoreline at an angle, sand or beach materials are also transported alongshore, either up- or downcoast. As each wave breaks, it stirs up the sand in the surf zone and carries the sand up the beach face. When a wave breaks at some angle to the shoreline, this path of the sand grains up the beach face is also at an angle that depends on the direction of wave approach. When carried back down the beach face by backwash, the individual particles move perpendicular to the shoreline (Fig. 5). So sand grains are transported in a zigzag or sawtooth pattern alongshore. This process is called littoral drift and is responsible for transporting thousands to hundreds of thousands of cubic meters alongshore each year.

So, although beaches may appear to be wide and stable environments in the middle of summer, they are constantly adjusting their size and shape in response to changing wave conditions. Sand moves on- and offshore seasonally as the beach is eroded and then rebuilt. Beach materials are also often moved alongshore by waves approaching the shoreline at an angle. Beach sand is transported many miles alongshore until it is ultimately transported offshore onto the continental shelf, carried down to the deep sea floor through a submarine canyon, or blown inland into a dune field.

Human activities have greatly altered beaches along many coastlines. The construction of dams and reservoirs in coastal watersheds, and sand and gravel mining along streams, have trapped or removed millions of cubic meters of sand that would formerly have been delivered to the shoreline to nourish beaches. Armoring coastal bluffs with seawalls and revetments temporarily halts bluff erosion but also eliminates these bluffs as a source of sand for the adjacent beaches. Large jetties, breakwaters, and groins all disrupt or trap littoral drift and serve to widen upcoast beaches but in many places have reduced the size of downcoast beaches. There are other locations where large volumes of sand have been added to the beaches, known as beach nourishment, to widen them or to make up for losses caused by upcoast human activities or construction.

SEE ALSO THE FOLLOWING ARTICLES

Coastal Geology / Sea Level Change, Effects on Coastlines / Surf-Zone Currents / Tidepools, Formation and Rock Types

FURTHER READING

Bascom, W. 1980. *Waves and beaches: the dynamics of the ocean surface.* New York: Doubleday.

Davis, R. A. Jr., and Fitzgerald, D. M. 2004. *Beaches and coasts.* Malden, MA: Blackwell.

Komar, P. D. 1998. *Beaches processes and sedimentation.* Upper Saddle River, NJ: Prentice Hall.

BENTHIC-PELAGIC COUPLING

JON D. WITMAN

Brown University

Benthic-pelagic coupling (BPC) refers to the linkage between the water column and the bottom. The process can be relatively passive, such as the sinking of particles, or highly energetic, such as upwelling or downwelling. This key process impacts all types of bottom communities including those on rocky shores, on coral reefs, and in soft bottom habitats. BPC delivers particulate food, organic and inorganic nutrients, and offspring in the form of larvae, spores, or propagules to the sea floor. BPC is an integral component of food webs, resulting in bottom-up regulation (control by nutrients, food resources) of marine webs. Furthermore, it can trigger a linkage between bottom-up and top-down (regulation by predation, grazing) control.

ECOLOGICAL IMPORTANCE

The ecological importance of BPC has been appreciated in deep-sea and coastal sedimentary habitats for decades, but it is only recently that the role of BPC in coral reefs, rocky intertidal, and subtidal habitats has been investigated. Earlier theories predicting the distribution and abundance of organisms in these habitats were focused on the roles of biotic interactions and environmental stress acting at local (0–20 km distance) spatial scales. The emerging recognition of the importance of oceanographic processes in marine benthic ecosystems has led to development of models of community structure applicable to larger landscape (20–200 km) and regional (200–4000 km) spatial scales. Major forms of BPC are sedimentation, upwelling, downwelling, and Langmuir circulation.

Sedimentation

Sedimentation is defined as the settling of solids out of a liquid. Sedimentation can be of inorganically or organically derived material. Inorganic sediments are either chemical precipitates or geologically derived material. They include a broad range of sedimentary particles—from sands, silts, and muds washing off the land, to volcanic material of all sizes, to dust deposited from the atmosphere. Major sedimentation of organic material occurs from the coastal zone to the deep-sea floor. Organic material sinking to the bottom includes phytoplankton and zooplankton of all sizes, macroalgae and marine angiosperms, flocculents (marine snow), woody debris, and animal carcasses. Reflecting this input, bottom sediments are a mixture of inorganic and organic matter.

Generally, sedimentation rates of small passive particles are inversely proportional to water velocity and particle size, as predicted by Stokes's law. BPC sedimentation forms soft-bottom habitats in places where water velocity is not high enough to keep particles in suspension. Sedimentation blankets the sea floor, creating a layer of soft-substrate habitat for invertebrates and fish that is, on average, 1.2 km thick.

Sediments are named after the type of material from which they were created. There are calcareous oozes composed of plant and animal skeletal material such as foraminiferal oozes (mostly *Globerigina*), pteropod shell oozes, and coccolith oozes composed of algae in the Coccolithophoridae. Siliceous oozes of radiolarians and diatoms are another major category of organic sediments. The biomass, distribution, and growth rates of invertebrates are correlated with the organic content of the sediment, which represents food for deposit feeders. The sedimentation of organic matter thus has a bottom-up influence on soft-bottom food webs.

Sedimentation occurs on rocky shores as well as soft-bottom habitats. On rocky shores it represents a major source of stress and disturbance for algae and invertebrates. By hindering settlement, growth, and survival of algae and epifaunal invertebrates, sedimentation reduces abundance and species diversity in rocky-shore communities. Many aggregation-forming species such as subtidal kelp beds, polychaete reefs, and mussel beds trap sediments, causing other direct and indirect ecological effects among the associated species and habitat formers (foundation species).

Development, pollution (sewage, industrial), and ocean dumping resulting from human activities have greatly increased the sediment load of many coastal waters. Consequently, negative ecological impacts of sedimentation are increasing and widespread. There is a need for more experimental studies of sedimentation effects on rocky shores and coral reefs.

Upwelling

Upwelling is a major type of BPC created by a variety of mechanisms including wind forcing, Coriolis deflection, and flow–topography interactions operating on a range of spatial and temporal scales. At the largest spatial scales, upwelling is driven by wind that blows parallel to a coastline, pushing a current of surface water in the same direction (Fig. 1). As a result of the Coriolis

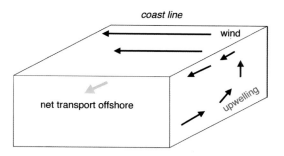

FIGURE 1. Diagram of wind-driven upwelling along a coastline in the southern hemisphere. Wind blowing along the coast creates a surface current in the same direction, as indicated by the parallel arrows. The Coriolis force deflects the current to the right in the northern and to the left in the southern hemisphere. This gradually causes the current to move perpendicular to the wind and offshore, as indicated by the net transport arrow. As the surface water is pushed offshore, it is replaced by upwelling currents from deeper areas that are often enriched in nutrients. When the upwelling currents relax or stop, the direction of net transport shown here is reversed so that it flows onshore instead of offshore. These onshore currents bring larvae of intertidal invertebrates to the shore.

force, the current deflects to the right in the Northern Hemisphere and to the left in the Southern Hemisphere. Eventually, this causes the current to move in a direction that is perpendicular to the wind, that is, offshore. This is called Ekman transport. The surface water that is pushed offshore is replaced by cool, deep water, often rich in nutrients. This form of wind-driven upwelling is common in eastern boundary currents on the western sides of continents. The Benguela Upwelling system off western South Africa, the California Current upwelling system, the Peru–Chile Current system, and the Canary Current upwelling system off the Iberian Peninsula and northwestern Africa are the principal examples of this type of wind forced upwelling. Major wind-forced upwelling also occurs in a western boundary current, the Somalia current off the coasts of Somalia and Oman in the northwest Indian Ocean.

The Coriolis effect also plays a major role in equatorial upwelling phenomena. Equatorial upwelling occurs when winds blowing westward along the equator drive currents from east to west. The Coriolis deflection causes the current on the Northern Hemisphere side of the equator to deviate to the right, whereas the currents moving parallel to the equator in the Southern Hemisphere diverge to the left. These deflections push the surface water away from the equator, which triggers upwelling as deep water replaces surface equatorial water. Moving deeper from westward-flowing equatorial surface currents, one finds equatorial countercurrents flowing to the east. For example,

the eastward-flowing equatorial countercurrent drives upwelling in the Galápagos Islands, when it runs into the western side of the archipelago.

One of the first ecological effects of upwelling noted was due to elevated nitrate levels in upwelled waters. It was hypothesized to stimulate the growth and production of macroalgae and explain the high abundance of macroalgae along shores bathed by the Benguela upwelling system. This hypothesis was tested and confirmed in the Chilean rocky intertidal, where the growth rates and biomass of a red alga were higher at sites of strong upwelling.

Researchers working along the coast of California discovered that wind-forced upwelling affects the supply of larvae as well as nutrients. When winds blowing along the central coast relax, upwelling currents stop and are replaced by onshore currents transporting invertebrate larvae entrained in surface waters to the coast. Pulses of high barnacle and crab settlement occur in the rocky intertidal as the water flows onshore during these upwelling relaxation events. The opposite situation occurs when upwelling resumes, as barnacle larvae are transported far offshore by surface currents (Fig. 1), explaining low barnacle settlement during upwelling periods.

Since intertidal recruitment varies with the timing and spatial extent of wind-forced upwelling, it is logical to hypothesize that species interaction strength and community structure varies with upwelling as well. This hypothesis has been tested along the West Coast of the United States and Chile. One of the effects is that cold water delivered by upwelling decreases the feeding rate of a keystone species, the predatory sea star, *Pisaster ochraceous*. Broad regions of the U.S. West Coast have been classified as the Intermittent Upwelling Region (IUR) in Washington and Oregon, the Persistent Upwelling Region (PUR) off central to northern California, and the Weak Upwelling Region (WUR) in southern California. Phytoplankton food concentration and the recruitment of mussels are highest in the IUR, indicating a bottom-up effect of upwelling. Interaction strength, measured as the per-capita and per-population rates of sea star predation on mussels, show regional-scale trends driven by upwelling. For example, per-population predation rates are higher in strong- than in weak-upwelling areas, lending support to the notion that BPC links bottom-up and top-down effects in intertidal food webs. To understand the causes of the often large variability in community structure observed from site to site, several authors have suggested that the upwelling regime should be measured at the same local scale as the study sites.

Rocky intertidal community structure and the recruitment of mussels and barnacles is sharply demarcated on a regional spatial scale along the coast of Chile. Recruitment is low in the strong upwelling region, north of 32° south latitude, and high south of it, where upwelling is weaker. High mussel recruitment in the weak-upwelling area leads to high cover of adult mussels, which typically outcompete other intertidal species. In this manner, upwelling changes intertidal community structure on large, regional spatial scales. In the Chilean intertidal system, the bottom-up effects of upwelling do not typically extend up the food web to herbivores of macroalgae and predators of barnacles and mussels, indicating that the extent of linkage between bottom-up and top-down control may be limited by factors other than prey availability such as predator recruitment, predator diet, and removal of predatory species by humans.

Internal Waves

Upwelling is often caused by the interaction of flow and topography. Internal wave upwelling and downwelling occurs in density-stratified waters worldwide as the tide flows over abrupt bottom topography (banks, ledges, reefs, continental slopes). Depending on the nature of the stratification, lee waves propagate forward obliquely or horizontally as a packet of internal waves as the tide changes (Fig. 2). Downward-flowing currents are followed by upwelling currents as the internal wave train moves away from the ledge. As observed on rocky shores of California, Chile, and on coral reefs, the train of internal waves breaks and becomes an internal tidal bore when it moves into shallow water. Internal tidal bores are well developed during periods of strong water column stratification. They deliver subthermocline water to the shore.

Similar to wind-forced upwelling, internal waves affect larval transport and recruitment and create bottom-up effects on benthic food webs. For example, internal waves can transport invertebrate larvae across the coastal shelf toward shore. In some cases, the cross-shelf larval transport has been linked to higher intertidal settlement and recruitment of bivalves, barnacles, and crustaceans. Two mechanisms have been proposed to explain larval transport by internal tides. One maintains that larvae are concentrated at or near the surface in convergence zones (slicks, Figs. 2, 3) above weak downwelling limbs of internal waves. The larvae and propagules are then carried along near the surface and toward the shore as the wave train propagates forward. Another mechanism involves larvae entrained in the downwelling phase of the internal tidal bore, which is conspicuous as a front of warm water. Larvae are transported to the intertidal as the front moves onshore.

FIGURE 3 Photograph of a slick from part of a large packet of internal waves on Cashes Ledge, Gulf of Maine. Downwelling by these internal waves pushes a midwater layer of dense phytoplankton to the rocky sea floor where filter-feeding invertebrates thrive. For scale, the dark object to the right is a 5.5-meter-long inflatable boat. Photograph by J. Witman.

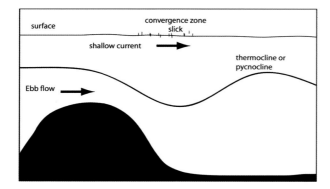

FIGURE 2 Diagram of an internal wave forming on the lee side of a bank. When the tide ebbs over the bank, water that was held on the lee side of the bank during the flood tide rushes forward forming an initial internal wave that depresses the thermocline and pycnocline. This is the downwelling phase. The parcel of water that was depressed by the initial tidal energy reaches a depth at which it is lighter (less dense) than water at the surrounding depth, so it rises. This is the upwelling phase of an internal wave. A convergence zone forms above the trough of the internal wave and is often visible as a slick. Repeated depression of the thermocline/pycnocline as the water flows over the bank generates a packet or train of internal waves.

Bottom-up effects of internal waves and bores are related to the strong vertical motion of internal waves that mixes nutrients and phytoplankton in the water column and pushes planktonic food and nutrients to the sea floor. During the stratified season, an energetic internal wave regime drives the subsurface chlorophyll maximum (SCM) layer 15 to 30 meters down to the bottom several times per day on subtidal pinnacles in the central Gulf of Maine. Enhanced food supply at the internal-wave downwelling zone at the tops of the pinnacles resulted in faster growth rates of mussels and sponges, indicating a bottom-up effect on subtidal food webs.

Internal tidal bores break and run up the reefs in the Florida Keys, delivering water that is rich in nitrate, phytoplankton, and zooplankton. The effect of breaking internal bores is well developed on the deep reef (>35 m). Upwelled water during the internal-bore phase is 10–40 times higher in nutrients (nitrate, phosphate) than during nonbore periods. Demonstrated biological responses to the internal bore regime on these coral reefs include use of nitrogen by benthic macroalgae, which are most abundant deeper on the reef, and elevated growth of a branched coral in the zone of internal-bore influence. In general, nutrients supplied to coral reefs by breaking internal waves may be a vitally important source of nitrogen for corals in the nutrient-limited environment of tropical oceans.

One of the few studies of upwelling in the subtidal zone suggests that the pace of community change may be faster at upwelling sites, because the diversity of epifaunal invertebrates doubled in a year at a strong subtidal-upwelling site in the Galápagos, as a result of high recruitment. Larvae are not necessarily advected offshore by localized upwelling in this subtidal system in which local physical regimes are often dominated by internal wave–generated upwelling and downwelling.

Langmuir Circulation

Like internal waves, Langmuir circulation results in convergence zones at the surface of the ocean, visible as slicks that aggregate biological material, but the mechanisms that produce them are completely different. For example, Langmuir circulation is driven by the wind, requiring wind speeds of greater than 3.0 meters per second to generate the characteristic rotating tubes of water aligned parallel to the wind. The slicks often contain lines of foam and are known as windrows. They are caused by wind shear acting on the surface of the water, which dissipates with water depth. Basically, the rotating tubes or vortices are caused by variation in the wind speed and resulting shear. Downwelling currents are formed in the convergence zone between the tubes; they typically exceed the velocity of the upwelling limbs. Small fish and jellyfish may reach greater abundances in Langmuir slicks than elsewhere. Although discovered in 1938, the biological potential of Langmuir circulation as a mechanism influencing benthic communities is only beginning to be realized. For example, in some cases the rotating Langmuir circulation can span the entire water column to a depth of 15 meters, as observed off the coast of New Jersey in the United States. The vertical flow in these supercells vigorously resuspended bottom sediments, apparently transporting sediment near the bottom. As downwelling phenomenona, the BPC implications of such supercells are that they are likely important mechanisms to supply larvae, propagules, and food to the shallow sea floor on an episodic basis.

SEE ALSO THE FOLLOWING ARTICLES

Food Webs / Near-Shore Physical Processes, Effects of / Nutrients / Upwelling / Waves, Internal

FURTHER READING

Airoldi, L. 2003. The effects of sedimentation on rocky coast assemblages. *Oceanography and Marine Biology: an Annual Review* 41: 161–236.

Farmer, D., and L. Armi. 1999. The generation and trapping of solitary waves over topography. *Science* 283: 188–190.

Gargett, A., J. Wells, A. E. Tejada-Martinez, and C. E. Grosch. 2004. Langmuir supercells: a mechanism for sediment resuspension and transport in shallow seas. *Science* 306: 1925–1928.

Leichter, J. J., H. L. Stewart, and S. L. Miller. 2003. Episodic nutrient transport to Florida coral reefs. *Limnology and Oceanography* 48: 1394–1407.

Menge, B. A., C. Blanchette, P. Raimondi, T. Fridenburg, S. Gaines, J. Lubchencho, D. Lohse, G. Hudson, M. Foley, and J. Pamplin. 2004. Species interaction strength: testing model predictions along an upwelling gradient. *Ecological Monographs* 74: 663–684.

Navarrete, S. A., E. A. Wieters, B. R. Broitman, and J. C. Castilla. 2005. Scales of benthic-pelagic coupling and the intensity of species interactions: from recruitment limitation to top down control. *Proceedings of the National Academy of Sciences USA* 102: 18046–18051.

Photographic atlas of internal waves. http://www.internalwaveatlas.com/Atlas_index.html.

Pineda, J. 1999. Circulation and larval distribution in internal tidal warm bore fronts. *Limnology and Oceanography* 44: 1400–1414.

Roughgarden, J., S. D. Gaines, and H. Possingham. 1988. Recruitment dynamics in complex life cycles. *Science* 241: 1460–1466.

Shanks, A. L., J. L. Largier, L. Brink, J. Brubaker, and H. Hoff. 2000. Evidence for shoreward transport of meroplankton by an upwelling relaxation front. *Limnology and Oceanography* 45: 230–236.

Witman, J. D., M. R. Patterson, and S. J. Genovese. 2004. Benthic pelagic linkages in subtidal communities: influence of food subsidy by internal waves, in *Food webs at the landscape level*. G. Polis, M. Power, and G. Huxel, eds. Chicago: University of Chicago Press, 133–153.

BIODIVERSITY, GLOBAL PATTERNS OF

MIRIAM FERNÁNDEZ, JUSTIN HOLL, AND SARA KIMBERLIN

Pontificia Universidad Católica, Santiago, Chile

Biodiversity, although present at all levels of organization of organisms, is most commonly measured by ecologists and biogeographers as species richness, or the number of species found at a particular point in space or time. By this definition, a section of rocky intertidal coast where 100 different species of fish, algae, and invertebrates can be found has greater biodiversity than a location of similar size where only 50 different species are found. Understanding patterns of global biodiversity, and the underlying processes, will continue to be a major priority as issues such as global climate change, conservation, and sustainable resource use increase in importance worldwide.

MEASURING BIOLOGICAL DIVERSITY

Although historically the number of species has been the most commonly used measure of biodiversity, scientists increasingly emphasize alternative methods of characterizing biological diversity, looking beyond the mere number of species to consider, for example, diversity in morphologies, genetics, or ecological relationships of organisms. Thus, although many of the hypotheses for global patterns of biodiversity use species richness as the measure of biodiversity, it is important to keep in mind that other ways of measuring the diversity of organisms may lead to important insights into the dynamics underlying biogeographic patterns.

GLOBAL PATTERNS OF BIODIVERSITY

Biodiversity varies spatially on the globe. The study of this variation has made rapid advances in recent decades through improvements in available datasets and analytical tools such as geographical information systems and remote sensing techniques. Interest in global patterns of biodiversity and the main factors determining species richness has also increased with the need to understand how biodiversity might change under different scenarios of global climate change, as well as to inform conservation and sustainable resource use efforts.

Although the total number of species on earth is yet unknown, it is clear that the tropics harbor many more species than colder environments. The latitudinal gradient in species richness, wherein species richness peaks near the equator and declines toward the poles, is a widely recognized phenomenon that holds true for many taxa, from marine to terrestrial organisms, and in habitats as diverse as the open ocean, coastal zones, rainforests, deserts, rivers, and lakes. In the ocean, the typical decline in species richness at high latitudes has been reported on continental shelves, in the open ocean, and surprisingly, in the environmentally stable deep sea. The pattern is so general that it has been observed from microscopic to macroscopic organisms, including the

most conspicuous and abundant groups such as molluscs (Fig. 1), crabs, and fishes. Moreover, diversity gradients not only exist for species but also for higher taxa (bauplans). However, some authors have claimed that there is not convincing evidence for the latitudinal gradient in species diversity across all marine taxa, comparable with that found for terrestrial taxa, based on exceptions to this pattern (e.g., infauna, certain peracarid crustaceans, southern oceans). Figure 2 shows one of these exceptions: the inverse latitudinal diversity pattern exhibited by molluscs (bivalves, gastropods, and placophora) in the southeastern Pacific.

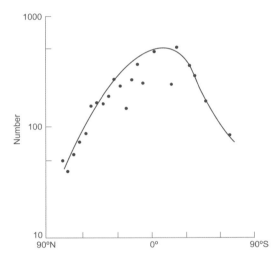

FIGURE 1 The relationship between bivalve diversity and latitude (after Stehli *et al.* 1967).

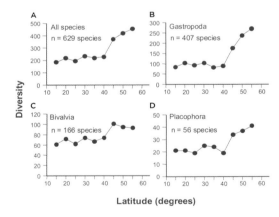

FIGURE 2 Relationship between diversity of molluscs and latitude in the southeastern Pacific (after Valdovinos *et al.* 2003).

In the ocean, species richness also varies with depth. The number of species peaks at about 2000 meters depth and then declines toward shallower and deeper waters in all taxa and oceans. Although our knowledge of the biodiversity of the deep ocean is particularly limited, and a remarkable number of new species are being discovered in the world ocean, the patterns reported seem unlikely to reverse even as more species are described.

Biodiversity patterns are scale dependent. It is interesting that the generalization of increased diversity with decreasing latitude applies to large-scale analysis (e.g., number of species per one-degree latitudinal band) but not necessarily to the number of species living within a small area (e.g., a square meter). For example, using a simple three-unit scale—local (patch dynamics), regional, and global—different biodiversity gradients can be obtained at different scales. The latitudinal species richness gradient is most detectable at broader scales. At a regional scale using latitudinal bins or blocks of 5–10 degrees latitude, strong gradients have been demonstrated in both hemispheres for bivalves, gastropods, coastal fishes, sea stars, planktonic foraminifera, benthic foraminifera, and pelagic copepods. These observed gradients are not uniform, but they do have highest diversity values between 0 and 30 degrees latitude and lowest values between 60 and 90 degrees.

A number of recent studies have suggested that if samples are taken on significantly smaller spatial scales, it can be far more difficult to detect a latitudinal diversity gradient. A famous early study by Thorson on a small scale found no evidence of an equatorial–polar gradient for macrobenthic taxa associated with soft substrates. One complication of gradients and scales is that diversity is highly variable among habitats. For instance, coral reefs, which are restricted to low latitudes, host higher marine biodiversity than any other habitat. This factor alone could explain the hotspots of diversity found at low latitudes. The question of diversity gradients on smaller scales will require further work in all habitats from the tropics to the poles. Explicit consideration of the relevance of scale may allow for a more nuanced view of global patterns of biodiversity, hopefully leading to insights into and improved understanding of the dynamics that cause these patterns.

PRIMARY CAUSES OF LATITUDINAL BIODIVERSITY GRADIENTS

Although some authors have stated that an almost universal pattern must have a common explanation, the processes behind one of the most universal and intriguing features of nature, the latitudinal biodiversity pattern, have not yet been elucidated.

A Common Pattern or Common Cause

Several authors have stated that the general latitudinal pattern in species numbers must be related to some climatic factor, or combination of factors, that change in a consistent manner with latitude. Several factors could serve as suitable candidates, from contemporary variation in average temperature, primary production, or seasonality, to historical perturbations of the physical environment.

The major risk of looking at a common cause is confounding correlation with causation. However, a strong correlation between diversity and some physical parameters may be a powerful indicator of a possible cause. Evidence relating temperature and species diversity of terrestrial and marine organisms has been accumulated for decades. Temperature is also an excellent predictor of the diversity of planktonic, benthic (crabs, bivalves, gastropods), and pelagic organisms (e.g., fish) in the world oceans. It is also remarkable that the rate of change in crab diversity with changes in sea surface temperature is exactly the same in the southern Atlantic and Pacific oceans (Fig. 3). In spite of the striking associations between temperature and diversity among marine organisms, ecologists have so far failed to find the primary cause linking biological diversity

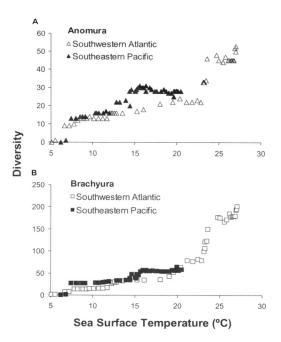

FIGURE 3 Plots of the relationship between annual mean sea surface temperature and diversity of anomuran (A) and brachyuran (B) crabs for species found between 10° and 56°S (after Astorga et al. 2003).

and temperature. The variation in diversity with depth, where there is no temperature difference, also remains unexplained.

Although most of the studies on causal factors have focused on latitudinal patterns of diversity, a factor that influences biodiversity on one spatial gradient, such as latitude, may also likely influence biodiversity along other spatial gradients for which the factor also varies (e.g., depth in the ocean, altitude on land). Thus, many of the hypotheses presented in the following subsections, developed to explain latitudinal gradients of diversity, may also inform our understanding of other types of spatial diversity gradients, such as gradients along depth in the marine environment. These hypotheses are broad, in that they apply to marine and terrestrial systems, but are also limited, in that they have been developed primarily using only species richness to measure biodiversity.

Identification of the Primary Cause of the Relationship between Temperature and Diversity

Considering the evidence linking temperature and diversity, the challenge is to identify the processes that link temperature and species diversity. A number of different hypotheses have been proposed to explain and understand the underlying causes of latitudinal biodiversity patterns, but only some of them have been tested in marine systems.

SPECIES-ENERGY HYPOTHESIS

A possible interpretation of the correlation between temperature and species diversity is that species richness is limited by energy supply. Solar energy is the most basic source of energy for photosynthesis and is also stored as the ambient temperature of the air or water in which organisms are immersed. Solar energy itself exhibits a strong latitudinal gradient, declining from the tropics to the poles. If productivity really represents the main constraint on biological diversity, then species richness might be strongly associated with the climatic factors that influence productivity, such as temperature. As has been shown previously, on a broad geographic scale, net primary productivity is higher at lower latitudes. It should be recognized, however, that other factors such as nutrients affect primary production and may affect diversity patterns, especially over smaller spatial scales. Therefore, the species–energy hypothesis proposes that the high biodiversity in the tropics compared to the poles is related to the fact that tropical regions exhibit greater primary productivity. Proponents of the species–energy hypothesis point to widespread evidence that productivity (or its climatic proxies) describes broad-scale diversity gradients better than alternative explanatory variables do.

According to the species–energy hypothesis, net primary productivity is critically important in determining biodiversity. In other words, somewhat simplified: Where there is more energy from the sun to allow more photosynthesis, there will be more algal matter, which will provide more food for herbivores, increasing their populations, thus providing more food for carnivores, increasing their populations, so that the total number of organisms will be greater than in areas with less primary productivity. The species–energy theory predicts that species richness for a given area is controlled by the population size of the total number of organisms of all species; therefore, the population-sizes of individual species are relatively unimportant. However, there is a logical limitation to this hypothesis: Why do a small number of species not monopolize the available energy? The hypothesis that species richness is limited by energy supply also assumes that habitats are saturated with species, which remains controversial.

As described previously, the species-energy hypothesis is partially based on the somewhat controversial idea that patterns in diversity observed at ecological and biogeographical scales can be explained by solely examining the population sizes of the total number of all species, without examining the specific biology of individual species. This approach requires the theoretical assumption that all individuals are identical and that all species are equal in competitive ability—a hypothesis known as the neutral or symmetric hypothesis. Because there is much support that species do, in fact, vary significantly in competitive ability, the neutral hypothesis is usually invoked more as a thought experiment and a structure for beginning to analyze biodiversity patterns, not as a stand-alone hypothesis. Its value is that it provides a simplified, unified structural approach that highlights factors long recognized as jointly influencing species richness: population density, area, and speciation rate.

NICHE-ASSEMBLY HYPOTHESIS

In contrast to the neutral hypothesis, the niche–assembly hypothesis incorporates from the start the idea that species are different in a variety of ways, including their niches and competitive abilities. The niche–assembly hypothesis of biodiversity views ecological communities as societies foremost, structured by species interactions as well as organisms' life histories, habitats, and trophic levels. The key concept in the niche–assembly hypothesis is adaptation of organisms to the climate and to each other. Classical niche–assembly hypothesis emphasizes the inherent uniqueness of all species in ecological communities. According to this perspective, competing species coexist in closed, stable assemblages by partitioning limiting resources through niche differentiation.

There are a number of ways that the niche-assembly hypothesis can be used to explain the latitudinal gradient in species richness. For example, the tropics might support more species than the higher latitudes because species in the tropics tend to occupy narrower niches, possibly because the resources in the tropics are more reliable and tend to be available year-round. This climatic stability allows for species to become increasingly specialized, so that one broad niche can be subdivided into multiple narrower niches, thus accommodating more species. In contrast, at higher latitudes, where significant climatic variation (e.g., seasonality in solar radiation, in temperature) is more common from season to season and year to year, such niche subdivision is not possible, because species that become too specialized for a specific narrow niche are at a higher risk of extinction when resource availability or climatic variables change. An important question, however, is what might drive this pattern: whether there are more species in the tropics because tropical niches tend to be smaller, allowing evolution of more species; or if niches are smaller in the tropics because the larger number of tropical species forces the subdivision of broader niches into narrower ones.

EVOLUTIONARY SPEED HYPOTHESIS

The difficulties encountered when trying to explain greater species richness in energy-rich environments, such as the tropics, disappear when energy (particularly temperature) is not related to species number but to evolutionary speed. Differences in evolutionary speed at different latitudes have been implicitly or explicitly assumed by several authors, and there is a considerable body of data that indicates higher evolutionary rates in the tropics. Several processes have been proposed to support the evolutionary speed hypothesis. One proposal is that since the climate in the tropics is warmer, generation times tend to be shorter, evolution tends to occur more rapidly, and therefore speciation is faster. The counterargument for this hypothesis, however, is that species' extinction rates, also tied to generation time, should increase as well, canceling out the effects of faster speciation rates, assuming speciation and extinction are in equilibrium. Higher mutation rates at higher temperatures can also favor higher speciation rates, and there is evidence supporting

this process. This may be accompanied by faster physiological processes at higher temperatures, which accelerates selection, leading to fixation of favorable mutants in populations. Although it can also be argued that the high diversity in tropical areas could be generated by reduced extinction rates, the limited fossil record does not provide evidence in this direction.

CENOZOIC RADIATION HYPOTHESIS

Finally, several authors have pointed out that the substantial increase in biodiversity during the Cenozoic Era, when a pronounced radiation event took place in the ocean, may explain why higher latitudes have lower biodiversity. Even though some polar marine taxa, especially in the Antarctic, went through intense radiation, most of the Cenozoic radiation took place in tropical and low-latitude regions. In fact, the latitudinal diversity patterns that we observe today can be traced back to the Cenozoic Era.

ALTERNATIVE WAYS OF MEASURING DIVERSITY

Although species richness has been the most commonly used measure of biodiversity, it is now frequently viewed as an inadequate metric taken by itself. Alternative definitions of biological diversity try to capture the true variety of life by looking beyond simply the number of taxa to include everything from genetic to ecological diversity. Species richness has been deconstructed, for example, to examine the morphologic, physiological, phylogenetic, and ecological relationship diversity among organisms.

The deconstructive approach to measuring biodiversity starts with the assumption that species richness does not sufficiently represent the true diversity of life, because it ignores the ways in which species differ from one another and in their responses to environmental changes. This approach emphasizes that it is critical to consider that the mechanisms controlling the number of species at a given location (migration, extinction, and diversification) are not independent of the life histories or physiological and ecological attributes of the species involved, because these attributes determine what environmental aspects are relevant to their survival. In other words, because the number of species in a given time and place depends on the interaction between the organisms' strategies and attributes and the characteristics of the environment in which they exist, any hypothesis that measures only numbers of species, while ignoring the similarities and differences in species' attributes, gives an incomplete picture of the complexity of the biodiversity pattern. To view the complete picture, one must consider biological attributes such as dispersal abilities, energetic demands, life history traits, physiological attributes, taxonomic distinctions (vertebrates vs invertebrates), ecological attributes such as food web position (decomposers, producers, primary consumers, and secondary consumers). For example, applying this deconstructive approach to the effects of energy availability on biodiversity would mean disaggregating mere species richness to examine the patterns in the attributes of species that are most relevant to the environmental variable of energy availability, for example, patterns in energy metabolism, endothermism versus ectothermism, body size, or diet.

An acknowledged characteristic of the deconstructive approach is that species richness can be disaggregated or deconstructed in many different ways and therefore achieve a better understanding of the environmental factor affecting biodiversity. With regard to the neutral or symmetric hypothesis, the deconstructive approach asserts that it is logical to first properly qualify the complexity of biodiversity patterns before attempting unification. Although some other approaches, such as the species–energy hypothesis, view environmental factors as causal and species as equivalent, the deconstructive approach claims that any hypothesis derived from environmental factors alone gives an incomplete view of the complexity of the biodiversity pattern. Although hypotheses driven by external factors may correlate to observed gradients of biodiversity, deconstruction allows researchers to explore in more complex detail how diversity patterns change depending on the criteria used to disaggregate richness, and it can be useful for explaining why certain groups with similar biological attributes, such as molluscs that show asymmetric patterns in the Northern and Southern Hemispheres, run counter to expected gradient patterns.

CONCLUSION

Global patterns of biodiversity, especially the latitudinal diversity gradient, have led to the development of numerous hypotheses to attempt to elucidate the underlying causes of this intriguing global pattern. Although each of the current hypotheses offers insights, no one mechanism seems to explain all instances of the observed patterns. Furthermore, many of the hypotheses are not mutually exclusive, and multiple hypotheses may play some role in determining the observed global patterns. Focusing on the complexities of biodiversity, including alternate methods of measuring diversity beyond number of species and the importance of scale in the patterns observed, may lead to a more nuanced view and improved understanding of the

dynamics underlying diversity patterns. Understanding patterns of global biodiversity, and the underlying processes, will continue to be a major priority as issues such as global climate change, conservation, and sustainable resource use increase in importance worldwide.

SEE ALSO THE FOLLOWING ARTICLES

Competition / Dispersal / Temperature Change

FURTHER READING

Astorga, A., M. Fernández, E. E. Boschi, and N. Lagos. 2003. Two oceans, two taxa and one mode of development: latitudinal diversity patterns of South American crabs and test for possible causal processes. *Ecology Letters* 6: 420–427.

Crame, J. A. 2004. Patterns and processes in marine biogeography: a view from the poles, in *Frontiers of biogeography: new directions in the geography of nature*. M. Lomolino and L. Heaney, eds. Sunderland, MA: Sinauer, 271–291.

Macpherson, E. 2002. Large-scale species-richness gradients in the Atlantic Ocean. *Proceedings of the Royal Society of London, Series B–Biological Sciences* 269: 1715–1720.

Marquet, P., M. Fernández, S. Navarrete, and C. Valdovinos. 2004. Diversity emerging: toward a deconstruction of biodiversity patterns, in *Frontiers of biogeography: new directions in the geography of nature*. M. Lomolino and L. Heaney, eds. Sunderland, MA: Sinauer, 191–209.

Rohde, K. 1992. Latitudinal gradients in species diversity: the search for primary cause. *Oikos* 65: 514–527.

Roy, K., D. Jablonski, J. Valentine, and G. Rosenberg. 1998. Marine latitudinal diversity gradients: test of causal hypotheses. *Proceedings of the National Academy of Sciences USA* 95: 3699–3702.

Stehli, F. G., A. L. McAlester, and C. E. Helsley. 1967. Taxonomic diversity of recent bivalves and some implications for geology. *Geological Society of America Bulletin*. 78: 455–466.

Valdovinos, C., S. A. Navarrete, and P. A. Marquet. 2003. Mollusk species diversity in the southeastern Pacific: why are there more species towards the pole? *Ecography* 26: 139–144.

BIODIVERSITY, MAINTENANCE OF

SERGIO A. NAVARRETE

Pontificia Universidad Católica, Santiago, Chile

One of the most important attributes of ecosystems on land and the ocean is the number of species that can be found coexisting in a local habitat. In the ocean, this essential component of biodiversity is first constrained by the ability of species to disperse and settle in appropriate habitat. Once there, a suite of local postsettlement biological and physical processes can maintain or alter biodiversity, favoring the establishment of some species and eliminating others.

SUPPLY OF INDIVIDUALS

The maintenance of biodiversity in marine environments depends critically on the successful arrival or supply of new individuals of different species to the local habitat. This is true of all systems, but it acquires special relevance in marine environments because the mere presence of healthy adults of a given species does not guarantee that new individuals will arrive and maintain that population over time. This is an obvious, but often times underappreciated, factor in the study and discussions of biodiversity of marine environments. In these systems, local communities are usually dominated by species with life histories that include a free swimming pelagic larva, which can spend from hours to months in the water column and have no means to remain close to their parents.

Let us assume that we are examining the number of species that occur in a small fraction of adult habitat, say a single rocky bench or a medium-size tidepool along a rocky shore. Let us also assume that we start with an empty tidepool and observe, over time, how this tidepool is colonized by individuals of different species. Clearly, the number of species that settle in our tidepool will be a function of the total number of species present in the area surrounding the tidepool and whose propagules—larvae and spores—have the potential to reach this pool. Note that in the case of invertebrates with long-lived pelagic larvae, the area supplying species to our pool can be several tens of kilometers up or down the coast. In the simplest case, our pool receives at least one individual of all the species within dispersal distance, in which case, at the time of settlement, our pool will be a faithful representation of the diversity in this region (Fig. 1A). However, this is highly unlikely. Our small pool will most likely receive only a fraction of those species within dispersal distance. Rare species or those producing few propagules or larvae at the time we start our observations will probably not reach the pool (Fig. 1B). In some cases, there might be sufficient individuals of one or a few species to completely fill the pool and this could prevent the arrival of new individuals. In these cases, the pool will no longer receive individuals of other species and will reach saturation (Fig. 1C). The later situation is common among sessile animals that require rock surface to settle and can quickly occupy the entire space.

POSTSETTLEMENT FACTORS: ONE TIDEPOOL

After our species colonize the pool, the individuals that settle grow in size and have to withstand the physical and biological conditions in the adult habitat. A number of

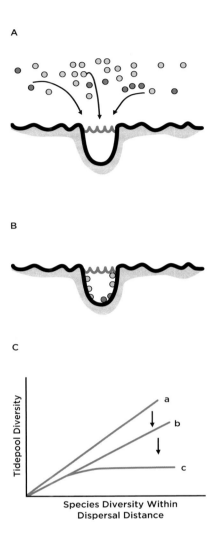

FIGURE 1 Colonization of newly open space or tidepool by individuals of different species found within dispersal distance. The top diagram illustrates how propagules of some species colonize and get established in the pool, setting the initial diversity of species. The bottom graph illustrates the relationship between the diversity of species in a region surrounding the pool, from which species can potentially colonize, and the number of species we can actually observe in the pool at the time of settlement. In (A) the pool receives one individual of each species within dispersal distance; in (B), representing conditions after some time has elapsed, only a subset of those species are found in the pool; in (C) the first ones to arrive monopolize the space and prevent the settlement of new species.

postsettlement processes can then cause differential mortality of individuals of different species and affect the number of species that we observe in the pool after some period of time. These processes can both decrease or help maintain the number of species that can coexist in the tidepool following settlement.

Competition for Limited Resources

EQUAL COMPETITIVE ABILITIES: THE LOTTERY MODEL

If the species that arrived at our pool have similar colonization and competitive abilities, as well as similar requirements and tolerances of abiotic conditions, then individuals will be able to hold their space without being overgrown, crushed, or displaced by others. When an individual dies, it will be replaced by an individual of the same or different species. Under these circumstances, species cannot secure the space beyond their lifetime, but can improve their chances of recolonizing the lost space by having more propagules available. In a way, this is like buying lottery tickets; buying more tickets improves your chances, but there is no certainty that you will win the prize and colonize the available space. Competition occurs in an indirect manner, through the appropriation and therefore pre-emption of space that otherwise will be available for colonization by self- and other species. In these cases, the final number of species will fluctuate somewhat over time, but it will be similar to the number of species that settled initially in the pool (Fig. 2A, curve a).

COMPETITIVE HIERARCHIES: THE EXCLUSION OF INFERIOR COMPETITORS

Much more common than species with similar colonization and competitive abilities is the existence of strong and persistent competitive hierarchies among algal and invertebrate species. One or a small subset of species is usually able to settle on, overgrow, crush, squash, suffocate, and eventually kill all other species. As species grow, these competitive hierarchies generate successional sequences from the time of settlement. Succession does not have to be linear, and multiple successional paths might be observed in the same community, but unless disturbed by other factors, the community will end up dominated by one or very few competitive dominant species. The number of species will therefore decline over time and in many cases the pool will end up with juts one single species (Fig. 2A, curve c). Numerous examples of this type of competitive structure abound in most rocky shores of the world. In many shores, mytilid mussels are dominant competitors for space at mid-intertidal elevations; in others, barnacles can take over this role and exclude other species, and in many shores different kelp species or surfgrass can form extensive monocultures that exclude all other algal and invertebrate species.

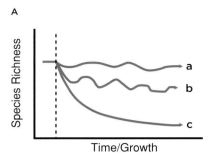

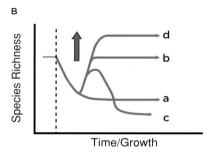

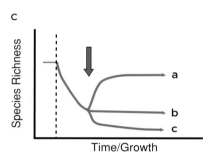

FIGURE 2 Processes maintaining or modifying the diversity of species after individuals have successfully settled into the new habitat and individuals grow. (A) Different forms of competition can maintain or alter biodiversity. In curve a, all species have equal competitive abilities and after settlement they hold the resource for life. In curve b, some species can outcompete and exclude others, but the process is not transitive: species A outcompeted a species B, which in turn dominates over C, but C can outcompete species A. In curve c, there is perfect competitive hierarchy, leading to monopolization of the resource by one or a small subset of competitively superior species. (B) Effect of predation on species diversity. Once predators start consuming prey (red arrow), they can increase species diversity by removing the dominant competitor as depicted in curve a, or can decrease diversity by removing inferior competitors (curve c), accelerating the effects of competitive hierarchies alone (curve b). (C) Effect of positive interactions among the established species on species diversity. In curve b, a settled species provides habitat for other species, diminishing the effects of competitive exclusion (curve a). In curve d, a settled species modifies local conditions for the settlement of other species, allowing the colonization of new species that otherwise could not even settle there. In curve c, the facilitator species favors the establishment of a superior competitor, which after some time accelerates the exclusion of other species.

COMPETITIVE NETWORKS: INTRANSITIVE INTERACTIONS

In intertidal systems there are few examples of coexisting species having equal colonization or equal competitive abilities. The most common pattern is the existence of competitively superior species that can overgrow competitively inferior ones. In some cases, however, competitive abilities might not be transitive; that is, species *A* might be able to exclude species *B* for the use of resource (e.g., space), species *B* might in turn exclude species *C,* but it is possible that *C* excludes *A* (Fig. 3). When all species are in contact and competition occurs simultaneously among them or in different sections of our pool, these intransitive networks can allow species to coexist. Species diversity in the pool should persist over time, amid continuous competitive interactions within the pool (Fig. 2A, curve b). It is unlikely, however, that the rates of competitive exclusion are similar among all pairs involved in the network; and therefore, some of the original species that settled will most likely be excluded from the pool. Lack of transitivity in competitive networks is usually produced by the existence of different mechanisms of competition between species. For instance, a species *A* could shade and kill a species *B*, which in turn might overgrow and suffocate a species *C;* this latter species could be shade resistant and outcompete species *A* for food. Few examples of intransitive networks have been documented in intertidal systems.

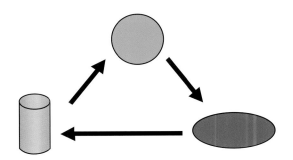

FIGURE 3 Nontransitive competitive networks among three species. A species depicted here as a green circle competitively excludes the red-oval species, as shown by the direction of the arrow. The red-oval species can in turn exclude the gray-cylinder species, but in contrast to a transitive hierarchy, this latter species can outcompete the green-circle species.

Facilitation and Other Positive Interactions

For years ecologists have emphasized the importance of negative interactions among species, such as competition, in discussions and theoretical models about the maintenance of species diversity. However, in real systems species not only compete for resources, having negative effects on each other, but they can also have positive effects on other species. Besides the positive effect that prey have on consumers, which are better treated in the context of predatory (trophic) interactions, species can indeed have a myriad of positive nontrophic effects on many species with which they interact. These positive effects might be long lasting or temporary, and they need not be reciprocal between species (i.e., mutualistic) to have dramatic effects on species diversity. Some definitions of positive interactions—"those that cause a positive effect on one species and cause no harm to neither"—are not particularly useful in the context of our tidepool community, because they underestimate the true importance of these interactions in the maintenance of species diversity and put an enormous burden on empirically demonstrating that the species recipient of the positive effect has no negative effects on the donor.

Among the most widespread and important biological interactions in rocky intertidal systems is the facilitation of settlement of invertebrate species by algal turfs and other invertebrates, or the refuge provided to mobile consumers by algae and invertebrates (e.g., limpets found under algal fronds, gastropods, crustaceans, and many other mobile species found inside beds of intertidal mussels, tunicates, or kelp holdfasts). This habitat enhancement can halt the negative effects of competition and help maintain species diversity (Fig. 2C, curve b). In many of these cases, the only means by which species can settle or survive is because another "facilitator" species is there. In our tidepool, this effect will lead to an increase in the total number of species as new ones can now settle in the habitat (Fig. 2B, curve). Now, it is entirely possible and actually very common that species have positive effects on their enemies or on the friends of their enemies; for example, algal turfs facilitate the settlement of competitively dominant mussels, which eventually overgrow the turf and all other species. Similarly, algal mats acting as bioengineers can prevent desiccation of many herbivore species, a positive effect on diversity, but these grazers can remove algal biomass, limit their growth, and potentially eliminate rare species, a negative effect. In these cases, the positive effect leads to temporary increase in the number of species in the pool, but eventually accelerates the decline of species diversity produced by competitive exclusion or over exploitation (Fig. 2B, curve c).

Predation

PREDATION ON COMPETITIVE SUPERIOR SPECIES

One of the most influential and striking experimental results in ecology has been the demonstration that predators can increase the local diversity of species. Experimental manipulations have repeatedly shown that carnivore and herbivore consumers—whelks, starfish, littorines, chitons, sea urchins—can control the abundance of a competitively dominant species, thus stopping the succession and preventing the exclusion of inferior competitors (Fig. 2C, curve a). The simplest way to observe this positive effect of consumers on species diversity is when they preferentially consume the competitively superior species over other prey. But even if they do not exhibit individual preferences for prey species, if the competitively superior species is relatively more affected by predation (e.g., slower to recover from predation), consumers can under some conditions still favor coexistence of species over local scales.

PREDATION ON SUBORDINATE COMPETITORS AND BEHAVIORAL EFFECTS

Of course, predators can also decrease species diversity. A common case of negative effects of predators on species diversity occurs when they preferentially consume competitively inferior species, thus accelerating the decline in diversity produced by competition alone (Fig. 2C, curve b). Even when predators prey preferentially on the dominant competitor, if the intensity of predation is too high, diversity of the prey assemblage will be reduced. A less well-studied negative effect of predation on diversity of intertidal systems occurs through behaviorally mediated indirect effects. In many cases, the mere presence of a predator (e.g., birds) can be perceived by its prey and cause a behavioral change that forces them to use and compete for refuges (e.g., crabs restricted to forage inside refuges during daytime). The intense competition for refuges produced by predators could lead to competitive exclusions and segregation by the mechanisms discussed in the section "Facilitation and Other Positive Interactions."

Disturbance and Species Diversity

Physical disturbances can kill or remove biomass from intertidal animals and plants. Physical disturbances include large waves that dislodge individuals or roll rocks inside tidepools, floating logs that crush sessile species when they land on the shore, the scouring action of ice sheets at high latitudes, and many other forms of death and destruction. In all cases, physical disturbance can alter the number of species that can coexist in the local habitat. The effect of disturbance on diversity, whether positive or negative, depends on a number of variables, including the intensity and frequency of disturbance, and the competitive and colonization abilities of the species that constitute the local community. A simple and intuitive hypothesis that relates all these factors is the Intermediate Disturbance Hypothesis, which has proven to be a useful framework in many empirical studies. The hypothesis predicts that over a gradient of increasing disturbance frequency, diversity follows a unimodal trend; number of coexisting species peaks at intermediate levels of disturbance (Fig. 4). When disturbance is infrequent, the dominant competitor excludes other species and diversity is low. When disturbance is too high, only those species with fast colonization and growth rates can get established in the habitat. At intermediate levels, removal of biomass by physical disturbances keeps the dominant competitor under checked and still allows for other species to colonize and grow.

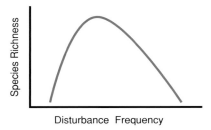

FIGURE 4 The Intermediate Disturbance Hypothesis predicts changes in species diversity across gradients of physical disturbance frequency or intensity. At low levels of disturbance (infrequent), dominant competitors exclude other species, and diversity is low. At very high levels of disturbance only a subset of species can get established in the habitat and diversity is again low. At intermediate levels, disturbance maintains the dominant competitor under check and still allows for a large number of species to be established in the habitat.

INCREASING THE SCALE OF OBSERVATION

So far we have focused on a given rocky bench or tidepool, where we followed the fate of the species that colonized that small area over time, as they undergo a suite of postsettlement processes that alter or maintain local species diversity. But in real situations, colonization by new individuals is occurring at the same time many of the postsettlement processes discussed here are taking place. As a predator removes biomass of a dominant competitor, the space or resource can be colonized by the same or a different species. If a tidepool or rocky bench is monopolized by a species, as we increase the spatial scale of observation, we are bound to find that predators and physical disturbances release resources (e.g., space) and allow the settlement and at least the temporal establishment of other species. Thus, processes maintaining biodiversity in marine systems are dynamic, can interact with each other, and their relative importance vary somewhat in different coasts of the world. Since the processes that determine the dispersal and colonization (settlement) of new individuals are fundamentally different from those taking place in the local habitat (e.g., species interactions), maintenance of diversity over larger scales, say a section of the coast, will be a function of the number of species in the region within dispersal distance, the rate of provision of new resources through natural death, predation and disturbance, and the balance between positive and competitive interactions. Moreover, plain counts of the number of species can confound these processes (colonization and postsettlement) and lead to wrong conclusions about the importance of local processes, such as species interactions in maintaining patterns of biodiversity. Studies of biodiversity in marine systems must therefore separate new settlers from those individuals that have "experienced" the local habitat.

SEE ALSO THE FOLLOWING ARTICLES

Competition / Disturbance / Facilitation / Predation / Succession

FURTHER READING

Huston, M. A. 1994. *Biological diversity. The coexistence of species on changing landscapes.* Cambridge: Cambridge University Press.
Johnson, K. H., K. A. Vogt, H. J. Clark, O. J. Schmitz, and D. J. Vogt. 1996. Biodiversity and the productivity and stability of ecosystems. *Trends in Ecology and Evolution* 11: 371–377.
Magurran, A. E. 1988. *Ecological diversity and its measurement.* Princeton, NJ: Princeton University Press.
Mooney, H. A., J. H. Cushman, E. Medina, O. E. Sala, and E. D. Schulze. 1996. *Functional roles of biodiversity: a global perspective.* Chichester: John Wiley & Sons.
Ricklefs, R. E. 1987. Community diversity: relative roles of local and regional processes. *Science* 235: 167–171.
Rosenzweig, M. L. 1995. *Species diversity in space and time.* Cambridge: Cambridge University Press.

BIODIVERSITY, SIGNIFICANCE OF

JOHN J. STACHOWICZ

University of California, Davis

Biodiversity refers to the numbers of different kinds of life forms present at a location. Biodiversity occurs at the species level (the number of species and their evenness), below the species level (genetic diversity within species), and above the species level (higher taxonomic or functional-group diversity). The aesthetic beauty of tidepools and rocky shores is in large part due to the wide variety and diverse coloration of the seaweeds and animals that inhabit them, and so the high biodiversity of these environments is also what often draws people to them. But there is growing evidence that biodiversity enhances the stability of marine systems by providing a form of biological insurance against changing conditions.

DEFINITIONS OF BIODIVERSITY

Measures of diversity often include not only the number (richness) of species in an area but the degree to which species are evenly represented versus dominated by a single species (evenness). There are many aesthetic reasons for maintaining biodiversity. For example, plants and animals of the rocky shores of the Pacific coast of North America are very diverse and colorful and attract many curious visitors to the shore each year. Most people would agree that the tidepool shown in Fig. 1A is somehow more appealing than the one shown in Fig. 1B and

FIGURE 1 Tidepools of similar size from Bodega Bay, California: (A) a pool with high diversity; (B) a pool that has lower diversity. Photographs by the author.

that this appeal has something to do with the fact that the former has a greater diversity of species. But does that diversity matter to the tidepool ecosystem?

In light of declining global diversity, scientists have become increasingly interested in whether biodiversity might have more quantifiable effects on communities. Do ecosystems with more species have greater productivity or stability than those with fewer species? As of mid-2006, relatively few papers had been published in this new, but rapidly expanding area of research, and a consensus had not yet emerged in marine systems. Thus some of the generalizations portrayed in this article may be subject to revision as the field matures. In discussing the significance of biodiversity, species-level effects are most easily ascertained, because available experimental data focus mostly on manipulations of richness. However, some examples of the effects of diversity at the genetic and functional-group level can also be considered.

POSSIBLE EFFECTS OF DIVERSITY ON ECOSYSTEM STRUCTURE AND FUNCTION

That diversity should somehow result in "higher" levels of ecosystem processes has some intuitive appeal and has long been part of conventional wisdom. The following are several mechanisms that have been posited to cause diversity effects on ecosystems. The reader should be aware, however, that there is considerable disagreement among scientists as to the relative importance of each of these mechanisms and indeed as to the definitions of some of the mechanisms themselves. More than one of these mechanisms may be in operation at any one time.

Complementarity

Resource levels or environmental conditions influence species' performance, such that different species often exhibit optimum performance at different resource levels or are limited by different resources. Because species have different ways of exploiting resources or use different parts of the resource spectrum, increasing diversity can lead to increased resource use and ultimately higher productivity. In some cases, species use resources in the same way as other species do but do so at different times of the year, or in different microhabitats, resulting in temporal or spatial complementarity.

Facilitation

Facilitation occurs when one species ameliorates some biological or physical stress, benefiting other species in the system. This can happen when, for example, a canopy-forming seaweed shades the substratum at low tide, permitting the growth of other species that are highly productive but less tolerant of drying out.

Sampling

The sampling effect occurs because more diverse communities may have a greater probability of containing a species that has a dominant effect on ecosystem functioning. This can result in an "apparent" effect of diversity in experiments that assemble communities of randomly selected species: Communities with more species have a greater chance of having the dominant species. This effect can also occur when increasing species richness increases the likelihood of including an important facilitator (see the preceding paragraph). Intertidal community structure and functioning often change dramatically depending on the presence of a single facilitator (such as a canopy-forming seaweed) or a single keystone predator (such as a sea star or sea otter).

Redundancy and Insurance

Often, species are redundant (i.e., they perform the same function or role in the community as some other species), but they may respond differently to environmental stress. For example, seaweeds may have different tolerances to changing temperature, such that the species that are the most productive under normal conditions die when temperatures are elevated. In these cases, less productive species that are tolerant of stresses can still persist, maintaining production. This observation implies that there may be a trade-off between achieving maximum performance at a single instant and consistent performance over time. Thus, diversity may have no significance in the short term but be essential for maintenance of ecosystem functioning in the longer term. Some think of this as similar to a "portfolio effect," analogous to a diversified investment strategy. Relying on a single stock can result in high payoff if the stock is up, but if it crashes, the investor loses everything. Similarly, spreading out resources over many species protects against wild fluctuations in the abundance of any one of those species.

Experimental studies on the functional consequences of biodiversity have been conducted in a wide range of marine environments, but relatively few have been conducted on rocky shores. Thus, the following examples also include examples from other shallow-water coastal habitats to illustrate mechanisms that will likely be shown to operate on wave-swept rocky shores.

SEAWEED AND SEAGRASS DIVERSITY

Most marine studies of diversity's consequences have focused on consumer diversity. What is known about the functional consequences of seaweed diversity on rocky shores is that effects of diversity on community stability can be overwhelmed by extreme physical forces or consumers. In fact, disturbances may affect diverse communities more than depauperate ones because more diverse communities have a greater biomass or cover of algae to begin with. However, communities with a diversity of functional types of algae are often able to recover from disturbances more quickly than those with lower diversity, consistent with the idea of diversity as biological insurance. The degree to which these effects are due to a single species that is resistant to stress rather than to species with complementary responses to stress is not yet clear. It is clear, however, that seaweeds respond differently to variation in the availability of different forms of nutrients, and thus having more species may allow greater and more consistent conversion of available resources into seaweed mass.

Seaweed diversity might also affect community function via bottom-up mechanisms that are independent of total production. For example, if particular invertebrates associate with certain seaweeds, then increasing seaweed diversity should lead to increasing animal diversity. Similarly, some grazing invertebrates grow, survive, and reproduce better on a mixed diet that includes a diversity of seaweeds. This kind of "diet mixing" can be favored because different foods provide complementary nutrients to grazers. Seaweed diversity could thus facilitate the health of grazing invertebrates in the same way that a balanced diet helps humans obtain proper nutrition and maintain health.

Similar effects of diversity below the species level has also been shown to stabilize communities against environmental change. For example, research on the seagrass *Zostera marina* suggests that genotypic diversity within monospecific *Zostera* meadows allows these important primary producers to persist in the face of natural (consumers) or anthropogenic (high temperatures) environmental change. Abundances of seagrass-associated invertebrates are also higher in more genetically diverse plots, perhaps because they offer a more reliable habitat. In these studies, the mechanisms appear to be very similar to those underlying the effects of species diversity previously discussed.

SESSILE INVERTEBRATE DIVERSITY

Although little is known about the functional significance of rocky-shore sessile invertebrate diversity, work on similar organisms on natural and artifical hard substrates in bays and harbors has been particularly informative regarding the role of diversity in determining the resistance of a

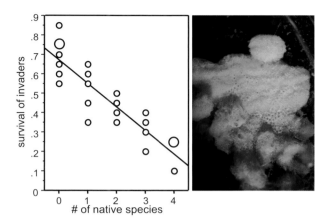

FIGURE 2 Species diversity enhances resistance of communities to invasion. Experiments manipulating native species richness find a strong negative effect on the survival and cover of introduced ascidians, *Botrylloides violaceus*, pictured at right. Redrawn from Stachowicz *et al.* (2002).

community to invasion by exotic species (see Fig. 2). This resistance occurs when native species are complementary in their use of resources so that as diversity increases, the amount of unused resources dwindles, leaving fewer "leftover" resources for invaders to use. In sessile invertebrate communities, decreasing native diversity increases the survival and overall abundance of invaders. High native diversity decreases the availability of open space, the limiting resource in this system, by buffering against fluctuations in the cover of individual species. Each species has a distinct seasonal pattern of abundance, so space is most consistently and completely occupied when more species are present. Diversity decreases invasion by providing insurance against fluctuations in the abundance of any one species.

However, field surveys tell us that diversity is only one of many factors affecting community resistance to invasion. Some surveys do find a strong and consistent negative relationship between native and non-native diversity, but many find a positive correlation between native and invader diversity. This may be because some locations are inherently capable of supporting more species, both native and exotic, perhaps because of high resource levels or high levels of habitat heterogeneity. Such areas both have more natives and more invaders because both native and exotic diversity are similarly affected by some third variable (for example, the rate of settlement or frequency of disturbance). Still, within a location (when all else is equal), the loss of species should lead to an increase in invasion by exotic species as a result of a freeing of resources.

Another possibility is that there is some positive causal relationship between native and exotic diversity. For example, increasing richness may increase the probability of including species that facilitate the colonization and establishment of new species. However, it is unlikely that, on balance, natives facilitate only invaders or vice versa. The invasive bryozoan *Watersipora subtorquata,* for example, enhances three-dimensional habitat structure, facilitating both native and non-native species similarly. Interestingly, where this bryozoan is abundant, the relationship between native and exotic richness is positive, whereas where it is absent, the relationship becomes negative, reflecting the influence of biotic resistance.

A study of genetic diversity in barnacles illustrates the potential effect of within-species diversity on intertidal community dynamics. Researchers compared settlement rates of barnacle larvae from a single brood (with a single pair of parents) to those of larvae from a mixture of two or three different broods and found that treatments with more broods had higher overall rates of settlement and metamorphosis. Although this study was conducted in the laboratory, in the absence of flow, it suggests that genetic diversity may enhance recruitment of an important intertidal species.

CONSUMER DIVERSITY

Mobile consumers play a large role in temperate coastal ecosystems, and single species such as sea otters, sea urchins, and sea stars can have dramatic effects that may overwhelm the bottom-up effects of diversity at the producer level. Additionally, because of the threats of overfishing and direct harvest, top predators are the most likely group of marine species to go locally extinct. An understanding of the significance of predator diversity is of great importance because of the cascading effects of predators on entire food webs. These "trophic cascades" appear to be most common in simple food chains, whereas diverse food webs are thought to be less likely to exhibit major shifts in community states when individual consumers are removed. For example, in the relatively simple food webs of Alaskan kelp forests, kelp biomass is maintained by sea otter predation on herbivorous sea urchins. In kelp forests containing a greater diversity of predatory and herbivorous fish and invertebrates, such as those of southern California, the extirpation of otters appears to have not had as large an impact on kelp. Both experimental and observational studies suggest that predator richness decreased grazing by herbivores, leading to a decrease in the biomass of the giant kelp *Macrocystis*. This occurs because predators have complementary effects on different species of herbivores such that total grazing is minimized only in a diverse predator assemblage. As kelps both serve as important habitat-providing foundation species and provide a significant energy source for much of the food web, declining predator diversity may have significant consequences for kelp forest ecosystem structure and function.

Other studies have pointed out that the presence of omnivores (consumers that eat both algae and herbivores) can overwhelm any effects of predator diversity. Given that many marine consumers are omnivores, these studies suggest that our ability to predict the consequences of changing diversity in food webs that include omnivores will be difficult and perhaps context-dependent. It is increasingly clear that the significance of diversity at any one trophic level cannot be understood without knowing the levels of diversity (and species composition) of other trophic levels.

BIODIVERSITY'S IMPORTANCE FOR ECOSYSTEMS

Although a growing number of examples suggest that diversity increases the absolute rates of processes such as production, many other examples indicate that a single dominant species (e.g., kelp) largely determines community productivity. However, diversity also buffers ecosystems from fluctuations in the environment or from the loss of particular species. Even if many species turn out to be functionally "redundant" because they perform similar ecological roles, having more species in a system may provide a critical insurance policy. Without the biological insurance provided by enhanced diversity, the ecosystem may collapse over the longer term and in the face of changing conditions.

Many of the community processes and ecosystem functions discussed in this article can be directly related to the benefits provided to humans by coastal ecosystems. It has been estimated that, globally, natural ecosystems provide humans with at least $16 trillion worth of goods and services per year (Costanza *et al.* 1997). If the effectiveness of these services is somehow linked to biodiversity (as evidence suggests that it is), then the loss of biodiversity will not only cost us our natural heritage but also cause us serious financial pain as we pay to develop technical means to provide the goods and services that the oceans currently provide us for free.

SEE ALSO THE FOLLOWING ARTICLES

Algae / Facilitation / Genetic Variation and Population Structure / Introduced Species / Predation / Surveying

FURTHER READING

Allison, G. W., B. A. Menge, J. Lubchenco, and S. A. Navarrete. 1996. Predictability and uncertainty in community regulation: consequences of reduced consumer diversity in coastal rocky ecosystems, in *Functional roles of biodiversity: a global perspective.* H. A. Mooney, J. H. Cushman, E. Medina, O. E. Sala, and E.-D. Schulze, eds. Chichester: John Wiley, 371–392.

Costanza, R., R. d'Arge, R. de Groot, S. Farber, M. Grasso, B. Hannon, S. Naeem, K. Limburg, J. Paruelo, R. V. O'Neill, R. Raskin, P. Sutton, and M. Van den Belt. 1997. The value of the world's ecosystem services and natural capital. *Nature* 387: 253–260.

Hooper, D. U., F. S. Chapin III, J. J. Ewel, A. Hector, P. Inchausti, S. Lavorel, J. H. Lawton, D. M. Lodge, M. Loreau, S. Naeem, B. Schmid, H. Setälä, A. J. Symstad, J. Vandermeer, and D. A. Wardle. 2005. Effects of biodiversity on ecosystem functioning: a consensus of current knowledge. *Ecological Monographs* 75: 3–35.

Loreau M., S. Naeem, P. Inchausti, J. Bengtsson, J. P. Grime, A. Hector, D. U. Hooper, M. A. Huston, D. Raffaelli, B. Schmid, D. Tilman, and D. A. Wardle. 2001. Biodiversity and ecosystem functioning: current knowledge and future challenges. *Science* 294: 804–808.

Naeem, S., F. S. Chapin III, R. Costanza, P. R. Ehrlich, F. B. Golley, D. U. Hooper, J. H. Lawton, R. V. O'Neill, H. A. Mooney, O. E. Sala, A. J. Symstad, and D. Tilman. 1999. *Biodiversity and ecosystem functioning: maintaining natural life support processes*. Washington, DC: Ecological Society of America.

Stachowicz, J. J., J. F. Bruno, and J. E. Duffy. 2007. Consequences of marine biodiversity for ecosystem functioning. *Annual Review of Ecology Evolution and Systematics* 38.

Stachowicz, J. J., H. Fried, R. B. Whitlatch, and R. W. Osman. 2002. Biodiversity, invasion resistance and marine ecosystem function: reconciling pattern and process. *Ecology* 83(9): 2575–2590.

Worm, B., E. B. Barbier, N. Beaumont, J. E. Duffy, C. Folke, B. S. Halpern, J. B. C. Jackson, H. K. Lotze, F. Micheli, S. R. Palumbi, E. Sala, K. A. Selkoe, J. J. Stachowicz, and R. Watson. 2006. Impacts of biodiversity loss on ocean ecosystem services. *Science* 314: 787–791.

BIOFILMS

RICHARD C. THOMPSON

Marine Biology and Ecology Research Centre, University of Plymouth, United Kingdom

Biofilms are assemblages of microorganisms that have colonized hard surfaces immersed in water. They cover rocks in the intertidal and even on the outer surface of marine plants and animals. These films are fundamental to the ecology of rocky shores, influencing settlement of invertebrates and algae and providing an important food resource for grazers. However, on man-made surfaces formation of a biofilm is described as "fouling" and is often seen as a nuisance; fouling on the hull of a ship, for example, reduces efficiency and increases fuel costs.

BIOFILM FORMATION

Biofilms begin to develop shortly after an object is immersed in water. Initial stages of biofilm formation are predictable and are driven by the physical and chemical properties of the surface, which influence adsorption of organic and inorganic chemicals from the surrounding water. Then, over longer time scales biological factors

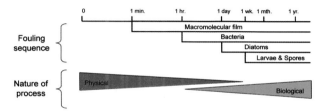

FIGURE 1 Schematic diagram showing sequence of colonization leading to biofilm formation on a hard surface immersed in water (redrawn after Wahl 1989, with permission).

become increasingly important as the surface becomes progressively colonized by bacteria, diatoms, protozoa, yeasts, and eventually juvenile stages of larger marine plants and animals (Fig. 1). Intertidal biofilms are predominantly composed of photosynthetic organisms, mainly diatoms and filamentous cyanobacteria, which are less than 20 μm in diameter (Fig. 2A). These organisms combine to form a film around 20 μm thick, and although not visible without a microscope, they can give rock surfaces a dark green appearance because of the chlorophyll and other photosynthetic pigments they contain. Organisms within these films secrete mucous and other extracellular polymeric substances (EPS), which bind the film together, and they can also make filmed surfaces slippery.

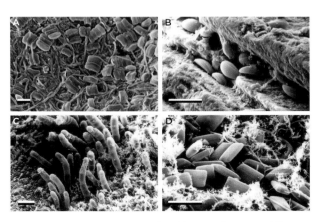

FIGURE 2 Electron micrographs of biofilms. (A) diatoms and filamentous cyanobacteria on the surface of intertidal rock, (B) diatoms among grooves on the shell of the mollusc *Gibbula cineraria* L., (C) cyanobacteria on the shell of the mollusc *Littorina obtusata* L., and (D) diatoms within a mucus film on the surface of the alga *Palmaria palmata* L. (scale bars = 20 μm). Reprinted from Thompson *et al.* (1996), with permission from Elsevier.

PHYSICAL FACTORS INFLUENCING BIOFILMS ON ROCKY SHORES

Despite the harshness of physical conditions in the intertidal, biofilms persist throughout the year from polar regions, where they are covered with ice during the winter to the tropics, where rock surfaces may exceed 50 °C. However, the type and abundance of organisms within these films can vary considerably according to physical conditions including temperature, solar radiation and wave action, and even according to the type and rugocity of the underlying rock. Because they are microscopic the organisms within these films are vulnerable to physical stresses during low tide and the species composition and abundance of organisms generally shows considerable seasonal variation. This is particularly evident on the mid and upper shore, where at temperate latitudes, the abundance of microorganisms can change dramatically, peaking in late winter and early spring and then declining markedly as insolation stress increases during the summer (Fig. 3).

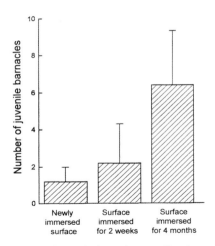

FIGURE 4 Number of juvenile barnacles recruiting to rock surfaces covered with biofilms of differing ages. Data were collected in laboratory choice chambers. (values are averages ± standard error). Reprinted from Thompson et al. (1998), with permission from Springer Science and Business Media.

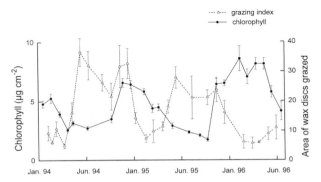

FIGURE 3 Seasonal changes in epilithic microbial standing stock (expressed using extracted chlorophyll to provide an index, n = 18 samples) and grazing intensity (measured using feeding marks on arrays of wax disks as an index, n = 3 arrays) on the midshore at Port St. Mary, Isle of Man (values are averages ± standard error). Reprinted from Thompson et al. (2000), with permission from Springer Science and Business Media.

EFFECTS OF BIOFILMS ON SETTLEMENT OF INVERTEBRATES AND ALGAE

Biofilms can influence settlement and colonization by larger organisms, most of which have a juvenile planktonic stage in their life cycle. The biofilm is the first point of contact between these juvenile stages and the underlying rock surface and can provide cues that influence the settlement of larvae (Fig. 4). Once settled these organisms grow and develop into adults, and as they do, their outer surface typically becomes colonized by a biofilm itself. Interestingly, these epibiontic biofilms are typically composed of different microorganisms than those on rock surfaces and so make an additional contribution to the microbial biodiversity of the shore (Fig. 2B, C, D).

BIOFILMS AS A FOOD RESOURCE FOR GRAZERS

Biofilms are also an important food resource for grazers, particularly molluscs such as limpets, but also for some urchins and fish. Because the organisms within these films are so small, they can multiply rapidly. This becomes apparent when grazers are removed from an area of shore, either experimentally or because of human exploitation or pollution. When this occurs the biomass of the microorganisms can double within days (Fig. 5). Although grazers directly remove organisms from biofilms, changes in feeding activity do not account for seasonal patterns in the abundance of photosynthetic microorganisms within biofilms (see Fig. 3). Hence, at some times of year, there can be a mismatch between the microbial productivity of biofilms, which is negatively correlated with insolation stress, and the demand for resources from the grazers that feed on them, which is positively correlated with temperature (Fig. 3). Grazers also consume the germlings

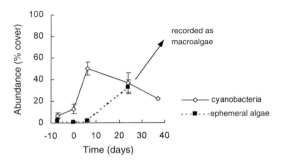

FIGURE 5 Changes in the abundance of cyanobacteria and ephemeral algae within intertidal biofilms as a consequence of excluding the principal grazer, limpets, from areas of shore at Port St. Mary, Isle of Man (values are averages ± standard error, recorded 10 days before grazer exclusion and for 40 days thereafter). Reprinted from Thompson et al. (2000), with permission from Springer Science and Business Media.

of macroalgae such as *Fucus* that have settled in the biofilm; in doing so, they effectively control the abundance of larger plants as well as microorganisms within the biofilm. Because large macroalgae provide an important habitat for many other species of intertidal organisms, the interaction between grazers and biofilms is a key factor influencing the ecology of many rocky shores.

METHODOLOGICAL ADVANCES IN THE STUDY OF BIOFILMS

Despite the importance of biofilms as food resource we know little about their rates of productivity and their contribution is often omitted from estimates of inshore primary productivity. We also know little of the full diversity of microbiota within these films. In part, this has been a consequence of inadequacies in methods for enumerating and identifying these organisms. Traditionally, the standing stock of organisms has been estimated destructively by removing samples of biofilm and counting the abundance of cells using microscopy or estimating photosynthetic biomass using extracted chlorophyll as an index. Considerable advances have recently been made using confocal microscopy to examine the three-dimensional structure of biofilms, using PAM fluorescence to quantify photophysiological responses of microorganisms within biofilms and using infrared photography to quantify photosynthetic microbial biomass *in situ* on the shore. Because of taxonomic and logistical problems with visualization and enumeration the nonphotosynthetic components of intertidal biofilms have received far less attention than the photosynthetic organisms. This is now being addressed using molecular approaches to help characterize these organisms. Collectively, these techniques should greatly advance our understanding of the role of biofilms in shallow water habitats.

SEE ALSO THE FOLLOWING ARTICLES

Herbivory / Larval Settlement, Mechanics of / Limpets / Microbes / Stromatolites / Succession

FURTHER READING

Bustamante, R. H., G. M. Branch, S. Eekhout, B. Robertson, P. Zoutendyk, M. Schleyer, A. Dye, N. Hanekom, D. Keats, M. Jurd, and C. McQuaid. 1995. Gradients of intertidal primary productivity around the coast of South Africa and their relationships with consumer biomass. *Oecologia* 102: 189–201.

Characklis, W. G., and K. C. Marshall, eds. 1990. *Biofilms,* Vol. New York: John Wiley & Sons.

Jenkins, S. R., F. Arenas, J. Arrontes, J. Bussell, J. Castro, R. A. Coleman, S. J. Hawkins, S. Kay, B. Martinez, J. Oliveros, M. F. Roberts, S. Sousa, R. C. Thompson, and R. G. Hartnoll. 2001. European-scale analysis of seasonal variability in limpet grazing activity and microalgal abundance. *Marine Ecology Progress Series* 211: 193–203.

Jesus, B., V. Brotas, M. Marani, and D. M. Paterson. 2005. Spatial dynamics of microphytobenthos determined by PAM fluorescence. *Estuarine Coastal and Shelf Science* 65: 30–42.

Murphy, R. J., A. J. Underwood, M. H. Pinkerton, and P. Range. 2005. Field spectrometry: new methods to investigate epilithic micro-algae on rocky shores. *Journal of Experimental Marine Biology and Ecology* 325: 111–124.

Norton, T. A., R. C. Thompson, J. Pope, C. J. Veltkamp, B. Banks, C. V. Howard, and S. J. Hawkins. 1998. Using confocal laser scanning microscopy, scanning electron microscopy and phase contrast light microscopy to examine marine biofilms. *Aquatic Microbial Ecology* 16: 199–204.

Thompson, R. C., P. S. Moschella, S. R. Jenkins, T. A. Norton, and S. J. Hawkins. 2005. Differences in photosynthetic marine biofilms between sheltered and moderately exposed rocky shores. *Marine Ecology-Progress Series* 296: 53–63.

Thompson, R. C., T. A. Norton, and S. J. Hawkins. 1998. The influence of epilithic microbial films on the settlement of *Semibalanus balanoides* cyprids—a comparison between laboratory and field experiments. *Hydrobiologia* 375/376: 203–216.

Thompson, R. C., T. A. Norton, and S. J. Hawkins. 2004. Physical stress and biological control regulate the producer-consumer balance in intertidal biofilms. *Ecology* 85: 1372–1382.

Thompson, R. C., M. F. Roberts, T. A. Norton, and S. J. Hawkins. 2000. Feast or famine for intertidal grazing molluscs: a mis-match between seasonal variations in grazing intensity and the abundance of microbial resources. *Hydrobiologia* 440: 357–367.

Thompson, R. C., B. J. Wilson, M. L. Tobin, A. S. Hill, and S. J. Hawkins. 1996. Biologically generated habitat provision and diversity of rocky shore organisms at a hierarchy of spatial scales. *Journal of Experimental Marine Biology and Ecology* 202: 73–84.

Underwood, A. J. 1984. Vertical and seasonal patterns in competition for microalgae between intertidal gastropods. *Oecologia* 64: 211–222.

Wahl, M. 1989. Marine epibiosis 1. Fouling and antifouling: some basic aspects. *Marine Ecology Progress Series* 58: 1–2.

Wieczorek, S. K., A. S. Clare, and C. D. Todd. 1995. Inhibitory and facilitatory effects of microbial films on settlement of *Balanus amphitrite amphitrite* larvae. *Marine Ecology Progress Series* 119: 1–3.

BIOLOGICAL MATERIALS

SEE MATERIALS, BIOLOGICAL

BIOLUMINESCENCE

JAMES G. MORIN

Cornell University

Bioluminescence is defined as visually detectable light generated by a chemical reaction originating from an organism. However, light production is largely the only unifying characteristic among luminescent organisms. Rather, *diversity* best describes bioluminescence. Bioluminescence is diverse taxonomically, biogeographically, ecologically, behaviorally, morphologically, physiologically, biochemically, and genetically. The available evidence indicates that ability to emit light has evolved many times and for a variety of overlapping functions.

TAXONOMIC AND BIOGEOGRAPHIC DIVERSITY OF LUMINESCENCE

Glowing waves crashing on rocky shores are a familiar sight to many who visit coastlines by night. This phenomenon is really an expression of light emission by planktonic organisms, most likely dinoflagellates, whose native habitat is the open sea rather than the shore. Most truly intertidal bioluminescence is restricted to relatively few cryptic species that dwell in and around the interstices of the benthic epifauna and a limited number of exposed colonial species, such as hydrozoans. Luminescence is most prevalent in dimly lit environments (the mesopelagic zone), somewhat less common where ambient light is completely

absent (the deep sea), and even less prevalent in periodically well-lit, photically heterogeneous environments, such as the rocky intertidal zone. The dominant luminescent intertidal organisms are found among the hydrozoa such as *Obelia* (Figs. 1A, B); the polychaetes, especially syllids (Fig. 1C), polynoids (Fig. 1D), and terebellids; the crustaceans (ostracods, rarely, Fig. 1E); echinoderms, especially ophiuroids (Figure 1F); molluscs (boring bivalves); and fishes (e.g., midshipmen during their mating period, Fig. 1G). However, the majority of the species in these higher taxa are not luminescent, and those that do luminesce are usually more abundant in subtidal environments; relatively few luminescent organisms are strictly intertidal.

Luminescence is found in less than 1% of all known genera worldwide, but there are still over 700 luminescent genera, which occur in 15 phyla. Seven major taxa (vertebrates, insects, molluscs, cnidarians, crustaceans, echinoderms, and annelids) account for over 90% of the known luminescent genera, and some of these occur in the intertidal. Nearly half of the luminescent genera are fishes and insects, while molluscs, cnidarians, and crustaceans add another third of the genera.

Over three-quarters of all luminescent genera are marine. Only about 1–3% of coastal (benthic and neritic) genera are luminescent, and of these only some occur in the intertidal zone. The open ocean represents the largest habitat on earth, and it is there that most of the luminescent genera occur. Light emission is highest among organisms dwelling in the mesopelagic zone of the open oceans, where well over 60% of the genera and almost every individual emits light. In the mesopelagic, luminescence is a way of life. Dominant luminescent organisms in this immense habitat are the fishes, squids, crustaceans, gelatinous zooplankton, and dinoflagellates. These forms occasionally wash into the intertidal zone, where their death throes may produce spectacular displays of light.

CHEMISTRY OF LUMINESCENCE

In all known luminescent systems an oxygenase (generically known as a luciferase) catalyzes the interaction of a substrate (generically known as a luciferin) with molecular oxygen or peroxide to form a peroxide intermediate, which breaks down to yield an oxidized oxyluciferin and light. However, both the luciferin substrates and the luciferases vary among taxa. It has been estimated that the ability to emit visible light has evolved over 30 times among organisms and with several distinctly different chemistries. Luciferases appear to be more diverse than their luciferin substrates. Accessory enzymes, as well as cofactors such as Ca^{2+} or H^+, are often involved in the

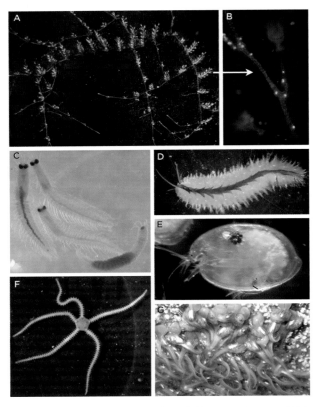

FIGURE 1 Luminescent organisms from shallow coastal waters. (A) A colony of the common luminescent hydrozoan *Obelia geniculata* from California. (B) Fluorescence micrograph of the photocytes in a colony of the hydroid *Obelia geniculata* from California. (C) Three male (left) and one female (lower right) luminescent syllid polychaetes *Odontosyllis parva* from British Columbia. (D) The luminescent polynoid polychaete (scale worm) *Harmothoe imbricata* from Maine. (E) The luminescent ostracod *Vargula nsp.* from Belize. (F) The luminescent brittle star *Amphipholis squamata* from Maine. (G) Developing young of the midshipman fish *Porichthys notatus* from a nest beneath an intertidal rock in California. Photographs by the author.

activation process, and some systems also involve a secondary emitter that changes the emission color or quantum yield of the light.

The diverse luciferins include (1) two systems that are composed of tripeptides that form an imidazolopyrazine nucleus (the coelenterazine and cypridinid ostracod systems), (2) a benzothiazole (fireflies), (3) a tetrapyrrole (dinoflagellates), and (4) several distinct aldehyde-related systems (bacteria, earthworms, a freshwater snail, and fungi). Only the firefly system uses ATP directly in the reaction. The specific luciferin is currently unknown or only partially characterized in about 30% of the known luminescent genera (representing at least 14 diverse higher taxa). Conversely, the coelenterazine system is very widespread. It occurs in about one-third of the luminescent genera, from at least eight phyla, and may have evolved independently several times using the same widespread tripeptide precursor. Other systems (e.g., earthworm, freshwater snail, firefly, and fungi systems) appear to be restricted to but one higher taxon (but appear in many species within that taxon). Both the cypridinid ostracod and the dinoflagellate systems occur in two different, distantly related taxa. The bacterial system, while being represented in five taxa (bacteria, nematodes, squids, thaliaceans, and fishes), occurs in the latter four taxa only by virtue of a symbiotic relationship between luminous bacteria and the host taxon.

STRUCTURE AND PHYSIOLOGY OF LUMINESCENCE

Light-emitting structures show extreme variation among taxa but generally can be classed as either simple emitters or light organs (complex emitters). Simple emitters show no or very little secondary structural manipulation of the light and include (1) single cells (e.g., dinoflagellates); (2) photocytes in cnidarians (see Fig. 1B), echinoderms, and polychaetes; and (3) photosecretory systems in cypridinid ostracods, copepods, bivalves, syllid polychaetes, and earthworms. Light organs (complex emitters) possess accessory structures that manipulate the light in complex ways with absorbing pigments, reflectors, lenses, diffusers, collimators, or filters. They include (1) glandular (usually bacterial) light organs (some fishes, crustaceans, cephalopods) and (2) photophores (fishes, cephalopods, crustaceans). Although the chemistry most often controls the color and intensity of the emitted light, and morphology determines to a large extent the spatial expression of the light (and sometimes intensity and color), temporal control of light emission is most often controlled by the nervous system or excitable cell membranes.

FUNCTIONS OF LUMINESCENCE

The majority of emitters luminesce only in response to contact stimulation. The resulting luminescent expression is usually either a slow glow lasting more than 5 seconds or a bright single pulse or burst of rapid intracellular flashes, each with a duration of less than about 2 seconds and often as short as 50 milliseconds. The bursts often propagate away from the point of stimulation as a traveling wave of luminescence and also away from vital structures. In order to be visually effective, bioluminescence must be produced when and where it can be physically detected. Maximum bioluminescent intensities reach about 10^{-2} μW cm^{-2} (with most in the range of 10^{-4} to 10^{-7} μW cm^{-2}), which is about equivalent to the boundary between nautical and astronomical twilight, well after the sun has set or before it rises. However, these dim to dark conditions occur virtually all of the time in crevices or under rocks.

Most bioluminescence almost certainly serves some form of communication. Communication is defined here as an exchange of a signal between a sender and a receiver to the benefit of at least the sender but not necessarily the receiver. Thus, whatever the incipient mechanisms were that gave rise to a visible chemiluminescent process, subsequent selection for enhancing the characteristics and expression of the light itself (e.g., intensity, color, kinetics, spatial distribution) was likely the result of evolutionary feedbacks between the emitter and the receiver.

There are four primary functions that bioluminescence might serve for an organism: defense (protection), offense (feeding), intraspecific communication (usually sexual courtship), and symbiosis (mutual benefit). Other potential functions are apparently relatively rare in nature. It is also important to note that most luminescent functions are not mutually exclusive, and more than one function is sometimes served by one light-emitting system.

Defense

Luminescent systems appear to have evolved most often for deterring predators or potentially damaging intruders that induce luminescence upon contact with the emitter. In these cases the emitter itself often does not have well-developed vision and often does not display complex behavior (see Figs. 1A, D, F), but the receiver, which is a different species, has both. These receivers—predators or intruders—are most likely to be fishes, crustaceans, cephalopods, or polychaetes. Many intertidal and subtidal predators visually forage in dim light, either nocturnally in the open or by day beneath rocks and epifauna. It is in both these situations where stimulated bioluminescence is most pervasive in the intertidal (or anywhere). In order to repel or deflect an intruder, nearly all the intertidal (and benthic) forms that luminesce appear to target the intruder's visual systems by startling ("boo" effect), blinding ("flashbulb" effect), warning (aposematic effect), decoying (via a sacrificed luminescent body part or deluding secretion), or attracting the intruder's predators (burglar alarm effect), all of which lead to a spatial separation of the emitter and receiver and also tend to direct the intruder away from vital regions of the emitter's body. These defensive tactics need not be mutually exclusive, and the signal could induce more than one of these effects both spatially and temporally. For instance, scaleworms (polynoid polychaetes) produce posteriorly directed flashes on initial contact, but they will drop one or more scales, which then rhythmically pulse, if the contact persists, which presumably detracts (decoys) the intruder away from the now-dark worm. An analogous situation occurs with the luminescent arms of brittle stars (ophiuroids). Any and all of these responses might well alert the intruder's predators to its presence as well, though this has not been experimentally tested in these cases.

Courtship

Another function, but probably a secondary effect co-opted from intruder deterrence, occurs in some intertidal (and benthic) forms where light is used for attracting mates. In these cases the receivers of the same species (and usually also the emitter) have well-developed eyes and complex behavior. Luminescence is usually displayed as a ritualized, species-specific courtship display pattern toward potential mates in the water column above the substrate, in a way analogous to that of fireflies on land. Cypridinid ostracods (see Fig. 1E) from the shallow waters of the Caribbean, syllid polychaetes (see Fig. 1C), and perhaps midshipmen fish, which come into the shallows and intertidal to breed (see Figure 1G), use species-specific light cues to attract mates. Male ostracod crustaceans use secreted patterns of light to attract females. Female syllid polychaetes produce a single or pulsed glow to attract rapidly flashing males. Midshipmen males may use light, along with sound, to attract females to potential nest sites, although this function has yet to be verified, and these fish probably use their lights differently in their deep-water foraging habitats.

Offense

Using luminescence for obtaining prey is known for glowworms at cave mouths on land and flashlight fishes in shallow tropical waters, where they use bacterial light to detect and attract prey, and is postulated (but not demonstrated) for the lures of deep-sea anglerfishes. In these cases, a controlled, usually continuous or long-lasting expression of luminescence can act as a flashlight of and/or a lure to visually orienting, actively motile prey in dim or dark light conditions. These conditions have not yet been demonstrated for any intertidal species.

Symbiosis

Finally, light could be involved in complex symbiotic relationships. For instance, mutualism has been demonstrated in which some fungi use light emanating from the gills of their fruiting body to attract insects, thus providing food for the developing insect larvae, which eat the hyphae in exchange for spore dispersal by the insect, much in the same way that flowers attract insects. Such mutualisms are unknown in the intertidal, however. On the other hand, parasitism of crustaceans by luminescent enteric bacteria has been suggested. Intertidal and subtidal caprellid and gammarid amphipods infected with species of luminescent enteric bacteria (genus *Vibrio*) have been documented. The hypothesis suggests that the glowing crustacean may increase the likelihood that it will be eaten by a nocturnal predator, thereby delivering the bacteria to a rich source of nutrients in the gut of the predator, but without harming the predator. However, this relationship has not been tested, and it remains only a tantalizing hypothesis.

Overall, what emerges from an examination of potential functions of luminescence found in the intertidal is that there are relatively few luminescent species in the intertidal, but their functions are roughly consistent with a broader comparison. Among all the luminescent genera, most use their light for defense, many also use light for intraspecific communication, some use light for obtaining food, and a very few use light for mutualism.

Finally there are the incidental pelagic species that are brought to the intertidal and sometimes adorn the shore with fascinating sparks and glows, which are really epiphenomena expressed by organisms usually in a moribund condition. Few survive their close encounter with what, to them, are foreign habitats. These incidental visitors include blooms of luminescent dinoflagellates or radiolarians, or, more rarely, clusters of deep-sea copepods, squids, or gelatinous zooplankters such as ctenophores, hydrozoan medusae, scyphozoan medusae, and thaliacean tunicates.

SEE ALSO THE FOLLOWING ARTICLES

Benthic-Pelagic Coupling / Camouflage / Foraging Behavior / Light, Effects of / Phytoplankton / Predator Avoidance

FURTHER READING

Hastings, J. W., and J. G. Morin. 1991. Bioluminescence, in *Neural and integrative physiology.* C. L. Prosser, ed. New York: Academic Press, 131–170.

Hastings, J. W., and J. G. Morin. 2006. Photons for reporting molecular events: green fluorescent protein and four luciferase systems, in *Green fluorescent proteins: properties, applications, and protocols.* 2nd ed. M. Chalfie and S. Kain, eds. New Jersey Wiley-Liss, 15–38.

Herring, P. J., ed. 1978. *Bioluminescence in action.* London: Academic Press.

Herring, P. J. 1987. Systematic distribution of bioluminescence in living organisms. *Journal of Bioluminescence and Chemiluminescence* 1: 147–163.

Morin, J. G. 1983. Coastal bioluminescence: patterns and functions. *Bulletin of Marine Science* 33: 787–817.

Morin, J. G., and A. C. Cohen. 1991. Bioluminescent displays, courtship and reproduction in Ostracodes, in *Crustacean sexual biology.* R. Bauer and J. Martin, eds. New York: Columbia University Press, 1–16.

Shimomura, O. 2006. *Bioluminescence: chemical principles and methods.* New Jersey: World Scientific.

BIRDS

J. TIMOTHY WOOTTON
University of Chicago

Birds are often overlooked as important components of coastal ecosystems because they are a land-based group of animals. However, they can play significant roles in shoreline communities by changing mortalities of coastal species via predation and other activities and by importing nutrients from other systems.

FIGURE 1 Glaucous-winged gulls *(Larus glaucescens)* resting on barnacles in the upper intertidal zone as the tide comes in. Gulls are conspicuous shoreline birds in most areas, and they have a stout bill with a slight hook at the end that facilitates a generalized diet and tearing apart food. Photograph by the author.

COASTAL BIRD GROUPS

Birds utilizing the shoreline span a remarkable range of taxonomic groups (Table 1). Birds primarily use shorelines as feeding habitats, although some take advantage of the adjacent seascape to establish nests, and many can be seen resting in these areas at low tide (Fig. 1). Coastal habitats such as offshore islands, flotsam-covered shores, and wave-carved cliffs often provide refuges from mammalian predators. The reasons shorelines make good resting areas are not well studied, but might include avoidance of territorial aggression by resident birds, a wide vista to scan for predators or offshore food sources, and a lack of physical habitat structure, which might facilitate social interaction or escape from predators. Birds divvy up the shoreline in different ways. Some species, such as oystercatchers *(Haematopus),*

TABLE 1
Bird Groups That Commonly Use Shoreline Habitats, and the Food Resources That They Consume

Common Name	Family	Food Resources
Cormorants	Phalcorcoraridae	Fishes
Herons	Ardeadae	Fishes
Ducks	Anatadae	Molluscs, crustaceans, echinoderms, annelids, fish
Geese	Anseridae	Algae.
Falcons, ospreys, and eagles	Falconiformes	Shorebirds, fishes, carrion
Oystercatchers	Haematopodiae	Large bivalves and gastropods, crabs, sea urchins, worms
Plovers	Charadriidae	Crustaceans, barnacles, small molluscs, annelids
Sandpipers	Scolopacidae	Crustaceans, barnacles, small molluscs, annelids
Gulls	Lariidae	Wide range of invertebrates, fishes, birds, and carrion
Alcids	Alcidae	Fishes
Kingfishers	Alcidinidae	Fishes
Flycatchers	Tyrannidae	Insects, amphipods
Ovenbirds [*Cinclodes*]	Funariidae	Molluscs, crustaceans, insects
Swallows	Hirudinidae	Insects
Crows	Corvidae	Wide range of invertebrates, fishes, and carrion
Pipits	Motacillidae	Small snails, insects
Starlings	Sternidae	Insects
Sparrows and longspurs	Emberizidae	Insects

establish and defend from other conspecifics territories, which include feeding areas. In contrast, some species, such as sandpipers (Scolopacidae), aggregate and forage in flocks. Finally, many birds, such as gulls *(Larus)*, defend an area only immediately around them, such as the space around a prey item on which they are currently feeding.

FEEDING

Because of the open vistas shorelines offer, the moderately large size of the prey, the high consumption rates of birds, and the frequent presence of hard parts such as shells and scales in many prey species, the diets of birds occupying shore habitats are readily observed, making them a model group for feeding studies. The diets of birds that feed on shorelines span a wide range of prey resources (Table 1). Gulls (Fig. 1) and crows (*Corvus* spp.) frequently consume a broad range of animals that includes bivalves, gastropods, barnacles, sea urchins, sea stars, fishes, sea cucumbers, isopods, crabs, polychaete worms, and ribbon worms. Other birds, such as sandpipers, plovers (Charadriidae), ducks (Anatinae), and pipits (Motacillidae) are more specialized, focusing on small amphipods, particular barnacle species, or molluscs. Some geese (Anserinae) feed on algae along the shore. Many groups concentrate just on fish or insects that utilize shoreline habitats. Even for those species that feed widely, diet observations or discarded hard parts in nesting territory middens often show individual specialization on particular prey species, termed "majoring." Feeding observations often reveal differences in prey between adults and nestlings, which may facilitate consumption by small nestlings, maximize rates of food delivery per unit time, or reduce predator detection of nests by minimizing the number of food delivery trips. Feeding by different groups of birds varies through the tidal cycle, depending on the activity patterns of prey and the morphological traits of the birds. Some birds, such as sandpipers and various passerine groups, feed when the shoreline is exposed to air at low tide. Others, such as fish-feeders (e.g., alcids, kingfishers [Alcedinidae], and mergansers [*Mergus*]) and invertebrate-eating ducks (e.g., eiders [*Somateria*], scoters [*Melanitta*], oldsquaws [*Clangula hyemalis*], and harlequin ducks [*Histrionicus histrionicus*]), feed when the shore is immersed. In many cases, birds forage right at the water edge, where the prey are adjusting their activities to meet the challenges of being immersed in water versus air.

Methods of prey capture and of prey consumption vary among bird species that use shorelines, and they depend strongly both on prey traits, such as structural defenses and use of refuges, and on the beak morphology of the birds. Many bird groups have skinny beaks, which they use as tweezers for extracting prey from crevices in the rocks or between sessile organisms. Bills of sandpipers are some of the most specialized for this task, and they range in both length and shape (straight, turned up, or turned down), which permits exploitation of different habitats. In general, sandpipers specializing on rocky shores tend to have shorter bills than those utilizing soft sediment habitats, because most prey cannot burrow deep into rock. Oystercatchers (Fig. 2) have long chisel-like beaks that are especially well suited for handling prey once they are captured, as well as for extracting prey individuals from refuges. The long beak permits probing of larger crevices for hiding prey such as crabs, or serves as scissors to snip the attachments of prey such as sea urchins and mussels, yet its sturdy chisel shape is often used to dislodge prey such as limpets with a sharp lateral blow. Once captured, oystercatchers extract the most edible portions of prey either by chiseling at shells, or more commonly by deftly probing between shell parts and snipping the meat from the shell by using the beak like a pair of scissors. Other birds, such as gulls (Fig. 1) and crows have more generalized bills that can be used both in capturing food by probing for prey or by tugging prey from the substrate, and in extracting the edible prey parts by prying apart or hammering shells and by ripping flesh from carcasses. The majority of shore-feeding birds do little processing of prey with their beaks and swallow prey whole. In these cases, structural prey defenses such as shells, exoskeletons, or scales either are crushed physically with a muscular gizzard (e.g., many mollusc-feeding ducks and sandpipers), digested chemically, or regurgitated undigested as pellets (many gulls and birds of prey), which facilitates diet investigations. Some species use behavior to circumvent prey defenses. When feeding on bivalves, oystercatchers

FIGURE 2 A group of American black oystercatchers *(Haematopus bachmani)* in the upper intertidal zone. These birds are efficient foragers on benthic invertebrates. Photograph by the author.

focus on situations in which bivalves are gaping open, such as when they are immersed in water and either feeding or rehydrating following extended exposure to air. Birds such as gulls and crows exhibit a notable behavior known as anvil use, where they drop prey from some height onto rocks below to crack shells. This procedure requires substantial skill because an anvil rock must be chosen that prevents the prey from ricocheting off to an inaccessible spot upon impact; the birds must be sufficiently accurate to hit the anvil from some height under challenging wind conditions; the birds must account for shell, weight, and shape properties of different prey species; and the prey must be dropped from an altitude that is high enough that the prey's shell will crack upon impact but low enough for the bird to reach the prey before lurking kleptoparasites steal it.

Because birds are derived from a terrestrial lineage, feeding on marine prey can pose special physiological problems for birds that live along the shore. Osmoregulation is an aspect that has received particular attention because many marine species have a higher salt content than terrestrial organisms and local sources of freshwater are often limited because of wave splash and local weather patterns. One solution is to feed on prey that are themselves low in salt content, such as many bony fish species and some crustaceans. As a further solution, many birds that inhabit shorelines have specialized salt excretory glands or exhibit reduced water loss from evapotranspiration.

Another problem posed by marine species is that they sometimes contain powerful toxins. For example, molluscs often filter planktonic dinoflagellates out of the water column and may sequester in their tissues the strong neurotoxins that these organisms produce, making them toxic to vertebrates. Yet there is little evidence of large-scale mortality of birds using shorelines following blooms of these dinoflagellates ("red tides"). How do birds avoid this hazard? One possibility is that they have evolved special physiological mechanisms to counteract the toxins, but to date there is little evidence of these. Instead, limited experiments on gulls suggest that they are adept at detecting and then either rejecting or regurgitating toxic food to avoid the hazard following detection. If no alternative food sources are available, the birds may be forced to move to other areas.

COASTAL IMPACTS OF BIRDS

Although birds are relatively rare compared to other coastal organisms, their energy requirements are great because of their large size and exceptionally high metabolic rate, so they can have substantial impacts on their prey populations in many situations. A variety of experimental manipulations and quantitative calculations have demonstrated that birds can significantly reduce populations and change the size structure of many benthic marine invertebrates, including bivalves, gastropods, crabs, sea urchins, barnacles, sea stars, and amphipods. Less is known about impacts on more mobile prey such as fish and insects. The temporal pattern of these impacts can vary as a consequence of migratory patterns; some birds such as oystercatchers, crows, and gulls tend to be resident and exert chronic impacts throughout the year, whereas migratory species such as sandpipers and pipits produce pulsed episodes of intense predation in an area. In the latter case, the effect of bird predation on prey populations depends on key life history events and points of regulation in the prey populations in relation to the timing of the predation pulse. For example, a pulsed bird presence is likely to have a much stronger effect on prey populations if it occurs just prior to the breeding season of prey than if it occurs after the population has been swelled by newly produced offspring. Birds often vary in abundance across space along the shore and consequently can generate significant spatial variation in prey populations. For example, birds such as oystercatchers, sandpipers, and passerines that feed at low tide usually require a horizontal surface on which to stand and hence cannot feed effectively on vertical walls. Many studies have found markedly higher abundance and individual sizes of avian prey on vertical walls compared to adjacent horizontal rock benches. This pattern has provided further evidence of the substantial role that birds can sometimes play on shorelines.

Bird impacts on the shoreline system are not limited to their feeding activities. First, birds can serve as prey resources for some species. Birds of prey, terrestrial mammals such as raccoons and otters, and the occasional shark will take adult birds. Furthermore, mammals, reptiles, and some birds such as gulls and crows regularly feed on the eggs and chicks of other birds. Additionally, because they often eat fish, molluscs and crustaceans, coastal birds can be a key host in the life cycles of parasites such as trematodes. In some cases these parasites can reduce the vitality of their avian hosts, but in others there seems to be little noticeable impact. Instead, the birds seem to be serving an important role as dispersal agents of the parasites to new areas, where they are released into the environment by defecation and go on to attack other invertebrate hosts in the life cycle, often severely.

Birds may also be a source of physical disturbance on the shore. As they feed, birds can sometimes cause considerable ancillary disturbance to nontarget species. Often prey hide on or under sessile organisms attached to the rocks, and birds may rip out or damage these sedentary organisms as they search for food. Because they are often structurally less robust, algae frequently sustain the greatest damage. In some situations, damage of algae by birds such as gulls is so ubiquitous that one wonders whether it has been done intentionally to maintain an area than can be easily searched for food. Gulls and other birds also disturb shorelines by harvesting benthic algae for use as nesting material.

Because birds are contained within a web of species interactions, strong impacts of birds on shore-dwelling organisms can have indirect effects on other species, and birds can themselves be indirectly influenced by other species. These indirect effects can arise in two general ways. First, chains of interactions can cause cascading effects through food webs. For example, experiments have shown that birds can reduce populations of grazing molluscs and sea urchins, which in turn can increase algal abundance and alter the food and shelter available for other grazers that are not susceptible to bird predators. Second, species can alter the intensity of interactions between individuals of pairs of other species. For example, experiments have shown that bird predation can change the relative abundance of different species of sedentary organisms attached to the rocks, and these species in turn can change interactions between predators and prey by making prey easier or harder for the predator to find. Similarly, subtidal predators such as large fish and sea stars often drive prey to the water surface or even out of the water, where they become easier for birds to catch.

Birds often can play an important role on the shoreline as conduits for material flow between ecosystems. Shoreline insect production, arising either from grazing on algae growing on the rocks or from scavenging algal and animal detritus washed up on the shore, can provide an important subsidy to insectivorous birds, allowing them to maintain higher populations in adjacent terrestrial habitats than would otherwise be possible. Similarly, piscivorous and zooplanktivorous birds often forage widely away from the shore and return to shoreline nesting areas, where they deliver both energy to the system in the form of dead prey and nutrients contained both within dead prey and in excreted guano. Through this mechanism, many groups of birds that do not utilize shoreline habitats for feeding, such as pelicans, alcids, and petrels, can still have substantial local impacts on the coastal zone. Several studies have investigated the consequences of nutrient inputs into costal zones and have found varying effects. In some cases, the added nutrients promote coastal production of algae. In other cases, the high nitrogen and salts contained in guano are toxic to shoreline plants and algae, shifting the balance of the system toward more resistant species.

EVOLUTIONARY IMPACTS

Because of the impacts they have on coastal systems, birds may be an important evolutionary force for shore-dwelling species. Many prey species of birds, such as certain limpets, fish, crabs, and sea stars, are remarkably cryptic in their environment, effective at finding or even making refuges, resistant to being dislodged, or armored heavily. Several studies have demonstrated that bird predation exerts strong selection against more visible individuals of a species, consistent with strong evolutionary impacts, but it has been hard to prove that birds are the primary driver of prey defensive traits, for several reasons. First, other agents of mortality are expected to have similar effects, including visually foraging predators such as fishes and octopuses, and waves that deliver strong forces that impact the shore. Additionally, the offspring of many prey organisms are highly mobile, which can counteract spatially variable bird abundance by introducing a stream of locally maladapted individuals that interbreed with locally adapted individuals and thereby counteract evolutionary change.

RELATIONSHIP WITH HUMANS

Because they occupy positions high in the coastal food chain, birds can serve as sensitive indicators of environmental impacts and change. Shoreline birds face a variety of anthropogenic threats. Coastal zones are some of the most heavily populated areas of the world. Human activity leads to high rates of habitat loss for nesting and feeding. Simply walking along the shore can disturb shore birds, increasing susceptibility of both adults and nestlings to predators or reducing feeding rates. Humans often harvest large quantities of coastal organisms used by birds, such as various mollusc species, and hence may compete with birds for food. Exotic species introduced by humans, particularly predatory mammals on islands, can lead to changes in the prey base affecting food resources and often produce catastrophic increases in nest predation. Elevated nutrients in coastal runoff from sewers, agriculture, and deforestation may be favoring blooms of toxic algae, which, if not directly poisoning birds, still restrict the food resources that birds can use. Finally, coastal spills of oil and chemicals have led to well-documented mortality events of birds using shorelines. Generally mortality is greatest for birds that forage on the water, which suffer both from the disruption to the critical insulating function of their feathers and from direct ingestion of the toxic chemicals while attempting to clean their feathers.

Mortality caused by consumption of polluted prey is also a possibility but has been less clearly documented.

SEE ALSO THE FOLLOWING ARTICLES

Ecosystem Changes, Natural vs. Anthropogenic / Food Webs / Foraging Behavior / Habitat Alteration / Predation / Vertebrates, Terrestrial

FURTHER READING

Bent, A. C. 1921. *Life histories of North American gulls and terns.* Smithsonian Institution, United States National Museum Bulletin 113. Washington, DC: U. S. Goverment Printing Office.

Bent, A. C. 1927. *Life histories of North American shore birds.* Smithsonian Institution, United States National Museum Bulletin 142. Washington, DC: U. S. Goverment Printing Office.

Croxall, J. P., ed. 1987. *Seabirds: feeding ecology and role in marine ecosystems.* Cambridge, UK: Cambridge University Press.

Ehrlich, P. R., D. S. Dobkin, and D. Wheye. 1988. *The birder's handbook: a field guide to the natural history of North American birds.* New York: Simon & Schuster.

Gaston, A. J. 2004. *Seabirds: a natural history.* New Haven, CT: Yale University Press.

Hori, M., and T. Noda. 2001. Spatio-temporal variation of avian foraging in the rocky intertidal food web. *Journal of Animal Ecology* 70: 122–137.

Olsen, K. M. 2004. *Gulls of North America, Europe, and Asia.* Princeton, NJ: Princeton University Press.

Paulson, D. R. 2005. *Shorebirds of North and Central America: the photographic guide.* Princeton, NJ: Princeton University Press.

Poole, A., and F. Gill, eds. 1993–2002. *The birds of North America.* Philadelphia: The Academy of Natural Sciences.

Schreiber, E. A., and J. Burger, eds. 2001. *Biology of marine birds.* Boca Raton, FL: CRC Press.

Wootton, J. T. 1997. Estimates and tests of per-capita interaction strength: diet, abundance, and impact of intertidally foraging birds. *Ecological Monographs* 67: 45–64.

BIVALVES

JONATHAN B. GELLER

Moss Landing Marine Laboratories

The Bivalvia is the second largest class of the phylum Mollusca and includes clams, mussels, and oysters. Bivalves differ greatly from other molluscs, with two hinged shells, a reduced head, and a narrow foot. These differences relate to an ancestral lifestyle spent buried in sand and mud. In most bivalves, the gills are used for filter feeding. From infaunal ancestors, many lineages of bivalves evolved mechanisms to attach to or bore into hard surfaces. Bivalves are thus very successful on rocky intertidal shores worldwide, with species on or in the rock substrata, and others nestled in protected microhabitats.

GENERAL FEATURES OF BIVALVES

Bivalves are ecologically important members of all benthic marine habitats, including the rocky shore, sand and mudflats, subtidal soft sediments, and hydrothermal vents. Visitors to the rocky shore are most familiar with mussels and oysters, which can be the major space occupiers and, through competition or the provision of secondary space, exert strong control over the abundance of other species. Less conspicuous bivalves also live in rocky intertidal communities but can be easily found when carefully looked for. The adaptations of both the obvious and the hidden bivalves of wave-beaten rocky shores make them superbly fit for life in this challenging environment.

Bivalves are members of the phylum Mollusca in the class Bivalvia, sometimes called the Pelycopoda. These names refer to characteristic features of most bivalves that set them apart from other molluscs: the possession of two shells ("valves") and a typically hatchet-shaped foot. Bivalves are also unlike familiar molluscs, such as snails and squids, by having a very reduced head that lacks obvious sensory structures such as eyes. Bivalves are the only major group of molluscs that are mostly filter feeders, using a modified gill as described subsequently. The shells are laterally arranged, with the anterior-posterior body axis running between them. A soft tissue, called the mantle, lines and secretes the shells. A hinge joins the two shells dorsally and may contain interlocking teeth; shell teeth may also be found adjacent to the hinge. A proteinaceous ligament joins the two shells and also provides the impetus for shell opening by virtue of its spring-like elasticity. Posterior and anterior adductor muscles traverse the two shells and counteract the opening force of the ligament; some species, as noted below, have only one adductor muscle. As in other molluscs, the dorsal region of the body contains most of the viscera (heart, gut, kidneys, and gonads). Two gills, one on each side, lie left and right of the main body in a space, called the mantle cavity, between the interior surface of the mantle and the viscera. The foot lies ventral to the visceral region and is primitively used for burrowing. The general features of the anatomy of bivalves are illustrated in Fig. 1.

SUSPENSION FEEDING

Feeding by collecting suspended particles from seawater is a common means by which marine animals acquire energy and essential nutrients. This mode of feeding, referred to as suspension feeding, takes advantage of the enormous amount of production of bacteria, phytoplankton, and protists in seawater, as well as particulate

A

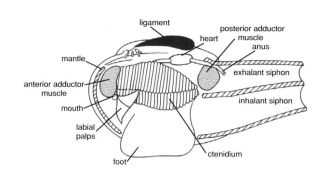

B

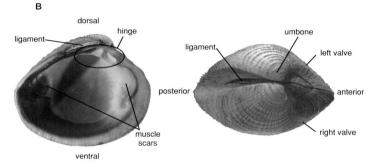

FIGURE 1 General features of bivalve morphology, from Carlton (2007).

detritus and smaller zooplankton. Animals that use sievelike structures to capture food particles are often called filter feeder. However, not all suspension feeders use sieves; food may also be captured by collision with mucus- and cilia-lined surfaces. Moreover, particles that can fit through sieve apertures might be collected nonetheless by collision and adherence to sieve surfaces. Among marine animals, feeding structures are often modifications of body parts that originally served another function. In bivalves, the gill has been modified from a strictly respiratory structure into a bifunctional respiratory and filter feeding apparatus.

The primitive bivalve had a pair of gills (called ctenidia) lying on each side of the body, as do modern bivalves. The primitive gill is a typical bipectinate molluscan ctenidium: a central axis from which a series of flat leaflike structures (lamellae) arise on either side. The lamellae are thin, to reduce the distance over which gas must diffuse, and broad, to provide ample surface area for gas exchange to occur. Beating cilia on the surfaces of lamellae pump water through the interlamellar spaces and ensure that the gills are constantly flushed with clean, oxygenated seawater. Because seawater is filled with particles that can foul gills and occlude water paths, specially

elongated cilia between lamellar edges serve as a barrier, while other cilia along the front edge carry potentially fouling particles into exhalant currents. This type of primitive gill is present in some extant bivalves and is termed a protobranch ctenidium.

The protobranch ctenidium has all the elements needed for an efficient filtering device: the ability to pump large volumes of water and a mechanism for removing particles from that water. In filter-feeding bivalves, the gill lamellae are elongated to increase the length of the frontal edge where filtering occurs. To contain the lengthened ctenidium within the mantle, the lamellae are folded, with tissue bridges holding the fold in place. Because each lamella is folded into a V shape, and each lamella lies on either side of an axis, the filter-feeding ctenidium has a W shape in cross section. Cilia carry food particles toward the mouth region where they are transferred to palps that transport the food to the mouth; other ciliary tracts carry inedible particles away. The simplest forms of these filter-feeding gills are termed *fillibranch*, whereas more complicated forms, in which tissue bridges connect adjacent lamellae, are called *eulamellibranch*.

ADAPTATIONS OF BIVALVES TO THE ROCKY SHORE

One of the greatest challenges of life on a rocky shore is resisting dislodgement caused by the great wave forces encountered there. Organisms that must move to feed, such as predatory sea stars or grazing gastropods, cannot protect themselves from waves by permanent attachment and have to find a compromise between mobility and a secure grip on the rock. As filter feeders, bivalves do not need to move to feed and therefore make no such compromise. Thus, all that was needed for ancestral infaunal bivalves in soft sediments to enter the rocky shore environment was a mechanism to avoid dislodgement. Bivalves have evolved three mechanisms. The first is to attach firmly to the rock surface, the second is to burrow into the substratum, and the third is to seek protected microhabitats.

Attaching with Threads: Mussels and Mussel-Like Bivalves

Several groups of bivalves use byssal threads to attach to rock surfaces or to other organisms. Byssal threads are proteinaceous strands produced by a byssal gland located near the base of the foot. The foot itself is highly specialized, no longer having a burrowing function and typically is used for locomotion only by juveniles. The thread is secreted as a liquid, molded by the foot into the thread shape, and attached by the foot to a hard surface with a disk-shaped plaque. Many threads are made, and the entire structure is called the byssus. While the most familiar of byssate bivalves are the mussels belonging to the family Mytilidae, byssal attachments have arisen independently in many other families (Fig. 2).

FIGURE 2 The mussel *Mytilus californianus,* found in the wave-exposed rocky intertidal zone from Baja California to the Aleutian Islands, attaches to rock surface with byssal threads that enable it to resist dislodgement. Photograph by the author.

The byssus varies in fine detail at several levels. First, the proteins constituting threads are not homogeneous but differ in the adhesive plaque, along the length of the thread, and in layers within the thread. Next, the number and thickness of threads in the byssus varies along the shore, seasonally, and among species in correlation to wave forces. As expected, tenacity of attachment depends on the thickness and number of threads in the byssus. Mussels do not resist waves by producing an overwhelming number of thick threads. Rather, they maintain a sufficient safety factor by adjusting thread numbers and thickness. For example, *Mytilus californianus,* living in the exposed rocky intertidal zone on the west coast of North America, have thicker threads than do *Mytilus trossulus,* which typically live in more protected sites. Further, mussels living outside of aggregations, and thus more fully exposed to waves, produce more threads than mussels living in aggregations. Energy saved by scaling the byssus to the actual risk of dislodgement can be devoted to reproduction, and fitness is maximal when the mussel has invested only the resources to the byssus as actually needed. Of course,

a mussel cannot "know" this fitness-maximizing value, and natural selection will continually track, with a generational lag, the optimum.

The byssus is an exemplary example of two evolutionary principles. The first is that the past and present adaptive significance of a structure can differ, and the second is that evolution tinkers with existing structures to perform new functions. The adult byssus has its origin as one or few threads used by newly settled larvae to attach to sedimentary particles while metamorphosing to juvenile clams. Larvae may also use byssal threads as a dragline to be carried by water currents. The byssal threads in ancestral and modern infaunal bivalves are quickly lost as the juvenile clam grows and gains the weight and burrowing ability needed to stay on the bottom. Mussels and several other groups of bivalves retain this larval feature into adulthood, an evolutionary change called neoteny. Instead of outgrowing the use of byssal threads, more threads are added to increase the strength of attachment, and this culminates in a strong byssus that can withstand wave shock on rocky shores.

Evolution has also changed shell shape and musculature in bivalves that adopt the byssal attachment strategy. These changes are most profound in the subclass Pteriomorphia in lineages in which both inhalant and exhalant water currents are located posteriorly. In these lineages, anterior tapering and posterior enlargement has resulted in a more or less triangular shell outline, in short the typical mussel shell shape. This posterior enlargement causes the elevation of the inhalant and exhalant water currents into the overlying water. This is thought to be an advantage in dense aggregations. The tapered anterior end leaves little interior room for soft body parts, and the anterior musculature is much reduced. This condition is called heteromyarian, referring to the asymmetry of anterior and posterior muscles, and is always associated with byssal attachment. In many mussels, there is also frequently a flattening of the ventral edges of each valve. This flattening lets the shell lie closely pressed to the rock surface, minimizing the space through which water can flow and dislodge the animal.

In byssate bivalves with anterior inhalant currents, restriction of the anterior end would be maladaptive, and these bivalves have limited success on wave-beaten hard surfaces. In the subclass Pteriomorphia, ark shells (family Arcidae) are an example of such bivalves, and one species found in the California fauna, *Acar bailyi,* is limited to calmer waters or rock undersides. Byssate bivalves from other subclasses are similarly unable to form mussel beds. Some examples from the Anomalodesmata are some species in the family Lyonsiidae; *Entodesma saxicola* and *Lyonsia californica* are common but limited to crevices or rock bottoms on the west coast of the North America.

The convergent evolution of the mussel form in distinct bivalve lineages is evidence of the survival value of these morphological changes. The most striking example is the freshwater genus *Dreissena*, which has strong resemblance to the marine genus *Mytilus* even though it belongs in a different subclass (Heterodonta). There is, however, disagreement on the evolutionary pathway leading to epibyssate bivalves, that is, bysally attached rock-dwelling bivalves. Some argue that retention of the byssus in adults first allowed attachment to rock surfaces, and this was followed by changes to shell shape and musculature. Others argue that an endobyssate stage, where infaunal byssate bivalves attached to buried stones or gravel, evolved first, and that the heteromyarian condition evolved before movement to hard surfaces. Updated phylogenetic analyses should be decisive in this debate.

Species of *Mytilus* are the most familiar of the byssally attached rocky shore bivalves, and these are among the most studied of all marine invertebrates. *Mytilus* spp. are found on most rocky temperate shores, and modern lineages appear to have arisen in the Northwest Pacific in the late Eocene or early Oligocene. Two lineages persist in the North Pacific, one leading to the modern *Mytilus californianus* on the Northeastern Pacific shore and *M. corsuscus* in the Northwest Pacific; the other led to members of the *Mytilus edulis* complex, a group of morphologically similar species *(M. trossulus, M. edulis,* and *M. galloprovincialis). Mytilus trossulus* is found sympatrically with M. *californianus* in the Northeastern Pacific. *M. trossulus* is usually found in more protected areas, including estuaries, than *M. californianus.* However, it is also found along the upper edges of *M. californianus* beds in Oregon and Washington. *Mytilus trossulus* expanded into the North Atlantic in the late Miocene or early Pliocene, and is now found in Nova Scotia and, curiously, in the Baltic Sea. *M. trossulus* is sister to a clade of *M. edulis* and *M. galloprovincialis.* *M. edulis* occupies the northeast and northwest Atlantic coasts. *M. edulis* expanded into the Mediterranean Sea where, upon the isolation of the Mediterranean Sea by the closure of the Gibraltar Strait, it diverged into *M. galloprovincialis.* Upon the reconnection of the Mediterranean Sea and the Atlantic Ocean, *M. galloprovincialis* expanded its range south and north of Gibraltar, coming into contact with *M. edulis* is southern Europe.

M. galloprovincialis also has proved to be a potent invader and is now established in California, Japan, South Africa, Hong Kong, and Australia. The biogeography of *Mytilus* in the southern hemisphere is not fully understood.

Cementation: Oysters, Scallops, and Similar Forms

In the previous section, we were introduced to heteromyarian bivalves that attach to rocks by a byssus along the ventral surface. Bivalves able to cement themselves to rocks have evolved from this template. These bivalves, including oysters, scallops, and some other similar forms, lay on their sides, with the lower valve cemented firmly to the bottom. This evolutionary transition was accompanied by further modifications to the soft anatomy, including the loss of the anterior adductor, enlargement and movement of the posterior adductor muscle to a more central position, and rearrangement of gills and viscera to surround the single adductor muscle. This body plan is called monomyarian. Permanent attachment by cementation has clear functional advantages. The permanence of attachment avoids the need to frequently renew byssal threads. Too, laying on their sides, oysters and scallops have a lower profile on rock surfaces, reducing the area against which waves can push. Cemented bivalves might also be better protected from some predators than byssally attached mussels: they are more difficult to lever off the rock surface and do not have a gap between the shell valves where the byssus passes through. This gap can provide access to the soft body for predatory sea stars or a point of leverage for the claws of crabs. The enlarged adductor muscle also allows the two shells to be more strongly closed.

Oysters (family Ostreidae) are monomyarian bivalves that are important bivalves on some rocky shores. On rocky temperate shores, oysters do not rival mussels in abundance, but they do achieve impressive densities in other areas. Of course, oysters are also very abundant in protected waters and estuaries, and are the basis of valuable fisheries worldwide. Oysters become ecologically important when they settle on each other to form mounds and reefs that provide habitat for other organisms, alter water flow, filter large amounts of phytoplankton, and stabilize underlying sediments. In the Pacific Northwest of the United States, we find *Ostrea conchalepa*, a relatively small oyster that is edible and was eaten by Native Americans despite its small size. *Ostrea conchalepa* is capable of forming reefs and in that state has been called *O. lurida.* Other oysters in this region are not native and have been introduced for mariculture. These include the European species *Ostrea edulis,* the Japanese species *Crassostrea gigas, C. ariakensis, C. sikamea*, and the American Eastern oyster, *C. virginica*. None of the introduced oysters are found on exposed rocky shores.

Scallops (family Pectinidae) are also monomyarian bivalves that lay on their right sides and can attach to hard surfaces by byssal threads while juvenile. In many species, adults are free living on soft sediments. Others, however, cement themselves to hard surfaces as they become adult. On rocky shores, high wave energy favors scallops that can attach firmly, such as *Crassoderma gigantea* (*Hinnites giganteus* in recent literature). This species attaches by byssal threads as juveniles, but can detach and actively swim to avoid predators. After reaching about 25 mm, the scallop cements itself by growing its right shell to low intertidal or subtidal rocks (Fig. 3A).

FIGURE 3 (A) The rock scallop *Crassoderma gigantea* and (B) the rock jingle *Pododesmus macrochisma* resist waves and currents by attaching one valve securely to rocks. In scallops, one shell is cemented to the rock surface, whereas in jingles an adductor muscle passes through the lower valve and attaches to the rock. Photographs by the author.

Other bivalves that cement themselves to rocks are jingle shells (family Anomiidae). These bivalves can be confused for true oysters, but in fact do not grow their shells to rock surfaces. Instead, they are attached by a modified byssus. The byssus extends through a large hole (actually a modified byssal notch extending from the shell's edge) in the left valve, which lies flat against a rock surface and is covered by the convex right valve. The byssus is calcified and cemented to rocks. The shell can be pulled tight against the rock by

muscles attached to the byssus; these muscles have the functional role of the enlarged posterior adductor of oysters and scallops. *Pododesmus macrochisma* (*P. cepio* in much of the literature) can be locally common on the rocky intertidal shore of the Northeastern Pacific shore (Fig. 3B).

Convergent evolution has given rise to yet another oyster-like group of bivalves, the family Chamidae, in the subclass Heterodonta.. These bivalves, called jewel box shells, differ greatly in anatomical detail from true oysters. For example, they are not monomyarian—they have two, albeit unequal, adductor muscles. However, they share the habit of cementing one valve to rock surfaces. Although chamid bivalves are mostly tropical or subtropical in geographic distribution, some species are found on temperate rocky shores. *Chama arcana* is common from Oregon to Baja California. Interestingly, different species in the Chamidae attach by different valves; *C. arcana* attaches by its left valve.

Boring Bivalves

Perhaps the most dramatically modified bivalves are those that bore into rock. Date mussels, such as those in the genera *Adula* and *Lithophaga,* use chemical secretions in mucus to dissolve calcareous substrata or mechanically abrade soft rocks (shale and sandstone) with their shell. In the later case, they attach byssal threads in forward and posterior positions, and alternately contract byssal retractor muscles to move the shell back and forth against the burrow wall. Date mussels are also reported from flint rock much harder than their shells, so the range of mechanisms for boring is not fully known. They can be quite common, yet are relatively unknown due to their inaccessible habitat. *Lithophaga* are also common in live or dead corals in tropical regions.

Bivalves in the family Pholadidae show elaborate morphological specialization for burrowing. These animals, called piddocks, have a sucker-like foot that extends through a large gape between the two valves and attaches to the bore hole. The anterior end of the shell bears filelike serrations on raised ribs. The shell is pushed against the end of the bore hole, rocked forward and back, and rotated to effect further excavation. Pholads have a long siphon, much exceeding shell length, that is used to reach the opening of their bore hole. The Wart-Necked Piddock (*Chaceia ovoidea*) has a siphon that reaches 1 m in length! Several species are common in the lower intertidal of the Northeast Pacific. These include *Zirfaea pilsbyri, Netastoma rostrata, Parapholas californica, Penitella conradi,* and *P. penita.* While piddocks are rarely seen alive, their bore holes are frequently found in eroded rocks and are evidence of their abundance. Pholads are evolved from ancestors that bored into compact sediments; modern examples of these include the common clam *Platyodon cancellatus* (family Myidae), found in packed clay in lower intertidal of bay mouths and the protected outer coast of the Northeastern Pacific.

Nestling Bivalves

Another strategy to avoid wave shock on the rocky shore is to nestle in protected microhabitats, such as sand pockets under rocks or in biogenic structures such as mussel beds, empty barnacle tests, or algal holdfasts. A taxonomically diverse array of bivalves have adopted this lifestyle, often by evolving a small body size that fits these small spaces. *Lasaea* is a cosmopolitan genus of tiny (to 3.5 mm), nestling clams. They are small enough to fit in spaces within and between empty barnacles shells, and can achieve densities of more than $275,000/m^2$. A tiny (to 4.5 mm) mytilid, *Musculus pygmaeus,* is also abundant in high rocky intertidal algal turfs, with reported densities of $10,000/m^2$ in California. Like many very tiny marine invertebrates, these bivalves brood their offspring, which do not disperse far. This may contribute to their extremely high local abundances. Another example from the U.S. west coast is *Kellia laperousii,* which is unusual among bivalves for its effective crawling on hard surfaces with its long foot. *Kellia* is common in mussel beds, kelp holdfasts, empty barnacle shells, and other protected spaces such as empty pholad holes. Although *Kellia* grows larger than the previous examples, to 25 mm, it too broods its offspring. File shells (family Limidae) improve the security of their nestling sites by making a nest from byssal threads on the undersides of rocks. Limids, such as *Limaria hemphilli* in California, can swim actively by clapping its valves together, as do true scallops. They have long, sticky tentacles of uncertain function on their mantle margin that are extended, medusa-like, from the gaping animal. These break away when the animal is disturbed.

Another place to find refuge from waves is the bodies of larger organisms. Larvae of *Mytilimeria nuttallii* (family Lyonsidae) settle on the tunics of compound ascideans in the lower rocky intertidal of the Northeast Pacific shore. They initially attach with byssal threads, but become overgrown and internalized in the tunic of their host. As internal symbionts, their shells are quite thin and fragile.

ECOLOGY OF ROCKY SHORE BIVALVES

The most important bivalves in rocky intertidal communities are mussels (Mytilidae) and, to a lesser extent, oysters (Ostreidae), and these dominate the middle intertidal region of rocky shores around the world. Their vertical distribution in the intertidal zone is often attributed to the stresses of exposure at low tide, which can include heat, desiccation, and reduced supply of waterborne food. It is often noted that the absolute height of mussels beds is higher in areas of greater wave splash, supporting the idea that aerial exposure sets the upper edge of mussel beds. However, the notion of physical upper limits is not necessarily universal, because consumers or competitors may exert pressure from above, as has been shown for some seaweeds. Physical conditions are benign for marine organisms at their lower tidal limits, and it is thought that predation or competition may prevent downward growth. For example, experimental removal of sea star predators resulted in downward growth of mussel beds in Washington State. In southern California, cages that excluded lobsters allowed growth of mussels below the natural lower limit. Suites of predators, including crabs, gastropods, sea urchins, and sea stars control the lower edge of *Mytilus edulis* on American and European Atlantic shores.

Bivalves are the dominant species at middle intertidal levels on most wave-exposed temperate rocky shores. *Mytilus californianus* dominates shores of western North America, while *Mytilus edulis* does the same in the northeast of North America and the northwest of Europe. *Perna perna* forms dense beds in wave-exposed south and eastern South Africa, although limpets dominate relatively protected areas. In the colder waters of western South Africa, and in rocky areas further north, *Choromytilus meridionalis* and *Aulacomya atra* (often reported as *A. ater*) are abundant. *Brachidontes rostratus* forms beds in the middle intertidal zone in Tasmania and Victoria, Australia. *Perna canaliculatus* fills this niche in New Zealand. The mid-intertidal zone of temperate regions of western and eastern South America, including parts of Peru, Chile, Argentina, and Brazil, contain beds of mussels *Perumytilus purpuratus* and *Aulacomya atra*. *Choromytilus chorus* may be locally abundant in Chile. Mussels may sometimes form beds in warmer waters: *Brachidontes exusta,* a species complex, can be abundant on intertidal rocks throughout the tropical western Atlantic. Although oysters are generally less abundant on rocky shores than mussels, the oyster *Saccostea cucullata* is common on intertidal rocks throughout the Indo-West Pacific.

When mussels are able to form continuous beds, they can compete with and exclude other species that must attach to the rock surface. Without predators or agents of disturbance, such as severe storms or wave-tossed logs, mussels can monopolize space and suppress the diversity of primary occupants of the rock surface. Indeed, multilayered mussel beds can contain enormous numbers of individuals: *Mytilus californianus* in Washington State contain between 459 and 11,098 individuals/m²; *Mytilus edulis* from the Bay of Fundy had 700–4,000 individuals/m²; subtidal beds of *M. trossulus* in the Baltic Sea achieved densities as high as 158,000 mussels/m²; and *M. galloprovincialis* in Italy reach 633–11,536 individuals/m². Despite this monopoly of primary rock space, mussel beds promote the diversity of secondary space occupiers. Barnacles, bryozoans, sea anemones, tube worms, algae, and many other organisms are able to live on mussel shells and so are not necessarily excluded from mussel beds. Furthermore, the internal matrix of multilayered mussel beds is habitable space for organisms that would otherwise be swept away by waves, such as sea cucumbers, free-crawling polychaete worms, and isopods. These internal spaces can also become filled with sediments, providing essential habitat for infaunal organisms. Consequently, mussel beds are quite rich in numbers of species: beds of *Mytilus californianus* contain at least 303 species, 69 species were found in *Mytilus galloprovincialis* beds in Japan, *Modiolus modiulus* and *Mytilus edulis* beds in Northern Ireland harbored 90 species and 34 species, respectively. Oyster beds can be similarly diverse: *Crassostrea virginica* reefs in South and North Carolina contained between 37 and 303 species in different regions of estuaries. These numbers vary greatly between studies because of different sampling methods and taxonomic precision; however, the general idea that mussel beds are rich in species is clear.

Bivalves are important not only as living space, but also as a trophic link between planktonic primary production and higher trophic levels. Bivalves are filter feeders that can remove particles as small as 2–3 μm at 80–100% efficiency. Particles as small as 1 μm are taken with less efficiency; thus a wide range of food sources, including bacteria, phytoplankton, and detritus are available. Most studies show that bivalves grow best on diets of phytoplankton. The amount of phytoplankton consumed by bivalves is enormous: bivalve populations have been estimated to require less time to filter 100% of the water in many bays worldwide than the water is resident in those systems. Bivalves on the rocky shore have no less filtering ability per capita; one estimate is 6–12 m³ of seawater cleared of phytoplankton/m² of mussel bed per hour. Coastal primary production, especially in areas of strong upwelling, is efficiently transferred into intertidal communities by filter-feeding bivalves, which in turn are consumed by gastropods, sea stars, crabs, lobsters, fishes, otters, and birds.

Bivalves are frequent biological invaders of marine waters. Athough less impacted than bays and estuaries, the rocky shore is no exception. *Mytilus galloprovincialis* has invaded the rocky coast of South Africa where it competes with native *Perna perna* and limpets. *Perna perna,* in turn, has invaded the Gulf of Mexico and Caribbean shores, and its congener *P. viridis,* an Indo-Pacific species, is now found in both the Gulf of Mexico and Australia. Because of their propensity to form beds, all mussels should be considered potential pests if introduced outside their native areas.

CLASSIFICATION

The classification of bivalves at higher levels has been based primarily on shell or gill characteristics, and the two systems have not been entirely compatible. Paleontologists have preferred to use shell characters, especially details of the hinge, as these are preserved in fossils. Anatomists have noted grades of gill architecture and proposed classification schemes based on those. Inclusion of more characters, including molecular data, has clarified the picture, and the most often used modern classification system recognizes the protobranch gill architecture as phylogenetically meaningful, whereas the eulamellibranch gill has evolved multiple times and reflects convergent evolution. The adaptations for life on the rocky shore discussed here are found in three of five subclasses of bivalves. An important conclusion of this chapter is that bivalves have evolved different solutions for life in the high wave energy of the rocky shores, and that each solution has evolved multiply and independently in these major groups.

Current schemes divide the Bivalvia into five subclasses: Protobranchia, Paleoheterodonta, Heterodonta, Pteriomorphia, and Anomalodesmata. The subclass Protobranchia is the only one that is based primarily on gill architecture. Protobranchs are considered the most primitive of bivalves. These bivalves possess a gill that most resembles a generalized bipectinate ctenidium, as described earlier. Protobranchs are deposit feeders, using oral palp proboscides to collect food from sediments, or harbor chemautotrophic bacterial symbionts that provide nutrition. Protobranchs are mostly found in the soft and often deep sea floor and not of importance on rocky shores. The Paleoheterodonta is comprised primarily of freshwater bivalves in the order Unionidae and are defined by their shells with interior nacre (mother-of-pearl) and details of their hinges.

Members of the subclass Heterodonta are very familiar to most students of marine life and to seafood eaters, and are the most specious of all classes of Bivalvia. This group includes the families that are known as clams and cockles. A unifying feature of heterodonts is interlocking teeth near or adjacent to the hinge. These bivalves usually have anterior and posterior adductor muscles, a eulamellibranch ctenidium, and siphons that direct water into and out of the mantle cavity. These siphons may be fused together and quite long, allowing some of these clams to burrow quite deeply in soft sediments and still maintain contact with the overlying water column. The mantle edges are often fused across the shell gape to prevent entry of sediment into the mantle cavity, with a gap only for the foot. The foot is typically muscular and wedge shaped, and used for burrowing. Heterodont bivalves discussed in this article include *Chama, Dreissena, Kellia, Lasaea,* and the family Pholadidae.

The subclass Anomalodesmata is a collection of ecologically and morphological diverse bivalves that occur in most marine habitats. Ecological specialization has driven considerable morphological divergence among families. Typical features include a nacreous shell, a toothless hinge (or secondary teeth only), and a eulamellibrach ctenidium. As in heterodonts, the mantle is usually fused along the shell edge, leaving only siphons and a gap for the foot as openings into the mantle cavity. A unique feature in this group is a fourth opening, located near the siphons. Another unique feature is a special gland (arenophilic radial mantle gland) that glues sand grains to the shell. Species may live in sediments, nestling in algal holdfasts, mussel beds, and in fouling communities, or as commensals with compound ascideans or sponges. This group also includes the only carnivorous bivalves. These employ a gill that is greatly modified into a muscular pumping septum to draw small prey items through the inhalant siphon. Such gill architecture is termed septibranch. Septibranch bivalves in the Northeastern Pacific are found in deep soft sediments. Bivalves in this subclass that are mentioned in this chapter are *Entodesma, Lyonsia,* and *Mytilimeria.*

Bivalves in the subclass Pteriomorpha are familiar as mussels, scallops, and oysters, among others. They have adapted to an epibenthic lifestyle through the evolution of various mechanisms to attach to hard surfaces and so are the most important bivalves on rocky shores. The shells of pteriomorphs are quite variable, some with similar left and right valves and others with distinctly different valves. In the latter, one valve is often flattened and lies close to the surface of the hard substratum. Pteriomorphs possess a byssal gland as larvae, juveniles, and often as adults. Most attach themselves to surfaces by byssal threads; some follow this by cementation. Pteriomorphs are heteromyarian or monomyarian. In the former case, they have two unequal

adductor muscles with the posterior muscle the larger, as in mussels. In the later condition, found in oysters and scallops, only the posterior muscle is retained and is located centrally between the valves. The gills are used for filter feeding and are filibranch or eulamellibranch in structure. Pteriomorphs generally do not burrow, and the foot, which is small and finger-like, is used by adults for the placement of byssal threads. Fouling of the mantle cavity by sediments is a lesser problem for epibenthic pteriomorphs than for infaunal heterodonts, and the mantle edges are not fused. Similarly, living at the substratum-seawater interface, siphons are unnecessary and these are not seen. Water flow can, however, be guided by pressing the mantle edge together on the posterior edge. Pteriomorphs mentioned here are *Acar, Adula, Aulacomya, Brachidontes, Choromytilus, Crassoderma, Crassostrea, Lima, Lithophaga, Mytilus, Modialis, Musculus, Ostrea, Perna, Perumytlilus, Pododesmus,* and *Saccostrea.*

SEE ALSO THE FOLLOWING ARTICLES

Adhesion / Molluscs / Stone Borers

FURTHER READING

Carlton, J. T., ed. 2007. *The Light & Smith manual: intertidal invertebrates from central California to Oregon,* 4th ed. Berkeley and Los Angeles: University of California Press.
Coan, E. V., P. V. Scott, and F. R. Bernard. 2000. *Bivalve seashells of Western North America. Marine bivalve molluscs from Arctic Alaska to Baja California.* Santa Barbara Museum of Natural History Monographs Number 2, Studies in Biodiversity Number 2. Santa Barbara, CA: Santa Barbara Museum of Natural History.
Gosling, E., ed. 1992. *The mussel* Mytilus: *ecology, physiology, genetics, and culture.* Developments in Aquaculture and Fisheries Science 25. New York: Elsevier.
Haderlie, E. C., and D. P. Abbott. 1980. Bivalvia; The clams and allies, in *Intertidal invertebrates of California.* R. H. Morris, D. P. Abbott, and E. C. Haderlie, eds. Stanford: Stanford University Press, 355–411.
Knox, G. A. 2001. *The ecology of seashores.* Boca Raton, FL: CRC Press.
Mathieson, A. C., and P. H. Nienhuis. eds. 1991. *Intertidal and littoral ecosystems. Ecosystems of the world,* Vol. 24. New York: Elsevier.
Morton, B., R. S. Prezant, and B. Wilson. 1998. Class Bivalvia, in *Mollusca: the southern synthesis. Fauna of Australia,* Vol. 5, Part A. P. L. Beesley, G. J. B. Ross, and A. Wells, eds. Melbourne: CSIRO Publishing, 195–234.

BLENNIES

ISABELLE M. CÔTÉ
Simon Fraser University, Burnaby, Canada

True, or combtooth, blennies belong to a large fish family (the Blenniidae) comprising 350 species found in temperate, subtropical, and tropical shallow marine waters around the world. Blennies exhibit remarkable variation in reproductive behavior, making them ideal models for behavioral studies. They are also ecologically important because most species are herbivorous and many are very abundant on rocky shores.

APPEARANCE

All blennies are relatively small (<15 cm), although a few species can reach more than 50 cm in length. They have scaleless, elongated bodies with large blunt heads (Fig. 1), often adorned with cirri, which can be large and extensively branched, particularly in males. The name "combtooth" is derived from the tightly packed, comb-like row of blunt teeth they possess on each jaw. Blennies are highly variable in color, with many species exhibiting spots, stripes, or bands. Some species change colors during breeding or during aggressive interactions. Cryptic coloration is widespread.

FIGURE 1 Two rock-pool blennies, *Parablennius parvicornis,* resting on the bottom of a tidepool at Île de Gorée, Senegal. Note the continuous dorsal fin, typical of the blenny family, and the small cirri above the eyes. Photograph © Dr. Peter Wirtz.

REPRODUCTIVE BEHAVIOR

Blenny reproduction occurs either seasonally, with a spring onset, or year-round with peaks of spawning around full moon. Males compete for access to small cavities in the substratum, empty shells, or barnacles, which they clean in preparation for nesting. As a result of this intense competition, male blennies are usually larger than females and more brightly colored during the breeding period. Males attract females to spawn in their nests with a series of characteristic head bobbing displays. Females lay a single layer of adhesive eggs on the nest wall and leave males to provide sole care of the eggs until hatching. Males may care simultaneously for eggs at various stages of development, which have been deposited by different females on different days.

In several species, alternative male mating tactics exist. Some males do not defend nests and guard eggs but mimic the females' morphology and behavior to approach nests, dart in, and parasitize fertilizations from nest-guarding males. This sneaky behavior is usually restricted to small males, who then adopt an egg-guarding strategy when larger.

Female blennies have fairly predictable mate preferences. Larger is usually better, although sometimes males with more developed cirri or nests that already have eggs are chosen. Larger males usually provide better care to the eggs in the nest, by fanning them assiduously with their pectoral fins and defending them from predators.

ECOLOGICAL IMPORTANCE

Most blennies are herbivores. Because of their sometimes phenomenal abundance, they exert strong grazing pressure, which is important in determining algal abundance and species diversity. Experimental exclusion of blennies from some rocky shore areas has been shown to result in drastically altered algal communities, which shifted from predominantly brown and red crustose macroalgae to green foliose species. Carnivorous blennies also have a large impact on algae. Blennies that rely on small invertebrates play a key role in controlling the abundance of herbivorous crustaceans, thus indirectly increasing the biomass and diversity of the algae upon which these crustaceans feed. Blennies therefore find themselves at the heart of many trophic cascades on rocky shores.

LIFE ABOVE THE WATER

Many species of blennies that live in the intertidal zone have the ability to withstand short periods of emersion. This ability is taken to extremes by a few species that have effectively become amphibious, living at the very top of rocky shores, where they are exposed to air for long periods. An amphibious lifestyle has required several adaptations, including providing internal support to the gills to prevent their collapse out of water, a flattened cornea to improve vision in air, various biochemical and physiological changes to deal with desiccation and ammonia excretion, and a phenomenal leaping ability to permit escape from aerial predators. For example, the small pearl blenny *Entomacrodus nigricans* (<10 cm) can jump more than 1 m in a single leap when threatened. This is approximately twice as far, relative to body length, as the currently held men's long jump world record!

SEE ALSO THE FOLLOWING ARTICLES

Camouflage / Desiccation Stress / Herbivory / Sculpins / Sex Allocation and Sexual Selection

FURTHER READING

Bruno, J. F., and M. I. O'Connor. 2005. Cascading effects of predator diversity and omnivory in a marine food web. *Ecology Letters* 8: 1048–1056.

Sayer, M. D. J. 2005. Adaptations of amphibious fishes for surviving life out of water. *Fish and Fisheries* 6: 186–211.

BODY SHAPE

STEVEN VOGEL

Duke University

The shapes of the inhabitants of tidepools and rocky shores must reflect the mechanical world in which they live—the hydrodynamic forces to which they may be subjected, the structural properties of their materials, and the arrangements with which they avoid dislodgement. But we commonly see them in tidepools at near low tide and under nearly calm conditions—mechanically the most benign and forgiving possible combination. It is as if we were listening to the instruments of an orchestra during intermission, not during the concert itself. To envision the mechanical challenges facing an organism, one must imagine a radically different situation. Intertidal areas of rocky, high-energy coastlines experience flows of rapidly changing speeds and directions together with the surge, splash, and plunge of breaking waves. Mechanically, then, a tidepool represents a particularly accessible bit of rocky intertidal, one stranded higher than the rest of the ocean when the tide recedes.

THE FORCES THAT MATTER

Drag

One first thinks of drag, a force on a body in the direction of the local flow that varies with the speed of flow, as in Fig. 1A. Older sources considered only drag, commonly with fanciful explanations of the relationship between form and drag. Applied to drag on rocky shores, our technological experience turns out to be less helpful than might be expected. In part, submarines and airplanes face directionally consistent flows and operate well away from substrata. Furthermore, both approach perfect rigidity, so shape and the forces of flow interact far less than they do in organisms. An airplane at the airport differs in shape only trivially from one going at top speed, while the shape of a macroalga in a storm surge will bear little resemblance to its shape in still water.

Acceleration reaction

Organisms face two other forces of comparable importance. The first, commonly termed the "acceleration reaction" (Fig. 1B), is a consequence not of the speed of flow per se but of the change of speed—the acceleration (and deceleration) of local flows. Accelerating or decelerating a body takes a force proportional to its mass. When a mass of fluid suddenly encounters and accelerates around a body, the body feels the effect as a force, one whose direction is that of the acceleration. One should note that whereas drag and acceleration reactions act concertedly for accelerating motion (the main present concern), for decelerating water (as when a swimming animal slows) they act in opposite directions. Drag typically depends (other things being equal) on the area of an object facing the flow; the acceleration reaction depends on the volume of the object. If both drag and attachment strength vary with area, then a large body is no worse off than a small one. But, given a volume-dependent acceleration reaction, the larger the body, the worse will be the force of flow relative to its attachment strength. So low volume has an advantage for organisms subject to severe flow surges beyond the advantage of low exposed surface for organisms in rapid but steady flows.

Lift

Yet another force is lift (Fig. 1C), defined as a force that acts perpendicularly to the direction of flow. Lift may be as hazardous for an organism clinging to a rock as it is for an automobile speeding down a highway. A low, rounded mound, something with lots of attachment area and low drag, turns out to be an especially effective lift producer. Shape changes that reduce lift often increase drag, which probably explains why the forms of some intertidal organisms (limpets in particular) either have more drag than expected or experience unexpectedly variable drag, individual to individual.

Impact Pressure

Something, whether liquid or solid, dropped on a body will make an impact. Breaking waves drop water on organisms, but the forces of impact rarely rise to significance relative

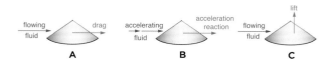

FIGURE 1 The main forces faced by coastal organisms exposed to waves: (A) drag; (B) acceleration reaction; (C) lift.

to the other forces at play, in part because compression of entrained air mitigates the suddenness of the event. But impacts of solids, mainly small rocks ("cobbles") and logs, are often significant and can even lead to temporarily or permanently bare patches.

As a result of the variability of flow magnitude and direction and of the diversity of operative forces, the classic drag-reducing streamlined form, an axisymmetric body rounded upstream and pointed downstream, is rare or absent among the sessile inhabitants of tidepools.

THE WAY THE FORCES IMPINGE

One might expect that encrusting organisms, which protrude only negligibly from their substrata (rocks or other organisms), would avoid flow forces entirely. In fact, they still experience shear, pulling them downstream, as in Fig. 2A. But even in very rapid flows and severe surges, shear forces are minimal; with large attachment areas, the resulting stresses (stress = force/area) will almost always be trivial.

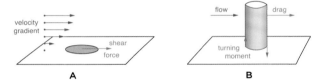

FIGURE 2 Less self-evident forces on intertidal organisms. (A) As a result of the velocity gradient adjacent to a surface and the shear that thus results from a fluid's viscosity, even an organism or colony that does not protrude will experience a downstream force. (B) A directly downstream force on a protruding organism will produce forces at right angles to the flow, tensile and compressive, between organism and substratum.

Pull on a rope and you impose a tensile load. Since the line of action of the force corresponds to the long axis of the object, it is both the easiest to envision and to resist. Macroalgae with long fronds may be subject to purely tensile loading; for organisms of most other shapes loading regimes are more complex and challenging. Neither the force parallel to flow (drag plus the acceleration reaction) nor that normal to flow (lift) will impinge directly; rather, these forces will produce turning moments about the attachment. These moments represent the product of the force and the perpendicular distance from its line of action to a rotation point, as in Fig. 2B. One sees the result of such a turning moment when a rearward push on the backrest of a chair makes its front legs rise.

Erect or protruding organisms thus face several problems beyond minimizing the forces of flow and resisting tensile loads. Since most flow goes around rather than over

it, a narrow, erect cylinder feels little lift. But making a cylinder taller increases the area facing the flow, the volume exposed to flow, and the length of the moment arm across which drag and acceleration reaction act. The turning moment of its drag will go up with the square of height, that due to the acceleration reaction with the cube of height. Few organisms of the rocky shore take the form of erect cylinders. Sea anemones, which do have that shape, avoid rapid flows by deflating when necessary down to a small fraction of both their extended heights and their volumes without changing attachment area. Cones do better than cylinders for the same volume and surface because the forces act on them with lower average moment arms; fairly low cones do better than high cones relative to both volume and surface. Thus tidepools contain nothing quite like branching corals except for tiny coralline algae and the like, but they host a great diversity of low, conical limpets, keyhole limpets, chitons, and such.

A purely tensile load imposes a nearly uniform stress on an attachment surface. By contrast, loads that produce turning moments create severe stress gradients. In the foregoing example of pushing the backrest of a chair rearward, its front legs rose while its back legs pressed down harder. If the chair were glued down, the push would produce upward tension in the front and increased downward compression in the back—a stress gradient from front to back. Similarly, an organism loaded in the direction of flow will feel an easily managed compressive stress at and near the downstream edge, but a tensile stress around the upstream edge. Lift, pulling outward from the substratum, will add to that tensile stress, its effect dependent on the location of the center of lift relative to the geometric center of the organism. Any glue or grapple will more likely fail at the upstream edge, where tension is maximal—the organism will face so-called peel failure, as in Fig. 3A.

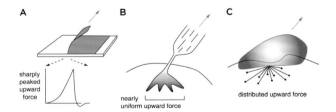

FIGURE 3 (A) Peel failure results when a slight directional asymmetry in a force normally distributed over an area makes that force act, with greatly increased stress, nearly on a line. (B) Applying a force through a stalk flexible enough that it resists mainly tension largely circumvents peel failure. (C) Using a large number of small attachments, each flexibly connected to the load, also defeats peel failure—among its other advantages.

What shapes can best deal with peel failure? A tapering margin of soft material peels with useful difficulty. The softness out near the edge converts much of the upward lift at the upstream edge to sideways sliding or shear, making a tensile force into a more easily withstood shearing force. The edges of the soft feet of gastropod molluscs, whether snails, limpets, or sea slugs, almost always taper outward. The feet of molluscs with shells may extend well beyond the shells' margins, both improving the conversion to shear and increasing contact area.

An additional device for minimizing the risk of peeling consists of using a tapered, flexible connection, a soft stalk, between an attachment pad and the rest of the organism, as in Fig. 3B. That ensures that almost any force on the organism will be taken by the pad in a more uniformly distributed combination of tension and shear, with little of the asymmetry that leads to peeling. Macroalgal holdfasts typically use such a pad-plus-joint, as do some sponges. Yet another approach involves multiple, tiny attachments, each flexibly connected to the rest of the organism (Fig. 3C). A large-area attachment will more likely peel than a small one; a very small one will simply be pulled straight or obliquely outward with more uniform stress over its contacting surface. A mussel attaching to a rock by numerous byssus threads (its "beard") uses the device. Starfish, sea urchins, and some other echinoderms do likewise with their tube feet, combining motility with good attachment to quite irregular substrata.

AVOIDING OTHER FORCE CONCENTRATIONS

Materials ordinarily fail at some level of stress, not of force per se. Stress, again, is force divided by area; more specifically, tensile stress is force divided by cross-sectional area. Thus, failure most often occurs at a narrowed region, say, of a shoe lace—same force but less area. A long algal frond, for instance, should be at greater risk where narrowed by the bite of some grazing animal. In fact, the risk may be far greater than one would expect from the relative reduction of cross section caused by the grazer. Stress may be especially high right at the base of an indentation, where force gets locally concentrated. A strip of aluminum foil, pulled lengthwise, will fail at much less force if it has a small crosswise nick in one edge. Even a small reduction in overall cross section can dramatically reduce effective tensile strength. In other words, cracks propagate.

How might either the occurrence or the danger of high local stress be avoided? One device widely used among the macroalgae consists of extending the length of edge beyond what would be needed for a planar structure, making what is termed an undulate margin. Such an

edge need withstand almost none of the tensile stress on the rest of the surface, with stress increasing with distance inward toward the long axis. Pressed and dried algal laminae almost always have folds near their edges because they are nontrivially three-dimensional, even if nowhere very thick. In addition or alternatively, algal laminae make use of material of low tensile stiffness—that is, material tolerant of stretching. Such stretching under load effectively reduces stress at the tips of sharp-ended indentations by rounding those sharp tips. Just this behavior underlies the greater difficulty of tearing a nicked piece of Saran Wrap compared with aluminum foil. An alternative to reducing the stress at the base of a nick is to mitigate its tendency to tear further. Grasses, in particular, do this with strong lengthwise fibers; one cannot easily tear a blade of grass transversely. But intertidal organisms seem not to make extensive use of this latter device.

Temporal as well as spatial force concentrations present hazards in the rocky intertidal. Consider the acceleration of a loosely tethered mass when hit by a wave. Its momentum (mass times velocity) will be converted to force on tether and attachments when the tether goes taut; how much force varies inversely with the suddenness with which it stops. Three devices among intertidal organisms mitigate such forces. First, concentrated masses on long tethers are rare. Second, tethers are made of extensible (low-stiffness) material, so stops are less sudden—macroalgae commonly do this. And third, tethers can be kept taut at all times—as are the byssus threads of mussels, where a specific muscle prevents slack.

FORCES, MATERIALS, FUNCTIONS, AND FORMS

Linking body form and environmental mechanics looks more complex than it did a decade or so ago. Flexibility—avoiding stiff, massy materials—provides no automatic benefit. Macroalgae, like the leaves on a tree, may curl and cluster in flow and thereby reduce their drag. But ordinary flags in air or water suffer several times the drag of rigid plates, so one cannot simply attribute their performance to flexibility. Furthermore, the mix of solidity and fluidity characteristic of almost all soft biological materials may make what appears soft to the touch to be anything but soft when suddenly loaded by a surge of water.

The variety of forms found in tidepools reflects both the complexity of the challenges and the diversity of tactics with which they can be met. An anemone stands erect in a pool but deflates into a small, limpet-like shape when waveswept. Macroalgae appear quite large, but their volumes remain relatively low, their materials tend to be extensible, and their holdfasts and stalks both taper to minimize peeling. Limpets and chitons have large attachment areas relative to their volumes, and they can adjust the attachment strength as circumstances change. Mussels use arrays of taut threads of sufficient extensibility to ensure good load sharing. Further many of the more mobile creatures, such as small whelks and crustaceans, take cover within the complex of larger organisms and substrata during stressful periods.

Beyond that, the shapes of the organisms of the rocky intertidal zone represent the interplay of the physical forces, the materials of which the organisms are made, and all other important biological functions. Thus, any advantage of an alteration in form that reduces force may be offset by excessive compromise of some other factor— feeding, reproduction, and so forth.

SEE ALSO THE FOLLOWING ARTICLES

Adhesion / Boundary Layers / Hydrodynamic Forces / Materials, Biological / Size and Scaling / Tidepools, Formation and Rock Types

FURTHER READING

Bascom, W. 1980. *Waves and beaches*, 2nd ed. Garden City, NY: Anchor/ Doubleday.
Denny, M. W. 1988. *Biology and the mechanics of the wave-swept environment*. Princeton, NJ: Princeton University Press.
Denny, M. W., and S. Gaines. 2000. *Chance in biology: using probability to explore nature*. Princeton, NJ: Princeton University Press.
Vincent, J. F. V. 1990. *Structural biomaterials*, rev. ed. Princeton, NJ: Princeton University Press.
Vogel, S. 1994. *Life in moving fluids*, 2nd ed. Princeton, NJ: Princeton University Press.
Vogel, S. 2003. *Comparative biomechanics*. Princeton, NJ: Princeton University Press.
Wildish, D., and D. Kristmanson. 1997. *Benthic suspension feeders and flow*. Cambridge, UK: Cambridge University Press.

BORING FUNGI

KATJA STERFLINGER

University of Natural Resources and Applied Life Sciences, Vienna, Austria

Fungi are primarily terrestrial organisms that developed in close association with higher plants. Although only few fungal species are commonly found in saline environments, most Ascomycetes and Zygomycetes (as well as the ascomycetous and basidiomycetous yeasts) easily tolerate the salt concentrations of the marine environment and are able to survive and to grow in seawater. Most of the so-called marine fungi in fact belong to the terrestrial soil microflora and are only secondary invaders of the marine environment. Although numerous fungal spores are distributed in marine water, colonies are formed only on solid surfaces. This means that fungi can be isolated from the water column as spores, but the adults are benthic organisms associated with marine organisms such as corals or bryozoans and also with mineral substrata.

BORING FUNGI IN INTERTIDAL ROCKS

Rock surfaces in extreme environments, such as high salinities, high temperatures, and intense UV radiation. are generally inhabited by black meristematic fungi, cyanobacteria, and lichens. Meristematic fungi are known from rock surfaces in arid and semiarid environments, from hot and cold deserts, and from rocky shores. The morphology of those fungi resembles that of colonies of boring cyanobacteria, and until 1981 they were erroneously identified as such. The meristematic fungi form darkly pigmented cell clusters with a diameter of up to 500 μm both on the surface and inside the rock. The surface colonies develop stolons and extremely thin hyphae, which penetrate the rock, leading the fungus to pores and cracks where new colonies are developed. As a result of the high turgor pressure of the cells and their extremely rigid cell walls, fungi growing in the rock push on the rock's structure, causing crystals to lose their adhesion to one another. This process finally leads to sloughing of material from the rock's surface. The depth of penetration into the rock can be up to 5 mm. Thus, boring fungi on rocky shores play a significant role in the weathering of the rocks. This is especially true for calcitic rocks (e.g., limestone, marble); boring fungi are found less often on gneiss and granite. From the analysis of field samples taken from rocky shores there is strong evidence that both mechanical action and acid attack play a role in rock penetration (Fig. 1A, B). Although meristematic fungi do not excrete organic acid under laboratory conditions, scanning electron microscopy has clearly shown the etching activity of fungal colonies on single calcite crystals (Fig. 1C).

TAXONOMY

The taxonomy of meristematic fungi is complex. Because the morphology, both on the rock and in culture, is similar for all species, identification is possible only on the basis of molecular data. Phylogenetic analysis based on

FIGURE 1 (A) A rocky shore with marble outcrop in the Mediterranean area showing (B) the typical growth pattern of meristematic fungi (width of image = 5 mm). (C) Etching of crystals by single colonies of fungi becomes visible under scanning electron microscopy (width of image = 300 μm).

the sequences of 18S rDNA, Internal Transcribed Spacers I and II, and 5.8S rDNA showed that the meristematic phenotype occurs in several orders of the Ascomycetes: Dothideales, Pleosporales, and Chaetothyriales. Many new species of meristematic fungi have been described in recent years; in saline environments the most important are *Hortaea werneckii, Coniosporium perforans,* and *Trimmatostroma salinum.*

The meristematic morphology found on intertidal rocks is interpreted as an ecotype that combines several adaptations to extreme environmental conditions. The sheltering effect of cell clusters (which have a small surface-area-to-volume ratio) is enhanced by the protective action of melanin, encrusting the cell walls. Each vegetative cell can provide propagation when the environment is benign and can also lie dormant when the environment is severe. In this way the production of spores is economized. With increasing osmotic stress, fungi induce the synthesis of high intracellular levels of glycerol and other compatible solutes. High trehalose levels, which protect enzymes under matrix and osmotic stress, seem to be present all the time. Energy consumption for synthesis of sugar alcohols and trehalose is high and occurs at the cost of growth rate. Meristematic fungi grow slowly, and

even under laboratory conditions colonies of some species do not exceed 5 mm in diameter within four weeks.

SEE ALSO THE FOLLOWING ARTICLES

Algal Crusts and Lichens / Heat Stress / Salinity Stress / Stone Borers / Ultraviolet Stress

FURTHER READING

de Hoog, G. S., ed. 2005. *Fungi of the Antarctic: evolution under extreme conditions.* Studies in Mycology 51. Utrecht: Centraalbureau voor Schimmelcultures.
de Hoog, G. S., ed. 1999. *Ecology and evolution of black yeasts and their relatives.* Studies in Mycology 43. Utrecht: Centraalbureau voor Schimmelcultures.
Sterflinger, K. 2000. Fungi as geologic agents. *Geomicrobiology Journal* 17: 97–124.

BOUNDARY LAYERS

LUCA A. VAN DUREN
Netherlands Institute of Ecology

A boundary layer is the transitional region from freely flowing water above a solid surface (e.g., the seabed) down to the stagnant water at the surface itself. The boundary layer is the result of friction exerted by the solid surface on the flowing water. The gradual increase in flow velocity from the bed to the free-flowing water above (the velocity gradient) determines the force the water exerts on the bottom, and on plants and animals living on the bottom. Strong velocity gradients require a very firm attachment to the bed, if the organism is to stay put. The velocity gradient also determines how quickly particles (e.g., food particles) are transported from the water column to the bed.

VISCOUS DRAG

Viscosity is a measure of a fluid's ability to resist deformation. In other words, it describes the "stickiness" of a fluid. Moving a teaspoon around in a cup of tea easily makes the tea swirl. Try the same in a cup of honey and one will find that it takes more effort to move both the teaspoon and the liquid because honey has a much higher viscosity than water. Although it is easier to stir water than it is to stir honey, water molecules nonetheless have a measurable tendency to resist movement relative to each other. This tendency is the basis for the formation of a boundary layer. Viscosity is not only a property of liquids. Gasses, such as air, also resist deformation, although gas molecules are much less tightly bound to each other than are water molecules. At a temperature of 20 °C seawater has a viscosity of 1.07×10^{-3} Pa s (pascal seconds), but air at the same temperature has a viscosity of only 18.08×10^{-6} Pa s.

Boundary layers occur on any solid surface in contact with a moving fluid. The fluid can be honey, tea, seawater or, as explained above, even air. The surface can be the inside of your teacup, the sea floor, a swimming fish, or the wing of an aircraft. For the sake of simplicity, in this article we deal with the boundary layer that exists between seawater and the rocky sea bottom; however, one should realize that the same principles apply to many different situations. The water immediately at the seabed has no motion relative to the surface of the bed. This is called the *no-slip condition,* and it is also an immediate consequence of the viscosity of seawater. As well as sticking to each other, the fluid molecules stick to the rocks on the bottom. Because it costs energy to move "parcels" of water relative to each other, water is slowed down near the bed. This is called *viscous drag.* Figures 1A and B illustrate two velocity profiles of water flowing over a smooth bottom. Theoretically the effect of viscous drag of the bed on the water column extends to infinity. However, at some distance away from the bed (a distance known as the *boundary layer thickness,* δ) the impact becomes negligible and the flow velocity here is defined as the *free-stream velocity.* Generally, the boundary layer thickness is defined as the height above the bed where the velocity reaches 99% of the free-stream velocity.

VELOCITY GRADIENTS AND SHEAR STRESS

For plants and animals attached to rocks, the velocity gradient, rather than the free-stream velocity, is often more important. The shear, that is, the velocity difference between layers of water, is what determines the force the water exerts on the bed and on the attached organisms. This *shear stress* is expressed in units of force per unit area, that is, N m^{-2}, or pascals (Pa). A very thick boundary layer with a very gradual velocity increase is much less likely to erode sediment or dislodge an object than is a thin boundary layer with a steep velocity increase. Consider the examples in Figs. 1A and B: both show a boundary layer over a smooth bed with a free-stream velocity of 20 cm s^{-1}, but one has a boundary layer thickness (δ) of 10 cm and the other has a thickness of 25 cm. The thinner boundary layer exerts a stress on the bed of 1.2 Pa, although for the thicker boundary layer the stress is about half as large (0.6 Pa). For larvae of bottom-dwelling animals that want to settle on

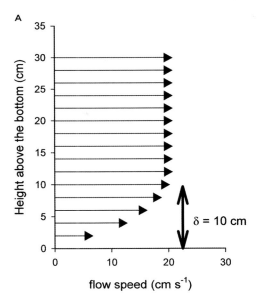

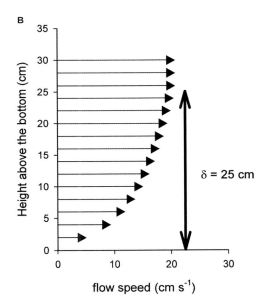

FIGURE 1 Two schematic examples of a boundary layer over a smooth flat bottom. Arrows represent the flow speed at a given height. Free-stream velocity in both examples: 20 cm s^{-1}, boundary layer thickness (A) 10 cm and (B) 25 cm.

the bed, this difference in stress can make the difference between being able to hold on to the bottom surface or being washed away.

A boundary layer does not appear instantaneously on a surface, it needs time to develop. The longer water flows with a particular speed over a surface, the thicker the boundary layer becomes. In deep, steadily flowing water, the boundary layer may have a thickness of several meters. In very dynamic environments such as rocky shores and tidepools, flow speed and direction constantly change as a result of waves and changing tides. In such environments, boundary layers are often thin and not fully developed. In shallow, wave-exposed sites the thickness of the boundary layer may be less than a centimeter.

ROUGHNESS AND FORM DRAG

The seabed is seldom smooth. Certainly, on rocky shores objects such as the rocks themselves, plants, and animals create a very complex, rough surface. This increased roughness increases the total drag on the water column and therefore influences the boundary layer. First, increased roughness increases the total surface area in contact with water, and thus increases viscous drag. Second, larger objects sticking up into the boundary layer (generally objects larger than about half a centimeter or so) create turbulent wakes in their lee, and the resulting eddies and vortices are mixed into the flow. The effect of this generated turbulence is an additional type of drag on the flow: *form drag*. The larger the area an object presents to flow and the higher the flow speed, the larger the form drag. In fact, form drag scales with the square of the flow velocity. A small increase in flow speed can therefore result in a large increase in form drag. Because in a boundary layer, flow speed increases with distance from the bottom and because drag increases with flow speed, reducing ones size is a means to reduce the drag one incurs. For example, many aquatic plants are very flexible. In still water they stand up straight, but in flowing water they bend with the flow (e.g., see Fig. 2). The resulting decrease in area presented to flow means that for these algae an increase in flow velocity may not involve a too drastic increase in drag.

The wake behind a single object attached to a rock may interact with objects standing downstream. If more than roughly one-twelfth of a surface is covered with objects (say, a rock densely covered in snails), eddies from the wakes cannot readily penetrate the space between the snails (Fig. 3). In between the snails the water is nearly standing still, while the boundary

FIGURE 2 *Caulerpa taxifolia* plants in an experimental flume tank. A picture in still water (A) and in flowing water (B). Photographs by the author.

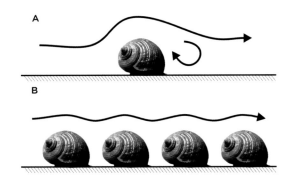

FIGURE 3 Schematic drawing of flow over a single snail (A) and over a group of snails (B) in close proximity, causing skimming flow.

layer forms above the snails. This phenomenon is called *skimming flow*. Sheltering in the interstices of tightly packed objects is thus another means of avoiding drag. In a turbulent, wave-exposed environment, a sub-merged tidepool can also provide protection against the full force of flow.

In water flowing slowly over a flat, smooth surface the flow resembles the profiles from Fig. 1: neatly organized layers of increasing velocity, without any crossing stream-lines and with little exchange between the layers. This type of flow is called *laminar flow*. Fast flowing water over rough objects, as well as the dynamic flow of break-ing waves, disturbs this neat pattern, causing the type of *turbulence* mentioned previously. Boundary layers still exist in turbulent flow, but they have different charac-teristics. At any instant, flow speed may vary in a ran-dom fashion near the seabed, but the *average flow speed* is lower near the bed than higher up in the water column. Turbulent boundary layers tend to be much thinner than laminar boundary layers. Because in turbulent flow water movement is not limited to horizontal motion, turbu-lence causes increased drag and mixing. As a result, turbu-lence in a boundary layer promotes exchange of water between animals and plants on the seabed and water in the main flow.

MARINE LIFE IN THE BOUNDARY LAYER

Rocky shores in general, and tidepools in particular, often support abundant life with an amazing variety of forms. The rapid flow in the turbulent boundary layers charac-teristic of this habitat are an important supplier of food, oxygen, and other necessities of life. Because of this flow, sessile creatures can simply sit and wait for food to come to them. Many creatures are very well adapted to withstand extremely large shear forces, without getting dislodged from their surface. For larval stages of bottom dwellers settlement in such a high flow environment does pose specific problems. Many of the bottom dwellers are filter feeders, animals that filter the water flowing past them. Flow speed is obviously important for the rate at which this food is delivered, and turbulent motion enhances food supply. Filter feeders such as mussels often occur in densities of thousands of animals per square meter (e.g., Fig. 4). In calm conditions, the upstream individuals can filter so much water that the boundary layer becomes depleted and downstream mussels get less food. As noted previously, turbulence enhances the exchange of water and food particles between the different water layers. The roughness of the mussels, and even the effects of their own

FIGURE 4 Dense aggregations of benthic filter feeders. Here: the blue mussel *Mytilus edulis*. Individual blue mussels can filter more than 5 liters of water per hour and occur in densities of several thousands of individu-als per square meter. One square meter of dense mussel bed can easily filter more than 10 cubic meters in one hour. Photograph by Jens Larsen.

feeding currents, promote turbulence and increase food supply. In general, boundary layers on wave-swept shores are thin enough not to pose a barrier to the delivery of food and the removal of wastes, but in sheltered estuaries this can be a serious limit to growth of organisms.

In principle, every organism living in flowing water interacts with the flow and changes it. Some species have such a profound effect on the structure of the boundary layer that they ultimately affect the whole physical regime near the bed. They can influence deposition and erosion of sediment, and can provide shelter from the flow for other plants or animals. Organisms that cause such a drastic change in the physical conditions of an ecosys-tem are called *ecosystem engineers*. Not only humans and beavers are capable of changing water currents. This type of engineering happens on many different levels and at different temporal and spatial scales.

SEE ALSO THE FOLLOWING ARTICLES

Hydrodynamic Forces / Larval Settlement, Mechanics of / Seawater / Size and Scaling / Turbulence

FURTHER READING

Denny, M. W. 1988. *Biology and the mechanics of the wave-swept environ-ment*. Princeton, NJ: Princeton University Press.
Massel, S. R. 1999. *Fluid mechanics for marine ecologists*. Berlin: Springer.
Van Duren, L. A., P. M. J. Herman, A. J. J. Sandee, and C. H. R. Heip. 2006. Effects of mussel filtering activity on boundary layer structure. *Journal of Sea Research*, 55: 3–14.
Vogel, S. 1994. *Life in moving fluids*. Princeton, NJ: Princeton University Press.
Wildish, D. J., and D. Kristmanson. 1997. *Benthic suspension feeders and flow*. Cambridge: Cambridge University Press.

BRACHIOPODS

GLENN JAECKS
Slovenian Academy of Science and Arts, Ljubljana

Brachiopods are a phylum of marine invertebrates that are bivalved, sessile, and use a ciliated, tentacled structure (called a lophophore) for feeding. Brachiopods are not major components of tidepool or rocky intertidal ecosystems, nor is there evidence that they have been a contributor to these environments in the geologic past. Nevertheless, brachiopods can be found in rocky intertidal environments, and they may occasionally be found inhabiting tidepools.

BRACHIOPOD NATURAL HISTORY

Most brachiopods attach to a hard substrate using a fleshy footlike pedicle or cement themselves directly. Brachiopod shells are either hinged (articulated) or unhinged (inarticulated); inarticulated brachiopod species have either a phosphatic or calcitic shell, whereas articulated brachiopods possess exclusively calcitic shells. Brachiopods are found from the intertidal to several thousand meters deep and in all of the world's oceans. It would seem that the aforementioned features would make brachiopods well suited to life in tidepools, but there are several reasons why they are not commonly found there.

First, compared to other marine macroinvertebrates such as molluscs, echinoderms, and cnidarians, little is known of brachiopods' biology, either because they are inaccessible, difficult to keep alive in the lab, or are unrecognized by the nonspecialist. This may lead them to be underreported in literature reports of tidepool communities. Additionally, many brachiopodologists are paleontologists, as brachiopods are among the most common fossils of the Paleozoic Era, and the majority of brachiopod genera are extinct. Further, rocky intertidal environments are extremely rare in the fossil record, because they tend to be erosional systems rather than depositional.

Second, brachiopods of the rocky intertidal tend to be cryptic, found under boulders and large, immobile cobbles or in deep crevices. This is related to the tendency of brachiopod larvae to settle in low light areas; hence many shallow water brachiopods are found in less accessible areas such as marine caves. This impedes direct observation of their behavior and also leads to brachiopods' being overlooked in the field.

Brachiopods may also be excluded from many rocky intertidal environments because they are likely incapable of surviving the large changes in temperature and salinity inherent to this ecological setting. Metabolism is positively correlated with temperature in marine invertebrates; brachiopods, for which there are data, are incapable of surviving short-term rises in temperature of more than 4–5 °C, although some brachiopods *(Lingula)* are found in waters with greater annual temperature variation (>20 °C). This may also contribute to shallow water brachiopods being more cryptic than their deeper water counterparts.

INTERTIDAL BRACHIOPODS

There are two groups of brachiopods that can be found relatively commonly in the intertidal: discinids, a group of inarticulated brachiopods with a limpet-like appearance because of their conical dorsal valve, and terebratulides, which can be found intertidally in the Pacific Northwest of North America.

Discninidae

Discinids are inarticulated brachiopods with black, phosphatic shells that attach to hard substrates, leaving only their conical dorsal valve visible (Fig. 1). They range in size from a few millimeters to a few centimeters. Similar

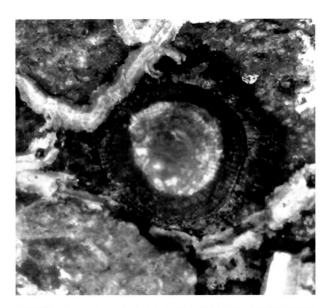

FIGURE 1 *Discaradisca strigata* from the intertidal zone, directly opposite NAOS marine station, Bay of Panama, Panama. This specimen is ~1.5 cm in diameter; note the other encrusters associated with the brachiopod and the halo of lophophore tentacles protruding from the margins. This specimen is housed in the Paleontological Department, the Natural History Museum, London, England. Photograph by the author (specimen does not yet have a number; collected by John Todd in 1998).

to other brachiopods, they usually live in clusters, so their distribution on a large scale tends to be patchy. Although brachiopods have been reported to live for a year without a food source, discinids tend to be found in areas with high nutrient availability, such as the tropical and subtropical Eastern Pacific. They may be found on the bottom of boulders, encrusting the substrate along with serpulids, bivalves, and bryozoans. Discinids may in turn be encrusted as well. As fossils, they can be found either encrusting a hard substrate or as free valves.

Terebratulida

Terebratulides are the dominant modern brachiopod group and are easily the most recognizable and best-known extant brachiopod group. They are found in virtually every marine environment, but are most common in the cooler, nutrient-rich waters of mid to high latitudes. Most reports of terebratulides inhabiting the intertidal refer to the Pacific Northwest, but they should be found elsewhere in similar environments. The most common intertidal terebratulide in the United States and Canada is *Terebratalia transversa*, which is very common from Puget Sound and further north. *Terebratalia* is morphologically quite plastic, in part because of the tight spaces in which it lives. Commonly, they are pink, more broad than long, and have longitudinal ribs. They attach with their pedicle, which allows them to alter their orientation, depending on flow characteristics. They live approximately 10–15 years, and little is known of their reproductive behavior. Typically, terebratulides are gregarious, so their spatial distribution is patchy. They are commonly attached to one another and, in turn, encrusted by bryozoans and annelids. Terebratulide predators include drilling gastropods, decapod crustaceans, and indiscriminate browsers such as regular echinoids and asteroids. It is not unusual to encounter *Terebratalia* whose shells exhibit signs of repaired damage.

SEE ALSO THE FOLLOWING ARTICLES

Bryozoans / Phoronids

FURTHER READING

Emig, C. 1997. Ecology of inarticulated brachiopods, in *Treatise on invertebrate paleontology, part H (revised) Brachiopoda*. A. Williams, C. H. C. Brunton, and S. J. Carlson, eds. Boulder, CO, and Lawrence, KS: Geological Society of America and University of Kansas.

Peck, L. S. 2001. Physiology, in *Brachiopods ancient and modern: a tribute to G. Arthur Cooper*. S. J. Carlson and M. R. Sandy, eds. New Haven, CT: Paleontological Society Papers.

Richardson, J. R. 1997. Ecology of articulated brachiopods in *Treatise on invertebrate paleontology, part H (revised) Brachiopoda*. A. Williams, C. H. C. Brunton, and S. J. Carlson, eds. Boulder, CO and Lawrence, KS: Geological Society of America and University of Kansas.

BRITTLE STARS

Sea stars and sea urchins are distinctive, easily recognized, and frequently encountered members of tidepool and rocky shore communities. Just as distinctive and easily recognized, brittle stars are the largest and most diverse of the major groups of echinoderms, yet they are rarely seen by casual observers. They usually are active at night and hide under rocks during the day. Most people will not find brittle stars unless they turn over rocks; and turning rocks is discouraged because it is destructive to habitat. In addition to under rocks in tidepools, brittle stars are very commonly found on the sea floor, often in very deep water. Brittle stars are very colorful and highly variable; some are luminescent.

IDENTIFICATION, CLASSIFICATION, AND FOSSIL RECORD

It is easy to distinguish brittle stars from other sorts of sea stars. The body is a central disk with five distinct long, thin, and flexible arms. The madreporite, or sieve plate, through which water enters the water vascular system is usually present, but it is on the oral surface instead of the aboral surface as in sea stars and sea urchins. Brittle stars also lack an anus and must consume food and expel wastes through the mouth on the underside of the central disk.

Brittle stars are members of the Ophiuroidea, the sister group of the Asteroidea (the conspicuous and familiar starfish or sea stars). Ophiuroideans comprise about 2000 species and are divided into two groups. The Ophiurina are the brittle stars and the serpent stars, which evolved from within the paraphyletic Eurylina (basket stars) (see Fig. 1).

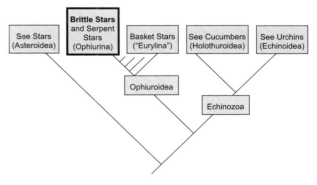

FIGURE 1 The evolutionary relationships of brittle stars. Adapted from Smith *et al.* (2004).

The fossil record for brittle stars is lengthy, extending back 490 million years to the Early Ordovician. During most of this long fossil history, the diversity of brittle stars has been relatively stable and much lower than current biodiversity. The relatively scant fossil record may be due to the delicate nature of brittle stars, exacerbated by low likelihood of fossilization in the habitats they may have occupied.

ECOLOGY, REPRODUCTION AND BEHAVIOR

Brittle stars feed on benthic detritus and small animals, either dead or living. Feeding is accomplished in different ways. Large food is moved toward the mouth by the arms. Smaller food particles can be aggregated by flagella, the action of spines, tube feet, and arms, and then transported toward the mouth for ingestion.

Little is known of the enemies of brittle stars. Most echinoderms avoid predation because they are spiny. In addition, brittle stars are photophobic and secretive in tidepools. Thus, natural observations of brittle stars are rare.

Many brittle stars are viviparous, retaining eggs inside the body and giving birth to tiny complete organisms. Among echinoderms, only the ophiuroids possess bursae—sacs inside the body that open onto the oral surface. Gonads line the walls of these five pairs of bursa (one pair for each leg), and gametes are released through genital slits into the water. Some brittle stars brood their young within these bursae.

Asteroid sea stars, sea urchins, and sea cucumbers are all slow-moving and sluggish. Brittle stars and their close relatives can be fast-moving in comparison. The arms of brittle and serpent stars are capable of only lateral, snakelike movements; basket stars can also move their arms vertically. Using these rapid arm movements, brittle stars are able to avoid predation by escaping. The arms of brittle stars lack an ambulacral groove and include numerous articulated ossicles called vertebrae. Covered by plates, muscles connect the vertebrae. Arms of brittle stars are easily lost or shed when attacked but are regenerated more rapidly than in other sea stars. The water vascular system functions in locomotion and respiration. This system of tubes and vessels is lined with ciliated cells and filled with water. Tube feet lack suckers and do not help in locomotion.

SEE ALSO THE FOLLOWING ARTICLES

Echinoderms / Sea Cucumbers / Sea Stars / Sea Urchins

FURTHER READING

Hickman, C. P., Jr., L. S. Roberts, and F. M. Hickman. 1988. *Integrated principles of zoology.* St. Louis, MO: Times Mirror/Mosby College Publishing.
Rosenfeld, A. W. 2002. *The intertidal wilderness,* rev. ed. Berkeley and Los Angeles: University of California Press.
Smith, A. B., K. J. Peterson, G. Wray, and D. T. J. Littlewood. 2004. From bilateral symmetry to pentaradiality: the phylogeny of hemichjordates and echinoderms, in *Assembling the tree of life.* J. Cracraft and M. J. Donoghue, eds. New York and Oxford: Oxford University Press, 365–383.

BRYOZOANS

MICHAEL TEMKIN

St. Lawrence University

Bryozoans are a group of sessile, colonial invertebrates that collect food particles with a retractable crown of ciliated tentacles called a lophophore (Figs. 1, 2). Bryozoans live in both marine and freshwater environments. In benthic communities, bryozoans may have important roles as competitors for space, as items of prey, and as habitat for a variety of protists, plants and animals.

ECOLOGY

Bryozoans grow on a wide variety of natural substrates, including rock, shell, and wood. Some species grow on brown or red macroalgae or some seagrasses. Other species grow on other invertebrates such as ascidians, gastropods, bivalves, barnacles, and other bryozoan colonies. In addition, many bryozoans will readily settle and grow on synthetic substrates such as glass, solid plastics, and plastic foam. Typically a biofilm, consisting of a matrix of

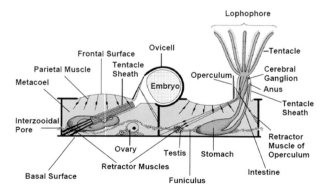

FIGURE 1 A diagram of two gymnolaemate zooids. The proximal zooid (on the left) has the lophophore retracted and is brooding an embryo in an ovicell. The distal zooid (on the right) has extended its lophophore through contraction of the parietal muscles that deforms the frontal surface to increase the intracoelomic pressure. Based on a diagram provided by C. Nielsen, University of Copenhagen.

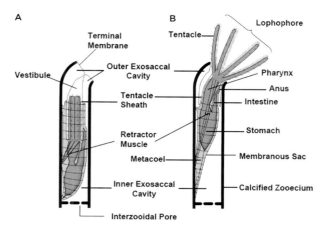

FIGURE 2 A diagram of two stenolaemate zooids. (A) Zooid with retracted lophophore, (B) zooid with lophophore extended. The areas shaded in light gray indicate the metacoel. Based on a diagram provided by C. Nielsen, University of Copenhagen.

microorganisms and their secretions, is usually required for bryozoan larval settlement. The larvae of some species, such as *Bugula neritina,* have been shown to preferentially settle adjacent to one another.

As members of benthic communities, bryozoans have been demonstrated to play important roles as competitors for space and as food for other organisms. Bryozoans compete for space with many other benthic organisms such as hydroids, ascidians, algae, and even other bryozoans. Some bryozoans, such as *Membranipora membranacea,* may almost completely cover fronds of kelp and will grow over other organisms that have settled on the frond. As prey items, bryozoans have been observed to be eaten by predators such as nudibranchs, caprellids, sea urchins, and fish. One nudibranch predator, *Doridella* (Fig. 3), has

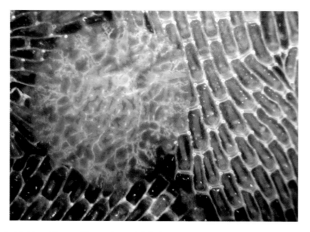

FIGURE 3 The nudibranch *Doridella* feeding on zooids of a *Membranipora membranacea* colony. Photograph by the author.

been shown to induce the gymnolaemate *Membranipora membranacea* to grow spines as a potential defense.

SYSTEMATICS

Traditionally, bryozoans have been grouped with phoronids and brachiopods into the Lophophorata based on morphological characters. Additional phylogenetic analyses using DNA sequences have suggested that the lophophorate phyla do not represent a monophyletic group and may be part of a larger group of animals that also includes the annelids and molluscs, called the Lophotrochozoa. Within the phylum Bryozoa, taxa have been organized into three classes: the Phylactolaemata, Gymnolaemata, and Stenolaemata. All phylactolaemates live in freshwater. The majority of the living marine taxa and a few freshwater species have been placed within the Ctenostomata and Cheilostomata, the two orders of the Gymnolaemata. Lastly, the Stenolaemata has only one remaining extant order, the Tubuliporata, that contains exclusively marine species. The separation of marine bryozoans into the Gymnolaemata and Stenolaemata is based on differences in zooid morphology and reproductive biology, including embryology and larval structure. Phylogenetic relationships both within and between classes of bryozoans remain largely unresolved.

COLONY STRUCTURE

Bryozoans may form colonies that are flat sheets, thick mats, creeping or erect stolons (stemlike filaments), or arborescent (treelike). Differences among colony growth forms have been proposed to be associated with adaptations for increasing feeding efficiency, structural rigidity, growth and regeneration, integration among members of a colony, and defenses against predation. Bryozoan colonies consist of numerous asexually budded individuals, or zooids (see Figs. 1, 2). Zooids are small (usually less than 1 mm long) and are typically cylindrical or boxlike in shape. Zooids in a phylactolaemate colony share a common body cavity, but gymnolaemate and stenolaemate zooids are largely separated from one another by partitions or septa. Gymnolaemate zooids are interconnected through the funicular system, a vascular-tissue homologue that serves as a circulatory system (see Fig. 1). In stenolaemates, small pores in the septa between zooids may allow for adjacent zooids to communicate with one another, including the sharing of nutrients. Polymorphism is common within bryozoan colonies, and zooids may be specialized for feeding, sexual reproduction, attachment of the colony to a substrate, or colony defense. The fundamental unit of a colony is the feeding individual, or autozooid. All other zooid types are called heterozooids. Heterozooids are typically derived by modifying autozooid body plans, usually resulting in anatomically incomplete zooids that are incapable of feeding. Consequently, nutrition for heterozooids must be provided by autozooids through connections between heterozooids and autozooids.

ZOOID STRUCTURE

The anatomy of an autozooid is organized into two functional subunits: the polypide and cystid. The polypide includes the lophophore, the U-shaped digestive tract and its musculature, the cerebral ganglion, and the tentacle sheath (see Figs. 1, 2). The tentacle sheath represents the protrusible portion of the body wall, which encloses the tentacles when the lophophore is retracted inside of the zooid. The cystid is the nonprotrusible portion of the body wall and consists of cellular and noncellular components. The cellular portion of the cystid is composed of an outer epidermal layer and a thin underlying mesodermal layer. The noncellular components of the cystid are secreted by the epidermal layer and form the zooecium or exoskeleton of the body wall. The zooecium consists of a thin mucopolysaccharide cuticle and a thicker matrix that may be gelatinous, chitinous, calcareous, or a combination of these substances. Increases in the calcification of zooecium and increases in the anatomical complexity of the frontal surface of zooids of different species have been proposed to be adaptations to predation.

During asexual budding, each new individual is formed first by the growth of the cystid of an existing zooid. The two typical patterns for budding process are intrazooidal, in which one zooid must complete its growth and development before budding new individuals, and zooidal, in which buds continually grow outward from the colony and new individuals become separated from one another by the construction of interior transverse walls. The process of zooidal budding unties the connection between colony growth and the development of individual zooids. Zooidal budding has been proposed as an adaptation for growth and regeneration. In both intrazooidal and zooidal budding, a localized thickening of the cystid of the developing zooid invaginates as a two-layered vesicle (epidermis and coelomic lining) and differentiates as the polypide of the new individual. In many species, the polypide periodically degenerates to form a "brown body" and a new polypide is regenerated from the cystid.

THE LOPHOPHORE AND FEEDING

The lophophore is the crown of tentacles that surround the mouth. The lophophores of phylactolaemates are horseshoe-shaped, whereas the lophophores of gymnolaemates and stenolaemates are circular. Tentacle number (typically between 8 and 35) and length vary among species and are frequently used as taxonomic characters. In addition, tentacle number has been correlated with lophophore size among species. Species with lophophores that have longer and more numerous tentacles capture more particles per unit time than species with lophophores that have shorter and less numerous tentacles.

The structure of tentacles is similar among bryozoans. Each tentacle contains a portion of the body cavity within the lophophore and has one frontal and two lateral tracts of cilia on its surface. The beat of the lateral cilia generate currents that bring water into the top of the lophophore and out between the bases of the tentacles. Water currents may transport food particles toward the mouth. Alternatively, food particles also may be transported to the mouth by the lateral and frontal ciliary tracts, a process that involves reversals in the effective stroke of the lateral cilia when they contact food particles.

Bryozoan zooids extend their lophophores by increasing the hydrostatic pressure within the main body cavity, or coelom, through muscular contraction. However, because of differences in zooid anatomy and degree of calcification, the mechanism by which hydrostatic pressure is increased within the body cavity varies among the three classes. The increase in hydrostatic pressure is due to the contraction of circular muscles in phylactolaemates and of parietal muscles in gymnolaemates, producing a deformation or change in shape of the body wall (see Fig. 1). In stenolaemates, the heavy calcification of the zooid prevents the deformation of the body wall by muscle contraction. Instead, stenolaemates are able to increase the hydrostatic pressure within their body cavities by redistributing fluid from a distal compartment (outer exosaccal cavity) to a proximal compartment (inner exosaccal cavity). These two compartments represent spaces that are pseudocoels, since they are located between the epidermal and mesodermal (membranous sac) layers of the cystid (see Fig. 2). In all bryozoan groups, the lophophore is retracted through the contraction of retractor muscles coordinated with the relaxation of the muscles that increase the hydrostatic pressure within the main body cavity.

REPRODUCTION AND DEVELOPMENT

Although variation in zooid sexuality exists among species, bryozoan colonies are typically hermaphroditic, containing individuals that may function as males, females, or both. Within a colony, zooids may remain sterile or function only as males, only as females, or as both males and females. Depending on the species, hermaphroditic zooids may produce sperm and eggs during the same reproductive period (simultaneous hermaphrodites), produce sperm before producing eggs (protandrous hermaphrodites), or produce eggs before producing sperm (protogynous hermaphrodites). Phylactolaemate zooids are exclusively hermaphroditic. Gymnolaemate and stenolaemate colonies are usually mosaics of sterile, sexually immature, and either hermaphroditic or male or female individuals. Sexual dimorphism is common, and both the cystid and polypide may become modified to support sexual reproduction. In gymnolaemates, most autozooids become sexually mature and are commonly hermaphroditic. In contrast, stenolaemate colonies are typically composed of mostly sterile autozooids, some male autozooids, and only a few female heterozooids called gonozooids. Fully developed stenolaemate gonozooids are much larger than either male or sterile autozooids. In some species, the gonozooid may include space that is actually external to the zooids. The actual brooding volume in all stenolaemate gonozooids is many times larger than any autozooid of the colony. Developing gonozooids may initially have a polypide, but they lose it as they grow and become sexually reproductive.

Internal fertilization has been reported for all three classes of bryozoans, although the process has been best described for gymnolaemates. In gymnolaemates, sperm are released through the tips of the tentacles and are transferred through the water column between functional male and female zooids of the same or different colonies. After entering maternal zooids, sperm fuse with eggs before they complete meiosis, usually while eggs are still primary oocytes. Typically, sperm–egg fusion occurs either within the ovary, while eggs are at early stages of development, or within the body cavity shortly after eggs have been ovulated. In stenolaemates, male zooids also release sperm into the water column through the tentacles. However, the exact timing of sperm–egg fusion is still uncertain in stenolaemates, but it appears likely to be while eggs are still inside of the ovary (intraovarian).

The development of gymnolaemate embryos usually begins after fertilized eggs have been spawned to a location outside of the maternal body cavity. The nature of the external location varies depending on whether the species freely spawns eggs into the water column or retains them. When gymnolaemates retain their eggs, embryos are most often brooded in either specialized chambers called ovicells or ooecia (see Fig. 1), invaginations of the body wall called embryo sacs, or spaces that form between the external body wall of the retracted polypides, called the introvert. The only known exception to this pattern of external development is *Epistomia bursia*, in which fertilized eggs

are brooded within the ovary rather than being spawned to an external location.

Three modes of development have been described for gymnolaemate bryozoans. In a few gymnolaemate species, maternal individuals produce many small, yolk-poor eggs within a reproductive season, which are spawned directly into the water column, where they grow and develop into feeding, or planktotrophic, larvae. In contrast, most gymnolaemates produce one large, yolk-rich egg at a time that they brood, releasing a nonfeeding, or lecithotrophic, larva that is about the same size as the original fertilized egg. These species often brood several embryos sequentially during a reproductive season. There are a few species in which individual zooids simultaneously brood several embryos at one time (e.g., *Alcyonidium gelatinosum*). The last developmental mode occurs in several genera, such as *Bugula* and *Celloporella*, in which the larvae are maternotrophic. In these species, maternal individuals produce one small egg at a time and supply their brooded embryo with nutrients through a simple placental system. Consequently, maternotrophic larvae are much larger in size than the fertilized egg cell from which they develop.

Stenolaemates have a very specialized reproductive biology called polyembryony. Polyembryony is a reproductive process in which a single zygote is cloned through a budding process to produce 100 or more genetically identical individuals. Early development of a single fertilized egg within a stenolaemate gonozooid establishes a large, multicellular primary embryo. The primary embryo increases in size and buds secondary embryos from lobate extensions. These secondary embryos may develop into larvae or may themselves bud to produce tertiary embryos that develop into larvae. As gonozooids lack a functional polypide, the nutrition to support polyembryony is thought to be supplied by the autozooids surrounding the gonozooid.

Larval forms differ among the three bryozoan classes. Gymnolaemates have the greatest larval diversity, with cyphonautes (shelled, feeding), pseudocyphonautes (shelled, nonfeeding), and coronate (nonshelled, nonfeeding) forms. The cyphonautes larva has been considered to represent the ancestral larval form for the group, since it contains a functional digestive tract. The cyphonautes larva of *Membranipora membranacea* uses a unique feeding mechanism in which cilia, arranged similarly to those on an adult tentacle, sieve particles from the water.

The free-swimming larval period varies from about four weeks for cyphonautes larvae to several hours for nonfeeding gymnolaemate larvae to under an hour for stenolaemate larvae. Settlement of bryozoan larvae is often associated with changes in the response of larvae to light and gravity.

Initially after their release, larvae typically move toward light (positive phototaxis) and away from the pull of gravity (negative geotaxis). At the time of settlement, larvae either move away from light (negative phototaxis) or no longer use light as a cue to orient their movement (neutral phototaxis) and swim toward the pull of gravity (positive geotaxis). Permanent attachment of bryozoan larvae to substrates occurs through the outward unfolding or eversion of an epidermal layer, called the internal sac, that was tucked up inside the larva, followed by the secretion of an adhesive.

In bryozoans, metamorphosis of the larva results in the formation of an ancestrula, the first member of a new colony. During their evolution, all bryozoans have apparently shifted the development of some adult characteristics into earlier life history stages, a phenomenon called heterochrony. One adult characteristic that has been shifted into larval stages is the ability of the cystid to form all other parts of the zooid. Consequently, all of the structures of the ancestrula, including the digestive tract, are formed from only epidermis, which is derived from ectoderm, and mesoderm. In phylactolaemates, many portions of the ancestrula, including the polypide, are preformed before the larva is released by the maternal zooid. In gymnolaemates and stenolaemates, the formation of an ancestrula requires extensive rearrangements of larval tissues to form an initial cystid from cells contained within invaginations (inward folds) of larval epidermis, and undifferentiated mesodermal cells within the larva.

SEE ALSO THE FOLLOWING ARTICLES

Biofilms / Brachiopods / Entoprocts / Nudibranchs and Related Species / Phoronids

FURTHER READING

McKinney, F. K., and J. B. C. Jackson. 1989. *Bryozoan evolution*. Boston: Unwin Hyman.
Ryland, J. S. 1970. *Bryozoans*. London: Hutchinson & Co.
Reed, C. G. 1991. Bryozoa, in *Reproduction of Marine Invertebrates*. Pacific Grove, CA: Blackwell Science/Boxwood Press, 85–245.

BUOYANCY

PAUL WEBB

University of Michigan

An organism's position in the water column, whether holding station on the bottom or swimming and floating at the water surface, depends first on the balance of vertical forces of weight acting downward counteracted by hydrostatic upthrust acting upwards. These forces alone are insufficient to fully regulate position, except at the bottom of the water column or floating at the surface. Animals control position within the water column using hydrodynamic forces generated by body and limb motions.

CONTROLLING LOCATION IN THE WATER COLUMN

Two hydrostatic forces interact with body density to determine the net weight of an organism in water. Gravity acts on body mass, resulting in the downward force of weight (Fig. 1A). This is counteracted by upthrust, equal to the weight of water displaced by the body volume (Archimedes's principle). Density (mass/volume) of tissues depends on amounts of materials such as fat, bone, and, in plants, cellulose. For example, densities of fish carcasses range from 1060 to 1090 kg m^{-3}. The density of seawater is typically about 1024 kg m^{-3}. As a result, organisms usually are negatively buoyant, with a positive net weight in water, and tend to sink to the bottom unless an additional force, lift, is created to offset that weight (Fig. 1B).

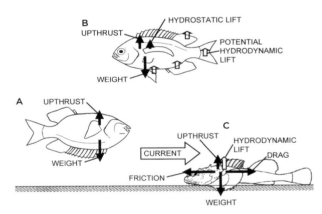

FIGURE 1 Controlling position in the water column, from the bottom to the surface, depends on the balance of hydrostatic and hydrodynamic forces that act on the body. (A) A carcass experiences only weight and upthrust, and because weight > upthrust, the body tends to sink. (B) The tendency to sink can be offset by the hydrostatic lift of low-density inclusions such as gas in the curved swimbladder in the abdominal cavity. No inclusions can make control of position or posture self-correcting, so hydrodynamic lift is always necessary for stability. (C) When there is a current, animals typically interact with the bottom to control position. They modify hydrostatic forces, hydrodynamic lift, and friction to avoid being displaced.

Hydrostatic lift is generated by incorporating low-density materials into the body, increasing upthrust without a commensurate increase in mass. Gas has a density around 1.2 kg m^{-3} and is commonly held in various structures such as lungs in marine iguanas, birds, and mammals (Fig. 2A), gas bladders such as those of inter-tidal algae (Fig. 2B), and swimbladders of bony fishes such as damselfishes (Pomacentridae), sea basses (Serranidae), and surgeonfishes (Acanthuridae).

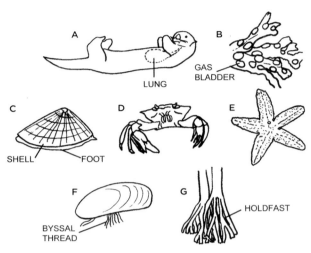

FIGURE 2 The various forces and adaptations to control position in the water column illustrated by fishes (Fig. 1) are found among other organisms. Gas inclusions are found in lungs of tetrapods such as reptiles, birds, and sea otters (A) and the bladders of marine algae (B). Body weight in water is increased in benthic organisms by the shells of molluscs such as limpets (C); the exoskeleton of crustaceans, such as crabs (D); and bony ossicles in starfishes (E). Friction is increased by passive rootlike threads, such as the byssal threads of bivalves (F) and holdfasts of algae (G), and actively by tube-feet of starfishes (E), the foot of molluscs (C), and legs with gripping clawlike structures (D). Drag and lift are minimized by the shapes of the shell of limpets (C), the body of crabs (D), and shapes of mussels (F). Hydrodynamic forces are produced by free-swimming animals using appendages, such as the legs of otters (A) and crabs (D).

However, gases follow the gas laws, according to which the volume varies directly with temperature and inversely with pressure. As a result, volume decreases by half for every doubling of pressure. Small differences in depth near the surface result in large pressure changes and hence volume changes, promoting instability. Some fishes, notably young stages of elasmobranchs in tidepools, avoid this problem by using a different low-density inclusion, lipid, accumulated in organs such as the liver, or in muscle. Lipids have densities around 850 to 930 kg m^{-3}, so that more of this material is needed to achieve a given density compared to a gas inclusion. Marine algae avoid potential effects of density changes near the surface by hyperinflating their gas bladders, ensuring that their fronds float.

Hydrostatic forces used for buoyancy are not self-correcting. That is, an organism with a gas inclusion does not return automatically to a position or posture when disturbed by inevitable external forces. As a result, all

animals supplement or even replace hydrostatic lift with hydrodynamic lift (see Fig. 1B). Hydrodynamic lift is created during swimming when water flows over the body and appendages, such as fins and legs.

CONTROLLING LOCATION IN CURRENTS

Special challenges are faced by tidepool animals when they are exposed to high currents. Organisms then control their positions to avoid being swept away, by interacting with the bottom in various ways. The principles can be illustrated for a sculpin holding position on the bottom in a flow. It still experiences the same hydrostatic forces as a fish in the water column (see Fig. 1C). In addition, current creates a drag force on the body, tending to push it downstream. Drag is resisted by friction with the bottom. Friction depends on the weight of the fish in water and a friction coefficient related to the roughness of both the bottom and the fish. However, flow over the body also creates hydrodynamic lift, which reduces the weight in water, hence also decreasing friction. The ability of a fish to remain in position therefore depends on maximizing both weight in water and friction and minimizing both hydrodynamic lift and drag.

Drag is reduced by a low profile, typical of limpet and mussel shells (Figs. 2C, F), flatfish such as dabs, and the bodies of many crabs (Fig. 2D). The lift force is minimized by a small body and fin area, as in cottids and blennies. The body must have sufficient volume for internal organs. As a result, a low profile with low drag is unavoidable without spreading out the body, increasing its area and hence lift. Conversely, compacting the volume and reducing lift increases drag.

The resulting inability to simultaneously minimize drag and lift are overcome by numerous adaptations. Many of these increase friction. Animals exposed to high currents often increase the proportion of heavier tissues in the body, especially the skeleton, to increase density and hence the weight in water and friction. This occurs in the skeleton of benthic bony fishes such as cottids, the shells of molluscs, the exoskeleton of lobsters and crabs, and bony ossicles in the arms of starfishes (Fig. 2E). The friction coefficient is very large for systems attached to the bottom, for example in the byssal threads of bivalves (Fig. 2F) and the holdfasts of algae (Fig. 2G). Other ways of increasing friction coefficients are through body roughness, especially scales of fishes, and behaviors such as grasping the bottom by the foot of molluscs, appendages such as the legs of crabs, the tube-feet on the ventral surface of starfish, and the fins of fishes. Hydrodynamic lift can be reduced by behaviors that create flow beneath the body, decreasing the current speed difference across the body. Some fish angle their paired fins upward, and crabs angle the body upward, to direct lift toward the ground. This adds to weight in water, increasing friction.

When currents overwhelm these mechanisms, animals living in the water column or on the bottom resort to avoiding flow. The forces sculpting habitats such as tidepools erode substrate and create numerous shelters for flow avoidance by animals.

SEE ALSO THE FOLLOWING ARTICLES

Body Shape / Fish / Hydrodynamic Forces / Locomotion: Intertidal Challenges / Size and Scaling

FURTHER READING

Alexander, R. McN. 1983. *Animal mechanics*. Oxford: Blackwell Scientific.

Arnold, G. P., and Weihs, D. 1978. The hydrodynamics of rheotaxis in the plaice *(Pleuronectes platessa)*. *Journal of Experimental Biology* 75: 147–169.

Gee, J. H. 1983. Ecological implications of buoyancy control in fish, in *Fish biomechanics*. P. W. Webb and D. Weihs, eds. New York: Praeger, 140–176.

Vogel, S. 1994. *Life in moving fluids*. Princeton, NJ: Princeton University Press.

Webb, P. W. 2002. Control of posture, depth, and swimming trajectories of fishes. *Integrative and Comparative Biology* 42: 94–101.

Webb, P. W. 2006. Stability and maneuverability, in *Fish physiology*. R. E. Shadwick and G. V. Lauder, eds. San Diego: Elsevier, 281–332.

CALCIFIED ALGAE

SEE ALGAE, CALCIFIED

CAMOUFLAGE

JOHN J. STACHOWICZ

University of California, Davis

Camouflage is a means by which animals avoid detection by other animals by blending in with the environment. As a result, the animal may either not be perceived at all or be perceived to be something it is not, and thus ignored or avoided. Although humans most often think of camouflage as visual, many marine organisms have poorly developed visual systems, so chemically mediated camouflage is common as well. Both chemical and visual camouflage protect animals from predators as well as facilitate predation by "sit-and-wait" predators. Camouflage can be a major reason why some organisms are able to live in areas with very high densities of predators.

GENERAL CONSIDERATIONS

Predators are numerous in the sea, and the intensity of their consumption of prey provides a strong selective force favoring adapations to avoid consumption. These generally fall into two categories: avoidance of detection and avoidance of consumption once detected. Camouflage deals largely with avoidance of detection and can involve general background-matching strategies or very specific matching of particular substrates or organisms. Most examples are inferred examples of camouflage because they appear difficult for humans to distinguish, but in a few cases careful

field experiments have demonstrated the protective value of camouflage. The following examples of camouflage are divided into two categories: those that are fixed throughout the life of an animal (innate) and those that can be changed as an organism's surroundings change (flexible).

Innate Camouflage

Among organisms that are of low mobility or are very faithful to a single site or habitat type, permanent structures or coloration may be adopted in order to match these backgrounds and avoid detection. For example, numerous limpet species excavate a "home scar" to which they return after foraging, spending the bulk of their life in the immediate vicinity of the scar. Such species often have shell colors, shapes, or textures that closely match the background of the home scars: the limpet *Notoacmea incessa* on the kelp *Egregia, Lottia digitalis* on the goose barnacle *Pollicipes,* and *Lottia scabra* on high intertidal rock surfaces are all excellent examples (Fig. 1A, B). Many nudibranchs are specialized feeders on one or a few closely related species of sessile invertebrates. Those that sequester toxic chemicals from their prey often adopt bright coloration to warn predators of their toxicity, while less toxic species may have morphologies that camouflage them on their host. For example, the nudibranch *Dendronotus frondosus* has finely branched cerata that are pigmented to look like the branches of *Obelia,* the hydroid on which it lives; *Rostanga pulchra* is completely orange and flattened and nearly impossible to distinguish from its host species, the encrusting orange sponge *Ophlitaspongia pennata.* Other species, such as the majid crab *Mimulus foliatus,* do not match their background per se but possess disruptive coloration such as large pigment blotches or alternating color patterns that break up the outline of the organism and make it difficult to distinguish.

FIGURE 1 Camouflaged limpets. Two species of *Lottia* that have coloration and shapes that help them blend into the background: (A) *Lottia* on gooseneck barnacles. Note the white, limpet-shaped bumps, which look almost exactly like the plates that constitute the barnacle's shell. (B) *Lottia scabra* has a home scar to which it returns after feeding. Both the shape and the color of the limpet's shell closely match those of the rock. Photographs by the author.

Absence of color altogether can also serve as camouflage for animals that live in the open water. Many larvae of benthic invertebrates, as well as holoplanktonic copepods, are transparent or nearly so, reducing the ability of visually oriented predators such as fishes to locate and capture them. Similarly, planktonic predatory chaetognaths may be another example of the lack of coloration enhancing the ability of a predator to capture food, because some of these "arrow worms" are known to prey on visually-oriented fish larvae.

Flexible Camouflage Strategies

Being well camouflaged in one environment might make an animal more conspicuous when it moves to another environment. Thus, animals that are mobile or inhabit variable environments may need to alter their camouflage depending on their current location.

A variety of crustaceans, molluscs, and fishes can change color in response to new surroundings, but the mode (and speed) with which they accomplish this feat varies dramatically. Some cephalopods, such as cuttlefish and octopuses, exert neuromuscular control over their color by contracting or expanding chromatophores—cells containing pigment—in their mantle tissue. This allows these animals to change color in a matter of seconds. Among fishes, a variety of sculpins, kelpfish, and gunnels are often well camouflaged. Some can rapidly change their coloration, while others do so more gradually as a result of the slow changing of the pigments that occur in their chromatophores. Because these pigments (usually carotenoids) are derived from their prey, which themselves are camouflaged to match the background, these fishes may passively track seasonal changes in environmental coloration. Camouflage may serve two functions in these fishes: simultaneously reducing their detection by visual predators such as birds and enhancing their ability to ambush and capture small fishes and shrimps as prey.

Many crustaceans also change color, but they usually do so only when they molt, or shed their shell, as they grow. Pigments from food are modified and sequestered in the carapace to achieve the desired coloration. Kelp crabs spend their entire juvenile life in intertidal red algae beds, where they have a deep red color, then they turn lighter amber as they migrate subtidally to amber-colored kelp plants as adults. Experiments raising crabs on red versus brown algal diets have shown that this color change is caused by changes in the diet. In *Pugettia producta*, color change occurs only during molting (Fig. 2). Molting occurs only when crabs grow to a large size; it is usually an infrequent process, with weeks to months between molts, so this color change does not occur in response to short-term change in background. This suggests that such camouflage will be most effective when animals remain within the same habitat for an extended period of time. Isopods in the genus *Idotea* can fine-tune their color using chromatophores that can expand or contract to darken or lighten the shade of the animal within a few minutes to an hour, enabling them to move among different-shaded habitats yet maintain the integrity of their camouflage.

Sequestering of diet-derived materials can also lead to chemical camouflage, as chemical compounds from the surfgrass *Phyllospadix* are incorporated into the shells of the limpet *Notoacmea paleacea*, rendering it indistinguishable from surfgrass to the sea star *Leptasterias*.

FIGURE 2 Flexible color camouflage. The crab *Pugettia producta*, shown here dark brown in color when living on the feather boa kelp *Egregia*. Smaller juveniles are found in red turfy algae and adopt a deep red color. Adults often live in giant kelp canopies and have a light amber morphology that matches the color of kelp. Photograph by K. Hultgren.

Other crabs, aptly called decorator crabs, have gained the ability to switch camouflage more rapidly by attaching living seaweed and marine invertebrates to hooked setae on their carapaces (Fig. 3). Some species have very strong decoration preferences, specializing on one or a few species, whereas others are quite generalized. Those that have specific preferences often choose chemically noxious species; this can reduce predation by making the crab appear to look like a distasteful species, causing predators to actively avoid it regardless of the background. The purple sea urchin *(Strongylocentrotus purpuratus)* and other urchins also attach drift seaweed, rocks, and shells on their body.

Other animals become decorated more passively, by the settlement of sessile organisms onto their shells. This

FIGURE 3 Decorator crabs. Shown here is *Pugettia richii*, which has covered itself with bladed red algae. Photograph by K. Hultgren.

occurs among molluscs: bivalves, gastropods, limpets, and chitons. In a few species this has been shown to protect the animals from nonvisual predators such as sea stars. Presumably the sea stars fail to recognize these animals for what they are, mistaking them for a rock overgrown by encrusting organisms in a case of chemical or tactile camouflage. However, in cases where the overgrowing organisms are chemically distasteful, avoidance by predators might be of a more active nature.

SEE ALSO THE FOLLOWING ARTICLES

Algal Color / Bioluminescence / Body Shape / Chemosensation / Predator Avoidance

FURTHER READING

Langstroth, L., and Langstroth, L. 2000. *A living Bay: the underwater world of Monterey Bay.* Berkeley: University of California Press.
Morris, R. H., D. P. Abbott, and E. C. Haderlie. 1980. *Intertidal invertebrates of California.* Stanford, CA: Stanford University Press.
Stachowicz, J. J. 2001. Mutualisms, positive interactions, and the structure of ecological communities. *BioScience* 51: 235–246.
Wicksten, M. K. 1980. Decorator crabs. *Scientific American* 242 (December): 116–122.
Wicksten, M. K. 1983. Camouflage in marine invertebrates. *Oceanography and Marine Biology Annual Review* 21: 177–193.

CEPHALOPODS

SEE OCTOPUSES

CHEMOSENSATION

RICHARD K. ZIMMER
University of California, Los Angeles

Essentially all living organisms possess membrane-associated proteins for detecting the chemical environment. Plants, algae, single-celled microorganisms, and animals share exquisite sensitivities to biologically relevant signal molecules. Within the tree of life, chemical sensing is far more widespread than other sensory modalities, including vision and hearing. Chemosensation thus plays a pivotal role in determining ecological interactions.

GENERAL CONCEPTS

Understanding the mechanisms by which environmental chemical signals mediate various life history processes can lead to important insights about the forces driving the ecology and evolution of marine organisms. For chemical signals released into the environment, establishing the principles that regulate chemical production and physical transport is critical for interpreting biological responses to these stimuli within appropriate natural, historical contexts. Knowledge of these factors makes it possible to analyze the constraints imposed by physicochemical phenomena on biological responses at individual, population, and community levels.

Chemistry indeed mediates a variety of critical ecological interactions within tidepools and rocky shores. Considerable information is now available on the types of biological responses to environmental chemical stimuli. Sensory perception of chemical stimuli, for example, strongly influences predation, predator avoidance, courtship and mating, aggregation and school formation, kin recognition, and habitat selection. Additionally, prey organisms (animals, plants, and microbes) often produce chemical defenses that render their tissues unpalatable to primary or secondary consumers. Outstanding examples are emerging in which either chemical signals or defensive compounds are known to regulate the behavioral or physiological responses of individuals at lower trophic levels. These regulatory effects are then transferred to consumers at higher trophic levels with profound impacts on the distributions and abundances of organisms.

The physical and chemical properties of habitats can determine the nature and success of ecological interactions. In terrestrial environments, for example, compounds with high vapor pressures (low molecular weights, hydrophobic) facilitate chemical transport in air. Because the requirement for gaseous volatility imposes strong constraints on molecular designs, the isolation and identification of signal molecules is often straightforward. By comparison, much less is understood about chemically mediated interactions in aquatic habitats. Aqueous solubility (imparted mainly by electronic charge or hydrophilicity), rather than gaseous volatility, may constrain the types of substances principally acting as waterborne chemical agents. Even insoluble compounds can, however, provide effective chemical signals when suspended and transported by fluid flow. Proteins, peptides, amino acids, organic nitrogen bases, sugars, complex carbohydrates, fatty and humic acids, and other types of chemicals are all putative agents mediating ecological processes in the ocean.

The sensory capabilities of an organism are important because they provide vital links between stimulus production/transmission and behavior, and between individual behavior and larger scale dynamics, in the abundance and spatial/temporal distributions of organisms. The nervous system represents a critical filter for translating environmental features into sensory stimuli,

and then into a behavioral task. Investigations on sensory physiology, particularly in conjunction with behavioral and population studies, establish linkages between stimulus space, behavior, and demographic consequences of decisions made by individual organisms.

Recent research into sensory biology has delved into how the properties of complex chemical stimuli are coded by the nervous system to modulate the motility patterns of organisms. Concomitant with these developments were increasing efforts to understand the connections between natural chemical and physical environments and the properties of animal perceptual systems. Taken together, these two lines of inquiry provide information on how biologically relevant natural stimuli are translated into specific strategies for navigation, orientation, and guidance during search for resources. Recent studies have documented the properties of chemosensory receptor cells (and occasionally other neural elements) in effective coding of chemical signal parameters that may specify the identity, direction, and distance to stimulus sources. Such efforts mostly have been applied to the physiological and behavioral strategies of large animals (e.g., crabs, snails, lobsters, and fish), living in physically benign habitats such as estuaries. In contrast, little is known about the navigational behavior of organisms in exposed, rocky-shore environments. Even though sensory and central neural mechanisms may not change much, chemical signals are more variable and quickly diluted below detection limits in wave-swept habitats.

EXAMPLES OF CHEMICALLY MEDIATED ECOLOGICAL INTERACTIONS

Provided here are examples of how chemical signals dictate ecological interactions for a wide variety of marine organisms. The examples are selected to represent an array of marine taxa and to summarize processes acting on many different temporal and spatial scales. The sea contains an impressive suite of chemical information—dissolved, in suspension, attached to substrates, and associated with organisms—that is critically involved in various aspects of faunal, floral, and microbial biology.

Animal Dispersal and Habitat Colonization

Habitat selection by larvae is one of the most important factors structuring marine communities. Critical interactions occur between substrate-adsorbed or waterborne chemical cues and hydrodynamics in establishing patterns of colonization. Recent studies have provided unequivocal experimental evidence that both particulate and dissolved chemical cues mediate settlement by invertebrate and fish

Peptide well   Control well

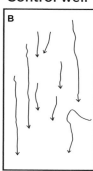

FIGURE 1 Oyster *(Crassostrea virginica)* larvae respond to a water-borne peptide cue associated with bottom-dwelling adults. In the laboratory, larvae were added to a large raceway flume, and transported by the flow (speed = 2 cm/s; shear velocity = 0.2 cm/s) over a well (i.e., depression) that contained clean oyster shell (as substrate) and either the (A) adult peptide signal, or (B) seawater (control). Direction of flow is downward in both figures. Each path of a single larva was generated using a computer-assisted video motion analysis system; arrowheads denote the direction of larval movement. Relative to seawater, the dissolved cue induces a change in the swimming behavior of suspended larvae, bringing them closer to the bed. Because of increased proximity to the bottom, the larvae are swept by turbulent eddies into contact with the bed much more frequently. (C) Ultimately, larval response to soluble adult cue enhances settlement and helps create large reefs. Scale bar = 2 m.

larvae in natural habitats. The soluble cues are, however, most effective in relatively benign estuarine flows, and under still-water conditions at peak flood, or ebb, tides within rocky-shore tidepools. Substrate-adsorbed cues, in contrast, are more resistant to hydrodynamic forces and, therefore, more important on wave-swept shores.

Small peptides (chains of amino acids) and large proteins have been isolated and identified as settlement and metamorphic inducers for oysters and barnacles (Fig. 1). Remarkably, the waterborne peptide cues are all structurally related to the carboxy-terminus sequence of mammalian C5a anaphylatoxin, a potent white blood cell chemoattractant. These results support an evolutionary link between the more primitive external receptors functioning in chemical communication of marine organisms and internal receptors for mammalian neuro- and immunoreactive agents. In barnacles, the substrate-adsorbed protein cues, called *arthropodins,* are surprisingly similar in structures to compounds described from insect and crustacean exoskeletons.

Predator–Prey Interactions

Previous studies on taxonomically diverse animals have established the general existence, and importance, of chemoreception in a wide variety of behavioral processes. Evidence suggests that the "sense of smell" mediates search in many organisms. Unexplored, however, are linkages between successful olfactory-mediated search and guidance mechanisms to the hydrodynamic environment (shear stress and turbulence). To establish interactions between hydrodynamics and chemoreception, recent laboratory and field studies were conducted on predatory success and search strategies of marine invertebrates (crabs and snails). Search ability was extremely sensitive to small changes in bottom boundary-layer structure. In estuaries, hydrodynamic regimes are transitional between smooth- and rough-turbulent conditions (low and high turbulence, respectively). Chemosensory systems of crabs, but not snails, extract information primarily from the more predictable hydrodynamically smooth flows (Fig. 2). These studies indicate that mechanisms governing the physical transport of odor signals can have profound influences on the development of sensory and behavioral mechanisms. Moreover, they also control biotic interactions such as predation, which, in turn, can mediate community structure.

FIGURE 2 Odor plume produced by the release of metabolites from the excurrent siphon of a hard clam, *Mercenaria mercenaria.* Flow was visualized by releasing fluorescein dye through the excurrent siphon and photographed using slit illumination. The metabolites act as chemical signals triggering search responses by predators, but the likelihood of predatory success depends on the dynamics of water flow. (Photograph is reprinted courtesy of the *Biological Bulletin,* Zimmer and Butman 2002.)

Courtship and Mating

Courtship and mating pheromones can be difficult to identify because breeding seasons are short and chemical and biological materials are hard to obtain. Specific behavioral acts of courtship are often troublesome to discriminate from other activities, thus making bioassay-guided fractionation impossible in some cases. Still, outstanding progress has been made toward elucidating the structures of mating pheromones in algae, worms, snails, and fishes. Whereas terpenes and other hydrocarbons appear to be the principal pheromones in worms and algae, steroid hormones and their metabolites produced by ovulating female fishes are potent attractants to mature males in some species. In comparison, small peptides function as mating pheromones within at least one family of opisthobranch mollusc: the large sea hares, *Aplysia* spp. (Fig. 3).

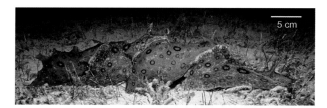

FIGURE 3 Sea hares (*Aplysia* spp.) release chemical signals that attract mates. Individuals are often attracted to a single site, thus forming large aggregations. Here, four individual sea hares *(Aplysia dactylomela)* are in a mating chain as a consequence of chemical communication. Photograph courtesy of Anne DuPont.

Chemical Communication and the Language of Sperm and Egg

Chemically mediated behavior is a key component of sperm-egg dynamics in habitats ranging from the turbulent ocean to a mammalian reproductive tract. Recent discovery of the sperm attractants, bourgeonal in *Homo sapiens* and L-tryptophan in abalone *(Haliotis rufescens)*, bridged the gap between molecular biology and fertilization ecology. In humans, a bourgeonal-sensitive, olfactory receptor protein (hOR17-4) is expressed exclusively on sperm membrane. Thus, sperm have a "nose." The receptor acts through a G protein–coupled cAMP transduction pathway in regulating calcium uptake and sperm chemotactic responses. In abalone (a large marine mollusc), sperm detect shallow concentration gradients that develop when tryptophan is released naturally by an egg (Fig. 4). The male gametes integrate chemosensory information over 200 ms and negotiate attractant gradients in

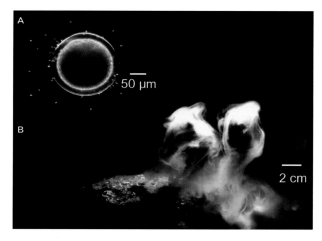

FIGURE 4 (A) Sperm of the red abalone (*Haliotis rufescens*) navigate toward a conspecific egg in response to a natural gradient of chemoattractant. (B) Male abalone releasing sperm. Like many marine animals, male and female abalone synchronously spawn their gametes into the surrounding seawater, where fertilization depends on contact between sperm and egg. The eggs release L-tryptophan, which attracts sperm to the egg surface. Photographs reprinted courtesy of the Society of Biologists, Limited, Riffell *et al*. 2002.

slow laminar-shear flows by ($\leq$ 2/s) by using helical klinotaxis to redirect swimming motions (rotational and translational velocities). Because human and abalone sperm navigate similarly, chemical communication apparently involves species-specific dialects of a common language at the molecular, cellular, and behavioral levels.

Biogeochemical Cycling in Microbes

Behavioral responses by microbes to chemical stimuli are critical for nutrient cycling in marine environments. Recent investigations have established the behavioral responses of protozoa to alternative nitrogen substrates, including amino acids, ammonium, and nitrate. Significantly, cells were attracted to these substrates at concentrations naturally occurring in seawater. The patterns of motility in response to amino acids and nitrate were differentially expressed by genetically identical cells depending on the nutritional environment. The observed chemosensory behavior was consistent with cells maximizing their use of alternative nitrogen substrates under contrasting conditions of nutrient limitation.

The behavior of bacteria also has been studied in response to biogenic sulfur compounds. Dimethylsulfide (DMS) is a predominant volatile substance emitted from the oceans to the atmosphere. It has been implicated as a major factor in regulating climate and in increasing the acidity of precipitation. DMS in seawater arises primarily via enzyme degradation of dimethylsulfoniopropionate (DMSP), a substance produced in high concentrations by phytoplankton. Following induction to synthesize the degradation enzyme, bacteria were attracted to DMSP at levels occurring in seawater near senescing phytoplankton cells. In contrast, genetically identical cells without enzyme induction were not attracted to DMSP. Investigators are now testing whether the degradation enzyme, or a protein coinciding with it, functions as a chemoreceptor that mediates chemical attraction. Alternatively, the enzyme (or coincidental protein) might serve as a transporter for DMSP uptake from the environment across the periplasmic membrane. When combined with enzyme activity, bacterial attraction to DMSP should substantially increase the rate of DMS production and therefore play a critical role in biogeochemical sulfur cycling between dissolved organic matter in seawater and the global atmosphere.

SEE ALSO THE FOLLOWING ARTICLES

Fertilization, Mechanics of / Foraging Behavior / Hydrodynamic Forces / Larval Settlement, Mechanics of / Predation / Turbulence

FURTHER READING

Derby, C. D. 2000. Learning from spiny lobsters about chemosensory coding of mixtures. *Physiology and Behavior* 69: 203–209.

Pawlik, J. R. 1992. Chemical ecology of the settlement of benthic invertebrates. *Oceanography and Marine Biology Annual Review* 30: 273–335.

Riffell, J. A., P. J. Krug, and R. K. Zimmer. 2002. Fertilization in the sea: the chemical identity of an abalone sperm attractant. *Journal of Experimental Biology* 205: 1459–1470.

Riffell, J. A., P. J. Krug, and R. K. Zimmer. 2004. The ecological and evolutionary consequences of sperm chemoattraction. *Proceedings of the National Academy of Sciences USA* 101: 4501–4506.

Sorensen, P. W., and J. Caprio. 1998. Chemoreception, in *The physiology of fishes*. D. H. Evans, ed. New York: CRC Press, LLC.

Susswein, A. J., and G. T. Nagle. 2004. Peptide and protein pheromones in molluscs. *Peptides* 25: 1523–1530.

Zimmer, R. K., and C. A. Butman. 2000. Chemical signaling processes in marine environments. *Biological Bulletin* 198: 168–187.

CHITONS

DOUGLAS J. EERNISSE

California State University, Fullerton

Chitons are an ancient lineage of molluscs, found only in the sea, that are classified as class Polyplacophora of phylum Mollusca. They can be recognized by their eight overlapping shell plates, known as valves (six intermediate valves and two terminal valves at the head and tail), which are firmly anchored in a tough muscular girdle (Fig. 1). The dorsal surface of the girdle is occasionally

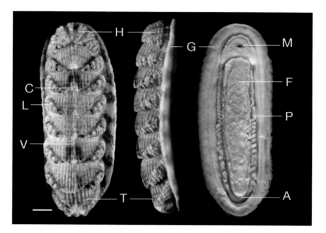

FIGURE 1 A chiton, *Callistochiton crassicostatus* (Monterey, California), in dorsal (left), lateral (center), and ventral (right) orientations. Abbreviations: V = one of eight valves, H = head (or first) valve, T = tail (or eighth) valve, C = central area, L = lateral area, G = girdle, M = mouth and oral region, F = foot, P = pallial groove and gills (ctenidia), A= anus, chiton in dorsal view, with labeled features. Scale bar = 2 mm. Photographs courtesy of A. Draeger.

nearly nude but is otherwise adorned with various scales, stout spines, elongate needles, hairs, branching setae, or dense microscopic granules (Fig. 2). Fossils of chitons, including some that are more than 500 million years old, suggest that throughout their history their lifestyle and appearance have not changed very much. Most of the more than 900 recognized species live on rocky habitats in intertidal to shallow subtidal habitats, where they are often common and ecologically important. Other species manage to inhabit the more sparse hard substrates found at greater depths, and some species have been dredged

FIGURE 2 Selected examples of different girdle ornamentation: (A) hairy and calcareous elements in *Stenoplax conspicua* (southern California); (B) clusters of spines at valve sutures in *Acanthochitona exquisita* (Gulf of California, Mexico); (C) imbricating scales in *Chiton virgulatus* (Gulf of California, Mexico); (D) setae with calcareous spicules in *Mopalia ciliata* (central California). Photographs A–C by the author; photograph D courtesy of A. Draeger.

from the deepest ocean trenches. Chitons usually attach firmly to hard substrates with a muscular foot, and they move by creeping with the aid of mucous secretions and by contractions of their foot.

FEEDING

Like many other molluscs, chitons feed with a thin strap bearing rows of teeth known as the radula. The anterior rows are used up and discarded or swallowed and replaced by new rows moving forward like a conveyor belt. Depending on the particular species, they scrape algal films off rocks, take bites of larger algal blades, eat encrusting colonial animals, or sometimes even ambush and eat mobile animals that come close enough to be trapped. The chiton radula is noteworthy because one pair of cusps in each row is hardened with magnetite, which provides these teeth with a coating harder than stainless steel. They are the only molluscs that have magnetite-coated teeth. In fact, they are the only organisms known to manufacture such vast quantities of magnetite.

The diet of many chitons consists of "diatom scuzz" scraped off rocks, but the largest chitons tend to take bites of large algal blades. Some chitons are specialists on particular marine plants (Fig. 3), scraping off the upper surface of coralline algal crusts (e.g., *Tonicella* spp.) or feeding on kelp (e.g., *Cyanoplax cryptica*, *C. lowei*, *Juvenichiton* spp., *Choriplax grayi*) or seagrasses (e.g., *Stenochiton* spp.). Even though chitons are important for their role as primary consumers of marine plants, many chitons feed predominantly on animals, for example, grazing on encrusting colonial animals in the low intertidal or on sponges or foraminifera in the deep sea or associated

with deep sunken wood or even deep-sea hydrothermal vents. Perhaps most striking is the highly convergent broad body shape and predatory behavior of three only distantly related chiton genera, all in separate families: *Placiphorella*, *Loricella*, and *Craspedochiton* (Fig. 4).

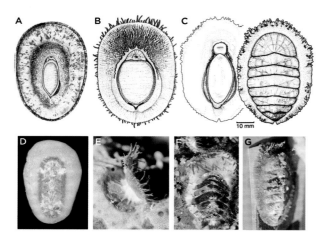

FIGURE 4 Convergent evolution of carnivorous feeding in three only distantly related chiton genera: *Placiphorella* (Mopaliidae), *Craspedochiton* (Acanthochitonidae), and *Loricella* (Schizochitonidae). (A) *Craspedochiton pyramidalis* (Japan), (B) *Placiphorella stimpsoni* (Japan); reprinted from Saito and Okutani 1992 (see Further Reading). (C) *Loricella angasi* (ventral and dorsal views, Western Australia); drawing by I. Grant, from Kaas *et al.* 1998 (see Further Reading). (D) *Craspedochiton productus* (South Africa), (E) *Placiphorella velata* (with oral hood raised, central California), (F) *P. velata* (central California), (G) *Loricella angasi* (Western Australia); photographs by the author. Members of all three genera have been shown to be ambush predators, with an expanded (especially anterior) girdle in comparison to their respective nearest nonpredatory relatives.

RESPIRATION

For respiration, most molluscs have a pair of gills (or ctenidia), sometimes reduced to a single gill, but chitons have entire rows of interlocking gills hanging from the roof of the pallial groove along each side of their foot. Members of the mostly deepwater Lepidopleurida (e.g., *Leptochiton*; Fig. 5, left) have a continuous semicircular arrangement of gills surrounding the anus, and this arrangement is likely primitive; however, more familiar chitons (Chitonida) have gill rows on either side of the foot, separated by an interspace between the ends of the rows (Fig. 5, right). The latter arrangement more effectively divides the outer and inner pallial groove into inhalant and exhalant spaces, respectively. Chitons use small hemoglobin proteins called myoglobins in tissues associated with feeding structures, but for delivering oxygen throughout their body they use a circulatory copper-based respiratory protein called hemocyanin. This

FIGURE 3 Chitons that are specialist grazers, feeding on particular algal species. (A) *Tonicella lineata* (British Columbia, Canada) feeds on crustose coralline algae. Photograph by the author. (B) *Cyanoplax cryptica* (southern California) lives and feeds on the southern sea palm kelp, *Eisenia arborea*. Photograph by R. N. Clark and the author.

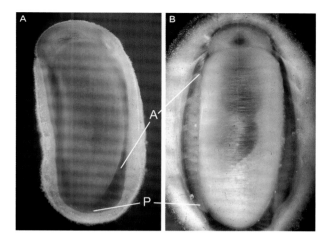

FIGURE 5 Ventral views of chitons from Washington, each about 1 cm in length, with contrasting arrangements of gill rows. Left: *Leptochiton rugatus* (Lepidopleurida: Leptochitonidae) with posterior gill rows that form a continuous semicircle of gills. Right: *Cyanoplax dentiens* (Chitonida: Lepidochitonidae) has lateral gill rows with an interspace at the posterior end. For each, the line points to the anterior (A) or posterior (P) end of one of the paired gill rows. Note: The red coloration of the foot and gills of *L. rugatus* is due to tissue hemoglobins, and this is the only chiton species for which these are known (Eernisse *et al.* 1988). Photographs by the author.

functional protein is unique to molluscs and is not related to a different copper-based respiratory protein found in arthropods, despite having the same name.

Underwater, an impressive respiratory current exits past a chiton's anus, generated by the numerous cilia on each gill. Oxygen is relatively more abundant in air than in water, so a large chiton sprawled with its gills partly exposed at low tide could be effectively involved in aerial respiration, provided that its gills do not dry out.

BEHAVIOR, NERVOUS SYSTEM, AND SENSORY ORGANS

A chiton's mouth is associated with the radula and a tonguelike subradular organ, but chitons really do not have a head. In this sense, they are typical molluscs; unlike the familiar subgroup of molluscs that includes snails and octopuses with a head, typically well equipped with a brain, tentacles, and eyes. The chiton has none of these, but even without a brain a chiton still manages to behave in an adaptive manner. For example, when touched, a chiton rapidly responds by clamping down on its attachment with powerful muscles in its foot and girdle and attached to its valves, and so resists being pried off a rock with amazing tenacity. To maintain such a tight grip indefinitely would be a waste of energy. Instead, a chiton chooses when and where to cling the tightest. Chitons will often move to feed or to seek shelter. For example,

a chiton can move with surprising speed when the rock it is on is overturned as it seeks to crawl back under the rock. When moving, it is at its most vulnerable to losing its grip when it might be surprised by an unusually large wave, a shorebird's beak, or human fingers. Even when dislodged, many species are able to escape from a would-be predator by rolling in a tight ball, like a common garden isopod (roly-poly, pillbug, sowbug). Such behavior allows them to be picked up by a passing wave and rolled out of harm's way, later to uncurl when conditions are safer. Some chitons (e.g., *Callistochiton* spp.) seem to spontaneously detach without provocation when a rock is overturned.

Many species show striking diurnal (daily cycle) patterns of activity, usually remaining hidden under a rock or wedged into a depression by day, and foraging by night when visual predators are not a threat. Tropical intertidal chitons of the genus *Acanthopleura* are well known to display homing behavior, possibly retracing the chemical cues in their own trails of mucus to return to the safe haven of their own home depression, which they also can aggressively defend, excluding other chitons. Particular members of *Nuttallina* (Fig. 6) are especially effective burrowers on soft sandstones and collectively can riddle the midshore with deep burrows. No one has studied how these chitons manage to make a home depression.

FIGURE 6 *Nuttallina fluxa* (southern California) creates a home depression in soft sandstones. Photograph by the author.

The most interesting aspect of a chiton's nervous system has to do with the many nerve bundles that innervate the upper (tegmentum) layer of each valve, leading to the primary complex sensory organs found in a chiton. These numerous shell organs and their supporting nervous tissue make the upper partly mineralized and partly living shell layer of chitons different from the shell of other molluscs, or from other animals with calcium carbonate skeletons in general. All chitons have esthetes, or small shell organs, and these were also present early in their evolutionary history during the Paleozoic Era (543 to 250 million years ago).

In addition to esthetes, some specific lineages of chitons also have considerably larger ocelli organs, and these are clearly photosensory. For example, some chitons normally clamp down to the rock when a potential predator (or scientist) creates a shadow over their body, but they lose this shadow response when their ocelli are covered with opaque material. Ocelli likely evolved separately in the families Schizochitonidae and Chitonidae.

REPRODUCTION

Chitons generally have separate sexes and spawn sperm or eggs from a simple gonad through paired gonopores near the posterior end of the pallial grooves alongside their foot. Spawning is often highly synchronous but is not necessarily exactly correlated with a particular stage of the lunar or annual solar cycle. Populations separated by some distance can be out of synchrony with each other. Chitons sometimes aggregate and simultaneously spawn (e.g., the giant gumboot chiton, *Cryptochiton stelleri*).

Normally gametes are free-spawned and exit past the anus, carried by normal respiratory currents into the plankton. Chiton embryos have typical spiral cleavage, leading to a trochophore larva (Fig. 7B) that hatches from the egg capsule normally within about two days. The trochophore is capped with a sensory plate with an apical tuft of cilia, and more flagella forming a band around the middle known as the prototroch. The rapidly beating prototroch propels the speedy larva through the plankton, but it is not involved in feeding as in some animals with a trochophore larva. Chiton trochophores depend entirely on the yolk supplied in the egg and are thus nonfeeding, or lecithotrophic.

Although free spawning is most common, females from about five percent of all chiton species instead brood their eggs (Figs. 7A, 8A–C), with embryonic and larval development completed within the pallial groove of the brooding mother, sometimes with embryos sticking together in rod-shaped broods (Fig. 8B). A few species (e.g., *Stenoplax heathiana*) are known to lay benthic strings of jelly-like egg masses. Unlike a free-spawned embryo, a brooded or benthic-egg-mass embryo hatches as a late-stage larva and already has a creeping foot (Fig. 7A). Such a "crawl-away" larva is at least potentially capable of remaining near its mother.

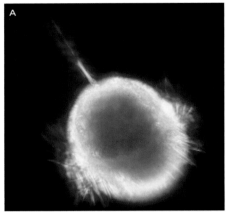
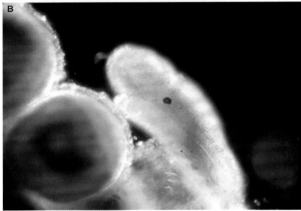

FIGURE 7 Contrast between stages of hatching in embryos of a free-spawner and a brooder. (A) Free-spawner *(C. hartwegii)* embryo hatches at early stage to a trochophore larva, topped with apical tuft and surrounded by prototroch flagella used for locomotion. (B) Brooder *(Cyanoplax thomasi)* larva just hatched and creeping over still unhatched embryos. In comparison to early-stage larva, this late-stage trochophore has already developed its foot so that it can crawl away; it has paired eyespots, and uncalcified shell precursors are visible on its dorsal surface. Photographs by the author.

Because most species do not seem to undergo metamorphosis spontaneously in cultures maintained in filtered seawater, the cues that promote larval settlement and metamorphosis are mostly unknown. Metamorphosis is not very dramatic but does involve some important changes, including the immediate start of biomineralization of the valves (Fig. 8D vs. 8E) and radula, the latter also apparent from the active feeding of newly metamorphosed juveniles. The prototroch and apical tuft are cast off, and the larva soon transforms from elongate to oval in body outline about 0.5 mm in length (Fig. 8F). At first there are only seven calcareous valves, with the tail (eighth) valve typically added up to a month or so later. One carryover from the larval stage is the retention of two bright red "larval" eyespots on the ventral surface (Fig. 8F); these do not correspond to the adult shell organs, and they persist for only about a month before they are lost.

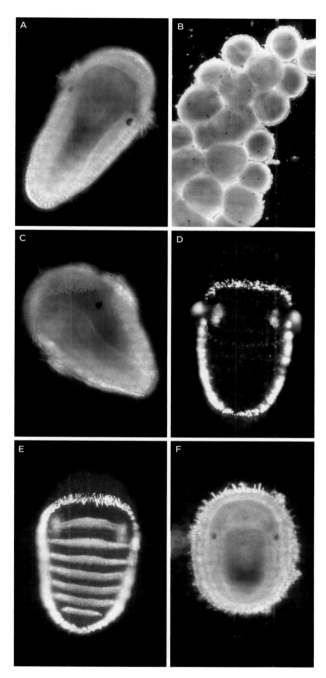

By the Late Cambrian Period (after 500 million years ago), chitons were already quite diverse and probably were important grazers on the then common cyanobacterial reefs. Particular fossils from earlier in the Cambrian or even the latest Precambrian (560 to 543 million years ago) that were previously considered as enigmatic "Problematica" or assigned to other phyla such as Annelida or Brachiopoda have recently been instead considered as close relatives of chitons. The range of fossils that are included within the chiton "crown group" has also been expanded by discoveries of exceptionally well-preserved articulated fossils, whereas most chiton fossil species are known only from separated valves. An Ordovician (approximately 450 millions of years ago) chiton, *Echinochiton dufoei,* is an example of an early chiton that had the normal eight valves but also had surprisingly gigantic spines on its girdle (Fig. 9). Other Paleozoic chitons probably had less dramatic smooth girdles.

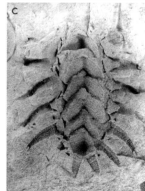

FIGURE 9 Ordovician fossil chiton from Wisconsin, *Echinochiton dufoei* (about 450 million years ago), which is remarkable for its especially prominent hollow spines as well as scutes along the margin of the valves. (A) Latex cast of *E. dufoei* holotype part, with 5 mm scale bar; (B) originally published reconstruction of the fossil; (C) external mold part of holotype, showing six posterior valves having attached lateral and posterior hollow spines and sediment filling of spines. From Pojeta *et al.* 2003.

FIGURE 8 Selected chiton developmental states through metamorphosis. (A) Ventral view of newly hatched late-stage trochophore larva of the brooder *Cyanoplax thomasi* (compare with Fig. 7A); (B) brooded embryos already at late trochophore stage before hatching, *C. fernaldi;* (C) newly hatched late-stage trochophore larva of brooder *C. fernaldi.* (D) Dorsal view of late-stage trochophore larva of free spawner *Mopalia lignosa,* using polarized light to emphasize already calcified girdle spicules (paired eyespots and more indistinct prototroch cilia also are bright in this polarized view); (E) dorsal view of recently metamorphosed *M. lignosa* juvenile as in (D) but now also with newly calcified valves; (F) ventral view of recently metamorphosed *C. fernaldi* juvenile, with prominent paired red larval eyespots. Photographs by the author.

PHYLOGENY AND CLASSIFICATION

There is general agreement that chitons are a monophyletic group. Most also agree that chitons are the sister taxon of a grouping that includes most other molluscan classes, known as Conchifera, including gastropods, cephalopods, bivalves, and scaphopods. Together, chitons plus conchiferans constitute a molluscan subgroup known as Testaria. The only molluscs that are generally not considered part of Testaria are two lineages of wormlike "aplacophoran" molluscs; however, the position of these is controversial. Most have considered aplacophorans as basal molluscs, usually as two separate lineages outside of Testaria, but some others have instead argued that aplacophorans are a monophyletic sister taxon of chitons.

Recent progress in morphological and molecular analysis of genealogical relationships within chitons has substantially improved understanding of living and fossil chiton relationships. Most familiar living chiton species belong to Chitonida, which share derived similarities in shell, gill, and egg hull, sperm, and molecular traits. Other living chitons are mostly restricted to deep water and belong to Lepidopleurida (e.g., *Leptochiton;* Figs. 5A, 10). Within Chitonida, combined shell, sperm, egg, and molecular analyses support the basal position of Callochitonidae within Chitonida (e.g., *Callochiton*). Particular derived sperm and egg hull features support the monophyly of all remaining members of Chitonida, which likewise is subdivided into two well-supported lineages: Chitonina and Acanthochitonina. This division corresponds to fundamental differences between both the pattern of sculpturing for an extracellular hull surrounding spawned eggs and also the particular arrangement of gill addition in the ontogeny of chitons. Members of Chitonina have spiny egg hulls and have retained a primitive "adanal" gill arrangement (Fig. 1C), with gills added both posterior and anterior of the first pair of gills to appear in ontogeny, which appear just posterior to paired renal openings. Acanthochitonina have cuplike or conelike egg hulls and derived abanal gill placement, with gills added only anterior to the first pair. In adults, if the last gill in each row is also the largest, the arrangement is abanal (Fig. 5B). In addition to these egg hull and gill arrangement features, both Chitonina and Acanthochitonina are strongly supported by molecular evidence. Within each of these taxa, most families and genera are well delimited, but relationships among these are not.

ECOLOGY

Chitons are often important members of rocky intertidal communities. For example, the "black Katy" chiton, *Katharina tunicata* (Fig. 11), is a large herbivore that is abundant and conspicuous on rocky shores from northern California to Alaska. With lower densities of grazers, the algal turf grows so profusely that those grazers that are present are less successful than if grazers were more common. At high densities, *K. tunicata* becomes a competitive dominant species in its community. *K. tunicata* keeps the kelp *Alaria marginata* at extremely low densities, but when it is excluded this alga forms a monodominant covering of the same habitat.

FIGURE 11 *Katharina tunicata* (foreground and background, Washington, about 8 cm length) is a competitive dominant grazer along much of the northwestern North American coast that is also a traditional subsistence food source for some Native Alaskans. Other grazers, such as the coralline alga specialist limpet, *Acmaea mitra* (center), benefit from the presence of high densities of *K. tunicata* (see text). Photograph by the author.

Other species of chitons occur in densities comparable to those of *K. tunicata* on various temperate and tropical shores but, unlike *K. tunicata,* they more normally retreat to under rocks or otherwise are hidden from sight during daylight hours.

FIGURE 10 *Leptochiton rugatus* (central California, about 1 cm length) is representative of the mostly deep-water order Lepidopleurida, although this species often occurs on the underside of rocks in the intertidal. Photograph by the author.

SEE ALSO THE FOLLOWING ARTICLES

Fossil Tidepools / Homing / Molluscs / Rhythms, Nontidal

FURTHER READING

Eernisse, D. J., and Reynolds, P. D. 1994. Polyplacophora, in *Microscopic anatomy of invertebrates*. Vol. 5, *Mollusca I*. F. W. Harrison and A. J. Kohn, eds. New York: Wiley-Liss Inc., 56–110.

Eernisse, D. J., N. B. Terwilliger, and R. C. Terwilliger. 1988. The red foot of a lepidopleurid chiton: evidence for tissue hemoglobins. *Veliger* 30: 244–247

Haderlie, E. C., and Abbott, D. P. 1980. Polyplacophora: the chitons, in *Intertidal invertebrates of California*. R. H. Morris, D. P. Abbott, and E. C. Haderlie, eds. Stanford, CA: Stanford University Press, 412–428.

Kaas, P., A. M. Jones, and K. L. Gowlett-Holmes. 1998. Class Polyplacophora, in *Mollusca: The Southern Synthesis*. Fauna of Australia, Vol 5. P. L. Beesley, G. J. B. Ross, and A. Wells, eds. Melbourne: CSIRO Publishing, 161–194.

Kaas, P., and R. A. Van Belle, eds. 1985–1994. *Monograph of living chitons (Mollusca: Polyplacophora)*. Vols. 1–5. Leiden: E. J. Brill/Dr W. Backhuys.

Kaas, P., R. A. Van Belle, and H. Strack. 2006. *Monograph of living chitons (Mollusca: Polyplacophora)*. Vol. 6. Leiden: E. J. Brill.

Okusu, Akiko, E. Schwabe, D. J. Eernisse, and G. Giribet. 2003. Towards a phylogeny of chitons (Mollusa, Polyplacophora) based on combined analysis of five molecular loci. *Organisms, Diversity, and Evolution* 3: 281–302.

Pearse, J. S. 1979. Polyplacophora, in *Reproduction of marine invertebrates*. Vol. 5. A. C. Giese and J. S. Pearse, eds. New York: Academic Press, 27–85.

Pojeta, J. Jr., D. J. Eernisse, R. D. Hoare, and M. D. Henderson. *Echinochiton dufoei*: a new spiny Ordovician chiton. *Journal of Paleontology* 77 (2003): 646–654.

Saito, H., and T. Okutani. 1992. Carnivorous habits of two species of the genus *Craspedochiton* (Polyplacophora: Acanthochitonidae). *Journal of the malacological Society of Australia* 13:55–63.

Schwabe, E., and A. Wanninger. 2006. Polyplacophora, in *The mollusks: a guide to their study, collection, and preservation*. C. F. Sturm, T. A. Pearce, and A. Valdés, eds. Boca Raton, FL: American Malacological Society and Universal publishers, 217–228.

Smith, A. G. 1960. Amphineura. in *Treatise on invertebrate paleontology*. Vol. I, *Mollusca 1*. R. C. Moore, ed. Lawrence: Geological Society of America and University of Kansas Press, 41–76.

CIRCULATION

IAIN MCGAW

University of Nevada, Las Vegas

Circulation is a general term that describes the movement of fluids within an animal. In most, but not all organisms, it refers to the movement of a transport medium (termed blood or hemolymph) through specialized conduits. At its most complex, the circulatory or cardiovascular system consists of a transport fluid, a series of conduits or vessels, and one or more pumping organs (hearts). Circulatory systems range in complexity from simple open systems to the high-pressure closed systems typical of the vertebrates. The circulatory system is a communication system providing a vital link between specialized organs and the tissues.

FUNCTIONS OF THE CIRCULATORY SYSTEM

Gases, nutrients, and wastes move in and out of cells by diffusion. In unicellular animals and some of the lower multicellular animals, this can be accomplished by diffusion directly across the body wall. In most multicellular animals the process of diffusion from the external environment into individual cells would be far too slow to maintain cellular activities. The circulatory system has evolved to transport substances from the external environment to individual body cells and vice versa.

The circulatory system carries oxygen from the respiratory organs (gills or lungs) to the cells and transports carbon dioxide from the cells to be expelled from the body. Oxygen is usually transported on specialized carrier pigments, which can be intracellular or extracellular; smaller amounts are carried in solution. Carbon dioxide can be carried in solution or bound up in various chemical forms such as bicarbonate. Once nutrients have been processed by the digestive system, they have to be transported to individual cells. Once inside the cell, oxygen is used to break down nutrients for energy and growth. Metabolic wastes produced by cellular activities are toxic to the system and are transported to specialized organs for excretion.

In complex multicellular animals the circulatory system acts as a conduit for hormone transport. Hormones are a control system allowing communication between various areas of the body. An array of hormones regulates ion levels and maintain the volume and pressure of the circulatory system. The blood itself is also an important buffer system, regulating pH of the extracellular fluids.

In homeothermic (warm-blooded) animals and some poikilothermic (cold-blooded) animals the circulatory system is important for distribution of heat and maintenance of body temperature. The circulatory system also serves as a proliferation and storage area for specialized cells. These cells function in the defense (immune system) and repair of the body (e.g., clotting, tissue regeneration).

Finally, in many invertebrates a fluid-filled system provides rigidity to the body, thus functioning as a skeletal system. Such hydrostatic skeletons are important for movement or extension of various structures. For example, echinoderms move by use of tiny tube feet. These tube feet are extended by pumping fluid into them via the water vascular system. During burrowing activity, annelids and bivalve molluscs use the hydrostatic skeleton to extend and contract the body as they bury in the sediment.

CIRCULATORY SYSTEM ORGANS

Circulatory systems have evolved more than once, in different animal taxa. Each is designed to best suit the needs of that animal. Although there are several recognized components, some may or may not be present, or have evolved to a greater degree in some groups. The circulatory systems of vertebrates are quite similar to each other. The wider array of circulatory systems in invertebrates reflects their diverse lifestyles.

Fluid

The fluid component of the circulatory system is typically an inert plasma (which is mainly water) with dissolved organic and inorganic substances. Many of these substances are carrier molecules, transporting nutrients, ions, and waste products. Specialized cells, or hemocytes, are suspended within this medium. These cells perform a variety of functions, including gas and nutrient transport and defense and repair of the body. The fluid is termed blood when this is held in a separate compartment from the rest of the extracellular fluids. In most invertebrates the blood and lymph system are not separate, and the fluid is termed hemolymph.

Pump

There must be some way of circulating the fluid around the body. The pump or pumps vary in complexity. In the simplest systems there is neither pump nor specialized vessels. Fluid circulates slowly through a low-pressure, open system. In acorn barnacles the hemolymph is contained within large vessels and sinuses surrounding the organs. The action of the cirri and thoracic muscles circulates fluid around the system (Fig. 1B).

In some organisms, in addition to distributing the fluid, the vessels themselves act as a pump. Here peristaltic waves of contraction pass along the vessel and force fluid along it. In the holothurians (sea cucumbers), pulsating vessels contract rhythmically to force fluid through the circulatory system.

The distinction between pulsating vessel and the tubular heart is not sharp. Tubular hearts are elongated organs that function solely for pumping rather than delivery of fluid, as in the pulsating vessels. Tubular hearts are found in the arthropods; some are very elongated and almost look like a vessel, others are more globular (Fig. 1A). Although single chambered, they are usually suspended within a pericardial sinus that acts as a primary receiving chamber.

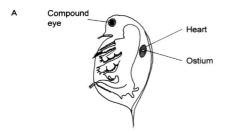

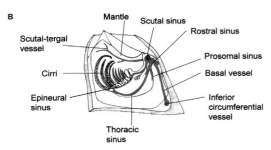

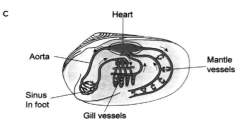

FIGURE 1 Representative examples of different types of open systems. (A) Cladoceran crustaceans possess a heart, but no distinct vessels. Hemolymph percolates between the tissues before returning to the heart via a pair of ostia. (B) Acorn barnacles have no heart *per se*. Hemolymph is circulated through a series of vessels by contraction of the thoracic muscles. (C) Bivalve molluscs have a complex open system. A muscular heart pumps hemolymph through a well-developed series of vessels.

Ampullar hearts are accessory hearts to the main one. These booster hearts help propel fluid into peripheral areas. The branchial hearts of cephalopods are situated at the base of the gills; these assist blood circulation through the gills (Fig. 2). The cor frontale of crustaceans is another

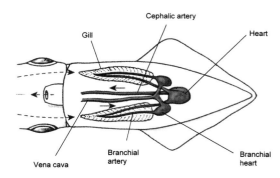

FIGURE 2 The branchial hearts of cephalopod molluscs aid circulation of blood into the gills, while the main heart circulates blood through the rest of the body.

example of an accessory heart. This helps propel hemolymph into the supraesophageal ganglion (brain) of the animal (Fig. 3).

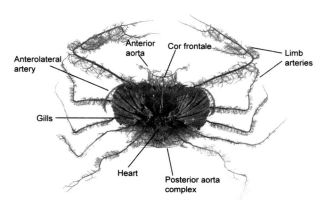

FIGURE 3 The circulatory system of the blue crab, *Callinectes sapidus*. The heart pumps hemolymph into five arterial systems. These split into fine capillary-like vessels that ramify within the tissues.

Chambered hearts are the most complex and are found in the vertebrates and the cephalopod molluscs. The number of chambers varies between two and four. They are typical of closed systems (see the section "Classification of Circulatory Systems") and thus generate higher pressures than simple tubular hearts. Chambered hearts can allow partial or complete separation of arterial (oxygenated) and venous (deoxygenated) blood.

Vessels

Vessels carry blood away from and back to the heart. Arteries are large vessels that carry (primarily oxygenated) blood away from the heart. Arteries in closed systems are surrounded by layers of smooth muscle. Contraction or relaxation of this muscle helps to re-distribute the blood.

Capillary networks are made up of small-diameter vessels (capillaries) having walls that are only one cell thick, consisting of flattened endothelium. Capillaries form dense networks within tissues, reducing the distance for diffusion. These vessels are the sites of gas, nutrient, and waste exchange between the blood and the tissues. Fluid exiting from the capillaries drains into the venous system (veins). The venous system carries blood, which is typically deoxygenated, back to the heart.

Sinuses and lacunae are typical of open systems (see the section "Classification of Circulatory Systems"). Hemolymph exiting from the heart is usually pumped into simple arteries, which open up into sinuses. The fluid in sinuses directly bathes the organs. Some sinuses branch into a network of smaller conduits known as lacunae. The

complexity and structure of sinuses varies considerably among the invertebrate phyla.

CLASSIFICATION OF CIRCULATORY SYSTEMS

Avascular Systems

Single-celled organisms such as the protozoans lack specialized organ systems, and exchange of nutrients and gases takes place over the body surface. In several larger, multicellular invertebrate taxa, components of a circulatory system are also absent. The sponges use flagellated cells to pump seawater through pores in the body, allowing nutrients and gases to reach individual cells. In the Cnidaria (jellyfish, corals, sea anemones) and the Platyhelminthes (flatworms), exchange of gases occurs over the body surface. Nutrients and wastes are exchanged with the tissues via the gastrovascular cavity, which may also function in delivery of gases to the tissues.

Open Systems

Most invertebrates have open systems (Fig. 4A); these are diverse in both form and function. Some of the simplest open systems are found in cladoceran crustaceans (water fleas). These small creatures possess a simple heart, but

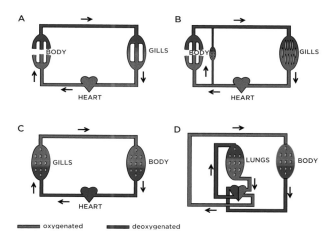

FIGURE 4 Schematic diagrams of the different types of circulatory circuit. (A) In an open system, hemolymph bathes the tissues in open sinuses. (B) Partially closed system of a decapod crustacean; here some of the organs may have closed, capillary-like vessels. The sinuses in the gills are well defined. (C) Closed system typical of the teleost fishes. The heart pumps deoxygenated blood through the gills, where it picks up oxygen. This oxygenated blood is pumped out to the tissues before returning to the heart. (D) The complex closed system typical of birds and mammals. The heart pumps deoxygenated blood to the lungs. The oxygenated blood returns to the heart and is pumped out via the systemic circulation to the body. Spent blood returns to the heart for reoxygenation. In birds and mammals there is complete separation of oxygenated and deoxygenated blood.

no vessels (Fig. 1A). Hemolymph percolates between the tissues in an open sinus and returns to the heart through small holes, or ostia. Acorn barnacles possess a series of interconnecting vessels and sinuses, but no heart. The contraction of the cirri and thoracic muscles circulate hemolymph around the body (Fig. 1B) The open circulatory system of bivalve molluscs is well developed. A muscular heart circulates hemolymph through a well-defined series of vessels (Fig. 1C).

Although the morphology of open systems differs considerably, they share a number of common characteristics. Although major distributing vessels are usually present, open systems lack a true capillary system. Instead, the fluid bathes the organs directly via sinuses. Because of this there is no separation of the blood from other extracellular fluids, and the resulting fluid is termed hemolymph. Open circulatory systems are characterized by larger blood volumes (20–40% of body volume) and lower pressures than closed systems. Efficient delivery of gases and nutrients and removal of wastes are essential to survival. Thus, although open systems appear more rudimentary than closed systems, their functioning is tightly regulated by neural and endocrine mechanisms.

Partially Closed Systems

The distinction between open and closed systems is not clear-cut. Although the absence of a true capillary system is one of the defining differences, several invertebrates with previously described open systems possess capillary-like vessels in some organ systems (Fig. 4B). Recently, the decapod crustacean system has been re-examined. The system is highly complex (Fig. 3). Well-defined arteries branch into capillary-like vessels, which may form closed circuits in the antennal gland (Fig. 5A, B) and the gills (Fig. 5C, D). Even the open components of the systems, the sinuses, are well defined. Thus, McGaw and Reiber (2002) have suggested another category that redefines the decapod crustacean circulatory system as one that is partially closed rather than open.

Closed Systems

All vertebrates have closed circulatory systems. Closed systems are also found in several invertebrate taxa—namely, the annelids and the cephalopod molluscs. In closed systems the extracellular fluids are divided into compartments: blood, lymph, and interstitial fluid. The blood is contained entirely within a series of vessels. Exchange of gases and nutrients takes place at the capillary level. Closed systems are high-pressure systems equipped with muscular

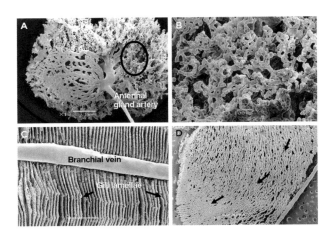

FIGURE 5 Electron micrographs to show the complexity of the partially closed circulatory system of decapod crustaceans. (A) Fine capillary-like vessels perfuse the antennal gland (kidney) of the Dungeness crab, *Cancer magister*. (B) Higher magnification of a the area circled in (A) shows a closed system of vessels that appear similar to the glomerulus in the vertebrate kidney. (C) In the gills the branchial vein delivers hemolymph to the gill lamellae. (D) Higher magnification of an individual lamella reveals a fine lattice network of vessels. Arrows show the direction of hemolymph flow. Images by the author.

hearts for distribution of blood. Because they are efficient systems, blood volume is comparatively low (5–10% of the body volume). Closed systems vary in complexity; fish have the simplest closed system. The heart pumps deoxygenated blood through the gills, where it picks up oxygen. This oxygenated blood is then delivered to the body tissues before returning to the heart (Fig. 4C). Birds and mammals have the most complex of the closed systems (Fig. 3D). Deoxygenated blood is pumped to the lungs, where it is oxygenated. The oxygenated blood returns to the heart and is pumped out to the body tissues. Spent blood then returns to the heart for reoxygenation in the lungs.

SEE ALSO THE FOLLOWING ARTICLES

Birds / Diffusion / Excretion / Fish / Sponges / Worms

FURTHER READING

Burggren, W. W., and C. L. Reiber. In press. Invertebrate and vertebrate cardiovascular systems. In *Evolution of cardiovascular systems*. W. C. Aird, ed. Cambridge, UK: Cambridge University Press, 24–44.

McGaw, I. J. 2005. The decapod crustacean circulatory system: A case that is neither open nor closed. *Microscopy and Microanalysis* 11: 18–36.

McGaw, I. J., and C. L. Reiber. 2002 Cardiovascular system of the blue crab *Callinectes sapidus*. *Journal of Morphology* 251: 1–21.

McMahon, B. R., J. L. Wilkens, and P. J. S. Smith. 1997. Invertebrate circulatory systems. In *Handbook of physiology. Section 13: Comparative physiology*. Vol. 2. Oxford, UK: Oxford University Press, 931–1008.

Randall, D., W. Burggren, and K. French. 1997. Circulation. In *Eckert animal physiology, mechanisms and adaptations*, 4th ed. D. Randall, W. Burggren, K. French, and R. Eckert, eds. New York: W. H. Freeman and Company, 467–516.

CLIMATE CHANGE, OVERVIEW

RAPHAEL SAGARIN

Duke University

Scientists are increasingly concerned about accelerating changes to the earth's climate caused by growing emissions of greenhouse gases from human activities. Greenhouse gases such as carbon dioxide and water vapor are essential to almost all life on earth because they prevent most of the solar radiation that reaches the earth from escaping back into outer space. Without our planet's greenhouse atmosphere the average surface of the earth would be around –18 °C instead of the comfortable average 15 °C we experience. But increases in the amount of greenhouse gases in the atmosphere—for example, through burning carbon-based fossil fuels that were previously trapped in underground reservoirs—are expected to increase the warming effect, a process often referred to as global warming. The physical changes brought by global warming, which include warming sea and air temperatures, melting glaciers, alterations of hydrological cycles, and increased intensity of tropical storms, are already apparent and expected to increase in the future. Likewise, these physical changes are already having effects on biological organisms and ecological communities, including those of wave-swept rocky shores.

HISTORY OF CONCERN ABOUT CLIMATE CHANGE

The Swedish chemist and Nobel prize winner Svante Arrhenius in the late 1800s was among the first scientists to draw a connection between greenhouse gases in the atmosphere and temperature at the surface of the earth. Through painstaking calculations Arrhenius estimated the potential rates of temperature warming associated with increased carbon dioxide in the atmosphere. Although Arrhenius's calculations of global warming in an atmosphere of doubled carbon dioxide were remarkably consistent with today's complex computer simulations, he expected a doubling of carbon dioxide concentrations in the atmosphere to take 2000 years. Today, although predictions of future emissions are complex, we expect a doubling of carbon dioxide from the already high 1990 levels in only 50–100 years. Our more complete (and more dramatic) understanding of the levels of carbon

dioxide in the atmosphere and its potential effects on climate is rooted in the 1950s with the work of chemists Charles Keeling and Hans Suess and oceanographer Roger Revelle. Keeling developed precise sensors of atmospheric carbon dioxide, which he first deployed on Mauna Loa in Hawaii in 1958. The Mauna Loa record, which shows a dramatic increase in atmospheric carbon dioxide since 1958, remains one of the most important records of the greenhouse gas increase that we have today. Revelle and Suess set the stage for keen scientific interest in global warming with their study (Revelle and Suess 1957) in which they showed that oceans would not be capable of permanently absorbing the added carbon dioxide in the atmosphere.

These pioneers in the study of climate change have been joined by thousands of scientists worldwide, in fields as diverse as biology, oceanography, physics, chemistry, mathematics, and economics, who are studying aspects of the climate change problem. Many of these scientists contribute, as authors or peer-reviewers, to the United Nations Intergovernmental Panel on Climate Change (IPCC) Assessment Reports. These reports attempt to give the most thorough evaluation of what is known and what is unknown about climate change and its effects on physical, socio-economic, and ecological systems on Earth. The most recent 2001 Assessment Report concluded that global average temperatures rose by 0.74° in the twentieth century, and it stated, "most of the observed increase in globally averaged temperatures since the mid-twentieth century is, very likely, due to the observed increase in anthropogenic greenhouse gas concentrations." Analysis shows that 11 of the 12 warmest years on record between 1850 and 2006 have all occurred since 1995. It is also important to recognize that temperatures may not warm gradually over many years; increases may be manifest in relatively sudden shifts to periods of much warmer average temperatures (as is seen during some El Niño years) or greater frequency of days with extremely warm temperatures.

CLIMATE CHANGE AND ROCKY SHORES: PREDICTIONS

Because rocky intertidal environments lie at the interface of the land and sea, and because their biological communities are so sensitive to environmental conditions, they are expected to be strongly affected by climate change. Changes in climate will affect intertidal environments by acting on individual organisms from the fine-scale molecular level, to the population and community level, to whole species ranges (Fig. 1). These effects may be

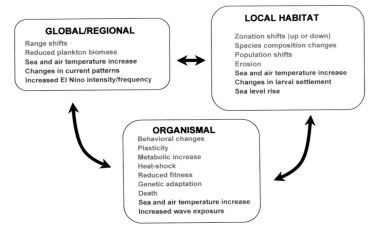

FIGURE 1 Climate change is expected to affect intertidal zones at a range of scales and levels of biological organization, from individual organisms, to local intertidal habitats and communities, to whole species ranges and global circulation patterns. Expected effects are listed in red; mechanisms driving those effects are listed in blue. Arrows indicate that changes at each scale are not independent but have cascading indirect effects on the other levels.

direct, in which changes in climate act on the physiology of individual organisms, or they may be indirect, in which case a cascade of events triggered by direct effects on a single species affects many species in a community, even those that are not particularly sensitive to climate change. Moreover, climate change may physically alter the intertidal environment.

The complexity of all the potential interacting factors and uncertainty about the intensity of future climate change makes it difficult to make precise predictions of how climate change will affect a given species or area. However, based on general knowledge of the natural history of intertidal ecosystems and predictions from global and regional climatic models, we can make some simplistic predictions about how future climate change will affect intertidal environments and organisms.

Direct Effects

Water and air temperature increases may have direct effects on the physiology of intertidal organisms, and this may ultimately result in changes in habitat use, greater susceptibility to predation or competition, and death. Warmer air temperatures, for example, may change intertidal zonation patterns by lowering the upper tidal height limits to plant and animal distributions, as these organisms will be less able to withstand conditions high on intertidal rocks. Warmer air and water temperatures have been shown to induce a "heat shock" response in intertidal organisms in which individuals must utilize heat shock proteins to restore proper metabolic function. The heat

shock response is believed to be very costly to the organism, because it cannot perform its usual activities such as feeding, digesting, or reproducing while it is repairing heat shock damage. Field and laboratory experiments have shown that some intertidal animals live in environments that reach temperatures close to their lethal limits. Thus, even slight increases in average temperatures may be expected to increase mortality in these species.

Even absent direct mortality, temperature changes may have sublethal effects that change distributions of intertidal plants and animals. Some species may have behavioral responses to temperature increases, such as moving to shaded crevices, or "gaping" in bivalve mussels. These responses may prevent their mortality but may exact a price in terms of reduced ability to acquire food or reduced energy allocated to reproduction. Organisms may also develop "plastic" responses to environmental change in which an organism changes color, morphology, or other traits within its lifetime. For example, some seaweeds develop different morphologies in areas of greater wave intensity. Finally, organisms may adapt over several generations through natural selection, although the pace of climate change may be faster than a species' ability to adapt genetically.

On a larger scale, species ranges may shift into environments that are cooler. Simplistic hypotheses suggest that ranges would shift poleward, where temperatures are cooler on average. However, detailed studies of temperature in intertidal sites show that high-latitude sites can have hotter conditions than low-latitude sites because of the timing of low tides, upwelling patterns, and weather patterns.

Sea level rise is also expected to occur with climatic warming, as a result of the thermal expansion of water and melting of glaciers and polar ice caps. Increased sea levels might allow some species to live higher on intertidal rocks, but this expansion of habitat might be curtailed by warmer air temperatures. Sea level rise might also change the dynamics of wave forces in the intertidal zone, which will change the distribution of species, depending on their tolerance to high wave action. The extent to which these effects change biological patterns in the intertidal zone will depend highly on the topography of individual rocky shores. In some shores with low relief, for example, there may be little habitat to move up into should sea level rise significantly.

The interaction between climate warming and El Niño events is uncertain, but it has been argued that El Niño events will increase in frequency and intensity, and some data suggest that this is the case since the 1970s. Based on past strong El Niño events, we would predict that rocky shores in some parts of the world (especially the west coast of North America) may experience higher water temperatures, invasions from low-latitude warm-water species, greater storminess, reduced upwelling, and changes in near-shore current patterns. El Niño temperatures and storms have resulted in die-offs of near-shore kelp forests as well as coastal erosion. Reduced upwelling reduces nutrients in near-shore waters and thus may affect the nutrition of planktonic stages of intertidal organisms as well as filter feeders on rocky shores. Changes in current patterns may bring species with planktonic larvae to locations where they are typically rare or absent, as was likely the case with shallow subtidal Kellet's whelk and sheephead fish, which expanded their range northward along the California coast during strong El Niño events.

Indirect Effects

Indirect effects, triggered by the direct effects just discussed, could occur through a number of pathways. In the most straightforward case, species that move into (or out of) a habitat or increase in numbers in response to warmer climate interact with another species that is not necessarily sensitive to climate change. For example, sea level rise may allow a barnacle species whose upper limit was caused by intolerance to emersion to move upward and outcompete other more emersion-tolerant species for space. In some cases climate-related species changes may benefit other species. For example, the tube snail *Serpulorbis squamigerus* has likely increased its numbers in central California because of climate warming. The hard, convoluted tubes of large colonies of this snail provide numerous hiding places for small organisms and vulnerable juvenile organisms such as recently settled purple sea urchins. The coupling of benthic and pelagic systems opens additional indirect pathways of change. Loss of kelp forests, through increased temperature or storm intensity for example, may allow greater settlement of larvae of intertidal species because kelp forests have been shown to act as "curtains" that prevent many planktonic larvae from returning to rocky shores. On the other hand, the cold water associated with upwelling has been shown to lessen the influence of keystone predators such as the sea star *Pisaster ochraceus,* so reduced upwelling overall might indirectly affect the prey of this species as well as the entire intertidal community.

CLIMATE CHANGE AND LIVING ORGANISMS: HISTORICAL COMPARISONS

In addition to simple predictions, we can look at historical records to see how past climate changes have affected biological organisms and communities. Because the twentieth century was a time of significant climate warming, studies that look at climate and biological records from this period are important proxy studies for what we might expect with future climate change.

Climate change in the twentieth and early twenty-first century has been shown to have affected the distributions, populations, and behaviors of living organisms throughout the planet. These effects include shifts in species ranges, increases in populations of warm-adapted species or decreases in populations of cold-adapted species, local extinctions of species, and changes in the timing of species migrations, flowering, and hibernation. Much of our understanding of these changes comes from historical studies, in which old observations made by scientists or amateur naturalists are compared to modern observations. These observations include records—such as those taken by writer Aldo Leopold—on the arrival dates of migratory birds or the flowering dates of plant species at a particular location, which can be used to determine long-term changes in biological indicators of the seasons. Other records include museum collections that document where and when specimens of butterflies and other species were made and thus provide insight into shifts in species ranges. Historical photographs (such as those of alpine areas) that show that the altitude of tree lines and even the condition of individual trees has changed with climate change, can be an invaluable source to document climate-related changes in biological systems.

However, very few historical data sets exist that cover a long enough time span or provide enough information to make inferences about connections between changes in intertidal environments and climate change. In one study from Pacific Grove, California, the exact location of a study in which a scientist in the 1930s counted all the tide-pool invertebrates he could find along a 105-square-yard transect line was relocated and resampled in the 1990s. The main pattern of change in the animal species was that, regardless of the type of species or its life history, southern species increased in numbers while northern species decreased in numbers. Water temperature records from this area showed a significant warming trend between the two studies. This suggests a link between the warming climate and community composition in the intertidal zone. However, this study leaves many open questions. As one example, we would like to know what happened in this community during the 60 years between studies to see whether changes in the community really tracked changes in climate. Another study from England that does have a more complete record of 70 years of data from the twentieth century on intertidal organisms showed that southern species increased and northern species decreased during warm periods, while the opposite occurred during cool periods.

HOW TO STUDY CLIMATE CHANGE IN INTERTIDAL HABITATS

The foregoing studies can show correlations between biological changes and changes in climate that occurred at the same time, but they are not controlled experiments that can more definitively isolate climate change as a mechanism causing the species changes. Experiments on climate change are difficult because many variables are changing with climate, including the concentrations of greenhouse gases, temperature, air and water circulation patterns, and water availability, as well as biological changes in entire species communities. Moreover, the predicted changes in these variables will differ in different environments and parts of the world. Trying to control for all these factors in an experimental design is often logistically impossible. Nonetheless, laboratory and field experiments that at least replicate some of the expected effects of climate change are being conducted. Such experiments are especially difficult to conduct in the wave-swept intertidal zone, where conditions are constantly changing throughout the tidal cycle and experimental apparatus used in terrestrial climate change experiments (such as heat lamps or free carbon emitters that simulate the rise in carbon dioxide) are likely to be dislodged or destroyed. One quasi-experimental approach that avoided these problems compared intertidal communities between two nearby areas, one of which was artificially warmed over ten years by the thermal outfall of a nuclear power plant. The study showed that warming-induced changes in intertidal communities were highly dependent on effects on key intertidal species, such as space-occupying algae. However, this approach also is problematic for several reasons. First, it is impossible to replicate this design without building another nuclear power plant! Second, the warming was brought on suddenly by the operation of the thermal outfall and maintained in a pattern that was not at all similar to the observed or expected warming caused by human-induced global climate change. Finally, the warming treatment was not accompanied by other expected effects of climate change, such as sea level rise or the arrival of new species from lower latitudes.

All of these studies only record changes in a limited area. If climate change is expected to act on large regions, we would ideally want to track changes in many sites. The only way to do this is through long-term monitoring of many intertidal areas. Several such studies are under way now, but most have not been conducted long enough to observe the effects of long-term climatic warming. Unfortunately, Revelle and Suess were prescient in their 1957 paper when they concluded, "Human beings are now carrying out a large scale geophysical experiment of a kind that could not have happened in the past nor be reproduced in the future."

SEE ALSO THE FOLLOWING ARTICLES

El Niño / Heat and Temperature, Patterns of / Monitoring: Long-Term Studies / Sea Level Change, Effects on Coastlines / Storms and Climate Change / Variability, Multidecadal

FURTHER READING

Christianson, G. E. 1999. *Greenhouse: the 200-year story of global warming*. New York: Penguin Putnam.

Intergovernmental Panel on climate change. 2007. *Climate change 2007: the physical science basis*. Summary for policymakers. Geneva.

Revelle, R., and H. E. Suess. 1957. Carbon dioxide exchange between atmosphere and ocean, and the question of an increase of atmospheric CO_2 during the past decades. *Tellus* 9: 18–27.

Root, T. L., J. T. Price, K. R. Hall, S. H. Schneider, C. Rosenzweig, and J. A. Pounds. 2003. Fingerprints of global warming on wild animals and plants. *Nature* 421: 57–60.

Sagarin, R. D., J. P. Barry, S. E. Gilman, and C. H. Baxter. 1999. Climate related changes in an intertidal community over short and long time scales. *Ecological Monographs* 69: 465–490.

Schiel, D. R., J. R. Steinbeck, and M. S. Foster. 2004. Ten years of induced ocean warming causes comprehensive changes in marine benthic communities. *Ecology* 85: 1833–1839.

Schneider, S. H., and T. L. Root, eds. 2001. *Wildlife responses to climate change: North American case studies*. Washington, DC: Island Press.

Southward, A. J., S. J. Hawkins, and M. T. Burrows. 1995. Seventy years' observations of changes in distribution and abundance of zooplankton and intertidal organisms in the western English Channel in relation to rising sea temperature. *Journal of Thermal Biology* 20: 127–155.

CNIDARIA, OVERVIEW

MARYMEGAN DALY

Ohio State University

On the one hand, the phylum Cnidaria is easily defined. Its members are alone among metazoans in having diploblastic bodies (i.e., only ectoderm and endoderm layers) and endogenous cnidae (intracellular capsules with highly folded tubules used for defense, prey capture, or attachment). On the other hand, the diversity of reproductive strategies, life histories, and anatomies seen in the Cnidaria defies easy generalization. Approximately 10,000 known species are included in the two major Cnidarian lineages, Anthozoa and Medusozoa. The Medusozoa include Cubozoa, Hydrozoa, and Scyphozoa, the classes whose members have (at least ancestrally) a free-living medusa as part of their lifecycle. The most important cnidarians on rocky shores are sea anemones (Anthozoa: Hexacorallia: Actiniaria, Corallimorpharia, Zoanthidea) and hydroids (Hydrozoa: Hydroidolina: Anthoathecata, Leptothecata).

THE BASIC CNIDARIAN

Architecture and organization

Describing a typical cnidarian is complicated, because the form and biology of medusae and polyps can be quite different, even when the two belong to the same species. A polyp is benthic, and limited in its locomotion to burrowing or to "inchworm" crawling by bending, attaching its oral or aboral region, then detaching the other region. In contrast, a medusa is pelagic, using muscular contractions of its bell to propel itself through the water; constrictions of the bell margin may enhance the propulsive capabilities. Medusae are oriented with the mouth facing downward, and the muscles and nerves are primarily circumferential, encircling the bell. Medusae tend to be more muscular than polyps, and the coronal muscles may be composed of striated fibers. Polyps are oriented with the mouth directed away from the substrate and have the muscles and nerves arranged in sheets that both ring the column and span its length.

The central element of cnidarian anatomy is a space rather than a structure: the coelenteron is the fluid-filled cavity enclosed by a bilayered epithelium (Fig. 1). The endodermal and ectodermal epithelial sheets that constitute the cnidarian body enclose a third layer, the mesoglea, which ranges from thin and acellular in hydroids to thick and fibrous in scyphozoans and many anthozoans. The epithelial sheet is extruded at one end into tentacles that

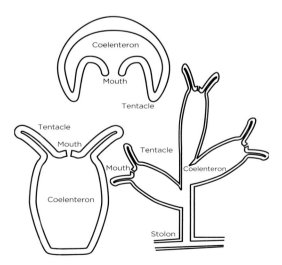

FIGURE 1 General anatomy and orientation of a cnidarian medusa, polyp, and colony. Internal anatomy, histology, and musculature of each morph differ across and within the cnidarian classes.

encircle the opening to the coelenteron, and closed on the opposite end. In medusozoans, the mouth is terminal; in anthozoans it leads to a short, ectodermally lined gullet, called the actinopharynx. The coelenteron is the gastrovascular cavity and serves as the primary hydrostatic skeleton. The coelenteron is subdivided by sheets of tissue in anthozoans and scyphozoans; anthozoan mesenteries are typically muscular and bear gametes and gastric filaments; scyphozoan gastric septae are nonmuscular and bear only gastric filaments. Biological functions such as respiration, circulation, and digestion are accomplished through diffusion, phagocytosis, and ciliary action. Although cnidarians lack true organs, the tissues of various regions of the body may be specialized and morphologically distinct. This regionalization and specialization is most extreme among cubozoans, in which photosensory structures may be elaborated into an image-forming eye. In all cnidarians, the nervous system is primarily netlike, although it may be concentrated in rings around the mouth or at the bell margin.

The hydrostatic skeleton provided by the coelenteron is bolstered in some species by an organic or mostly mineral skeleton. Most notable among these are the calcium-carbonate skeletons of scleractinian corals (Anthozoa: Hexacorallia: Scleractinia) and the hydrocorals (Hydrozoa: Anthoathecata: Sylasteridae, Milleporidae). Calcium carbonate is the principal component of endoskeletal elements called sclerites that are embedded in the flesh of many octocorals. Proteinaceous skeletons range from the tough but flexible axial skeletons of gorgonian soft corals and antipatharian black corals to the cuticular covering of hydroids. The perisarc (periderm) is a flexible, chitin-based exoskeleton secreted by the ectoderm; in the thecate hydroids (order Hydroidolina, suborder Leptothecata) it is enlarged distally into a cuplike theca, out of which extends the tentacle crown. The perisarc may be fused and stiffened into a more rigid skeleton that gives the colony a permanent and characteristic shape, such as the feather-like colonial form of *Agalophenia*. Nonskeletalized anthozoans gain additional support and protection from exogenous material such as sand grains: ceriantharians use cnidae and mucus to agglutinate sediments into a flexible but tough tube, and zoanthideans may incorporate sand grains into their ectodermal epithelium.

Cnidae

Cnidae are the most notable attribute of Cnidaria. These highly complex intracellular structures function in defense, prey capture, locomotion, and attachment. A cnida is produced by the Golgi apparatus within a specialized sensory cell called a cnidoblast, and all types have the same basic structure: an eversible tubule continuous with a capsule into which it is packaged. The size of the capsule varies in length from 3 to 120 μm, depending on the type of cnida and the animal bearing it. The tubule everts through a lidded operculum or a tripartite flap at the apex upon reception of appropriate mechanical, chemical, or electrical stimulation. Mechanical stimuli are mediated through a specialized ciliary complex on the cnidocyte surface. Although at least some cnidae act as independent effectors, the number of cnidae responding to a stimulus may be controlled by the animal. Such control of discharge is effected through synapses between the cnidocyte and adjacent nerves. Discharge occurs at speeds in excess of 2 meters per second with an acceleration force greater than 40,000 g. The mature cnidoblast, or cnidocyte, is anchored to the surrounding tissues to prevent it from being ripped from the epithelium by the explosive force of discharge. Once discharged, the cnidocyte degenerates and is replaced by a new one that differentiates from multipotent interstitial cells (I-cells).

Three major kinds of cnidae have been described, based on the morphology of the capsule and tubule (Fig. 2). A spirocyst (Fig. 2A) has a single-walled capsule in which a

tubule bearing tiny, hollow threads is coiled. Spirocysts are found only in members of the anthozoan subclass Hexacorallia. They are primarily adhesive and are characteristic of tentacle ectoderm; the threads solublize in seawater and form a meshwork that enwraps prey or aids in attachment. Like spirocysts, ptychocysts (Fig. 2D) are known only from hexacorallians, but ptychocysts are more restricted in their distribution, being found only in members of order Ceriantharia. The single-walled capsule of a ptychocyst contains a smooth, closed tubule that is packaged into the capsule through pleating and folding rather than coiling. Ptychocysts are found in the column of ceriantharians, and their everted tubules are the principal component of the unique tube of these burrowing anemones.

Nematocysts are the most variable type of cnida in terms of structure and function and are the most widespread, occurring in all body regions of members of all classes. Approximately 30 categories of nematocysts are recognized. A nematocyst has a double-walled capsule, and the coiled tubule bears spines or barbs that facilitate penetration and anchor the capsule in prey tissue (Fig. 2B, C, E). The tubule may be differentiated distally into a more slender thread; the tip of the thread may be closed, as in the tubule of spirocysts and ptychocysts, or be open. The open-tipped nematocysts (stomocnidae), capable of delivering toxins through the hollow shaft, are more diverse and taxonomically widespread than the closed-tipped astomocnidae, which are found only among Hydrozoa. Only one kind of nematocyst is known to occur across all cnidarian classes; this nematocyst, variously called a holotrichous isorhiza, an atrichous isorhiza, or simply a holotrich, has a thread of uniform diameter adorned with minute spines along its entire length (Fig. 2B). Electron microscopy has revealed that atrichous isorhizae, so called because they were formerly believed to be spineless, do bear minute spines along the length of the tubule, thus rendering atrichous and holotrichous isorhizae synonymous. Holotrichs are integral in aggressive and defensive interactions.

The types of nematocysts and their location in the body change temporally during development, during morphogenesis of new polyp types, and in response to outside stimuli. The cnidom of adult scyphozoans is almost invariant across taxa, but those of planulae, ephyrae, or scyphistomae may be diagnostic at the level of species. Particular kinds of nematocysts may be associated with specific functions: in sea anemones, for example, microbasic p-mastigophores are found in the regions of the body concerned with digestion, and holotrichs are found in regions concerned with intraspecific aggression.

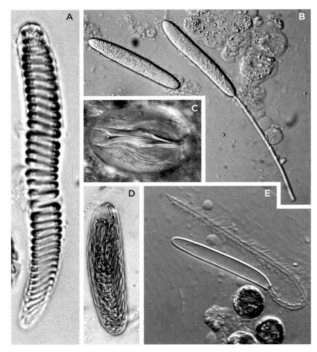

FIGURE 2 Cnidae are the most distinctive attribute of cnidarians. Thin-walled spirocysts (A) are common in the tentacles of hexacorallians. Nematocysts are the most diverse kind of cnidae, including holotrichs (B) with a tubule of uniform diameter and spination, harpoon-like stenoteles (C) and mastigophores (E). Ptychocysts (D) are the least diverse and most taxonomically restricted type of cnida, being found only in the column ectoderm of ceriantharians. Photographs by author.

COMPLEXITY THROUGH OTHER MEANS

Complexity through Specialization

Cnidarians are necessarily anatomically simple, being limited in terms of their diploblastic, tissue-grade bodies. However, they may achieve physical complexity through iteration into colonies and subsequent differentiation of individuals. Coloniality is important in polypoid cnidarians such as hydroids, corals, and soft corals and in the large "superorganisms" of the hydrozoan order Siphonophora. In all instances, the colonies represent a single genetic individual, with the constituent polyps connected by a shared gastrovascular system. Coloniality allows a single genotype to maximize its use of space and provides opportunities for anatomical specialization and complexity. In polymorphic colonies, individuals differ in anatomy, often to achieve functional specialization. Specialization is achieved through differential gene expression and manifests in distinct anatomies and complements of nematocysts. Polymorphism is most developed in Hydrozoa, where particular individuals may be specialized for feeding (gastrozooids), defense and prey capture (dactylozooids), reproduction (gonozooids), or locomotion (nectophores). In Anthozoa, polymorphism is seen only among octocorallians, in which there is a distinction between feeding polyps and the nonfeeding axial polyp on which the former are situated.

Complexity through Symbiosis

Symbiotic interactions, like polymorphism, serve to increase the biological complexity of cnidarians. Photosymbiosis is known in all major cnidarian lineages. The relationship between the host cnidarian and the photosynthetic microorganism contributes positively to growth and nutrition of both partners: the cnidarian receives carbon fixed by the photosymbiont, which in turn receives a steady supply of nitrogen and carbon dioxide. Partnerships in which sessile cnidarians gain motility are common: the juveniles of many species of burrowing sea anemones parasitize pelagic jellyfish or ctenophores, which allows them to traverse greater distances than they could cover as planktonic larvae. Hydroids of the genus *Hydractinia* and members of several genera of sea anemones, most notably *Adamsia, Stylobates,* and *Calliactis,* live on shells inhabited by hermit crabs. The hermit crab provides substrate, motility, and food for the cnidarian and may deter some of its predators. The benefit to the hermit crab differs somewhat depending on the cnidarian it has partnered with. For example, *Hydractinia* seems to provide only modest protection from predators, acting primarily as camouflage, whereas *Calliactis* and *Adamsia* provide demonstrable protection from predators.

Diversification of Life History and Development

Cnidarians are among the most complex of the free-living metazoans in terms of their life history and reproductive biology. As the incidence of coloniality indicates, cnidarians are unusually proficient at asexual reproduction. In addition to colonial growth, asexual reproduction can result in physically separate, genetically identical individuals called clones; in Cnidaria, clones arise through many pathways, ranging from regeneration of a complete animal from autotomized tissue or tentacles to transverse and longitudinal fission. Clonal reproduction plays an especially important role in space-limited habitats and is a critical element in the intraspecific aggressive interactions in both actiniarian sea anemones and hydroids. In these groups, individuals have historecognition mechanisms through which they distinguish clone-mates from other conspecifics, and they have stereotyped aggressive behaviors involving nematocysts to compete for space.

The early phases of sexual reproduction are highly variable across Cnidaria. The site of maturation for the gametes differs among lineages: in Anthozoa, Cubozoa, and Scyphozoa the gametes mature in the mesoglea, whereas gametes of most hydrozoans mature in the ectoderm. In Hydrozoa, Cubozoa, and many Scyphozoa, the gametes are produced by pelagic medusae (Fig. 3); because anthozoans are exclusively polypoid, gametes are necessarily made by the polyp. Sexuality, mode of fertilization, and cleavage pattern may vary among even closely related species. Nonetheless, the end product of sexual reproduction remains remarkably similar: the planula, a ciliated, egg-shaped larva that develops directly from the gastrula, is common to the vast majority of all cnidarian lineages. A

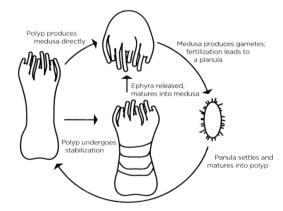

FIGURE 3 Metagenesis, in which a sexually reproducing medusa is produced by an asexually reproducing polyp, is characteristic of the superclass Medusozoa. The pathways and particulars of metagenesis may vary within groups, but the general pattern is remarkably uniform across classes, as illustrated in this diagram (not drawn to scale). Hydrozoans and cubozoans typically follow the outside path, whereas scyphozoans follow the inside path.

planula swims or creeps using cilia and typically moves with the oral end (blastopore) trailing rather than leading. In anthozoan planulae, the apical region, opposite the blastopore, has a tuft of especially long cilia and a dense concentration of sensory cells, which may be instrumental in evaluating substrates for settlement. In all cnidarians with a polyp phase, the primary polyp develops directly from the planula: the blastopore becomes the adult mouth, and tentacles are extruded from the oral tissue surrounding it.

In medusozoans, the path of postplanular development varies across lineages. In stauromedusans, as in anthozoans, the sexually mature adult is a polyp, and it develops directly from the planula. In all other Scyphozoa, the primary polyp, called a scyphistoma, produces many pelagic medusae through serial transverse fission (strobilation); the juvenile medusae, called ephyrae, mature into the sexually reproducing adults (Fig. 3). In Cubozoa, the primary polyp produces a single medusa. Typically in Hydrozoa, medusae bud from the sides of the polyp as gonophores; the gonophores may be produced on the body wall, within the oral region, on the stolons of a colony, or on a specialized polyp called a gonozooid. In some hydroids, such as the hermit crab epibiont *Hydractinia,* rather than resulting in free-swimming medusae, the gonophores remain attached to the parent colony and release eggs and sperm directly. In these hydroids and in some of the "direct-developers" that have lost the polyp stage entirely, the planula develops into a pelagic feeding larva called an actinula. In direct-developing trachyline hydrozoans such as *Liriope,* the actinula develops into a medusa; in hydroids such as *Tubularia,* it becomes a polyp.

EVOLUTIONARY HISTORY

Precedence of Polyp or Medusa

The debate over cnidarian evolution has historically focused on metagenesis and its implications for cnidarian and metazoan evolution. Most debate focused on the historical primacy of polyps or medusa and broader hypotheses about the early evolution of animals. Embedded within this larger question are debates about the origins of each of the major groups, and arguments about the ancestral symmetry of the phylum and the origin of coloniality. Although modern data and methods have been unable to fully resolve these long-standing controversies, substantial progress towards resolving the polyp-or-medusa debate has been made. The ancestral form of Cnidaria seems almost certainly to have been a polyp; the medusa is a derived life history stage that characterizes a subset of superclass Medusozoa.

The conjecture that the medusa is the ancestral form for Cnidaria is integral to hypotheses of a close evolutionary link between Cnidaria and Ctenophora. These two groups share similar gross morphology, as members of both have gelatinous, organless, radially symmetrical medusiform bodies and historically have been considered together as Coelenterata. This proposed relationship is a component of a broader hypothesis of metazoan phylogeny that draws a distinction between the bilaterally symmetric triploblasts and the supposedly radially symmetric diploblasts. Although deeply rooted in metazoan phylogenetics, this hypothesis seems to have little scientific merit. No modern study has demonstrated a close affinity between Cnidaria and Ctenophora. Cladistic analyses of morphology have suggested that Cnidaria is the sister group to Ctenophora + Bilateria; molecular and developmental data have either been inconclusive or favored the view that Cnidaria is more closely related to Bilateria than to Ctenophora. Furthermore, studies of developmental gene expression have demonstrated that radial symmetry is secondary rather than the ancestral condition for Cnidaria.

Although identifying their sister group has been contentious, there has been little debate about the monophyly of Cnidaria, because of the seemingly irrefutable synapomorphy of cnidae. Although molecular sequence data concur that Cnidaria is monophyletic, there is uncertainty as to whether Myxozoa and the enigmatic *Polypodium hydriforme* are cnidarians. Although *P. hydriforme* cannot be readily placed in any group, it is cnidarian-like in its biology and anatomy and has long been considered a hydrozoan of unknown affinity. Phylogenetic analyses of 18S sequences indicate a close affinity between *P. hydriforme* and the endoparasitic Myxozoa, which have nematocyst-like polar capsules. The inclusion of *Polypodium* within Cnidaria is called into question because the affinity of Myxozoa is unclear: some analyses have found that Myxozoa lies within Cnidaria, whereas others have asserted that myxozoans are triploblastic.

New Interpretations of Old Groups

It is now widely accepted that Cnidaria comprises two major clades, Anthozoa and Medusozoa, the latter being characterized by the possession of a linear mitochondrial genome and a life history that, at least primitively, includes a medusa. Virtually all analyses of molecular sequence data and all modern analyses of anatomical or developmental data support this dichotomy. The resolution of these two major lineages represents a significant advance in our understanding of cnidarian relationships but does not resolve the controversy over whether the medusa or the polyp is the ancestral form for the phylum. The primacy of the polyp is justified based on its relative commonness across Cnidaria. This hypothesis is bolstered by phylogenetic analyses of ribosomal genes, which suggest that the polypoid "stalked jellyfish" of the scyphozoan order Stauromedusae is the most basal lineage within Medusozoa and that a metagenetic life history characterizes a more inclusive group comprising Cubozoa, Hydrozoa, and non-stauromedusan Scyphozoa. Although it is often described as the basal lineage, Anthozoa is not necessarily more primitive than Medusozoa; in fact, the anthozoan polyp is far more complex than that of medusozoans in terms of its anatomy and cnidom and certainly represents a significant divergence from the ancestral cnidarian body plan.

The primary divergence within Anthozoa is between Hexacorallia (anemones and corals) from Octocorallia (soft corals and gorgonians). Octocorals have endoskeletal sclerites, a fused axial skeleton, or both; are exclusively colonial; and have eight-part symmetry and pinnately branched tentacles. Relationships among octocorals remain contentious: molecular sequence data suggest relationships that are not supported by traditional, morphologically based taxonomy. Hexacorals are more diverse in their anatomy than octocorals, having either six-, eight-, or (less commonly) ten-part symmetry and being colonial, solitary, or clonal. Hexacorals all have spirocysts and muscular coupled mesenteries. In contrast to the octocorallian orders, most major groups within Hexacorallia are monophyletic, but relationships among and within them are enigmatic. In particular, relationships among the solitary actiniarian sea anemones, antipatharian black corals, clonal corallimorpharian anemones, scleractinian corals, and colonial zoanthidean anemones remain problematic. Resolution of these relationships is relevant to the inference of the ancestral state for coloniality among the Anthozoa: if one of the primitively colonial groups (Scleractinia, Zoanthidea, or Antipatharia) is basal, then the coloniality would be interpreted as the ancestral condition for anthozoan polyps.

Relationships among the three classes included in Medusozoa are more controversial. Monophyly of cubozoans is widely assumed based upon morphology; in particular, the structural complexity of the eye and molecular analyses that include multiple representatives of Cubozoa support this hypothesis. Sequence data also provide strong support for the monophyly of Hydrozoa, the most diverse group of medusozoans. In contrast, Scyphozoa is paraphyletic: Stauromedusae appears to be the sister group to the remaining Medusozoa. The absence of a medusa in stauromedusans and their apparent phylogenetic separateness has led to their proposed elevation as a fifth class, Staurozoa. Morphological data

alone, and in combination with 18S ribosomal sequence data, support a clade comprising the three scyphozoan orders (Coronatae, Rhizostomeae, Semaeostomeae) in which a small polyp generates medusae through strobilation. Morphology and life history suggest that Hydrozoa is sister to a clade containing Cubozoa and all scyphozoan groups. The medusa is inferred to be homologous across Medusozoa, but certain resolutions of relationship among these major groups allow for the possibility that medusae have evolved multiple times.

SEE ALSO THE FOLLOWING ARTICLES

Corals / Genetic Variation and Population Structure / Hydroids / Octocorals / Sea Anemones / Symbiosis

FURTHER READING

Arai, M. N. 1996. *Functional biology of Scyphozoa.* London: Chapman and Hall.
Bridge, D., C. W. Cunningham, R. DeSalle, and L. W. Buss. 1995. Class-level relationships in the phylum Cnidaria: molecular and morphological evidence. *Molecular Biology and Evolution* 12: 679–689.
Fautin, D. G., and S. L. Romano. 1997. *Tree of life: Cnidaria.* http://tolweb.org/tree?group=Cnidaria&contgroup=Animals.
Fautin, D. G., J. A. Westfall, P. Cartwright, M. Daly, and C. R. Wyttenbach. 2004. *Coelenterate biology 2003: trends in research on Cnidaria and Ctenophora.* Amsterdam: Kluwer Academic.
Harrison, F. W., and J. A. Westfall. 1991. *Microscopic anatomy of the invertebrates,* Volume 2: *Placozoa, Porifera, Cnidaria, and Ctenophora.* New York: Wiley-Liss.
Shick, J. M. 1991. *A functional biology of sea anemones.* London: Chapman and Hall.

COASTAL CURRENTS

SEE CURRENTS, COASTAL

COASTAL ECONOMICS

SEE ECONOMICS, COASTAL

COASTAL GEOLOGY

GARY GRIGGS
University of California, Santa Cruz

The diverse landforms that characterize any particular area of coastline reflect that region's geologic history and the interplay among tectonic processes, geology, climate, and the sea. The recent geologic history of the coastal areas of California, Oregon, Washington, and Alaska, for example, is strikingly different from that of either the Atlantic or the Gulf coast of the southeastern United States. Even a casual visitor to the coastline will recognize the obvious differences between the coastal mountains and sea cliffs that characterize much of California's and Oregon's coastal zone and the broad, low relief coastal plain, sand dunes, and barrier islands of Virginia, North Carolina, or Louisiana.

PLATE TECTONICS AND COASTAL GEOLOGY

The concept of plate tectonics provides a comprehensive explanation of the origin and distribution of the large-scale features on the Earth—the mountain ranges, volcanoes, trenches, faults, and earthquakes—and how they all fit together, and it also provides a useful framework for our understanding of the diversity of coastal landforms and geology. The coastal mountains, rocky seacliffs, and intertidal zone that characterize much of the coastline of California, Oregon, and Alaska are evidence of a youthful coastline, where plates are colliding or have collided in the recent past, the land is often being uplifted, and the landscape is being worn down by wave attack. In striking contrast, the coastline of the central and south Atlantic coast of the United States or the Gulf Coast is much older geologically; the terrain has been eroded or worn down over millions of years and is predominantly a depositional environment. There are no sea cliffs or rocks exposed, but rather sandy barrier islands, dunes, and back bays.

THE DISTRIBUTION OF COASTAL LANDFORMS

Based on geology and tectonic history, coastlines can be broken down into perhaps four general geomorphic environments: (1) steep coastal mountains and seacliffs with hundreds of feet of relief (Fig. 1); (2) uplifted marine terraces and low seacliffs, tens to perhaps several hundred feet in height (Fig. 2); (3) low-relief coastal plains, deltas,

FIGURE 1 High, near vertical cliffs eroded into volcanic rocks on the north side of the Hawaiian island of Moloka'i. Photograph by the author.

FIGURE 2 Terraced coastline eroded into sedimentary rocks along the central coast of California, south of San Francisco. Photograph by the author.

or lowlands with beaches, sand dunes, bays or lagoons, sand bars, and often barrier islands (Fig. 3); and then there are (4) a variety of organic coasts or shorelines formed primarily by organisms such as coral reefs and mangroves (Fig. 4). Much of the geologically young California coast, for example, consists of coastal mountains and actively eroding seacliffs that would include (1) and (2) in the foregoing list. The geology of Florida, on the other hand, consists almost entirely of coastal plain, reef, and mangrove environments.

FIGURE 3 The low-relief sandy coastline along the Gold Coast of eastern Australia. Photograph by the author.

The tectonic history is important in determining the general appearance and large-scale landforms of any coastline. What alters, modifies, and influences how any local coastline appears, however, is the underlying geology, or

FIGURE 4 The reef coastline of Moorea in the Society Islands of the South Pacific. Photograph by the author.

the type of rock and the weaknesses within the rocks; all of the physical processes that impact the coast, such as the climate, rainfall, wave energy and tidal range, wind, ice and glaciers; as well as organic activity.

On a sandy or low-relief depositional coastline, for example, sand dunes, marshes, mud flats, sand bars, lagoons, tidal channels, ebb tide deltas, or some combination of these may be present. Sediments are much finer-grained than those of rocky coasts; they may be fine or very fine-grained sand, or even mud. These coastlines can erode or change shape rapidly during hurricanes or strong storms because there is no rock to resist wave attack, only sand or mud to be easily moved around. Cape Hatteras, for example, which is on a sandy barrier island, has been retreating about 11 feet per year over the past 18,000 years in response to the gradual sea level rise accompanying global warming and ice melting. The lighthouse was recently moved 2000 feet inland at a cost of $9 million.

There are a number of different landforms or geomorphic features, commonly observed along rocky, mountainous, terraced, or cliffed coastlines, that are distinct from those of depositional coastlines. On a rocky or tectonically active coast, we typically see raised marine terraces, sea cliffs, a rocky intertidal zone with tidepools, and often a rich assemblage of flora and fauna. (Fig. 5). The beaches in these areas may consist of medium to coarse-grained sand, pebbles, or even cobbles, much different from the beaches of a depositional coast. It is the specific characteristics of the rocks exposed along the coastline, however—their type (granite, sandstone, or shale, for example), their layering or bedding; their internal weaknesses, such as joints or faults; and how resistant they are to erosion—that will determine how any individual stretch of coastline will look. Where the

rocks are hard, where they are layered or eroded in such as way as to leave benches or outcrops in the intertidal zone, tidepools may develop. Many of the points and capes that are landmarks for sailors or sites for lighthouses consist of very hard granitic or volcanic rocks that are very resistant to erosion.

FIGURE 5 The rugged rocky Yehliu coastline of Taiwan. Photograph by the author.

The older, more subdued coastlines just described have been eroded down to and below sea level by millions of years of weathering, wave attack, glaciation, and other agents. As a result, depositional landforms and sand and mud predominate. There are no tidepools along these coastlines, because there is no rock exposed.

Years of human activity have produced some highly altered coastlines today, where seawalls, revetments, and breakwaters have completely replaced and covered the natural coastal materials (Fig. 6). In some highly developed

FIGURE 6 The completely altered coastline of Hong Kong, where revetments and breakwaters have covered most of the natural materials. Photograph by the author.

FIGURE 7 A revetment along the San Mateo County, California, coast has produced a new coastal environment and replaced the weaker sedimentary rocks with resistant granitic boulders. Photograph by the author.

areas, 30–50% of the coastline has now been so armored that completely new landforms and habitats have been created in efforts to protect human development in the coastal zone (Fig. 7).

FURTHER READING

Davis, R. A. Jr., and D. M. Fitzgerald. 2004. *Beaches and coasts*. Malden, MA: Blackwell.

Bird, E. C. F. 1984. *Coasts—an introduction to coastal geomorphology*. Oxford: Blackwell.

Carter, R. W. G., and C. D. Woodroffe, eds. 1994. *Coastal evolution*. Cambridge, UK: Cambridge University Press.

Woodroffe, C. D. 2003. *Coasts*. Cambridge, UK: Cambridge University Press.

COLD STRESS

CAREN E. BRABY

Monterey Bay Aquarium Research Institute

Physiological stress can be broken down into two categories—lethal and sublethal stress—and both have been correlated with biogeographic distributions of intertidal organisms. Although cold temperatures universally decrease the rates of metabolism in organisms (and thus decrease the feeding, growth, and reproduction of that organism), there is no known lethal stress associated with cold until the temperature drops below the freezing point. Once below this point, the organism must find methods to avoid or tolerate ice crystal formation and the associated stresses. These physiological strategies of mitigating cold stress are costly and divert cellular energy away from other processes such as growth and reproduction.

COLD STRESS IN THE INTERTIDAL

The intertidal zone can experience extreme temperatures—both hot and cold—and the organisms that live there must be able to cope with the extremes as well as rapid changes in temperature, as the tide ebbs and flows. Most marine organisms are ectothermic (their body temperature is not regulated but rather is dependent on many environmental and behavioral factors) and thus experience the temperature fluctuations in the air or water around them. Although most of the ocean is buffered from temperature stress by the high heat capacity of water, intertidal habitats experience periods of aerial exposure during which body temperature of ectotherms can vary greatly. Polar and temperate latitude coastal zones can experience extreme cold during freezing winter storms that coincide with low tide events, and the organisms that are exposed during these events must cope with the physiological threats associated with cold stress. Although the more temperate regions may experience fewer days each year of extreme cold than the poles, they nonetheless experience a great deal of fluctuation in temperature, which is a challenge to the physiological mechanisms used by that organism to survive and thrive.

LETHAL COLD STRESS

Lethal cold stress occurs as body fluids reach their freezing temperature and ice crystals develop. Ice crystals mechanically damage cell membranes and increase the osmotic concentration of the remaining fluid. To understand the importance of freezing on physiological stress, we have to understand the nature of water freezing. The freezing point of pure water is 0 °C. However, the addition of solutes (such as salts, proteins, sugars, or other components) changes the behavior of that fluid by decreasing the freezing point, because the solutes interfere with the crystal formation of the water molecules. As water freezes, the solutes are excluded from the crystal lattice, increasing the osmotic concentration in the surrounding liquid water. If there are no nucleation sites for ice to form, fluid may become supercooled (the temperature falls below the freezing point). When supercooling occurs, the addition or presence of an ice nucleator initiates abundant and rapid ice formation, which can easily tear through delicate cellular structures. In the context of an organism, both ice crystals and changes in osmotic concentration pose physiological hurdles to overcome. By investing cellular energy into making molecules that mitigate freezing stress (such as antifreeze or ice-nucleating proteins), marine organisms that face freezing conditions can avoid or tolerate ice crystal formation.

Freeze Avoidance

Although mobile ectotherms can mitigate habitat temperature through behavior (such as migrating to areas that are less stressful—for example, a snail crawling into a moist rock crevice during low tide), sessile ectotherms cannot change their temperature by migrating and thus can only develop nonmigratory behavioral adaptations (such as mussels gaping their shells) or physiological solutions to cope with temperature. Teleost fish have developed antifreeze proteins, which prevent ice formation in the blood despite body temperature falling below the freezing point. Because teleost fish have a much lower solute concentration in their body fluids relative to seawater (Fig. 1), they are especially susceptible to freezing in the absence of antifreeze proteins. Increased concentration of small osmolytes (such as glycerol) in body fluids can also decrease the freezing point and help avoid lethal cold stress.

Freeze Tolerance

Another strategy to deal with cold stress is to control—rather than avoid—ice crystal formation so that it is not damaging to the cells. Marine organisms have developed ways to control both the speed with which ice crystals form and the exact location. Intracellular ice formation appears to be lethal; however, controlled extracellular ice

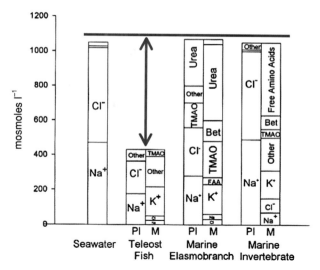

FIGURE 1 Solute composition of the body fluids of seawater and three categories of marine organisms: teleost fish (bony skeleton), elasmobranch fish (cartilaginous skeleton), and invertebrates (in milliosmoles-L^{-1}). Teleost fish have a much lower solute concentration than elasmobranchs and invertebrates (emphasized by the arrow), which makes teleosts more susceptible to freezing stress. Bet = glycine betaine; FAA = free amino acids; M = intracellular fluid of muscle tissue; Pl = extracellular fluid of the plasma or hemolymph; TMAO = trimethylamine-N-oxide. From Hochachka and Somero (2002).

formation can be tolerated. The marine mussel *Mytilus* produces ice-nucleating proteins that are localized in the extracellular spaces and prevent supercooling of body fluids. The extracellular ice-nucleating proteins encourage ice crystal formation at temperatures at which the ice crystals form slowly and have a smaller final size, allowing the crystals to coexist with the delicate cellular membranes (Fig. 2). As ice crystals form and exclude solutes, the extracellular fluid increases its osmotic concentration and causes the cells to lose water, increasing the intracellular osmotic concentration and further depressing the freezing point of the intracellular space, where ice crystal formation is lethal.

SUBLETHAL COLD STRESS

Because there is no obvious metric for sublethal stress, it is difficult to define and difficult to demonstrate. We do know that a decrease in temperature is accompanied by a slowing of metabolic processes in ectothermic organisms. This slowing down of physiological function has an effect on an organism's ability to forage, grow, and reproduce, and thus it can affect the overall fitness of the individual organism. There is evidence that cold stress can alter behavior and physiological function. For example, the low-intertidal sea star *Pisaster* forages in

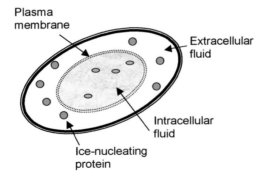

A

Plasma membrane

Extracellular fluid

Intracellular fluid

Ice-nucleating protein

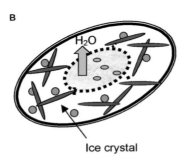

B

H₂O

Ice crystal

FIGURE 2 A schematic diagram of an organism (the mussel, *Mytilus*) containing a cell. The first panel (A) shows the cell above its freezing point, and the second panel (B) shows the same cell below its freezing point, when ice crystals begin to form in the extracellular space. Ice crystal formation in *Mytilus* is initiated by the production of ice-nucleating proteins to control the timing and extent of ice crystal formation. Ice formation in the extracellular fluid leads to higher solute concentration in both the extracellular and the intracellular fluids, which depresses the freezing point, draws water out of the cell, and causes the cell to shrink. In addition to causing osmotic stress, the ice crystals can cause mechanical damage by breaking up the cell membrane and denaturing proteins. From G. N. Somero (unpublished).

the mid-intertidal on the mussel *Mytilus* when the tide is high. During lengthy cold upwelling periods, *Pisaster* foraging activity in the intertidal dramatically decreases and resumes only when the temperature rises above a threshold of about 10 °C. This is an example of habitat temperature regulating an important trophic relationship (a keystone predator and its prey), such that the fitness (feeding rate and growth) of the predator is likely affected in a sublethal manner. *Mytilus* mussels themselves provide another example. As temperature decreases, *Mytilus* heart rate also decreases and will cease if a critical low temperature is passed. In fact, there is evidence that periods of quiescence in temperate-latitude intertidal organisms can be followed by periods of higher activity

and metabolic efficiency once conditions improve. So the question of whether cold is stressful—in the absence of ice formation—remains to be answered.

SEE ALSO THE FOLLOWING ARTICLES

Circulation / Heat Stress / Temperature Change / Upwelling

FURTHER READING

Braby, C. E., and G. N. Somero. 2006. Following the heart: temperature and salinity effects on heart rate in native and invasive species of blue mussels (genus *Mytilus*). *Journal of Experimental Biology* 209: 2554–2566.

Hochachka, P. W., and G. N. Somero. 2002. *Biochemical adaptation: mechanism and process in physiological evolution*. New York: Oxford University Press.

Loomis, S. H. 1995. Freezing tolerance of marine invertebrates. *Oceanography and Marine Biology: An Annual Review* 33: 337–350.

Sanford, E. 1999. Regulation of keystone predation by small changes in ocean temperature. *Science* 283: 2095–2097.

COLLECTION AND PRESERVATION OF TIDEPOOL ORGANISMS

GARY C. WILLIAMS

California Academy of Sciences

The careful collection, preservation, and processing of tidepool organisms are all essential for accurate biological studies. The usefulness of biological specimens as material for study is related to how well preserved the material is regarding degree of shrinkage, retraction, contraction, autotomy, damage, and color loss.

BACKGROUND AND HISTORY

Wherever in the world there are rocky shores, tidepooling, or the exploration of the rocky intertidal zone and its marine life, can take place. Arguably, two of the best tidepooling regions in the world are along the coasts of California and South Africa. The heyday of tidepool collecting in California was no doubt between the 1920s and the 1940s, when the legendary exploits of marine biologist Ed Ricketts (Fig. 1) were used as a prototype for a fictional character ("Doc") in two novels of the Nobel Prize–winning writer John Steinbeck. This was an era of tidepool collecting for commercial purposes through the Pacific Biological Laboratories, located in Pacific Grove, California. Contemporaneously, research conducted in South African

FIGURE 1 Ed Ricketts collecting in a tidepool near Carmel, California, 1930s. Photograph by Jack Calvin, from Ricketts and Calvin (1939), courtesy Stanford University Press.

intertidal regions resulted in several significant scientific publications and eventually to guides and descriptions aimed at a broader readership.

Many noteworthy books have been written about the endeavor of tidepooling, beginning in England with Philip Henry's 1853 volume, *A Naturalist's Rambles on the Devonshire Coast.* From the 1920s onward, published accounts were based more on original observations at the seashore and with a greater emphasis on the detailed natural history and ecology of organisms. Included here are several well-known works dealing with the Atlantic Coast of North America, central California, and the Cape Province of South Africa, such as Rachel Carson's *The Sea Around Us, Between Pacific Tides* by Ricketts and Calvin, and *Life Between Tidemarks on Rocky Shores* by Stephenson and Stephenson.

TIDEPOOL COLLECTING AND PHOTOGRAPHY

Beforehand planning is important to maximize the time spent at the field site. Tide tables need to be consulted for the time and extent of low tide. It is best to arrive at an intertidal site before maximum low tide. Start working at the waters and follow the tide going out (ebb tide), work the water's edge at slack tide, then move in with the incoming tide (flood tide). Keep an eye on the sea at the water's edge in case of rogue waves, and take care not to be trapped on rocky outcrops during flood tide.

Film photography is rapidly giving way to digital photography as the medium of choice for researchers as well as almost everyone else. Important potential dangers with tidepool photography involve salt spray and moisture with electronic camera equipment. An underwater camera can be used in deep tidepools. For many photographs of field research quality specialized equipment is necessary,

such as a macro lens and an external flash. A small tray or aquarium is useful for photographing specimens in the field that may lose vitality during transport.

SUGGESTIONS FOR COLLECTING

Depending on several factors—the nature of the field study intended, the amount of time one needs to stay in a rocky intertidal area at one time, the duration of time one needs to wade in tidepools, and the amount of submergence needed in particular tidepools—various modes of clothing should be considered, as well as a hat and sunscreen. Temperate regions with high productivity and copious lithophytic algal growth are usually in regions with upwelled, nutrient-rich water, which is cold. Tropical or warm temperate regions, or regions with warm nearshore currents usually provide more benign conditions. Three options to consider are tennis shoes or water shoes or rain boots with long or short pants; hip boots or waders; or wet suit and booties.

Some of the best areas for collecting are on the shady side of deeper tidepools at the low tide level, rock crevices and cavities, under loose rocks, and on seaweed or in kelp holdfasts.

COLLECTING PERMITS

A scientific collecting permit or research permit may be required, depending upon regional governmental requirements. Inquiries should be made to relevant governing agencies as part of initial research planning.

As an example, a scientific collecting permit from the California State Department of Fish and Game is required for all collecting in intertidal regions of California. Also, in areas under jurisdiction of the National Park Service on the west coast of the United States (such as Point Reyes National Seashore, Golden Gate Recreational Area, and Channel Islands National Park), or in any one of the many coastal California State Parks, written permission for collection must be acquired from the relevant park superintendent office. Additionally, if a particular species to be collected is considered a game species (such as fish or shellfish), a hunting permit is required from the California State Department of Fish and Game.

CONSERVATION

Regarding the burgeoning human populations of states like California and the ever-increasing pressures on coastal areas, it is critically important, now more than ever, to use the intertidal regions with a conservation consciousness. Environmental impact such as trampling, littering, and overcollecting can result from a variety of sources—food gatherers, student classes, scientific collectors, and hordes of tourists and casual visitors (many of which collect with no purpose other than souvenir taking). Also, pollution can result from runoff and siltation from adjacent land areas and coastal development projects.

Four of the seven principles of the *Leave No Trace* outdoor ethics code are immediately applicable here: plan ahead and prepare, leave what you find, respect wildlife, and be considerate of other visitors. It is wise to remember the saying, "take only pictures – leave only footprints." Of course, if the goal is to find specimens for scientific study, the intention will obviously be to remove living material from the tidepools, but minimal impact can be achieved by taking only what is necessary for a well-planned and coherent scientific study, and to collect over a wide intertidal region so that one small area or even a single tidepool is not overly impacted. Watch where you step in order to prevent damage to marine life, make sure to replace rocks that have been overturned to their original positions, and take only those specimens necessary for the study at hand.

EQUIPMENT

The necessity of some of the tools of the trade will vary with the nature of the study intended and local conditions. A day pack or plastic bucket can be used to carry most items—forceps, putty knife, pocket knife, closable plastic bags, various sizes of plastic jars and vials with lids, notebook and pencil. A good hand lens (6–10× magnification) is indispensable; it can be worn around the neck on a cord. Other items that can prove important in field work include a small flashlight for peering into dark crevices, small shovel and sieves, binoculars, and a camera with flash and macro capabilities. Also to consider are a water bottle, a light lunch, sunscreen, hat, coat or sweatshirt, and sunglasses.

FIELD NOTES

When collecting for scientific study and especially for permanent collection, it is important to keep careful field notes of all organisms in their undisturbed setting that are to be collected, *before* specimens are preserved. Color in life should be described in a notebook and can be recorded with color photography as well. Measurements can be made with a plastic metric ruler. Ambient conditions such as water temperature, degree of exposure to air or submergence, sky conditions (amount of sunlight and degree of cloudiness, etc.) are often important regarding the appearance and behavior of some organisms. Notes on habitat and natural history, such as tidal level, associations with other organisms, and behavior, can also be made.

Particular features of the organism, such as surface texture and morphological characteristics, distinctive odors and appearance, and other features, should be noted.

For collection labels and subsequent databasing, the following essential information must be recorded:

Exact Location

A Global Positioning System (GPS) unit is useful for providing exact coordinates when recording collection data in the field.

Date

Time and nature of the tide, such as tide table data, may also be useful here.

Collector's Name

It is best to use the full name of the collector or collectors (not just initials), to avoid confusion.

Specimen Details

This information includes the number of specimens collected and whether a whole or partial specimen was collected.

PRESERVATION

A recent article by Williams and Van Syoc provides a synopsis and table of recommended preservatives regarding field and lab collecting for various groups of marine organisms. Ethanol (75–95%) is adequate for most invertebrates, with the exceptions of most jellyfish, anemones, flatworms, and tunicates. Buffered formalin is normally used as a fixative for members of these groups and for fishes. Bouin's solution is used for histological or dissection purposes for nudibranchs and other molluscs and soft-bodied invertebrates. Any specimens collected for molecular analysis should be preserved in 75–95% ethanol. Acidic compounds such as unbuffered formalin and Bouin's solution are destructive to delicate structures composed of calcium carbonate such as spicules, sclerites, ossicles, or caryiophyllidia. Seaweeds are often collected in buckets or jars of seawater and are taken to the lab, where most specimens can be floated in pans of water, settled on herbarium sheets, dried, and labeled.

Before actual preservation, specimens should be relaxed as much as possible. Good relaxants include crystals of magnesium chloride ($MgCl_2$) or Epsom salts ($MgSO_4$), which can be added directly to pans of seawater containing living organisms. Preservative can subsequently be slowly dripped into the pans, and then the specimen placed in a glass jar of final preservative.

STUDY AND CURATION

The transportation of living material in hot weather presents a problem. Bottles for very small specimens can be transported in buckets of cool seawater packed with wet seaweeds. Large specimans can be kept directly in the bucket. A portable, battery-powered aquarium aerator can be used to keep a bucket of seawater oxygenated. In the laboratory, material should be separated and placed in dishes of well-oxygenated seawater or in aquaria (Fig. 2). Observation of living material under the microscope and the making of careful sketches and lab notes are desirable when feasible.

FIGURE 2 Field study conducted at a marine laboratory with outdoor aquaria from recently collected material, Christensen Research Institute, Madang, Papua New Guinea. Photograph by the author.

Museum curation involves the long-term preservation of material, which can be wet (usually bathed in ethanol) (Fig. 3A) or dry (dried specimens in separate, properly labeled plastic bags or specimen boxes in dustproof cabinets) (Fig. 3B). Overall, the best

FIGURE 3 Marine invertebrate curation (long-term storage). (A) Wet collection, South African Museum, Cape Town. (B) Dry collection, California Academy of Sciences, San Francisco. Photographs by the author.

environmental conditions for collections are indoors with cool, dry ambient air and low light levels (darkness with short-term periods of artificial illumination is best). Wet material is isolated into individual species in separate glass jars with tight-fitting lids and filled to the top with liquid preservative; the material is properly identified, and a label is written on special label paper (such as PolyPaper®, Nalge Nunc International, Rochester, NY), containing collection data and a catalog number that is placed in the bottle with the specimen. Subsequently, all information is entered in a computerized database, and specimen jars (lots) are systematically arranged on storage shelves in cool, dry, low-light environments.

SEE ALSO THE FOLLOWING ARTICLES

Habitat Alteration / Museums and Collections / Photography, Intertidal / Ricketts, Steinbeck, and Intertidal Ecology

FURTHER READINGS

California Department of Fish and Game. *Permit Directory.* http://www.dfg.ca.gov/html/permits.html.
National Park Service. 2001. *Obtaining scientific research and collecting permits for Bay Area National Parks.* http://www.nps.gov/pore/science_permits.htm.
Ricketts, E. F., and J. Calvin. 1939. *Between Pacific tides.* Stanford, CA: Stanford University Press.
Smith, G. M. 1944. Introduction, in *Marine algae of the Monterey Peninsula.* Stanford, CA: Stanford University Press, 3–16.
Smith, R. I., and J. T. Carlton. 1975. Introduction, in *Light's manual: intertidal invertebrates of the Central California coast,* 3rd ed. R. I. Smith and J. T. Carlton, eds. Berkeley: University of California Press, 1–16.
U.S. Antarctic Program/National Museum of Natural History. *Museum collection management terms and invertebrate processing procedures: methods of fixation and preservation.* http://www.nmnh.si.edu/iz/usap/usapspec.html.
Williams, G. C., and R. Van Syoc. 2007. Methods of preservation and anesthetization of marine invertebrates, in *The Light and Smith manual: intertidal invertebrates from Central California to Oregon* 4th ed., J. T. Carlton, ed. Berkeley: University of California Press, 37–41.

COLLECTIONS

SEE MUSEUMS AND COLLECTIONS

COMPETITION

JOSEPH H. CONNELL, SUSAN L. SWARBRICK, AND STEPHEN C. SCHROETER

Marine Science Institute, University of California, Santa Barbara

Competition occurs when organisms use common resources that are in short supply and negatively affect one another as a result. As the density of organisms increases, every individual has access to a smaller share of the limited resource. This reduction in resources available to individuals can curtail their growth, reproduction, and survival, which in turn can negatively affect population growth rate, abundance (density), and spatial distribution. Competitive interactions may also occur when resources are not in short supply if individuals harm one another when using a common resource.

RESOURCES IN INTERTIDAL HABITATS

Space on rocky shores is often a limiting resource for sessile organisms, which use it in many different ways. Algae and sessile invertebrates (e.g., barnacles, mussels, anemones, tube-building worms, bryozoans, sponges, and tunicates) require hard substrate on which to settle and permanently attach. These attachment sites must provide access to sunlight and nutrients or food from the surrounding water, and space on which to grow and reproduce. Mobile animals such as snails, limpets, and sea stars also compete for food resources. Many studies have shown that grazing gastropods compete for algae. Food and space resources are often closely linked because algal abundance is limited by space. Intertidal organisms may also compete for many other resources inextricably related to space on the shore, including protected sites that provide refuges from predators or physical disturbances, and sites for laying eggs. Intertidal animals may also compete for resources, such as mates, that might not be directly related to space.

MECHANISMS OF COMPETITION

There are two general ways in which organisms can compete for common resources: exploitation and interference competition. Exploitation competition (also called resource or scramble competition) is an indirect interaction that occurs when individuals consume a common limited resource and thus reduce the amount available to others. This kind of competition requires no physical contact between individuals. For example, limpets grazing in a patch of rocky shore on the same species of algae can negatively impact one another by reducing the abundance of the algae when it is in short supply.

Interference competition occurs when individuals interact directly and exclude others in the course of using a resource. This often involves some directly damaging behavior. For example, mussels overgrow and smother other species, some limpets actively defend a territory by butting up against nearby grazers and forcing them to leave, and faster-growing barnacles undercut or overgrow and kill slower-growing neighbors. Figure 1 shows interference competition for space between sponges, tunicates, and coralline algae on the undersides of intertidal boulders. Sponges are the dominant competitiors, overgrowing the tunicates (white) and coralline algae (purple). Unlike exploitation competition, interference competition may occur even if the resource is not in short supply.

FIGURE 1 Interference competition for space between sponges (red, orange, tan, and blue), tunicates (white), and coralline algae (purple) on the undersides of intertidal boulders.

Competitive interactions are often complex and may involve both indirect exploitation of a resource and direct interference. For example, one species may be able to occupy all of the rocky substrate in a patch of the shore, thus preventing recruitment of larvae or juveniles that need bare space to settle. The dominant competitor exploits all of the resource and at the same time excludes other individuals from the space through direct interference.

INTRASPECIFIC AND INTERSPECIFIC COMPETITION

When competitors belong to the same species, the interaction is called intraspecific competition; if they belong to different species, it is called interspecific competition. Intraspecific competition may be very intense because individuals of the same species usually have completely overlapping requirements for resources and thus exploit them in the same way. There is ample evidence for intraspecific competition in rocky intertidal habitats in a wide range of species including algae and mobile and sessile animals. One example that has been well documented by G. M. Branch occurs on the shores of South Africa, where the limpet *Patella cochlear* can reach very high densities. As limpet density increases, the growth rate, survival, maximum biomass, and reproductive output of individuals decline. There are many other examples of intraspecific competition in limpets worldwide. When intraspecific competition is intense, the population of a species can be self-regulating. High population density will cause a decrease in recruitment and survival, which will result in the decline of the population growth rate toward zero.

Competition between different species will vary from slight to intense depending on how much overlap there is in the use of limiting resources by the two species. The sum total of a species's abiotic and biotic resource requirements is called its fundamental niche. It is usually very difficult to completely describe the fundamental niche because it can have many dimensions (i.e., composed of many different kinds of resources). The fundamental niche of a species may be very different from its realized niche, which describes its actual use of resources in the presence of the other organisms in the habitat. Because interspecific competition can restrict a species ability to utilize resources, one indication of competition might be that when species 1 is absent, species 2 can exploit more of its fundamental niche and thus can occur over a wider area or range of habitats. However, factors other than competition can also produce this result, so this type of evidence must include a careful analysis of the contested resource and verification that it is limited.

Interspecific competition commonly occurs on rocky shores. There are many examples of competition for food between gastropod species, although the profusion of examples for this group compared to other taxa may simply reflect the relative ease of studying gastropods; they are abundant on many shores, easy to manipulate, and their algal resources can be readily altered for experiments. Studies have also shown that species of predators sometimes compete for prey. For example, predatory whelks compete for barnacles, and sea stars vie for a wide variety of species of invertebrates.

RELATIVE STRENGTHS OF INTRA- AND INTERSPECIFIC COMPETITION

A species may experience the effects of both intraspecific and interspecific competition simultaneously if its population growth is limited by the density of its own population as well as by the density of other species. *Cellana tramoserica*, a limpet found on rocky shores in southeastern Australia, shows a negative response to increases in density of both its own and other species. W. J. Fletcher and R. G. Creese have shown that as the density of *Cellana* increases, the growth rate of individuals and the survival rate both decline, and there is a negative correlation between the density of recruiting juveniles and adults. A second species of limpet, *Patelloida alticosata*, co-occurs with *Cellana tramoserica*; their distributions overlap and

they graze on the same microalgae. Experimental manipulations of the densities of the two species reveal that interspecific competition occurs and that *Cellana* is the superior competitor; it has a larger negative effect on the growth and mortality of *Patelloida* than *Patelloida* has on *Cellana*. However, the same study reveals that *Cellana* does worse at high densities of its own species than at high densities of *Patelloida*. Thus for *Cellana*, the negative impact of intraspecific competition is greater than that of interspecific competition. Conversely, *Patelloida* does better in the presence of its own species than when surrounded by *Cellana*, so interspecific competition is more important than intraspecific competition for *Patelloida*.

Ecological communities are typically composed of many species that can interact in complex ways. Organisms on the shore may be competing with a number of different species simultaneously, which can make it difficult to unravel the relative effects and strengths of the competitive interactions among the various species. Finally, species do not need to be closely related to compete for resources; they simply need to use the same resources and live in the same places. For example, algae and sessile animals often compete for space on rocky shores (see Fig. 1), and grazing limpets sometimes compete for food with herbaceous seastars.

COMPETITION AND INTERTIDAL ZONATION

Competition between species for space can affect local distributions of populations. One of the most striking biological features of rocky shores worldwide is the distribution of organisms in horizontal bands along the gradient from low to high tide zones. Competition for space can be an important factor in setting the lower limits of a number of sessile species. In an experiment that was one of the first field studies of the demographic consequences of interspecific competition on the shore, J. H. Connell manipulated the densities of two species of barnacles found in Millport, Scotland. *Chthamalus montagui* occurred in the upper intertidal just above the zone occupied by *Semibalanus balanoides*. Connell found that neither the larvae of *Chthamalus* that settled below the zone occupied by adults nor adult *Chthamalus* transplanted down into the *Semibalanus* zone survived unless *Semibalanus* were removed. When *Semibalanus* were not removed, they undercut or crushed *Chthamalus*, eliminating them from the lower shore. In the 45 years since this study, a number of experiments have demonstrated that interspecific competition plays a major role in setting the lower limits of zone-forming algae. For example, on Scottish shores, when *Fucus spiralus* was removed, *Pelvetia canaliculata* extended its range downshore; on the northeast coast of the United States, red turf-forming algae excluded *Fucus distichus* from the low intertidal zone; and in Chile, when the kelp *Lessonia nigrescens* was removed, *Codium and Gelidium* extended into the low zone.

Some experiments suggest that competition can also set the upper limit of distribution of some species. When *Fucus spiralis* was removed from the mid-intertidal zone on the British coast, the distribution of *F. vesiculosus*, which normally occurs just below the mid-intertidal zone, extended higher up the shore. Similarly, when the high-shore limpet *Collisella digitalis* was removed, the distribution of the limpet occupying the zone directly below it, *C. strigatella*, expanded up into the high zone.

COMPETITION AND SUCCESSION

Competition can play an important role in the succession of species that occurs after patches of rocks are cleared by some type of disturbance that kills or damages the residents. After an area is cleared, there is often a sequential replacement of species until a climax assemblage of one or more species is formed that persists in the absence of disturbance. Competition may or may not play a role in the replacement of species during course of succession. However, the species that persist in the climax stage of succession are dominant competitors for space because they are able to prevent other species from invading.

Mussels are strong competitors for space and often dominate large areas of the shore on the northwest coast of North America. When bare patches are created in mussel beds, competition can play a key role throughout the successional sequence as the patches are recolonized. In larger patches, once mussel larvae settle into the patch, they relentlessly overgrow other species that use the rocky substrate, eliminate them, and then continue to exclude them. In small patches, mussels may simply immigrate into the patch from the edges and smother and exclude other colonizing species.

In contrast, the succession of algae on stable boulders in southern California is not driven by competition. W. P. Sousa has shown that the green alga that colonizes first cannot hold the space, because it is eaten by crabs. The group of red algal species that form the mid-successional stage do not persist, because they are vulnerable to desiccation and overgrowth by epiphytes. The red alga *Chondracanthus canaliculatus*, which dominates the climax stage of succession, takes longer to recruit but is able to occupy the space vacated by earlier species and

prevent other species from invading, because it is hardy and has vigorous vegetative growth.

COMPETITIVE ABILITY AND LIFE HISTORY TRAITS

The competitive abilities of different species, particularly sessile species on rocky shores, are often correlated with life history features. In general, species such as the late-successional red alga, *Chondracanthus canaliculatus*, are good competitors for space because they have traits that enable them to capture substrate, persist, and resist invasion. These species tend to invest more energy into growth than reproduction. They may be capable of vigorous vegetative growth, resistant to both herbivores and overgrowth by other species, and tolerant of physical stress. By contrast, species that are good reproducers and dispersers such as the early-successional green alga, *Ulva*, often tend to be poorer competitors. They invest more of their energy and resources into reproduction at the expense of traits that promote persistence, such as resistance to physical stress and grazers. Such species maintain themselves by constantly colonizing newly created open space and reproducing before better competitors arrive and become established.

APPARENT COMPETITION

When two species compete, they negatively affect one another. If one species is a better competitor than the other, when resources are limited, an increase in the density of the better competitor will result in a decline in the density of the inferior competitor. However, this pattern of an increase density of one species linked to the concurrent decline in a second species can also arise from interactions that do not involve competition. If two prey species share a common predator, an increase in the density of prey 1 could result in an increase in the predator population, which could in turn eat more of prey 2, causing its population to decline. Since the population of prey 2 declines when prey 1 increases, it may seem that they are competing when they are not. This type of indirect negative interaction between species is called apparent competition.

An example of apparent competition occurs in Australia, where it can seem that barnacles compete for space with the limpet *Patella granularis*, because the growth of adult limpets declines when barnacles are present. The barnacles actually compete for space with algae. When barnacle density increases, algal abundance declines. Adult limpets eat the algae, and when their food supply declines, their growth rate slows. Thus, barnacles and limpets are not competing for space; rather the impact of barnacles on adult limpets is mediated through the algae.

COMPETITIVE EXCLUSION

Mathematical models of interactions between species predict that two species sharing a limited resource cannot coexist if one has any advantage over the other in its ability to exploit the resource. The superior competitor will eventually exclude the inferior. This has been called the competitive exclusion principle.

Competitive exclusion has rarely been demonstrated in communities on rocky shores. Determining whether competitive exclusion occurs in nature is difficult. There are many examples of competition impacting small-scale patterns of distribution such as the zonation of species on the shore. However, we know of no examples of competitive exclusion of a species over an entire shore. By its very nature, competitive exclusion will be hard to detect through direct evidence. Evaluating whether competitive exclusion has occurred requires documenting the competitive interaction before the species has been excluded, subsequently demonstrating that the species has disappeared from the habitat, and eliminating all other factors, such as predation or climate change, that might have caused the species to disappear.

COEXISTENCE OF COMPETING SPECIES

Competing species often coexist; if they did not, competitive interactions would rarely be observed. Mechanisms that promote coexistence of potential competitors can be divided into three types: (1) those that operate when competition is intense and invariant, (2) reduction of competitive intensity by predation or disturbance, and (3) reduction of competitive intensity by spatial or temporal variability in the recruitment of competitors or resources.

Coexistence When Competition Is Intense

A possible outcome of intense interspecific competition over an extended time is that the two species could coevolve to partition their use of resources. The assumption that adaptations that allow species to coexist in the present are the result of intense competition in the past is very difficult to test. Coevolution of two species requires that they occur in close proximity over a long time. J. H. Connell has suggested that temporal and spatial variability in the physical and biological environment on a patch of shore will make it unlikely that two species will be linked to one another for periods long enough for competition to cause the evolution of divergence in their use of resources. Coevolution seems more probable between pairs of species

that are closely linked through predator–prey, plant–herbivore, or host–parasite interactions. It is more likely that species that share similar resources evolved separately and were able to coexist when they came together because they had already become adapted to different resources or parts of the habitat.

Two species competing through the exploitation of shared, limited resources may coexist if the negative effects of intraspecific competition are greater than those of interspecific competition. Competition between individuals of the same species could be more intense because their use of resources is more similar than use by individuals of a different species. When the density of one of the species increases, the resulting negative effects on conspecifics will reduce the growth rate of the population and prevent it from reaching a density that would exclude the other species from the patch. In the case of *Cellana* and *Patelloida* discussed previously, *Cellana* is the superior competitor. However, competitive exclusion of the inferior competitor, *Patelloida*, is unlikely because of the strong intraspecific effect that *Cellana* has on itself.

When the distributions of two competing species overlap completely and competition is intense, the superior competitor will eventually exclude the inferior competitor from the area. However, when the distributions of the species do not overlap completely, offspring produced by the inferior competitor in areas where the superior competitor is absent can disperse and continually recolonize areas of co-occurrence, thus promoting coexistence.

Competitive Intensity Reduced by Predation or Disturbance

Species compete when their densities become high enough that common resources become limiting. Predation and natural disturbances may reduce the densities of potential competitors on rocky shores well below the limits set by their resources, reducing the intensity of competition, and thereby allowing species to coexist.

R. T. Paine has shown that sea stars are key predators in some areas of the low zone on rocky shores on the northwest coast of the United States. They consume the dominant competitor, *Mytilus californianus,* and prevent it from competitively excluding a wide range of species. When sea stars were removed from large patches of the lower shore, mussel larvae settled into the patches and eventually dominated the substrate, excluding a number of macroalgae and invertebrates that occupy rock surfaces. Herbivores can have similar impacts on communities by removing dominant algae that would otherwise monopolize space.

Disturbances can cause the death, displacement, or damage of resident organisms. Common agents of disturbance on rocky shores include waves, waveborne objects such as logs, ice, and sand, and trampling by large mammals. In areas where space is limited, disturbance can reduce the intensity of competition by creating patches of open substrate and so allow coexistence of inferior competitors. Experiments in intertidal boulder fields have shown convincingly that diversity of species is highest when disturbance occurs with intermediate frequency and severity.

Competitive Intensity Reduced by Temporal or Spatial Variation in Recruitment of Competitors and Resources

Variation in recruitment of species to an area is a common feature of rocky intertidal habitats. For many species, the number of larvae settling on the shore can vary dramatically between seasons and is often unpredictable from year to year. The supply of propagules or offspring of species that do not have a larval stage can also be highly variable in time. Even if the supply of propagules is relatively constant, the density of settlers on the shore can vary greatly among patches. This temporal and spatial variability in recruitment can generate large fluctuations in the density of populations of competing species and result in variation in the intensity of competition. Since it is unlikely that competition will be intense at all sites simultaneously, this variation can allow the inferior competitors to persist. In addition, variation in recruitment is frequently not synchronous among different species. A good recruitment year for an inferior competitor may be a poor year for the dominant. Relative gains in density of inferior competitors during good recruitment years can promote coexistence.

The supply of resources can also vary over space and time, which can produce variation in the intensity of competition if the resource is limiting. If the density of competitors remains constant, the intensity of competition will decline when the supply of the resource is high and increase when the supply is low. Space on which to settle and grow is often a limited resource. The rate of creation of patches of open space by predators and disturbances (as discussed previously) varies with spatial and temporal variation in the density of predators and the seasonal occurrence of storms. Variation in local recruitment or mortality of food organisms can cause large fluctuations in the supply of food for competitors. Although these temporal and spatial fluctuations in resources clearly have the potential to periodically reduce the intensity of competition and consequently promote coexistence, there are relatively few studies that have measured how the spatial and temporal variability in resources affects the intensity of competition.

IMPORTANCE OF COMPETITION ON ROCKY SHORES

It is generally recognized that competition for limited resources of space and food is a widespread feature of communities on rocky shores and can influence the population sizes and local distributions of species. Whether competition plays a dominant role in determining the structure (the presence and relative abundances of species) of communities on rocky shores is subject to debate. Many of the observed patterns of distribution and abundance of species, such as the distinct bands of species along the gradient of elevation in the intertidal zone, or the observation that species 1 occupies a wider range of habitats in the absence of species 2 than when they co-occur, are consistent with the expected outcomes of intense competitive interactions. However, competition is only one of several mechanisms that can account for such observed distributions. Many other processes, including variation in time and space in the physical characteristics of habitats or the supply of recruits, and other biotic interactions such as predation or parasitism, could produce these patterns. Evaluating the importance of competitive interactions in structuring communities on rocky shores requires identifying contested resources, determining when and where they are limiting, and studying the mechanisms that influence resource use by potential competitors. The most direct approach for measuring the importance of competitive interactions is to conduct carefully designed field experiments that can test alternative explanations for observed patterns of distribution and abundance.

SEE ALSO THE FOLLOWING ARTICLES

Biodiversity, Maintenance of / Disturbance / Predation / Recruitment / Succession / Zonation

SUGGESTED READING

Branch, G. M. 1984. Competition between marine organisms: ecological and evolutionary implications. *Annual Review of Oceanography and Marine Biology* 22: 429–593.

Connell, J. H. 1961. The influence of interspecific competition and other factors on the distribution of the barnacle *Chthamalus stellatus*. *Ecology* 42: 710–723.

Fletcher, W. J., and R. G. Creese. 1985. Competitive interactions between co-occurring herbivorous gastropods. *Marine Biology* 86: 183–191.

Menge, B. A., and G. M. Branch. 2000. Rocky intertidal communities. Chapter 9, in *Marine community ecology.* M. D. Bertness, S. D. Gaines, and M. E. Hay, eds. Sunderland, MA: Sinauer, 221–251.

Sousa, W. P. 1979. Experimental investigations of disturbance and ecological succession in a rocky intertidal algal community. *Ecological Monographs* 49: 227–254.

Underwood, A. J. 2000. Experimental ecology of rocky intertidal habitats: what are we learning? *Journal of Experimental Marine Biology and Ecology* 250: 51–79.

CONDENSATION

SEE EVAPORATION AND CONDENSATION

COPEPODS

SEE AMPHIPODS, ISOPODS, AND OTHER SMALL CRUSTACEANS

CORALS

PETER W. GLYNN
University of Miami

Corals are multicellular marine animals, members of the phylum Cnidaria that produce calcareous or horny skeletal structures. There are several kinds of corals, for example, hydrocorals, octocorals, black or thorny corals, and hexacorals (stony or true corals). The members of many of these groups are flexible and fleshy with skeletons composed of minute calcareous sclerites embedded in an organic matrix or of horny proteinaceous material. Other kinds of corals, such as hydrocorals (fire corals), blue corals, and stony corals, have rigid and often massive limestone skeletons. The focus here is on the stony or true corals, which are especially prominent in shallow tropical waters where they can form biogenic structures (coral reefs) of enormous dimensions.

THE CORAL ANIMAL

While corals are the quintessential builders of modern reefs, they are members of only one of many groups of organisms that contribute to the rich biodiversity found in coral reef ecosystems. Coral reef communities are composed of a mix of species, some beneficial and some detrimental to reef-building activities. The composition of coral communities is controlled by physical and biotic factors whose influences vary over space and time. The physical structure or morphology of coral reefs is also shaped by a subtle interplay of physical and biotic

conditions that demonstrate interocean differences. Of first consideration are the corals themselves, their biology, life history traits, and susceptibilities to various disturbances. An awareness of these attributes will offer some understanding of the adaptations that allow certain species to occur in the intertidal environment.

Corals are members of the order Scleractinia in the class Anthozoa of the phylum Cnidaria. This phylum also includes hydroids, jellyfish, sea anemones, and sea fans. Corals construct a rigid calcareous skeleton over which lies a thin and delicate layer of tissue (Figs. 1A, B, 2). Millions of single-celled dinoflagellate protists, called zooxanthellae, reside as obligate symbionts within the gastrodermal cells of reef-building corals. Approximately 1400 species of corals are presently recognized and are about equally divided between zooxanthellate and azooxanthellate species. Azooxanthellate corals do not contain zooxanthellae, generally have lower rates of calcification, and often occur in cryptic shallow habitats or in deeper waters below the euphotic zone. Recent exploration of deep (50–2000 m

FIGURE 1 (A) A pocilloporid coral of branching colony morphology, Galápagos Islands (-10 m depth, May 2006), colony -60 cm maximum diameter. Photograph courtesy of Graham Edgar. (B) Expanded polyps of *Montastraea cavernosa*, Bonaire (-10 m depth, December 2003), each polyp is -1 cm in diameter. Photograph courtesy of Michael C. Schmale.

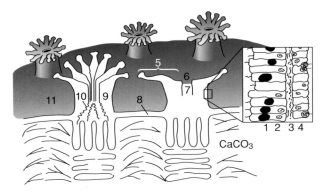

FIGURE 2 Schematic cutaway view of a massive coral colony illustrating relationship of polyps and exoskeleton: (1) gastrodermis (ciliated), (2) zooxanthellae (spherical opaque cells), (3) mesoglea, (4) epidermis (with nematocysts), (5) tentacle, (6) mouth, (7) pharynx, (8) gastrovascular canal, (9) mesentery, (10) mesenterial filament, (11) epidermal surface.

and deeper) offshore areas has revealed numerous coral assemblages that form large reef-like structures.

Reef-building corals are typically colonial and are composed of hundreds to thousands, and sometimes millions, of interconnected polyps. These polyps may contract and withdraw deeply in the skeleton or remain exposed near the colony surface, depending on the species. The internal cavity serves as a gut, where captured particles are digested. Wastes exit through a single opening, the mouth. Each tentacle (typically in multiples of six) is armed with thousands of stinging cells, the nematocysts, which assist in the capture of minute zooplankton and particulate organic carbon of animal origin. The captured food is moved to the mouth by the tentacles and is sometimes assisted by ciliary mucoid action. The saclike interior contains mesenterial filaments, which secrete digestive enzymes to break down ingested food. The mesenterial filaments can also be extruded through the mouth and other openings and are capable of digesting organic matter or other organisms outside the polyp. Carbon photosynthetically fixed by the zooxanthellae is also an important food source for corals. Some coral species satisfy as much as 90–100% of their energy needs from this source. Nitrogen and phosphorus sources from zooplankton, however, are needed for coral tissue growth. Recent molecular genetic studies have revealed that the zooxanthellae that inhabit corals are extraordinarily diverse and have different susceptibilities to solar radiation and temperature stressors. Some reef flat corals host zooxanthellae that are resistant to the harsh conditions that often occur in the intertidal environment. Another defensive mechanism that helps protect shallow-living corals involves the production of mycosporine-like

amino acid pigments that act like a sunscreen in blocking ultraviolet radiation.

Although there are many variations on the theme, the majority of corals reproduce in two ways: by asexual fragmentation and sexually through the fusion of gametes. The fragmentation of colony parts is common in corals, especially in environments with high rates of physical or biotic disturbances. The genetic diversity in some coral populations is very low, because of a predominance of asexual fragmentation and the formation of clones. In sexual reproduction, fertilized eggs may be retained and brooded by the maternal polyp or released into the water column where they spend varying periods developing into larvae (planulae) that eventually settle and metamorphose into juvenile corals. Brooded planulae typically settle close to their colony of origin, whereas planulae that spend days to weeks in the water column may disperse great distances.

During periods of low tidal exposure, physical conditions often reach stressful levels (Fig. 3). At midday, low tides lead to intense solar radiation, desiccation, and the heating of shallow pools, which can exceed the tolerance limits of corals. Such stresses are exacerbated during rainstorms with sudden changes in salinity. Corals often bleach when subject to extreme conditions. Bleaching involves the loss of zooxanthellae or their photosynthetic pigments or both. The tissues of bleached corals remain intact but become transparent, revealing the stark white underlying calcareous skeleton. If bleaching persists for several weeks, coral death often results. However, if conditions return to normal zooxanthellae can increase pigment levels and repopulate coral hosts, thus leading to recovery.

Cyclonic storms and tsunamis that generate unusually high waves can cause significant physical damage to coral reefs. Extreme water motion causes sediment scouring

FIGURE 3 Bleached corals, extreme low tidal exposure, Uva Island, Panamá, February 1992. Photograph courtesy of Susan B. Colley.

and burial of coral colonies as well as impacts by projectiles such as dislodged reef blocks, logs, and debris.

The great majority of coral species occupying shallow reef environments have branching morphologies (Figs. 1A, 4). Species of *Acropora, Pocillopora,* and *Stylophora* are particularly common on Pacific reef flats. On western Atlantic reef flats, calcifying hydrocorals occur as platy colonies *(Millepora complanata)* on exposed seaward reef margins or as finely branching colonies *(Millepora alcicornis)* in protected leeward habitats. Thickets of branching *Porites* spp. colonies also occur on leeward reef flats. Colonies at high energy sites often develop thick, robust skeletons that grow parallel to the direction of water flow. Encrusting species that offer little resistance to water motion also inhabit wave-swept areas (Fig. 4). Laminar *(Acropora* spp.) and massive or mounding colonies *(Porites* spp.) can be found where water motion is reduced (Fig. 4). The "blue coral" *Heliopora,* an octocoral with a dense skeleton of sturdy plates, is also abundant on many Pacific reef flats. *Porites* and *Heliopora* often form microatolls, along with more than 40 other species (Fig. 5A, B). Microatolls form when tissues at the summits of corals are killed during extreme tidal exposures. Colony growth continues laterally where tissues are still largely submerged. The feeding activities of sea urchins, fishes, and other herbivores result in the erosion of central colony surfaces (Fig. 6). Resulting rings of live tissue mark the level of extreme low

FIGURE 5 Microatolls on a Pacific Costa Rican reef flat at low water (A) and a *Porites* microatoll showing erosion on the upper colony surface (B). Photographs by the author.

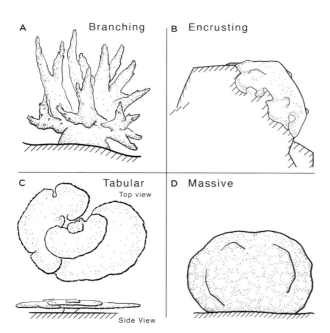

FIGURE 4 Some common coral colony morphologies occurring in the intertidal zone: (A) branching, (B) encrusting, (C) tabular, and (D) massive.

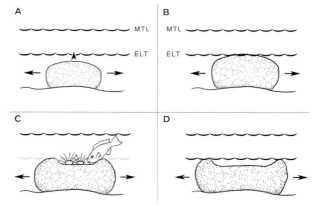

FIGURE 6 Schematic of a microatoll illustrating mode of formation. MTL, mean tide level; ELT, extreme low tide level.

spring tides. This dead/live interface denoting sea level shifts has been employed in documenting tectonic activity and El Niño–Southern Oscillation events.

BIOGEOGRAPHIC PATTERNS

Zooxanthellate corals are mainly restricted to warm tropical waters. Several factors contribute toward high reef coral abundance and rapid growth at low latitudes, notably high sea water temperature, elevated aragonite saturation state, and high year-round light levels. Under such conditions, provided that sedimentation is not too great and oceanic salinity is stable, zooxanthellae promote tissue growth and rapid calcification in their coral hosts. The latitudinal occurrence of coral reefs is generally greater along western ocean margins where favorable warm currents (e.g., Kuroshio, East Australian, and Gulf Stream currents) reach into higher latitudes. In like manner, the distribution of coral reefs is restricted in eastern ocean basins because of cold ocean currents (e.g., Peru and Benguela currents) and upwelling centers.

It is also likely that some sort of biological control limits the occurrence of zooxanthellate corals to tropical seas. For example, symbiotic zooxanthellae appear to be intolerant of the physical conditions at higher latitudes, even though potential host corals do exist. Numerous species of scleractinian corals occur in shallow temperate and polar regions, but these species are not found in the tropics and they do not form a symbiotic relationship with zooxanthellae. By the same token, it may also be that the tropical corals that are able to host algal symbionts are intolerant of conditions at high latitudes and are responsible for these latitudinal limits.

Within the tropics, coral faunas form five more or less homogeneous species assemblages or provinces that are spatially coherent. The largest and most diverse of these is the Indo-Pacific province, which extends from the central and southern Pacific to the western Pacific and across the Indian Ocean to eastern Africa, the Red Sea, and Arabian Gulf. Its center of diversity is in the Philippine and Indonesian archipelagos. In this region, nearly 600 coral species are known. The coral fauna of the eastern Pacific, extending from the Gulf of California (Mexico) to Ecuador, is considerably less diverse. About 40 zooxanthellate coral species are presently known in this region. This relative impoverishment of fauna is due in large measure to its isolation from the central and south Pacific. Ekman's barrier, a 5000- to 6000-km expanse of open ocean waters, separates these regions. In the tropical Atlantic Ocean, some 65 species make up the coral fauna of the Caribbean, Gulf of Mexico, and Bahamian regions. South of the Orinoco and Amazon River drainages, the Brazilian coral fauna consists of about 20 species, and six of these are endemic or unique to this region. The West African coral fauna is also species poor, consisting of only about 15 species and one endemic genus. These present-day faunas are a result of deep evolutionary divergences, dispersions, and extinctions following the fragmentation

of the circumtropical Tethys Seaway (30–15 million years ago), the closure of the Central American Isthmus (3.5–3 million years ago), and more recent Pliocene and Pleistocene climates and their effects on ocean circulation.

CORAL COMMUNITIES

Some of the most arresting and provocative images of coral reef biodiversity were first revealed in W. Saville-Kent's (1893) photographic images of reef-building corals along Australia's Great Barrier Reef tract and the Torres Straits region between Northeast Australia and Papua New Guinea. Great varieties of coral species, colony morphologies and sizes, as well as many associated reef organisms are shown exposed at low spring tidal stands in these early pictures. Before the advent of scuba diving, much of the work on coral reefs was directed toward accessible intertidal areas. The interest in shallow and intermittently exposed reef zones has continued apace, with more recent studies in subtidal reef areas and to the greatest depths of occurrence of zooxanthellate, reef-building corals. Here the shallowest assemblages of zooxanthellate corals are considered, those species with symbiotic algae that occur on wave-swept tropical reef shores.

While corals contribute importantly to the building of reefs, the complex living spaces thus formed provide a multitude of microhabitats for numerous organisms other than coral. Many sessile organisms, such as algae, sponges, other cnidarians, bryozoans, mollusks, and tunicates, share habitat space with corals and are therefore potential competitors. Since many of the neighboring animals are suspension feeders, they can capture coral larval stages and thus play a role as predators. The unusually high inter- and intraphyletic diversity on coral reefs raises the question: How is this elevated biodiversity generated and maintained? Certain species possess adaptations that would seem to work to their advantage in monopolizing space, for example, (1) an ability to resist various physical stresses, (2) high recruitment rates, (3) high growth rates, (4) overtopping capabilities, (5) ability to produce toxic or noxious substances for resisting competition and predation, and so on. Why do these species not win out and dominate coral reef communities?

A much-celebrated study by J. H. Connell (1978) on coral reef flat assemblages in Australia demonstrated that community succession was interrupted by numerous sorts of disturbances that promoted diversity. For example, if a potentially dominant competitor were to be accidentally killed by storm-tossed debris, this would create a patch that could be invaded and occupied by a weaker competitor. Reef habitats with low rates of disturbance would potentially support a few strong competitors. At high rates of disturbance, only a small number of species might be expected to survive. Maximum diversity would be expected at some intermediate level of disturbance. This model is known as the intermediate disturbance hypothesis and has been called upon to explain diversity patterns in many different marine communities as well as tropical rain forests.

Despite the many physical challenges that shallow-living corals experience in wave-swept zones, they do avoid certain biotic hazards in this environment that often affect subtidal corals. For example, mollusc and echinoderm corallivores that prey on coral tissues generally do not invade the intertidal zone because they cannot maintain a hold on the substrate under the turbulent conditions that prevail there. Since the late 1960s several major crown-of-thorns sea star outbreaks have devastated enormous areas of coral in the Indo-Pacific region. However, most coral populations inhabiting reef flats are spared because this corallivore is unable to maintain a hold on wave-swept substrates. Corals surviving in this spatial refuge can therefore serve as source populations to replace those killed by the sea star.

An important indirect effect involving fishes that invade the intertidal zone at high tide can also benefit corals. Coralline, filamentous, and macroalgae are abundant and frequently compete with corals for space. Where overfishing has not occurred, at high tide large schools of fish herbivores (e.g., parrotfishes and surgeonfishes) flood onto reef flats and consume algae, thus indirectly limiting competition with corals.

Due to the high topographic complexity of shallow reef substrates, water-filled channels, pools, and depressions often form at low tide. Where these occur, they can provide refuges for organisms during periods of emersion. Diverse communities of algae, invertebrates, and fishes find shelter in such settings, providing solar heating is not too intense. Such patches also contribute toward the high biotic diversity of reef flat coral communities.

CORAL REEF MORPHOLOGY

The kinds of species inhabiting shallow reef zones are very much influenced by wave action. On some Pacific atolls that experience a marked gradient in wave energy from seaward to leeward exposures, as many as 11 coral zones have been recognized. Crustose coralline algae generally have a greater tolerance than corals do to windward reef exposures subject to high wave assault. Under such conditions certain species of calcareous algae predominate and can build impressive algal ridges that may reach 1 to 2 meters above

sea level (Fig. 7A). Algal ridges are prominent on many reefs in the Pacific and Indian Oceans where wave action is generally high and seasonally constant. In areas with diminished wave energy, for example, near the equator (in the doldrums belt) or on the leeward sides of islands, coralline algae buildups are less pronounced (Fig. 7B). Finally, the seaward exposures of reefs that receive little wave action are devoid of calcareous algal buildups (Fig. 7C). While many coral reefs in the Caribbean Sea are lacking algal ridge structures, some reefs located at trade wind latitudes with persistent high seas can have well-developed algal ridges similar to those in the Indo-Pacific region.

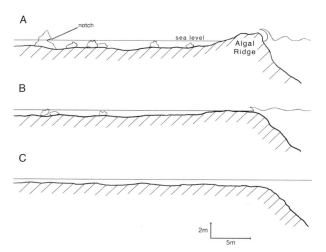

FIGURE 7 Profiles of seaward reef margins illustrating the buildup of an algal ridge under conditions of high to moderate wave assault.

On reef shores with emergent limestone structures, such as cliffs, terraces, and large storm-deposited blocks, the erosive action of numerous organisms that feed on and bore into these substrates near sea level produce notches, nips, and variously sculpted undercut features (Fig. 7A). The organisms responsible for this erosion—for example, boring algae, sponges, and peanut worms, and herbivorous chitons, snails, and sea urchins—are collectively referred to as bioeroders. Their effects are most strongly felt along leeward or backreef zones where they are most abundant. Wave action is reduced in these areas, thus allowing animal herbivores to maintain a purchase on rock surfaces. The rasping action of their mouth parts removes both algae and the underlying coral rock substrate.

OUTLOOK

In a global warming scenario, rising sea level, if not too rapid, would favor vertical coral growth and an upward expansion of reef structures. However, if sea temperatures continue to rise and solar radiation levels remain high or increase, one could expect increasing incidences of coral bleaching and mortality. Also declining aragonite saturation state caused by rising atmospheric CO_2 will reduce the reef-building capacity of corals. With the disappearance of zooxanthellate corals, other benthic taxa resistant to such changes could become more abundant and lead to phase shifts in community structure. With the loss of calcifying, reef-building organisms and continuing bioerosion, unique reef habitats would be lost. This would depress biodiversity. Additionally, with the exponential increase in human populations there will be a greater demand for seafood products, higher levels of coastal development, and decreased coastal water quality, and therefore overall negative impacts on nearshore coral communities. Conditions favoring coral growth on tropical shores are presently in decline. Whether these inimical human-induced trends can be mitigated or reversed remains to be seen.

SEE ALSO THE FOLLOWING ARTICLES

Biodiversity, Maintenance of / Cnidaria / Octocorals / Salinity Stress / Storm Intensity and Wave Height

FURTHER READING

Birkeland, C. 1997. *Life and death of coral reefs.* New York: Chapman & Hall.

Connell, J. H. 1978. Diversity in tropical rain forests and coral reefs. *Science* 199: 1302–1310.

Done, T. J. 1983. Coral zonation: its nature and significance, in *Perspectives on coral reefs.* D. J. Barnes, ed. Manuka, Australia: Brian Clouston, 107–147.

Dubinsky, Z. 1990. *Coral reefs.* Ecosystems of the World 25. Amsterdam: Elsevier Science.

Karlson, R. H. 1999. *Dynamics of coral communities.* Dordrecht, Netherlands: Kluwer Academic.

Sebens, K. P. 1994. Biodiversity of coral reefs: what are we losing and why? *American Zoologist* 34: 115–133.

Stoddart, D. R. 1969. Ecology and morphology of recent coral reefs. *Biological Reviews* 44: 433–498.

Veron, J. E. N. 2000. *Corals of the world,* Vols. 1–3. Townsville, Qld: Australian Institute of Marine Science.

CRABS

JONATHON STILLMAN

San Francisco State University

Crabs are decapod crustaceans that are common inhabitants of intertidal rocky shores in tropical and temperate regions of the world. Although some types of common rocky intertidal-zone crabs are easily recognized, a great diversity of cryptic and obscure species inhabit rocky shores. Across crab species there is huge diversity of body forms, behaviors, physiologies, and life histories.

GENERAL MORPHOLOGY AND PHYSIOLOGY

External Anatomy

Crabs have an external skeleton (exoskeleton) made up of layers of cuticle composed of the polysaccharide chitin, proteins, lipids, and minerals such as calcium carbonate. Dorsally the "body" of the crab (cephalothorax) is covered by the carapace, which is the largest single piece of exoskeleton in most crab species. The five pairs of main appendages (pereopods) of most crabs are one pair of claws (chelipeds), and four pairs of walking legs. The pereopods are numbered 1 to 5, 1 being the most anterior (claws) and 5 being the most posterior legs. Between the two chelipeds at the anterior end of the crab are the compound eyes (on stalks of varying length depending on species), the antennae, and the antennules. In general the antennae are longer than the antennules.

On the ventral surface of the crab are the abdomen and telson. The end of the digestive tract is at the tip of the telson. In males, the abdomen is narrower (often dramatically) than in females, because females brood their developing embryos under the abdomen. Underneath the abdomen are the pleopods, paired segmented appendages. Males usually have one (in anomuran crabs) or two (in brachyuran crabs) pairs of pleopods, one pair of which is very long. The long pleopods are called gonopods and are the organs used to deliver sperm into the gonopores (openings that connect to the seminal vesicles) of the female during reproduction. Females have large pleopods on four of the six abdominal segments, and these often have long setae or hairs. They function to hold the developing brood of embryos (from hundreds to millions of embryos, depending on the species) when the crab is ovigerous, or "in berry." At the anterior end of the ventral surface is the mouth, the most exterior appendage of which is the third maxilliped.

Internal Anatomy (Conspicuous Structures Only)

Underneath the exoskeleton is a thin layer of epidermis (skin) that separates the hemocoel (viscera) from the innermost layers of cuticle. When crabs grow, they molt, or shed their exoskeleton. In preparation for molting, the crab forms a new epidermis and cuticle underneath the old one and simultaneously digests and resorbs much of the material from the old exoskeleton. During molting the old exoskeleton splits open between the cephalothorax and abdomen, the carapace lifts up, and the soft crab backs out. The soft-shelled crab quickly uptakes large amounts of water to swell the soft exoskeleton to its new size, and it deposits the resorbed minerals from its old shell to mineralize and harden the new exoskeleton.

The gills, lamellar and often "spongy" structures, lie on either side of the hemocoel within the branchial chamber. This chamber, although underneath the carapace, is functionally outside of the body of the crab so that the gills may be bathed with external water for exchange of gases and ions across their highly lamellar surface. Within the hemocoel, the heart is located centrally on the dorsal surface just posterior to the anterior–posterior midline. Because crabs have a copper-based respiratory pigment (hemocyanin), their blood and highly vascularized tissues are not red. Thus, it is difficult to discern the arterial systems, although these are well developed in most species. In the central anterior region of the hemocoel is the cardiac stomach, where sorting and processing of ingested food occurs. Small food particles are shuttled into the digestive glands (hepatopancreas), paired large yellow-brownish structures at the left and right sides of the anterior region of the hemocoel. In female crabs, developing ovaries will be observed lying on top of the hepatopancreas. The color of the ovaries ranges from yellow-orange to dark purple depending on the crab species and the stage of development of the ovaries.

CRAB DIVERSITY

Crabs fall into two taxonomic orders: Anomura and Brachyura. Anomuran crabs, often called half-crabs or false crabs, include hermit crabs, king crabs, umbrella crabs, mole crabs, and porcelain crabs (Fig. 1). Brachyuran crabs, often called true crabs (Fig. 2), include shore crabs, stone crabs, kelp crabs, sheep crabs, spider crabs, decorator crab, rock crabs, snow crabs, tanner crabs, and many others. Gross anatomical differences between anomuran and brachyuran crabs make it easy to distinguish these two groups. Decapod crustaceans are so named because they have "ten feet," or five pairs of leg and claw appendages. Brachyuran crabs always possess five fully developed paired appendages, four pairs of legs and one pair of claws (Fig. 2). In anomuran crabs, generally one or more pairs of legs are dramatically reduced in size and may be adapted for alternative functions. For example, the porcelain crabs have three pairs of walking legs used for locomotion, and the fourth (most posterior) pair is reduced in size and is used as a cleaning appendage (Fig. 2E), and the hermit crabs have only two pairs of walking legs used for locomotion, because the third and fourth pairs are used to hold the crab into its snail shell. Anomuran and brachyuran crabs also differ in the relative location of the

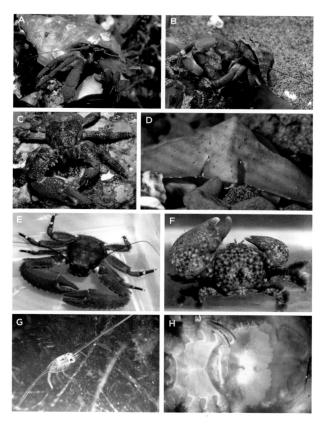

FIGURE 1 Anomuran crabs encountered in the rocky intertidal zone: (A) hermit crab, *Pagurus granosimanus*, (B) hermit crab, *Pagurus hirsutiusculus*, (C) granular-claw crab or soft-bellied crab, *Oedignathus inermis*, (D) umbrella crab, *Cryptolithodes sitchensis*, (E) porcelain crab, *Petrolisthes coccineus*, (F) porcelain crab, *Pachycheles grossimanus*, (G) *Petrolisthes cinctipes* zoea larvae (planktonic), (H) porcelain crab *Petrolisthes cabrilloi* with abdomen pulled back to reveal infection of the rhizocephalan barnacle parasite *Lernaeodiscus porcellanae*. Photographs A–D courtesy of Dr. Gregory Jensen.

eyes and the antennae. In anomuran crabs, the antennae are posterior to the eyes (e.g., Fig. 1A), whereas in brachyuran crabs, the antennae are anterior to the eyes and are thus in-between the pairs of eyes (e.g., Fig. 2E). In addition, anomuran crabs tend to have very long antennae, whereas the antennae of brachyuran crabs are very short and thus are not visible from many of the photographs in Fig. 2. Following are short descriptions of groups of anomuran and brachyuran crabs commonly encountered in rocky intertidal ecosystems, along with some general information about their ecology, physiology, and behavior. Although the majority of these examples are drawn from the North American Pacific coastline to highlight the diversity found at one rocky shore, the taxonomic groups listed here are also representative members of rocky-shore communities in other regions of the world.

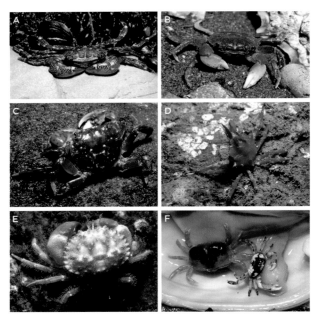

FIGURE 2 Brachyuran crabs encountered in the intertidal zone: (A) striped shore crab, *Pachygrapsus crassipes*, (B) yellow shore crab, *Hemigrapsus oregonensis*, (C) purple shore crab, *Hemigrapsus nudus*, (D) graceful kelp crab, *Pugettia gracilis*, (E) pigmy rock crab, *Cancer oregonensis*, (F) female (left) and male (right) mantle pea crab, *Pinnixa faba*, in their host, a gaper clam *(Tresus capax)*. Female *P. faba* are similar in appearance to *Fabia subquadrata*, found occasionally in the mantle cavity of the intertidal mussel *Mytilus californianus*. Photographs courtesy of Dr. Gregory Jensen.

Anomuran Crabs of Rocky Intertidal Shores

Decapod crustaceans in the infraorder Anomura that are commonly observed in rocky intertidal zone and tidepool habitats include species from the following three families: Paguridae (right-handed hermit crabs), Lithodidae (stone crabs and king crabs), and Porcellanidae (Porcelain crabs).

PAGURIDAE (RIGHT-HANDED HERMIT CRABS)

These crabs (Fig. 1A, B) are perhaps the best known and most common of the crabs found in tidepools. These crabs inhabit evacuated shells of snails (gastropod molluscs). When one observes a snail shell moving rapidly in a tidepool, this shell is very likely inhabited by a hermit crab. Paguridae have claws that are dissimilar, with one claw being much larger than the other. This larger claw can be used by the crab as a blockade to the opening of the snail shell when the crab has fully retracted within the shell, which these crabs often do when handled by humans or other potential predators. Hermit crabs tend to have only two pairs of walking legs that are used for locomotion; the other two pairs are diminutive and are used by the crab to move within the snail shell. The crabs

have a long and soft abdomen that is curled into a spiral to match the snail shell spiral.

Hermit crabs can be found in both marine and terrestrial habitats, and the terrestrial forms have highly developed adaptations for reduction of water loss (desiccation) and for aerial respiration. In general, the only intertidal-zone hermit crabs are found in tidepools. Hermit crabs eat algae that they graze from rocks, tiny benthic invertebrates (e.g., newly settled mussels), and carrion that they scavenge.

Hermit crabs must molt in order to grow, just like any other decapod crustacean. For hermit crabs, growth is complicated by the fact that they must find a suitably larger snail shell to inhabit. While not actively feeding, hermit crabs spend much of their time searching for snail shells that are larger or of better quality than their current shell. Although not commonly observed on rocky shores, where shells are common, on snail-shell-limited tropical sandy beaches it is not uncommon to observe long "trains" of 10–20 hermit crabs, each gripping its sequentially bigger neighbor. At the "engine" of the train is a hermit crab with an empty snail shell that it will move into. When this "engine" crab switches shells, all of the other individuals in the train also trade up shells with their neighbor so that there is an empty snail shell (presumably the smallest and lowest-quality shell) at the "caboose" when the crabs have finished the trades.

Hermit crabs are very diverse, and the hermit crab "form" is seen in multiple families of Anomuran crabs, including Coenobitidae (land hermit crabs), Diogenidae (left-handed hermit crabs), Paguridae (right-handed hermit crabs), and Parapaguridae (deep-water hermit crabs). According to genetics-based analyses by Morrison and others (2002) in decapod crustaceans, focusing on anomuran crabs, a single origin of the hermit crab "form" evolved only once, and the family Lithodidae (stone and king crabs) group evolved within the various "hermit crab" families. Thus, we can consider king crabs and other lithodid crabs to be hermit crabs as well, but hermit crabs that have stopped inhabiting snail shells and have adopted a more typical "crab" form. Although Lithodid crabs may actually be hermit crabs, they are listed separately next, in accord with most systematists.

LITHODIDAE (STONE AND KING CRABS)

These crabs (Figure 1C, D) include genera with a typical crab form, such as the granular-claw or soft-bellied crab, *Oedignathus inermis* (Fig. 1C), as well as oddities with very unusual forms, such as the umbrella crab, *Cryptolithodes sitchensis*. The most recognizable lithodid crab is the Alaskan king crab, *Paralithodes camtschatica*, although this crab is rarely encountered on rocky intertidal-zone shores.

O. inermis is a common inhabitant of the low intertidal zone of wave-swept rocky shores in the northern Pacific rim. These crabs can be found in crevices, although all that would be observable to most is their purple claw. These crabs are predators on mussels, which they crush with their large claws, or amphipods and worms, which they catch with their small claws. Like many lithodid crabs, the abdomen of *O. inermis* is not hardened, although it is held tucked under the body.

C. sitchensis is often found in the low intertidal zone of wave-sheltered rocky shores. These crabs are very unusual in that their carapace nearly completely covers the three pairs of walking legs; only the distal tips of the third pair is visible from under their shells. These crabs have a well-formed exoskeleton around all of their appendages, including the abdomen. The carapace can be highly variable in color, including dazzling shades of orange, red, and pink.

PORCELLANIDAE (PORCELAIN CRABS)

Porcelain crabs (Fig. 1E–H) are found throughout the world's oceans, from deep-water habitats to coral reefs, to intertidal zones, where species can be found living nearly as high on the shore as the high-intertidal brachyuran crabs. These crabs are generally small, with carapace widths ≤1.5 cm in most cases, but they can be very abundant. Porcelain crabs are most commonly found living underneath stones in intertidal wave-sheltered boulder fields, or living on wave-swept rocky shores underneath mussel beds or deep in crevices. Some species can be found living in very high densities: *Petrolisthes cinctipes* has been reported as high as 3800 individuals per square meter of mussel bed in the northeastern Pacific, and *Petrolisthes armatus* has been reported in densities three times that high in oyster beds of the U.S. Atlantic coast.

Porcelain crabs have very large claws, but they do not use these claws for feeding, because they are filter feeders and use their fanlike mouthparts to filter particles out of the water. The large claws may be used to secure space for the mouthpart filters to function, or they may be used to scrape food up from the substratum into the water column. Some species of porcelain crab are very fast-moving, aggressive, and "delicate." If these crabs are even gently handled, they will autotomize or "drop" claws or legs, presumably an escape response in nature. Other species are slower and more robust to handling and will not drop their appendages so readily.

The two most common genera in the family Porcellanidae are *Petrolisthes* (Fig. 1E) and *Pachycheles* (Fig. 1F). The genus *Petrolisthes* is by far the most physiologically diverse as well as speciose (with approximately 110 species,

approximately one-third of the species in the family). One remarkable feature of the diversity across species of *Petrolisthes* is patterns of vertical zonation of species. On wave-sheltered rocky shores throughout the eastern Pacific, there are at least two, but as many as 12 or more (in the tropics), sympatric species of *Petrolisthes* living in different vertical zones. The high diversity, widespread distribution across temperate and tropical habitats, and vertical zonation patterns make these crabs fascinating subjects for studies of physiological adaptation to the harsh environment of the intertidal zone.

Crabs in the genus *Pachycheles* are remarkable because they are almost always found in size-matched male–female pairs. These crabs are very sedentary and likely live most of their lives in the same small rock crevice or hole. Thus, it is likely that *Pachycheles* are monogamous and that they pair-bond for life in nature. Of note is that often there is a small shrimp that lives along with the pair of *Pachycheles* in their rock hole or crevice.

Porcelain crabs are also remarkable because of their planktonic zoea larvae (Fig. 1G). Although these larvae will not be seen in the rocky intertidal shore, they are commonly caught in near-shore plankton nets. They are remarkable because of their long stiff spines, thought to provide buoyancy as well as escape from predation, as the long spines make it difficult for predators to eat them.

Porcelain crabs, along with many other groups of anomuran and brachyuran crabs, are subjected to parasitic castration by a rhizocephalan barnacle parasite. This barnacle infects the host crab and spreads throughout its body. The parasite will feminize male crabs, causing them to grow a larger abdomen upon subsequent molts. When the barnacle is ready to reproduce, it extrudes a reproductive structure where the crab would have held a brood of embryos (Fig. 1H). The crab aerates and protects broods for the barnacle, hatching out larval barnacle parasites into the water column, just as it would have done for its own larvae.

Brachyuran Crabs of Rocky Intertidal Shores

Brachyuran crabs, or true crabs, are found throughout the vertical stretches of rocky intertidal shores and occupy nearly every ecological niche. Many families of brachyuran crabs can be commonly found on rocky shores and in tide pools, including the Grapsidae (shore crabs), Majidae (spider crabs), Cancridae (rock crabs), Xanthidae (mud crabs), and Pinnotheridae (pea crabs).

GRAPSIDAE (SHORE CRABS)

On rocky shores throughout temperate and tropical regions, the crabs observed moving over rocks and hiding in crevices in the high intertidal and splash zone are in the family Grapsidae (Fig. 2A–C). *Pachygrapsus* and *Grapsus* are the most common genera of crabs that live highest on the shore, in the splash zone. These crabs are usually visible in crevices during the day but are principally active at night, when they emerge to scavenge and graze algae off the rocks. They are extremely fast and agile crabs, and are usually dorsoventrally flattened, which likely assists them from being swept away underneath each crashing wave. The high intertidal grapsid crabs are principally air breathers, and they have reduced gills and enlarged branchial chambers, the interior lining of which functions as a respiratory structure similar to a lung. The crabs must keep this structure moist, but they do not keep it full of water.

MAJIDAE (SPIDER CRABS)

These crabs are very diverse, but relatively few spider crabs are found in the rocky intertidal zone. In tidepools, it is possible to find decorator crabs and kelp crabs (Fig. 2D). These crabs generally have long legs and are cryptic, their carapaces either blending in with the kelp or being well decorated with other intertidal organisms such as algae and bryozoans. Some species of majid crabs actively decorate themselves with pieces of kelp or other organisms and thus are often referred to as "decorator crabs."

CANCRIDAE (ROCK CRABS)

All of the crabs in this family (Fig. 2E) are in the genus *Cancer*, including *Cancer magister*, the Dungeness crab, which although not found in rocky intertidal habitats, is an important economic species in the northeastern Pacific Ocean. On wave-swept rocky intertidal zones, *Cancer oregonensis* is often encountered living in rock holes, its round carapace often fitting precisely within the rock hole. *Cancer productus* is commonly encountered underneath stones in wave-sheltered intertidal boulder fields. These crabs are aggressive, active predators and will crush other crabs, snails, and bivalve molluscs with their powerful claws.

XANTHIDAE (MUD CRABS)

Although often encountered under rocks in wave sheltered boulder habitats as well as in estuarine habitats, these crabs (not pictured) are principally absent from wave-swept rocky shores. These crabs generally have large powerful claws, which they use to crush prey items such as snails and barnacles.

PINNOTHERIDAE (PEA CRABS)

Pea crabs (Fig. 2F) are sometimes found living within the mantle cavity of bivalve molluscs (e.g., *Mytilus californianus*) living on wave-swept rocky shores. Depending on the species, pea crabs are also found living in other locations, including burrows of shrimps and worms, as well as on chitons, sea cucumbers, and other organisms. The life cycles of pea crabs include two stages: one inside the host, where crabs have extremely reduced appendages and very thin membranous exoskeletons (as in the female on the left in Fig. 2F), and one in the water column, either as larvae or as adults during mating, depending on the species. In some species, only adult females are found in the bivalve host (e.g., *Fabia subquadrata* in *M. californianus*), whereas in other species both female and male forms are found in the host (Fig. 2F).

SEE ALSO THE FOLLOWING ARTICLES

Camouflage / Hermit Crabs / Locomotion: Intertidal Challenges / Predation / Rhizocephalans

FURTHER READING

Abbott, D. P. 1987. *Observing marine invertebrates.* Stanford, CA: Stanford University Press.

Brusca, R. C. 1980. *Common intertidal invertebrates of the Gulf of California.* Tucson: University of Arizona Press.

Jensen, G. C. 1995. *Pacific Coast crabs and shrimps.* Monterey, CA: Sea Challengers.

Kozloff, E. N. 1983. *Seashore life of the northern Pacific coast.* Seattle: University of Washington Press.

Light, S. F., R. I. Smith, F. A. Pitelka, D. P. Abbott, and F. M. Weesner. 1954. *Intertidal invertebrates of the central California coast.* Berkeley and Los Angeles: University of California Press.

Morris, R. H., D. P. Abbott, and E. C. Haderlie. 1980. *Intertidal invertebrates of California.* Stanford, CA: Stanford University Press.

Morrison, C. L., A. W. Harvey, S. Lavery, K. Tieu, Y. Huang, and C. W. Cunningham. 2002. Mitochondrial gene rearrangements confirm the parallel evolution of the crab-like form. *Proceedings of the Royal Society Series B Biological Sciences* 269: 345–350.

Ricketts, E. F., J.Calvin, and J. W. Hedgpeth. 1968. *Between Pacific tides,* 4th ed. Stanford, CA: Stanford University Press.

Stillman, J. H. 2003. Acclimation capacity underlies climate change susceptibility. *Science* 301: 65.

Waterman, T. H. 1960. *The physiology of crustacea*, Vol. I and II. New York: Academic Press.

CURRENTS, COASTAL

LIBE WASHBURN
University of California, Santa Barbara

Coastal currents flowing along the boundaries of land masses disperse larvae of many coastal species to distant habitats. Larvae colonizing new habitats arrive by coastal currents. Many urban areas rely on coastal currents to dilute pollutants and transport them to deeper offshore waters. In contrast to open-ocean currents, the dynamics of coastal currents are strongly affected by bathymetry, coastline features, and tides.

COASTAL CURRENT FORCING

Because coastal currents flow over the rotating earth, they experience a sideways force, called the Coriolis force, acting 90 degrees to the right of the current direction in the Northern Hemisphere (Fig. 1A). In the Southern Hemisphere the Coriolis force acts to the left of the flow direction (Fig. 1B). The Coriolis force acts over the entire ocean depth and is therefore a "body force." For short-duration water movements such as surface gravity waves approaching the seashore, the Coriolis force is not significant, but for more persistent flows such as coastal currents it is a dominant force. The strength of the Coriolis force is proportional to flow speed and varies with location on the earth; it is strongest at the poles and zero along the equator. Its strength increases with increasing current speed. The Coriolis force acts on existing water movements and does not produce coastal currents by itself.

Another important body force, and one that does produce coastal currents, is the tide-raising force arising from the gravitational attraction of the moon and sun. Although the tide-raising force is very weak compared with other forces such as the earth's gravity, it often dominates the acceleration of ocean currents because it acts horizontally. The tide-raising force of the sun is similar to the moon's but only about half as strong.

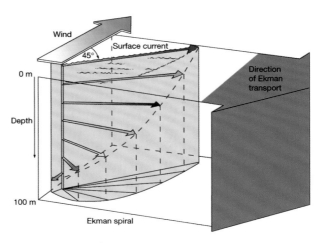

FIGURE 2 The spiraling pattern of currents (Ekman spiral) produced as wind (green arrow) blows over the sea surface in the Northern Hemisphere. The current direction at the surface is 45 degrees to the right of the wind direction. With increasing depth, the pattern of surface currents spirals counterclockwise (colored arrows). The depth of the spiral pattern (Ekman layer) is up to 100 meters. The net transport of water in the Ekman layer is 90 degrees to the right of the wind direction (brown arrow). In the Southern Hemisphere, the surface current is 45 degrees to the left of the wind, the spiraling pattern of currents turns clockwise with increasing depth, and the net transport is to the left of the wind direction. (Pearson Prentice Hall, Inc., 2004.)

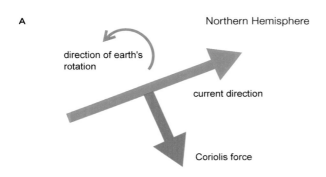

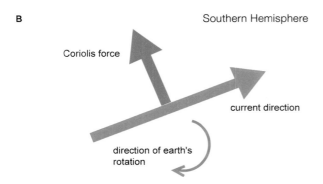

FIGURE 1 (A) A coastal current (blue arrow) in the Northern Hemisphere and the resulting Coriolis force (red arrow). (B) As in (A) but for the Southern Hemisphere. Curved arrows show the directions of the earth's rotation in a horizontal plane in each hemisphere.

Wind is an important driving force for many coastal currents. Because the wind acts directly only on the sea surface, it is a "boundary force." As wind blows over the ocean it exerts a stress on the sea surface that accelerates surface waters downwind. Turbulence generated by the wind transfers momentum downward, accelerating successively deeper ocean layers. If the wind persists for several hours, the resulting ocean currents are affected by the Coriolis force. Together with turbulence, the wind stress and Coriolis force produce a right-turning (in the Northern Hemisphere) spiral, over which the current speed diminishes with depth (Fig. 2). At the surface the current direction is 45 degrees to the right of the wind. (In the Southern Hemisphere the current spiral is left-turning and the surface current is 45 degrees to the left of the wind.) The depth range of the spiraling currents is called the Ekman layer and is up to ~100 meters deep. Within the Ekman layer, net water transport (Ekman transport) is to the right of the wind direction in the Northern Hemisphere (large brown arrow, Fig. 2); in the Southern Hemisphere the Ekman transport is to the left of the wind direction. Ekman layers also form above the sea floor in coastal currents as a result of the frictional stress at the sea floor acting opposite the current direction above the Ekman layer. Close to shore in shallow water the surface and bottom Ekman layers can merge to produce a single, vertically well-mixed water column.

Another important driving force of coastal currents is horizontal pressure gradients, the variation in pressure from one location to another at the same depth. Because pressure forces act on boundaries of flowing water masses, they are a boundary force. Horizontal pressure gradients result from hydrostatic pressure variations caused by differences in sea surface elevation and by internal differences in seawater density. Typical sea level differences are on the order of a few centimeters over horizontal distances of several kilometers, so the resulting sea surface slopes are very small. Pressure gradients arise both along and across coastal boundaries and produce flow components in both directions. For persistent coastal currents, pressure gradients are typically balanced by the Coriolis force in the so-called geostrophic balance.

The geostrophic balance is a fundamental force balance controlling most ocean currents. Figure 3 shows a cross section through a coastal current in geostrophic balance in the Northern Hemisphere. The sloping sea surface generates a horizontal pressure gradient force that is balanced by the Coriolis force acting in the opposite direction. The resulting current direction is oriented so that the Coriolis force is 90 degrees to its right (i.e., for a current flowing south, the Coriolis force acts to the west). Coastal currents often have salinity and temperature differences with respect to offshore ocean waters. These produce horizontal density gradients and contribute to the horizontal pressure gradient in the geostrophic balance.

UPWELLING COASTAL CURRENTS

In coastal ocean regions where the wind's stress on the surface water has a component parallel to the coastline, the resulting Ekman transport has a component perpendicular to the coastline. Since no ocean flow may cross the shore, a vertical flow must develop to compensate for water moved cross-shore within the Ekman layer. For winds blowing parallel to shore, with the coastline to the left of the wind direction in the Northern Hemisphere, the Ekman transport is offshore; in the Southern Hemisphere this occurs if the coastline lies to the right of the wind direction. In each case this produces upwelling near shore. For coastlines to the right of the wind direction in the Northern Hemisphere, or to the left of the wind direction in the Southern Hemisphere, the Ekman transport is onshore. This produces downwelling near shore. Figure 4A and B show examples of downwelling and upwelling, respectively, for winds blowing parallel to a coastline in the Northern Hemisphere.

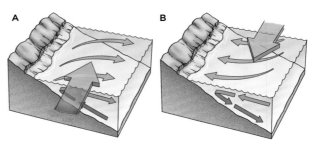

FIGURE 4 Wind (green arrows) blowing over the sea surface produces Ekman transport (dark blue arrows) to the right of the wind direction in the Northern Hemisphere. (A) When winds blow with the coastline to the left of the wind direction in the Northern Hemisphere, the Ekman transport is offshore and upwelling (purple arrow) occurs near shore. (B) When winds blow with the coastline to the right of the wind direction in the Northern Hemisphere, the Ekman transport is onshore and downwelling (purple arrow) occurs near shore. (Pearson Prentice Hall, Inc., 2004.)

In addition to the near-surface Ekman transport, upwelling- and downwelling-favorable winds generate deeper coastal currents flowing along shore. During upwelling, as water is moved offshore in the Ekman layer, coastal sea level declines and the sea surface slopes upward offshore (e.g., Fig. 2). This sets up a shoreward pressure gradient force that, over the course of about a day, comes into geostrophic balance with a seaward Coriolis force. A coastal current in the downwind direction arises such that the seaward Coriolis force acting on it balances the

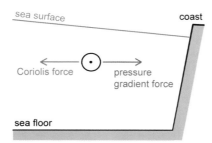

FIGURE 3 A cross section through a coastal current in the Northern Hemisphere. Because the sea surface slopes upward offshore, the pressure gradient force (green arrow) points onshore. The flow is in geostrophic balance, so the Coriolis force (red arrow) balances the pressure gradient force. The black circle with a dot in the center indicates a current vector out of the page.

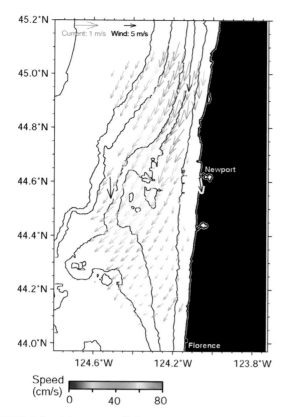

FIGURE 5 Coastal currents off Oregon (USA) during summertime upwelling. Black arrows indicate wind speed and direction; colored arrows indicate current speed and directions. Speed scales for the wind and current arrows are given at the top of the figure. Current speeds are also indicated by the color bar. Solid black lines are depth contours at depths of (moving offshore) 50, 80, 100, 150, 200, 500, and 1000 meters. Adapted from Kosro (2005).

effect by transporting nutrient-poor, offshore waters to coastal ecosystems. Downwelling during storms adds to coastal flooding by producing onshore Ekman transport and higher coastal sea levels. Onshore flows resulting from downwelling deliver larvae to nearshore habitats.

BUOYANCY-DRIVEN COASTAL CURRENTS

Coastal currents can also form in response to "buoyancy forcing" resulting from strong horizontal density gradients in the coastal ocean. The discharge of lower-density waters into the coastal ocean from freshwater river outflows or tidal flows from brackish estuaries are important sources of buoyancy forcing. Other sources occur in fiordic systems such as along the coasts of Norway and Alaska, where numerous creeks and rivers act as distributed sources of freshwater, creating extensive regions of horizontal density gradients.

Major rivers such as the Yangtze in China or the Columbia in the United States are examples of point sources of low-density runoff to the coastal ocean. These rivers form large, low-salinity plumes in coastal waters. Often the rivers transport sediments, so their plumes are visible in satellite imagery. Figure 6 shows the plume of the Yangtze River as it flows into the East China Sea. The plume is visualized by high sediment concentrations in surface waters.

Near the mouth of a river outflow, low-density river waters flow offshore over the underlying denser ocean

shoreward pressure gradient force. Figure 5 shows a strong coastal current of this sort resulting from upwelling-favorable winds along the Oregon coast. Surface coastal currents, shown by arrows in Fig. 5, are derived from high frequency radar measurements.

During downwelling the directions of forces and currents reverse: Ekman transport is onshore, causing elevated coastal sea level; a seaward pressure gradient develops balanced by a shoreward Coriolis force. The resulting coastal current flows in geostrophic balance, and (because the direction of the wind has reversed) the current still flows downwind. Coastal currents resulting from cycles of upwelling and downwelling are common around the world's coastlines. Upwelling is critical for replenishing nutrients lost to phytoplankton growth in surface waters of coastal ecosystems. The resulting coastal currents distribute these nutrients along shore in coastal waters. In contrast, downwelling can have the opposite

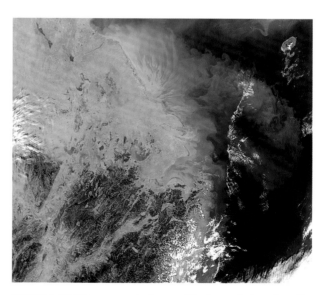

FIGURE 6 Satellite ocean color image of the buoyant plume from the Yangtze River as it discharges into coastal waters of the East China Sea. Brown and green areas indicate high sediment concentrations. White areas are clouds. The rightward turning of the plume offshore may result from the Coriolis force. (Courtesy of NASA.)

waters but are soon deflected by the Coriolis force to the right of the flow direction in the Northern Hemisphere (in the Southern Hemisphere they are deflected to the left of the flow direction). The rightward turning of the offshore portion of the Yangtze River plume in Fig. 6 may result from deflection by the Coriolis force. Often this deflection causes a river flow to reattach to the coastline and form an alongshore, low-salinity coastal current.

The strength of coastal currents resulting from river discharge is often highly variable because of changes in seasonal rainfall. Wind direction also strongly affects the coastal currents produced by river plumes. Upwelling-favorable winds disperse and mix the plumes into offshore waters; downwelling-favorable winds push the plumes onshore and concentrate coastal currents near shore.

TIDALLY DRIVEN COASTAL CURRENTS

In seas on continental shelves, tides produce strong currents that oscillate on diurnal and semidiurnal time scales. Tidal amplitudes over continental shelves are typically much larger than in the open ocean because of these seas' shallower depths. Tide-raising forces of the moon and sun accelerate ocean waters to form tide waves, which propagate in the deep ocean and move over continental shelves as long-wavelength, shallow-water waves. (Note that these tide waves are different from the tsunamis caused by earthquakes, which formerly and popularly were misnamed "tidal waves.")

As sea level rises and falls as a result of the passage of tide waves, horizontal pressure gradients produce strong coastal currents whose speeds typically increase toward shore. In regions where tidal currents are very strong, such as the continental shelves of the North Sea, turbulence generated at the sea floor can mix away density gradients to produce well-mixed water columns. Near river outflows, input of low-density water forms strong vertical density gradients (low-density freshwater on top, high-density seawater below) that can inhibit vertical mixing caused by tidal flow over the sea floor, producing a density-stratified water column. Coastal currents often form along the boundaries between these mixed and stratified regions.

In many coastal regions the lunar and solar tides travel along coastlines as Kelvin waves. In the alongshore direction, the dynamics of Kelvin waves are similar to gravity waves commonly observed at the seashore. In the cross-shore direction the force balance is geostrophic such that the sea surface slopes up toward the coast during high tide and down during low tide. The height of Kelvin waves and the speeds of the resulting coastal currents drop off rapidly away from shore. At high and low tide, the alongshore

slope of the sea surface is small so coastal currents generated by the Kelvin wave are also small. Midway between high and low tide, the alongshore slope of the sea surface is maximum, which creates strong alongshore pressure gradients and strong coastal currents.

COASTAL TRAPPED WAVES

Kelvin waves are an example of a broader class of phenomena called coastal trapped waves, which produce variable coastal currents on time scales as long as a few weeks. Coastal trapped waves arise because of periodic variations in winds along coastlines that cause corresponding variations in cross-shore Ekman transport. As water moves offshore at the surface during upwelling-favorable winds, sea level drops near the coast and deeper waters move onshore. During downwelling-favorable winds the surface flow is onshore, elevating coastal sea levels with deeper waters moving offshore. These regions of elevated and lowered coastal sea level propagate as waves along shore even after the winds that produced them have ceased. Alongshore pressure gradients occur between the regions of elevated and lowered sea level that drive coastal currents. Many factors alter the propagation of coastal trapped waves, including variable bathymetry, changes in coastline shape and orientation, and density stratification. Coastal trapped waves are important for producing variations in coastal currents on time scales of days to a few weeks. In many locations they are the dominant source of variability for coastal currents on these time scales.

SEE ALSO THE FOLLOWING ARTICLES

Buoyancy / Ocean Waves / Surf-Zone Currents / Tides / Upwelling / Wind

FURTHER READING

Brink, K. H., and A. R. Robinson. 1998. *The global coastal ocean*, Vol. 10, *The sea*. New York: John Wiley & Sons.

Colling, A., and the Open University Course Team. 2001. *Ocean circulation*, 2nd ed. Oxford, UK: Butterworth-Heinemann.

Cushman-Roisin, B. 1994. *Introduction to geophysical fluid dynamics*. Englewood Cliffs, NJ: Prentice Hall.

Kosro, P. M. 2005. On the spatial structure of coastal circulation off Newport, Oregon, during spring and summer 2001 in a region of varying shelf width. *Journal of Geophysical Research* 110: C10S06; doi:10.1029/2004JC002769.

Pond, S., and G. L. Pickard. 1983. *Introductory dynamical oceanography*, 2nd ed. New York: Pergamon.

Thurman, H. V., and A. P. Trujillo. 2004. *Introductory Oceanography*, 10th ed. Upper Saddle River, NJ: Pearson Prentice Hall.

Tomczak, M. 2005. Shelf and coastal oceanography. http://www.es.flinders.edu.au/~mattom/ShelfCoast/index.html.

Wright, J., A. Colling, D. Park, and the Open University Course Team. 1999. *Waves, tides, and shallow water processes*, 2nd ed. Oxford, UK: Butterworth-Heinemann.

DESICCATION STRESS

STEVEN C. HAND AND MICHAEL A. MENZE

Louisiana State University

The threat of desiccation for organisms inhabiting the intertidal zone occurs during emersion at low tides or when organisms are positioned in the high intertidal zone, where wetting occurs primarily by spring tides, storm waves, and spray. Drying due to evaporative water loss is the most common mechanism for dehydration, although during winter in northern temperate regions freezing can also occur, which reduces the liquid water in extracellular fluids and can lead to intracellular dehydration in multicellular organisms. Freezing tolerance has been reported and characterized for a number of intertidal invertebrates, including gastropods such as an air-breathing snail and a periwinkle, and bivalve genera including the common and ribbed mussels.

BIOLOGICAL RESPONSES TO THE THREAT OF DESICCATION

Behavioral defenses against stress and injury from the loss of water from cells, tissues, or body fluids are seen across all organisms resident in the rocky intertidal zone: microorganisms, animals, and plants. For organisms that are highly mobile, the first response to water stress is generally behavioral: to leave the area or to seek microhabitats that afford some degree of protection against dehydration (Figs. 1, 2). Such refugia include crevasses in the substratum, animals aggregated into clumps, cover and shade underneath macroalgae, and accumulated organic detritus. Other, less mobile organisms (Figs. 1, 2) restrict

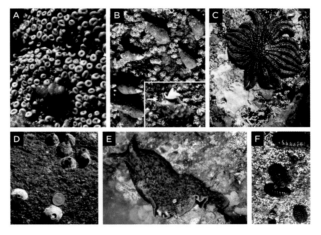

FIGURE 1 Selected inhabitants of the intertidal zone in the northern Gulf of California, where tidal amplitudes are among the largest in the world (6–9 m of vertical displacement). (A) The colonial zoanthid anemone, *Palythoa ignota,* tightly constricted during emergence and pictured with the brown bubble gum alga *Colpomenia;* (B) a field of acorn barnacles, *Chthamalus,* occupied by the prosobranch snails *Cerithium* and (inset) *Acanthina angelica,* whose apertural spine (arrow) is used to pry open opercular plates of barnacles on which it feeds; (C) the gulf sun star, *Heliaster kubiniji,* tightly adhered to rock surface and with restricted locomotion during emersion; (D) the thatched barnacle, *Tetraclita stalactifera,* from the higher mid intertidal zone; (E) the sea hare, *Aplysia californica,* trapped at low tide in a rock crevasse; and (F) the green chiton *Chiton virgulatus.* The behavioral and physiological mechanisms possessed by these organisms for avoiding or tolerating water loss during emersion are varied (see text). Photographs by the author.

various activities such as filter feeding and irrigation of respiratory epithelia (e.g., ectoprocts, barnacles, bivalves), constrict to reduce surface area and attach gravel/shell debris to body wall (sea anemones), and adhere more tightly to the rocky substratum (e.g., chitins, limpets, snails)—all behaviors that retard water loss. Behavioral responses to emersion may also include the synchronization of gamete release or hatching of embryos.

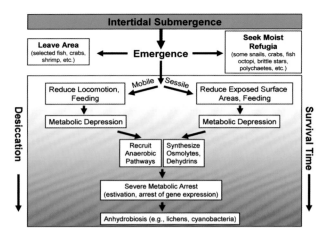

FIGURE 2 Metabolic depression is a natural consequence of assuming postures that restrict water loss; reduction in gas exchange across respiratory surfaces promotes a decrease in energy metabolism and, commonly, a greater reliance on anaerobic metabolism. In more extreme cases, states of estivation (snails) and anhydrobiosis (lichens, cyanobacteria) can be entered.

Physiological and biochemical features important for resisting or tolerating water loss are numerous in organisms of the intertidal zone. These include (a) the deployment of desiccation-resistant egg cases for embryonic development, (b) reduction in water permeabilities of body coverings and epithelia (with the unavoidable disruption of gas exchange), (c) accumulation of metabolic end products as a result of concomitant oxygen limitation, (d) short-term, facultative depression in metabolic and developmental rates, (e) maintenance of intracellular osmolytes for water retention and macromolecular protection at low-water activities, (f) differential gene expression for the production of protective macromolecules, and, in extreme cases, (g) global downregulation of gene expression and metabolism during estivation and anhydrobiosis. Energy conservation, as in options (d) and (g), is critical during extended periods when nutrient acquisition is offset by the potential for severe water loss.

DEPRESSION OF METABOLISM

Water loss during emersion is increased by active processes such as locomotion, feeding, and ventilation of respiratory surfaces. Sessile organisms, which often dominate significant areas of the intertidal zone, respond to desiccation stress by reducing the exposed surface areas across which water loss takes place. Thus, when these processes and surface areas are restricted, overall metabolism is commonly depressed, either by active downregulation or "automatically" by impeded gas exchange (e.g., collapsed gills). Different strategies in response to aerial exposure and water loss can be distinguished across species.

Depending on the species, some sea anemones store significant seawater within the body cavity (coelenteron) during emersion. Intertidal anemones become quiescent during exposure, and overall metabolism declines as a function of time in air. For intertidally acclimatized individuals of some species, aerobic metabolic pathways (ones requiring oxygen) predominate, as evidenced by the lack of an oxygen debt upon reimmersion and by the observation that rates of respiration and heat release by the animals are consistent with the complete conversion of energy fuels to carbon dioxide and water. However, subtidally acclimatized specimens rely partly on anaerobic pathways (those that do not require oxygen) as length of experimental emersion is extended. Anaerobic pathways are apparently recruited in certain other anemones as well during aerial exposure. End products, such as the amino acids alanine and glutamate, have been measured.

A rich literature exists concerning physiological responses of intertidal molluscs to emersion. For example, the common mussel withstands the physical, chemical, and biotic factors during low tide by trapping water in the mantle cavity and tightly closing its valves to restrict water loss. Long-term closure is accomplished using catch muscles, which substantially reduces the energetic cost of keeping the valves tightly closed. When valves are closed, gas exchange is greatly depressed, oxygen supply limited, and metabolism downregulated; the mussel relies generally on anaerobic ATP generation during these periods. The oxygen debt that builds up during the anaerobic phase is paid off by aerobic respiration on the next incoming tide. Water loss from body tissues is minimal during air exposure if valves are kept closed. However, in some individuals of the common mussel, and more typically in other species such as the common cockle and the ribbed mussel, the phenomenon of "gaping" of valves can occur during emersion, which allows for a greater rate of oxygen consumption with the concomitant tradeoff of greater tissue desiccation.

Snails that live above the high-tide mark in the intertidal zone greatly minimize locomotion and feeding during emergence, and as a result, evaporative water loss from tissues is reduced. Nevertheless, water loss can reach 30% or more after many hours of air exposure at elevated temperatures. In some species, entry into a state of estivation is a common occurrence, which involves withdrawing the body deeply into the shell, occluding the shell aperture, and even cementing with mucus the edge of the aperture to the substratum. Estivation in snails is commonly associated with major metabolic depression. In nonestivating snails in the upper intertidal, desiccation is normally associated with varying degrees of metabolic depression, either active or passive. Compared with other intertidal gastropods with shells, limpets exhibit much greater water losses during emersion, apparently in part because of the wide aperture of their dish-shaped shells. Aerobic metabolism accounts for the vast majority of overall ATP turnover during emersion in snails. Although anaerobic metabolism and accumulation of end products (alanine and the organic acid succinate) have been documented in at least one gastropod species during emersion, the contribution to overall energy production is very low. Many species of gastropods are reported not to recruit anaerobic pathways at all during air exposure.

Opisthobranch gastropods (sea slugs, sea hares, nudibranchs) can become stranded occasionally during low tide in the intertidal zone and exposed to air. Some sea hares (Fig. 1E) tolerate emersion for many hours, during which time some loss of body water occurs and oxygen uptake drops dramatically compared to the aquatic value. Evidence indicates the metabolic depression is accompanied by anaerobic energy production; whole-body lactate and alanine are elevated severalfold.

As early as the 1950s, studies on crustacea underscored that intertidal barnacles ceased filter feeding and became quiescent upon emersion, in contrast to subtidal barnacles, which remained active and were more quickly desiccated. Aerobic metabolism in barnacles is the primary energy source, even though anaerobic production of lactate occurs to a limited extent. Similarly, intertidal crabs often show a depressed aerobic metabolism soon after exposure that is thought to arise from impeded gas exchange due to clumping of gills, absence of ventilation, and depleted oxygen in unventilated water surrounding the gills. The presence or absence of lactate accumulation varies across species.

Most intertidal fishes do not emerge voluntarily, in contrast to marine amphibious fishes such as mudskippers and rockskippers. Consequently, the duration of air exposure is very limited in intertidal fishes, and thus desiccation is an infrequent concern. Still, it has been established that numerous families of intertidal fishes, including sculpins, gobies, pricklebacks, gunnels, and sea chubs have the capacity for aerial gas exchange. In species for which data exist, anaerobic pathways are not recruited to supplement routine metabolism during brief bouts of emergence.

Water loss during emersion can be very fast among species of intertidal seaweeds, ranging from 10% to over 90% loss in water content after only a few hours under moderately desiccating conditions. Tolerance to water loss is key to survival of seaweeds in the intertidal zone, rather than the possession of efficient mechanisms for the prevention of water loss. Species with high rates of water loss also exhibit large inhibitions of photosynthesis and dark (mitochondrial) respiration. The degree of water loss is a better predictor of metabolic depression than is the height in the intertidal zone at which various species reside. Similarly, within a given species (as for selected brown algae), vertical distribution in the intertidal zone is not tightly correlated with photosynthetic rates during air exposure. Desiccation as a result of emersion has also been shown to retard embryonic development in at least one intertidal alga.

Very high in the intertidal zone (supralittoral fringe), one can find blackish lichens and cyanobacteria that are quite tolerant to the dry environmental conditions that exist, where seawater spray can often be the only mechanism for hydration. Some cyanobacteria are embedded in a gelatinous mass that helps retard water loss and can produce a stress protein that is similar to plant dehydrins (see the next section). Other cyanobacteria experience massive water loss and tolerate such air-dried states for prolonged periods. The severe desiccation observed for lichens over extended periods (days, weeks) is an example of anhydrobiosis. Lichens can survive multiple cycles of anhydrobiosis. Other anhydrobiotic organisms, as well as species experiencing milder forms of water stress, are afforded protection by small organic solutes and protective proteins found intracellularly.

ORGANIC OSMOLYTES AND PROTECTIVE MACROMOLECULES DURING DESICCATION

When cellular water is lost as a result of desiccation or some other form of water stress (freezing, fluctuating salinity), intracellular osmolytes become concentrated. The beneficial effect of intracellular osmolytes is to retard water loss from the cell. While some contribution to the intracellular osmotic pressure of cells comes from inorganic ions, a substantial fraction of the osmotically active solutes are organic osmolytes. The term "compatible solute" was originally coined in 1972 by A. D. Brown and colleagues and applied to small carbohydrates termed polyols (i.e., polyhydric alcohols). When accumulated to high levels, these compounds do not disturb macromolecular structure and function, yet they serve to provide osmotic balance with the external environment. In addition to organic osmolytes that can be considered nonperturbing or compatible, other organic osmolytes can actually stabilize macromolecules during water stress. Protective effects of osmolytes have been documented for proteins and for phospholipids that form membrane bilayers. An evaluation of the dominant types of organic osmolyte systems and their evolution later suggested that a common feature of many osmolytes was that they exerted influences on macromolecules by acting on the solvent properties of water. It is appropriate to note that in the case of anhydrobiosis, disaccharides such as trehalose are technically not serving the role of an osmolyte, because virtually all cellular water is lost in this state. However, trehalose is considered an osmolyte for other nonanhydrobiotic species. Table 1 lists some representative organic osmolytes that are found in various groups of organisms inhabiting the intertidal zone. The presence of organic osmolytes across these diverse organisms reflects the common threat of water stress that is faced.

TABLE 1
Representative Organic Osmolytes Accumulated in
Various Groups of Intertidal Organisms

Osmolytes	Organisms
Polyhydric alcohols	
Glucosylglycerol	Cyanobacteria
Mannosidomannitol	Lichens
Mannitol	Multicellular brown algae
Amino acids and derivatives	
Various amino acids*	Invertebrates (all phyla)
Octopine	Mollusca, Cnidaria
Methylamines	
Glycine betaine,	Invertebrates (multiple phyla)
Proline betaine,	
Trimethylamine oxide	

*Amino acids commonly accumulated as osmolytes: alanine, β-alanine, serine, glycine, taurine, proline.

Late embryogenesis abundant (LEA) proteins were first identified in land plants, and their expression is associated with desiccation tolerance in seeds and anydrobiotic plants. One family of proteins within this group is dehydrins, which are thought to confer dehydration tolerance in plants. The precise mechanism by which dehydrins act is unclear. As discussed in the preceding section, brown algae display remarkable tolerances to water loss. Some of these algae constitutively express proteins that are related to dehydrins, which can be specific for certain embryonic stages and certain species. It has not been demonstrated whether these dehydrin-like proteins protect the algae against intertidal desiccation during emergence. Upon osmotic challenge, dehydrin-like proteins are also inducible in certain cyanobacteria represented in the intertidal zone.

APPLICATION OF FINDINGS TO CELL STABILIZATION

The comparative physiology of animals inhabiting the intertidal zone and other environments in which water stress is frequent has led to the appreciation of mechanisms, such as those summarized in the preceding section, that can limit damage incurred during desiccation. One approach that appears to improve survivorship during desiccation stress is the accumulation of low-molecular-weight carbohydrates. Trehalose is a prime example of a sugar utilized by a broad range of species that naturally cope with periods of severe water loss. The lessons learned from organisms that are naturally desiccation-tolerant are being applied to cell stabilization problems in the biomedical field. The ultimate goal is to prepare desiccated human cells for storage at ambient temperature. Trehalose has been shown to be effective in improving desiccation tolerance for a variety of macromolecular assemblies and cells including liposomes, enzymes, retroviruses, platelets, fibroblasts, hematopoietic stem cells, and macrophages (Fig. 3). For biostabilization of cells, trehalose has its greatest impact when present on both surfaces of the membrane. To succeed in preparing dried cells that retain high viability upon rehydration would offer tremendous economic and practical advantages over traditional cryopreservation protocols.

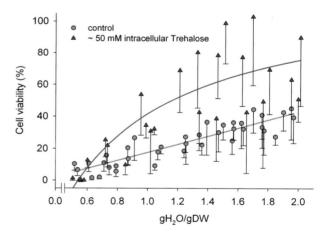

FIGURE 3 Naturally occurring osmolytes promote cell stabilization during drying. Loading an organic osmolyte such as the disaccharide trehalose into mammalian cells increases survival post-drying. Mouse macrophage cells were air dried to a range of final water contents in an intracellular-like medium containing 250 mM trehalose. Upon rehydration, cells that had been previously permeabilized in order to load approximately 50 mM intracellular trehalose showed higher survival at all water contents, compared to control cells that were not loaded with trehalose. Values are means ± SD (n = 10). Modified from Elliot et al. (2006).

FURTHER READING

Crowe, J. H., L. M. Crowe, W. F. Wolkers, A. E. Oliver, X. Ma, J.-H. Auh, M. Tang, S. Zhu, J. Norris, and F. Tablin. 2005. Stabilization of dry mammalian cells: lessons from nature. *Integrative and Comparative Biology* 45: 810–820.

Elliott, G., X.-H. Liu, J. L., Cusick, M. Menze, J. Vincent, T. Witt, S. Hand, and M. Toner. 2006. Trehalose uptake through P2X$_7$ purinergic channels provides dehydration protection. *Cryobiology* 52: 114–127.

Hand, S. C., and I. Hardewig. 1996. Downregulation of cellular metabolism during environmental stress: mechanisms and implications. *Annual Review of Physiology* 58: 539–563.

Hoekstra, F. A., E. A. Golovina, and J. Buitink. 2001. Mechanisms of plant desiccation tolerance. *Trends in Plant Science* 6: 431–438.

Sleator, R. D., and C. Hill. 2001. Bacterial osmoadaptation: the role of osmolytes in bacterial stress and virulence. *FEMS Microbiology Review* 26: 49–71.

Yancey, P. H. 2005. Organic osmolytes as compatible, metabolic and counteracting cytoprotectants in high osmolarity and other stresses. *Journal of Experimental Biology* 208: 2819–2830.

Yancey, P. H., M. E. Clark, S. C. Hand, R. Bowlus, and G. N. Somero. 1982. Living with water stress: the evolution of osmolytes systems. *Science* 217: 1214–1222.

DIFFUSION

GEORGE JACKSON

Texas A&M University

Diffusion is the name given to the process in which cumulative random motions result in the net movement of material. Diffusion theory has been used to describe the movement of a variety of properties, including heat through walls, molecules through membranes, eddies throughout the ocean, and bacteria swimming around a leaky cell.

MOLECULAR DIFFUSION

Life is full of random motions that combine to form predictable patterns, particularly when large numbers of objects are involved. One of the best known patterns is molecular diffusion, the net movement of molecules in air or water as a result of random thermal motions. The biological importance of molecular diffusion to marine systems results from the need of aquatic organisms to gather those molecules needed for life from solution and to be rid of those that are toxic. Because water currents are damped down to nothing next to a surface, the last

micrometer that must be traversed to get to the surface, at least, is controlled by the random process of molecular diffusion.

Diffusion Coefficient

Molecular diffusion results from thermal vibrations pushing at random on molecules, propelling them along, and opposing retardation provided by viscosity, slowing them down (Fig. 1). The cumulative effect of these movement spurts is to move a molecule along a random walk. With all molecules subjected to their own random walks, there is a net movement of molecules called diffusion. The rate of diffusion is described using a diffusion coefficient D with units of length^2/time: cm^2 s^{-1} or m^2 s^{-1}. Typical values are those for O$_2$ molecules in water: 1.0×10^{-5} cm^2 s^{-1} at 0 °C increasing to 2.7×10^{-5} cm^2 s^{-1} at 30 °C. In air, values are about ten-thousandfold larger: 0.18 cm^2 s^{-1} at 0 °C and increasing only slightly to 0.21 cm^2 s^{-1} at 30 °C. Much of the difference in the temperature effect between the atmosphere and water results from how temperature changes the viscosities of water and air.

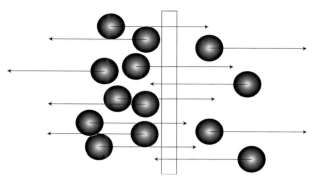

FIGURE 1 If there are more molecules on one side of a surface, then there will be an average movement of molecules to the region of lower concentration even though each molecule has an equal chance of moving toward or away from the surface. In this case, there is a net movement of two molecules to the right.

Dimensional Analysis

A rough idea of the distances and times involved in a diffusive process can be gained by using the dimensions of the process as a guide. In a given time t, there is only one way to calculate the distance L that diffusion will spread a substance without invoking other processes described by their own dimensional constants: $L \sim \sqrt{Dt}$. The square root implies that diffusion is very rapid for small distances and slow for larger ones: in 1 s, a patch of O$_2$ molecules in water will spread 30 µm, 300 µm = 0.3 mm in 100 s, 0.9 cm

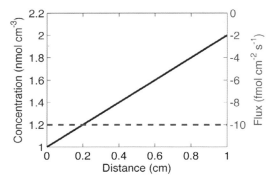

FIGURE 2 The time that it takes diffusion to spread increases dramatically with distance. Movement is rapid for short distances and slow for long ones. $D = 10^{-5}$ cm^2 s^{-1}.

in 1 day, and 18 cm in 1 year (Fig. 2). If other processes, such as water flow, are also involved, the time scales change.

Flux

An important property of diffusion is that it acts to decrease differences in concentration, because the net movement of material is always from high- to low-concentration regions. We usually describe the material movement in terms of a flux, the net rate at which the substance passes through a surface normalized by the area of the surface. A flux is, thus, mass per area per time, with typical units such as moles m^{-2} s^{-1} or g cm^{-2} d^{-1}. The flux resulting from diffusion is encapsulated in Fick's first law, which describes the net flux in terms of spatial concentration differences using the concentration gradient. The concentration gradient is the rate at which concentration changes over distance and is calculated using a derivative. For example, if the oxygen concentration C changes in the x direction, then the oxygen gradient is dC/dx. Fick's first law states that the flux in the x direction is given by

$$\text{Flux}_x = -D \, dC/dx \qquad \text{(Eq. 1)}$$

The minus sign indicates that the flux is in the opposite direction from the increase of oxygen; that is, the oxygen flows from high to low concentrations (Fig. 3).

Total Flow

The rate at which diffusion delivers material to or from a spherical cell with a radius r_o is given by

$$\text{Flow to cell} = 4\pi r_o D \, \Delta C \text{ (mass time}^{-1}, \text{ such as} \\ \text{mole s}^{-1} \text{ or g d}^{-1}) \qquad \text{(Eq. 2)}$$

where ΔC is the difference between the concentration far away from the cell (the background concentration in

solution, C_o) and the concentration next to the cell. If the difference is negative, the concentration is higher next to the cell and material diffuses away from the cell; if positive, it diffuses toward the cell. A cell increases the flow of material by taking material through its membrane, decreasing the concentration next to the cell. Because the lowest a cell can take that concentration is to 0, the *maximum* rate that material can diffuse to a cell is then $4\pi r_o DC_o$, despite the best efforts of the cell. That is, the maximum rate that material can reach the cell by diffusion is controlled by external properties, D and C_o. In addition, the rate at which material arrives is proportional to the cell's radius, increasing much slower than a cell's surface area (which increases as radius squared) and volume (which increases as radius cubed). Metabolic activity cannot increase this, without increasing relative water movement.

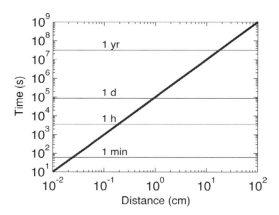

FIGURE 3 Example of net diffusive flux as a function of distance for a constant concentration gradient. A linearly increasing concentration has a constant gradient. Diffusion moves material down the gradient, in the opposite direction to the direction of concentration increase. In this case, the flux is constant because the slope of the concentration is constant.

The effect of the cell on the concentration in the surrounding water drops off rapidly with distance from the cell: the concentration a distance r from the cell center = $C_o - \Delta C \, r_o/r$. Any depression or elevation in concentration next to the cell is decreased by half at one radius from the surface. This relationship tells us that large cells or other organisms influence the concentrations away from their surfaces farther out than do small ones. The molecules right next to a cell can be renewed only by molecular diffusion, but further away they are also moved by water motion, such as caused by waves. When we stand in front of a fan during a hot day, we are using the breeze to speed up the movement of heat away from our bodies, which, without the breeze, would be slower because it would occur only by thermal diffusion.

Effect of Water Motion

The relative importance of water motion depends on the relative sizes of the organism, the speed of the water motion, and the diffusion coefficient. The Peclet number Pe = $r_o v/D$, where v is the water velocity, is a useful index of the relative importance of diffusion versus water motion in controlling transport to the cell: when Pe is larger than 1, water motion becomes the dominant control of mass movement, while diffusion dominates when Pe is less than 1. For example, a 1-μm-radius bacterial cell swimming at 10 μm s^{-1} has a Pe = 0.01 for a molecule with D = 10^{-5} cm^2 s^{-1}, indicating that swimming has little effect overcoming diffusive transport limitation; a 1-cm-diameter macroalga sitting in a tidal channel with water rushing by at 1 m s^{-1} has a Pe = 10^7 for the same diffusivity, indicating that water movement has a large effect enhancing transport to the plant. The Peclet number does not actually describe how much of an enhancement there is of transport; it is only an index describing the importance of water motion. Calculations and laboratory experiments can be used to actually measure diffusive transport for different shapes and water motions. The resulting rates can be divided by the transport rate for the case when only diffusion moves material. This value is a dimensionless ratio known as the Sherwood number, Sh. Its value depends on the value of Pe as well as on the object shape.

The diffusion coefficient depends on temperature (as noted above), the size of the molecule, and the viscosity of solution. Organic compounds released by seaweeds can increase the viscosity of water in a tidepool, thereby reducing the diffusion coefficient and the rate of diffusive transport.

DIFFUSION USED TO DESCRIBE OTHER PROCESSES

Other transport processes can be thought of as being composed of a sequence of random motions, leading to the same mathematical formulation to describe them as to describe molecular diffusion. For example, the movement of material in a turbulent intertidal regime is the result of random eddies and can be described in terms of an eddy diffusion coefficient. It is important to remember, though, that Fick's laws for molecular diffusion describe the results of average motions of large numbers of molecules. Using eddy diffusion coefficients to describe

dispersal of material in a turbulent environment works best when describing the average pattern for large numbers of events, in what is called the ensemble average. The concentration pattern associated with a single dispersal event does not always mimic the average and can actually be quite different.

Other examples of aquatic systems that are well described as diffusive processes include heat conduction and bacterial movement in the absence of chemical cues. The colonization rate of particles by bacteria in solution has been predicted from the theory of diffusion, and these predictions match what is observed.

SEE ALSO THE FOLLOWING ARTICLES

Boundary Layers / Hydrodynamic Forces / Turbulence

FURTHER READING

Berg, H. C. 1983. *Random walks in biology*. Princeton, NJ: Princeton University Press.
Csanady, G. T. 1973. *Turbulent diffusion in the environment*. Dordrecht, Holland: Riedl.
Denny, M. W. 1993. *Air and Water*. Princeton, NJ: Princeton University Press.
Kiørboe, T., and G. A. Jackson. 2001. Marine snow, organic solute plumes, and optimal chemosensory behavior of bacteria. *Limnology and Oceanography* 46: 1309–1318.
Okubo, A., and S. A. Levin. 2001. *Diffusion and ecological problems: modern perspective*, 2nd ed. New York: Springer.
Turchin, P. 1998. *Quantitative analysis of movement*. Sunderland, MA: Sinauer.

DISEASES OF MARINE ANIMALS

KEVIN D. LAFFERTY

United States Geological Survey

Disease is an identifiable group of symptoms that corresponds to a pathological condition of an organ or system. Diseases can result from various causes, such as an infectious agent, genetic defect, or environmental stress such as temperature or a toxin. Compared to many habitats, we know very little about disease in the intertidal zone. This is surprising, given how many scientists have studied the intertidal zones around the world. Perhaps sick organisms get eaten, or the tides wash them away. But disease should prevail in the intertidal zone. Intertidal organisms are often at the limits of their physiological tolerance to temperature and desiccation.

Being near shore, the intertidal zone is increasingly exposed to poisons and pathogens that come from human activities. Finally, intertidal creatures are often packed cheek by jowl; such high densities facilitate the transmission of infectious agents.

NONINFECTIOUS DISEASE

The physical stress of living along the sea–land interface can contribute to mortality through noninfectious disease. For instance, exposure to high temperatures can lead to the expulsion of symbiotic algae from some invertebrates such as corals and sea anemones. Coral bleaching is the most notable form of this response to heat stress. Ultraviolet light, which can be intense in shallow waters, can damage cells and promote cancerous growths. Toxic chemicals, whether natural (e.g., from "red-tide" dinoflagellates) or from human sources, can also impair living organisms.

INFECTIOUS DISEASE

Only rarely are infectious agents visually apparent. A cyanobacterium is associated with shell wear in mussels (perhaps changing them from shiny black to chalky blue). Each marine species is usually host to several infectious agents, from viruses and protozoans to metazoans. Most of these are not well known. Collectively, we call these infectious agents pathogens and parasites. Such infectious agents are usually a normal part of an organism's experience, and most do not have noticeable effects. For example, parasitic flatworms infect sea urchins but do not appear to cause disease. In other cases, the effects can be dramatic. For instance, some fly larvae enter barnacles and devour them from the inside.

Some infectious agents do have effects on hosts, and these may limit host populations. One "unseen" strategy used by infectious agents is parasitic castration. In a single tidepool, parasitic barnacles may castrate hermit crabs and shore crabs, trematode worms may castrate periwinkles, and parasitic isopods may castrate acorn barnacles. These animals look perfectly healthy, but they do not reproduce. Bacterial diseases have led to quite noticeable mass mortalities in the intertidal zone. A rickettsial (intracellular bacterium) disease suddenly appeared in black abalones in California in the mid-1980s. This disease spread rapidly and reduced abalone populations by 99%. A bacterial epidemic caused similarly high mass mortalities of a multi-armed sea star in the Sea of Cortez (Mexico). Such bacterial epidemics seem to spread faster when host densities are high. For example, in California, dense sea urchin populations are more likely to suffer epidemics of bacterial disease. The high temperatures in the intertidal zone may favor bacteria. When temperatures increase above normal, bacterial infections can cause sea stars to literally disintegrate (Fig. 1).

FIGURE 1 An orange bat star, *Asterina miniata*, from the California Channel Islands, with one disintegrating arm, most likely the result of a bacterial infection linked to warm water. Photograph by the author.

People can accidentally move infectious agents from one place to another. For example, a sabellid polychaete worm was accidentally introduced from South Africa to abalone farms in California. The worm, which has a negligible effect on South African abalone, stunted California abalone by interfering with shell deposition. These worms soon escaped into the intertidal zone, where they were able to infect other intertidal snails (especially turban snails). Luckily, an effort to remove millions of infected and susceptible snails from the area eradicated the worm from the wild.

Some pathogens make their way from land to the ocean. These pathogens are essentially a form of pollution rather than a true infectious disease, because they do not spread from host to host once in the water; the term "pollutogens" has been coined for these organisms. Unlike a true pollutant, they do grow and multiply in this new host. For instance, bacteria from human sewage can infect and kill corals, and fungi from soil runoff can infect and kill sea fans. Sea otters are susceptible to deadly infectious agents that come from land, such as toxoplasmosis and valley fever. The infectious stages of these organisms probably enter the intertidal zone with runoff from land and then may float out to kelp beds, where sea otters contact them.

Although marine diseases are a normal and important part of marine ecosystems, changes may be occurring. Global change, toxic pollution, and the direct input of pollutogens from land to sea may increase the role of some infectious agents in the ocean. In the last three decades, scientists have noted apparent increases in the reports of diseases of intertidal organisms such as corals, mammals, sea urchins, and molluscs. Further study of disease in intertidal organisms may help us to better protect this resource.

SEE ALSO THE FOLLOWING ARTICLES

Heat Stress / Introduced Species / Parasitism / Rhizocephalans / Ultraviolet Stress

FURTHER READING

Harvell D., R. Aronson, N. Baron, J. Connell, A. Dobson, S. Ellner, L. Gerber, K. Kim, A. Kuris, H. McCallum, K. Lafferty, B. McKay, J. Porter, M. Pascual, G. Smith, K. Sutherland, and J. Ward. 2004. The rising tide of ocean diseases: unsolved problems and research priorities. *Frontiers in Ecology and the Environment* 2: 375–382.
Lafferty, K. D. 2004. Fishing for lobsters indirectly increases epidemics in sea urchins. *Ecological Applications* 14: 1566–1573.
Lafferty, K. D., and A. M. Kuris. 1993. Mass mortality of abalone *Haliotis cracherodii* on the California Channel Islands: tests of epidemiological hypotheses. *Marine Ecology Progress Series*, 96: 239–248.
Lafferty, K. D., and A. M. Kuris. 1999. How environmental stress affects the impacts of parasites. *Limnology and Oceanography* 44: 925–931.
Lafferty, K. D., and A. M. Kuris. 2002. Trophic strategies, animal diversity, and body size. *Trends in Ecology and Evolution* 17: 507–513.
Lafferty, K. D., J. Porter, and S. E. Ford. 2004. Are diseases increasing in the ocean? *Annual Review of Ecology, Evolution and Systematics* 35: 31–54.
McCallum, H. I., A. M. Kuris, C. D. Harvell, K. D. Lafferty, G. W. Smith, and J. Porter. 2004. Does terrestrial epidemiology apply to marine systems? *Trends in Ecology and Evolution* 19: 585–591.
Torchin, M. E., K. D. Lafferty, and A. M. Kuris 2002. Parasites and marine invasions. *Parasitology* 124: S137–S151.

DISPERSAL

STEVEN D. GAINES

University of California, Santa Barbara

All organisms move. To an observer walking along a rocky shore and seeing so many plants and animals glued to the rock, this definitive statement may seem clearly false, yet it is correct. As immobile as some species may seem, at some point in their life cycle they move. In broad terms, the active or passive movement of individuals from one place to another is called dispersal. This movement plays fundamental roles in many ecological and evolutionary processes, and it has important implications for how a variety of human activities impact rocky shores.

REASONS TO MOVE

Species move for many reasons. Some move to find and capture food. Others move to avoid becoming food for someone else. Some move to seek more favorable microclimates (e.g., moving into or out of the shade) that may enhance their performance. Others move to find a mate. Examples of each of these types of movement can be found on nearly any rocky shore as sea urchins forage from a crevice to seek algae, limpets "run" when contacted by the tube foot of an approaching sea star, periwinkles seek cracks at low tide that hold moisture, or sea slugs aggregate in large mounds of individuals to mate. Often these forms of movement occur over relatively short distances on a given rocky shore. When they do, they can play important roles in creating patterns in the local distribution of species across a shore. Such limited movement of many species on rocky shores has made them a favorite target for scientific studies.

Movement can also occur across great distances. Elephant seals that haul out and mate at the same California beach year after year, leave to forage across the entire North Pacific. Rock lobsters that forage on New Zealand and Australian shores can migrate as much as 1000 kilometers to deep offshore waters to spawn. Oystercatchers, key predators on mussels and cockles, seasonally migrate between Britain and Iceland to feed. Movements across such large distances connect ecosystems that are widely separated. As a result, processes that occur in one location may drive changes on distant shores through the ecological coupling that occurs via long-distance movement. Without knowledge about the pattern of these connections, it may be impossible to interpret the cause of changes observed at a given place.

These types of movement can all redistribute individuals locally, regionally, or even globally, depending on the scale of movement. Yet another form of movement is the focus of most ecological attention on dispersal: the movement of young away from their parents. There are many potential reasons for offspring to flee the site of their birth. Dispersing can reduce competition with an organism's parents. For example, kelps and other seaweeds can form dense canopies that shade the rocky surface below. If algal spores were released right under their parents, their growth could be severely compromised. Similarly, dispersing young can reduce the likelihood that siblings will compete with each other for scarce resources or mate with each other and incur the costs of inbreeding. Dispersing young can also escape the pathogens and predators that are attracted or accumulated by their parents, and dispersal reduces the chance that a single disturbance (e.g., a large storm, an oil spill, a log rolling in the waves along a rocky shore) will kill all members of a genetic family. By spreading young across a variety of settings, the risks of shared catastrophe is reduced.

These arguments all focus on dispersal as providing direct benefits to young. Dispersal can also occur as a byproduct of choices that have nothing to do with the benefits of moving. This is especially true of animals in the sea. Since seawater is commonly a rich soup laden with plankton, one potential reproductive strategy available to marine animals is to produce enormous numbers of very small offspring (called larvae in invertebrates and fish) who fend for themselves finding food in the plankton. Rather than having to provision each young with sufficient food to reach a large size, marine species can produce far more offspring with little nutritional investment in individual offspring. If the added number of young that can be produced offsets the added mortality they experience fending for food on their own, this reproductive strategy can enhance an individual's fitness.

This type of life history is extremely common in animals of rocky shores. The great majority of all invertebrates and nearly all fish produce young that are microscopic and grow by feeding in the plankton. Unlike the atmosphere, which has limited food and requires an animal to expend considerable energy to stay suspended, the sea commonly provides a rich potential food supply. Larvae of invertebrates can spend days, weeks, or months, drifting, eating, and growing in the plankton. Larvae commonly grow by an order of magnitude in size even with minimal nutritional investments from their mother. Such benefits from being able to produce minute young that can forage on their own, however, have a key side effect: larvae are dispersed away from their natal site as they drift and feed. Thus, some patterns of dispersal could be an unintended consequence of feeding more than something providing direct evolutionary benefit from dispersal *per se*.

DISPERSAL DISTANCES FOR MARINE LARVAE AND SPORES

Because the spores of seaweeds and larvae of invertebrates and fish are typically minute and can drift in the plankton for hours, days, weeks, or months (depending on the species), tracking their movement and quantifying

their patterns of dispersal have proven to be daunting challenges. Using a diverse array of approaches, marine ecologists are beginning to get a handle on patterns of the dispersal of young.

A recent synthesis of more than 100 marine species showed that dispersal of young varies enormously among marine species (Fig. 1). Average dispersal distances vary by more than seven orders of magnitude, ranging from meters to thousands of kilometers. Although the number of species for which we have data is typically small within many taxonomic groups, there are also strong patterns among taxa. Seaweeds, on average, disperse far less than marine invertebrates and fishes, with average dispersal distances of a few kilometers or less. Some disperse only meters from their parent because they are released at low tide and never enter the plankton. At the other taxonomic extreme, fishes tend to be consistently broadly dispersed with average dispersal distances of tens to hundreds of kilometers. Invertebrates span nearly the entire range between seaweeds and fishes, reflecting the enormous taxonomic diversity found within the multiple phyla in this group.

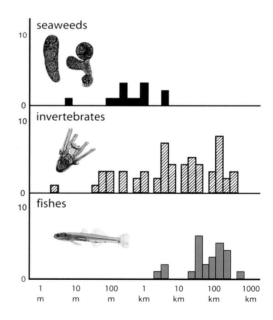

FIGURE 1 Frequency distribution of estimated dispersal distances for marine algae, invertebrates, and fishes. Estimates of average dispersal distances come from several forms of evidence: direct observations, rates of spread of exotic species, and genetic estimates from patterns of population genetic structure. Figure courtesy of PISCO, the Partnership for Interdisciplinary Studies of Coastal Oceans.

These scales of propagule dispersal are enormous relative to comparable patterns on land. Focusing just on plants, comparisons between dispersal of seaweeds and marine angiosperms versus terrestrial plants show that propagule dispersal in the sea is typically greater than dispersal on land. The potential for long-distance dispersal of the spores of seaweeds and seeds of marine angiosperms may be greater than that of many of their terrestrial counterparts just as it is for marine animal larvae, because it is far easier to stay suspended in water than in the air, particularly if the propagule is large. Marine plants do not have to rely on dispersal by animals to transport offspring large distances.

DISPERSAL AND INVASIONS

The scale and pattern of dispersal plays a fundamental role in species invasions. Species have a range that bounds where they occur on the planet. These ranges are not static, however. If the factors that control a species' range change, the range can expand. This can happen when climate changes make previously inhospitable areas hospitable, or when barriers to movement break down. Over evolutionary time scales, species ranges can change frequently (e.g., as a result of glacial cycles). In recent decades, the pace of changes in species ranges has accelerated dramatically as a variety of human activities have moved species to parts of the planet where they previously did not occur. Such exotic introductions have occurred in all habitats, and rocky shores are no exception. The means of introductions are diverse, including fouling on the bottom of ships, larval transport in the ballast water of ships, aquaculture, and the aquarium trade. Often such exotic introductions occur at a single location. If the introduction is successful, the exotic species can spread from the location in its new home. This subsequent spread of an exotic often depends more on its natural dispersal abilities rather than continued transport by human patterns.

The rates of spread of marine exotic species can be rapid, since their natural dispersal distances can be large. Expanding range edges that move many tens of kilometers per year are not uncommon. Surprisingly, the rates of spread of seaweeds and other species with relatively modest average dispersal distances can also be quite large—far in excess of their average dispersal distance. These patterns highlight the importance of the extremes of dispersal distances rather than the average. A seaweed may have an average dispersal distance of only a few hundred meters, but some spores may be able to disperse tens of kilometers in a single jump. These

individuals are the extreme tails of the distribution of dispersal distances, and they can be orders of magnitude larger than the average. For seaweeds, one of the reasons that the extreme tails can be such large distances is the presence of alternative forms of dispersal. Spores can disperse directly by drifting. They can also be dispersed when adult plants are ripped from the shore and raft in surface currents. If these rafting plants eventually land on a distant rocky shore, they can release spores that have been effectively dispersed great distances from where the parental plant grew. Invertebrates that grow on seaweeds can disperse by such rafting as well. Rates of spread of exotic seaweeds are often as fast as those of invertebrates with long-lived planktonic larvae. This suggests that the rare long-distance dispersal events can often play as important a role as the average.

SEE ALSO THE FOLLOWING ARTICLES

Algal Biogeography / Biodiversity, Global Patterns of

FURTHER READING

Highsmith, R. C. 1985. Floating and algal rafting as potential dispersal mechanisms in brooding invertebrates. *Marine Ecology Progress Series* 25: 169–180.

Kinlan, B., and S. D. Gaines. 2003. Propagule dispersal in marine and terrestrial environments: a community perspective. *Ecology* 84: 2007–2020.

Palumbi, S. R. 2003. Population genetics, demographic connectivity and the design of marine reserves. *Ecological Applications* 13: S146–S158.

Shanks, A. L., B. Grantham, and M. H. Carr. 2003. Propagule dispersal distance and the size and spacing of marine reserves. *Ecological Applications* 13: S159–S169.

Zacherl, D. C., S. I. Lonhart, and S. D. Gaines. 2003. The limits to biogeographical distributions: Insights from the northward range extension of the marine snail, *Kelletia kelletii* (Forbes, 1852). *Journal of Biogeography* 30: 913–924.

DISPERSAL, MEASUREMENT OF

DANIELLE C. ZACHERL

California State University, Fullerton

The apple may not fall far from the tree, but it is an open question just how far marine offspring move away from their parents. Thus, measurement of dispersal involves determining the extent to which individual organisms move away from a starting population to a destination. In marine systems, dispersal often takes place during the larval or juvenile phase of an organism's life cycle. Determination of larval and juvenile dispersal distances and determining the percentage of a particular cohort that have dispersed away from their birthplace rather than being retained are important factors in understanding how particular populations are replenished with the next generation of young. It also helps to build an understanding of how separate populations might be connected by larval dispersal (Fig. 1). Despite the fact that dispersal distances remain unmeasured for most marine species, quantifying the extent of population connectivity is vitally important in conservation and management of marine fishery resources, because planners need to identify which populations are most important in promoting the local persistence of marine species. In this way, reserve planners can identify those areas that require maximal protection and can map out an effective network of marine reserves that are connected by larval dispersal and thus are capable of supporting one another.

FIGURE 2 This barnacle larva (nauplius stage) is microscopic and nearly transparent, as is common for most marine larvae. Directly tracking this individual larva from its site of production to its settlement location would be a formidable challenge. Copyright Wim van Egmond/Visuals Unlimited.

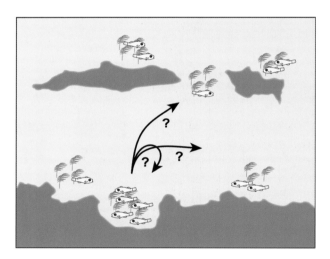

FIGURE 1 In this hypothetical shoreline with offshore islands, individual fish larvae are released from their birth population. Their dispersal trajectory is unknown and the connectedness of local populations, therefore, is unknown.

CHALLENGES AND APPROACHES

The challenges associated with measuring larval dispersal distance and percent retention are formidable. Larvae are typically microscopic, making direct observation impractical (Fig. 2). Further, they can exhibit complex behaviors (swimming, vertical migration) and can be advected away by oceanic currents and turbulent mixing processes. Scientists have employed artificial tagging techniques to track them directly but more commonly use indirect methods such as natural tagging (e.g., genetics, calcified tags) and modeling to estimate dispersal distances and extent of retention.

ARTIFICIAL TAGGING

Artificial tagging is the only way to unequivocally measure dispersal distance when larvae move further than they can be followed directly. Larvae can be immersed in solutions containing marker chemicals such as fluorescent compounds (tetracycline, calcein), elemental tags (strontium, rare earth elements), or radioactive isotopes (^{65}Zn, ^{85}Sr) that typically tag calcified structures. They can also be artificially marked with novel DNA sequences using transgenic methods, thus producing a genetic tag. Larvae are then released from a source population and recaptured at a destination of interest, where they are screened for presence of the artificial tag. Recapture at the destination location is difficult because the probability of recovering a tagged larva can be exceedingly small as a result of high mortality rates and diffusive processes that can dilute their concentrations substantially. These factors make artificial tagging costly, time-consuming, and of limited effectiveness in most situations.

Rather than measuring dispersal distances, artificial tags have more typically been used to estimate rates of larval retention to their source population. Scientists combine the number of tagged larvae recaptured, relative to the total number of tagged and untagged larvae captured at the release site, with estimates of the percentage of larvae tagged at the release site to estimate the rate of retention. Because of the exceedingly low recapture rates of tagged larvae, the accuracy of such estimates is typically quite limited.

NATURAL TAGGING

Natural tagging approaches take advantage of population-specific tags generated either by naturally occurring genetic variation among populations or by variation in environmental conditions. Every larva is effectively tagged, and this eliminates the problems arising from diffusion of tagged larvae and low recapture rates.

One natural tagging approach censuses differences and similarities in allele frequencies of particular genes among populations. When allele frequencies are different among populations, the aggregate of populations is said to exhibit genetic structure. The magnitude of genetic structure can provide an estimate of the extent of larval exchange among populations and allow estimates of dispersal distances, but there is a large amount of uncertainty in interpreting data. Because even a small amount of gene flow can homogenize genetic structure among populations, estimates of per-generation connectivity can be impossible to assess. When dispersal distances are large, genetic approaches typically cannot determine variability in the extent of larval exchange over the short time scales relevant to resource management but instead can provide information about long-term average gene flow over many generations. Genetic estimates of dispersal have nonetheless provided very useful information about general trends in larval dispersal. For example, genetic estimates of dispersal distance typically support a correlation between realized dispersal distance and dispersal potential (measured as pelagic larval duration).

An alternative genetic approach that can provide information on a per-generation time scale is paternity analysis. This technique relies upon the ability to sample most potential parents in a sampling area, and thus are useful over small spatial scales where all adults can be readily located.

Environmentally induced tags rely upon variation in environmental conditions, such as gradients in temperature or salinity or in metal concentrations in seawater to generate site or region specific tags. This variation in environmental factors is thought to generate variation in the chemical composition of calcified structures formed by dispersing larvae, such as otoliths (ear stones) of teleost fish, statoliths of molluscs, and larval shells of molluscs (Fig. 3). These calcified structures, then, potentially record environmental history in discrete time slices for the entire time period that the structure is forming. In some species, calcified structures are formed before larvae are released

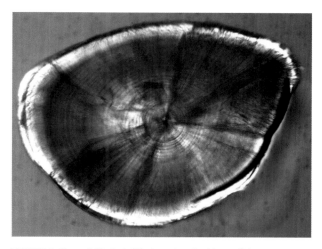

FIGURE 3 The calcified otolith (ear stone) of bony fish can act as a "flight recorder" of the environmental history experienced by larval fish at their birth location and during their dispersal phase. Otoliths can thus act as naturally induced tags of birth location. Photograph by Michael Sheehy.

into the ocean currents, and so these structures can act as natural tags of birth location. Fish scientists have examined trace elements present in the otolith to reconstruct migration and dispersal patterns and to identify spawning grounds and juvenile nursery habitats.

To take advantage of either genetic tagging or environmentally induced tagging, scientists must first determine the spatial scale at which tags are generated. For example, if each population along a stretch of coast were genetically distinct from one another and therefore produced larvae that were uniquely tagged with their birth population's tag, a scientist could easily track larvae collected at any particular location back to its birthplace by simply analyzing the unique genetic tag. On the other hand, if distinctive genetic tags were present only in groups of populations separated by distances greater that 100 km, a researcher might, at best, be able to assign a larva's birthplace within a 100-km range. The ability to answer questions about larval exchange is therefore limited to that spatial scale at which variation in the tags occurs. Thus, successful application of any natural tag to identify source population information typically requires a complete characterization of the spatial and temporal variability in the tags. For both genetic and environmentally induced tagging approaches, there can be a large amount of uncertainty in interpreting the variation in tags among locations.

MATHEMATICAL MODELING

Using mathematical modeling, scientists can predict the trajectory and distance traveled by larvae released from a hypothetical location. Modelers can incorporate into their mathematical equations a large number of variables influencing larval dispersal, such as oceanography (current direction/speed, presence of eddies/complex flow fields), larval duration, mortality in the plankton, and larval competency duration. Estimating dispersal outcomes using models can be particularly valuable to determine the relative importance of these variables on dispersal outcomes.

Some models assume that larvae act as passively floating particles despite the knowledge that larvae can exhibit complex behavior (such as vertical migration and swimming). Estimating dispersal using models relies upon the acceptance of a number of assumptions about the parameter values of variables influencing dispersal.

SEE ALSO THE FOLLOWING ARTICLES

Genetic Variation, Measurement of / Larval Settlement, Mechanics of / Monitoring: Techniques

FURTHER READING

Kinlan, B. P., S. D. Gaines, and S. E. Lester. 2005. Propagule dispersal and the scales of marine community process. *Diversity and Distributions* 11: 139–148.

Levin, L. A. 2006. Recent progress in understanding larval dispersal: new directions and digressions. *Integrative and Comparative Biology* 46: 282–297.

Mora, C., and P. F. Sale. 2002. Are populations of coral reef fish open or closed? *Trends in Ecology and Evolution* 17: 422–428.

Thorrold, S. E., G. P. Jones, M. E. Hellberg, R. S. Burton, S. E. Swearer, J. E. Neigel, S. G. Morgan, and R. R. Warner. 2002. Quantifying larval retention and connectivity in marine populations with artificial and natural markers. *Bulletin of Marine Science* 70: 291–308.

DISTURBANCE

WAYNE P. SOUSA

University of California, Berkeley

The term "disturbance" refers to the displacement, damage, or death of organisms caused by an external physical force or condition or incidentally by a biological entity. Physiological or mechanical stress that does not result in tissue loss or death would not be considered a disturbance, although such stress is a common precursor to disturbance. The force or condition that causes disturbance is referred to as the agent of disturbance. Disturbance affects the structure and dynamics of intertidal populations and communities in a variety of ways. By displacing, damaging, or killing resident organisms, disturbance may (1) free up limiting resources, particularly space, for

exploitation by colonists or survivors and thereby reset the successional state of the assemblage, (2) promote or hinder the coexistence of competitors, and (3) disrupt or enhance the influence of positive interspecific interactions. The nature and consequences of these effects depend on characteristics of both the disturbance regime and the affected organisms and assemblages.

COMMON AGENTS OF DISTURBANCE ON ROCKY SEASHORES

Common agents of physical disturbance on rocky seashores include wave forces; impact or abrasion by wave-borne objects such as cobbles, logs, or ice; extremes of air or water temperature; and desiccation associated with long periods of exposure at low tide. Abrasion by suspended sand or burial under deposited sand is an important agent of disturbance in areas where sandy beaches are contiguous with areas of hard substrate.

Biological entities also cause disturbance on rocky seashores. Biological disturbance occurs when organisms (other than targeted prey) are damaged, displaced, or killed by activities of animals or by algal fronds whiplashing rock surfaces. Examples of disturbance caused by animals include the bulldozing of sessile invertebrates or algae from the interior of territories maintained by limpets (Fig. 1) and the crushing and abrasion of invertebrates and algae by seals as they haul out onto emergent rocks to rest. Some authors also refer to the negative impacts of predation, herbivory, and parasitism as biological disturbance. It is, however, useful to distinguish between these trophic interactions and the phenomena just described, because the patterns and consequences of the two can be quite different.

FIGURE 1 Defended territory of an owl limpet *(Lottia gigantea)*. The limpet (at arrow) has bulldozed barnacles, smaller limpets, and other sessile space competitors from its territory (lighter-colored central area). This behavior maintains open space, promoting the recruitment of diatoms and early successional algae on which the owl limpet grazes. Photograph by the author.

Some agents of disturbance permanently alter physical characteristics of the habitat, thereby causing irreversible changes in community structure. Examples include lava flows that cross the intertidal zone and sudden vertical displacement of the shore by seismic activity. Such events are less common than less severe forms of disturbance from which intertidal populations are able to regenerate.

CHARACTERISTICS OF DISTURBANCE

Defining Features

To study the effects of disturbance and compare them among different habitats, it is important to adopt a common set of terms and metrics for describing and quantifying its characteristics. Disturbances can differ in a variety of ways, including their severity (i.e., degree of damage caused) and size (i.e., areal extent). They also vary in duration; some are discrete, short-term events that last for a fraction of the life time of an organism ("acute" disturbances), while others exert their impact over much longer periods ("chronic" disturbances). A floating log crushing a stand of barnacles would be considered an acute disturbance. Cobbles battering and abrading the sides of a tidepool or surge channel on every rising tide would represent a chronic disturbance to most macroscopic organisms. These defining features of a disturbance (severity, size, and duration) strongly influence the rate and mechanisms by which the affected population or community regenerates.

The Regime of Disturbance

Seashores vary in the sizes, severities, and durations of disturbance they experience. The distributions of one or more of these characteristics often differ in mean or variance among shores or sections of a shore. The disturbance regime is described by these distributions of disturbance properties, together with the spatial and temporal patterning of their occurrence. Frequency and predictability of disturbance are especially important temporal components of the disturbance regime. Frequency refers to the number of disturbances that occur over an interval of time. The predictability of disturbance is inversely related to the variation in time between successive events. The more regular a disturbance, the more predictable it is. This patterning of disturbance has an important effect on the abundance and persistence of intertidal populations and, thus, the species composition and diversity of intertidal communities (see "The Impact of Disturbance," later in this article).

Spatial and temporal variation in the regime of disturbance is a key source of heterogeneity in intertidal environments. Within a geographic region, some sites suffer more damage from drift logs, ice scour, the impact of cobbles, or wave forces than others. At larger spatial scales, the importance of ice damage or heat stress varies with latitude. The intensities and frequencies of disturbing forces vary in time as well; strong seasonal variation in disturbance rates is typical in most regions. More biomass is lost and open space generated during months when large storm waves or floating ice is common. For example, on the outer coast of the state of Washington, the mean rate at which bare patches are cleared in beds of the mussel *Mytilus californianus* is more than an order of magnitude greater in the winter, when large storm waves strike the shore, than during the summer, when waves are much smaller. Similarly, in southern California intertidal boulder fields, boulders are more likely to be overturned during stormy winter months than in the summer, when wave energy is low. Ice scour is obviously a very seasonal phenomenon at temperate latitudes. Large interannual variation in disturbance rates has been documented in all these systems.

Interactions between Properties of Organisms or Populations and Disturbing Forces

The regime of disturbance is determined not only by the patterning and intensity of external agents of disturbance (e.g., wave forces, fluctuations in air or water temperatures) but also by intrinsic variation in the susceptibility of the target organisms. Consider, for example, the forces associated with breaking waves on the seashore. As a wave breaks against the shore, intertidal organisms in the path of the flowing water experience drag, lift, and accelerational forces. An organism's size, shape, and flexibility influence the absolute and relative strengths of these forces. The first two forces, drag and lift, result from the unequal pressures that develop on different sides of an organism as water flows past it. In the case of drag, greater pressure develops on the upstream side of the organism than downstream in the turbulent wake; the net force tends to push the organism in the direction of flow. All else equal, the force of drag increases with the size of an organism, or more specifically, the area of the organism that is projected in the direction of flow. Organisms that are inherently more streamlined in shape, or are flexible enough to assume a streamlined shape in flow, will experience less drag than those that have a fixed shape and a relatively larger area projected into flow. Organisms in flow may also experience a vertical force that tends to lift them off the substrate. A net upward force develops because water cannot flow under an attached organism and only slowly around its base, potentially resulting in relatively high pressure beneath it, while pressure above the organism drops because the flow of water is forced to accelerate over it. Size and shape have similar effects on the magnitude of lift as they do on drag. The third kind of hydrodynamic force, accelerational force, develops when flow around an organism is accelerating. In contrast to lift and drag forces, which are proportional to the area exposed to flow, accelerational force is proportional to the volume of fluid displaced by the organism.

Our understanding of the combined impact of these different forces on intertidal organisms is a work in progress, but recent theory and empirical studies are providing insight to the manner in which disturbance from wave forces can limit organism size and density. Generally speaking, as an organism grows larger, the hydrodynamic forces it experiences increase, making adhesive failure or breakage more likely. However, accurate predictions concerning the effects of wave forces on sessile organisms are made more challenging by the fact that neighboring organisms or substrates intercept and modify flow, thereby influencing the forces that target organisms experience. As a result, laboratory measurements on isolated individuals may not translate well to the situation in the field.

An organism's size and shape also affect its susceptibility to damage or death by climatic extremes. Small individuals or organisms have larger surface-to-volume ratios, making them more susceptible to drying out or heating up during prolonged exposures to the air at low tide. This pattern has been documented for common intertidal organisms such as barnacles, limpets, sea anemones, and various species of macroalgae. On the other hand, small individuals are better able to fit into cracks and crevices, where they can escape the extreme conditions. Surrounding organisms of the same or different species can also modify local physical conditions, often retaining moisture, blocking the wind, or insulating against extreme temperatures and thus ameliorating potentially lethal conditions. For example, at higher levels on the shore, where high temperatures and long exposures to the air create highly stressful conditions, the barnacle *Semibalanus balanoides* suffers less mortality when growing at high density than at low or medium density. Dense stands of barnacles shade the rock surface, so that it remains cooler than areas supporting lower barnacle densities, and, as a consequence, barnacles in dense aggregation experience less thermal stress.

Under more benign conditions, high population density and species interactions can increase the likelihood of disturbance by an external force. For example, when barnacles recruit in high numbers to sites with plentiful food, few predators, and moderate physical conditions, they quickly grow to fill the open space. As they come into contact with neighboring individuals, they begin to grow upward instead of expanding at the margin. This process leads to the formation of hummocks of weakly attached, elongate individuals that are more prone to being torn loose by wave forces than are barnacles of the shorter, conical shape characteristic of individuals growing at lower densities. Similarly, as mussel beds *(Mytilus californianus)* develop, individuals coalesce into aggregations that form a single-layer over the rock surface, to which the byssal threads of most individuals are attached. As the mussels grow larger and new individuals recruit to the bed, primary space becomes limiting. The bed gradually becomes multilayered, with an increasing proportion of mussels being attached to the shells of other mussels rather than to the rock itself. Multilayered beds have a higher profile than single-layered beds and develop hummocks, where densely packed individuals have been lifted away from the substratum. As this bed morphology develops, a lifting force is generated by the pressure difference between the water that slowly flows through the interstices of the bed and the rapidly flowing water within waves that break over the top of it. This force pulls weakly attached mussels out of the bed, increasing the roughness of the bed surface. As edges of disrupted sections of the bed are projected into the flow, drag increases, and the fabric-like matrix of mussels begins to flap in flow and to peel away from the substratum, creating patches of cleared space.

This feedback between the growth and development of an assemblage and its vulnerability to one or more agents of disturbance has the potential to generate endogenous cycles of disturbance within some communities. For example, long-term monitoring of *Mytilus californianus* beds at a wave-exposed site on the outer coast of the Pacific Northwest of the United States has shown that any given area of mussel bed is redisturbed by wave forces at an average interval of 7–8 years. This coincides with the duration of succession from bare rock to a mature, multi-layered, wave-vulnerable mussel bed. Thus, it appears that the rate of succession and associated changes in community vulnerability to disturbance strongly influences the disturbance recurrence interval. At less wave-exposed sites, the interval between successive disturbances is longer. This is because succession to a more vulnerable state is slowed by a lower supply of propagules (i.e., lower recruitment), lower concentrations of suspended food and dissolved nutrients (i.e., slower growth), and more stressful abiotic conditions (i.e., more desiccation), and because forceful waves are less frequent.

THE IMPACT OF DISTURBANCE

The impact of disturbance is "in the eye of the beholder." For example, the loss of several fronds from an alga as a result of moderate drag forces can markedly affect its subsequent growth or reproduction but may have little or no effect on the dynamics of its population or the structure of the community in which it lives. In contrast, if intense forces associated with a large wave break an alga's stipe or tear its holdfast from the rock, not only is the individual alga affected, but so are the density and age/size structure of its population. Furthermore, the space opened by the disturbance becomes available for colonization, which is likely to alter assemblage structure.

From a population or community dynamics perspective, a disturbance has occurred when an external force or physiological stress kills one or more resident organisms or damages them sufficiently to indirectly affect the abundance of other organisms in the assemblage. The indirect effect of this damage may be positive or negative. When damage frees up limiting resources, other organisms may benefit. In rocky intertidal habitats, bare space is the most important limiting resource for most sessile and some mobile species (e.g., limpets). It allows for secure attachment, access to resources (light and suspended and benthic food), and room for growth and reproduction. On the other hand, damage to an organism by disturbance may disrupt a mutualistic or predator–prey interaction, adversely impacting nontarget species.

Effects of Disturbance Characteristics on Succession

The severity, size, and location of a disturbance can markedly affect the patterns and mechanisms of regeneration. A very intense, acute disturbance may kill all residents in the affected area, in which case the site can be recolonized only via the settlement and recruitment of juvenile stages (larvae, spores, or zygotes) dispersing in from outside source populations or from the lateral movement of older individuals living just outside the disturbed area. In contrast, reoccupation of open space generated by less severe disturbances can also be from vegetative regrowth of damaged survivors. The relative contribution of vegetative regrowth obviously depends on the growth form of the affected species. Hard-shelled, "unitary" organisms, such as barnacles and mussels, cannot regenerate from severe damage, while "modular" organisms, such as

many red and brown macroalgae and clonal invertebrates such as sponges, bryozoans, colonial tunicates, and some species of anemones, are capable of vigorous regrowth from surviving tissues. Generally speaking, a disturbed assemblage will regenerate more rapidly and with greater fidelity to its original composition when reestablishment can occur by vegetative regrowth than when recruits must come from outside the affected area.

Succession is the sequence of species replacements that occurs during recolonization of a disturbed area. Early successional species colonize shortly after a disturbance; characteristically, such species produce many small propagules (e.g., spores or larvae) that are dispersed over large distances. In contrast, the later successional species, which replace earlier colonists, produce fewer, larger propagules that are not dispersed as far from the parent. Because of these life history differences, the size of a disturbed area can influence the rate of successional replacement, particularly following a severe disturbance, in which recruits must come from outside the patch. When the disturbed area is small, both early- and late-arriving species can colonize the entire area, which usually results in a faster sequence of species replacement than occurs in large cleared patches. In contrast, only better-dispersing early successional species may be able to immediately colonize the centers of large disturbances; the more poorly dispersing species that typically dominate later in the sequence must slowly colonize their way into the center over multiple generations. This slow encroachment by late successional species may be by propagule dispersal, vegetative growth, or both.

Disturbance size can also indirectly affect successional dynamics by influencing the impact of consumer species. The influence of selectively grazing limpets on algal succession in patches cleared by disturbance in intertidal mussel beds is a good example. When wave forces or drifting logs create patches of bare space in mussel beds, limpets inhabiting the surrounding intact bed forage only a limited distance (about 10–15 cm) into the patch. Thus, the entire interior of small clearings (<20–30 cm in diameter) will be grazed, while the centers of large clearings remain ungrazed (Fig. 2) until the edges of the mussel bed encroach or a protective canopy of large algae develops in the clearing. Consequently, the centers of large disturbed patches provide temporary refuges for algal species fed on by limpets. This interaction between the size of a disturbed area and grazing pressure affects the rate and pattern of algal recolonization and the species composition of the assemblage that develops. Large clearings develop dense canopies of species whose juvenile stages are vulnerable to

FIGURE 2 Recolonization of a patch of bare space created by wave forces in a bed of the mussel *Mytilus californianus*. A dense turf of the early successional green alga *Ulva* dominates the center of the clearing; the bare zone around the perimeter of the patch is maintained by limpet grazing. Limpets forage only 10–15 cm from the protective environment of the surrounding mussel bed. Photograph by the author.

grazers, whereas small clearings are dominated by algal species that grow as grazer-resistant turfs or crusts.

Effects of Disturbance Characteristics on Population Dynamics

Discussions by non-ecologists of the impact of disturbance in natural communities commonly emphasize the losses resulting from disturbances: Organisms are damaged or killed, and their population density is reduced, sometimes resulting in local extinction. However, as is true for other ecosystems, the abundances of certain intertidal organisms are enhanced by particular regimes of disturbance; in fact, some appear to have evolved a dependence on disturbance for their persistence. Two particularly compelling examples, one an invertebrate and the other an alga, come from seashores along the Pacific coast of North America. Recruitment of the sabellid polychaete worm *Phragmatopoma californica*, which can be found growing in dense aggregations in rocky intertidal (Fig. 3) and shallow subtidal habitats from central California to Panama, is strongly enhanced by disturbance. Larvae of this species settle in dense aggregations on the rock surface. Individual worms live in tubes constructed of cemented sand grains; neighboring tubes fuse to form large, honeycomb-like structures that can be 50 cm thick and cover several square meters of substrate. These aggregations are often disturbed during winter storms by strong wave forces or wave-borne projectiles. When the tubes are broken, large numbers of eggs are spawned and then fertilized in the water column. The larvae are competent to settle 2–8 weeks later, but they can delay settlement and remain in the plankton for several months. A study in southern California found that

recruitment of *Phragmatopoma* was highly correlated with wave power measured at the monitored sites 2.5–5 months earlier, with the highest correlation at a 5-month lag interval. In addition, recruitment of the worm was highest during years with the greatest storm activity.

FIGURE 3 Aggregations of the sabellid polychaete worm *Phragmatopoma californica*, growing along the edge of a rocky intertidal bench. Photograph by the author.

Not only are populations of the sea palm *Postelsia palmaeformis* enhanced by a certain regime of space-clearing disturbance, but their persistence may depend on it. This brown alga has an annual life history: The macroscopic sporophyte phase of its life cycle recruits most densely in disturbance-generated clearings in beds of the mussel *Mytilus californianus* (Fig. 4). The mussel is a superior space competitor; as it gradually reoccupies the patch by larval recruitment or the inward movement of edge individuals, fewer and fewer sea palm sporophytes are able to establish each generation. Eventually, the local density of sporophytes falls below that which can produce sufficient gametophyte stages to maintain the population, and the local population goes extinct. For *Postelsia* to persist across

FIGURE 4 Stands of the sea palm *Postelsia palmaeformis*. The clumps of sporophytes occupy patches of cleared space that were opened by wave forces within a bed of the mussel, *Mytilus californianus. M. californianus* is the dominant space competitor at this tidal height. Photograph by the author.

an intertidal landscape that is dominated by mussels, new disturbances must create cleared space within the dispersal distance of the alga's spores (roughly 1–3 m from the edge of an adult stand). Its spores must then successfully colonize this open space within the time it takes for the source population to go extinct or fall to such low density that successful dispersal is precluded. Therefore, regional extinction will occur when either the average disturbance rate (area cleared per unit time) is too low or the interval between successive disturbances is too long. The term "fugitive species" has been coined for organisms that, like *Postelsia*, experience frequent population extinction at small spatial scales but nonetheless persist at larger scales by virtue of being able to rapidly establish new populations in recently disturbed sites where resources are plentiful.

Effects of Disturbance Characteristics on Community Structure

LOCAL SCALE

The disturbance regime has a marked influence on the kinds, numbers, and relative abundances of species in a local community. When disturbance is chronic and severe, the time between events is so brief that few if any invertebrates or macroalgae can colonize and grow to maturity; under such conditions the habitat will remain relatively barren, except for rapidly colonizing films of diatoms or persistent, disturbance-resistant crusts of some coralline algal species. When disturbance is slightly less frequent and severe, the assemblage will be dominated by early successional species (e.g., green algae such as *Ulva*) that can colonize and grow rapidly during the short intervals between successive disturbances. As the interval between disturbances increases or their severity declines, a more

diverse suite of longer-lived, more slowly colonizing species are able to establish and gradually replace early successional species. Where disturbance is rare, succession is able to continue uninterrupted, leading to dominance of the assemblage by one or a small number of long-lived late successional species and a decline in species diversity. The mechanisms of successional replacement can be several, including competitive exclusion of early by later species or differential resistance of later species to consumers (e.g., predators or grazers) or pathogens. Such observations inspired the Intermediate Disturbance Hypothesis, which predicts that local species diversity will be maximal when disturbance occurs with intermediate frequency and severity. Such a disturbance regime allows both sufficient time for species to accumulate and a sufficient rate of resource renewal to prevent dominance by one or a few late successional species. The community is maintained in a diverse, mid-successional state. The faster the rate of successional replacement, the higher the disturbance rate must be to maintain diversity. The role of intermediate disturbance rates in maintaining rocky intertidal diversity has been most clearly documented in boulder fields, where wave forces overturn the boulders at frequencies inversely related to their size and mass. When a boulder is flipped over, organisms on what was formerly the upper surface are abraded, suffocated, or shaded to death. When the boulder is later righted to it original position, the denuded surface is available for recolonization. Boulders of intermediate size are overturned by storm waves less frequently than smaller boulders but more frequently than large boulders. As a consequence, the algal assemblages on their upper surfaces are maintained in a middle successional stage that is more diverse than the early or late successional stages that characterize the surfaces of smaller and larger boulders, respectively.

An alternative mechanism by which disturbance can maintain high local diversity is called the Compensatory Mortality Hypothesis. If damage or mortality caused by disturbance falls disproportionately on the species that dominates late in succession, it will be prevented from excluding earlier successional species, and diversity will be maintained. In other words, the selective impact of disturbance on the potential dominant compensates for its competitive or other advantage over earlier successional species. Although some selectively feeding intertidal predators have been shown to maintain diversity by this mechanism, there do not appear to be any examples of compensatory mortality caused by physical agents. The increased vulnerability of older, multi-layered *Mytilus californianus* beds to wave forces is a possible example. In this case, however, competitive exclusion of earlier successional species has already occurred long before disturbance intercedes, so diversity is not being maintained, strictly speaking.

REGIONAL SCALE

The regime of disturbance also affects the species composition and dynamics of assemblages over larger areas comprising multiple local patches of habitat; this is often referred to as the regional or landscape spatial scale. As describe earlier, the occurrence of disturbance varies in space and time and often transforms rocky intertidal landscapes into mosaics of different successional stages that vary in species composition and diversity as a function of the time since the last disturbance event. This is true of continuous rocky shores (Fig. 5) as well as inherently patchy habitats such as intertidal boulder fields (Fig. 6).

FIGURE 5 Intertidal mosaic of algal- and mussel-dominated patches. These two phases of the mosaic represent, respectively, mid- and late stages of succession following space-clearing wave disturbance. Photograph by the author.

The richness and diversity of species that occupy such a landscape will be highest when the variance in patch age (i.e., time since last disturbed) across the landscape is maximal. Typically, this will occur at sites with intermediate rates of space-clearing disturbance. If the rate is very high, most clearings will be in an early stage of succession, dominated by a small number of rapidly colonizing species. If the rate of disturbance is low, most clearings will be in a late stage of succession, dominated by competitively dominant or consumer-resistant species that have replaced earlier colonists. At intermediate rates of disturbance, most patches will be in a diverse middle successional stage composed of a mixture of surviving earlier colonists and establishing later colonists.

FIGURE 6 Boulders supporting different stages of algal succession following overturning of the substrate by wave forces. The recently disturbed boulder in the center of the photo is covered with the early successional green alga *Ulva*. The surfaces of surrounding, undisturbed boulders are covered with late successional red algae. Photograph by the author.

The regime of disturbance that maximizes species diversity depends on myriad characteristics of the initial disturbance (e.g., distributions of size, shape, and severity), its spatial and temporal pattern of occurrence, and the life histories and other aspects of the biology of the affected organisms and other species that might newly colonize the system from outside. Precise predictions are made all the more challenging because these factors rarely operate independently. Interactive effects are commonplace; some were mentioned earlier, and the interested reader can find a more detailed discussion of these complex but important phenomena in the Further Reading.

Effects of Anthropogenic Disturbance

Rocky intertidal habitats are increasingly affected by various forms of disturbance caused by human activities. Anthropogenic agents of disturbance include increased rates of sedimentation associated with shoreline development or the installation of breakwaters, oil spills, trampling, and harvesting of edible or bait species. Some, such as exposure to oil, are qualitatively novel agents of disturbance for most rocky intertidal communities, to which member species cannot have evolved morphological resistance, physiological tolerance, or compensatory life history attributes. Other kinds of human disturbance, such as trampling and sedimentation, are qualitatively similar to natural agents of disturbance but differ in their patterns of occurrence and severity of effect. They often occur more frequently and at times that do not match the seasonality of comparable natural disturbances. In addition, their impacts are often more severe and chronic. Consequently, we expect the changes resulting from anthropogenic disturbance to be more extreme and persistent than those of natural disturbance events.

SEE ALSO THE FOLLOWING ARTICLES

Biodiversity, Maintenance of / Ice Scour / Projectiles, Effects of / Size and Scaling / Succession / Wave Forces, Measurement of

FURTHER READING

Barry, J. P. 1989. Reproductive response of a marine annelid to winter storms: an analog to fire adaptation in plants? *Marine Ecology Progress Series* 54: 99–107.

Connell, J. H. 1978. Diversity in tropical rain forests and coral reefs. *Science* 199: 1302–1310.

Dayton, P. K. 1971. Competition, disturbance, and community organization: the provision and subsequent utilization of space in a rocky intertidal community. *Ecological Monographs* 41: 351–389.

Dayton, P. K. 1973. Dispersion, dispersal, and persistence of the annual intertidal algal, *Postelsia palmaeformis* Ruprecht. *Ecology* 54: 433–438.

Denny, M., and D. Wethey. 2001. Physical processes that generate pattern in marine communities, in *Marine community ecology.* M. D. Bertness, S. Gaines, and M. E. Hay, eds. Sunderland, MA: Sinauer, 3–37.

Paine, R. T. 1979. Disaster, catastrophe, and local persistence of the sea palm *Postelsia palmaeformis. Science* 205: 685–687.

Paine, R. T. 1988. Habitat suitability and local population persistence of the sea palm *Postelsia palmaeformis. Ecology* 69: 1787–1794.

Paine, R. T., and S. A. Levin. 1981. Intertidal landscapes: disturbance and the dynamics of pattern. *Ecological Monographs* 51: 145–178.

Sousa, W. P. 1985. Disturbance and patch dynamics on rocky intertidal shores, in *The ecology of natural disturbance and patch dynamics.* S. T. A. Pickett and P. S. White, eds. New York: Academic Press, 101–124.

Sousa, W. P. 2001. Natural disturbance and the dynamics of marine benthic communities, in *Marine community ecology.* M. D. Bertness, S. Gaines, and M. E. Hay, eds. Sunderland, MA: Sinauer, 85–130.

E

ECHINODERMS, OVERVIEW

RICHARD B. EMLET

University of Oregon

The phylum Echinodermata (from Greek *echino-* "spiny" and *derma* "skin") contains exclusively marine animals with five-part (pentamerous) radial symmetry, distributed among five living classes: Asteroidea, sea stars (starfish) (Fig. 1A); Echinoidea, sea urchins and sand dollars (Fig. 1B); Ophiuroida, brittle stars or serpent stars (Fig. 1C); Holothuroidea, sea cucumbers (Fig. 1D); and Crinoidea, sea lilies and feather stars. Sea stars and sea urchins are the most likely echinoderms to be found in wave-swept and rocky intertidal habitats, including tidepools. Brittle stars and sea cucumbers may also be present but less obvious, because they will usually be associated with mussel beds or algal holdfasts, under boulders, or burrowed in sediments beneath exposed rocks and cobbles. Feather stars (Crinoidea) occur in shallow waters of tropical reefs but are usually restricted to deeper water in temperate regions. In shallow, subtidal reef environments, all five echinoderm classes may be encountered. Depending on the species, some are hidden or aggregated by day and out grazing or suspension feeding at night. Sea stars are important predators, and sea urchins are important grazers. Brittle stars and sea cucumbers will be either suspension feeders or deposit feeders, and all crinoids are suspension feeders.

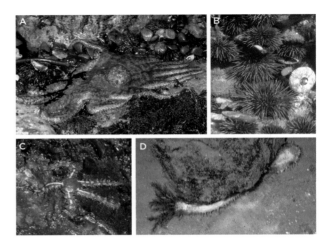

FIGURE 1 (A) The sunflower star, *Pycnopodia helianthoides,* with a purple sea urchin in its stomach. (B) Red and purple urchins, *Strongylocentrotus franciscanus* and *S. purpuratus,* found in tidepools and shallow subtidal habitats along the temperate coast of western North America. (C) The brittle star *Ophiopholis aculeata* is usually under rocks or in kelp holdfasts, but it raises its arms into the water column to suspension-feed. (D) The sea cucumber *Cucumaria miniata* nestles under rocks and uses its oral tentacles (left) to suspension-feed. Photographs by the author.

PHYLUM CHARACTERISTICS

In addition to the adult pentamerous radial symmetry, echinoderms share a unique endoskeleton (because the hard parts are made within a cellular cover) composed of porous plates called ossicles, made of calcium carbonate in the form of calcite. Ossicles can be microscopic, measuring tens of micrometers across, or many centimeters in length, such as the spines of sea urchins. While often complex in overall shape and appearing like a porous fabric at high magnification (Fig. 2), each separate ossicle behaves optically as a single crystal. This porous calcite is called stereom. The porous microscopic structure and

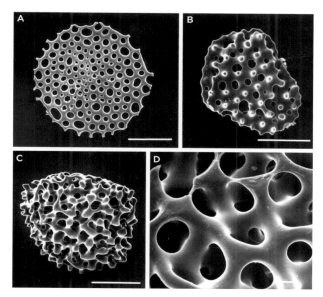

FIGURE 2 Echinoderm ossicles made of calcium carbonate in the form of calcite. The unique porous structure is known as stereom. All images are from sea cucumber, *Psolus chitonoides*. The scale bars in A, B, and C are 100 μm, and the scale bar in D is 10 μm. From Emlet (1982).

on sea urchins. These structures are all connected via a plumbing system of internal pipes. Tube feet connect to a radial canal that is associated with each ray or arm in sea stars, brittle stars, and crinoids, or to one of five radial canals that run along the inside of the body wall from the mouth to the anus of sea urchins and sea cucumbers. The radial canals are interconnected by a ring canal that encircles the oral region of the digestive system. In most sea stars and sea urchins, the tips of tube feet are shaped like tiny suction cups and can adhere to surfaces and are used to hold onto the bottom and to prey.

FIGURE 3 A sea star with tube feet showing. Photograph by the author.

the unicrystalline behavior of most echinoderm calcite permit immediate recognition of living or fossil skeletal material as belonging to an echinoderm.

The ossicles are usually embedded in the outer body wall (the dermal layer), which contains numerous cell types and varying amounts of extracellular fibers made of the protein collagen. A unique feature of echinoderms is that some of the tissues with these collagen fibers have changeable mechanical properties. The tissue can be made soft and pliable or stiff and is apparently under control of the organism's nervous system. This so-called mutable collagenous tissue (MCT) allows sea urchins to hold their spines rigidly in place, the body wall of a sea star to be come rigid and act was a braced framework during feeding, the body wall of a sea cucumber to soften to permit fission or the release of the digestive system, and brittle stars to release (autotomize) all or part of an arm to escape predation. The change in stiffness of the MCT is controlled by nerve cells that terminate in the extracellular collagen matrix. Increased concentration of calcium ions (Ca^{2+}) increases stiffness, and decreased calcium concentration reduces stiffness of the MCT.

A unique organ system, the water vascular system, found in echinoderms can be seen externally in the form of water-filled tube feet, or podia, used for locomotion and feeding (Fig. 3). Tube feet are usually concentrated on the underside of arms of sea stars and in distinct rows

Echinoderms lack a centralized brain, but they have numerous sensory capabilities including the ability to perceive light and chemicals, and they show clear behaviors during predation or escape from predators. They have large nerve tracts associated with the main internal canals of the water vascular system and a diffuse nerve net that occurs throughout the dermal layer and in internal organ systems. They do not have a muscular heart and a dedicated circulatory system, but they are able to exchange oxygen and carbon dioxide via diffusion across thin epithelial regions in the epidermis or the water vascular system. They excrete nitrogenous wastes, primarily as ammonia, by diffusion across the same surfaces. The thin epithelia and an inability to regulate internal ion concentrations restrict echinoderms to fully marine waters. They are intolerant of lower salinities found in estuarine and deltaic environments.

Like other marine invertebrates, most echinoderms have complex life cycles, with planktonic larvae (Fig. 4) and benthic adults. Sexes are usually separate and males and females shed their gametes into the water column where fertilization occurs. Species that produce tiny eggs (diameter 100–200 micrometers) that develop into feeding larvae can produce millions of eggs on single spawning events. These larvae take weeks to develop and must acquire energy and material from the microscopic algal cells they consume. Feeding larvae are bilaterally

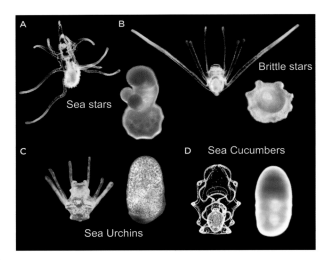

FIGURE 4 Echinoderm larvae from four classes. For each class the form on the left is a feeding larva, and the form on the right is a non-feeding larva. The form of feeding larva is diagnostic for each class. Images by the author.

symmetrical, about 1 millimeter long, and usually transparent except for pigment cells and a pigmented gut. Other species produce larger, energy-rich eggs (diameter 400 micrometers to greater than 1 millimeter) that develop in days to weeks without feeding. The larvae are often over a millimeter long and usually opaque but brightly colored. Still other species may brood their eggs on or within their bodies or protect them on the substratum until they develop enough to crawl away from the parent.

CLASSES OF ECHINODERMS

The sea stars (Asteroida) are recognized by their star shape, with five (or more) arms tapering gradually from the central disk, and their gliding movement produced by tiny tube feet under the arms (or rays). Arms can be so short as to join the central disk to form a pentagon or so long as to exceed the central disk diameter by tenfold or more. Many sea stars are voracious predators on barnacles, mussels, snails, sea anemones, and even other echinoderms. The concept of keystone predator, one that has a major impact on the composition (abundance and diversity) of the community, was first demonstrated with studies on a sea star *(Pisaster ochraceus)*. While most sea stars consume other animals, some feed on algae and seagrasses and others ingest sediments and digest the organic components therein. Sea stars can extrude their stomachs onto a prey item and digest it, or they may swallow a prey whole and digest it internally (see Fig. 1A).

The sea urchins (Echinoida) are globose with a rigid body wall (called a test) and movable spines that vary in length and strength within and between species (see Fig. 1B). The test is made of many thin calcite plates sewn together at the edges by collagen fibers. The spines are attached to projections called tubercles on the outer surfaces of the test plates by layer of collagen and muscle. The tube feet are attached to the epidermis of the test, and at each of these attachment points two holes are present in the test plates to allow continuity of the fluid in the tube feet with the rest of the water vascular system. Sea urchins have an elaborate jaw apparatus called an Aristotle's lantern that is composed of calcite ossicles and muscles that support five teeth used for biting prey or pieces of algae or sea grasses. Sea urchins are dominant grazers in many shallow-water marine environments. They can consume large quantities of kelp and other macroalgae and sea grasses (see Fig. 5). When populations of sea urchins are high, they can reduce algal abundance to a minimum and create what are called barren grounds by their effective consumption of algae. Worldwide, many shallow-water sea urchins are the subject of fishery for their "roe" (actually the gonads rather than the eggs).

The brittle stars or serpent stars (Ophiuroida) are usually smaller than sea stars and sea urchins, and they occur in more protected microhabitats, such as under cobbles and boulders or in the holdfasts of macroalgae. They have a central (circular) body disk covered with

FIGURE 5 Purple urchins in a tidepool have caught drifting bull kelp. So many urchins are holding onto separate blades that the kelp cannot be easily dislodged. Each urchin will eat as much of the blades as possible and possibly even the stipe. Photograph by the author.

scales or small spines and containing the mouth, gut, and gonad. Narrow, bony, segmented arms, often with lateral spines, extend out from the body disk (see Fig. 1C). Brittle stars usually move with their arms and use their tube feet to catch food and move it toward the mouth. As suspension feeders they usually have little obvious impact on the shallow-water community. However, they are prey for fishes, crabs, and other echinoderms.

The sea cucumbers (Holothuroida) are also rather inconspicuous members of wave-swept shores but are more abundant in shallow subtidal habitats and where refuge from waves and predators can be found. True to their name they are wormlike, extended along the oral-anal axis, and often show secondary bilateral symmetry including an upper surface with fleshy protuberances or bumps and a lower surface with most of the tube feet. Tiny, centimeter-long individuals can be extremely abundant (hundreds per square centimeter) but unseen suspension feeders in mussel beds on temperate shores. Others seek safety under rocks or in crevices and hold their branched oral tentacles in the water to remove suspended materials (see Fig. 1D). These animals are prey for sea stars. A number of sea cucumbers are deposit feeders in temperate and tropical shallow water environments. Often large (tens of centimeters to one meter in length), these organisms ingest surface sediments and digest organic and algal cells adhering to the sediments. Several of these species are also subject to fisheries that collect and dry the body wall for consumption (called *bêche-de-mer* or *trepang*), primarily in Asian countries.

SEE ALSO THE FOLLOWING ARTICLES

Herbivory / Kelps / Predation

FURTHER READING

Emlet, R. B. 1982. Echinoderm calcite: a mechanical analysis from larval spicules. *Biological Bulletin* 163: 264–275.

Hendler, G., J. E. Miller, D. L. Pawson, and P. M. Kier. 1995. *Sea stars, sea urchins, and allies: echinoderms of Florida and the Caribbean*. Washington, DC: Smithsonian Institution Press.

Lawrence, J. M. 1987. *A functional biology of echinoderms*. Baltimore: Johns Hopkins University Press.

Lawrence, J. M. 2001. *Edible sea urchins: biology and ecology*. Amsterdam/New York: Elsevier.

Ruppert, E. E., R. S. Fox, and R. D. Barnes. 2004. *Invertebrate zoology*, 7th ed. Chapter 28, *Echinodermata*. Belmont, CA: Brooks Cole.

University of California Museum of Paleontology. *Introduction to the Echinodermata . . . from starfish to sea cucumbers . . . .* http://www.ucmp.berkeley.edu/echinodermata/echinodermata.html.

Wilkie, I. C. 1996. Mutable collagenous tissue: extracellular matrix as mechano-effector, in *Echinoderm studies*, Vol. 5. M. Jangoux and J. M. Lawrence, eds. Rotterdam: A. A. Balkema. 61–102.

ECHIURANS

DAVID JULIAN

University of Florida

Echiurans, commonly known as "spoon worms" or "innkeeper worms," are nonsegmented, sausage- or sac-shaped marine worms typically characterized by a highly extensible, sometimes "spoon-shaped" proboscis that cannot be retracted into the body. Echiurans vary in body length from a few to 50 cm and may live up to 25 years. Most echiurans are found in shallow-water habitats, where they excavate permanent, U-shaped burrows that harbor a variety of commensal animals. Echiurans are probably annelids with a sister relationship to polychaetes (therefore the lack of anatomical segmentation is a derived condition), and are grouped into two orders: the Echiuridae (*ca.* 150 species) and the Xenopneusta (four species, all in the genus *Urechis* in the family Urechidae).

HABITAT AND DISTRIBUTION

Echiurans occur in polar, temperate, and tropical seas, although the families differ in typical habitat: Echiuridae are found in sublittoral to abyssal depths, and Urechidae are found in shallow habitats (from littoral to several meters depth). Although most echiurans create their burrows in fine sediments of shallow-water habitats including mud flats, mangroves, and sea grass beds, others live in crevices and cavities of rocks and deep-water corals; create burrows inside buried, silt-filled bivalve shells and sand dollar tests; burrow into limestone (possibly with the aid of an acidic mucus secretion); or remain exposed on the sea floor in shallow depressions.

ANATOMY AND PHYSIOLOGY

The muscular body wall of echiurans surrounds an undivided coelomic cavity, within which are suspended the long alimentary canal, nerve cord, two or more paired nephridia, and a pair of anal sacs, the latter of which function as excretory structures and are unique among the bilateria. Most species have a closed vascular system, but in some the coelomic fluid contains abundant coelomocytes with intracellular hemoglobin for O_2 transport and storage. Regular peristaltic contractions of the muscular body wall produce burrow irrigation. In the Xenopneusta ("strange breathers"), the highly distensible, thin-walled hindgut is tidally ventilated through the anus with large volumes of water and is the site of 50% or more of the animal's O_2 uptake, with the remainder occurring across the body wall. The sexes are superficially similar in echiurans other than the Bonellidae, which show extreme sexual dimorphism and environmental sex determination. Sexually undifferentiated *Bonellia viridis* larvae that settle away from females usually become females, typically reaching 8 cm in body trunk length, whereas larvae that settle on a female become 1–3 mm long dwarf males. Up to four dwarf males reside completely within the female's modified nephridium, which stores eggs for fertilization. Some intertidal echiurans are periodically exposed to the toxic chemical hydrogen sulfide, particularly during low tide, and a number of detoxification mechanisms have been described in the urechid *Urechis caupo*.

FEEDING

Echiurid species are surface detritivores that extend the proboscis through one burrow opening and trap particles from the sediment surface on extruded mucus. When fully loaded, the proboscis is withdrawn and the skimmed particles are ingested. Each proboscis withdrawal leaves a narrow track on the sediment surface, producing characteristic lines in the sediment radiating out from the burrow opening. Following digestion, the fluidized sediment is expelled from the opposite burrow opening, forming a large conical mound. When fully extended for feeding, the proboscis may be over 100 times the length of the worm's body trunk. Since each worm is sedentary, its feeding area is limited to what is within reach from the burrow openings.

In contrast, Urechid species secrete a mucus net from glands near the short, comparatively nonextensible proboscis (Fig. 1). As the burrow is irrigated, the net captures suspended matter, including bacteria and macromolecules as small as 4 nm in diameter, after which the net and its contents are consumed.

COMMENSALS

The common name "innkeeper worm" refers to the commensal association between some echiurans and smaller infaunal animals that also reside in the burrow. The most common inhabitants include small bivalves, polychaetes (e.g., polynoids or scale worms), crabs (e.g., pea crabs), shrimp, and fish (gobies); and some echiurans can host eight or more species from five phyla. Although some commensals are obligate symbionts that have not been found outside echiuran burrows, most are facultative. However, the interactions and trophic relations between the host and commensals—and indeed whether some interactions are actually mutualistic or parasitic—are mostly unknown.

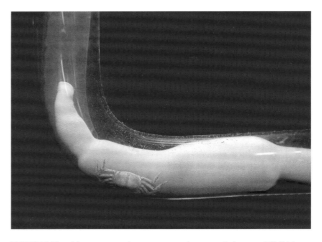

FIGURE 1 *Urechis caupo* and a commensal pea crab in an artificial burrow. The worm has created a mucus net, through which it is pumping water to trap suspended food particles. Photograph by Peter Macht, Seymour Marine Discovery Center at Long Marine Laboratory, University of California, Santa Cruz.

ECOLOGICAL AND COMMERCIAL IMPACT

Population-level studies of echiurans are largely lacking, so whether species abundances are changing is unknown. A variety of marine vertebrates feed regularly on echiurans, including leopard sharks, rays, sea otters, and gulls. Urechids are harvested as fishing bait, and as seafood in China, Japan, and Korea. In addition to providing habitat for commensals, the feeding and metabolic activities of echiurans are likely to substantially modify the local habitat through bioturbation and bioirrigation. For example, in the filter-feeding *U. caupo*, which occurs in mud flats at 60 worms per m^2, burrow irrigation exceeds 350 liters per day per worm (and therefore 21,000 liters per day per m^2), most of which may be filtered through the mucus net.

SEE ALSO THE FOLLOWING ARTICLES

Phoronids / Sipunculans / Worms

FURTHER READING

Anker, A., G. V. Murina, C. Lira, J. A. V. Caripe, A. R. Palmer, and M. S. Jeng. 2005. Macrofauna associated with echiuran burrows: a review with new observations of the innkeeper worm, *Ochetostoma erythrogrammon* Leuckart and Ruppel, in Venezuela. *Zoological Studies* 44(2): 157–190.

Arp, A. J., J. G. Menon, and D. Julian. 1995. Multiple mechanisms provide tolerance to environmental sulfide in *Urechis caupo*. *American Zoologist* 35: 132–144.

Hessling, R., W. Westheide. 2002. Are Echiura derived from a segmented ancestor? Immunohistochemical analysis of the nervous system in developmental stages of *Bonellia viridis*. *Journal of Morphology* 252(2): 100–113.

Osovitz, C. J., D. Julian. 2002. Burrow irrigation behavior of *Urechis caupo*, a filter-feeding marine invertebrate, in its natural habitat. *Marine Ecology-Progress Series* 245: 149–155.

ECONOMICS, COASTAL

JUAN CARLOS CASTILLA

Pontificia Universidad Católica, Santiago, Chile

Man has extracted fish and shellfish from coastal zones from millennia. Presently there are approximately 50 million coastal small-scale fishers (commercial, artisanal, traditional, subsistence) and some 400 million household dependents or persons linked to this activity. They depend directly on shoreline and inshore fisheries for food consumption, economic revenue, and well-being. These systems contain marine resources extracted mainly by small-scale artisanal and subsistence intertidal food-gatherers (hand gathers, hook, trap, netting and skin divers), who are important components of shore ecosystems. The small-scale annual marine catch ranges between 20 and 30 million metric tons.

SUBSISTENCE SHORELINE/ INNER-INSHORE FISHERY ECONOMICS

Small-scale subsistence food-gathering and skin diving are part- or full-time activities occurring at the shoreline and in inner-inshore coastal systems, usually encompassing lone, family, community, or tribal group operators who in many cases follow tidal cycles. Past and present maritime traditional societies have practiced food-gathering subsistence without using boats, or using simple hand-made rafts, such as the artisan totora reef rafts, called *caballitos de mar*, still used in southern Peru, or the sea lion inflated rafts used in northern Chile by the *changos*. These societies have invented fishing tools and practiced sustainable coastal resource management operations successfully for centuries. Their investment is small or non-existent, and fishing gear (hooks, nets, traps, boxes, spears, mollusc or algae detachers; Fig. 1) are usually locally made. Their catches are consumed primarily by operators, families, friends, and tribal groups, and they are occasionally used for exchanges by barter (Fig. 2). Sales occur occasionally, and therefore income from this endeavor is small (Fig. 3) to nonexistent.

Generally speaking, these activities are informal and lack legal regulations. Under management options such as open-access fishery scenarios, and with a continuously growing shoreline population, if the fishery is not duly controlled by the users themselves (based on traditions, local, community, or tribal rules), shoreline and shallow-inshore-water resources are easily overexploited.

FIGURE 1 Intertidal food-gatherer detaching kelp *Lessonia nigesrcens* with an iron bar during low tide (northern Chile). Photograph by Natalie Godoy.

FIGURE 2 Inner-inshore rock fish speared by skin divers in northern Chile for local consumption or barter. Photograph by Natalie Godoy.

Exploitation is exacerbated if external market forces press for these resources that have been traditionally managed exclusively for community needs. Worldwide, there is only very limited information regarding coastal food-gathering or skin-diving noncommercial fishery activities.

FIGURE 3 Accumulation and transport of kelp *Lessonia nisgrescens* extracted by a group of four intertidal food-gatherers in northern Chile. Photograph by Natalie Godoy.

The collection of information is difficult, because usually there are many small operative fishery units distributed over wide spatial sectors. Further, shoreline and inshore-shallow-water food gatherers and skin divers, within societies showing long sea-going traditions and dependence on coastal resources for subsistence (such as Pacific Island cultures), show a high level of self-organization and have developed unique sustainable marine resource management schemes. In other shoreline small-scale fishery societies, where traditional organization is lacking, the operators respond to a sociological "resource-hunter" behavior: operators are individualistic, lack community root-ties, and are averse to following rules.

ECOSYSTEM SERVICES AND CO-MANAGEMENT

The aim of management is to promote a sustainable use of resources over time to (1) prevent resource extinction (commercial and biological, including ecological damages); (2) optimize benefits derived from extraction. Overall, a fishery may be seen as an ecosystem service for humans; thus, multiple direct and indirect drivers may be used to control and operate the activity with the objective of achieving specific management goals and adequate services regarding human well-being.

For small-scale subsistence shoreline/inner-inshore fisheries, which differ substantially from large-scale commercial operations, the "fishery optimization of benefits" does not refer, or only peripherally refers, to economic optimization drivers: Instead, optimization of benefits refers to political and social drivers related to the specific resources to be managed, as much as to the users directly

linked to them. In this case, with a goal of rational management of resources, referring mainly or exclusively to economic drivers or to the use of traditional economic models (i.e., maximum economic yield; profit maximization) is an inexcusable mistake. In short, in small-scale subsistence marine shoreline/shallow inshore fisheries, the fishers are more important than the fish, and economical considerations are less important than societal values, including governance.

In managerial planning, the key elements are the roles of decision makers, stakeholders, and the control system used. In the case of marine fisheries, where resources belong to the "people" (represented by the government) and property rights do not exist, this interaction is a key management driver. Fishery managerial strategies with maximum top-down control mean that the government is the exclusive decision maker. At the other end of the scale lies the fully community-based management system: a management scheme in which the community (tribe, association, users), based on traditions, has 100% control of resource management. What lies between is termed co-management, a managerial resource system in which there is some degree of shared responsibility between the government and the resource users in making management decisions. A fundamental character of co-management strategy is that governments provide the general legal framework for the user organization, while the organizations must be able to regulate the actions of their members. Co-management is expected to increase the efficiency of fisheries management, because compliance and self-regulation are assumed to be better than in top-down approaches. According to the different levels of devolution of power to local communities there are three types of co-management: (1) delegated co-management, in which the government allows formally organized users (traditional or legal organizations, communities, tribes) to make all the decisions and has little or no control; (2) collaborative co-management, in which the government and the users work closely and share decisions; (3) consultative co-management, in which the government interacts often with users but makes all the decisions.

CO-MANAGEMENT AND SUBSISTENCE SHORELINE/INSHORE FISHERIES

Based on the foregoing considerations, it appears that subsistence shoreline/inner-inshore waters fisheries are best suited to be managed by delegated and/or collaborative co-management governance arrangements. Both models have been tested in the field, and in many cases their implementation has proven successful. Critical aspects for success are teamwork; minimum and essential biological-ecological fishery knowledge; social and economic understanding of the resources/systems to be co-managed; use of adaptive strategies; and mutual trust among players (not only between government and users but also, importantly, among users). Delegated co-management is welcomed by small-scale fishers, and the system has shown success in shoreline/inshore fisheries, where strong and long-standing fishery traditions on marine resources extraction and management are deeply rooted in the culture. Collaborative co-management has proven successful in small-scale commercial benthic fisheries, where explicit Territorial User Rights for Fisheries (TURFs) have been assigned to organized small-scale fishery communities.

SMALL-SCALE SUBSISTENCE FISHERY ECONOMICS AND HUMAN WELL-BEING

Worldwide, the shoreline/inner-inshore subsistence fishery is a key, though as yet unquantified activity, and as such it is impossible to be economically evaluated. In shoreline subsistence fishery resource sustainability is amenable to regulations via co-management schemes, in which fishers, still un-technologized, are the central management drivers. Subsistence fishery management models should be sociologically, rather than economically, oriented, and fishers' well-being must be central. Older, single-species fishery management strategies, as well as purely biophysical ecosystems-integral prescriptions for fishery sustainability, appear to be less amenable, or unrealistic, for these fisheries. In small-scale subsistence fisheries, humans have played a central role for thousand of years, as opposed to large-scale fisheries, in which mechanization, technology, and subsidies appear as the managerial drivers to be controlled. Therefore, it is unacceptable to approach the rational management of shoreline/inshore small-scale subsistence fisheries using the same tools as for commercial fisheries.

SEE ALSO THE FOLLOWING ARTICLES

Algal Economics / Food Uses, Modern / Management and Regulation

FURTHER READING

Berkes, F., R. Mahon, P. McConney, R. Pollnack, and R. S. Pomeroy. 2001. *Managing small-scale fisheries: alternative directions and methods.* IDRC publication 1-320. Ottawa, Canada: International Development Research Centre. http://www.idrc.ca/en-ev-9328-201-1-DO_TOPIC.html.
Caddy, J. F., and O. Defeo. 2003. Enhancing or restoring the productivity of natural populations of shellfish and other marine resources. FAO Fisheries Technical Paper No. 448. Geneva: UN Food and Agricultural Organization.
Castilla, J. C. 1994. The Chilean small-scale benthic fisheries and the institutionalization of new management practices. *Ecology International Bulletin* 21: 47–63.
Castilla, J. C., and M. Fernández. 1998. Small-scale benthic fisheries in Chile: on co-management and sustainable use of benthic invertebrates. *Ecological Applications* 8: S124–S132.
Castilla, J. C., and O. Defeo. 2001. Latin American benthic shellfisheries: emphasis on co-management and experimental practices. *Reviews in Fish Biology and Fisheries* 11: 1–30.
Castilla, J. C., and O. Defeo. 2005. Paradigm shifts needed for world fisheries. *Science* 309: 1324–1325.
Gelcich, S., G. Edwards-Jones, and M. J. Kayser. 2005. Importance of attitudinal differences among artisanal fishers towards co-management and conservation of marine resources. *Conservation Biology* 19: 865–875.
Johannes, R. E. 2002. The renaissance of community-based marine resource management in Oceania. *Annual Review in Ecology and Systematic* 33: 317–340.
McClanahan, T., and J. C. Castilla. 2007. *Fisheries management: progress toward sustainability.* Oxford, UK: Blackwell Publishers.
McConney, P. A., R. Pomeroy, and R. Mahon. 2003. *Guidelines for coastal resource co-management in the Caribbean: communicating the concepts and conditions that favor success.* Caribbean Conservation Association and University of West Indies.

ECOSYSTEM CHANGES, NATURAL VERSUS ANTHROPOGENIC

KAUSTUV ROY

University of California, San Diego

Intertidal communities are incredibly dynamic, exhibiting natural fluctuations on many different spatial and temporal scales. These habitats are also easily accessible and subject to increasing human impacts as more and more people inhabit coastal areas. Such impacts can directly or indirectly affect intertidal species and communities. Separating natural from anthropogenic changes is important both from a scientific as well as a management perspective. Yet as more and more people use resources from the tidepools, it is becoming increasingly difficult to separate these effects.

NATURAL CHANGES

Natural changes in the distribution and abundance of intertidal species can result from a number of abiotic as well as biotic factors ranging from changes in temperature, coastal upwelling, and circulation to variations in larval supply and recruitment. Temperature and coastal circulation patterns are important determinants of the geographical distributions of many intertidal species and can influence compositions of intertidal communities on a variety of time scales. A common biological response to changes in temperature is a shift in the distributions or abundances of species or both. During episodes of climatic warming, many species extend their ranges northward, and others show an increase in abundance near their northern distributional limits. The reverse is true during cooling events. Depending on the nature and magnitude of the climatic change, such distributional shifts can separate co-occurring species or lead to formation of new species associations, thereby changing the composition and diversity of local intertidal communities. These changes can happen over a variety of time scales, from short-term fluctuations to decades to geological time. For example, during El Niño Southern Oscillation (ENSO) events, populations of some warm-water species can be found much farther north of their normal distributional limits, but such shifts are ephemeral in that these populations rarely persist much beyond the duration of the ENSO event. Over longer time scales, warming of coastal waters over multiple decades has led to increases in the abundances of warm-water species in temperate intertidal communities such as in Monterey Bay. On even longer time scales, major fluctuations in global climate during the Pleistocene and Holocene (the last 1.8 million years) led to large changes in the compositions of intertidal communities. For example, during warm interglacial periods, such as around 125,000 years ago, intertidal habitats in California harbored many species that today are only found much farther south; the resulting communities have no modern analogs. Such climate-driven changes in community compositions are well documented from many parts of the world. In addition to changes in community composition, such range shifts can also lead to changes in the morphology and genetic population structures of intertidal species. In the Northern Hemisphere, more northerly populations of many species exhibit lower genetic diversity compared to more southerly ones. This is because the northern populations went extinct during Pleistocene glaciations and these regions were recolonized by individuals from southern refugia only after the glacial period was over. Thus these northern populations are too young (less than 20,000 years old) to have accumulated much genetic diversity.

Although changes in species distributions and abundances are a common response to changes in the ambient environment, it is important to note that not all species show these responses; distributions and local abundances of many species can be remarkably stable even in the face of substantial environmental change. Similarly, species that do respond to changes in climate differ in terms of the magnitude of the responses. The ecological and life history characteristics that drive such individualistic responses remain poorly understood. What is clear is that although responses of some species to climate change are predictable from their thermal physiology alone, those of others reflect more complex interactions between biotic and abiotic factors. For example, even small changes in water temperature (such as those resulting from changes in coastal upwelling patterns) have been shown to influence the rate of predation by the sea star *Pisaster ochraceus* on its mussel prey. Since *Pisaster* is a keystone predator that controls the local abundance of mussels, thereby maintaining a diverse intertidal assemblage of algae and invertebrates, in this case change in temperature has the potential to influence community composition by affecting predator–prey interactions.

HUMAN ACTIVITIES AND INTERTIDAL ECOSYSTEMS

Human activities can impact the easily accessible habitats at the land–sea interface in a number of different ways. Our impacts on intertidal ecosystems range from sewage discharge and industrial pollution to more episodic disturbances such as trampling of the intertidal by foot traffic; harvesting of intertidal organisms for food, fish bait, aquariums, and other needs; and moving of rocks and other material such as dead shells that serve as habitats for many invertebrates. Each of these activities, by itself or in conjunction with other impacts, can substantially change species compositions and the nature of intertidal ecosystems. In addition, introduced species can also change the nature and composition of intertidal communities.

Harvesting

Plants and animals living in intertidal habitats are not only a source of food for people, but they are also used in a variety of other ways. For example, many species of molluscs provide the basis for a thriving ornamental shell trade. Because of such diversified use, a large variety of intertidal species ranging from fish and shellfish to algae are harvested by humans. Such harvesting ranges from localized recreational or subsistence collecting to more organized and widespread fisheries, and the species involved varies from one part of the world to another. Harvesting of intertidal organisms for food and cultural use has been going on for thousands of years in many coastal areas, although as coastal human population densities continue to increase rapidly, so does the harvesting pressure and its impact on intertidal species. In effect, humans insert themselves into the intertidal ecosystem as top predators, leading to both direct and indirect impacts on other species.

Human harvesting of intertidal species is highly selective; not all species that live in intertidal habitats are directly targeted for harvesting and those that are harvested are not all harvested to the same extent. Furthermore, harvesting of many species tends to be strongly size selective, with larger individuals preferentially taken. The ecological effects of such removal are complex and vary from one region to another, depending on which species are taken and the harvesting methods. Nonetheless, some general trends are clear on a global scale.

Size-selective harvesting changes the population size structures and abundances of the targeted species; large individuals of harvested species become very rare in areas under heavy exploitation. Such changes have been documented over large stretches of coastlines covering tens to hundreds of kilometers and over many decades (Fig. 1). Removal of large individuals also leads to a decrease in the biomass of the species and can negatively affect its reproductive output. For some species, decreases in spawning biomass as a result of overexploitation have been shown to be responsible for dramatic declines of local populations. Changes in the size distribution and abundance of individual species also lead to changes in the structure of intertidal ecosystems. For example, grazers such as limpets control the abundance of algae and help maintain bare space on the rocks. Removal of these species, a common food item in many parts of the world, can lead to a rapid increase in algal cover and thus loss of space that can be used by other species. Similarly, harvesting of important intertidal predators can lead to a dramatic increase in the abundance of their prey, which in turn can change the nature of the local ecological community. In Chile, *Concholepas concholepas*, a large gastropod, is a prized local food item and heavily harvested. But it is also a keystone species that preys on mussels and creates bare space on the rocks, some of which are then colonized by barnacles. Since the removal of *Concholepas* favors mussels, human harvesting of this key species changes an intertidal community characterized by open space and few mussels into one with a thick cover of mussels (Fig. 2). Mussels themselves are commonly harvested along many coasts, and removal of clumps of mussels indirectly kills many smaller species that live in the complex structures created

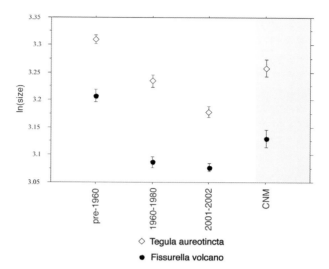

FIGURE 1 Declines in body sizes of two intertidal gastropod species, a turban snail and a small limpet, over time in Southern California in response to human harvesting. Each data point represents the average size (log transformed) across multiple populations in Southern California. For each species, largest individuals were found before 1960. The shaded region of the plot shows data from Cabrillo National Monument in San Diego, one of the few intertidal reserves in Southern California with a human-exclusion zone. That today the largest individuals of these species are found in the only reserve where human harvesting is prevented indicates that size-selective harvesting rather than natural changes have caused the temporal decline in body size. Modified from Roy *et al.* (2003).

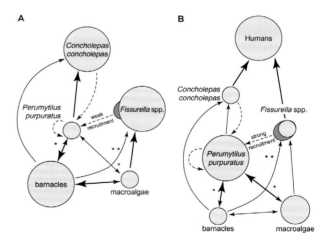

FIGURE 2 Changes in the intertidal community structure in Chile as a result of human harvesting. The size of the circles are proportional to the density of each species shown. Single-headed arrows indicate predation, with the direction of the arrowhead indicating the direction of energy flow. Double-headed arrows indicate that the species involved compete with each other. Broken lines indicate settlement of juveniles with the arrowhead pointing toward the species benefiting. Panel (A) shows the natural condition with no harvesting when *Concholepas* and *Fissurella* dominate. Panel (B) shows the effects of human harvesting of *Concholepas* and *Fissurella*. The mussel *Perumytilus* and various macroalgae increase in abundance at the expense of the barnacles and the harvested snails. Reprinted from Castilla (1999), with permission from Elsevier.

by the mussels. Despite region-specific differences in harvesting practices and species compositions, it is clear that human harvesting of intertidal species leads to predictable changes in community compositions; exploited localities tend to have higher dominance by algae and consequently higher abundances of species associated with algae, but reduced open space and lower abundance and biomass of harvested species as well as many grazers and animals that use the primary substrates.

Trampling

Trampling by humans is another significant yet underappreciated threat to the health of intertidal ecosystems. People can change the composition of intertidal assemblages simply by walking on them. The force exerted by human footsteps crushes many species of algae as well as small invertebrates (e.g., barnacles and small molluscs) and so the direct effect of trampling is a reduction in or a loss of various types of algae including species that form thick turfs and reduced densities of animals associated with the turf. Susceptibility of algal and invertebrate species to trampling does vary, as does the ability of species to recover if and when trampling is discontinued. In places with high human visitation, trampling can transform an intertidal area with a thick cover of algal turf (e.g., coralline turf) to one that is essentially bare rock. Trampling also has indirect effects; densities of limpets and other grazers can increase as algal turf disappears and more bare rock becomes available. Like harvesting, impacts due to trampling are increasing as densities of coastal populations increase on a global scale and more people use the intertidal habitats for recreational or other usage.

Other Impacts

Many other anthropogenic activities ranging from pollution, eutrophication, and oil spills to the presence of introduced species affect the health of intertidal assemblages. Episodic disturbances such as oil spills generally have localized effects, but recovery can be slow, and not all species may come back. Pollution is becoming an increasing problem for intertidal habitats near rapidly growing coastal cities, but the effects of various pollutants on the health of intertidal ecosystem remain poorly known. Some pollutants impair immune response and make intertidal species more vulnerable to diseases, whereas endocrine disrupters (e.g., tributyl tin, TBT) associated with antifouling paints can cause female sterility in some species, leading to reduced reproductive output and the potential for local extinctions.

A number of plants and algae have been intentionally or accidentally introduced by humans into intertidal habitats that are outside the natural distributions of these species. Such introductions have the potential to change the nature and composition of the recipient assemblage, although the effects vary depending on the nature of the introduced species. In some cases introduced species simply increase local diversity without displacing the native species, but in other cases such invasions affect the abundances of native species either directly or indirectly.

SYNERGIES AND THE FUTURE OF INTERTIDAL ECOSYSTEMS

Few ecosystems today are buffered from anthropogenic impacts, and tidepools, being easily accessible, are particularly vulnerable. Also in a world dominated by human activities, the distinction between natural and human-mediated change is increasingly becoming blurry; even changes in global climate are now driven by anthropogenic emissions of greenhouse gases, and the dynamics of global climate in the future are likely to be very different from the conditions under which the species living in the tidepools evolved. Of particular concern is the problem of synergistic interactions between many of these stressors; as populations of many species decline as a result of direct or indirect effects of human harvesting, climate warming, or both, they may become particularly vulnerable to diseases or other disturbances. The dramatic decline of the black abalone (*Haliotis cracherodii*) in California, where overexploitation, climatic warming, and disease have all been implicated, already provides one example of such synergistic interactions. How tidepool ecosystems will look in the future depends largely on our abilities to understand and manage these threats.

SEE ALSO THE FOLLOWING ARTICLES

Climate Change / Economics, Coastal / Habitat Alteration / Introduced Species / Predation / Upwelling

FURTHER READING

Castilla, J. C. 1999. Coastal marine communities: trends and perspectives from human-exclusion experiments. *Trends in Ecology and Evolution* 14: 280–283.

Hellberg, M. E., D. P. Balch, and K. Roy. 2001. Climate-driven range expansion and morphological evolution in a marine gastropod. *Science* 292: 1707–1710.

Roy, K., A. G. Collins, B. J. Becker, E. Begovic, and J. M. Engle. 2003. Anthropogenic impacts and historical decline in body size of rocky intertidal gastropods in southern California. *Ecology Letters* 6: 205–211.

Thompson, R. C., T. P. Crowe, and S. J. Hawkins. 2002. Rocky intertidal communities: past environmental changes, present status and predictions for the next 25 years. *Environmental Conservation* 29: 168–191.

EDUCATION AND OUTREACH

ALI WHITMER

University of California, Santa Barbara

The rocky intertidal habitat offers many opportunities for students of any age to learn science and math in a living laboratory. In addition, learning extensions in language arts are easily accommodated through topical readings and creative writing. Undergraduate and graduate students have long been taken into the rocky intertidal habitat as part of their education. Interpretive centers and docents with agencies such as the National Park Service and the National Marine Sanctuaries programs provide high-quality public education programming. The biggest strides in recent years have been made in the development of K–12 educational opportunities. For example, advances in the development of ocean literacy principles for K–12 education will guide the application of marine studies in the classroom. Following are a few examples of how the rocky intertidal environment can enrich teaching science and mathematics standards along with some exemplar materials and programs. In addition, undergraduate textbooks in marine science and ecology are excellent resources for background information and relevant scientific literature.

BIOLOGICAL SCIENCES

The biodiversity of intertidal organisms is amazing. Careful searching will reveal numerous invertebrates, vertebrates, algae, and fungi. Invertebrates are, by far, the most diverse macroscopic organism, with nearly every phyla represented in this environment. Intertidal organisms are marvelous examples of physiological tolerance. The intertidal is an extremely stressful environment. The daily tidal cycle results in large shifts in temperature, osmotic balance, and water balance. When this is combined with seasonal climate change, from blazing summer heat to winter ice scour, one can see how these organisms provide good models of physiological adaptations. Some of the earliest experimental ecology was conducted in the rocky intertidal. These studies include predation, competition, and how biological versus physical factors set species distributions. The concept of the keystone species was advanced in the rocky intertidal. Many of these experiments can be repeated by K–12 students to ask questions about the ecology of local organisms (Fig. 1).

FIGURE 1 University of California Santa Barbara graduate student Mackenzie Zippay collects egg capsules with San Marcos High School student Veronica Pessino for her biology science project at Coal Oil Point, Santa Barbara. Photograph by Monica Pessino.

EARTH AND PHYSICAL SCIENCES

Seawater is a solution with variable solute concentrations depending on locale, tidal cycle, substrate, and other parameters. Proximity to freshwater sources will alter water chemistry and enables students to study density and buoyancy, including saltwater wedges and density-driven circulation. The intertidal is also an excellent location to study motion and forces as well as transfer of energy using the example of waves. Simple wave meters or dissolution blocks can be made and used in these studies. The rocky intertidal is always a fascinating place to discuss geology and climate. The substratum of the intertidal may reveal variations in geomorphology as well as provide opportunities to examine earth's history through uplifted sections. And the tides provide one of the very best ways to study the earth in the solar system.

MATHEMATICS

Scientists use math as a way to describe, explain, and predict elements of the environment. Students can learn and apply math skills while engaged in study of the rocky intertidal. For example, students can count, estimate, determine the mean, identify outliers, and learn to sample populations of marine organisms. More advanced students can learn basic statistics and compare methods of estimation and probability in the intertidal.

SCIENTIFIC METHOD

Much of what has been described previously has direct correlations with the national science education standards for science as inquiry. The rocky intertidal is a terrific environment in which to propose hypotheses, conduct experiments, collect and analyze data, and develop explanations.

NOTABLE MATERIALS AND PROGRAMS

Educators can turn to local institutions and agencies, such as parks, marine laboratories, universities, and nonprofit groups, for educational materials and activities that focus on intertidal environments. Following are three examples of high-quality programming that has been developed to give students authentic marine science experiences that connect to national and state education standards.

The Tidepool Math lesson series was developed by the Pacific OCS Region Minerals Management Service of the Department of the Interior. The lessons were developed for K–8 and high school and are aligned with California State Mathematics Education Standards. In the K–8 activities, students use a standardized plot of a mussel bed to examine the scientific applications of counting and estimating. The high school lesson plans focus on sampling methods, comparing counts and percent covers to estimate abundance, and understanding basic statistical concepts and methods.

The Long-term Monitoring Program & Experiential Training for Students (LiMPETS) program is a collaborative effort among the five West Coast National Marine Sanctuaries: Olympic Coast, Cordell Bank, Gulf of the Farallones, Monterey Bay, and Channel Islands. One goal of the program is to establish a long-term, quantitative intertidal monitoring program that can be used to assess the health of local rocky intertidal habitats. The program includes direct involvement of K–12 teachers and their students and provides professional development for teachers, monitoring equipment, and a LiMPETS Web site and database.

The Center for Ocean Science Education Excellence (COSEE) program is a national network of centers funded jointly by the National Science Foundation (NSF) and the National Oceanic and Atmospheric Administration (NOAA). The goals of COSEE are as follows: to promote the development of effective partnerships between research scientists and educators; to disseminate effective ocean sciences programs and the best practices that do not duplicate but rather build on existing resources; and to promote a vision of ocean education as a charismatic, interdisciplinary vehicle for creating a more scientifically literate workforce and citizenry. Centers are located across the nation and include locally based and nationally distributed programs for educators, scientists, and the public. The national Web site provides an overview of current centers and links to each COSEE Web site.

SEE ALSO THE FOLLOWING ARTICLES

Aquaria, Public / Collection and Preservation of Tidepool Organisms / Marine Stations / Museums and Collections

FURTHER READING

Centers for Ocean Science Education Excellence Network. www.cosee.net.
National Marine Sanctuaries LiMPETS. http://limpets.noaa.gov/.
Ocean Literacy Network. www.coexploration.org/oceanliteracy.
Tidepool Math lesson plans. www.mms.gov/omm/pacific/kids/Tidepool_Math/tidepool.htm.

EL NIÑO

STEVEN D. GAINES

University of California, Santa Barbara

Climate varies in ways that are random and in ways that are relatively predictable. The predictable components of variation are typically associated with conditions that repeat over different intervals of time. Day–night and seasonal cycles are the most familiar, because we have experienced their consequences so many times and because their physical causes are readily obvious. Climate also varies over longer time scales, with repeated cycles on intervals of years, decades, centuries, and millennia. At the longer end of this list are the glacial–interglacial cycles that have played significant geological and evolutionary roles for hundreds of thousands of years. Climate cycles on these long time scales have played crucial roles in shaping the patterns we see today on rocky shores, even though they occur over time intervals that span hundreds to thousands of generations of even long-lived species. Individuals would never realize such cycles exist, because they spend their entire lifetime in only one part of the cycle. At the other end of the spectrum are interannual climate cycles that recur every few years. In the ocean, the most prominent of these climate cycles is the El Niño/Southern Oscillation (ENSO).

THE BIRTHPLACE OF EL NIÑO

Climate describes the pattern of weather at a place—both its average conditions and its pattern of variation over time. Although people typically think of climate in terms of conditions in the atmosphere, oceans have climates too. Indeed, the climates of oceans and of atmospheres are closely coupled because their physics are tightly intertwined. These linkages are clearly evident in the ENSO.

To understand El Niño, we must start in the atmosphere over the tropical Pacific. The tropics are characterized by trade winds that blow from east to west. When they blow over the ocean, they push warm surface waters to the west. In the tropical Pacific, the trade winds can push for an enormous distance—from the coast of South America to Indonesia (Fig. 1). The surface currents driven by the trade winds eventually reach the diverse land masses of Southeast Asia and Australia. As a result, the winds pile up warm surface water in the western Pacific, and the sea level tilts up from east to west. In a typical year, the trade winds can raise the sea level of the western Pacific a half meter higher than it is in the eastern Pacific.

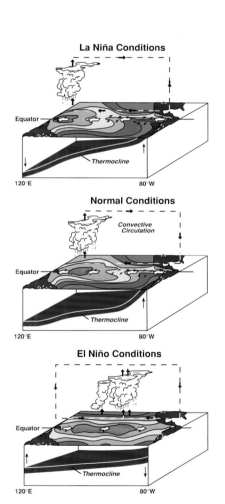

FIGURE 1 Schematic diagram of ocean and atmospheric conditions in the tropical Pacific during La Niña, normal, and El Niño years. Shifts in the strength of the trade winds dramatically alter the distribution of warm sea water along the equator. These shifts in the distribution of warm water also feed back to affect atmospheric conditions by changing the convective circulation in the atmosphere, which has large effects on the distribution of rainfall and intensity and frequency of storm events. Image courtesy of NOAA/PMEL/TAO Project Office, Dr. Michael J. McPhaden, Director.

The equatorial Pacific does more than tilt under these wind conditions. The trade winds also drive a large gradient in sea surface temperature. Along the equator, the westerly push of surface waters is associated with the upwelling of cold, deeper waters to the surface. The resulting gradient in temperature across the equatorial Pacific is typically 8 °C. In addition, the upwelling in the eastern Pacific pumps nutrients into sunbathed tropical surface waters, creating ideal conditions for rapid plankton growth. Since warm tropical seas are typically quite low in nutrients, upwelling in the eastern Pacific provides the fuel for an unusually rich food web in the tropics that has historically supported abundant fisheries.

Now imagine you could throw a switch and shut off or greatly reduce the trade winds. Without the strong westward push of surface waters, the tilt in sea surface height could not be maintained, and waves of warm water driven by gravity, called Kelvin waves, would move from the western to the eastern Pacific. The gradient in temperature across the Pacific would be reduced. Moreover, in the absence of the trade winds, the upwelling along the coast of South America would diminish or even cease. The productivity of the eastern equatorial Pacific would plummet, and fisheries would collapse. These phenomena constitute an El Niño event.

WHY IT IS CALLED EL NIÑO

The name arises from the timing of the most obvious signal of El Niño—anomalously warm water off the coast of equatorial South America around the time of Christmas. Fishermen from Ecuador and Peru long called a warm current around the time of Christmas *El Niño*, "the little boy," for the Christ child. This seasonal warming is much stronger in years when the trade winds relax, a harbinger of a poor fishing season to come. Such years with unusually warm water temperatures are now known elsewhere as El Niño.

EL NIÑO AND HIS SISTER

The strength of the trade winds can vary greatly among years in an irregular cycle. Years with anomalously weak winds are called El Niño, but there are also years with anomalously strong trade winds relative to long-term average conditions. These years have a strengthened temperature gradient across the Pacific and have come to be called La Niña, "the little girl," to signify their contrast to El Niño (Fig. 1). The variation in the strength of the trade winds is driven by an oscillation in atmospheric pressure between the central and western Pacific that was noted a century ago by Sir Gilbert Walker, who called it the Southern Oscillation. Now we know that the changes in the atmosphere and the ocean are coupled, and the cycle through years with normal, El Niño, and La Niña conditions has been termed the El Niño/Southern Oscillation or ENSO.

ENSO is not a regular cycle that alternates predictably between different states. Rather, in recent decades, El Niño has tended to recur every 3 to 7 years. Not all ENSO cycles include both El Niño and La Niña years. At the moment, scientists are unable to predict when an El Niño or La Niña will occur, but they can now detect their onset early in their infancy. Given the dramatic effects these phenomena can have on climate, such early detection is of great value.

Studies of ancient corals in the Huon Peninsula of New Guinea have revealed the history of ENSO over the last 130,000 years. The intensity of El Niño was reduced during glacial periods, but ENSO occurred throughout this entire interval. One of the most striking findings from this study was that El Niño intensity has never been higher than during the past 100 years. The patterns observed during the last century are atypical and raise the question whether human activities that have altered the composition of the atmosphere are changing the frequency and intensity of the El Niño phase of ENSO.

THE IMPACT OF EL NIÑO

The signature of El Niño depends on where you are. Because it is a large-scale phenomenon generated by changes occurring across the entire equatorial Pacific, El Niño has large impacts globally, but the specifics differ markedly from place to place. For example, on land, El Niño years commonly bring greatly elevated rainfall and floods to Peru and the southwestern United States, whereas much of the western Pacific experiences drought. In extreme cases, the latter is associated with much more extensive brush fires in Australia. At sea, El Niño and La Niña have equally large and geographically diverse effects. The main features altered are ocean temperatures, nutrients, currents, and storms.

Along the temperate and tropical coasts of South and North America, sea surface temperatures can be greatly elevated during El Niño, even well away from the equator. These elevated temperatures are often associated with poleward range expansions of warm-water species. In some settings, the warm waters are also associated with dramatic disease outbreaks, as with the echinoderm disease outbreaks in the Channel Islands off the coast of California in the United States (Fig. 2). In addition, productivity frequently plummets in coastal habitats as nutrients decline when

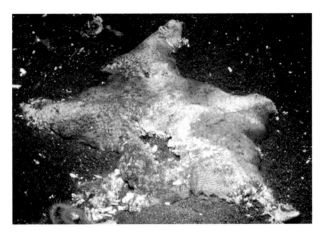

FIGURE 2 A disintegrating bat star (*Asterina miniata*) attacked by a bacterial disease that appears strongly linked to warm water. Outbreaks of the disease are common during El Niño years and cause extensive mortality of sea stars off the coast of Southern California, in the United States. Photograph courtesy of R. Hermann.

water temperatures rise. Kelps can decline precipitously in South and North America as they become physiologically starved for nutrients. Upwelling-favorable winds occur less frequently, and even when they do occur, the benefits for coastal productivity are limited. Since the thermocline is deeper, upwelling commonly only brings warm, nutrient-poor water to the surface. Marine iguanas in the Galápagos shrink in size during El Niño because of the declines in algal productivity as average temperatures rise as much as 14 °C. Even higher up the food chain, marine mammals along the West Coast of the United States commonly have abysmally poor pupping success in El Niño conditions because of food shortages.

Although the decline in upwelling has negative consequences for a wide range of species, one positive effect is commonly seen—higher recruitment rates of some species with planktonic larvae. The reduction in upwelling can keep larvae of fish and invertebrates closer to shore, which may enhance their likelihood of returning to adult habitat after they complete recruitment.

Shifting the distribution of warm water around the Pacific also has dramatic effects on storms. In some areas, El Niño events greatly increase the intensity and frequency of storm activity (e.g., the South Pacific, the Central North Pacific near Hawaii, the Northwest Pacific, Tasmania, and the west coasts of temperate North and South America). In other areas, storm activity tends to decrease (e.g., the Atlantic and the Australian region). In general the opposite trends occur during La Niña. These changes can greatly affect shoreline habitats by increasing or decreasing the normal level of wave-generated disturbances. For example, along the coast of Southern

California in the United States, wave disturbance increases dramatically, on average, in El Niño years. Kelp forests can be entirely removed, with large amounts of seaweed wrack deposited on the shore. When this large increase in disturbance is coupled with elevated temperatures and low nutrients, the structure of coastal ecosystems can be changed dramatically during El Niño.

SEE ALSO THE FOLLOWING ARTICLES

Climate Change / Iguanas, Marine / Temperature Change / Upwelling / Variability, Multidecadal

FURTHER READING

Dayton, P. K., and M. J. Tegner. 1984. Catastrophic storms, El Niño, and patch stability in a southern California kelp community. *Science* 224: 283–285.

Trenberth, K. E. 1997. The definition of El Niño. *Bulletin of the American Meteorological Society* 78: 2771–2777.

Tudhope, A. W., C. P. Chilcott, M. T. McCulloch, E. R. Cook, J. Chappell, R. M. Ellam, D. W. Lea, J. M. Lough, and G. B. Shimmield. 2001. Variability in the El Nino-Southern Oscillation through a glacial-interglacial cycle. *Science* 291: 1511–1516.

Vinueza, L. R., G. M. Branch, M. L. Branch, and R. H. Bustamante. 2006. Top-down herbivory and bottom-up El Niño effects on Galapagos rocky-shore communities. *Ecological Monographs* 76: 111–131.

Wikelski, M., and C. Thom. 2000. Marine iguanas shrink to survive El Niño. *Nature* 403: 37–38.

Wootton, J. T., M. F. Power, R. T. Paine, and C. A. Pfister. 1996. Effects of productivity, consumers, competitors, and El Niño events on food chain patterns in a rocky intertidal community. *Proceedings of the National Academy of Sciences USA* 26: 13855–13858.

Zacherl, D. C., S. I. Lonhart, and S. D. Gaines. 2003. The limits to biogeographical distributions: insights from the northward range extension of the marine snail, *Kelletia kelletii* (Forbes, 1852). *Journal of Biogeography* 30: 913–924.

ENTOPROCTS

TOHRU ISETO

Kyoto University, Japan

Entoprocts are small benthic animals with a tentacular crown at the top of the body. They superficially resemble hydroids and bryozoans. Entoprocts are characterized by a U-shaped gut and an anus positioned so that it opens inside the tentacular crown, as well as by the active bending movement of their flexible stalk. They may be common in the rocky shores over the world. However, their small size makes it difficult to find entoprocts in the field; thus, their taxonomy, distribution, and ecology have been poorly documented.

MORPHOLOGY

Entoprocts may be either colonial or solitary. Their generally globular body (calyx) contains a U-shaped gut, gonads, protonephridia, and brain, and it has a tentacular crown on top. Usually, a long stalk supports the calyx and attaches the body to a substratum. In the colonial species (Fig. 1), individuals are connected by thin tubes (stolons) that fix the colony to the substratum. Solitary species have a variety of attaching apparatus (feet) that enable the animal to attach to, crawl on, or glide over the substratum; in many species the foot degenerates in the adult stage and cements on the substratum.

FIGURE 1 Part of a colony of colonial entoprocts, *Barentsia discreta*, living on a stone found at a rocky shore at Shimoda, middle Japan. Bar = 1 mm. Photograph by the author.

TAXONOMY AND PHYLOGENY

To date approximately 140 solitary and 50 colonial species have been described in worldwide seas from tropic to polar regions and in seas ranging from shallow to more than 500-m-deep. Only two colonial species has been known to live in freshwater. The taxonomic survey of this animal is still in an early stage, and a number of species are currently undescribed. Recent molecular studies imply that entoprocts are a member of the Lophotrochozoa, an animal group that is characterized by trochophore larvae and spiral cleavage in early development.

REPRODUCTION

Entoprocts reproduce both sexually and asexually. Most entoprocts are believed to be hermaphroditic, of which some are protandrous. Fertilized eggs attach to the mother body by strings and develop into trochophore-like larvae, which finally swim away from the mother. Budding occurs in all species and is quite vigorous. In solitary species, grown buds detach from the parent and then crawl or glide on the substratum using the foot; otherwise they swim using ciliated tentacles.

ECOLOGY

Entoprocts are suspension feeders. They generate a water current by means of ciliary tentacles and catch phytoplankton and organic particles suspended in the current. Most solitary species have been found with specific association to larger animals. They are often found on the body or tube of polychaetes (Fig. 2) and on the bodies of sponges, bryozoans, and sipunculids. Such commensal relationships with larger host animals are believed to provide entoprocts safe habitats and fresh water that brings food particles. Some solitary species (Fig. 3) and almost all colonial species do not have specific hosts but have been found on rocky walls, stones, algae, and artificial objects.

SEE ALSO THE FOLLOWING ARTICLES

Bryozoans / Hydroids / Phoronids / Sipunculans / Sponges

FURTHER READING

Brusca, R. C., and G. J. Brusca. 2003. *Invertebrates,* 2nd ed. Sunderland, MA: Sinauer.

Iseto, T. 2003. Entoprocta: Entoprocts, in *Grzimek's animal life encyclopedia,* 2nd ed. Vol. 1, *Lower metazoans and lesser deuterostomes.* M. Hutchins, D. A. Thoney, and N. Schlager, eds. Farmington Hills, MI: Gale Group, 319–325.

Wasson, K. 2002. A review of the invertebrate phylum Kamptozoa (Entoprocta) and synopses of kamptozoan diversity in Australia and New Zealand. *Transactions of the Royal Society of South Australia* 126: 1–20.

Nielsen, C. 1989. Entoprocts. *Synopses of British Fauna, New Series* 41:1–131.

Nielsen, C. 2002. Phylum Entoprocta, in *Atlas of marine invertebrate larvae.* C. M. Young, ed. San Diego: Academic Press, 397–409.

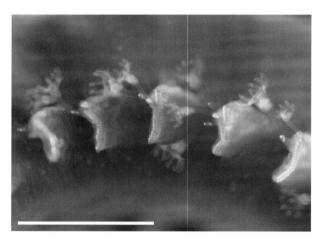

FIGURE 2 A solitary entoproct, *Loxosomella akkeshiensis*, living on parapodia of a polychaete found at a rocky shore at Shizugawa Bay, northern Japan, collected by Dr Katsuhiko Tanaka. Bar = 1 mm. Photograph by the author.

FIGURE 3 A solitary entoproct, *Loxosomella shizugawaensis*, living on an alga found at a rocky shore at Shizugawa Bay, northern Japan. Bar = 1 mm. Photograph by the author.

EVAPORATION AND CONDENSATION

MARK W. DENNY

Stanford University

As water evaporates from a plant or animal, a surprising amount of heat is lost, and the organism is cooled. Although evaporative cooling can help intertidal organisms avoid overheating during low tide, this advantage is inherently linked to the problem of desiccation. To remain cool, organisms must dry out, and the resulting tradeoffs play important roles in the physiology of intertidal plants and animals. The rate of evaporation depends on the size, shape, and temperature of the organism, the relative humidity of the air, and the speed of the wind. The converse of evaporation is condensation: if air is cooled to the dew point, water condenses and in the process releases heat that warms the organism. Condensation may help intertidal organisms to avoid freezing.

HYDROGEN BONDS

Water—H$_2$O—is an asymmetrical molecule. Contrary to one's intuition, the two hydrogen atoms in a water molecule are not attached at opposite ends of the oxygen atom, 180° apart. Instead they are attached only 105° apart (Fig. 1), and this arrangement has important consequences.

Oxygen is much more electronegative than hydrogen is. As a result, the electron of each hydrogen atom in the

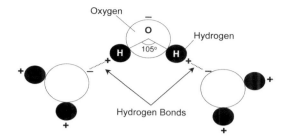

FIGURE 1 A water molecule consists of two hydrogen atoms attached to an oxygen atom.

water molecule is attracted toward the oxygen atom, leaving each hydrogen with a net positive charge due to the proton in its nucleus. By the same reasoning, the oxygen atom has a net negative charge from the proximity of the hydrogen atoms' electrons. This separation of charges makes the molecule an electric dipole, and the molecule is therefore said to be polar.

Water's polar nature gives it unusual properties. The net positive charge of each hydrogen atom in one water molecule is attracted to the net negative charge of an adjacent oxygen atom from another water molecule, forming a hydrogen bond (see Fig. 1). The hydrogen bonds between water molecules keep water in a liquid state to an exceptionally high temperature. Water boils at 100 °C, whereas nonpolar methane (CH_4), which has a similar molecular weight, boils at a frigid −162 °C. The hydrogen bonds among water molecules give water the highest surface tension of any room-temperature liquid, and they play a key role in evaporation and condensation.

TEMPERATURE AND MOLECULAR MOTION

Kinetic energy is the energy associated with mass in motion, and temperature is a measure of the average kinetic energy of individual molecules. The higher the temperature, the faster each molecule moves and the more kinetic energy it has. This thermal energy has a variety of effects. For instance, at temperatures below the freezing point of water, the kinetic energy of each water molecule is sufficient to cause it to vibrate in place but is insufficient to allow it to break the hydrogen bonds that hold it to its neighbors. As a result, ice is a solid. Above the freezing point, the thermal energy of water molecules is sufficient to allow them occasionally to break from their neighbors and move to new positions. This motion allows water to change its shape but not its volume, and as a consequence, at temperatures between 0 °C and 100 °C, water behaves as a liquid. In the transition from ice to water, approximately 15% of hydrogen bonds are broken. If the temperature of

water is increased to the boiling point, the kinetic energy of individual molecules is sufficient to break the remaining hydrogen bonds, and molecules become free to move independently. This freedom of motion converts the liquid to a gas.

If, at a given temperature, all molecules in liquid water moved at the same speed, no individual molecule could escape into the air ahead of any other. In this case, until liquid water reached its boiling temperature, no molecule would have sufficient energy to become a gas. In reality, however, temperature simply reflects the average molecular speed, and at any given temperature, molecules in water have a wide range of velocities (Fig. 2). For present purposes, it is the small fraction of molecules in the right-hand tail of this distribution—the molecules that have exceptionally high speeds—that are important. Because of their higher-than-average speed, these molecules have higher-than-average kinetic energy, sufficient energy to break the hydrogen bonds holding them to their neighbors. In the bulk of the liquid, the presence of these energetic molecules simply acts to make water a bit more fluid than it would otherwise be. If by chance, however, one of these energetic molecules arrives at the water's surface, it can break free and enter the air as a molecule of water vapor. This is evaporation.

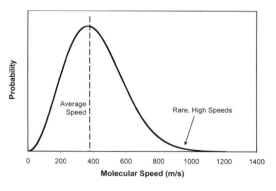

FIGURE 2 At any given temperature, water molecules possess a range of speeds. The distribution shown here is for 20 °C.

There are two important consequences of evaporation. First, it cools the water left behind. Because it is only the most rapidly moving molecules that are capable of escaping into the air, their exit reduces the average speed of the molecules remaining in the liquid. By analogy, if you removed all the world-class runners from the Boston Marathon, the average speed of the remaining field would be reduced. As noted in the preceding paragraph, temperature is directly related to the average speed of molecules, so when this speed is reduced as water molecules evaporate, the water left behind during evaporation is cooled. Furthermore, in escaping from the liquid surface into the air, evaporating molecules break the bonds holding them to other water molecules. This breakage requires energy, which is drawn from the remaining liquid, thereby cooling it.

This cooling effect is substantial. For each kilogram of water that evaporates, approximately 2,500,000 joules of heat energy are removed from the remaining liquid (see Table 1). To put this in a familiar context, we first note that the specific heat capacity of water, S, is approximately 4000 joules per kilogram per °C. That is, it takes the removal of approximately 4000 joules to cool one kilogram of water (the mass of one liter) by one degree Celsius. Next, we note that a large man has a mass of about 100 kilograms. If he were to lose just 1% of his body mass (one liter of water) by evaporation, his temperature would be cooled by $1 \text{ kg} \times 2,500,000 \text{ J kg}^{-1} / (4000 \text{ J kg}^{-1} °C^{-1} \times 100 \text{ kg}) = 6.25 °C$ (11.25 °F), sufficient to make him seriously hypothermic.

TABLE 1
The Latent Heat of Vaporization for Pure Water

Temperature °C	Latent Heat of Vaporization J/(kg °C)
0	2,513,000
10	2,489,000
20	2,465,000
30	2,442,000
40	2,394,000

NOTE: Data from Denny 1993.

The second important effect of evaporation concerns the location of the energy removed from the liquid. Where does that heat go? As noted above, most of the energy required to move water molecules from the water's surface into the air is needed to break the hydrogen bonds that hold molecules to their neighbors. A small amount of additional energy is required to break other weak bonds between water molecules (van der Waals attractions) and to power the volume increase as liquid becomes a gas. The sum of these bond-breaking and volume-increasing energies is stored as potential energy in the newly created water vapor. In other words, a water vapor molecule contains latent ("hidden") within itself the energy required to separate it from its neighbors when it evaporated. The 2,500,000 joules of energy removed from water when one kilogram evaporates is known as the latent heat of vaporization, Q. The latent heat of vaporization of pure water is the highest for any known liquid. The Q of seawater is also high, essentially that same as that of freshwater.

The latent heat of vaporization is made manifest when water vapor condenses. A water-vapor molecule that by chance strikes the water's surface and is captured by the liquid, reforms the bonds broken during evaporation. The latent heat of the vapor is then converted to sensible heat, and as a result, heat energy is released into the water. This heat of condensation is exactly equal to the latent heat of vaporization: approximately 2,500,000 joules per kilogram of vapor condensed.

It should be noted that evaporation and condensation occur at the same time. While some molecules escape from liquid into air, water vapor molecules collide with the liquid's surface and are captured. Thus, when we refer to evaporation or condensation, we are referring to the net difference between the two processes at any time.

EVAPORATIVE COOLING

If a bottle is partially filled with water and then tightly stoppered, water vapor accumulates in the enclosed air until the rate of evaporation equals the rate of condensation. At that point, the air is saturated with water vapor. The concentration of saturated water vapor (measured in moles of vapor per cubic meter of air) varies with temperature: the higher the temperature of the air, the more water vapor it can contain (Fig. 3). In nature, the concentration of water vapor in air is usually less than this saturation value, and the ambient concentration of water vapor, expressed as a fraction of the saturation concentration, is the relative humidity. Note that relative humidity is sensitive to temperature. For air with a fixed concentration of water vapor, relative humidity decreases with increasing temperature because at high temperatures the constant concentration of vapor is compared to a larger saturation concentration. This effect can be seen by examining a horizontal line—a line of constant

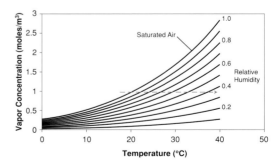

FIGURE 3 The concentration of water vapor in air depends on temperature and relative humidity. At constant vapor concentration, an increase in temperature results in a decrease in relative humidity (as shown by the dashed arrow).

vapor concentration—in Fig. 3. Moving along this line to higher temperatures (as shown by the arrow) brings you to lower values of relative humidity.

The relative humidity of air is commonly measured using a sling psychrometer, a device consisting of two thermometers. The bulb of one thermometer is surrounded by a wet wick that freely evaporates water, and thus its temperature may be lower than that of the air, while the second thermometer remains dry and measures air temperature directly. When the device is swung through the air, evaporation from the wick cools the wet-bulb thermometer to the degree allowed by the relative humidity, and the difference in temperature between the wet- and dry-bulb thermometers is known as the wet-bulb depression. The wet-bulb depression can then be used to calculate relative humidity (Fig. 4).

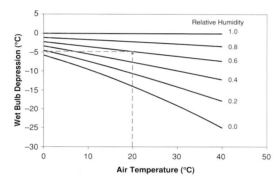

FIGURE 4 The wet-bulb depression depends on air temperature and relative humidity. In the case highlighted by the dashed lines, a measured wet-bulb depression of × 4.9 °C and a measured air temperature of 20 °C indicate that the relative humidity is 0.6.

However, it is not the measurement of relative humidity that concerns us here; rather it is the magnitude of the wet-bulb depression itself. Many intertidal algae (and a few intertidal animals, such as sea anemones) evaporate water sufficiently freely that when the wind blows, their body temperatures are cooled by an amount approaching the wet-bulb depression. This cooling effect can be substantial (see Fig. 4). For example, for the case shown by the dashed lines in the figure, we see that if air temperature is 20 °C and the relative humidity of the local air is 0.6 (a common value for the intertidal environment), the wet-bulb depression is approximately −5 °C. In other words, under these environmental conditions, algae on the shore can be at 15 °C, 5 °C cooler than the air around them. In an extreme case (air temperature of 40 °C and zero relative humidity), the wet-bulb depression is 25 °C, and algae

could maintain the same cool body temperature (15 °C) they would experience when the air was considerably cooler. However, the proximity of the ocean makes it rare for intertidal sites to experience very low relative humidity, so extreme wet-bulb depressions are unlikely.

There are some constraints on this process. First, a plant or animal realizes the wet-bulb depression only if there is no heat influx to the organism. For example, if the organism is heated by the sun, its temperature is higher than one would calculate from the wet-bulb depression. Second, the wet-bulb depression is achieved only when water freely evaporates, and in contrast to algae, this is rarely seen in animals. Because evaporation leads to desiccation, and desiccation can be injurious, many animals have evolved mechanisms to reduce the rate of evaporation. In these cases, the magnitude of evaporative cooling is controlled by the rate of evaporation, which in turn is controlled by the factors in Equation 1:

$$M = h_m A(C_{sat} - C_{air}) \qquad \text{(Eq. 1)}$$

Here, M is the flux of water out of the animal (measured in kilograms per second), h_m is the mass transfer coefficient (measured in kilogram meters per second per mole), and A is the area (in square meters) available for evaporation. C_{sat} is the saturation concentration of water vapor at the temperature of the animal's body, and C_{air} is the water vapor concentration in the local air. Both concentrations are measured in moles of water vapor per cubic meter of air. From Equation 1 we see that the larger the mass transfer coefficient and the greater the area available for evaporation, the more rapidly water is lost. Similarly, the greater the difference in concentration between water-saturated air next to the organism and air farther away, the more rapidly water evaporates.

Three of these four variables can potentially be controlled by the organism, vapor concentration in air being the sole exception:

Mass Transfer Coefficient

The mass transfer coefficient depends to a minor degree on the shape of the organism—oganisms whose evaporative areas have many small projections tend to have higher mass transfer coefficients than do organisms with smooth surfaces—but h_m is more dependent on size. The larger the organism, the smaller the mass transfer coefficient. Using L, the length of the organism, as our measure of size, the mass transfer coefficient is typically proportional to approximately $\sqrt{1/L}$.

The mass-transfer coefficient also depends on wind speed, U. The faster the wind, the higher the mass-transfer coefficient and the more rapidly water evaporates. Typically, h_m is proportional to approximately $\sqrt{U}$.

These general rules governing the behavior of h_m are unlikely to apply exactly to any specific organism, and in practice, mass transfer coefficients are best measured empirically. This can be accomplished by placing the organism in question in wind of a known velocity (in a wind tunnel, for instance) and weighing the animal periodically, thereby measuring the rate at which water is lost. The evaporative area, A, can be measured directly, as can C_{air}, the water vapor concentration of the air (see the preceding discussion about sling psychrometers). C_{sat} can be computed by measuring the temperature of the organism and using that to look up the corresponding saturated vapor concentration (see Fig. 3). With all of these values measured, h_m can be calculated from Equation 1.

Evaporative Area

Algae do not have the same sort of waxy cuticle found on terrestrial plants, and therefore they have their entire surface area exposed to evaporation. In contrast, the evaporative area of animals is under both evolutionary and behavioral control. For example, the calcareous shells of acorn barnacles, mussels, and snails are nearly impervious to water. Thus, through the evolution of shells, the fraction of these organisms' bodies area exposed to the air has been reduced.

Behavioral control takes a variety of forms. For instance, when the tide is out, periwinkles (littorine snails) retreat into their shells, tightly sealing the shell's aperture with a "trap door" (the operculum) attached to the foot. This minimizes the area subject to evaporation, and these animals can survive for weeks without contact with water. Many limpets clamp their shells tightly to the rock surface at low tide, thereby reducing evaporative area, and some species enhance the effect by constructing a mucus "curtain" at the edge of the shell to prevent air flow past the body. Mussels commonly close their shells at low tide, and some anemones attach bits of broken shells to their bodies, reducing the area open to evaporation.

Saturation Vapor Concentration

Air in contact with the evaporative area of an animal can usually be assumed to be saturated with water, and for any given temperature this saturation vapor concentration is purely a matter of physics—it is outside the control of the organism. Saturation vapor concentration is, however, a function of temperature (Fig. 3), and the temperature during evaporation is at least potentially under animals' control. For example, many shelled

organisms can avoid evaporation by "clamming up" but must emerge from their shells for some period each day to feed and respire. The choice of when to emerge is up to the animal. The saturated vapor concentration at a typical night-time temperature of 10 °C is only about a third that during a hot day (30 °C). Thus, if an animal exposes its evaporative area at night when it is cool, it evaporates water much more slowly than if it exposes the area during the day.

The concentration of salts in water can affect the saturation vapor pressure in the adjacent air—the higher the concentration, the lower the vapor pressure. By concentrating salt at evaporative surfaces, organisms could reduce the saturation concentration and thereby reduce the rate of evaporative loss (see Equation 1). The concentration of salt would need to be quite high to have appreciable effect, however.

Rate of Cooling

Once the rate of water loss from an organism is known, the rate of cooling can easily be calculated. The rate of heat loss, H, is equal to the rate of water loss, M, multiplied by the latent heat of evaporation, Q:

$$H = MQ \qquad \text{(Eq. 2)}$$

For example, if 0.1 gram of water is lost each second ($M = 0.0001 \text{ kg s}^{-1}$), the heat flux is $0.0001 \text{ kg s}^{-1} \times 2,500,000 \text{ J kg}^{-1} = 250$ joules per second. Now, one joule per second is one watt, so the amount of heat lost when 0.1 gram of water evaporates each second is equal to the heat produced by a 250-watt light bulb.

Given H, we can calculate T, the rate at which the organism is cooled. T (measured in degrees per second) depends on m, the mass of the organism (measured in kilograms) and S, the specific heat capacity of the material from which it is constructed:

$$T = \frac{H}{mS} \qquad \text{(Eq. 3)}$$

The specific heat capacity of animal tissue is close to that of water, approximately 4000 joules per kilogram per °C.

For example, the animal noted above loses 250 joules per second. If this animal's mass is 1 kilogram, its temperature changes by 250 J s^{-1} / ($1 \text{ kg} \times 4000 \text{ J kg}^{-1} \text{ °C}^{-1}$) = 0.06 °C per second. By losing water at this rate, the animal could quickly cool itself, losing 6 °C in 100 seconds.

It is important to note, however, that this cooling comes at a cost. In the 100 seconds required to lower its temperature by 6 °C, the animal loses 10% of its body mass in water. For many intertidal organisms, this rate of desiccation would be deleterious, if not fatal.

CONDENSATION

As noted previously, heat energy imparted to water as it evaporates is returned to liquid water as vapor condenses. During low tide, the rate of evaporation typically exceeds the rate of condensation, and it is this net rate of evaporation that we notice. At times, however, the rate of condensation exceeds that of evaporation. This is most commonly seen when the temperature of an object is lowered. For example, water condenses on a glass of iced tea or a can of cold beer. Similarly, if the surface of an intertidal organism is below the dew point (defined subsequently), water condenses on it, and the organism is warmed.

How cold must a surface be to accumulate condensation? This question can be answered through the use of Fig. 5. Consider the following example. The afternoon air temperature at an intertidal site is 30 °C and the relative humidity is 0.6. We can use Fig. 5 to tell us what the vapor concentration is under these conditions by following arrow 1 up from 30 °C to the line for 0.6 relative humidity. We then desire to know how much we must lower the temperature before this vapor concentration becomes saturating. We follow arrow 2 horizontally left until it contacts the line of saturation (relative humidity = 1.0) and then arrow 3 down to find the corresponding temperature. In this case, the temperature required to have net condensation is 21 °C. This is the dew point for these conditions. Different conditions have different dew points. The lower the relative humidity, the greater the drop in temperature required to reach the dew point.

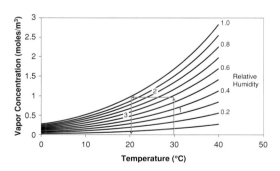

FIGURE 5 The information of Fig. 3 allows one to estimate the dew point corresponding to a given current air temperature and relative humidity (see text).

Condensation may be important in helping intertidal organisms to avoid freezing. Just as evaporation cools organisms, condensation warms them, and the condensation forming on an organism as air temperature approaches 0 °C can help to maintain the organism's body temperature several degrees above that of the air. Given the high relative humidity typical of intertidal environments, this mechanism may be particularly effective.

When water condenses onto a surface that is below 0 °C, it condenses as frost. The latent heat released under these conditions (approximately 2,830,000 joules per kilogram) is 16% larger than that released by vapor condensing into liquid water, augmenting the warming effect.

The biological role of condensation has been extensively studied for terrestrial plants (in particular for crop plants, for which frost damage has monetary consequences), but the effect has not been studied in detail for intertidal organisms.

SEE ALSO THE FOLLOWING ARTICLES

Desiccation Stress / Heat Stress / Seawater

FURTHER READING

Campbell, G. S., and J. M. Norman. 1998. *An introduction to environmental physics*, 2nd ed. New York: Springer-Verlag.
de Podesta, M. 2002. *Understanding the properties of matter*, 2nd ed. New York: Taylor and Francis.
Denny, M. W. 1993. *Air and water: the biology and physics of life's media*. Princeton, NJ: Princeton University Press.
Nobel, P. S. 1991. *Physicochemical and environmental plant physiology*. New York: Academic Press.

EXCRETION

MATTHEW E. S. BRACKEN

University of California, Davis

All animals produce nitrogen-containing compounds as byproducts of their metabolic processes. These compounds can be highly toxic and are actively excreted by intertidal invertebrates, usually in the form of ammonium ion (NH_4^+). In temperate marine ecosystems, the growth of seaweeds and phytoplankton is largely determined by nitrogen availability, and invertebrate-excreted ammonium can play an important role in ameliorating this nitrogen limitation and thereby influencing the recruitment, growth, and diversity of seaweeds on rocky shores.

NITROGEN EXCRETION BY INTERTIDAL INVERTEBRATES

Nitrogen is an essential component of many biological molecules, including amino acids, proteins, and nucleic acids. Nitrogen is therefore a somewhat unusual element, as it is both produced by and toxic to animals. Like all animals, invertebrates on rocky shores must consume proteins and other nitrogen-containing compounds, and the digestion and metabolism of these nitrogenous compounds produces waste nitrogen that must be excreted. In marine invertebrates, the primary excretory product is ammonium (NH_4^+), though small amounts of urea and organic nitrogen (amino acids) are also excreted by some species. Because ammonium is soluble in water, most excreted ammonium readily diffuses across the body surface (often gills) and into the surrounding water. In coelomate invertebrates, excretion of other compounds, including nonammonium nitrogenous wastes (urea, amino acids), salts, and byproducts of metabolism, occurs *via* ultrafiltration and reabsorption in specialized structures called filtration nephridia.

Nitrogen excretion rates vary over tidal and seasonal cycles as a result of changes in invertebrates' food supply (quantity and quality) and the status and use of their internal nitrogen reserves. Because nitrogen excretion is linked to protein metabolism, increases in protein in the diet (e.g., consumption of bacteria instead of phytoplankton) or the use of protein reserves as a substrate for respiration during starvation typically lead to increases in ammonium excretion. Site-specific differences in the quality and quantity of particulate organic material (phytoplankton and detritus) available to filter feeders can result in large differences in the biomass-specific ammonium excretion rates of filter-feeding invertebrates at different sites. Within a site, temporal variation in food availability and quality (e.g., summer versus winter or upwelling versus relaxation) can lead to similarly large differences in ammonium excretion rates. Other factors that influence excretion rates include tide height (which influences immersion time and thereby food availability), temperature (which affects physiological rates, including ammonium excretion), and salinity.

INVERTEBRATE-EXCRETED AMMONIUM AS A NITROGEN SOURCE FOR INTERTIDAL SEAWEEDS

Nitrogen is typically the most important growth-limiting nutrient for autotrophs in temperate coastal ecosystems, so the productivity of these systems is closely tied to nitrogen availability. In the near-shore waters of the northeastern Pacific Ocean, the amount of particulate organic nitrogen (associated with phytoplankton and detritus) can equal or exceed the amount of inorganic nitrogen (nitrate or ammonium). However, seaweeds and other marine primary producers cannot utilize particulate

nitrogen, and they rely largely on nitrate and ammonium, which they assimilate into amino acids. By consuming particulate nitrogen, which is unavailable to seaweeds as a nitrogen source, and excreting inorganic nitrogen (ammonium), which is readily taken up and assimilated by seaweeds, invertebrates play an important role in the nitrogen dynamics of intertidal ecosystems.

For decades, biological oceanographers have differentiated between new production, or primary production associated with upwelled nitrate, and regenerated production, which is fueled by local-scale ammonium excretion by animals. However, it is only comparatively recently that benthic marine ecologists have begun to consider the potential role of nitrogen regeneration in fueling primary production in near-shore and intertidal ecosystems. Measurements of ammonium excretion and uptake by amphipods and seaweeds suggest that benthic macroalgae can potentially obtain approximately 50% of the nitrogen needed for growth from associated epifaunal invertebrates in subtidal seaweed beds. Excreted nitrogen has also been shown to influence seaweed growth, recruitment, and tissue carbon-to-nitrogen ratios (an indication of nitrogen limitation) in intertidal seaweeds.

Tidepools are a convenient system for quantifying the effects of invertebrate-excreted ammonium on seaweed growth and diversity. Intertidal pools, especially those high on the shore, are isolated from the ocean for substantial periods of time, and the seaweeds living in those pools (which rapidly deplete available nitrogen from the water) are therefore subjected to long periods without any external nitrogen. Adding controlled-release fertilizer pellets (inorganic nitrogen and phosphorus) to tidepools enhances seaweed growth and diversity, suggesting that seaweeds in mid- and high-zone pools are nutrient limited.

Can invertebrate-excreted ammonium ameliorate this limitation? Sessile invertebrates, especially mussels and sea anemones, excrete substantial amounts of ammonium into tidepools, so that inorganic nitrogen concentrations in invertebrate-dominated pools can equal or exceed those in the adjacent ocean, even during upwelling events. Seaweeds take up and assimilate this excreted ammonium; the rate of nitrogen incorporation into seaweed tissues increases with the rate of nitrogen loading by invertebrates into tidepools.

The excretion of ammonium by tidepool invertebrates and its uptake by seaweeds is associated with increased seaweed growth and diversity. For example, in Oregon-coast tidepools, a fourfold increase in the rate of ammonium loading by invertebrates into tidepools resulted in a doubling of the number of seaweed species present in

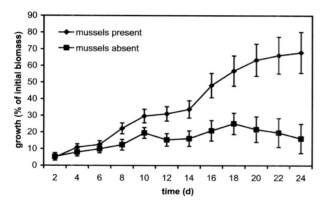

FIGURE 1 Growth of intertidal seaweeds in the presence and absence of mussels. When mussels are present in tidepools, seaweeds grow >40% more than when mussels are absent, because ammonium excreted by mussels enhances seaweed growth. From Bracken (2004), with permission of Blackwell Publishing.

those pools. Growth of the red alga *Odonthalia floccosa*, a common seaweed in high intertidal pools, is more than 40% higher when mussels are present (Fig. 1). Positive interactions such as this one tend to occur when conditions are stressful. Nitrogen limitation in tidepools is ameliorated by local-scale ammonium excretion, which promotes algal diversity and growth in an otherwise inhospitable environment.

SEE ALSO THE FOLLOWING ARTICLES

Algae / Facilitation / Nutrients / Water Chemistry

FURTHER READING

Bayne, B. L. 1976. *Marine mussels: their ecology and physiology.* Cambridge, UK: Cambridge University Press.

Bracken, M. E. S. 2004. Invertebrate-mediated nutrient loading increases growth of an intertidal macroalga. *Journal of Phycology* 40: 1032–1041.

Campbell, J. W., ed. 1970. *Comparative biochemistry of nitrogen metabolism:* 1. *The invertebrates.* New York: Academic Press.

Carpenter, E. J., and D. G. Capone, eds. 1983. *Nitrogen in the marine environment.* New York: Academic Press.

Dame, R. F. 1996. *Ecology of marine bivalves: an ecosystem approach.* Boca Raton, FL: CRC Press.

FACILITATION

MARK D. BERTNESS

Brown University

Facilitation occurs when an organism benefits from the presence of another organism that is not itself negatively impacted. Two types of facilitation, or positive interactions, are common in rocky intertidal communities: habitat-ameliorating positive interactions and associational defenses.

TYPES OF FACILITATION

Habitat amelioration and associational defenses occur under predictable conditions. Habitat amelioration occurs when an organism reduces potentially limiting physical stresses, such as heat, desiccation, and wave forces, on other organisms. Habitat-ameliorating positive interactions are common in rocky intertidal habitats because (1) they are physically stressful habitats for most of the organisms that live in these habitats, and (2) many of the key stresses, such as heat, desiccation, and waves, can be ameliorated by sessile intertidal neighbors. Associational defenses, a second common type of positive interaction in intertidal communities, occur when organisms are protected from their enemies when living in association with other organisms. They are common in intertidal communities because (1) consumer pressure is intense in many shoreline habitats, and (2) common sessile organisms are often chemically or physically defended from consumers and provide refuge for less-defended neighbors.

Facilitations can be facultative, meaning that although beneficial they are not necessary for the survival and reproduction of facilitated species, or obligatory, meaning that they are absolutely necessary for facilitated species' survival and persistence. Facilitation can also occur intraspecifically, when individuals of the same species benefit one another (also often called group benefits), or interspecifically, when a facilitator species positively impacts other species. Mutualisms, in which two species benefit each other, are a specific type of reciprocal facilitation or positive interaction that can be either facultative or obligatory and highly evolved.

FOUNDATION SPECIES

Facilitation is a particularly important process in rocky intertidal communities because they are commonly dominated by foundation species. Foundation species are common and abundant species in a community that create and maintain habitats that other community members are dependent on. Most common rocky intertidal foundation species, such as canopy-forming seaweeds and bed-forming mussels, live in dense groups that occur because of strong intraspecific facilitation or group benefits of physical stress amelioration or associational defense, and the same amelioration of physical and biotic conditions that drive intraspecific group benefits typically facilitate other species. Species that modify environmental conditions and provide suitable habitats for other organisms have also often been recently referred to as ecosystems engineers. High-intertidal algal canopies are a good example of an intertidal foundation species or ecosystem engineer. On New England rocky shores, *Ascophyllum nodosum* forms a dense algal canopy on wave-protected rocky shores (Fig. 1). At high-intertidal

FIGURE 1 Intertidal seaweed canopy in the Gulf of Maine that buffers understory organisms from thermal, desiccation, and wave stress and buffers many understory organisms from predators. Photograph by the author.

elevations virtually all of the mobile and sessile organisms that live in the understory are dependent on habitat amelioration by the algal canopy to live in the intertidal habitat. Other common foundation species that are responsible for providing habitats for rocky-intertidal organisms around the world include other seaweed canopies, mussel beds in many temperate-zone rocky intertidal habitats, and tunicate beds on the Pacific coast of South America (Fig. 2).

FIGURE 2 Intertidal mussel bed on the Patagonian coast of Argentina that protects interstitial organisms from thermal, desiccation, and wave stress as well as predators. Photograph by Caitlin Crain.

WHERE FACILITATIONS ARE FOUND

Although facilitation is a common process in many rocky intertidal communities, the types of facilitation common in specific communities vary predictably as a function of the limiting stresses in a given community. In particular, consumers generally are not important in physically

stressful habitats, while physical stress ameliorating facilitations are common in physically stressful habitats. Conversely, in physically benign habitats, physical stress ameliorating facilitations are unimportant, whereas consumer pressures are elevated and associational defenses are of increased importance. These general patterns lead to predictable variation in the occurrence and importance of facilitation in rocky intertidal communities at local, regional, and global spatial scales (Fig. 3).

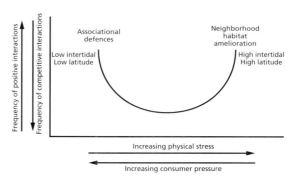

FIGURE 3 Conceptual model showing the predicted spatial distribution of facilitations in intertidal systems.

Heat- and desiccation-ameliorating positive interactions are one of the most common facilitations in rocky-intertidal communities. They occur when algal canopies, mussel beds, tunicate beds, algal turfs, and dense aggregations of barnacles and other sessile organisms in the intertidal provide relief from high temperatures and desiccation to organisms that live under their cover. Because the importance of this type of interaction in rocky-intertidal communities varies as a function of physical stress, these types of facilitation are more important at high-intertidal heights than at low-intertidal heights and more important at lower latitudes than higher latitudes. At high-intertidal heights many intertidal organisms are dependent on living under algal canopies or in mussel beds to limit physical stresses, but at lower tidal heights intertidal organisms are less dependent on habitat amelioration by foundation species. The importance of these habitat-ameliorating interactions, however, varies with latitude. On the east coast of North America, for example, intertidal algal canopies facilitate understory organisms in southern New England but not in northern New England, where temperatures and desiccation stress are reduced. Similarly, on the west coast of North America, physical stress–ameliorating positive interactions have been found to be a dominant process on rocky shores in Baja California and the Patagonian shores of Argentina, but not on the Oregon or Washington coasts.

Wave stress–ameliorating positive interactions are also common facilitations, but only in wave-swept habitats. Seaweed canopies and mussel beds are the most common examples of foundation species facilitators found in wave-swept rocky intertidal habitats. In high-wave-energy rocky shores intertidal seaweeds and mussels often facilitate members of their own species, resulting in dense seaweed canopies and mussel beds, and these canopies and beds, in turn, facilitate other organisms by buffering them from wave stress and dislodgement.

In contrast to physical stress–ameliorating facilitations, associational defenses are more common and important processes in physically benign rocky-intertidal habitats where consumer pressure is the greatest. At local spatial scales on rocky shores this leads to associational defenses being more common and important at lower tidal heights than at high tidal heights. Low-intertidal and subtidal algal turfs and mussel beds have been shown to protect associated organisms from large mobile consumers such as fish and sea urchins. The importance of associational defenses on rocky shores also varies regionally and with latitude as a function of variation in consumer pressure. On tropical shores consumer pressure by prey-crushing fish and crustaceans that are unique to lower latitudes leads to rocky shores that are devoid of most intertidal organisms except for those that find refuge from predators in cracks and crevices or associated with physically or chemically defended neighbors.

SEE ALSO THE FOLLOWING ARTICLES

Desiccation Stress / Heat Stress / Mutualism / Predator Avoidance

FURTHER READING

Bertness, M.D., and P. Ewanchuk. 2002. Latitudinal and climate-driven variation in the strength and nature of biological interactions. *Oecologia* 132: 392–401.

Bertness, M.D., G. Leonard, J.M. Levine, P. Schmidt, and A.O. Ingraham. 1999. Habitat modification by algal canopies: testing the relative contribution of positive and negative interactions in rocky intertidal communities. *Ecology* 80: 2711–2726.

Bruno, J., and M.D. Bertness. 2001. Positive interactions, facilitations and foundation species, in *Marine community ecology.* M.D. Bertness, S.D. Gaines, and M. Hay, eds. Sunderland, MA: Sinauer Associates, 201–218.

Bruno, J., J.J. Stachowiz, and M.D. Bertness. 2003. Including positive interactions in ecological theory. *Trends in Ecology and Evolution* 18: 119–125.

Stachowicz, J.J. 2001. Mutualisms, positive interactions and the structure of ecological communities. *BioScience* 51: 235–246.

Stachowicz, J.J., and R.B. Whitlatch. 2005. Multiple mutualists provide complementary benefits to their seaweed hosts. *Ecology* 86: 2418–2427.

FERTILIZATION, MECHANICS OF

KRISTINA MEAD

Denison University

The phrase "mechanics of fertilization" refers to the physical processes that affect sperm and egg interaction prior to and during fertilization. Although physical processes will affect all types of reproduction, they are especially important in reproductive modes in which gametes are exposed to the external environment. Many marine organisms (most invertebrates, some fish, some algae) reproduce by releasing gametes into the water column, where they mix and unite to form the zygotes that will become the next generation. Exposure to gentle water movement can enhance mixing and contact rates between eggs and sperm, increasing fertilization success. Intense turbulence can still increase the likelihood that eggs and sperm will be near each other, but it can limit fertilization success by decreasing sperm concentration or interfering in contact between eggs and sperm. Surge channels may provide habitats that limit sperm dilution and therefore increase fertilization.

REQUIREMENTS FOR FERTILIZATION

Fertilization occurs when the nucleus of the sperm fuses with the egg nucleus. For fertilization to succeed, the sperm must contact the egg. The first sperm that is recognized as belonging to the correct species is allowed to enter the egg; rapid changes in the egg's surface inhibit entry by other sperm. Subsequently, the genetic material of the egg and sperm fuse, and zygote development is initiated.

Sperm–egg recognition is commonly mediated by rapidly evolving, species-specific compatibility molecules in the sperm, the egg, or both. There can be several recognition steps. For instance, factors in the egg jelly of sea urchins activate the sperm and cause the sperm to initiate the acrosome reaction. When the activated sperm has moved through the egg jelly and contacts the outer layer of the egg's vitelline envelope, an acrosomal protein called bindin binds to sperm receptors on the vitelline envelope. This process is followed by lysis of the envelope near the sperm head and fusion of the sperm membrane with the cell membrane of the egg. Then the male pronucleus migrates through the egg to fuse with the female pronucleus, and development is initiated.

FACTORS AFFECTING FERTILIZATION

There are many factors that affect fertilization success. These include gamete properties, features of the organism, and population- and environmental-level aspects. Although all of the following factors have been studied in the laboratory or in the field, and many have been incorporated into mathematical models of fertilization, it is not well known how these factors interact under natural conditions.

Gamete Properties

Sperm characteristics that are important include cell shape, swimming behavior and speed, and longevity. The longer and faster the sperm can swim, the greater the probability that they will bind to and then fertilize an egg. Egg size, presence of accessory layers or cells, and sperm receptor density are relevant egg features. In both cases, a viscous extrusion medium may help to minimize gamete dilution.

Organismal Properties

Relevant organismal features include size and shape, reproductive output, and factors related to spawning. For successful fertilization to occur, the concentration of sperm must be within a particular species-specific working range. Too few sperm limits fertilization, especially since highly dilute sperm appear to age more rapidly than concentrated sperm solutions. Too many sperm can enable more than one sperm to fertilize an egg (polyspermy), which is often fatal. Because sperm dilution appears to critically affect fertilization, factors that affect sperm concentration, such as spawning rate, synchrony among individuals, and reproductive output, are likely to heavily influence fertilization success.

Population-Level Factors

Important population characteristics include population density, size, and sex ratio, especially as they affect sperm dilution. Recent theory suggests that the spatial distribution of males and females within the population is probably important as well.

Environmental Characteristics

Environmental features affecting fertilization include the site's topography, wave-driven water velocity and turbulence, water depth, and aspects of water quality including salinity, temperature, and pH. These factors are probably especially important to the extent that they affect sperm longevity and dilution. In addition, wave-generated turbulence may affect egg–sperm interactions, either positively or negatively.

EFFECTS OF SURF ZONE TURBULENCE ON FERTILIZATION

Surf zone turbulence has several potential effects on fertilization, depending on the time and spatial scale under study, and the intensity of the turbulence. At large time and spatial scales, turbulent mixing dilutes gametes, decreasing fertilization success. At intermediate time and spatial scales, turbulent mixing can actually enhance the ability of eggs and sperm to interact. At the smallest time and spatial scales, the behavior of the gametes within vortices can limit egg–sperm binding and thus fertilization.

Long and Intermediate Temporal and Spatial Scales

The rocky intertidal zone can be subject to intense wave energy. As each wave breaks, its ordered motion is transformed into a collection of vortices. The energy of the wave is dissipated in part by straining processes within and between the vortices. Until recently, most studies on the effects of turbulent mixing have considered average effects. For instance, a person on shore can easily see how a patch of dye rapidly gets larger and less intense as turbulent mixing and molecular diffusion act to spread out and dilute the dye. When considering eggs and sperm, however, the dilution effects may occur at long time scales relative to those associated with fertilization. When instantaneous values are considered at intermediate spatial and temporal scales, a different picture emerges. As gametes are released from the parent organisms, turbulence first stirs the eggs and sperm into well-defined filaments surrounded by "clean" water. Turbulent vortices stretch and fold the filaments, increasing the surface area over which eggs and sperm can interact (Fig. 1). Turbulent structures tend statistically to cause sperm and egg filaments to overlap, generating higher fertilization rates than those predicted by time-averaged concentrations. The extent to which turbulence can cause egg and sperm filaments to coalesce before blurring into a cloud depends on the diffusivity of the gametes and the nature of the turbulent stirring. Gametes that are extruded in a medium that is viscous enough to inhibit rapid dilution but does not inhibit filament formation may be able to extend the time during which overlap can develop, and fertilization can occur before the filaments blur and dilution effects limit further fertilization.

Small Temporal and Spatial Scales

In addition to affecting filament overlap and dilution at intermediate and long temporal scales, turbulent mixing can affect eggs and sperm at even smaller scales, resulting

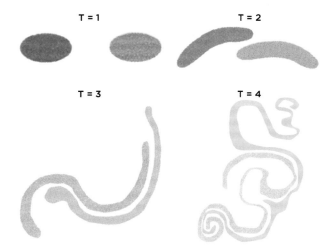

FIGURE 1 Turbulent mixing. At intermediate time scales, turbulent stretching can at least theoretically increase the contact area between concentrated filaments of eggs and sperm. This increase in encounter rate could enhance fertilization success.

in diminished egg–sperm binding, in gamete damage, and in abnormal development.

INHIBITION OF EGG-SPERM BINDING

If we spatially simplify and temporarily ignore the dynamic nature of a vortex in turbulent flow, a vortex can be thought of as a collection of constantly changing velocity gradients. Eggs, like any spherical object exposed to a velocity gradient, are likely to rotate; rotation speeds may be many revolutions per second. Sperm swimming speeds are much smaller than average velocity fluctuations, so sperm are likely to line up passively in the direction of flow, parallel to the egg's surface (Fig. 2). This could theoretically impede the sperm's ability to bind to the egg. Under conditions of intense turbulence, interference with egg–sperm binding could outweigh the increase in encounter rate caused by overlap of sperm and egg filaments.

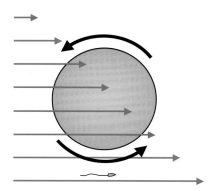

FIGURE 2 Eggs exposed to velocity gradients rotate. Sperm are likely to line up in the direction of the flow, parallel to the egg's surface. This may interfere with egg–sperm contact.

GAMETE DAMAGE

In addition to interference with egg–sperm contact, experimental exposure to velocity gradients in a Couette cell suggests that exposure to shear stress damages eggs and sperm. The amount of damage experienced by gametes subjected to velocity gradients prior to fertilization varies immensely among species. Some marine invertebrate eggs are able to repair injury by releasing exocytotic vesicles (granules), which fuse to the plasma membrane. In some species, these vesicles are involved in the slow block to polyspermy and are thus depleted shortly after fertilization.

EFFECTS OF SHEAR STRESS ON DEVELOPMENT

Many of the eggs fertilized while exposed to velocity gradients showed diminished developmental success. The mechanism is unclear, but it may involve damage to the egg membrane combined with a decreased ability to repair membrane damage.

FERTILIZATION RATES IN TIDEPOOLS AND SURGE CHANNELS

Early models focusing on time-averaged effects of turbulent mixing on gametes (i.e., dilution) predicted low rates of fertilization in the rocky intertidal zone (<1%) even during synchronous spawning. However, even though the few field studies of natural spawning that exist suggest that only a small percentage of organisms spawns within a given tidal cycle, the average fertilization success rate in the field is often more than 5% and can be greater than 90% when conditions are optimal (synchronous spawning of many nearby individuals at low tide). There is a huge variety in observed fertilization success, sometimes attributable to the environment or to characteristics of particular taxa (see the discussion of "Factors Affecting Fertilization"). Variation can be high even within a single spawning event in an isolated population.

Because surge channels are exposed to wave action, abundant mixing can occur within a surge channel. However, water exchange between the surge channel and the surrounding ocean can be very slow. Depending on the volume of the surge channel and the number of males spawning, fertilization rates inside surge channels have the potential of being much higher than outside the surge channel for equivalent conditions. If this is true, the greater contribution to the next generation of surge channel organisms may have important consequences for the conservation biology and population genetics of the species.

In sum, the fertilization success of free-spawning marine invertebrates is highly variable and is affected by a complex suite of biological, chemical, and physical factors ranging from gamete and organism characteristics to turbulence and topography.

SEE ALSO THE FOLLOWING ARTICLES

Boundary Layers / Diffusion / Reproduction / Surf-Zone Currents / Turbulence

FURTHER READING

Crimaldi, J. P., and H. S. Browning. 2004. A proposed mechanism for turbulent enhancement of broadcast spawning efficiency. *Journal of Marine Systems* 49: 3–18.

Levitan, D. R. 1995. The ecology of fertilization in free-spawning invertebrates, in *Ecology of marine invertebrate larvae*. L. McEdward, ed. Boca Raton, FL: CRC Press, 123–156.

Marshall, D. J. 2002. *In situ* measures of spawning synchrony and fertilization success in an intertidal, free-spawning invertebrate. *Marine Ecology Progress Series* 236: 113–119.

Mead, K. S., and M. W. Denny. 1995. The effects of hydrodynamic shear stress on fertilization and early development of the purple sea urchin *Strongylocentrotus purpuratus*. *Biological Bulletin* 188: 46–56.

Swanson, W. J., and V. D. Vacquier. 2002. The rapid evolution of reproductive proteins. *Nature Reviews Genetics* 3: 137–144.

FISH

MICHAEL H. HORN

California State University, Fullerton

LARRY G. ALLEN

California State University, Northridge

Fishes that live in the rocky intertidal environment are subjected to one of the harshest marine environments in the world. During high tides they live in a highly turbulent zone buffeted by both wave and tidal surge. At low tide, they are literally fish out of water, and desiccation is a major problem. Tidepools and under-rock habitats offer the only refuge from the exposure to air. Even the fish seeking refuge in such areas must tolerate a wide range of temperature, salinity, pH, and dissolved oxygen (DO). Low tides during daylight hours typically raise temperatures, salinities, and DO and lower pH dramatically. With all of these challenges, one might ask, then, why fish live there at all. Rocky intertidal areas are typically heterogeneous habitats that offer a variety of hiding places in the form of crevices and under-rock habitat. Also, being at the interface of the sea and land, these areas tend to be very productive with large stocks of macroalgae and eelgrass that thrive in the well-lit, nutrient-rich waters. These advantages have, in some fish species, outweighed the disadvantages and led to the evolution of resident species of intertidal fishes that are well suited to living in the harsh environment.

INTERTIDAL FISH SPECIES

What kinds of fishes commonly inhabit the rocky intertidal zone? In temperate regions of the world, intertidal fish assemblages tend to be dominated by sculpins (family Cottidae), clingfishes (Gobiesocidae), kelpfishes (Clinidae), snailfishes (Liparidae), pricklebacks (Stichaeidae), and gunnels (Pholidae) (Fig. 1). Gobies (Gobiidae), blennies (Blenniidae, Labrisomidae), clingfish, triplefins (Tripterygiidae), wrasses (Labridae), sea chubs (Kyphosidae), and pipefish (Syngnathidae) are commonly found on rocky shorelines throughout the tropical regions of the world (Fig. 2).

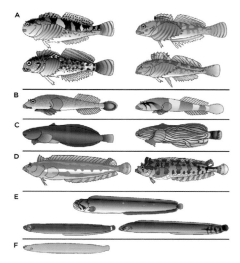

FIGURE 1 Representative intertidal fishes of temperate latitudes: (A) sculpins (Cottidae), (B) clingfishes (Gobiesocidae), (C) snailfishes (Liparidae), (D) kelpfishes (Clinidae), (E) pricklebacks (Stichaeidae), and (F) gunnel (Pholidae).

How have these fishes adjusted to life in such a harsh environment? Various shallow-water fishes may have evolved specialized anatomical, physiological, and behavioral features that allow them to cope with aforementioned conditions encountered in the rocky intertidal habitat.

BODY PLANS

Fishes that spend most or all of their lives in rocky intertidal habitats are small, usually less than 150 mm in length. The most common species are frequently even smaller, less than 100 mm in length. These fishes exhibit a variety of colors and pigment patterns—mostly muted grays, greens, reds, and browns—in the heterogeneous rocky intertidal habitat,

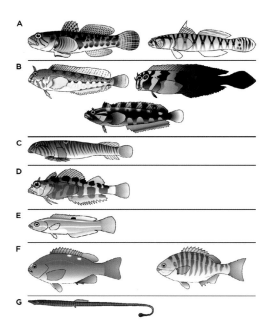

FIGURE 2 Representative intertidal fishes of tropical latitudes: (A) gobies (Gobiidae), (B) blennies (Blennidae, Labrisomidae), (C) clingfish (Gobiesocidae), (D) triplefin (Tripterygiidae), (E) wrass (Labridae), (F) sea chubs (Kyphosidae), and (G) pipefish (Syngnathidae).

FIGURE 4 Fluffy sculpin *(Oligocottus snyderi)* representing the roughly cylindrical and tapered body shape.

FIGURE 5 Striped kelpfish *(Gibbonsia metzi)* representing the laterally compressed, slender body shape.

FIGURE 6 Underneath view of a northern clingfish showing the pelvic (ventral) fins fused to form a sucking disk.

whether they occur in tidepools, beneath boulders, within crevices, or associated with the larger, branching forms of algae on the shore. Resident species represent only a few of the bewildering array of body shapes and paired fin arrangements found among fishes in general. Three basic body plans largely encompass those of most intertidal species: (1) flattened top to bottom (Fig. 3), (2) roughly cylindrical and tapered (Fig. 4), or (3) compressed side to side and shallow or deep bodied (Fig. 5). All three body plans show some degree of variation among intertidal fishes. The flattened body shape is either accompanied by small pelvic (ventral) fins or by these fins formed into a suction

FIGURE 3 Northern clingfish *(Gobiesox maeandricus)* representing the flattened body shape.

disk (Fig. 6). Cylindrical species possess either fused pelvic fins and pectoral fins positioned on the side of the body or separate pelvic fins and large, low-set pectoral fins (see Fig. 4). The latter arrangement is perhaps the most widely represented body plan of fishes in intertidal habitats. Slender-bodied (compressed) fishes usually possess moderately sized paired fins with the pectorals positioned on the side of the body (see Fig. 5). Fishes with an elongate, eel-like body (Fig. 7) are rare in most rocky intertidal habitats. Some of these species may have small paired fins or lack one or both sets (Fig. 8).

FIGURE 7 High cockscomb *(Anoplarchus snyderi)* representing the elongate, eel-like body shape.

FIGURE 8 Rockweed gunnel *(Apodichthys fucorum)* representing the elongate, eel-like body shape with reduced or absent paired fins.

Resident intertidal species are usually negatively buoyant, meaning that they are heavier than seawater and thus can remain on the bottom without effort. With this weight, most are labored swimmers and spend little time moving about in a habitat that often features strong surge and wave action. Movement occurs almost entirely when the fish are submerged at high tide.

All of the characteristics mentioned so far are anatomical features typical of fishes that live in shallow, turbulent waters, whether intertidal or subtidal or both. They are not traits exclusive to resident intertidal species. This point raises the question of whether intertidal fishes possess traits that can be labeled unequivocally as adaptations to the rocky intertidal environment. Studies show that intertidal fish assemblages are largely distinct in species composition from adjacent subtidal assemblages, which implies that resident species, at least, ought to show adaptations to the demands of intertidal existence.

Assuming that resident intertidal fishes evolved from subtidal ancestors, a trait possessed both by an intertidal species and its closest subtidal relatives can be interpreted as having evolved before colonization of the intertidal habitat. Such a trait is not an adaptation to intertidal life but an exaptation—that is, a trait that originated for some other selective or historical reason unrelated to current function. On the other hand, if the intertidal species possesses the potentially adaptive trait but its subtidal relatives do not, then the trait possibly can be interpreted as an adaptation to the pressures of intertidal existence. Sorting out this evolutionary question requires that robust phylogenies exist for the lineages containing both intertidal and subtidal species. The case would be strengthened further if distinct lineages showed the same pattern, that is, convergent adaptation among unrelated species to the similar selective forces of geographically separated intertidal habitats. In most cases, neither the phylogenetic requirement nor the test of unequivocal evidence for convergent evolution has been met. Nevertheless, the three kinds of functional traits discussed next are generally thought to be among such adaptations, and they offer the opportunity for rigorous testing of the adaptational hypothesis.

DESICCATION TOLERANCE

Resident intertidal fishes that dwell mainly in boulder fields or among algae may spend several hours each day out of water (Fig. 9). Most spend their time exposed in the damp environment beneath boulders or among blades of algae, but a few, such as the salariine blennies, may sit completely exposed on sunny tropical shores during a low-tide period although returning frequently to the water to dampen their bodies. Two abundant intertidal fish groups, clingfishes and pricklebacks, can tolerate high evaporative water loss and survive long periods exposed to the air. For example, certain clingfishes in the Gulf of

FIGURE 9 Monkeyface prickleback *(Cebidichthys violaceus)* partially out of water at low tide.

California can tolerate water losses as great as 60% of their total water content in a low-humidity (5%) environment and 93 hours under high-humidity (90%) conditions. Depending on body size, the monkeyface prickleback can sustain losses of as much as 27% of its body water and survive as long as 18 hours out of water in a damp environment.

The vertical distributions of many intertidal fish species appear to be limited mainly by tolerance to desiccation. Among four prickleback species and a related gunnel (Pholidae) that live in the same central California rocky intertidal habitat, the monkeyface prickleback occurs highest on the shore, has the highest initial water content, and tolerates the greatest water loss. In contrast, two species that live lower on the shore, the black prickleback and rockweed gunnel, show less tolerance to water loss and even continue to lose water after reimmersion. Fishes, however, that spend most of their lives in tidepools rather than beneath boulders or within algae seem to be less tolerant to desiccation, as might be expected in fishes that rarely experience exposure to air. In support of this contention, five species of sculpins that occur along a vertical intertidal gradient in Puget Sound, Washington, show no differences in rate of water loss.

With regard to adaptation to the intertidal habitat, the tolerance to desiccation exhibited by the clingfishes and pricklebacks may represent convergent adaptation. The necessary experiments on survival and water loss capacities, however, have not been conducted on the nearest subtidal relatives of these fishes to determine whether the intertidal and subtidal species actually differ in these traits. Until those studies have been undertaken, we cannot be certain that intertidal species have evolved adaptations to the threat of desiccation exposed shores.

AIR BREATHING

The ability to breathe air has been demonstrated in a variety of intertidal fishes and is expected in many other still unstudied species. As a result, air breathing is often regarded as an adaptation evolved by fishes in response to aerial exposure in the rocky intertidal habitat. Air-breathing ability may have evolved independently in amphibious members of families such as the Blenniidae, Cottidae, Gobiesocidae, Gobiidae, Stichaeidae, and Tripterygiidae as an adaptive response to periods of severely depleted oxygen levels that occur in tidepools during nighttime low tides (when respiration but not photosynthesis occurs in the algae). Alternatively, air breathing may have arisen as an outcome of low metabolic rate and sedentary habit typical of many intertidal fishes. If this scenario is true, then air breathing can be considered as an exaptation, not an adaptation. This interpretation means that air breathing has occurred as a trait associated with limited activity and low energy demand and therefore not unique to intertidal species.

A study of air-breathing ability in five species of sculpins occurring along an environmental gradient from the highest intertidal zone to deep subtidal waters supports the adaptation hypothesis rather than the exaptation hypothesis. The two intertidal species exchange respiratory gases (oxygen and carbon dioxide) at high rates in air and do not rely on anaerobic metabolism. In contrast, the three other species, representing shallow subtidal and deeper-water habitats, do not exhibit these traits. Nevertheless, the phylogenetic relationships among these five species have not been established; thus, abilities associated with the intertidal habitat cannot be claimed to be traits evolved in response to intertidal conditions. A rigorous test of air breathing as a convergent adaptation awaits future studies.

PARENTAL CARE

Resident intertidal fishes of a variety of lineages show a narrow range of reproductive behavior and life history patterns. These few kinds of patterns suggest that convergent adaptation in traits such as parental care (Fig. 10) has evolved among these distantly related species. The question in this regard is whether the behavioral patterns associated with parental care arose in different groups by convergent evolution in the process of colonizing the intertidal habitat or whether the patterns are possessed as exaptations that made these groups successful in colonizing the habitat. In other words, did parental care evolve after or before colonization of the intertidal habitat? If it

FIGURE 10 Male plainfin midshipman *(Porichthys notatus)* guarding his egg mass.

occurred in the intertidal species after colonization, thus contrasting with the behavior of the subtidal species, then the behavior can be considered as a true adaptation to intertidal life.

The patterns of parental care in blennioid fishes (members of the Blenniidae and related families) of the northeastern Atlantic provide the opportunity to weigh these alternatives because about half of the species reside in the intertidal habitat and the other half in the subtidal environment. A survey of parental care in these blennioids largely supports the exaptation hypothesis because in most cases the intertidal species and its subtidal relative both exhibit parental care. The investigation, however, does show that male courtship displays are modified in intertidal species compared to subtidal species, probably to reduce swimming time and loss of contact with the surface of the bottom in response to the wave action and turbulence characteristic of the intertidal habitat. These displays, therefore, may constitute intertidal adaptations. As with desiccation tolerance and air breathing, definitive conclusions on reproductive behavior, including parental care, as adaptations of resident intertidal fishes rest with the availability of resolved phylogenies and completion of detailed ecological and physiological studies.

SEE ALSO THE FOLLOWING ARTICLES

Air / Body Shape / Buoyancy / Camouflage / Desiccation Stress / Size and Scaling

FURTHER READING

Almada, V. C., and R. S. Santos. 1995. Parental care in the rocky intertidal: a case study of adaptation and exaptation in Mediterranean and Atlantic blennies. *Reviews in Fish Biology and Fisheries* 5: 23–37.

Horn, M. H. 1999. Convergent evolution and community convergence: research potential using intertidal fishes, in *Intertidal fishes: life in two worlds.* M. H. Horn, K. L. M. Martin, and M. A. Chotkowski, eds. San Diego: Academic Press, 356–372.

Horn, M. H., and K. L. M. Martin. 2006. Rocky intertidal zone, in *Ecology of marine fishes: California and adjacent waters.* L. G. Allen, D. J. Pondella II, and M. H. Horn, eds. Berkeley: University of California Press, 205–226.

Martin, K. L. M., and C. R. Bridges. 1999. Respiration in water and air, in *Intertidal fishes: life in two worlds.* M. H. Horn, K. L. M. Martin, and M. A. Chotkowski, eds. San Diego: Academic Press, 54–78.

FLIES

BENJAMIN A. FOOTE

Kent State University

Although insects are supremely successful in terrestrial and freshwater habitats, they have been unable to penetrate marine habitats to any great extent. Only a handful of species, mostly belonging to the group known as water striders, are truly marine in the sense that they can be encountered at considerable distances from shorelines. On the other hand, several orders of insects have been able to adapt to the physical and chemical factors that dominate shoreline habitats. Among these, flies of the order Diptera are particularly successful in exploiting the food resources available in coastal regions, although most of the species are restricted to salt marshes, sandy beaches, tidal mud flats, and mangrove swamps. Only 16 families contain species that are regularly encountered on rocky shores and around tidepools.

LARVAL FEEDING HABITS

Most of the rocky-shore species of flies are considered to have larvae that are scavengers of decaying organic matter. Species of Anthomyiidae *(Fucellia intermedia),* Canaceidae *(Canace macateei),* Coelopidae *(Coelopa frigida),* Ephydridae *(Lamproscatella dichaeta),* Helcomyzidae *(Helcomyza ustulata),* Sepsidae *(Orygma luctuosa, Saltellaspondylli),* and Sphaeroceridae *(Thoracochaeta brachystoma, T. seticosta, T. zosterae)* consume the decaying remains of seaweeds (wrack) that has been washed up onto rocky shores by wave action. Larvae of another species of Ephydridae, *Hecamede albicans,* are known to attack the decaying remains of crabs and other shellfish.

Larvae of a few species of Chironomidae *(Paraclunio* spp., *Saunderia* spp.), Tipulidae *(Limonia* sp.), Canaceidae *(Nocticanace arnaudi),* Ephydridae *(Scatella obsoleta, S. picea, S. stagnalis)* are somewhat more specialized, as their larvae are consumers of microalgae growing on intertidal or supratidal rocks.

Undoubtedly, the most unusual feeding habit among the rocky shore flies is that of *Oedoparena glauca* (Dryomyzidae), whose larvae are predators of intertidal barnacles along the West Coast. At low tide, females of this species seek out exposed, living barnacles for egg laying. Larvae feed on the softer tissues of the barnacle, eventually killing their prey. Infestation rates can be as high as 35% during late spring and early summer. There is one generation a year.

ECOLOGICAL SIGNIFICANCE

Ecologically, larvae of Diptera play several important roles on rocky shorelines. Scavenging larvae are significant in the recycling of organic matter in that they speed up decomposition of wrack deposits by burrowing through the accumulated mass of vegetation. This burrowing activity, in turn, creates passageways that allow greater access to oxygen for the aerobic microorganisms that are essential in the decomposition process. The larvae of Chironomidae, Tipulidae, Ephydridae, and Canaceidae can affect the abundance, species composition, and microdistribution of rocky-shore algae because some species are rather selective in their choice of algal food. There are few studies documenting the effect of larval feeding on the population dynamics of intertidal barnacles, but if *Oedoparena* larvae become unusually abundant, they definitely can effect the population density of their prey. Finally, larvae of Diptera, because of their abundance and diversity, undoubtedly are important food sources for shoreline birds.

SEE ALSO THE FOLLOWING ARTICLES

Birds / Food Webs / Predation / Surface Tension

FURTHER READING

Burger, J. F., J. R. Anderson, and M. F. Knudsen. 1980. The habits and life history of *Oedoparena glauca* (Diptera: Dryomyzidae), a predator of barnacles. *Proceedings of the Entomological Society of Washington* 82: 360–377.

Cheng, L. 1976. *Marine insects.* Amsterdam: North-Holland.

Harley, C. D. G., and J. P. Lopez. 2003. The natural history, thermal physiology, and ecological impacts of intertidal mesopredators, *Oedoparena* spp. *Invertebrate Biology* 122: 61–73.

Marshall, S. A. 1982. A revision of the Nearctic *Leptocera* (*Thoracochaeta* Duda) (Diptera: Sphaeroceridae). *The Canadian Entomologist* 114: 63–78.

Robles, C. 1984. Coincidence of agonistic larval behaviour, uniform dispersion, and unusual pupal morphology in a genus of marine midges (Diptera: Chironomidae). *Journal of Natural History* 18: 897–904.

Robles, C. D., and J. Cubit. 1981. Influence of biotic factors in an upper intertidal community: dipteran larvae grazing on algae. *Ecology* 62: 1536–1547.

FOG

WENDELL A. NUSS

Naval Postgraduate School

Fog is the occurrence of clouds near the ground, consisting of small droplets of liquid water suspended in the air. The development and occurrence of fog is controlled by a variety of processes in the atmosphere that concentrate moisture near the ground, produce cooling, and condense moisture into fog. Fog alters the near-surface environment by reducing solar radiation and moderating air temperatures.

PHYSICAL PROCESSES OF FOG FORMATION

The basic ingredient for fog is water vapor molecules that occur in the air. Water vapor is put into the air through evaporation from liquid water sources on the earth's surface, such as the ocean, lakes, rivers, and moist soil. The amount of water vapor that can occur in the air depends upon the temperature of the air. Warmer air holds larger amounts of water vapor than colder air. The amount of moisture in the air is expressed as the specific humidity or mixing ratio, which is the mass in grams of the water vapor contained in a kilogram of air (g/kg). Typical values for air near the surface range from 2 to 30 g/kg, with the higher values associated with very warm tropical air. The maximum amount of water vapor that can occur in air at a given temperature and pressure is the specific humidity of saturation or saturation mixing ratio.

Common methods for expressing the amount of water vapor present in the air are the relative humidity and dew point temperature. The relative humidity is the ratio of the mixing ratio to the saturation mixing ratio, expressed as percentage of saturation (0–100%). The saturation mixing ratio of a given volume of air will change as the temperature or pressure changes; however, the mixing ratio for the same volume of air changes only through evaporation and condensation processes. Consequently, an 80% relative humidity in the cool of morning may only be 40% when the peak afternoon temperature occurs, because the actual moisture content does not change but the saturation mixing ratio rises with the temperature. Another common method of expressing the moisture content of the air is the dew point temperature, which is the temperature to which a volume of air must be cooled before condensation begins. The dew point simply represents the temperature for which the mixing ratio (actual moisture content) is the saturation mixing ratio. When the temperature is equal to the dew point, then the mixing ratio and the saturation mixing ratio are equal and the relative humidity is 100%.

The basic mechanism by which fog, or any cloud in the atmosphere, forms is through cooling the air to its dew point, or a relative humidity of 100%, at which point the water vapor begins to condense. Cooling of air can occur through three primary methods: lifting, diabatic processes, and mixing. Although the lifting mechanism is typically most important for cloud formation above the surface, diabatic cooling in the form of heat exchange by radiation and conduction is most important for fog formation. To cool the air adjacent to the ground, the ground or underlying water surface must be colder than the air just above so that heat flows from the air into the colder ground or water. If the ground cools enough to lower the air temperature to its dew point, fog forms in a shallow layer near the ground. Radiation fog forms at night when the ground cools by radiating thermal energy, absorbed from the sun (solar radiation) during the day, back to space at night. Advection, or dynamically driven fog, occurs when the air near the surface moves over a colder surface, such as cold upwelled ocean water.

ATMOSPHERIC CONDITIONS THAT PRODUCE FOG

Atmospheric conditions that are favorable for fog formation depend upon the type of fog but must promote cooling of a moist layer near the surface. For radiation fog, low wind and clear skies allow strong surface cooling by radiation to space and limit vertical mixing of the moisture-laden air away from the ground. These conditions are typically associated with regions of high atmospheric pressure at the surface. High-pressure regions occur when air in the atmosphere descends toward the surface and are generally characterized by fair weather and clear skies. High-pressure centers are also usually characterized by rather low winds, which is also conducive to fog formation. Wind impacts fog and fog formation by producing vertical mixing of the air near the ground. Because the air above the ground tends to have less moisture, vertical mixing of this air toward the surface tends to delay or prevent fog formation. Critical to the formation of radiation fogs is a supply of surface moisture. Moisture is supplied by the ocean in coastal regions and is carried inland through the afternoon sea breeze. Additional moisture may be supplied through rain that moistens the ground to subsequently be evaporated back into the air near the ground. Since strong radiational cooling occurs from the land and not the ocean, radiation fogs are most prevalent over the land areas in coastal environments.

For advective or dynamic-type fogs, winds that blow relatively warm, moist air over cooler regions of the ocean or adjacent coastal land areas produce the required surface cooling. Winds also tend to produce vertical mixing, which can be unfavorable to fog formation. However, the cooling from below tends to produce a cool layer that produces fog or low-level stratus clouds as a result of vertical mixing with strong winds. These types of fogs are very common along the U.S. West Coast or other regions of the world characterized by coastal upwelling of colder water (Fig. 1).

FIGURE 1 Bank of coastal fog at the mouth of the Little Sur River, Monterey County. Photograph by Thomas H. Mikkelsen.

The actual weather conditions under which advective or dynamic-type fog forms varies from location to location and season to season. For the West Coast of the United States during the summer, air subsides in the subtropical high-pressure center that is located offshore. Anticyclonic flow around the high results in northerly along-coast winds and coastal upwelling produces cold surface water. The near-surface air that is trapped in the low levels by a strong temperature inversion is cooled by this cold water to result in a cloud-filled layer near the ocean surface. Because of the wind and other processes that control the depth of this layer, the clouds are often elevated above the surface by a few hundred meters to produce marine stratus as opposed to true fog. The distribution of fog or low clouds can vary along the coast depending on how the air interacts with coastal mountains. For the East Coast of the United States, advective fogs occur when warm moist air moves northward ahead of a low-pressure system, which brings the warm moist air over colder ocean water to produce fog if enough cooling occurs.

INTERTIDAL ENVIRONMENT AND THE EVOLUTION OF FOG

Fog impacts the intertidal environment by limiting sunshine and moderating both daytime and nighttime temperatures. Both these effects depend on the tendency for fog to persist through the day and recur multiple days in a row. Once fog has formed, the fog may evaporate during the daytime because of warming of the ground and near-surface air by the sun. This process is referred to as burn-off, and the degree to which fog will clear depends primarily on the depth of the fog layer. Deeper layers require longer to clear because the sunshine, which is absorbed at the ground, must heat the entire fog layer to a temperature above the dew point temperature. A shallow layer can heat quickly even with limited sunshine so that fog will not persist. The depth of the fog layer depends on the winds, strength of subsidence, degree of surface cooling, and various other factors. Typically, radiation fogs tend to be shallow and burn off more readily than dynamic-type fogs, which occur in deeper layers of moist air. The number of hours of sunshine is determined by the persistence of the fog through the day, and this in turn limits the amount of warming observed on a given day. The development of fog at night effectively stops or slows the nighttime cooling by radiation to maintain warmer temperatures at night, which reduces the diurnal temperature range in foggy regions.

Fog will usually re-form if it burns off unless the characteristics of the near-surface air for a given region change. The characteristics of the low-level air change as weather systems pass, lowering or raising the moisture content and altering the vertical structure of the lower atmosphere. These changes are very significant for the maintenance and evolution of fog over periods of days and produce more extended periods of clear or foggy conditions.

SEE ALSO THE FOLLOWING ARTICLES

Air / Evaporation and Condensation / Temperature Change

FURTHER READING

Fog and low stratus. http://www.meted.ucar.edu/topics_fog.php.
Houze, R. A., Jr. 1993. *Cloud dynamics.* New York: Academic Press.
Wang, Binhua. 1985. *Sea fog.* New York: Springer-Verlag.

FOOD USES, ANCESTRAL

MICHAEL A. GLASSOW
University of California, Santa Barbara

Because of their visibility and ease of acquisition, rocky-intertidal fauna may have been the first marine foods used by humans and their hominid ancestors. Archaeologists have demonstrated that humans were acquiring, and presumably eating, a variety of fauna from rocky intertidal zones by 150,000 years ago, when Neanderthal peoples living along the shores of the Mediterranean were collecting mussels and other shellfish. Evidence from Europe as well as South Africa and Australia is more certain for periods beginning about 75,000 years ago. Fish remains also occur at coastal habitation sites of this age, but whether fish were acquired from tidepools is unclear.

ANTIQUITY

Between 8000 and 10,000 years ago, peoples in diverse coastal areas of the world began to accumulate substantial shell middens at their coastal habitation sites. Middens along open coasts typically contain dense concentrations of shells of rocky-intertidal molluscs (Fig. 1). Particularly after 5000 years ago, some middens became mounds several meters tall largely because of the accumulation of shellfish remains at relatively stable residential bases.

FIGURE 1 An example of a shell midden, exposed along the edge of a sea cliff on western Santa Cruz Island. Mussel shell is the dominant constituent in the midden, which dates between approximately 2300 and 200 years ago. Photograph by the author.

Problems of preservation make it difficult for archaeologists to document use of rocky-intertidal fauna during the late Pleistocene and early Holocene. The remains of these fauna, in the form of shells and bones, survive from such ancient times only under certain circumstances (e.g., in cave deposits), and fluctuations in sea level resulting from glacial cycles surely have destroyed or inundated many coastal habitation sites.

PREHISTORIC CONTEXTS

Rocky intertidal shorelines provided a wide variety of invertebrates and vertebrates collected and consumed as foods. Molluscs such as mussels, abalones, limpets, and turbans, as well as clams residing in gravels between intertidal rocks, were collected. Less popular but sometimes important were sea urchins, a variety of crabs, and tidepool-dwelling fishes. The popularity of particular taxa is related to their ease of collection, abundance, and rate of replenishment after collection. Along the North American Pacific coast, for instance, the abundant and prolific California mussel *(Mytilus californianus)* frequently constitutes more than 75% of the quantity of shellfish remains in middens.

Shells of easy-to-acquire molluscs often are nearly the exclusive marine food remains present in the earliest coastal sites. With the development of new technology, however, other marine foods were added to the diet, and fully maritime adaptations developed. Nonetheless, even the most sophisticated prehistoric maritime peoples continued to include rocky-intertidal shellfish in their diets if these resources were abundant.

In their efforts to understand the changing context of rocky-intertidal marine resources in the evolution of prehistoric subsistence systems, archaeologists have been concerned with two major issues. First, they have attempted to distinguish the effects of environmental change from the role of increasing cultural complexity related to population growth. Lack of sufficient knowledge of such factors as the effects of sea level rise on coastal environments and fluctuation in marine productivity, however, has given rise to varying opinions and interpretations. Second, the energetics of marine resource utilization, particularly of shellfish, in relation to terrestrial foods have been the subject of sharp differences of opinion. Some archaeologists have argued that shellfish would not have been important unless terrestrial food resources were scarce, whereas others have argued that the abundance and reliability of marine foods, particularly rocky-intertidal shellfish, were always important because of ease of collection, predictability of location, and reliability.

SEE ALSO THE FOLLOWING ARTICLES

Abalones / Fossil Tidepools / Molluscs / Sea Level Change, Effects on Coastlines / Symbolic and Cultural Uses

FURTHER READING

Bailey, G., and J. Parkington, eds. 1988. *The archaeology of prehistoric coastlines.* Cambridge, UK: Cambridge University Press.
Claassen, C. 1998. *Shells.* Cambridge, UK: Cambridge University Press.
Erlandson, J. M. 2001. The archaeology of aquatic adaptations: paradigms for a new millennium. *Journal of Archaeological Research* 9: 287–350.
Yesner, D. R. 1980. Maritime hunter-gatherers: ecology and prehistory. *Current Anthropology* 21: 727–750.

FOOD USES, MODERN

KERRY SINK AND JEAN M. HARRIS

University of Cape Town, South Africa

Rocky-shore fauna and flora are popular, ubiquitous items on modern menus, offering key ingredients for gourmet feasts, yet remaining the mainstay of simple meals that nourish the impoverished. The narrow fringe of shoreline habitat is highly accessible, and the species prized as food are generally easy to find and harvest. Rocky shores support subsistence, recreational, and commercial fisheries, and important mariculture species originate from these intertidal ecosystems. This, coupled with a norm of common-property status for rocky shores, renders their biodiversity and the natural resources they yield extremely vulnerable. Although well-regulated and managed examples of intertidal fisheries do exist, sustainability of the harvest is often exceeded, biodiversity compromised, and stocks of target organisms are severely depleted on countless shores. Many of the dominant visible organisms targeted are keystone species, and dramatic ecosystem shifts and cascade effects have been documented as a result of overharvesting by humans, who have taken the role of an unregulated top predator. Permit systems, monitoring, research, and marine-protected areas are valuable management measures to ensure that sustainability of the harvest and conservation of biodiversity are achieved in near-shore habitats.

FISHERS AND THEIR HARVEST

In many parts of the world, particularly in underdeveloped countries, indigenous and poor people depend on the use of natural marine resources, readily accessible without the use of sophisticated gear, to meet basic livelihood needs. These are termed subsistence fishers. Although historical culturally based fisheries, including collection for traditional medicines, account for a significant proportion of this use, there is also a tendency for rocky-shore resources to provide a last resort to modern poverty. In general, the resources used by this group of fishers as a direct food source are of relatively low value; they are nevertheless also often used for barter or local sale. Important taxa include mytilid mussels, limpets, whelks, chitons, sea urchins, barnacles, crabs, and ascidians (*Pyura* spp.).

Species that are harvested for traditional medicinal and magical uses include sponges, seaweeds (particularly for the treatment of skin ailments), chitons and other molluscs, as well as echinoderms such as urchins and starfish. In Asia, sea cucumbers are believed to have aphrodisiac properties, whereas in Africa medicine made from sea cucumbers is believed to make an unfaithful wife's stomach fall out! This is linked to the holothuroid habit of expelling a portion of their intestine in response to a predatory attack.

Recreational use—that is, collecting for sport or to gain gourmet food "free" from the sea—can account for a large portion of the total harvest from the intertidal and can contribute significantly to economic activity through associated industries (such as purchase of permits, collecting implements, or protective clothing) and by stimulating tourism. Mussels, oysters, abalone, limpets, cockles, winkles, stalked barnacles, and crabs are prized as gifts from the sea by recreational harvesters all over the world.

The potential for commercial activities that are based on direct harvest of intertidal organisms is limited compared to that of offshore fisheries, but it does underpin the local economy in some coastal areas. High-value resources such as rock lobster, abalone, sea cucumbers, and oysters are targeted, although in many cases these taxa are no longer accessible in the intertidal. Seaweed is harvested in such operations throughout the world as a food source as well as an export commodity for the production of agar and carrageenan products. Small-scale commercial operations that generally have a traditional or cultural basis, and can be referred to as artisanal fisheries, provide the economic basis for some coastal communities. In Chile, the murcid snail or *loco (Concholepas concholepas)* is economically the most important artisanal resource, although sea urchins *(erizo)* and keyhole limpets *(lapas)* are also important targets of shellfish gatherers.

FIGURE 1 Mussels are one of the most popular seafoods worldwide and are enjoyed smoked, dried, steamed, cooked on the grill or open fire, and in stews or soups, such as these *Mytilus galloprovincialis* cooked in white wine and garlic.

FIGURE 2 Subsistence harvesters in the Greater St. Lucia Wetland Park, South Africa, depend on rocky-shore species such as mussels, limpets, chitons, and tunicates as a primary protein source.

Although there are few options for *in situ* farming, numerous rocky-shore species are the mainstay of many lucrative mariculture ventures. Globally the most important group of invertebrates in mariculture that originate from rocky shores is mytilid mussels. They are cultured in approximately 40 countries around the globe, with four species accounting for the bulk of world production (the European or blue mussel, *Mytilus edulis*; the Mediterranean mussel, *Mytilus galloprovincialis*; the Thailand green mussel, *Perna viridis*; and the New Zealand green-shell mussel, *Perna canaliculus*). Farming of oysters (*Crassostrea* and *Ostrea* species) is also an important commercial activity with considerable economic benefits. The taste of oysters is strongly influenced by the characteristics of local waters, and species tend to be marketed by origin. Seaweed has also been cultured traditionally for decades and probably for centuries in several Asian nations such as China, Korea, and Japan. Modern small-scale seaweed farming is a growing industry often considered to uplift the socioeconomic status of small-scale fishers and to reduce fishing pressure on overexploited fisheries. Species grown for harvesting include varieties of *Eucheuma*, *Gracilaria*, *Sargassum*, and *Porphyra*, among others, with farming now even extending into Africa and several island states.

SUSTAINABLE USE AND IMPACTS ON BIODIVERSITY

The productivity of rocky-shore organisms is dependent on the productivity of the local marine environment. Upwelling areas support rich intertidal communities with extensive biomass, whereas less nutrient-rich waters tend to have low potential for supporting intertidal fisheries. Nevertheless, even in the poor to moderately productive subtropical and tropical areas, rocky shores do provide an important primary protein source to coastal communities. Because intertidal resources are easily visible and highly accessible at low tide, they are vulnerable to overexploitation. Monitoring population trends using catch-per-unit effort often fails to detect problems within an intertidal fishery because many species are prone to hyperstability, such that although the resource may be declining, because collectors can actively search for these visible species, overexploitation is often detected only when the resource has all but disappeared. Abalone are one of the rocky-shore organisms with the highest economic value, and the high economic incentive of these species has resulted in extensive illegal and unregulated fishing, compromising the future of several taxa.

There is a rich source of literature documenting the ecological implications of intertidal harvesting on rocky shores, and this has provided a key contribution to the international debate on community regulation and ecosystem impacts of fisheries. Harvesting of rocky-intertidal resources not only impacts directly on the population of target species but can have serious consequences for intertidal biodiversity through habitat destruction or transformation, recruitment failure, cascading ecosystem impacts, and impacts on energy flow.

Large-scale modifications of rocky-shore communities have been attributed to human exploitation of intertidal organisms. In Chile removal of grazing molluscs caused increased abundance of the macroalgae they feed on, whereas harvesting of predatory gastropods modified community structure by increasing the abundance of prey, particularly small, inedible mussels and barnacles. These impacts can be considered as cascading ecosystem effects because the removal of one species ripples through the food chain and impacts on prey or food species. In South Africa, studies on the southeast coast indicated that harvesting of intertidal invertebrates by subsistence collectors modified community structure, with filter-feeding communities switching to stable algal communities dominated by articulated coralline turfs. There, changes are driven by competitive release, which sees the algae able to gain occupancy of space on the shore when their competitors (filter-feeding mussels) are removed from the shore. This can further impact on sustainability through recruitment failure. The settlers, or spat, of several mussel species prefer to settle among established mussels, and therefore reduced cover of adult mussels can impact on recruitment. On the west coast of South Africa, the spat of *Mytilus galloprovincialis* settle among adult mussels at densities 20–100 times greater than they do on bare rock or primary colonizing algae. Similarly, the recruitment success of several oyster species and a tunicate species have been shown to rely on sufficient abundance of adult stocks.

Removal of filter feeders can also have impacts for other species because these taxa play a pivotal role in shallow-water food webs. Mussels and oysters capture phytoplankton and particulate matter that is inaccessible to other trophic groups. They are therefore responsible for a large share of the energy flow from pelagic to benthic systems. Many intertidal taxa including mussels, oysters, and tunicates form dense three-dimensional matrices that support a diverse infauna. Overharvesting or harvesting with destructive methods (such as the use of wide-bladed implements) can destroy the habitat of many infaunal species.

Commercial culture of intertidal species has had serious impacts on many coastal ecosystems. Many cultured molluscs have escaped from mariculture operations, and several species have become truly invasive, competing with indigenous species, transforming intertidal communities, and threatening local biodiversity. In some cases these taxa also cause problems in other mariculture operations and foul bridges, docks, and boats. Farming of indigenous species can also pose a significant risk to marine biodiversity, with genetically manipulated animals impacting on local stocks when they escape and breed with wild populations. Mariculture can also suffer from similar problems normally associated with intensive agriculture on land, such as local pollution, habitat destruction, disease, and contamination by toxins and stimulants.

MANAGEMENT AND CONSERVATION

Because intertidal species are easily accessible and vulnerable to overexploitation and their harvesting can impact on many other species, management systems to ensure sustainable harvesting are appropriate. Carefully determined harvesting offtakes, permit systems, allocation procedures, research, and monitoring have been effective in the regulating of intertidal fisheries and are considered the best practice.

The effects of small-scale artisanal, subsistence, and recreational fisheries should be examined and measures implemented to ensure sustainability. Unlike other fisheries, catch per unit effort may not provide an accurate indication of the stock status. Human impacts in the intertidal should also be assessed in light of their effect on the entire ecosystem, not just on the resource species. Analyses of whole ecosystems are also important in assessing how much any change matters in terms of biodiversity conservation. Some studies have used an experimental approach to assess optimal offtake for intertidal fisheries. Controlled harvesting at predetermined levels can effectively evaluate the effects on target species as well as the impacts on biodiversity and ecosystem processes. An interesting example of this can be found in South Africa, where subsistence harvesters were recently recognized as a new category of fishers and specific management schemes have been developed to ensure that they be given preference in certain areas and that their harvesting is conducted at a sustainable level.

Another important consideration in the management of intertidal fisheries is the methods used for harvesting. It can be advantageous to impose gear limitations. Control of the implements used in shellfish collection is advantageous because wide-bladed tools create larger areas of bare rock than narrow-bladed implements (e.g., a screwdriver) and remove a considerable unwanted bycatch of inedible juveniles and infaunal species. Smaller or more selective implements cause less habitat damage than those with wide blades and are therefore less likely to impact on recruitment success. They encourage collectors to be more selective, reducing the bycatch. Restricting the use of apparatus that facilitate exploitation in adjacent shallow subtidal waters, such as mask and snorkel, or self-contained underwater breathing apparatus (SCUBA), is often advisable for those intertidal species that extend into the subtidal, because intertidal recruitment success and sustainability of these species may depend on these subtidal stocks remaining inaccessible to collectors.

Effective resource management is particularly critical to the long-term success of small-scale and subsistence fisheries. Comanagement arrangements have been identified as a potential way forward for in-shore fisheries that have a strong common-property nature and where community-based ownership rights are strong. Furthermore, in most parts of the world, governments do not have the resources or will to effectively police the entire shoreline, and community-based solutions need to be sought. The essence of comanagement is that user groups and government share the responsibility for managing a resource. The benefits of comanagement include greater participation of the user groups in decision making and resources monitoring and research, better relationships between fishers and authorities, and improved information sharing and capacity building among fishers and law enforcement staff. Documented outcomes include greater acceptance of regulations, improved compliance, restoration of access rights for traditional fishers, and the hope of more efficient, equitable management systems and sustainability of the harvest. There are now quite a few examples of comanagement systems for intertidal fisheries (e.g., in Chile the turf system is a unique leading example for a commercial fishery, and in South Africa local comanagements systems for subsistence harvesting are being implemented on the east coast), and there is a need to develop indicators to measure success of these models and to export lessons learned from case studies to other areas.

Fully protected representative marine-protected areas contribute significantly to the sustainability and maintenance of intertidal biodiversity. Such areas can seed recruits to adjacent exploited areas and help to achieve resilience to overexploitation. They also serve as valuable reference areas needed to measure and monitor harvesting impacts. One management option that has been considered for intertidal fisheries is rotational cropping. However, rotational cropping is not a viable management strategy for fisheries in which harvesting can transform community structure. Evidence suggests that such changes may take long time periods (if ever) to be reversed. A more viable management strategy is to use zonation to protect representative sections of the coast in the long term and to institute controlled harvesting at predetermined levels at other sites and conduct *in situ* monitoring of stocks and community structure to allow an adequate evaluation of optimal harvesting.

As the global demand for intertidal resources escalates, wild stocks alone cannot be relied on. The cultivation of intertidal species is advancing, and such initiatives play an important role in providing protein and sustaining coastal and even inland communities. However, because mariculture can have serious impacts on marine and coastal biodiversity, it is critical that responsible farming techniques are used with implementation of lessons learned and best practices in all operations.

FURTHER READING

Castilla, J. C. 1999. Coastal marine communities: trends and perspectives from human-exclusion experiments. *Trends in Ecology and Evolution* 14: 280–283.

Charles, A. T. 1994. Towards sustainability: the fishery experience. *Ecological Economics* 11: 201–211.

Harris, J., G. Branch, S. Sibiya, and C. Bill. 2003. The Sokhulu Subsistence Mussel Harvesting Project—a case study for fisheries comanagement in South Africa, in *Waves of change. Coastal and fisheries co-management in South Africa*. M. Hauck and M. Sowman, eds. Cape Town: University of Cape Town Press, 61–98.

Hillborn, R., and C. J. Walters. 1992. *Quantitative fisheries stock assessment: choice, dynamics and uncertainty*. New York: Chapman and Hall.

Hockey, P. A., A. L. Bosman, and R. W. Siegfried. 1988. Patterns and correlates of shellfish exploitation by coastal people in Transkei: an enigma of protein production. *Journal of Applied Ecology* 25: 353–364.

Kingsford, M. J., A. J. Underwood, and S. J. Kennelly. 1991. Humans as predators on rocky reefs in New South Wales. *Australia. Marine Ecology Progress Series* 72: 1–14.

Lasiak, T. A. 1999. The putative impact of exploitation on rocky intertidal macrofaunal assemblages: a multiple area comparison. *Journal of the Marine Biological Association of the UK* 79: 23–34.

McClanahan, T., ed. 2007. *Fisheries management: progress towards sustainability*. Oxford, UK: Blackwell.

Pomeroy, R. S., and F. Berkes. 1997. Two to tango: the role of government in co-management. *Mar. Pol.* 21: 464–480.

Siegfried, W. R., ed. 1994. *Rocky shores: exploitation in Chile and South Africa*. Berlin: Springer-Verlag.

Underwood, A. J. 1993. Exploitation of species on the rocky coast of New South Wales (Australia) and options for its management. *Ocean and Coastal Management* 20: 41–62.

FOOD WEBS

BENJAMIN S. HALPERN

University of California, Santa Barbara

All biological communities are composed of species that eat or are eaten by other species. This network of species interactions is called a food web and is one of the most common means of describing what a community, or collection of species, looks like. No two food webs are identical, whether measured across space or through time, even if they might have the same basic structure. Similarities and differences among food webs have motivated a long history of research in community ecology to understand how to effectively and efficiently describe and compare food webs, what causes these differences and similarities in food webs, and what are the ecological, evolutionary, and management consequences of food web structure.

DESCRIBING FOOD WEBS

If one draws the connections between and among species in a community, indeed a weblike picture emerges (Fig. 1). However, many communities have hundreds of species, or more, making such web pictures cumbersome and difficult to analyze or understand (Fig. 2). Since species often provide similar functions to a community—two species may share a common food or provide a similar resource to the community—several approaches have been developed for assigning species to functional groups. In other words, if species provide redundant services to a community, they can be grouped into a single category, and these

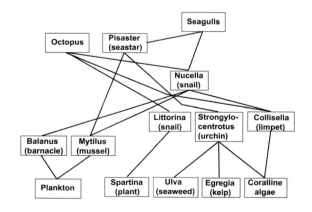

FIGURE 1 A very simple rocky intertidal food web from the Pacific Northwest shores of North America.

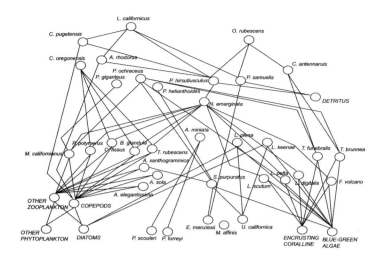

FIGURE 2 A more complete food web for the entire rocky shore. Only a small percentage of all possible species is depicted, and only some of the key trophic links are shown. Species complexes are indicated with all-caps labels.

groupings in turn simplify our view of what the food web looks like.

The most common approach to lumping species into functional groups is to assign them to a trophic level. In this view of a food web, we have autotrophs (plants and algae such as seagrass and kelp), herbivores, predators, and perhaps omnivores, secondary predators, and tertiary predators. All of a sudden, a very complex food web (Fig. 2) becomes very simple (Fig. 3). More recently a variety of approaches have been developed for subdividing these trophic classifications while still preserving the simplicity and usefulness of functional groups. These approaches focus on more accurately describing the niche of different groups of species, and include both categorical and continuous methods for defining niches. For example, plants and algae can be classified as separate groups because they

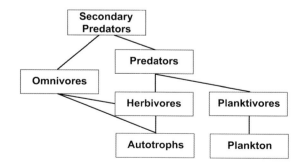

FIGURE 3 The food web with species grouped into trophic levels.

have different palatability and resource needs, mobile and territorial herbivores can be separated because their spatial impact on community dynamics is fundamentally different, or predators can be defined by the different proportions of their diet made up of a variety of potential prey. These approaches can even put different life stages of a single species (e.g., juvenile vs adult) into different functional groups if they align more closely to other species than to their own life stages. For example, some prickleback fishes in California tidepools shift from being carnivores to herbivores as they grow. Such views of food webs are still fairly simple but capture more of the nuances in how species relate to each other, partition niche space, and affect whole communities (Fig. 4).

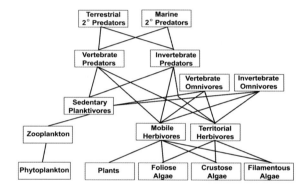

FIGURE 4 The food web with species lumped into 15 typical functional groups.

PATTERNS IN FOOD WEBS

The number and identity of species in a community changes, sometimes dramatically, as one moves along a coastline or down the coastal slope, and in some cases the number and type of functional groups changes as well. The horizontal zones of species groups in rocky intertidal areas are some of the best studied and most notable examples of such species-level changes (Fig. 5). Clear differences in species composition exist in these different zones, yet they are very similar in their functional group composition—barnacles replace mussels as one moves higher up the slope, but both food webs have space-filling filter feeders. In contrast, moving down the slope into the subtidal zone, space is often filled by macroalgae and other invertebrate species, sometimes to the exclusion of these filter feeders.

Food webs also change dramatically in composition as one moves from the tropics to the poles and from the Atlantic to the Pacific to the Indian Ocean. It is particularly interesting that these changes tend to vary among functional groups. So although overall species richness typically increases as one moves towards the

FIGURE 5 Distinct zones of intertidal communities at the Olympic Coast National Marine Sanctuary. Each zone has different species composition, whereas functional group composition among zones can differ or be the same. Photograph courtesy of www.photolib.noaa.gov.

tropics and into the Pacific, very different patterns exist for particular functional groups. For example, herbivores and structural species (e.g., corals) generally follow the previous patterns, but some suspension feeders (e.g., polychaetes) and detritivores (e.g., nematodes) show uniform or opposite patterns. In some cases these changes lead to novel functional groups being added to the community, but in other cases existing functional groups are simply enriched with more species.

Recently there has been interest in describing and understanding food web structure based only on the pattern of connections among species rather than on the identity or functional role of the species. This approach uses graph theory to compare and analyze food webs, with particular interest in understanding emergent properties of a food web, such as its stability. In this frame of mind, food webs are thought of as networks (if one squints at the food web in Fig. 1B, it could just as easily represent the network of airplane routes in North America or the pattern of Internet connections in London), and analyses focus on the relative importance of particular nodes (species) in maintaining the network structure (see "Consequences of Food Web Structure"). Keystone species—single species that regulate the abundance of multiple other species (typically through predation)—are examples of hub nodes (i.e., important, well-connected species). *Pisaster* sea stars are the classic example of a keystone species in rocky shores.

CAUSES OF FOOD WEB STRUCTURE

There are essentially two core questions about the causes of food web structure—how are communities initially constructed, and what maintains or influences community structure once it is in place—and these questions have

occupied ecologists since the inception of the field. Colonization and community succession in previously barren areas (e.g., lava fields, new seamounts) is a classic example of the construction of a community. It is of course impossible to have an herbivore without a plant, and similarly impossible to have a predator without both an herbivore and plant, and so there is an inherent order to which species are added to a food web. But why would a second herbivore establish before a predator in some cases and a predator show up first in other cases? In more species-rich communities, what leads to, for example, omnivores versus secondary predators versus new herbivores being added next to the community? And why are there rarely tertiary predators in communities, and almost never any higher trophic levels?

Two primary competing theories that provide insight into the first two of these questions. Robert MacArthur first described how species partition resources to create unique niches—if a niche is already filled, a new species must either occupy a novel niche or somehow subdivide the existing niche with the species already present in the community. When sufficient resources are not available to support a novel niche (e.g., the predators in a community have insufficient biomass to support secondary predators), then a new species in that niche cannot invade the community. In contrast, Stephen Hubbell later proposed that species are added to communities randomly, such that the addition of new species to a community is simply a function of which ones show up first. Both theories have proven capable of predicting some aspects of community structure, and so the debate remains active. Answers to why tertiary predators are unusual invoke a different set of theories, but ones that are equally unresolved. Higher trophic levels may be rare or impossible because energy transfer up a food chain is far from efficient (a relatively small portion of prey biomass gets converted into predator biomass), yet there are examples of marine communities with huge amounts of biomass in the top trophic level. The other explanation for limitations to food chain length focuses on the stability (or instability) of longer food chains. In this line of argument, direct and indirect species interactions up and down the food chain become more unstable with longer food chains, creating an upper stability threshold for the number of additional trophic levels that can be added to a food web. In other words, longer food chains have more predator–prey interactions that can have potentially destabilizing effects on each other. However, there is still much debate about whether or not longer food chains are less stable.

Once communities are in place, the question turns to what processes change and maintain community structure.

Drivers of community structure can be classified into four broad categories: top-down effects (predators), bottom-up effects (nutrients and primary productivity), recruitment, and abiotic control. The first two represent two sides of a long-debated and still unresolved question in ecology—are food webs driven more by bottom-up or by top-down forces? Nutrient enrichment can lead to greater plant growth, which can in turn fuel biomass transfer up the food chain. In contrast, predator abundance has been shown to lead to trophic cascades—herbivores are eaten or displaced by predators, in turn releasing plants from herbivore grazing pressure—in a wide range of systems and locations. Humans are, of course, the ultimate top-down control of communities and have had profound impacts on entire communities through their omnivorous consumption or extermination of species, as with the extensive harvest of loco in Chile. Yet we have also driven many species and systems to local extinction through extreme eutrophication of coastal waters, which is a bottom-up effect, as seen in the dead zone off the mouth of the Mississippi and the rocky shores of the Baltic Sea. In reality, both bottom-up and top-down forces are important at different times and different places, and so research has begun to focus on understanding where, when, and why one type of control is more important than the other.

A recent and interesting twist on the bottom-up hypothesis is the idea that community structure is regulated by the stoichiometry, or ratio of carbon to nitrogen to phosphorus (C:N:P), of the species present in the community. Plants are commonly limited in their growth and survival by the availability of nitrogen and phosphorus (e.g., we fertilize our gardens to increase growth rates), and animals are also often nutrient-limited in a similar fashion (e.g., potassium deficiencies in humans): these species will only be present (or abundant) in a community that has prey species that are rich in P or N.

Recruitment patterns can also greatly influence the development and maintenance of food web structure by controlling how many of which species show up at a location. Indeed, intertidal areas in regions with consistent upwelling, where larvae are transported offshore and therefore unable to recruit (e.g., the coasts in northern Chile and Namibia), have very different community structures compared to nearby areas that do not experience consistent upwelling, where species composition and abundance are largely driven by patterns of recruitment. When recruitment is particularly high (i.e., no longer limiting), then other factors, such as the top-down and bottom-up drivers discussed previously and abiotic stresses discussed herein, become more important in the control of food web structure.

The importance of all of these drivers of food web structure can be greatly modified by the abiotic conditions (i.e., level of abiotic stress) at a location. In highly stressful environments (e.g., high salinity or temperature stress), positive interactions among species, such as facilitation, can become far more important than predation (top-down control) or nutrient availability (bottom-up control) in allowing species to exist and coexist in a community. For example, conspecific and interspecific aggregation on rocky shores helps mediate thermal and desiccation stress, and the presence of these species then affects the entire community structure. The intensity and frequency of natural disturbances such as large storms (and associated storm surge) can also act to shift food web structure, sometimes into alternate stable states, by removing dominant or key species (such as beds of mussels) from the community.

CONSEQUENCES OF FOOD WEB STRUCTURE

Regardless of the causes, food web structure can have important consequences for the succession and stability of the community. There is growing evidence that more complex food webs—those with more species linked to each other through a greater number of connections—are more resistant to disturbance. This greater stability is conferred in part by having more redundancy within functional groups, such that the loss of a single or few species does not lead to the loss of an entire functional group, as well as by the presence of numerous weakly linked species that have a small effect on the whole food web if they are removed. The complexity of food webs may also influence its invasibility, whether by natural species (i.e., succession) or alien species (i.e., human introductions), because existing species or functional groups may already fill the niche of the invading species.

SEE ALSO THE FOLLOWING ARTICLES

Biodiversity, Global Patterns of / Herbivory / Predation / Recruitment / Succession / Zonation

FURTHER READING

Barabasi, A. 2002. *Linked: the new science of networks.* Cambridge: Perseus.

Elser, J. J., R. W. Sterner, E. Gorokhova, W. F. Fagan, T. A. Markow, J. B. Cotner, J. F. Harrison, S. E. Hobbie, G. M. Odell, and L. W. Weider. 2000. Biological stoichiometry from genes to ecosystems. *Ecology Letters* 3: 540–550.

Hooper, D. U., F. S. Chapin, J. J. Ewel, A. Hector, P. Inchausti, S. Lavorel, J. H. Lawton, D. M. Lodge, M. Loreau, S. Naeem, B. Schmid, H. Setälä, A. J. Symstad, J. Vandermeer, and D. A. Wardle. 2005. Effects of biodiversity on ecosystem functioning: a consensus of current knowledge. *Ecological Monographs* 75: 3–35.

McCann, K. S. 2000. The diversity-stability debate. *Nature* 405: 228–233.

Menge, B. A. 1991. Relative importance of recruitment and other causes of variation in rocky intertidal community structure. *Journal of Experimental Marine Biology and Ecology* 146: 69–100.

Menge, B. A. 2000. Top-down and bottom-up community regulation in marine rocky intertidal habitats. *Journal of Experimental Marine Biology and Ecology* 250: 257–289.

Navarette, S. A., E. A. Wieters, B. R. Broitman, and J. C. Castilla. 2005. Scales of benthic-pelagic and the intensity of species interactions: from recruitment limitation to top-down control. *Proceedings of the National Academy of Sciences of the USA* 102: 18046–18051.

Pimm, S. L. 2002. *Food webs.* Chicago: University of Chicago Press.

FORAGING BEHAVIOR

GRAY A. WILLIAMS
University of Hong Kong

COLIN LITTLE
University of Bristol, United Kingdom

All consumers must have some strategy to obtain energy—whether they feed on dead organic matter, plants, and algae, or capture and eat animal prey. To do this some species actively search for their prey, killing and ingesting them, whereas others simply remain stationary and wait for their food to come to them. Many species, such as anemones and barnacles, adopt this second strategy and are reliant on tidal currents or the chance wanderings of prey items to bring their food within reach. Most consumers, however, adopt the first strategy and actively move to seek and ingest their food—that is, they show some form of foraging behavior. Understanding the timing, location, and intensity of such behavior is important in helping to interpret and predict the organization and dynamics of intertidal communities.

GENERAL PATTERNS OF FORAGING BEHAVIOR

The success of a consumer depends on the trade-off between the energy gained from foraging (i.e., the energy ingested and assimilated) and the energy expended to obtain the food item. The distribution of food, whether sessile animals such as barnacles and mussels, clumps of algae, or mobile species such as slow herbivorous snails or rapidly moving crabs, is rarely uniform in any rocky intertidal habitat: food items are patchily distributed, both in time and space. Foraging animals therefore need to search to find their food. The physical conditions a foraging animal will face when searching and feeding are also not constant and will vary with location and related environmental changes (such as time of day, tidal cycle, season, etc.). As a result, the rewards of foraging, in terms of finding and gaining energy from feeding, must be offset against the costs incurred due to physical stresses (e.g., desiccation and osmotic stresses), the risk of being eaten, or the energy required to find and ingest the food items. Foraging species, therefore, adopt strategies that either maximize their rewards, or minimize their losses, to achieve a net energy gain or profit. They can do this either by varying their time and place of feeding, or by choosing between food items to achieve the highest rewards. This series of strategies is described under the theory of optimal foraging.

THE TIMING OF FORAGING

General Patterns

The two main factors affecting the timing of foraging are related to diurnal and tidal cycles. Many intertidal foragers move to feed only when they are under water or out of water, or when they are awash by the rising or falling tides. Some feed with no regular pattern in relation to night and day, but most have distinct diurnal patterns. These tidal and diurnal patterns are affected by a number of factors, but principally by physical stress when out of water, the chance of dislodgement by waves, and predation risk. The most common pattern on temperate shores appears to be foraging while submerged (Fig. 1), irrespective of the diurnal cycle. This strategy minimizes physical stresses that may be experienced while out of water. Some species are active when emersed, but usually these species only forage at night, when the shore is moist and cool, and physical stress is reduced (Fig. 1). On tropical shores the risk of foraging while submerged is high, due to the large numbers of predatory crabs and fish, and because physical stress is too high to allow foraging during emersion, most species forage while awash, often migrating with the rising and falling tide. On shores with wide tidal ranges, some foragers are unable to maintain themselves within the awash phase of the tide, and when immersed will become inactive in refuges until they are washed by the ebbing tide.

Rhythmic Patterns of Foraging

The timing of foraging in many animals appears to be rhythmic and in some cases (e.g., crabs, nerites, and limpets) has been shown to be controlled by an endogenous

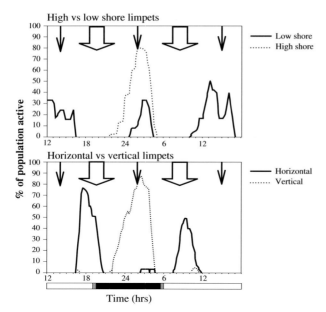

FIGURE 1 Variation in activity patterns of the limpet *Patella vulgata* at Lough Hyne, Ireland. Top: Variation between high-shore and low-shore populations on a vertical surface, both of which are active out of water, but the high-shore population is only active in the nighttime, whereas the low-shore population is active at day and night. Bottom: Variation between populations on vertical surfaces that are active out of water and those on horizontal surfaces that are active while submerged. Thin arrows represent the times of low water, whereas open arrows represent the time and approximate duration of submersion. Horizontal bar represents day time (clear bar), dusk and dawn (stippled bar), and night time (filled bar) (redrawn from Little *et al.* 1988 and Williams *et al.* 1999).

clock, most likely set by the tidal cycle and diurnal zeitgebers. In most species this clock tells animals when to initiate activity, although there is evidence that some species such as limpets have a stop clock as well to help maintain their tidal height. Fixed predictions of the timing of foraging are, however, difficult to make as the endogenous rhythms can be overridden by changes in environmental conditions. Wave splash due to storms and typhoons can influence foraging, in some cases stimulating movement principally in high-shore species (e.g., limpets such as *Patella* and *Cellana*). Other stimuli such as rain or strong sunlight can inhibit movement (e.g., the limpets *Patella* and *Helcion*, the chiton *Acanthopleura*, and the whelk *Nucella*).

As a result, the timing of foraging is labile and can vary between populations of the same species and even individuals within that population. For example, in Lough Hyne, southern Ireland, the limpet *Patella vulgata* showed spatial variation when it foraged between, and even within, shores. Low-shore limpets on vertical surfaces were active when out of water during both night and day whereas high-shore individuals, on the same shore, were only active during nighttime and tracked the tide, shifting their activity period from early morning low tides approaching daylight, back to late evening low tides. Populations on vertical and horizontal rock surfaces less than 200 m apart foraged while emersed at night and when submerged by day, respectively (Fig. 1). The timing of individual foraging activity can also vary with the state of the animal. Individual *Nucella* that have just ingested a large prey item, for example, remain inactive for a number of days while they digest their last meal.

SPATIAL PATTERNS OF FORAGING

The spatial patterns of foraging involve the directions and distances of movement from the start of the foraging cycle until the animals become inactive. These can be broadly defined according to the spatial availability of food supply and refuges. Free-range foraging movements are made by grazers surrounded by their food supply; such species often move randomly and have no fixed resting area. In situations where food distribution is patchy, grazers must locate their food source and therefore often have very directed foraging movements. Where grazers are limited in their foraging movements by the distribution of refuges or shelters, which can include the algal food source, grazing is often localized around these areas. It should be noted, however, that many species change between these categories with season, age, or variation in food availability.

Most molluscan grazers that feed on the biofilm fall within the free-ranging category, although the majority remain within a certain shore level and return to some form of shelter after foraging (Fig. 2). Some limpets and

FIGURE 2 Limpets, *Cellana toreuma*, return from their foraging excursions while awash to take refuge during the emersion period. Photograph by Gray A. Williams.

chitons home—that is, they return to a fixed location on the shore. These species feed on resources surrounding their homes and return home using chemical cues in their mucus trails to orientate. In some cases, the spatial pattern of foraging by these herbivores has been suggested to stimulate or garden their food supply. Selectively cropping or fertilizing their algal food via mucus and excretory products has been shown to enhance the productivity of the alga, or in some cases, prevent the alga from being overgrown, suggesting some degree of mutualism. These species (e.g., *Patella* and *Lottia*) have small foraging ranges and will often defend their territories.

Many species move from a refuge to find their food and then must return back to the refuge before emersion. This is especially true for predators, such as starfish (e.g., *Pisaster*) and whelks (*Nucella*, *Morula*, and *Thais* species), which spend a long time handling their prey and are inactive while digesting their previous meal. These species are strongly limited in their foraging range by the distribution of damp, cool refuges where they can hide when they are not searching for or capturing their prey. The patchy distribution of both food and shelter, therefore, plays a large role in how far an animal must forage and when it must return to seek shelter.

There are, however, only a limited number of studies that have quantified the spatial foraging patterns of intertidal grazers. Many studies simply measure displacement during feeding, or after foraging excursions, whereas others record large-scale spatial patterns (e.g., movement between tidal heights). Time-integrated movement patterns, although difficult to attain, have revealed spatial patterns of variation within a foraging excursion, even in free-ranging grazers. In many herbivorous species foraging can be clearly divided into three phases: rapid outward and inward movement away and back to their refuges and a slower middle phase to the cycle. This triphasic pattern has been recorded in limpets such as *Patella* and *Cellana* and has been interpreted as rapid movement from the refuge to a feeding area, an intense, slow feeding period, and then rapid return. The triphasic pattern may, in some cases, simply reflect the tidal cycle. On tropical shores where many grazers forage while awash, animals become active and rapidly follow the rising tide, after which there is a period of reduced movement as the tide turns (when animals slow their movements and remain in one area), followed by rapid movement back down shore on the ebbing tide.

VARIATION IN SPATIAL AND TEMPORAL PATTERNS

When species have been studied over a number of tidal cycles, and individuals tracked, a great deal of variation in timing and spatial movements is revealed. Many species show longer foraging excursions and greater activity as neap tides change to spring tides (e.g., *Patella*). Foraging duration and frequency can also vary with season, according to levels of physical stress, tidal conditions, and physiological state of the animal. The chiton *Acanthopleura*, for example, is more active in summer than winter and individuals high on the shore are active for twice as long as those found lower down.

The direction of movement also varies greatly between, and within, species. Populations of the limpet *Patella vulgata*, found on low-shore vertical rock faces, move short distances laterally. Limpets higher on the shore, however, move greater distances, vertically up the shore. This difference has been interpreted as a form of habitat selection, the limpets avoiding movement within a band of barnacles found in the mid shore. Movement patterns are thus closely constrained by the habitat over which the grazers must move. Movement patterns are more complex as habitat topography increases, and there is evidence that some species will actively avoid areas of high complexity (e.g., *Nodilittorina* and *Cellana*).

CHOICES MADE DURING FORAGING AND THEIR CONSEQUENCES

Food Choice

When they move, many foragers start feeding immediately, especially herbivorous molluscs, which rasp the rock surface with their radulae as they move. In most cases these foragers are not selecting different food items from the epilithic biofilm but rather passively scraping materials from the rock that their teeth can dislodge. Other species, however, move to find their food items. To do so, many species are able to search visually for their prey (e.g., birds) or follow chemical cues exuded by their food (e.g., urchins). Once they have located their food, consumers either capture or choose one item from those available. This selection can be based on a number of factors: the nutritional value of the food item, the ease of capture and ingestion, or minimizing the time required to feed. Herbivorous crabs such as *Grapsus*, for example, select filamentous algae over foliose algae, which are more difficult to handle and ingest. Predatory crabs (e.g., *Carcinus* and *Callinectes*) often choose mussels of a size that they are able to crack and feed on with minimal energy expenditure so that they can attain the maximum profit. Many predators, such as starfish, have a range of preferences and can vary their diet, choosing one species when it is present but switching to others when it is absent. Food items that are too big, of low energy value or unpalatable, or that are found outside the range of the foraging animals are not, therefore, eaten, and escape from consumers.

Models of Foraging Behavior

Optimization models for crabs, dogwhelks, and limpets have been used to try to predict optimal strategies for foraging, based on energetic costs and benefits. These models have tried to incorporate food choice in terms of size and nutritional quality including costs of finding, handling, and energy to be gained. In the laboratory, crabs and whelks will indeed choose optimally sized mussels and barnacles, but this does not always apply on the shore. Optimal choices of food items often appear to be overridden in the natural environment when animals seem to adopt time minimization strategies due to the risks of physical stresses or predation.

Individual physiological condition and morphological constraints also affect foraging behavior. Glycogen storage capacity and use during starvation have, for example, been linked to patterns of behavior of differently zoned *Patella* in the Mediterranean, where feeding rates varied with gut fullness and standing crop availability. These studies have culminated in more refined models involving individual condition in state-dependent models of foraging patterns. In such models individual energy costs and reserves are combined with mortality risks to produce different individual patterns with reference to the tidal cycle and short-term organization within a foraging period. Understanding such individual constraints on the timing and organization of foraging will lead to improved interpretation of the link between species behavior and subsequent community organization.

SEE ALSO THE FOLLOWING ARTICLES

Biofilms / Chitons / Homing / Limpets / Mutualism / Rhythms, Nontidal / Rhythms, Tidal

FURTHER READING

Burrows, M.T., G. Santini, and G. Chelazzi. 2000. A state-dependent model of activity patterns in homing limpets: balancing energy returns and mortality risks under constraints on digestion. *Journal of Animal Ecology* 69: 290–300.

Chapman, M.G., and A.J. Underwood. 1992. Foraging behaviour of marine benthic herbivores, in *Plant-animal interactions in the marine benthos.* D.M. John, S.J. Hawkins, and J.H. Price, eds. Systematics association special volume no. 46, Oxford: Clarendon Press, 289–317.

Hughes, R.N., and M.T. Burrows. 1994. An interdisciplinary approach to the study of foraging behaviour in the predatory gastropod, *Nucella lapillus* (L.). *Ethology and Evolution* 6: 75–85.

Little, C. 1989. Factors governing patterns of foraging activity in littoral marine herbivorous molluscs. *Journal of Molluscan Studies* 55: 271–284.

Little, C., G.A. Williams, D. Morritt, J.M. Perrins, and P. Stirling. 1988. Foraging behaviour of *Patella vulgata* in an Irish sea-lough. *Journal of Experimental Marine Biology and Ecology* 120: 1–21.

Santini, G., M.T. Burrows, and G. Chelazzi. 2004. Bioeconomics of foraging route selection by limpets. *Marine Ecology Progress Series* 280: 189–198.

Williams, G.A., C. Little, D. Morritt, P. Stirling, L. Teagle, A. Miles, G. Pilling, and M. Consalvey. 1999. Foraging in the limpet *Patella vulgata*: the influence of rock slope on the timing of activity. *Journal of the Marine Biological Association of the UK* 79: 881–889.

FOSSIL TIDEPOOLS

DAVID R. LINDBERG

University of California, Berkeley

Rocky shores and their associated tidepool faunas have been features of coastlines since the first marine metazoans appeared on earth over half a billion years ago. However, because tidepools themselves are erosional features (cracks and cavities) of the intertidal bedrock and typically occur in high-energy environments, they are rarely found intact in the fossil record, and the animals that would have inhabited them are often preserved in different, albeit close–by, depositional environments.

SUBSTRATES

Approximately 33% of all the world's coastlines are currently rocky shores, and while these habitats have been well studied by biologists, few paleontologists have investigated rocky shore habitats in the fossil record. Tidepools are typically found on rocky shores, where they are created when weaker portions of the bedrock erode, thereby producing cracks and cavities that retain water at low tide. Tidepools form in all types of rocks, including metamorphic, igneous, and sedimentary, and can be recognized in the fossil record by both their geomorphology and associated fossil organisms. For example, while features similar to the pothole shown in Fig. 1 can be found in both freshwater and marine habitats, the infilling of the cavity by marine limestone clearing identifies this structure as a Middle Ordovician

FIGURE 1 Middle Ordovician tidepool near Quebec City, Canada. Tidepool is formed in Precambrian gneiss and filled with marine limestone. Reproduced from Johnson (1988).

fossil tidepool (458 million years ago). Figure 1 also highlights another problem in recognizing and interrupting fossil rocky shores and tidepools. Note that the Ordovician limestone lies unconformably on the Precambrian gneiss, which is over one billion years old. That is, a complete sequence of the rock record is not present—the Middle Ordovician limestone lies in direct contact with the Precambrian bedrock, and the intervening Cambrian rock record is missing. In most paleontological studies this is seen as a problem because there is a gap of missing rock record. However, rocky-shore substrates will always be older than the animals that lived on them, and therefore geological structures containing fossil rocky shore will often be considered to be incomplete and lacking information, rather than being recognized as a signal associated with the preservation of a fossil rocky shore.

Another bias associated with fossil rocky shores is the association of rocky shores with high-energy environments. The same strong erosional processes responsible for the formation of tidepools are also highly destructive to the remains of the organisms that inhabited them. Forces associated with high wave action grind and alter both substrates and organismal remains. In addition, the almost continuous wave action of rocky shores generates waterborne rock debris that impacts the habitat at different scales. Add to this multiple sea level rises and falls during the Neogene, and it is remarkable that any remains of these habitats and their taxa are found intact. However, in New Guinea, Chile, California, and elsewhere, eustatic sea level change and uplift have preserved numerous rocky shores and tidepools, as described in the following paragraphs.

Fossil rocky shores are found worldwide (e.g., Canada, Australia, Mexico, United States, Africa, Russia, Poland, Great Britain) and are continuously present from the Precambrian to the Recent, with the possible exception of the Triassic. They are more common along the active margins than on passive margins of the continental plates. This difference was recognized and contrasted as "Atlantic" and "Pacific" coast types in the late nineteenth century and in the 1970s was reframed in the context of plate tectonics. Simply stated, active margins and the dynamic properties associated with them (subduction, faulting, accretion, etc.) have resulted in both a higher abundance of rocky shores as well as a greater diversity of rock types along these shores than are found along the quieter and less dynamic passive margins of continents. Not unexpectedly, living biodiversity patterns of rocky-shore species generally reflect this "active" and "passive" dichotomy as well.

Many fossil rocky shores reflect major sea level changes, such as the Miocene emptying and refilling of the Mediterranean Sea from the Atlantic Ocean (6–5.4 million years ago) and the multitude of Pleistocene glaciations and interglacials (1.8 million years to 11,000 years ago). Some of the best-preserved fossil rocky shores are found in the Pleistocene of southern California, where a combination of recency, eustatic sea level change, and uplift has preserved as much as 77% of the shelled bivalve and gastropod fauna. Many of the ecologically important taxa in these communities today can also be traced through this record, which extends back about a million years.

BIOTAS

Fossil rocky shores and associated tidepools are often characterized by distinctive taxa. During the Paleozoic (542–251 million years ago), rocky shores were characterized by encrusting brachiopods, tabulate corals, and chitons. In Mesozoic rocks (251–65.5 million years ago) rocky shores are characterized by boring and encrusting bivalves, scleractinian corals, terebratulid brachiopods, and coralline algae. Cenozoic rocky shores (65.5 million years ago–Recent) faunas are identified by several of the former taxa, including chitons and boring and encrusting bivalves, as well as the presence of patellogastropod limpets, byssate bivalves, and barnacles.

Most of the characteristic taxa are sessile and either attach to the substrate or burrow into it. Calcium carbonate–producing taxa (e.g., oysters, barnacles, corals, coralline algae) that affixed themselves to the substrate generally have a higher probability of being preserved, not only as attached body fossils but also as an attachment scar on the substrate. Burrowing taxa such as bivalved molluscs and sea urchins leave distinctive cavities in the substrate. The boundary between the tidepool and adjacent aerial substrate is often well demarcated by the presence of these encrusting and burrowing taxa and therefore provides a distinctive signal in the fossil record (Fig. 2).

FIGURE 2 Recent tidepool at Arroyo de los Frijoles, San Mateo County, California. Note sharp demarcation between tidepool and adjacent rock habitat in foreground. Encrusting coralline algae, burrowing annelids, and bivalves are common in the tidepool but are absent from the adjacent exposed substrate. Photograph by the author.

Motile taxa such as chitons, limpets, whelks, and littorines, while closely associated with rocky shores, are much less likely to be found *in situ*. Instead they are typically transported offshore, where they form death assemblages in which relative abundance measures of different species can be used to robustly infer local habitat and substrate type (Fig. 3). Many of these taxa

FIGURE 3 An Oligocene rocky shore death assemblage consisting of patellogastropod limpets (arrows) among mussel shell fragments. Scotts Mill Formation, Oregon. Photograph by the author.

provide direct evidence of the presence of rocky shores, other taxa with which they interacted, as well as the longevity of these ecological interactions. Examples include fossil epibionts whose distribution of fossil gastropod shells are identical to colonization patterns seen in gastropod shells occupied by living hermit crabs; blisters and burrows in fossil bivalves that are identical to those produced by parasites in living clams; shell borings and distinctive shell breakage patterns of that reflect specific predators; and taxa with distinctive morphologies associated with obligate lives on marine plants. All of these taxa and markers are readily discernable and informative from past rocky shores.

SEE ALSO THE FOLLOWING ARTICLES

Algae, Calcified / Chitons / Coastal Geology / Stone Borers / Stromatolites / Wave Exposure

FURTHER READING

Johnson, M. E. 1988. Hunting for ancient rocky shores. *Journal of Geological Education* 36: 147–154.
Johnson, M. E., and B. G. Baarli. 1999. Diversification of rocky-shore biotas through geologic time. *Geobios* 32: 257–273.
Lindberg, D. R., and J. H. Lipps. 1996. Reading the chronicle of temperate rocky shore faunas, in *Evolutionary Paleobiology*. D. Jablonski, D. H. Erwin, and J. Lipps, eds. Chicago: University of Chicago Press, 161–182.

G

GENETIC VARIATION AND POPULATION STRUCTURE

STEPHEN R. PALUMBI
Stanford University

Measurement of genetic variation within and between species living on rocky shores shows a wide variety of different patterns that depends on population size, dispersal prowess, recent climate history, and other features of the physical and biological environment. In addition, DNA comparisons have been very useful in helping distinguish species that are extremely similar morphologically. For example, the Northeast Pacific barnacles *Chthamalus dalli* and *C. fissus* are similar enough visually to require identification by an expert (Fig. 1) but are easily distinguished by a short sequence from a mitochondrial gene.

POPULATION STRUCTURE: DIVERSITY WITHIN AND BETWEEN POPULATIONS

Rocky-shore species can be very abundant and have large geographic ranges, leading to enormous population sizes—in the hundreds of millions or billions. Overall genetic diversity within any species varies with population structure but is determined by two fundamental forces: the mutation rate of genes from generation to generation and the rate at which genetic variation is lost through genetic drift. Drift occurs when alleles in a population are lost because individuals by chance leave different numbers of offspring. Generally speaking, small populations have high amounts of drift

FIGURE 1 The rocky shore barnacle *Chthamalus fissus* or *C. dalli*. These species are visually so similar that it is often difficult to tell them apart. However, genetic differences between the two species are significant, and they can be easily distinguished using this approach. Photograph by Chris Patton.

and subsequently low amounts of genetic diversity. A consequence is that rocky-shore species with large populations have low rates of drift and very large amounts of genetic diversity. In fact, genetic studies of rocky shore species often show very large numbers of alleles within populations, and a high degree of DNA sequence divergence among alleles. Exceptions occur in species with low dispersal and low population size, in species that have experienced recent population bottlenecks, or for genetic loci under strong natural selection (see also Measurement of Genetic Variation).

Much of the work on the genetics of rocky shores involves comparison of different populations within a species from one place to another. The major motivation for this kind of work is to understand the spatial scale of population differentiation. Populations of a species diverge from one another through the action of natural selection and through random genetic drift from one generation to the next. Populations also experience immigration from other populations; this individual movement (of spores, eggs, larvae, juveniles, or adults) can result in movement of alleles from one population to another and causes populations to remain genetically similar. By examining the alleles in a population at a particular genetic locus and comparing the frequencies of these alleles among populations, we can ask which populations have so little immigration that they have become genetically distinct, and which populations are genetically homogenized by dispersal.

MODES OF DISPERSAL: CRAWLERS AND FLOATERS

Although rocky-shore species are in many different animal and algal phyla, many of them are firmly attached to the rock surface as adults and have poor powers of adult dispersal. Some of them, such as many algae, are permanently attached to one place. Others, such as starfish and snails, move slowly along rock surfaces. Other species, such as many crabs, amphipods, and worms, are mobile as adults but do not move large distances. For these species, migration among populations occurs primarily during the larval stage of development. Differences in the dispersal potential of larvae of different species can be a prime determinant of differences in genetic differentiation from place to place.

For example, some species of intertidal snails in the genus *Nucella* lay egg capsules on rock surfaces, from which hatch out tiny juvenile snails that crawl away from the egg capsule. These species have poor powers of dispersal at both adult and larval stages, and populations tend to be genetically distinct from one another over small spatial scales. In such cases, DNA sequences from mitochondrial genes of individuals are much more similar among individuals collected from the same population than among individuals collected in different populations. The ratio of genetic variation within populations (π_{within}), compared to average genetic variation between populations ($\pi_{between}$), is used to estimate Wright's fixation index, $F_{ST} = 1 - \pi_{within}/\pi_{between}$, a measure of the fraction of genetic variation that is distributed geographically in a population. For snails with crawl-away larvae, F_{ST} is high even over short spatial scales of a few kilometers, showing that dispersal over these scales is infrequent. Because of this large amount of genetic differentiation, these populations can show a surprising amount of local adaptation, becoming adapted to environmental conditions that differ from

place to place along the shore. As an example, different populations of *Nucella lapillus* are genetically different and show local adaptation to prey abundances along the west coast of the United States. Other species that have low dispersal as adults and young also show strong genetic differences among populations. Many amphipods, isopods, and algae show this pattern.

On the other extreme, rocky-shore species with larvae that spend a long time floating planktonically before settlement can show low levels of geographic genetic differentiation. Across the world's rocky shores, sea urchins tend to have larvae that spend weeks or months in the plankton and show only slight genetic differentiation across hundreds of kilometers. The level of geographic genetic differentiation, F_{ST}, is often as low as 1–2% and can be statistically indistinguishable from zero.

In these species, high immigration pressure from outside populations keeps a local population from differentiating significantly by genetic drift. Only strong natural selection can counteract this high immigration rate. When genes that are potentially under natural selection are examined, they can change abruptly across locations. One example of this pattern is the strong shift in allele frequencies at the mannose phosphate isomerase locus in barnacles of the northwest Atlantic coast. Individuals with different alleles at this locus are differentially susceptible to heat and desiccation, and gene frequencies can change from the sunny side of a rocky channel to the shady side. Another example from European shores is the strong differentiation of the snail *Littorina saxatillis* from the upper to lower parts of the intertidal. This genetic shift signals a change at many genetic loci and may reflect an ongoing process of differential selection across a steep physiological gradient countered by gene flow across the shore. This kind of genetic differentiation across a genetic gradient may also be visible in the steep genetic cline of the barnacle *Balanus glandula* along the west coast of the United States (Fig. 2). In all these cases, an interaction between gene flow (due to dispersal) and natural selection (due to environmental gradients) is probably the major determinant for the geographic scale of genetic differentiation.

In between the extremes of crawl-away juveniles and long-duration planktonic larvae live thousands of rocky-shore species with different powers of dispersal that depend on the life history of the individual species and the environment in which they live. Some genera have species that differ remarkably in dispersal potential, with one species producing planktonic larvae and another producing crawl-away young. In these cases, such as the intertidal

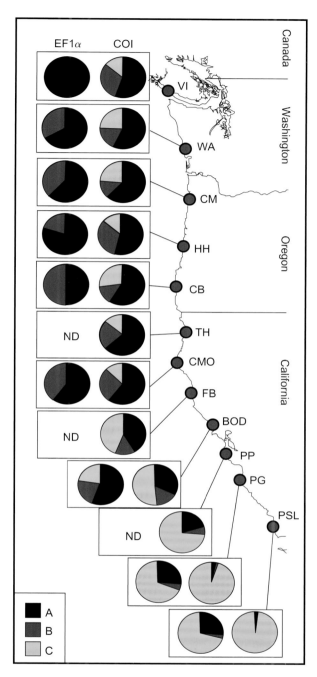

FIGURE 2 Strong genetic differences among populations of the North American barnacle *Balanus glandula*. Pie diagrams show the relative proportion of three major classes of alleles (denoted A, B, or C) at two loci: the nuclear elongation factor alpha and the mitochondrial COI. From Sotka *et al.* (2004).

copepods such as *Tigriopus californicus* have such strong genetic differences from one rocky outcrop to another that their dispersal at this scale must be very limited. In another case, the tropical rock-burrowing mantis shrimp *Haptosquilla pulchella* has a long duration planktonic larva but shows dramatic genetic breaks across Indonesian islands separated by less than 100 km. Other species have very different dispersal mechanisms. For example, rafting of adult algae can create gene flow far higher than achievable by their spores, but rafting depends on dislodgement rates, current flow, and other environmental factors. Such rafts can transport other species as well, from rocky-shore amphipods to seahorses to burrowing invertebartes living in algal holdfasts. In a few cases, individual species can produce larvae with variable dispersal potentials, but such variation is considered rare.

STEPPING STONES: GENETIC VARIATION AT DIFFERENT SPATIAL SCALES

Across this wide array of species, it is possible to use the geographic buildup of genetic differentiation along a coastline to estimate the dispersal distance. A genetic model of population differentiation called the steppingstone model envisions the movement of individuals from one adjacent population to the next along a coast. Genetic differentiation increases steadily with increasing geographic distance among populations in this model, and the slope of the genetic vs geographic line is steeper for species with low migration rates or low population size (Fig. 3). A recent summary of data using this isolation-by-distance approach shows that genetic differences accumulate over short spatial scales for algae, over moderate but variable spatial scales for

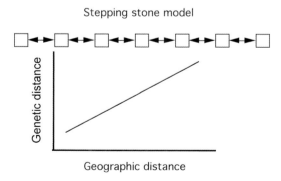

FIGURE 3 A steppingstone model of population exchange is set up to allow movement between the closest areas. In this framework, populations at greater geographic distance accumulate larger genetic distances, and the slope of a line relating genetic and geographic distance is related to the average dispersal distance moved by individuals.

snails of the genus *Littorina*, geographic differentiation is typically far higher over the same spatial scale for the low-dispersal species. In some cases, species that appear to have high powers of dispersal turn out to have strong population structure. For example, temperate tidepool invertebrates, and over larger spatial scales for fish. The degree of genetic differentiation suggests that larvae move up to 50–100 km per generation in many invertebrate and fish species with planktonic larvae—though this result depends strongly on the species and location. Algae and invertebrates that lack planktonic larvae can show dispersal distances of a few kilometers or less.

Some studies of rocky-shore species have failed to find this comforting increase of genetic distance with geographic distance. Such cases tend to fall into three categories of explanation. The first is simply that there is no geographic pattern because dispersal or population size is so high that the slope of a line relating genetic distance to geographic distance is indistinguishable from zero. The second is that the population may have only recently invaded the habitat and may not have had enough time to develop a steady-state relationship between genetic and geographic distance. This explanation may be particularly true for temperate rocky shores that have been affected by recent glaciation. Along the west coast of North America, for example, the genetics of some snail species suggests a steady movement from south to north, presumably as northern climate warmed after the last glacial maximum 18,000–20,000 years ago. These snails have also changed shape as their populations expanded northward, suggesting the impact of local adaptation as the population expanded.

The third explanation for a lack of isolation by distance is that planktonic larvae may be carried along in clouds of related individuals produced by a small number of parents. If these clouds of larvae settle together, the genetics of the cloud—determined largely by chance because of the genetics of the parents that happened to produce that cloud—will dominate the genetics of the local population in the next generation. This could cause significant genetic differences between populations without an increase in genetic distance with increasing geographic distance. This pattern of local genetic structure without isolation by distance is called chaotic patchiness and has been seen in a number of rocky shores species studied at the protein level. Few DNA-based studies have shown this pattern, suggesting that it may sometimes be due to selection on protein variants within populations, not larval clouds.

GENETIC TOOLS: SEQUENCING

Studies of the genetics of rocky shore organisms have been limited by low access to genes that can be studied. Often researchers must invest considerable time and effort to find genes of interest before studies of genetic diversity or structure can be conducted. This limitation has been

FIGURE 4 The purple sea urchin *Strongylocentrotus purpuratus* is one of the most recent species for which the entire genome is now available. Complete genomes allow a more complete understanding of the regulatory machinery that allows genetic control of development, provide a bank of genes for comparison among species, help enormously in understanding physiological relationships to environment, and facilitate understanding of genetic structure. Photograph by Chris Patton.

dramatically reduced by the sequencing of the complete genomes of four rocky-shore species. Two are ascidian species in the genus *Ciona;* the third is the common rocky-shore sea urchin *Strongylocentrotus purpuratus* (Fig. 4). The fourth is the large owl limpet *Lotiia gigantea* from the California coast. Complete genome sequences for the sea urchin, for example, show a large number of coding genes—about 20,000 to 27,000. A particularly large number of these appear related to genes conferring innate immunity to pathogens. Outside coding regions, the sea urchin genome has a plethora of single sequence repeats, especially repeats of the dimer CACACACA. Genome sequences from these and other species can accelerate investigation of patterns of genetic structure and can help reveal differences inherent in the genome organization of different phyla.

SEE ALSO THE FOLLOWING ARTICLES

Algal Biogeography / Dispersal / Sex Allocation and Sexual Selection

FURTHER READING

Hellberg, M.E., R.S. Burton, J.E. Neigel, and S.R. Palumbi. 2002. Genetic assessment of connectivity among marine populations. *Bulletin of Marine Science* 70(1): 273–290.

Kinlan, B., and S.D. Gaines. 2003. Propagule dispersal in marine and terrestrial environments: a community perspective. *Ecology* 84: 2007–2020.

Palumbi, S.R. 2003. Population genetics, demographic connectivity, and the design of marine reserves. *Ecological Applications* 13: S146–S158.

Palumbi, S.R. 2004. Marine reserves and ocean neighborhoods: the spatial scale of marine populations and their management. *Annual Review of Environmental Resources* 29: 31–68.

Sotka, E. E., J. P. Wares, J. A. Barth, R. K. Grosberg, and S. R. Palumbi. 2004. Strong genetic clines and geographical variation in gene flow in the rocky intertidal barnacle *Balanus glandula*. *Molecular Ecology* 13: 2143–2156.

GENETIC VARIATION, MEASUREMENT OF

JOHN WARES

University of Georgia

Genetic variation is simply a description of how many distinguishable versions of a gene (or genetic marker) exist in a given population. There are two primary motivations for measuring genetic variation in marine populations. First, a researcher may be interested in the size and reproductive dynamics (known as demography) of a particular population and will use genetic variation, such as the number of distinct alleles (distinguished by the size or composition of a fragment of DNA or protein) at a given gene region, as a descriptor of population size and reproductive life history. Second, a researcher may be interested in the patterns of individual movements between populations across a geographic range (known as connectivity) and will use comparisons of allele frequencies (and similarity of alleles) within and among populations to infer the rate of movement of individuals from one sampled geographic location to another. In some cases, these measurements are made at different developmental stages of an organism to see how larval/juvenile and adult population dynamics differ.

DEMOGRAPHY AND GENETIC VARIATION

Genetic variation can describe the overall demographics and reproductive patterns of marine populations. The greater the number of individuals that contribute offspring to the following generation (known as the effective population size), the greater the amount of genetic variation there may be in a population. Effective population size (often symbolized N_e) is typically measured as though a population is at equilibrium between the gain of alleles through mutation and the loss of alleles through genetic drift. This assumption is violated if populations have passed through dramatic or cyclic fluctuations in population size. Thus, when the ratio of N_e to actual census size is extremely low, the cause may be high rates of inbreeding, high variance in reproductive success, or large stochastic variation in population size over time.

Different methods of measuring genetic variation can, to an extent, separate such effects. A single temporal sample of genetic variation tends to integrate all of these long-term demographic characteristics of the population over recent evolutionary time. The estimated genetic diversity reflects the history of the sample, dating back to approximately the reciprocal of the mutation rate for that genetic marker. For example, given sequence data from a mitochondrial protein-coding gene from a sample of individuals, the observed variation typically incorporates overall demographic changes that have occurred over the past $\sim 10^4$–10^5 years (the allelic substitution rate μ at these genes is typically around 1×10^{-5} substitutions per fragment per generation). These estimates are based on our understanding of the neutral model of molecular evolution, and simple estimators—such as the average number of pairwise sequence differences between individuals—are expected to be proportional to the product of the effective population size (N_e) and the substitution rate (μ). However, data must be tested for the possibility of non-neutral evolutionary patterns resulting from natural selection and other mechanisms of molecular evolution.

More contemporary demographic patterns can be resolved using multiple temporal samples of genetic variation in a population. The variance in allele frequencies from one generation to the next indicates what proportion of individuals actually succeeded in reproducing. If the variance is quite high, then the effective population size—using this class of estimators—is low. Comparisons between the long-term and short-term methods of estimating N_e may provide the most information for understanding the interaction between contemporary environmental change and reproductive life history on the stability of a population, because it becomes apparent when N_e has been relatively constant over evolutionary time periods and when it has recently changed.

CONNECTIVITY AND GENETIC VARIATION

To analyze the connectivity of marine populations, there are two primary approaches to determining the number of migrants between regions. One is based on spatial variation in allele frequencies, and the other method involves the genealogical relationship of the sampled alleles themselves. Frequency-based approaches assume that the variance in allele frequencies from one region to another should be inversely related to the frequency of migration between the regions. These methods are generally based on Sewall Wright's *F* statistic, and the components of variance can be measured between individuals within a population, between populations within a region, or between regions. These methods can be used when there is no known evolutionary relationship among alleles (in other words, the exact DNA mutations responsible for the observed difference between two alleles is unknown), and randomization procedures can be used to assess the significance of differences among populations.

When the evolutionary relationship among alleles can be inferred, as with DNA sequence data or some classes of fragment-size markers such as microsatellites, genealogical-based methods can reconstruct the relationship between populations more quantitatively in terms of the effective number of migrants and the time scale over which migration may be taking place. The analysis of migration requires careful consideration of how these patterns vary with scale and with geographic location. For example, some species may appear to be panmictic (fully mixed genetically) through much of their geographic range but exhibit very low gene flow across particular boundaries because of environmental or oceanographic variation. Genealogical methods for quantifying gene flow are often better at assessing statistical confidence intervals on an estimate, because the estimates are based on a large number of possible genealogical relationships between sampled individuals. Gene trees are inferred using statistical models just as a species phylogeny is, and many equivalent genealogies may be examined using likelihood-based searches of all possible relationships among individuals. On any given gene tree, migration is inferred by reconstructing the number of changes of state (in this case, sample location; Fig. 1). These methods may, in some cases, be able to quantify asymmetric gene flow as well.

SAMPLING EFFORT AND THE MEASUREMENT OF GENETIC VARIATION

Measurement of genetic variation for a given population involves certain assumptions about the spatial limits of the population; the census size and geographical extent of a population can be difficult to establish empirically. Within a defined population, sampling effort will depend on the types of genetic variation being measured and the analytical needs of the researcher. Some measures of genetic variation do not require large sample sizes. For example, estimates of average pairwise sequence differences between sampled individuals will scale up at a rate of $2/n$, where n is the sample size. Thus, an estimate of

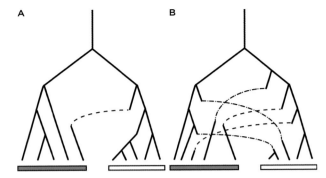

FIGURE 1 Examples of genealogies for samples from two populations. Genealogy A represents very low migration, with only a single change of state (location) reconstructed on the tree; genealogy B represents high migration between the two regions, and individuals from one population may be genetically more similar to those from the other population than to those from their own.

overall genetic variation of an unstructured population can be obtained with only 10–15 individuals (with each additional sample, there are diminishing returns for this estimator). However, this does not mean that most of the alleles at a given locus are recovered with such a small sample size. The Ewens sampling distribution shows that increasing sample size translates to an increased number of alleles found in a population. If the estimate for overall genetic variation ($\theta \propto N_e\mu$, easily estimated using the number of polymorphic sites, or the average number of pairwise sequence differences in a population) is low, the effect of sample size is small, but can translate into larger effects as θ increases (Fig. 2). For this reason, when allelic

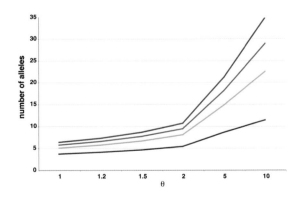

FIGURE 2 The number of alleles recovered in a population is more dependent on sample size in highly diverse populations. Ewens' sampling distribution shows that at low levels (e.g., less than 1) of θ, which is a function of N_e and the substitution rate μ, there is little benefit to increased sample sizes; at larger values, however, θ it can make a large difference in the number of alleles found. Plots shown are for population samples of 10, 40, 80, and 150 individuals, from bottom to top. Data taken from Nei (1987, table 8.3).

diversity itself is an important component of a population study, iterative approaches to determining the most effective sample size may be necessary.

Within a single uniform, panmictic population, estimates of N_e at selectively neutral markers will be scale-independent. However, if there is additional population structure (geographic variation caused by low levels of migration among regions, habitat preferences, or other physical barriers), then levels of genetic variation will be dependent on the geographic scale of the sample. One of the first hypotheses that must be investigated is whether migration and gene flow are limited by mechanisms common to all sampled regions, or whether there are areas in which migration or gene flow is more limited than in others. If gene flow is restricted by the natural dispersal mechanisms of an organism, a pattern of isolation by distance may be found, in which the genetic divergence of samples increases linearly with the geographic distance between them. This pattern represents an equilibrium between the loss of alleles over evolutionary time by genetic drift, and the influx of alleles from neighboring regions due to migration. If there is apparent population structure but this linear relationship between geographic and genetic distance is not recovered, it may suggest that other types of interactions—including poor habitat or physical barriers—are limiting dispersal between one or more populations.

CLASSES OF MARKERS USED FOR MEASURING GENETIC VARIATION

The type of genetic markers used in a study will affect the sampling for a study because of cost, reliability, substitution rate, and the questions being addressed. Measures of genetic variation may be extremely basic, such as the number of alleles recovered in a population, but they often require consideration of the most appropriate statistical model of substitution for a given marker (for example, we know that nucleotide substitutions in protein-coding genes occur at unequal rates among sites of a DNA sequence, dependent on the effect of a DNA substitution on the overall amino acid structure of the protein; an appropriate model will account for these different rates).

Electrophoretic markers involve alleles that are scored based on the migration of a protein or DNA fragment in a gel matrix under strong electrical current. Variation at these markers is scored by measuring the distance that a fragment migrates, which differs because of differences in the charge and size of the molecule, relative to known allelic standards. This class of markers includes allozymes, microsatellites, and a variety of other nonspecific-fragment-length polymorphism approaches. Data based on variation at allozyme loci—reflecting differences in ionic charge at the amino acid level of proteins—are inexpensive to collect, though fresh or frozen specimens are necessary, but the evolutionary relationship of one observed allele to another is not known. The advantage of using these markers is that multilocus, codominant genotypes (meaning that the full genotype is apparent at a given locus) can be obtained for any organism with very little optimization or expense, and they are good options for analyses of kinship and gene flow. However, variation at these markers may be limited to only a few alleles because of a low substitution rate at the amino acid level of these proteins. Other markers, such as microsatellites (rapidly mutating loci that include short nucleotide repeat motifs, such as AGTAGTAGT . . . or ACACACAC . . .) are often much more variable, and statistical models can be used to infer the relationship of particular alleles. These markers are appropriate for both frequency-based and genealogy-based analytical approaches, but the cost of developing these species-specific markers can be prohibitive, and anecdotal evidence suggests that they can be more difficult to develop for many marine invertebrates than they have been for other taxa. Some methods are available to obtain broad multilocus genotypes of organisms without prior genome-specific screening procedures, but these fragment-length polymorphism protocols (including RAPDs, AFLPs, etc.) generate only dominant markers, meaning that the full genotype cannot be determined for all individuals and the mechanism for gain or loss of an observable allele is unknown. This is a problem for many types of subsequent analyses, because exact allele frequencies are unknown.

The most complete information can be extracted from DNA sequence data polymorphisms, because of the detailed substitution models and analyses that can be used to reconstruct the genealogical relationships among individuals. However, the cost of DNA sequencing, and the limited number of variable loci that are characterized for most organisms, can be limiting factors. Nevertheless, sequences of particular genes are now being used as "DNA bar codes" for identifying many marine organisms, particularly when species groups include morphologically indistinct but evolutionarily diverged lineages. The advantages to using such markers as taxonomic tools can be great, because efficient screening methods may help researchers limit their study to a single cryptic species in a region, and these tools can aid in fully characterizing the biodiversity of a region.

SEE ALSO THE FOLLOWING ARTICLES

Biodiversity, Significance of / Monitoring: Techniques

FURTHER READINGS

Avise, J. C. 1994. *Molecular markers, natural history, and evolution.* New York: Chapman and Hall.

Ferraris, J. D., and S. R. Palumbi, eds. 1996. *Molecular zoology: advances, strategies, and protocols.* Wilmington, DE: Wiley-Liss Publishing.

Grosberg, R. K., and C. W. Cunningham. 2001. Genetic structure in the sea: from populations to communities, in *Marine community ecology.* M. D. Bertness, M. E. Hay, and S. D. Gaines, eds. Sunderland, MA: Sinauer, 61–84.

Hudson, R. R. 1990. Gene genealogies and the coalescent process. *Oxford Surveys in Evolutionary Biology* 7: 1–44.

Nei, M. 1987. *Molecular evolutionary genetics.* New York: Columbia University Press.

Wakeley, J. 2005. *Coalescent theory: an introduction.* Denver: Roberts and Company.

GEOLOGY, COASTAL

SEE COASTAL GEOLOGY

GULLS

J. TIMOTHY WOOTTON

University of Chicago

Gulls are the birds perhaps most frequently associated with coastlines, as they soar in coastal updrafts and walk on the shoreline in search of food. Some species also frequent large bodies of water inland. They are relatively large birds in the family Laridae with characteristic white, gray, and black colors, and they feed on a wide range of foods derived both naturally and through human activities. Because of their active lifestyles and relatively high densities, gulls can be important components of food webs along shorelines. As a group, they are fairly tolerant of human presence, which has facilitated many fascinating observations of their behavior, ecology, and life history.

APPEARANCE

Gulls are notable for their generally uniform appearance. They are typically colored white on the underside and on the tail, with their back and wings ranging in shade from white to black. The two major groups of gulls are differentiated by having either an all-white head or a head covered all or in part in black feathers. Black-headed species usually replace most of their black head feathers

with white feathers in the nonbreeding season. A few species (e.g., Heerman's gulls and slaty gulls) are exceptional in being colored a fairly uniform gray on the entire body. The beaks of most species are usually some shade of yellow with species-specific black or red markings, although a few species have solid yellow, black, or red beaks. Young gulls lack the distinctive coloration of adults, but instead exhibit dull gray or brown plumages for one to three years in most cases. The often-nondescript juvenile plumage contributes to the difficulty of identifying gull species in the field. Gull identification is further complicated because there are several geographic areas where different species hybridize, creating individuals that look like neither parent species. For example, on the Pacific coast of North America, the glaucous-winged gull *(Larus glaucescens)*, a northern species characterized by a light gray back and wing tips, hybridizes with the Western gull *(Larus occidentalis)*, a southern species characterized by a black back and wingtips, to produce offspring that look superficially like species such as the herring gull *(Larus argentatus)*, that has gray backs with black wing tips. The study of gull hybrid zones provides biologists with an opportunity to better understand the processes of species formation.

The function of the uniformity of adult gull coloration is not well understood. One possibility is countershading, with the white underside making it hard for fish prey or swimming predators to see them from below, whereas darker shading above may make it harder to be seen by aerial predators. Alternatively, the strong contrast between light and dark colors may facilitate long-distance signaling to other gulls that a source of food is present. When fish become locally abundant near the water surface, gulls often congregate in large groups, sometimes referred to as "herring balls." There is evidence that such group feeding may be beneficial to individual gulls by disrupting the defensive schooling behavior of fish, thereby increasing feeding efficiency. Other evidence suggests a role in communication for the color differences. The age-dependent plumage pattern exhibited by gulls indicates that the coloration pattern may play a role in social interactions and dominance. Experiments exposing young birds to elevated levels of testosterone, a hormone often related to maturity and social interactions, have caused early molting into adult plumage patterns in some species.

Aside from coloration, most gulls also have a similar morphology. The beaks of most species tend to be rather robust with a slight hook at the end, well suited

FIGURE 1 An adult herring gull *(Larus argentatus)*, illustrating the coloration and morphology typical of many gulls. Photograph by the author.

fish, sea birds, invertebrates, and marine mammals that accumulate along the drift line of the shore.

As human populations have expanded, some gull species have been observed feeding extensively on garbage at dumps and in plowed fields where invertebrates such as earthworms have been exposed. They are also well known to follow fishing boats, feeding on the excess bait and fisheries by-catch that is dumped back into the ocean. The role that these human-generated food resources contribute to gull population dynamics is a current area of interest. Some studies indicate that garbage dump use is not extensive in relation to other food resources, and they also find that the gulls using garbage resources tend to be males. Several studies report lower reproductive success for gulls that feed on garbage than for those that feed on natural food sources. Hence, the role that garbage feeding plays in gull population dynamics is unclear, as it could either provide a low-quality supplemental food supply that augments populations to some extent, or trigger population reductions if gulls preferentially choose to focus their foraging activity on highly concentrated "junk food" sources.

Gulls also deploy several interesting behaviors to facilitate food capture. First, detailed diet observations reveal individual specialization on particular food types, which may increase foraging efficiency because gulls learn specialized techniques for prey capture. Second, individuals that feed on hard-shelled prey often learn to drop prey from the air onto anvil rocks to crack the shells and gain access to the enclosed meat. Third, some individuals engage in kleptoparasitic behavior, often forcing sea birds or other gulls to drop the prey they have captured. In many cases, adults focus such piracy on young individuals, who are less-skilled fliers and have smaller body sizes.

Many food items of gulls contain high salt concentrations, and freshwater availability can be low in many coastal situations. To deal with the osmoregulatory problems this situation creates, gulls possess in their head specialized salt glands, which actively pump salt out of their blood and into their nasal passages, where it is exuded into the environment.

ENEMIES

As fairly large and agile birds, gulls are at relatively low risk of succumbing to predation. Nevertheless, several groups of animals will feed on gulls and in some circumstances may have a large effect on gull populations. Immature gulls seem particularly vulnerable, probably because of a combination of inexperience in detecting and evading predators and because of their smaller body size relative

to a generalized diet. Gulls also have webbed feet to facilitate swimming on the water surface, but they do not have the streamlined shape to permit underwater swimming or diving, as do some other seabirds. Males and females look very similar, but males tend to be slightly larger.

FEEDING

Most gulls are dietary generalists. Aside from being fairly tolerant of direct human observation when feeding, analysis of feeding is facilitated because gulls regurgitate the indigestible shells, bones, and scales of their prey as pellets, which can be regularly seen both along shorelines and in middens at nesting areas. Fish usually make up the largest component of the diet, but a range of benthic and pelagic invertebrates regularly contributes to the diet of many species. Invertebrate prey includes gastropods, squid, bivalves, crabs, shrimp, goose barnacles, starfish, sea urchins, krill, isopods, polychaete worms, and earthworms. Gulls also opportunistically capture and eat small mammals and birds, particularly burrow-nesting seabirds such as auklets and storm petrels. Additionally, gulls will often consume the eggs and young of other birds, including other gulls, during the nesting season. There is even one report of extensive grazing on algae by herring gulls in Japan. Because they often occur in high densities, gulls can significantly deplete prey resources along shorelines. Aside from capturing live prey, gulls also function as scavengers, feeding on dead to adults. Predators such as sharks and pinnipeds attack from underwater while gulls are floating on the surface. Eagles and other large birds of prey attack gulls while they are either flying or standing on the shore. Some land mammals may also take gulls when the opportunity arises. Such predation effects can be extreme when land mammals are introduced to previously predator-free offshore nesting colonies. Crows and other gulls can also have a substantial impact on survival of gull eggs and nestlings.

FIGURE 2 A recently hatched glaucous-winged gull *(Larus glaucescens)* chick hiding just outside of its nest, illustrating typical nestling coloration. Photograph by the author.

LIFE HISTORY

Gulls typically nest in colonies on offshore islands and on steep cliffs, although some smaller species may nest inland in trees or on buildings. The reasons for usually nesting in dense aggregations are an area of debate, but several explanations may contribute, including preferential nesting in areas with low predation risk, group defense against predators, nesting at densities that facilitate finding acceptable mates, and living in situations where information on food resources can be transmitted socially. For example, gulls actively mob larger animals that enter their breeding colony, including clawing with their feet, pecking with their beaks, and defecating on the intruder, which can be effective at discouraging predation in the colony. On islands, gulls often nest in a range of microhabitats from vegetated edges of the island to cliffs on the sides to drift lines on open sandy beaches. Older, more experienced individuals tend to use vegetated flat areas at the edges of the island, which appear to be beneficial by providing shelter from both flying predators and climate extremes, by giving individuals a better vantage point for spotting feeding aggregations, and by being in locations where nests are least likely to be washed away by waves. Nests are made out of a variety of materials including grasses, forbs (nongrass herbs), and seaweeds. Nest material collection can often have strong local effects on the cover of terrestrial plants and intertidal algae as substantial amounts of plant material are removed. This impact can subsequently affect the species composition of communities of terrestrial plants and sessile intertidal organisms, favoring species that are poor competitors for light or space but are resistant to being removed.

The high-density lifestyle of most gulls often leads to squabbles among neighbors. Perhaps correlated with the regular aggression observed on both nesting and feeding areas, gulls maintain high levels of the hormone testosterone in their bodies throughout the year, which contrasts with the pattern typically exhibited by birds, wherein testosterone levels increase at the onset of the breeding season and then drop back to low baseline levels.

Gulls exhibit a curious pattern of clutch size, with the majority of pairs having three eggs, a moderate number having two eggs, a few having one egg, but virtually none having four or more. The absence of larger clutches is not an absolute energetic constraint, because females will lay a total of four or more eggs if the first egg is experimentally removed from the nest during the egg laying period. Despite this pattern, several lines of evidence indicate a role of food limitation in egg production. Reduced clutch sizes have been observed in low-food (e.g., El Niño) years, and feeding supplementation studies show that the weight of the last egg laid, and hence the size of the chick produced, may depend on food, with runt chicks produced when food is more scarce. The strategy of variably provisioning the final egg is probably advantageous, because a strong size hierarchy among the chicks may ensure that at least one chick receives sufficient food to survive, rather than having all chicks starve on a food supply that could have supported fewer. Gulls also engage in courtship feeding, in which males provide food to females, and so clutch size patterns may be influenced by male decisions on how to allocate courtship feeding relative to other activities such as mate guarding over the course of the breeding season. Although individuals are not known to lay more than three eggs under nonexperimental circumstances, nests with four eggs are occasionally observed. Careful study has revealed that these nests usually result from female–female pairs, which may develop because gull populations often have female-biased sex ratios. Environmental pollutants may be partially to blame for these skewed sex rations, because feminization of gulls exposed to pollutants has been documented. Hence gulls are a useful indicator of environmental impacts that may be of concern to humans. The existence of successful female–female nests also indicates that gulls will mate outside their social pair bond. The strongly spotted eggs are laid every other day and hatch after 3–4 weeks. Because gulls will start incubating their eggs before the clutch is complete, chicks hatch somewhat asynchronously, which can further contribute to a size hierarchy among chicks. The chicks, which hatch with tan to gray down with heavy dark spots that make them harder to see, can walk after several days, but they are often viciously attacked if they stray away from the nest onto a neighboring territory. Adults feed chicks on the nesting territory for 1–2 months, and the chicks follow adults around for at least several more weeks while being fed by the adults and learning to capture prey through direct observation of their parents and trial-and-error pecking at potential food items.

MOVEMENT PATTERNS

Gulls vary in their migration patterns with the seasons. Adults of many coastal species are relatively sedentary, with only northernmost populations moving south for the winter, presumably because ocean temperatures significantly buffer temperature variation along coastlines. Juveniles are much more nomadic. Species that nest in interior habitats, such as California gulls (*Larus californicus*) and most black-headed species, are much more migratory, typically moving from interior to coastal habitats following the breeding season, with some (e.g., Franklin's gull, *Larus pipixcan*), migrating between the Northern and the Southern Hemisphere. Some unique migratory patterns also exist. For example, Heerman's gulls *(Larus heermanni)* breed in late winter and then migrate north during the summer nonbreeding season.

SEE ALSO THE FOLLOWING ARTICLES

Birds / Food Webs / Foraging Behavior / Oystercatchers / Penguins / Predator Avoidance

FURTHER READING

Bent, A. C. 1921. *Life histories of North American gulls and terns.* Smithsonian Institution. United States National Museum Bulletin 113. Washington, DC: U.S. Goverment Printing Office.

Ehrlich, P. R., D. S. Dobkin, and D. Wheye. 1988. *The birder's handbook: a field guide to the natural history of North American birds.* New York: Simon and Schuster.

Grant, P. J. 1986. *Gulls: a guide to identification.* Friday Harbor, WA: Harrell Books.

Hand, J., W., Southern, and K. Vermeer, 1987. Ecology and behavior of gulls: proceedings of an international symposium of the Colonial Waterbird Group and the Pacific Seabird Group, San Francisco, California, 6 December 1985. *Studies in Avian Biology* 10: 1–140.

Olsen, K. M. 2004. *Gulls of North America, Europe, and Asia.* Princeton, NJ: Princeton University Press.

Poole, A., and F. Gill, eds. 1993–2002. *The birds of North America.* Philadelphia: The Academy of Natural Sciences.

Schreiber, E. A, and J., Burger, eds. 2001. *Biology of marine birds.* Boca Raton, FL: CRC Press.

Tinbergen, N. 1971. *The herring gull's world.* New York: HarperCollins Publishers.

H

HABITAT ALTERATION

STEVEN N. MURRAY

California State University, Fullerton

Rocky shore landscapes and the habitats contained therein are shaped by physical and biological disturbances that vary in their magnitude and duration. Many agents of disturbance act over broad regional scales, but others have more localized effects and can alter shore habitats and have significant effects on the structure of intertidal populations and communities.

FORMS OF DISTURBANCE

Disturbances contribute to the high degree of spatial heterogeneity and patchiness characteristic of rocky shores and can be categorized as being of physical or biological origin. Physical disturbance is caused, for example, by the dislodging forces of storm waves, inputs of low-salinity water following precipitation events, sand scour and inundation, contact with drifting logs or other objects, effects of ice, and low tide emersion during hot, dry, or extremely cold days. Biological disturbance is defined here as damage, mortality, or displacement caused by living organisms. Human visitors to the rocky intertidal zone can be viewed as agents of disturbance because their activities often have an impact on shore habitats and organisms. The intertidal zone is a rarity among marine habitats because it is readily accessible without the use of special underwater gear during low tides, and humans can create localized disturbances by their exploratory activities, including the collection of organisms.

PHYSICAL DISTURBANCES

Storm Waves and Substratum Alteration

One of the most obvious environmental variables affecting shore habitats is wave energy. Headlands and promontories generally receive higher wave forces than do more sheltered locations, such as those in the lee of islands or peninsulas or in estuaries. Temperate shores receiving stronger wave energy are generally characterized by having higher abundances of filter-feeding, sessile invertebrates and seaweeds with wave-resistant morphologies. Even at a single site, wave energy varies greatly over time, and periodic storm events can create strong, localized disturbances. These disturbances are manifest in substratum breakouts; the overturning and movement of large, otherwise stable rocks and boulders; the ripping of sessile invertebrates and seaweeds away from the substratum; and the displacement of mobile animals that cannot maintain their positions during extraordinary water flows.

Storm influences on rocky shores often can be seen in the high abundances of more short-lived, opportunistic seaweeds and invertebrates that are able to quickly colonize newly available substratum. These seaweeds usually include rapidly growing green ulvoids (leafy green algae) and small filamentous and filament-like brown and red algae. Early invertebrate colonizers often consist of small acorn barnacles and some tube-forming worms. Hence, storms function as disturbances that contribute to the biotic patchiness of rocky shores, altering habitats, increasing mortality in established populations, and offering increased recruitment opportunities for new colonists.

Logs and Ice

In certain parts of the world, coastal ecosystems are subject to drifting logs or ice formation. These disturb rocky intertidal habitats by crushing, abrading, and dislodging intertidal organisms and excavating rock surfaces. Damage to mussels, barnacles, and other intertidal populations resulting from drifting logs is particularly common on the Pacific Northwest coast of America and other shores adjacent to landscapes supporting terrestrial logging activity. Whereas damage from drifting logs can create more localized effects on intertidal communities, ice damage will usually be regional in nature. Ice formation is a common seasonal phenomenon on high-latitude shores, where freezing and thawing events mechanically damage rocky intertidal organisms and scour rock surfaces.

Sand Influence

Although rocky shores are by definition eroding shores, sedimentation can have a strong influence on shores proximal to sandy beaches or shallow, offshore sediment deposits. Sand is typically transported on- and offshore with longshore current flow and variations in the magnitude and direction of wave forces. A consequence of this transport is that rocky shore substratum is regularly scoured by sediments and at times can be buried for days when sediments are left cast ashore (Fig. 1). Rocky habitats receiving high sand influence often are characterized by high abundances of certain species, such as coralline and other algal turfs in southern California, that have morphologies and biologies that enable them to resist the mechanical effects of sand abrasion and burial. Other species common on sand-influenced rocky shores include ulvoids and small acorn barnacles, which can opportunistically colonize freshly scoured substratum. Although poorly understood, some intertidal species are characterized as psammophilic, meaning that they preferentially occupy sand-influenced shores, perhaps because of reduced

FIGURE 1 Fronds of the lower intertidal feather boa kelp *(Egregia menziesii)* on the surface of rocky substrata completely inundated by sand. Photograph by the author.

competition or predation. Other organisms, including some mobile invertebrates and delicate seaweeds, are less tolerant of sand influence and occur in low abundances on sand-disturbed rocky shores.

Precipitation Events and Tidal Emersion

PRECIPITATION EVENTS AND LOW SALINITY

Shores proximal to the mouths of rivers and drainage channels or gullies will receive inputs of low-salinity water following precipitation events. Where a river input is large and occurs throughout the year, the effects of low salinity will be spread over a wide region and represent a relatively constant environmental feature. However, if freshwater inputs are sporadic and highly seasonal, effects on intertidal marine life will be more localized and temporal. During rain events, tidepools and rocky channels can collect freshwater during periods of tidal emersion. Marine organisms differ considerably in their tolerance of low-salinity water. For example, in southern California unusually high mortalities have been found in purple urchins and octopuses following winter rains coinciding with low tides.

DESICCATION AND FROST

During periods of tidal emersion, intertidal organisms are removed from bathing ocean waters and exposed to the atmosphere. This results in the loss of water from their emersed surfaces, an event known as desiccation, which is accelerated at warmer temperatures and at higher air flows. In addition to water loss, emersed organisms are no longer buffered by ocean waters from temperature extremes and will heat up or even freeze in response to hot and cold air temperatures. The impacts of temperature extremes and desiccation events on marine life are generally regional in nature and can be seen in bleaching seaweed fronds or fragile barnacles that easily break off, leaving white scars behind on the rocks. Extreme cold events, resulting in frost, can have a major impact on shore life, particularly at higher latitudes.

More localized effects of temperature and desiccation on marine life also occur and can contribute to shore patchiness. For example, in the Northern Hemisphere, organisms living on more sun-exposed south-facing shores and rock surfaces are more vulnerable to daytime low tides during warmer seasons.

BIOLOGICAL DISTURBANCES

The activities of organisms can create disturbances, which affect rocky intertidal populations and communities by displacing, injuring, or causing mortality. Biological disturbances, in the broadest sense, include activities associated with the actual consumption of prey by predators or grazers, as well as such incidental disturbances such as the whiplash of algal fronds across the substratum, dislodgement of organisms or substratum by foraging predators, burial by seaweed and plant wrack, and the mechanical effects of the hauling out of pinnipeds. For example, the wave-induced whiplash disturbance by fronds of upper shore rockweeds can create halos of unoccupied surface on temperate shores by increasing mortality and reducing the recruitment success of acorn barnacles (Fig. 2). For most rocky shores, these incidental biological disturbances are more localized and less significant than physical disturbances are in structuring rocky-intertidal communities.

FIGURE 3 A large group of young people walking on the rocks and exploring the intertidal zone as participants in an educational field trip in the Dana Point Marine Life Refuge in southern California. Photograph by the author.

FIGURE 2 Whiplash effect of fucoid fronds, resulting in a halo of substratum cleared of most barnacles and other organisms. Modified and redrawn from Carefoot (1977).

HUMAN VISITOR DISTURBANCES

Many shores around the world are heavily influenced by the activities of human visitors (Figs. 3, 4). These activities include the collecting of intertidal organisms for food, souvenirs, or bait; the handling and manipulation of organisms while observing them; and the trampling of organisms by people walking across the shore. All of these activities can have significant and long-lasting direct and indirect effects on rocky-shore populations and communities.

Collecting

Intertidal species have long been gathered for food, and in Asia and other parts of the world where human foraging in the intertidal zone occurs on a daily basis, targeted

FIGURE 4 Collectors filling a bag with mussels in the Laguna Beach Marine Life Refuge in southern California, a marine protected area where such collecting is not allowed. Photograph by the author.

species are characteristically low in abundance. Research has demonstrated that humans tend to preferentially collect the largest animals, so human exploitation not only reduces the abundances of harvested organisms but also has size-related effects on their populations. Because larger individuals are selectively removed, exploited populations tend to be characterized by low abundances of larger-sized individuals and to be dominated by smaller individuals. This has been shown for many exploited intertidal invertebrates, including owl limpets, which have been harvested for food for hundreds of years along the western coast of North America (Fig. 5). Today, owl limpet populations rarely include larger individuals in areas readily accessible to people, presumably a reflection of past exploitation events.

For most invertebrates, reproductive output increases with size, so removal of the larger individuals from a population reduces the numbers of its most fecund members.

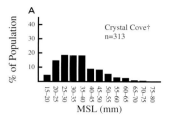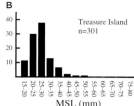

FIGURE 5 Size–frequency plots for owl limpets from two southern California sites. (A) Crystal Cove is located within the Irvine Coast Marine Life Refuge, which was designated a marine protected area in 1971 and is currently a state park with active patrol by park rangers. (B) Treasure Island is a heavily visited site without protection from collectors, except for state bag limits, until January 1999. The mean size of owl limpets is much larger at Crystal Cove (35.2 ± 0.6 SE) than at Treasure Island (26.6 ± 0.4 SE), and the former site has population profiles that include higher percentages of larger-sized limpets. Modified and redrawn from Kido and Murray (2003).

Although not well explored for intertidal invertebrates, size-selective removal of larger individuals significantly reduces reproductive success in fish populations, which similarly show a strong positive relationship between fecundity and larger body size. Moreover, some intertidal invertebrates, including owl limpets, change their sex as they age and become larger in size. For these species, size-selective human predation not only can eliminate the more fecund members of a population but also can alter the ratios of males and females.

In addition to foraging for food, humans also exploit intertidal organisms for use as souvenirs and bait. These cases are less well studied, but it appears that collecting also is size-selective and larger individuals are preferentially taken. Sea stars and shelled snails are among the most common species collected for souvenirs, whereas mussels, snails, chitons, and worms are the most likely candidates for use as bait. Like the exploitation of intertidal organisms for food, souvenir and bait collection can reduce the abundances of targeted species, affect the size structures of their populations, and affect intertidal community structure by altering important species interactions such as competition and predation.

Tidepooling

Because of their accessibility and high biological diversity, rocky shores have long been attractive sites for "tidepoolers" interested in viewing marine life. These visitors frequently move substratum, turn over rocks, and pick up organisms to examine them. Although mostly unintentional in their effects, these manipulations serve as disturbances that alter the habitats of cryptic, under-rock organisms, increasing their exposure to desiccation, and injure or kill displaced animals by handling them and

failing to return them to the protection of their natural habitats. On heavily visited shores, frequently overturned rocks appear like "monks' heads" because of the bald patches of dead and bleached organisms that appear on their flipped surfaces and the fringes of algal growth remaining on their sides.

Trampling

The mere act of walking on rocky shores can injure marine organisms and create localized disturbances. On temperate coasts, upper-shore rockweeds are particularly susceptible to visitor foot traffic, but mussels, barnacles, tube-forming worms, and most larger, fleshy seaweeds also are vulnerable to trampling. For seaweeds, trampling disturbance compresses thalli and breaks branches, resulting in more turflike and shorter, bushy morphologies. Taller barnacles, isolated or one-to-two deep mussels, and reef-forming tube worms such as *Phragmatopoma californica* can suffer significant damage after being walked on. The degree of trampling disturbance is directly related to the magnitude of shore visitation and varies spatially across intertidal habitats, with upper, flatter rocks and benches and natural intertidal paths receiving the most foot traffic. Barren rock patches can be observed where shore visitation is very high and people repetitively walk across the same surfaces (Fig. 6). However, where people disperse widely across intertidal habitats and where visitation is moderate, most trampling damage will be subtle and difficult to distinguish from the natural heterogeneity so characteristic of rocky shores.

FIGURE 6 A worn path through populations of the upper mid shore rockweed *Silvetia compressa* caused by human foot traffic on a southern California shore in San Diego. Photograph courtesy of Jack Engle.

SEE ALSO THE FOLLOWING ARTICLES

Beach Morphology / Desiccation Stress / Ice Scour / Projectiles, Effects of / Salinity Stress / Storm Intensity and Wave Height

FURTHER READING

Addessi, L. 1994. Human disturbances and long-term changes on a rocky intertidal community. *Ecological Applications* 4: 786–797.

Airoldi, L. 2003. The effects of sedimentation on rocky coast assemblages. *Oceanography and Marine Biology, An Annual Review* 41:161–236.

Carefoot, T. 1977. Pacific seashores: a guide to intertidal ecology. Seattle: University of Washington Press.

Dayton, P. K. 1971. Competition, disturbance, and community organization: the provision and subsequent utilization of space in a rocky intertidal community. *Ecological Monographs* 41: 351–389.

Kido, J. S., and S. N. Murray. 2003. Variation in owl limpet *Lottia gigantea* population structures, growth rates, and gonadal production on southern California rocky shores. *Marine Ecology Progress Series* 257: 111–124.

Knox, G. A. 2001. *The ecology of seashores.* Boca Raton, FL: CRC Press.

Raffaelli, D., and S. Hawkins. 1999. *Intertidal ecology.* 2nd ed. Dordrecht: Kluwer.

Scheil, D. R., and D. I. Taylor. 1999. Effects of trampling on a rocky intertidal algal assemblage in southern New Zealand. *Journal of Experimental Marine Biology and Ecology* 235: 213–235.

Smith, J. R., and S. N. Murray. 2005. The effects of bait collection and trampling on a *Mytilus californianus* mussel bed in southern California. *Marine Biology* 147: 699–706.

Sousa, W. P. 1979. Disturbance in marine intertidal boulder fields: the non-equilibrium maintenance of species diversity. *Ecology* 60: 1225–1239.

Sousa, W. P. 2001. Natural disturbance and the dynamics of marine benthic communities, in *Marine community ecology.* M. D. Bertness, S. D. Gaines, and M. E. Hay, eds. Sunderland, MA: Sinauer, 85–130.

HABITAT RESTORATION

RICHARD F. AMBROSE
University of California, Los Angeles

Ecological restoration is the process of assisting the recovery of a damaged ecosystem. As human impacts on natural ecosystems have increased, so has the importance of finding ways to restore damaged habitats. Restoration ecology is a nascent field in general, but restoration efforts are much more prevalent in terrestrial ecosystems, where the science of restoration ecology is better understood. In the marine world, techniques for restoring seagrass beds and kelp forests are fairly well developed, and methods for restoring damaged or degraded coral reefs are being refined. In temperate regions, the restoration of subtidal rocky reefs has focused mainly on artificial reefs, which have been constructed extensively for reasons other than restoration (mainly to increase fishing opportunities).

Being on the interface between terrestrial and marine environments, rocky shores have been degraded by a wide variety of different factors, so efforts to restore them have relied on a diversity of approaches.

CAUSES OF ROCKY-INTERTIDAL DEGRADATION

As a first step toward restoring a habitat, we need to consider *why* the habitat needs to be restored. The cause may be obvious if degradation of the habitat was caused by a distinct event, such as an oil spill. More often, a habitat is degraded over a period of time by a multitude of stressors, and it may not be simple to disentangle the relative influences of the different stressors. It is unwise to initiate expensive restoration efforts before understanding why the habitat was degraded, especially if the sources of degradation cannot be removed and an ongoing stressor is going to degrade it again. Nonetheless, in some cases it may be worth restoring a habitat even when all stressors cannot be eliminated. If the restoration efforts can substantially improve the natural resource values of a habitat, restoring the habitat may be worthwhile even if the full potential level of ecosystem function cannot be achieved, as long as the improvement is judged to be worth the cost and effort of restoration.

Rocky intertidal habitats are affected by a wide variety of human influences. Two main types of impacts can be distinguished: short-term (acute) effects and long-term (chronic) effects. Many short-term effects are accidental. The most dramatic examples are oil spills, such as the *Exxon Valdez* spill in 1989. In some spills, dispersants or other efforts to control the oil cause as much damage as the oil itself. Other accidental impacts include chemical spills, landslides, and boat groundings. In addition to these accidental impacts, some short-term impacts are planned, typically as the result of coastal construction. For example, construction activities might temporarily increase the sediment load in coastal waters.

Although accidental spills, especially oil spills, receive the most press coverage, and planned impacts (such as pipeline construction through the intertidal) garner the most regulatory attention, the majority of rocky-intertidal degradation is probably caused by chronic, often inconspicuous, impacts. Worldwide, the largest chronic impacts to rocky shores are probably (1) harvesting of organisms for consumption or other purposes, and (2) impacts from visitors to rocky intertidal habitats. Poor water quality also undoubtedly affects a number of rocky shores, although the data on actual impacts is sparse and so the importance of this impact is not well understood.

RESTORATION GOALS

A critical step in the restoration of a degraded habitat is establishing the goal of the restoration. Restoration goals determine how a restoration should be designed and provide the basis for determining the success of the project. In most cases, the general goal will be to restore the habitat to some predisturbance condition (often termed the "original ecosystem"). In North America, the predisturbance condition is often idealized as the condition of the habitat before European settlement. This goal is not practical in most cases. Data on the condition of rocky intertidal habitats 500 years ago is nonexistent, and there have been so many changes in the general coastal conditions since then that returning to the centuries-old state is not possible. More realistically, the goal of restoration is often to return to an earlier state considered more desirable, such as the condition of the rocky intertidal habitat 50 years ago, or to restore particularly valuable species, such as abalone.

Despite its difficulty, defining clear restoration goals is essential to the success of a restoration project. Without clear goals, the nature and extent of the effort needed to restore a habitat cannot be determined, nor can the success of the restoration project be measured. The restoration goals need to be clear, simple, and attainable. They should address any particularly problematic situations. For example, extirpated species can raise various difficulties for a restoration project, so the goals should indicate whether or not it will be important to re-establish extirpated species or ecologically similar species, since re-establishing them could greatly increase the complexity and cost of the restoration effort.

Restoration goals may be established by a variety of different procedures. Goals may be determined by an independent group promoting the restoration, by the involvement of various stakeholder groups, or by a sponsoring agency. In the case of restoration in response to an event that caused natural resource damages, such as an oil spill, the goal is established by the legal and regulatory framework guiding the Natural Resources Damage Assessment. Restoration is also required as mitigation for damages permitted by different legislation, such as the California Coastal Act.

RESTORATION METHODS

The choice of the best restoration approach depends on the goal of the restoration, the cause of the degradation, and the specific details about the habitat to be restored.

The ecology of rocky shores provides distinct advantages and disadvantages for restoration. Organisms inhabiting rocky shores are already adapted to a harsh and dynamic environment. Unlike many terrestrial or wetland restoration projects, where extensive site manipulation may be required followed by active planting of vegetation or other direct biotic manipulation, rocky-intertidal restoration generally does not require any direct manipulations of species. Because sessile organisms attach directly to rock, there is no need to be concerned about appropriate soil characteristics (e.g., grain size, organic or nutrient content) or preparing the soil by amendments, which are important issues for restoring wetlands and other habitats. Because many rocky-intertidal species have excellent dispersal capabilities, restoration can rely on natural recruitment rather than having to employ manual transplantation and attachment (but see the following discussion), another major task for most terrestrial habitat restoration projects. These characteristics can simplify rocky-intertidal restoration.

On the other hand, the area available for rocky-intertidal restoration is extremely limited and not easily expanded. As challenging as wetland restoration can be, it is relatively simple to expand wetland habitat by adjusting topographic relief (e.g., excavating areas) and connecting to existing water bodies. The presence of rocky habitat in the intertidal zone reflects particular geological and hydrodynamic processes that are not easily manipulated. In addition, the seaward extent is limited by the level of the ocean and the bottom slope, and the landward extent is often constrained by bluffs. Thus, in contrast to wetland restoration, where new wetland habitat can be created from nonwetland habitat, the opportunities for creating new rocky intertidal habitat are extremely limited. Rocky intertidal habitat could conceivably be created by placing rocks on top of a sandy beach, but this would destroy the biological value of the sandy beach and would be confounded by the dynamic nature of that habitat (e.g., the seasonal movement of sand off- and onshore would likely sink and then bury rocks placed on a beach), and there are apparently no instances where this has been attempted as a way to restore degraded rocky-intertidal resources. Breakwaters and other man-made structures do generate new rocky intertidal habitat, but the vertical nature of the habitat limits the extent of the intertidal zone, and the species occurring on these habitats are typically a specialized subset of all rocky-intertidal organisms. As a consequence of these difficulties, nearly all rocky-intertidal restoration efforts focus on improving conditions in existing rocky intertidal habitat.

FIGURE 1 Workers spraying rocky intertidal beach with high-pressure hoses to remove oil spilled from the *Exxon Valdez*. Hot water was originally used to remove oil, but workers switched to cold water after discovering that the hot water was killing organisms. Photograph courtesy of the *Exxon Valdez* Oil Spill Trustee Council.

Much of the experience with rocky-intertidal restoration has come from oil spills. An oil spill is an acute impact, and in many cases there is a rapid response to remove the oil in order to speed up recovery. The need for rapid and visible action is partly based on public perception and political pressure, and the resulting "restoration" efforts can actually be more damaging than leaving the oil would have been. This was the case for the *Exxon Valdez* oil spill, where washing rocks with hot water caused significant damage to organisms in the intertidal (Fig. 1). After initial efforts to remove excess oil from the area, two main restoration approaches have been used: (1) natural recovery and (2) bioremediation through nutrient enrichment. Natural recovery relies on the inherent resilience of intertidal communities, with the assumption that either enough oil was removed to enable the community to recover or that the remaining oil will degrade or disperse quickly enough to allow the community to recover within a reasonable time. Recovery of the main species in a community does appear to be relatively rapid in many cases, occurring within a few years after the spill. However, studies after the *Exxon Valdez* spill have demonstrated that effects of a spill can persist for more than a decade. The long-term consequences of a spill can be due to ongoing effects of oil remaining in the environment, altered community dynamics, or both. Although the most visible oil on rock surfaces in exposed rocky intertidal habitats weathers away in a few years, oil in interstices of rocks may persist to affect organisms for many years afterwards. In addition, even though rocky-intertidal organisms may return to oiled shores fairly quickly, the community composition and dynamics may differ from nonoiled shores for many years. There is a potential for an alternative state to develop at a site, precluding (or at least delaying) the return of the original community.

Bioremediation through nutrient enrichment attempts to speed up recovery by enhancing biodegradation of oil remaining on a site. Nutrient enrichment relies on the fact that bacteria occurring naturally in the environment can degrade oil, but their populations are limited by nutrient availability; adding the appropriate nutrients can lead to increased biodegradation activity. Because they work on subsurface oil, the bacteria have the potential to remove oil from less-exposed areas of the rocky shoreline, such as between small rocks. The efficacy of nutrient enrichment over large regions is not well established, and in addition it can be costly (e.g., $50 million for treating 71,000 kg of subsurface oil from the *Exxon Valdez* oil spill). "Restoration" of rocky intertidal habitats damaged by other acute events generally entails natural recovery alone, since there is usually no need for enhanced biodegradation.

Restoration of rocky intertidal habitats following degradation from chronic impacts focuses on removing the source(s) of impacts. In most cases, the restoration of community structure and function relies on the natural resilience of rocky intertidal communities. As previously mentioned, the main impacts to rocky intertidal communities stem from collecting organisms and trampling. Collecting is relatively easy to eliminate, at least in theory. In places where collecting has been prohibited, and the prohibition enforced, there have often been dramatic and relatively rapid changes in intertidal communities. Although trampling is also easy to eliminate in theory, in practice it can be difficult to balance the competing goals of restoring natural resources with encouraging people to enjoy those resources. Unlike collecting, which involves an active removal of organisms and is easily perceived as impacting the community, trampling is a side effect of passive enjoyment of the rocky intertidal community. The number of visitors walking through an intertidal area can be remarkable, with more than 20,000 people visiting many intertidal sites in southern California each year. Elevated boardwalks, designated paths, and visitor education could reduce the effects of trampling, but the most effective way to restore an area degraded by trampling is to close the area. To be effective, such a closure needs either strong enforcement (as can be possible in a national or state park) or strong support from the user community. The exclusion of humans can result in changes not only to rocky-intertidal organisms, but also to birds using intertidal habitats.

There is one exception to the general approach of relying on natural recovery in rocky intertidal habitats.

Although most marine species have fairly wide dispersal, some do not. As a group, many of the seaweeds have relatively limited dispersal, but some invertebrates (such as abalone) also do not disperse far. If the degradation of the habitat was not so severe as to eliminate these species from the site, then natural recovery is possible, although it may take time for the species to spread throughout the site. On the other hand, if a limited-dispersal species has been extirpated at the site, active management such as transplantation will likely be necessary. Although this has rarely been attempted in the rocky intertidal (a notable exception is surfgrass), experience in other habitats indicate that it could be relatively easy for some species, but that unexpected difficulties often arise. Active transplantation is likely to be worthwhile only for particularly valued species such as abalone.

SEE ALSO THE FOLLOWING ARTICLES

Biodiversity, Maintenance of / Coastal Geology / Ecosystem Changes, Natural vs. Anthropogenic / Nutrients / Recruitment

FURTHER READING

Crowe, T. P., R. C. Thompson, S. Bray, and S. J. Hawkins. 2000. Impacts of anthropogenic stress on rocky intertidal communities. *Journal of Aquatic Ecosystem Stress and Recovery* 7: 273–297.

Hackney, C. T. 2000. Restoration of coastal habitats: expectation and reality. *Ecological Engineering* 15: 165–170.

Hawkins, S. J., and A. J. Southward. 1992. The *Torrey Canyon* oil spill: recovery of rocky shore communities, in *Restoring the nation's marine environment*. G. W. Thayer, ed. College Park, MD: Maryland Sea Grant, 583–631.

Palmer, M. A., R. F. Ambrose, and N. L. Poff. 1997. Ecological theory and community restoration ecology. *Restoration Ecology* 5: 291–300.

Suding, K. N., K. L. Gross, and G. R. Houseman. 2004. Alternative states and positive feedbacks in restoration ecology. *Trends in Ecology and Evolution* 19: 46–53.

HEAT AND TEMPERATURE, PATTERNS OF

KIMBERLY R. SCHNEIDER AND
BRIAN HELMUTH

University of South Carolina

The temperature of an organism's body can affect almost every aspect of its physiology, and small changes in climate can translate into dramatic ecological consequences. Intertidal organisms have long served as a model system for examining the impacts of climate on the distribution of species in nature, and may additionally serve as an early warning for the impending impacts of human-induced global climate change. Therefore it is important to explore patterns of temperature on the shore—over large geographic scales that are more relevant to populations and communities, as well as over smaller spatial scales more relevant to individual organisms.

DETERMINANTS OF BODY TEMPERATURE

Rocky-intertidal ectotherms experience particularly large fluctuations in body temperature compared to organisms in other ecosystems because they must contend with both terrestrial and marine conditions. During submersion, the body temperature of an intertidal organism closely mimics the temperature of the surrounding seawater. In contrast, body temperature during low tide is driven by multiple climatic factors, including wind speed, air and surface temperature, relative humidity, solar radiation, and cloud cover. Characteristics of the organism's body, such as its shape, color, and its ability to evaporatively cool, also significantly affect the temperature of its body. As a result, two organisms exposed to identical climates can have very different temperatures, and intertidal ectotherms can have temperatures that are very different from the temperature of the surrounding air or rock surface. Scientific studies have shown that both submerged temperatures (when animals or algae are most physiologically active) as well as aerial temperatures (when extreme thermal events are most likely to occur) are important to the survival and physiological performance of intertidal organisms. Therefore, quantifying spatial and temporal patterns of temperature at the interface of these two worlds, and predicting patterns of thermal stress in relation to changes in climate in this model ecosystem, can be quite complex.

The difference between habitat temperature and organism temperature is an important distinction. Although the temperature of an organism's microhabitat (such as the rock that it adheres to) can have a significant influence on the temperature of its body, the two are not always identical. It is therefore convenient to consider microhabitats (e.g., different angles of exposure to sunlight or differing intertidal zonation heights) separately from the influence of organism size and shape on body temperature.

Intertidal Microhabitats

The amount of direct solar radiation received by an organism or surface is a major determinant of heat flux, and in the Northern Hemisphere north-facing (shaded) surfaces can be 10–15 °C cooler than adjacent, unshaded surfaces.

This factor alone is a good indication that it is not possible to define a particular rocky bench using a single temperature measurement. Instead, an intertidal bench includes a complex mosaic of temperatures because of the interaction of substrate complexity, emersion time, and the unique interaction of each species of organism with its local microclimate. As a result, temperatures can often vary over surprisingly small spatial and temporal scales, as can be seen by a thermal image of the intertidal bench at Botany Bay, New South Wales, Australia (Fig. 1).

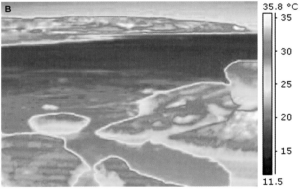

FIGURE 1 An intertidal bench during low tide at Botany Bay, New South Wales, Australia (A). A thermal infrared image (B) shows high thermal variability, with temperatures changing by more than 10 °C over the scale of centimeters. Images by Brian Helmuth.

Not surprisingly, temperatures also vary with tidal height and wave exposure. Although lower intertidal zones and wave-exposed regions are typically much cooler than upper intertidal or wave protected regions due to decreased exposure time, during extreme low tides or periods of calm waves, these sites may experience rare instances of temperature extremes. Because

organisms in these habitats are acclimated to less extreme conditions, they may be harmed during rare events of high or low temperature exposure.

Tidepools create a unique thermal microhabitat for organisms, which, at first glance, appear to be ideal environments to avoid extreme temperatures during aerial exposure. The temperature of a tidepool during low tide, however, depends strongly on the size and depth of the pool. Although seawater has a high specific heat, the amount of heat (in joules) required to raise the temperature of 1 kg of a substance by 1 K, small bodies of water can heat and cool very quickly, depending on climatic conditions. Therefore, a shallow tidepool on a warm sunny day can rapidly increase in temperature, reaching lethal or near lethal thermal limits of the organisms inhabiting the pool, particularly because animals often remain active during submersion (Fig. 2).

Influence of Organism Morphology on Body Temperature

During low tide, the rate of heat transfer between an organism and its environment is affected by characteristics of the organism, such as its mass, shape, size, and color. As a result, extremes in organism temperature during low tide can far exceed those experienced during submersion, often by 20 °C or more. In general, heat flux is divided into six categories (metabolic heat production by intertidal ectotherms is usually considered negligible): short-wave (visible) solar radiation, long-wave (infrared, IR) radiation to and from the sky, IR radiation to and from the ground, conduction to and from the ground, heat convected between the animal and the surrounding air, and heat lost through the evaporation

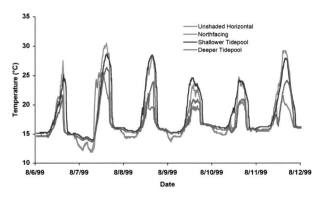

FIGURE 2 Mussel body temperature changes on an unshaded, horizontal surface (in red), an adjacent, vertical (north-facing) surface (in green), and in tidepools of two different depths (in blue and purple) in Monterey, California.

of water. Subsequently, two species exposed to identical environmental conditions can experience markedly different body temperatures. Moreover, different species of organisms appear to be affected by different aspects of their physical environment; the intertidal zone has a high diversity of organisms with different shapes and colors. For example, barnacles that have a large portion of their body in contact with the underlying substratum have body temperatures that are tightly coupled with rock temperature. Rock temperature in turn is strongly affected by the interaction of water temperature and solar radiation. Mussels, in contrast, have body temperatures that are largely decoupled from the temperature of the underlying rock and are often several degrees warmer than the surrounding air. These animals are most strongly affected by the amount of solar radiation that they receive. Organisms with wetted areas exposed to moving air, such as sea stars and algae, cool via evaporative water loss, provided relative humidity is not too high. For example, hydrated intertidal algae are often several degrees cooler than air temperature, but when desiccated can be much warmer than the surrounding air.

GEOGRAPHIC PATTERNS OF ORGANISM TEMPERATURE

Because both aquatic and aerial temperatures have been shown to have significant physiological impacts on intertidal organisms, it is important to quantify patterns in both types of potential thermal stress in intertidal habitats. Interestingly, spatial patterns in aquatic and aerial temperature can show contrasting patterns over a cascade of spatial scales.

Sea Surface Temperatures

Ocean environments respond differently to solar radiation than terrestrial environments. The ocean has a high heat capacity, allowing large amounts of heat to be absorbed and released with relatively little change in temperature. The ocean's annual change in temperature is therefore smaller than its terrestrial counterpart. Temperate seas in the northern hemisphere display the greatest annual change in sea surface temperature (SST), 8–9 °C, whereas tropical areas close to the equator can experience only 1–2 °C annual changes in SST.

Variability in SST also occurs within any given latitude. Ocean currents on the western sides of oceans carry warm water from lower, tropical latitudes (e.g., the Gulf Stream in the western Atlantic), and currents on the eastern side of the ocean carry water from higher

latitudes. These currents cause the SST temperatures on the eastern sides of the ocean to be lower than the SST on the western side. Additionally, along the eastern boundaries of oceans low SSTs are also caused by upwelling of subsurface cool water along the coastline. For example, the west coast of America (eastern Pacific) has cooler water than the east coast (western Atlantic), at the same latitude.

In general, body temperatures of intertidal invertebrates are assumed to be similar to the temperature of the surrounding water during submersion. In some cases, depending on thermal stratification and heating from the rock during the return of the tide, near-shore measurements of SST may not always serve as an effective proxy for submerged body temperatures. *In situ* measurements of body temperature are therefore preferable to reliance on offshore measurements of SST for predictions of intertidal water temperature.

Aerial Temperatures

MICROHABITAT (WITHIN-SITE) PATTERNS

The numerous microhabitats found within an intertidal habitat due to substrate angle, the presence of a tidepool, or tidal height can lead to large differences in the body temperatures between and within species. A good example of these differences has been shown through comparing mussel body temperatures (see "Measuring Temperature") on both horizontal surfaces and north-facing surfaces. Mussel body temperatures can be up to 10 °C cooler on a north-facing vertical slope than on a horizontal exposed surface, despite that the mussels are only a few centimeters apart (Fig. 2). Additionally, organisms that live in beds (e.g., mussels can live in beds, two to three layers deep) may experience different thermal environments as a result of position within the bed. Temperature differences of up to 8 °C have been documented in the top versus bottom of mussel beds. Therefore, minimal movement within a bed can drastically change an individual's microhabitat and thus its thermal environment. Some more mobile species (e.g., sea stars) often alter their location during high tide and may experience different thermal habitats with each low tide.

These differences in microhabitat thermal regimes may have significant effects on ecological interactions. A good example can be seen in the competitive interactions between two barnacle species (*Chthamalus fragilis* and *Semibalanus balanoides*) wherein the competitive dominant is dependent on the substrate angle of the rock that the barnacles inhabit. One species wins when the rocks are cooler (e.g., north-facing vertical slopes) and the other is the competitive dominant in a more thermally stressful environment (e.g., horizontal slopes).

Intertidal organisms must alternately contend with terrestrial and aquatic climatic regimes, and the timing of low tide (coupled with the effects of wave splash) determines the exposure and duration of these two alternate lifestyles. When low tide occurs midday in summer, temperatures experienced by intertidal species can be extremely hot. In contrast, when low tides occur just before dawn in winter, then the risk of freezing and cold stress may be at a maximum. At least along some coastlines, such as the west coast of the United States, the timing of low tide varies seasonally and with latitude. For example, in Washington State, and particularly in Puget Sound, low tides in summer frequently occur in the middle of the day. In contrast, in Southern California the chances of midday exposure in June and July are minimal, especially in lower intertidal regions. The timing of low tide results in body temperatures of intertidal organisms during aerial exposure that do not always increase with decreasing latitude. Research with the competitive dominant mussel *(Mytilus californianus)* along the west coast of North America suggests that instead of an increase in temperature with decreasing latitude, mussels show peaks in temperature at a series of hot spots, where the timing of low tide at midday coincides with regions of hot climate and episodes of low wave splash. These results suggest the effects of climate change may have impacts throughout species ranges rather than just at northern and southern boundaries. Predictive approaches to understand where climate is most likely to impact intertidal species are being developed by researchers to further understand these patterns.

SEE ALSO THE FOLLOWING ARTICLES

Body Shape / Climate Change / Desiccation Stress / Evaporation and Condensation / Heat Stress / Temperature, Measurement of

FURTHER READING

Denny, M. W. 1993. *Air and water: the biology and physics of life's media.* Princeton, NJ: Princeton University Press.

Helmuth, B. 2002. How do we measure the environment? Linking intertidal thermal physiology and ecology through biophysics. *Integrative and Comparative Biology* 42: 837–845.

Helmuth, B., C. D. G. Harley, P. Halpin, M. O'Donnell, G. E. Hofmann, and C. Blanchette. 2002. Climate change and latitudinal patterns of intertidal thermal stress. *Science* 298: 1015–1017.

Hochachka, P. W., and G. N. Somero. 2002. *Biochemical adaptation.* New York: Oxford University Press.

Schmidt-Nielson, K. 1997. *Animal physiology.* Cambridge, UK: Cambridge University Press.

HEAT STRESS

GEORGE SOMERO

Stanford University

Heat (expressed in calories or joules) is a measure of the total thermal energy of a system. Temperature (expressed in Kelvins [K] or degrees Celsius [°C]; 1 K = 1 °C) is the intensity of this thermal energy. Temperature affects every aspect of biological function and structure and, therefore, plays a pivotal role among abiotic environmental factors in governing the distributions and activities of organisms. During emersion at low tide, intertidal organisms may be confronted with high levels of heat stress, the severity of which varies with the vertical position of the organism. Extreme heat stress may be rapidly lethal if physiological processes are strongly perturbed. Sublethal stress may require significant amounts of repair of heat-damaged cellular structures, and these costs of restoring the cell may strongly impact energy budgets and thereby reduce growth rates and fecundity. Heat stress thus plays a pivotal role in establishing the upper limits of vertical distribution and the physiological status of intertidal species.

THE NATURE OF HEAT STRESS:
WHY CHANGES IN TEMPERATURE HURT

Addition to or removal of thermal energy (heat) from a system leads to alterations in the kinetic energy of the molecules of the system. Because rates of biological activity are determined largely by the kinetic energy of the reacting molecules, changes in temperature cause large alterations in all biological rates. Commonly, an increase in temperature of 10 °C leads to an approximate doubling of the rates of physiological processes such as metabolism. Conversely, a 10 °C decline in temperature causes rates to decrease by approximately 50%. Intertidal organisms are ectotherms, species whose body temperatures are strictly governed by the environment (by water temperature during immersion and by air temperature, wind speed, and solar radiation during emersion), so their physiological rates are apt to vary sharply during a tidal cycle. Changes in body temperature of 10–20 °C are common for intertidal species during a tidal cycle, so the rates of their physiological activities may vary by two- to fourfold or more. Heat stress may accelerate rates of metabolic processes to levels that cannot be sustained because of limitations in oxygen transport to cells or in provision of substrates such as sugars for fueling metabolism.

Temperature's effects on biological structures such as proteins, cellular membranes (which are about one-half lipid and one-half protein), and nucleic acids (DNA and RNA) may present even greater challenges to intertidal species than perturbation of physiological rates. Effects on proteins, lipids, and nucleic acids are pervasive because the intricate three-dimensional structures of these large molecular systems, which are essential for their functions, are stabilized by noncovalent (weak) chemical bonds whose energies are of the same order of magnitude as the thermal energy of the system. Thus, all of the structures on which life depends are vulnerable to perturbation by temperature. It follows that, during evolution under different thermal conditions, adaptive changes in the properties of these diverse molecular systems lead to large differences in thermal tolerance limits and thermal optima for physiological activities such as metabolism, locomotion, feeding, and growth. Phenotypic adaptations during the course of an individual's lifetime also are critical in adjusting thermal optima and tolerance limits. Species that experience wide variations in body temperature on a seasonal basis alter (acclimatize) their physiologies in manners that offset thermal perturbation. In the case of intertidal organisms that experience large changes in temperature on a daily basis, adaptive changes in physiological systems may occur over periods of only a few hours, in concert with the tidal cycle.

ADAPTIVE VARIATION IN THERMAL TOLERANCE AMONG INTERTIDAL ANIMALS

Because of the strong and pervasive effects of temperature on biological structures and processes, it should come as little surprise to learn that organisms from different latitudes and from different heights along the subtidal-to-intertidal axis differ in tolerance of high temperatures. Figure 1A illustrates this commonly seen relationship for LT_{50} values (the LT_{50} is the temperature at which one-half of the study population dies following thermal exposure) of congeneric porcelain crabs (genus *Petrolisthes*) from temperate, subtropical, and tropical habitats. Heat tolerances of tropical (Panama) and subtropical (Gulf of California) crabs are much higher than those of congeneric species from temperate regions (California and Chile). Species of porcelain crabs that occur highest in the intertidal zone at a given latitude have heat tolerances that significantly exceed those of congeners that occur low in the intertidal zone or subtidally. The biogeographic patterning of species' distributions and the vertical patterning seen along the subtidal-to-intertidal axis at a single site are reflections of pervasive adaptive changes in the physiological tolerances of organisms, as discussed here.

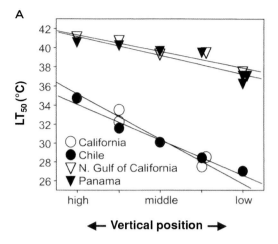

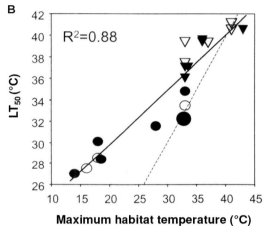

FIGURE 1 The thermal tolerance limits of porcelain crabs reflect the latitudes and vertical positions of the species' habitats. (A) Upper thermal tolerance limits (LT_{50} in °C) of congeneric porcelain crabs (genus *Petrolisthes*) native to eastern Pacific habitats in temperate (California and Chile), subtropical (Gulf of California), and tropical (Panama) latitudes. At each latitude, species occur at different vertical positions along the subtidal–intertidal axis. Each symbol represents a different species. LT_{50} increases with rising adaptation temperature: Tropical/subtropical species are significantly more heat tolerant than temperate species and, at any latitude, species found highest in the intertidal zone have the highest LT_{50} values. (B) Thermal tolerance versus maximal habitat temperature for congeners of *Petrolisthes*. The dotted line (line of unity) can be used to gauge the proximity of species to their upper thermal limits. Species adapted to the highest temperatures have LT_{50} values that lie closest to current maximal habitat temperatures. Thus, the most warm-adapted species appear most threatened by further increases in habitat temperature, which might result from climate change. Data are from Stillman and Somero (2000) and Stillman (2002).

To gauge the threat that increases in temperature may pose to survival of intertidal organisms during emersion, it is necessary to compare the thermal tolerance of a species to the maximal temperatures it may experience in its habitat during low tide. As shown by the dashed line in Fig. 1B, which is a line of unity where lethal temperature

(LT$_{50}$) equals maximal habitat temperature, those species that are most tolerant of high temperatures also are the most threatened by increases in temperature beyond the current thermal maxima of their habitats. Thus in temperate and subtropical or tropical habitats, the species of porcelain crabs that occur highest in the intertidal zone are more likely to encounter temperatures that are near (or even reach) the lethal limit. Similar trends have been observed in several intertidal molluscs. The greater threats that may confront higher-occurring species are exacerbated because some of these species have lesser abilities to increase their heat tolerance through acclimation to higher temperatures than their low-intertidal and subtidal congeners. One can therefore predict that, other things being equal, the effects of global warming will be in direct proportion to the vertical position of animals along the subtidal-to-intertidal gradient.

WEAK LINKS IN THE PHYSIOLOGICAL CHAIN

Because temperature affects the structures and rates of activity of all cellular processes, it is likely that thermal limits and, therefore, biogeographic and vertical distribution patterns may be set by a number of weak links in the physiological chain. Especially in heat-tolerant species currently living close to their thermal limits, several physiological systems have been shown to collapse at temperatures near the LT$_{50}$ of the species. Physiological systems of more cold-adapted species that dwell in low-intertidal or subtidal habitats may not be as vulnerable to increases in habitat temperature, but these systems still play a critical role in vertical patterning because they are not adapted to withstand the temperatures commonly experienced by mid- to high-intertidal species. These generalizations are illustrated in the following discussion by effects that have been observed at several levels of biological organization, ranging from the organ level (heart), to the organelle (mitochondrion), to complex biochemical reaction systems (protein synthesis), to the individual protein molecule (heat shock proteins and proteolytic systems).

Heart Function

The survival of animals with circulatory systems reliant on some type of heart for propelling the internal fluids is apt to be compromised when this pumping system is damaged by temperature. For marine invertebrates, including porcelain crabs and several species of molluscs, the temperature at which heart failure occurs reflects the species' vertical distribution and may lie close to its LT$_{50}$. These relationships are illustrated in Fig. 2 for temperate zone congeners of the genus *Tegula* (turban snails) with

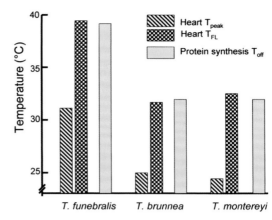

FIGURE 2 Heat tolerance of heart function and protein synthesis differs among congeneric turban snails (genus *Tegula*) that occupy different vertical positions along the subtidal-to-intertidal axis: *T. funebralis* (mid-intertidal), *T. brunnea* (low intertidal/subtidal), and *T. montereyi* (subtidal). For heart function, increases in measurement temperature led to a steady rise in heart rate until a species-specific temperature of maximal rate (T_{peak}) was reached, after which activity decreased rapidly and eventually ceased at the flatline temperature (T_{FL}). Exposure to T_{FL} was lethal for all *T. brunnea* and *T. montereyi* and 92% of *T. funebralis*. In their low-intertidal and subtidal habitats, neither *T. brunnea* nor *T. montereyi* is likely to experience temperatures that impair cardiac function. However, *T. funebralis* commonly experience temperatures of 32 °C and higher on hot days. Data are from Stenseng et al. (2005). Rates of protein synthesis showed a pattern similar to that seen for heart rate: a steady rise in rates of protein synthesis occurred with increasing temperature up to a species-specific optimal temperature, beyond which the rate of synthesis fell rapidly and soon ceased entirely (T_{off}) (see Tomanek 2002). For the two subtidal species, neither heart function nor protein synthesis is able to persist at temperatures that the intertidal species *T. funebralis* commonly encounters. These differences in heat tolerance of key physiological processes thus may be important in establishing the vertical limits of subtidal and intertidal species.

different vertical distributions. Increases in experimental temperature led to a rise in heart rate, as expected, but this increase in frequency of beating occurred only up to a species-specific high temperature (designated as T_{peak}). Above T_{peak} a rapid drop in heart rate occurred with further heating and, within a few degrees Celsius, the heart stopped (the flatline temperature, T_{FL}). T_{peak} and T_{FL} vary adaptively among species with different vertical distributions. Both temperatures are higher in *Tegula funebralis*, a mid-intertidal species, than in two low-intertidal/subtidal congeners, *T. brunnea* and *T. montereyi* (Fig. 2). *Tegula funebralis* encounters temperatures near 30–35 °C in its habitat, so its cardiac function may be impaired by heat stress on hot days during low tides. The two lower-occurring species rarely see temperatures higher than 20 °C, so cardiac function is unlikely to be impacted by heat stress. However, the greater thermal sensitivities of their cardiac function may contribute to their absence from the intertidal habitat where *T. funebralis* is abundant; neither subtidal species could carry out cardiac activity at the higher temperatures that *T. funebralis* encounters. Acclimation of the three congeners of *Tegula* to increased temperatures led to larger increases in thermal tolerance of cardiac function (about 6 °C) in the subtidal species than in *T. funebralis*, which increased tolerance by only approximately 1 °C. Similar relationships involving thermal tolerance and acclimation capacity of heart function have been discovered in porcelain crabs. Thus, as seen for whole organism thermal tolerance, the congeners of *Tegula* and *Petrolisthes* found highest in the intertidal zone seem in greatest danger from further increases in environmental temperature.

Mitochondrial Respiration

The cell's ability to generate adenosine triphosphate (ATP), its energy currency for doing work, depends on the integrity of the membranes of the mitochondrion, the organelle serving as the primary site of ATP synthesis. The lipid and the protein constituents of membranes are highly sensitive to changes in temperature, and the integrity of membrane function is often regarded as a key weak link under conditions of thermal stress. Adaptive variation in thermal tolerance of mitochondrial function is an important element in coping with heat stress. Comparisons of congeners of the genus *Haliotis* (abalone) found at different latitudes and at different vertical positions along the subtidal-to-intertidal axis have revealed interspecific and acclimation-induced variation in the heat tolerance of mitochondrial function and membrane lipid structure. The temperatures at which membrane damage occurs are a reflection of adaptation and acclimation temperatures. The ability to modify the membrane of the mitochondrion through altering the types of lipid molecules inserted into the membrane—a process termed homeoviscous adaptation, which adjusts the physical state (viscosity) of the membrane in a temperature-compensatory manner—is lost when animals are brought to acclimation temperatures slightly above the maximal habitat temperature. Mitochondrial failure, which results in a suboptimal production of ATP, thus is another weak link in the physiological chain. And, as in the case of cardiac function, species living near the upper limits of their thermal tolerance range may have minimal abilities to increase heat tolerance of mitochondrial function when exposed to higher temperatures.

Protein Biosynthesis

A large fraction of the ATP produced in mitochondria is used in the biosynthesis of proteins. Proteins have life spans that range from minutes (e.g., proteins that regulate the cell division cycle) to the lifetime of the entire organism (e.g., eye lens proteins). Commonly, proteins survive in the cell for periods of several days. Their lifetimes are governed in part by temperature because increases in temperature disrupt (denature) the folded conformations of proteins. If repair (renaturation) of heat-damaged proteins is not possible, the irreversibly damaged proteins must be degraded and new proteins synthesized to replace those that are lost.

Protein biosynthesis is a complex biochemical process that exhibits a sharp dependence on temperature and, similarly to cardiac and mitochondrial function, a distinct upper thermal limit (T_{off}) that reflects adaptive variation among species. *Tegula* snails again provide a clear illustration of the extent to which temperature adaptive differences characterize congeneric species from different thermal habitats (Fig. 2). *Tegula funebralis* exhibits a higher thermal tolerance of protein biosynthesis than the two lower occurring congeners, *T. brunnea* and *T. montereyi*. Neither of the subtidal species could synthesize proteins at the upper range of temperature experienced by *T. funebralis*. Thus, we see another illustration of how thermal effects on physiology may help to establish vertical patterning in the intertidal zone.

Thermal Damage to Proteins: The Heat Shock Response and Proteolysis

The unfolding of protein structure caused by heat stress may lead to two distinct types of restorative response: repair of the damaged protein or, if this cannot be achieved, degradation of the protein through a process termed proteolysis. Repair of unfolded proteins may be achieved through the activities of molecules termed heat shock proteins (Hsps). These stress-induced proteins assist in the restoration of the native three-dimensional structure through a process known as molecular chaperoning. Heat shock proteins typically are synthesized quickly when body temperature rises to levels close to the organism's thermal tolerance limits. The interspecific variation in Hsp induction temperatures noted between subtidal and intertidal species is a reflection of differences in protein thermal stability between cold- and warm-adapted species.

When Hsps are unable to repair an unfolded protein, the irreversibly damaged protein must be removed from the cell. Protein degradation, like the Hsp-mediated refolding of damaged proteins, requires considerable amounts of ATP. Thus, the repair of unfolded proteins, the removal of irreversibly heat-damaged proteins, and the replacement of degraded proteins through protein biosynthesis represent substantial energy costs to heat-stressed intertidal organisms.

EFFECTS OF LETHAL AND SUBLETHAL HEAT STRESS ON BIOGEOGRAPHIC AND VERTICAL PATTERNING

Heat stress is one of the most critical abiotic environmental factors in governing where species can live and how well they can function in their habitats. Severe heat stress may cause rapid, lethal perturbation of physiological activity, as seen for cardiac function (Fig. 2). Even if heat stress is not immediately lethal to an organism, the high costs of repairing cellular damage may be a major determinant of the upper limits of distribution in the intertidal zone. The costs of coping with thermal stress, for instance, by manufacturing new proteins and rebuilding cellular membranes, will involve redirection of ATP and building blocks such as amino acids away from growth and reproduction toward repair of heat-induced cellular damage. These complex and pervasive effects of heat stress not only play pivotal roles in the establishment of current biogeographic and vertical patterning of intertidal species but need to also be considered when the potential effects of climate change on marine intertidal ecosystems are evaluated.

SEE ALSO THE FOLLOWING ARTICLES

Cold Stress / Monitoring: Long-Term Studies / Rhythms, Tidal / Temperature Change / Zonation

FURTHER READING

Hochachka, P. W., and G. N. Somero. 2002. *Biochemical adaptation: mechanism and process in physiological evolution.* New York: Oxford University Press.

Somero, G. N. 2002. Thermal physiology and vertical zonation of intertidal animals: optima, limits, and costs of living. *Integrative and Comparative Biology* 42: 780–789.

Stenseng, E., C. E. Braby, and G. N. Somero. 2005. Evolutionary and acclimation-induced variation in the thermal limits of heart function in congeneric marine snails (genus *Tegula*): implications for vertical zonation. *Biological Bulletin* 208: 138–144.

Stillman, J. 2002. Causes and consequences of thermal tolerance limits in rocky intertidal porcelain crabs, genus Petrolisthes. *Integrative and Comparative Biology* 42: 790–796.

Stillman, J., and G. N. Somero. 2000. A comparative analysis of the upper thermal tolerance limits of eastern Pacific porcelain crabs, genus *Petrolisthes*: influences of latitude, vertical zonation, acclimation and phylogeny. *Physiological and Biochemical Zoology* 73: 200–208.

Tomanek, L. 2002. The heat-shock response: its variation, regulation and ecological importance in intertidal gastropods (genus *Tegula*). *Integrative and Comparative Biology* 42: 797–807.

Tomanek, L., and B. Helmuth, eds. 2002. Physiological ecology of rocky intertidal organisms: from molecules to ecosystems. *Integrative Comparative Biology* 42: 771–775.

HERBIVORY

PATRICIA M. HALPIN

University of California, Santa Barbara

Herbivory is the consumption of algae or plants by animals. Herbivory, also referred to as grazing, is a common mode of feeding, seen in a wide variety of marine animals. Equally diverse are the types of algae and plants consumed. Herbivory can play a major role in determining species abundance and ecosystem diversity.

IMPORTANCE

Herbivory is found in nearly all the animal phyla, a testament to its ubiquity in natural ecosystems. Where photosynthetic organisms such as seaweed or phytoplankton occur, herbivores are typically found as well. Seaweeds, phytoplankton, plants, and other photosynthesizing organisms are referred to as primary producers because by converting carbon dioxide into sugars and complex molecules they "produce" the bulk of organic materials that are passed into the food web. The majority of primary producers in the sea are single celled phytoplankton and algae, neither of which are technically "plants," although all share the ability to photosynthesize. Herbivores are also referred to as primary consumers, reflecting their trophic role (secondary consumers eat primary consumers, and so on).

Like their terrestrial counterparts, larger seaweeds and seagrasses provide habitat as well as food for herbivores. One of the major effects of herbivory is not just the consumption of the primary producer, but alterations in the habitat. Herbivores play an important ecosystem role as prey in addition to their role as consumers.

Research on herbivory frequently focuses on trophic cascades, a dynamic linkage among predator, herbivore, and seaweed populations. For example, oystercatchers feeding in intertidal areas can consume large numbers of herbivorous limpets, resulting in an increase in seaweed abundance. In central California, oystercatchers prefer to feed on the giant owl limpet, *Lottia gigantea*. Predation by oystercatchers confines the giant owl limpet to highly sloping rock faces, where they are harder to catch. Experimental removal of the giant owl limpets *(Lottia gigantea)* resulted in an increase in seaweeds. However, this system is more complex than a simple trophic cascade of oystercatchers, owl limpets, and sea weed. In the areas where owl limpets were removed, small herbivorous limpets gradually became more abundant and seaweed abundance declined again, indicating that owl limpets exclude small limpets from potential feeding areas through competition. This example shows the most dramatic type of herbivory, active foraging, where herbivores consume or dislodge growing seaweeds. Two other important types of herbivory occur through food subsidies and mesograzers. All three are described in the following sections.

FOOD SUBSIDIES

Because of the ocean's fluid nature, marine herbivores do not always depend upon local algal growth for their food supply. This contrasts sharply with terrestrial systems, where local plant productivity strongly influences the local populations of herbivores. In the sea, fronds or whole seaweeds may be torn loose to travel on ocean currents. Such "drift algae" may travel long distances from their original source to be eaten as food in an entirely different ecosystem. These food subsidies are one of many ways that seemingly separate marine ecosystems are in reality deeply connected.

Dense populations of sea urchins are often shown to be dependent on drift kelp. Sea urchins (Fig. 1) feed by using their powerful five-part jaws to eat a wide variety of seaweeds. Though mobile and seemingly well protected from predators, urchins prefer have seaweeds come to them as drift algae, rather than move about to feed. In many instances, urchins can gain a remarkable amount of food simply by waiting for these stray pieces of kelp to float by. Drift algae are captured by the sticky tube feet covering the urchin and passed on to the mouth.

FIGURE 1 The purple sea urchin, *Strongylocentrotus purpuratus*, is an important grazer on Pacific seashores. Urchins can survive on drift algae when the supply is plentiful, but they will actively forage to create urchin barrens when it is not. Photograph by the author.

This important phenomenon was first observed in California in the early 1980s, when it was demonstrated that, in areas with abundant kelp, large numbers of urchins could survive simply by sitting in crevices and feeding on drift algae. When drift algae supplies fell off, urchins would leave their crevices to feed on seaweed growing nearby. Similarly, in Chile, nearly 70% of the diet of the intertidal urchin *Tetrapygus niger* was provided by drift algae.

Drift kelp can also form an important part of the diet of other marine species. Intertidal populations of the South American limpet *Patella granatina* survive by capturing and eating pieces of subtidal kelp that float by as the tide rises and falls. In this way, they escape competing for food with another limpet species, *Patella argenvillei*, that feeds directly on the algae growing on nearby rocks.

ACTIVE FORAGING

As mentioned in the preceding section, sea urchins will move out of crevices to actively forage when supplies of drift algae decline. Urchins can be voracious foragers, and when densities are high enough, they create urchin barrens: areas devoid of all seaweed. Urchin barrens are typically covered with a coating of pink, encrusting coralline algae—the only algal life that survives the sea urchins' jaws. Urchin barrens are observed all over the world in tropical, temperate, and polar waters.

In addition to urchins, other important grazers on rocky shores include molluscs such as snails, limpets, and chitons. All feed by scraping their radula (basically a tongue with a rasping surface like a file) across the surface of rocks to remove algae. Active foragers often have a home site that they return to when the tide is falling and leave to graze when the tide is high. In some cases, herbivores create hollows in the rock, called home scars, to which they return. As a result of their homing behaviors, a grazing halo can often be observed surrounding the grazers at low tide, where algae has been eaten (Fig. 2).

Active foragers have important effects besides removing algae from the rocks. Grazers can also remove small invertebrates that have recently settled out of the plankton and metamorphosed. Limpets kill newly settled barnacles by eating them, bulldozing them off the rock, or otherwise preventing them from attaching to the substrate. In areas where limpets have been experimentally removed, barnacles recolonize bare space at far greater densities than in areas where limpets are allowed to continue foraging (Fig. 3).

FIGURE 2 The giant owl limpet, *Lottia gigantea,* in its home scar. The limpet has grazed a halo of bare rock around the scar. The faint ripples that can be seen on the surface of the rock are marks from the limpet's radula in a faint film of microscopic algae. The small green spot above the quarter and to the right of the owl limpet is another small limpet covered in green algae. Note that the owl limpet also has a patch of red crustose algae on its surface. Shelled animals often have grazers living on them that feed off of the algae growing on their shell. Photograph by the author.

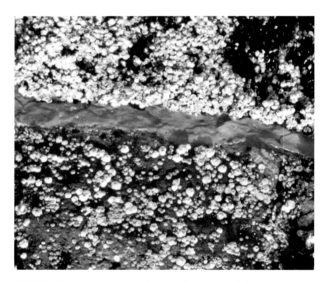

FIGURE 3 Grazers can keep rocks free of some invertebrate species as well as the algae they eat. Limpets, in particular, are known for eating or bulldozing barnacles off the rock as the latter settle out of the plankton from their larval stage. This photograph shows an experimental plot where limpets and other molluscs have been excluded with paint containing copper (top of photo) and a control area where they were allowed to graze freely. Notice the much higher density of barnacles within the experimental plot than in the control. Photograph by the author.

MESOGRAZERS

Though large herbivores such as sea urchins and abalones garner more attention from both scientists and tidepoolers, close examination of seaweeds reveals they host a wealth of small animals, including worms, isopods, amphipods, crabs, snails, and limpets. Many of these animals use the seaweed as habitat, without feeding on their host seaweed. The small herbivores inhabiting seaweed, or mesograzers, survive by feeding either on fouling organisms that settle on their seaweed host, on the host itself, or both. As a result, the relationship between seaweed and mesograzers has aspects of both mutualism and parasitism.

Large brown kelps on rocky shores often host small limpets (genus *Lottia*) that create shelter for themselves by eating a small hollow into the kelp stipe (Fig. 4). They stop short of eating through the stipe; instead they feed by grazing on ephemeral algae that grow on the surrounding kelp stipe. Though they depend on their host kelp for shelter and food, the creation of their home scar can also weaken the host plant, causing breakage. A grazing halo, akin to the one described for active grazers, can often be seen around the home scar. A similar pattern is seen in other seaweeds, where mesograzers may benefit their host by removing fouling organisms but also harm them by feeding upon them and increasing their risk of breakage.

Mesograzers frequently camouflage themselves to resemble their host seaweed and escape predation. The isopod *Idotea montereyensis* uses pigments from its host to change color. Young *I. montereyensis* settle onto red algae into intertidal areas. By feeding on their host, they turn a reddish-brown color by depositing carotenoid pigments from their host into their cuticle. As they grow in size, adult isopods move into beds of surfgrass (*Phyllospadix,* a

FIGURE 4 Limpets in the genus *Lottia* can live upon large kelp species. They create a home scar by consuming part of the kelp stipe. Their main food source is algae and other organisms that settle on the kelp and not the kelp itself. The limpet–kelp relationship has aspects of both parasitism, where the limpet eats part of the plant and weakens the stipe, and mutualism, where the limpet helps keep the kelp free of overgrowth. Photograph by the author.

green plant) and take up residence on the blades. Eventually, the isopods molt and form a new cuticle colored green with pigments from their new host.

SEAWEED DEFENSES

Seaweeds exhibit an array of defenses against herbivory. At the most simple level, algae find refuges in space and time, thereby avoiding contact with grazers. In Panamanian rocky intertidal areas, only seaweeds growing in crevices and holes escape consumption by a suite of fish and invertebrate grazers. Seaweeds may also avoid grazing by associating with other species of seaweeds that produce chemical defenses.

Seaweeds produce a variety of chemical compounds, some of which act to decrease herbivory by either rendering the alga unpalatable or harming the herbivore. In addition to deterring herbivores, these chemicals may perform other functions: preventing fouling of the seaweed by other organisms, poisoning neighboring organisms, killing microbes, and protecting against ultraviolet (UV) damage. For example, the phlorotannins produced by brown algae both protect against UV damage and deter herbivores. It has been hypothesized that herbivore feeding strategy selects for patterns of chemical defense production in seaweeds. The chemicals involved in plant defense are costly to produce. Therefore, only seaweeds susceptible to herbivory that either occurs frequently or quickly kills the seaweed should always produce defensive chemicals. In contrast, mesograzer damage is slow and not always large. This could select for inducible defenses, in which antiherbivore chemicals are produced only when mesograzer damage becomes severe.

Algal defense can also be based on morphology. One type of morphological defense involves a curious aspect of the life cycle in many marine algae: the alternation of morphological types. Many algae have complex life histories that involve different morphologies, for example, a crustose and an upright form. Crustose morphologies are harder for grazers to feed upon and are associated with high grazing pressure, whereas upright morphologies are associated with maximizing growth and reproduction when grazing pressure is low.

Lastly, coralline algae defend themselves against consumption through calcification. By depositing calcium carbonate throughout their tissues, coralline algae become quite hard and resistant to the jaws and radulas of herbivores. Littorine snails forced to feed on crustose coralline algae exhibit broken teeth, a testament to the effectiveness of this defense. The crustose coralline algae that persist in urchin barrens are utilizing two forms of defense: a crustose growth form and calcification.

SEE ALSO THE FOLLOWING ARTICLES

Algal Life Cycles / Foraging Behavior / Kelps / Limpets / Sea Urchins

FURTHER READING

Bustamante, R. H., G. M. Branch, and S. Eekhout. 1995. Maintenance of an exceptional intertidal grazer biomass in South Africa: subsidy by subtidal kelps. *Ecology* 76: 2314–2329.

Dayton, P. K. 1971. Competition, disturbance, and community organization: the provision and subsequent utilization of space in a rocky intertidal community. *Ecological Monographs* 41: 351–389.

Duffy, J. E., and M. E. Hay. 1990. Seaweed adaptations to herbivory. *BioScience* 40(5): 368–375.

Harrold, C., and D. C. Reed. 1985. Food availability, sea urchin grazing, and kelp forest community structure. *Ecology* 66: 1160–1169.

Lindberg, D. R., J. A. Estes, and K. I. Warheit. 1998. Human influences on trophic cascades along rocky shores. *Ecological Applications* 8: 880–890.

Viejo, R. M., and P. Aberg. 2003. Temporal and spatial variation in the density of mobile epifauna and grazing damage on the seaweed *Ascophyllum nodosum. Marine Biology* 142: 1229–1241.

HERMIT CRABS

DANIEL RITTSCHOF

Duke University

Hermit crabs are highly mobile decapod crustaceans that are distinguished by their practice of occupying used gastropod shells (Fig. 1). Hermit crabs are found in most estuarine, intertidal, and marine ecosystems. In benthic communities, hermit crabs are detritivores and the center of shell habitat webs. Shell habitat webs are groups of invertebrates that live with hermit crabs (Fig. 2).

FIGURE 1 A colorful tropical hermit crab. Six legs are visible. The eyes are on the ends of the two stalks between the largest legs. The antennules (noses) are between the two eyestalks. Photograph by Ming-Shiou Jeng.

FIGURE 2 A hermit crab and some members of the shell habitat web. The antennae are the long white striped appendages next to the eyes. Members of the shell habitat web that are visible include barnacles (two bumps left top of shell), a filter-feeding snail (white on top of shell), and worms (circular tubes below barnacles) attached to the outside of the shell. There is a symbiotic crab on the outside of the shell between the hermit crab's eyes. On the inside of the shell, next to the crab's large left claw is a symbiotic anemone. Photograph by Scott Taylor Photography, Inc.

MORPHOLOGY AND LIFE CYCLE

Hermit crabs are common in tidepools. They are the creatures that you take for snails but that move too fast and that drop to the bottom of the pool when you move. Claws of the two kinds, pagurid and dioginid hermit crabs, are different. Paguridae are right handed, Diogenidae are left handed. Out of their shells, hermits have soft curved abdomens with modified appendages that help them hold onto the shell. Females have feathery appendages on the outside curve of their abdomen where eggs are attached and brooded until egg hatching and release of planktotrophic larvae.

After spending weeks to months growing through planktonic larval stages, naked larvae metamorphose to benthic juveniles and occupy snail shell chips and eventually intact snail shells. Snail shells provide protection, control growth and fecundity, and are important to sexual success. Shells wear out, are damaged by predators, become overgrown, and must be replaced as hermits grow. Usually, any intact snail shell in a tidepool will house either the snail responsible for growing the shell or a hermit crab. Shells are usually a limiting resource, and most crabs occupy shells that are too small. If you sit quietly by a tidepool, hermit crabs will entertain you with aggressive social interactions centered upon trading shells, obtaining a new shell, or in sexual activity.

BEHAVIOR

The most common behaviors displayed by hermit crabs are interactions involved with shell trading. While new shell acquisition may involve tens of crabs, shell trading usually involves just two crabs. The crabs grasp each other's shells and perform behaviors such as shell probing, shell rocking, and shell rapping. Behaviors enable the crabs to assess the size and quality of each other's shells. Often, when crabs trade shells, both crabs benefit.

Finding the Perfect Shell

Hermit crabs spend their lives looking for perfect shells. What we think is perfect and what crabs think is perfect are often different. Each hermit crab species has a preferred type of shell and fit. If you give hermit crabs a lot of extra shells and about 24 hours to try them out, you will find what each hermit crab chooses as a perfect shell. Hermit crabs do not kill snails for their shells. Hermit crabs are omnivorous, usually eating microscopic food scraped off surfaces. Hermit crabs acquire a new shell when a snail dies.

Dead and dying snails release odors that cause hermit crabs in good shells to run away and attract hermit crabs in poorly fitting shells. The species of hermit crabs that usually occupy the shell of the dead snail are attracted. Crabs arrive within minutes when a predatory snail begins eating a prey snail. Attracted hermit crabs locate the potential new shell and then usually assess the shell of every other hermit crab in the area. Many species of crabs line up from largest to smallest and wait for the release of the new shell.

Some species of hermit crabs are attracted by the blood of other hermit crabs that are being eaten. Sometimes more than one species is attracted. That hermit crabs are attracted to dangerous sites such as those where another hermit crab died a violent death is testimony to the importance of shells.

So, hermit crabs in poorly fitting shells are attracted to odors of dead snails, odors of snails being eaten by predators, and the odor of hermit crab blood. Crabs of all sizes are attracted by these odors, interact, and prepare to occupy the new shell or a shell vacated when another crab switches. Tens of crabs can all change shells in less than a minute. After entering a new shell, crabs run away. After about a half hour, crabs can tell whether the new shell still does not fit, and the quest resumes.

Mating and Reproduction

If the hermit crabs you are watching are different in size, or if there are several crabs all approaching one crab in an old shell, it may be breeding season. Males and females produce attractive sex pheromones. Breeding males are aggressive. They fight with each other and drag females around by their shells. Mating is a complex ritual. Each species has its own set of stereotyped behaviors that culminate in lining up gonopores on the legs and transferring sperm from the male to the female. Gonopores on the male are at the base of the fifth walking leg. Gonopores on the female are on the base of the third walking leg.

COMPLETING THE LIFE CYCLE

Days to weeks or months after mating, the female extrudes a clutch of fertilized eggs. Each egg is covered by glue, which hardens about 30 minutes after egg extrusion. During that time, the female positions the eggs so that they become glued to the feathery appendages on her abdomen. Eggs are brooded inside the shell until embryo development is complete. At egg hatching, eggs release a pheromone that stimulates the female to perform larval release behaviors. Female movements help hatch the eggs and generate currents that deliver the larvae to the plankton. Larval release is often at high tide after dark, around the time of the new and full moon. After weeks to months in the ocean and several swimming larval stages, the last larval stage finds a shell chip in a tidepool, transforms to a tiny juvenile hermit crab, and begins its quest for the perfect shell.

SEE ALSO THE FOLLOWING ARTICLES

Arthropods / Crabs / Snails

FURTHER READING

Asakura, A. 1995. Sexual difference in life history and resource utilization by the hermit crab. *Ecology* 76: 2295–2313.

Brooks, W. R., and R. N. Mariscal. 1985. Protection of the hermit crab *Pagurus pollicaris* Say from predators by hydroid-colonized shells. *Journal of Experimental Marine Biology and Ecology* 87: 111–118.

Elwood, R., and H. Kennedy. 1988. Sex differences in shell preferences of the hermit crab *Pagurus bernhardus* L. *Irish Naturalists' Journal* 22: 436–440.

Elwood, R. W., N. Marks, and J. T. A. Dick. 1995. Consequences of shell-species preferences for female reproductive success in the hermit crab *Pagurus bernhardus*. *Marine Biology* 123: 431–434.

Elwood, R. W., and S. J. Neil. 1992. *Assessments and decisions: a study of information gathering by hermit crabs*. London: Chapman and Hall.

Fotheringham, N. 1976. Population consequences of shell utilization by hermit crabs. *Ecology* 57: 570–578.

Hazlett, B. A. 1981. The behavioral ecology of hermit crabs. *Annual Review of Ecology and Systematics* 12: 1–22.

Rittschof, D., C. M. Kratt, and A. S. Clare. 1990. Gastropod predation sites: the role of predator and prey in chemical attraction of the hermit crab *Clibanarius vittatus*. *Journal of the Marine Biological Association of the UK* 70: 583–596.

Rittschof, D., D. W. Tsai, P. G. Massey, L. Blanco, G. L. Kueber, Jr., and R. J. Haas, Jr. 1992. Chemical mediation of behavior in hermit crabs: alarm and aggregation cues. *Journal of Chemical Ecology* 18: 959–984.

Williams, J. D., and J. J. McDermott. 2004. Hermit crab biocoenoses: a worldwide review of the diversity and natural history of hermit crab associates. *Journal of Experimental Marine Biology and Ecology* 305: 1–128.

HOMING

ROSS A. COLEMAN

University of Sydney, Australia

Some animals inhabiting the rocky intertidal show a remarkable ability to return to the same spot time after time. This behavior is commonly known as homing. Individuals of a species are described as homing if, after moving away, usually to feed, they consistently return to the same exact location in the environment from which they left. This location could be a place on the substratum, a crevice in the shore, or a tidepool; the location is set by the scale of the organism.

HOMING vs RANGING

Most animals home in that they return to some feature of their environment that offers some resource value, such as shelter. This is particularly important for less mobile intertidal organisms, which need to survive the next unfavorable phase in the tidal cycle.

Homing is normally described as the proportion of a sampled population returning to the same location on repeated sampling occasions. For homing to exist, this proportion must be substantially greater than would be expected if animals moved randomly. This is in contrast to ranging behavior, in which an organism remains within a given area but does not necessarily return to the same exact location. There are two aspects to homing behavior: spatial resolution and temporal consistency. Spatial resolution is a function of the size of the animal—the animal returns to either the same site on the substratum or to some other piece of spatial resource (crevice, crack, pool, etc.) that is roughly 1–2 times its body size. Homing must also be temporally consistent; an organism considered as homing must demonstrate home site fidelity for a substantial proportion of its life.

Examples of Homing Organisms

Many organisms from wave-swept rocky shores have been described as homing. Some useful examples are given here. Most attention has been directed towards the Mollusca. Perhaps the most familiar examples are limpets, particularly *Patella vulgata* from the Northeast Atlantic, which is known to return to a home scar (Fig. 1). These are not the only patellid limpets to home; others include *Scutellastra argenvillei* from South Africa, *P. flexuosa* from Japan, and *Lottia gigantea* from North America. The siphonariid limpets have

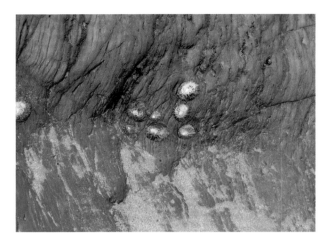

FIGURE 1 Limpets *(Patella vulgata* and *P. depressa)* in a group. Note the close fit of the larger limpets to the rock surface topography. These individuals are highly likely to return to their home scar after foraging and are good examples of organisms exhibiting homing behavior. Photograph by the author.

also been shown to have homing behaviors, as have a few chitons. Other members of the Gastropoda, such as whelks (e.g., *Nucella lapillus*) and snails (e.g., *Littorina obtusata*), have been suggested as homing, but on closer examination of the reports it is apparent that the scale of spatial resolution noted is too large for these animals to be considered homing. Homing has been noted in crabs (e.g., *Cancer magister*) as well as in other decapod crustacea. Fish have also been cited as homing, notably those that use tidepools as a resource, for example *Lipophrys pholis* from the United Kingdom, *Bathygobius cocosensis* from Eastern Australia, *Oligocattus maculosus* from northwestern America, and *Sebastes inermis* from Japan. The unifying theme is that for all of these examples, homing represents returning to a resource feature of the habitat that has a high value to the organism.

Mechanisms of Homing Behavior

A variety of mechanisms have been suggested that enable animals to find their way back to home sites. In the most frequently given examples of limpet homing, surface-derived cues have been ruled out, as have reports of animals using the position of celestial bodies. The most likely mechanism for homing behavior in limpets is that of using chemical cues associated with mucus trails. *Patella vulgata* and other limpets have been shown to correctly identify trail direction as well as determine whether the trails belonged to the individual or to conspecifics. There is some evidence that learning by limpets of microtopographic features can contribute to the homing process. Other taxa seem to rely more on learning of visual cues. Some crabs and lobsters are able to use visible features of the habitat to relocate their home site.

This has been most commonly shown in fishes. Two good examples of this are *Bathygobius cocosensis* and *Lipophrys pholis*, which have been shown to be able to construct spatial maps. The use of cognitive landscape maps to navigate and, in particular, to home has also been suggested for decapod crustaceans. Evidence from field and laboratory experiments in which components of the habitat are modified or moved have shown that crabs and lobsters use the positions of prominent objects for navigation marks. These marks are explicitly georeferenced in the memory of the animal, as it is the distance relationship between the landmarks that is of importance. New technology and instrumentation have demonstrated that some animals (e.g., birds, and a notable marine example, the spiny lobster *Palinurus argus*) use magnetic cues as part of the homing process. It is possible that other organisms utilize this sensory field, but experimental evidence is not yet available.

Benefits of Homing

Organisms return to a site because the site has a particular value. The site may be beneficial in reducing the risk from the biotic environment. For example, the sea urchin *Paracentrotus lividus* bores holes in the rock, which reduces its vulnerability to fish predators; also the fit of the shell of the limpet *Patella vulgata* to the substratum means it is harder for bird and crab predators to remove the limpet. Equally, there may be benefits in reduction of risk from the physical environment such as desiccation or wave exposure, although in some cases the evidence for this benefit is circumstantial at best. For an organism such as an intertidal fish that relies on a cognitive map of its environment (e.g., *L. pholis*), homing is the most efficient means of maximizing information use. If the individual had to relocate a substantial distance, it would have to learn a new set of landmarks to achieve maximal foraging efficiency.

SEE ALSO THE FOLLOWING ARTICLES

Birds / Chemosensation / Limpets / Predator Avoidance / Size and Scaling

FURTHER READING

Branch, G. M. 1981. The biology of limpets: physical factors, energy flow and ecological interactions. *Oceanography and Marine Biology: an Annual Review* 19: 235–280.

Davies, M. S., and Hawkins, S. J. 1998. Mucus from marine molluscs. *Advances in Marine Biology* 34: 1–71.

Gibson, R. N. 1982. Recent studies on the biology of intertidal fishes. *Oceanography and Marine Biology: an Annual Review* 20: 363–414.

Johnsen, S., and Lohmann, K. J. 2005. The physics and neurobiology of magnetoreception. *Nature Reviews Neuroscience* 6: 703–712.

Vannini, M., and Cannicci, S. 1995. Homing behaviour and possible cognitive maps in crustacean decapods. *Journal of Experimental Marine Biology and Ecology* 193: 67–91.

HYDRODYNAMIC FORCES

BRIAN GAYLORD

University of California, Davis

Hydrodynamic forces represent the tendency of water to push on organisms as it flows past them. On rocky shores, these forces result primarily from fluid motions associated with ocean waves that break on the shore. Hydrodynamic forces can act in the direction of water motion, perpendicular to it, or even against flow, depending on the specific causal mechanism. Drag and lift constitute the dominant forces if the pattern of flow surrounding but outside the immediate vicinity of the organism is constant over time and space, whereas additional forces arise when patterns of flow vary. In all cases, magnitudes of hydrodynamic force depend on organism shape, size, and properties of the tissues from which a plant or animal is constructed. These forces can act as important agents influencing ecological processes in coastal marine communities.

GENERALITIES OF SHORELINE FLUID FORCES

Both gases and liquids are fluids and can impart forces when flowing past objects and organisms. In coastal marine habitats, the relevant fluids are largely air and seawater. However, because fluid forces scale in proportion to fluid density, and because the density of seawater is more than 800 times greater than that of air, hydrodynamic forces due to flowing seawater usually dominate over aerodynamic forces associated with wind.

The seawater flows that result in the largest hydrodynamic forces are produced by ocean waves as they approach and break on the shore, and to a lesser extent by surf zone currents, themselves often tied to wave conditions as well as changes in tidal elevation. Such flows can vary substantially in both space and time such that organisms may be subjected to multiple types of hydrodynamic force in short succession or even simultaneously.

DRAG

The most familiar hydrodynamic force is that of drag. Drag acts in the direction of flow and therefore tends to push organisms downstream. It arises due to a combination of two factors: skin friction and an upstream–downstream pressure difference. The skin friction component emerges as a consequence of the no-slip condition, which dictates that seawater in contact with the surface of an organism does not move relative to that organism. Because

seawater at other locations far from the plant or animal flows unimpeded, this means that in intervening regions closer to the organism, fluid layers must move relative to one another. Skin friction results from the fact that the viscosity of seawater resists such relative motion.

In many cases, particularly when seawater is flowing past a nonstreamlined organism, a wake may also be created behind a plant or animal. Wakes arise when the downstream contour of an organism is curved too sharply for the flow to follow along it, such that the fluid stops tracking the shape of the organism (it separates from it) and heads more or less directly downstream. This effect in turn creates a downstream region where fluid recirculates in vortices of a range of sizes. In such wake regions, pressures are typically lower, and in combination with higher pressures generated on the upstream side of the organism, lead to a net force directed downstream. This component of force is pressure drag (Fig. 1). For organisms with a sizable wake, pressure drag can greatly exceed the accompanying skin friction.

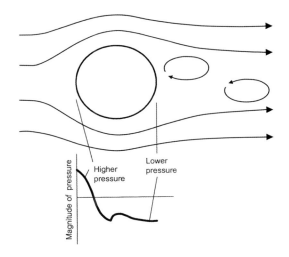

FIGURE 1 The pressure distribution around a circular cylinder produces pressure drag due to differences between above-ambient pressures upstream and below-ambient pressures downstream in the wake.

The difference between upstream and downstream pressures varies roughly with the square of the seawater's velocity relative to the organism, and because the pressure component often dominates, the total drag typically operates similarly. Pressure itself has units of force per area, which means that the total drag (F_D) also tends to scale in proportion to the area of the organism that faces into flow. Usually these relationships are expressed by means of the drag equation: $F_D = 1/2 \, C_D \, \rho \, S \, U_R^2$, where ρ is the mass density of the fluid, S is typically the frontal area of the organism, U_R is the fluid's velocity relative to the organism, and C_D is the drag coefficient. A primary complication is that for streamlined organisms, an alternative convention for S is used (the wetted area = total surface area) because the lack of an appreciable wake means that friction over the full surface of the organism is more relevant for dictating the force than the area facing flow. The drag coefficient itself varies widely, depending on the shape of the organism as well as properties of the flow, as is discussed further herein.

LIFT

Lift, unlike drag, acts perpendicular to flow. It is the force that holds birds aloft, but it can also act horizontally or downward; its line of action simply depends on details of the flow pattern around an organism. Lift arises from differences in pressure between two sides of an organism as induced by differences in flow speed. Fluid traveling around an organism (or an organism's body part, for that matter) often must travel faster around one side than the other, to rejoin smoothly at the organism's downstream edge. Due to a physical rule called Bernoulli's principle, regions of flow characterized by high velocities tend to be accompanied by low pressures, and vice versa. As a consequence, whenever there is asymmetry in the split paths that fluid takes in passing around an organism, there is a capacity for a lift force to be produced, directed laterally toward the side of the organism that experiences the faster flow. Everyday examples include bird wings or fish tails, where fluid travels at a more rapid rate around their convex sides than their flat sides, as shown in Fig. 2.

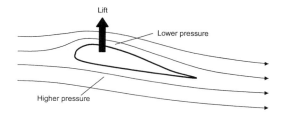

FIGURE 2 Top versus bottom asymmetries in flow around a bird wing or fish fin (represented schematically in cross section here) produce lift, directed perpendicular to the arriving flow.

Although such lift forces can act on the fins of fishes and appropriately shaped sedentary marine organisms, there is another, somewhat different class of lift that may operate more routinely in shoreline habitats. This force is near-wall lift. In the case of organisms that live

attached to the substrate, seawater cannot flow readily around all sides of their bodies because of their positioning against the rock. Nonetheless, there can still be mechanisms by which pressure gets transmitted to the substrate side of the organism, which can enable a net lift to be induced. Flow past the limpet shown in Fig. 3 provides an example: As seawater speeds up in passing over the elevated shell of the animal, this causes a reduction in pressure above it. At the same time, beneath the limpet, the imperfect seal of the shell against the substrate enables seawater to seep under its edge, whereby it moves inward against the animal's body and upward against the shell's underside. The net result is a relatively high internal pressure that is not fully counteracted by the lower pressure outside. The pressure mismatch can lead to a tendency for the limpet to be pulled away from the rock.

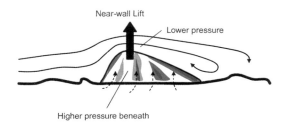

FIGURE 3 Near-wall lift acts on a limpet as higher-pressure seawater seeps under the shell and lower pressure, faster flowing seawater passes over it.

Both classes of lift forces (F_L) are related to flow speed in much the same way as drag and can be expressed via an analogous lift equation: $F_L = 1/2\ C_L\ \rho S\ U_R^2$. Thus, lift, like drag, increases with the square of the speed of the fluid relative to the organism. In this case, however, S is typically taken to be the planform area: the area one would see if viewing the organism along the line of action of the lift force, oriented perpendicular to the incident flow. C_L is the lift coefficient, another index of shape that usually does not equal C_D.

FORCES TIED TO CHANGES IN VELOCITY

Drag and lift are potentially present in all flows. Additional hydrodynamic forces, however, arise if flow fields vary in time or space. The first of these forces results from the pressure gradient that intrinsically accompanies an accelerating parcel of fluid. The second derives from a mass of fluid immediately adjacent to the organism that alters the organism's interaction with the rest of the flow. A third force arises when an air–sea interface, for example, that associated with the leading edge of a breaking wave, impinges on an emergent plant or animal.

Virtual Buoyancy

Basic physics dictates that a parcel of seawater (which has mass) can accelerate only if a larger force is imposed on one side than the other. The effects of such a difference in force, however, are also transferred through the interior of the parcel such that a pressure gradient arises within it. An organism immersed in this pressure gradient will experience a net force due to seawater pushing harder on one side than the other (Fig. 4). This force is often termed virtual buoyancy because it is related to the familiar buoyancy force that arises in a stationary column of fluid as a result of the vertical pressure gradient induced by gravity. Virtual buoyancy acts in the direction of fluid acceleration and is quantified using the expression $F_{VB} = \rho V A$, where V is the volume of the organism and A is the acceleration of the fluid relative to the earth. Note that this relationship means that if the fluid is decelerating, virtual buoyancy acts—nonintuitively—opposite to the direction of fluid movement. Virtual buoyancy is independent of the shape of an organism.

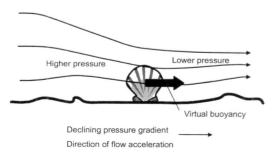

FIGURE 4 Virtual buoyancy as imposed on a scallop attached to the rock. The relationship of this force to buoyancy proper can be observed by mentally rotating the page 90 degrees counterclockwise. Then the acceleration-associated pressure gradient becomes directly analogous to the standard gravitationally induced one characteristic of a stationary body of water, where pressure increases with depth and produces an upward force on any immersed organism (realized as flotation).

Added Mass Force

Another force arises in association not with how a flow changes in an absolute sense but with how it changes with respect to an organism. Seawater in the vicinity of a plant or animal is influenced by the organism and thus moves differently than fluid farther away. Indeed, there is a mass of fluid (the added mass) that behaves in

a physical sense as if it were attached to the organism. The consequences of this added mass can be evaluated from two perspectives. In considering situations where seawater accelerates past a stationary organism, the added mass can be understood in terms of its tendency to cause the organism to displace more of the surrounding fluid than it would otherwise. This effect results in the imposition of an added mass force (F_A) that, in direct analogy to virtual buoyancy, arises as a consequence of the pressure gradient tied to the accelerating flow (Fig. 5A). A second perspective pertains to situations where the organism itself accelerates (Fig. 5B). Under these circumstances, assuming the fluid does not accelerate at the same rate as the organism, the added mass acts like an additional mass over and above the organism's own body mass. This mass provides extra resistance to acceleration, thereby functioning effectively as an opposing force. In both of these two acceleration scenarios, the added mass force is expressed as $F_A = C_A \rho V A_R$, where C_A is the added mass coefficient, another shape factor, and A_R is the acceleration of the fluid relative to the

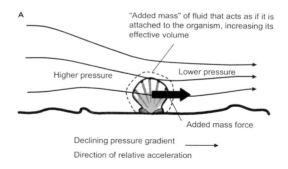

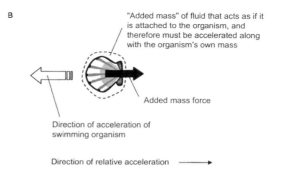

FIGURE 5 The added mass force as imposed on an attached or free-swimming scallop. (A) When seawater accelerates past a stationary organism, the added mass increases the effective volume of the organism, which results in a supplementary virtual buoyancy-type force. (B) If the same organism were to itself accelerate relative to flow, the added mass acts like extra mass to retard the acceleration, functioning in the same way as a force directed opposite to the acceleration.

organism. It may also be noteworthy that the added mass force is sometimes termed the acceleration reaction, whereas on other occasions the sum of the virtual buoyancy and added mass forces are lumped together under this name.

Impingement Force

A third force associated with changes in velocity arises when the air–sea interface of the leading edge of a breaking wave directly impacts a plant or animal. In this case, there is a sudden need for moving seawater to shift its trajectory (i.e., decelerate and shift laterally to establish a new flow pattern) to pass around the organism. This deceleration requires a force, which is provided by the presence of the organism. Naturally, the flow pushes back, and this response produces the impingement force. Recordings on rocky shores suggest that this force, although often lasting only very briefly, can be among the largest imposed on surf zone organisms.

EFFECTS OF SIZE AND FLOW SPEED— THE REYNOLDS NUMBER

One of the major complications in estimating magnitudes of hydrodynamic force derives from the difficulty of determining the three force coefficients: C_D, C_L, and C_A. There are a number of reasons for this difficulty, but paramount among them is that these coefficients are not constants but depend on aspects of the flow.

The drag coefficient, for example, is a function not only of an organism's shape, but also of a parameter called the Reynolds number. The Reynolds number is defined as $Re = \rho U_R L / \mu$, where ρ is again the mass density of the fluid, μ is its viscosity, U_R is the velocity of the fluid relative to the organism, and L is a length term that characterizes the size of the organism, usually its maximal length along the axis of flow. The Reynolds number represents the relative importance of fluid inertia versus viscous effects in a flow: At high Re, the flow has a tendency to maintain its original trajectory and frictional effects are relatively minor, whereas at low Re, the flow is less resistant to directional or speed changes and frictional processes become more important. In the surf zone where seawater is the primary fluid of interest, differences in Reynolds number are equivalent to differences in flow speed for a given organism of fixed size.

In general, skin friction is a greater fraction of the total drag at low Re than at high Re. Furthermore, skin friction depends more on the relative velocity than it does on the square of relative velocity. To account for this feature, the drag coefficient varies essentially as $1/U_R$ at low Re. This pattern can be seen in Fig. 6, which depicts the C_D of a smooth sphere as a function of Reynolds number. This graph also serves as a reminder that identical organisms in different fluids (each characterized by a ρ and μ), or identically shaped organisms of different sizes (as indexed by L), can have distinct drag coefficients because of the consequent change in Re.

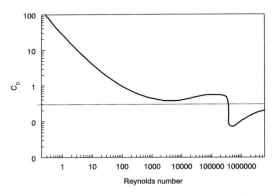

FIGURE 6 The drag coefficient for a smooth sphere as a function of Reynolds number. The sudden decrease (the drag crisis) at Re ~ 3×10^5 reflects an abrupt downstream shift in the location where flow separates from its rear side.

In some cases, shifts in the drag coefficient with Reynolds number can be dramatic. For instance, a drag crisis occurs with smooth spheres and cylinders at around Re ~ 10^5 such that the flow very near their surfaces, in what is called the boundary layer, switches abruptly from a smooth state in which mixing is nearly absent to a turbulent state in which tiny swirls, vortices, and eddies are produced. These turbulent motions enable the point of separation along the contour of the sphere or cylinder to slide further around to the rear (Fig. 7). This process in turn dramatically reduces the size of the wake and thus the drag coefficient. Although most organisms have surfaces that are sufficiently rough that analogous drag crises do not arise in nature, there is at least one curious example of a situation where it does (Denny 1989).

The lift coefficient and added mass coefficient depend, as well, on aspects of flow. The lift coefficient, in particular, becomes increasingly minuscule at low Reynolds numbers. Lift coefficients are also strongly influenced by the orientation of an organism or its body part with respect to the arriving flow. A fish fin tilted moderately relative to the incident flow, for example, can have a large lift

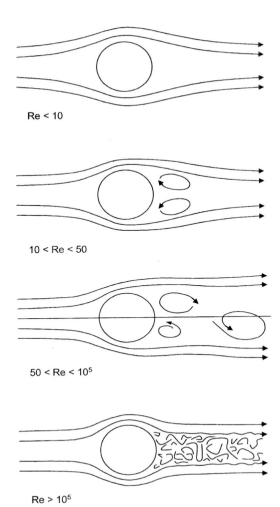

Re < 10

10 < Re < 50

50 < Re < 10⁵

Re > 10⁵

FIGURE 7 Representative flow patterns past a circular cylinder oriented perpendicular to the direction of fluid movement, as a function of Reynolds number. At low Re, flow passes smoothly around the cylinder and, although there is upstream–downstream asymmetry, no obvious wake is apparent. At somewhat higher Reynolds numbers (10–50), stable recirculating vortices form behind the cylinder. For $50 < Re < 10^5$, these vortices are shed alternately from one side then the other, creating a vortex street. At higher Re yet, the wake narrows and becomes turbulent, characterized by disorganized fluid motion with considerable mixing. Note that the numerical values given are approximate and can vary within a factor of 2–5 as a function of conditions in the incident flow.

coefficient, whereas the same structure inclined at either a lower or sharper angle may exhibit a smaller or even zero lift coefficient. The added mass coefficient can vary in bidirectional flow as a function of the distance of travel of the fluid past the organism before the flow reverses.

TISSUE PROPERTIES AND FORCE

There is a further complication that arises when estimating the hydrodynamic forces imposed on actual shoreline organisms. Many, if not most, intertidal plants and animals are not entirely rigid. Flexible seaweeds, for instance, readily reorient and reconfigure in flow, with fronds compressing together as velocities increase. As a consequence, their drag coefficients become strong functions of flow speed when computed—as is the convention—relative to constant reference areas (usually the maximal frontal area that could face flow if the organism were held upright). Stiffer organisms naturally do not exhibit the same degree of conformational change as compliant ones and so show less of a decline in drag coefficient with flow speed. In this regard, the tissue properties of plants and animals influence their interaction with flow and thereby the hydrodynamic forces they experience.

The fact that flexible organisms such as seaweeds move passively in response to seawater motion has two other implications as well. First, in longer organisms that not only compact in flow but also sway or flop back and forth, a tendency to move with the fluid can reduce the speed at which seawater translates relative to an organism. This behavior decreases the velocity term in the drag equation and thus the applied force. Second, as an organism moves, it acquires momentum. This momentum can cause an organism attached to the rock to impose a force on itself as it reaches the end of its range of motion and is jerked to a halt. Interactions among these various effects indicate that flexibility has both advantages and disadvantages. In some cases, flexibility may result in a reduction in drag coefficient and decreased relative velocities, whereas in other cases it may elevate an organism's vulnerability to whiplash-type effects.

HYDRODYNAMIC FORCES ACROSS TIME AND SPACE

A number of the hydrodynamic processes identified previously can operate in concert on rocky shores and will sum together to impose an overall force. The relative magnitudes of the total and individual forces, however, can change as the tide rises and falls. At low tide, hydrodynamic forces are entirely absent over much of the shore. At intermediate tidal levels, waves begin to arrive at locations where organisms were previously emergent, and can crash directly onto them to impose impingement forces. At high tide, organisms often become completely immersed such that drag and lift become the major forces, and those from impingement disappear. It is also the case that magnitudes of total force are modulated over longer time scales. Storms, for instance, produce large waves that chronically increase the severity of the flows faced by organisms. Both major and minor storms may be more likely during certain seasons of the year, or during specific years characterized by unusual weather patterns.

Hydrodynamic forces are also linked to geometrical features of the shore. At small scales, crevices and holes can provide protection from rapid water motions. Similarly, organisms low against the rock or behind upstream protrusions can exist in regions where average velocities are slower. Such plants and animals exploit the fact that velocities are reduced within and immediately adjacent to the roughness elements that make up the rugosity of the substrate. At the same time, at the scale of meters, velocities are often increased where topographical features accelerate flows, such as within converging channels aligned with the direction of wave travel, or along the sides of boulders where the flow speeds up to pass around them. At larger scales, coastal features such as headlands can focus waves and increase their sizes, leading to faster wave-generated water velocities and larger hydrodynamic forces.

COMMUNITY IMPLICATIONS OF HYDRODYNAMIC FORCE

Patch Formation in Space-Limited Habitats

A primary motivation for understanding hydrodynamic forces derives from the important roles they play in coastal communities. Lift, for example, can act to create new open patches in shoreline mussel beds. It arises because faster flows and lower pressures above the bed go unmatched by slower velocities and higher pressures in interstices within the bed (Fig. 8). The initiation of

open patches is particularly relevant in coastal areas of the Northeast Pacific where mussels are often a dominant space occupier, growing in such densities that they form extensive beds composed of multiple layers of individuals. Because unclaimed rock substrate is commonly a limiting resource in these habitats, without the removal of mussels the amount of free space decreases and the abundance of other plants and animals declines. By removing mussels, therefore, hydrodynamic forces enable inferior competitors that would otherwise be excluded from a population to persist within it. A classic example is the sea palm (Fig. 9). This seaweed operates as a "fugitive" species in that although any specific individual lives only a transient existence in its own slowly disappearing patch, the species as a whole can reliably maintain a presence in the community by exploiting a continually changing assortment of open patches within the mussel bed.

FIGURE 9 Open patches in mussel beds provide substrate for the sea palm, a Northeast Pacific seaweed species found only on outer rocks subjected to large hydrodynamic forces imposed by breaking waves. Photograph by the author.

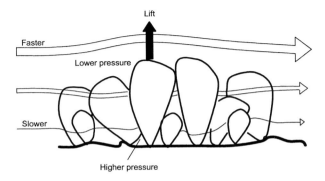

FIGURE 8 Lift can function as an important agent initiating patch removal in mussel beds, arising due to the combination of higher pressures in slower flow regions within the interstices of the bed and lower pressures in faster flow areas above the bed.

Effects on Food Supply and Consumption Rates

Hydrodynamic forces and the rapid flows that produce them can also carry dislodged organisms to nearby locations where they may become food for other shoreline animals. For instance, sedentary sea anemones that live on the bottom of surf zone pools acquire a majority of their sustenance from mussels, snails, and other animals that are knocked off the rocks and fall within the grasp of their tentacles. In much the same way, the fronds of many tattered or dislodged seaweeds are washed onto the beach where they provide fodder for a variety of sand-dwelling grazers (Fig. 10).

FIGURE 10 Dislodged seaweed individuals and fragments, washed onto the beach near Santa Barbara, California, following a storm accompanied by large waves. Such material provides food for many beach-associated grazers. Photograph by the author.

In other situations, hydrodynamic forces alter rates of acquisition or consumption of the food items that become available. For instance, barnacles may retract their feeding appendages when hydrodynamic forces exceed a given threshold. Other organisms that cannot hide their feeding structures may instead be bent over in response to hydrodynamic forces. Such changes modify the orientation of body structures that may be used for food collection and can thereby alter rates of food acquisition and consumption. Hydrodynamic forces can also affect the feeding strategies of animals with nonsedentary lifestyles. For instance, sea stars, voracious predators on many rocky shores, reduce their foraging activity when subjected to large waves.

Scaling Considerations

Different hydrodynamic forces also have the potential to be more or less important for small or large organisms. Both drag and lift depend on an area term, either the frontal area, the planform area, or the wetted area. In contrast, virtual buoyancy and the added mass force depend on an organism's volume. Because factors that vary in proportion to volume increase more rapidly with increases in size than do factors that vary with area (that is, as L^3 vs L^2, where L is a characteristic length of the organism), one would expect that the latter two forces would become increasingly important relative to drag or lift as plants or animals get bigger.

Virtual buoyancy and added mass forces may indeed be the dominant ones that act on exceptionally large organisms such as massive corals growing in deeper shoreline habitats exposed to nonbreaking waves (Massel and Done 1993). By contrast, however, volume-dependent flow forces do not appear to be more important for most surf zone organisms subjected to breaking waves. In these latter habitats, the spatial dimensions over which velocities vary are sufficiently small that individual accelerating parcels of fluid are unable to encompass the full bodies of larger organisms. Because virtual buoyancy and added mass forces each depend on the volume of the organism enclosed in the accelerating parcel of fluid, this characteristic limits the magnitude of force that can be imposed. As a consequence, acceleration-dependent forces in surf zone habitats do not appear to become large relative to drag and lift, even in bigger organisms.

On the other hand, momentum-related forces produced when flexible organisms reach the ends of their ranges of motion tend to vary in proportion to an organism's mass. Mass itself also increases with size in much the same way as volume. This relationship suggests that momentum-related forces should become disproportionately important relative to area-dependent fluid forces as flexible organisms get bigger. The capacity of such forces to outweigh drag and lift in large individuals, however, has not been fully explored.

SEE ALSO THE FOLLOWING ARTICLES

Body Shape / Buoyancy / Disturbance / Size and Scaling / Surf-Zone Currents / Wave Forces, Measurement of

FURTHER READING

Carrington, E. 1990. Drag and dislodgment of an intertidal macroalga: consequences of morphological variation in *Mastocarpus papillatus* Kutzing. *Journal of Experimental Marine Biology and Ecology* 139: 185–200.

Denny, M. W. 1988. *Biology and the mechanics of the wave-swept environment*. Princeton, NJ: Princeton University Press.

Denny, M. W. 1989. A limpet shell shape that reduces drag: laboratory demonstration of a hydrodynamic mechanism and an exploration of its effectiveness in nature. *Canadian Journal of Zoology* 67: 2098–2106.

Denny, M., and D. Wethey. 2001. Physical processes that generate patterns in marine communities, in *Marine community ecology*. M. D. Bertness, S. D. Gaines, and M. E. Hay, eds. Sunderland, MA: Sinauer Associates, 3–37.

Gaylord, B. 2000. Biological implications of surf-zone flow complexity. *Limnology and Oceanography* 45: 174–188.

Koehl, M. A. R. 1982. The interaction of moving water and sessile organisms. *Scientific American* 247: 124–134.

Massel and Done. 1993. Effects of cyclone waves on massive coral assemblages on the Great Barrier Reef: meteorology, hydrodynamics and demography. *Coral Reefs* 12: 153–166.

Vogel, S. 1994. *Life in moving fluids*. Princeton, NJ: Princeton University Press.

HYDROIDS

LEA-ANNE HENRY

Scottish Association for Marine Science,
Argyll, United Kingdom

Hydroids are the benthic phase of the hydrozoan cnidarians, a group of gelatinous suspension-feeding or carnivorous invertebrates. Most hydroids are composed of polyps integrated into a single colony, although solitary species also exist. Hydroids alternate between this benthic polyp phase and a more mobile planktonic stage that is often a swimming medusa (Fig. 1), but the mobile phase may also be represented by a crawling planula larva. This life cycle strategy has historically resulted in the independent classification of polyps and medusae, culminating in much present-day confusion regarding hydroid systematics.

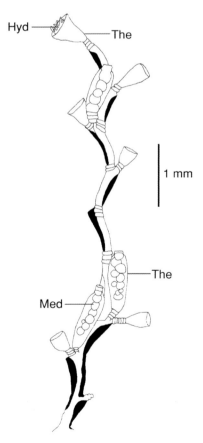

FIGURE 1 General hydroid morphology and life cycle demonstrated by *Obelia geniculata*. The: Thecate cover, which protects both the hydranths and developing medusae; Hyd: a hydranth with its tentacles emerging from its theca; Med: a developing medusa inside its protective theca.

DISTRIBUTION AND DIVERSITY

Hydroids are often cryptic but ubiquitous components of rocky intertidal ecosystems. Although most hydroids found in the littoral zone are not wholly restricted to the intertidal, many species exhibit morphological and reproductive adaptations to the environmental conditions in these ecosystems. Life histories of intertidal hydroids are often linked with the timing of events on wave-swept rocky shores. Hydroids are associated with other rocky fauna and flora, as prey, competitors, and as microhabitats themselves. These associations, their ubiquity, and at times their dense abundance make hydroids an integral part of rocky-shore ecosystems.

Most hydroids on rocky shores are not restricted to the intertidal zone. Rather, predominantly shallow sublittoral species extend their upper vertical ranges onto the shore by inhabiting microhabitats that rarely emerge except during spring tides or storms, for example, crevices, channels, tidepools, or under the macroalgal canopy. These shady microhabitats reduce desiccation and protect delicate and gelatinous hydroids from the shearing forces of waves and ice scour, while ensuring adequate immersion times for these suspension feeders to acquire food from the overlying water column. As is common in many sessile rocky shore organisms, strictly intertidal hydroids have their upper distribution limits set by physical forces and their lower limits restricted by biological factors, for example, competition and predation.

Diversity of hydroids on rocky shores tends to increase with proximity to the sublittoral, because optimal habitats for hydroids are shady current-swept and submerged hard substrata. But although diversity increases in the sublittoral, hydroids still comprise much of the species diversity of sessile fauna in rocky shore ecosystems, particularly in tidepools (Fig. 2) and macroalgal epiphytic assemblages.

The environmental conditions imposed on sessile organisms on rocky shores often impart a strongly seasonal aspect to the life histories of the organisms inhabiting these ecosystems. Consequently, hydroids tend to exhibit one of three life history strategies to cope with these conditions: (1) a highly seasonal cycle of activity and regression (which may involve complete coenosarc resorption and even encystment), (2) a generally sporadic presence but with some seasonal tendencies, and (3) a fairly constant level of activity throughout the cycle.

MORPHOLOGICAL AND REPRODUCTIVE ADAPTATIONS

Few hydroids can withstand the range of conditions experienced on wave-swept rocky shores. Yet some species

FIGURE 2 Population of the hydroid *Ectopleura* (*=Pinauay*) *crocea* (shown by arrows) coinhabiting a tidepool with chlorophytes, anemones, and mussels on a rocky shore in Mar del Plata, Argentina. Photograph courtesy of Dr. Gabriel N. Genzano.

feature morphological characters that permit them to thrive on rocky shores. Some hydroids have protective thecae covering hydranths and reproductive structures and special ringed outer coverings of the polyps themselves that confer rotation and flexibility, for example, in *Dynamena pumila*. These traits permit this hydroid to feed longer despite occasional emersion, reduce effects of desiccation and drag, and prevent detachment from its substratum. Rocky-shore hydroids also tend to exhibit vertical and horizontal zonation of phenotypes, with increased height and degree of branching in hydroids closer to the sublittoral and more hydranths in sheltered areas, although strong shearing wave action and rapid currents may restrict the most robust colonies to slightly higher zones on the shore.

Vertical zonation of sexually fertile hydroids is also evident in some species, with the lowermost hydroids being more fertile (e.g., number of gonophores, percentage of fertile colonies). Although approximately one-half of all hydroid species release a swimming medusa as the sexual phase of the life cycle, many intertidal hydroid species extrude a planula larva with short distance dispersal and rapid settlement and metamorphosis. These larvae are often highly adapted to ensure optimal habitats are colonized. For example, larvae released from *Clava multicornis* and *Dynamena pumila* exhibit positive schizo- and phototaxis. These responses, combined with metamorphosis induced by microbial films on macroalgae, help to ensure rapid settlement and metamorphosis on optimal habitats such as along the stipes of large fucoid macroalgae.

SPECIES INTERACTIONS

Many animals, both sublittoral and littoral, consume intertidal hydroids. Nudibranchs, pycnogonids, and fish are notorious hydroid predators on rocky shores, sometimes limiting the lower vertical range of hydroids to the infralittoral.

Although not the best competitors for space, hydroids are often the earliest recruits to open areas and interact with other sessile species for occupation of those spaces. Alternatively, hydroids successfully colonize other intertidal organisms such as mussels and macroalgae and live as epibionts and use asexual proliferation to rapidly colonize and spread over these organisms.

Hydroids are themselves microhabitats for an array of other intertidal organisms, including other hydroids, gammarid amphipods, mussel recruits. Mostly these epibionts colonize the hydroid stem without harming the host, or are predators that feed on other epibiont while browsing along the host, or are associated with sediments that clumps of hydroids often attract. Occasionally these relationships impart negative contramensalism effects on the hydroid host, for example, in the case of the polychaete *Procerastea halleziana* feeding on the hydranths, or in the case of pycnogonid larvae parasitizing and developing in hydranths.

SEE ALSO THE FOLLOWING ARTICLES

Cnidaria / Metamorphosis and Larval History / Mutualism / Nudibranchs and Related Species / Pycnogonids / Zonation

FURTHER READING

Brinckmann-Voss, A. 1996. Seasonality of hydroids (Hydrozoa, Cnidaria) from an intertidal pool and adjacent subtidal habitats at Race Rocks, off Vancouver Island, Canada. *Scientia Marina* 60: 89–97.

Calder, D. R. 1991. Vertical zonation of the hydroid *Dynamena crisioides* (Hydrozoa, Sertulariidae) in a mangrove ecosystem at Twin Cays, Belize. *Canadian Journal of Zoology* 69: 2993–2999.

Genzano, G. N. 1994. La comunidad hidroide del intermareal de Mar del Plata (Argentina). I. Estacionalidad, abundancia y periodos reproductivos. *Cahiers de Biologie Marine* 35: 289–303.

Genzano, G. N. 2001. Associated fauna and sediment trapped by colonies of *Tubularia crocea* (Cnidaria, Hydrozoa) from the rocky intertidal of Mar del Plata, Argentina. *Biociências* 9: 105–119.

Henry, L. A. 2002. Intertidal zonation and seasonality of the marine leptothecate hydroid *Dynamena pumila* (Cnidaria: Hydrozoa). *Canadian Journal of Zoology* 80: 1526–1536.

Hughes, R. G. 1992. Morphological adaptations of the perisarc of the intertidal hydroid *Dynamena pumila* to reduce damage and enhance feeding efficiency. *Scientia Marina* 56: 269–277.

Orlov, D. 1996. Observations on the settling behaviour of planulae of *Clava multicornis* Forskål (Hydroidea, Athecata). *Scientia Marina* 60: 121–128.

Rossi, S., J. M. Gili, and R. G. Hughes. 2000. The effects of exposure to wave action on the distribution and morphology of the epiphytic hydrozoans *Clava multicornis* and *Dynamena pumila*. *Scientia Marina* 64 (Suppl. 1): 135–140.

HYDROMEDUSAE

CLAUDIA E. MILLS
University of Washington

YAYOI M. HIRANO
Chiba University, Japan

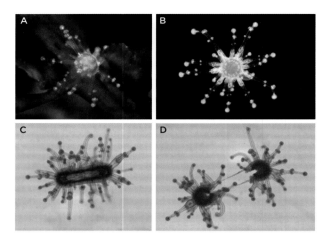

FIGURE 1 (A) *Staurocladia oahuensis* on seaweed. (B) *Staurocladia bilateralis*, isolated from seaweed. (C and D) *Staurocladia oahuensis* medusa undergoing asexual reproduction by binary fission: time series showing initial elongation of medusa followed by pulling apart to form two medusae, nearing completion. All photographs by Yayoi M. Hirano of medusae in Kominato, Boso Peninsula, Japan; central disk diameters approximately 0.5 mm. C and D reproduced with permission from *Scientia Marina*.

Hydromedusae are jellyfishes in the phylum Cnidaria (class Hydrozoa), most of which are born from bottom-living hydroids and are then set free for a relatively short existence in the plankton during which they feed and reproduce sexually. A few hydromedusae, after being released from their hydroids, are adapted to remain attached to the bottom by their tentacles, which they use to crawl slowly around on seaweeds or other substrates along rocky shores, even in places of moderate wave action. These are very tiny species, the largest of which is no more than a few millimeters across the flattened central disk. They are sometimes known as the crawling or creeping hydromedusae and belong to the genera *Staurocladia* and *Eleutheria*. In the rocky intertidal, crawling hydromedusae are perhaps most often encountered in tidepools, but some may also be found by the careful observer to be hanging on tightly to rock or seaweeds along some open rocky shores.

HABITAT, BIOGEOGRAPHY, AND PHYSIOLOGY

Crawling (or creeping) hydromedusae are quite widely distributed globally, although they are not often seen because of their very small size. In some locations these little jellyfishes can be quite common. Seaweeds may require inspection by using a low-power microscope to discover these tiny jellyfish crawling on the surface, and individual blades of algae may bear numbers of these minute animals upon their upper or lower surfaces, even in areas of substantial wave action. These little jellyfish tend to be present in summer through late fall; their parent hydroids are likely present for a longer season, but may be even more cryptic and difficult to find in the field.

There are approximately 15 species of *Staurocladia* (Fig. 1A, B) known so far—from Japan, Hawaii, New Zealand, Australia, Papua New Guinea, Seychelles, Kerguelen Island in the southern Indian Ocean, several locations in Antarctica, South Georgia, South Africa, Chile, Brazil, Falkland Islands, Bermuda, and the Mediterranean. *Eleutheria* is generally considered to have two species that are known from the shores of the European North Atlantic, the Mediterranean, the Black Sea, and perhaps also a single sighting from the Caribbean. *Eleutheria* has recently also been found in Australia, where a morphological and molecular analysis indicates that it has been introduced from the North Atlantic or Mediterranean (Fraser *et al.* 2006).

Species in both genera seem exceptionally well adapted to living in tidepools, and some of the temperate and warm-water species show wide ranges of temperature tolerance from the teens to more than 30 °C.

MORPHOLOGY AND LIFE HISTORY

The various species of crawling hydromedusae have 5 to 60 tentacles extending out from the central disk. There is a red eyespot facing upward at the base of each tentacle. The tentacles are bifurcated, with one branch terminating in a small adhesive sucker that attaches to the substrate and the other branch terminating with a knob of stinging cells known as "cnidocysts", which are used to capture prey. The distribution, or pattern, of additional cnidocyst clusters along the tentacles is one of the important features for distinguishing species. The mouth opens on the center of the bottom side of the central disk, toward the substratum.

These crawling hydromedusae reproduce both sexually and asexually. Asexual reproduction by species of *Staurocladia* is generally accomplished by a process known as fission (also used by many sea anemones that live on the rocky shore), dividing their small bodies in half by adhering with their tentacles to the bottom and then pulling themselves apart into two approximately equal halves (Fig. 1C, D). *Eleutheria* reproduces asexually by budding tiny new jellyfish from the edges of the central disk between existing tentacles, or over the canals within the subumbrella; at least two species of *Staurocladia* are also capable of asexual budding from the edges of the bell margin.

Both *Staurocladia* and *Eleutheria* medusae can also reproduce sexually, which usually occurs after a period of asexual reproduction by each individual. Each jellyfish can produce eggs or sperm; where sexual reproduction is known, sexes are separate in *Staurocladia* medusae, meaning that each jellyfish is either a female or a male, whereas some *Eleutheria* are reported to be hermaphroditic, which means that each jellyfish is both male and female, producing both sperm and eggs, often sequentially. In the case of some *Eleutheria* and some species of *Staurocladia*, the fertilized embryos are protected for a short while within the central disk before they are then released; in other cases the gametes are fertilized externally in the sea without maternal protection, as is typical for most other hydromedusae. The fertilized embryos develop into free-swimming planula larvae and then settle to the bottom where they form tiny, inconspicuous hydroids, which will eventually produce more medusae. The hydroids of these species have only rarely been found in nature, but for a few of these species, are quite easily raised in the laboratory.

It is difficult to estimate age of individuals in populations where asexual reproduction can be so prolific at times and where individuals are so tiny that it is nearly impossible to keep track of them. It is thought that most of these crawling hydromedusae live less than about one month. However, some individual crawling hydromedusae may successfully overwinter in the field and medusae of at least one species of *Staurocladia* are known to persist as long as one year in the laboratory.

There are a few other species of benthic hydromedusae, which live at least partially attached to the bottom in shallow water, including species of *Cladonema* and *Gonionemus*, but these are characteristic of quieter water including bays with abundant sea grass, rather than exposed rocky shorelines.

ECOLOGY

These tiny crawling jellyfish feed primarily on harpacticoid copepods, which also move over the surfaces of seaweed blades and rock. The jellyfish creep slowly along by raising one tentacle off the surface at a time and throwing it forward in the direction of movement. Some species have been observed to release from the surface and swim for brief periods, but most of these crawling species of jellyfish are unable to swim.

Little is known about the predators of these little hydromedusae. At least one species of aeolid nudibranch feeds on a *Staurocladia*, and it is likely that other such predator–prey associations exist between various species of these hydromedusae and other nudibranchs in the field.

Eleutheria dichotoma has recently been found to be well established as an introduced species along approximately 400 km of the New South Wales (Australia) shoreline. Although these medusae, with central disk diameters of less than 0.5 mm, occur in densities up to about 100 individuals per 10 cm^2 of algal surface (especially *Ulva*) in some places along the rocky shore (Fraser *et al.* 2006), we still do not know what impact they might have in the intertidal communities in which they live. *Staurocladia* species in Japan have been shown to have high asexual reproductive rates when conditions are good, allowing rapid increase of the microscopic medusa populations at times; the same is likely true for *Eleutheria*.

TAXONOMY AND PHYLOGENY

Both *Staurocladia* and *Eleutheria* are anthomedusan hydromedusae and produce athecate hydroids. They are usually now placed in the family Cladonematidae, although some authors still recognize the Eleutheriidae as a separate family for these two genera. Recent morphological comparison studies indicate that this is a polyphyletic group (Schuchert 2006); distinction of the two genera is problematic and has varied widely between authors.

SEE ALSO THE FOLLOWING ARTICLES

Amphipods, Isopods, and Other Small Crustaceans / Hydroids / Nudibranchs and Related Species / Sea Anemones / Stauromedusae

FURTHER READING

Bouillon, J., M. D. Medel, F. Pagès, J. M. Gili, F. Boero, and C. Gravili. 2004. Fauna of the Mediterranean Hydrozoa. *Scientia Marina* 68 (Suppl. 2): 1–449. (Full text and illustrations available online at http://www.icm.csic.es/scimar/vol68s2.html.)

Fraser, C., M. Capa, and P. Schuchert. 2006. European hydromedusa *Eleutheria dichotoma* (Cnidaria: Hydrozoa: Anthomedusae) found at high densities in New South Wales, Australia: distribution, biology and habitat. *Journal of the Marine Biological Association of the UK* 86: 699–703.

Hadrys, H., B. Schierwater, and W. Mrowka. 1990. The feeding behaviour of a semi-sessile hydromedusa and how it is affected by the mode of reproduction. *Animal Behaviour* 40: 935–944.

Hirano, Y. M., Y. J. Hirano, and M. Yamada. 2000. Life in tidepools: distribution and abundance of two crawling hydromedusae, *Staurocladia oahuensis* and *S. bilateralis*, on a rocky intertidal shore in Kominato, central Japan. *Scientia marina* 64 (Suppl. 1): 179–187.

Kramp, P. L. 1961. Synopsis of the medusae of the world. *Journal of the Marine Biological Association of the UK* 40: 1–469. (Full text available online at http://www.mba.ac.uk/nmbl/publications/jmba_40/jmba_40.htm.)

Mills, C. E. Hydromedusae. http://faculty.washington.edu/cemills/Hydromedusae.html.

Mills, C. E., and J. T. Rees. 2007. Hydrozoa, in *Light and Smith's manual: intertidal invertebrates of the central California coast*, 4th ed. J. T. Carlton, ed. Berkeley: University of California Press, 118–168.

Russell, F. S. 1953. *The medusae of the British Isles*. Cambridge, UK: Cambridge University Press. (Full text and illustrations available online at http://www.mba.ac.uk/nmbl/publications/medusae_1/medusae_1.htm.)

Schuchert, P. 2006. The European athecate hydroids and their medusae (Hydrozoa, Cnidaria): Capitata part 1. *Revue Suisse de Zoologie* 113: 325–410.

Schuchert, P. The Hydrozoa. http://www.ville-ge.ch/musinfo/mhng/hydrozoa/hydrozoa-directory.htm and http://www.ville-ge.ch/musinfo/mhng/hydrozoa/antho/staurocladia-wellingtoni.htm.

ICE SCOUR

LADD E. JOHNSON

Université Laval, Quebec, Canada

Ice scour is the physical disturbance of intertidal and shallow subtidal benthic communities caused by the mechanical abrasion of the rock surface by sea ice or the removal of organisms frozen into ice that forms on the shoreline and is subsequently transported away.

GENERAL ATTRIBUTES

The widespread influence of ice on rocky shores—15% of the planet's shoreline, principally in arctic regions—goes largely unnoticed because few people live in the polar and subpolar regions where marine ice occurs, and thus it is generally underappreciated as an ecological force structuring intertidal communities. Whereas ice disturbance in subtidal benthic habitats has been relatively well examined (especially on soft bottoms), its influence in intertidal zones, particularly on rocky shores, is still poorly known and appears much more complex than a simple "titanic" scenario of ice colliding with rocks. Indeed, the effect of ice varies over large spatial and temporal scales, depending ultimately on both the physical environment (climate, circulation, geomorphology, etc.) and species characteristics (recruitment, growth, mobility, etc.).

Although ice can occur in a variety of forms, only two usually influence the intertidal zone: free-floating sea ice and the ice foot (Fig. 1). (The better known icebergs rarely affect intertidal zones as they usually run aground before reaching shore. Likewise, anchor ice normally only affects subtidal habitats.) Free-floating sea ice (the ensemble of

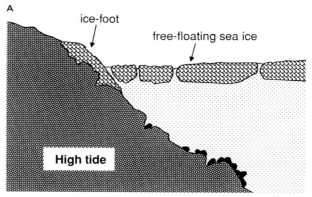
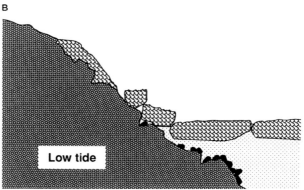

FIGURE 1 Position of the ice foot and free-floating sea ice at high (A) and low (B) tide. Note distribution of organisms and the importance of refuges in the intertidal zone.

ice floes and smaller pieces of ice) that comes into contact with the shore has the potential to crush or scrape organisms off the surface, in other words, scour the shore (Fig. 2). The effect of the ice foot (the accumulation of ice frozen fast to the rock surface) is less obvious, because it can benefit organisms by buffering temperatures and shielding them from the abrasion of sea ice but can also cause

FIGURE 2 Ice scour in action along the shoreline in the Gulf of St. Lawrence (Nova Scotia), Canada. Photograph by Ricardo Scrosati.

harm by reducing access to resources in the water column (food, oxygen, etc.) and eventually lifting or tearing them off the substratum. Despite the more static nature of the ice foot, this combination of effects appears to be more disruptive than scouring by sea ice because the substratum underlying the ice foot is often completely denuded. The relative importance of these two kinds of ice disturbance appears to depend on the relative magnitude of the tidal amplitude and the local sea ice thickness—when the former exceeds the latter, an ice foot can form in the upper intertidal zone. The action of free-floating sea ice is then restricted to lower zones and is driven by water motion resulting from tides, currents, wind, and wave action.

TEMPORAL AND SPATIAL VARIATION

As with any form of disturbance, the impact of ice varies temporally, depending on both its frequency and intensity. Frequency of ice disturbance is relatively easy to estimate, at least over longer temporal scales. Seasonally, the development of the ice foot generally follows the annual temperature cycle. In contrast, the seasonality of free-floating sea ice depends on its production and persistence, and it can even occur year-round in locations near ice shelves or glaciers. Annually, ice disturbance can vary from being chronic to rare "once-a-century" events. Intensity is much more difficult to estimate and ultimately may be "in the eye of the beholder," because the force needed to crush or dislodge a mussel will differ from that required to remove completely the holdfasts or crustose parts of algae.

Effects vary over spatial scales from latitude to local topography. At large scales, ice scour will obviously have its greatest effects at northern latitudes. In contrast, ice scour can be unpredictable at meso-scales due to variation in winds and waves that influence the onshore transport and movement of sea ice—headlands may be more affected by wave-driven ice, whereas the accumulation of ice can be greater in bays. Finally, at smaller scales, local topography has a large influence because only larger pieces of sea ice will have the mass necessary to damage or dislodge organisms. Thus, certain microhabitats, such as crevices and areas between boulders, can provide refuge to both sessile and mobile organisms. Low-lying areas are also less affected, and tidepools in particular appear relatively untouched by ice.

EFFECT ON COMMUNITIES

The overall effect of ice on intertidal communities is typically a reduction in invertebrate and algal biomass and diversity. (Care must be taken in attributing any such pattern to ice disturbance because other environment factors, such as the low salinity associated with ice melt and low temperatures, may also play a local and regional role.) Sessile, long-lived organisms are the most affected, and in chronically disturbed environments, dominant organisms are typically ephemeral algae (filamentous or foliose forms) and mobile herbivorous invertebrates (amphipods and gastropods). The perennial algae (rockweeds, kelps) and invertebrates (barnacles, mussels) that dominate temperate zones are either rare or restricted to spatial refuges. More ecologically interesting situations arise in the transition zone between polar and ice-free seas. Here an ice foot never forms and scouring by sea ice happens with varying annual predictability, both in its occurrence and timing. Community consequences can be dramatic because the resulting disturbance resets the successional clock, often over widespread areas. Diversity can increase in situations in which dominant species are removed (*sensu*, the intermediate disturbance hypothesis), and the irregular occurrence of ice disturbance may be responsible for persistent patchiness in certain communities.

Given the current interest in global warming, the signals of climate change should be nowhere as evident as in the influence of ice on intertidal communities. Clearly, temporal and spatial variation in the frequency and intensity of ice formation and scouring will change as the climate is altered. Unfortunately, our understanding of these changes is still based on just a handful of studies, representing just the tip of the proverbial iceberg of knowledge necessary to predict ecosystem responses to environmental changes.

SEE ALSO THE FOLLOWING ARTICLES

Climate Change / Cold Stress / Disturbance / Succession / Temperature Change / Tides

FURTHER READING

Barnes, D. K. A. 1999. The influence of ice on polar nearshore benthos. *Journal of the Marine Biological Association of the UK* 79: 401–407.

Bergeron, P., and E. Bourget. 1986. Shore topography and spatial partitioning of crevice refuges by sessile epibenthos in an ice disturbed environment. *Marine Ecology Progress Series* 28: 129–145.

Ellis D. V., and R. T. Wilce. 1961. Arctic and subarctic examples of intertidal zonation. *Arctic* 14: 224–235.

Gutt, J. 2001. On the direct impact of ice on marine benthic communities, a review. *Polar Biology* 24: 553–564.

McCook, L. J., and A. R. O. Chapman. 1997. Patterns and variation in natural succession following a massive ice scour of a rocky intertidal shore. *Journal of Experimental Marine Biology and Ecology* 214: 121–147.

Scrosati, R., and C. Heaven. 2006. Field technique to quantify intensity of scouring by sea ice in rocky intertidal habitats. *Marine Ecology Progress Series* 320: 293–295.

Wethey, D. S. 1985. Catastrophe, extinction and species diversity: a rocky intertidal example. *Ecology* 66: 445–456.

IGUANAS, MARINE

RODRIGO H. BUSTAMANTE

CSIRO Marine and Atmospheric Research, Cleveland, Australia

LUIS R. VINUEZA

Oregon State University

The endemic Galápagos marine iguana, *Amblyrhynchus cristatus*, is the world's only seagoing iguanid lizard that feeds on seaweeds along the rocky shores, tidepools, and shallow subtidal rocky reefs of the Galápagos Archipelago. Marine iguanas have behavioral, morphological, and physiological adaptations to cope with their unique feeding habits, including a rounded snout, serrated teeth to mow algae, and special glands to excrete salt. Females are normally half the size of the males, which can reach up to 1.5 meters in length and weigh as much as 12 kilograms (Fig. 1). The iguanas' body size has evolved in response to sexual selection, food availability and temperature. While bigger males have access to more females, they are negatively selected (suffer higher mortalities) during periods of food shortage such as El Niño/Southern Oscillation (ENSO). The marine iguana population around the islands is estimated as 700,000 individuals, but they are considered vulnerable by the red list of the World Conservation Union (IUCN 2004) because of their endemic status, susceptibility to anthropogenic (pollution, fisheries, introduced species, global warming) and environmental (ENSO) perturbations.

FIGURE 1 A large male (front) with very vibrant reproductive colors for mating (grey-green-reddish) and a smaller female marine iguana (back), basking in the sun. Photograph by Luis R. Vinueza.

FROM LAND TO SEA

Evolution

Three species of iguanas coexist in the Galápagos Islands: the two land iguanas *Conolophus subcristatus* and *C. pallidus* and the marine iguana *Amblyrhynchus cristatus*, all endemic species of this unique archipelago. It is believed that these iguanas evolved from a common ancestor of the genus *Iguana*, now widely distributed in mainland South America. For many years it was assumed that the iguanas (and much of the other extant biota) arrived at the Galápagos Islands from mainland South America in floating rafts transported by winds and currents. However, genetic studies have shown that the two sister taxa *Amblyrhynchus* and *Conolophus* differentiated between 10 and 20 million years ago. This has puzzled scientists, because the present islands emerged from the bottom of the ocean less than 5 million years ago. Nevertheless, the accepted current hypothesis for the origin of these much older lizards is that they originally inhabited the former, now sunken, Pacific–Caribbean island arcs from the early Tertiary period (~40 million years ago) off the coast of South America.

Reproductive Traits

The breeding season of marine iguanas appears to be synchronized with peaks of food availability and quality in a strategy to cope with the costs involved in egg production and to maximize their breeding success. Breeding begins in December and extends through March. Marine iguanas have a lekking mating system, in which males cluster during the mating season and defend small territories to attract females. These leks (clusters of territorial males) occur high in the rocky coastline, where females also aggregate. The much larger males defend their mating leks by continually signaling their aggression by head-nodding to intimidate potential transgressors. Males are visited by females who select them as mates based on their size, the level of activity toward females, and lek attendance. Meanwhile, smaller males in the periphery of the leks will attempt to copulate forcibly with females. Following copulation, and approximately four weeks after mating, females will leave the colony to find a spot in the soft sand high up the shore to dig a nest. There, they will bury a clutch of between one to six eggs, whose incubation will take between 89 and 120 days.

Hybridization

Hybrids between the marine iguana *(Amblyrhynchus)* and the land iguana *(Conolophus)* have been recognized by their unusual morphology and genetic analyses. These hybrid iguanas typically behave like the land iguanas, but they have darker colors and exhibit a dorsal striping, both characteristic of juvenile marine iguanas. The fact that there is crossbreeding between the two genera of iguanas raises interesting questions about the evolutionary relationship between the two species. It is not known whether the hybrids are fertile.

Physiological Adaptations

Because marine iguanas lack temperature regulation (ectotherms), they must warm up their bodies with the equatorial sun before and after feeding in the cold sea. Most iguanas feed on rocky shores around low tide. Thus, foraging time is constrained by cold water, the duration and height of the low tide, and the height of the swell. Iguanas stop feeding when full or when cold, whichever happens first. The iguana's body temperature falls steeply in the cool season and more slowly in the hot season. When algal biomass is high, the iguana suffices with a single feeding excursion, but if algal biomass in the warm season is low, the iguana may require two or more feeding excursions during a low tide.

ECOLOGICAL ROLE

Grazing Up and Down

Typically, the marine iguanas feed during diurnal low tides (Fig. 2A). Yearlings (recently born iguanas) feed on tidepools and higher on the rocky shores to avoid wave action. As marine iguanas get older and bigger, they will venture lower on the shore, synchronizing their feeding bouts with the spring low tides. Only large individuals, mostly males, will feed subtidally on the shallow reefs of the islands down to 1.5–5 meters depth, where food is more

FIGURE 2 (A) Juvenile marine iguana feeding among tidepools. Photograph by Luis R. Vinueza. (B) Male adult grazing underwater on the abundant algal communities. Photograph by Angel I. Chiriboga.

predictable and abundant (Fig. 2B). Dive times are only a few minutes long, but it is known that marine iguanas can be submerged for more than half an hour. The preferred food for the marine iguanas includes several species of ephemeral algae, such as *Ulva* sp. (green sea lettuce) and a combination of different species of filamentous red algae that includes species of the genera *Gelidium*, *Ceramium*, *Centroceras*, *Hypnea*, and *Spermothamnion*, among others, which normally form lawns of patches or algal turfs.

Mowing the Algal Lawns

Marine iguanas share their food with other grazing herbivores, including crabs, fish, and molluscs (Fig. 2A). The effect on the algal communities of this diverse group of grazing herbivores varies according to wave action, levels of nutrients, and temperature. At wave-protected sites, the iguanas and other grazers normally decrease the diversity to a few or even single dominant species (mostly green or red algae) or to grazing resistant, crust-forming species. At other locations, particularly in the western islands, the effects of these herbivores give way to more diverse communities because rocky shores are exposed to stronger wave action or higher productivity as a result of the upwelling of cold and nutrient-rich waters.

COPING WITH CLIMATE VARIATION AND EL NIÑO

Boom-and-Bust Life Histories

The climate and oceanographic conditions of the Galápagos Islands are all about change and dynamics. The unique mix of tropical and temperate seas makes finding tropical penguins and diving lizards possible. The southwestern islands harbor cold water and subtropical biota, whereas the northeastern islands have truly warm, tropical ecosystems. In addition, the El Niño/Southern Oscillation (ENSO), a large-scale oceanographic/atmospheric disruption in the tropical Pacific, has dramatic impacts on the marine iguanas and the ecosystem as whole. The variability of the sea and the alternating cycles between El Niño (wet and warm) and La Niña (dry and cold) conditions have shaped the evolutionary traits of the marine iguana. Thus, differences in food quality, availability, abundance, composition, and temperature between biogeographic regions seem to underlie the almost tenfold difference in body size and mass between the iguanas of the northern (up to 0.41 kg) and western islands (up to 12 kg). The consequences of El Niño on the intertidal and reef ecosystems of the Galápagos are dramatic. The warmth and low productivity of the otherwise rich seas starve most of the marine biota to death, including the marine iguanas. The abundance of *Ulva* sp. and filamentous red and brown algae (the most frequent species in marine iguana diet, Fig. 3A) is drastically reduced and the normally lush algae lawns and turfs are then replaced by the opportunistic *Giffordia mitchelliae*, a brown slimy filamentous alga that has been observed only during El Niño years to cover all the rocky shore habitats. *Giffordia* has a high content of tannins and a very low energetic value for the grazers, and it appears also to be difficult to digest by the now-starving iguanas (Fig. 3B). The signs of famine are evident and cause an elevated mortality among marine iguanas (up to 90% on some islands), especially among small and large individuals (Figs. 3C, 3D). After El Niño passage, marine iguanas recover rapidly because of resumed high algal production, reduced competition, and increased fecundity.

Shrinking to Survive

To cope with these stressful conditions, marine iguanas will change their diet to alternative sources of food, including coastal shrubs and carrion, or could even reduce their body size. This is in repose to the fact that smaller iguanas outcompete larger ones when food availability declines. During particularly lean (El Niño) years,

FIGURE 3 Starving during El Niño. (A) In non-ENSO years, a lush and abundant algal growth of green, brown, and red algae occurs in the low and exposed rocky shore platforms of the western islands of the Galápagos archipelago. Photograph by Luis R. Vinueza. (B) A weakened marine iguana grazing on the invasive filamentous brown alga *Giffordia mitchelliae*. Photograph by Luis R. Vinueza. (C) An emaciated female marine iguana during El Niño 1997–1998. Photograph by Linda J. Cayot. (D) Dried carcass of a marine iguana that died of starvation during El Niño 1997–1998. Photograph by Luis R. Vinueza.

some animals shrink—a net reduction in body length as well as in body mass—and those that shrink have better chances of surviving (Fig. 3C). The survival and recovery of marine iguana populations might be hindered by the presence of introduced predators, particularly dogs and cats, and by pollution, such as a recent oil spill that indirectly killed 65% of the population of marine iguanas on Santa Fe Island, where scientists had been studying them for the last 25 years. Therefore, the frequency and strength of environmental perturbations such as ENSO and climate change, together with the intensity of anthropogenic perturbations, will dictate the fate of this unique species.

SEE ALSO THE FOLLOWING ARTICLES

Cold Stress / El Niño / Herbivory / Rhythms, Tidal / Size and Scaling / Vertebrates, Terrestrial

FURTHER READING

Laurie, W., and D. Brown. 1990. Population biology of marine iguanas II. Changes in annual survival rates and the effects of size, sex, age, and fecundity in a population crash. *Journal of Animal Ecology* 59: 529–544.

Rubenstein, D. R., and M. Wikelski. 2003. Seasonal changes in food quality: a proximate cue for reproductive timing in marine iguanas. *Ecology* 84: 3013–3023.

Shepherd, S. A., and M. W. Hawkes. 2005. Algal food preferences and seasonal foraging strategy of the marine iguana, *Amblyrhynchus cristatus*, on Santa Cruz, Galápagos. *Bulletin of Marine Science* 77: 51–72.

Vinueza, L. R., G. M. Branch, M. L. Branch, and R. H. Bustamante. 2006. Top-down herbivory and bottom-up El Niño effects on Galapagos rocky-shore communities. *Ecological Monographs* 76: 111–131.

Wikelski, M. 2005. Evolution of body size in Galapagos marine iguanas. *Proceedings of the Royal Society of London B* 272: 1985–1993.

Wikelski, M., L. M. Romero, and H. L. Snell. 2001. Marine iguanas oiled in the Galápagos. *Science* 292: 437.

Wikelski, M., and C. Thom. 2000. Marine iguanas shrink to survive El Niño. *Nature* 403: 37–38.

INTERNAL WAVES

SEE WAVES, INTERNAL

INTRODUCED SPECIES

JAMES T. CARLTON

Williams College

Introduced species are important members of many open-coast rocky-intertidal communities around the world. Also known as nonnative, nonindigenous, exotic, alien, or invasive species, introduced species are those organisms that have been transported by human action to a region in which they did not previously occur in historical time. For most regions and species, scientific records of the movement of marine organisms by humans typically date back less than 200 years, although introductions may have occurred over the past 1000 or more years. This lack of deep history may influence our understanding of the evolutionary and ecological history of many rocky shores.

VECTORS

Multiple human-mediated transport vectors have been available to move rocky-shore species across and between oceans. An early vector was the movement of intertidal rocks for ships' ballast. A wide variety of animals and plants (such as seaweeds, snails, mussels, barnacles, other crustaceans, mites, and insects) were transported for centuries in the damp holds of sailing vessels on and among ballast rocks. Open-coast rocky-shore organisms also colonize the buoys, floats, pilings, seawalls, and jetties of marine ports and harbors, and thus find themselves adjacent to ocean-going vessels, whose hulls they may colonize as fouling organisms or into whose ballast water systems they may be entrained. Edible rocky-shore animals and plants (such as snails, mussels, and seaweeds) have also been moved intentionally by people. These mechanisms, and others, have provided ample opportunity for the larvae, juveniles, and adults of many taxa to be successfully translocated around the world.

ECOLOGY OF INVADED SHORES

Studies in the Northeast Atlantic, South Africa, South America, and the Northwest Pacific provide experimental evidence for the role of invasions in rocky-shore communities. Additional observations elsewhere in the world further suggest the global nature of the impact of invasions on rocky coastlines.

Northeast Atlantic: The Canadian Maritimes and New England

The region from Nova Scotia to Long Island Sound represents one of the best-studied shores in terms of introduced species. Here scientific records of the larger and more conspicuous species date back to the early 1800s; the efforts of seashell collectors contributed importantly to our early knowledge of this coastline as well. These shores were only deglaciated in the past 10,000 years: Natural recolonization in this rigorous continental climate resulted in a low-diversity biota that greeted the species that began to arrive on and in ships.

The periwinkle snail *Littorina littorea*, well known on European shores, was first found at Halifax, Nova Scotia, in 1854, and from there (in one of the best documented invasion scenarios in the sea) spread south to New Jersey in just 30 years; its earlier history in the Gulf of St. Lawrence remains uncertain. *L. littorea* is one of the largest, most abundant, and most important consumers on all but the most exposed rocky shores of eastern Canada and New England. In the winter, *L. littorea* moves down shore, partially relieving the intertidal of predation pressure and permitting the development of extensive mats of ephemeral filamentous and leafy green and red algae on boulder tops. In the spring, vast numbers of *L. littorea* return to the middle and upper shore and reduce these winter algal populations to nearly zero percent cover. At times, algal patches a meter or more in diameter may be surrounded by hundreds of nibbling snails, with the patch shrinking rapidly over the following days. The aboriginal role of the native snail *Littorina saxatilis*, perhaps displaced by *L. littorea* and consumed by introduced crabs (following discussion), in this regard remains to be determined. The growth of *L. saxatilis* is markedly depressed in the presence of *L. littorea* (which grows to three times the size of *L. saxatilis*), the two species overlap broadly in habitat and potential diet, and nineteenth-century naturalists noted that *L. saxatilis* became less common as *L. littorea* moved down the coast. Growth and survival of the native limpet *Lottia testudinalis* are also depressed in the presence of *L. littorea*.

The removal of most edible algae by *L. littorea* leaves most shores covered by monocultures of the large fucoid seaweeds *Ascophyllum nodosum* and *Fucus vesiculosus*. These plants are unpalatable as adults because of their concentrations of polyphenolic compounds, but they can be eaten by snails in the sporeling and germling stages, and thus predation by *L. littorea* may locally impact the recruitment of these seaweeds as well. Dense populations of snails—hundreds to thousands per square meter—leave many rocks bare. *Littorina* also acts as a generalized omnivore, consuming in season vast numbers of settling barnacle cyprids. In many regions, *L. littorea* also provides the major—and sometimes the only—shell resource for native adult hermit crabs.

Adding to the complexities of understanding the restructuring of these shores has been the introduction of two predatory crabs that have become important community regulators. The European green shore crab *Carcinus maenas* is a eurytopic omnivore that is, or was, the most abundant crab on rocky shores from Canada to New Jersey. It is a voracious consumer of molluscs, barnacles, polychaetes, and a wide variety of other prey. Joining it—and in some cases displacing it—in the 1990s is the equally omnivorous Asian crab *Hemigrapsus sanguineus*, now the most abundant crab on many shores north and south of Cape Cod. Interactions among introduced crabs and native and introduced snails form a portion of the intricate mosaic of these reshaped communities. Snail morphology, particularly shell thickness, in the native seaweed periwinkle *L. obtusata* and the native carnivorous whelk *Nucella lapillus* changes through selection and induction in the presence of *Carcinus*. *Carcinus* also indirectly influences algal populations by reducing snail abundance through predation and by significantly suppressing snail grazing via waterborne stimuli emanated by the crab and detected by the snails.

In the rocky sublittoral, only a few meters below the intertidal zone on these same shores, the green alga *Codium*

fragile tomentosoides, the ascidians *Styela clava, Botrylloides violaceus, Diplosoma listerianum*, and *Didemnum* sp., and the bryozoan *Membranipora membranacea*—all introduced—are community engineers that have significantly altered the previous communities dominated by native kelps (*Laminaria* spp.) and sea urchins. Direct and indirect interactions play out on this stage: For example, the European bryozoan *Membranipora*, by growing epiphytically on *Laminaria*, reduces the growth and survival of kelp, resulting in plant defoliation and gap formation in the kelp beds. The Japanese alga *Codium* colonizes these gaps, preventing kelp recolonization: The demise of these New England kelp beds thus appears to be the result of one introduced species facilitating the spread of a second, with communitywide consequences.

South Africa

In the 1970s the Mediterranean mussel *Mytilus galloprovincialis* appeared on the west coast of South Africa; by 1984 it had become the dominant intertidal mussel in many areas (Fig. 1), and by 2005 it occupied 2000 kilometers of the South African coast, radically altering the appearance and structure of rocky shores. Quantitative and experimental studies have elucidated a complex network of interactions resulting from this invasion. Indigenous mussels, including *Aulacomya ater*, have been displaced wherever *Mytilus* is abundant, and as a result of *Mytilus*'s higher growth rate, greater fecundity, and greater desiccation tolerance compared to native mussels, there has been an up-shore movement of mussel beds. *Mytilus* also generally forms

FIGURE 1 (A) Intertidal reef of the Mediterranean mussel *Mytilus galloprovincialis* on the exposed rocky shore of South Africa with close-up of individual shells (B). Photographs by Charles Griffiths.

multilayered beds on the west coast, as opposed to the single-layered beds of the indigenous mussels: As a result there has also been a massive increase in mussel biomass, with a simultaneous increase in the density of mussel bed infauna. *Mytilus galloprovincialis* dominates primary surface space; as a result, the native limpet *Scutellastra granularis* is largely excluded from the rock surface, although the mussel shells themselves have led to an increase in overall density of this limpet by providing a recruitment substrate for juveniles (resulting in turn in a decrease in mean limpet size, dictated by the size of the host shell). In an interesting development, the threatened African black oystercatcher, *Haematopus moquini*, has shifted its diet and now feeds predominantly on *M. galloprovincialis;* as a result, there has been an increase in the breeding success of the oystercatcher as a result of this increased food supply.

South America: Chile

The medium-sized (35 cm tall) Australian ascidian *Pyura praeputialis* was introduced by ships to Antofagasta Bay, Chile, perhaps in the nineteenth century, where it occurs in dense intertidal aggregations. *Pyura* is an aggressive competitor for primary space and outcompetes the native mussel *Perumytilus purpuratus*, reducing the latter's fundamental niche by constraining it to the upper mid-intertidal, when it previously occurred in the lower intertidal as well. *Pyura*'s broad belts and dense three-dimensional matrices have in turn created a novel intertidal habitat, increasing local species richness.

Northwest Pacific: British Columbia and Washington

In the 1940s the Japanese brown alga *Sargassum muticum* appeared in Washington and British Columbia; it has since spread south to Baja California and is one of the few invaders of the open rocky shore of the Pacific Coast of North America. In experimental studies on British Columbia rocky shores, *Sargassum* dominates new bare patches and excludes the native alga *Neorhodomela larix* (which, however, once it regains space, can prevent recolonization by *Sargassum*). The impact of *Sargassum* is greater in the shallow sublittoral, where it reaches greater densities and overshadowing heights. In Washington, on shallow subtidal rocky habitats, *Sargassum* has effects at multiple trophic levels: As a result of shading, *Sargassum* significantly reduces the native canopy (brown) and understory (red) algae. *Sargassum* also has a strong negative indirect impact on native sea urchins *(Strongylocentrotus droebachiensis)* by reducing the abundance of native kelp on which the urchin feeds.

Additional Invaders and Regions

Nonnative species have become characteristic members of additional rocky shores around the world. In Southern California, the Mediterranean mussel *Mytilus galloprovincialis* displaced the native mussel long known as *Mytilus edulis* (now called *M. trossulus*). This invasion went unnoticed for decades: *M. galloprovincialis* was misidentified as the morphologically similar *M. edulis*, so that the disappearance of the latter, and the transition from native to alien, was overlooked. The interactions of *M. galloprovincialis* (under the name *M. edulis*) and the native *M. californianus* were the subject of a series of classic studies in Southern California in the 1960s.

The Asian algae *Undaria pinnatifida, Sargassum muticum,* and *Codium fragile tomentosoides* have invaded many rocky shores of the northern and southern hemisphere. In North America, Europe, and Australia, native specialized herbivores (sacoglossan sea slugs) have assumed the introduced *Codium* as a new prey, in a process described as emerging associations. Nonnative barnacles are conspicuous elements in many communities: The New Zealand barnacle *Elminius modestus* is now one of the commonest species of western and northern Europe; the American Pacific Coast *Balanus glandula* now forms the predominant band of high intertidal barnacles in Argentina, and the Caribbean barnacle *Chthamalus proteus* has turned the upper intertidal zone of the Hawaiian shores white.

DEEP HISTORY AND THE SCALE OF UNDETECTED INVASIONS

The examples of the invasions described here, and all other invasions that might be mentioned on rocky shores of the world, date only from the 1800s. A rich recorded history of human exploration, colonization, and global commercialization, combined with our knowledge of transport vectors, indicates that unrecorded rocky-shore (and many other) species must have been carried by human activity across and between oceans over many earlier centuries. Future work will reveal that some of the most characteristic if not iconic species of other rocky shores arrived only in the 1500s, 1600s, or 1700s, and, in the process of short-term integration in ecological time, challenge our ability to distinguish them from species that have been present over evolutionary time.

SEE ALSO THE FOLLOWING ARTICLES

Algal Biogeography / Bryozoans / Ecosystem Changes, Natural vs. Anthropogenic / Food Webs / Monitoring: Long-Term Studies / Snails

FURTHER READING

Britton-Simmons, K. H. 2004. Direct and indirect effects of the introduced alga *Sargassum muticum* on benthic, subtidal communities of Washington State, U.S.A. *Marine Ecology Progress Series* 277: 61–78.

Carlton, J. T. 2003. Community assembly and historical biogeography in the North Atlantic Ocean: the potential role of human-mediated dispersal vectors. *Hydrobiologia* 503: 1–8.

Castilla, J. C., R. Guinez, A. U. Caro, and V. Ortiz. 2004. Invasion of a rocky intertidal shore by the tunicate *Pyura praeputialis* in the Bay of Antofagasta, Chile. *Proceedings of the National Academy of Sciences of the USA* 101: 8517–8524.

Geller, J. B. 1999. Decline of a native mussel masked by sibling species invasion. *Conservation Biology* 13: 661–664.

Harris, L. G., and M. C. Tyrrell. 2001. Changing community states in the Gulf of Maine: synergism between invaders, overfishing, and climate change. *Biological Invasions* 3: 9–21.

Levin, P. S., J. A. Coyer, R. Petrik, and T. P. Good. 2002. Community-wide effects of nonindigenous species on temperate rocky reefs. *Ecology* 83: 3182–3193.

Lohrer, A. M., and R. B. Whitlatch. 2002. Interactions among aliens: apparent replacement of one exotic species by another. *Ecology* 83: 719–732.

Robinson, T. B., C. L. Griffiths, C. D. McQuaid, and M. Rius. 2005. Marine alien species of South Africa—status and impacts. *African Journal of Marine Science* 27: 297–306.

Trowbridge, C. D. 2004. Emerging associations on marine rocky shores: specialist herbivores on introduced macroalgae. *Journal of Animal Ecology* 73: 294–308.

Trussell, G. C., P. J. Ewanchuk, and M. D. Bertness. 2002. Field evidence of trait-mediated indirect interactions in a rocky intertidal food web. *Ecology Letters* 5: 241–245.

Trussell, G. C., and M. O. Nicklin. 2002. Cue sensitivity, inducible defense, and trade-offs in a marine snail. *Ecology* 83: 1635–1647.

INVERTEBRATES, OVERVIEW

JAMES M. WATANABE
Stanford University

Look in any habitable place on earth and you will find invertebrates living there. They range in size from microscopic mites to giant squid that grow to more than 20 meters and weigh a couple of tons. Some invertebrates rival open-ocean fishes as pelagic wanderers, while others spend their entire lives attached permanently to rocks or pilings, never going anywhere at all. There are invertebrates whose adult lives are over in less than a day, while others have lifetimes that span decades or even centuries. Some invertebrates, such as mosquitoes or houseflies, are irritatingly familiar to everyone; others, such as spiders, evoke irrational fear and loathing despite their beneficial role in nature's economy. Still others, such as crustaceans, sea urchins, squid, and

bivalves, provide important sources of human food. Some invertebrates such as fleas have even affected the course of human history as vectors of disease, while many others have wrought havoc as agricultural pests. But most invertebrates go quietly about their day-to-day lives unnoticed by all but a mere handful of recondite specialists. On an evolutionary scale, invertebrates are tremendously old. Representatives of several present-day groups have been found among the earliest clear fossils of multicellular animals, nearly 600 million years old. In short, no matter what yardstick you use to measure life on earth, you will find at least a few invertebrates at every extreme and many, many more everywhere else in between.

INVERTEBRATES, PRESENT AND PAST

With over a million animal species described so far (and probably millions more insects and mites still to go), it is rather arbitrary to divide the animal kingdom into vertebrates (<3%) and *in*vertebrates (everything else) simply because the organisms that do the classifying happen to have a backbone. It would be far more sensible to divide the animal kingdom into arthropod and nonarthropod, since arthropods (crustaceans, insects, spiders, and their kin) comprise 85–90% of all living animal species, or even into mollusc (8%) and nonmollusc. But regardless of how we categorize them, the invertebrates are as varied as the animal kingdom itself in form, function, and life style.

Invertebrates are most diverse in the sea. Life first evolved in the sea, and most groups have never left. Echinoderms (which include those quintessential tidepool invertebrates, the starfish), are exclusively marine. Many other major groups are predominantly marine, with only a few species living in freshwater and none at all on land. The only animal groups to make terrestrial habitats truly their own are the arthropods (insects and arachnids) and the vertebrates. In order to prevent the small puddle of the sea that they still harbor inside their bodies as blood or body fluid from evaporating away into the surrounding air, these groups have evolved a specialized array of structures and functions. Most all other terrestrial invertebrates are restricted to damp habitats, having been less successful in solving the problem of water loss to the environment during their transition from sea to land. So when we speak of biodiversity on earth, we are speaking of mostly of the sea.

Today there are roughly 35 phyla of multicellular animals. Each phylum embodies a fundamentally different body plan, comprising a combination of derived characteristics whose presence implies a common evolutionary ancestry of all its component species. Some of these body plans, such as those of sponges or jellyfish, are comparatively simple and have persisted nearly unchanged for hundreds of millions of years. Others are complex and highly derived; that is, they have added a great number of modifications to their ancestral form (e.g., mammals or cephalopod molluscs). But whether simple or complex, all of these body plans succeed in meeting the ever-changing challenges of life, and that is what matters most. The simpler phyla (sometimes referred to as lower or primitive invertebrates) have not remained that way because the organisms are too simple or primitive to change. They have persisted because they work. Natural selection smiles on things that work, but eventually eliminates things that do not. Contrary to the popular conception, natural selection is not survival of only the fittest; rather, it is elimination of those whose fitness falls below some minimum threshold for a given time and place.

To consider how invertebrates evolve, it is important to recall that adaptations do not arise simply because they are needed. Rather, the process of adaptation occurs through incorporation of heritable traits that directly or indirectly affect the reproductive success of individuals possessing those traits. Just because a certain trait would be an improvement does not necessarily mean it will evolve. Constraints placed on organisms by their evolutionary history or by the absence of sufficient genetic variability limit the degree of adaptation that natural selection can produce. Organisms are not always (or perhaps ever) adapted perfectly to their environment, since their environment changes constantly. The most maladapted ones just do not persist. The environment, through natural selection, sifts and sorts blindly, without a predetermined design for the final product. The myriad, sometimes convoluted, ways in which various invertebrates have solved the basic problems of life attest to the opportunistic, but nevertheless effective, nature of natural selection.

The state of invertebrates today is inextricable from their evolutionary past. The fossil record tells us that by the time animals acquired hard body parts that are easily fossilized, many of the present-day phyla were already thriving and quite diverse. This great flowering of invertebrate diversity marks the beginning of the Cambrian period (543 million years ago) and is referred to as the Cambrian explosion. Over a period perhaps as short as 13 million years, the fossil record shifts from only traces of comparatively few, mostly small and presumably soft-bodied forms to conspicuous, larger-bodied representatives of nearly all modern groups.

Since the Cambrian, the history of animal life on earth is an interwoven tapestry of expansion and contraction of various groups. Some groups, such as crinoid echinoderms or brachiopods, have blossomed and expanded, then contracted to just a few present-day species, while others, such as trilobites, dominated Paleozoic seas for hundreds of millions of years and then declined and disappeared forever. Others such as crustaceans, insects, and vertebrates diversified more or less continuously and dominate the earth today. Still other phyla have remained modest in size but surprisingly persistent. For example, wormlike priapulids comprise about a dozen living species and only a dozen fossil species, but they have been present for half a billion years since the Cambrian.

These contrasting histories have been mediated by events on spatial and temporal scales that range from day-to-day, one-on-one ecological interactions to global-scale extinctions spanning millions of years. At least five times in the last 500 million years, mass extinctions have altered the face of life on earth. At the end of the Permian (248 million years ago), 80–95% of shallow-water (continental shelf) marine species are thought to have gone extinct. The Cretaceous–Tertiary extinction 65 million years ago ended of the age of dinosaurs. After each event, life recovered, but often with deeply altered patterns. Formerly dominant groups were replaced by others that for one reason or another were more successful (or just fortunate) in recolonizing the altered ecological landscape depopulated by extinction.

So the history of animal life on earth has not been a steady, stately progression from low diversity to high diversity or from simple to complex body plans. Biodiversity has had a bumpy trajectory, full of peaks and valleys, albeit with a general upward trend. Tracing this history is a major goal of many who study invertebrates. The task is difficult, in part because the vast majority of species that have ever lived are extinct and most left few traces of what they looked like. The living species that we see today are the merest twig-tip buds of a vast genealogy of evolutionary relationships whose earliest branches are hidden forever in the distant past by the one-way flight of time's arrow. Aided by an incomplete, though sometimes spectacularly preserved fossil record, we can only make inferences about those relationships based on the current states of characters provided by morphology, embryology, biochemistry, and, most recently, DNA sequences and patterns of gene expression: the sometimes ambiguous or seemingly contradictory end points of a process more than half a billion years old. Despite its frustrations, the detective work involved in solving this biological jigsaw puzzle makes the study of invertebrates a fascinating challenge.

CLASSIFICATION

Before delving into the diversity of the invertebrates, we must consider how this vast amount of information is organized. Everything needs a name. Without names for things, communication would be impossible. But names have to be consistent from place to place, otherwise they are not very useful. This is the major problem with common names for plants and animals: a lobster from southern California is a very different species of crustacean from a lobster from New England, yet people in both areas refer to the local species as a lobster. In order to avoid such confusion, we have a standardized system for classifying and naming organisms (and groups of organisms), originated by Linnaeus in the eighteenth century.

Under this scheme, each species's name consists of two parts: the genus name and the species name (e.g., *Pisaster ochraceus* or *Homo sapiens*). The genus name is always capitalized and the species name is never capitalized, even when it is derived from a proper noun. The rules for naming species ensure that each species has one and only one valid scientific name. A species name may change for a variety of reasons. If it turns out that the species has been described under a different name at an earlier time, the earlier name has priority. If a species was originally placed in an inappropriate genus, its name can (and should) be changed. Names may also change when a taxonomist reworks a group and decides that the relationships of its species are sufficiently different from what was previously thought to revise the classification of the entire group.

The scheme of classification is hierarchical, with each nested subdivision based on increasingly detailed criteria. From the larger, more inclusive categories to the smaller, more exclusive ones, the divisions are domain, kingdom, phylum, class, order, family, genus, species. The defining characteristics of each descending division are sets of unique, derived (rather than primitive) traits that are shared by all members of that subdivision to the exclusion of members of all other subdivisions at the same level. Thus, members of the class Gastropoda within the phylum Mollusca are more similar to each other than they are to members of any other class of molluscs (Bivalvia, Cephalopoda, etc.). Larger phyla such as the arthropods or the molluscs have additional layers of subdivisions to accommodate their much larger numbers of species.

Taxonomists use a wide variety of characters to classify organisms and to infer their evolutionary relationships. These characters may be morphological, embryological, or biochemical. The fossil record, although incomplete, also provides a wealth of information. Comparison of DNA sequences is the most recent tool to be applied to questions of phylogeny, and it has revitalized interest in phylogeny of the animal kingdom since the late 1980s. Even behavioral traits can be used to differentiate closely related living species. Regardless of the approach, it is important to remember that organisms are classified by evidence from a large number of sources, not just by one or two characteristics by themselves.

To a certain degree, all the categories just described are arbitrary. However, that does not diminish their utility in the least. A scheme of classification serves to organize and make accessible a vast amount of information. By following such a scheme of classification, we can determine the identity of any creature we find. In doing so, we also learn a considerable amount about its biology, simply by knowing to whom it is closely related. But taxonomy is more than simply pigeonholing organisms into rigid categories or collating vast amounts of information. Anyone with a careful eye can accomplish the feat of Cuvier or Linnaeus of simply grouping like with like. It was only through the perception of Darwin and his intellectual kin and the concept of mutability of species by natural selection that taxonomy was given true biological meaning. The notion of evolution changed taxonomy from a classification by simple *similarity* to a classification by *relatedness*. In doing so it revealed not only the vast history of life on our planet but that such a history even existed.

THE PHYLA

There are ~35 animal phyla. One of the current, though not unanimous, views of the phylogenetic relationships among these phyla is depicted in Fig. 1. A dozen or so of these phyla have many representatives living on rocky shores, and you can find them with little effort (Table 1). Many of the rest, except for a few that live wholly offshore in pelagic habitats, are no doubt also present, but they are either rare, transient, very small, or very cryptic.

Despite the bewildering array of names and characteristics, there are patterns among the animal phyla. They fall into four broad groups: the so-called lower invertebrates (sponges, placozoans, cnidarians, ctenophores), deuterostomes (echinoderms, chordates, and hemichordates), and protostomes, which can be divided further into two groups: ecdysozoa (arthropods, nematodes, and

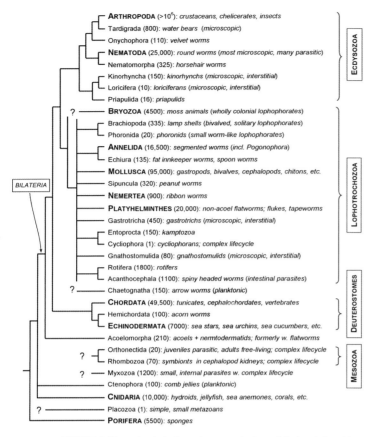

FIGURE 1 The animal phyla and a current view of their relationships. Numbers in parentheses are estimated numbers of living species. Phyla in boldface type are ones likely to be encountered on rocky shores.

other molting phyla) and lophotrochozoa (annelids, molluscs, flatworms, bryozoans, and a number of small-bodied phyla; Fig. 1). A degree of debate still centers around subdivisions within the protostomes, but the protostomes as a whole are supported by a broad array of evidence.

These groups of phyla are distinguished by suites of both specific and general characteristics, which include the number and complexity of tissue layers (or germ layers) that appear during embryonic development, presence and type of body cavity, patterns of early embryonic development, and DNA sequences (see Table 2 for others). It is important to realize that what follows are very broad generalizations and that none of the suites of characters is universally consistent. Some groups are so large and vary so greatly in the nuances of these features that they defy honest classification. Other groups have only a handful of species for which these features have been studied. However, when taken together, the distribution of these characters provide a framework for viewing the phylogenetic organization of the animal phyla. This framework should

be viewed as a testable set of phylogenetic hypotheses that must be probed with additional data.

Tissues: Ectoderm, Endoderm, and Mesoderm

The fundamental organization of an invertebrate body appears early in embryonic development with the establishment of tissues or germ layers. Tissues are integrated groups of cells organized into a functional unit (more in the following paragraphs). The number and complexity of germ layers that arise during embryonic development distinguishes the lower invertebrates from the other, more complex phyla. We will first digress and discuss several phyla whose cell layers are not sufficiently well organized to qualify as tissues, then go on to describe how these germ layers develop in the other invertebrates.

The sponges are the simplest of multicellular animals and in many ways stand apart from all the others. Sometimes they are referred to as the Parazoa to reflect these differences. They have no true tissues, organs, or nervous system and previously were thought to be closer to complex colonies of quasi-independent cells rather than integrated multicellular animals. However, evidence from molecular, genetic, and cellular sources indicate that sponges are basal on the main line of metazoan evolution rather than an early evolutionary detour into multicellularity that went nowhere. As such, they merit status as full-fledged, albeit simple, metazoans.

Four additional small phyla of simple but highly enigmatic multicellular animals also lack true tissues. These include orthonectids and rhombozoans (sometimes grouped together as mesozoans), myxozoans, and placozoans (Fig. 1). The relationship of these phyla to the rest of the invertebrates is unclear and has generated considerable debate. Placozoans are simple, free-living metazoans composed of two cell layers. These layers are not considered true tissues because they lack an extracellular basement membrane (discussed subsequently). The most recent molecular data suggest that placozoans may be basal to all other metazoans and thus represent the simplest grade of multicellularity among the animal phyla. However, not all the data are consistent with this conclusion, so the debate continues. Mesozoan orthonectids and rhombozoans also possess only two cell layers and are parasites or endosymbionts with complex life cycles. As in sponges and placozoans, their cell layers are not considered to be true tissues. One current view is that mesozoans may have evolved from more advanced groups (possibly platyhelminth flatworms) and become smaller and simpler in association with their parasitic way of life. Lastly, the myxozoans are very small, structurally simple endoparasites of both vertebrates and invertebrates. They have

TABLE 1
Phyla with Representatives on Rocky Shores

PHYLUM PORIFERA: Sponges. Simple, sessile animals; porous bodies with internal spaces lined with flagellated collar cells (choanocytes) that pump water & capture tiny particles for food. Mostly low on the shore or on protected undersides of boulders; often brightly colored. Most sponges do not tolerate prolonged exposure to air and are more diverse in subtidal rather than intertidal habitats.

    CLASS CALCAREA: Skeletal spicules of $CaCO_3$; color usually white/cream.

    CLASS DEMOSPONGIA: Largest class (almost all living species); skeletal spicules of SiO_2; wide variety of colors & growth forms.

PHYLUM CNIDARIA: Bodies are functionally radial in the form of either a vase-shaped polyp or bell-shaped medusa, both with a ring of tentacles surrounding mouth; possess unique stinging cells (nematocysts) used for prey capture.

    CLASS HYDROZOA: Hydroids, hydrocorals, siphonophores, etc.; both polyp (asexual) & medusa (sexual) stages alternate in many (but not all) species; colonial forms often highly polymorphic. Abundant but inconspicuous at low levels on rocky shores, under overhands or in pools.

    CLASS ANTHOZOA: Sea anemones, corals, gorgonians, soft corals, black corals, etc; exclusively polypoid; largest class of phylum. Predators. Conspicuous inhabitants of mid- to low levels on rocky shores.

    CLASS SCYPHOZOA: Jellyfish. Sometimes wash ashore or are stranded in tide pools.

PHYLUM ACOELOMORPHA: Small acoelomate flatworms formerly classified in Platyhelminthes, but probably basal bilaterians. Hermaphroditic. Very small and cryptic, but can be abundant on some rocky shores.

Deuterostomes: Triploblastic, bilaterally symmetrical phyla w. radial, indeterminate egg cleavage; blastopore becomes anus; enterocoelous coelom.

PHYLUM ECHINODERMATA: 5-part radial symmetry derived from a bilateral ancestor; unique water vascular system drives tube feet used for locomotion, prey capture, gas exchange; endoskeleton of distinctive $CaCO_3$ ossicles; exclusively marine.

    CLASS ASTEROIDEA: Sea stars. Important predators, especially on many rocky shores; also deposit feeders and scavengers.

    CLASS OPHIUROIDEA: Brittle stars. Circular central disk with long slender snake-like arms. Lack anus due to compression of central disk; numerous species are hermaphroditic brooders; predators & scavengers. Under rocks or in sandy patches along rocky shores.

    CLASS ECHINOIDEA: Sea urchins, sand dollars, heart urchins. Sea urchins are important algal grazers in intertidal and subtidal rocky habitats. Often abundant at mid- to low levels on rocky shores.

    CLASS HOLOTHUROIDEA: Sea cucumbers; skeletal ossicles reduced; functionally most bilateral group of phylum; most species are benthic suspension & deposit feeders. Most diverse in subtidal habitats; rocky shore inhabitants are small and inconspicuous, mostly in pools.

PHYLUM CHORDATA: Possess a notochord, dorsal nerve cord, pharynx w. gill slits, post-anal tail at some stage in life history.

    SUBPH. UROCHORDATA: Ascidians (benthic "sea squirts"), salps & larvaceans (both are planktonic); solitary or colonial suspension feeders; most possess unique tunic made of cellulose-like tunicin that covers body; most chordate features only obvious in tadpole larval stage. Abundant, usually at lower levels on rocky shores; sometimes brightly colored.

    SUBPH. VERTEBRATA: Possess vertebral column & skull. Fishes at high tide and birds at low tide are important intertidal predators.

Protostomes, Lophotrochozoa: Triploblastic, bilaterally symmetrical. Most phyla w. spiral, determinate egg cleavage, blastopore becomes mouth; schizocoelous coelom (when present).

PHYLUM BRYOZOA: Moss animals; only wholly colonial phylum; colonies composed of numerous individuals each < 0.5mm & enclosed in chitinous box or tube (zooecium). Suspension feed with specialized crown of ciliated tentacles (lophophore) that surrounds mouth; colonies & individuals hermaphroditic. Abundant but inconspicuous at lower levels on rocky shores, often growing as encrusting sheets under overhangs or in pools.

PHYLUM ANNELIDA: Segmented worms with bodies built on a repeated, modular plan with spacious subdivided coelom.

    CLASS POLYCHAETA: Marine segmented worms; each segment w. lateral lobe-like appendages (parapodia) w. extensible bristles (setae); may be free-crawlers, burrowers, tube-dwellers. Abundant but often inconspicuous in algal holdfasts, pools, or other cryptic microhabitats. Some species of tube-dwellers build extensive reefs of sand-grain tubes on some rocky shores. Mid- to low shore levels.

    CLASS CLITELLATA

        SUBCL. OLIGOCHAETA: Earthworms, sludgeworms, etc; mostly freshwater & damp terrestrial; possess setae but no parapodia; hermaphroditic.

PHYLUM MOLLUSCA: Bodies unsegmented, divided into head, foot, & visceral mass; typically possess shell(s) secreted by mantle, a strap-like toothed feeding structure (radula); main body cavity a hemocoel, but technically a coelomate phylum.

    CLASS GASTROPODA: Marine snails, nudibranchs, terrestrial snails & slugs. Abundant and diverse at all levels on rocky shores.

    CLASS BIVALVIA: Mussels, clams, scallops, etc.; filter feed using expanded gills. Abundant on rocky shores at mid- & low levels.

    CLASS POLYPLACOPHORA: Chitons; 8 shell plates surrounded by tough girdle; grazers on hard substrate. Can be abundant on rocky shores.

    CLASS CEPHALOPODA: Squid & octopus; most highly advanced nervous system of any invertebrate. Octopus very cryptic at low levels on rocky shores.

PHYLUM SIPUNCULA: Peanut worms; bulbous bodies with elongate tubular snout. Small, nestling deposit feeders often abundant in sandy patches under cobbles and boulders.

PHYLUM NEMERTEA: Ribbon worms; elongate thin bodies. Possess tubular proboscis nearly as long as body that can be everted explosively to capture prey. Arguably coelomate: possess a fluid-filled body cavity lined with mesoderm, but it surrounds the proboscis instead of the gut (as is typical for eucoelomates). Abundant but most are small and cryptic on rocky shores.

PHYLUM PLATYHELMINTHES: Flatworms. Bodies very flat and thin; acoelomate; gut lacks anus; complex hermaphroditic reproductive systems. Three of the four classes within the phylum are parasitic but free-living forms can be abundant gliding on undersides of cobble and boulders or in pools. Predators, scavengers.

(Continued)

TABLE 1

(Continued)

CLASS TURBELLARIA: Free-living flatworms (excluding acoel & nemertodermatids); mostly predatory.

Protostomes, Ecdysozoa: Triploblastic, bilaterally symmetrical paracoelomates. Possess an external cuticle or exoskeleton that is periodically molted as individuals grow.

PHYLUM ARTHROPODA: Segmented bodies w. rigid, jointed exoskeleton of chitin reinforced w. protein (terrestrial) or $CaCO_3$ (aquatic); pair jointed appendages per segment; largest phylum (more arthropod spp. than all other phyla together).

SUBPH. CRUSTACEA: Aquatic arthropods; bodies divided into 3 body regions (head, thorax, abdomen); number of segments per region vary; head usually with 2 pr. antennae.

CLASS MAXILLIPODA: Barnacles, copepods, ostracods, and kin. Acorn & gooseneck barnacles are abundant & ecologically important on rocky shores; benthic & planktonic copepods are also abundant but inconspicuous (small size).

CLASS MALACOSTRACA: Higher crustaceans; most with 8 thoracic & 6 abdominal segments.

SUPERORDER PERACARIDA: Isopods, amphipods, etc.; brood pouch under thorax. In tufts of algae, under cobbles and boulders, in holdfasts. Abundant and diverse at most levels on rocky shores, but small in size.

SUPERORDER EUCARIDA: Decapods (crabs, shrimp, hermit crabs, porcelain crabs, etc) and planktonic euphausids (krill); brood eggs under abdomen. Abundant, but often cryptic at low tide on rocky shores.

SUBPH. HEXAPODA: Insects and kin; head w. 1 pr. antennae, insects w. thorax of 3 segments; abdomen w. 11 segments (no appendages); abundant a low tide on rocky shores (larvae of some flies are parasites on adult barnacles).

SUBPH. CHELICERIFORMES: Bodies w. two regions: cephalothorax (=prosoma) & abdomen (=opisthosoma); most are terrestrial (scorpion-like arachnids are some of the oldest fossils of terrestrial animals), but ancestors were marine.

CLASS PYCNOGONIDA: Sea spiders. Enigmatic marine arthropods. Inconspicuous, but can be common, living on hydroid colonies at low levels on shore & in low tide pools.

CLASS CHELICERATA: First head appendages fang-like chelicerae followed by pincer-like pedipalps; no jaws; 4 pr. walking legs.

SUBCL. MEROSTOMATA: Largely extinct except for horseshoe crabs (*Limulus*).

SUBCL. ARACHNIDA: Spiders, scorpions, mites, ticks, daddy longlegs, etc.

PHYLUM NEMATODA: Round worms; extremely abundant & diverse; free-living & parasitic. Most are microscopic, but very numerous on rocky shores.

NOTE: Within each phylum, only those classes that are likely to occur in intertidal habitats are listed. Shore levels refer to vertical height on the shore at low tide.

complex life cycles and multiple hosts. The simple structure of myxozoans is also thought to be derived secondarily from a more complex ancestor. Myxozoans have often been allied with hydrozoan cnidarians, but recent evidence suggests that they may be descended from a more advanced, but as of yet unidentified bilateral ancestor.

All the other animal phyla (Fig. 1, Table 2) possess true tissues: integrated groups of cells organized into a functional unit. These units may be a sheet of cells held together by an extracellular basement membrane, a matrix of cells organized into connective tissue, a bundle or sheet of muscle cells, or elements of the nervous system. Organs are still more highly organized units, usually composed of more than one type of tissue. As mentioned previously, the complexity of embryonic tissue (or germ) layers separates the lower non-sponge invertebrates from the other, more complex phyla. To see how this is so, we must delve briefly into early embryonic development.

After fertilization, cell division typically produces a hollow, fluid-filled ball of cells called a blastula. The enclosed fluid-filled space is the blastocoel. The next stage in development is gastrulation, which produces the embryonic gut (or archenteron). Gastrulation occurs by a variety of mechanisms but frequently produces a saclike gut, with just a single opening to the exterior called the blastopore. Gastrulation also establishes the first two germ layers: the outer, enclosing ectoderm and internal endoderm. Ectoderm eventually gives rise to epidermis, the nervous system, sensory organs, some types of excretory organs, and other external structures. Endoderm gives rise to the digestive tract and other structures involved with nutrition. The third and final germ layer to develop is mesoderm, which lies between the ectoderm and endoderm. Typically, mesoderm does not form until later in development. Of the three embryonic germ layers, mesoderm varies most in its degree of development among invertebrate phyla, but it gives rise to important structures such as muscles, some types of body cavities, and various internal organs, especially gonads.

The complexity of these germ layers differentiates the lower invertebrates from the deuterostomes, ecdysozoans, and lophotrochozoans. In the lower nonsponge invertebrates (cnidarians and ctenophores), the layer of cells in between the ectoderm and endoderm is not considered to be a true mesoderm. It is relatively poorly organized and varies a great deal in its ratio of cellular to extracellular material. As such,

TABLE 2

Summary of Some Basic Characteristics of the Invertebrate Phyla

Phylum (Number of Living Species)	Tissue Layers[a]	Body symmetry[b]	Body Cavity (Type, Mode of Formation)[c]	Gut[d]	Circulatory System[e]	Primary Skeletal Support[f]	Reproductive Mode[g]	Egg Cleavage[h] (Mode, Cell Fate)	Fate of Blastopore[i]	Typical Larval Type[j]
Porifera (5500)	—	obscure	acoelomate	none	none	spicules, spongin fibers	hermaphroditic, dioecious, asexual	radial, indeterminate	—	various
Placozoa (1)	—	obscure	acoelomate	none	none	none	asexual	radial?, ?	—	?
Cnidaria (10,000)	2	radial	acoelomate	no anus	gut	hydrostatic	dioecious, hermaphroditic, asexual	radial, indeterminate	(mouth)	planula
Ctenophora (100)	2	radial	acoelomate	mouth, anus	gut	gelatinous mesenchyme	hermaphroditic, (dioecious)	idiosyncratic, determinate	(mouth)	cydippid, but ±direct
Orthonectida (20)	—	bilateral	acoelomate	none	none	none	asexual, dioecious	mod. spiral, ?	—	various
Rhombozoa (70)	—	bilateral	acoelomate	none	none	none	dioecious, asexual	spiral, ?	—	various
Myxozoa (1200)	—	bilateral?	acoelomate	none	none	none	—	—	—	—
Acoelomorpha (210)	3	bilateral	paracoelomate (functionally acoelomate)	no anus	none	none	hermaphroditic	idiosyncratic, determinate	(mouth)	direct
Deuterostomes										
Hemichordata (100)	3	bilateral	coelomate, enterocoelous	mouth, anus	open	hydrostatic	dioecious, asexual	radial, indeterminate	(anus)	tornaria[j]
Echinodermata (7000)	3	radial (2°)	coelomate, enterocoelous	mouth, anus	water vascular system, hemal system, and coelom	endoskeletal ossicles	dioecious, (asexual)	radial, indeterminate	anus	dipleurula[j]
Chordata (49,500) **Urochordata** (3000)	3	bilateral	coelomate (urochordates functionally acoelomate)	mouth, anus	open	cellulose tunic (urochordates)	hermaphroditic, asexual	radial, determinate	(anus)	tadpole
Protostomes: Lophotrochozoa										
Chaetognatha (150)	3	bilateral	coelomate?, mod. enterocoelous	mouth, anus	open	hydrostatic with cuticle	hermaphroditic	radial, ?	(anus)	direct
Rotifera (1800)	3	bilateral	paracoelomate	mouth, anus	none	lamina	parthenogenetic, dioecious	mod. spiral, determinate	(mouth)	direct

(*Continued*)

TABLE 2
(Continued)

Phylum (Number of Living Species)	Tissue Layers[a]	Body Symmetry[b]	Body Cavity (Type, Mode of Formation)[c]	Gut[d]	Circulatory System[e]	Primary Skeletal Support[f]	Reproductive Mode[g]	Egg Cleavage[h] (Mode, Cell Fate)	Fate of Blastopore[i]	Typical Larval Type
Acanthocephala (1100)	3	bilateral	paracoelomate	none	none	lamina	dioecious	mod. spiral, determinate	no gut	acanthor
Gnathostomulida (80)	3	bilateral	paracoelomate? (functionally acoelomate)	no anus	none	none	hermaphroditic	spiral, ?	?	direct
Entoprocta (150)	3	bilateral	paracoelomate (functionally acoelomate)	u-shaped	none	cuticle (calyx)	hermaphroditic, dioecious?, Asexual	spiral, determinate	(mouth)	trochophore-like
Cyclophora (1)	3	bilateral	paracoelomate? (functionally acoelomate)	u-shaped	none	cuticle	dioecious, asexual	?; ?	?	trochophore-like
Gastrotricha (450)	3	bilateral	paracoelomate	mouth, anus	none	hydrostatic	hermaphroditic, parthenogenetic	mod. radial, determinate	neither	direct
Platyhelminthes (20,000)	3	bilateral	paracoelomate (functionally acoelomate)	no anus	none	hydrostatic	hermaphroditic, (dioecious)	spiral, determinate	(mouth)	müllers
Nemertea (900)	3	bilateral	coelomate, schizocoelous	mouth, anus	closed	hydrostatic	dioecious, hermaphroditic, (asexual)	spiral, determinate	(mouth)	~direct (pilidium)
Sipuncula (320)	3	bilateral	coelomate, schizocoelous	u-shaped	none	hydrostatic, NS cuticle	dioecious, (asexual)	spiral, determinate	mouth	trochophore, pelagosphera
Annelida (16,500)	3	bilateral	coelomate, schizocoelous	mouth, anus	closed	hydrostatic, NS cuticle	dioecious, asexual, hermaphroditic	spiral, determinate	mouth	trochophore
Echiura (135)	3	bilateral	coelomate, schizocoelous	mouth, anus	closed	hydrostatic, NS cuticle	dioecious	spiral, determinate	mouth	trochophore
Mollusca (95,000)	3	bilateral	coelomate, schizocoelous	mouth, anus	hemocoel	shell and hydrostatic	dioecious, hermaphroditic	spiral, determinate	mouth	trochophore, veliger cyphonautes[j]
Bryozoa (4500)	3	bilateral	coelomate, mode unclear	u-shaped	none	exoskeletal zooecium	hermaphroditic, asexual	radial, indeterminate	—	"lobate"
Brachiopoda (335)	3	bilateral	coelomate, mod. enterocoelous coelomate, mod.	mouth, anus or no anus	open	shell	dioecious	radial, indeterminate	(mouth)	"lobate"
Phoronida (20)	3	bilateral	coelomate, schizocoelous	u-shaped	closed	hydrostatic (with external tube)	dioecious, hermaphroditic, (asexual)	biradial, indeterminate	mouth	actinotroch
Protostomes: Ecdysozoa										
Kinorhyncha (150)	3	bilateral	paracoelomate (functionally acoelomate)	mouth, anus	none	jointed cuticle	dioecious	?; ?	?	direct

Loricifera (10)	3	bilateral	paracoelomate	mouth, anus	none	exoskeletal lorica	dioecious	?, ?	?	higgins
Priapulida (16)	3	bilateral	paracoelomate?	mouth, anus	none	hydrostatic, NS cuticle	dioecious	radial, indeterminate	not mouth	unnamed
Nematomorpha (325)	3	bilateral	paracoelomate	reduced or none	none	thick cuticle	dioecious	idiosyncratic, ?	(anus)	parasitic
Nematoda (25,000)	3	bilateral	paracoelomate	mouth, anus	none	thick cuticle	dioecious	idiosyncratic, determinate	mouth, (anus)	direct
Onychophora (110)	3	bilateral	paracoelomate	mouth, anus	hemocoel	hydrostatic, NS cuticle	dioecious	idiosyncratic, ?	mouth, anus	direct
Tardigrada (800)	3	bilateral	paracoelomate	mouth, anus	hemocoel	jointed cuticle	dioecious	idiosyncratic, ?	neither	direct
Arthropoda ($>10^6$)	3	bilateral	paracoelomate	mouth, anus	hemocoel	jointed exoskeleton	dioecious, hermaphroditic, parthenogenetic	idiosyncratic (mod. radial?), determinate	neither	nauplius, zoea, etc.

NOTE: Phyla in **boldface** are likely to be present on rocky shores (some are inconspicuous or mostly microscopic); "?" = not known; "—" indicates that a character is not applicable to that phylum.

[a]Porifera (sponges) have an outer layer of cells (pinacoderm) and a layer of feeding cells (choanoderm) that line portions of internal canals, but both lack basement membranes. Placozoans have upper and lower cell layers considered equivalent to ectoderm and endoderm, but both lack basement membranes. In ctenophores, muscle fibers develop within the mesenchyme (middle layer), making the phylum technically triploblastic. Rhombozoans and orthonectids have two cell layers but lack basement membranes. Most life cycle stages of myxozoans are unicellular, and the multicellular spore stage lacks clear tissue layers.

[b]Technically, many cnidarians and ctenophores are biradial or even bilateral when examined in detail. There are few if any purely radial phyla, but the distinction between radial and bilateral is still meaningful at the gross anatomical/functional level.

[c]"mod." = not typical for category; "?" = status unclear. In urochordates (tunicates), fluid-filled body cavity is secondarily reduced or lost, but the ancestral condition is presumed to be coelomate, since other chordates are eucoelomate. In molluscs the coelom is reduced to the fluid-filled pericardial cavity surrounding heart (as well as the intestine), and the main body cavity of noncephalopod molluscs is a hemocoel (cephalopods have an atypically spacious coelom). In bryozoans, drastic rearrangement of tissues at metamorphosis obscures mode of coelom formation. In arthropods, tardigrades, and onycophorans, mesodermally lined schizocoelous coelomic spaces (enterocoelous in tardigrades) appear during development, but all eventually either disappear, are restricted to specific organs (gonads or excretory ducts), or fuse with pseudocoelomic hemocoel.

[d]"None" = gut absent in adults. Porifera have a system of internal canals (aquiferous system) through which external seawater is pumped; microscopic food particles are captured by flagellated collar cells (choanocytes); digestion is intracellular. In articulate brachiopods U-shaped gut lacks anus, but anus is present in inarticulate brachiopods.

[e]"Hemocoel" = open system with very spacious sinuses derived from blastocoel. "Closed" = circulation (mostly) in enclosed vessels. "Open" = circulation through spacious sinuses of various origins (some enclosed vessels may also be present). "None" = no separate circulatory system, usually associated with small body size. In Porifera aquiferous systems (see note d) provides internal circulation. In echinoderms, circulation occurs by combination of hemal system (spongy vessels, derived from coelom, through which fluid is circulated mostly by ciliary action), water vascular system (hydraulic system that drives tube feet, derived from coelom), and coelomic cavity. The complexity of the hemal system varies among different classes and is most well-developed in holothuroid sea cucumbers. Cephalopod molluscs possess a closed circulatory system, while all other molluscs have an open system.

[f]"Cuticle" = extra-epidermal cuticle that provides some degree of body support; "NS cuticle" = extra-epidermal cuticle, usually nonsupporting; "None" = skeletal support absent (mostly due to small body size). In cnidarians other skeletal types include gelatinous mesenchyme between ecto- and endoderm (various medusae), chitin-protein exoskeletal perisarc (colonial hydroids), CaCO$_3$ exoskeleton (e.g., corals), proteinaceous endoskeleton (e.g., gorgonians, black corals). In rotifers and acanthocephalans the skeletal lamina is a unique intraepidermal skeletal structure.

[g]Parentheses indicate that a mode occurs relatively rarely within the phylum. In Placozoa, asexual modes include fission and budding (producing multicellular, flagellated swarmers); sexual reproduction also occurs but is poorly known. Adult orthonectids are free-living, but juveniles are parasitic; they possess complex life cycles in which asexual modes predominate but sexual reproduction also present. Rhombozoans are obligate symbionts (parasites or just commensals?) of cephalopod kidneys and possess complex life cycles with alternating sexual and several asexual phases. Myxozoans are parasites with complex life cycles; most stages are unicellular, but the infective stage is a multicellular spore that encloses gametic cells. The phylogenetically enigmatic, bilateral, wormlike *Buddenbrockia* now appears to be a myxozoan, strengthening the myxozoans' ties to the Bilateria rather than the Cnidaria. Cycliophorans have a complex, multistage life cycle, including dwarf males and sexual and asexual phases.

[h]"Mod." = highly modified from typical pattern; "?" = not known.

[i]Parentheses indicate that blastopore closes but mouth/anus opens nearby. "?" = not known. "Neither" = mouth and anus open secondarily at sites remote from site of blastopore, or blastopore closes, or gastrulation occurs by method other than invagination. In bryozoans, drastic rearrangement of tissue at metamorphosis obscures fate of blastopore. In nematodes, blastopore elongates and closes in the middle with mouth and anus forming at either end. In some onychophorans blastopore elongates and closes in the middle, with mouth and anus forming at either end; other onycophorans have large yolky eggs with superficial cleavage; only one onychophoran species lays eggs (oviparous); all others are viviparous (young born alive).

[j]Tornaria of hemichordates is a dipleurula-type larva (pterobranch hemichordates have a brooded larva that differs from a tornaria). Echinoderm classes each have distinctive larvae; early stage of all is a dipleurula-type, but later stages are brachiolaria for sea stars, ophiopluteus for brittle stars, echinopluteus for echinoids, and a doliolaria for sea cucumbers. In bryozoans the cyphonautes is a feeding larva typical of broadcasters, but most groups brood and release ciliated, nonfeeding larvae that vary in form from group to group. Priapulid larvae resemble an adult loriciferan.

these phyla are referred to as diploblastic or two-layered. The deuterostomes, ecdysozoans, and lophotrochozoans all have well-defined mesodermal tissues and are referred to as triploblastic or three-layered (Table 2).

Body Symmetry

Another characteristic that distinguishes the lower invertebrates from the higher is body symmetry. The two basic types of body symmetry are radial and bilateral. Bodies of radially symmetrical animals can be divided into two roughly equal halves in many ways by planes running through and parallel to their central longitudinal axis. The lower invertebrates, including cnidarians and ctenophores, are functionally radial (but see the footnote in Table 2).

Most of the higher invertebrate phyla are bilaterally symmetrical. Bilateral bodies can be divided into two equal halves in only one way, usually by a vertical plane running down the longitudinal axis and through the dorsal and ventral midlines. This type of symmetry is typical of motile organisms. Bilaterality juxtaposes parts of the body such that locomotion becomes more effective. Obviously, the ability to move about and hunt for food and mates has advantages over sitting and waiting for things to come to one or drifting randomly at the mercy of one's surroundings. In keeping with this greater motility, bilateral symmetry is usually accompanied by elaboration and concentration of sensory organs and nervous tissue at the functionally anterior end of the body (cephalization).

All the bilaterally symmetrical invertebrates (deuterostomes, ecdysozoans, lophotrochozoans) are often referred to collectively as the Bilateria. The nature of the ancestral bilaterian is unclear at present. Pre-Cambrian trace fossils have been interpreted as small-bodied, shallow burrowers, suggesting a paracoelomate body plan (see subsequent discussion of body cavities). The most basal living group of bilaterians appears now to be the Acoelomorpha (Fig. 1), comprising the most primitive groups, formerly considered within the platyhelminth flatworms but now apparently a separate and considerably less-derived phylum.

Body symmetry is also associated with mode of existence, regardless of taxonomic relationship. Several bilaterian groups such as bryozoans, tunicates, or tube-dwelling polychaete annelids have secondarily assumed functional radial symmetry in association with a sessile existence. Radial symmetry is advantageous for sessile organisms in that it allows them to face all points of the compass at once. Despite being functionally radial, most of these groups reveal their true bilateral nature on close examination. Echinoderms are unusual among bilaterians in being both motile and radially symmetrical. The

apparent contradiction is resolved once it is realized that echinoderms passed through a stage in their evolutionary history where they were sessile (and radial), then resumed a motile existence later.

Body Size

In addition to tissue layers and body symmetry, a third trend that arose as animals diversified through the Cambrian is an increase in body size. A larger body has obvious advantages. An organism with a larger body can eat larger things (and also can be eaten by fewer things). It can also produce more gametes, since a larger body supports more reproductive tissue. However, larger body size also raises two serious challenges: adequacy of internal transport (of oxygen, nutrients, wastes) and efficiency of locomotion.

Simple invertebrates have a more-or-less solid body, a condition referred to as acoelomate (that is, lacking a body cavity). Small-bodied acoelomates can do perfectly well using simple diffusion for exchange and transport of oxygen and wastes. However, as body size increases, the body's internal volume increases disproportionately faster than its surface area, rendering diffusion inadequate to maintain cells deep within the body. This in turn helps to limit the maximum size of acoelomate organisms. Options for locomotion are also limited with a solid body, placing additional constraints on body size of acoelomates. Although many acoelomate groups are sessile as well as small-bodied, motile acoelomates move about by gliding on mucus sheets using cilia or by simple muscular undulation of the entire body. As body size increases, these modes of locomotion become less effective.

All the lower invertebrates (sponges, placozoans, cnidarians, ctenophores) are acoelomate, as are some of the higher groups, notably the flatworms (Acoelomorpha and Platyhelminthes, Table 2). The only fluid-filled internal space that these groups possess is their gut. Still, acoelomates have several adaptations that permit larger body sizes. Acoelomate sea anemones, for instance, have a large gut cavity whose interior wall is folded to form radially arranged sheetlike longitudinal mesenteries. These mesenteries increase internal surface area and facilitate diffusion. Their gut cavity thus serves double duty, both as a digestive tract and a simple circulatory system. Jellyfish and ctenophores evolved another strategy. They support larger bodies by packing gelatinous extracellular material between the ectoderm and endoderm. This gelatinous layer harbors only a sparse population of cells and keeps total metabolic demand low relative to body volume. An extensive, branched digestive tract supports cells within the gelatinous layer. Yet other acoelomates such as flatworms have grown in only two dimensions, keeping their bodies thin and flat so that diffusion is still adequate. The larger flatworms also have a highly branched gut that ramifies throughout the body, facilitating internal transport of food and wastes. Still, even with these adaptations, acoelomate body size is ultimately limited, and the majority of acoelomate animals are small-bodied or even microscopic.

Another adaptation that may facilitate increased body size is the addition of an anus to the digestive tract. Such an innovation permits one-way passage of food through the gut as well as regional specialization. In an animal equipped with both a mouth and an anus, food capture, ingestion, preliminary breakdown, digestion, absorption, and waste formation can all take place serially along the length of the gut. So a one-way linear gut promotes more efficient energy processing, but it does little to solve the problem of internal transport and distribution.

Body Cavities: Hydrostatic Skeletons, Coeloms, and Pseudocoeloms

A far more widespread solution to the problems of internal transport and more efficient locomotion is a fluid-filled body cavity between the outer body wall and the gut. Such a space acts as a primitive circulatory system, permitting a larger body without any cell being too far away from either the exterior or the sluggishly circulating internal body fluids. In addition, suspending the gut in this internal cavity frees it from direct attachment to the body wall, permitting peristaltic transport and greater digestive efficiency. With a more or less solid body, food can only be pushed through the gut by ciliary action. Lastly, a fluid-filled body cavity can be used to store gametes.

A spacious body cavity also enhances locomotion by providing a hydrostatic skeleton. A hydrostatic skeleton comprises layers of circular and longitudinal muscles in the body wall that work against an incompressible, fluid-filled internal space to change the body's shape. If the circular muscles contract while the longitudinal muscles relax, the body becomes long and thin. Alternatively, if longitudinal muscles contract and circular muscles relax, the body becomes short and fat. By selective contraction and relaxation of different muscles in different parts of the body, an elaborate repertoire of movements can be generated from a relatively simple set of structures. New forms of locomotion such as peristaltic burrowing, more efficient crawling, and even swimming are all augmented by a hydrostatic skeleton. Comparable changes in shape are much more limited with a solid body.

Many different invertebrates use a hydrostatic skeleton for body support and locomotion, including sea anemones, burrowing polychaetes, and crawling sea cucumbers (Table 2). The hydrostatic skeleton probably evolved independently multiple times, wherever a fluid-filled space surrounded by opposing muscles was present. Although more advanced groups have abandoned it in favor of more durable or functionally sophisticated skeletons such as the exoskeleton of arthropods or the endoskeleton of vertebrates, the hydrostatic skeleton was probably the earliest evolutionary solution to the problem of how to support a larger body.

So a fluid-filled body cavity permits larger body size by solving two potentially limiting problems: internal transport and locomotion. But where do body cavities come from? They are of two basic types: the coelom or the pseudocoelom. Although the two types function in a similar manner, they differ a great deal in how they arise during embryonic development. This in turn has been used to infer phylogenetic relationships among the phyla that possess each type.

A pseudocoelom is essentially a blastocoel that persists into adulthood (Fig. 2). Recall that in a developing

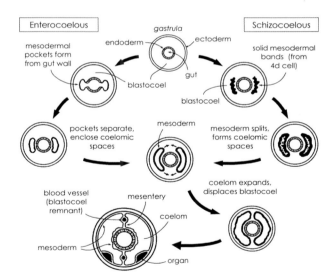

FIGURE 2 Transverse cross-sections of a developing embryo illustrating the difference between enterocoelous (left) and schizocoelous (right) coelom formation. Both schemes begin with a gastrula whose structure is similar (top center): ectoderm surrounding a fluid-filled blastocoel with the blind-ended embryonic gut running through it. Origin of mesoderm and formation of coelomic cavities differ between the enterocoely and schizocoely, but the final outcome is similar (bottom, left center). In pseudocoelomates, the structure of the gastrula (top center) persists, maintaining the blastocoel as the body cavity and mesoderm (mostly musculature) developing on the inner side of the ectoderm and, to a lesser extent, around the wall of the gut.

embryo, the blastocoel is the fluid-filled space that lies between the outer ectoderm and the embryonic gut. The blastocoel is thus the primary body cavity of most all animals, as it is the first fluid-filled space to appear in an embryo. As development proceeds in pseudocoelomate phyla, this fluid-filled space persists. Mesoderm arises from ectoderm, forming musculature of the body wall and other organs (if present). Other mesodermal tissues may also arise from endoderm and form an incomplete layer around the tubular gut. The most important, if somewhat obscure, distinction is that a pseudocoelom is *not* fully lined with mesodermal tissue. Even more obscure, but significant, is that the immediate lining of the body cavity is not cellular epithelium, but rather the extracellular basement membrane that holds the mesodermal layers of the body wall together.

The Ecdysozoa and roughly half the lophotrochozoan phyla are predominantly pseudocoelomate (Table 2). Many of these phyla have bodies in the millimeter or submillimeter size range (e.g., most nematodes, kinorhynchs, rotifers, gastrotrichs). Arthropods have larger bodies and a spacious pseudocoelom that functions as a blood circulatory space, but in other phyla the pseudocoelom may be reduced to only small spaces or filled in entirely by cells or organs. The latter groups are functionally acoelomate, but still more complex than sponges or cnidarians. Because the presence of a spacious pseudocoelom varies so greatly amongst adults of these phyla even though they are all of a comparable grade of complexity, the term *paracoelomate* has been suggested as a general, all-encompassing term for this grade of organization.

In arthropods (the dominant ecdysozoan phylum), the pseudocoelom develops into large sinuses through which blood is circulated. This type of spacious circulatory pseudocoelom is referred to as a hemocoel. How the hemocoel develops varies considerably from group to group within the arthropods. In some, transient spaces open up from within solid mesodermal bands but disappear before adulthood. In others these spaces eventually fuse with the hemocoel or persist within excretory or reproductive organs. But in all arthropods, the principal body cavity is the hemocoel, traceable to the blastocoel rather than arising wholly from within mesodermal tissue.

In contrast to a pseudocoelom, a true coelom (or eucoelom) is a fluid-filled body cavity that arises within the mesoderm itself. It is thus a wholly new, secondarily derived body cavity, completely lined with mesodermal epithelium. It expands into the blastocoel during development, eventually reducing the blastocoel to the circulatory spaces contained within blood vessels or slightly more spacious sinuses (Fig. 2). A eucoelomate body cavity is fundamentally more derived than a pseudocoelomate body cavity.

A coelom forms in one of two ways (Fig. 2). In deuterostomes (e.g., echinoderms, hemichordates), coelom formation is enterocoelous. That is, mesoderm typically forms as lateral outpockets of the embryonic gut. During development these small pockets separate from the gut, forming wholly new, fluid-filled vesicles completely surrounded by mesoderm and lying within the blastocoel. The mesoderm proliferates, expanding these new cavities like water balloons filling up inside a fluid-filled barrel. Eventually these growing vesicles encircle the gut from either side with two fluid-filled cavities. Where the two compartments meet along the embryo's midline dorsal and ventral to the gut, a vertical mesentery forms and suspends the gut within the coelom. The blastocoel is reduced to vessels or sinuses of the blood circulatory system located around the gut, against the body wall, or suspended in the mesenteries between two layers of mesoderm.

In lophotrochozoans (annelids, molluscs, and several smaller phyla), coelom formation is schizocoelous (Fig. 2). That is, instead of originating as outpockets of the embryonic gut, the mesoderm forms as solid bundles of cells located on each side of the gut. These cells are not derived from endoderm tissue but instead are the descendants of a single specific cell, known as the 4d blastomere or mesentoblast, that can be identified very early in development. The bundles of mesoderm are initially solid masses of cells aligned on either side of the gut, but as they grow, they hollow out and expand to encircle the gut, forming a fluid-filled space lined completely with mesodermal epithelium, with the gut suspended by vertically oriented dorsal and ventral mesenteries (Fig. 2). Thus, the end result looks just the same as that of enterocoelous coelom formation, but the origin of the mesoderm and the pattern of development are quite different. As with enterocoelous coeloms, the only remnants of the blastocoel are spaces within blood vessels around the gut, against the body wall, or suspended in the mesenteries (Fig. 2).

In the past, the presence and type of body cavity was thought to reflect phylogenetic history, with radial acoelomates being most primitive, followed by bilateral acoelomates (flatworms and nemerteans), then paracoelomates, and finally the two major groups of coelomate phyla (deuterostomes and protostomes). A great deal of debate centered on how enterocoelous and schizocoelous modes of coelom formation could have arisen from a common ancestor. Views have changed a great deal since the late 1980s. Mostly on the basis of DNA sequences, but also as a result of reexamination of morphological characters, the paracoelomate phyla and most of the more advanced acoelomate flatworms now appear to be distributed among the protostomes (Lophotrochozoa and Ecdysozoa; Fig. 1, Table 2). Some are grouped with the arthropods in the Ecdysozoa and the others with the annelids and molluscs in the Lophotrochozoa. Enterocoelous and schizocoelous coeloms probably evolved independently in the deuterostome and lophotrochozoan lines, respectively. In addition, recent evidence suggests that at least some of the small-bodied paracoelomates may have evolved from coelomate ancestors, having lost their coelom or its mesodermal lining during evolution to smaller body size. So, our current view of invertebrate phylogeny encompasses a more convoluted set of relationships than the stately transitions from acoelomate to pseudocoelomate to eucoelomate as previously thought.

In summary, then, the three major bilaterian groups can be distinguished in part by the nature of their body cavities: deuterostomes (echinoderms, hemichordates, and chordates) are enterocoelous eucoelomates, lophotrochozoans (annelids, molluscs, etc.) are schizocoelous eucoelomates, and ecdysozoans (arthropods, nematodes, and kin) are paracoelomates.

Other Distinguishing Characteristics

In addition to differences in their body cavities, other characteristics that unite deuterostomes, lophotrochozoans, and ecdysozoans include developmental features such as mode of egg cleavage, fate of the blastopore, and type of larvae (Table 2, Fig. 3).

Egg cleavage may be radial or spiral. The distinction refers to the orientation of daughter cells to one another during early development (Fig. 3A). Deuterostome zygotes exhibit radial cleavage, in which the mitotic spindles are aligned parallel to or at right angles to the main axis of the embryo. The resulting daughter cells are stacked directly above or below one another. Radial cleavage is thought to be the ancestral mode of egg cleavage among animals, because cnidarians and ctenophores also exhibit radial cleavage. Coelomate lophotrochozoans possess spiral cleavage, in which the mitotic spindles are oriented at an oblique angle to the embryo's axis, producing a spiral arrangement of daughter cells. Each daughter cell is nestled between the daughter cells above or below it (Fig. 3A). Egg cleavage of ecdysozoans varies from radial to idiosyncratic (i.e., neither classically radial or spiral).

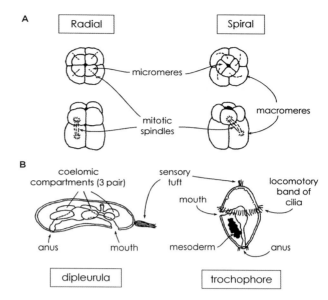

FIGURE 3 Embryonic cleavage and larval types. (A) Radial and spiral cleavage at the 8-cell stage viewed from top (polar) and lateral sides. (B) Dipleurula (left) and trochophore (right) larvae. The dipleurula is typical of deuterostomes, especially echinoderms and hemichordates. The top-shaped trochophore is typical of many coelomate protostomous lophotrochozoans such as annelids, primitive molluscs, sipunculids, and echiuroids.

Arthropod cleavage is often obscured by large amounts of yolk and is traditionally described as a highly modified spiral type. However, recent analyses suggest that the ancestral mode of arthropod cleavage is radial. Nematodes, the other major ecdysozoan phylum, have a pattern of cleavage that is unique, differing greatly from that of lophotrochozoans, deuterostomes, or other ecdysozoans (Table 2).

In addition to differences in the orientation of daughter cells in typical deuterostome and protostome embryos, the timing by which the fate of cells is determined also differs (Table 2). In deuterostomes, many cells remain relatively undifferentiated until later in development. If the cells of a deuterostome embryo such as a sea urchin are separated at the four-cell stage, each cell can develop into a whole, albeit smaller, embryo. This type of cleavage is called indeterminate. In contrast, the fate of cells in a coelomate lophotrochozoan embryo is determined very early in development. If an annelid embryo is divided at the four-cell stage, none of the cells can form a complete larva and all will eventually die. As early as the fifth or sixth round of cell division, certain cells can be identified that will give rise to specific tissues or structures in the larva. This type of cleavage is called determinate or mosaic. Cleavage among ecdysozoans also tends to be determinate, but the fates of specific cells during early cleavage are not the same as in coelomate lophotrochozoans.

In deuterostomes the blastopore (the first external opening to the embryonic gut) becomes the anus and the mouth opens secondarily at another location later in development. The opposite is true for most coelomate lophotrochozoans: the blastopore becomes the mouth and the anus opens secondarily (Table 2). In ecdysozoan arthropods, the blastopore closes and both mouth and anus open later at some distance from the site of the blastopore. In nematodes, the blastopore is often elongate and closes at its center. The anterior portion subsequently becomes the mouth. The posterior portion closes, but the anus opens later at the same site as the closed posterior end of the blastopore (Table 2).

With regard to larval types, a top-shaped trochophore larva, with a conspicuous equatorial band of locomotory cilia, is the primitive larval form of molluscs, annelids, and several other small, wormlike coelomate lophotrochozoans (Fig. 3B). Noncoelomate lophotrochozoans have a wide variety of different larval types that are difficult to generalize (Table 2). Larvae of ecdysozoans likewise vary a great deal. Arthropod larvae themselves range from a variety of swimming, feeding larvae of crustaceans to wormlike caterpillar larvae of insects to direct development without a larval stage at all. Nonparasitic nematodes have direct development and lack a distinct larval phase. The dipleurula larva is typical of deuterostomes. It has three pairs of enterocoelous coelomic compartments and extensive bands of locomotory cilia (Fig. 3B). The details of its shape and pattern of ciliation differ between echinoderms and hemichordates. The term was also formerly used for a hypothetical composite adult ancestor of echinoderms and hemichordates.

Other differences between deuterostomes, lophotrochozoans, and ecdysozoans are not restricted to embryology. Evolution of photoreceptors and surface ciliation differ between the groups. Many lophotrochozoans and ecdysozoans use chitin, a nitrogen-containing polysaccharide, as a structural material (e.g., the bristle-like setae of polychaetes and the exoskeleton of arthropods). In contrast, chitin is found rarely among the deuterostomes. Lophotrochozoans and ecdysozoans typically use arginine phosphate as their primary means of storing phosphates, while creatine phosphate is more typical of deuterostomes. Lastly, DNA sequences separate deuterostomes, lophotrochozoans, and ecdysozoans as distinct phylogenetic groups.

As mentioned at the beginning of this section, these suites of characters are generally consistent, but not every species whose phylum fits one category or another necessarily possesses *all* the requisite features. Exceptions can be found easily within every phylum, especially with regard to developmental features. Simple differences such as the amount of yolk in an egg or the degree of development that occurs before hatching can alter the expression of many of these distinguishing features. These generalizations oversimplify the vast diversity and biological richness in the way that invertebrates body plans arise. The point is that, when considered all together, the distribution of these features is sufficiently consistent to distinguish the deuterostomes, lophotrochozoans, and ecdysozoans as major branches of the tree of higher animal life on earth (Table 2).

INVERTEBRATE ECOLOGY

All organisms must meet a common set of fundamental challenges in order to survive and grow. First, they must obtain and process energy. They must find, handle, ingest, and digest food, then respire and dispose of metabolic wastes. Second, they must find a suitable place to live and deal with physiological constraints imposed by the physical environment and biological constraints imposed by ecological interactions with other organisms (competition, predation, parasitism, etc). Lastly, they must reproduce successfully. Although this list is deceptively short, the myriad ways that invertebrates accomplish these tasks are as diverse as the animal kingdom.

Feeding

Invertebrates obtain food in many ways that nevertheless fall into a few broad categories or modes. They can be predator/grazers, deposit feeders, or suspension feeders. Each mode requires a suite of specialized structures and functions.

Predators and grazers represent the most familiar mode of feeding. These animals (for the most part) actively seek out their food. A sea star captures prey with its tube feet, then turns its stomach inside out through its mouth and slides it into its prey, digesting it in place. Cone snails use poisoned harpoons to subdue prey as big as small fish, then swallow them whole. Sea anemones and corals capture prey using thousands of microscopic stinging cells that inject toxins and fasten prey to their tentacles. Snails, limpets, nudibranchs, and chitons rasp small pieces out of their food using a tonguelike strap studded with transverse rows of renewable teeth. Many different groups, such as sea urchins, cephalopods, crustaceans, and insects, have hardened jaws that tear and dismember prey. Some invertebrate consumers search diligently for cryptic prey.

Others, such as octopus, use stealth and camouflage to stalk their prey. Still others lie hidden in waiting, ambushing unsuspecting prey that venture too near. Many invertebrate predators also have distinct food preferences and employ elaborate foraging strategies to fulfill their diets. Some spiders can even alter their diet to balance fats and protein depending on the nutritional composition of recent meals.

Deposit feeders ingest sediment and particulate detritus that settles to the sea floor, eking out what seems to be a meager living on the small quantities of digestible organic material that the detritus contains. Deposit feeders must ingest a large volume of material in order to meet their energetic needs. A moderate-size sea cucumber swallows and processes 70 kg (150 lbs) of sediment a year in order to feed itself. The low food value of detritus is offset by its ready availability in all parts of the sea. Deposit feeders include a wide variety of polychaete annelid worms (burrowers, nestlers, tube dwellers), myriad crustaceans and gastropods, sea cucumbers, sea stars, brittle stars, and many of the small-bodied paracoelomates. Nearly every motile phylum has species that deposit-feed.

Suspension feeders capture tiny particles suspended in the water. Some feed on larger phytoplankton such as diatoms (~0.1 mm in diameter), while others feed on things as small as bacteria (0.001 mm). Methods of catching particles range from simple passive structures such as mucus-coated tentacles (e.g., some sea cucumbers) or even gossamer nets of mucus alone (e.g., some corals or tube snails) to very elaborate filtration systems with biological pumps, filters, and ciliary conveyor belts (e.g., tunicates or bivalve molluscs). Suspension feeders harvest food nearly continuously in order to meet their energy needs.

Because of their basic body plans, some invertebrate phyla feed using just a single mode, while others have diversified to use several modes. Sponges, bryozoans, brachiopods, phoronids, bivalve molluscs, and tunicates are typically suspension feeders, but even among these very consistent groups, there are rare exceptions, such as a predatory sponge or a predatory tunicate, to take the investigator by surprise. Cnidarians, ctenophores, flatworms, and nemerteans are mostly all predators. Arthropods, gastropod molluscs, and echinoderms are much more varied in the ways they feed. The broad array of specializations that invertebrates use to gather food is truly remarkable and has been a rich area of research.

From the prey's point of view, predation is also a very potent selective force, and marine invertebrates have a wide array of defenses against their predators. Some have evolved heavy armor such as thick exoskeletons or massive shells. Others can detect their predators at a distance and have extremely active escape responses. Even such unlikely groups as bivalves and anemones have species that swim actively to escape their predators. Yet others have developed distasteful secondary chemical compounds that make them unpalatable to their predators. The evolutionary arms race between predators and the defenses of their prey is another well-studied area of invertebrate ecology.

Habitat

In the sea, habitats where invertebrates live are defined by whether the organism swims actively (nekton) or drifts passively (plankton) in the water column above the sea floor, or whether it lives on the sea floor itself (benthic). Benthic dwellers can be divided further into those that live on the surface of the substrate (epibenthic) and those that burrow into soft sediment (infauna). Infaunal organisms are often categorized by size. Meiofauna are the smallest microscopic animals that live interstitially on or among sand grains. As is evident from Fig. 1, many of the less conspicuous paracoelomate ecdysozoans and lophotrochozoans are interstitial.

Many factors affect where an organism can live. Physiological constraints, such as extreme temperature or salinity, and ecological constraints imposed by competition, predation, disease, or availability of food and space are all important. Every species responds differently to fluctuations in these challenges, and no single species is best at everything all the time. The abundance and distribution of each species are set by a complex compromise of numerous processes all acting simultaneously. So the location where a given species is most abundant may not be where it does best in relation to any single factor acting alone, no matter how important we naïvely assume that factor to be. Determining what actually limits a species's distribution requires careful, painstaking experimental research. Unfortunately, a species's ideal habitat cannot be discerned by just walking around and looking to see where the species can be found.

A rich diversity of behaviors is associated with finding and maintaining a place to live. Some rocky-shore limpets are territorial and sometimes bulldoze other limpets off their patch of grazing space. Others learn the lay of the land, crawling off in search of food at high tide, but returning faithfully as the tide drops to a specific site where their shell fits precisely to the shape of the rock. The snug fit helps to minimize evaporative water loss during low tide and enables these limpets to live on exposed rock faces that are intolerable to other species. Even the microscopic larvae of many sessile invertebrates can test and reject possible settlement sites until the right cue is encountered, which then triggers them to metamorphose and fasten themselves down to that site for the rest of their lives. Time and again, research has shown that we should not underestimate the behavioral capabilities of these so-called simple animals.

Symbiosis: Mutualism, Commensalism, and Parasitism

Symbiosis refers to species living in intimate association with one another. It is a strategy by which organisms can secure both habitat and food at the same time. In the broadest sense, symbiotic relationships may be mutualistic (both members benefit from the interaction), commensal (one member benefits and the other is not affected), or parasitic (one member of the interaction is harmed to the benefit of the other). These categories are not discrete, isolated types of interactions, but rather points along a very broad continuum.

Parasitism is widespread among the invertebrates. Some phyla, such as flatworms and nematodes, contain both free-living and parasitic groups. Others, such as the paracoelomate acanthocephalans, are wholly parasitic. Nearly every phylum has at least a few species that make their living as parasites. Some parasitic species have a simple life cycle with only a single host, while others have extremely complex life cycles with one to many intermediate hosts. Parasitic species are often very specific in their choice of host species, in part because of the specializations required to overcome the host's defenses. Parasites are also usually much more fecund than their nonparasitic relatives, since the majority of their offspring will perish before finding and infecting another host. At first glance, the complex parasitic life cycles of some species appear to be unnecessarily baroque. Sexual reproduction can occur only in the final, definitive host, so why have so many more in between? The advantage of adding one to several intermediate hosts, in which only asexual reproduction occurs, is that it permits a great deal of biological amplification of the initial output from sexual reproduction. In some platyhelminth flukes, a single sexually produced larva that infects a secondary host can lead to release of 250,000 larvae of the next stage. This amplified reproductive effort in turn increases the likelihood of eventually reinfecting the definitive host. Complex parasitic life histories do not evolve all at once but are built up incrementally. Natural selection acts to add additional intermediate hosts only when the ecological interactions among those hosts increase the likelihood of successful transmission of larvae from one host to the next and, ultimately, to the definitive host.

Mutualism and commensalism are also widespread among invertebrates. Some of these relationships simply involve two species living in close association. The fat innkeeper worm (an echiuroid) lives in mud flats in a U-shaped burrow, which is also inhabited by a small polychaete scale worm, a clam, a pea crab, and a small fish. The commensals benefit from shelter provided by the worm's burrow but probably have little if any impact on the worm itself. Keyhole limpets, giant gumboot chitons, and sea cucumbers often have a different species of scale worm living on their bodies. The keyhole limpet provides shelter for the worm, and the worm helps to defend its host against predatory sea stars by biting the sea star's tube feet when it attacks. Slipper limpets hitchhike on the shells of larger snails, suspension feeding with their enlarged gill and probably not having much impact on their host at all. A small polychaete annelid lives among the tube feet of bat stars, obtaining shelter and probably scavenging bits of food, again without impacting its host seriously.

Other forms of symbiosis are much more intimate and biochemically complex, such as endosymbiosis of unicellular algae with anemones and corals. Here, unicellular zooxanthellae (a kind of dinoflagellate) live within the tissue of the coral or anemone. There they capture sunlight and carry on photosynthesis, releasing as much as 30% of the carbohydrates they produce to their host. The algae not only obtain a place to live but also obtain important raw materials (carbon and nitrogen) from their host's metabolic by-products. Both partners benefit from this mutualistic arrangement. That this is not an easy or conflict-free interaction is evidenced by coral bleaching, in which corals actively expel their symbionts when they become temperature stressed.

One of the most spectacular examples of invertebrate endosymbiosis is the hydrothermal vent worms or vestimentiferans (formerly in their own phylum but now a family with the polychaete annelids). These large tube worms lack a mouth or gut and instead have a large internal space packed with symbiotic bacteria. Through a series of elaborate adaptations, the worm takes up hydrogen sulfide (an energy-rich but highly toxic compound produced by volcanic activity at these deep-sea vents) from the surrounding water and transports it to its bacterial symbionts. The bacteria then use the energy to produce carbohydrates that support not only the bacteria but the worm host as well. This type of ecosystem is one of the few, if not the only one, on the face of the earth that do not rely ultimately on sunlight for energy. These systems are made possible only by these symbiotic relationships.

Reproduction

Reproduction is a prerequisite for ecological and evolutionary survival. The problems that all organisms face include finding a mate, ensuring successful fertilization, and having sufficient numbers of offspring survive until they reach reproductive age.

In most phyla, individuals are either male or female (dioecious). However, hermaphroditism has arisen many times in many different invertebrate groups (Table 2). Some phyla, such as flatworms or bryozoans, are composed almost entirely of simultaneous hermaphrodites. Others have species that are sequential hermaphrodites, being one sex when they first reach maturity, then changing to the other with increasing age or body size. The largest phyla (arthropods and molluscs) are difficult to categorize with regard to reproductive mode. In the arthropods, most of the terrestrial groups tend to have separate sexes (insects and chelicerates), whereas marine groups vary a great deal (e.g., barnacles are simultaneous hermaphrodites, while most decapods have separate sexes). Amongst the gastropod molluscs (snails and slugs), the more primitive groups are dioecious, but the more derived groups became hermaphroditic (e.g., nudibranchs and terrestrial snails and slugs). As a rule, hermaphroditic species do not self-fertilize but must exchange sperm with a mate. Groups in which self-fertilization probably does occur include ctenophores, bryozoans, and urochordates (tunicates, salps, and larvaceans).

The simplest mode of reproduction of marine invertebrates is just to shed gametes into the sea (broadcast spawning). Fertilization is external, and development proceeds without any further investment from the parents. Groups that reproduce in this manner include many sponges, corals, jellyfish, echinoderms, chitons, most of the more primitive gastropods, most bivalves, and many polychaete annelids. Despite its widespread occurrence, this mode of reproduction is not very efficient. Many eggs are probably never even fertilized, let alone able to complete development. So huge numbers of gametes must be spawned to ensure that even a single offspring survives to reproductive age.

In many groups of broadcasters, particularly those that are sessile or suspension feeders, males broadcast sperm but females retain eggs. Females take up sperm (often in the process of feeding), and fertilization is internal. This strategy reduces the waste of female gametes suffered by pure broadcasters. Embryos are then brooded internally to varying stages of development and then released. A wide variety of sponges, sea anemones, tunicates, and bryozoans reproduce in this manner.

Many of the more derived groups in many phyla have evolved mechanisms for copulation and internal fertilization. This further decreases waste of gametes, but its evolution is constrained by the morphology of a given group's reproductive organs. Some groups lack structures that can be modified by natural selection for copulatory appendages or for storing sperm internally. Others, such as primitive gastropods, have gonoducts that share an excretory function with the kidneys, making it difficult to modify them for specialized reproductive roles. Still, many divergent groups have evolved various means of copulation. Many of these groups also provide a greater degree of protection for the resulting embryos by either brooding them on their body (e.g., many higher crustaceans) or building protective egg cases (e.g., many advanced gastropods).

One of the unique aspects of most marine invertebrates' life histories is that their early life is spent with a completely different body form in a very different habitat from where they live as adults. The majority of marine invertebrates have microscopic planktonic larvae that hatch from fertilized eggs and swim off into the water column for periods ranging from just a few hours to several weeks or even months. Some larvae feed while swimming (planktotrophic), whereas others depend wholly on internal yolk reserves (lecithotrophic). After this free-swimming period, the larvae undergo an abrupt metamorphosis into their juvenile form and take up a benthic, or bottom-dwelling, existence. This metamorphosis is usually accompanied by a dramatic reorganization in body structure. During a single individual's lifetime, then, it is subjected to selection pressures in two very different environments. There are no truly analogous equivalents to a planktonic larval stage among terrestrial invertebrates.

For benthic invertebrates, the ecological consequences of these indirect life histories are many. A great deal of research has focused on discerning the best reproductive strategy in an uncertain world. Planktonic larvae suffer tremendous mortality rates, but they can disperse across distances much greater than an adult could move in its lifetime (especially if it is sessile). Dispersal permits discovery and colonization of new habitats, as well as an escape from an unfavorable benthic habitat. Wholly benthic development bypasses planktonic mortality but restricts dispersal severely.

Planktotrophic larvae can feed themselves and are energetically cheaper to produce, because their eggs require only sufficient yolk to support pre-hatching development. However, they are more subject to starvation if larval food supplies are unpredictable. Lecithotrophic larvae carry their energy supply with them and do not have to be concerned with food availability, but their larger yolk supply requires a greater amount of energy per egg, which means that fewer larvae can be produced for a given amount of adult energy. The length of a lecithotrophic larva's planktonic phase is fixed by its supply of yolk. If appropriate habitat is not found within that time, it will perish. Planktotrophic larvae are somewhat more flexible, as long as food is available. Obviously, the trade-offs and compromises incurred by different strategies are mediated by environmental conditions and their variability in space and time. The best strategy is always a moving target.

Clearly, life histories respond to these selection pressures. The evolutionary shift from planktotrophic to lecithotrophic modes of reproduction has occurred many times independently in many marine invertebrates. A more extreme strategy of bypassing a planktonic stage completely and reproducing by brooding or using benthic egg cases to produce crawl-away young has also evolved independently in many groups. These strategies reduce losses to planktonic mortality but curtail potential dispersal. Curiously, the transition from planktotrophy to lecithotrophy or benthic development appears to be easier to evolve than the reverse. At least the reattainment of planktotrophy from lecithotrophic development appears to have occurred much less often than the converse. Regardless, such shifts suggest that planktonic mortality may be a potent selective force.

In favor of planktotrophy, indirect evidence suggests that planktotrophic species suffered disproportionate extinction rates during at least some of the global extinctions in the geologic past. Despite this, planktotrophy is still the most common mode of reproduction amongst marine invertebrates, especially those that inhabit shallow benthic and intertidal environments. This suggests that it too has an enduring advantage.

The evolution of life histories of marine invertebrates is clearly a complex but fascinating topic. The need to find and study microscopic larvae out at sea, often miles from shore, makes field work difficult. Determining where the larvae that arrive at a given location actually came from has been one of the most daunting, but crucial, challenges of invertebrate larval ecology. Molecular genetic markers are providing important solutions to some of these problems. The role that larval dispersal and life histories play in maintenance and re-establishment of populations has obvious implications for conservation and habitat restoration.

CONCLUSION

To study marine invertebrates is to study the animal kingdom. The topic is truly vast, and no one can hope to master all its fine details, even with several lifetimes of study. Each path to understanding, however expert or novice, begins with curiosity and a good eye for detail. It takes time and effort to learn how to *see* things rather than just look at them. It takes practice and time spent out in nature, delving into natural things. Rocky shores are an ideal place to get acquainted with invertebrates, since the seashore abounds with all the major phyla and many representatives of each. But do not take what you see for granted. Question your initial assumptions about how a given structure works and how the animal uses it. Think about why a species lives where it lives and what you need to learn to choose between competing explanations. Do not always believe your intuition. Ask hard questions of the organisms you study, and learn to understand their answers. The answers an organism gives you are never wrong, but they may be difficult to understand or easy to misinterpret. So be patient, but be persistent.

SEE ALSO THE FOLLOWING ARTICLES

Arthropods / Bryozoans / Cnidaria / Echinoderms / Molluscs / Sponges / Worms

FURTHER READING

Giese, A. C., and J. S. Pearse, eds. 1974–. *Reproduction of marine invertebrates.* Vols. 1–5, 7, 8. New York: Academic Press. Vols. 6, 9. Pacific Grove, CA: Boxwood Press. Vol. 9. Palo Alto, CA: Blackwell Scientific (vol. 9).

Halanych, K. M. 2004. The new view of animal phylogeny. *Annual Review of Ecology, Evolution, and Systematics* 35: 229–256.

Harrison, F. W., and E. E. Rupert, eds. 1991–. *Microscopic anatomy of invertebrates.* Vols. 1–15. New York: Wiley Liss.

Morris, R., D. P. Abbott, and E. C. Haderlie. 1980. *Intertidal invertebrates of California.* Stanford, CA: Stanford University Press.

Ricketts, E. F., J. Calvin, J. W. Hedgepeth, and D. W. Phillips. 1985. *Between Pacific tides,* 5th ed. Stanford, CA: Stanford University Press.

Rosenfeld, A. W. 2002. *The intertidal wilderness.* Berkeley: University of California Press.

Valentine, J. W. 2004. *On the origin of phyla.* Chicago: University of Chicago Press.

ISOPODS

SEE AMPHIPODS, ISOPODS, AND OTHER SMALL CRUSTACEANS

KELPS

DAVID R. SCHIEL

University of Canterbury, Christchurch, New Zealand

The large brown algae that dominate marine algal beds and forests of the world come from two major taxonomic groups. True kelps belong to the order Laminariales, which contains the largest marine plants and comprises the greatest mass of seaweeds in the coastal zone. They occur in both hemispheres, but most of them by far are northern temperate or boreal. Fucoids belong to the order Fucales, are most abundant in the Southern Hemisphere, and are usually confined to the shallower portions of reefs from the high intertidal zone to around 10 meters depth, depending on light quality.

ECOLOGICAL SIGNIFICANCE

Charles Darwin, in his voyages on HMS *Beagle*, was impressed by the great aquatic forests of the Southern Hemisphere, comparing them in complexity to terrestrial forests. He wrote that, "the number of living creatures of all Orders whose existence intimately depends on the kelp is wonderful." He also recognized the importance of kelp communities and their resident organisms to the peoples who relied on them for food. Comparably to land forests, large marine algae provide vertical structure to the near-shore ecosystem, alter the physical and biological environment, and provide habitat, food, and nurseries for a wide range of other species.

Large brown algae inhabit rocky shores from the low intertidal zone to depths of around 50 meters in most temperate areas of the world, as well as the boreal regions of the Northern Hemisphere and the sub-Antarctic regions of the Southern Hemisphere. They can even extend into the tropics at depths where the water is cool and clear. The often dense canopies of these algae, some extending 30 meters or more to the sea surface, reduce light to areas below (Fig. 1)

FIGURE 1 Giant kelp *(Macrocystis pyrifera)* forest along the coast of southern California. The multiple fronds rising from the sea floor and the dense canopy floating on the sea surface produce dense shade to areas below. Photograph courtesy of Jay Carroll.

and dampen the effects of turbulence and waves. Large brown algae interact with myriad invertebrates, fish, mammals, birds, and other seaweeds that rely on them for critical resources. Numerous species of molluscs and echinoderms feed on kelp, and the turnover of the great biomass of coastal seaweeds contributes much of the productivity of nearshore waters, with flow-on trophic effects throughout the entire nearshore ecosystem.

LIFE HISTORY AND MORPHOLOGY

The order Laminariales includes species that form floating canopies on the sea surface, such as the giant kelp *Macrocystis pyrifera*, and the stipe-bearing kelps that form canopies up to 2 meters high above the sea floor. The order Fucales includes relatively small species of the genus *Fucus*, reaching only about 30 centimeters in length and widely distributed in the intertidal zone along the shorelines of the North America and much of Europe, and large subtidal forms, such as *Cystoseira osmundacea*, found along central and southern California, that can extend their canopies many meters above the bottom. The two orders have distinct differences in their life histories. Laminarians are diplohaplontic, with an alternation of generations consisting of a separate attached haploid gametophyte (Fig. 2), which is microscopic in size and can live for several months, and a diploid sporophyte, the large visible structures that can persist for many years in most species. Fucoids, in contrast, have direct development from eggs released from adult plants. Another group, the southern "bull kelps" of the genus *Durvillaea*, is widely distributed in the Southern Hemisphere and is considered by some to be in its own group (order Durvillaeales) and by others to be a fucoid. It consists of robust species that grow exceptionally large, such as *D. antarctica*, which reaches 10 meters in length and thrives on some of the most turbulent, wave-exposed shores of the world.

REPRODUCTION AND RECRUITMENT

The microscopic phases from reproduction through settlement and recruitment are crucial to the population biology of large algae. Vast numbers of propagules are usually necessary to produce successful recruitment in kelps, and only a tiny fraction of them survive. For example, a study in Nova Scotia on the kelp *Laminaria longicruris* found that nine billion spores per square meter produced nine million microscopic gametophytes, from which only one juvenile plant developed. Spores are tiny, around 7 micrometers in diameter, and all kelps produce many millions of them. In contrast, fucoids produce far fewer eggs, in the order of hundreds of thousands, that are around 200 micrometers in diameter.

Delivering reproductive propagules to the sea floor is a considerable challenge. The processes and structures necessary to achieve this in sufficient quantities for successful fertilization and development vary considerably among the kelps and fucoids. All kelps produce spores within specialized regions of their blades called sori. These appear as darkened ovoid areas when held up to the light. Some species, such as giant kelp, have specialized blades, called sporophylls, that bear the reproductive sori, but many others have sori throughout their blades. Spores are bi-flagellated, so they have some ability to swim. However, they are so tiny that water is viscous to them, and they are largely carried by water movement. Giant kelp have their sporophylls at the base of plants, and spores are released near the sea floor. Another float-bearing kelp, *Nereocystis luetkeana*, has no blades at all except at the top of plants, usually at the sea surface. This species releases the sorus itself, which drifts to the substratum as it releases spores. Once spores settle, they can develop over a few days into male or female gametophytes. These are single cells or lightly pigmented filaments that develop over 20 days or so until becoming fertile. Male gametophytes produce flagellated sperm that fertilize the egg extruded at the end of the female gametophyte. The density of settlement is crucial here because males cannot swim far. It is estimated that

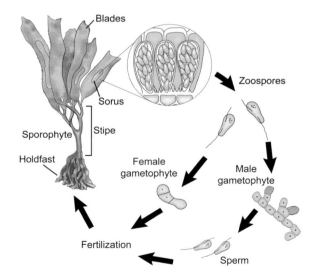

FIGURE 2 Kelp (laminarian) life history, featuring an alternation of generations with microscopic spores (about 7 μm long) developing into male and female gametophytes (around 1 mm long), fertilization, and then the growth of macroscopic sporophytes (the large visible plants). Graphic courtesy of Erika MacKay, National Institute of Water and Atmospheric Research Ltd, New Zealand.

males and females can be no more than around 1 millimeter apart for successful fertilization.

For fucoids, the timing of fertilization is mostly unknown but seems to occur rapidly as eggs are released. Fertilization is undoubtedly aided by the high densities at which most fucoids occur. However, this process varies considerably because there are both monoecious (both sexes in the same plant) and dioecious species of fucoids. Where separate sexes occur, there is an obvious requirement for having close proximity of males and females, as well as synchronous release of gametes. Some fucoids also have the equivalent of parental care of their propagules, avoiding some of the hazards of early life on the substratum. One notorious example is *Sargassum muticum*, a highly invasive fucoid native to Japan that has established along much of Europe and the west coast of North America. This species (and many others) extrudes eggs that stick by mucilage to ephemeral structures called receptacles, the part of the plant that produced them. Here they are fertilized and grow to a germling stage up to around a millimeter long, after which they detach and spiral down to the substratum, adhering quickly by the sticky rhizoids that have already developed.

Recruitment periods in kelps are often referred to as "windows," in which appropriate conditions coincide. These involve critical levels of light, temperature and nutrients, which affect the germination process and production of sporophytes and undoubtedly vary among species. In southern California, for example, irregular episodes of good recruitment of giant kelp occur when temperatures are between 12.5–16.3 °C and irradiance is between 0.4 and 4 einsteins per square meter per day (one einstein = one mole of photons). Few sporophytes are produced above 18 °C or below 0.4 einsteins per square meter per day. The deep shade of dense sea surface canopies can reduce irradiance to as little as 0.2% of surface light. When levels reach below around 1% of surface light, successful reproduction and recruitment do not occur. The relationships among all factors can be complicated, because temperature in many places is negatively correlated with nitrate levels, and the interaction between light and nutrients may be of primary importance.

GEOGRAPHIC DISTRIBUTIONS

Generalizing about kelp communities is difficult because of the wide variation in forms, oceanographic conditions, and wave climates in which they occur and the distinct assemblages of organisms associated with them worldwide. One of the most obvious differences in kelp

FIGURE 3 Stand of the stipe-bearing kelp *Pterygophora californica*, about 1.5 m high, with fronds streaming in the current, beneath a canopy of giant kelp; the sparse understory on the rocky bottom results from dense shading caused by both canopies of kelp. Photograph courtesy of Dan Reed, University of California, Santa Barbara.

communities globally is between those dominated by forests of kelps reaching to the sea surface and those with beds of stipe-bearing smaller kelps, rarely more than 2 meters high. These produce different tiers of canopy layers above the sea floor and therefore different levels of structural complexity and biomass in their communities (Fig. 3). There are relatively few large kelps, but they vary considerably in morphology. *Macrocystis pyrifera* is the largest of all marine plants, often dominating coastal reefs in both hemispheres with as few as 1 plant per ten square meters. Plants can have hundreds of fronds reaching upwards from the holdfast, and over half of the mass of plants may float on the sea surface, producing deep shade to areas below. With individual fronds elongating up to tens of centimeters daily, *Macrocystis* can quickly grow above other kelps throughout most of its range. Individual plants can live up to around 10 years. *Nereocystis luetkeana*, an annual kelp, has a single long stipe and dominates cooler waters along the west coast of North America from California to Alaska and the islands of the Bering Sea. Another annual kelp, *Alaria fistulosa*, reaches around 10 meters in length and dominates many shallow areas in Alaska and the Pacific coast of Asia. *Ecklonia maxima* is a Southern Hemisphere analog of *Nereocystis* with a single robust stipe extending up to 15 meters long with a dense canopy on the sea surface. It dominates nearshore subtidal reefs around the Cape Province of South Africa. Other areas of the world are dominated by smaller kelps and often only a few species of these. These include members of the most speciose genus, *Laminaria*, which dominate the shores of Europe and northeastern America, *Ecklonia radiata* of New Zealand and much of temperate Australia, and the

Lessonia species of Chile. Still other laminarians have a prostrate form that rather flops on the substratum. These include some species of *Laminaria* in the high Arctic and understory species such as *Dictyoneurum californicum* and *Costaria costata* along California.

Fucoids can be extremely abundant in the intertidal zone and shallow subtidal areas of temperate reefs, often reaching hundreds of plants per square meter of reef. In the most diverse area for laminarians, the coast of California with around 19 species, there are relatively few fucoids, and these are confined to very shallow water and the intertidal zone. In contrast, the mainland coastlines of New Zealand and southeastern Australia have only about four laminarian species but at least thirty species of fucoids. Some of these dominate shallow subtidal areas to the local exclusion of stipitate kelps (Fig. 4).

FIGURE 4 A dense fucoid forest (*Cystophora* species) at 7-m depth off southern New Zealand. Densities or plants reach hundreds per square meter, and fronds can extend over 2 m in height. Photograph by Steve Mercer, National Institute of Water and Atmospheric Research Ltd, New Zealand.

It is generally considered that the effects of light, temperature, and nutrients acting on one or more life stages of seaweeds restrict their distributions geographically. Although some kelps occur in cold Arctic and sub-Antarctic waters, they are restricted by low light levels and short summers, which affect growth and reproduction. Conversely, few laminarians can survive in sustained temperatures above around 23 °C because reduced photosynthesis, nutrient depletion, and the prevalence of fouling organisms cause severe deterioration of plants. This is illustrated over long contiguous coastlines such as the west coast of North America, where few kelps occur in the warmer waters of Mexico. At their southern limit around central Baja California, Mexico, seasonal upwelling of cool, nutrient-rich water supports kelp populations. These types of special conditions allow kelp to persist in other warm-water regions, such as the northern coast of Chile to Peru and the central coast of western Australia. A curious example of this involves the kelp *Ecklonia radiata*, which co-occurs with corals in the Gulf of Oman. Despite Oman being well within the tropics, there is seasonal upwelling generated by monsoon winds in late summer, and seawater temperatures drop below 20 °C. These lower temperatures combined with extremely high nitrogen levels allow rapid growth and reproduction of *Ecklonia*, which will overgrow corals.

Fucoids are generally more tolerant of high temperatures than laminarians are, especially the large number of *Sargassum* species that extend into the tropics and intertidal taxa such as *Pelvetiopsis* species that commonly withstand temperatures over 35 °C. Because each species has optimal conditions for growth and reproduction, the interaction of temperature, light, nutrients and wave exposure can affect geographic distributions and abundances, tipping the balance in favor of particular species. An example is *Macrocystis pyrifera*, which can be found from the Mexican border to Alaska but is rarely the dominant species north of Point Santa Cruz in central California, north of which *Nerocystis luetkeana* dominates.

The tropical zone is considered to be a major biogeographic barrier to temperate species, and relatively few large brown algae have made the jump between hemispheres. Chief among these is the giant kelp genus *Macrocystis*, which dominates much of the shoreline of central and Southern California, as well as shores along southern Chile, Argentina, South Africa, southeastern Australia, southern New Zealand, and the string of sub-Antarctic islands extending across the Southern Ocean. *Macrocystis* can form large floating rafts, which provide a means for long-distance dispersal. Anomalous distributions of large brown algae occur where appropriate conditions coincide, as in the case of *Ecklonia radiata*. It is only speculative, however, how a species that has no ability to float can transcend hemispheres.

LOCAL DISTRIBUTIONS

There is a complex interplay of biological, ecological, and physical factors that dictate the local distribution and abundance of kelp and fucoids. Their habitats consist of many gradients that affect the structure and physiology of plants operating over spatial scales ranging from microscopic to hundreds of kilometers and over time scales from seconds (as in wave impacts) to decades. These include light, salinity, temperature, availability of suitable hard rock, nutrients, sedimentation, water motion, interactions with other species, and herbivory, particularly by sea urchins.

FIGURE 5 Intertidal kelps in southern California. The sea palm *Postelsia palmaeformis* and three juvenile *Alaria marginata*, surrounded by numerous purple sea urchins (*Stongylocentrotus purpuratus*), wedged into individual pits by their spines. Photograph courtesy of Mike Foster.

Within any given site, these gradients and their effects are readily seen. Large algae tend to be zoned in bands from the high intertidal to deep subtidal zones. Turbulence-adapted species, such as the feather boa *Egregia menziesii*, the sea palm *Postelsia palmaeformis* and the kelp *Laminaria setchelli*, occur in wave-exposed inshore areas (Fig. 5). Deeper areas of reefs, generally between around 5 and 20 meters, where wave force dissipates and light is sufficient, have a wide range of kelps, often including a canopy of giant kelp, a subcanopy of stipe-bearing species, and an understory of corticated and corallinaceous seaweeds. Some species do well beyond 20 meters depth, such as the elk kelp *Pelagophycus*, and *Agarum cribosum*. The lower depth limit is usually set either by light limitation or a lack of hard substrata. Both the quantity and the spectral distribution of light change with depth, affecting the distribution and abundance of algae, which need particular wavelengths for reproduction and growth. These processes are complicated, however, by the ability of some species to cope with low light levels, for example during winter, by storing organic reserves during favorable periods and using them when conditions are poor.

Wave force and physical stresses greatly affect the upper boundary of kelp and fucoids as well as their coastal patterns. The eggs and developing germlings of many species of fucoids are unable to attach securely where wave forces are high, with considerable differences among species. Exposed, sheltered, and intermediate sites usually have different characteristic dominant species, even within single-reef systems that have a gradient of wave exposures. These effects may translate to large spatial scales, such as the geographic shift in dominance between *Nereocystis* and *Macrocystis*.

Most physical factors are affected by very large-scale events, such as El Niño episodes. During these periods along the west coast of North America, temperatures increase, nutrients are low, and waves from storms are severe. El Niños have caused major loss to kelp communities over very wide areas, sometimes taking years to recover, especially near the warm-water limits of populations. The loss of kelp canopies often results in greater recruitment of understory algae and competitive interactions among species for space.

COMMUNITY AND ECOSYSTEM EFFECTS

There is considerable debate about the influences of human-induced impacts on kelp communities worldwide. There is no doubt that many commercially valuable species, such as large predatory fishes, abalone, sea urchins, and lobsters, have been fished to the point where they are functionally unimportant in many communities. The loss of "apex" predators and potential cascading effects through food webs have caused much concern about changes to the functioning of kelp communities. The consequences of historical overfishing of marine mammals, such as sea otters and seals, may have given us a distorted view about the natural state of communities and their functional relationships. Localized deforestation of kelps by sea urchins occurs worldwide (Fig. 6), but it is still unknown whether the demise of sea urchin predators in some areas is having greater, long-lasting effects on the dynamics of kelp communities. The kelp themselves mostly respond to the physical environment within their ranges. They flourish where

FIGURE 6 The tough, leathery Southern Hemisphere kelp *Lessonia variegata* under grazing pressure by the sea urchin *Evechinus chloroticus* at the Chatham Islands, New Zealand. Note the barren grazed understory and presence of abalone *(Haliotis iris)*. Photograph by Steve Mercer, National Institute of Water and Atmospheric Research Ltd, New Zealand.

upwelling brings cool, nutrient-rich water to the coast and wane during warm-water, nutrient-poor events such as El Niño/Southern Oscillation. As trophic elements are impacted by human activities, however, a major concern is that the resilience of kelp-dominated communities to disturbances will be lessened because the buffering in the community is compromised. Increased sedimentation through coastal development and land use practices has already taken a toll on algal communities in many areas, such as the Baltic Sea.

Kelp communities continue to be intimately linked with human habitation of the coastal zone. Fortunately, many of the activities that affect them can be managed on a local scale, which is the underlying basis for the increasing number of marine protected areas. Darwin's observations on the mingled relationship between kelp communities and people still remain true, but in many areas the strength of the interaction has shifted to the effect of people on kelp.

SEE ALSO THE FOLLOWING ARTICLES

Algal Biogeography / El Niño / Food Webs / Light, Effects of / Recruitment / Zonation

FURTHER READING

Abbott, I.A., and G.J. Hollenberg. 1976. *Marine algae of California.* Stanford, CA: Stanford University Press.

Foster, M.S., and D.R. Schiel. 1985. The ecology of giant kelp forests in California: a community profile. *U.S. Fish and Wildlife Biological Report* 85(7.2).

Branch, G., C. Griffiths, M. Branch, and L. Beckley. 1994. *Two oceans: a guide to the marine life of southern Africa.* Cape Town: David Philip Publishers Ltd.

Steneck, R.S., M.H. Graham, B.J. Bourque, D. Corbett, J.M. Erlandson, J.A. Estes, and M.J. Tegner. 2002. Kelp forest ecosystems: biodiversity, stability, resilience and future. *Environmental Conservation* 29: 436–459.

LARVAL SETTLEMENT, MECHANICS OF

CHERYL ANN ZIMMER

University of California, Los Angeles

Most intertidal invertebrates have a complex life cycle involving a sedentary adult preceded by a planktonic larval stage (Fig. 1). Larvae typically colonize new habitat and recruit into existing populations. Settlement occurs when a competent larva touches down onto the substratum. If it then undergoes metamorphosis, that site has become its home. Larval settlement mechanisms are both passive and active. Passive processes include larval transport and

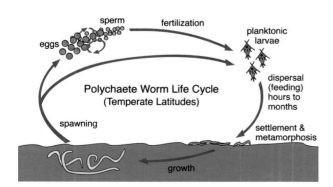

FIGURE 1 Complex life cycle of a typical bottom-dwelling invertebrate. Drawing by J. Doucette.

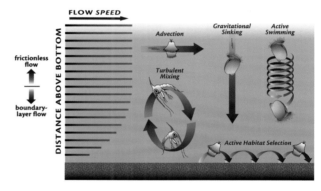

FIGURE 2 Larval movements within near-bottom flows. Drawing by J. Doucette. Vertical velocity profile is shown at the left (flow speed on abscissa, distance above bottom on ordinate). Flow speed is zero at the bed. Reprinted with permission from the *Biological Bulletin*.

deposition due to advection, turbulent mixing, or gravity (Fig. 2). Once passively deposited over a relatively broad region, larvae may differentially survive in favorable adult habitat. Alternatively, larvae use chemical, tactile, or visual cues to actively (via swimming or sinking) select their settlement sites. Active and passive processes are not mutually exclusive. Passively deposited larvae could, for example, reject their touchdown site, return to the near-bottom flow, and test another site downstream.

CONCEPTUAL FRAMEWORK

To provide a synoptic overview of the mechanics of larval settlement, passively transported larvae are tracked from inner-shelf waters to the shore and from a shallow water column to the rocky intertidal. Within each flow environment, larval populations can be dispersed or concentrated, and their paths directed, by physical processes.

In wave-swept rocky habitat, the mechanics of settlement—when and how competent larvae make physical contact with and attach to the substratum—is the least studied and quantified of the physical/biological couplings. It is difficult, if not impossible, to catch larvae in the act of settling on intertidal surfaces. Simulating complex intertidal flows in the laboratory is challenging, at best. Thus, settlement processes have been inferred largely from studies of more organized, unidirectional flows offshore, from patterns of intertidal recruitment, and from field or flume experiments. Most examples provided herein are for barnacles, arguably the most extensively studied group of intertidal invertebrates.

LARVAL DISPERSAL IN INNER-SHELF WATERS

Planktonic larvae spawned intertidally often are transported offshore to feed and grow and then moved back onshore to settle and metamorphose. Length of larval life balances the risks of planktonic mortality and settlement in an unsuitable locale, with the benefits of growth and dispersal. Evolution has reconciled these factors in several ways, and thus intertidal invertebrates have multiple larval types. At one extreme, long-distance dispersal decouples larval source and sink populations, enhancing fitness over the species's range. Alternatively, nondispersing larvae maintain localized patches by recolonizing natal habitat, with potentially severe population-genetic consequences.

The inner-shelf flow regime—from just outside the surf zone to several kilometers offshore (5–30 meters depth)—is unaffected by physical characteristics of the shore. Wind-driven currents are strong in the shore-parallel direction and weaker perpendicular to the coast. In addition, there are flows driven by tides and by differences in the density of water masses. Although several types of currents may occur simultaneously, the dominant flow tends to be relatively unidirectional, steady, and predictable.

Larvae can exploit reasonably reliable circulation patterns that would carry them to shore. For example, upwelling and downwelling flows are cross-shelf circulations caused by intense and prolonged alongshore winds (Fig. 3). Upwelling currents move surface water (and larvae) out to sea. Years of poor barnacle (several species) recruitment have been correlated with strong upwelling flows along the west coast of the United States. In contrast, downwelling currents transport offshore surface water shoreward and down at the coast. Relaxation of upwelling has the same effect. Larvae concentrated (e.g., via upward swimming or buoyancy) at the surface would ride downwelling currents to shore.

On the inner shelf, horizontal currents (of order 10–100 cm/s) are much faster than swim speeds (of order 0.1–1.0 cm/s) of most invertebrate larvae, resulting in passive transport. In contrast, velocities in the vertical are at least 1–2 orders of magnitude smaller than in the horizontal, and thus larvae can swim up and down in the water column, navigating via salinity, temperature, light, or turbulence gradients. Enhanced near-bed mixing, for example, caused snail *(Ilyanassa obsolete)* larvae to cease

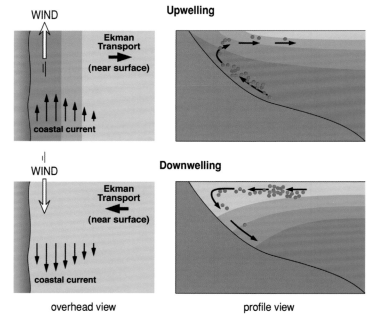

FIGURE 3 Cross-shelf larval transport along the east coast of a continent in the Northern Hemisphere. Ekman transport occurs where surface water moves to the right of the wind stress as a result of the Coriolis force (an apparent force associated with the earth's rotation). Thus, a northward-blowing wind moves surface water offshore; cold, deep water upwells at the coast to replace it. A southward-blowing wind transports surface water shoreward, where it downwells at the coast. Larvae may track with the water masses. Shades of blue depict stratification in water temperature, where the deepest color is the coldest water. Red dots are larvae transported by the flow. Drawing by J. Doucette.

swimming and actively sink to the bottom, presumably facilitating settlement.

In a classic salt-wedge estuary, less-dense ocean water from offshore rides on top of warmer, more-saline (denser) estuarine water. On the flood tide (influx of ocean water), larvae actively move up from the bottom in response to a gradient in salinity or turbulence and are transported into the estuary. On the ebb tide, larvae swim down to the sea floor and thus avoid going back out to sea. This mechanism retains late-stage blue crab (*Callinectes sapidus*) larvae within the estuary for settlement.

Soluble chemical cues, as emitted by the gregarious oyster *Crassostrea virginica*, cause conspecific larvae to swim down and actively explore adult habitat. For an oyster bed in 3 m of water, and slow (of order 10 cm/s) horizontal currents, larvae with vertical swim speeds of ~0.5 cm/s would be advected 60 meters before hitting bottom. Thus, larvae cued to swim down could easily land on a conspecific reef (hundreds of meters long). Near-bottom turbulent eddies also should facilitate larval contract with an oyster bed. Together, chemical, behavioral, and physical processes play critical roles in this system.

LARVAL SUPPLY TO THE INTERTIDAL

Once transported by the regional-scale flow to the surf zone, larvae encounter stronger, oscillatory, and less-predictable flows. Breaking waves (velocities as high as 25 m/s) pummel the shore. Onshore wave-driven flow bathes exposed rocky surfaces and, seconds later, moves back out to sea, leaving isolated puddles (tidepools) behind. In addition, weaker tidal flows oscillate over diurnal or semidiurnal time scales.

Larval settlement is challenging in this highly chaotic flow region. Rare are predictable flow patterns at scales that would be useful for navigation. Moreover, there are no persistent gradients in those environmental variables (e.g., temperature and salinity) that afford successful navigation offshore.

Swimming larvae make little headway in such disorderly flows. Specific settlement patterns can result, however, because of larval supply alone. "Supply-side ecology" asserts that physically dispersed and deposited larvae determine community structure. That is, variation in passively transported larvae is responsible for recruitment variation. Such patterns also could result from downward swimming or active sinking, especially in more tranquil flows. For example, vertical stratification of planktonic barnacle larvae (*Balanus glanula* and *B. crenatus*), probably due to active movements, ultimately resulted in two vertically distinct settlement horizons on shore. In fact, supply-side mechanisms may span a continuum, anchored at either end by passive physical processes and active behaviors.

MECHANICS OF LARVAL SETTLEMENT

Intertidal currents usually are very strong at the water surface. Near the sea floor, however, is a vertical gradient in current speed, known as the bottom boundary layer, resulting from the frictional drag of the bed on the flow (Fig. 2). Flow speed decreases to zero at the bottom, which is clearly defined on smooth surfaces and a moving target on the highly uneven topography of the rocky intertidal. Still, relatively slow boundary-layer flows can facilitate settlement. The thickness of this layer is proportional to the period of flow steadiness. For example, in a typical wave boundary layer with a 6 s period, the boundary-layer thickness is ~2 cm—not a very expansive region for active maneuvering. Moreover, a wave boundary layer forms and reforms every few seconds, when the flow changes direction. In contrast, offshore boundary layers due to tidal forcing (period of 6 h) are ~8 m thick. Thus, swimming behaviors are more likely to be effective in near-bed inner-shelf flows than in the surf zone. Once settled, active perusal of the substratum for meaningful cues appears to be common in both habitats.

To reach the bed, larvae can swim down, sink under gravity, or be mixed by turbulent eddies (Fig. 2). Turbulence is generated at the bottom. A bumpier bed generates more turbulence than a smooth one. The intensity of turbulence is maximal just above the bed and decreases with height above bottom, to zero at the top of the boundary layer. In contrast, eddy size scales with distance from the bed; small eddies are generated by small roughness, and the largest eddies circulate throughout the entire boundary layer. Eddies exchange fluid between the upper and lower portions of the water column, often homogenizing gradients in velocity, salinity, temperature, and larvae. In turbulent flow, velocities are characterized by a mean and a fluctuating (instantaneous turbulent) component, defined as the standard deviation of the mean. The force per unit area on the bottom (the boundary shear stress) results from the turbulent fluctuations and is roughly proportional to velocity squared.

The moment a larva attempts to settle, a sufficiently strong instantaneous shear stress would dislodge it. Thus, a competent larva may be bounced around near the bed until hydrodynamic conditions permit settlement. Higher bed roughness enhances turbulent fluctuations on the bottom, making touchdown increasingly more difficult.

The average distribution of a settled larval population integrates over temporal and spatial variability in shear stress. Like bottom sediments, settled larvae largely are in equilibrium with the average shear stress regime. Just as fine sediments persist under low mean shear stress, higher mean shear stress erodes fine particles, leaving coarser sands behind. Likewise, larvae should be rare in shear stress regimes that are, on average, sufficiently energetic to dislodge them. In fact, laboratory flume studies have shown that barnacle *(Balanus amphitrite)* larvae were supplied to a smooth plate surface in regions of maximal vertical velocity but settled only where local shear stress was minimal.

Although strong adhesives have evolved for attachment in the high-energy intertidal, larvae may "hide" from the flow by settling in cracks, crevices, under overhangs, and among other organisms. Moreover, settlement can be restricted to specific hydrodynamic regimes. The gregarious honeycomb worm *Phragmatopoma californica* builds conspecific aggregations of closely packed sand tubes cemented to rocks in the low intertidal. Larvae settle in response to tube cement secreted by the adults. In laboratory flume experiments, settlement was maximal at intermediate velocities (and bed shear stresses) (Fig. 4). Larvae actively rejected slow flows that might not transport sufficient food to suspension-feeding adults, and they were physically prevented from attaching in high flows.

The rocky intertidal consists of convoluted surfaces with bumps spanning a wide range of spatial scales. These irregularities direct the flow, resulting in local accelerations and decelerations that could affect where and when larvae settle. A spatial mosaic of near-bed turbulence also dictates touchdown and attachment sites. At very small scales (tens of micrometers), for example, distributions of barnacle larvae *(Semibalanus balanoides)* were determined by the locations of minuscule pits on the substrate. Oriented head-down, barnacles attach to a surface by anterior antennules. These structures have tiny (40 μm in diameter) terminal pads that adhere most effectively within pits of the same size.

At intermediate scales (millimeters to centimeters), the distribution of local shear stresses can determine the spatial distribution of settlers. For example, when replicate plaster casts of a natural rocky surface were oriented in the same direction, newly settled barnacles *(Chthamalus fragilis and Semibalanus balanoides)* occurred in nearly identical positions on all of the casts. Larvae were distributed in and among microtopographic features according to local hydrodynamics.

Patterns of flow at larger scales (tens of centimeters to meters) also can determine larval distributions. Flow approaching an intertidal boulder, for example, splits and accelerates

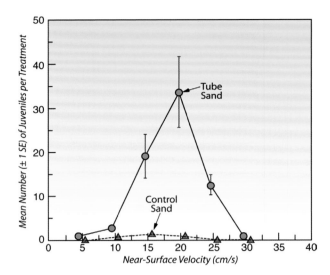

FIGURE 4 Larval settlement as a function of near-surface velocity for the honeycomb worm *(Phragmatopoma californica)*. Experiments were conducted in a unidirectional flume flow. Larvae were offered tube sand (green circles)—particles on which adults had secreted cement for tube building and that induce metamorphosis—and control sand (red triangles), with no cement or inductive activity.

to round the front face and then separates into recirculating eddies on the lee (shoreward) side of the rock. On the back face, eddy velocities are slow relative to the oncoming flow, and thus, larvae may be passively retained within the recirculation zone. This mechanism may explain the more abundant and larger honeycomb worm *(P. californica)* aggregations on shoreward than on seaward faces of boulders (Fig. 5).

FIGURE 5 Intertidal rock showing asymmetrical distribution of honeycomb worm aggregations *(Phragmatopoma californica)*. Larger and more numerous aggregations were observed on the shoreward than on the seaward face of boulders located on intertidal beaches along the California coast. Photograph by S. Simmons, with permission from the *Journal of Marine Research*.

In summary, larval settlement is a challenging proposition in the rocky intertidal. Given the highly energetic, unpredictable flows and intricate, convoluted surfaces, the odds are against larvae landing in hospitable adult habitat. But, because of behavioral interactions with local near-bed flow regimes, sometimes they do.

SEE ALSO THE FOLLOWING ARTICLES

Boundary Layers / Dispersal / Metamorphosis and Larval History / Recruitment / Surf–Zone Currents / Turbulence

FURTHER READING

Butman, C. A. 1987. Larval settlement of soft-sediment invertebrates: The spatial scales of pattern explained by active habitat selection and the emerging role of hydrodynamical processes. *Oceanography and Marine Biology: An Annual Review* 25: 113–165.

LeFevre, J., and E. Bourget. 1992. Hydrodynamics and behaviour: transport processes in marine invertebrate larvae. *Trends in Ecology and Evolution* 7: 288–289.

Underwood, A. J., and M. J. Keough. 2001. Supply-side ecology: the nature and consequences of variations in recruitment of intertidal organisms, in *Marine Community Ecology*. M. D. Bertness, S. D. Gaines, and M. E. Hays, eds. Sunderland, MA: Sinauer, 183–200.

LICHENS

SEE ALGAL CRUSTS AND LICHENS

LIGHT, EFFECTS OF

MOLLY CUMMINGS

University of Texas

SÖNKE JOHNSEN

Duke University

Light is the portion of the electromagnetic spectrum that stimulates visual and photosynthetic pigments, comprising radiation with wavelengths in air between 300 and 700 nanometers (ultraviolet to red). The spectral distribution of light energy is important for both vision and photosynthesis, and can be affected by the medium in which the light travels, particularly by water. The absorption of light by water molecules is high and depends strongly on wavelength, with the least absorption occurring at 460 nanometers (blue light). Light underwater is also strongly attenuated by interactions with other constituents, such as absorption and scattering by phytoplankton, dissolved organic matter, and suspended solids. Because of this variation, the aquatic near-shore environment is a challenging world for organisms that see, as well as for organisms trying to communicate or hide.

THE INTERTIDAL LIGHT FIELD

Three primary factors alter the light field in the intertidal zone: (1) tides, (2) waves, and (3) dissolved and suspended substances. The tides modify the benthic intertidal light field by changing the path length that light travels in this strongly absorbing medium. At low tide, when the intertidal zone is exposed, organisms encounter spectral irradiances similar to terrestrial environments. As the tide moves in, however, the light passes through a medium that changes both its intensity and its spectral distribution. Because water molecules absorb light most strongly in the short (blue-violet) and long (red) wavelength regions of the spectrum, increases in tidal depth selectively attenuate these wavelengths. Figure 1A illustrates how increasing water depth alters the ambient light field. For example, in the near-shore water column of temperate California, light at 650 nm is present at 1 m and nearly absent at 10 m. In contrast, green light ($\lambda = 500$–550 nm) is most prevalent at depth.

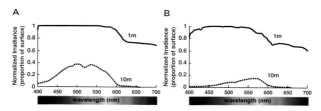

FIGURE 1 Downwelling irradiance measurements, $I_D(\lambda)$, normalized to surface spectral irradiance, at 1 (—) and 10 (⋯) m depths collected in Hopkins Marine Life Refuge, Pacific Grove, California, with a diver-operated spectroradiometer. Measurements were collected under sunny skies between 12 and 1 PM on two days that differed in phytoplankton concentrations: (A) minimal concentration, and (B) a dense phytoplankton bloom. In B, most of the blue and green light has been absorbed.

As tides rise in the intertidal zone, they generally are accompanied by waves. The effect of waves on the intertidal light field is difficult to estimate because waves complicate the underwater light field immensely. First, waves significantly change the depth of the water column and do so over short time scales. Whereas tidal cycles change in average depth over a matter of hours, waves can alter the water column depth in the intertidal in seconds. Even relatively modest waves substantially change both the intensity and the color of the light in the benthic environment.

A second effect of waves is that they act as lenses. Wave peaks act as positive lenses, creating bright regions at their foci. Wave troughs act as negative lenses, darkening the region below them. Given the almost limitless variety of shapes the ocean surface takes near the shore, the light field underneath has enormous temporal and spatial variability. The differences between this effect and that of the increased height of the waves are that it is spectrally neutral (affecting all wavelengths equally) and that it occurs over smaller spatial scales.

Last, wave breaking alters the intertidal light field. Turbulence, particularly in protein-laden seawater, creates large numbers of bubbles of all sizes. These scatter light, deflecting it from its original direction. Areas directly below the bubbles are darker; areas to the side of the bubbles can be much brighter. In addition to their effect on illumination levels, bubbles dramatically affect underwater image propagation. Even a relatively thin layer of bubbles can render any object behind it invisible. Similar to the effects of waves and light focusing, the formation and persistence of bubbles is unpredictable and highly variable on small spatial and temporal scales.

Light in the intertidal is affected not only by tides and waves but also by the filtering effects of particles within the water column. At low tide, organisms experience the full spectrum of daylight. However, as tides and waves roll in, they usually bring in particles that interact with light more intensely than do the water molecules themselves. Microscopic algae known as phytoplankton are a major group of light-interacting particles that can be present in near-shore and intertidal waters. Many marine algal species differ from terrestrial plants by having a range of accessory pigments that are able to absorb light more effectively in the middle wavelengths. These accessory pigments are important in the aquatic environment, because water differentially filters out the wavelengths that the typical terrestrial plant and green algae pigments (chlorophylls *a* and *b*) absorb (400–450 nm; 680 nm). A concentration of phytoplankton in the water column removes much of the short and middle wavelengths of light leaving only green-yellow wavelengths (e.g., 530–580 nm) for benthic organisms (Fig. 1B).

ADAPTATIONS TO THIS VARIABLE WORLD: ANIMAL VISION

In general, visual systems adjust to the average background light levels and respond only to changes in it. This process, known as adaptation, is useful in that it greatly increases the range of response of the receptor cells. In the terrestrial world, changes in illumination levels are often slow, and consequently visual systems usually adjust to these changes over a matter of several minutes. However, in intertidal habitats, the changes can be abrupt, making it difficult to adapt over the appropriate timescale. The bright flashes and shadows of a flickering surface make it difficult for visual species to focus on small objects or adjust to the diverse levels of background light. As anyone who has spent time directly below a wavy surface can attest, the variation of intensity in this region approaches and can surpass one's ability to adapt, particularly when looking up.

Although wave focusing and bubbles produce spectrally neutral challenges for intertidal organisms, the wavelength-dependent effects of tidal range, wave depth, and phytoplankton change the color of the ambient light, which creates new challenges. Both intertidal and shallow-dwelling organisms address these using visual systems characterized by broadened visual sensitivity (e.g., greater number of visual receptor cells known as photoreceptors sensitive to different parts of the spectrum) as well as visual sensitivity shifted to longer wavelengths relative to deeper or open-ocean environments.

Photoreceptors' properties differ by species according to the light found in their habitats. In general, receptors shift to be sensitive to shorter wavelengths as depth increases, because the longer wavelengths have been attenuated (Fig. 1B). This shifting is a common property of both invertebrate and vertebrate marine and aquatic organisms. Crustaceans such as the intertidal *Gonodactylus* (stomatopod) species have λ_{max} values (wavelength of peak sensitivity) that are long wavelength shifted relative to their deeper dwelling congeners. Vertebrate fish show a similar trend, as shown by the temperate marine surfperch and aquatic Lake Baikal cottoids, which have long wavelength–sensitive photoreceptors exhibiting a shift toward longer λ_{max} values at shallower depths. The intertidal peacock blenny, *Salaria pavo*, living in Mediterranean mudflat regions has a visual system that also exhibits a shift to long wavelengths in shallow, intertidal environments. The intertidal peacock blenny differs from most other shallow-water fishes in that it lacks a short wavelength–sensitive cone entirely. It is also unique among fish in containing a screening pigment in some of its photoreceptor outer segments, which shifts the absorption of the long wavelength–sensitive photoreceptor to even longer wavelengths.

ANIMAL SIGNALS IN THE INTERTIDAL

The challenge of communication in the intertidal zone arises from two different properties of the intertidal light field: (1) high tide conditions, which reduce the number of wavelengths available (the full spectrum is reduced to just middle wavelengths), particularly if the water contains phytoplankton; and (2) spectrally neutral effects of waves—such as lens effects and bubbles—that produce dynamic changes in brightness.

The optical variability and the inability to see well in the intertidal combine to make reliable signaling very difficult. Animals using color for sexual advertisement, warning coloration, or other purposes find it difficult to send a constant color signal, and the relevant viewers find it difficult to see anything at all, particularly in the surf zone. There is little that can be done about the latter problem, but the former can be mitigated by a judicious choice of colors. The constancy of the perceived color (or hue) of an animal's color signal is affected by whether its reflectance is saturated or unsaturated. Saturated colors are those that reflect light in a concentrated portion of the spectrum, so that only a narrow bandwidth of light is reflected from the animal. On the other hand, unsaturated colors, such as silver or white, have a high degree of reflectance across a wide range of wavelengths. The hue of an animal with a broad, or unsaturated, reflectance curve (with some reflectance at all wavelengths) is significantly changed by any changes in the spectral quality of the light striking it. But the hue of an animal with a narrow or saturated reflectance curve (with high reflectance only at a few select wavelengths) remains relatively constant.

Narrow reflectance curves are difficult to produce using pigments, however, because natural pigments generally have very broad reflectance curves. There are some solutions to this problem. One is to have a broad reflectance curve that appears narrow because most of it occurs outside the range of vision. For example, an animal using a pigment that reflects at very long or short wavelengths at the right concentration can produce an intense orange or red (at long wavelengths), or violet or blue (at short wavelengths). In both cases, the majority of the reflectance curve is found outside the visual realm, in the infrared for the red case, in the ultraviolet for the violet case. Consequently, the color will change little under varying illumination. Another solution is to use structural colors. These are not bound by the restrictions of natural pigment absorption curves and can produce narrow reflectance curves if they are constructed in a sophisticated layering system in which interference restricts the bandwidth of light reflected. Therefore, one would predict that intertidal animals sending visual signals would use either interference

colors to produce nearly monochromatic signals or pigments concentrated on either end of the spectrum (violet, blue, orange, or red) so that only a small portion of the pigment's reflectance occurs within the animal's visual spectrum, producing visually saturated coloration.

An intertidal fish demonstrates how coloration may follow the general principles outlined above. The peacock blenny has sexually dimorphic coloration, with males that signal to females using orange-yellow crests and blue-ring eyespots against a dark brown body (Fig. 2A). The blues and oranges of the peacock blenny reflect approximately equal amounts of light under

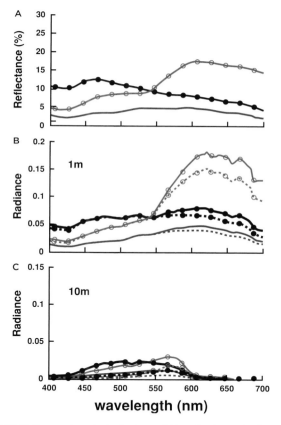

FIGURE 2 Measured spectral reflectance, $R(\lambda)$, and estimated radiance π of a territorial male peacock blenny, *Salaria pavo*. (A) Mean spectral reflectance measurements of *S. pavo* dark flank (—), blue ring (●), and orange crest (○). (B) Estimated color pattern radiances of these three color elements at 1-m depth under clear optical conditions (—) and high plankton conditions (---) using spectral irradiance measurements from Fig. 1. These colors are easily distinguished from each other at this depth, regardless of phytoplankton density. (C) Estimated color pattern radiances at 10 m under clear optical conditions (—) and high plankton conditions (---). These colors are less distinguished from each other at this greater depth, where the bandwidth of available light is diminished.

full-spectrum conditions, and each color reflects wavelengths at opposite regions of the spectrum. Both the blue- and orange-colored spots exhibit greater saturation than the unsaturated brown background coloration. The brown body coloration reflects similar amounts of light across the visual spectrum, reflecting 33% of its total reflectance in the orange (>600 nm) end of the spectrum and 26% in the blue (<500 nm). Meanwhile, the orange and blue color elements have more concentrated reflectance, with the orange crest reflecting 48% of its total reflectance in the longwave portion of the visual spectrum (>600 nm) and the blue rings reflecting 42% of their total reflectance in the shortwave bandwidth (<500 nm). These saturated colors are likely to be less mutable across the dynamic changes in intertidal ambient light spectra. We can investigate how these colors change with light conditions by estimating radiance (see Glossary) of these color patterns under different light fields. The estimated average radiance $L_{avg}(\lambda)$ from a diffusely reflective cylindrical target (e.g., fish) viewed horizontally in a light field dominated by direct downward light is given by

$$L_{avg}(\lambda) = \frac{I_D(\lambda)R(\lambda)}{2\pi}$$

where $I_D(\lambda)$ is the downwelling light spectrum representing the light falling upon the surface of the signaler, $R(\lambda)$ is the diffuse reflectance of that particular color element on the animal's body.

Calculating the estimated radiances for the intertidal peacock blenny shows that its richly saturated color patterns maintain a degree of color constancy (they retain their spectral shape) despite very different ambient light conditions (e.g., presence and absence of phytoplankton; Fig. 2B). The dynamically changing optical environment of the intertidal, therefore, may favor saturated and dark colors that are not greatly affected by rapid changes in the ambient spectrum. As species inhabit deeper waters in the near-shore environment where phytoplankton drastically reduces the spectral bandwidth, however, these dark colors become less distinguishable (Fig. 2C), suggesting that such dark and saturated colors would not be favored in deeper environments.

CRYPSIS IN THE INTERTIDAL

The optically dynamic intertidal environment also has special implications for crypsis. For a benthic animal, crypsis is achieved when its spectral reflectance matches that of the background (or of some random sample of the background). If the match is good at all wavelengths where vision occurs, then changing illumination does not affect crypsis. Note, for crypsis to occur, it need only match body and background wavelengths in the spectral region to which its predators can see. This means for animals whose predators have simple, single-photoreceptor vision (such as crabs), crypsis can be crude. However, for animals in the intertidal with bird predators that have multiple photoreceptor cell types, color crypsis must be a much better match, as these predators can see across a wide spectral range (300–700 nm).

Animals can be cryptic by matching the color of their background, as well as matching the brightness (total reflected light) of their background. The dynamic change in incident brightness over small spatial scales is likely to make brightness matching against a background very difficult in the intertidal zone. The lens effect of waves produces focal regions small enough that many animals more than a few centimeters in diameter have multiple regions of differing brightness. However, it may also provide a means to hide, as viewers cannot discern the entire form of an animal with multiple bright spots of illumination. The dark pigmentation found in intertidal animals, such as seen in the peacock blenny (Fig. 2), may make it difficult for these animals to match a brightly flickering background. Dark pigmentation reflects back only a small fraction of the incident light. If the total reflectance of the animal coloration is much lower than the rock, sand, or algal background, then there will be a mismatch between animal and background brightness. Hence, as the light flickers on dark-pigmented animals (such as the blenny), these spots will differ from those of the possibly brighter background (granite rock, sand).

The spectral quality of the illuminating light in the intertidal zone is highly variable, for all the reasons described in previous sections. Therefore, animals must match their reflectance to that of the background for all wavelengths relevant to vision if they wish to be continuously cryptic. Given that the rock background is made of molecules quite different from those found in biological pigment, this may be a challenging task.

SEE ALSO THE FOLLOWING ARTICLES

Bioluminescence / Blennies / Camouflage / Photosynthesis / Tides / Vision

FURTHER READING

Johnsen, S. 2002. Cryptic and conspicuous coloration in the pelagic environment. *Proceedings of the Royal Society of London: Biological Sciences* 269: 243–256.

Lythgoe, J. N. 1975. Problems of seeing colours under water, in *Vision in fishes*. M. Ali, ed. New York: Plenum Press, 619–634.

Partridge, J. C., and M. E. Cummings. 1999. Adaptation of visual pigments to the aquatic environment, in *Adaptive mechanisms in the ecology of vision*. S. N. Archer, M. B. A. Djamgoz, E. R. Loew, J. C. Partridge, and S. Valerga, eds. Dordrecht: Kluwer Academic.

LIMPETS

GEORGE M. BRANCH

University of Cape Town, South Africa

Limpets are perhaps the best known animals on rocky shores. They are defined as gastropod molluscs with a conical cap-shaped shell, but this definition covers species from at least three radically different and only distantly related groups. The first and largest group is the Patellogastropoda, which includes what most people would regard as true limpets: the Lottiidae in the Americas and Pacific Ocean; the Patellidae in Africa, Europe, and the Indo-Pacific; and the Nacellidae in the sub-Antarctic. The second group is the keyhole limpets or Fissurellidae, which are characterized by shells with a hole at the apex, through which wastewater and feces are passed, thus avoiding fouling the gills. Third, the family Siphonariidae includes the so-called false limpets, which are related to land snails and have lost their original gills and acquired lungs, only to develop secondary false gills so that they can now breathe in air using their lungs and respire under water using their gills.

Despite being only distantly related in evolutionary terms, all three groups share features that result in their having a common lifestyle. All have a cap-shaped shell that covers and protects the body, all have a large foot that allows tenacious attachment, and all have a radula that allows them to rasp algae off the rock surface.

HANDLING PHYSICAL STRESSES

Compared with most other animals, limpets are superbly adapted to wave action. Their shells are low and flat, reducing drag to water movement, and they have disproportionately large feet, increasing their tenacity. Attachment is not by means of suction, as popularly believed. If they were attached by suction, they would not be able to exceed one atmosphere of pressure in terms of the forces they can resist. In reality, they use an adhesive-like mucus, which is thinly spread between the foot and the rock face, and the powers of attachment of some species exceed the equivalent of five atmospheres of pressure. Part of their effectiveness is due to the relative rigidity of the foot. The more rigid it is, the less easy it is to dislodge a limpet. Take a limpet by surprise, and it is easy to dislodge it with a quick tap; but give it warning and its foot tenses up, increasing rigidity and strength of attachment.

The shape and texture of the shell also influence the relative resistance of the shell to passing waves. Theoretically, there are two avenues open to limpets to reduce drag. First, the shell may be low rather than tall. Most limpets achieve this, although, as seen herein, there are good reasons why some have relatively tall shells. Second, the shell may be streamlined. This can be accomplished by having the apex of the shell about one-fourth of the distance from the front end of the shell. However, few species conform to this theoretically desirable shape. Perhaps this is not unexpected: Most limpets will face waves rushing up the shore one moment, only to be confronted soon after by the same waves in reverse flow down the shore. As limpets cannot alter their positions at the rate that waves march up and down the shore, a streamlined shell that would be perfect for water movement in one direction would achieve the opposite effect when water movement is reversed. There are a few species of limpet that do have a guaranteed direction of water flow, namely, those that live their entire lives attached to the fronds of kelp plants. *Lottia insessa* in North America and the kelp limpet *Cymbula compressa* in South Africa are two examples. Because kelp sways back and forth with the waves, these species always experience water flow from the base of the kelp plant toward to its tips. In these species, the shells are strongly streamlined, to capitalize on the predictable flow of water.

Limpets may have a wonderful design in terms of wave action; but they are not ideal in terms of other physical stresses. Heat stress and water loss are major issues, especially on tropical shores, where the rock face may reach 60 °C. Limpets gain heat from solar radiation, and the flatter they are, they greater the planar area exposed to the sun. They also potentially gain heat by conduction though the rock face via their feet. Flat shells and large feet may be brilliant for dealing with strong wave action, but they are a disaster for reducing heat uptake. Shell color and texture can help avoid overheating. Pale shells are reflective, and textured shells increase the area from which heat can be lost by reradiation. Water loss is also a problem for limpets. Having a flat shell that covers the body inevitably means a large circumference to the mouth of the shell, and hence a large perimeter from which water can be lost. Again, the flatter the shell is, the greater the problem. Limpet shells are thus a compromise between different selective pressures. What is good for wave action is not necessarily good for avoiding heat and water loss.

As a solution to some of these problems, many limpets return faithfully to a particular position on the rock after each bout of feeding, so their shells grow to fit the rock face exactly, or they may even erode the rock to increase the fit, forming home scars. These serve multiple purposes. They allow the limpet to seal itself off more effectively, reducing water loss and osmotic stress, and to attach itself more strongly, diminishing the risk of being swept away by waves or detached by a predator. Moreover, most limpets have strong rhythms of activity. Some are active only when awash with the rising or falling tides, presumably the time when predation poses least threat. Others leave their scars only during low tide, when wave action is nonexistent. Some are even more particular and forage only at night during low tide. Yet others are active when submerged, and for them desiccation is presumably a greater problem than wave action.

How do limpets find their way back to their home scars after foraging? Many simply retrace their outward trails. However, experimental elimination of outward trails (or even elimination of the rock surface over which the limpets have traveled) fails to prevent limpets from relocating their home scars, so it has been suggested that they have memory of local topography or may even be using external clues to navigate, such as polarized light or the position of the sun or moon. In an ingenious series of experiments, the German scientist Funke reared limpets in aquaria where he could remove and replace portions of the floor after limpets had moved off their home scars to feed. He showed that if he replaced a portion of the floor with an equivalent piece that another limpet had traveled over, the returning limpet would fail to locate its scar. The implication is that limpets can recognize their own individual homeward trails and are not fooled by a section of trail belonging to another limpet.

EFFECTS OF LIMPETS ON ALGAE

Because limpets are grazers, they have the capacity to profoundly influence the development of algae (Fig. 1). As early as 1946, Jones and Lodge demonstrated this when they experimentally removed limpets from a 10 × 100-m stretch of shore on the Isle of Man: A lush growth of algae developed on what was previously bare rock, extending

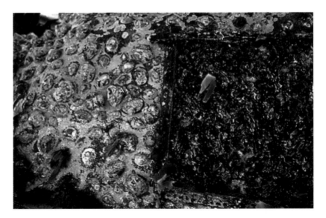

FIGURE 1 Experimental exclusion of limpets (right) leads to the development of dense growths of foliose algae, even in zones where such algae are normally absent. Conversely, some encrusting coralline algae rely on grazing by limpets to avoid being overgrown by foliose algae (left).

up the shore to cover zones where algae are not normally found. This simple but trend-setting experiment graphically demonstrated that limpets are continually controlling the growth of algae, primarily by consuming microscopic algal sporelings before they have a chance to become established. However, not all algae are negatively affected by limpet grazing. Some species of encrusting coralline algae, for example, rely on limpets to prevent their overgrowth by foliose algae (Fig. 1).

Limpets leave a trail of mucus when they move. This is energetically costly but does have the advantage that it traps settling algal spores, enhancing the supply of food in the vicinity of the limpet.

Not all limpets are equally effective at controlling algae. So-called true limpets (the Patellogastropoda) have relatively few but very large teeth on their radulae, and these teeth are tipped with iron oxide to strengthen them. Moreover, the iron oxide is concentrated on the front edge of each tooth, whereas the trailing edge is predominantly made up of softer elemental silicon. The net effect is that as the teeth are worn down through use, they are self-sharpening, retaining a sharp edge rather than being blunted. For all of these reasons, patellogastropods have a decided advantage over the siphonariid false limpets, which have multiple, soft teeth and cannot rasp away algae nearly as effectively. Whenever the two compete, the false limpets are at a disadvantage.

Although limpets can control the growth of algae, if beds of adult seaweeds do develop, many limpets are at a disadvantage because they feed less effectively on upright, foliar algae than they do on sporelings that can be scraped off the rock face. Some species of limpets will even die in the middle of apparent plenty if they are introduced into established algal beds.

The control that limpets exert over algae has spin-off effects on other organisms. For example, when algae proliferate after the removal of limpets, they smother barnacles, which indirectly suffer as a consequence. On the other hand, herbivorous winkles that live on algae benefit from the increase in algae, and even predatory whelks may benefit because the algae reduce desiccation stress. Because whelks feed on barnacles, this further increases the plight on barnacles. The effects of limpets are thus not straightforward. Their influence on algae may be simple to predict, but they set in action a chain of events that are often difficult to forecast.

OTHER WAYS OF FEEDING

Not all limpets act as simple grazers of algae. Two other modes of feeding have been developed by some species, with profound consequences for their population dynamics and for the operation of rocky shores as a whole. The first can be described as gardening. Gardening limpets have evolved specific associations with particular algae, which they tend and defend in a remarkably intimate manner. Each limpet is associated with a small patch of algae that it territorially defends against other grazers by lifting its shell and thrusting it against intruders until they retreat. This behavior was first described for the Californian limpet, *Lottia gigantea*, but it also takes place in South African limpets (Fig. 2), in the Kermadec Islands, and in the monstrous 20-cm-long Mexican limpet, *Scutellastra mexicana*. The famous naturalist William

FIGURE 2 Two South African territorial gardening limpets, the long-spined limpet, *Scutellastra longicosta* (center, with a large surrounding garden of the brown encrusting alga *Ralfsia expansa*), and the pear limpet, *Scutellastra cochlear* (surrounding individuals with narrow fringing gardens of filamentous red algae). Juveniles of *S. cochlear* often occur on shells of adults, as on the left.

Beebe (1942) described battles of the latter species as "a struggle between living tanks."

Territorial limpets not only defend their gardens against intruders, they weed the gardens by grazing away other species of algae, fertilize them with their ammonia-rich excreta, and water them at low tide by the slow release of water from beneath their shells. All these processes increase the productivity of the gardens, which are small but exceptionally productive. The limpets also graze the gardens in a manner that keeps them in a youthful, fast-growing phase in which their tissues are highly nutritious and low in herbivore-repellent chemicals. An intimate association develops, truly justifying the use of the word gardening.

Territorial gardening allows limpets to dominate particular zones on the shore and to achieve remarkable densities. In South Africa, the pear limpet, *Scutellastra cochlear*, attains densities of up to 1600 per square meter, with juveniles piled in layers on top of adult shells, and it excludes most other species from a characteristic cochlear zone low on the shore.

A second novel method of feeding is by trapping drift algae rather than grazing. Two species of limpet that feed this way are Argenville's limpet, *Scutellastra argenvillei*, and the granite limpet, *Cymbula granatina*, both occupants of the west coast of South Africa. Adults of Argenville's limpet feed by lifting their shells in a mushrooming stance and smashing them down if a blade of kelp washes beneath the shell. The edge of the shell is uniquely edged with sharp teeth to grip the kelp. A tug-of-war then ensues, with waves tugging the blade until it snaps, leaving the limpet in possession. Although a single limpet is responsible for capturing a piece of kelp, others may then share in the spoils, and groups of up to 24 limpets may cluster on a single kelp blade. This cooperative feeding means that, in contrast to the situation for other limpets, high densities are an advantage, and *Scutellastra argenvillei* characteristically lives in dense assemblages. *Cymbula granatina*, the other limpet that feeds by trapping, behaves quite differently. It lives in calm-water bays where swathes of drift kelp accumulate. It traps detached drift kelp under its sticky foot in a relatively lethargic manner that contrasts with the energetic reactions of *Scutellastra argenvillei*.

Because of the subsidy of food that they receive from outside the intertidal system, drift-trapping species attain exceptionally high biomasses—greater than have been recorded for any other grazer anywhere else in the world. As a consequence, *S. argenvillei* reduces the flora to a thin coating of grazer-resistant encrusting corallines; and

FIGURE 3 The granite limpet, *Cymbula granatina*, attains exceptionally high densities and biomass by trapping drift kelp that subsidizes intertidal shores.

where *C. granatina* reigns, the rock is denuded of most other life (Fig. 3).

PREDATORS

Limpets are consumed by many predators, including starfish, whelks, fish, birds, and mammals such as racoons and baboons. Some species specialize on limpets, including the suckerfish *Chorisochismus* in South Africa and *Sicyaces* in Chile. The shells of limpets are also prone to boring by lichens and blue-green algae, eroding them away faster than they can be replenished and causing death, particularly of older, slow-growing individuals. Human consumption is a serious problem in various parts of the world, including Chile, the east coast of South Africa, and the Canary Islands. In the latter case, depletion of limpets may have been a contributory cause of the extinction of the Canarian black oystercatcher, *Haematopus meadewaldoi*.

Most of the siphonariid false limpets are relatively immune to predators because they produce toxic chemicals. The true limpets do not have chemical defenses and depend on their strong attachment and shells for protection. Large individuals are safe against most predators and may even retaliate against whelks and starfish by smashing their shells down on would-be attackers.

The effectiveness of predators depends partly on how large and how fast limpets can grow. For example, on the west coast of South Africa, islands are home to large numbers of seabirds, and their guano fertilizes algae, enhancing productivity and thus increasing the growth and size of limpets. The African black oystercatcher, *Haematopus moquini*, congregates on these islands and feeds extensively on limpets, reducing their densities and further aiding algal growth. Fast growth to a large size is

essential for the survival of the limpets. Thus, a complex web of interactions exists, with the limpets occupying center stage as consumers of algae and a source of food for oystercatchers.

ALIEN SPECIES

The introduction of alien species constitutes a major threat to marine biodiversity worldwide. Limpets are not immune to this threat. For example, the Mediterranean blue mussel, *Mytilus galloprovincialis*, has now invaded Hong Kong, Japan, Korea, Hawaii, the west coasts of the United States, Canada, and Mexico, and southern Africa. In South Africa it now contributes 75% of the biomass of mussels, having displaced local mussels. Its effects on limpets depend on their sizes and their preferences for different intensities of wave action. The granular limpet, *Scutellastra granularis*, is almost completely displaced from primary rock space, but is small enough to survive and reproduce on the secondary substrate provided by the mussels themselves. The situation for a larger species of limpet, *S. argenvillei*, is more perilous because it is too large to achieve reproductive maturity while living on mussel shells, and is becoming extinct in areas of strong wave action that favor the Mediterranean mussel.

Much has been learned about rocky shores by using limpets as models to explore the factors that control the distribution and abundance of intertidal organisms. They regulate algae by grazing on them, compete with other sessile organisms such as barnacles, and constitute an important source of food for higher tropic levels. Clearly, they play a pivotal role in the ecology of rocky shores.

SEE ALSO THE FOLLOWING ARTICLES

Body Shape / Herbivory / Homing / Introduced Species / Predator Avoidance / Territoriality

FURTHER READING

Beebe, W. 1942. *Book of bays.* New York: Harcourt, Brace and Co.
Branch, G. M. 1981. The biology of limpets: physical factors, energy flow, and ecological interactions. *Oceanography and Marine Biology: An Annual Review* 19: 235–380.
Branch, G. M. 1985. Limpets: their role in littoral and sublittoral community dynamics, in *The ecology of rocky coasts.* P. G. Moore and R. Seed, eds. London: Hodder & Stoughton Educational, 97–116.
Bustamante, R. H., G. M. Branch, and S. Eekhout. 1995. Maintenance of an exceptional grazer biomass in South Africa: subsidy by subtidal kelps. *Ecology* 76: 2314–2329.
Lindberg, D. R. 1988. The Patellogastropoda. *Malacological Review* 4 (Suppl.): 35–63.
Peschak, T. P., and C. Velasquez Rojas. 2005. Limpets—a life between the tides, in *Currents of contrast.* Cape Town: Struik Publishers, 104–117.

LOBSTERS

CARLOS ROBLES
California State University, Los Angeles

Lobsters are a polyphyletic group of crawling decapod crustaceans that have a cylindrical, usually laterally compressed carapace and prominent abdomen with tail fan. The evolutionary origins of lobster-like crustaceans can be traced back at least 245 million years to the early Triassic Period. Of the many extant families in the group, the Nephropidae and Paniluridae stand out as rich in species, geographically widespread, and functionally important in near-shore and intertidal environments.

TAXONOMY AND NATURAL HISTORY

The family Nephropidae, or clawed lobsters, consists of approximately 118 species in 18 genera. They are characterized by claws on the first three pairs of walking legs, with the first pair often greatly enlarged relative to the other two. The antennae are small and located lateral to a prominent rostrum (forward projection of the carapace between the eyes). The Nephropidae includes the commercially and ecologically important Atlantic clawed lobster, *Homarus americanus*.

The family Paniluridae, or spiny lobsters, consists of approximately 45 species and occurs in all major oceans. Spiny lobsters lack both a rostrum and claws on the anterior walking legs. They defend themselves with long whip-like first antennae covered with sharp spines. A pair of forward projecting horns protect the stalked eyes. Many species emit harsh rasping sounds (stridulation) by rubbing the plectrum, a finely ribbed scale at the base of the antennae, against smooth ridges on the adjacent surface of the carapace. Stridulation apparently functions to startle predators and warn conspecifics. The most species-rich genera, *Jasus* with 9 species and *Panulirus* with 19 species, are found in shallow depths (<100 meters) of cool temperate latitudes (approximately 30–45°) and warm temperate to tropical latitudes (35–0°), respectively.

Homarus americanus in the Atlantic and *Panulirus* spp. in the Pacific play an important role in communities of the shallow subtidal zone (depth <35 meters). Acting in concert with others consumers, they hold populations of sea urchins and other herbivorous invertebrates in check, helping to maintain kelp beds and the diverse species assemblages that inhabit them. In some locations, lobster fisheries have severely reduced stocks, incurring outbreaks of grazing sea urchins and the subsequent loss of the kelp bed community.

Perhaps as a consequence of heavy exploitation reducing their numbers, as well as their secretive habits, lobsters are seldom thought of as intertidal species. However, early accounts of the colonists made reference to Native Americans collecting copious amounts of clawed lobsters from New England tidal flats during low tide. Early twentieth century naturalists on the southwest west coast of North America remarked on the shore-loving tendencies spiny lobsters and their frequent occurrence in tidepools. Contemporary observations on shores of California Channel Islands protected from sport and commercial exploitation reveal that *Panulirus interruptus* remain during the day in subtidal shelters and emerge at night to forage, and if the tide is high, they may roam into the upper levels of the intertidal zone (Fig. 1). Spiny lobsters crush shelled prey with large shearing mandibles on the ventral surface. Intertidal prey include polychaetes, urchins, barnacles, turbin snails, limpets, littorines, and most frequently, mussels (*Mytilus* spp.).

FIGURE 1 *Panulirus interruptus* foraging in the intertidal zone at night. Barnacles, anemones, mussels, and algae comprise the carpet of sedentary species. While foraging, spiny lobsters creep along tapping the substratum with the terminal segments (dactyls) of the forward walking legs. The dactyls pry up the prey, positioning it in front of the mouth on the ventral surface.

SPINY LOBSTERS AS KEYSTONE PREDATORS IN THE ROCKY-SHORE COMMUNITY

Keystone predation is confirmed experimentally when removal of the purported keystone predator species allows a prey species to increase sufficiently to eliminate its competitors. On rocky shores of the California

Channel Islands, dense populations of mussels *(Mytilus californianus)* and the other shelled invertebrates are restricted to wave-beaten headlands. Areas of shore that lie in the lee of the headlands are covered by a dense algal turf comprised of fleshy red algae (principally *Chondrocanthus* [= *Gigratina*] *canaliculata* and *Gelidum coulteri*) and articulated coralline algae *(Corallina pinnatifolia)*. Exclusion of *Panulirus interruptus* by cages over plots arrayed on midshore levels along the horizontal wave exposure gradient caused changes in the composition of attached life to varying degrees. Plots on a wave-exposed headland that were initially covered with mussels showed further increases in this competitively dominant prey species. Plots in the moderately wave-exposed (middle) segment of the array showed the rapid replacement of the algal turfs by a continuous cover of mussels; those in the most wave-protected areas showed a continuing dominance of the algal turfs, with only slight increases in mussels. The spiny lobsters are, therefore, important intertidal predators in the Channel Islands shore community, but they exert keystone predator effects only in a portion of the horizontal wave exposure gradient.

INFLUENCE OF HYDRODYNAMIC FACTORS ON THE IMPACT OF PREDATION BY SPINY LOBSTERS

The differences in the apparent effect of lobsters along the shoreline of the California Channel Islands result from the influence of hydrodynamic factors on both lobster foraging and the recruitment of the mussels. As waves pass over, the oscillatory flow across the rock surfaces of the intertidal zone subject the lobsters to the hydrodynamic stresses of drag and lift, which can damage or dislodge them. These hydrodynamic stresses are proportional to the square of the flow speeds. In turn, the flow speeds are proportional to the wave energy reaching the site and inversely proportional to the depth of the tide. Thus, for any point along the shoreline, maximum bottom flow speed traces a U-shaped curve over the course of a high tide, a consequence of the changing water levels: Water velocity is high when the tide first covers the rocks, but velocities decrease as the tide rises to its maximum height (Fig. 2). The values of the maximum speeds reached during the cycle of tidal immersion increase from wave-protected (relatively low energy) sites to seaward (relatively high energy) sites (Fig. 2). The differing time courses of flow speed, and hence stresses, over the horizontal gradient of wave exposure allow the lobsters prolonged access to wave-protected shores, but only a relatively brief window of opportunity to forage on wave-exposed shores, as

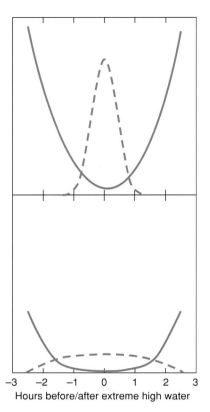

FIGURE 2 Diagram of idealized relationship between maximum bottom flow speed (solid lines) and lobster densities (dashed lines) as water levels change over the course of a high tide. Curves depict relationships for midshore levels from the beginning of submergence on the rising tide to reemergence on the falling tide. (Top) Time courses for wave-exposed site. (Bottom) Time courses for wave-protected site. The duration of foraging is longer on wave-protected sites, but the peak abundances of lobsters are lower in these prey-sparse areas. These relationships obtain in the calm summer season. Little foraging occurs on exposed sites in the stormy winter months.

evidenced by the shifting abundances of intertidal foragers (Figs. 2, 3).

Mussel recruitment (accumulation of juveniles on the shore following their settlement from plankton stocks) increases over the wave exposure gradient from wave-protected to wave-exposed sites. Therefore, the mussels' capacity to replenish their populations increases from protected to wave-exposed sites, as the accessibility of these populations to the predatory lobsters decreases over the same span. Hydrodynamic factors influence relative levels of these crucial processes such that one observes predatory elimination of sparse mussel recruits along midshore levels in protected areas, keystone predation at midshore levels of moderately exposed sites, and predatory reduction of otherwise dominant mussel beds on wave-exposed headlands.

The axis label below the figure reads: −3 −2 −1 0 1 2 3 — Hours before/after extreme high water

FIGURE 3 Time-lapse photographic sequence: four successive frames of a fixed plot on a midshore level recorded over the course of a high tide in summer. (A) During late afternoon, fishes enter midshore levels on the rising tide. (B) As night falls and the water levels rise still higher, maximum flow speeds decrease, and lobsters move onto the algal turfs. (C) Peak lobster abundances occur at extreme high water. (D) As water levels fall with the receding tide, maximum flow speeds and turbulence mount, and the lobsters leave the site well in advance of emergence. Photographs by the author.

SEE ALSO THE FOLLOWING ARTICLES

Algal Turfs / Hydrodynamic Forces / Predation / Recruitment / Tides / Wave Exposure

FURTHER READING

George, R. W., and A. R. Main. 1967. Evolution of the spiny lobsters (Palinuridae): a study of evolution in the marine environment. *Evolution* 21: 803–820.

Patek, S. N., R. M. Feldmann, M. Porter, and D. Tshudy. 2006. Phylogeny and evolution of lobsters, in *Lobsters: biology, management, aquaculture and fisheries.* B. F. Philips, ed. Ames, IA: Blackwell Publishing.

Robles, C. D. 1997. Changing recruitment rates in constant species assemblages: implications for predation theory in intertidal communities. *Ecology* 78: 1400–1414.

Robles, C. D., M. A. Alvarado, and R. A. Desharnais. 2001. The shifting balance of littoral predator-prey interaction in regimes of hydrodynamic stress. *Oecologia* 128: 142–152.

Shears, N. T., and R. C. Babcock. 2002. Marine reserves demonstrate top-down control of community structure on temperate reefs. *Oecologia* 132: 131–142.

LOCOMOTION: INTERTIDAL CHALLENGES

MARLENE M. MARTINEZ

American River College

The rocky-intertidal zone poses a unique set of challenges for locomotion. An animal is subject not only to periodic immersion and emersion, high water velocity, and uneven substrata but also to an extraordinary level of heterogeneity in water flow conditions—factors with the potential to assist, constrain, or even prevent locomotion. Furthermore, for animals using a form of pedestrian locomotion, in which thrust is generated by pushing or pulling on the substratum, the characteristics of the substratum will affect the animal's ability to hold on in waves or to gain purchase for forward motion.

ENVIRONMENTAL CONDITIONS ENCOUNTERED

An animal in the intertidal zone may encounter three basic types of environmental conditions that present different sets of forces affecting locomotion. Above the water line, but not in a tidepool, an animal experiences terrestrial conditions for locomotion, where buoyancy and fluid dynamic forces (due to the relative movement of air) are minimal and weight of the animal is the defining force. Within a tidepool that is above the reach of waves,

FIGURE 1 A rocky intertidal shelf in Hawaii at low tide. Note the many sea urchins in tidepools above the water line. Photograph by the author.

an animal is immersed in water but not subject to ambient water flow; thus the only fluid dynamic forces experienced by the animal are due to its movement through the water (Fig. 1). At or below the water line, an animal experiences immersion in water and hydrodynamic forces due to both locomotion and ambient water flow.

EFFECT OF IMMERSION IN WATER

A defining characteristic of the intertidal zone is the periodic immersion and emersion of the substratum. As the tide recedes, many organisms experience a temporary terrestrial habitat in which they must support the entire weight of their bodies without the assistance of buoyancy from the water. In terrestrial conditions, weight is the primary destabilizing force on an animal, as exemplified by an animal sprinting on land, when its weight is more than two orders of magnitude larger than any fluid dynamic force it experiences. An exception, however, occurs if the animal is both small and very fast, in which case it may generate fluid dynamic forces that approach the magnitude of its weight. The challenge of supporting one's body weight in terrestrial conditions is most severe for animals, such as fish, that lack strong supportive structures such as a crab's legs or a limpet's muscular foot. Even legs, however, do not ensure locomotion without difficulty in terrestrial conditions: Larger crabs such as *Cancer magister* (Fig. 2) and *Carcinus maenas* glide with apparent ease in water but, without the assistance of buoyancy, resort to dragging their bodies across the substratum on land.

Another locomotory hazard posed by terrestrial conditions is the extra force imposed by breaking waves. Should an emersed animal be hit directly by a breaking wave, it would be subject to severe impact forces not seen by an immersed animal, which is cushioned by the surrounding water.

Tidepools, an aquatic refuge when the tide is out, can provide a still-water environment with the most permissive locomotory conditions for swimmers and pedestrians alike. The main advantage of still-water conditions is the release from the requirement of holding up one's body weight. Because of this effect, aquatic pedestrian locomotion has been compared to human locomotion in reduced gravity. Because animal bodies commonly have a specific gravity near 1, immersion in water generates significant buoyancy, decreasing the apparent weight of an animal to near one-tenth its weight on land. Because of this smaller apparent weight, animals can use smaller, weaker structures to support the body during locomotion underwater than they must use when not submerged. This increased buoyancy may even translate into a decreased cost of transport in water compared to on land, especially for slower locomotory speeds in which fluid dynamic forces are small.

FIGURE 2 *Cancer magister*, the Dungeness crab, experiencing emersion in the rocky intertidal. Photograph by the author.

In still-water environments, animals have more freedom to use different gaits with shorter and less-regular periods of contact with the ground such as in underwater punting by shore crabs. In punting, an animal pushes off the substratum and glides; at a slow speed, the animal can achieve periods of noncontact with the substratum that are typical of faster speeds on land and that would not be possible in chaotic or high-velocity ambient water flow conditions.

Increased buoyancy in water compared to on land, however, imposes a different set of challenges. For nonswimming animals in the tidepools, force must be exerted against the substratum to generate thrust for movement. Although relieving the necessity of generating high vertical forces to support body weight, the added buoyancy in water may also cause difficulties for a pedestrian animal by decreasing the vertical force with which it can gain traction on the substratum. Furthermore, although buoyancy offsets an animal's weight, it does not decrease the animal's mass, thus disrupting the pendulum-like action that characterizes terrestrial walking gaits. With the resultant mismatch of kinetic and potential energy fluctuations, aquatic pedestrians may not be able to use the energy-saving walking gaits commonly used on land. Finally, by reducing an animal's apparent weight in water, buoyancy decreases the stabilizing moment that resists overturning. This effectively reduces the speed the animal can attain underwater, because as animals attain higher speeds, the hydrodynamic forces generated by their own movement can destabilize them even in the absence of ambient water flow.

EFFECTS OF WATER MOTION: HYDRODYNAMIC FORCES

Intensity of water motion varies with position on the shore, with mid and low intertidal positions experiencing faster flows than the highest positions on the shore that are not subject to breaking waves. Local topography will also influence the flow in microhabitats; for example, water will flow more slowly in the lee position behind a boulder and faster in a surge channel. Water motion from both waves and currents, in addition to that caused by an animal's own motion, generates hydrodynamic forces on the animal that sum to either assist or work against an animal's locomotion. These hydrodynamic forces can limit the speed, direction, gait options, and conditions under which an animal can locomote in the intertidal zone by forcing an animal to actively grasp the substratum to avoid dislodgement.

The main hydrodynamic forces a macroscopic animal will encounter are pressure drag (D) and lift (L), which can be described by the following equations, respectively:

$$D = 0.5 \, \rho U^2 A C_D$$

$$L = 0.5 \, \rho U^2 A C_L$$

where ρ = water density (kg per m^3), U = velocity (m/s, due to locomotion or ambient water motion), A = reference area of the animal (m^2, such as the area projected to the oncoming flow), C_D = coefficient of drag, and C_L = coefficient of lift. These equations illustrate the effects of size and shape on hydrodynamic forces. All else being equal, a larger animal has a larger reference area, thus it experiences larger drag and lift forces. An animal's shape affects forces as well and is loosely characterized by the drag and lift coefficients; lower drag and lift coefficients indicate lower hydrodynamic forces.

Drag, the force acting parallel to the direction of water motion, acts downstream. An animal locomoting in still water will generate hydrodynamic drag that acts against the animal's forward motion. An animal in moving water will experience drag in the instantaneous downstream direction, with the effects depending on how the animal is oriented relative to the flow at the moment. Water flow from behind will cause drag that enhances an animal's locomotion, similar to running with the wind at one's back. Drag in a direction not aligned with the direction of motion could perturb an animal's gait, potentially destabilizing the animal.

Lift is the hydrodynamic force that acts perpendicular to water flow. For an animal moving forward through still water, lift could act to either draw it up off the substratum or push it down toward the substratum. Like drag, lift can act in different directions with an animal's change in orientation relative to the water flow. Low-profile, streamlined shapes (i.e., with a lower C_D) minimize hydrodynamic drag but may actually increase the hydrodynamic lift, potentially pulling the animal off the substratum.

Another component of hydrodynamic force that might impact an animal's locomotion is the acceleration reaction force. Acceleration reaction opposes changes in water velocity. Thus, as an animal accelerates or as water accelerates over the animal, acceleration reaction will augment the drag force, but it will counteract the drag as the water flow relative to the animal decelerates. Acceleration reaction forces scale with the magnitude of acceleration, shape of the animal, and volume of the animal. Although the water accelerations in the intertidal zone can reach exceedingly high values, it is now recognized that their effect on organisms is minimal, because an animal that is small enough to fit within the spatial scale of these extreme accelerations has little volume on which the force can act. If acceleration reaction forces are important, they will be due to the acceleration of the body itself during locomotion.

EFFECT OF THE VARIABILITY OF WATER MOTION

In addition to high velocity, the variable nature of water flow in the rocky intertidal poses an extreme challenge to locomotion in this environment. Water flow may differ substantially in locations that are mere centimeters apart on the shore; water velocity commonly changes on a range of time scales, from once every few seconds to many times per second. The constantly changing magnitude and direction of water flow produce a constantly changing set of forces on an animal. An animal must therefore change the forces it generates on the ground or on the surrounding water to maintain control of its locomotion. Because the time scale for these changes can be small and the variation so large, responding in real-time may not be possible for many animals. However, there is evidence that some animals can sense information about oncoming waves and respond appropriately by anchoring or limiting locomotion during either specific waves or times of high wave action.

Some intertidal animals can manipulate the hydrodynamic forces imposed on them by adopting specific postures relative to the flow, such as a negative angle of attack (orienting downward into oncoming flow) to generate lift that pushes the animal onto the substratum. While their postural strategies can assist with locomotion in still water or steady flow, the extreme variability of flow direction in the surf zone renders this behavior ineffective. Indeed, this postural adjustment has only been reported for animals locomoting through still water, not for animals locomoting through the surf.

EFFECT OF SUBSTRATUM ROUGHNESS

The complexity of the substratum, which varies at multiple spatial scales, affects not just the local water flow but also an animal's ability to adhere to the surface. For animals that use creeping or pedestrian locomotion, the type of substratum present can affect the ability to gain purchase for generating thrust or the ability to grip for stabilization against being overturned or washed away. Gastropods such as limpets use adhesive mucus and suction with a single muscular foot to cling to the substratum while locomoting (Fig. 3). Echinoderms such as urchins and sea stars use mucus with multiple suckered tube feet that move like hundreds of little legs. Animals relying on suction may experience reduced tenacity on more rugose substrata. Due to the surface irregularity, animals may have difficulty maintaining the integrity of the seal between a sucker and the substratum. Furthermore, gripping roughness elements may require shape changes of suckers that decrease the projected area of the sucker and therefore reduce the adhesive force it can generate. The grasping abilities of legged animals, such as crabs, are sensitive to the scale of substratum roughness and tend to show a decreased ability to grasp a substratum when the roughness elements are smaller than the pointed ends of their appendages. Larger roughness elements also allow crabs to use more segments of a leg and therefore stronger muscles to attain tenacity. Spacing of roughness elements

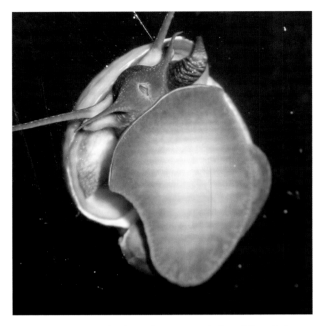

FIGURE 3 The muscular foot of a gastropod viewed from below as the animal crawls across the glass of an aquarium. Photograph by the author.

is also a factor, because if the elements are too far apart, a crab cannot reach from one foothold to another and is limited to locomoting under conditions in which actively grasping the substratum is not necessary, such as slow locomotion in still water.

SEE ALSO THE FOLLOWING ARTICLES

Adhesion / Buoyancy / Hydrodynamic Forces / Ocean Waves / Surf-Zone Currents / Turbulence

FURTHER READING

Full, R. J. 1997. Invertebrate locomotor systems, in *The handbook of comparative physiology*. W. Dantzler, ed. Oxford: Oxford University Press, 853–930.

Lau, W. W. Y., and M. M. Martinez. 2003. Getting a grip on the intertidal: flow microhabitat and substratum type determine the dislodgement of the crab *Pachygrapsus crassipes* (Randall) on rocky shores and in estuaries. *Journal of Experimental Marine Biology and Ecology* 295: 1–21.

Martinez, M. M. 1996. Issues for aquatic pedestrian locomotion. *American Zoologist* 3: 619–627.

Martinez, M. M. 2001. Running in the surf: hydrodynamics of the shore crab *Grapsus tenuicrustatus*. *Journal of Experimental Biology* 204: 3097–3112.

Martinez, M. M., R. J. Full, and M. A. R. Koehl. 1998. Underwater punting by an intertidal crab: a novel gait revealed by the kinematics of pedestrian locomotion in air versus water. *Journal of Experimental Biology* 201: 2609–2623.

MANAGEMENT AND REGULATION

JOHN UGORETZ AND TONY WARRINGTON
California Department of Fish and Game

The intertidal zone is rich with biodiversity and contains many fragile habitats and species. At the same time, the intertidal zone is the most accessible marine habitat to humans. The coast provides a scenic setting for recreation on sandy beaches and exploration of rocky-intertidal habitats. Marine intertidal resources also attract recreational anglers and commercial fishermen. Because of these characteristics, the marine intertidal zone is particularly vulnerable to human impacts. Effective management of the marine intertidal zone is essential to protect and preserve the functional and aesthetic characteristics of this important transition between land and sea.

MANAGEMENT GOALS

In the United States, most activity in the marine intertidal zone is managed by individual states, which generally have authority over the resources within three nautical miles seaward of their mainland and island shores. Individual states may delegate the responsibility for management of the marine intertidal zone to a resources agency, which contains a department of wildlife or fisheries. These departments tend to focus on regulating particular human activities such as hunting and fishing and, in special cases, access to important areas. More recently, the departments have broadened their missions to include conservation of ecosystems and the services they provide. In California, for example, the Department of Fish and Game's mission for managing marine ecosystems is "to protect, maintain, enhance, and restore California's marine ecosystems for their ecological values and their use and enjoyment by the public." To achieve their goals, state agencies and departments maintain a diverse staff, including policy experts, biologists, and enforcement personnel.

REGULATORY FRAMEWORK

Both state and federal laws and regulations govern marine intertidal resources in the United States. Although individual states govern most resource use, some regions are designated as national seashores, national parks, national marine sanctuaries, or national estuarine research reserves, which have additional federal regulations. State regulations must also be consistent with and may be more conservative than federal laws. In some regions, federal agencies have worked together with state partners to increase protection for marine intertidal habitats. For example, the Monterey Bay National Marine Sanctuary, which overlaps state waters in central California, worked with state and other federal agencies to reduce the input of agricultural and other pollutants that were carried in rivers, streams, and runoff to the ocean. Another example of partnership in marine management between state and federal agencies is the Channel Islands National Park and Marine Sanctuary, which overlap with state waters around California's five northern Channel Islands. The state Department of Fish and Game worked with these federal agencies to form a recommendation on marine protected areas around the Channel Islands. After the state established these protected areas, the agencies continued to partner on enforcement and monitoring.

State legislatures create laws that govern marine intertidal habitats or delegate regulatory authority to appointed commissions. State marine resource regulations generally focus on individual species or species groups. Recent legislation (such as California's Marine Life Management Act of 1998 and Marine Life Protection Act of 1999), however, has shifted this focus of management to an ecosystem approach that considers all species and their habitats.

In the United States, state and federal regulatory processes encourage public participation. Proposed new regulations and proposed changes to existing regulations must be announced to the public, and regulatory authority must allow ample time for public comment. Existing regulations must be reviewed regularly, and the review process must include opportunities for public participation.

The state departments of wildlife must gather data on targeted species and habitats for several purposes. The departments must understand how human activities are impacting the resources they manage. The departments must be able to provide useful scientific data to decision makers who regulate resource use.

REGULATION OF INTERTIDAL RESOURCES

The intertidal zone must be defined to create and enforce regulations on resource use. In California, for example, the intertidal zone is usually defined for the purpose of regulation as the area between the high tide mark (Mean Higher High Water) and 1000 feet seaward and lateral to the low tide mark (Mean Lower Low Water). In some municipalities, regulations are not specific to the intertidal zone, and those species found in the intertidal are managed under the same regulations for other areas.

Regulations can be subdivided into recreational regulations and commercial regulations. Regulations generally protect intertidal organisms with exceptions for those that are used for commercial or recreational purposes. The exceptions are intended to allow for the reasonable use of recreationally and commercially important species. This regulatory framework also protects species that are targets of new and emerging fisheries.

For example, in California, recreational fishing for intertidal invertebrates is limited to the following species and species groups, except where otherwise prohibited (such as in marine protected areas or closed fishing zones): chiones, clams, cockles, crabs, limpets, lobster, moon snails, mussels, native oysters, octopus, red abalone (north of San Francisco only), rock scallops, sand dollars,

FIGURE 1 A recreational angler fishes from the intertidal zone in Carmel Bay, California. Anglers frequently use tidepools as a source of bait species, in addition to being an access point to catch game species farther offshore.

sea urchins, shrimp, squid, turban snails, and worms (as incidental take to mussels) (Fig. 1). Regulations for recreational use may include minimum size and daily bag limits for some species. Invertebrate species without specific bag limits fall under a general bag limit of 35 of any species per day.

Commercial fishermen generally are required to purchase a special permit to take invertebrates from the intertidal zone. For example, in California, commercial fisherman must have a Tidal Invertebrate Permit to gather intertidal invertebrates. However, lobster, sea cucumber, squid, crab, and sea urchin fishermen are exempt from this special permit.

Scientific collection permits are also issued to individuals to take intertidal or other organisms for bona fide scientific, educational, or propagation purposes. These permits are issued by the appropriate government agencies on a case-by-case basis and require the reporting of specimens taken.

ENFORCEMENT

The enforcement of marine resource management laws and regulations varies from jurisdiction to jurisdiction, and many different agencies bear overlapping responsibilities. Enforcement duties include all commercial and sport fishing statutes and regulations, other restrictions, marine water pollution incidents, homeland security, and general public safety. State wardens and rangers who are deputized may be cross-deputized to enforce federal laws and regulations within the entire 200-mile span of the Exclusive Economic Zone.

Enforcement of marine intertidal and offshore regulations requires wardens to use a variety of enforcement tools, including large and small patrol boats, skiffs, ground vehicles, and aircraft. Technology such as vessel monitoring systems (VMS) may be used effectively to enforce commercial regulations offshore, but this approach cannot be transferred easily to use of intertidal habitats, which are accessible by foot. Special Operations Units (SOU) are needed to investigate large-scale poaching operations, which may severely impact intertidal resources. For example, special operations are used to track illegal take of abalone and lobster in California.

Enforcement personnel from various state and federal agencies work together on matters of mutual interest. Enforcement officers from state departments of wildlife, and parks and recreation, and federal fisheries, sanctuaries, parks, and the Coast Guard work together to achieve regional management goals. In spite of concerted efforts by state and federal agencies, the overall level of enforcement in the United States is barely adequate to protect marine intertidal habitats and state and federal waters.

Enforcement officers are frequently the primary point of contact for the public. For example, each year, wardens contact between 250,000 and 500,000 individuals fishing or hunting in California. Enforcement concerns within the intertidal zone range from educating the public to addressing illegal recreational and commercial take of the many marine creatures found there. Wardens routinely contact individuals within the intertidal zone to determine whether they are properly licensed or permitted and to ensure that no laws or regulations have been violated. Wardens strive to make these contacts informational and impress upon individuals the importance of protecting state marine resources for future generations.

Fishing regulations and license and permitting requirements are subject to change and should be checked regularly. The easiest and best way to protect intertidal resources may be to look and not touch when visiting the intertidal zone.

SEE ALSO THE FOLLOWING ARTICLES

Economics, Coastal / Education and Outreach / Habitat Alteration / Marine Reserves

FURTHER READING

California Department of Fish and Game, Marine Region, Fishery Regulations. http://www.dfg.ca.gov/mrd/index_regs.html.

MANTIS SHRIMPS

ROY L. CALDWELL

University of California, Berkeley

Mantis shrimps are a group of predatory marine crustaceans characterized by a pair of greatly enlarged and powerful feeding appendages. Specialized for either impaling or smashing prey, these weapons not only reflect the type of prey eaten but also determine fighting ability. Most spearing species excavate their own burrows in soft sediments, but those species that smash their prey are more likely to live in preexisting cavities in rock or coral. It is the possession of these lethal hammers that allows smashing mantis shrimps to compete effectively against other cavity-living specialists in the rocky intertidal.

HISTORY

During the Lower Carboniferous approximately 350 million years ago, an obscure group of crustaceans in the order Hoplocarida began evolving a novel pair of raptorial feeding appendages. This set them on a unique evolutionary trajectory that would eventually make them into mantis shrimp (stomatopods), well suited for inhabiting tropical and subtropical marine rocky and coral tidepools. Before this time, Paleozoic proto-stomatopods possessed five pairs of subchelate mouth appendages of approximately equal size. These appendages (maxillipeds) were probably used to sift detritus or to collect small, relatively nonmobile prey from soft substrates. However, some species in the family Tyrannophontes evolved the second pair of these maxillipeds into much larger, more elongate appendages capable of quickly reaching out to seize bigger, more mobile quarry. By the late Paleozoic the first true stomatopods can be recognized. They are characterized by greatly enlarged raptorial second maxillipeds, large complex eyes set on mobile stalks, a pair of triramus antennules, a pair of antennae bearing large antennal scales, a carapace covering just the anterior portion of the cephalothorax, three pairs of walking legs, five pairs of abdominal flaplike pleopods bearing gills and also used for swimming, and a well-developed tail fan with large, armed uropods and telson (Fig. 1). All seven modern stomatopod superfamilies (Stomatopoda) were in place by the end of the Cretaceous. Today we recognize more than 500 species in 17 families. All are marine predators. Most occur in shallow tropical or subtropical seas. Species range in size from less than 2 to more than 38 cm in total length.

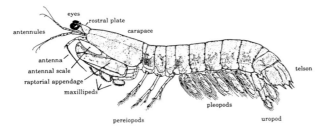

FIGURE 1 *Pseudosquilla ciliata* with the major body parts and regions labeled.

FEEDING APPENDAGES

The raptorial appendage of stomatopods has evolved into an effective feeding device. Normally carried folded beneath the head and thorax (Fig. 2), the terminal segments can be extended rapidly with their sharp spines impaling and seizing soft-bodied prey such as shrimp and fish (Fig. 3).

FIGURE 2 The largest of all stomatopods, *Lysiosquillina maculata*, has a typical spearing raptorial appendage. Photograph by the author.

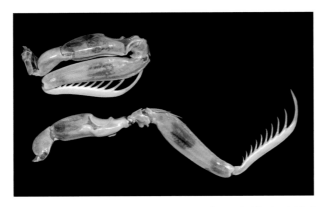

FIGURE 3 The spearing appendage is normally carried in the folded position (top), rapidly extends to impale soft-bodied prey (bottom). The pair of appendages photographed are from a 30-cm male *Lysiosquillina maculata*. Photograph by the author.

The strike is analogous to that of a praying mantis, which gives stomatopods their common name of mantis shrimp. There are two fundamental differences, however. First, the strike of a mantid's forelimbs is overhanded; that of mantis shrimp is underhanded. Second, the strike of a mantid is fast—about 100 milliseconds. This is about the time it takes a person to blink an eye. The strike of the mantis shrimp is up to 50 times faster, occurring in just a few milliseconds. Although the strike of mantis shrimp with a spearing appendage is effective at capturing unarmored prey, the slender dactyl spines are useless against armored animals such as snails, hermit crabs, and barnacles. A few times in the evolution of stomatopods a modification to the raptorial appendage has occurred, making it a more effective weapon used to capture and process heavily armored prey. The base of the dactyl heel is enlarged and highly calcified, turning it into a functional hammer used to smash armored prey to bits (Fig. 4). Rather than strike with an open dactyl as spearers do, smashers hit with the dactyl closed, delivering a crushing blow. Almost all modern smashers occur in the superfamily Gonodactyloidea. The power of such a strike is impressive. The force generated by the strike of a large smasher, the 14-centimeter-long *Odontodactylus scyllarus*, has been measured in excess of 1500 newtons (337 pounds of force). In fact, the forces recorded when a smasher hits a hard target occur as two very powerful but brief impulses. The first is generated when the dactyl impacts the target. The second, about 0.5 millisecond later, is caused by cavitation bubble collapse and can be about 50% of the initial impact force. This hammering mechanism allows smashing mantis shrimp to produce peak forces that far surpass those produced by other similarly sized predators such as fish or crabs that typically crush or peel armored prey. Even though these high forces are delivered over very brief periods (about 50 microseconds), repeated blows by smashing stomatopods enable them to fracture even heavily armored shell.

HABITAT

The evolution of the spearing and smashing appendage has had a profound impact on almost every aspect of stomatopod biology. For example, spearers typically excavate burrows in soft sediments such as sand or mud, and adopt an ambush mode of hunting. Burrows are not vigorously defended, in part because they are easy to construct and in part because the raptorial appendages of spearers are not effective weapons against armored competitors. Smashers, on the other hand, typically live in preexisting cavities in rock, coral, or coralline algae. These homes provide excellent protection from predators such as large fish, but

FIGURE 4 *Gonodactylus chiragra* is a 9-cm-long smashing stomatopod that lives in tidepools in the Indo-Pacific. Its enlarged dactyl heel (arrow) serves as an effective hammer. Photograph by the author.

cavities are often in short supply, and competition for them can be intense. The strike of a smashing species is potentially lethal to opponents of similar size. Smashing stomatopods fight vigorously to secure and defend their cavities.

These two factors, that smashers are able to feed on armored prey and that they have the fighting ability to obtain and defend scarce cavities, are primarily responsible for why most tidepool species are gonodactyloid smashers. In the harsh world of the rocky tidepool, many potential prey are heavily armored and are able to clamp down tightly to the substrate. Only smashing stomatopods with their lethal sledgehammer blows are able to dislodge and process them.

Some species, such as the common Indo-Pacific species *Gonodactylus chiragra*, are generalists, taking a variety of prey such as nerite and other armored snails, hermit crabs, limpets, and crabs. They forage for only a few minutes during daylight when the tide is low and when large predatory fish have been driven from the shallow tidepools. Roaming up to several meters from their home cavity, they select a suitable prey item and carry it back to the safety of their cavity, where they can safely break it apart as high water and predators return. The solid wall of the cavity serves as an anvil against which to smash prey.

Other smashers are more specialized. *Neogonodactylus festae*, a 7-cm-long smashing species from the Pacific coast of Panama, feeds heavily on barnacles that settle near its cavity. It quickly darts out of its cavity at most a few body lengths, smashes in the crowns of barnacles, picks out the soft tissues using its dexterous third to fifth maxillipeds, and rapidly returns to the safety of its cavity. By so doing, the mantis shrimp clears an area around its cavity entrance, providing suitable space for recruiting juvenile barnacles—in effect, farming a renewable crop.

COMPETITION FOR CAVITIES

Rocky tidepools are often worn smooth by surf. They lack the nooks and crannies that provide shelter for many animals on coral reefs. The few flask-shaped cavities that offer shelter to stomatopods are typically formed by rock-burrowing organisms such as razor clams or sipunculids or are built by the stomatopods themselves from natural crevices. Pounding pebbles and shell into a crack, these stone masons are able to construct a cavity. Over time, coralline algae cement the walls together, yielding a strong cavity. Typically, the cavities are shaped such that they retain at least some water during exposures, allowing the stomatopods to keep their gills moist. With time, cavities in rock or coral are gradually enlarged through the constant activity of the occupants, particularly through the smashing of prey in the cavity. Or, as residents molt and grow, they attempt to enlarge the cavity by chipping away at the entrance. There is a tight correlation between the size of a stomatopod and the diameter of the entrance of the cavity it occupies. If the entrance is too large, bigger stomatopods can force their way in, evicting the resident. If the entrance is too small, the resident can barely squeeze in, let alone bring prey inside for processing.

Most species of tidepool stomatopods live four to six years. Even though they slow their growth after reaching sexual maturity, they still need to move to larger quarters every few months to obtain an optimal defensible fit. Fights for suitable cavities are frequent and often escalate into potentially lethal confrontations. A first line of defense from an opponent's blows is heavy body armor, particularly on the telson. The telson can be wedged in the cavity opening to prevent the intruder from entering, or the intruder can coil near the entrance using its tail as a shield to block the resident's strikes before launching its own attack (Fig. 5). Since cavity occupants may remain in the same cavity for weeks or even months, it pays to learn the identity of their neighbors and their fighting abilities. Gonodactylid smashers have been shown to learn the individual odors of their neighbors and avoid attacking those who have defeated them while trying to evict animals they have defeated. This ability to associate an animal's identity with its fighting prowess is retained for at least several days.

FIGURE 5 *Haptosquilla trispinosa* is a 4-cm-long smasher that uses its armored telson as a shield when trying to evict another stomatopod from its cavity. Photograph by the author.

PHYSIOLOGY

Tropical tidepools often experience extreme fluctuations in temperature and salinity. Heavy rain during low tide can dramatically reduce salinity. The midday sun can heat pools to in excess of 40 °C. Although most subtidal stomatopods are relatively intolerant of such changes, tidepool-inhabiting species can withstand more dramatic exposures. On the Atlantic coast of Panama, *Neogonodactylus bredini* and *N. oerstedii* frequently inhabit tidepools, whereas *N. austrinus* occurs on the fringing reef crest and is rarely exposed. When exposed to 39 °C water for 90 minutes, most *N. bredini* and *N. oerstedii* survived, but not *N. austrinus*. Similarly, the tidepool species were more tolerant of immersion in freshwater than were reef crest and subtidal species, reflecting their ability to withstand a sudden cloudburst at low tide.

Neogonodactylus species brood their eggs for three weeks, and the first three stages of larvae for another week, before they enter the plankton for a month. The eggs and early larval stages of the tidepool species of *Neogonodactylus* are just as tolerant of heat and hyposaline water as are their mothers. However, once the larvae enter the open water, their resistance to environmental stresses decreases dramatically. Even so, brooding tidepool mothers can suffer dramatic losses of their clutches. For example, immediately prior to extreme spring low tides in May, 40% of large *N. bredini* and *N. oerstedii* females were brooding eggs and early-stage larvae. After five consecutive midday low tides with bright sun, not a single female in the intertidal was found with surviving offspring. Most of the females survived by temporarily abandoning their cavities and moving to deeper water, but they did not transport their eggs and larvae, which perished.

The evolution of the specialized smashing raptorial appendage has allowed stomatopods to occupy tidepools by competing effectively for scarce cavities, modifying those cavities when necessary and feeding on the armored prey found there. For their size, there are few nonvenomous predators as formidable as a mantis shrimp.

SEE ALSO THE FOLLOWING ARTICLES

Competition / Heat and Temperature, Patterns of / Predation / Stone Borers / Territoriality

FURTHER READING

Caldwell, R. L. 1987. Assessment strategies in stomatopods. *Bulletin of Marine Science* 41: 135–150.
Caldwell, R. L. 1991. Variation in reproductive behavior in stomatopod crustacean, in *Crustacean sexual biology*, R. Bauer and J. Martin, eds. New York: Columbia University Press, 67–90.
Caldwell, R. L., and H. Dingle. 1976. Stomatopods. *Scientific American* 1976 (Jan.): 80–89.
Patek, S. N., and R. L. Caldwell. 2005. Extreme impact and cavitation forces of a biological hammer: strike forces of the peacock mantis shrimp (*Odontodactylus scyllarus*). *Journal of Experimental Biology* 208: 3655–3664.
Reaka, M. L., and R. B. Manning. 1987. The significance of body size, dispersal potential, and habitat for rates of morphological evolution in stomatopod Crustacea. *Smithsonian Contributions to Zoology* 448: 1–46.
University of California Museum of Paleontology. Introduction to the Stomatopoda. http://www.ucmp.berkeley.edu/arthropoda/crustacea/malacostraca/eumalacostraca/stomatopoda.html.

MARINE IGUANAS

SEE IGUANAS, MARINE

MARINE RESERVES

STEVEN D. GAINES

University of California, Santa Barbara

Marine reserves are areas of the sea set aside for special protection with very simple rules—you cannot remove any animals or plants. Often called no-take marine reserves, they are a special type of Marine Protected Area (MPA) that prohibits all fishing and other forms of collecting. Other types of MPAs, which provide less extensive protection from fishing, have been called by many different names—marine parks, marine sanctuaries, marine conservation areas, marine refuges, and so on. Although most marine reserves target subtidal habitats, some focus primarily on rocky-intertidal shores.

GOALS OF MARINE RESERVES

Marine reserves are a form of spatial management of the ocean. By applying specific rules to specific places, they are similar to the more common zoning regulations that dominate the management of terrestrial landscapes. There are a wide range of goals that have driven the establishment of marine reserves. They include the protection of unique places of special biological, historical, or cultural significance, the recovery of threatened and endangered species, the protection or restoration of fragile habitats, wilderness sites for diving and snorkeling, reference sites for scientific study of ecosystems with more limited human impacts, and the more effective management of fisheries.

One might expect vast areas of the ocean would be *de facto* marine reserves simply as a result of their isolation from ports. Although this was true many decades ago, advances in technology for capturing and processing fish at sea have left few areas of the world's oceans untouched by humans. The oceans, like the land, are human-dominated ecosystems. Marine reserves, therefore, are rarely placed in settings where they protect pristine habitat with little history of human impact. Rather, they reduce human impacts and set the stage for ecosystem responses to reduced fishing.

Collecting marine animals and plants from rocky intertidal shores at low tide has a rich history in many cultures. Although most people imagine scenes with boats, nets, and hooks when they think of fishing in the sea, harvesting from the shore is a common practice in many regions of the world (Fig. 1) and has been a subsistence source of food for millennia.

CHANGES WITHIN RESERVE BORDERS

Only a minute fraction of the world's oceans are protected in marine reserves. Yet marine reserves are not rare. There are well over 200 marine reserves around the world, spread across more than two dozen countries. Studies from many of these reserves have been published in the scientific literature. Therefore, we have been able to learn a lot about how marine ecosystems change when fishing is banned. Some marine reserves were created many decades ago; others are in their infancy. Establishing a marine reserve starts an experiment in which the impacts of humans change. By comparing how populations inside the reserve respond after the reserve is implemented or by comparing patterns inside reserves to comparable unprotected habitats nearby, we can tease apart the complex ways that people alter marine ecosystems.

FIGURE 1 Subsistence harvesting from South African shores has an incredibly long history. In the top panel, South African women scour the shore at low tide in search of mussels, ascidians, chitons, and other food. Many areas of South Africa's coastline have now been zoned into different types of marine protected areas. The bottom panel shows the welcome sign for the Maputaland Marine Reserve with warnings in English, Afrikaans, and Zulu. Photographs by the author.

A synthesis of roughly 100 scientific studies of existing marine reserves from around the world clearly demonstrates that fishing has substantial impacts on marine populations (Fig. 2). Compared to protected areas outside the reserves, on average the density of individuals increased 200%, their size increased 82%, total biomass increased more than 400%, and the number of species increased by 70% inside reserves. Although the great majority of these studies came from subtidal reefs, changes on intertidal shores were consistently in the same direction. Marine reserves typically harbor more, larger individuals of a greater diversity of species.

Although, on average, individuals are large and more abundant inside marine reserves, some species become substantially rarer in areas that exclude fishing and collecting. This is a predictable consequence of ecological

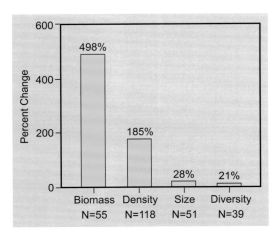

FIGURE 2 Marine reserves tend to increase the biomass, density, size, and diversity of species relative to unprotected areas. These data are average responses from scientific studies of reserves from around the world that were synthesized by B. Halpern in 2003. Most of the studies are from subtidal habitats. Bars represent average increases from all studies. Figure courtesy of S. Lester and the Partnership for Interdisciplinary Studies of Coastal Oceans (PISCO).

interaction between species. Since fishing is often concentrated on predators and herbivores within marine food webs, when they become more abundant their prey may become rarer. For example, a predatory whelk, commonly known as the *loco*, is one of the most sought-after animals from the rocky shores of Chile. As a result, outside of marine reserves it can become quite rare. As a result, one of its favored prey, a small mussel, becomes extremely abundant. When a marine reserve was established along Chilean shores, the abundant beds of mussels disappeared as *locos* returned to the shore in abundance. Such cascading effects of removal of fishing are quite common. On many subtidal reefs, the cessation of fishing for lobster and predatory fish can lead to dramatic declines in the abundance of their common prey—sea urchins. Since urchins are voracious consumers of seaweed, increased predation on urchins within marine reserves can lead to dramatic increases in the abundance of kelps and other seaweeds. Such trophic cascades through multiple trophic levels (↓ human fishing ⇒ ↑ lobster and predatory fish ⇒ ↓ sea urchins ⇒ ↑ kelp and seaweed) have been seen repeatedly on reefs in South Africa, New Zealand, and the United States (Fig. 3).

Because marine reserves generally ban only fishing and collecting, they do not remove all human impacts. This is especially true on rocky intertidal shores, where people can have large negative impacts even when they only walk around and look. Exploring the shore involves stepping on invertebrates and seaweeds, and this trampling has

been found to have dramatic effects on the fate of some species. Similarly, turning over rocks to find the biological treasures hiding beneath can expose species to conditions they cannot tolerate. Therefore, if marine reserves attract many more visitors to the shore to observe the rich biological diversity, they may substitute one human impact for others without thoughtful planning and responsible behaviors.

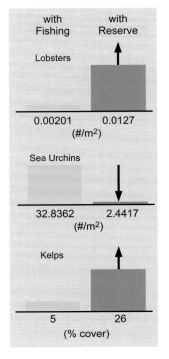

FIGURE 3 Reduced fishing at one level of marine food webs can have cascading effects on lower trophic levels. Lobster are a favorite fishery target. They are also important predators themselves. Under the protection of a marine reserve, lobster abundances increase more than sixfold. As a result, the abundance of their prey, sea urchins, declines more than tenfold. Such large reductions in the abundance of these key herbivores greatly enhance the success of kelps. Banning fishing on lobster indirectly benefits kelps by reducing the density of herbivores. Data are from Anacapa Island off of the coast of California, courtesy of M. Behrens and K. Lafferty.

CHANGES BEYOND RESERVE BOUNDARIES

Marine reserves restrict fishing only within their borders, yet they can alter areas outside in several ways. First, animals protected within reserves do not know where the boundaries of the reserve are. As a result, their random movement can lead to the spillover of adults to areas where they can be caught. If a species moves too much, there will be little benefit of the reserve at all because individuals will regularly move beyond the protection of its borders. This will preclude the buildup of dense populations of older, larger individuals within the reserve boundary. If movement distances are more modest relative to the size of the reserve, the resulting dense populations inside the reserve borders can generate substantial spillover of adults at the border. As a result, the edge of the reserve can be an extremely successful place to fish. Moreover, since the fish inside the reserve tend to be larger and older, reserve spillover can be an important source of large, trophy fish. Off the coast of Florida in the United States, a marine reserve, established in the 1960s because its waters were beneath the flight path of rockets launched from nearby Cape Canaveral, is now the source of a remarkably large number of world record catches of black and red drum, two fish popular with recreational fishermen.

The spillover of adults and its support of fisheries beyond the reserve boundary depends on both the high densities within reserve borders and the movement of individuals. For rocky-intertidal habitats, most of the common species are relatively immobile or even permanently attached to the rock. As a result, spillover of adults across the reserve boundary will generally be trivial. You may wait a long time for an abalone to randomly crawl across the border.

The second way that marine reserves can affect areas beyond their border is through the export of young. If individuals are older and larger inside reserves, they likely produce many more young. In general, the number of eggs produced by marine invertebrates and fishes or the number of spores produced by seaweeds increases exponentially with the size of the adult. Therefore, reserves can produce far more young from a given area than comparable unprotected habitat. Since biomass increases on average more than 400% inside reserves and reproductive output generally increases more than biomass, the potential added production can be enormous.

If all this added reproduction stayed within the reserve borders, the benefits to protected populations could be large, but the benefits to fished areas and fishermen would be minimal. For most marine species, however, even large marine reserves cannot contain their young. The great majority of marine fish and invertebrates produce young, called larvae, that are microscopic and drift in the plankton as they develop. They may drift from hours to months depending on the species. While they are developing, they are moved by ocean currents. Similarly, seaweeds produce eggs and spores that can also be moved by currents before they settle to the shore. Therefore, for any given reserve, the young of many species are likely to be exported well beyond the reserve's borders. If these exported larvae survive the perils of planktonic development, they can grow up to be a part of some fisherman's catch. As a result, the benefits of the added production of young within reserves can be transferred to fisheries beyond the reserve's borders. Given that most intertidal animals are slow-moving or immobile as adults, this export of young is the primary way that reserves may benefit unprotected populations.

Finally, reserves can also potentially have negative impacts on unprotected areas nearby. The source of these impacts is the displacement of fishing. Once a reserve is implemented, the fishing that previously occurred within the reserve is displaced. If the total amount of fishing does not decrease, then unprotected areas will experience a greater fishing intensity. By squeezing fishing into a smaller area, the negative impacts may increase in unprotected areas. Few studies have adequately studied how this displaced fishing affects unprotected areas after marine reserves are implemented. A few studies, however, have monitored comparable areas inside and outside of marine reserves before and after implementation. These kinds of studies provide unique insight into how the establishment of the reserve affects unprotected populations. Because there are biological ways that reserves can benefit unprotected areas (adult spillover and larval export) and other ways that reserves can harm unprotected areas (displacement of fishing effort), what are the likely net effects? Although the number of studies where this question can be addressed is small, nearly all cases showed increases in abundance and biomass outside reserve boundaries after reserves were established. This suggests that at least in these cases, the benefits of reserves can compensate for the displacement of fishing effort.

NETWORKS OF MARINE RESERVES

In the past, marine reserves focused on protecting single places. Today, there are global efforts to create larger networks of multiple marine reserves that collectively provide more benefits than the sum of the individual marine reserves. Large networks of marine reserves and other MPAs have been or are being implemented in Australia, South Africa, New Zealand, and along the coast of California in the United States. Each of these efforts includes dozens of MPAs. Two broad goals are driving the focus on networks. First, the ocean contains diversity of habitats that often harbor different suites of species. As a result, any attempt to conserve the rich biological diversity of the sea must capture the rich diversity of habitats. Because marine reserves placed at an individual site will include, at most, a few habitats, creating protection that represents the diversity of marine species must include multiple reserves and MPAs. Networks can represent a much larger fraction of marine biological diversity than individual marine reserves.

The second push for marine reserve networks is more functional. To have a substantive impact on a population, a sizable fraction of individuals must be protected. Imagine a 100-meter stretch of shore set aside as a marine reserve to protect abalone. The reserve could be wildly successful, with ten times the biomass of abalone inside relative to comparable stretches of unprotected shore. Yet, the contribution of this single marine reserve to abalone as a species would be minimal. Even with high densities, the total number of abalone protected is small. The only way to increase the total impact is to increase the total area protected.

Creating very large marine reserves would undoubtedly have great conservation benefits, but they impose an important added cost—they reduce the fisheries benefits to areas outside their borders. The rationale is simple. As the size of the reserve increases, spillover of adults and export of larvae across the reserve boundaries do not increase proportionally. For intertidal reserves, the relevant boundaries (edge of protected area along the shore) do not increase at all. As reserves grow in size, more of the reserve is far from the boundary. Networks of multiple reserves help reduce some of this inherent conflict between conservation value and fisheries.

Considerable scientific effort is now being focused on design guidelines for MPA networks to achieve both conservation and fisheries benefits where possible. The emerging findings focus guidelines for the size of individual MPAs on the average movement distances of target species. By contrast, the spacing of MPAs is focused more on movement of larvae to simultaneously connect MPAs through dispersal of young and spread the export of reserve larval production across all unprotected areas between reserves. Since different species move different distances both as adults and larvae, these guidelines inevitably include compromises in any attempt to achieve benefits for the diverse array of species common to most marine habitats.

SEE ALSO THE FOLLOWING ARTICLES

Biodiversity, Maintenance of / Collection and Preservation of Tidepool Organisms / Dispersal / Disturbance / Food Uses, Modern / Monitoring, Overview

FURTHER READING

Behrens, M. D., and K. D. Lafferty. 2004. Effects of marine reserves and urchin disease on southern California rocky reef communities. *Marine Ecology Progress Series* 279: 129–139.

Castilla, J. C. 1999. Coastal marine communities: trends and perspectives from human-exclusion experiments. *Trends in Ecology and Evolution* 14: 280–283.

Gaylord, B., S. D. Gaines, D. A. Siegel, and M. H. Carr. 2005. Marine reserves exploit population structure and life history in potentially improving fisheries yields. *Ecological Applications* 15: 2180–2191.

Gell, F. R., and C. M. Roberts. 2003. Benefits beyond boundaries: the fishery effects of marine reserves. *Trends in Ecology and Evolution* 18: 428–434.

Halpern, B. S. 2003. The impact of marine reserves: do reserves work and does reserve size matter? *Ecological Applications* 13: S117–S137.

National Research Council. 2001. *Marine protected areas: tools for sustaining ocean ecosystems*. Washington, DC: National Academy Press.

MARINE SANCTUARIES AND PARKS

WILLIAM J. DOUROS AND
ANDREW P. DEVOGELAERE

Monterey Bay National Marine Sanctuary
Monterey, California

Most nations set aside sites that protect nationally significant areas. Many include intertidal communities. In the United States, such nationally protected sites are called National Parks, National Monuments, and National Marine Sanctuaries. Protected areas provide distinct challenges to managers. Although they typically have a mandate to protect intertidal resources, managers often must also promote human access and visitation. Many human impacts to intertidal communities come from human access. Management issues related to rocky shores include public benefits and human uses.

NATIONAL MARINE SANCTUARIES AND NATIONAL PARKS

National Parks, National Monuments, and National Marine Sanctuaries (Sanctuaries) in the United States protect nationally significant areas, including intertidal communities. The respective management agencies for these federal protected areas have mandates to protect and interpret them for the public in terms of natural and cultural resource values. This includes optimizing components of ecosystem health, such as historic biodiversity, and protection of cultural resources, such as shipwreck artifacts. Moreover, managers of Sanctuaries and National Parks are mandated to understand population dynamics and ecosystem trends, then interpret the information for effective resource management decisions and public education. Perhaps most difficult in the management of intertidal communities in National Parks and Sanctuaries is that although the areas are to be protected, public access is also to be provided, and in some cases promoted, as long as the primary goal of protection can be maintained. This can lead to a challenging balance of human expectations.

Although intertidal areas are found throughout the coastal states of the United States, they are most prominent in National Parks and Sanctuaries along the West Coast, in the states of California, Washington, and Alaska. National Parks most typically manage terrestrial areas whereas Sanctuaries manage marine resources, with intertidal communities being a habitat sometimes along the edge of their jurisdictional boundaries. Managers of National Parks and Sanctuaries whose boundaries abut will often work collaboratively to manage intertidal areas by sharing information, staff, and equipment (e.g., at the Channel Islands and Pt. Reyes Seashore in California and the Olympic Coast in Washington). This optimizes government resources for the public benefit.

RELEVANCE OF INTERTIDAL COMMUNITIES IN MARINE MANAGED AREAS

After beaches, intertidal communities are the next most accessible marine habitat to humans. With no special equipment such as boats or SCUBA gear, humans routinely access rocky shores for recreation, education, and research (Fig. 1). For instance, in the nation's largest National Marine Sanctuary, Monterey Bay, rocky shores make up 56% of the 280 miles of shoreline; however, the width of these shores is narrow (rarely over 30 meters), and so the percentage of total habitat of this same Sanctuary is actually very small (<1%). Although some intertidal areas in National Parks are remote, such as Glacier Bay in Alaska or the Channel Islands National Park/National Marine Sanctuary, other areas receive huge visitation from the public. Approximately 8 million people live within 50 miles of the Monterey Bay Sanctuary's rocky shoreline, and millions more visit communities along that

FIGURE 1 Rocky shores are one of the most accessible marine habitats, and they are used for recreation, education, research, and commercial purposes. Photograph by Andrew DeVogelaere, MBNMS/NOAA.

shoreline. Thus, the very accessible habitat is geographically small yet can potentially be overwhelmed by human activities.

Intertidal communities are the best-studied marine habitat because of their accessibility and the relative ease of setting up manipulative experiments on rocky platforms. Results from research on intertidal communities are often used to help understand general ecological paradigms for other habitats that are more difficult to manipulate. Moreover, habitats in the intertidal zone are linked to other habitats and cannot be managed in isolation. Terrestrial organisms such as rats can be significant predators on intertidal species. Birds will feed on barnacles and limpets as well as graze on algae. Bird guano seems to enhance the growth of *Prasiola*, an alga found in the splash zone. When harbor seals haul out on intertidal sites, their waste products bleach algae, and the disturbance from their movement affects the local assemblage of species. Sea otters feed on intertidal mussels, creating patchy distribution patterns. Links with offshore systems include tidepools that act as nursery grounds for subtidal fishes, kelp forests and associated fishes that filter larvae of intertidal species, black abalone beds with large numbers of individual abalone that move between intertidal and subtidal habitats, and strong upwelling currents that prevent the offshore larvae of intertidal species from returning to the shore.

Intertidal communities can be considered as cultural resources in terms of their historical use and the presence of artifacts. In California, Native Americans such as the Ohlone and early immigrants such as the Chinese used the shoreline as a source of food; middens and historic photographs remain for each of these groups, respectively. Today, some Native American tribes in the Pacific Northwest, in and around Olympic National Park and Olympic Coast National Marine Sanctuary, continue their heritage of harvesting intertidal invertebrates. Literary work by John Steinbeck has memorialized the scientific biological collection and natural history work of Edward Ricketts (Fig. 2). Managers can use this information to interpret National Parks and Sanctuaries to a much broader public audience than those simply interested in natural resources. Moreover, the presence of cultural remains that indicate past human intertidal harvesting shapes human activities, such as response efforts to an oil spill. Finally, events such as television images showing shoreline impacts from the 1969 Santa Barbara oil spill significantly affected public opinion on coastal oil drilling, influencing national policy such as the designation of Sanctuaries with prohibitions on exploring for oil, gas, or minerals.

FIGURE 2 Author John Steinbeck (left) memorialized the rocky-shore natural history research of Edward Ricketts (right). Monuments of these men are tourist attractions in Monterey, California. Note the sea star in Ricketts's hand. Photographs by Andrew DeVogelaere, MBNMS/NOAA.

HUMAN USES OF THE INTERTIDAL COMMUNITIES

Many human uses commonly found in unprotected areas are prohibited, or their effects otherwise mitigated, in National Parks and Sanctuaries. Typical human uses in these federal marine managed areas can be grouped as follows:

- Research and education—usually allowed in both National Parks and Sanctuaries
- Harvesting and collecting—usually allowed only in Sanctuaries
- Passive recreation—usually allowed in both National Parks and Sanctuaries

More important, the threats to intertidal communities come not only from the typical activities noted above, but also from human activities onshore or offshore that lead to impacts to the intertidal habitats. It is these activities that have received extensive scientific study, and for which a sizable body of scientific literature exists.

OIL SPILLS

Oil spills are the most dramatic potential human impact to intertidal communities in National Parks and Sanctuaries as well as outside these protected federal areas, and they have received considerable popular and scientific attention. Oil spills can cause large-scale disturbances, but it is difficult to generalize about their damage because spills vary greatly in volume, chemical composition, and degree of weathering before reaching the shore. Moreover, species and habitats (e.g., kelp versus surfgrass, and exposed versus protected shores) are also affected differently by oil. Ongoing research in central California National Marine Sanctuaries indicate that the upper intertidal zone can recover from a disturbance in as little as one year but that mid-intertidal mussel beds may not recover for well over a decade. Recovery patterns vary greatly between sites, and their rates are not related to a latitudinal gradient. Moreover, recovery rates will vary with the intensity of the disturbance and with methods chosen for subsequent cleanup efforts. Although decisions for managing past oil spills have been based in large part on politics, research suggests that oil should be kept off shorelines through various techniques and that cleaning oil from the shoreline can slow intertidal community recovery.

HUMAN TRAMPLING AND HARVEST

Increasingly, humans visit the rocky shore as naturalists and sightseers, to collect bait and souvenirs, and to harvest food (Fig. 3). Trampling from foot traffic is unavoidable during these activities. In many, but not all, cases, an overall pattern of higher diversity and density at less-accessible sites has been documented. For species with short dispersal distances such as the sea palm, *Postelsia*, one 15-minute collecting spree could eliminate the population from a site for the foreseeable future. Harvesting can also interfere with scientific experiments. In one case near Long Marine Laboratory, dozens of painstakingly marked owl limpets disappeared overnight, when they were apparently collected for a bouillabaisse repast. Federal and state enforcement officers provide some control on collecting in tidepools, but innovative measures to minimize trampling impacts will be needed at sites

FIGURE 3 Sea stars, important in determining the community structure on rocky shores, are collected and dried to sell as souvenirs. Photograph by Andrew DeVogelaere, MBNMS/NOAA.

near areas of high human density. In some protected tide-pool areas, experiments are taking place with access or no-access areas on rocky benches and quasi-trail systems more commonly found in other terrestrial park habitats. Increasingly, no-fishing areas (often termed marine reserves) are being planned for intertidal areas in several California Sanctuaries and National Parks. In the coming years, considerable science will address this expanding management technique.

PHYSICAL DISTURBANCE FROM SHIP GROUNDINGS

Ships periodically run aground, and their initial impact and subsequent removal can damage rocky-shore assemblages. In 1996, the fishing vessel *Trinity* ran aground at Point Piños in the Monterey Bay National Marine Sanctuary (Fig. 4). When the boat was pulled from the rocks, it left gouged and broken rock, with these patches totaling 251 square meters. Although this damage was dramatic, intertidal systems are adapted to recover quickly from physical damage. In addition, ship grounding events cause very localized damage and are infrequent. Therefore, they can be considered a relatively minor resource management issue. As discussed previously, spilling the contents of a ship is of more concern than the physical scraping.

FIGURE 4 The fishing vessel *Trinity*, a 51-foot steel-hull seiner, grounded on the granite shores of Pacific Grove, California. The ship was salvaged by flipping it up the shore with cables, using tractor tires to effectively limit rock breakage and crushing of intertidal organisms. Photograph by Andrew DeVogelaere, MBNMS/NOAA.

COASTAL DEVELOPMENT AND ROAD MAINTENANCE

An interesting contrast between intertidal protection in National Parks and that found in Sanctuaries is that National Parks typically own the land on which the park resides. Thus, a National Park manager is generally better able to control immediate coastal development that may impact tidepools than a Sanctuary manager, who does not own the land and must work with myriad agencies to protect from human activities. This is particularly pronounced when considering coastal development.

Human-made structures to counter coastal erosion can increase and decrease intertidal habitats (Fig. 5). In California, for instance, roughly 10% of the coast is structures made by humans. Rocky habitat has increased in Elkhorn Slough, the largest wetland in the Monterey Bay Sanctuary, as landowners have armored their shorelines against tidal erosion. Of course, adding rocky shores means an equivalent loss of some other habitat.

FIGURE 5 Rocky-shore habitat is generated by rip rap (center of image) and seawalls (along the distant shore) installed to protect property from coastal erosion. Natural rock is in the foreground. Photograph by Becky Stamski, MBNMS/NOAA.

Especially along the Big Sur coastline in California, maintenance of the coastal highway by pushing construction material over the road shoulder toward the ocean creates periodic disturbances that simulate landslides (Fig. 6). These events can immediately eliminate rocky-shore habitat or have long-term impacts related to sand movement and burial on adjacent sites. Studies on the frequency of natural slides versus road-related slides are in development and may help coastal managers determine what additional steps are needed to protect tidepools in as natural state as is possible.

OTHER LARGE-SCALE IMPACTS

Intertidal areas in central California, within the Monterey Bay National Marine Sanctuary, have been and will continue to be used to detect human impacts on a global scale.

FIGURE 6 Natural and human-caused landslides can bury rocky shore habitat. Photograph by Steve Lonhart, MBNMS/NOAA.

Studies have indicated increasing numbers of warmer-water species in Monterey and concluded that the pattern is consistent with available water temperature data and the hypothesis of global warming. Nuclear fallout from the 1986 Chernobyl power plant accident in Ukraine was detected on Canadian and U.S. shores by measuring iodine 131 in the intertidal alga *Fucus*. Increased and improved monitoring efforts will facilitate future studies on global scales and more local studies on human-induced and natural changes.

MONITORING STUDIES TO ASSESS OVERALL INTERTIDAL HEALTH

One of the benefits of protecting intertidal communities in National Parks and Sanctuaries is that these agencies are mandated to monitor these resources to detect natural versus human-caused changes and to assess the long-term health of these habitats. Intertidal monitoring and research can be incorporated in base funding for managing a National Park or Sanctuary, or the mandate can provide justification for funding from other sources. Some very effective studies on long-term health have occurred and are continuing in intertidal communities protected in National Parks or Sanctuaries, and this existing monitoring as well as developing programs (e.g., for invasive species) are critical information for National Park and Sanctuary management. The management agencies can also provide mechanisms for sharing information among scientists and with resource managers through synthetic programs such as the Sanctuary Integrated Monitoring Network. The intertidal zone contains important habitats in National Parks and Sanctuaries, and partnerships between scientists, the public, and resource management are key to protecting these resources for future generations.

SEE ALSO THE FOLLOWING ARTICLES

Biodiversity, Maintenance of / Collection and Preservation of Tidepool Organisms / Dispersal / Disturbance / Food Uses, Modern / Monitoring, Overview

FURTHER READING

Barry, J. P., C. H. Baxter, R. D. Sagarin, and S. E. Gilman. 1995. Climate-related, long-term faunal changes in a California rocky intertidal community. *Science* 267: 672–675.

DeVogelaere, A. P., and M. S. Foster. 1994. Damage and recovery in intertidal *Fucus gardneri* assemblages following the *Exxon Valdez* oil spill. *Marine Ecology Progress Series* 106: 263–271.

Douros, W. J. 1987. Stacking behavior of an intertidal abalone: an adaptive response or a consequence of space limitation? *Journal of Experimental Marine Biology and Ecology* 108: 1–14.

Foster, M. S., A. P. DeVogelaere, C. Harrold, J. S. Pearse, and A. B. Thun. 1988. Causes of spatial and temporal patterns in rocky intertidal communities of central and northern California, in *Memoirs of the California Academy of Sciences No. 9*. San Francisco, CA: California Academy of Sciences.

Foster, M. S., E. W. Nigg, L. M. Kiguchi, D. D. Hardin, and J. S. Pearse. 2003. Temporal variation and succession in an algal-dominated high intertidal assemblage. *Journal of Experimental Marine Biology and Ecology* 289: 15–39.

Foster, M. S., J. A. Tarpley, and S. L. Dearn. 1990. To clean or not to clean: the rationale, methods, and consequences of removing oil from temperate shores. *Northwest Environmental Journal* 6: 105–120.

Gaines, S. D., and J. Roughgarden. 1987. Fish in offshore kelp forests affect recruitment to intertidal barnacle populations. *Science* 235: 479–480.

Ricketts, E. F., J. Calvin, J. W. Hedgpeth, and E. W. Phillips. 1985. *Between Pacific tides*. Stanford, CA: Stanford University Press.

Sousa, W. P. 1985. Disturbance and patch dynamics on rocky intertidal shores, in *The ecology of natural disturbance and patch dynamics*. S. T. A. Pickett and P. S. White, eds. New York: Academic Press, 101–124.

MARINE STATIONS

KEVIN J. ECKELBARGER

University of Maine

Marine stations are specialized research and educational facilities located on or near the seashore that serve as windows to the sea for students, faculty, and research scientists interested in marine science (e.g., marine biology, oceanography, marine engineering, or marine archaeology). They offer direct access to a wide variety of salt-water habitats of unusual beauty and scientific interest where research investigations and educational programs can be conducted in relatively unspoiled natural environments.

NATURAL HISTORY STUDIES AND THE RISE OF MARINE STATIONS

Marine stations did not appear until the nineteenth century, following two centuries of growing interest in natural history. Natural history studies were originally rooted in the natural theology concept that celebrated the handiwork of the Creator, so interest in nature was viewed as a morally worthy and appropriate activity, especially during the Victorian era. In practice, natural history consisted of collecting, describing, preserving, and displaying many objects of nature including rocks, plants, fossils, and both invertebrate and vertebrate animals. A major emphasis was placed on plants, insects, birds, and shells, because they were easily preserved and required minimal storage space. During the Enlightenment movement of the eighteenth century, interest in natural history swept through Europe and Britain before reaching North America in the late eighteenth and early nineteenth centuries. Until the early nineteenth century, no universities in the world offered formal training in natural history, so most naturalists were self-trained amateurs (often physicians and clergymen) who conducted their studies as hobbies. The collection of marine animals from the seashore became especially popular in England and postcolonial America, when the majority of the population lived near the coast.

During the late eighteenth century, and prior to the widespread construction of museums, the natural-history-cabinet craze, in which amateur naturalists displayed specimens (including marine animals) and artifacts in their homes, spread from Italy to other parts of Europe and Britain. In the mid-nineteenth century, the Wardian case was invented to display living plants in sealed glass terrariums, and this later evolved into the home aquarium for holding seashore animals. The aquarium craze was further fueled by the publication of seashore guides to common marine animals encountered on the beaches of Europe, Britain, and America, stimulating the construction of large public aquaria. At the same time, many nations were funding ocean expeditions to explore new regions and to collect marine organisms to extend their knowledge of natural history as well as their political influence. The specimens collected from these expeditions stimulated the formation of natural history museums, and marine stations soon followed, in part, to provide further animal and plant specimens for the museums. Marine stations took on further importance in the 1870s and 1880s, when universities began establishing graduate programs in biology that benefited from seaside facilities where faculty and students could study marine organisms firsthand. The early marine stations also grew into summer resorts for academics who wished to escape insects and the oppressive heat so prevalent in large cities of the nineteenth century. It soon became fashionable for professors to take their families and students with them each summer to study living organisms in rustic settings by the shore. Many marine stations became second homes for generations of marine scientists whose professional lives alternated between bustling, urban university campuses during the nine-month academic year and the quiet, collegial communities that grew around the marine stations.

EARLY MARINE STATIONS

The first marine stations appeared in Sweden in the 1830s, when small, temporary facilities were constructed. The first permanent facility was established in 1859 in Concarneau, France, followed by seven more French stations in rapid succession. By 1880, sixteen additional marine stations were established between Sweden and the Black Sea. Three principal types of marine stations appeared in the second half of the nineteenth century: (1) the seaside laboratory and public aquarium first seen in Italy, (2) the seaside school of natural history for teacher education first seen in America, and (3) the seaside marine station for college-level research and instruction as exemplified by most present-day marine stations. In 1973, Harvard professor Louis Agassiz established the Anderson School of Natural History on Penikese Island near Cape Cod, Massachusetts, the world's first marine station dedicated to the education of teachers in natural history. Although the facility only operated for two years, its educational innovations, including the participation of women in the program and the emphasis on living marine organisms ("Study nature, not books"), encouraged the creation of other summer schools and summer stations that evolved into the marine stations seen today. In 1874, the Naples Zoological Station and aquarium in Italy opened as the first permanent marine station. Its purpose was to provide research space to marine scientists from any nation so it was the first international station. In 1877, Agassiz's son, Alexander, established a unique, private marine station in Newport, Rhode Island, that was dedicated primarily to graduate student research. A year later, the Chesapeake Zoological Laboratory was established by Johns Hopkins University professor William Keith Brooks as a novel moving marine station in which graduate students and faculty conducted research at different North American locations each summer.

In the 1880s and 1890s German scientific innovations, including improvements in microscopes and tissue-cutting microtomes, resulted in a surge in morphological and embryological work, much of it conducted at marine stations. In America alone, 38 marine stations were founded between 1873 and 1940, primarily on the coasts of New England, in the Mid-Atlantic States, and in California. Following World War II the number of marine stations in America more than doubled, and many other nations established new facilities. Today there are nearly 200 marine stations located in at least 55 countries that range in size from small seasonal (usually summer) facilities staffed by a few workers, to large complexes employing hundreds of personnel throughout the year. Woods Hole Oceanographic Institution in Massachusetts (Fig. 1) is one of the largest in the world. The University of Maine's Darling Marine Center (Fig. 2) is a smaller station that is more typical of a year-round facility.

FIGURE 1 Aerial view of Woods Hole Oceanographic Institution (WHOI), Woods Hole, Massachusetts, United States. Photograph by Doug Weisman, ©Woods Hole Oceanographic Institution.

FIGURE 2 Research vessel and waterfront pier at the University of Maine's Darling Marine Center, Walpole, Maine, United States. Photograph by Linda Healy, Darling Marine Center.

THE SPECIALIZED FACILITIES OF MARINE STATIONS

Most modern marine stations have office and laboratory space, housing and dining services, a library, and classrooms, as well as specialized facilities unlike those found on traditional university campuses. They include a variety of specially designed coastal or oceanic research vessels (Fig. 2) for transporting students and faculty to remote habitats offshore where they can collect organisms in deeper water; waterfront facilities to support the vessels (piers, floating docks, fueling stations, engine repair shops); SCUBA support facilities (air tank–filling stations, equipment repair shops, showers, and dive equipment storage areas); and complex flowing seawater systems that pump clean, fresh seawater from the ocean to indoor research and classroom facilities. Flowing seawater systems are critical to the operation of any marine station because they allow researchers to study living organisms and to conduct experiments under controlled environmental conditions. The seawater pumping systems must be free of any metal parts because of their toxicity, and the seawater must have high levels of oxygen and low sediment loads to keep the organisms healthy. Many stations are located in remote coastal areas where population densities and levels of coastal water pollution are relatively low so that investigators can conduct research with few distractions.

THE EDUCATIONAL AND RESEARCH MISSIONS OF MODERN MARINE STATIONS

The research conducted at marine stations is often fundamental, and several Nobel Prizes have been awarded to investigators conducting biomedical research on marine organisms at these facilities. Studies of the nervous systems of the squid and sea slug have led to a better understanding of the molecular mechanisms of learning, memory, and nerve repair. Studies of sharks, stingrays, and skates have resulted in a greater insight into the human immune system and its resistance to carcinogens. Biomedical research has resulted in the discovery of numerous medically useful drugs and other pharmaceuticals derived from both shallow-water and deep-sea marine organisms. Through the use of large-scale experimental manipulations and long-term environmental change studies, marine stations have also been important to our understanding of ecosystem processes and the effects of population growth and subsequent increases in coastal pollution. Many ecological theories have been developed through research conducted at marine stations along with assessments of ecological damage caused by exotic species invasions. Marine stations have also been leaders in conservation biology by conducting basic research on endangered plant and animal populations.

Most marine stations offer many educational programs for K–12 (elementary and secondary) and university-level students that emphasize the study of living organisms in their natural habitats (Fig. 3). The marine station experience can be a career-defining one for young scientists who get to use research vessels and SCUBA (Fig. 4) to collect and

FIGURE 3 Marine ecology class studying living marine organisms along the shoreline of the Gulf of Maine. Photograph by Linda Healy, Darling Marine Center.

conduct research on living organisms under natural field conditions. These unique educational opportunities often expose students to important global issues such as marine biodiversity, fisheries exploitation and restoration, and marine conservation biology. International marine stations are recognized as national resources for environmental and biological research and education, and with the increasing impact of humans on our biosphere, these unique facilities are likely to become even more important in the future.

SEE ALSO THE FOLLOWING ARTICLES

Aquaria, Public / Ecosystem Changes, Natural vs. Anthropogenic / Education and Outreach / Museums and Collections

FURTHER READING

Jack, H. A. 1945. Biological field stations of the world. *Chronica Botanica* 9: 1–73.
Kofoid, C. A. 1910. The biological stations of Europe. *U.S. Bureau of Education Bulletin* 4: 1–356.
Lillie, F. R. 1944. *The Woods Hole Marine Biological Laboratory.* Chicago: University of Chicago Press.
Lohr, S. A., P. G. Connors, J. A. Stanford, and J. S. Clegg, eds. 1995. *A new horizon for biological field stations and marine laboratories.* Miscellaneous Publication No. 3. Crested Butte, CO: Rocky Mountain Biological Laboratory.
Terwilliger, R. C. 1988. Teaching and research at marine stations: present priorities, future directions. *American Zoologist* 28: 27–34.

MATERIALS, BIOLOGICAL

DOUGLAS FUDGE

University of Guelph, Canada

JOHN M. GOSLINE

University of British Columbia, Vancouver, Canada

Just as humans manufacture a wide diversity of materials for various uses, organisms construct themselves, and sometimes structures outside of themselves, out of materials that are adapted to their particular ecological and physiological roles. Marine organisms are no exception, and they manufacture an impressive diversity of materials that help them survive and reproduce in ocean habitats. These materials include slimes, rubbers, fibers, and crystalline composites.

MATERIAL PROPERTIES

Material properties describe how a material behaves when subjected to mechanical force. Humans have a natural feel for the material properties of objects, and this is reflected by the rich vocabulary of adjectives, such as "rigid," "stiff," "soft," "hard," "strong," "weak," "tough," "brittle," "resilient," "elastic," "viscous," and "plastic," that we use to describe how materials behave when they are acted upon by external forces.

Many of the material properties that are vital to the work of engineers and architects are also important to organisms, and it is worthwhile to define some of these terms here. One

FIGURE 4 Students diving into shallow water to collect marine animals using SCUBA. Photograph by Linda Healy, Darling Marine Center.

of the most fundamental and informative mechanical tests one can perform is called a force-extension test, in which the deformation of a material is measured as it is subjected to an external force (Fig. 1A). The magnitude of a material's deformation is referred to as its strain, which is defined as the change in a material's dimensions divided by its original dimensions. Dividing by the original dimensions standardizes the strain so that samples of different sizes can be directly compared. The force can be similarly standardized by dividing it by the cross-sectional area over which the force is applied to the sample. This standardized measure of force is called the stress and is expressed in units of N/m^2 or Pa. The relationship between stress and strain is defined as the stiffness, or modulus, which is calculated as the slope of the stress–strain curve, and it also has units of Pa. Materials with steep stress–strain curves, such as steel, possess high stiffness, whereas materials with shallow stress–strain curves, such as rubber, have low stiffness. Other information can be extracted from stress–strain curves, such as the maximum stress a material can withstand before it ruptures, which is referred to as its strength, and the strain at which a material ruptures, which is called its extensibility. The amount of energy that a material absorbs when it is deformed is referred to as the strain energy, and this is measured as the area under the

stress–strain curve (Fig. 1B). Materials that can absorb large amounts of energy are considered tough, whereas those that fail at low strain energies, such as glass, are considered brittle. While the stress–strain curves for ideal solids are linear and insensitive to the rate of deformation, biological materials rarely have linear stress–strain curves, and their mechanical deformation is typically strain rate dependent. In this way, these materials exhibit properties of both solids and fluids, a property known as viscoelasticity. In mechanical terms, the difference between a solid and a fluid has to do with the time dependence of their response to external loads. For an ideal solid, stress is determined entirely by the amount of strain and has nothing to do with how fast the material was deformed or how long it was held at that strain. For fluids, stress is determined by how fast the material is strained, whereas the total amount of strain is less important for fluids because of their ability to flow, which disperses strain energy. The resistance of a material to flow is referred to as its viscosity, which is defined as the slope of the stress-strain rate curve. When viscoelastic materials are deformed and held, the resultant stress decays over time, a phenomenon referred to as stress relaxation. Similarly, subjecting a viscoelastic material to a constant stress causes the strain to increase over time, a phenomenon referred to creep.

From Table 1, one can see that organisms make materials with a wide diversity of properties, with stiffness values spanning eleven orders of magnitude. Based on their material properties, it is possible to classify biomaterials into roughly four classes—slimes, rubbers, fibers, and crystalline composites—each of which is suited to the particular functions they perform in life.

SLIMES

Slimes are gel-like materials that are composed of swollen networks of crosslinked glycoproteins and are similar to secretions such as saliva and gastric mucus. Seawater makes up the vast majority of the volume of marine slimes and dominates their material properties. When slimes are deformed, they respond with a very low-stiffness behavior that is mediated by a network of glycoprotein molecules, and a viscous behavior mediated primarily by the viscosity of water. Slimes are secreted by a wide range of marine organisms as a protective barrier against the chemical, physical, and miocrobiological insults of the outside world. Such slime coatings are produced by fishes and most invertebrate groups such as molluscs, annelids, cnidaria, and echinoderms. Arthropods are a notably unslimy exception.

Slimes are used by marine organisms for a diversity of functions including feeding and digestion, locomotion, defense, desiccation resistance, and navigation. All animals

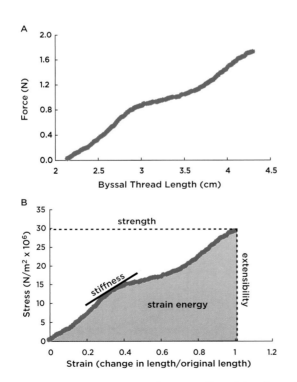

FIGURE 1 (A) Force-extension curve for a mussel byssal thread tested in seawater. (B) Same data as in (A), but expressed as stress and strain. Data for this figure kindly provided by Emily Carrington.

TABLE 1
Mechanical Properties of the Four Groups of Biological Materials and Some Human-made Materials for Comparison

	Strength Pa × ·10^5	Extensibility	Breaking Strain Energy J/m³ × ·10^5	Stiffness Pa ·10^6
Slimes				
Gastropod pedal mucus[a]	0.001–0.003	1–5	0.1–0.5	0.0001–0.0005
Barnacle cement[b]	1–3	——	——	——
Rubbers				
Resilin	40–60	2–3	400–900	2
Abductin	80–120	2–3	800–1,800	4
Elastin	40–60	2–3	400–900	2
Octopus arterial elastomer[c]	40–60	2–3	400–900	2
Fibers				
Silks	5,000–10,000	0.2–0.35	5,000–18,000	5,000–10,000
Collagen	500–1,000	0.08–0.10	200–500	2,000
Cellulose	5,000–10,000	0.02–0.10	500–5,000	20,000–80,000
Chitin	1,000	0.01–0.02	50–100	40,000
Keratin	1,000–2,000	0.3–0.8	1,500–3,000	4,000
Mussel byssal threads[d]	320	0.9	150	62
Fiber-reinforced composites				
Wood, parallel to grain	1,000	0.01	5	10,000
Wood, perpendicular to grain	50			500
Locust tibial cuticle	950	0.02	1.3	10,000
Locust intersegmental cuticle				0.01
Jellyfish mesoglea[e]				0.001
Bull kelp stipe[f]	30	0.4	6	12
Crystalline composites				
Coral skeleton[g]	400	0.0003	0.6	60,000
Mussel shell[h]	560	0.0018	5	31,000
Bone	1,900	0.01	950	18,000
Human-made materials				
Steel	30,000	0.015	2,000	200,000
Glass	1,000	<0.001	<5	100,000
Cement	40	<0.001	<2	4,000
Fiberglass	3,000–10,000	0.01	150–500	30,000–100,000

NOTE: Based on Table 12.2 in Denny 1988.
[a]Denny and Gosline 1980. All parameters measured in shear, not tension.
[b]Denny et al. 1985.
[c]Shadwick and Gosline 1983.
[d]Bell and Gosline 1996.
[e]Megill et al. 2005.
[f]Koehl and Wainwright, 1977.
[g]Vosburgh 1982.
[h]Currey 1980.

line their digestive tract with slimy mucus secretions of some kind. These materials are used to lubricate food and protect the underlying cells from being digested. Mucus is used in other capacities related to food capture and processing as well, especially among the invertebrates. Polychaete worms, tunicates, limpets, and pteropods all possess members that use mucus nets for capturing or concentrating food particles. The slimes used to construct feeding webs are typically stiffer and more elastic than other slimes. The gastropod molluscs employ a mode of locomotion in which they crawl on a single foot that adheres to the substrate via a layer of "pedal" mucus. Forward progress is made possible by the mechanical properties of the mucus, which behaves as an elastic solid at low strains but flows like a fluid at higher strains. By passing waves of contraction over the foot, the mucus provides elastic resistance to stationary bits (which is good for pushing), and only viscous resistance in areas that are moving forward (which is good for sliding). Using slime for locomotion has a fortuitous side effect in navigation. Limpets are known to return to a "home scar" before every low tide, and it has been shown that limpets and other gastropods use their slime trails to help them navigate the topologically complex rocky intertidal habitat and find their homes.

Some animals coat themselves in a thick layer of slime when threatened by predators. This strategy makes the animal difficult to handle or eat and is used commonly among gastropods and echinoderms. The "ink" ejected by cephalopods when they are startled contains mucus-like glycoproteins that lend the dark cloud coherence, which is presumably more distracting to pursuing predators than ink diluted in seawater alone. Hagfishes, or "slime eels," secrete slime in great quantities when they are startled by a predator. Hagfish slime differs from other slimes in that it is permeated by thousands of fine protein fibers that lend the slime coherence. Ecologically, the slime likely functions to thwart predatory attacks by fishes by catching on and clogging the gills.

RUBBERS

Rubbers are soft, extensible, and highly elastic materials that are generally produced by organisms as a means of storing energy. Rubbers consist of a crosslinked network of protein molecules that resist deformation not by the stretching of chemical bonds, but instead by a thermodynamically unfavorable rearrangement of the polymer chains in the rubber. These rearrangements of the polymer molecules are highly reversible and can be repeated many times with minimal damage to the material. A classic example of a rubber made by marine organisms is "abductin," which is found within the hinge that connects the two shells in bivalve molluscs. This rubber pad is compressed when the adductor muscle closes the shells, and it causes the shells to passively reopen when the adductor muscle relaxes. Because abductin is so good at storing energy, the animal can clam up for long periods of time and be confident that almost all the energy stored in the abductin pad will be available when it is safe to open up again.

FIBERS AND FIBER-REINFORCED COMPOSITES

Marine organisms make a wide range of fibrous materials, which are generally three to four orders of magnitude stiffer than rubbers. At the molecular level, these fibers

consists of polymer chains that are arranged in both crystalline and amorphous configurations. It is the crystalline structures that give fibrous materials their high strength and stiffness, whereas the amorphous domains generally make them more extensible and therefore tend to increase toughness. Because large-scale deformations can disrupt the bonds within the crystalline domains, fibers are generally not as well suited for energy storage as rubbers are. Fibrous materials may be molded into complex three-dimensional structures, such as the keratin of whale baleen, but fibers often have high length-to-width ratios, which makes them ideal for bearing tension, but not compression, because of the tendency of fibers to buckle. Mussels stay put in wave-swept environments via a system of byssal threads that they glue to the substrate. Byssal threads consist of proteins resembling the collagen fibers found in vertebrate tendons and ligaments, but they also show some structural features characteristic of elastin (a rubber) and silk (an exceptionally tough fiber). As a consequence, byssal threads are considerably more extensible than tendons. The byssal threads are typically laid down onto the substrate in a circular pattern (Fig. 2). When the mussel is acted upon by wave action, some threads will invariably buckle, but others will be loaded in tension, and in most cases these will keep the animal attached. The extensibility of byssal threads allows them to stretch far enough that other threads can be recruited to share the load.

Other fibers found in the marine environment are collagen, which is found in tendons and ligaments and the body wall of many soft-bodied marine invertebrates;

FIGURE 2 Mussels use fibers called byssal threads to attach themselves to the substrate. In this photo, the mussel in the center has attached itself to its neighbor via several byssal threads. Photograph courtesy of Laura Coutts.

chitin, which is found in the cuticle of crustaceans such as lobsters and crabs; and cellulose, which makes up much of the cell wall of plants and lends stiffness to a variety of structural elements in plants and algae. In most cases these fibers are found in the form of fiber-reinforced composites, where they are embedded in a matrix that links the fibers together to create a hybrid material that combines the properties of the fiber and matrix. The benefit of composite designs is that they are able to resist compression and bending loads in spite of the flexibility of the thin fibers they contain. Because collagen, chitin, and cellulose are all stiff fibers, the maximum stiffness of these composite materials is limited by the fibers, but when the fibers are combined with softer matrix materials, these composites can have properties that span a huge range. For example, jellyfish mesoglea (the very soft, elastic material that makes up the animal's bell-shaped swimming structure) contains both collagen fibers and rubber-like elastic fibers, and these fibers are embedded in a very dilute matrix of slimelike glycoproteins. Because the matrix makes up most of the volume of this material, the mesoglea has a very low stiffness (approximately 10^3 Pa), and it functions as a soft spring that powers the refilling of the bell during a jellyfish's swimming cycle. The body wall connective tissues of many soft-bodied invertebrates and the internal tissues of other animals that have more rigid skeletons are also constructed as composites of collagen, elastic fibers, and glycoproteins. In these tissues, however, the collagen and elastic fiber content dominates and the composites are considerably stiffer, with modulus values in the range of 10^5 to 10^7 Pa. In marine algae the stiff fiber cellulose is found in association with gels made from charged polysaccharides (agar, a thickening agent used in foods, is a gel-forming polysaccharide derived from marine algae), and these form a soft composite with stiffness values in the range of 10^6 to 10^7 Pa. These reasonably strong but remarkably deformable materials allow intertidal algae to survive the extreme flow velocities created by breaking waves by allowing the plants to deform and largely avoid the huge forces that they would experience if they were designed to rigidly resist waves. Interestingly, the same fiber, cellulose, when embedded in a rigid matrix, can form the high-strength rigid composite that forms the secondary cell wall of most land plants, a material that we use in its dry form (i.e., wood) to make furniture and houses. This material is a composite of cellulose and lignin, with stiffness in the range of 10^9 to 10^{10} Pa. Similarly, chitin fibers in association with a rigid matrix of protein are found in insect cuticle, a rigid composite with stiffness reaching

about 10^{10} Pa. The interaction of the fibers with a strong but more deformable matrix in these rigid composites provides mechanisms that control the growth of cracks and allow these materials to be remarkably stiff, strong, and tough.

CRYSTALLINE COMPOSITES

Whereas the presence of crystalline polymer fibers in organic composites can result in materials with stiffness up to about 10^{10} Pa, materials that are based on inorganic crystals can be as stiff as 10^{11} Pa. Calcium carbonate is by far the most common form of inorganic crystal used by marine organisms, and it occurs in the hard shells of molluscs, brachiopods, and barnacles. Materials such as the skeleton of hard corals resist high stresses with very little deformation. The drawback of such high stiffness is that the area under the stress–strain curve is relatively small, which means that these materials can be quite brittle. This problem is mitigated in materials such as mollusc shell by the presence of microstructures that resemble fiber-reinforced composites. That is, small elements of the brittle, crystalline material, typically having at least one dimension on the scale of about one micron, are surrounded by a protein-rich "matrix" with high extensibility and toughness. The matrix in a composite serves to isolate the individual brittle elements, making it difficult for cracks to propagate through the composite. Nacre, the crystalline material found in pearls and in many mollusc shells, is an exceptionally tough biocomposite (Fig. 3).

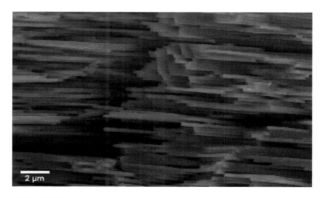

FIGURE 3 Scanning electron micrograph of the nacre layer from the mollusc *Nautilus pompilius*. The "brick wall" structure comes from the arrangement of inorganic aragonite tablets, which are separated by a thin layer of organic matrix. Image courtesy of Fabio Nudelman.

SEE ALSO THE FOLLOWING ARTICLES

Adhesion / Bivalves / Corals / Hydrodynamic Forces / Kelps / Limpets

FURTHER READING

Wainwright, S. A., W. D. Biggs, J. D. Currey, and J. M. Gosline. 1976. *Mechanical design in organisms*. London: Edward Arnold.

REFERENCES

Bell, E. C., and J. M. Gosline. 1996. Mechanical design of mussel byssus: material yield enhances attachment strength. *Journal of Experimental Biology* 199: 1005–1017.

Currey, J. D. 1980. Mechanical properties of mollusk shell. *Symposia of the Society for Experimental Biology* 34: 75–97.

Denny, M. W., and J. M. Gosline. 1980. The physical properties of the pedal mucus of the terrestrial slug, *Ariolimax columbianus*. *Journal of Experimental Biology* 88: 375–393.

Denny, M. W., T. Daniel, and M. A. R. Koehl. 1985. Mechanical limits to size in wave-swept organisms. *Ecological Monographs* 55: 69–102.

Koehl, M. A. R., and S. A. Wainwright. 1977. Mechanical design of a giant kelp. *Limnology and Oceanography* 22: 1067–1071.

Megill, W. M., J. M. Gosline, R. W. Blake. 2005. The modulus of elasticity of fibrillin-containing elastic fibres from the hydromedusa *Polyorchis penicillatus*. *Journal of Experimental Biology* 208: 3819–3834.

Shadwick, R. E., and J. M. Gosline. 1983 The structural organization of an elastic fibre network in the aorta of the cephalopod *Octopus dofleini*. *Canadian Journal of Zoology* 61: 1866–B1879.

Vosburgh, F. 1982. *Acropora reticulata*: structure, mechanics and ecology of a reef coral. *Proceedings of the Royal Society of London. Series B, Biological Sciences* 214: 481–499.

MATERIALS: STRENGTH

LORETTA ROBERSON

University of Puerto Rico

Strength is a mechanical property of materials defined as the amount of force required to break a material of a given cross-sectional area. The measurement of strength is given in pascals (Pa), units of force per area, where $1\ \text{Pa} = 1\ \text{N/m}^2$. Of the various mechanical properties of materials (strength, stiffness, hardness, etc.), strength is of particular interest to intertidal biologists because the failure of material has direct consequences on the survival of organisms. For example, strength can determine whether or not an organism will be dislodged or damaged by breaking waves, currents, moving objects such as ice, or predators. Strength therefore plays an important role in the fitness, survivorship, and evolution of organisms that live in the dynamic and often stressful environment of wave-swept shores.

INTERTIDAL STRESSES

There are three main mechanical stresses experienced by organisms in the intertidal: tension, compression, and shear. Tension occurs when forces pull on a material such as an algal stipe (Fig. 1A) in opposite directions along the same axis (Fig. 1B), as when a rope is pulled taut. Compression occurs when a force crushes a material (Fig. 1C). Compressive stress can be found in concrete pier pilings that support a structure above. Shear stresses occur when forces are applied in parallel but in opposing directions in different planes (Fig. 1D). For example, skidding car tires experience a stress from the force of the street opposing the forward momentum of the vehicle. In addition, multiple types of stresses often occur in a material at the same time. When a material is bent, for instance, both compressive and tensile stresses are applied to the material simultaneously (Fig. 1E).

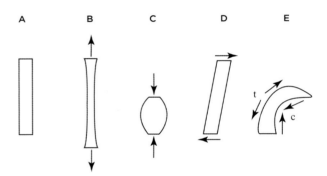

FIGURE 1 Force diagram of algal stipe. Arrows indicate direction of force. (A) Portion of algal stipe at rest. (B) Algal stipe in tension. As the stipe is stretched, the tissue narrows, thereby reducing the cross-sectional area. (C) Algal stipe in compression. As the stipe is flattened, the tissue bulges outward, increasing the cross-sectional area. (D) Algal stipe in shear. Note that as the stipe is sheared, the tissue is deformed (the rectangle approximating the stipe becomes a parallelogram) but the cross-sectional area remains the same. (E) Algal stipe bending, with opposite sides undergoing tension (t) and compression (c).

The most persistent mechanical stresses in the intertidal are the result of waves (Fig. 2). Waves break on a given shoreline more than 8000 times each day, moving water as the wave impacts the shore (Fig. 2A), as well as after the wave breaks when water is pushed up on the shore (Fig. 2B) and when that water then returns to the sea (Fig. 2C). This water movement creates hydrodynamic forces such as drag or lift that directly push, pull, and bend organisms. Indirectly, wave action can mobilize objects in the intertidal such as rocks or ice that consequently crush or scour organisms. Currents generated by waves (i.e., longshore currents) and tides (Fig. 3) generate hydrodynamic forces as well, but these are generally much less than those experienced in a breaking wave and not as frequent. Exceptions occur in areas with large tidal ranges, such as Puget Sound in Washington State, where tidal velocities can approach those seen in medium-sized waves. In addition to forces generated by waves, intertidal organisms are also continually subject to forces imposed by predators. Crabs, for example, apply compressive forces with their claws to break the protective shells of their gastropod prey. Other predators such as birds, sea stars, and fishes, impose a whole range of forces, from the impact of beaks (e.g., oystercatchers) or

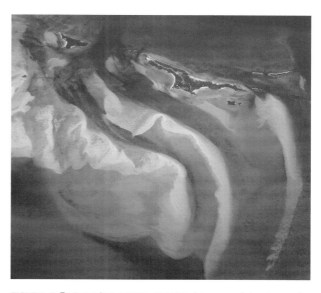

FIGURE 3 Fast-moving waters carrying heavy particles large distances are reflected in the sand banks of the Bahamas. In addition to waves, strong currents can produce large hydrodynamic forces in the near-shore region. Strong currents are produced by extreme tidal exchanges and longshore currents as well as when slower-moving water is forced through a restriction. Photograph by the author.

FIGURE 2 Waves with a 10-s period break 6 times every minute. Waves affect the intertidal not only while breaking (A) but also during the run-up onto shore after the wave breaks (B) and as the waters rush back out to sea in the opposite direction (C). Punta Cerro Gordo, Puerto Rico. Photographs by the author.

hammering (e.g., sea otters) to peeling (e.g., box crabs) and drilling (e.g., dog whelks).

Organisms can respond to forces in several ways. Mobile creatures such as crabs and snails can move to the safety of depressions or cracks where water movement is reduced and where they are less conspicuous to predators. Kelps, barnacles, and other sessile organisms with limited mobility, however, must withstand all waves and predators to survive. By maintaining a relatively small size or streamlined profile, organisms can effectively hide from hydrodynamic forces or the eyes of predators. Larger organisms such as kelps (Fig. 4) effectively do this by being flexible so that a large surface area, good for collecting sunlight, can be folded into a bundle when exposed to large hydrodynamic forces, thereby decreasing drag. Some organisms simply get stronger to withstand forces. The mere presence of predators results in the increase of shell strength in some gastropods. More frequently, however, organisms are damaged or removed completely from the substratum (Fig. 5). This does not necessarily mean the death of the organism; organisms such as algae and sea stars can regenerate damaged tissue, and some corals can eventually reattach to the substrate. It is thought that many algae and corals actually take advantage of "weakened" structures to increase

FIGURE 4 Sessile intertidal organisms such as the southern sea palm (*Eisenia arborea*) are at the mercy of waves and currents. Santa Catalina Island, California. Photograph by the author.

FIGURE 5 A battered holdfast is all that remains of the kelp *Eisenia arborea* after herbivores introduced flaws in the stipe. The arrow denotes the base of the missing stipe. Santa Catalina Island, California. Photograph by the author.

dispersal and asexual reproduction. A striking example of where high strength is not always preferred is in the specialized feeding apparatus of sea urchins, known as Aristotle's lantern. Most of the hard structural elements of the Aristotle's lantern are very strong, but the teeth exhibit a gradient such that the center of the tooth is quite strong but becomes gradually weaker toward the exterior of the tooth. The weakness of the outer layer allows the teeth to be self-sharpening, an important

quality for an organism that feeds by scraping algae off rocks.

CAUSES OF MATERIAL STRENGTH

Organisms are subjected to forces, but the strength of the organism depends on the materials with which they are constructed and how those materials are organized. A wide variety of materials are found in the intertidal but can be grouped according to broadly similar mechanical properties or performance. These are gels, elastomers, polymers, ceramics, and composites.

Gels generally are amorphous with a high water content and are relatively weak (strength << 1 MPa). Examples include pedal mucus secreted by gastropods, mucus feeding nets (e.g., ascidians), sea anemone mesoglea, and agarose extracted from red algae. Some gastropods, such as limpets and periwinkles, can alter their pedal mucus viscosity thereby enhancing the attachment strength by several orders of magnitude, causing it to act more like an adhesive instead of an aid to locomotion.

Elastomers are a subset of polymers that are very stretchy and extensible, like rubber, with medium strength (1–100 MPa). Examples include elastin found in skin and arteries, abductin found in the ligament of bivalve hinges, and mussel byssal threads. Instead of being strong, elastomers are extensible, which may reduce their susceptibility to tearing, and they readily store energy, which makes them more efficient for doing certain kinds of work.

Polymers are long chains of molecules, such as cellulose or proteins, and can be quite strong (100–1000 MPa). Examples of polymers include chitin, collagen, and keratin. Some of the strongest known materials (Kevlar® and carbon nanotubes) are polymers with strengths of 2000–4000 MPa.

Ceramics are strong, generally highly structured materials that include coral skeletons, mollusc shells, and glass. Strengths of ceramics are comparable to those of polymers (100–1000 MPa).

Composites, as the name implies, are mixtures of different material types. To a certain degree, all natural structures are composites. Good examples include bone, wood, and fish scales. Composites exhibit a wide range of strengths (0.01–1000 MPa). The strengths of tidepool materials and various artificial materials (for comparison) are listed in Table 1.

The way that materials are put together can greatly affect the strength and performance of the overall structure. For example, the nacre of mollusc shells is a highly specialized, ordered composite that exhibits a higher strength and performance (i.e., resistance to cracking)

TABLE 1
Intertidal Materials: Mechanical Properties

Material	Species	Common Name	Test	Strength (MPa)	Stiffness, E (MPa)	References
Gels						
	Lottia limatula	limpet mucus	s	0.05 to 0.21		Smith 2002
		pectin	s	0.04 to 0.09		Smith 2002
		pedal mucus	s	0.0001 to 0.0003	0.0001 to 0.0005	Denny 1988
	Metridium senile	sea anemone mesoglea	t	3 to 4.5	0.0003 to 0.08	Chapman 1953; Gosline 1971
	Botrylloides sp.	colonial ascidian	t	0.1	0.3	Edlund *et al.* 1998
	Botrylloides sp.	colonial ascidian	a	0.004		Edlund *et al.* 1998
Elastomers						
	abductin	bivalve hinge ligament	t	8 to 12	4	Denny 1988
	elastin	skin, cartilage, arteries	t	2	1	Gosline *et al.* 2002
	Mytilus californianus	California mussel byssal thread	t	35 to 75	16 to 870	Gosline *et al.* 2002
	Mytilus galloprovincialis	bay mussel byssal thread	t	20 to 150	50 to 500	Waite *et al.* 2002
	Busycon canaliculatum	whelk egg capsule	t	5	4 to 88	Rapoport *et al.* 2002
	Busycon carica	whelk egg capsule	t	5	3 to 66	Rapoport *et al.* 2002
Polymers						
	cellulose	plant cell wall	t	500 to 1000	20000 to 80000	Denny 1988
	chitin	arthropod exoskeleton	t	100	40000	Denny 1988
	collagen	tendon, bone, teeth, skin	t	120	1200	Gosline *et al.* 2002
	keratin	skin, scales, hair, baleen	t	100 to 200	4000	Denny 1988
		kevlar ©	t	3600	130000	Gosline *et al.* 2002
		carbon-nanotubes	t	1800	80000	Dalton *et al.* 2003
Ceramics						
	Pinctada margaritafera	pearl oyster nacre	c	140 to 210	34000	Currey *et al.* 2001
	Haliotis rufescens	abalone	c	180 to 540	60000 to 70000	Menig *et al.* 2000
	Fragilariopsis kerguelensis	diatom frustule	c	700	20000	Hamm *et al.* 2003
	glass	(artificial)	t	100	100000	Denny 1988
	Strombus gigas	conch shell	c	70 to 200		Hou *et al.* 2004
Composites						
	Cucumaria frondosa	sea cucumber	t	15	7300	Thurmond *et al.* 1996
	Dreissena polymorpha	zebra mussel attachment	a	0.2 to 1.5 N		Ackerman *et al.* 1996
	Mytilus edulis	common mussel shell	c	56	31000	Denny 1988
	Mercenaria mercenaria	surf clam shell	c	50 to 200 N		Zuschin *et al.* 2001
	Anadara ovalis	blood ark clam shell	c	100 to 1000 N		Zuschin *et al.* 2001
	Littorina littorea	common periwinkle shell	c	250 N		Cotton *et al.* 2004
	Gibbula cineraria	grey topshell shell	c	300 N		Cotton *et al.* 2004
	Osilinus lineata	flat topshell shell	c	230 N		Cotton *et al.* 2004
	Gibbula umbilicalis	toothed topshell shell	c	300 N		Cotton *et al.* 2004
	Acropora reticulata	coral skeleton (CaCO$_3$)	t	40	60000	Denny 1988
	Strongylocentrotus purpuratus	urchin blastula	c		0.0005 to 0.0025	Davidson *et al.* 1999
	Paracentrotus lividus	urchin tooth (MgCO$_3$)	c	1370		Wang *et al.* 1997
	Asterias rubens	sea star arm	t	0.1 to 4	0.5 to 15	Marrs *et al.* 2000
	Phyllospadix scouleri	surfgrass	t	13	155 to 300	Hale 2001
	Egregia menziesii	featherboa kelp	t	5	8 to 25	Hale 2001
	Fucus distichus	rockweed (brown alga)	t	5	20	Hale 2001
	Endocladia muricata	red alga	t	8	10 to 25	Hale 2001
	Calliarthron cheilosporoides	coralline alga, genicula	t	25	8 to 30	Hale 2001
	Calliarthron cheilosporoides	coralline alga, whole	t	17	90 to 150	Hale 2001
	Codium fragile	dead man's fingers	t	0		Hale 2001
	Nemalion helminthoides	red alge	t		0.16	Hale 2001
	Prionitis lanceolata	red alga	t		50	Hale 2001
		algae break force	a	1 to 600 N		Thomsen *et al.* 2005
	Rhinoptera bonasus	stingray jaw trabeculae	c		0.01	Summers *et al.* 1998

(*Continued*)

TABLE 1
(*Continued*)

	Intertidal materials		Test	Strength (MPa)	Stiffness, E (MPa)	References
Material	**Species**	**Common Name**				
Composites	*Balanus c.f. variegatus*	barnacle	a	0.05 to 2		Becker 1993
(continued)	*Pomatoleios kraussii*	polychaete worm	a	0.03 to 2		Becker 1993
	Pagrus major	red Sea bream scales	t	30		Ikoma *et al.* 2003
	Ursus maritimus	polar bear adult femur	t	140 to 150	11000 to 22000	Currey 1999
	Aptenodytes patagonica	king penguin humerus	t	175	23000	Currey 1999
	Homo sapiens	human adult femur	t	166	17000	Currey 1999
	Fagus sylvatica	beech wood	t	75		Burgert *et al.* 2001
		concrete pilings	c	10 to 70	18000 to 30000	http://www.matweb.com/
		stainless steel	c	1500	300000	Gosline *et al.* 2002
		PVC	c	34 to 59	1000 to 5000	http://www.matweb.com/

NOTE: Types of stress applied: s = shear; t = tensile; c = compressive; a = attachment strength. Note that some strength data were published as force (N = newtons).

than the individual materials (Fig. 6A). This is accomplished in two ways: (1) by an elegant ordering of interlocking tiles of crystalline calcium carbonate (Fig. 6B) and (2) by an organic matrix binding the tiles comprised of long polymers looped into a series of domains, each domain filled with a high degree of within-chain interactions (i.e., intermediate-strength bonding)(Fig. 6C).

Of course, any damage or flaws in the materials or tissues can dramatically decrease the overall strength from the values listed in Table 1. Partial predation, repeated bending, or abrasion can introduce cracks or weak points in the material (see Fig. 5). In addition, factors such as temperature or hydration can affect material strength. Lastly, the strength of the substratum can override the strength of the organism in terms of survivorship if the substratum itself is unable to withstand intertidal forces. For example, spores of the giant kelp *Macrocystis pyrifera* settle on hard substratum, but some of these surfaces may be on small cobbles. The kelp can survive until the buoyancy generated by its pneumatocysts approaches or exceeds the weight of the cobble.

MEASUREMENT OF STRENGTH

Strength can be measured with a device called an extensometer or tensometer (Fig. 7). The tensometer consists of three main parts: (1) a driving arm to exert an extension, (2) a force transducer that records how much force is consequently applied to a sample, and (3) clamps to secure the sample between the driving arm and the force transducer. The driving arm is used to extend or compress the sample, and the stress applied when the sample fails or breaks is the breaking stress or strength. In practice, the proportional extension of the sample (strain)

is recorded simultaneously to create a stress–strain curve (a graphical presentation of stress values plotted against strain values); the resultant curves are unique to each material type. Engineers use stress–strain curves to understand the behavior of materials when a force is applied and to predict under what conditions the material is likely to fail. Stress–strain curves also are used to calculate a property related to strength: the stiffness, or modulus of elasticity. Stiffness is often confused with strength, but stiff materials such as glass are not necessarily very strong (see Table 1).

Intertidal ecologists can use stress–strain curves and strength in a manner similar to engineers to predict the risk of dislodgement or breakage for organisms exposed to a specific intertidal force, such as forces generated by waves or the force generated by the crushing jaws of a lobster. To predict the probability of dislodgement by waves for instance, a minimum of three measurements is required: (1) the wave height offshore, (2) the shape coefficient of drag or lift, and (3) the strength of the organism. Significant wave heights can be obtained for many areas from offshore wave buoys used to monitor and predict weather. Shape coefficients are not as readily available, but many have been published for common intertidal organisms (e.g., mussels, barnacles, limpets, and some algae) and simple engineering shapes such as a cylinder or plate. The attachment and breaking strengths of organisms can be measured relatively easily using inexpensive and readily available materials. These three pieces of information can then be analyzed using statistical methods such as the random sea (to predict the maximal wave height), wave theory (to calculate breaking wave height and resultant water velocities),

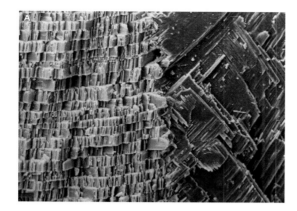

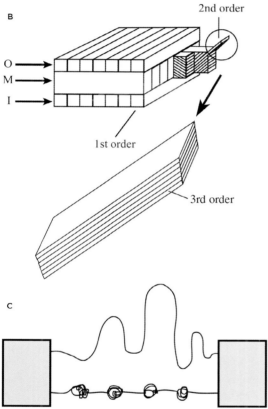

and the statistics of extremes (to predict the probability of dislodgement by drag or lift forces generated by water of a given velocity).

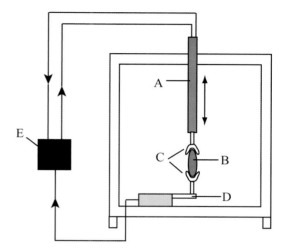

FIGURE 7 Tensometer used for measurement of mechanical properties. The driving arm (A) tugs on a sample (B) held in place by a set of clamps (C). A fixed force transducer (D) measures the force applied to the sample and sends the data to a computer (E) for recording. The computer also can be used to control the driving arm and simulate more natural, repeated tugging on a sample. Strain can be measured using calipers, digitized video images, sonomicrometers, or anything that will allow the measurement of material length before and after breakage. Redrawn from Hale (2001).

FIGURE 6 The complex microstructure of nacre increases the survivorship of the material beyond that for CaCO₃ alone. (A) Electron micrograph of the fracture surface of *Conus litteratus* shell. Scale bar = 100 μm. Reprinted by permission from Currey (1999). (B) Crystalline aragonite (calcium carbonate) hierarchical organization of *Strombus gigas* shell. Adapted by permission of Macmillan Publishers Ltd from Kamat *et al.* (2000). The outer layer (O) represents the exterior shell surface and the interior (I) is closest to the animal, with the middle layer (M) in between. These layers are further subdivided into smaller and smaller tiles or lamellae called first-, second-, and third-order layers, each rotated approximately 45° from the adjacent layer. An organic matrix binds each lamellar layer. (C) Organic matrix polymer organization. Long molecules can be either looped (top) or straight with looped and folded domains (bottom) to take up the slack between two surfaces. Adapted by permission of Macmillan Publishers Ltd from Smith *et al.* (1999).

SEE ALSO THE FOLLOWING ARTICLES

Adhesion / Disturbance / Hydrodynamic Forces / Wave Forces, Measurement of

FURTHER READING

Alexander, R. M. 1983. *Animal mechanics*, 2nd ed. Oxford: Blackwell Scientific Publications.

Denny, M. W. 1988. *Biology and the mechanics of the wave-swept environment*. Princeton, NJ: Princeton University Press.

Denny, M. W. 1995. Predicting physical disturbance: Mechanistic approaches to the study of survivorship on wave-swept shores. *Ecological Monographs* 65(4): 371–418.

Niklas, K. J. 1992. *Plant biomechanics: an engineering approach to plant form and function*. Chicago: University of Chicago Press.

Vogel, S. 2003. *Comparative biomechanics: life's physical world*. Princeton, NJ: Princeton University Press.

REFERENCES

Ackerman, J. D., C. M. Cottrell, C. R. Ethier, D. G. Allen, and J. K. Spelt. 1996. Attachment strength of zebra mussels on natural, polymeric, and metallic materials. *Journal of Environmental Engineering* 122: 141–148.

Becker, K. 1993. Attachment strength and colonization patterns of two macrofouling species on substrata with different surface tension (*in situ* studies). *Marine Biology* 117: 301–309.

Burgert, I., and D. Eckstein. 2001. The tensile strength of isolated wood rays of beech (*Fagus sylvatica* L.) and its significance for the biomechanics of living trees. *Trees* 15: 168–170.

Chapman, G. 1953. Studies on the mesogloea of coelenterates II. Physical properties. *Journal of Experimental Biology* 30(3): 440–451.

Cotton, P. A., S. D. Rundle, and K. E. Smith. 2004. Trait compensation in marine gastropods: shell shape, avoidance behavior, and susceptibility to predation. *Ecology* 85: 1581–1584.

Currey, J. D. 1999. The design of mineralised hard tissues for their mechanical functions. *Journal of Experimental Biology* 202: 3285–3294.

Currey, J. D., P. Zioupos, P. Davies, and A. Casinos. 2001. Mechanical properties of nacre and highly mineralized bone. *Proceedings of the Royal Society London* 268: 107–111.

Dalton, A. B., S. Collins, E. Muñoz, J. M. Razal, V. H. Ebron, J. P. Ferraris, J. N. Coleman, B. G. Kim, and R. H. Baughman. 2003. Supertough carbon-nanotube fibres. *Nature* 423: 703.

Davidson, L. A., G. F. Oster, R. E. Keller, and M. A. R. Koehl. 1999. Measurements of mechanical properties of the blastula wall reveal which hypothesized mechanisms of primary invagination are physically plausible in the sea urchin *Strongylocentrotus purpuratus*. *Developmental Biology* 209: 221–238.

Denny, M. W. 1988. *Biology and the mechanics of the wave-swept environment.* Princeton, NJ: Princeton University Press.

Edlund, A. F., and M. A. R. Koehl. 1998. Adhesion and reattachment of compound ascidians to various substrata: weak glue can prevent tissue damage. *Journal of Experimental Biology* 201: 2397–2402.

Gosline, J. M. 1971. Connective tissue mechanics of *Metridium senile*. *Journal of Experimental Biology* 55: 775–795.

Gosline, J., M. Lillie, E. Carrington, P. Guerette, C. Ortlepp, and K. Savage. 2002. Elastic proteins: biological roles and mechanical properties. *Philosophical Transactions of the Royal Society London* 357: 121–132.

Hale, B. B. 2001. Macroalgal materials: foiling fracture and fatigue from fluid forces. Ph.D. dissertation, Stanford University.

Hamm, C. E., R. Merkel, O. Springer, P. Jurkojc, C. Maier, K. Prechtel, and V. Smetacek. 2003. Architecture and material properties of diatom shells provide effective mechanical protection. *Nature* 421: 841–843.

Hou, D. F., G. S. Zhoua, and M. Zheng. 2004. Conch shell structure and its effect on mechanical behaviors. *Biomaterials* 25: 751–756.

Ikoma, T., H. Kobayashi, J. Tanaka, D. Walsh, and S. Mann. 2003. Microstructure, mechanical, and biomimetic properties of fish scales from *Pagrus major*. *Journal of Structural Biology* 142: 327–333.

Kamat, S., X. Su, R. Ballarini, and A. H. Heuer. 2000. Structural basis for the fracture toughness of the shell of the conch *Strombus gigas*. *Nature* 405: 1036–1040.

Marrs, J., I. C. Wilkie, M. Skolde, W. M. Maclaren, and J. D. McKenzie. 2000. Size-related aspects of arm damage, tissue mechanics, and autotomy in the starfish *Asterias rubens*. *Marine Biology* 137: 59–70.

Menig, R., M. H. Meyers, M. A. Meyers, and K. S. Vecchio. 2000. Quasistatic and dynamic mechanical response of *Haliotis rufescens* (abalone) shells. *Acta Materialia* 48: 2383–2398.

Rapoport, H. S. and R. E. Shadwick. 2002. Mechanical characterization of an unusual elastic biomaterial from the egg capsules of marine snails (*Busycon* spp.). *Biomacromolecules* 3: 42–50.

Smith, A. M. 2002. The structure and function of adhesive gels from invertebrates. *Integrative and Comparative Biology* 42: 1164–1171.

Smith, B. L., T. E. Schäffer, M. Viani, J. B. Thompson, N. A. Frederick, J. Kindt, A. Belcher, G. D. Stucky, D. E. Morse, and P. K. Hansma. 1999. Molecular mechanistic origin of the toughness of natural adhesives, fibres and composites. *Nature* 399: 761–763.

Summers, A. P., T. J. Koob, and E. L. Brainerd. 1998. Stingray jaws strut their stuff. *Nature* 395: 450–451.

Thomsen, M. S., and T. Wernberg. 2005. Miniview: What affects the forces required to break or dislodge macroalgae? *European Journal of Phycology* 40: 139–148.

Thurmond, F. A., and J. A. Trotter. 1996. Morphology and biomechanics of the microfibrillar network of sea cucumber dermis. *Journal of Experimental Biology* 199: 1817–1828.

Waite, J. H., E. Vaccaro, C. Sun, and J. M. Lucas. 2002. Elastomeric gradients: a hedge against stress concentration in marine holdfasts? *Philosophical Transactions of the Royal Society London* 357: 143–153.

Wang, R. Z., L. Addadi, and S. Weiner. 1997. Design strategies of sea urchin teeth: structure, composition and micromechanical relations to function. *Philosophical Transactions of the Royal Society London* 352: 469–480.

Zuschin, M., and R. J. Stanton. 2001. Experimental measurement of shell strength and its taphonomic interpretation. *Palaios* 16: 161–170.

METAMORPHOSIS AND LARVAL HISTORY

NICOLE PHILLIPS

Victoria University, Wellington, New Zealand

Most marine invertebrates have a complex life cycle, including a larval stage that is drastically different from the adult stage in both form and function. Metamorphosis is the physiological process by which larvae transform into juveniles. Larval history, including stresses experienced by larvae, can have an influence on the ability of larvae to metamorphose and also have lingering effects on the success of juveniles and adults after metamorphosis.

POSSIBLE ENVIRONMENTAL STRESSES ENCOUNTERED BY LARVAE

Planktonic larvae may encounter a variety of environmental stresses in the water column, including variability in salinity, temperature, or oxygen concentration; exposure to pollutants, toxicants, or suspended sediment; and (for feeding larvae) food availability. For nonplanktonic larvae developing in benthic egg masses in the intertidal zone, there may be additional environmental stresses associated with exposure to air at low tide, including extremes of air temperature, desiccation, and exposure to ultraviolet light or waves. Any or all of these stresses may influence the ability of larvae to undergo metamorphosis or may influence the success of later juvenile or adults.

The effects of environmental stress on larval survival and development have been investigated in the laboratory for a wide variety of stressors and a relatively large number of species. To date, however, evidence for the effects of larval exposure to environmental stress on postmetamorphic success is limited to experimental studies on only some of these potential stressors, and for only a few species. Of those listed in the preceding paragraph, the best examined to date is probably larval nutritional stress. Reduced food availability for larvae of a variety of species (including mussels, polychaete worms, a slipper limpet, and a sea star) has resulted in decreased ability to undergo metamorphosis, smaller initial juvenile size, decreased juvenile growth, and increased juvenile mortality compared to juveniles fed abundant food rations as larvae.

DELAY OF METAMORPHOSIS

Many, if not most, planktonic larvae have the capacity to delay metamorphosis for variable periods of time (e.g., days to weeks or months) while remaining competent to undergo metamorphosis (i.e., physiologically capable of doing so). From experimental studies it has been shown that, for some species, delaying metamorphosis can have strong effects on metamorphic and postmetamorphic success (examples include species of bryozoans, sponges, ascidians, sand dollars, polychaete worms, abalone, oysters, crabs, and barnacles). Delaying metamorphosis may result in reduced initial juvenile size or structures, slower juvenile growth, increased juvenile mortality, increased time to reproduction, and reduced adult fecundity; but these effects and their severity are species-specific and also often related to the length of the delay.

In general, it appears that delaying metamorphosis can be particularly important for species with nonfeeding larvae. It is thought that because nonfeeding larvae are dependent on internal energy stores for development, delaying metamorphosis forces them to use energy that is otherwise necessary for undergoing metamorphosis and beginning early juvenile life.

MECHANISMS

The mechanisms that allow larval history to influence the process of metamorphosis, or the success of juveniles or adults, are not well understood. One often cited hypothesis, which may account for the effects of delaying metamorphosis (for nonfeeding larvae) and larval nutritional stress (for feeding larvae), is that these result in reduced energetic reserves necessary for fueling metamorphosis and early juvenile performance. For some species there is also evidence that a relatively small body size at metamorphosis (which may be the result of a variety of larval experiences or stresses) may be disadvantageous for early juveniles. Relatively small body size may be

correlated with smaller initial juvenile feeding structures, for example, or increased vulnerability to predation. Finally, because larvae develop juvenile structures, or the precursors to juvenile structures, over the course of the larval period, there are many potential physiological pathways for stresses that occur during larval life to influence postmetamorphic performance.

SEE ALSO THE FOLLOWING ARTICLES

Desiccation Stress / Larval Settlement, Mechanics of / Recruitment / Reproduction / Size and Scaling

FURTHER READING

Marshall, D. J., J. A. Pechenik, and M. J. Keough. 2003. Larval activity levels and delayed metamorphosis affect post-larval performance in the colonial ascidian *Diplosoma listerianum. Marine Ecology Progress Series* 246: 153–162.

Pechenik, J. A., D. E. Wendt, and J. N. Jarrett. 1998. Metamorphosis is not a new beginning: Larval experience influences juvenile performance. *Bioscience* 48: 901–910.

Phillips, N. E. 2002. Effects of nutrition-mediated larval condition on juvenile performance in a marine mussel. *Ecology* 83: 2562–2574.

MICROBES

PATRICIA A. HOLDEN AND ALLISON M. HORST

University of California, Santa Barbara

Microbes are organisms too small to see without the aid of a microscope, but they are the most numerous organisms on earth and are hugely important to the functioning of the entire biosphere. At the ocean's rocky edge, just as elsewhere on the planet, microbes serve as food for higher organisms, contribute to disease, and catalyze nutrient-cycling reactions.

CLASSIFICATION OF MICROBES

Microbes are traditionally classified into prokaryotes and eukaryotes, but the modern taxonomic system defines three broad domains: the Archaea and the Bacteria, which together comprise the prokaryotes, and the Eukarya, which includes all eukaryotes. Bacteria, as the domain name implies, include hetero- and autotrophic, single-celled organisms commonly known as bacteria. Archaea are also commonly called bacteria and share similar metabolisms with Bacteria, but they differ biochemically, are often associated with high-temperature and other extreme environments, and evolved earlier than the Bacteria. Eukarya are higher, more structurally complex organisms such as algae and diatoms.

Heterotrophic bacterio-plankton are major recyclers of organic carbon made by autotrophic (CO_2-eating) phytoplankton, the marine algae. Marine algae are important primary producers and also recycle nutrients. In the rocky intertidal, the main microbes present are the (prokaryotic) bacteria and cyanobacteria and the (eukaryotic) microalgae and diatoms.

Viruses are also important microbes, but, although numerous and diverse, they are not cells and thus are not classified as either prokaryotes or eukaryotes. Rather, viruses are genetic material infecting specific prokaryotic and eukaryotic cells; in that sense, they occur with other microbes and contribute to their turnover in the rocky intertidal.

HABITATS AND METHODS OF STUDY

In the rocky intertidal, just as in any aquatic system, microbes can be found in three general compartments: attached at the air/water interface (neuston) at the top of the water column, suspended in the water column (limnos) as either freely living or particle-attached cells, and attached to rocky or sediment surfaces at the bottom (benthos). In addition, microbes colonize the exterior and interior surfaces of higher organisms, such as molluscs.

However, microbes in nature are mostly attached to surfaces, where they form biofilms composed of cells and hydrated extracellular polymeric substances (EPSs). In theory, biofilms begin once microbial cells adhere to a surface that is precoated with an organic conditioning film. Cells feast on the film, then grow and manufacture EPSs, the glue that helps microbes stick to surfaces and to each other. To a degree, EPSs also protect microbes from external threats such as toxins, desiccation, and predators. EPSs, composed of polysaccharides, protein, and DNA, are also food sources when biofilm microbes are starved.

In the rocky intertidal, microbial biofilms contain layers of bacteria, diatoms, cyanobacteria, microalgae, and encrusting algae. The type of surface a biofilm colonizes (that is, whether it is rough or smooth, and whether it is rock or organic material) makes a difference to which organisms are there and their abundance. Biofilms in the rocky intertidal are studied by outplanting synthetic surfaces for colonization; they are also studied by chipping natural rock surfaces. Samples are examined for biofilm microbe shape and size using high-resolution microscopy. To quantify phototrophic microbial biomass, chlorophyll is extracted from samples and analyzed. Heterotrophs are cultured and quantified by washing biofilms into suspension, then plating onto solid media (agar) in the laboratory, and counting colonies. Total microbes are also counted via staining and epifluorescent microscopy. Determining exactly which organisms are there requires surveying microbes by culture-independent (DNA-based) methods, and this is an exciting frontier for future researchers interested in the microbial communities of the rocky intertidal.

MICROBIAL ECOLOGY OF THE ROCKY INTERTIDAL

When microbes colonize surfaces in the rocky intertidal, there is a succession (Fig. 1) that starts with bacteria and is later followed by diatoms, then additional phototrophic bacteria and microalgae. The accumulation of EPS can be seen along this progression. But where do biofilms begin? One idea is that mucus trails from molluscs serve as the conditioning films that are feasted upon by heterotrophic bacteria that make depolymerizing, extracellular enzymes. In this way, molluscs feed initial microbial colonizers. However, molluscs, in that they are grazers, are also the predators of microbial biofilms in the rocky intertidal. Thus, there is a complex interaction between biofilms as prey, and molluscs as predators, with both being subject to influences of environmental factors.

The rocky intertidal is a dynamic physical environment: High-energy waves break, shores alternate between wet and dry with the tides, and there are diurnal and seasonal variations in temperature and shading, or reduced insolation. It is also a dynamic chemical environment where salinity can shift with varying terrestrial freshwater inflow, and pollutants can arrive from either terrestrial (e.g., pesticides and fertilizers) or oceanic (e.g., oil spills) sources. Microbes and their predators (macrograzers such as molluscs, and micrograzers such as protozoa and copepods) are all affected by these environmental and anthropogenic factors. When in the surf, microbial biofilms can be eroded by water, leaving only the most firmly attached fractions behind. At low tides, biofilms may dry out, killing some microbes, but others are protected by the spongy, hydrated EPSs. In warm seasons, microbial biofilm population sizes decline, perhaps due to increased feeding by grazers. In the winter, populations increase because grazing decreases. Across seasons, due to selective feeding by grazers, the relative abundance of different microbes changes within biofilm communities. The abundance of microbes during the summer may even control the overall abundance of grazers.

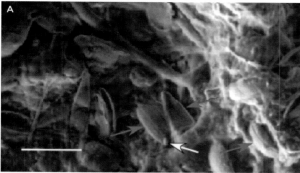

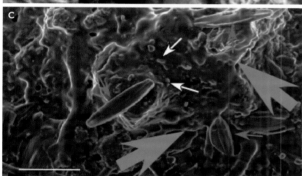

and they release copious amounts of nitrous oxide (N_2O), a potent greenhouse gas. Sulfur and silicon are also cycled rapidly by intertidal biofilms. Other elements, including toxic metal pollutants, are transformed by biofilms but are also trapped in EPSs, where they can then be transferred up the food chain by grazers.

Microbes respond rapidly to changes in their environment. Predicted large-scale changes in the global climate are expected to change the rocky intertidal physical and chemical environment. Understanding the overall responses of microbes to changing environmental factors such as temperature, salinity, and inundation frequency will improve the ability to predict specific responses of rocky-intertidal microbes including community composition and function shifts.

SEE ALSO THE FOLLOWING ARTICLES

Biofilms / Diseases of Marine Animals / Nutrients / Temperature Change

FURTHER READING

Decho, A. W. 2000. Microbial biofilms in intertidal systems: an overview. *Continental Shelf Research* 20: 1257–1273.

Magalhaes C. M., W. J. Wiebe, S. B. Joye, and A. A. Bordallo. 2005. Inorganic nitrogen dynamics in intertidal rocky biofilms and sediments of the Douro River estuary (Portugal). *Estuaries* 28: 592–607.

Thompson, R. C., T. P. Crowe, and S. J. Hawkins. 2002. Rocky intertidal communities: past environmental changes, present status and predictions for the next 25 years. *Environmental Conservation* 29: 168–191.

Whitman, W. B., D. C. Coleman, and W. J. Wiebe. 1998. Prokaryotes: the unseen majority. *Proceedings of the National Academy of Sciences of the USA* 95: 6578–6583.

FIGURE 1 Microbial biofilms cultivated on a sandpaper substrate suspended in an intertidal estuary over a 35-day period. Images were captured with an environmental scanning electron microscope (ESEM) operating in wet mode at 5 kV accelerating voltage and 2500× magnification. Scale bar = 10 μm. Red arrows = diatoms; white arrows = bacteria; blue arrows point to EPS-rich regions. (A) Fourteen-day-old biofilm where bacteria and their EPSs dominate with few, small diatoms appearing. (B) Twenty-five-day-old biofilm with larger diatoms and increased EPS around and on diatoms. (C) Thirty-five-day-old biofilm with more EPS and other surface growth covering diatoms and bacteria underneath.

Nutrient cycling by rocky-intertidal microbial biofilms is a fascinating area of research. Primary producers in rocky-intertidal biofilms fix carbon and release copious amounts of oxygen. Relative to nearby sedimentary biofilms, rocky-intertidal biofilms rapidly cycle all inorganic and organic nitrogen-bearing compounds such as proteins,

MICROMETAZOANS

JÖRG OTT

University of Vienna, Austria

Many higher taxa of metazoans never exceed microscopic size. Together with specialized dwarf forms from taxa of normally larger size (macrofauna), they constitute the meiofauna of benthic habitats. Because of their small size, these animals live under conditions often radically different from those of the macrofauna and have evolved peculiar biological features. Although inconspicuous in terms of biomass, they may act as an important regulatory link in the microbial component of tidepool systems.

Most marine invertebrates pass through a stage where they would be classified as meiofauna. The majority, however, outgrow this size range over time and may be considerably larger upon reaching adulthood. The term temporary meiofauna has been introduced for these juvenile stages of macrofauna species—in contrast to the permanent meiofauna, which complete their whole life cycle at a small size.

In several phyla, all species belong to the meiofauna, for example, the Gnathostomulida, Loricifera, Kinorhyncha, Gastrotricha, or Tardigrada. Among them are the smallest invertebrates known, with some adult forms not exceeding 100 micrometers in length. Most marine free-living Nematoda and Platyhelminthes are also of microscopic size.

Within other phyla (for example, the Arthropoda), higher taxa such as classes and orders may be entirely meiofaunal. This includes the mites (Acari) and, among the crustaceans, the Ostracoda and Copepoda. In addition, many taxa that, as a rule, belong to the macrofauna are represented among the meiofauna with dwarf forms, for example, the Cnidaria, Gastropoda, Annelida, or Tunicata. These are treated elsewhere in this volume.

Benthic micrometazoans are especially abundant and diverse in subtidal algae and sediments. In particular, sandy bottoms are the prime habitat of the most highly adapted meiofauna, which inhabit the interstitial spaces between the sand grains (interstitial fauna or mesopsammon). Only a limited number of micrometazoan species has been reported from the intertidal of rocky shores. This is partly due to the physiologically and ecologically demanding environment, but also partly to the lack of attention that has been given to these small creatures in this habitat, in which taxonomic and ecological studies have largely focused on macroscopic organisms.

Platyhelminthes (Flatworms)

This group of acoelomate animals (Fig. 1A) is positioned at the basis of bilaterian evolution. All have a ciliated epidermis and a digestive tract with a ventrally located mouth. The gut may be equipped with a muscular pharynx and lacks an anus. It is saclike in small species but may be highly branched in larger forms, such as the Tricladida or Polycladida. Although the latter are macroscopic animals that are often confused with slugs, most other free-living flatworms are microscopic, rarely exceeding 5 mm in length. They may be drop-shaped, ribbon-shaped, or filiform. A system of fine longitudinal and circular muscles beneath the epidermis allows for an extremely flexible body shape. They move by ciliary swimming or gliding

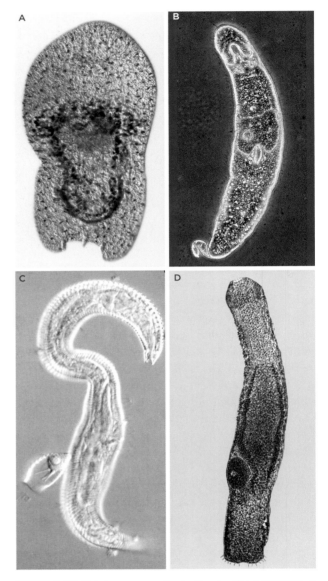

FIGURE 1 (A) *Polychoerus carmelensis* (Acoela, Platyhelminthes), a common tidepool inhabitant. Length, 1.4 mm. Photograph courtesy of M. Hooge. (B) *Chirognathia dracula*, from the reduced sediment among surfgrass (*Phyllospadix* ssp.) roots, U.S. and Canadian West Coast. Length, 700 μm. From Sterrer and Sorensen (2006). (C) Epsilonematid nematode from finely branched calcareous algae, Adriatic Sea. Length, 400 μm. Photograph courtesy of R. Novak. (D) *Paradasys* sp. (Macrodasyoidea, Gastrotricha) from the Oregon coast. Length, 700 μm. Photograph courtesy of M. Hooge.

and are, by far, the fastest among the micrometazoa. Marine Catenulida, for example, are currently known only from sheltered sediments. The Acoela and the small Rhabditophora occur in all marine habitats and may be the dominant micrometazoans in marine algae or tidepools. The larger rhabditophoran Polycladida are mainly found under boulders and among sedentary animals.

Gnathostomulida (Jaw Worms)

These small sluggish acoelomate worms (Fig. 1B) are distinguished from the Platyhelminthes by a monociliated epidermis and cuticularized, paired jaws with an unpaired basal plate in the muscular pharynx. They are either elongate with a long rostrum anterior to the ventral mouth (Filospermoidea) or bowling pin–shaped with a slightly swollen anterior end bearing a characteristic set of sensory cilia (Scleroperalia). So far they have been found only in sheltered intertidal and subtidal sediments, in or close to the anoxic and sulfidic layers below the redox potential discontinuity (RPD). Species of both Filospermoidea and Scleroperalia have been reported from the sediment among surfgrass roots.

Nematoda (Roundworms)

Together with platyhelminthes and copepods, the nematodes (Fig. 1C) are the most abundant micrometazoans inhabiting both algae and sediments. Free-living marine nematodes range from less than 100 µm (some species of Desmoscolecidae and Epsilonematidae) to about 10 or even 20 mm in length. Nematodes lack cilia and have only a limited ability to swim. The larger species are thread-shaped and move by snakelike writhing through sand or algae. The tiny desmoscolecids and epsilonematids crawl caterpillar- or inchworm-like over surfaces. The body surface is covered with a cuticle that is molted four times during the life span of the worms. In most cases the cuticle is devoid of microbial fouling. In the families Epsilonematidae, Draconematidae, and Desmodoridae, however, epigrowth of bacteria, diatoms, or suctorians is frequently observed in some species. In the desmodorid subfamily Stilbonematinae, all representatives have a species-specific coat of symbiotic sulfur bacteria covering all or certain parts of the animal's surface.

Small species (such as most Chromadoridae) are typical for algal turf or fine epiphytes; larger species are found in coarse algae and assemblages of sedentary animals, where families such as the Enoplidae, Oncholaimidae, or Phanodermatidae dominate.

Gastrotricha

These tiny aschelminthes (Fig. 1D) only rarely exceed 1 mm in length. In the purely marine Macrodasyoidea the body is ribbon-shaped. Adhesive tubules connected to epidermal glands are distributed over the whole body and concentrated in certain body regions, such as the long threadlike tail in the genus *Urodasys*. In the Chaetonotoidea, which also occur in freshwater, the body is shaped like a bowling pin, and only two adhesive glands (toes) are present at the posterior end. The body is covered by a multilayered cuticle, often forming scales or spines. They move by means of cilia on their ventral side. Whereas the freshwater Chaetonotoidea live in the periphyton and are nimble swimmers, in the sea gastrotrichs are exclusively mesopsammic and—in the rocky intertidal—confined to sediment accumulations. Both macrodasyoids and chaetonotoids have been reported from the sediment among surf-grass roots.

Loricifera

With an adult size of <0.5 mm, loriciferans (Fig. 2A) are the smallest metazoans thus far described. Their body lacks cilia and is encased by a vase-shaped cuticular shell (lorica). The minute terminal mouth is surrounded by symmetrically arranged spines (scalids), which can be withdrawn together with the front end. The Loricifera develop through several larval stages whose appearance can differ considerably from that of the adults. Most species have been found in deep water, and none has been reported from the rocky intertidal yet. Findings in shallow, even intertidal sediments, however, indicate that the tiny creatures may have been overlooked in past studies of rocky-shore sediment accumulations.

Kinorhyncha

The barrel-shaped body of kinorhynchs (Fig. 2B) is superficially segmented (zonites). They move by means of a retractable anterior end (introvert) bearing several circles of spines (scalides). The body surface lacks cilia and is covered by a stiff, jointed cuticle, often armed with spines. The cuticle of the zonites forms a closing structure when the introvert is retracted. Most kinorhynch species inhabit muddy sediments from the intertidal to the deep sea. Species of the genus *Echinoderes*, however, may be abundant in finely branched coralline algae.

Tardigrada (Water Bears)

The Tardigrada (Fig. 2C) are considered to be a sister group of the Arthropoda within the Articulata. Their body is divided into four segments, each of which bears a pair of legs. The mouth is terminal, leading to a sucking pharynx equipped with two protrusive stylets. They inhabit terrestrial, freshwater, and marine habitats. The marine Arthrotardigrada have extensible legs with four to six toelike projections ending in claws or suckers. Many genera and species have been described from sediments, from beaches to the deep sea. The Echiniscoidea bear several claws on short stubby legs. They are almost exclusively terrestrial and limnic. The cosmopolitan marine species, *Echiniscoides sigismundi*, however, may reach high

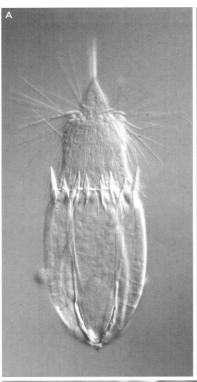

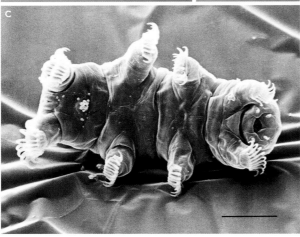

FIGURE 2 (A) *Nanaloricus* sp. (Loricifera) from shell gravel. Length, 350 μm. Photograph courtesy of M. Bright. (B) *Echinoderes* sp. (Cyclorhaga, Kinorhyncha) from finely branched calcareous algae, Adriatic Sea. Length, 400 μm. Photograph courtesy of M. Bright. (C) *Echiniscoides sigismundi*, a heterotardigrade common in intertidal algae and mussel beds. Length, 300 μm. Photograph courtesy of R. Kristensen.

densities in turfs of algae and cyanobacteria among barnacles or mussels in the rocky intertidal.

HABITATS

Although micrometazoans may be found swimming in the water of tidepools, almost all regular members of the tidepool and rocky-shore microfauna are benthic and associated with structures of the sea bottom. Because of

their small size the animals depend on small-scale structures such as crevices and cracks in the rock, algae, seagrass, or sessile animals. Such structures create shelter against strong water movement, retain moisture during periods of emersion, and collect sediment and organic matter.

Erosional processes fracture the rocks and create a complicated relief. Especially on limestone coasts, where biocorrosion and bioerosion play important roles in coastal geomorphology, the surface of the rock may be structured in alternating sharp crests and deep furrows. Boring cyanobacteria and lichens weaken the rock surface and are, in turn, grazed by chitons and gastropods, leading to profiles with high fractal dimension. In the lower intertidal the abandoned internal cavities created by boring sponges and bivalves provide additional habitats for micrometazoans.

In large and deep tidepools, sediment may collect on the bottom. Because of the generally wave-exposed situation of rocky shores, this is commonly gravel or coarse sand with a high proportion of biogenic material (debris from coralline algae and mollusc shells). This yields larger interstitial spaces suitable for colonization by larger meiofauna species.

Many rocky shores, especially in cool temperate climates, are covered by intertidal algae. Finely branched macroalgae or those with a cover of epiphytes are an especially favorable habitat for micrometazoans, whereas smooth, fleshy thalli are only sparsely colonized. Even algal turfs that are only a few millimeters high can harbor high densities of nematodes and platyhelminthes. A special algal habitat is the interstitial space created by coralline algae. Such algae may form dense banks of porous limestone in the lower intertidal (the trottoir), which is at the same time the habitat of specialized arthropods of terrestrial origin (pseudoscorpions, myriapods, insects).

On temperate Pacific rocky shores the root mats of surf grass, *Phyllospadix* spp., create a special habitat for meiofauna. The roots, which are attached to the rock surface, accumulate rather fine sediment. Due to the high organic matter content of the sediment, anoxic and sulfidic conditions develop in the interstitial space. A special micrometazoan community, the sulfide system, adapted to low oxygen/high sulfide conditions, inhabits this sediment (including gnathostomulids and stilbonematid nematodes with symbiotic sulfur bacteria); this community is otherwise found only in sheltered tidal flats or subtidal sand.

Similar conditions are created within banks of intertidal sessile animals, such as mussels, sabellarid worms, or barnacles. The feces and pseudofeces of these suspension-feeding animals enrich the sediment trapped within the aggregations with organic matter, leading to a depletion of oxygen and the development of sulfide.

PHYSIOLOGICAL PROBLEMS

Large temperature amplitudes, rapid salinity changes, large and rapid variations in oxygen concentrations, and desiccation and freezing are the main physiological stresses of all intertidal and rockpool animals. In some cases the meiofauna can avoid the more extreme conditions by retreating into crevices and interstitial spaces; in others, however, it is subject to even higher stresses when confined in water films or in the boundary layer. Because the severity and frequency of stress increases with height above low tide, only a few micrometazoan species—highly adapted to physiological stress—are present in the high intertidal and supratidal.

Temperature

Little is known about temperature tolerances in intertidal micrometazoans. In the platyhelminth *Procerodes littoralis*, oxygen consumption rates are independent of temperature between 5 and 20 °C and elevated above 25 °C.

Salinity

Because of their small size, micrometazoans are mostly osmoconformers but can keep their physiological functions more or less constant under changing salinities.

Oxygen

Oxygen saturation may reach more than 500% in the light under stagnant conditions in the first millimeter on macroalgal surfaces to which most of the meiofauna animals are confined. Because of the small size of the meiofauna, diffusion will rapidly equilibrate the internal milieu of the organisms with the surrounding water; formation of reactive oxygen species could lead to hyperoxic stress. Within fine algal turf, anoxic and sulfidic conditions are rapidly established in the darkness. Meiofauna react to anoxic conditions by leaving the algal mat. The transition between hypoxic, normoxic, or hyperoxic conditions can be abrupt, for example, when the tide reaches a previously stagnant pool.

Sulfide

The high sulfide concentrations and permanent anoxic conditions found in the interstitial space of sediment collected among seagrass roots or between mussels selects for a special meiofauna adapted to at least temporally anaerobic life. The dominant micrometazoans here are nematodes and gnathostomulids, which have extremely low respiration rates and may tolerate long periods of anoxia. Sulfide detoxification in the animal tissue has been reported for nematodes in sulfidic environments and hypothesized as the role of symbiotic sulfur bacteria in stilbonematid nematodes, but it is unlikely to play a major role because of the small mass of both animals and microbes.

Desiccation

Splash pools and high intertidal pools periodically dry out. Meiofauna may avoid complete drying by retreating into moisture-conserving structures. Under unfavorable conditions Tardigrada can enter a cryptobiosis stage, with a water content of only 3%, for years. In this stage the animals can also endure extremes of temperature and radiation.

ECOLOGY

Reproduction and Dispersal

The small size of these animals limits the number of eggs produced by a female. Therefore, adaptations to ensure fertilization (e.g., hermaphroditism, sperm storage) or development without fertilization (parthenogenesis), brood protection, or vivipary are common. Direct development without planktonic stages is the rule in meiofauna. Dispersal operates via drift of mainly adults during submersion, and new and insular habitat patches are rapidly colonized.

Feeding

Little is known about the food of meiofauna. Gnathostomulids, gastrotrichs, and many nematodes probably feed on bacteria and may play a role in stimulating bacterial productivity. On algae, nematodes equipped with buccal teeth and some platyhelminthes feed on epiphytic diatoms. Tardigrada are thought to pierce algal cells and suck the contents. Some of the larger nematodes and many platyhelminthes are predators on other meiofauna.

Predation

Whether predation of macrofauna on meiofauna plays a role in ecosystem function is still open to debate. Meiofauna populations, however, may be regulated by predation, and refuge from predation has been put forward as an explanation for the colonization of physiologically demanding habitats such as supratidal pools.

SEE ALSO THE FOLLOWING ARTICLES

Amphipods, Isopods, and Other Small Crustaceans / Microbes / Protists / Worms

FURTHER READING

Barnes, R. S. K. 2001. *The invertebrates*. Oxford: Blackwell Science.
Giere, O. 1994. *Meiobenthology*. New York: Springer.
Higgins, R. P., and H. Thiel, eds. 1988. *Introduction to the study of meiofauna*. Washington, DC: Smithsonian Institution Press.
Sterrer, W., and M. V. Sorensen. 2006. *Chirognathia dracula* gen. et. spec. nov (Gnathostomulida) from the west coast of North America. *Marine Biology Research* 2: 296–302.

MOLLUSCS, OVERVIEW

DAVID R. LINDBERG
University of California, Berkeley

Molluscs are bilaterally symmetrical Lophotrochozoa that range in size from giant squids more than 20 meters in length to some adult snails and bivalves that barely reach one-half millimeter (0.5 mm). The molluscs are often cited as the second largest phylum next to Arthropoda, with about 200,000 living species, of which about 75,000 living and 35,000 fossil have been described, thereby making them one of the better-known marine invertebrate groups. In addition to size variation, molluscs exhibit a broad range of physiological, behavioral, and ecological adaptations that have enabled them to diversify into most of the earth's biomes. Molluscs also have an excellent fossil record extending back over 500 million years to the Early Cambrian and perhaps into the Precambrian (>542 million years). Three major classes—Gastropoda (snails, slugs, limpets), Bivalvia (scallops, clams, oysters, mussels), and Cephalopoda (squid, cuttlefish, octopuses, nautilus)—are well known, whereas the remaining five classes are less morphologically and ecologically diverse and often less abundant. These include the former aplacophoran taxa (spicule worms) Chaetodermomorpha and Neomeniomorpha, the Polyplacophora (chitons), the Scaphopoda (tusk shells), and the Monoplacophora (a small group of deep-sea limpets with a long fossil history).

MORPHOLOGY

The molluscan body commonly consists of a head, foot, and visceral mass, covered with a mantle that typically secretes the shell (or, more rarely, spicules); the shell can be secondarily lost in some groups (e.g., nudibranchs, octopuses). Generally, there are one or more pairs of gills (ctenidia), which lie in a posterior mantle cavity or in a posterolateral groove surrounding the foot. The kidneys, gonads, and anus open into this cavity, which typically contains a pair of sensory structures known as the osphradia. The mouth opens into the buccal cavity, which contains a radula—a ribbon of teeth supported by a muscular odontophore. The radula and associated structures have been lost in the bivalves. There is a ventral foot used for locomotion by using muscular waves or cilia, or both, in combination with mucus. Molluscs are coelomate animals, although the coelom is reduced and represented by the kidneys, gonads, and pericardium; the main body cavity is a hemocoel. Molluscs lack segmentation and undergo spiral cleavage during development. Trochophore or veliger larvae are found in many aquatic taxa, but direct development is also common in the phylum.

The earliest undoubted molluscs are found in the Early Cambrian (approximately 540 million years ago), from which several major and minor groups (e.g., gastropods, bivalves, monoplacophorans) are known. Cephalopods are found from the Middle Cambrian, polyplacophorans from the Late Cambrian, and scaphopods from the Middle Ordovician. Studies on molluscan evolution are often able to use this rich fossil diversity and can be particularly useful when combined with morphological, anatomical, embryological, and molecular studies of recent taxa. Studies of the genetics, diversity, phylogeny, and ecology of molluscs have also provided important contributions to the more general fields of evolutionary biology, biogeography, and ecology.

SYSTEMATICS

There are seven class-rank taxa of molluscs (Fig. 1): Polyplacophora, Monoplacophora, Bivalvia, Chaetodermomorpha, Neomeniomorpha, Scaphopoda, Cephalopoda, and Gastropoda. The Chaetodermomorpha, Monoplacophora,

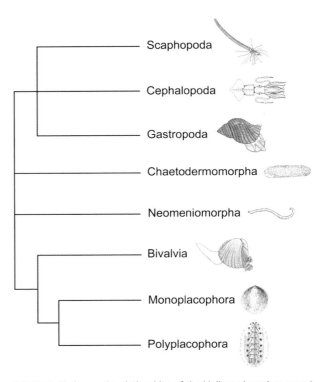

FIGURE 1 Phylogenetic relationships of the Mollusca based on recent morphological and molecular studies.

Neomeniomorpha, and Scaphopoda do not occur in rocky-intertidal habitats or tidepools.

Polyplacophora

Chitons (Fig. 2) are dorsal-ventrally flattened molluscs, elongate-oval in outline, and with eight overlapping dorsal shell plates or valves. The valves are typically bordered by a thick girdle that may be covered with spines, scales, or hairs. The mantle cavity contains multiple pairs of small gills and surrounds the foot, with which the animal tenaciously clings to hard surfaces. The shell valves can be reduced or even internal, sometimes to the point of giving the chiton a wormlike body. Most chitons are relatively small animals (0.5–5 cm), but one Californian species *(Cryptochiton stelleri)* can reach more than 30 cm in length.

FIGURE 2 The chiton *Katharina tunicata* in the low intertidal zones at Arroyo de los Frijoles, San Mateo County, California. Photograph by the author.

All chitons are marine or estuarine and members of the group are found worldwide. This relatively small group of molluscs is estimated to include between 650 and 800 living species. High diversities of chitons are found on the rocky shores that encircle the Pacific Rim. Most live in the intertidal zone or shallow sublittoral, but some live in water deeper than 7000 meters. Most species are found on hard rocky substrates, although a few species are associated with algae and marine plants; in the deep sea, waterlogged wood is a common habitat for some species. Numerous species include tidepools as part of their habitat, but they are most common under rocks

resting on hard substrates or resting in sand or gravel. The rocky-shore genus *Tonicella* is common on coralline algae.

Chitons feed on encrusting organisms such as sponges and bryozoans and nonselectively on diatoms and algae that are scraped from the substrate with their radula. The radular teeth of chitons are hardened by the incorporation of metallic ions into the tooth matrix. One group—the *Placiphorella*—captures small crustaceans by trapping them under the anterior part of their body. Most chitons exhibit nocturnal behavior and show homing behavior—returning to a specific spot on the substrate, where they sometimes form depressions (Fig. 2).

Chitons are generally dioecious, with sperm released by males into the water. In most species fertilized eggs are shed singly or in gelatinous strings and, once fertilized, develop into a trochophore larva that soon elongates and then directly develops into a juvenile chiton with no veliger stage. In brooding species the eggs remain in the mantle cavity of the female, where they are fertilized by sperm moving through with the respiratory currents. Upon hatching from the brooded eggs the offspring may remain in the mantle cavity until they crawl away as young chitons or exit the mantle cavity as trochophores for a short pelagic phase before settling.

Bivalvia

Bivalves (Fig. 3) are named for their shell, composed of a pair of laterally compressed hinged valves in which an expansive mantle cavity surrounds the body. The shell can be internal, reduced, or even absent; some bivalves are even wormlike, such as the shipworms *(Teredo)*. Bivalves lack a head, radula, and jaws. The bivalve foot is modified as a powerful digging structure in many groups, but in those that live attached to the substrate (e.g., oysters) the foot is reduced. A few groups of bivalves, such as oysters, are cemented permanently to the substrate.

Bivalves are the second largest group of molluscs. Most are marine, but there are also substantial radiations into brackish and freshwater habitats. Bivalves may be infaunal or epifaunal, and epifaunal taxa may be either sessile (cemented or byssally attached) or motile. The bivalves are an extremely diverse group, with about 10,000 described species that range in adult size from 0.5 millimeter to giant clams that reach 1.5 meters.

Five major groups of bivalves are recognized: Protobranchia, Pteriomorphia, Palaeoheterodonta, Heterodonta, and Anomalodesmata. Only two of these groups—Pteriomorphia and Heterodonta—are typically represented along rocky shores and in tidepools.

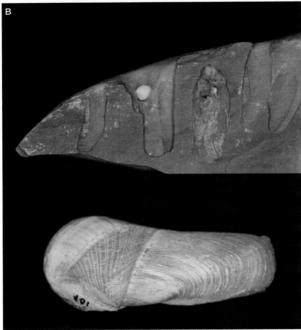

FIGURE 3 Rocky-shore bivalves. (A) Mid-intertidal aggregation of the mussels *Mytilus californianus* and *Mytilus galloprovincialis*. The lower limit of the mussels is limited, in part, by the predatory sea star *Pisaster ochraceus* (Dillon Beach, Marin County, California, USA). (B) Burrows created in sedimentary substrate by rock-boring bivalves of the family Pholadidae (Santa Cruz, Santa Cruz County, California, USA). Photographs by the author.

The Pteriomorphia include many of the familiar bivalves—the scallops (Pectinina), oysters (Ostraeidae), mussels (Mytilidae), arks (Arcidae), and others. Many pteriomorphs are epifaunal whereas others are shallow burrowers, including the arcid *Adula* and the mytilid *Lithophaga*, which burrow into soft shales. Rocky-shore pteriomorphs include the ubiquitous mussels (e.g., *Mytilus* spp.), oysters, rock scallops, spondylids, and the Anomiidae, all of which

attach to the rocky substrate by a bysuss or are permanently cemented to the substrate. All are filter feeders, and the high densities of mussels often make them an important ecological habitat in the rocky intertidal (Fig. 3A).

The Heterodonta includes the majority of burrowing bivalves. There are more than 40 families, including the Veneridae, Cardiidae, Mactridae, and Tellinidae. Most of these groups are shallow burrowers in soft sediments, but this group also includes the rock-boring Pholadidae, the Chamidae (which are cemented to rocky substrates), and a host of often small, rocky-shore nestlers such as *Lasaea*, *Petricola*, *Semele*, and *Hiatella*. However, of all rocky-shore bivalves, the rock-boring pholads are perhaps the most amazing (Fig. 3B). They use their anterior shell valves as rasps; the shell is rocked by muscular contractions while held against the substrate with the foot. Repeated rotations of the animal produce the tunnel-like burrow. Boring rates range between 4 and 50 mm per year, depending on the hardness of the rock.

Bivalves are hermaphroditic or have separate sexes. The eggs are typically small and not very yolk-rich. Fertilization is usually external except in brooding species. Embryos developing in the water column go through both trochophore and veliger larval stages. Although these larval stages are morphologically similar to the gastropod veliger larva, phylogenetic analyses have suggested that they are convergent and not shared by common ancestry. The initial uncalcified shell grows laterally in two distinct lobes to envelop the body. Many bivalve species retain their eggs in the mantle cavity and bring in sperm via the inhalant water current. In these brooding bivalves the larvae develop in special pouches in the gills in some taxa or, in others, simply lie in the mantle cavity. Many brooding bivalves release their young as swimming veliger larvae, whereas others retain them longer and release them as juveniles.

Gastropoda

Gastropods (Fig. 4) are characterized by the possession of a single (often coiled) shell and a body that has undergone torsion such that the mantle cavity faces forward; the shell may be lost in some sluglike groups. Gastropods have a well-developed head, which bears eyes and a pair of tentacles (cephalic tentacles), and a muscular foot, which is used for locomotion in most species whereas in others it is modified for swimming or burrowing. The foot typically bears an operculum that seals the shell aperture when the head-foot is retracted into the shell.

Externally, gastropods appear to be bilaterally symmetrical; however, they are one of the most successful clades of asymmetric organisms known. The ancestral state of

FIGURE 4 *Nucella ostrina* feeding on barnacles on the Bodega Bay headlands, Sonoma County, California, in the United States. Photograph by Jacqueline L. Sones.

this group is clearly bilateral symmetry (e.g., chitons, cephalopods, bivalves), but during development, organ systems can be twisted into figure-eights by torsion, rotated, or differentially developed or lost on either side of their midline, and they produce shells that may coil to either the right or the left.

Gastropods are extremely diverse in size, body and shell morphology, and habits and occupy the widest range of ecological niches of all molluscs, being the only molluscan group to have invaded the land. Snails, such as the whelk *Syrinx aruanus*, can reach lengths in excess of 600 mm, but there is also a very large (and poorly known) fauna of micro-gastropods (0.5–4 mm) that live in marine, freshwater, and terrestrial environments. Gastropods are, by far, the largest group of molluscs, with more than 62,000 described living species, comprising about 80% of all living molluscs. Estimates of total extant species range from 40,000 to more than 100,000 but may be as high as 150,000, with about 13,000 named genera for both recent and fossil species. They have a long and rich fossil record, from the Early Cambrian on, that shows periodic extinctions of groups, followed by diversification of new species.

Gastropods occupy all marine habitats, ranging from the deepest ocean basins to the highest reaches of the intertidal, as well as freshwater habitats and salt lakes. Although most aquatic gastropods are benthic and mainly epifaunal, a few groups are planktonic (e.g., Janthinidae, Heteropoda, and Gymnosomata). Gastropods are also found in the terrestrial biomes, occurring in lowlands and high mountains, in deserts and rain forests, and from the tropics to high latitudes.

Gastropod feeding habits are extremely varied, although most species use the radula in some aspect of their feeding behavior. Gastropods, by feeding habits, include grazers, browsers, suspension feeders, scavengers, detritivores, and carnivores. Carnivory in some taxa may simply involve grazing on colonial animals, whereas others engage in active hunting of their prey (Fig. 4). Some gastropod carnivores drill holes in their shelled prey, a method of entry that has been acquired independently in several groups (e.g., Muricidae and Naticidae). Some gastropods feed suctorially and have lost the radula.

Traditionally the gastropods were divided into three sub-classes: the Prosobranchia, the Opisthobranchia, and the Pulmonata. Modern classifications, confirmed from several morphological and molecular studies, recognize five mono-phyletic groups: (1) Patellogastropoda, (2) Vetigastropoda, (3) Neritimorpha, (4) Caenogastropoda, and (5) Hetero-branchia. The Patellogastropoda, Vetigastropoda, and Neritopsina make up the former Archaeogastropoda, and the Caenogastropoda includes the former Mesogastropoda and Neogastropoda. All of these constitute the former Prosobranchia. The Heterobranchia is made up of the former Opisthobranchia and the Pulmonata. At present, there is no general agreement regarding the ranks applied to these major groups. Members of all these groups are often common along rocky shores and in tidepools.

The Patellogastropoda or true limpets (Patellidae, Acmaeidae, Lottiidae, Nacellidae, and Lepetidae) are often abundant and characteristic species of rocky shores. All are marine and limpet-shaped, and most live in the intertidal zone. The shell is never nacreous, and the oper-culum is absent in adults. Most patellogastropods are nonspecific grazers and their radula has only a few teeth per row, some of which are strengthened by the incorpo-ration of metallic ions such as iron.

The Vetigastropoda contains the keyhole and slit lim-pets (Fissurellidae), abalones (Haliotiidae), top shells (tro-chids), and about 10 other families. All are marine and have coiled to limpet-shaped shells. The shell is nacreous in many of these taxa (e.g., abalones), and an operculum is usually present. The radula has many teeth in each row and is not mineralized.

The Neritimorpha are found in marine, freshwater, and terrestrial habitats. They have coiled to limpet-shaped shells, and one group has lost its shell and become sluglike (Titiscaniidae). Although uncommon in most temperate environments, nerites can be common and important grazers in tropical and subtropical habitats. Their radula is similar to that of the vetigastropods.

The Caenogastropoda are a very large and diverse group containing about 100, mostly marine, families. They include many rockpool and rocky-habitat groups such as the periwinkles (Littorinidae), whelks (Buccinidae), and muricids (Muricidae). Caenogastropod shells are typically coiled, a few are limpet-like (e.g., the slipper limpets, Calyptraeidae), and one family (Vermetidae) has shells resembling worm-tubes. There are several common radular configurations in this group, reflecting a remarkable specialization for a diverse array of food types.

The Heterobranchia is also a very large group and includes marine and one predominately freshwater group (Valvatidae) that were previously included in the Mesogastropoda, as well as two very large groups that were previously given subclass status—the Opisthobranchia and Pulmonata (collectively known as the Euthyneura). The basal members of the Heterobranchia include about a dozen families that are mostly small sized and poorly known. One of these groups is the Pyramidellidae—a diverse group of small-sized ectoparasites that are commonly associated with other rocky-shore molluscs, especially abalone *(Haliotis)*.

The opisthobranchs comprise about 25 families and 2000 species of bubble shells (Cephalaspidea) and sea slugs (Nudibranchia) as well as the sea hares (Anaspidea), all of which are common on rocky shores and associated tidepools. Virtually all opisthobranchs are marine, with the majority showing shell reduction or shell loss. Only a few of the primitive shell-bearing taxa have an operculum as adults.

Most gastropods have separate sexes; but some groups (mainly the Heterobranchia) are hermaphroditic, although most hermaphroditic forms do not normally engage in self-fertilization. The basal gastropods release their gametes into the water column in which they undergo development, but others use a penis to copulate or exchange spermatophores and produce fertilized eggs surrounded by protective capsules or jelly. The first gastropod larval stage is typically a trochophore that transforms into a veliger and then settles and undergoes metamorphosis to become a juvenile snail. Although many marine species undergo larval development, there are also numerous marine taxa that have direct development, a mode that is more common in freshwater and terrestrial taxa. Brooding of developing embryos is widely distributed throughout the gastropods, as are sporadic occurrences of hermaphrodism in the nonheterobranch taxa. The basal groups have nonfeeding larvae, but veligers of many neritopsines, caenogastropods, and heterobranchs are planktotrophic. Egg size is reflected in the initial size of the juvenile shell or protoconch, and this feature has been useful in distinguishing feeding and nonfeeding larvae in both recent and fossil taxa.

Cephalopoda

Cephalopods (Fig. 5) are dorso-ventrally elongated molluscs with well-developed sense organs and large brains. They are often argued to be the most intelligent of all invertebrates. Nearly all are predatory and most are very active swimmers. A few taxa are benthic, drifters, or medusa-like, and some are detritus feeders. All are active carnivores in marine benthic and pelagic habitats from near-shore to abyssal depths. Giant squid *(Architeuthis)* are the largest invertebrates, and the cephalopods include the largest living as well as largest extinct molluscs: Ammonite shells range to more than 2 meters across, and body sizes of living squid range up to 8 meters with tentacles exceeding 21 meters in length. The smallest cephalopods are about 2 centimeters in length.

Cephalopods are found worldwide, all are marine, and only a few can tolerate brackish water. All are found in

FIGURE 5 *Octopus bimaculatus* moving between tidepools at night (Pigeon Point, San Mateo County, California, USA). Photograph by the author.

benthic and pelagic habitats from near-shore to abyssal depths. Cephalopods were once one of the most dominant animal groups in the ocean, but today there are only about 700 living species. In contrast, more than 10,000 fossil species are known.

Cephalopods are much more variable in their diversity through time than other molluscan groups. Cephalopods have experienced numerous extinctions (e.g., the terminal Permian, Triassic, and Cretaceous events), but typically showed rapid replacement and subsequent radiation by the survivors. Three major clades (usually treated as subclasses) are recognized: the Nautiloidea, Coleoidea, and Ammonoidea. Only one cephalopod group is active in intertidal rocky-shore communities—the coleoid Octopoda.

Coleoids have eight or ten suckered or hooked tentacles and a single pair of gills, and an ink sac is often present. There are two main groups of coeloids—the Octobrachia (or Octopodiformes), which includes the octopuses, paper argonauts, and the pelagic cirrate octopods (Octopoda), and the vampire squid (Vampyromorpha); all have four pairs of tentacles and no internal shell.

Octopuses are often common in rockpools, where they retreat into crevices during the day but lie in wait at the edges of the pools at night, from where they pounce on arthropods venturing onto the open pool bottom (Fig. 5). Octopuses also feed on a variety of gastropods, which they drill with their radula to inject a poison, which is derived from the octopuses' salivary glands, into the snail's body.

Cephalopods have a single gonad and separate sexes, with males transferring spermatophores to females following typically complex courtship. The spermatophore is transferred by the male using a penis (some squid, vampire squids, and cirrate octopuses) or, in nearly all others, with a modified tentacle (the hectocotylus). Some taxa are highly sexually dimorphic. Fertilization is internal, with egg capsules laid on the substrate, and development is direct; the eggs are large and yolk-rich. There is no larval stage, just direct development into juveniles, although the juveniles may undergo a pelagic phase.

HABITS

Molluscs occur in almost every habitat found on earth, where they are often the more conspicuous and sometimes predominant organisms. Although most are found in the marine environment, where they extend from the highest intertidal habitats to the deepest oceans, several major gastropod clades live predominantly in freshwater or terrestrial habitats. Marine diversity is highest nearshore and becomes reduced as depth increases beyond the shelf slope. Like many other organisms, marine molluscs reach their highest diversity in the tropical Western Pacific and decrease in diversity toward the poles.

Marine molluscs occur on a large variety of substrates including rocky shores, coral reefs, mud flats, and sandy beaches. Gastropods and chitons are characteristic of hard substrates, and bivalves are commonly associated with softer substrates, where they burrow into the sediment. However, there are many exceptions. The largest living bivalve, *Tridacna gigas*, nestles on coral reefs; many bivalves (e.g., mussels, oysters) are attached to hard substrates; and microscopic gastropods live interstitially between sand grains. Because they retain water, tidepools often maintain subtidal molluscan species in intertidal settings. This is especially true of tidepools sheltered by rock overhangs or those that occur in sea caves (Fig. 6). Often it is the presence of their food (e.g., sponges and tunicates) that enables these deeper-water species to maintain themselves in these sheltered tidepools.

FIGURE 6 Tidepools located deep in sea caves often include species that are more common in subtidal habitats (Low Arch sea cave, Southeast Farallon Island, San Francisco County, California, USA). Photograph by the author.

The adoption of different feeding habits appears to have had a profound influence on molluscan diversification. The change from grazing to other forms of food acquisition is one of the major features in the adaptive radiation of the group. The earliest molluscs were carnivores or grazed on encrusting animals and detritus. Such feeding may have been selective or indiscriminate and would have included algal, diatom, or cyanobacterial films and mats, or encrusting colonial animals. Truly herbivorous grazers are relatively rare and are limited to some polyplacophorans and a few gastropod groups. Most chaetodermomorphs, monoplacophorans, and scaphopods feed on protists and bacteria, and neomeniomorphs feed on cnidarians. Cephalopods are mainly active predators as are some gastropods, whereas a few chitons and septibranch bivalves capture microcrustaceans. Most bivalves are either suspension or deposit feeders that indiscriminately take in particles, but then elaborately sort them based on size and weight.

Shell morphology is often thought to be correlated with lifestyle and habitat, and some substantial changes in body form are clearly associated with major adaptive changes. Frequently, however, morphology is not readily correlated with habitat, and similar shell morphologies do not necessarily indicate similar habits or habitats. For example, limpet taxa occur on wave-swept platforms, on various substrates in the deep sea including hydrothermal hot vents, in fast-flowing rivers, in quiet lakes and ponds, and as parasites on oysters and starfish. It is often suggested that strong wave action selects for limpet morphology, but it is obvious from their wide range of habitats that molluscs with limpet-shaped shells do very well in a wide range of settings.

SEE ALSO THE FOLLOWING ARTICLES

Bivalves / Chitons / Fossil Tidepools / Limpets / Octopuses / Snails

FURTHER READING

Lindberg, D. R., W. F. Ponder, and G. Haszprunar. 2004. The Mollusca: relationships and patterns from their first half-billion years, in *Assembling the tree of life*. J. Cracraft and M. J. Donoghue, eds. New York: Oxford University Press, 252–278.
Vermeij, G. J. 1993. *A natural history of shells*. Princeton, NJ: Princeton University Press.
Wilbur, K. M., editor-in-chief. 1983–1988. *The Mollusca*. Vols. 1–12. San Diego, CA: Academic Press.

MONITORING, OVERVIEW

A. J. UNDERWOOD

University of Sydney, Australia

Monitoring is the measurement of the numbers, types, sizes, condition, and so forth, of animals and plants and their habitat, usually at several times or in several places. Monitoring should be carefully designed to ensure that the information collected is reliable and allows logical interpretations and conclusions. The data are variable, so statistical procedures are needed to describe and display the information and to analyze the samples to detect patterns and changes. Statistical analyses of samples are univariate (single measures) or multivariate (many simultaneous measures), needing a range of procedures for description and analysis.

GENERAL ISSUES ABOUT MONITORING

One of the great things about the habitats (such as algal mats, mussel beds, tidepools, and boulder fields) that we can see on rocky shores is that they are different from each other and from other areas, such as mangrove forests or sandy beaches. Common sense tells us that intertidal animals and plants may not enjoy being out of the water during low tide, so that many of them can be expected to take advantage of the more frequently wet conditions lower on a shore. Common sense sometimes matches reality in ecology, and often there are more or larger animals (such as sea urchins or snails) in lower than in higher areas, or in pools rather than around them.

So, for some animals and seaweeds, it is obvious that there are differences in presence or absence or in their numbers from place to place. If things are obvious, why should measurements be necessary? This raises an important issue—measurements are needed only when the information cannot be obtained by some simpler method, such as just by looking. So, it is important to sort out what you want to know before going to measure. This guides what sort of data to collect, how to collect the data, and importantly, how to interpret or use the information.

Monitoring is the process of measuring the types, the number, the sizes, and so forth, of the animals and plants or their environment (temperature, salinity, depth, area, etc.). There are always thousands of things we could measure, so some serious thought is needed about what to measure and why to spend time and energy collecting information. Monitoring is useful only if there are clear questions that need answers and that need quantitative information to supply the answers. In the scientific jargon of ecology, the questions are usually stated as predictions or hypotheses about rocky shores, based on our understanding and knowledge of ecological processes on these shores (in other words, based on our theories or models about what is going on). Ecologists argue about the details of how important predictions might be, but very few people argue that monitoring should be done without advance understanding of what the information is for. This, in turn, influences (and usually decides) what to measure, where and when to measure it, and how to interpret the answers.

Once a prediction (question) or series of predictions has been sorted out, we usually know what we need to measure. The whole business of how, when, where to measure is called the experimental design or sampling design. Because the information is usually quantitative, interpretation depends on detailed analysis. Because the information is usually variable (explained later), analyses are usually statistical. It turns out that it is sometimes difficult to collect information that can be analyzed sensibly, allowing proper and believable interpretation. As a result, designing how to monitor and how to analyze data go hand in hand.

EXPLORATION AND DISCOVERY

The interesting part of ecological monitoring is when the purpose is to find out about patterns and understanding processes that cause patterns. For example, a pattern in space would be seeing more large crabs in tidepools than elsewhere on a shore. It is a pattern of difference—things are not the same in the two habitats. A pattern in time would be an increase from summer to winter in the cover of large seaweeds on the bottoms of pools. This is a pattern of change.

Usually, very careful sampling is needed to demonstrate that patterns exist. Often we get fooled into believing we are seeing a pattern when, in reality, there is no real pattern and we are being conned by variability in the observations.

There are also patterns of association of relationship between things. For example, we might see lots of seaweeds in some pools in which there are few sea urchins. If there were other pools with many urchins, but few seaweeds, we would identify a pattern. There would be evidence of negative association between the urchins and the plants (negative because many of one are found with few of the other). The statistical analysis of patterns is an extremely important component of ecology.

Having found patterns in space or time, the challenge is to explain them. What causes there to be more urchins in pools than elsewhere, or more seaweeds in winter than summer? Why are there more urchins where there are few seaweeds? Determining processes and the relative importance of different processes almost always requires experiments to sift out correct ideas from ideas that are wrong. Designing, monitoring, and interpreting the experiments is often more fun (although often more difficult) than recording patterns.

Both types of study (patterns and processes) share the need for care in designing the monitoring.

DECISIONS AND APPLICATIONS

The other type of study of habitats on rocky shores is often called applied ecology (as opposed to ecology for discovery, which is often called pure or basic ecology). Monitoring in applied ecology is to get information to help answer questions about management or conservation of animals, plants, or habitats. For example, in many parts of the world, marine reserves have been made where catching fish or harvesting of animals or plants for bait or food is not allowed. It is important to monitor the ways people use the areas, to determine whether the bans are working. The reason for stopping people from fishing or harvesting is not just to control them, but because it has been predicted that biodiversity will be enhanced (there will be more species or larger numbers of some species), breeding populations of fish will be sustained, and so forth. It is therefore very important to monitor the areas to test these predictions. Are there more species or more fish?

The other type of applied monitoring is to detect environmental impacts. For example, the harmful effects of sewage or industrial waste or the consequence of oil spills (or cleaning up after an oil spill) can be known only by monitoring the fauna and flora.

Clearly, monitoring in these cases is not just to find out how the ecology of the shore works. The information is needed to help decision makers (governments, regulatory agencies, community groups) make correct decisions about what to do to manage reserves or to prevent or fix environmental problems.

It makes no difference whether information is gained for basic or for applied reasons—the information must be collected in ways that are appropriate and interpretable.

SEE ALSO THE FOLLOWING ARTICLES

Habitat Alteration / Management and Regulation / Marine Reserves

FURTHER READING

Clarke, K. R. 1993. Non-parametric multivariate analyses of changes in community structure. *Australian Journal of Ecology* 18: 117–143.
Downes, B. J., L. A. Barmuta, P. G. Fairweather, D. P. Faith, M. J. Keough, P. S. Lake, B. D. Mapstone, and G. P. Quinn. 2002. *Monitoring ecological impacts: concepts and practice in flowing waters.* Cambridge, UK: Cambridge University Press.
Green, R. H. 1979. *Sampling design and statistical methods for environmental biologists.* Chichester/New York: Wiley.
Helmuth, B., N. Mieszkowska, P. Moore, and S. J. Hawkins. 2006. Living on the edge of two changing worlds: forecasting responses of rocky intertidal ecosystems to climate change. *Annual Review of Ecology, Evolution & Systematics* 37: 373–404.
Schmitt, R. J., and C. W. Osenberg. 1996. *Detecting ecological impacts: concepts and applications in coastal habitats.* San Diego, CA: Academic Press.

MONITORING: LONG-TERM STUDIES

STEVE HAWKINS, KEITH HISCOCK,
PATRICIA MASTERSON, PIPPA MOORE,
AND ALAN SOUTHWARD

Marine Biological Association of the United Kingdom

To assess human impacts, long-term studies are often needed. These studies can be used to separate natural fluctuations from human-driven changes. The most extensive long-term and broad-scale datasets for rocky shores are from Europe. There is also a great heritage of amateur natural history in this region. Long-term monitoring studies have shown the broad-scale influences of climate change and nonnative species, plus regional and local impacts and subsequent recovery from pollution on marine biodiversity.

HISTORICAL BACKGROUND

What a delight it is to scramble among the rough rocks that gird this stern iron-bound coast, and peer into one after another of the thousand tide-pools that lie in their cavities! . . . we are indebted to the pools, which make the coast so rich and tempting a hunting ground to the naturalist.

P. H. GOSSE (1853)

In the nineteenth century, polite Victorian society was enthused by the radical findings of Darwin on the origin of species. There was an explosion of interest in the seashore and its easily accessible diversity of life. This demand generated a plethora of books for the serious and merely curious amateur. Thus, there is a long tradition of study and recreational enjoyment of the seashore and rockpools in Europe, pioneered by the work of Philip Henry Gosse, who was preeminent among marine naturalists in this field. His forays to the shore led to the writing of enthusiastic descriptions from places such as Torbay, Ilfracombe, and Tenby. Some of these descriptions provide a basis for comparison today. Figure 1 captures the spirit of the age.

Writing of his excursions on the Devonshire coast during 1852, Gosse observed:

> Towards Oddicombe . . . I found a pretty tide-pool, a delightful little reservoir, nearly circular, a basin about three feet wide and the same deep, full of pure sea-water, quite still, and as clear as crystal. From the rocky margin and sides, the puckered fronds of the Sweet Oar-weed

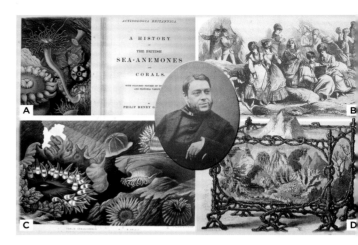

FIGURE 1 A montage of mid-Victorian images of rock pooling. (Inset) Author Philip Henry Gosse. (A) The front page of one of his master works. (B) Polite society collecting on the shore. (C) An illustration of a rock pool with diverse sea anemones. (D) An aquarium providing a rock pool for the drawing room.

(*Laminaria saccharina*) sprang out, and gently drooping, like ferns from a well, nearly met in the centre; while other more delicate sea-weeds grew beneath their shadow. Several sea-anemones of a kind very different from the common species, more fat and blossom-like, with slender tentacles set around the fringe, were scattered about the sides: when touched they contracted, more and more forcibly, into a whitish grey tubercle.

These observations can prove useful today; for instance, the detailed description of anemones and the beautiful plates in the books referred to previously allows us to identify the species as probably being the gem anemone, *Aulactinia verrucosa*.

As science became professionalized in the latter part of the nineteenth century, marine stations were established throughout Europe: Concarneau (1859, the oldest in the world) and Roscoff (1872) in France; by Austrians in Naples (1872), Italy; and Helgoland (1892) in Germany. Laboratories were founded in the United Kingdom (e.g., Gatty, St. Andrews, 1884; Scottish Association for Marine Science in the Firth of Forth and then the Firth of Clyde, 1885; The Marine Biological Association of the UK; Plymouth, 1888; and Port Erin Marine Laboratory, 1892—closed in 2006). Similar laboratories were founded in the United States (e.g., Woods Hole Marine Biological Laboratory, 1888; Scripps Institute of Oceanography, 1903; and Hopkins Marine Station, 1892).

These marine stations spawned both local and broader scale studies of marine flora and fauna, much of it on easily accessible seashores. At first these were qualitative lists of species such as the flora and fauna of the area in the vicinity of marine stations (e.g., Roscoff, St. Andrews, Plymouth, Isle of Man)—cataloguing and classifying the diversity of life in the Darwinian context. With time, distributions began to be mapped and individual species or groups of species counted and quantified.

CLIMATE-RELATED CHANGES IN THE BRITISH ISLES AND EUROPE

The legacy of these pioneering studies provides a baseline against which global changes (human-driven climate change, and introduction of nonnative species) and localized impacts (pollution, coastal development and habitat loss, overexploitation of shellfish, and trampling from recreational activities) can be judged. The need for a long-term perspective is emphasized by the recent changes in sea temperature off Plymouth, England, over the twentieth century. Although in the past there have been warmer (1930–1950s) and cooler (1900–1920s, 1963–1987) periods, present-day temperatures have accelerated to their current high levels since the late 1980s (Fig. 2).

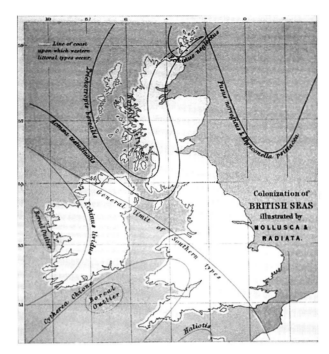

FIGURE 3 Forbes 1859 map of the British Isles and Ireland showing biogeographic boundaries (from Godwin-Austen, 1859).

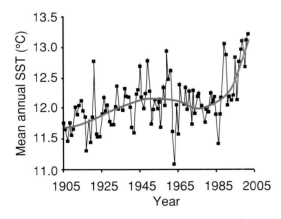

FIGURE 2 Annual mean sea surface temperature trends from square 50° to 51°N, 4° to 5°W, off Plymouth (Hadley Centre for Climate Research).

Extensive mapping of species distributions, particularly conspicuous and ecologically important animals and seaweeds, was undertaken by the French biologist Fischer-Piette, from the Paris Museum, from the 1930s until the 1960s. Similar work was continued in Britain by Crisp and Southward in the 1950s. They were interested in the boundaries of northern and southern species around the British Isles and France—long known to be a boundary zone following the pioneering work of Forbes (Fig. 3).

Fischer-Piette was also interested in boundaries further south in Europe, particularly the southern limits of many cold-temperature species in Spain and Portugal.

The baseline of species distributions around the British Isles, French, and Iberian coastlines was established during a warm period in the 1950s. Crisp, Southward, and colleagues showed how the extremely cold winter of 1962–1963 trimmed back the distribution of many species at their northern limits in the British Isles. Recovery was slow because the next 20 or so years were typified by generally cooler weather. Recently, however, the range of several species has been shown to extend beyond the limits of the 1950s. The barnacle *Balanus perforatus* has extended on both coasts of the English Channel through the Eastern Basin. Warm-water chthamalid barnacles have extended around Scotland and penetrated further south into the colder North Sea. Trochid snails (topshells) have extended their ranges along the North Coast of Scotland (*Gibbula umbilicalis*) and in North Wales and eastward along the English Channel (*Osilinus lineatus*) toward the colder eastern basin of the Channel. The rock pool–dwelling seaweed, *Bifurcaria bifurcata*, has also advanced eastward along the coast of the English Channel. The changes on rocky shores have, however, occurred slowly compared to widespread changes in zooplankton and fish in European waters.

Species such as the kelp *Alaria esculenta* retreated in the warm period of the 1950s and never recovered. The species now appears to be suffering further contraction. The circumpolar tortoiseshell limpet, *Tectura testudinalis*, has also retreated; where once it was easy to find on the Isle of Man, none have been seen by the lead author, despite much searching, since the mid-1990s.

In the 1950s, the northern barnacle, *Semibalanus balanoides*, was shown to be rarer in the Plymouth area than in the 1930s by Southward and Crisp. This observation spawned a remarkable 40-year quantitative time series from the 1950s to the 1980s, charting fluctuations in the abundance of the southern *Chthamalus* species (occurring from North Africa to Scotland) and the more northerly *S. balanoides* (from the Arctic to northern Spain), which compete for space on European shores. During the warm 1950s *Semibalanus* became rarer. It prospered after the extremely cold winter of 1962–1963, which heralded a regime shift to colder conditions. From 1987 onward the influence of global climatic warming driven by greenhouse gas emissions became apparent with a series of warm years, including the warmest on record in Europe in 2003. The barnacle time series was restarted in the 1990s by the lead author, and by 2004 the warm temperate *Chthamalus* were now doing well and were at higher levels than in the 1950s—the previous warm period (Fig. 4).

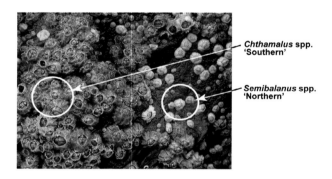

Chthamalus spp. 'Southern'

Semibalanus spp. 'Northern'

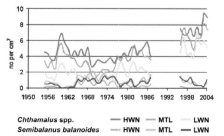

FIGURE 4 (Top) The barnacles being studied. (Bottom) Long-term changes in northern (*Semibalanus balanoides*) and southern (*Chthamalus* spp.) barnacles averaged for several shores on the south coast of Devon and Cornwall. From Southward and Hawkins, unpublished; Southward (1991); see also Southward *et al.* (2005).

NONNATIVE SPECIES

In addition to climate-related change, introduction of nonnative species by human agency has gathered pace with globalization of trade (including by ship fouling and as larvae in ballast water) and with deliberate introductions via aquaculture. The Japanese seaweed *Sargassum muticum* has spread throughout the Pacific coast of America and in Europe from Scandinavia to Spain since the 1970s. This seaweed, which seems to grow far faster abroad than at home, was probably introduced via oyster cultivation. In Europe, the Australasian barnacle *Eliminius modestus* first took hold in the 1940s and has now spread throughout much of Europe. Figure 5 illustrates the interaction of two kinds of global change: fluctuations in warm temperate (*Chthamalus*) and cold temperate (*Semibalanus balanoides*) barnacle species—plus the addition and establishment of a nonnative species. The European shore crab (known as the green crab in the United States) has spread to the eastern and western seaboards of North America, South Africa, and Australia. Long-term datasets are required to measure not only the advances of these species but also the impacts on the native fauna and flora.

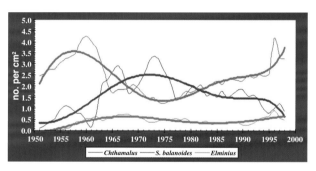

FIGURE 5 Cellar Beach: Smoothed abundances of intertidal barnacles and polynomial trend lines. From Southward (1991) and unpublished observations.

CHANGES OBSERVED IN NORTH AMERICA

Similar changes in the relative abundance of intertidal organisms with northern and southern biogeographic distributions were found following a resurvey of a rocky shore, 60 years after the first sampling effort, along a transect in California by Sagarin and co-workers, published in 1999. The resurvey found an increase in the abundance of southern species (10 of 11 species), a decrease in the abundance of northern species (5 of 7 species), and no clear trend being exhibited by cosmopolitan species (12 increased and 16 decreased). Barry and co-workers also observed a decrease in the abundance of canopy-forming macroalgae and an increase in small turf-forming algae, which are characteristic of warm temperate latitudes. These changes in the relative abundance of these species occurred during a period when surface seawater temperatures close to the shore had risen by 0.8 °C.

OVERVIEW

For a whole suite of rocky-shore species it has become apparent that warm temperate species are more common and have extended farther north than in the previous warm period of the 1950s. Conversely, cold-water species have declined with some retreats.

Long-term datasets are difficult to maintain. In the past, all too frequently, funding agencies were dismissive about monitoring as not being hypothesis-driven, process-oriented research. Such studies are, however, essential for discriminating the noisy local natural dynamics of systems from the longer-term and lower-amplitude signal typical of global environmental change. Throughout the world, rock-pool and other shore species are moving poleward, as has been shown on land. Unfortunately, some northern species may have nowhere to go in a rapidly warming world.

SEE ALSO THE FOLLOWING ARTICLES

Ecosystem Changes, Natural vs. Anthropogenic / Heat and Temperature, Patterns of / Introduced Species / Marine Stations

FURTHER READING

Beaugrand, G., P. C. Reid, F. Ibanez, J. A. Lindley, M. Edwards. 2002. Reorganization of North Atlantic marine copepod biodiversity and climate. *Science* 296: 1692–1694.

Crisp, D. J., and A. J. Southward. 1958. The distribution of intertidal organisms along the coasts of the English Channel. *Journal of the Marine Biological Association of the United Kingdom* 37: 157–208.

Godwin–Austen, R., ed. 1859. *The natural history of Europe's seas.* London: John Van Voorst.

Gosse, P. H. 1856. *Tenby: a sea-side holiday.* London: John Van Voorst.

Hawkins, S. J., A. J. Southward, and M. J. Genner. 2003. Detection of environmental change in a marine ecosystem—evidence from the western English Channel. *Science of the Total Environment* 310: 245–246.

Helmuth, B., N. Mieszkowska, P. Moore, and S. J. Hawkins. 2006. Living on the edge of two changing worlds: forecasting responses of rocky intertidal ecosystems to climate change. *Annual Review of Ecology, Evolution & Systematics* 37: 373–404.

Hiscock, K., A. Southward, I. Tittley, and S. Hawkins. 2004. Effects of changing temperature on benthic marine life in Britain and Ireland. *Aquatic Conservation, Marine and Freshwater Ecosystems* 14: 333–362.

Lewis, J. R. 1964. *The ecology of rocky shores.* London: English University Press.

Murray, S. N., R. F. Ambrose, and M. N. Dethier. 2006. *Monitoring rocky shores.* London: University of California Press.

Parmesan, C., and G. Yohe. 2003. A globally coherent fingerprint of climate impacts across natural systems. *Nature* 421: 37–42.

Raffaelli, D. G., and S. J. Hawkins. 1996. *Intertidal ecology.* London: Chapman & Hall (reissued by Kluwer 1999).

Southward, A. J. 1991. 40 years of changes in species composition and population density of barnacles on a rocky shore near plymouth. *Journal of the Marine Biological Association of the United Kingdom* 71: 495–513.

Southward, A. J., O. Langmead, N. J. Hardman-Mountford, J. Aiken, G. T. Boalch, P. R. Dando, M. J. Genner, I. Joint, M. Kendall, N. C. Halliday, R. P. Harris, R. Leaper, N. Mieszkowska, R. D. Pingree, A. J. Richardson, D. W. Sims, T. Smith, A. W. Walne, and S. J. Hawkins. 2005. Long-term oceanographic and ecological research in the western English Channel. *Advances in Marine Biology* 47: 1–105.

MONITORING: STATISTICS

A. J. UNDERWOOD

University of Sydney, Australia

Information (or data) collected in ecological sampling or in experimental studies is usually quantitative—the number of species present, the numbers of each species, their sizes, how far they move or grow, and so forth. This information is usually (and probably should be always) collected to answer particular questions, or, in other words, to test predictions about patterns and processes regulating the ecology of intertidal habitats.

Nearly everything we might want to measure on rocky shores, however, is variable. As an example, the number of urchins per square foot of surface on the bottom of a tide-pool is unlikely to be the same over every square foot of a whole pool. Even if it were, it is quite unlikely that the numbers in two adjacent pools would be the same. Even if they were, it is unlikely that the numbers would be the same in three months' time. So, most things we want or need to know vary from place to place or time to time.

WHY WE NEED STATISTICS

Describing Variability

The variability makes it difficult to know how many there really are or to identify differences from place to place or changes from time to time. Consider a small example. You would like to know the number of limpets in some area of a shore. You do not have time to count all the limpets in all of the area. So, you can only look in a subset of the area—you will take a sample of quadrats, that is, small areas of chosen size scattered over the whole area. Your sample will be a set of replicate sample units (in this case, the quadrats). Which quadrats to include can be very complex, but, in principle, you need to ensure that the ones you choose are typical of all the possible quadrats you might have examined in the area. You must not choose quadrats only in the highest or the lowest parts of the shore—or the prettiest, or most weed filled, and so on, quadrats—or the results will not be accurate. A very common way to pick which things to sample is to make a grid of all possible quadrats, to number them all, and then to pick the ones to include, using a random-numbers table so that you have no way of choosing them because of certain features. Alternatively, you could choose, in advance, to sample every tenth quadrat you came across and then move systematically back and forth across the whole area, counting limpets in every tenth quadrat.

You count all the limpets in the ten chosen quadrats (as in one of the samples in Table 1). There is an average of 18.2 limpets per quadrat. If your sampled quadrats really are typical of the whole area of interest, it is reasonable to expect that the true (but unknown) number in all of the area would also be somewhere around an average of 18.2. Because the numbers of limpets vary from quadrat to quadrat, they will also vary from one set of ten quadrats to another set of ten quadrats. The average number in your sample is therefore affected by sampling error—if you took another sample of ten quadrats, the result would not be exactly the same, but should be somewhere around 18.

TABLE 1
Sampling and Comparing the Numbers of Limpets in Areas of a Shore

Area	Number of limpets	
Replicate area	With fishermen	Without fishermen
1	15	22
2	21	15
3	19	28
4	26	19
5	17	26
6	12	19
7	19	24
8	22	27
9	18	31
10	13	22
Average	18.2	23.3
Sample variance	17.9	23.6
Sample standard deviation	4.2	4.9
Standard error	1.3	1.5

NOTE: t-statistic = 2.51; probability < 0.025.

As a result, when describing the results of sampling (in this case, the average number of limpets in each area), it is important also to describe the sampling error. A very common method is to calculate the variance of the data—a measure of the variability of the numbers in each sample. The formula is

$$\text{Sample variance} = \frac{\sum_{i=1}^{n}(X_i - \overline{X})^2}{(n-1)} \qquad \text{(Eq. 1)}$$

where X_i is the number of limpets in any quadrat in a sample, n is the number of quadrats in that sample, and $\overline{X}$ is the average number. Because this measure involves squared numbers, it is usual to take its square root, called the standard deviation (SD), which also measures how variable are the numbers of limpets from quadrat to quadrat.

An alternative, widely used measure is to divide the variance by the size of the sample (n). The square root of this number is called the standard error (SE) and describes how variable samples are. It gives an estimate of how different the average number of limpets would be if you had counted limpets in a different sample of ten quadrats in the same area. Its accuracy depends on some assumptions about how the numbers of limpets vary from quadrat to quadrat, but this is not explored here.

Describing the data then consists of describing the average or mean number in the sample, the size of the sample (how many quadrats were counted), and how variable are the data (using SD) or sample means (using SE), as shown in Fig. 1, for the numbers of urchins in samples of pools at three heights on a shore.

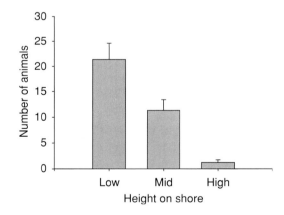

FIGURE 1 The average number of sea urchins in tidepools at different heights on a shore; ten pools were sampled at each height. Error bars are standard errors.

Examining Differences and Changes

As an example, consider the previous example of limpets in some area of a shore. The following example would, in reality (see later discussion), require more complicated spatial replication than described here, but will illustrate the point. Perhaps people think that the areas of shore where fishing occurs and the numbers of limpets in these areas are affected by fishermen, who use them as bait or disturb them when trampling around the area where they fish. If the first sample of quadrats (as described previously) had been near a favorite place for rock fisherman, you now need another sample from an area where there are no fishermen. You take a second sample of ten quadrats away from fishermen and get the counts in Table 1. The numbers certainly look a bit bigger in the second area, but some quadrats had fewer limpets than were found in some of the quadrats in the first area. Are they really different?

A statistical answer to this recognizes that the difference we see (an average of 18.2 per quadrat for the first area and 23.3 in the second area) may be real—there really are more limpets in the second area. Alternatively, the numbers are only different because of sampling error. A second sample from either area might give just as big a difference. In that case, the two samples would really be measuring the same average number (about 20), but differing because of sampling error.

A statistical procedure tries to measure how much difference between two samples there might be just because of sampling error. The procedure then compares how much difference we actually get (in this case, 18.2 to 23.3, a difference of 5.1 limpets per quadrat) with the amount of difference we might expect due to sampling error.

There are several ways of working out the sampling error. One would be to count limpets in many sets of ten quadrats in each area. The difference in averages from these samples is obviously sampling error—each sample of ten quadrats would be estimating the numbers of limpets in the same area. This is hard work, however, and statistical tests have been invented to reduce the work.

There are many types of statistical test. They all require important assumptions about the data to be true, to calculate how much difference there would be just due to sampling error. For example, a t-test assumes that if we took hundreds of samples, their averages would have a particular distribution of values called a normal distribution, often referred to as a bell-shaped curve. If the data might match this assumption, the standard errors of the two samples can be used to work out how likely the difference between two samples is just due to sampling error.

Other procedures are called randomization tests. These assume that the data from the samples in each area are the only data—no other numbers could happen. They then shuffle the 20 numbers into two new sets of 10 and calculate how different are their averages. Each time the data

are shuffled, a new pair of samples is made. If the quadrats in the two areas really have the same average number of limpets, the shuffling produces an estimate of sampling error and can be compared with how much difference exists between the two real samples.

The data from the two samples just described are tested by a t-test in Table 1. The difference (23.3 − 18.2 = 5.1) is quite large compared with sampling error. The probability given is the chance that a difference of 5.1 would happen if the two samples were really measuring the same average number of limpets and only differed because of sampling error. It is a small probability, and therefore it is reasonable to assume that the difference is more likely to be due to the two areas having different numbers—there are more limpets per unit of area (i.e., quadrat) where there are no fishermen.

ENSURING MONITORING IS DESIGNED PROPERLY

Assumptions about Data

It is very important to design sampling programs carefully, or there will be serious problems for interpreting the data. Problems can occur because of two different features of any study—statistical issues and logical ones. Statistical issues occur because all analytical procedures require assumptions about the data. Assumptions that apply quite generally are requirements that data are unbiased and, usually, that data are independent.

Unbiased data can come only from samples that are typical or representative of the thing we are trying to measure. Suppose samples are taken to estimate diversity of species in areas of shore on a stretch of coast. Because of shortage of time, only the higher areas of the shores actually got sampled (because they are uncovered for a longer time by the tide and can therefore be sampled more easily). These sampled areas do not represent all of the relevant parts of the shores—the information obtained will be biased if the upper areas do not have the same numbers of species as areas of rock lower on the shore.

Independence of data is much more complicated, but it is a requirement for many statistical procedures. A simple example is the situation in which crabs hide in large tidepools, but forage for snails to eat in surrounding smaller pools. The numbers of snails in the smaller pools are influenced by how many crabs are in nearby large pools. They are not independent; where there is a large number of crabs, all nearby small pools will have few snails because most have been eaten. Sampling would have to be designed to ensure that the pools sampled for number of snails are sufficiently far apart not to be influenced by the same nearby large pool.

Problems for Interpretation

Problems in logic occur when sampling is badly designed so that the data do not really measure the correct property—the one you wanted to know about. In the previous example about fishermen, the conclusion that numbers of limpets were different in two areas, one with and one without fishermen, is correct. It would not be correct to conclude that the smaller numbers were caused by the fishermen. There are many reasons why the number of limpets might vary from one place to another—for example, there may be bigger waves in one area, which is why fishermen like it. If large waves decrease the numbers of limpets, the pattern seen is not due to the fishermen.

To demonstrate an effect of fishermen requires an experiment to prevent fishing in some areas to test the prediction that numbers of limpets will then increase. If the numbers do not increase, it must be because of some feature of the areas that is not the fault of the fishermen. If the numbers do increase to match those in the areas that are naturally without fishermen, that would be strong evidence that the fishermen are causing the decreased numbers of limpets. Monitoring on its own can never resolve such problems.

Appropriate Replication

As another sample of problems for logic when sampling is not designed properly, suppose we want to understand seasonal changes in the numbers of juvenile fish in tidepools. We can representatively sample them by choosing random pools in the area to be studied and then searching through them with small dip nets, catching and counting the small fish. When we believe we have caught and counted them all, we can release them back into the pool. So, the data will be replicated counts of the number of fish in a set of pools.

If we propose that there will be differences across the four seasons of the year, we need to take samples in each season. Commonly, this is done by counting fish at one time in each season, using the replicate pools to measure natural variability in each season. This produces data, as in Fig. 2A. The top part of the figure shows what is really happening to the average numbers of fish throughout the year. There is a lot of variation from week to week, because of changes in weather, storms, and so on. Remember that we do not know this pattern; our only data come from the one time of sampling in each season. Our data, however, show a strong seasonal pattern (the middle graph). Statistical analyses of the data would probably reveal that this is unlikely to be chance, so we would conclude the data demonstrate a strong seasonal pattern.

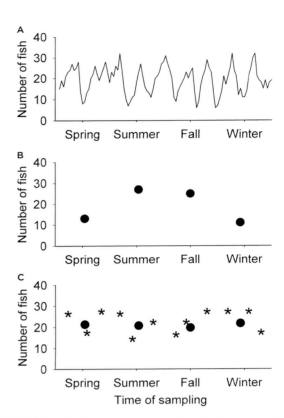

FIGURE 2 Sampling to detect seasonal patterns in the number of fish in tidepools. (A) The real number of fish, averaged over all the pools throughout the year—which can never be known. (B) The average number if sampled once in each season at the times indicated by the filled circles. (C) The average number in each season (filled circle), from three samples in each season (asterisks).

But the real pattern is not changing seasonally (the top graph). What has gone wrong? The problem is that we have not designed the study to measure how much temporal change there is at shorter intervals within each season. By bad luck, the times we picked to go sampling happened to show larger values in summer and fall than in spring and winter (Fig. 2B). We need temporal replication. Measuring the spatial variation among replicate pools at one time cannot measure this.

Instead, we must take several samples in each season, as shown in Fig. 2C. We now sample a random set of pools at (in this example) three times picked at random in each season. These will now show the noise, or variation, from time to time in each season. Averaging them will give a much more accurate measure of the number of fish in each season—because it takes into account the temporal variation. These averages do not show any seasonal pattern—which is what is really happening to the numbers of fish.

Ecological research programs must be carefully designed to ensure that the questions asked (predictions made) are properly considered. This involves care about choosing which replicates to include in samples, ensuring that samples are arranged in space and at appropriate times to allow proper interpretation. Analyses will often be complex, and it is important to be very clear about how to interpret the information to reach valid conclusions. Most statistical analyses will be much more complex than indicated here because of variability in the data.

BIODIVERSITY AND MULTIPLE MEASUREMENTS

Multivariate Data

So far, only single measures (univariate data) have been considered. Often, though, we need to understand what is happening for multiple measures (multivariate data). For example, if there is concern about changes in the entire set of species in pools—the assemblage of everything living there—we need to examine simultaneous changes in numerous quantitative sets of data.

There are many ways of approaching this, but they share some common features. The first step is to choose a method that measures all of the differences between the replicates in the samples. As an example, consider the effect of large boulders on the fauna on the sides of pools. We propose that large boulders should affect the animals, because they cause disturbances during storms, provide shelter for predators, and so forth. We therefore predict that assemblages in pools with a large boulder should be substantially different from those without a boulder.

Dissimilarity between Replicates

Suppose the data are the numbers of different species of limpets, anemones, urchins, and so on., on the sides of four randomly chosen areas of shore with boulders and four without (the data for five species are shown in Table 2A). Inspection of the numbers suggests fairly large differences for species 2 (more where there is no boulder) and species 4 and 5 (more where there is a boulder). Typically, however, data are much more variable, and there may be hundreds of species—so we need a statistical approach.

There are many measures of difference between the individual samples. One commonly used is the Bray–Curtis measure of dissimilarity:

$$d_{j,k} = \frac{\sum_{i=1}^{s} |X_{ij} - X_{ik}|}{\left(\sum_{i=1}^{s} X_{ij}\right) + \left(\sum_{i=1}^{s} X_{ik}\right)} \times 100 \quad \text{(Eq. 2)}$$

where X_{ij} is the number of species i in sample j and X_{ik} is the number of the same species (i) in sample k. So, $\sum_{i=1}^{s} X_{ij}$

TABLE 2
Multivariate Measurements: Biodiversity

(A) Numbers of each of 5 species in 4 areas with and 4 areas without boulders

Area	With boulders				Without boulders			
	1	2	3	4	5	6	7	8
Species								
1	3	4	7	8	1	0	3	2
2	5	8	12	15	16	29	35	22
3	11	9	4	7	3	9	12	6
4	28	22	25	34	11	16	14	9
5	19	15	17	14	4	8	6	2
Total	66	58	65	70	35	62	70	41

(B) Matrix of Bray–Curtis dissimilarities

Area	1	2	3	4	5	6	7
2	13						
3	18	14					
4	21	20	19				
5	52	42	38	35			
6	41	32	37	30	30		
7	43	38	42	37	33	12	
8	55	45	45	39	18	24	26

(C) Matrix of ranked dissimilarities

Area	1	2	3	4	5	6	7
2	2						
3	4	3					
4	8	7	6				
5	27	22	19	15			
6	21	13	16	12	11		
7	24	18	23	17	14	1	
8	28	26	25	20	5	9	10

adds up all the numbers for all species in sample j (for example, when j is 3, this total is 7 + 12 + 4 + 25 + 17 = 65). $\sum_{i=1}^{s} X_{ik}$ is the total of everything in sample k. $|X_{ij} - X_{ik}|$ measures the absolute difference in the number of one species (i) between the two samples. For example, if $j = 1$ (the first sample) and k is 5 (the fifth area, or the first without a boulder), $|X_{ij} - X_{ik}|$ for the first species ($i = 1$) is (3 − 1) = 2; when $i = 2$ (the second species), it is (5 − 16) = −16, but 16 in absolute value. These differences are added up over all the species for the two samples and multiplied by 100 to give the percentage dissimilarity between samples j and k, called $d_{j,k}$. The results for each pair of samples are shown in Table 2B.

The results are written as a dissimilarity matrix—a triangle of values describing how different the all possible pairs of samples are. Large numbers represent pairs that are quite different. Small numbers represent pairs of samples that are quite similar. 0 would mean that two areas had exactly the same number of every species; 100 means that the two areas had no species in common.

The dissimilarity matrix is a triangle, with three parts. The top left triangle represents differences between pairs of areas that have boulders (1 vs. 2, 1 vs. 3, . . . 3 vs. 4), that is, natural variation among areas of the same type. The bottom right triangle represents differences among areas without boulders (5 vs. 6, 5 vs. 7, . . . 7 vs. 8). The remaining rectangle is all the differences between each area with boulders and every area without boulders (1 vs. 5, 1 vs. 6, . . . 4 vs. 7, 4 vs. 8).

If having a boulder makes a difference (as predicted), the differences between the two sets of areas (between samples) should be greater than the differences in each type (within samples). If having a boulder makes no difference to the animals in an area, the differences between pairs of areas with boulders or pairs of areas without boulders will be about the same as the differences between an area with boulders and one without them. Thus, it is useful to examine the average dissimilarity within the set of areas without boulders (the average of the top left triangle in Table 2B) and within the set with boulders (the average of the bottom right triangle) in comparison with the average dissimilarity between areas with and those without boulders (the rectangle in Table 2B). These are 17.3, 24.0, and 40.7, respectively (Table 2). The latter is much larger than the former, suggesting very strongly that the assemblages are different in areas with boulders from those without boulders.

There are formal statistical tests of such hypotheses (predictions), such as ANOSIM (analysis of similarity tests) and PERMANOVA (permutation tests). There are also procedures for detecting which species are most different, how the differences in the animals might be correlated with differences in a set of other data, for example, physical data (height on the shore, cover of algae, numbers of small pools in each area, etc.) about the areas.

Displaying the Data

It is impossible to illustrate the data using simple graphs. One method of displaying results is called MDS (multidimensional scaling). This can be done with the dissimilarities or, more commonly, using their ranks.

In Table 2B, the smallest number identifies which two areas have the most similar set of animals (this is 12%, for the dissimilarity between areas 6 and 7). This is therefore given rank 1 (the most similar). The next most similar pair are areas 1 and 2, with dissimilarity 13%, which are ranked 2. Ranking continues until the last pair, that is, the most different is found. This is areas 1 and 8 (55% dissimilar and ranked 28th). Now, we have a dissimilarity matrix (Table 2C) of ranks. The ranks identify increasingly different pairs of areas.

These can be illustrated in an MDS (technically, because they are ranks, a nonmetric MDS, or nMDS). Every area is shown by a symbol. The procedure plots the most similar areas closely together. The next most similar areas are then plotted, then the next most similar, keeping them the right distances apart to maintain the pattern of differences between all pairs. This continues until all areas are fitted into the picture that best preserves how far apart they are. Areas that are similar in their biota are close together.

The data shown in Table 2 are plotted in Fig. 3. It is now easy to see how scattered the areas are that have boulders (the empty circles) or that do not (the filled circles). It is also very clear how different the two sets of areas are. Those with boulders are quite separate (i.e., have large differences in their assemblages of species) from those without boulders.

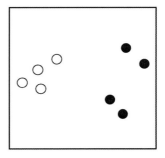

FIGURE 3 Displaying multivariate data in an nMDS. Empty circles represent areas without a boulder; filled circles represent areas with a boulder. The closer two areas are plotted, the more similar are the types and numbers of animals and plants in them.

SEE ALSO THE FOLLOWING ARTICLES

Dispersal, Measurement of / Monitoring, Techniques / Surveying

FURTHER READING

Clarke, K. R. 1993. Non-parametric multivariate analyses of changes in community structure. *Australian Journal of Ecology* 18: 117–143.
Manly, B. F. J. 1986. *Multivariate statistical methods: a primer.* London, UK: Chapman & Hall.
Quinn, G. P., and M. J. Keough. 2002. *Experimental design and data analysis for biologists.* Cambridge, UK: Cambridge University Press.
Underwood, A. J. 1994. On beyond BACI: sampling designs that might reliably detect environmental disturbances. *Ecological Applications* 4: 3–15.
Underwood, A. J. 1997. *Experiments in ecology: their logical design and interpretation using analysis of variance.* Cambridge, UK: Cambridge University Press.

MONITORING: TECHNIQUES

MEGAN N. DETHIER

University of Washington

Scientists, the public, and resource managers are increasingly aware of the value of monitoring natural habitats and communities to enable us to assess change over time or quantify impacts from a given event. Rocky-shore communities are susceptible to a wide variety of impacts, from global warming to being "loved to death" by visitors. How can they be monitored effectively? This is a complex subject with no easy answers.

HOW TO MONITOR

Sampling Units and Their Arrangement

Most monitoring programs use some combination of transects, quadrats, and sampling points. Transects are laid out either within a zone of interest (e.g., through a zone where most human trampling occurs) or across the whole shore, perpendicular to the water line. Along these transects, organisms are counted or collected either using quadrats or under discrete points (Fig. 1). Random location and replication of sampling units are critical elements of a sampling design; such issues are described in detail in books on statistics and sampling design. If a given site is to be monitored repeatedly, some investigators use fixed plots (e.g., marked with bolts in the rock), whereas others use random plots each time. In special cases, the sampling unit might be a whole area. For example, timed searches might be done across a whole site, either for generating a species list or for counting numbers of a very patchy or rare species of interest.

FIGURE 1 A transect line being sampled in the mid-upper shore on the coast of Washington state. Sampling units along this line could be quadrats or point-contacts. Photograph by the author.

Methods to Quantify Density and Percent Cover

The actual data collected in monitoring programs usually include estimates of abundance—most often, densities of mobile organisms and percent covers of sessile organisms. For some programs, simply listing presence/absence of species might suffice. A somewhat more quantitative method is to put abundances into categories (e.g., predefined scales such as "abundant, common, rare"); such

categories can be used either in quadrat-based sampling or in searches over a whole area. Very rapid surveys can sometimes be accomplished by taking detailed video images of a whole area or transect and then analyzing these images later in the laboratory.

Density is usually straightforward to measure, although quadrat sizes may vary with how abundant and how "clumped" different species are. Quantifying percent cover is more difficult. The most common method is point sampling, using many (usually random) points either within each quadrat or laid out along a line (Fig. 2), and recording the species found under each point. If enough of these points are sampled, good estimates of cover can be generated. If organisms are not multi-layered (for example, with seaweeds overlying barnacles and mussels), such data can be gathered later from photographs or videos taken during low tide. Visual estimates of percent

FIGURE 2 Point-contact sampling along a horizontal transect line in the low intertidal zone on the coast of Washington state. Points are chosen randomly; the sampling "tool" for precise observation of organisms lying under each point is a shish kebob stick. Photograph by the author.

cover can sometimes be made accurately, especially when quadrats are subdivided into smaller visual units, and when observers are trained in this method.

Methods to Quantify Biomass

Biomass data may be useful for detailed comparisons of areas or times; such data can be gathered for all organisms (mobile and sessile, plants and animals) and thus may be useful for community-based analyses and may better represent "abundance" than two-dimensional measures such as percent cover. Biomass values also can be converted into other units such as carbon or calories. Unfortunately, biomass sampling is destructive and time-consuming.

Methods to Quantify Individual-Based Parameters

For a target species of particular value or known to be a good indicator of environmental stressors, quantifying individual-based parameters is labor-intensive and often destructive but can provide more information about population status than simple abundance measures. Examples of measured parameters include size or age structure of a population, growth rates over time, sex ratios, and reproductive output. These data can demonstrate what is happening to a species over time; for example, whether individuals are growing less fast in a polluted area or whether only old individuals are present because of poor recruitment. If program goals allow or suggest such focus on one or a few species, individual-based data may provide a stronger signal about ecosystem conditions than density data can.

WHY MONITOR

Monitoring programs are set up for a wide variety of reasons. Often the goal is simply to try to detect change from any source on any time scale; unfortunately, this broad goal is the hardest to reach, because it is difficult to do sufficiently focused sampling with such a broad question. When goals are measurable (e.g. "detect declines in mussel abundance"), then it is easy to design a monitoring program. When they are not directly measurable (such as "improve ecosystem health"), then monitoring is difficult. The forces that might cause change to rocky shore ecosystems are diverse: oil spills, trampling or overharvesting by visitors, sea level change, global warming, pollution, invasive species, nearby land-based development, and so forth. The optimal methods for monitoring for change from each of those sources might be different. If we knew that the biggest problem in one area was, for example, an invasive algal species, then we could target our monitoring in the habitat type most likely to be invaded, and focus our methods on

the local algal species with which the invader might compete. If we knew that we wanted to quantify the impacts of an oil spill about to hit the shore, we could gather information quickly on the species known to be most susceptible to oil, and sample both in vulnerable and relatively "safe" sites. Groups considering setting up a monitoring program would thus be well advised to carefully consider key local stressors and possible impacts before proceeding with a design. A sample decision tree is illustrated in Figure 3.

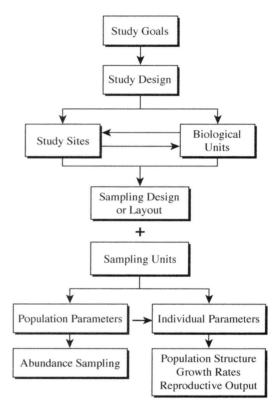

FIGURE 3 Decision tree for designing rocky intertidal sampling programs. Study Goals (at the top) must first be determined, as they will determine Study Design. The goals and design will focus sampling on specific sites (e.g., near a sewage outfall) and/or biological units (e.g., sensitive species). Selection of sampling units (such as quadrats or transects) and how they are laid out on the shore should flow from both the biological units (above) and the type of data (below: population-based or individual-based parameters) to be collected. From Murray et al. (2006), reprinted with permission from the authors.

WHERE TO MONITOR

This question is wholly dependent on program goals. If the goal is to characterize the "health" of shoreline communities of a region, then numerous sites spread out over the region will need to be monitored; sampling must be *extensive*. If, in contrast, we simply want to know how a sewage outfall is affecting local shorelines, we can monitor communities at several impacted and several control sites, or along a gradient away from the source of pollution; sampling can be *intensive*.

A critical factor, unless only one site is being studied, is site-selection. Sites must be as similar as possible in all aspects except for the factor being studied. In rocky shore habitats, physical features such as wave exposure, slope and roughness of the rock, and compass direction often have a huge effect on the biological communities found there; thus, to avoid confounding natural site-to-site differences with impacts of the variable being monitored, matching these physical features is essential.

Monitoring programs also must select the habitat to be sampled; should it be the whole rocky shore, or just one zone, or just tidepools, or just mussel beds? Focusing program goals on habitats that are of most interest (e.g., the most diverse ones, or the tidepools that attract visitors) or most at risk from a given stressor (e.g., mussel beds where oil spills often concentrate) may be more efficient—again, by allowing sampling to be more intensive.

WHAT TO MONITOR

Many monitoring programs quantify the abundances of one or more species in target habitats, but this is not the only choice (Fig. 3). Monitoring all the species in a community is intensive and requires much taxonomic expertise; however, it does allow examination of the community as a whole, including the possible loss of rare species. Monitoring "target" or "indicator" species is a valid alternative if there are particular species known to be sensitive to the stressor of interest; unfortunately, we often lack this information. Usually there is no way to know ahead of time what components of an ecosystem will change. Some studies monitor functional groups, such as grouping animals by trophic level and algae by growth form; this is likely to be quicker than monitoring at the species level and may allow detection of ecologically relevant changes.

At the other end of the spectrum, monitoring sometimes focuses on particular individual-based parameters for target species, such as growth rates or reproductive success. In some cases, these parameters are more sensitive to environmental change than abundances of species are, but considerable prior work in a system is necessary to demonstrate this.

SEE ALSO THE FOLLOWING ARTICLES

Climate Change / Ecosystem Changes, Natural vs. Anthropogenic / Introduced Species

FURTHER READING

Ellis, J. I., and D. C. Schneider. 1997. Evaluation of a gradient sampling design for environmental impact assessment. *Environmental Monitoring Assessment* 48: 157–172.

Murray, S. N., R. F. Ambrose, and M. N. Dethier. 2006. *Monitoring rocky shores.* Berkeley: University of California Press.

Raffaelli, D., and S. Hawkins. 1999. *Intertidal ecology.* Dordrecht, Boston: Kluwer.

Schoch, G. C., and M. N. Dethier. 1996. Scaling up: the statistical linkage between organismal abundance and geomorphology on rocky intertidal shorelines. *Journal of Experimental Marine Biology and Ecology* 201: 37–72.

Underwood, A. J. 1994. On beyond BACI: sampling designs that might reliably detect environmental disturbances. *Ecological Applications* 4: 3–15 .

Underwood, A. J. 1997. *Experiments in ecology: their logical design and interpretation using analysis of variance.* New York: Cambridge University Press.

MULTIDECADAL VARIABILITY

SEE VARIABILITY, MULTIDECADAL

MUSEUMS AND COLLECTIONS

RICH MOOI
California Academy of Sciences

Partly through historical constraints, and partly by expediency, museums apportion their biological collections by taxonomic group, not by habitat. Therefore, museums generally do not have a "tidepool collection." Nevertheless, the value of museum collections to understanding biodiversity of the rocky shore is without rival. New technologies confer added significance upon collections at a time when extinction rates and habitat destruction, especially along shorelines, are reaching unparalleled levels.

THE ESSENCE OF A MUSEUM COLLECTION

Natural history collections are sometimes compared with a library of books and journals. However, library materials almost always exist in multiple identical copies, whereas a museum specimen is completely unique. Each specimen records a unique set of traits and data, which must be recorded and disseminated. This informs research and policy decisions that rely on knowing what an organism is, where it occurs, and when it occurred there. Museums are in the business of preserving and providing these data.

Because museums are usually organized by taxonomic group, and not by habitat, collections of rocky-shore organisms are usually queried from a taxonomic standpoint. Algae can be found in botanical collections, fish in ichthyological collections (or in vertebrate zoology holdings, depending on the organizational structure of a given museum), and noninsect invertebrates in a "general invertebrate" collection, usually held separately from the entomological (insect) collection. Groups strongly represented by marine forms will be found in such "general invertebrate" collections. Not surprisingly, museums with the strongest collections of tidepool organisms are at or near coastlines (the Smithsonian's National Museum of Natural History, located in Washington, D.C., is a notable exception).

There is a wide diversity of preservational methods (distinguished from fixation), each suited to characteristics of the specimens and the uses to which they might be put. Plants are stored in herbaria, often laid out on herbarium sheets. Many animals with hard parts, such as sea urchins, shells, or corals, can be dried. Soft-bodied organisms, notably fishes, worms, and many molluscs, are preserved in wet collections, usually in alcohol, although sometimes in weak formalin. Increasingly, museums are storing at least some material in 95% ethanol to facilitate extraction of DNA for molecular analysis. Tissue samples are also kept in subzero freezers, thereby greatly enhancing future extraction of DNA and other molecules.

The single most important unit in a museum collection is the specimen itself. However, divorced of ancillary data, the specimen's value is greatly diminished. The most important piece of information about a specimen—arguably more important than the organism's name—is where it came from, along with some indication of when the specimen was collected. Much of a museum's resources are aimed at connecting these data with the specimen in perpetuity. Before computers, searchable card catalogs and ledgers recorded museum holdings, usually assigning each specimen lot a unique number keyed to other data. Computerized databasing permits connections through nearly limitless cross-indexed searches.

Geographic information systems (GIS) have enhanced these properties of natural history collections. Increasingly, devices recording precise coordinates for the location of the collecting event on the Earth's surface immediately integrate these data with a museum's catalogs. This type of system is a powerful tool for studying the distributions of organisms, correlating these in space and time with factors that cause biogeographic change.

THE NEED FOR COLLECTIONS

Anyone who thinks that museum collections are dusty, cloistered repositories of the dead is ascribing to a view that was never accurate. Museums, most of which have active research staff, fill an important educational role through public programs that are augmented by comprehensive collections of real objects that illuminate curricula and displays. Many museums, particularly those near the shore, have living collections in artificial tidepools and aquaria.

Museum collections are a permanent record of what the world was like at specific points in time, and they can be used to develop ecological baselines. Human-mediated global change, such as the introduction of organisms and the subsequent establishment of nonnative populations with a variety of effects on native inhabitants, is a significant issue in coastal, intertidal habitats. Some introduced taxa (invasives) can dominate the environment to which they have been introduced, often with deleterious effects on the ecology. Museum records constitute the primary way of determining whether a species was introduced.

Museum collections contribute directly to developing pictures of environmental health. Analysis of specimens from a time series can reveal instances in which human activities have influenced physiology. Direct measurements of specimens can show that certain organisms, such as shelled molluscs, no longer occur at the larger sizes exhibited by older specimens in museum collections. New molecular techniques and GIS mapping can extract information from even the oldest material in a museum's collection. By these means, natural history collections themselves evolve to enhance their already great value to researchers in many disciplines.

SEE ALSO THE FOLLOWING ARTICLES

Aquaria, Public / Biodiversity, Significance of / Collection and Preservation of Tidepool Organisms / Education and Outreach / Introduced Species / Invertebrates

FURTHER READING

Suarez, A. V., and N. D. Tsutsui. 2004. The value of museum collections for research and society. *BioScience* 54: 66–74.

Shaffer, H. B., R. N. Fisher, and C. Davidson. 1998. The role of natural history collections in documenting species declines. *Trends in Ecology and Evolution* 13: 27–30.

Ponder, W. F., G. A. Carter, P. Flemons, and R. R. Chapman. 2001. Evaluation of museum collection data for use in biodiversity assessment. *Conservation Biology* 15: 648–657.

Krishtalka, L., and P. S. Humphrey. 2000. Can natural history museums capture the future? *BioScience* 50: 611–617.

University of Maine, Department of Biological Sciences. 2001. WWW Sites on invertebrates. http://www.umesci.maine.edu/biology/inv/inverts.htm.

University of California Museum of Paleontology. 2000. Other invertebrate collection catalogs. http://www.ucmp.berkeley.edu/collections/otherinv.html.

MUTUALISM

JOHN J. STACHOWICZ
University of California, Davis

Mutualism is an interaction between two species that benefits both and causes harm to neither. These benefits are varied and include directly provided benefits such as food or living space, or indirect benefits that are mediated through the presence of another species (e.g., protection from natural enemies).

GENERAL IMPORTANCE

Mutualisms are ubiquitous in the natural world and are at the root of such key phenomena as the evolution of green plants (a mutualism between a eukaryotic cell and a photosynthetic bacterium) and the accretion of coral reefs (a mutualism between the coral and microscopic algae). While these "obligate" mutualisms capture our immediate attention, facultative mutualisms, in which both partners benefit but are capable of living independently, may be much more common in the natural world. In these interactions, species are able to expand their distribution into new habitats or grow faster and become more abundant in old habitats than they could without the mutualism. It is important to realize that even obligate mutualisms are not altruistic: organisms act "selfishly" and usually benefit others incidentally rather than intentionally. In reading the examples in this article, rather than thinking about how one species acts to benefit another, think of how each species exploits the activities of the other for its own benefit without inflicting harm on its counterpart.

NUTRITIONAL BENEFITS

Tidepool sea anemones such as *Anthopleura xanthogrammica* (Fig. 1A) and *A. elegantissima* contain microscopic algae (zooxanthellae, which are dinoflagellates; and zoochlorellae, which are green algae) living as symbionts in their tissues (Fig. 1B). The algae photosynthesize and contribute sugars toward their host anemone's nutritional requirements. Meanwhile, the algae obtain a reliable source of nitrogen from anemone excretions, and anemones help regulate the exposure of the algae to light, possibly protecting them from damage due to overexposure. This arrangement is similar to the well-known relationship between zooxanthellae and reef-building corals in the tropics. For tropical corals the mutualism may be obligate—they often derive 100% of their nutrition from the food produced by

FIGURE 1 (A) The great green anemone, *Anthopleura xanthogrammica*, owes its bright green coloration, as well as part of its food supply, to microscopic mutualistic algae: zoochlorellae and zooxanthellae. The anemone gains food produced by the algae by photosynthesis, while the algae are protected from extreme light levels and gain a steady supply of nitrogen from anemone wastes. Photograph by the author. (B) Picture of a squashed tentacle of *A. elegantissima* using fluorescence microscopy to highlight the zooxanthellae (yellow disks). Photograph courtesy of Matt Bracken.

the algae and die if they go without their mutualists for extended periods. In temperate rocky-shore anemones, the association is more facultative; anemones can live in the dark in the lab or in crevices or underhangs in the field, but they are often pale in color, lacking mutualistic algae. Such anemones do survive well by eating animal matter trapped by their tentacles. Algal symbionts do provide a significant amount of fixed carbon to tidepool anemones, and this can account for up to 100% of the anemone's metabolic needs, so anemones without their mutualists may grow at a slower rate. Anemones also differ from corals in that they may have zoochlorellae or zooxanthellae. The dominance of *A. xanthogrammica* by zoochlorellae contributes to its characteristic bright green color. Zooxanthellae are thought to provide more food to anemones than zoochlorellae, but the latter are more tolerant of cold temperatures and also reduce the palatability of anemone tentacles to foraging tidepool sculpins.

A more diffuse mutualism supplies tidepool algae with nutrients. Small, turf-forming algae such as *Cladophora* or *Endocladia* are teaming with small invertebrates from many phyla, collectively called "meiofauna." These meiofauna use the seaweed as a refuge from drying out at low tide, and they graze microscopic algae living on the seaweeds or consume decomposing seaweed in the sediments trapped in the algal filaments. The waste excreted by these invertebrates contains nutrients (ammonium) that can account for the entire nitrogen demand of their seaweed hosts, allowing the seaweeds to colonize and dominate intertidal pools where nitrogen is scarce. In some ways the ammonium regeneration abilities of these invertebrates is analogous to the mutualistic bacteria in the root nodules of legumes, in that both transform nitrogen from a form that is unusable by their host to a form that is usable.

PROTECTION FROM NATURAL ENEMIES

In the web of complex interactions among species, indirect mutualisms can arise because "the enemy of one's enemy is one's friend." For example, robust branching seaweeds or calcified colonial invertebrates are often overgrown by faster-growing competitors. Such overgrowth is often called fouling because of the negative effects it can have on the host: shading, smothering, increased drag, and increased risk of dislodgement by waves. However, these slow-growing species are often structurally complex and provide a stable refuge to mobile invertebrates, such as snails, crabs, polychaetes, and amphipods, that consume fouling species, indirectly benefiting their host. As one example, the branching red seaweed *Chondrus crispus*, which dominates low-intertidal shores of New England, becomes overgrown by sessile invertebrates in deeper subtidal locations. It hosts a diverse array of species within its branches, including several small gastropod species that consume newly settled sessile invertebrates living on their host's branches (Fig. 2A). In the absence of these snails, *Chondrus* becomes overgrown and smothered (compare Figs. 2B, C). Note that these mutualisms require no altruism or intentional provision of a benefit to the partner on the part of either species. Structurally complex species, simply by growing, provide nooks and crannies for other species to exploit as shelter; those species must then find a source of food, and sessile invertebrates growing on or near their host often provide a convenient source. Some mutualists may be more effective than others at cleaning fouling organisms from the surface

FIGURE 2 The snails *Anachis* (larger) and *Mitrella* (smaller) live on the branches of the red seaweed *Chondrus crispus*, where they consume newly settled sessile invertebrates (A). In the absence of either snail, the seaweed becomes overgrown and smothered by sessile invertebrates; compare panels B (with snails) and C (without snails). Photographs by the author.

of their hosts. The most effective ones will be those that consume the specific species that overgrow their particular host. In some cases one mutualist species is not enough. For example, two different mutualistic snails are required to keep *Chondrus* clean, because no one snail consumes a broad enough range of encrusting organisms to completely clean the host (Fig. 2).

Mobile species also protect sessile organisms from their own predators. For example, caprellid amphipods that inhabit hydroids are known to actively defend them from predation by harassing specialist sea slugs (nudibranchs) that consume hydroids. The caprellids gain a suitable substrate that may protect them from predation by fish, either by camouflage or by fish actively avoiding the stinging hydroids. This type of protective mutualism can be very important in helping maintain lush kelp forests. Sea otters benefit giant kelps by consuming herbivores (sea urchins), preventing overgrazing; meanwhile, by remaining in the kelp forest, otters likely gain a refuge from larger oceanic predators that cannot penetrate the kelp canopy.

FIGURE 3 A number of sea star species harbor commensal polychaetes, some of which may function mutualistically by cleaning debris and fouling organisms from their host surface. Pictured is the underside of the sea star *Asterina*, with a close-up (panel B) of a commensal polychaete called *Ophiodromus pugettensis*. The effect of this particular worm on its host is not known. Photograph by the author.

Other mutualisms may superficially appear less consequential but are nonetheless fascinating bits of natural history. Several sea star species harbor polychaete worms that live within their rows of tube feet (Fig. 3). Some of these polychaetes can locate their hosts from a distance by following waterborne chemical cues and presumably gain protection from predators, as well as scraps of food from their hosts' messy eating habits. But there is some suggestion that these interactions may be mutualistic, as some of these worms have been observed removing organisms from the dorsal surface of their host. One such example involves the sea star *Dermasterias* and the scale worm *Arctonoe vittata*. Especially for sea stars that lack pedicellariae (the pinching structures that clean their surfaces), this could be an important means of removing fouling organisms. *Arctonoe* is by no means restricted to *Dermasterias* as its host, and it seems to exhibit a generalized aggression toward the natural enemies of other hosts it inhabits. For example, worms inhabiting the keyhole limpet *Diodora aspera* have been observed to bite the tube feet of the sea star *Pisaster* when it tries to attack the worm's host, causing the sea star to retreat.

TURNING COMPETITORS INTO ACCOMPLICES

Although many species host mutualists to remove fouling organisms (also called epibionts) because of the harm they can cause, in some cases being overgrown may be beneficial for hosts. For example, bivalve molluscs such as scallops, oysters, clams, and mussels often become overgrown by colonial invertebrates (sponges, bryozoans, sea squirts) or algae. In some cases the hard shell provides the fouling organism (also called an epibiont) a refuge from competition for space on the primary substrate (rock), or access to a better location from which to remove food from the water. Overgrowth can benefit the host by protecting it from predators. In some cases, this is simply because predators no longer recognize the bivalve as prey and ignore it, whereas in others predators may actively avoid the prey because the overgrowing organism produces noxious chemical defenses. As one example, the jewel box clam *(Chama pellucida)* grows attached to rock, but it is protected from predation by foraging sea stars by the dense growth of sessile plants and invertebrates on its shell. The benefits to the epibionts are unclear, but these do settle more rapidly on the rough surface of the shell than on smooth shell, suggesting that there is some benefit. Similarly, several sponges grow on the shell of scallops in the genus *Chlamys*. The mobility of the scallop allows the sponge to escape from a slow-moving predatory sea slug. The sponge protects the scallop from predation by some sea stars because it prevents firm attachment of the tube feet, allowing the scallop to flee more readily once contacted. Sessile organisms can often gain a refuge from predators or herbivores by settling and growing on top of mobile species. Indeed, chitons, limpets, and snails on rocky shores often support epibionts that are less abundant on the primary substrate, but the costs and benefits to each species in these associations are usually unknown (Fig. 4).

FIGURE 4 A limpet with shell overgrown by coralline algae. The outcome of this interaction is unknown, but similar overgrowth interactions are known to be mutualistic, with the overgrowth disguising or camouflaging the host from predators, and the mobile host's shell providing a competition or grazer-free refuge to the epibiont. Photograph by the author.

VARIATION IN THE OUTCOME OF INTERACTIONS

It is difficult to discuss mutualism without also mentioning commensalism, an interaction in which one species benefits and the other is unaffected. Mutualisms may grade into commensalisms as the benefits derived by one of the partner species becomes exceedingly small. This can happen quite easily when the physical or biological stress from which one species is gaining a refuge disappears. Consider the example of the benefits of sessile invertebrate growth on clams and scallops. There may be some costs to fouling overgrowth in terms of increased weight or drag or decreased feeding efficiency, but where predators are numerous, these costs are outweighed by the benefits of protection from predators. However, in areas where predatory sea stars are rare or absent, the benefit is small, so the costs equal or exceed the benefits, making the interaction commensal or even parasitic. Similar scenarios can be constructed for all the mutualisms discussed: each interaction is a combinations of costs and benefits that must be tallied on the ecological ledger. Mutualisms become commensal or antagonistic as aspects of the physical or biological environment change. In fact, it is quite common for one species to benefit from the presence of another under physically stressful conditions in the high intertidal, then be harmed by competition with that same species in more benign low intertidal or subtidal habitats.

SEE ALSO THE FOLLOWING ARTICLES

Excretion / Facilitation / Parasitism / Sea Stars / Sea Urchins / Symbiosis

FURTHER READING

Caine, E. A. 1998. First case of caprellid amphipod–hydrozoan mutualism. *Journal of Crustacean Biology* 18: 317–320.

Hay, M. E., J. D. Parker, D. E. Burkepile, C. C. Caudill, A. E. Wilson, Z. P. Hallinan, and A. D. Chequer. 2004. Mutualisms and aquatic community structure: the enemy of my enemy is my friend. *Annual Review of Ecology, Evolution and Systematics* 35: 175–197.

Martin, D., and T. A. Britayev. 1998. Symbiotic polychaetes: review of known species. *Oceanography and Marine Biology Annual Review* 36: 217–340.

Muller-Parker, G., and S. K. Davy. 2001. Temperate and tropical alga-sea anemone symbioses. *Invertebrate Biology* 120(2): 104–123.

Stachowicz, J. J. 2001. Mutualisms, facilitation and the structure of ecological communities. *BioScience* 51: 235–246.

Stachowicz, J. J., and R. B. Whitlatch. 2005. Multiple mutualists provide complementary benefits to their seaweed host. *Ecology* 86: 2418–2427.

Wahl, M. 1989. Marine Epibiosis. I. Fouling and antifouling: some basic aspects. *Marine Ecology Progress Series* 58(1-2): 175–189.

N

NEAR-SHORE PHYSICAL PROCESSES, EFFECTS OF

MARGARET ANNE MCMANUS AND
ANNA PFEIFFER-HOYT

University of Hawaii

Oceanographers have long known that organisms are not distributed homogeneously in the water column. Distributions of plankton vary both horizontally and vertically across a continuum of temporal and spatial scales. These distribution patterns result from a combination of physical circulation patterns as well as planktonic growth, behavior, and mortality. In the coastal environment, some of the physical processes important for structuring distributions of planktonic organisms include water column stratification, fronts, internal waves, and tidal bores.

VERTICAL STRATIFICATION

The Pycnocline

A pycnocline is a region of the water column with a rapid vertical change in density (Fig. 1). In coastal waters, the pycnocline separates water at the sea's surface, which is mixed by wind and waves, from water near the sea floor, which is mixed by tidal action. When a pycnocline exists, the water column is vertically stratified. The depth and intensity of the pycnocline depend on factors such as wind strength, the input of low-salinity water, and the degree of tidal mixing. When the density difference between the layers of water on either side of the pycnocline is great, the pycnocline can act as a barrier to vertical transport of

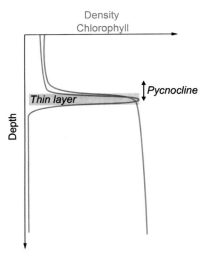

FIGURE 1 Density of a stratified water column (red line) and variation in chlorophyll concentration (green line) with depth. Chlorophyll is a light-harnessing pigment that is common to phytoplankton. A thin layer, indicated by the peak in chlorophyll concentration, is most often found at the base of the pycnocline.

dissolved and particulate materials. This barrier has significant consequences for living organisms.

Thin Layers

Recent advances in optical and acoustical technology have allowed oceanographers to examine the distribution of plankton at fine vertical scales. These advances have led to the discovery of thin layers of plankton below the sea surface (Fig. 1). A thin layer is an aggregation of plankton in which the concentration is at least three times greater than the concentration of these organisms outside the layer, with a vertical thickness of a few centimeters to a few meters and a horizontal extent on the scale of kilometers. Over 71% of thin phytoplankton layers observed in the coastal ocean have been found within the pycnocline of a stratified water column. Phytoplankton require both light and nutrients to grow. Light is often ample in the upper reaches of the water column above the pycnocline, whereas nutrients are often ample in the lower reaches of the water column below the pycnocline. The combination of suitable light and the diffusion of nutrients at the pycnocline may create an optimal location for the growth of phytoplankton. The pycnocline is also an area where currents are generally low and turbulence is suppressed. It has been shown that organisms in thin layer structures decrease their transport distances because they are associated with these regions of reduced current flow.

NEAR-SHORE FRONTS

Fronts are features in the ocean where physical, chemical, and biological properties change rapidly over a narrow horizontal area. Fronts often occur where two water masses with distinct properties meet (Fig. 2). The spatial scale of oceanic fronts ranges from 1 to 1000 meters in depth, from 10 meters to 100 kilometers across the front, and from 100 meters to 10,000 kilometers along the front. Fronts that are commonly found in the coastal zone include tidal fronts, upwelling fronts, fronts associated with geomorphic features (such as headlands, islands, and canyons), shelf break fronts, plume fronts, and estuarine fronts.

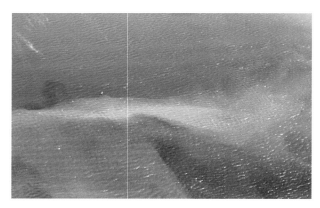

FIGURE 2 Photograph of a river plume front in Kaneohe Bay, Hawaii. A sharp boundary is apparent on the sea surface between the orange river plume water and the blue water of the bay. Photograph by Chris Ostrander.

Aggregation of Plankton at Fronts

Most fronts are regions of enhanced plankton biomass, which leads to higher production all the way up the food chain. Several mechanisms have been proposed to explain the increased productivity at fronts.

One mechanism involves the advection of plankton and accumulation in the region of convergent flow in the front. In the convergent zone, plankton will stay at the surface if their buoyancy or swimming speed is great enough to overcome the downward currents, which are typically less than 0.2 millimeters per second. Indeed, many plankton can swim faster than vertical currents in the ocean, allowing them to maintain a preferred depth in the water column by responding to cues such as light and pressure.

A second mechanism for enhanced biomass at fronts is enhancement of phytoplankton growth within the frontal region. As previously mentioned, phytoplankton need both light and nutrients to grow. Consider an example of a front between a water mass that is well mixed throughout the water column and a second water mass that is highly stratified. In the well-mixed water mass, although nutrients will be adequate for growth, phytoplankton productivity will be limited by light because phytoplankton in the well-mixed water mass are mixed from the surface to the bottom, spending reduced amounts of time in the photic zone. In the stratified water mass, phytoplankton will stay within the photic zone in the surface layer, but growth will be limited by low nutrient concentrations. Horizontal mixing by vertical eddy diffusion at the front will transfer nutrients from the well-mixed side to the stratified side. This introduction of nutrients increases the productivity of phytoplankton adjacent to the front.

TRANSPORT OF PLANKTON INTO THE INTERTIDAL

The delivery of planktonic larvae is critical for the survival of intertidal marine populations. Researchers propose that plankton can be transported into the intertidal zone by several physical mechanisms, including nonlinear internal waves, tidal bores, and shoreward-moving fronts.

Internal Waves

Internal waves are waves generated on the interface between two water masses of different densities (i.e., the pycnocline). Internal waves typically have much longer periods and higher amplitudes than surface waves, because the density difference between two water masses is much less than the density difference between water and air. Internal waves often propagate from the edge of the continental shelf, where they can be generated, into the near-shore region. The passage of internal waves may be evident on the ocean surface by long bands of foam called slicks. Slicks, which develop where surface water converges over the passing internal wave, are convergent zones that have been shown to collect floating material, including planktonic organisms. The concentration of plankton in a slick can be several times greater than concentrations outside the slick. As a linear internal wave passes through the water, the net transport of particles in the water is generally negligible. If the wave travels into an area that is shallow in comparison to the wave's length, however, it becomes a nonlinear wave. A nonlinear wave creates a current that surpasses the speed of the wave, and thus is capable of significantly displacing matter. When internal waves are nonlinear, the plankton collected in the slicks can be transported toward shore.

Tidal Bores

A tidal bore is a phenomenon in which the leading edge of the incoming internal tide forms one or more breaking waves that propagate into the near-shore environment. Internal tidal bore fronts, which are regions of increased biological activity, have been shown to transport plank-tonic material in a direction perpendicular to the coastline. If larvae are concentrated in the shoreward-propagating front of a tidal bore, these larvae can be transported into the intertidal zone.

Relaxation Events.

When upwelling-favorable winds blow along a coastline and upwelling commences, a convergent front forms between cold upwelled waters near shore and warmer sur-face waters offshore. This front is an area of increased bio-logical activity. When upwelling-favorable winds decrease in strength or reverse direction, the upwelling front is advected toward shore. This is called a relaxation event. Plankton that accumulated at the upwelling front can be transported shoreward with the onshore movement of the front. Evidence to support this hypothesis comes from dramatic increases in the number of planktonic larvae that settle in the intertidal following relaxation events.

SEE ALSO THE FOLLOWING ARTICLES

Tides / Upwelling / Waves, Internal / Wind

FURTHER READING

Belkin, I. 2002. New challenge: ocean fronts. Preface. *Journal of Marine Systems* 37: 1–2.

Farrell, T. M., D. Bracher, and J. Roughgarden. 1991. Cross-shelf transport causes recruitment to intertidal populations in central California. *Limnology and Oceanography* 36: 279–288.

Franks, P. J. S. 1992. Sink or swim: accumulation of biomass at fronts. *Marine Ecology Progress Series* 82: 1–12.

Le Fevre, J., and E. Bourget. 1992. Hydrodynamics and behaviour: transport processes in marine invertebrate larvae. *Trends in Ecology and Evolution* 7: 288–289.

Mann, K. H., and J. R. N. Lazier. 1996. *Dynamics of marine ecosystems: biological-physical interactions in the ocean.* Malden, MA: Blackwell.

McManus, M. A., O. M. Cheriton, P. T. Drake, D. V. Holliday, C. D. Storlazzi, P. L. Donaghay, and C. E. Greenlaw. 2005. The effects of physical processes on the structure and transport of thin zooplankton layers in the coastal ocean. *Marine Ecology Progress Series* 301: 199–215.

Mullin, M. M. 1993. *Webs and scales: physical and biological processes in marine fish recruitment.* Seattle: Washington Sea Grant Program/University of Washington Press.

Pineda, J. 1999. Circulation and larval distribution in internal tidal bore warm fronts. *Limnology and Oceanography* 44: 1400–1414.

Shanks, A. L. 1995. Mechanisms of cross-shelf dispersal of larval inverte-brates and fish, in *Ecology of Marine Invertebrate Larvae.* L. McEdward, ed. Boca Raton, FL: CRC Press, 323–368.

NUDIBRANCHS AND RELATED SPECIES

JAMES NYBAKKEN

Moss Landing Marine Laboratories

The nudibranchs and related species are all simultane-ous hermaphroditic members of the molluscan subclass Opisthobranchia of the class Gastropoda, and most are marine in distribution. All nudibranchs are char-acterized by lacking shells, opercula, mantle cavity, and ctenidia as adults. They have the genital openings on the right side of the body, and a pair of unique sensory organs, called rhinophores, on the head. The naked mantle surface often has projections of various kinds on its dorsal side (Fig. 1). Nudibranchs range in size

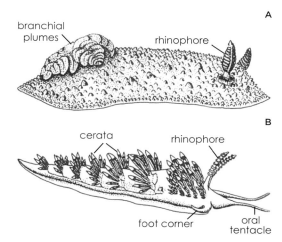

FIGURE 1 (A) A typical dorid nudibranch; (B) a typical aeolid nudi-branch. Figures used with permission from University of California Press.

from a few millimeters up to 30 centimeters, but most are in the 1-to-5-cm range.

ECOLOGY

One of the most remarkable features of most of the opisthobranchs is their striking colors and color patterns. It has been said that they are to the molluscs what the orchids are to the angiosperms or the butterflies are to the arthropods. Certainly no other group of marine invertebrates displays such dramatic colors. Although the colors and color patterns are conspicuous when the animals are seen against a uniform or plain background, when these animals are in their natural environment they often blend in so well that they are difficult to pick out. This seems to be especially true for those species that inhabit the colorful coral reef areas of the world. Some species have a pattern that effectively duplicates their food source. These cryptic colors and patterns suggest that they are significant in hiding the animals from potential predators.

The number of described species of of opisthobranch molluscs is somewhat over 3000, but some students of the taxon have estimated that the total number of species could be as high as 10,000. Thus, they are considerably less numerous than the prosobranch gastropods, with perhaps 70,000–100,000 species. However, the geographical distribution of opisthobranchs encompasses all oceans and seas of the world. In terms of depth, they can be found down to several thousand meters, but the majority of species are found in shallow water.

The life cycles of most opisthobranchs are not well known or studied. Data for the nudibranchs seem to indicate that there are two types of life histories. Those nudibranchs that feed on stable and abundant prey items such as sponges and ascidians seem to have life spans of about one year. Those that feed on transitory prey items such as hydroids have much shorter life spans and may have several generations per year.

All nudibranchs are predators, feeding primarily on sessile or sedentary marine invertebrates such as sponges, tunicates, bryozoans, and cnidarians. However, there are exceptions, with some species known to feed on barnacles and others that feed on planktonic organisms, which they trap in the water column. A few nudibranchs and sacoglossans are specialized egg predators. Most sacoglossans and anaspideans are herbivores feeding on various species of algae. The sacoglossans feed on single green algal cells, which they puncture with their characteristic radula and suck out the cell contents, whereas the anaspideans ingest larger pieces of macroalgae. The cephalaspideans are primarily burrowing carnivores that feed on various polychaete worms, prosobranch gastropod molluscs, and bivalve

molluscs. Most notaspideans appear to be predators on ascidians, although the species of *Pleurobranchaea* are predaceous on other opisthobranchs and even cannibalistic.

The opisthobranchs have a number of defense mechanisms against potential predators. Perhaps the most common defense is chemical. Notaspideans and many nudibranchs produce acid secretions when disturbed, and other nudibranchs produce repugnatorial secretions that are observed to deter potential predators. The aeolid nudibranchs concentrate nematocysts from their cnidarian prey in the cnidosacs at the tips of their cerata and employ them against predators. Certain groups of nudibranchs that feed on sponges have large numbers of spicules in their mantle that, some have suggested, reduce their attractiveness as food to potential predators. As noted previously, the cryptic coloration of many of the opisthobranchs suggests that the color and pattern serve to conceal them from visual predators.

The ecological role that opisthobranchs play in tidepools and other marine communities is very poorly understood. There have in fact been very few scientific studies worldwide that have centered on how these animals might affect the communities in which they occur. This lack of information may well be due to the general feeling that these animals are short-lived and transitory members of the marine communities in which they are found, so meaningful ecological studies cannot be made.

ANATOMY

Opisthobranchs have an anatomical structure that varies among the different orders. In all opisthobranchs the larval form, the veliger, has a shell, which may be completely lost at metamorphosis or can be further developed into an external or internal shell in the adult. In most opisthobranchs with an internal shell or no shell in the adult, the mantle tissue is often enlarged and covers the dorsal surface of the animal. It also may develop various other structures such as cerata, secondary gills, and secretory glands.

No opisthobranch has more than a single ctenidium or gill, but many have lost the original gill and have developed various kinds of secondary gills.

The head may have an expanded oral shield or an oral veil with one or more sets of tentacles. The dorsal surface of the head always contains a pair of sensory rhinophores that vary considerably in shape among the different taxa. Eyes, if present at all, are usually sunk beneath the skin and lie on the surface of the brain.

The digestive tract has a terminal or subterminal mouth opening and may or may not have jaws in the buccal cavity. With few exceptions, behind the buccal cavity is a complex organ called the radula, which is a ribbon of tissue bearing few to many chitinous teeth and moved by muscles. The radula is employed by the animal to obtain food. Some opisthobranchs have a crop to store food items, and some may also have a gizzard to grind food. The stomach is often small, and the intestine usually terminates in an anal opening that is on the right side of the animal or posteriorly. Associated with the digestive tract is a large digestive gland, which in some nudibranchs and all sacoglossans branches extensively into finger- or platelike projections of the mantle surface called cerata (singular ceras). As mentioned previously, in one group of nudibranchs (the aeolids) the cerata have sacs at their distal ends in which the nudibranch sequesters the undischarged nematocysts from the cnidarian prey and uses them for its own protection.

Because most of the opisthobranchs are simultaneous hermaphrodites, the reproductive systems are large and complex. They include not only the male and female gonads but also an array of glands for treating the ova and receptacles for storing sperm. Although they are simultaneous hermaphrodites, the reproductive system usually has mechanisms that prevent self-fertilization, and the animals usually undergo cross-copulation. Fertilization is internal, and the eggs are laid in gelatinous masses rather than in capsules as is the case in most gastropods.

The nervous system is concentrated into a circumesophageal ring, and there are no dorsal or ventral nerve cords. The major sense organs are the rhinophores, which are chemical sense organs.

SYSTEMATICS

The subclass Opisthobranchia comprises eight orders, of which the order Nudibranchia is by far the largest in number of species. The most primitive relatives of the nudibranchs are in the order Cephalaspidea (Fig. 2).

FIGURE 2 *Bulla gouldiana*, an externally shelled cephalaspidean opisthobranch. Note the paired rhinophores on the head and the tiny sunken black eyes. Photograph by the author.

These animals often have shells, either external or internal, as well as a single ctenidium in a lateral, open mantle cavity. Others lack a shell as adults. Members of this order are mainly burrowers in soft sediments, but some are capable of swimming using expanded parapodial lobes of the foot.

The order Anaspidea includes some of the largest opisthobranchs, such as *Aplysia californica*, which can reach weights of 10 kilograms (Fig. 3). All these animals lack an external shell, but they may have an internal one. They have large parapodial lobes which fold over the dorsal mantle cavity and enclosed gill. These lobes may also be used for swimming. The head has both rhinophores and paired rolled tentacles. These animals are herbivores, feeding on algae, and hence restricted to shallow water.

FIGURE 3 *Aplysia californica*, an internally shelled anaspidean opisthobranch. Note the paired oral tentacles and the paired rhinophores on the head. The visceral mass is partially covered with the parapodial lobes. Photograph by the author.

The order Sacoglossa includes a group of small opisthobanchs that appear externally very similar to some nudibranchs (Fig. 4). They may be separated from nudibranchs in that their rhinophores are rolled and hollow rather than solid and they have a distinctive radula consisting of a single row of piercing teeth, which they use to puncture algal cells on which they feed. Some species are capable of sequestering the chloroplasts from the algae, which they maintain as functioning entities in a symbiotic relationship in their bodies. Sacoglossans with such symbionts absorb the photosynthetically derived organic molecules.

The order Notaspidea is a small group of generally large animals with a head having an oral veil and lateral tentacles in addition to rhinophores (Fig. 5). These animals may have the shell external, internal,

FIGURE 4 *Elysia hedgepethi*, a sacoglossan opisthobranch. Note the large parapodial lobes and the single set of paired rolled rhinophores on the head. Photograph by the author.

FIGURE 5 *Pleurobranchaea californica*, a notaspidean opisthobranch with an internal shell. Note the broad oral veil and large rhinophores. Photograph by the author.

or absent. The mantle tissue is able to secrete a strong acid (pH = 1) when the animal is disturbed. These animals are carnivores, feeding on sponges and ascidians or other opisthobranchs.

The remaining opisthobranch orders (Gymnosomata, Thecosomata, and Acochlidiacea) are not found in tidepools.

SEE ALSO THE FOLLOWING ARTICLES

Bryozoans / Camouflage / Chemosensation / Snails / Sponges / Tunicates

FURTHER READING

Morton, J.E. 1967. *Molluscs*, 4th ed. London: Hutchinson & Co.

Solem, A. 1974. *The shell makers.* New York: John Wiley & Sons.

Thompson, T.E. 1976. *Biology of opisthobranch molluscs*, Vol. 1. London: The Ray Society.

Thompson, T.E., and G.H. Brown. 1984. *Biology of opisthobranch molluscs*, Vol. 2. London: The Ray Society.

NUTRIENTS

FRANCIS CHAN

Oregon State University

Nutrients are fundamental inorganic compounds essential for the production of new biomass by phytoplankton, benthic microphytes, and macrophytes. This process of carbon fixation or primary production underpins the size and structure of coastal food webs. Essential nutrients for primary producers comprise a large suite of elements. These nutrients are commonly divided into macro- and micronutrients. Macronutrients such as nitrogen (N), phosphorus (P), and silicon (Si), in conjunction with carbon, oxygen, and hydrogen, make up the basic building blocks of primary producer biomass. Micronutrients are trace elements such as iron, molybdenum, and zinc that individually make-up a minor fraction of primary producer biomass but play crucial roles in enzymatic processes such as photosynthesis, inorganic carbon acquisition, and uptake of macronutrients.

NUTRIENT LIMITATION

In contrast to the large suite of elements essential to the growth of marine primary producers, a much smaller subset of elements in marine environments typically occur at such low concentrations that they serve as limiting nutrients to the productivity of whole ecosystems. Nutrient limitation can occur over multiple temporal scales and scales of biological organization. This range has contributed to some confusion as to the nature and identity of limiting nutrient(s) in the sea. For example, a nutrient can limit the growth of a particular species of phytoplankton, gross rates of photosynthesis, or net rates of ecosystem production. For many ecologists, a limiting nutrient is defined as a nutrient that limits the potential rate of net primary production (NPP). Because NPP represents the amount of energy available to herbivores and other consumers, this definition is particularly useful when examining the effects of nutrient availability on food web dynamics. By explicitly recognizing the effects of nutrients on potential rates of NPP, this definition also allows for species turnover in response to changes in nutrient supply. Thus, NPP reflects the interaction between the supply rate of limiting nutrient(s) and the potential suite of primary producers that can colonize a system in response to changes in nutrient supply.

Since the supply of limiting nutrient(s) can impact the structure and dynamics of aquatic food webs, characterizing the identity of limiting nutrient(s) and the processes that control their supply has been an important focus of research. In marine ecosystems, three macronutrients (nitrogen, phosphorus, and silicon) are most commonly in the shortest supply relative to demands by primary producers. In surface waters, nitrate, phosphate, and silicic acid represent the inorganic pools of N, P, and Si. Relative demands for these three nutrients ultimately reflect the elemental makeup of primary producers. For phytoplankton, average cellular C:N:P ratio for bulk communities is described by the Redfield ratio of 106:16:1. This canonical ratio was described by Alfred Redfield in 1934. Redfield's observations of the surprisingly close correspondence between the elemental ratio of phytoplankton and inorganic nutrients in the sea led him to propose that the chemistry of nutrient elements in the sea was in fact regulated by biological processes. For phytoplankton communities, elemental ratios for C:N:P:Si converges toward mean values of 106:16:1:1. For benthic marine macrophytes, mean elemental ratios for C:N:P of 550:30:1 have also been described. It should be noted that for both pelagic and benthic primary producers, such elemental ratios represent global means, and composition of individual species can deviate markedly as a consequence of phylogenetics and environmental conditions. In a closed system in which primary producer biomass is completely remineralized and is available for subsequent bouts of primary production, no single nutrient will be in short supply relative to others. In actuality, a number of processes in the coastal ocean can decouple elemental ratios of phytoplankton from those of inorganic nutrients. Preferential loss or additions of individual nutrients can arise during remineralization, transient storage of N or P in consumer biomass, long-term burial in sediments, nitrogen fixation, and denitrification. In addition, atmospheric and hydrologic inputs of nutrients from coastal watersheds can play important roles in modifying the supply ratios of nutrients to coastal primary producers.

Along eastern boundary currents, coastal upwelling typically dominates the supply of inorganic nutrients to inner-shelf waters, including intertidal systems that lie at the water's edge. For example, along the Oregon coast of the United States, summertime upwelling can result in surf zone nitrate concentrations in excess of 30 micromoles per liter (μM). Such values are among the highest surface values that will be seen on the upwelling shelf, making such intertidal systems some of the most naturally nutrient-rich communities in the sea. This fertility reflects the efficient wind-driven transport of deep, nutrient-laden water across the shelf into the very edge of the shore.

Because upwelling-driven nutrient flux dominates nutrient supply, nutrient limitation on the inner-shelf tends to strongly reflect the nutrient characteristics of offshore sourcewater pools from depth. Upwelled water commonly experiences a deficit of N relative to P and Si. As evident in Fig. 1, the N:P ratio in upwelled water begins below the Redfield ratio. Such patterns reflect the preferential loss of nitrogen via denitrification in the deep ocean or as water transits over shelf sediments. Although nitrogen fixation has the potential to make-up for nitrogen deficits relative to other nutrients, blooms of planktonic nitrogen-fixing cyanobacteria are largely absent from temperate coastal seas. Because of preferential losses, nitrogen commonly emerges as the limiting nutrient to coastal production. Initial nitrogen deficit in upwelled water is accentuated as nutrients are taken up by primary producers, such that nitrate is completely depleted while excess P and Si remain in surface waters. It is worth noting that although nitrogen limitation is widespread, it is by no means universal. Phosphorus and silicon can be limiting or colimiting nutrients in systems where nitrogen fixation is sufficiently active or where relative demands for silicon are particularly strong.

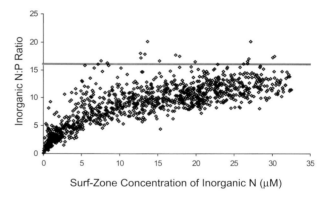

FIGURE 1 Changes in the ratio of inorganic N:P with decreasing concentrations of inorganic N in surf-zone water samples collected along the Oregon and Northern California coasts. Note the Redfield ratio line (red) for an N:P ratio of 16:1. Illustration by the author.

ACQUISITION OF NUTRIENTS FROM THE ENVIRONMENT

Marine organisms live in an aqueous environment where nutrients occur in concentrations that are many orders of magnitude more dilute than those found inside the cell. To overcome this unfavorable concentration gradient, marine primary producers must actively transport nutrient ions across the cell membrane. The kinetics of nutrient transport ultimately control the growth rate of an organism and the concentration at which growth becomes nutrient-limited. Nutrient transport is mediated by enzymes that vary in activity rates as a function of nutrient concentration outside the cell. This functional relationship can be described by Michaelis–Menten uptake kinetics, in which uptake rate increases with nutrient concentration up to an asymptote at which the enzymes responsible for nutrient uptake become saturated (Eq. 1).

$$\text{Nutrient uptake rate} = V_{max} \qquad \text{(Eq. 1)}$$

$$\times \frac{\text{Nutrient concentration}}{K_m + \text{Nutrient concentration}}$$

Two central parameters describe such uptake kinetics: K_m, the half-saturation, and V_{max}, the maximum uptake rate. The half-saturation constant is the concentration at which half of the maximum uptake rate is reached. As nutrient concentration increases beyond K_m, further increases in concentration result in smaller and smaller increases in uptake rate. Under balanced growth, the Michaelis–Menten equation is directly analogous to the Monod equation (Eq. 2), where organismal growth rate increases as a saturating function of nutrient concentration.

$$\text{Growth rate} = \text{Maximum growth rate} \qquad \text{(Eq. 2)}$$

$$\times \frac{\text{Nutrient concentration}}{K_s + \text{Nutrient concentration}}$$

In the Monod equation, maximum growth rate replaces V_{max}, and K_s, the half-saturation constant for growth, replaces K_m. Understanding nutrient uptake and growth kinetics is important, because a nutrient ceases to be limiting for NPP when nutrient concentrations become much larger than K_m and uptake rates approach V_{max}. Although K_m values vary widely among phytoplankton taxa, reported K_m values for coastal phytoplankton of 1 μM or lower indicate that nutrients may be often be replete for pelagic production in inner-shelf waters during the upwelling season.

For benthic microphytos and macrophytes, nutrient uptake and growth kinetics can also be described by Michaelis–Menten and Monod equations. However, since uptake rates decrease as reductions in surface-area-to-volume ratios impose further constraints on the diffusion of nutrients across boundary layers, benthic primary producers typically exhibit higher values of K_m and higher nutrient concentrations required before V_{max} is approached. Although the absolute flux of nutrients to benthic habitats by wave action can be high, high K_m values relative to average nitrate concentrations suggest that benthic primary producers may experience nutrient-limited growth more often than their pelagic counterparts

do. For long-lived macrophytes that experience inter- and intraseasonal changes in nutrient conditions, luxury uptake can decouple short-term growth rates from nutrient concentrations. Luxury uptake is nutrient uptake in excess of current nutrient demands for growth by organisms. Such excess nutrients are stored and mobilized for growth at later dates.

PATTERNS IN NUTRIENT SUPPLY

The supply of nutrients to intertidal systems can be highly variable in time and across space. Even for upwelling systems where average nutrient flux is high, day-to-day changes in nitrate concentration between saturating and limiting levels can occur in response to time-varying winds (Fig. 2). On interannual time scales, the linkages between nutrient supply and ecological interactions can be seen in changes in macrophyte population dynamics in response to El Niño events. During strong El Niño events, the supply of nutrients is curtailed as upwelling winds weaken and the surface layer of nutrient-poor water increases in thickness and further restricts the upwelling of nutrients from deeper layers in the water column. Such climate-mediated reductions in nutrient supply can lead to widespread depressions in kelp forest abundance, for example. Because the strength and seasonality of upwelling is geographically variable, intertidal primary producers also experience differences in nutrient regimes along broad latitudinal clines and among sites generated by local-scale topographic features such as offshore banks and coastal headlands. The extent to which such broad-scale patterns in nutrient supply interact with biogeographic patterns in

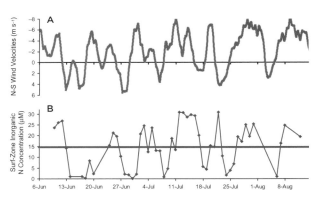

FIGURE 2 (A) Changes in the strength of upwelling-favorable winds (negative values) recorded on the Oregon coast. (B) Corresponding increases and and decreases in surf zone nutrient concentration at a nearby rocky intertidal bench. The blue horizontal line is a reference line for the mean inorganic N concentration encountered in this time series. Note the rapid changes in inorganic N concentration as winds shift between upwelling- and downwelling-favorable conditions. Illustration by the author.

intertidal community structure and processes remains an active area of research. Since nutrient concentrations can range from predominantly saturating to predominantly limiting over broad spatial scales, differences in nutrient supply may be important determinants of geographical patterns in the productivity of intertidal primary producers and their interactions with herbivorous consumers.

While nutrient supply from coastal ocean can dictate ecological structure and dynamics on benthic substrates, feedback interactions between pelagic and benthic processes can also shape intertidal communities. The residence times of water over intertidal systems can span from seconds over wave-swept rocks to days and weeks in tidepools that lie at the upper edges of the high zone. This range in residence times determines not only the supply rate of nutrients to primary producers but also the extent to which local benthic processes can modify water column nutrient dynamics. For tidepools that are transiently isolated from coastal waters, microbes and macroinvertebrate consumers can serve as locally important sources of remineralized nutrients. Conversely, primary producers such as benthic diatoms and macroalgae can act as active local nutrient sinks. The activities of tidepool organisms can thus give rise to spatial mosaics in nutrient availability that contribute to the maintenance of diversity at local scales.

EUTROPHICATION

For many intertidal systems along the open coast, the ocean serves as the dominant source of nutrients to primary producers. However, for intertidal sites situated within estuaries or downstream of river plumes or sewage discharge points, nutrient supply can be greatly elevated by anthropogenic inputs. For systems in which nutrient limitation is strongly expressed, nutrient overenrichment, known as eutrophication, can have particularly important impacts on intertidal communities. Symptoms of eutrophication include enhancement of algal productivity, reductions in algal diversity, and increased dominance by fast-growing ephemeral algal taxa. Increased nutrient

loading has also increased prevalence of phytoplankton blooms that reduce light availability to benthic primary producers and, in the case of harmful algal blooms, have resulted in the introduction of phycotoxins into intertidal food webs. Watershed processes can also alter not only the magnitude but also the ratio of nutrient supply. Human activities tend to mobilize the loss of P from watersheds at higher rates than that of N. For example, sewage typically has low ratios of N:P, and such inputs can reinforce N limitation in coastal waters. Watershed activities can also alter Si fluxes. Sediment trapped behind reservoirs serves as a long-term sink for Si and has reduced Si fluxes to coastal margins. This change can increase the dominance of nonsiliceous phytoplankton at the expense of diatoms and their zooplankton consumers. The wide-ranging ecological impacts of eutrophication provide stark illustrations of the sensitivity of food web interactions in marine ecosystems to human alterations of biogeochemical cycles in the coastal landscape.

SEE ALSO THE FOLLOWING ARTICLES

Algae / Diffusion / Food Webs / Near-Shore Physical Processes, Effects of / Photosynthesis / Upwelling

FURTHER READING

Bracken, M. E. S., and K. J. Nielsen. 2004. Diversity of intertidal macroalgae increases with nitrogen loading by invertebrates. *Ecology* 85: 2828–2836.

Dayton, P. K., M. J. Tegner, P. B. Edwards, and K. L. Riser. 1999. Temporal and spatial scales of kelp demography: the role of oceanographic climate. *Ecological Monographs* 69: 219–250.

Howarth, R. W. 1988. Nutrient control of net primary production in marine ecosystems. *Annual Review of Ecology and Systematics* 19: 89–110.

Menge, B. A., B. A. Daley, P. A. Wheeler, E. Dahlhoff, E. Sanford, and P. T. Strub. 1997. Benthic-pelagic links and rocky intertidal communities: bottom-up effects on top-down control? *Proceedings of the National Academy of Sciences of the USA* 94: 14530–14535.

National Research Council. 2000. *Clean coastal waters.* Washington DC: National Academy Press.

Redfield, A. C. 1958. The biological control of chemical factors in the environment. *American Scientist* 46: 205–221.

Worm, B., H. K. Lotze, and U. Sommer. 2000. Coastal food web structure, carbon storage, and nitrogen retention regulated by consumer pressure and nutrient loading. *Limnology and Oceanography* 45: 339–349.

OCEAN WAVES

PAUL D. KOMAR
Oregon State University

Wind-generated ocean waves are generally the most significant physical force involved in the erosion of coasts and experienced by wave-swept plants and animals that live along the shore. The heights of the waves generated by storms over the ocean commonly reach 10 to 15 meters, with the highest reliably measured wave having achieved 34 meters (measured in 1933 in the South Pacific to the east of the Philippines). The energy carried by the waves as they cross the expanse of the ocean is eventually delivered to the coast, where they break on beaches or crash against cliffs. There the waves expend their energy as they wash across the intertidal plants and animals and, over the long term, etch out the natural weaknesses in the rocks to locally form tidepools.

CHARACTERISTICS OF OCEAN WAVES

Waves are generated on the ocean whenever the wind blows across the water's surface, the process involving the transfer of energy from the wind to the waves. The greater the speed of the wind, the greater the heights and energies of the waves that can be formed. The highest waves are generated by major storms, since they generally have the strongest winds and can persist for several days during which energy is transferred to the waves. Therefore the heights of the waves depend on the duration of the storm as well as on the speed of its winds, and also on the storm's fetch: the area or length over which the energy is transferred from the winds to the waves. Techniques have been

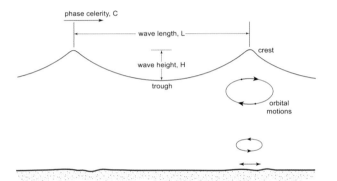

FIGURE 1 Regular ocean waves, characterized by the wave height, length, and wave period, the time interval of the passage of successive wave crests.

developed by researchers to predict the heights of waves generated by storms, depending on the wind speeds and on the storm's fetch and duration.

The characteristics of relatively simple ocean waves are depicted in Fig. 1, with the waves consisting of a series of crests and troughs. The wave height, H, is the vertical distance from the trough to its crest, while the wavelength, L, is the horizontal distance between successive crests. The motion of waves is periodic, repeating through a relatively fixed interval of time: the wave period, T. From the geometry of the waves it is apparent that they will move a horizontal distance L during the period T, so their speed, or wave celerity, is calculated as

$$C = L/T \qquad \text{(Eq. 1)}$$

As the waves pass it can be observed that a floating object (such as a cork) rises and falls as the crests and troughs alternately affect its movement, with some to-and-fro horizontal movement as well; the paths of the floating objects are thereby circular to elliptical. This pattern of movement within the wave extends downward with depth beneath the water's surface (see Fig. 1), but with the diameters of the circular movements progressively decreasing and flattening, so that at the seafloor the water movement is a simple horizontal orbital motion. The water itself makes no net advance in the direction of wave movement, unless there is also a net ocean current in that direction. The waves thereby transfer energy across the ocean's surface, in the form of the wave itself and its internal orbital water movements, but with a negligible net drift of the water. Having derived their energy from the storm's winds, the waves are capable of carrying that energy for hundreds to thousands of kilometers across the expanse of the sea, ultimately delivering it to the coasts that form the ocean's boundaries.

The *energy* of the waves, *E*, is directly related to their heights:

$$E = \frac{1}{8} \rho g H^2 \qquad \text{(Eq. 2)}$$

where ρ is the density of the water and *g* is the acceleration of gravity. It is seen from this relationship that if the wave height is doubled, the energy is increased by a factor of 4. Important is the rate of transfer of this energy across the ocean to the coasts, which is calculated by multiplying *E* by the wave celerity *C* to obtain the *wave power*,

$$P = ECn \qquad \text{(Eq. 3)}$$

The factor *n* has been included because waves travel across the ocean in groups, those formed by a particular storm. It turns out that although individual waves travel at their celerity, *C*, in deep water the group as a whole travels at half that speed (that is, $n = 1/2$). The observation is that an individual wave progressively moves to the front of the group as a whole, where it then disappears, while at the same time a new wave is formed at the back edge of the moving group. Fundamentally this occurs because the individual waves are traveling faster than their energy. This is the condition only in the deep water of the ocean. As the waves reach the shallower depths of the ocean basin and approach the coast, the factor *n* progressively increases until, in shallow water close to the coast, it achieves its maximum value of $n = 1$, at which point the energy of the waves is advancing at the celerity *C* of the individual waves.

In deep water the wave celerity depends on its period *T* according to the relationship

$$C = gT/2\pi \qquad \text{(Eq. 4)}$$

That is, the longer the period, the faster the wave's speed and the more rapidly the waves cross the ocean basin. In shallow water close to the coast *C* depends instead on the local water depth, causing the waves to slow down as they approach the shore. In intermediate water depths, *C* depends on both the wave period and water depth. Researchers have developed analyses that employ the mathematical relationships in Eqs. 1–4 for the wave energy and power and for the celerity *C* and factor *n*, which make it possible to track the paths and rates of movement of the waves from their area of generation by a storm, across the ocean width, and ultimately to the coast where they deliver their energy.

The foregoing consideration of regular waves is primarily applicable after they have left their area of generation by the storm. In the storm area itself the waves are highly irregular and are termed *sea*, while the regular waves are referred to as *swell*. The irregularity of the waves within the fetch of the storm is due to the fact that the winds generate waves having ranges in both periods and heights. When the wind blows over initially still water, it forms small ripples, but with time those ripples grow to waves of progressively longer periods and greater heights, while at the same time new ripples are formed, hence the resulting ranges of periods and heights. With this mix of waves in the storm area, it is difficult to follow the movement of an individual wave for any length of time, as the observed highest crests generally occur as the chance summation of several waves that individually have different heights and periods; in this case the crest exists for only a few seconds during which these several waves have combined. In the extreme this may produce what is termed a *rogue wave*, one of such exceptional height that it can be a danger to ships; fortunately, this degree of chance summation is rare. As this mixture of waves leaves the storm area, they sort themselves out by period, a process that is termed *wave dispersion*, since the longer-period waves travel away from the storm faster than the shorter-period waves. This dispersion eventually yields the more regular swell, but even then there will be waves having a narrow range of periods, not yielding waves of perfectly uniform periods and heights.

As a result of these processes, whenever waves are measured in the ocean, one finds ranges of wave heights and periods, distributions that need to be analyzed statistically. One can, for example, define an average wave height from those that have been measured, but it is more difficult to establish the maximum wave height, because this tends to depend on how long the measurements are made (which typically is on the order of 20 minutes to make the measurements representative). A commonly used measure of waves is the significant wave height, the average of the highest third of the measured waves—"significant" in the sense of having ignored the smallest waves in calculating the average, focusing instead only on those that have the greatest energies. Many measurements of waves under a variety of conditions have shown that the maximum height of the waves is on the order of 1.5 to 2.0 times greater than the significant wave height. Therefore, when you hear predictions of wave conditions, for example by the Coast Guard, it generally is the significant wave height that is being reported. You therefore need to realize that if you go to the shore, you can expect to actually experience waves that are potentially twice as high as that prediction.

THE WAVE CLIMATE

The *wave climate* for a specific location in the sea refers to its ranges of wave conditions experienced over the years, including the wave heights, periods, and directions of travel. Waves are most commonly measured with buoys that are tethered to the sea floor and contain instruments that continuously record the pitch and heave of the buoy, which can then be analyzed to decipher the waves that produced those motions. The recorded data can then be transmitted to a satellite, which relays it to a laboratory, where the wave data is further analyzed and compiled to form the wave climate. Lacking such direct measurements, the waves at the site can be calculated from the storm systems present on that body of water, using the analysis techniques described in the preceding section, which depend on the wind speeds of the storms and on their fetches and durations. This requires a number of years of wave measurements or calculations to adequately define the wave climate for a particular site, primarily because of the importance of establishing the occurrence of the extreme but rare events, the highest wave conditions, because they are most relevant to ship hazards and to the erosion of coasts.

Figure 2 is an example of the distribution of significant wave heights measured by a buoy in the North Pacific located off the coast of Oregon, representing a partial documentation of the wave climate for that coast. This graph is the product of 30 years of hourly wave measurements by this buoy. It is seen in the bar-histogram for the number of observations that the dominant significant wave height is on the order of 2 meters, with fewer occurrences of large waves because their generation depends on the comparatively rare major storms; those more extreme storm waves are more easily identified

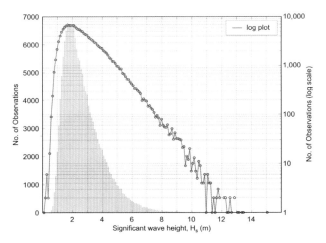

FIGURE 2 The histogram distribution of the hourly measurements of significant wave heights measured by a buoy in deep water off the Oregon coast from 1975 through 2005. The graph with the logarithmic scale provides a better identification of the rare extreme storm waves, which have reached a maximum of 15 meters significant wave height. Graph courtesy of Dr. Jonathan Allan.

where the numbers of observations are plotted using a logarithmic scale, also included in Fig. 2. It is the wave conditions that occur during these major events that are of greatest interest in the wave climate, and it is seen in this example that they commonly exceed 10 meters, with the highest measured significant wave height having been 15 meters, which occurred during a storm in March 1999 that resulted in considerable erosion along the Oregon coast. Statistical analyses are applied to the extreme-wave conditions in order to project what might be expected to be the most severe storm event and wave heights: the so-called 100-year storm, which actually is the event that has a 1% probability of occurring during any particular year. In this example for the Oregon coast, the 100-year projected significant wave height is 16 meters. Recalling that the highest waves generated by a storm are 1.5 to 2.0 times as high as these significant wave heights, the most extreme storms off the Oregon coast can be expected to generate individual waves as high as 25 to 30 meters, roughly the height of a 10-story building.

In that the highest waves are generated by extreme storms, their occurrence depends on the earth's climate, its seasonality during a year, and any long-term changes. Here it is also necessary to distinguish between tropical and extratropical storms. Tropical storms (the most severe of which are known as hurricanes and typhoons) develop close to the equator during the summer to fall in the hemispheres they affect, since their energy is derived from the warm surface water in the tropics. Extratropical storms form at higher latitudes by the interactions of cold and warm air masses and are associated with the paths of the jet streams. There have been long-term changes, spanning decades, in the frequencies and magnitudes of tropical storms, but the connection of such changes with the earth's changing climate is not fully understood. It has been suggested that the recent increase in hurricanes in the Atlantic Ocean, associated with warmer water temperatures (particularly in the Gulf of Mexico), is associated with human-induced global warming resulting from our emissions of greenhouse gases. Hurricane Katrina in the summer of 2005, a category 5 hurricane when it was offshore, generated significant wave heights of 19 meters, with its maximum wave heights likely having been on the order of 30 to 35 meters. The strongest extratropical storms occur during the winter, so the highest wave conditions seen in Fig. 2 for the Oregon coast begin in the late fall and end in the early spring. It also has been found that higher waves are formed along the U.S. West Coast by storms during El Niño winters, a change that is most dramatic along the California coast, because the El Niño storms follow tracks that take them further to the south than during normal years. Furthermore, the storm intensities and wave heights in both the North Pacific and North Atlantic have increased markedly during the past 25 to 50 years (Carter and Draper 1988; Allan and Komar 2006); researchers remain uncertain whether this has resulted from a natural cycle in the earth's climate or is associated with global warming.

THE APPROACH OF THE WAVES TO THE COAST

Wave climates generally are developed for deep-water locations, that is, before the wave heights and directions of travel are modified by crossing the shallower water of continental shelves. In wave analyses the term *deep water* has a technical definition: the water depth is greater than half the wave length (L). Because L is shorter for lower-period waves,

$$L = CT = gT^2/2\pi,$$ (Eq. 5)

it follows that "deep water" extends to shallower depths closer to shore for the lower-period waves (for example, waves of $T = 5$ seconds are still in deep water at a depth of 20 meters, while waves of 15 seconds period are in deep water at a depth of 175 meters). It follows that the modifications of these waves begin shoreward of those water depths, seen in their changing heights and directions, but not in their periods, which remain unchanged during shoaling.

Part of the change in the waves results directly from the progressively decreasing water depths, which slows their rate of advance and causes the heights of the waves to generally increase. This transformation results because the power of the waves remains relatively constant during the shoaling process; that is,

$$P = ECn \approx \text{constant}$$ (Eq. 6)

so that as C decreases as the water depths decrease, the wave energy E must increase, resulting in an increase in the wave heights H. Actually, the transformation is more complex in that while C decreases with water depth, n increases from $\frac{1}{2}$ in deep water to 1 in shallow water, such that the rate of energy advance (Cn) initially increases slightly, resulting in a small decrease in E and H, but this is followed by a more dramatic decrease in Cn so that E and H undergo significant increases; it is this latter increase that is most easily seen as the waves approach the coast and eventually break on the beaches and cliffs. This transformation of the waves in shallow water is further complicated by the frictional drag of the waves on the sea floor, so that some of their power is lost rather than being constant as previously assumed, and especially by the changes in the wave directions as they approach the coast, which can either concentrate or spread out (defocus) the energy of the waves.

This latter process is termed *wave refraction*, seen in the photograph in Fig. 3, where the crests of the waves become more closely parallel to the shoreline as they progressively reach shallower water depths (the increase in the wave heights is also apparent in this photograph). This change in crest orientation and direction of wave advance also results from the dependence of the celerity C on the water depths in shallow water. As illustrated in the diagram in Fig. 3, this results in a rotation of the wave crests with respect to the depth contours and shoreline. The portion of the wave crest at B is in deeper water than at A, and accordingly moves faster and therefore reaches B′, which represents a greater shift in position than the movement from A to A′, where the wave celerity is slower. This trend continues such that the wave crest progressively rotates and becomes more nearly parallel with the shore, just as seen in the photograph.

The refraction of water waves in the ocean is analogous to the bending of light rays when they pass through glass, as occurs when a convex lens focuses the energy of the light to a degree that it can cause

WAVE REFRACTION

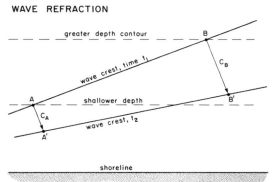

FIGURE 3 (Upper) The refraction of swell as the waves approach the California coast, rotating so that their crests become more closely parallel to the shore. Photograph by. (Lower) The refraction is caused by their more rapid movement while in greater water depths, so that in moving faster at B than at A, a few seconds later the shifts have respectively been to B′ and A′. Photograph from the University of California, Berkeley, library archives.

paper to burst into flame. In contrast, a concave lens spreads the energy of the light passing through it. The similar patterns for the refraction of ocean waves are depicted in Fig. 4 for the concentration of wave energy along a rocky headland on the coast or its spreading due to refraction over the deeper water of a submarine canyon. In this diagram, rather than considering the bending of the wave crests due to their refraction, the analysis is presented as the changing directions of the wave energy, the *wave rays* that are perpendicular to the wave crests (the directions AA′ and BB′ shown in Fig. 3). In the case of the waves approaching a headland, the rays are bent by refraction so that they become concentrated along the headland, the shallow water depths offshore from the headland having much the same effect as the convex lens in focusing the energy of light rays. Ignoring wave energy losses due to friction, the wave power between adjacent wave rays in Fig. 4 is constant, so it becomes more concentrated as the rays converge on the headland, locally increasing

the energy and wave heights. This would tend to increase the rates of erosion of the headlands, compared with the shores of the adjacent bays, except for the fact that the headlands likely exist because of the greater resistance of their rocks compared with those backing the shores to the side of the headland. It is apparent that the refraction of the waves must exert a strong control on the wave intensities and hydrodynamic forces experienced by plants and animals living in tidepools along the shores of rocky headlands.

Any irregularities in the bottom topography and water depths on the continental shelf can affect the refraction of the waves, resulting in significant variations in wave energies and heights along the shore. This is seen in Fig. 5 for the shoreline at La Jolla (San Diego), California, where both the wave crests and wave rays are included in the refraction diagram, and the water depths (in fathoms) are shown by the dashed contours. The pronounced spreading of the rays where the waves cross the deeper water of the two submarine canyons is readily apparent, resulting in reduced wave energies

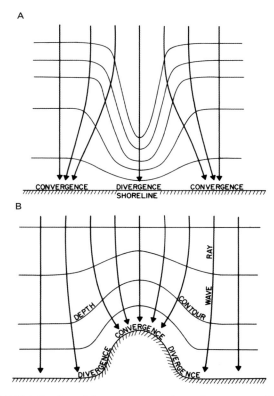

FIGURE 4 Bending of the wave rays as they refract over a submarine canyon or over the shallow water depths offshore from a headland, with the latter resulting in a convergence of the wave rays and increased wave energies along the headland.

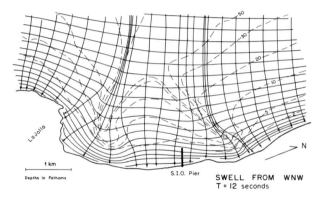

FIGURE 5 Wave refraction over the submarine canyons offshore from La Jolla, California, showing a divergence of the wave rays and a decrease in wave heights shoreward of the canyons but a convergence along the La Jolla headland. Modified from Munk and Traylor (1947).

and heights along their lee shores, while there is a degree of concentration or focusing of the wave energy between the submarine canyons, and also along the La Jolla headland. These longshore variations in the wave conditions are well known to surfers who frequent this beach.

WAVE FORCES AND THE EROSION OF ROCKY SHORES

The heights and energies of the waves that reach a coast depend on the severity of the storms that generated those waves. This determines the coast's wave climate, which can vary from site to site depending on the location relative to the occurrence of the most severe tropical or extratropical storms and on the subsequent paths of those generated waves as they cross the ocean. The waves in deep water can be considerably modified as they cross the shallower water depths of the continental shelf, generally increasing in their heights compared with deep water, but altered in complex ways by wave refraction that can either concentrate (focus) the wave energy or defocus it. All of these processes ultimately determine the energy and power of the waves once they reach the shore and their capacity to erode rocky cliffs to form tidepools and to disturb the plants and animals living there.

SEE ALSO THE FOLLOWING ARTICLES

Climate Change / Hydrodynamic Forces / Rogue Waves / Storm Intensity and Wave Height / Tidepools, Formation and Rock Types / Wave Forces, Measurement of

FURTHER READING

Allan, J. C., and P. D. Komar. 2006. Climate controls on U. S. West Coast erosion processes. *Journal of Coastal Research* 22, 511–529.
Bascom, W. 1980. *Waves and beaches.* Garden City, NY: Anchor Books.
Carter, D. J. T., and L. Draper. 1988. Has the northeast Atlantic become rougher? *Nature* 332: 494.
Dean, R. G., and R. A. Dalrymple. 1984. *Water wave mechanics for engineers and scientists.* Englewood Cliffs, NJ: Prentice-Hall.
Komar, P. D. 1998. *Beach processes and sedimentation*, 2nd ed. Upper Saddle River, NJ: Prentice-Hall.
Munk, W. H., and M. A, Traylor. 1947. Refraction of ocean waves: a process linking underwater topography to beach erosion. *Journal of Geology* 55: 1–26.

OCTOCORALS

CATHERINE S. MCFADDEN

Harvey Mudd College

Octocorals are cnidarians belonging to the subclass Octocorallia (also known as Alcyonaria), one of the two major taxonomic divisions within the class Anthozoa. The Octocorallia include all of the organisms known familiarly as soft corals, sea fans, and sea pens. Soft corals are the octocorals that are most likely to be encountered in tidepools; sea fans and sea pens typically are found only in deeper, subtidal habitats.

DISTINGUISHING CHARACTERISTICS

As their name implies, octocorals are distinguished from members of the other anthozoan subclass, Hexacorallia (Zoantharia), by their basic eight-part radial symmetry. The feeding polyps of octocorals always have eight tentacles and eight internal mesenteries, a number that is invariant within the subclass. In contrast, the hexacorals (e.g., sea anemones and stony corals) have variable, often large numbers of tentacles and mesenteries, typically in multiples of six. The polyps of octocorals are further distinguished from those of hexacorals by the presence of pinnules on the tentacles, small lateral projections that give each tentacle a feather-like appearance (Fig. 1). With only a single known exception, all octocorals have a colonial growth form.

MAJOR TAXONOMIC GROUPS

The major taxonomic groups of octocorals are distinguished from one another by the form of the colony and

FIGURE 1 A small soft coral colony (Family Xeniidae). Each feeding polyp has eight pinnate tentacles, a defining characteristic of all octocorals. Photograph by the author.

the type of supporting skeleton produced, if any. Soft corals (order Alcyonacea) range from small, encrusting forms in which polyps are connected basally by a simple stolon or membrane to large, massive or branched colonies in which the polyps can retract into a fleshy mass of coenenchyme (Fig. 2). Soft corals lack a supporting skeleton, although calcareous sclerites embedded in the tissues may impart some rigidity to the otherwise soft colony. Sea fans, also commonly called gorgonians, typically form highly branched, upright colonies that resemble fans, trees, or shrubs. The branches of sea fans are supported by an internal skeletal axis composed of calcium carbonate and gorgonin, a hornlike proteinaceous material; the polyps are

FIGURE 2 A colony of *Sarcophyton*, a soft coral whose large, leathery colonies are common on shallow near-shore reefs throughout the tropical Indian Ocean and western Pacific. This colony was exposed at low tide on a reef in northern Australia. The polyps are retracted within the fleshy mass of coenenchyme, and the polyp apertures appear as small dots on the colony surface. This species harbors microalgal symbionts (zooxanthellae) whose photosynthetic pigments give the soft coral tissue its greenish-gold hue. Photograph by the author.

housed in a thin layer of coenenchyme surrounding this axis. Sea pens (order Pennatulacea) have a highly specialized colony growth form in which a single, very large axial polyp (oozooid) develops into a basal peduncle, used to anchor the colony in mud or sand, and a polyp-bearing rachis that is supported by an internal rod of calcium carbonate. The entire colony often resembles an old-fashioned quill pen, giving the group its common name.

DISTRIBUTION AND ECOLOGY

Octocorals are found throughout the world's ocean habitats, ranging from tropical to polar waters and from the intertidal zone to the deep sea. All octocorals feed by passively filtering small particulate matter, including both zoo- and phytoplankton, from the water column. As a result they are restricted to living in areas where water movement is sufficient to deliver an adequate food supply. Many tropical species also harbor microalgal symbionts (zooxanthellae) and are therefore limited to shallow, clear-water habitats where photosynthesis can occur. The octocorals that are most commonly encountered in intertidal habitats are soft corals with encrusting or low, unbranched growth forms that are not easily dislodged by waves (see Fig. 2). Upright, branched colonies, including most sea fans and sea pens, are unable to withstand wave forces and as a result are usually found only in deeper, wave-protected habitats. Octocorals are generally long-lived and undergo sexual reproduction annually. Most species are gonochoric and broadcast spawn eggs and sperm into the water column; the nonfeeding planula larvae typically spend a few days to weeks in the plankton. Many octocorals produce unique secondary metabolic compounds that they use as predator deterrents or to compete for space allelopathically.

SEE ALSO THE FOLLOWING ARTICLES

Cnidaria / Corals / Sea Anemones

FURTHER READING

Fabricius, K., and P. Alderslade. 2001. *Soft corals and sea fans: a comprehensive guide to the tropical shallow-water genera of the central-west Pacific, the Indian Ocean and the Red Sea*. Townsville, Australia: Australian Institute of Marine Science.

Grasshoff, M., and G. Bargibant. 2001. *Coral reef gorgonians of New Caledonia*. Paris: IRD Éditions.

Williams, G. C. 1993. *Coral reef octocorals: an illustrated guide to the soft corals, sea fans, and sea pens inhabiting the coral reefs of northern Natal*. Durban, South Africa: Durban Natural Science Museum.

Williams, G. C. 1995. Living genera of sea pens (Coelenterata: Pennatulacea): illustrated key and synopsis. *Zoological Journal of the Linnean Society* 113: 93–140.

Williams, G. C. The Octocoral Research Center. http://www.calacademy.org/research/izg/orc_home.html.

OCTOPUSES

ROLAND C. ANDERSON
The Seattle Aquarium

F. G. HOCHBERG
Santa Barbara Museum of Natural History

Octopuses are a group of animals with eight boneless arms having one or two rows of suckers running along their undersides. The mantle contains the internal organs. The head, which has a mouth underneath, two perceptive eyes on top, and a brain, is located between the mantle and the arms. Octopuses are strictly marine, living under intertidal rocks down to 5000 meters deep in all the world's oceans. They feed mainly on crabs and are eaten in turn by shorebirds, large fish, and marine mammals.

MORPHOLOGY

Octopuses are placed in the phylum Mollusca, which comprises such familiar soft-bodied animals as clams, oysters, and snails as well as lesser-known marine creatures such as chitons, chambered nautiluses, and monoplacophorans. They belong to the class Cephalopoda, which means "head-footed," because their eight "feet" emerge from their heads. The eight appendages for which octopuses are named are, however, traditionally called arms by biologists.

Octopuses (not "octopi," because the word *octopus* is Greek, not Latin; the pedantically correct plural would be *octopodes*) typically live in the subtidal or in deeper water offshore. However, they are sometimes found in tidepools, under boulders, or crawling across a tidal flat or coral reef. They are extraordinarily adapted to live and survive on the shore or in shallow water.

A notable feature of these animals is their large eyes. An octopus eye has a lens, a pupil, and a retina, surprisingly similar to those of humans, though the octopus's ancestors diverged from those of the vertebrates more than a billion years ago. Such similarity is obviously a case of convergent evolution.

Despite the apparent similarities, the eyes of octopuses differ from those of humans in several ways. For instance, humans focus their eyes by thickening or thinning the lens, while an octopus retains the shape of the lens and actually moves the lens back and forth within the eye in order to focus. Also, an octopus cannot see colors. That

FIGURE 1 This giant Pacific octopus (*Enteroctopus dofleini*) demonstrates the eyes, eight arms, and suckers common to all octopuses. Photograph by Leo Shaw, Seattle Aquarium.

octopuses are colorblind has been proven by both experimentation and the absence of color-receptive cells in its retina. Because many octopuses are nocturnal or crepuscular (active at dawn and dusk) or live at great depths where there is little light, there may be little need for them to see colors.

One of the key features of these animals are the eight prehensile appendages armed with suckers (Fig. 1). Because octopuses are invertebrates, the arms are free to move in all directions and are more limber than a monkey's tail. The suckers, on the oral surface or underside of each arm, are used to adhere to the substrate to hold the animal in place, to move the whole animal by sucking on to the bottom while an arm retracts, for capturing food, and for moving food to the mouth, which is located on the underside of the animal at the center where the arms converge. The suckers are well supplied with chemosensory cells on their surfaces, so they are also able to "taste" food and other chemicals in the water. Mature males of some species of octopus have several distinctly enlarged suckers in the middle of one or more pairs of arms. The function of these enlarged suckers is not clearly understood. They may be displayed or flashed at other males or to females, as a demonstration of their sex and size, possibly used to grasp a female during mating, or to locate (by chemosensing) a female for possible mating.

All near-shore species of octopus are capable of changing their color. Color changes are achieved by tiny pigment-filled sacs in the skin, called chromatophores. Each chromatophore is capable of expanding or contracting to expose a color such as yellow, red, brown. or black.

Octopuses have structures in the skin, termed iridophores, which reflect blue, green, and metallic colors such as gold and silver. Leucophores, which reflect white, are also present in their skin. So an octopus has an amazing palette to work with in changing colors or producing patterns. Octopus color cells are directly controlled by nerves from the brain; an individual has the ability to change colors in a fraction of a second, unlike vertebrates such as chameleons or fishes, which may take several minutes to change their color. Octopuses change color to match their background for camouflage in defense against predators and even possibly to show their emotions. (A common myth is that a red octopus is angry or frustrated.) They also can produce a diversity of contrasting color patterns to startle a potential predator or prey organism or to communicate with another octopus.

DIET

Crabs and their relatives (shrimps, amphipods, etc.) are favorite food of octopuses, which are very well adapted to catch and eat prey animals of this type. An octopus typically captures its prey by pouncing on a crab and wrapping its body and arms around the victim. In another type of feeding behavior, called webover, an octopus envelopes a rock or coral head with its arms and spreads the webs of skin between its arms like a net to capture any potential prey living on or under the rock. It then feels around the rock with the arm tips for any crabs, shrimps, or small fishes it may have trapped. When a crab or other prey species is caught, the octopus holds it with its arms while it drills a tiny hole either in the eye or carapace (shell) of the crab, using a small proboscis armed with fine teeth in combination with a chemical substance secreted by a pair of salivary glands located in the buccal region (near the mouth). It next injects venom from a second pair of salivary glands into the hole, paralyzing or killing the crab. This second chemical also helps to dissolve the muscle attachments so that the "meat" can easily be extracted. All octopuses also have a hard, parrot-like beak inside the mouth, which is used to cut the meat into small pieces that can easily be swallowed. Because the esophagus of an octopus goes from the mouth right through the middle of the brain, the animal needs to eat and swallow very small pieces of its food so as not to damage the brain.

Many octopuses also drill and eat clams and snails, and some even feed on fishes and fresh carrion. They have been known to catch and eat small sharks in captive situations such as public aquariums. During a shark research event in Puget Sound outside the Seattle Aquarium, an octopus carried a salmon carcass 100 meters to its den from a bait box used to attract six-gill sharks.

PREDATORS

Many different kinds of predators eat octopuses. Large fishes, such as groupers in tropical water and ling cod in colder water, readily attack and eat octopuses, as will sharks and eels. Marine mammals such as seals, sea lions, sea otters, and even killer whales also eat octopuses. On the shore or in shallow water, seabirds and shorebirds eat small octopuses whenever they find them. Larger octopuses have also been known to catch and eat sea gulls and other seabirds in turn.

Octopuses use skin texture and color to camouflage themselves so that they can blend in with the bottom or background in order to be invisible from predators. They often flash displays of enlarged dark eye rings or false eye spots in their skin, called ocelli, to startle predators so as to escape. Some species of long-armed octopuses are capable of dropping an arm, as a lizard does with its tail, as a predator decoy. If an octopus loses an arm to a predator, it is able to regrow or regenerate the appendage. An octopus can bite and envenomate a predator that has tried to capture and eat it. In some species the body slime may taste bad so it will be spit out by a predator. Another common defense strategy of an octopus that has been disturbed by a predator is to shoot out a cloud of ink. This serves not only to hide the octopus but to confuses the predator's sense of smell. Once an ink cloud is present in the water, the octopus immediately turns pale and jets away along the bottom. When the animal lands, it instantly camouflages itself to blend in with the background so that it is difficult to see.

REPRODUCTION

Reproduction is similar in most octopus species. However, exactly how males find females is not known for sure. They may simply encounter or bump into them by chance or may locate them by "smell" or chemosensory means. Once a male encounters a female, he needs to assess somehow whether she is the same species and whether she is ripe and ready to mate. Mature females typically allow mating to occur, but occasionally they will chase off the amorous male.

During mating, the male places a sperm packet, or spermatophore, into one of the female's two oviducts. The male's third arm (either right or left depending on the species) is modified for copulation. The tip of the modified copulatory arm, termed the ligula, lacks suckers and has a groove on its underside for grasping a spermatophore and inserting it into a female's oviduct. Depending on the species, octopuses assume one of two positions during mating. In mounted mating, the male grasps the female from above, whereas in distance mating the male extends his modified, or hectocotylized, third arm toward and into the body cavity of the female without mounting her. In distance mating the two animals can be as much as a body width apart, and hence the male is able to mate with a receptive female located in a den.

Octopuses are semelparous; that is, like salmon, they reproduce once in a lifetime and die after reproducing. Males typically die within a month or two after mating, although they can mate with other females during that time period. Males of the giant Pacific octopus have only about ten spermatophores ready at any one time for transfer to females, whereas males of the common octopus may produce several hundred spermatophores. Small species of octopuses may take only a few minutes to an hour to consummate mating. In contrast, up to four hours may be needed for spermatophore transfer in the giant Pacific octopus.

After mating, females often take several months to "bulk up" and to find a suitable den in which to lay their eggs. The female typically blocks up the entrance to her small den with rocks. Small-egg species of octopus most often lay eggs in strings that are attached to the ceiling of the maternal den. In contrast, large-egg species either lay a cluster of individual eggs in a den or carry a relatively few eggs clutched in the ventral arms and web or under the body. In tropical waters the female broods her eggs without feeding for a month or two before the young hatch. In cold Alaskan waters it may take 8 to 10 months for a giant Pacific octopus female to brood eggs. The female does not eat while aerating and guarding the eggs. Because octopuses have no fat, she lives off protein metabolism, usually shrinking to about half her starting weight, during the brooding process and dies shortly after the eggs hatch. Studies on brooding females have determined that if the female is eaten or is taken away from her clutch of eggs, they will all die before hatching, victims of egg predators such as sea stars.

The hatchlings of large-egg species are small juveniles that resemble adults. They crawl away and take up a benthic existence just like the adults. In contrast, hatchlings of small-egg species are planktonic and are adapted for swimming and feeding in the water column. They are called paralarvae because they do not go through a metamorphosis like other invertebrate larvae. The tiny paralarvae have shorter arms and bigger eyes than the adults, and they swim all the time via jet propulsion. They may

FIGURE 2 Occasionally, octopuses are found dead on the shore. Death occurs after mating, and these stranded animals are inevitably spawned-out individuals at the ends of their lives. Photograph by Leo Shaw, Seattle Aquarium.

spend several months in the plankton before settling out onto the bottom as a small juvenile. Depending on body size, adult females of large-egg species typically lay only 10 to 100 eggs, while small-egg species may lay more than 100,000 eggs.

Octopuses do not live very long by human standards. Most near-shore tropical species live less than a year; the largest octopus species, the giant Pacific octopus, normally lives three to four years and may live to five years in cold water off Alaska.

Depending on the species, males die shortly after mating. Before they die, the males may crawl out of the water onto the beach (Fig. 2). As their organs, especially the brain, start to deteriorate, they begin to behave abnormally, appearing to crawl around aimlessly. During this senescent period, their state is quite a bit like that of humans with Alzheimer's disease. They stop eating and lose weight. They often develop white sores on their bodies. They leave the shelter and protection of their dens and do not camouflage themselves to hide from predators. They are especially vulnerable at this time to being eaten by large, relatively slow predators such as six-gill sharks or killer whales.

INTELLIGENCE

Octopuses are the most intelligent of all invertebrate animals. However, there are no criteria for measuring octopus intelligence or for comparing intelligence between octopus species (there are over 150 described species). If brain size is compared to body size, octopuses should be as intelligent as some birds, and certainly more intelligent than fishes, amphibians, and reptiles, but their brains are really not very large. The common octopus has a brain with about 70 million neurons; for comparison, a human's has about 100 billion. New findings have shown that more than half of an octopus nerves are in its arms, whose movements are largely self-directed. To demonstrate this phenomenon, early scientists did experiments in which they would cut off an octopus arm and then observe the results (an experiment that would not be condoned today). The detached arms continued to crawl around and were even able to sense prey, such as a piece of food, and pass it correctly along the arm in the direction of the phantom mouth. Because the arm movements do not need to be controlled by the brain most of the time, brain functions may be available for higher purposes.

Octopuses display their intelligence by a number of complex learned behaviors. In laboratory settings, they can be trained, in classical conditioning experiments, to go to specific targets for food. They can learn to navigate simple mazes or open jars to get at the food inside. In the wild they have been observed to use simple tools or landmarks to find their way back to their dens. They have proven to have individual personalities and even exhibit simple play behavior by blowing floating bottles back and forth in their tanks, like a child bouncing a ball.

Based on their intelligence, octopuses are the only invertebrates in public zoos and aquariums that are given environmental and behavioral enrichment. Such enrichments provide a more natural environmental setting, present new experiences, and constantly keep the animal occupied. A typical octopus enrichment session may consist of placing food inside a puzzle, which the animal has to figure out how to open or access in order to feed. The Association of Zoos and Aquariums, the accrediting agency for public zoos and aquariums, now requires enrichment for all mammals and birds and recommends it for such other intelligent animals as octopuses.

VENOMOUS BITES

Live octopuses encountered on a beach, on an exposed coral reef, or in a tidepool should never be picked up or handled. Handling often removes the protective mucus coating from the skin, thus exposing the animals to bacterial infections. In addition, many near-shore species may have a venomous bite, which in some cases can be deadly. The blue-ringed octopuses of Australia and Indonesia are known to kill humans, and at present there is no known antidote to the venom. When an octopus in this group bites, it injects a potent neurotoxin into the wound. Death is usually from suffocation as the victim's diaphragm is paralyzed by the toxin and the lungs cease to function. To escape death, a person bitten by a blue-ringed octopus has about 30 minutes to get to a hospital that has artificial breathing equipment.

The bites of other kinds of octopuses, although typically not as severe as those of blue-rings, can still be quite serious. They usually involve fiery pain, swelling, and paralysis of the bitten area. The bite from even a small (50-gram) common red octopus may cause these symptoms. Immediate first aid for someone bitten by an octopus is to bathe the area in hot water, as hot as the victim can tolerate. Although such treatment may alleviate the local symptoms, there may be systemic symptoms, such as dizziness, for up to a week after a bite. In addition, the site of the bite can fester and become an open, necrotic wound or ulcerous lesion that may require medical treatment. The best protection is simply to leave the octopus alone on the beach and not handle it, even if it appears to be dying.

SEE ALSO THE FOLLOWING ARTICLES

Camouflage / Chemosensation / Molluscs / Predator Avoidance / Vision

FURTHER READING

Hanlon, R. T. and J. B. Messenger. 1996. *Cephalopod behaviour.* Cambridge, UK: Cambridge University Press.
Hochberg, F. G., and W. G. Fields. 1980. Cephalopoda: the squids and octopuses, in *Intertidal invertebrates of California.* R. H. Morris, D. P. Abbott, and E. C. Haderlie, eds. Stanford, CA: Stanford University Press, 429–444.
Lane, Frank W. 1957. *Kingdom of the octopus.* London: Jarrolds Publishers.

OSCILLATION, MULTIDECADAL

SEE VARIABILITY, MULTIDECADAL

OTTERS

SEE SEA OTTERS

OYSTERCATCHERS

SARAH E. A. LE V. DIT DURELL

Centre for Ecology and Hydrology, Dorchester, United Kingdom

Oystercatchers are a group of shorebirds, forming the family Haematopodidae, with a single genus, *Haematopus.* They are very distinctive birds with all black or pied

FIGURE 1 Adult Eurasian oystercatcher *Haematopus ostralegus*. Photograph by the author.

plumage, long red bills, and loud piping calls (Fig. 1). All eleven recognized species are similar in size and shape, but vary in their coloration. In particular, American species have yellow eyes, whereas Eurasian, African, and Australasian species have red eyes. Oystercatchers are primarily birds of coastal regions and are found world-wide, except for the polar regions.

HABITAT

Oystercatchers nest on the ground in a variety of coastal habitats such as sand, shingle or rocky beaches, dunes, saltmarsh, and cliff tops. Eurasian oystercatchers, which are the most widespread species, also breed inland along shingle river valleys, along lake edges, in arable and grass fields, and even on rooftops. At the end of the breeding season, Eurasian oystercatchers migrate to coastal overwintering sites further south. Other species tend to have more limited dispersal or are sedentary, remaining close to their breeding grounds throughout the year. Migratory oystercatchers show a high rate of site fidelity, returning to their breeding and wintering sites year after year.

Oystercatchers on the coast feed primarily in the intertidal zone, following the tide out as it recedes. In the winter, large numbers of oystercatchers can be found, along with other shorebird species, resting at high-water roost sites while their low-water feeding grounds are covered by the tide (Fig. 2).

FIGURE 2 Oystercatchers jostling for position at a high-water roost. Photograph by the author.

DIET

Oystercatchers feed on a wide range of intertidal invertebrates but, in spite of their name, not oysters. They can also feed on terrestrial invertebrates such as earthworms and insect larvae. Oystercatchers are particularly well known for eating shellfish, including cockles, mussels, limpets, winkles, and clams. Oystercatchers are unusual in that they open shellfish to eat the flesh, rather than consuming the prey whole. This means that they can eat much bigger sizes of shellfish than other shorebirds can. This also means that, in some situations, they can be in competition with man for fishable shellfish stocks.

Extensive long-term studies have been carried out on the foraging behavior of the Eurasian oystercatcher. These studies have shown that individual birds specialize in particular prey throughout their adult life. Birds also specialize in the feeding method that they use when opening shellfish: "Stabbers" use their bills to prize bivalve molluscs open without breaking the shell, whereas "hammerers" use their bills to break the shells to gain entry. These feeding specializations are related to a bird's age and its sex: Young birds and females are more likely to be worm feeders or shellfish stabbers, whereas adult males are more likely to be shellfish hammerers.

Oystercatcher feeding specializations affect the shape of their bill tips: Worm feeders have pointed bill tips, shellfish stabbers have chisel-shaped bill tips, shellfish hammerers have blunt bill tips, and winkle feeders have a notch near the bill tip. If a bird changes its diet, its bill tip shape will also change.

BREEDING

Oystercatchers are essentially monogamous birds, with pair bonds lasting many years. Breeding pairs return to the same nesting territory year after year, and territories are defended by vigorous piping displays. Both male and female birds play a part in defending the territory, incubating the eggs and tending the young. One clutch of two to four eggs is laid each year, with one to two young being successfully reared to fledging. Both eggs and young are highly cryptic. Chicks are capable of leaving the nest after one or two days, but they are fed by their parents until they are fully fledged.

AGE AND SEX

Oystercatchers are long-lived birds, with recorded ages of over 30 years. Birds do not breed until they are at least four or five years old. Juvenile (in their first year) and immature (two to four years old) birds differ from adults in their coloration, generally having duller plumage and paler bills, eyes, and legs. Immature Eurasian oystercatchers do not migrate to the breeding grounds during the summer months but remain on their overwintering sites. Young birds are also subdominant to adults and are forced to move from preferred feeding sites by adult birds when the latter return in the autumn.

Male and female oystercatchers do not differ in their coloration and are difficult to tell apart. However, female birds tend to be slightly heavier and have longer and thinner bills.

SEE ALSO THE FOLLOWING ARTICLES

Birds / Foraging Behavior / Gulls / Penguins

FURTHER READING

Blomert, A. M., B. J. Ens, J. D. Goss-Custard, J. B. Hulscher, and L. Zwarts, eds. 1996. *Oystercatchers and their estuarine food supplies. Ardea* 84A.

Cramp, S., K. E. L. Simmons, eds. 1983. *Handbook of the birds of Europe, the Middle East and North Africa.* Vol. 3, *The birds of the western Palearctic.* Oxford: Oxford University Press.

Goss-Custard, J. D., ed. 1996. *The oystercatcher: from individuals to populations.* Oxford: Oxford University Press.

Hayman, P., J. Marchant, and A. J. Prater, 1986. *Shorebirds: an identification guide to the waders of the world.* London: Christopher Helm.

PARASITISM

ARMAND M. KURIS

University of California, Santa Barbara

A parasitic association is an intimate and durable relationship (symbiosis) between a smaller individual consumer and its larger prey (host). Each stage in the consumer's life cycle can be a different type of association. Major types of consumers in these trophic interactions include macroparasites (typical parasites), trophically transmitted parasites, pathogens, parasitoids, parasitic castrators and micropredators. Historically, bacteria and viruses causing infectious diseases have not been termed parasites, even though those interactions fully satisfy the definition given here.

INCIDENCE OF INTERTIDAL PARASITES

In contrast to many other habitats, there are relatively few studies of parasites in the rocky intertidal zone. The default perception is that parasites are not important in this habitat. This lack of study probably derives from two reasons. The first is that few potential hosts (other than fishes) have been investigated. Of the 110 decapod crustaceans listed in the *Light and Smith Manual*, only 15% have been investigated for more than one type of parasite, and 68% have not had even a minimal examination (say 25 individuals) for even one kind of parasite. Most of the parasitological investigations that have been conducted were restricted to a few taxa (mostly parasitic crustaceans and symbiotic egg predator nemerteans). Many invertebrate groups (e.g., polychaetes, sipunculans, chitons, anthozoans) are hardly ever examined for parasites because few parasitologists even consider their potential as hosts.

The other reason for the absence of studies of parasitism in the rocky intertidal zone is that many trophic interactions among intertidal invertebrates are not considered parasitic even though the nature of the interaction with their food items is fully consistent with various types of parasitic interactions. For example, although a tick feeding on a dog is parasitic, few would consider a nudibranch feeding on a sponge to be so, even though the degree of intimacy and the reduction in host fitness are at least comparable.

MACROPARASITES

Macroparasites include most monogeneans, nematodes and copepod parasites of fishes, notodelphyoid copepod parasites of ascidians, as well as gregarines and some coccidian parasites of polychaetes and crustaceans (if the parasites' asexual generations are self-limited). Although there are few studies of the ecology of these creatures in rocky intertidal habitats, they are well studied in other marine environments.

TROPHICALLY TRANSMITTED PARASITES

Trophically transmitted parasites are generally larval stages of tapeworms, trematodes, acanthocephalans, many nematodes, and some parasitic protozoans. They are widespread, and most of the adult stages are in fishes or birds. Well-studied examples include acanthocephalan cystacanths and trematode metacercarial cysts in crabs. These likely play a significant role in the ecology of their hosts because many studies indicate that they make their intermediate hosts significantly more susceptible to predatory final hosts.

PATHOGENS

Many pathogens are well controlled by the immunological defenses of their hosts; for example, the nearly ubiquitous rhombozoan (mesozoan) infections in the kidneys of octopuses. However, some can overcome host defenses and kill their hosts, sometimes causing epidemic mortality. These include microsporidioses of crustaceans, orthonectid infections in nemerteans, and bacterial epidemics that sometimes devastate sea star populations.

PARASITOIDS

Parasitoids—always lethal, infectious natural enemies—appear to be relatively uncommon in rocky intertidal zones. The best studied example is the flatworm *Fecampia erythrocephala* along the Atlantic coast of Europe (Fig. 1). These parasitoids only infect very small juvenile crabs. They appear to be a major mortality cause for the common European green crab. Driomyzid flies also act as parasitoids of barnacles. Since very small juvenile hosts are rarely examined for parasites, there may be other unrecognized but ecologically important intertidal parasitoids.

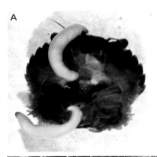

FIGURE 1 The flatworm *Fecampia erythrocephala.* (A) Parasitoid worms in a 6-mm (carapace width) juvenile European green crab, *Carcinus maenas,* from Poyll Vaish, Isle of Man, Great Britain. (B) Cocoon adhering to the underside of a rock in the mid-intertidal zone at the same location. Orange eggs are being expressed through the aperture of the cocoon. Photographs by Kevin D. Lafferty.

PARASITIC CASTRATORS

Parasitic castrators, which hijack their hosts' reproductive resources, are often common in rocky intertidal habitats. They include larval trematodes in littorine snails in the upper intertidal zones around the world. Others are bopyrid and entoniscid isopods parasitic on or in shore

crabs (Fig. 2) and rhizocephalan barnacle parasites of on a variety of intertidal crabs. Although less commonly investigated, many barnacles are also parasitically castrated. Experiments and theoretical models indicate that these often have a controlling impact on host populations.

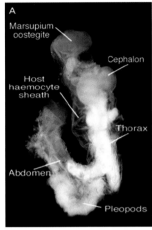
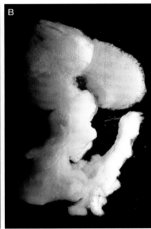

FIGURE 2 The entoniscid isopod parasitic castrator *Portunion conformis,* removed from the body cavity of an adult shore crab, *Hemigrapsus oregonensis,* from Santa Barbara, California. (A) Juvenile female, structures and host response. (B) Gravid adult. Photographs by Todd Huspeni.

MICROPREDATORS

These consumers take a small bite out of their hosts (like a parasite), but move from host to host (like a predator). This is a common lifestyle in the intertidal zone. It includes many pycnogonids, caprellid amphipods, pyramidellid snails, and nudibranchs. These are the mosquitos of the sea. Their feeding impacts the success of sessile and clonal organisms such as tunicates, bryozoans, sponges, and hydroids.

CONCLUSION

Overall, parasites are a natural part of rocky intertidal ecosystems and clearly play a role in the distribution and abundance of some species. Their general impact on the communities of this habitat remains to be studied.

SEE ALSO THE FOLLOWING ARTICLES

Diseases of Marine Animals / Food Webs / Introduced Species / Predation / Rhizocephalans / Symbiosis

FURTHER READING

Kuris, A. M. 2007. Intertidal parasites and commensals, in *The Light and Smith manual: intertidal invertebrates from central California to Oregon,* 4th ed. J. T. Carlton, ed. Berkeley: University of California Press, 24–27.
Kuris, A. M., and K. D. Lafferty. 2000. Parasite-host modeling meets reality: adaptive peaks and their ecological attributes, in *Evolutionary*

biology of host-parasite relationships: theory meets reality. R. Poulin, S. Morand, and A. Skorping, eds. Amsterdam: Elsevier Science B.V., 9–26.

Kuris, A. M., M. E. Torchin, and K. D. Lafferty. 2002. *Fecampia erythrocephala* rediscovered: prevalence and distribution of a parasitoid of the European green crab, *Carcinus maenas. Journal of the Marine Biological Association of the UK* 82: 955–960.

McCallum, H. I., A. Kuris, C. D. Harvell, K. D. Lafferty, G. W. Smith, and J. Porter. 2004. Does terrestrial epidemiology apply to marine systems? *Trends in Ecology and Evolution* 19: 585–591.

Rohde, K., ed. 2005. *Marine parasitology.* Collingwood, Australia: CSIRO Publishing.

PARKS

SEE MARINE SANCTUARIES AND PARKS

PENGUINS

RORY P. WILSON

University of Wales Swansea, United Kingdom

Of the some 300 species of birds that exploit the sea as a primary food source, penguins are the most highly specialized for an aquatic existence. They have secondarily lost the power of flight, with their wings being transformed into highly efficient flippers that, together with their fusiform shape, high body density, and extremely low drag coefficient, enable them to exploit water depths that are inaccessible to other birds. Penguins alternate time between their mating and nesting grounds on land and their feeding grounds at sea and, being unable to fly, are therefore obligate, regular transients of intertidal zones.

ANCESTRY AND TAXONOMY

Penguins are believed to have evolved during the Cretaceous (140–65 million years ago) and are most closely related to the albatross group (Procellaridae) and the divers (Gaviidae). Today all penguins belong to the family Spheniscidae, the only family within the order of the Spheniscidae. There are six genera, containing a total of 17 species.

SPECIES AND DISTRIBUTION

The most species-rich genus is *Eudyptes*, comprising the six crested penguins, distinguished by the striking, long, yellow feathers in the head crest and probably best known for the rockhopper penguins. All eudyptid penguins frequent sub-Antarctic zones, primarily ranging from 45° to 60° S. Four species from only two genera exploit continental Antarctica or adjacent regions: the three species of the genus *Pygoscelis* (the Adelie, gentoo, and chinstrap penguins) and

the emperor penguin, which is the largest penguin (weighing some 30 kilograms) and is one of two members of the genus *Aptenodytes*. The other member of this genus is the king penguin, which, at 12 kilograms, is the second largest penguin and breeds on sub-Antarctic islands. The temperate or tropical genus *Spheniscus* contains four species, one of which breeds in southern Africa (the African penguin; Fig. 1), two breed on continental South America (the Magellanic and Humboldt penguins), and one occurs only in the Galapagos (the Galapagos penguin). The remaining two genera contain one species each: the little penguin *(Eudyptula)*, the smallest penguin at 1 kilogram or less, which occurs in southern Australia and New Zealand, and the yellow-eyed penguin *(Megadyptes)*, which lives in and around New Zealand. The distribution of penguins is thus limited to the Southern Hemisphere and is dictated by their association with waters with enhanced biological productivity. This association arises because their inability to fly does not allow them to travel as fast, and therefore as far, to forage as many other seabirds, and this shortfall must be compensated by high densities of prey.

FIGURE 1 African penguin, *Sphensicus demersus*, swimming rapidly through a tidepool (Saldanha Bay, South Africa) in transit between its breeding colony and its foraging site several miles out to sea. Photograph by the author.

ECOLOGY AND BEHAVIOR

All penguins are gregarious, with extremes ranging from yellow-eyed penguins, which often travel in groups at sea but tend to nest spaced in thick vegetation, to macaroni and king penguins, which form dense breeding colonies that may number hundreds of thousands of birds. Such colonies impact the terrain substantially, both by the trampling and scratching of the birds' feet (even on rock) and by the guano runoff, which affects the ecology of tidepools adjacent to the colonies. Emperor and king penguins lay only one egg, while all others lay two, although, interestingly, the crested penguins never rear more than one chick to fledging. Eggs take between 30 and 68 days to incubate, depending on species, and at hatching chicks are covered in down but are semi-altricial. The young are normally fed, by regurgitation, by both parents for between 50 and 80 days, an exception being king penguin chicks, which do not achieve independence until they are many months old. Since hours, or even days, might elapse between prey ingestion and chick feeding, penguins have developed a remarkably complex system to ensure that the swallowed food neither is digested nor rots in the stomach. This includes a variable pH (enabling digestive enzymes to be activated or deactivated as appropriate), bird-regulated stomach churning, and chemical mechanisms for stopping bacterial growth.

Penguins feed primarily in the water column on pelagic shoaling fish, although squid are sometimes eaten (by *Spheniscus*, *Eudyptula*, and *Aptenodytes*) or swarming crustaceans such as krill (by *Pygoscelis*, *Eudyptes*, and *Aptenodytes*). Some species may also hunt for benthic species (e.g., *Megadytes*). Prey may be captured at speeds up to 5 meters per second, although normal cruising speeds are of the order of 2 meters per second. The larger species are capable of diving deeper and longer; for example, the maximum depth and duration record for the little penguin is about 70 meters and 90 seconds, but emperor penguins have exceeded depths of 530 meters and durations of 1300 seconds.

ADAPTATIONS TO DIVING

Penguins have relatively little plumage air compared to volant birds and so are less subject to the extreme buoyancy of many diving seabirds, which means that they can swim underwater expending correspondingly less energy. Since energy expenditure equates with oxygen use, this, and large quantities of oxygen-carrying myoglobin, explains why dive durations can be so extreme. Recent studies also suggest that penguins can extend dive durations by having a flexible body temperature, allowing it to decrease during the dive so as to reduce the metabolic rate, and thus oxygen use, of many of the body tissues. Excessive heat loss in water is, however, hindered by having countercurrent heat exchange systems in the flippers and feet; thick skin; and some subcutaneous fat.

SEE ALSO THE FOLLOWING ARTICLES

Birds / Gulls / Habitat Alteration / Oystercatchers

FURTHER READING

Ainley, D. G. 2002. The Adelie penguin: bellwether of climate change. New York: Columbia University Press.
Davis, L. S., and M. Renner. 2003. *Penguins.* London, UK: Christopher Helm/A & C Black Publishers Ltd.
Davis, L. S., and J. T. Darby, eds. 1990. *Penguin biology.* San Diego, California: Academic Press.
Dann, P., I. Norman, and P. Reilly, eds. 1995. *The penguins: ecology and management.* Chipping Norton, New South Wales, Australia: Surrey Beatty and Sons Pty Ltd.
Williams, T. D. 1995. *The penguins.* Oxford: Oxford University Press.

PHORONIDS

RUSSEL L. ZIMMER

University of Southern California

Phoronids are small, tube dwelling suspension feeders. They are found most commonly in subtidal areas, but they may also be found burrowed into soft rocks in tidepools.

MORPHOLOGY

Phoronids are tube-dwelling marine animals. The smallest species is about 5 millimeters in length and 0.35 millimeters in diameter; the largest, 200 millimeters by 5 millimeters. Phoronids are suspension feeders, using a crown of ciliated tentacles that encircle the mouth, and they contain a cavity called a coelom that is separated from the visceral (trunk) coelom by a septum. Phoronids have a U-shaped gut, a closed circulatory system containing erythrocytes, and paired excretory organs called metanephridia. Both hermaphroditic and dioecious species occur. All but one species have a characteristic planktotrophic larva, the actinotroch; the exception, *Phoronis ovalis*, produces a slug-shaped, demersal, lecithotrophic larva.

CLASSIFICATION AND ECOLOGY

In older classifications, phoronids were allied with both Bryozoa and Brachiopoda as Lophophorata or Tentaculata, but their broader relationships with other phyla were unresolved. Recent molecular evidence identifies phoronids as Protostomia and supports their alliance with brachiopods but weakens that with bryozoans.

Of the twelve species of Phoronida, all occur subtidally, and all have been reported to occur intertidally in at least part of their range. However, documentation that they occur in rocky tidepools has not been published. In part, this is not unexpected, since many species are known to burrow only into soft substrates or

FIGURE 1 *Phoronis vancouverensis* individuals within an aggregation collected at Garrison Bay, San Juan Island, Washington. Photograph by the author.

are commensals with hosts that live in soft substrates. Interestingly, of the four species known to burrow into calcareous substrates *(Phoronis hippocrepia, P. ijimai, P. ovalis,* and *P. vancouverensis),* three have been found in tidepools, although rarely *Phoronis vancouverensis* (Fig. 1) and *P. ovalis* have been found burrowing into limestone outcropping in tidepools, and *P. hippocrepia* has been found in rocky tidepools. Additional records of phoronids in tidepools will probably be forthcoming, but they are seemingly better adapted to relatively quiet waters in shallow bays or subtidal habitats.

SEE ALSO THE FOLLOWING ARTICLES

Brachiopods / Bryozoans

FURTHER READING

Cohen, B. L. 2000. Monophyly of brachiopods and phoronids: reconciliation of molecular evidence with Linnean classification (the subphylum Phoroniformea nov.). *Proceedings of the Royal Society of London Biological Sciences* B267: 225–231.

Emig, C. C. 1974. The systematics and evolution of the phylum Phoronida. *Zeitschrift fur zoologische Systematik und Evolutionsforschung* 12: 128–151.

Emig, C. C. 1982. The biology of the Phoronida. *Advances in Marine Biology* 19: 1–89.

Halanych, K. M., J. D. Bacheller, A. M. A. Aquinaldo, S. M. Liva, D. M. Hillis, and J. A. Lake. 1995. Evidence from 18S ribosomal DNA that the lophophorates are protostome animals. *Science* 267: 1641–1643.

Johnson, K., and R. L. Zimmer. 2001. Phylum Phoronida, in *Atlas of marine invertebrate larvae.* C. Young and M. Rice, eds. London: Academic Press, 477–487.

Peterson, K. J., and D. J. Eernisse. 2001. Animal phylogeny and the ancestry of bilaterians: inferences from morphology and 18S rDNA gene sequences. *Evolution & Development* 3: 1–35.

Santagata, S., and R. L Zimmer. 2002. Comparison of the neuromuscular systems among actinotroch larvae: systematic and evolutionary implications. *Evolution & Development* 4: 43–54.

PHOTOGRAPHY, INTERTIDAL

ANNE WERTHEIM ROSENFELD

Author and Photographer, Kentfield, California

The vicissitudes of nature—inspiring, invigorating, and often frustrating—present a unique set of technical challenges to photography in the intertidal. Several creative and technical issues are germane to taking successful pictures in this environment. The following discussion assumes a fundamental working knowledge of photography.

THE ART OF SEEING

Everyone has a point of view on the world, a prevailing arc along which the eye travels. A particularly meaningful exercise when photographing the intertidal is to take a break and look up, even briefly, to try to see on a different scale. Think about the gradient from close-up photography to mid-views to larger overviews. Time spent just looking around and observing things—an edge of something, a scrap of color next to another scrap of color—may inform your approach. You may also spot some fascinating organism or behavior that would otherwise have escaped your view.

The art of seeing while photographing is a balancing act: the photographer is trying to be at the ready, in terms of equipment, yet relaxed and flexible in terms of his or her frame of mind and frequently equivocal conditions.

The more you can mix it up, and play the combination of visual thinking and technical proficiency, the more interesting the results will be.

ENVIRONMENTAL CHALLENGES

In any given year, opportunities to photograph the intertidal are relatively few. The most important variable, of course, is the rhythm of the tides. A good low tide during daylight hours may permit five solid working hours. Another tide, either less low or perhaps clipped by water, weather, or light conditions, may cut that time by as much as half. It is essential to know the tide schedule and to become familiar with the specific site. By observing how the water goes in and out at that site, and how it is affected by daily differences in swells, currents, and wave action, it becomes easier to estimate how long you can spend in parts of an exposed rocky shore without having to swim back.

Another difficult variable is wind. Whether the camera is on or off a tripod, strong winds can buffet it; even if you want to convey some sense of wind motion, camera shake must be avoided in order to prevent blurry photographs. One solution is to shoot faster film, or increasethe ISO speed on a digital camera. Another solution is to keep yourself and your equipment low and sheltered. With the camera on a tripod, hang a sand bag full of sand or rocks from the center axis; the weight will steady the camera.

Water clarity is a crucial variable. Turbidity in the water can interfere with getting a good photograph. Particles in water will absorb and reflect light from both sun and flash; light reflected off suspended particles can cause an objectionable problem called backscatter.

Contrary to what might be expected, rain need not be a deterrent, short of a torrential downpour. Coastal habitats and rocky intertidal outcroppings may provide some degree of shelter, and rain shields, plastic bags, and umbrellas will protect the gear. Although rain can make your day more laborious and uncomfortable, clearing storms are often accompanied by gorgeous light.

Fog reduces the amount of ambient natural light but also evens it out, reducing contrast. Although one's eye adjusts constantly to ambient light and is able to naturalize colors, film cannot. Therefore, your photographs may benefit from warming filters to balance the color of the light. In both fog and rain, the front element of your lens should be dry before photos are taken.

Cloud cover diffuses light and, like fog, reduces contrast. Moving clouds and mixed light can be challenging,

not just in terms of exposure but in using the right accessories to help light a scene before it quickly changes again.

Full sun, particularly harsh, midday top light, is hard to work with. All the details and the great contrast between intensities of light in a scene may be difficult for film and sensors to recognize.

TECHNIQUE AND EQUIPMENT

There are two major technical challenges in intertidal photography: photographing through water and coping with endless reflective surfaces and highlights. The level of illumination of the light source and the degree of reflection, transmission and absorption are all related.

Film and digital sensors record light as levels of brightness. Human vision discriminates approximately twice as many levels and much more detail, particularly in the highlights and shadows. Light varies in intensity, color, direction, and texture, and it changes constantly. Both internal and external light meters are meant to interpret scenes as if they were average, a mixture with no extreme dominant areas of lights or darks, and will make everything a middle tone. Pronounced problems arise when the subject is not average, as is often the case in the intertidal, because water has peculiar effects on light. Light rays are bent or refracted, colors are absorbed, and light is scattered by particles in the water. The depth of the water, and the absorptive qualities of the water and the subject, will affect light and therefore exposure. Daylight may need to be supplemented, and the tools to do this come in a variety of forms.

Flash, an important source of supplemental light in the intertidal, must be off-camera. Using flash off-camera and away from the lens at a 45-degree angle to the water will allow flash output to cut through the surface of the water rather than bounce off it. Moisture in the air or on your hands as you handle flashes, cords, and mounting shoes can cause them to short out, if only temporarily; thus when using flash, it is prudent to carry a set of spare equipment.

Through-the-lens (TTL) metering has made electronic flash more practical, especially for shooting through water. In general the intensity of electronic flash falls off quickly with distance: a close background will be lit in a similar to way to the subject, whereas a distant background usually will be darker. The problem is confounded when shooting through water, where the effects of water become difficult to gauge with any certainty.

Even with the automatic feature of TTL, bracketing shots is recommended.

There are specific mathematical formulas related to the flash power output, referred to as guide numbers. Guide numbers may not apply to all natural situations, particularly the intertidal. Exposure is also affected by the focal length of the lens. When using suggested guide numbers to calculate exposure with manual flash, varying the flash-to-subject distance can be an effective method of bracketing.

Photographs can also be lit with multiple flashes, aimed at the water at 45 degrees or slightly more acute angles. A device that holds multiple flashes (typically two) on flexible arms can be attached to the camera. Each flash can be directed independently, and, if needed, a third flash can be handheld. An external power pack will keep the flash units ready to fire.

Preflashes integrate data for determining the flash output level for optimal exposure. To avoid their being misled by reflective surfaces, it may be best to turn them off.

A circular polarizing filter will be needed, whether shooting with film or digitally. Polarizers remove glare and reflections. These filters can reduce or increase the amount of specular reflection, creating richer, deeper colors. Unfortunately, their filter factor reduces the exposure anywhere from 1/2 to 2 full stops. This is most easily handled with TTL metering systems, because they are reading light that is passing through the lens, including the filter on the lens.

A black umbrella for close-up photography is quite useful. Before heading to the intertidal, turn the umbrella upside down and spray its inside with a couple of coats of matte black paint. The umbrella can make rainy days more workable; it can also blow away if you are not attached to it.

Photography manuals discuss how to get good exposures, but exposure in the intertidal has special problems. First, the Auto and Manual metering systems in cameras are based on the concept of averaging and are easily fooled in the intertidal. If there is a lot of black or a color and value close to it, stop down to prevent overexposure. Conversely, for a lot of white, or something close to it in value, open up about two stops.

The second issue involves the exposure for photographing moving water. A correct exposure with a fast shutter speed will stop the motion. These choices are highly subjective. Shutter speeds slower than those stopping the dynamic force of moving water produce a wide variety of impressionistic results.

There are several tools to help expose a photograph so that light in the scene is balanced. Graduated Neutral Density Filters, or split NDs, are especially useful for high-contrast scenes. Split ND filters are half clear and half a neutral gray, which will block light from part of a scene without affecting the other part or distorting the colors. Available in one, two or three f-stops, the soft-step models are better suited to the intertidal, where you rarely have a distinct boundary between light and dark zones. The filters mount in a special holder attached to the front of the lens. To use them to deliver the desired exposure control, meter the scene, evaluating the difference between its lightest and darkest parts. Using the depth-of-field preview button, position the selected split ND slightly below the density shift and then take the picture. The easiest example is a landscape that includes sky. Without some amount of ND filtration, the correct exposure for the intertidal will yield a washed-out sky. Conversely, the correct exposure for the sky will most likely yield an underexposed foreground. The split NDs may help you get good results without having to shoot far-flung brackets that you will have to recompose digitally.

One or more disk-shaped reflectors and diffusers are helpful. These come in many sizes; they are portable, flexible, lightweight and collapsible; and they are essential for close-up work.

If the light is extremely bright, a translucent diffusing disc will produce a broader light source and soften the effects of high-contrast light Concurrently, this makes the diffuser helpful at tempering reflective surfaces, especially wet ones. A reflective white disc can be positioned to catch and throw natural light in an area of the photo, especially an area in shadow. When taking long exposures, wiggle the disc a bit, just enough to create a more natural effect in the final photo. It is important to do the best possible job lighting on scene; editing programs cannot bring out detail that either was not there to begin with or was not captured adequately.

It is never practical or rewarding to wait around for good lighting conditions. An alternative is to poke around looking at smaller subjects, and then use one of these diffusing dics to modify and improve the lighting conditions.

Tripods are indispensable if you are using low ISO speed numbers, because the exposures will be longer than manageable with hand-held equipment. For long exposures, a cable release is also necessary. The best tripods for the intertidal are those made of light, long-lasting, sturdy materials. Key features include adjustable leg lengths, and legs designed so that the lateral movement of each one operates independently. Interchangeable center columns are important: a short one will make it possible to get close to ground level without the column being in the way. Ideally, the camera should be close to the tripod's center of gravity no matter what position the tripod is in. Any tripod used to photograph the intertidal also has to withstand salt water. After photographing in the intertidal, take the appropriate steps to maintain your tripod.

UV filters on all lenses are important prophylactic accessories (in addition to blocking ultraviolet light). They will protect the front lens element from salt spray, blowing sand, and any abrasive accidents.

The more equipment you take, the more cumbersome your trip, and the harder it will be to move quickly. The following items are important when photographing in the intertidal zone:

- A watch, to know what time it is in relation to the predicted time of the tide
- Spare parts and backup for any equipment that may go down in the field
- The appropriate batteries for your equipment
- A supply of chamois cloth to help keep photography equipment dry from both humidity and rain
- Lens cleaning cloths and a strong air-blowing bulb for removing particles from the lens before using the cloth
- A supply of extra plastic bags in various sizes
- Any small tool that might come in handy when you are far from other resources
- A small bit of strong tape
- Paper and pen, preferably moisture-proof, for taking notes
- Any necessary notes or technical information, pared to just a few moisture-proof sheets or laminated for protection
- A pair of fisherman's gloves, for both warmth and nimble fingers.
- A few bandages for "barnacle bites" and the like

Clean your equipment immediately upon returning from the intertidal. After cleaning, you might place all equipment except the tripod into a large plastic bucket, add packets of a drying agent, and tightly seal the bucket. After a day or two the deleterious effects of the salt water will be deterred or at least delayed.

GOING DIGITAL

While opinions on this new technology vary widely, the digital camera has much to offer for intertidal work, especially because of the playback feature of the LCD monitor. Photos can be reviewed as they are taken, histograms checked, and exposures fine-tuned. Or, if the scene exceeds the dynamic range of the camera, an exposure can be made for good shadow detail and another for good highlight detail; they can later be combined in a single image using editing software. With the playback feature you can also confirm that the selected equipment is doing its job without creating untoward effects of its own. In addition, digital cameras can reduce the number of times a photographer will need to reload film in the field. In the moist and salty conditions of the intertidal, the less often you open a camera, the better. A single memory card has the capacity to hold many more photographs than a roll of film.

As digital quality and performance increase, the arguments against shooting with this new technology will lessen. A substantive drawback is that there is a lot more work to be done following a digital shot than a film shot, and the digital workflow relies on ever-changing software and equipment, as well as quite a bit of discipline. Still, considering the difficulty of photographing in the intertidal, there are many times when the benefits of digital photography will outweigh the problems.

Perhaps the most important argument in favor of digital photography is a simple truth known to all photographers: "You cannot put your foot in the same river twice." Once lost, the visual conditions of a scene will never be exactly the same again. With the playback capabilities of digital photography and the opportunity to resolve both esthetic and technical problems as you work, you have a better chance of preserving the content and relationships in the scene, not only as you first discovered it but exactly as you hoped to capture it.

SEE ALSO THE FOLLOWING ARTICLES

Collection and Preservation of Tidepool Organisms / Education and Outreach / Light, Effects of

FURTHER READING

American Society of Media Photographers. http://www.asmp.org.
Grimm, T., and M. Grimm. 2003. *The basic book of photography*, 5th ed. New York: Plume Books.
Koehl, M., and A. W. Rosenfeld. 2006. *Wave-swept shore: the rigors of life on the rocky coast.* Berkeley: University of California Press.
North American Nature Photography Association. http://www.nanpa.org
Rosenfeld, A., and R. T. Paine. 2002. *The intertidal wilderness—a photographic journey through Pacific Coast tidepools*, rev. ed. Berkeley: University of California Press.
Rotenberg, N., and M. Lustbader. 1999. *How to photograph close-ups in nature.* Mechanicsburg, PA: Stackpole Books.

PHOTOSYNTHESIS

RICHARD C. ZIMMERMAN

Old Dominion University

Photosynthesis is the biological conversion of light energy into chemical energy stored in the form of organic material. Marine organisms are responsible for approximately 40% of all photosynthesis on earth, and 25–30% of the ocean photosynthesis is accomplished by marine macrophytes— the seaweeds and seagrasses—which occupy less than 10% of the ocean surface area along the submerged and tidal margins of the continents and islands.

DEFINITION AND ORIGIN OF PHOTOSYNTHESIS

At its most fundamental level, photosynthesis represents an oxidation-reduction reaction in which light energy is used to drive the transfer of chemical energy, via electrons and hydrogen ions from a donor molecule (H_2A) to carbon dioxide, creating water and organic carbon in the process:

$$2H_2A + CO_2 \xrightarrow{\downarrow light} (CH_2O) + H_2O + 2A \quad \text{(Eq. 1)}$$

Photosynthesis evolved early in the history of life on earth, and the original substrate (A) represented a variety of atoms, such as sulfur. Oxygen, however, is the preferred substrate for most modern photosynthetic organisms, which means the photosynthesis reaction is usually written as

$$2H_2O + CO_2 \xrightarrow{\downarrow light} (CH_2O) + H_2O + O_2 \quad \text{(Eq. 2)}$$

The oxygen molecule (O_2) released in this reaction is produced by splitting water (H_2O). The organic compounds produced by photosynthesis (CH_2O, a carbohydrate) can be oxidized to generate chemical energy in the form of adenosine triphosphate (ATP), which is capable of fueling cellular work. Oxidation (also called respiration) and photosynthesis can be summarized as a reversible reaction that ultimately yields chemical energy (in the form of ATP) from light energy:

$$H_2O + CO_2 \underset{\downarrow ATP}{\overset{\downarrow light}{\rightleftarrows}} (CH_2O) + O_2 \quad \text{(Eq. 3)}$$

Oxygenic photosynthesis probably evolved in relatively shallow regions of coastal seas that allowed for the growth of microbial mats. The proliferation of photosynthetic

O_2 began a long and dramatic alteration of the earth's atmosphere that changed forever the course of evolution. Because photosynthetic organisms consume light energy for growth, they are frequently referred to as photoautotrophs or photolithotrophs; the latter name recognizes that they consume inorganic carbon in the photosynthetic process. Organisms that consume other organisms are known as heterotrophs. There are approximately 30,000 living species of photoautotrophs. Their evolutionary history is complex and highly branched, with deep divisions that extend back to the origins of life on earth. As a consequence, many photosynthetic organisms are as distantly related to each other as they are to heterotrophs.

Marine tidepools and wave-swept shores provide a particularly rich habitat for a wide diversity of species from virtually all major groups of photosynthetic organisms. In contrast, terrestrial photoautotrophs are all derived from the chlorophytes, a single lineage of photoautotrophs characterized by two photosynthetic pigments: chlorophylls *a* and *b*.

THE BIOCHEMISTRY AND PHYSIOLOGY OF PHOTOSYNTHESIS

The essential oxidation–reduction reaction described in Eqs. 1–2 represents a highly condensed summary of a rather complex biochemical process. To appreciate the environmental constraints to photosynthesis placed on organisms inhabiting wave-swept shores, a little more detail is necessary. The chemistry of photosynthesis is usually compartmentalized into light reactions and dark reactions. The light reactions involve (1) the absorption of photons by photosynthetic pigments, (2) the transfer of that light energy from the pigments to the photosynthetic reaction centers, (3) the extraction of electrons and protons from water by the reaction centers, and (4) the transport of electrons, which ultimately generates energy and reducing power (ATP and NADPH) for photosynthetic carbon fixation. The dark reactions use the reducing power of ATP and NADPH generated by the light reactions to assimilate carbon dioxide into organic matter.

Photosynthetic Pigments, Photosynthetic Units, and the Light Reactions

The light reactions are initiated when photosynthetic pigments absorb packets of light energy called photons. The pigments, or chromophores, can be classified into three basic types: the chlorophylls, the carotenoids, and the phycobilins. All photosynthetic organisms contain chlorophyll *a* (Chl *a*), which serves to harvest light energy and transfer it to electron acceptors in the two reaction centers. The chlorophytes (green algae and seagrasses) contain Chl *b* in addition to Chl *a*. In contrast, the chromophytes (diatoms, dinoflagellates, and brown seaweeds) contain Chl *c* and the carotenoid fucoxanthin (discussed in a subsequent paragraph). Although chlorophylls absorb light with wavelengths between 400 and 700 nanometers, they absorb light most efficiently in the blue (wavelengths from 400 to 500 nm) and red (650 to 700 nm). Photosynthetic organisms are often green in color because green and yellow photons (500 to 600 nm) are poorly absorbed by chlorophyll and are therefore transmitted and reflected.

The major light-harvesting chromophores in red algae are the linear tetrapyrroles called phycobilins. Chlorophylls, by comparison, are condensed, closed-ring tetrapyrroles. Unlike chlorophylls, the phycobilins absorb strongly in the green region (500 to 600 nm). The red algae also contain Chl *a* (which absorbs strongly in the blue), and the two pigment types combine to make the algae appear red.

The carotenoids are a diverse group of chromophores characterized by two six-carbon rings separated by an 18-carbon unsaturated chain, in which some carbons are bound by double rather than single bonds. Subtle differences in ring structure, side chain length, and degree of unsaturation (the number of C=C double bonds) produce a wide range of optical properties. In general, the carotenoids absorb light in the blue-green part of the spectrum, which partially overlaps the Soret (blue) absorption band of the chlorophylls. Some carotenoids (e.g., fucoxanthin, peridinin) play a fundamental role in photosynthetic energy transfer, whereas other carotenoids (e.g., β-carotene, violaxanthin, antheraxanthin, and zeaxanthin) protect the photosynthetic apparatus from damage caused by oxygen radicals and high light intensity.

The photosynthetic pigments are bound to specific proteins that control their orientation and distribution within the thylakoid membranes of the chloroplasts (Fig. 1) in photosynthetic eukaryotes, and they function within a larger photosynthetic unit, defined as the minimal number of pigment–protein structures that must undergo a photochemical reaction to produce one molecule of oxygen (Fig. 1). The light-harvesting pigments, also called antenna pigments, are coupled to two photochemical reaction centers, called Photosystem I (PSI) and Photosystem II (PSII). The reaction centers are connected to each other via an electron transport chain. This chain transfers energy from PSII to PSI, thereby generating electrochemical gradients and reducing power needed to

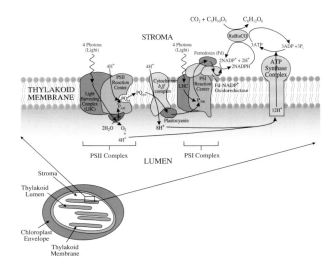

FIGURE 1 The structure of the chloroplast, showing the arrangement of the photosynthetic apparatus within the thylakoid membrane. The thylakoid membrane encloses the lumen and separates it from the stroma. The light reactions of photosynthesis are accomplished by protein-based structures in the two photosystems (PSII and PSI) and the electron transport chain that generates NADPH in the stroma and drives protons (H⁺) across the membrane into the lumen. The resulting electrochemical gradient is used by the ATP synthase complex to generate ATP in the stroma. Carbon is fixed in the stroma by the enzyme ribulose 1,5-bisphosphate carboxylase oxygenase (RuBisCO), in conjunction with other enzymes responsible for the C3, or Calvin/Benson, cycle. Carbon fixation consumes NADPH and ATP.

produce NADPH and ATP used for carbon fixation and, ultimately, cell growth. Much of the adaptive physiology to different light environments (discussed in the next section) involves changes in the abundance of the antenna pigments (and their associated proteins) and reaction centers within the membrane.

Chloroplasts probably arose when a photosynthetic prokaryote developed a symbiotic relationship with a nonphotosynthetic cell. They are found in the cells of all eykarotic phytoplankton. In seaweeds and seagrasses, however, chloroplasts are found primarily within the epidermal cell layer, which facilitates gas exchange with, and nutrient uptake from, the water medium. The interior cells (mesophyll) of seaweeds and seagrasses are virtually unpigmented. In contrast, the leaves of terrestrial plants are covered by a waxy cuticle that helps prevent water loss, and the chloroplasts are located primarily in the mesophyll.

The Dark Reactions and Photosynthetic Carbon Assimilation

The photosynthetically driven fixation of inorganic carbon into organic matter is accomplished by the enzyme ribulose 1,5-bisphosphate carboxylase/oxygenase

(RuBisCO). This soluble protein is located in the stroma of the chloroplast in close proximity to PSI (see Fig. 1). Photosynthetic carbon assimilation is accomplished by all marine macrophytes using RuBisCO and the Calvin/Benson cycle, or C3 pathway, which ultimately leads to the production of sugars, sugar alcohols, and starch.

CO_2 is the obligatory inorganic carbon substrate for RuBisCO. The high diffusion coefficient of CO_2 in air provides a ready supply of this important substrate for terrestrial plants directly from the atmosphere. Dissolved inorganic carbon is highly abundant in seawater (approximately 2.5 millimolar), and 80% of it exists as bicarbonate (HCO_3^-), a hydrated form of CO_2. Less than 1% exists as free dissolved CO_2 at the typical pH range of seawater (7.8 to 8.2), and the diffusion coefficient for CO_2 is much lower in water than in air. Consequently, most marine macrophytes possess mechanisms for extracting CO_2 from the abundant pool of dissolved bicarbonate. Some photoautotrophs possess enzymes that mediate the active transport of bicarbonate into the cell. Many utilize enzymes knows as carbonic anhydrases that dehydrate bicarbonate, allowing free CO_2 to cross the cell membrane by diffusion. These carbonic anhydrases are localized in the periplasmic space between the cell membrane and cell wall. Other photoautotrophs exploit the process of calcification to reduce the alkalinity of seawater, thereby extracting CO_2 from the bicarbonate pool.

The high concentration of bicarbonate in seawater, combined with high activity of the transport/dehydration mechanisms just described, typically prevent carbon limitation of marine photosynthesis. In addition to utilizing dissolved bicarbonate in seawater, macrophytes exposed to air are capable of taking up CO_2 directly from the atmosphere as long as the tissue remains sufficiently hydrated. Seaweeds growing in the upper intertidal zone, as well as subtidal photoautotrophs capable of forming extensive surface canopies, such as the giant kelp *Macrocystis pyrifera*, will take up a significant amount of CO_2 directly from the atmosphere. Carbon utilization pathways of some chlorophytes (e.g., the sea lettuce *Ulva*) are capable of environmental acclimation with respect to the utilization of dissolved inorganic carbon. When grown fully submerged, *Ulva* exhibits high periplasmic activity of carbonic anhydrase and utilizes bicarbonate as the primary inorganic carbon source. When grown in an energetic surf zone, *Ulva* exhibits very little carbonic anhydrase activity and relies primarily on diffusion of CO_2 from the air.

In contrast to the marine algae, seagrasses (aquatic flowering plants) exhibit limited capacity for bicarbonate utilization. The resulting chronic CO_2 limitation restricts light-saturated photosynthesis rates to less than half the metabolic capacity of these plants under CO_2 saturation. The reduced photosynthetic capacity leads to characteristically high light requirements for seagrass survival—on the order of 10–20% of the light present at the sea surface—whereas algae are generally capable of survival with 1% of surface light, or less.

ENVIRONMENTAL CHALLENGES TO PHOTOSYNTHESIS ON WAVE-SWEPT SHORES
Desiccation

The most important environmental (as opposed to biological) factor limiting the upward distribution of photosynthetic organisms along the shore is desiccation. Unlike terrestrial plants, marine photoautotrophs have few effective mechanisms to prevent water loss and no capacity to extract water from the rocky substrate to which they cling. The highest, and driest, intertidal regions are colonized by algal films (benthic diatoms) and multicellular algal crusts that can exploit small depressions, minimize the amount of surface area exposed to the air, or both. The algae assume turflike morphologies at slightly lower (and wetter) elevations. Multistory canopies begin to emerge in the extremely low intertidal, as algae with three-dimensional fronds extend above the smaller turfs and crusts.

Although the short, compact forms of algal crusts and turfs serve to reduce water losses by desiccation, many of these algae (particularly the turflike red algae) are capable of losing up to 80% of their fully hydrated water content without suffering significant impairment of either photosynthesis or respiration. Most of the water in these fully hydrated algae is located within the relatively thick and hydrophilic cell walls, which provides a buffer against the loss of metabolically important water inside the cells. Evaporative water loss from the cell wall also provides a natural air conditioning that helps maintain temperatures within physiologically tolerable ranges.

Those macrophytes growing in low intertidal and fully submerged habitats are much less tolerant of desiccation, primarily because their cell walls provide a much smaller hydration buffer. Constant wave motion and wind-blown spray from the sea surface prevent desiccation of floating surface canopies of the large kelps (e.g., *Macrocystis pyrifera*, *Nereocystis lutekeana*), and these algae are less tolerant of water loss than algae from the high intertidal zone. Ironically, the seagrasses, which are derived from terrestrial plants, grow best when fully submerged, because they have an extremely thin leaf cuticle, which retains very little extracellular water to buffer the effects of desiccation.

Temperature

Although marine photoautotrophs inhabit all seas from the equator to the poles, they experience a much narrower, and more temporally stable, temperature range than their terrestrial counterparts. The high heat capacity of water limits the daily temperature range of the ocean surface caused by solar heating and nighttime cooling to about 1 °C. Annual variations in mean water temperature rarely exceed 20 °C and are typically less than 10 °C along most rocky shores. Consequently, many marine photoautotrophs exhibit a narrow range of temperature tolerances matched to their environmental conditions.

Like all chemical reactions, the metabolic processes that drive photosynthesis and respiration rely on the rate and intensity of molecular collisions. The speed and frequency of the collisions increases with temperature, causing metabolic rates to increase with temperature until enzymes and structural proteins begin to break down, a process called denaturation. Further temperature increases cause metabolic rates to decline as the protein pool becomes increasingly denatured.

Clearly, exposure to high temperatures can be lethal to marine photoautotrophs. Intertidal algae experience large potential daily fluctuations in ambient temperature because of their periodic exposure to air. Evaporative losses of periplasmic water (see the previous discussion of desiccation) help prevent the overheating of intertidal seaweeds when exposed to afternoon low tides.

Temperature affects the chemistry of seawater in ways that may reduce photosynthesis well before protein denaturation occurs. The solubility of dissolved CO_2 decreases as temperature increases. Thus, photosynthesis of organisms that rely primarily on dissolved CO_2 for photosynthesis (e.g., seagrasses) may become increasingly carbon limited as temperature increases. Under such conditions, the photosynthetic response to temperature may be suppressed, preventing the capture of enough light energy to offset the respiratory demand. Under such conditions, stored energy (sugars and other carbohydrates) must be mobilized to support the respiratory demand, depleting the stored reserves.

The carbonate saturation state of seawater increases with temperature. Consequently, calcification is more common among tropical and warm temperate species (e.g., the coralline algae and hard corals with their photosynthetic algal symbionts) than among coldwater plants. The calcification process generates CO_2

for photosynthesis and provides structural rigidity that may deter grazing.

Physical Isolation

Tidepools provide a unique and challenging environment for intertidal photoautotrophs. The presence of water eliminates desiccation as a significant limiting factor, but the physical isolation of the tidepool from the larger ocean permits dramatic changes in ambient temperature and dissolved gas content that are not experienced by subtidal photoautotrophs. In tidepools with dense populations of macrophytes, high rates of photosynthesis can reduce the dissolved inorganic carbon concentration to the point where it limits photosynthesis. In addition, large increases in the concentration of dissolved O_2 may reduce photosynthesis by inducing photorespiration, in which RuBisCO functions as an oxygenase that consumes oxygen and generates CO_2 rather than fixing it.

Light

The light environment of wave-swept shores is highly variable. Photoautotrophs growing in the intertidal zone are frequently exposed to full sunlight that can be as bright as any terrestrial environment. On the other hand, photoautotrophs growing at the base of dense, multistory canopies of marine algae and seagrasses may be exposed to extremely low, and spectrally altered, light environments (Fig. 2). Photosynthetic organisms possess two mechanisms that help cope with this dramatic

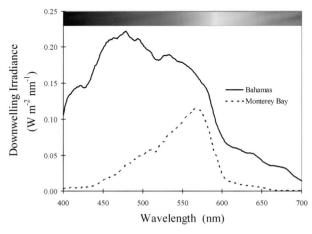

FIGURE 2 Examples of submarine irradiance spectra measured at 5-m depth beneath the clear tropical waters of the Great Bahamas Bank and at about 8-m depth beneath the plankton-rich waters of Monterey Bay, California. Shifts in the spectral quality of the light can have significant impacts on the distribution and abundance of marine photoautotrophs.

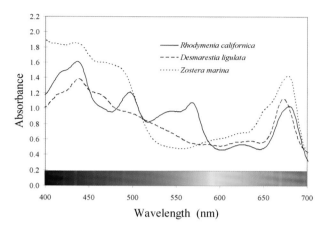

FIGURE 3 *In vivo* absorption spectra of representative marine plants *Rhodymenia californica* (a red alga), *Desmarestia ligulata* (a brown alga) and *Zostera marina* (eelgrass, a green plant) reveal differences in their ability to utilize light for photosynthesis. The red alga absorbs green light more efficiently than the brown alga or the green plant.

range of light environments—chromatic adaptation and photoacclimation.

Chromatic adaptation describes the evolution of different pigment systems capable of exploiting different parts of the submarine light environment. Although all photoautotrophs contain Chl *a*, the accessory pigments (primarily carotenoids and phycobilins) confer significant advantages on the algae in different light environments (Fig. 3). The strong absorption bands of chlorophylls and carotenoids in the blue, combined with the ability of water to absorb red light, mean that the submarine light environment of productive coastal waters becomes dominated by green light as water depth or algal abundance increases (see Fig. 2). Consequently the chlorophytes (green algae), which contain only Chls *a* and *b*, are typically restricted to the shallowest habitats, where blue light is more abundant. The carotenoid fucoxanthin broadens the blue absorption band of chromophytes (brown algae) significantly into the green, and the depth distribution of these algae generally exceeds that of the green algae. Phycobilin pigments of the rhodophytes (red algae) are very efficient at absorbing green light and, when combined with chlorophyll, can make these algae appear almost black—an indication that they are absorbing most of the light that hits them. The ability to exploit green light enables red algae to grow beneath dense canopies of other photoautotrophs, and to greater depths than their green or brown counterparts. Examples of depth distributions based on chromatic adaptation are best illustrated from temperate and boreal environments, where benthic plants and phytoplankton combine to produce green-light environments.

In clear tropical waters, where the light environment is predominantly blue, the depth distribution of green algae can match or exceed that of the red algae.

Photoacclimation to different light environments involves physiological adjustments in the abundance of light-harvesting pigments and photosynthetic reaction centers within an organism that optimize the photosynthetic process (Fig. 4). As just noted, red algae are generally capable of growing to greater depths than the green algae. Both groups, however, can exploit extremely high-light environments of intertidal shores. When grown in high light, rich in blue photons, red algae reduce the abundance of phycobilin pigments, appearing more green as Chl *a* becomes the dominant pigment. In

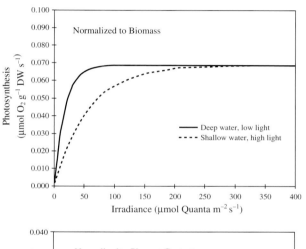

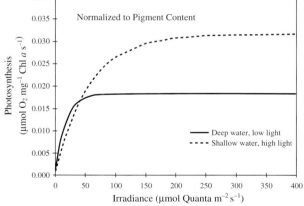

FIGURE 4 Photoacclimation alters the photosynthesis–irradiance response function exhibited photoautotrophs. The high pigment content of plants acclimated to low light increases the efficiency of photosynthesis, as indicated by the steeper slope of the initial part of the response normalized to biomass (e.g., grams dry weight), but it does not necessarily change the light-saturated (maximum) rate of photosynthesis. When normalized to pigment content (e.g., chlorophyll *a*), the photosynthetic efficiencies (initial slopes) of photoautotrophs acclimated to low light are often greater than those of their high-light counterparts, but rates of maximum photosynthesis are lower.

general, acclimation to high light involves a reduction in the abundance of light-harvesting pigments and an increase in the number of photosynthetic reaction centers capable of processing the captured photons into chemical energy. Plants adapted to low-light environments generally contain more pigments, which increase the rate (or efficiency) of light harvesting. In dim-light environments the plants need fewer reaction centers per molecule of antenna pigment to process the absorbed light.

MEASURING PHOTOSYNTHESIS

Energy captured by photosynthetic organisms represents the base of the marine food web. Consequently, the ability to estimate the rates of gross and net photosynthesis, as well as respiration, is critical to understanding the energy balance of marine food webs. There are four common ways to estimate primary productivity of photoautotrophs growing on wave swept shores.

Growth

A simple way to estimate net productivity is to measure growth over time. Growth represents the net production accomplished after photosynthesis has provided the energy required for respiration. In plants with intercalary meristems (e.g., kelps) or basal meristems (e.g., seagrasses) that exhibit linear patterns of growth, productivity can be determined by measuring the migration of a mark (hole punch, needle prick, tag, etc.) along the thallus or leaf over time. Growth of algae with apical meristems can be measured by marking plants with fluorescent stains (e.g., calcifluor or tetracycline) that bind to cell walls. New growth can be distinguished from old tissue because it lacks stain. The hole punch method is fast, accurate. and easy. Staining procedures require microscopic examination of the tissue and can be relatively tedious to perform. Although both measurements of growth can provide accurate estimates of net productivity in many instances, they usually require a significant time lag (days to weeks to months) between the initial mark and the growth measurement, they assume that growth rate was constant during that period, and they do not account for the release of dissolved organic matter or losses due to grazing. Thus, they provide a lower bound on net primary productivity. Finally, they do not provide any insight into rates of respiration or gross photosynthesis.

O_2 Evolution

Changes in the oxygen content of seawater are relatively easily measured using a variety of electrode systems and incubation chambers that allow precise control over water temperature, gas and dissolved solute composition, flow rate, and light environments. Measuring oxygen consumption in the dark provides an estimate of respiration, while measurement of net O_2 evolution across a range of light intensities permits a quantitative evaluation of the photosynthesis–irradiance response function. This method is extremely effective and fast for measuring photosynthesis of seaweeds and seagrasses when relatively large amounts of tissue can be placed in small, controlled volumes of water. This method requires that the tissue be isolated within an incubation chamber that may alter the flow, light, and temperature, but all these conditions can be controlled precisely. It can, however be difficult to interpret the steady-state photosynthesis–irradiance response function derived from precise laboratory measurements within the context of a dynamic natural light environment, in which intensity and spectral composition of the light can change rapidly. Estimating rates of carbon fixation from O_2 measurements requires a photosynthetic quotient (= moles O_2 evolved/moles CO_2 fixed). Although the respiratory quotient is reliably 1 (see Eq. 3), the photosynthetic quotient can be as high as 3 or 4 if large amounts of photosynthetic energy are diverted to nitrate assimilation or anabolic metabolism rather than carbon fixation. Accurate measurement of photosynthetic quotients requires simultaneous measurement of dissolved inorganic carbon flux, as well as oxygen, between the plant and the water medium.

CO_2 Fixation

The high concentration of dissolved inorganic carbon in seawater precludes the use of physiologically driven bulk changes in dissolved inorganic carbon to measure metabolic activity. Instead, aquatic physiologists and production ecologists typically rely on the use of bicarbonate tagged with ^{14}C, a radioactive isotope of carbon. The assimilation of ^{14}C into organic matter can be measured with great precision and ease by modern scintillation counters. This method is used by oceanographers to study populations of marine phytoplankton that routinely occur in densities that are too low to permit routine measurements of bulk oxygen flux. Like the O_2 flux measurement, the ^{14}C method require the photoautotrophs to be isolated in an incubation chamber in which environmental conditions can be carefully controlled. Although short-term incubations with ^{14}C can provide estimates close to gross photosynthesis, they cannot provide estimates of net photosynthesis or respiration, which are readily available from measurements of oxygen flux. Further, because

^{14}C incorporation is accomplished by RuBisCO (see the discussion of dark reactions), these measures provide little insight into the dynamics of light harvesting or energy transfer processes mediated by the photosynthetic reaction centers.

Variable Fluorescence

The light energy captured by photosynthetic pigments is dissipated in the photosynthetic unit via three competing pathways. Most of the energy (~85%) is dissipated as heat. About 10% is captured as photosynthetic energy. The remaining 5% is re-emitted from chlorophyll *a* as fluorescence. Recent advances in computerized fluorometry enable measurements of variable fluorescence to be made of intact plants in their natural light environments using pulsed amplitude modulation (PAM) or fast repetition rate (FRR) fluorometers. Although the two measurement procedures differ, they both rely on the inverse relationship between the rate of photosynthesis and the amount of fluorescence emitted under illumination, assuming heat losses are constant. That is, when the alga is actively photosynthesizing, little fluorescence is given off. The ability to measure variable fluorescence of intact tissues in the natural environment and to rapidly construct light–induction curves offers several advantages over the more traditional O_2 or ^{14}C techniques, and these measurements can provide important insights into the process of energy transfer between the reaction centers. However, because the PAM or FRR fluorometers do not measure photosynthesis directly, the estimation of photosynthetic carbon fixation rates requires a number of assumptions or careful calibration against the O_2 or ^{14}C incubation methods previously described.

SEE ALSO THE FOLLOWING ARTICLES

Algal Color / Diffusion / Heat Stress / Light, Effects of / Seagrasses / Water Chemistry

FURTHER READING

Falkowski, P., and J. Raven. 1997. *Aquatic photosynthesis*. New York: Blackwell.

Kirk, J. T. O. 1994. *Light and photosynthesis in the sea*. Cambridge, UK: Cambridge University Press.

Larkum, A, R. J. Orth, and C. Duarte. 2005. *The seagrasses—their biology, ecology and conservation*. New York: Springer.

Margulis, L. 1998. *Symbiotic planet: a new look at evolution*. New York: Basic Books.

Meyers, J. 1980. On the algae: thoughts about physiology and measurements of efficiency, in *Primary productivity in the sea*. P. Falkowski, ed.. New York: Plenum, 1–16.

Zimmerman, R. 2003. A bio-optical model of irradiance distribution and photosynthesis in seagrass canopies. *Limnology and Oceanography* 48: 568–585.

PHYTOPLANKTON

MARK A. BRZEZINSKI

University of California, Santa Barbara

The term *phytoplankton* is used to describe the myriad of microscopic species of photoautotrophs in the sea. In addition to growing in suspension in the sunlit surface waters, where sunlight reaches the bottom phytoplankton can also be found attached to rocks, to other organisms, and even inhabiting the spaces between grains of sand. Phytoplankton cells range in size from under one micron to just over two millimeters, but most are between 0.5 and 200 microns. Despite their small size, phytoplankton account for half of the photosynthetic carbon fixation on planet Earth. Phytoplankton has been called the grass of the sea, because the vast majority of marine life depends on phytoplankton for energy and minerals. In the intertidal and subtidal zones phytoplankton share their role as primary producers with the macroalgae or seaweeds.

DISTINGUISHING CHARACTERISTICS

Phytoplankton have traditionally been referred to as microalgae to distinguish them from the larger seaweeds, or macroalgae. They are unicellular, but many species form chains of individual cells. The phytoplankton include both eukaryotic and prokaryotic forms. Although some eukaryotic phytoplankton were once regarded as being related to higher plants, most are now considered to be protists with more animal-like characteristics. Prokaryotic forms are predominantly cyanobacteria. Thus, phytoplankton represent a phylogenetically diverse group of organisms that arose from multiple evolutionary lines.

The ecological classification of phytoplankton is in part based on their size. In near-shore environments like the intertidal, the larger microalgae, or net phytoplankton (so called because these forms can be captured in fine-meshed plankton nets) dominate. These are typically eukaryotic organisms. Smaller plankton, the ultraplankton, are less than 2 microns in size, with some being less than one micron. Ultraplankton include picophytoplankton as well as heterotrophic bacteria, archea, and viruses. Picophytoplankton consists of the prokaryotic cyanobacteria just mentioned. Pico- and net phytoplankton show characteristic distributions, with picophytoplankton dominating the phytoplankton biomass of the open sea while net phytoplankton tend to be the dominant forms in near-shore environments.

The larger eukaryotic phytoplankton include the diatoms, coccolithophores, and dinoflagellates. The diatoms are some of the most beautiful of phytoplankton, because they literally live in an intricately sculptured house of glass (Fig. 1). The cell wall of diatoms contains approximately 20% silica and is called the frustule. Variations in the patterning of the silica within the frustule form the basis for diatom taxonomy. The frustule is composed of two valves that resemble the lid and bottom of a Petri dish. Between the two valves there exist one or more siliceous girdle bands. Diatoms are among the most prolific of phytoplankton, forming large blooms in coastal oceans that fuel food webs and coastal fisheries. In areas where diatoms are routinely abundant, their frustules can accumulate on the sea floor, forming sediments called siliceous oozes. Some ancient diatom deposits have been uplifted onto land and are mined as diatomaceous earth, to be used in many filtration and purification processes. Several species of diatoms of the genus *Pseudo-nitzschia* produce domoic acid, which acts as a neurotoxin in mammals. Blooms of *Pseudo-nitzschia* can result in mass mortality of seabirds and marine mammals.

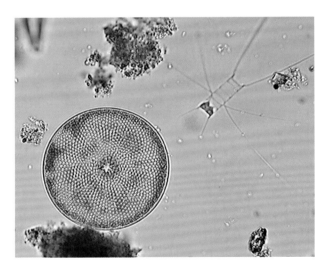

FIGURE 1 Light micrograph of the diatoms *Coscinodiscus* (lower left) and *Chaetoceros* (upper right). Photograph courtesy of T. Villareal.

The term *dinoflagellate* means "whirling flagella." Dinoflagellates have two flagella that propel the cells through the water. One flagellum lies in the sulcus, which is a groove running from the midpoint of the cell to the posterior end. A more ribbon-like flagella lies on a groove that encircles the midsection of the cell, called the cingulum. The action of these two flagella causes the cells to rotate on their longitudinal axis while moving forward, and this helical motion is the characteristic embodied in the name of this group. Dinoflagellates are referred to as being armored or naked. Armored forms are covered with thecal plates made of cellulose (Fig. 2), while naked forms lack theca.

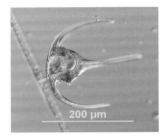

FIGURE 2 Light micrograph of the thecate dinoflagellate *Ceratium*. Photograph courtesy of T. Villareal.

Dinoflagellate metabolism is very diverse. They can be photosynthetic, some are heterotrophic, others are mixotrophic, some are bioluminescent, and others are parasitic or produce toxins. Dinoflagellates produce red tides, which sometimes consist of toxin-forming species. A variety of species produce several neurotoxins that accumulate in the food web affecting humans that consuming contaminated fish or shellfish. It should be emphasized that not all dinoflagellate blooms, or red tides, are toxic.

Coccolithophores are covered with small plates of calcium carbonate called coccoliths. The coccoliths cover the cell and vary widely in morphology among species (Fig. 3). During intense coccolithophore blooms the sea can turn milky white because of the high reflectivity of the coccoliths. Coccolithophore blooms have been detected from satellites in space by this reflectivity. Coccoliths can be a major component of marine sediments forming calcareous oozes. When these deposits are uplifted onto land, they can form large calcite deposits such as the White Cliffs of Dover in Great Britain.

The smaller prokaryotic phytoplankton are referred to as picophytoplankton. These cyanobacteria were largely unknown before the 1980s. Picophytoplankton

FIGURE 3 Scanning electron micrograph of the coccolithophore, *Emiliama huxleyi*. Photograph courtesy of J. Fritz.

are now known to be extremely abundant, and they account for a large fraction of global marine primary production. Two genera, *Prochlorococcus* and *Synechococcus*, are among the most numerous. One species, *Prochlorococcus marinus*, may be the most abundant organism on Earth.

DISTRIBUTION AND ECOLOGY

Phytoplankton is ubiquitous in the sea, having adapted to every condition in the surface ocean from the -2 °C temperatures of the polar seas to the >30 °C temperatures in the tropics. Because they are photoautotrophs, they are confined to depths where sufficient sunlight penetrates to support photosynthesis. This well-lit zone is referred to as the euphotic zone. In the clear waters of the mid-ocean, the euphotic zone can reach 150 meters, whereas in the more turbid coastal ocean it is typically less than 25 meters deep.

Like all autotrophs, phytoplankton require a source of mineral nutrients to supply carbon, nitrogen, phosphorus, trace metals, and other essential elements. Nutrients are most abundant in deep waters, and mixing processes that bring nutrient-rich deep water to the surface ocean stimulate phytoplankton growth. Close to shore the input of nutrients from rivers can stimulate phytoplankton blooms. Phytoplankton can double their biomass in a day or less, creating populations over 100,000 cells per liter in a matter of days. In coastal regions, blooms are most frequent in areas where local winds drive the upwelling of deep waters to the surface, fertilizing the euphotic zone with nutrient from below. Upwelling is common along many coasts and it is the resulting blooms of phytoplankton that make these areas some of the richest habitats for marine life in the world.

The carbon dioxide fixed into organic matter by phytoplankton photosynthesis is the main pathway by which ocean biota affect carbon dioxide levels in the Earth's atmosphere. Phytoplankton photosynthesis consumes carbon dioxide molecules in the surface ocean, which is replenished by the diffusion of carbon dioxide from the atmosphere into the sea. Some of the carbon fixed into organic matter by phytoplankton is passed up the food web. A portion of this carbon sinks to the deep sea in the form of corpses, fecal matter, detritus, and living phytoplankton. This so called "biological pump" transfers carbon from the atmosphere to the deep sea, where it is sequestered for long periods of time. If the biological pump were eliminated, the carbon dioxide concentration in the atmosphere would double, leading to significant warming.

Phytoplankton growth in the near-shore environment feeds many of the filter feeders in the intertidal, and the relative abundance of phytoplankton food may partially determine the abundance of filter feeders in intertidal communities. Some diatoms are specially adapted to live amongst the waves breaking along the shore and are referred to as surf zone diatoms. Blooms of surf zone diatoms occur as dark golden-brown discoloration within the surf off sandy beaches. Occasionally the number of diatoms in a patch has been so high that the accumulation is mistaken for an oil slick. Only a handful of diatom species form dense surf zone blooms. Their accumulation results from their ability to ride wave-generated bubbles to the surface. At night the patches disappear as the diatoms descend into the sand, only to reappear the next morning. These patches contain several hundred thousand cells per milliliter and provide food for many surf and beach invertebrates. Their importance to the rocky intertidal is less, because their distribution is restricted to sandy beaches.

Along sandy beaches and mudflats, microalgae, including diatoms, dinoflagellates, and cyanobacteria, inhabit the benthos. Among the best studied are the benthic diatoms. Many benthic diatoms are motile, gliding along grains of sediment. The benthos is a challenging environment for diatoms. If they remain on the surface, turbulence caused by tidal currents can scour them from the bottom. If they remain below the surface, the overlying sediment blocks most of the light, severely inhibiting photosynthesis. Many benthic diatoms have adapted to this environment by migrating vertically within the sediment, spending time at the surface during low tide and migrating downward during high tide. They do not move far, often only a few millimeters, but migrations of a few centimeters have been observed.

Other benthic phytoplankton can be found attached to rocks, seagrasses, macroalgae, and many other surfaces in the intertidal and subtidal regions. These encrusting forms are an important food source for invertebrates.

SEE ALSO THE FOLLOWING ARTICLES

Algal Blooms / Food Webs / Nutrients / Photosynthesis

FURTHER READING

Round, F. E. 1990. *Diatoms: biology and morphology of the genera.* Cambridge, UK: Cambridge University Press.
Reynolds, C. S. 2006. *Ecology of phytoplankton.* Cambridge, UK: Cambridge University Press.
Tomas, C. R., ed. 1997. *Identifying marine phytoplankton.* San Diego, CA: Academic Press.
Williams, P. J. Le B., D. N. Thomas, C. S. Reynolds, eds. 2002. *Phytoplankton productivity: carbon assimilation in marine and freshwater ecology.* Ames, IA: Blackwell Publishing Professional.

POLYPLACOPHORES

SEE CHITONS

PREDATION

ROBERT T. PAINE
University of Washington

The key characteristic of predation in marine intertidal environments is that the victim rarely survives. Unlike other major biological processes—competition, parasitism, herbivory, mutualism—in which both members of the relationship may survive an encounter, predation as defined here usually ends in victim mortality. This generally unilateral outcome reduces ecological ambiguity and has enhanced exploration by controlled manipulation or experimentation of predation's role in structuring marine rocky shores.

THE TROPHIC LEVEL CONCEPT

Predation (as opposed to herbivory, another consumer–victim relationship, but one in which plants are consumed) usually involves pairwise interactions between individuals. Mass predatory attacks (e.g., wolves on elk or killer whales on other whales) are virtually unknown on marine rocky shores. In an overly simplistic sense, consumers of animal tissue are embedded in whole *trophic level* descriptions of communities as "carnivores." Thus, a consumer feeding on a species whose primary diet is composed of plant material, for instance an oystercatcher eating limpets or a snail drilling mussels, becomes a primary carnivore. When the oystercatcher also consumes that snail, it becomes a secondary-level carnivore, thereby blurring the concept of discrete trophic levels. Tertiary and even quaternary predator levels are possible. The point is that these observations are based on animals attacking and consuming living animal prey. When prey exist at more than one trophic level, the act is described as omnivory; if dead items are eaten, the consuming individual becomes a scavenger.

THE ACT OF PREDATION

An attribute of intertidal predation is that exposure at low tide often freezes an act of consumption while in progress. Although there are difficulties in translating such observations into rates of predation or food preference, little doubt exists about the identity of consumer and victim.

Ease of observation, coupled to relevant natural history, has provided the cornerstone on which the ecological contribution of intertidal predation and its many nuances has been developed.

An act of successful predation, as defined by death and consumption of the prey, can be considered as a set of serially connected processes. All require time (t) and involve the expenditure of energy (E), producing the convenient and common currency, E/t, or energy per unit time. The prey must be searched for and, once found, handled. The latter term embodies such activities as pursuit, catching, subduing, and consumption. Predators on rocky shores can be roughly partitioned into useful categories depending on the allocation of time to each of these endeavors. Mammals (e.g., sea otters, rats, bears, and raccoons) and birds (e.g., oystercatchers, sandpipers, gulls, and some ducks) actively search, as do flatworms, nemertean worms, gastropods, octopuses, starfish, and true fishes. Other species, however, fall into an ambush or "sit and wait" category in which little or no time or energy is dedicated to searching, and ultimate success depends on attracting, catching, and subduing the victim. At least one carnivorous chiton and the globally ubiquitous hydroids and sea anemones provide examples.

All intertidal predators express some choice, or food preference, in their prey selection; none consume all available prey types. Many sea slugs (nudibranchs) are prey specialists; that is, they attack a single prey taxon or a group of closely related species. The same appears characteristic of some nemertean worms that can be differentiated by their diets (for instance, either barnacles, annelid worms, or amphipods). On the other hand, many starfish tend to be generalists; that is, their diet is drawn from a taxonomically diverse group. The starfish *Pisaster ochraceus* (Fig. 1) is known to consume almost 50 prey species distributed among four taxonomic classes. Caution, however, is advised when applying these terms. A carnivorous whelk, *Nucella*, has been shown at the population level to eat many different species. However, individually marked whelks within the broader study group at the same site were shown to be specialists, eating a taxonomically restricted prey subset.

PREY RESPONSES TO PREDATION

Intertidal prey illustrate a variety of responses to predation. Prey species may be limited to tidal heights at which their predators cannot forage effectively or are physiologically challenged. This feeding limitation often determines lower limits to mussel beds on eastern Pacific shores, in New Zealand, and both sides of the Atlantic, and it generates a similar pattern for tube worms in eastern Australia

FIGURE 1 The starfish *Pisaster ochraceus* attacking mussels, *Mytilus californianus*. Photograph courtesy of A. W. Rosenfeld.

and barnacles at many sites globally. The resultant zonation is a striking feature of many rocky shores. A less obvious process also permits prey and predator to coexist: simply by becoming too large to be handled, mussels, barnacles and some sea slugs, for instance, cannot be consumed by their major enemies and thus coexist with them.

Other responses of prey to predators are still more subtle. One predatory snail attacks its barnacle prey by jamming a spine on its shell's lip between the plates protecting the prey's mouth. However, a single contact with the predator's mucus is sufficient to alter the barnacle's subsequent growth patterns, resulting in "bent" barnacles with their aperture facing downward and hence much less accessible to the predator. Similarly, the presence of predatory crabs leads to thicker shells in their bivalve or snail prey. Such prey responses, termed induced defenses, can enhance prey survival.

Finally, a rich variety of behaviors and morphologies have evolved in response to the threats posed by predation. Cryptic and warning coloration characterize many intertidal species: sea slugs often match the background color of their sponge prey; some amphipods mimic, by coiling, the shell morphology of small snails; some crabs can flash brightly colored chelae at potential enemies. Escape behaviors are readily observed when prey are exposed to chemical cues from or a brief contact with a potential predator. For instance, some anemones, scallops, and large sea slugs launch themselves into the water column

and swim away, tube worms withdraw their crowns if a shadow passes over, and limpets and other gastropods move directionally away when touched by a starfish tube foot. The specificity of many of these responses suggests a deep evolutionary history in the sense that specific predators elicit predictable responses from their prey.

THE MULTISPECIES PERSPECTIVE

When these one-on-one (pairwise) interactions are assembled into a multispecies perspective, ecologists can construct a food web. Food webs are road maps to specific interactions and therefore the flow of energy in a natural community based solely on feeding relations; they provide the scaffolding around which the ecology of whole communities has been developed. Rocky intertidal shores have played an important role in the development of this perspective, for two basic reasons. First, many of the acts of predation can be directly observed and measured: for instance birds eating molluscs and sea urchins, nemertean worms tracking down and capturing amphipods or worms, starfish consuming a wide range of species, tide pool fishes eating copepods. Second, the solid surface on which these interactions take place has facilitated exploration of the relationships and their consequences by controlled manipulations (experiments). That is, areas with and without predators can be established, and changes in the biotic assemblages that develop compared.

INSIGHTS BASED ON EXPERIMENTAL MANIPULATION

The simplest, though not the earliest, of these manipulations involved the manual removal of one predator species, thereby producing a "predator-free" area, and a comparison with a control or unmanipulated site of the resultant changes in the identity and spatial distribution of the resident species. This work, involving the removal of the generalist predatory starfish *Pisaster*, contributed to an understanding of the dynamics underlying zonation, helped explain why species richness might vary significantly from shore to shore, and led directly to the keystone species hypothesis. The basic result has been repeated anywhere keystone predators are sufficiently abundant to have an influence: in New Zealand with another starfish, in Chile with a carnivorous snail, and in Southern California, where the dominant consumer is a spiny lobster. The keystone concept has matured since its introduction (1969), and its relevance to management and conservation is well established. Only a few species in any ecosystem are recognized as keystones, and their identification has proven challenging. Keystones are species whose ecological impact is disproportionately great when compared to their abundance or mass. Thus in certain forests, no one doubts the importance of, say, oaks, but their production scales with their abundance. On the other hand, rinderpest virus and *Pisaster* play major roles in the appearance and function of their respective communities despite their comparatively low biomass. The rocky intertidal examples offer a further refinement: Any species capable of controlling the distribution and abundance of a competitively superior prey is potentially a keystone species, because the consequences of this "control" will be expressed as increased abundance of the many other species requiring space whenever intertidal real estate is limiting.

The spiny lobster manipulation employed a technique characteristic of rocky intertidal experiments: exclusion barriers. Among the earliest of these were 100-cm² steel mesh cages designed to protect two barnacle species from predatory snails. The experiment revealed that the less preferred species flourished in the predator's presence, but lost when the dominant competitor was protected. In the spiny-lobster-as-predator example, larger cages protected the algal turf and recently settled mussels from predation, resulting in a dramatic switch in community composition from the turf to an expanding bed of mussels. A previously underappreciated predatory impact of foraging birds (oystercatchers, gulls) has been revealed by protecting prey with vinyl-covered letter baskets attached to the rocky surface: Birds can accelerate the success and dominance of mussels by preferentially consuming one of their competitors.

A variety of other manipulations illustrate the experimental benefit of working on solid, rocky surfaces. Rocks with established barnacles can be transplanted from secure-from-predator heights to lower zones, and barnacle survival measured. Barriers, positioned around refuges, establish that small crevices that protect predators at low tide also lead to adjacent bands in which their prey are absent. Prey can be transplanted to areas of varying predator density, and once establishment is ensured and a protective plastic mesh removed, relative survival can be measured. Prey also can be "tethered"—that is, attached by monofilament line to specific spots—with or without their primary predator, and their relative survival quantified. All these manipulations or procedures require a solid surface on or to which the experimental organism can be attached to or excluded from. The conclusions derived from such experiments have been of fundamental significance to understanding the role predation has played in the dynamics and organization of natural communities.

SEE ALSO THE FOLLOWING ARTICLES

Competition / Food Webs / Herbivory / Oystercatchers / Vertebrates, Terrestrial / Zonation

FURTHER READING

Bertness, M. D. 1999. *The ecology of Atlantic shorelines*. Sunderland, MA: Sinauer.

Branch, G., and M. Branch. 1981. *The living shores of southern Africa*. Cape Town: Struik.

Connell, J. H. 1961. Effects of competition, predation by *Thais lapillus*, and other factors on natural populations of the barnacle *Balanus balanoides*. *Ecological Monographs* 31: 61–104.

Lewis, J. R. 1964. *The Ecology of Rocky Shores*. London. The English Universities Press.

Menge, B. A., and G. M. Branch. 2001. Rocky intertidal communities, in *Marine community ecology*. M. D. Bertness, S. D. Gaines, and M. E. Hay, eds. Sunderland, MA: Sinauer, 221–251.

Paine, R. T. 1966. Food web complexity and species diversity. *The American Naturalist* 100: 65–75.

Paine, R. T. 1994. *Marine rocky shores and community ecology: an experimentalist's perspective*. Oldendorf/Luhe, Germany: Ecology Institute.

Power, M. E., D. Tilman, J. A.Estes, B. A. Menge, W. J. Bond, L. S. Mills, G. Daily, J. C. Castilla, J. Lubchenco, and R. T. Paine. 1996. Challenges in the quest for keystones. *BioScience* 46: 609–620.

Rosenfeld, A.W., and R. T. Paine. 2002. *The intertidal wilderness*. Berkeley: University of California Press.

Tollrian, R., and C. D. Harvell. 1999. *The ecology and evolution of inducible defenses*. Princeton, NJ: Princeton University Press.

PREDATOR AVOIDANCE

GEOFFREY C. TRUSSELL

Northeastern University

PATRICK J. EWANCHUK

Providence College

Predator avoidance comprises the various strategies that prey utilize to reduce their risk of consumption by predators. Predators expend great effort to catch their next meal and, it has been theorized that prey should try inexorably harder to avoid being eaten (the "Life/Dinner Principle"). The term *predator avoidance* thus embodies the failure of prey to submit passively to a fate on the dinner plate. Both plants and animals have evolved a wonderfully diverse suite of adaptations that reduce their risk of being eaten. Such strategies of evasion are widespread in nature, and their prevalence in marine habitats such as the rocky intertidal zone is refining our views of how predators may shape the dynamics of communities and populations in this system.

HOW PREY DETECT PREDATORS

The ability to detect predators is critical to the success of prey in avoiding predation. In marine environments, prey often utilize tactile or visual cues that alert them to the presence of predators, but growing evidence suggests that chemoreception may be the most common approach to detect predators. In the context of predator avoidance, chemoreception is the detection of waterborne chemicals released by predators that disclose their presence to prey in the local environment. The number of organisms known to rely on chemoreception to assess predation risk is rapidly growing as scientists continue to explore the nature of predator–prey interactions in marine environments. Several predatory species, including sea stars, turtles, crabs, snails, fish, and lobsters, are known to release chemical cues that induce predator avoidance strategies in an equally diverse array of prey (barnacles, bryozoans, sea anemones, snails, scallops, mussels, sea urchins, sand dollars, sea cucumbers). Of equal interest are studies showing that such signaling need not involve what we typically view as predators and prey. For example, marine algae exposed to chemical cues released by herbivorous snails initiate their own set of chemical defenses that act to make their tissues less palatable to potential grazers.

Despite the numerous studies that document prey detection of predator risk cues, we know very little about the chemical composition of these cues, how long they reside in the environment after being released, and over what distances they can influence prey populations. These questions must be a focal area of future research to provide a better understanding of the influence of chemical signaling on marine population and community dynamics.

PREY AVOIDANCE STRATEGIES

After sensing predator risk cues, prey may adopt either behavioral or morphological strategies that reduce their risk of being consumed. The first lines of defense are behavioral responses, such as reducing conspicuous feeding activities; fleeing risky habitats containing predators; or, if this is not possible, seeking refuge (cracks and crevices) within risky habitats. These responses are quite common simply because they can be activated quickly, thus producing immediate benefits.

Equally important are those responses involving relatively rapid morphological changes by prey species that are exposed to predator risk cues. After detecting predator risk cues, prey can modify their developmental or growth trajectories to augment their level of morphological defense. Such morphological changes are a form of phenotypic plasticity known as inducible defenses. Compared to behavioral responses, morphological plasticity often requires more time to produce adequate levels of defense. The effectiveness of morphological plasticity is evidenced by the many species that utilize it. For example, several species of rocky intertidal snail (*Littorina littorea, L. obtusata, Nucella lapillus*) thicken their shells when exposed to seawater containing risk cues from predatory crabs (e.g., *Carcinus maenas*, Fig. 1), and this thickening is magnified if crab cues are accompanied by alarm cues from damaged conspecifics. Because thicker shells are stronger and thus more difficult for crabs to crush, these thickened shells are adaptive in reducing the risk of crab predation. Mussels—a culinary favorite of snails, crabs, and humans—also employ shell thickening and changes in shell shape when exposed to cues from predators. In addition to thickening, mussels also produce more and thicker byssal threads, which they use to attach to rocky surfaces. Presumably, reinforcement of these threads makes it more difficult for crabs to pull mussels away from the rock.

A final, elegant example of predator-induced morphological plasticity comes from studies on the rocky shores of the Gulf of California. Barnacles (*Chthamalus anisopoma*) on these shores typically produce the familiar volcano-shaped

FIGURE 1 Waterborne risk cues released by the predatory green crab, *Carcinus maenas* (A) can induce strong changes in the behavior and morphology of snails such as *Littorina obtusata* (B), *L. littorea* (C), and *Nucella lapillus* (D). Photographs by the author.

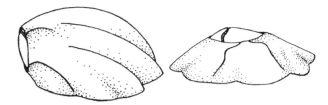

FIGURE 2 Shell morphs of the barnacle *Chthamalus anisopoma*. The "bent" morph (left) is induced by cues released by the predatory snail *Acanthina angelica*, whereas the "conic" morph (right) is produced in the absence of snail cues. Drawing courtesy of Curt Lively.

morphology ("conic" morph) in the absence of the predatory snail *Acanthina angelica* (Fig. 2). However, if juvenile barnacles are exposed to cues from this snail early in their development, they undergo a radical shift in their morphology that leads to a barnacle that is best described as bent over (the "bent" morph). Like the increases in shell thickness induced by crabs, this plasticity is adaptive because bent morphs, which are oriented parallel, rather than perpendicular, to the rock that the barnacles are attached to, are more difficult than conic morphs for *A. angelica* to consume (Fig. 2). This plasticity has evolved as a solution to survivorship and reproductive trade-offs that often exist in environments with variable predation risk. Although the bent morph is less vulnerable to snail predation, this shape also reduces the number of offspring produced by bent barnacles because it reduces the volume of the brood chamber in which eggs are kept before their release into the water column.

THE ECOLOGICAL CONSEQUENCES OF AVOIDANCE

The trophic cascade concept has been one of the most influential ideas in the field of ecology. According to this idea, predators can have strong indirect effects on the abundance of primary producers (plants, algae, phytoplankton) that form the base of most marine food webs, by regulating herbivore density via consumption. For example, in subtidal habitats of the North Pacific, sea urchin grazing can produce large expanses of habitat called urchin barrens, which are devoid of kelp and other macroalgae. However, when sea otters are present, their consumption of sea urchins releases kelp and other algae from urchin grazing. Thus, the cascading effects of sea otters on kelp, which are mediated by reductions in urchin density, can promote the establishment of kelp forest that serves as vital habitat for many species. Recent work reveals that these cascading trophic interactions also operate in four-tiered food chains. In Alaska, killer whales feeding on sea otters reduce the impact of otters on urchins, thereby allowing urchins unfettered access to kelp. The outcome of the cascade therefore

reverses, and urchin barrens become more prevalent despite the presence of sea otters.

Trophic cascades emphasize the density-based consequences of species interactions on overall community dynamics. This concept has been applied to algal community dynamics and diversity on rocky shores where algal diversity in tide pools is strongly influenced by the density of resident herbivorous snails *(Littorina littorea)*. When *L. littorea* density is low, algal diversity is low and dominated by competitively superior ephemeral species that are the preferred food of this snail. At high snail densities, algal diversity is also low because snail grazing rapidly eliminates ephemeral species and only tough, less preferred species persist. Only at intermediate snail densities is algal diversity high because snail grazing is sufficient to prevent ephemeral species from competitively excluding perennial species. Hence, the central question becomes, "What regulates *L. littorea* density in rocky-shore tidepools?" Although many factors are operating, it has been suggested that a trophic cascade caused by crab predation might be the principal driving agent: The direct consumption of snails by crabs indirectly leads to an algal community dominated by ephemeral algae.

Although density-mediated cascades are clearly important to the dynamics of many marine communities, recent research has focused on the importance of predator avoidance strategies in initiating this phenomenon. Trait-mediated cascades arise after predators alter prey traits, such as behavior, instead of prey density. Hence, this perspective emphasizes the "ecology of fear" rather than the ecology of density. As discussed in the previous section, when prey detect predators, they may leave the local habitat or seek local refugia within the habitat that are more difficult for predators to access. Regardless of the avoidance strategy, the end result is the same: The feeding rates of prey species are greatly reduced.

Recent work in tidepools provides empirical support that trophic cascades can be generated by green crab risk cues, which reduce snail density via avoidance behaviors. Because of this strong behavioral response, the abundance of ephemeral green algae was similar for pools having either direct predation or predation risk. Hence, snail emigration caused by fear of predation appears to be just as important as the actual consumptive effects of crabs in regulating snail density and producing the cascade within tidepools. Although this cascade is still ultimately driven by snail density, it is important to note that trait-mediated cascades are likely to be stronger and operate faster, because the entire prey population can simultaneously respond to risk. In contrast, the consumptive effects of crabs are expected to take more time to develop and may be relatively weak, especially in the presence of alternative, preferred prey.

The risk of predation can clearly exert a profound influence on the behavior, morphology, and life history of prey species inhabiting rocky shores. Moreover, because predator avoidance can initiate trophic cascades by strongly modifying the intensity of interactions between prey and their food resources, one must consider how such avoidance influences the structure and dynamics of ecological communities before using density-based cascade theory to manage natural ecosystems.

SEE ALSO THE FOLLOWING ARTICLES

Barnacles / Camouflage / Chemosensation / Food Webs

FURTHER READING

Dawkins, R., and J. R. Krebs. 1979. Arms races within and between species. *Proceedings of the Royal Society of London, B* 205: 489–511.

Estes, J. A., M. T. Tinker, T. M. Williams, and D. F. Doak. 1998. Killer whale predation on sea otters linking oceanic and nearshore ecosystems. *Science* 282: 473–476.

Lively, C. M. 1986a. Predator-induced shell dimorphism in the acorn barnacle *Chthamalus anisopoma*. *Evolution* 40: 232–242.

Lively, C. M. 1986b. Competition, comparative life histories, and maintenance of shell dimorphism in a barnacle. *Ecology* 67: 858–864.

Lubchenco, J. L. 1978. Plant species diversity in a marine intertidal community: importance of herbivore food preference and algal competitive abilities. *American Naturalist* 112: 23–39.

Kats, L. B., and L. M. Dill. 1988. The scent of death: chemosensory assessment of predation risk by prey animals. *EcoScience* 5: 361–394.

Werner, E. E., and S. D. Peacor. 2003. A review of trait-mediated indirect interactions in ecological communities. *Ecology* 84: 1083–1100.

Trussell, G. C., P. J. Ewanchuk, and M. D. Bertness. 2002. Field evidence of trait-mediated indirect interactions in a rocky intertidal food web. *Ecology Letters* 5: 241–245.

Trussell, G. C., P. J. Ewanchuk, and M. D. Bertness. 2003. Trait-mediated indirect effects in rocky intertidal food chains: Predator risk cues alter prey feeding rates. *Ecology* 84: 629–640.

Trussell, G. C., P. J. Ewanchuk, M. D. Bertness, and B. R. Silliman. 2004. Trophic cascades in rocky shore tide pools: distinguishing lethal and nonlethal effects. *Oecologia* 139: 427–432.

PROJECTILES, EFFECTS OF

ALAN SHANKS

University of Oregon

In the eyes of intertidal ecologists, one of the primary variables defining the intertidal environment is the degree of wave exposure. Organisms living at sites with strong wave action experience higher current speeds, greater accelerative forces, and higher shock pressures. These forces help to define the community of organisms found in an intertidal habitat. What is not well appreciated is that breaking waves can hurtle projectiles, greatly increasing the lethality of waves. The most common projectiles are woody debris (e.g., logs and driftwood) and rocks (e.g., pebbles, cobbles, and even boulders). Few studies have addressed the role of projectiles in shaping intertidal communities.

WOODY DEBRIS

The destructive role of woody debris, in the form of logs and driftwood, has been studied in the Pacific Northwest in North America. This is an area of dense forests, the source of the logs and driftwood. The landward sides of beaches in this part of the world are often covered with large piles of logs and driftwood. Logs with squared-off ends have come from logging operations, while logs with root balls have been washed into the sea or a river by erosion. Logs and driftwood, because they float, enter the intertidal zone on the tops of waves as if they were surfing toward the shore. Driftwood, because of its smaller size, tends to be swept rapidly through the intertidal zone by the waves. Logs tend to run aground in the intertidal zone, and each passing wave drives the log forward like a battering ram.

Because of their size, logs propelled by even small waves can be highly destructive. In areas where log battering is common, one can often find masses of large wood splinters jammed into cracks in the rocks (Fig. 1). Researchers have measured the frequency of log battering by making thickets of nails embedded partway into the rock substrate of intertidal zones and following their destruction (bending) over time. On a small spatial scale, logs tend to accumulate along some sections of

FIGURE 1 Wood splinters embedded in cracks in an intertidal boulder. Photograph by the author.

coast and be pushed away from others as a result of local topography and currents. Thus, on the scale of hundreds of meters to kilometers, sites with similar wave exposure may have very different exposure to log damage. Within a given stretch of shore, there appears to be a tendency for logs to ultimately wash into surge channels, where the damage from their battering may be concentrated.

On a very large spatial scale (hundreds to thousands of kilometers) the amount of log damage will vary with the availability of logs. For example, along the coast of Washington State in the United States, a heavily forested area, log damage is a common feature of the intertidal zone. In contrast, in southern California forests are nearly absent, logs are uncommon on beaches, and log damage is rare.

There may also be temporal variability in the amount of log damage. A log at sea ultimately becomes water-logged, sinks, and is no longer a threat to intertidal organisms. Alternatively, a log may be washed ashore and become stranded at the high-tide line. These latter logs can reenter the ocean either at extreme high tides (spring tides) or during storms. This temporal variability in the abundance of floating logs may lead to a temporal variability in log damage to the intertidal zone.

The force of log battering is extreme. Organisms in the path of a battering log are smashed unless they are protected within cracks and crevices in the rock. The destruction caused by log damage is a form of disturbance. Where the battering occurs, the organisms are destroyed, and the bare rock surface is available to larvae settling from the plankton. In this way, competitively dominant organisms, which can tend to control space in the intertidal zone, are removed, opening space to other organisms that might otherwise be excluded. A good example of this is the effect of log damage on intertidal mussel beds. Mussels tend to dominate space at certain heights in the intertidal zone. Log bashing removes patches of mussels, and these patches can then be enlarged by wave action. The longevity of a patch varies with its size. Large patches last long enough that one can find unique assemblages of organisms that otherwise would be competitively excluded from the mussel zone. In this way, the disturbance from log damage can increase the diversity of the intertidal community.

ROCKS

Logs are not the only projectiles the ocean can "throw" at the shore. Pebbles, cobbles, and even boulders can become projectiles. Pebbles and cobbles traveling at even moderate velocities can be lethal projectiles to small organisms. Currents within a breaking wave or bore are quite high, high enough to cause the mobilization of pebbles and their saltation (i.e., movement by "hopping" or "leaping" across the bottom). On a pebble beach, if one watches waves rush up the shore, one will often see pebbles tossed up into the air out of the wave. This will happen even in relatively small waves (0.5-m height). The intense turbulence of a breaking wave may throw pebbles into the air, or pebbles embedded in wave-generated currents may strike other pebbles and, like a bad shot in a game of pool, propel the latter pebbles out of the water. Pebbles have been recorded with the momentum at impact ranging from 0.2 to 4.0 N s (newton seconds). An impact with a momentum of 1.25 N s can be generated by a 0.5-kg rock traveling at 2.5 m s^{-1}. Collisions by pebble-sized projectiles embedded in the flow or thrown out of the water and falling back to the ground are energetic enough to dent aluminum targets set in the intertidal zone (Fig. 2). Experimentally produced impacts, which produced similar-sized dents in the targets to those made by pebble impacts, were energetic enough to kill intertidal organisms (e.g., barnacles and limpets). Impacts from even marble-sized pebbles (1 to 2 cm in diameter, about 10 g) can be energetic enough to kill organisms. By placing targets at a number of locations within an intertidal zone and following the production of impacts over time, researchers gained insight into the spatial and temporal production of wave-borne rock damage.

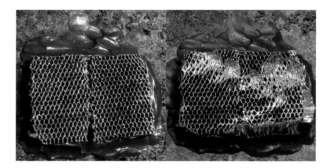

FIGURE 2 An aluminum honeycomb target shortly after placement in the intertidal zone (left) and after a week of exposure to wave-borne projectiles (right). Photograph by the author.

Impacts are uncommon in rocky shelf habits, where small rocks that could become projectiles are uncommon, but impacts are frequent in boulder fields, where projectiles are abundant. Impacts are more common near the high-tide line, perhaps because larger hydrodynamic forces tend to occur higher in the intertidal zone. By looking carefully at boulders high in the intertidal zone, one can see the effects of impacts on the surface of the boulders. As a result of oxidation and the growth of microscopic algae, the surfaces of boulders are usually darker than the underlying rock. When a projectile strikes a boulder, it chips off a bit of this darker outer layer. The surfaces of boulders high in the intertidal zone are often covered with these chip marks, indicating that they are often hit by projectiles (Fig. 3).

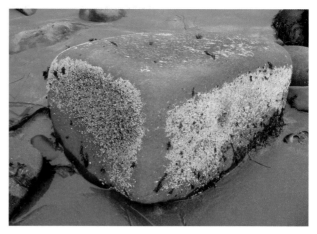

FIGURE 4 Distribution of barnacles growing on an intertidal boulder. The distribution of barnacles is probably due to the distribution of projectile impacts. Where impacts are frequent (along sharp edges and the top of the boulder), the rock is bare, but where impacts are less frequent, the barnacle cover is complete. Photograph by the author.

FIGURE 3 Close-up photograph of the surface of a boulder high in the intertidal zone. The dark coloration of the rock is due to either oxidation or the growth of microalgae. The light patches are due to impacts by projectiles chipping off pieces from the surface of the boulder. Photograph by the author.

On a yet smaller scale, within a boulder field, impacts were least common on the face of boulders, far more common on the tops of boulders, and are extremely common at sharp edges (Fig. 4). This distribution of impacts is mirrored by the distribution of damage to organism growing on rocks; the cover of organism is often continuous on the face of boulders. On the tops of boulders, numerous gaps are often present in the cover, and on the edges of boulders there are often few or no organisms. These observations suggest that in habitats with abundant potential projectiles (e.g., boulder fields), impacts from wave-borne projectiles are probably a major factor molding community structure and composition and the small-scale distribution of organisms.

In habitats where impacts are common, it is easy to find animals that appear to have experienced sublethal

impacts—that is, impacts that have only damaged their shell (Fig. 5). Sublethal impacts to limpets and barnacles appear to generate species-specific patterns of damage. For example, impacts to *Lottia scabra* tend to crush the apex of the shell, whereas impacts to *Lottia gigantea* and *Lottia persona* tend to either break off chips from the margin of the shell or produce cracks from one side of the shell to the other (Fig. 5).

FIGURE 5 Close up photograph of the shell of a *Lottia gigantea*. Note the large chip in the shell, which has been repaired with new growth. Chips like this one are produced by impacts from projectiles. Photograph by the author.

If waves are large and energetic enough, even large rocks and boulders can become projectiles. One can, at times, see the consequences of this in the field. After large storms one can sometimes find an abundance of large cobbles and rocks scattered above the high-tide line. In the right geologic setting, one can at times find boulders in the intertidal zone of a different rock type from that of

FIGURE 6 The intertidal zone on the north side of Cape Blanco, Oregon. The large, smooth boulders are made of serpentine, while the surrounding rock from which the intertidal platform is formed is sandstone. The smooth boulders are from a source outside the intertidal zone and were carried into the intertidal zone by waves. Photograph by the author.

the surrounding shore (Fig. 6). These errant boulders were transported by waves from a source outside the intertidal zone. One can imagine the destruction they must cause as waves propel them through the intertidal landscape.

Engineering studies of the failure of stone breakwaters describe the conditions under which large rocks and even boulders can be mobilized. What engineers report is that breakwaters composed of large boulders fail (i.e., the boulders are washed away) not when a storm is approaching shore and the waves are large, often the largest of a storm event, and with a long period. Rather, the breakwater fails as the storm arrives at the coast and the breakwater is subjected to large waves with short periods. As the storm approaches shore and the breakwater is hit by very large, long-period waves (swells), it is subjected to the highest wave-generated current velocities. The boulders remain in place during this phase of a storm, suggesting that the drag forces generated by the high current velocities are inadequate to mobilize boulders. As the storm hits the coast, the breakwater is impacted by large, short-period waves

(seas), and these waves can mobilize boulders. During this phase of the storm the rapid changes in current direction caused by these large, short-period waves generate high accelerative forces, and it is these forces that ultimately can move even large boulders.

Traditionally, most rocky-intertidal ecological studies have been conducted on rocky platforms where the substrate is composed of stable basement rock. Projectiles tend to be uncommon in these types of habitats, and damage from projectiles is usually light. Perhaps for this reason the role of projectiles in intertidal ecology has received little attention. Boulder-field intertidal zones are as common as, if not more common than, rock platforms. In boulder fields, projectiles are abundant, and the evidence of damage due to projectiles is obvious. Here projectiles may be one of the most important defining physical forces in the habitat.

SEE ALSO THE FOLLOWING ARTICLES

Coastal Geology / Habitat Alteration / Hydrodynamic Forces / Wave Exposure

FURTHER READING

Carstens, T. 1968. Wave forces on boundaries and submerged bodies. *Sarsia* 34: 37–60.

Dayton, P. K. 1971. Competition, disturbance, and community organization: the provision and subsequent utilization of space in a rocky intertidal community. *Ecological Monographs* 45: 137–159.

Levin, S. A., and R. T. Paine. 1974. Disturbance, patch formation, and community structure. *Proceedings of the National Academy of Sciences of the USA* 71: 2744–2747.

Shanks, A. L., and W. G. Wright. 1986. Adding teeth to wave action: The destructive effects of wave-borne rocks on intertidal organisms. *Oecologia* 69: 420–428.

Sousa, W. P. 1979. Experimental investigations of disturbance and ecological succession in a rocky intertidal algal community. *Ecological Monographs* 49: 227–254.

PROTISTS

GERARD M. CAPRIULO

Saint Mary's College of California

JOHN J. LEE

City College, City University of New York

Prokaryotic (bacteria and cyanobacteria) as well as eukaryotic (protists; micro "algae" and "protozoa") microorganisms abound in all of earth's habitats from the benign to the extreme. With their fast growth and high metabolic and respiration rates, they are important components of all ocean systems and subsystems. Protists (as well as bacteria and cyanobacteria) are integral components of marine food webs as primary producers and consumers, and as nutrient recyclers of biomass and detrital materials.

HABITATS, ECOLOGICAL CONSIDERATIONS, AND THE PHYSICAL ENVIRONMENT

In the rocky intertidal zone and tidepools, protists are found as plankton; in relationship with organic and inorganic suspended particles, sediments, and rock surfaces; and as exobionts and endobionts of other organisms. In response to the dynamic and continually changing nature of the intertidal zone and tide pools, attached and trapped microorganisms must be adapted to the associated large periodic fluctuations in salinity, temperature, desiccation levels, and solar irradiance. When these habitats are marked with robust detrital decay or are isolated and characterized by high photosynthetic rates, adaptations to wide changes in pH are also necessary. Because of diffusion dynamics and their small sizes relative to the viscosity of water, protists experience the rocky intertidal zone and tidepools as a series of subdivided—but interconnected microhabitats, each one of which promotes their success or demise based on the microhabitat's individual microscale characteristics. The success of protists in intertidal rocky shores and their tidepools thus depends on their physiochemical tolerances, abilities to compete for resources, nutrient availability, the organic chemistry and molecular properties of the habitat, the presence of prokaryote microbes, and, to a major extent, substrate and sediment characteristics. For their part, tidepools can be considered islands ready to recruit "r–selected" organisms (those with high rates of reproduction and growth) as they are formed. Since colonization is stochastic, very often nearby different tide pools have blooms of different protists and different food webs of protists and micrometazoa.

Shorelines are born of past and present geological forces acting on above- and below-sea-level land masses, which are also altered by air, sea, and limnologically mediated physical and chemical (e.g., erosion, depositional forces, and chemical reactions) and biological interactions. Each shoreline has its own history and unique set of sediment properties, including particle chemical characteristics, crystalline structures, and size distribution. The latter ranges from the finest sediments (i.e., clays and muds), to sand (fine to coarse), and rocks (from gravel to smooth and jagged boulders and mountain ranges).

As tides advance and retreat, the rocky intertidal zone undergoes rhythmic variations in exposure to extreme physical, chemical, geological, and biological factors. Because each point is exposed differently depending on its position and the slope of the shore grade, the littoral zones of rocky shores generally show more pronounced elevation gradients and associated permanent horizontal banding patterns of life forms than are found in other marine habitats. The specific distribution dynamics of life forms in these habitats vary with their respective sizes and are primarily driven by physiochemical size-scale parameters.

The rocky intertidal world inhabited by microbes is very different from that which we experience. Because of their small size, bacteria and protists live life at very low Reynolds numbers. This means that viscosity plays a key role in their world. Living in the water to them is the equivalent of living in a sea of cold molasses syrup for humans. The forces the protists exert to move often are barely able to overcome the viscous forces of the water in which they live. Also, their diminutive sizes means they are much closer in size to that of the molecules they encounter. This influences their environmental directional world and their perceptions and responses to stimuli. For this reason we must carefully consider questions of scale, diffusion, micro-level habitat structures, and environmental patchiness when engaging ecological questions related to bacteria, cyanobacteria, and the protista. From this perspective, a rocky shore is, for microbial eukaryotes, much more than it appears, at first glance, to the human eye. It contains a host of unique habitats that include spray; midlittoral and infralittoral zones; tidepools; soft, hard, and porous rock zones and their associated cracks and crevices; sand and mud zones; seaweed surfaces; algal and bacterial mats; planktonic zones; detritus and dead organism remains; bird and other animal droppings; and the bodies of live organisms.

THE PROTIST PLAYERS OF ROCKY SHORES

In the late twentieth century, detailed electron microscopic and molecular genetics studies changed scientists' concepts of the phylogenetic relatedness of protist groups from earlier conceptualizations, which divided protists into "protozoa" and "algae." At present many protist phylogenetic relationships are not neatly resolved, and many discrete, monophyletic, and often seemingly unrelated groups of protists are recognized that could be raised to the level of phyla.

Associated with rocky shoreline habitats are a plethora of both prokaryotic and eukaryotic marine microbes, many of which can be sampled relatively easily (Table 1). The protist microorganisms are an extensive, taxonomically

TABLE 1

Some Simple Techniques for Sampling the Protists of Rocky Shores and Tidepools

1. Gently grasp a small piece of seaweed, place into a test tube with filtered seawater, vigorously shake to dislodge epiphytic protist community.

2. Place a test tube close to a submerged rock and scrape a sample from the rock into the test tube.

3. For sandy or muddy benthos use a small piece of plastic tubing attached to a 5-mL or 10-mL syringe (made more quantitative by pushing a plastic ring into the benthos and then sucking the surface with the tube and syringe).

4. Recruit protists to microscope slides using a grooved rubber stopper that will hold several slides. Place on the benthos for a few days to a week, then cover one side of slide with a cover-glass, clean the other side, observe via microscopy.

5. For the water column (a) tie polyfoam sponges to a fishing line and weight the other side with a fishing sinker. The floating sponge will be colonized by the microorganisms in the water column, which are then squeezed into a beaker, or (b) pass the sample through a membrane filter (0.2 or 0.45 μm for bacteria and protists). Filters with larger pore sizes (e.g., 3 μm, 5 μm, 8 μm) can be used to selectively capture by size (note, some protists are too fragile for sampling via filtration). A hand-held vacuum pump with pressure gauge, 5 psi or less, and stopping the filtering before the membrane is dry, is recommended.

6. For special protist collections additional preparation may be necessary.

FIGURE 1 Scanning electron micrographs of a representative sampling of protist groups that inhabit rocky intertidal zones and tidepools: (A) centric diatom, *Asterophalus*, size 30 μm; (B) pennate diatom, *Nitzschia*, size 10.5 μm, (C) pennate diatom, *Amphora*, size 20 μm; (D) Ciliophora, *Suctoria*, size 80 μm; (E) Ciliophora, Spirotrichea, Tintinnida, *Codonellopsis* lorica, size 133 μm; (F) pennate diatom, *Cocconeis*, size 9 μm; (G) centric diatom, *Cyclotella*, size 18 μm; (H) Ciliophora, Heterotrichea, *Condylostoma*, size 115 μm; (I) Ciliophora, *Euplotes* (ventral side), size 40 μm; (J) Prymnesiida, coccolithophorid, *Cruciplacolithus* (coccosphere with interlocking placoliths), size 10 μm; (K) ramicristate amoeba, *Vexillifera*, size 34 μm; (L) Volvocida, *Dunaliella*, size 8 μm; (M) Ciliophora, Scuticociliatia, *Pseudocyclidium*, size 10 μm; (N) Granuloreticulosa (Foraminifera), *Elphidium*, size 350 μm; (O) Granuloreticulosa (Foraminifera), *Quinqueloculina*, size 160 μm; (P) Dinozoa, *Ceratium*, size 150 μm; (Q) Dinozoa, *Dinophysis*, size 75 μm; (R) Granuloreticulosa (Foraminifera), *Textularia*, size 800 μm; (S) Granuloreticulosa (Foraminifera), *Ammonia*, size 1300 μm; (T) Granuloreticulosa (Foraminifera), *Millammina*, size 850 μm; (U) Dinozoa, *Amphidinium*, size 20 μm. Images by J. J. Lee.

diverse group of photosynthetic and heterotrophic flagellates, nonmotile photosynthetic cells, gliding photosynthetic cells, testate and naked amoebae, and ciliates. It is beyond the scope of this article to detail all of them. However, it is important to stress that rocky shores and their tidepools support a rich community of these protists (Fig. 1).

PROTISTS TOO BIG TO PLACE BETWEEN A SLIDE AND A COVER GLASS

A few groups of protists (Granuloreticulosa, Ciliata, and Dinozoa) that are common in tidepools and the intertidal zone include organisms that cannot be compressed between ordinary microscope slides and cover glasses. These organisms should be examined using a good-quality dissecting microscope or, without cover glasses, under a low-power compound microscope.

Phylum Granuloreticulosa (Foraminifera)

Most members of this phylum of testate amoebae (see Fig. 1N, O, R, S, T) are larger than 0.1 mm. Some of the largest found in tidepools reach >3 mm. Commonly called foraminifera, this group has anastomosing (spider web–like) granular pseudopods with bidirectional streaming. Most of the common ones build either calcareous tests (shells) or tests made by agglutinating particles. Agglutinated tests are easiest to detect in a dissection microscope when they are dry. Foraminifera with agglutinated tests are common in the littoral and supra-littoral zones and in the deep sea. Test composition (calcareous or agglutinated) is a major taxonomic criterion. Calcareous tests include those with or without pores. Most foraminifera are multichambered, adding chambers as they grow. The pattern of test growth is a character used to separate genera. Chambers can be added linearly to produce monoserial, biserial, or triserial tests, in a flat spiral, or in a raised spiral. Some grow to look like a coil or a paper clip. Location of apertures, places where the pseudopods emerge from the test, and ornamentation are other important characters. Foraminifera are usually very patchy in their distribution in the field. As is true for many protists, their distribution may be restricted to particular zones, regions, or habitats. Two large (1–3 mm) foraminifera, *Amphistegina* and *Heterostegina*, which bear orange-brown symbionts, are common in Hawaiian tidepools and the shallow subtidal zones of many well-illuminated tropical and subtropical seas. Star sands (calcarinid foraminifera) are found in similar habitats in the Pacific.

Phyla Ciliata and Dinozoa

Only a fraction of the members of these phyla found in tide pools and the littoral zone are too large to be examined in the compound microscope, so they will be treated in the next section.

PROTISTS EASILY IDENTIFIED USING A COMPOUND MICROSCOPE

Before the wide employment of electron-microscopic and molecular techniques, protists were identified by placing them in groups based on whether they were photosynthetic or not and whether they were motile or not. If they were motile, the type of motility (flagellated, amoeboid, ciliated, or gliding) became an important separating factor. The nature of the cell covering (tests, frustules, scales, pellicle) or internal skeletal elements (spines, axonemes) are also major factors separating groups of protists in tidepools and littoral zones. Following are the characteristics of some of the major groups of protists found in these habitats.

Phylum Bacillariophyta (Diatoms)

Diatoms (Fig. 1A, B, C, F, G) have exquisitely beautiful silica glass shells (frustules) that are best seen when the rest of the organism is digested away. They are extremely abundant primary producers in all aquatic habitats, including the ones of present interest, and are important food sources for many organisms. Diatoms may be round and radially symmetrical (Centric Bacillariophyta) or elongate with a bilateral plane of symmetry (Pennate Bacillariophyta). The frustule is made of two valves, which fit together like a Petri dish and its cover. The centric forms tend to dominate the plankton in the water column, while the pennate forms (some of which attach and others actively glide) are found in the benthos. Some of the centric diatoms are globelike, while others are flattened. Some of the most common genera of diatoms found in tidepools include *Cyclotella* (Fig. 1G), *Coccinodiscus, Navicula,*

Nitzschia (Fig. 1B), *Amphora* (Fig. 1C), and *Synedra.* Other golden-pigmented (Chromista) photosynthetic protists with silica skeletons include the Prymnesiophyta and Silicoflagellata.

Phylum Dinozoa (Dinoflagellates), Also Known by Phycologists as Division Dinoflagellata

Dinoflagellates (Fig. 1P, Q, U) are very common in most aquatic environments and quite abundant in tidepools and the intertidal. Most dinoflagellates are photosynthetic and important in food webs. The nucleus of dinoflagellates is a unique feature of the group, because it contains chromosomes that are usually condensed during all phases of the cell cycle. Most dinoflagellates contain two flagella that are quite different in their structure. These produce a corkscrew-style locomotion. The flagella emerge from the ventral side of the cell. The flat ribbonlike flagellum, which causes the cell to rotate on its axis, wraps around the body in a groove known as a cingulum or girdle. The longitudinal flagellum lies in a groove (sulcus) that runs posteriorly from the flagellar pore to the posterior end of the cell. The position of the cingulum (anterior, central, posterior, or spiraling) is an important character used to separate genera of some groups. Dinoflagellates have a distinct theca, the structure of which is seen only by transmission electron microscopy (TEM). This complex cell covering, known as the amphiesma, consists of inner and outer membranes, between which lie a series of flattened vesicles. At the light microscopic level, it is possible to recognize thecal plates, which are formed in the vesicles of armored species. Dinoflagellate groups are separated from each other by whether they have recognizable plates (Gonyaulacales, Peridinales, Dinophysiales, Prorocentrales) or not (Syndales, Phytodiniales, Noctilucales, Bastidinales, Desmocapsales, Suessiales, and Gymnodiniales), and by the plate patterns of those dinoflagellates that have them. One might expect almost any of the >100 common genera to be found in the waters of tidepools and littoral zones. Species of the dinoflagellate *Amphidinium* (Fig. 1U) are found swimming between the sedimentary particles of all intertidal habitats and on the benthos of every tidepool. Other unarmored common genera are *Gymnodinium* and *Gyrodinium.* Most common among the armored genera that might be found are *Peridinium, Gonyaulax,* and *Dinophysis* (Fig. 1Q). Often a tidepool will have a reddish tinge caused by a bloom of dinoflagellates. Of additional note are the dinoflagellates often referred to as the zooxanthellae. These are the photosynthetic, mutualistic endosymbionts (e.g., of the genus *Symbiodinium*) found in certain rocky intertidal and subtidal sea anemones (e.g., the aggregating anemone, *Anthopleura elegantissima*, Fig. 2, and the giant green anemone, *Anthopleura xanthogrammitica*), as well as within reef-building corals, certain marine sarcodines (e.g., Foraminifera and Radiolaria), flatworms, and the tropical Indo-Pacific giant clam *Tridacna*.

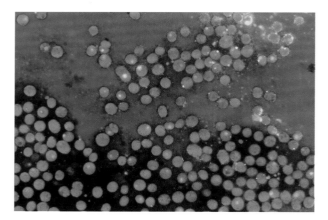

FIGURE 2 Epiflorescence micrograph (under blue light excitation) of *Symbiodinium*, a dinoflagellate endosymbiont extracted from the aggregating anemone *Anthopleura elegantissima*. Cells average about 10 μm in diameter. Image by G. M. Capriulo.

Phylum Ciliophora

Ciliates (Fig. 1D, E, H, I, M) are abundant in every aquatic habitat and in soils. All but a few have cilia on their surface at one time in their life cycle. Most have two kinds of nuclei; the larger one (or more) the macronucleus regulates metabolism; the smaller one (or more) is primarily involved in sexual recombination. Most ciliates are phagotrophic, ingesting nutrients through their cytostomes into a cytopharynx, which may be surrounded by specialized ciliary structures. The organization of the oral and somatic regions (the somatic region functions in locomotion, attachment, forming protective coverings, and sensing the environment) are the main characters used to subdivide the phylum into 10 classes. Identifying ciliates involves careful study of the distribution of cilia and associated fibrils. Ciliate specialists use a silver staining technique and TEM for identification. The nonspecialist can identify certain groups of common ciliates by some superficial observations, such as attachment to the sides or bottom of tide pools. One of the easiest groups to identify is the subclass Peritrichia. These may have long or short contractile stalks. Their cytostome is surrounded by a crown of prominent peristomal ciliation. Related to the peritrichs, the chonotrichs have noncontractile

bodies. Another common group of attached ciliates are the Suctoria. Adults lack cilia and a cytostome. They have tentacles that capture and digest prey. Some of the ciliates that crawl along surfaces are flattened and ribbon-like (class Litostomatea). Another group of crawling ciliates likely to be found in the littoral and every tidepool is the subclass Hypotrichia. These small, ovoid, dorsoventrally flattened ciliates have composite tufts of somatic cilia (cirri) that function as a single unit. The cirri occur in definite patterns on their ventral surfaces. Another group of ciliates, class Karyorelelecta, order Protostomatida, resemble nematodes crawling and swimming on and in the benthos. Their cytostomes are anterior. Members of the class Oligohymenophorea are abundant in the water column. Some of them have a prominent velum (sail-like structure), which acts to funnel food into the cytostome. The proper identification of a large variety of other small ciliates requires more sophisticated methods.

SMALL FLAGELLATED GROUPS

There is a diversity of nonrelated, free-living small flagellates (Fig. 1J, L) that include marine representatives (e.g., class Choanoflagellata (collored flagellates), order Volvocida (biflagellated oval green flagellates), straminophiles (nonpigmented and orange- or red-pigmented flagellates with chlorophylls *a* and *c*; order Bicoecida, class Chrysomonadida, class Pelagiophyae, order Raphidomonadida, class Silicoflagellata), order Cryptomonadida, phylum Euglenozoa, class Pedinophyceae, order Prymnesiida, and some residual heterotrophic flagellates with too few representatives to be grouped above the family level. A few of these flagellates are easily recognized because of their unusual skeletons, a cleft in their anterior, or scales that are observable in the light microscope, but most require observation under TEM for definitive identification. Except for the very small (5 μm) flagellates, these groups are very rarely abundant in rocky-shore and tidepool habitats, most being brought in the wash of plankton in each tidal cycle. Some genera, such as *Dunaliella* (Volvocida), are always present in small numbers.

NAKED AMOEBAE

Naked amoebae (Fig. 1K) are always present in tide pools and in littoral habitats. They are usually small, inconspicuous, and easily overlooked. Stellate forms of many genera can be found in the water column, but most of them crawl on vegetation, debris, or benthic substrate. To find them it is often better to place a sample in a covered aquarium or on a Petri plate with or without a non-nutrient agar. As soon as the surface is colonized by bacteria and diatoms, the amoebae become abundant. Sometimes the explosive bursts of amoebae are dramatic, with a resultant dramatic clearing of a bacterial or diatom layer where the amoebae have eaten. Common amoebae found in tidepools can be recognized by their shape, possession of a uroid (posterior end, which may be bulbus, globular, papillate, etc.), the nature of their pseudopods (monopodal, polypodal, slender, radiating, etc.) and whether locomotion is smooth or eruptive. Some marine amoebae have scales or spicules, which are not always obvious to the untrained eye.

NUTRITION AND FOOD WEB INTERACTIONS

Marine organisms sustain themselves as autotrophs (strictly photosynthesis), mixotrophs (chemosynthesis or photosynthesis supplemented with other outside nutritional sources or symbioses), or heterotrophs (outside food obtained from some combination of absorption and ingestion). A plethora of algae sustaining themselves entirely or primarily by photosynthesis live both in the euphotic zone of the water column as well as in, on, or associated with shallow-water coastal marine sediments and seamounts. Which ones are to be found is determined by many physiochemical and biological factors. Key to this are sediment type and size (rocks to sand, muds, organic oozes) and living surfaces. Well represented in association with the sediments are diatoms, certain dinoflagellates, and small flagellates as well as other microalgae and seaweeds (Figs. 1 and 3).

In general, primarily heterotrophic protists gain nutrition from some combination of absorbtion of dissolved organic materials, photosynthesis, chemosynthesis, phagotrophy, or symbiosis (including mutualism, commensalisms, and parasitism). Phagotrophy might involve bacteria, cyanobacteria, microalgae, other protists, and, for some (e.g., foraminiferan and radiolarian solitary and colonial amoeboids), even multicellular invertebrates and small vertebrates. For parasitic protists, large hosts often serve as a nutritional source.

Because of limitations imposed by the physics of diffusion relative to organism size and surface-area-to-volume ratios, dissolved organics as a sole source of nutrition are generally adequate only for bacteria. In turn, because of normal water column bacterial concentrations, physical hydrodynamic filtration limitations, and metabolic needs, planktobacteria cannot support the growth of most larger protists such as planktonic ciliates but do support small flagellate predators. Many of the flagellates also can eat other protists their own size and larger. The planktonic ciliates generally eat microalgae and other protists, including ciliates. Many of the water column amoeboid forms feed on larger prey, including various planktonic crustaceans and small fish, in addition to smaller food items. Surface-to-volume ratios increase with decreasing particle size. So, in the sediments, depending on particle sizes and associated surface-to-volume ratios, concentrations of organics are generally higher than in the water column. For that reason, in addition to bacterivorous flagellates in sediments, many of the ciliates found there are bacterivorous in nature, and thus taxonomically distinct from the water column ciliates. Many amoeboid forms are also found in the sediments feeding on a host of prey, including bacteria, cyanobacteria, diatoms and other algae, larger protists, and certain metazoans. Taxonomic variations follow the sediment size gradients as well as physiochemical and biological stresses.

FURTHER READING

Capriulo, G.M. ed. 1990. *Ecology of marine protozoa.* New York: Oxford University Press.
Jahn, T. L., E. C. Bovee, and F. F. Jakn. 1979. *How to know the Protozoa.* Dubuque, IA: Wm. Brown Co..
Knox, G. A. 2001. *The ecology of seashores.* Boca Raton, FL: CRC Press.
Lee, J. J., G. F. Leedale, and P. Bradbury, eds. 2000. *An illustrated guide to the protozoa,* 2nd ed. Lawrence, KS: Society of Protozoologists/Allen Press..
Lee, J. J., and A.T. Soldo. 1991. *Protocols in protozoology.* Lawrence, KS: Society of Protozoologists/Allen Press.
Reid, P. C., C. M. Turley, and P. H. Burkill, eds. 1991. *Protozoa and their role in marine processes.* NATO ASI Series 25. Berlin: Springer-Verlag.
Sieburth, J, M. 1979. *Sea microbes.* New York: Oxford University Press.

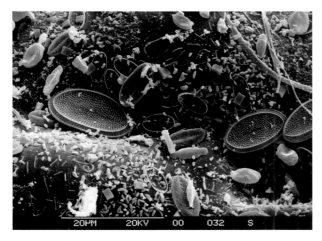

FIGURE 3 Scanning electron micrograph of a small tidepool rock surface showing colonization by bacteria and diatoms, some typical rocky-shore and tidepool food for grazing protists. Image by J. J. Lee.

PYCNOGONIDS

BONNIE A. BAIN
Southern Utah University

Pycnogonida (Pantopoda) are spider-like marine organisms found in nearly all types of marine habitats, including the intertidal zone. These interesting animals, overlooked by most biologists, have recently become the focus of attention of several different groups of researchers, with interests ranging from molecular systematics to the evolution of paternal care.

CLASSIFICATION AND ECOLOGY

Pycnogonida are a group of understudied marine chelicerates (7–9 families, 88 genera, 1,400+ species) found worldwide in all marine habitats (intertidal zone to abyssal depths). They range in size from less than 1 millimeter to over 70 centimeters in leg span and can occasionally be present in very large numbers. Most larger pycnogonids (leg span greater than 3–4 cm) (e.g., *Colossendeis, Decolopoda, Dodecolopoda*) are found in shallower waters in the Antarctic, and some (e.g., *Colossendeis*) can also be found in cold deep waters worldwide. Intertidal and shallow-water pycnogonids in other areas are much smaller than their Antarctic and deeper-water cousins (1 mm to several centimeters) (Fig. 1). In addition, intertidal pycnogonids are

FIGURE 2 (A) *Pallenoides* sp. on a bryozoan. (B) *Pallenoides* sp. removed from the bryozoan. Photographs by Fredric Govedich, Southern Utah University.

usually very well concealed, either hiding under rocks when the tide is out, or perfectly camouflaged (Figs. 2A, B).

MORPHOLOGY

Pycnogonids have a reduced body, four pairs of walking legs, and paired chelicerae, pedipalps, and ovigerous legs or ovigers. In overall appearance, they resemble a daddy-longlegs spider, giving rise to their common name of "sea spider." Depending on family association, some pycnogonids lack various combinations of chelicerae, pedipalps, and ovigers. For example, *Anoplodactylus* (family Phoxichilidiidae) is characterized by well-developed chelicerae, no pedipalps, and only males have ovigers, whereas others such as *Nymphon* have chelicerae, five-segmented pedipalps, and well-developed ovigers in both sexes. Several pycnogonid genera, across three families, have additional pairs of walking legs: *Decolopoda, Pentacolossendeis, Pentanymphon*, and

FIGURE 1 Pycnogonids collected from South Channel Fort, Port Philip Bay, Melbourne, Australia. Centimeter scale at bottom of photograph for size comparison. Photograph by Fredric Govedich, Southern Utah University; specimens collected by Bruce Weir, Monash University, Clayton, Victoria, Australia.

Pentapycnon have five pairs, whereas *Dodecolopoda* and *Sexanymphon* each have six. Possible explanations for this situation include polyploidy; metameric instability; and that extra pairs of legs are the primitive condition. The most likely explanation, in light of current molecular advances, is that these extra pairs of legs are due to a duplication event in the segmentation genes.

Ovigers, located between the pedipalps and first pair of walking legs, are unique to pycnogonids. Oviger functions include grooming; food handling; courtship, mating, and egg transfer; and transport of eggs and larvae. In more primitive genera, male and female ovigers are identical in size and structure, and the last four segments of the oviger have either single or multiple rows of species-specific spines. More derived genera either have ovigers in which the structure of the last four segments has been highly modified, or have lost some or all of these segments; in extreme cases they have lost the entire structure.

REPRODUCTION, LARVAL DEVELOPMENT, AND PATERNAL CARE

Pycnogonids are one of the few animal groups with exclusive paternal care. Because of this, they have attracted the attention of evolutionary biologists studying the evolution of sexual selection, mating systems, and different aspects of parental care. Male and female pycnogonids undergo a series of courtship and mating behaviors in which the ovigers play a major role. Then the female begins to lay her eggs, and the male picks the eggs up with his ovigers, dips them in glue from his cement gland, and attaches them to his ovigers. He then carries the eggs around until they hatch. Males ventilate the egg masses and provide them protection from predators. Depending on species and phylogenetic position, the eggs hatch into one of two possible larval types. The protonymphon, a free-living, feeding larva, is the most common type and is found in several families (Ammotheidae, Nymphonidae, Phoxichilidiidae, Pycnogonidae). When the eggs hatch, the protonymphon larvae leave the male and either take up a free-living existence (typical protonymphon), form

a cyst in a hydroid or other cnidarian (encysted larva), or live in or on another marine invertebrate (atypical protonymphon). The attaching larva (Callipallenidae, some Nymphonidae), unlike the protonymphon, remains on the parent after hatching and undergoes several more molts before it leaves. The longest duration of paternal care has been found in several Arctic species *(Nymphon, Boreonymphon)*, in which males were observed carrying juveniles which were nearly adult size.

FEEDING BEHAVIOR

Very little is known about pycnogonid feeding behaviors. What little we do know indicates that many of them feed on hydroids, medusae, sea anemones, bryozoans, polychaetes, brine shrimp, and various molluscs (including clams, mussels, snails, and nudibranchs). Some taxa (e.g., *Anoplodactylus*) are generalized predators and have been observed eating everything from medusae to brine shrimp, while others (e.g., *Pycnogonum*) feed only on sea anemones.

SEE ALSO THE FOLLOWING ARTICLES

Bryozoans / Camouflage / Hydroids / Sex Allocation and Sexual Selection

FURTHER READING

Arnaud, F., and R. N. Bamber. 1987. The biology of the Pycnogonida, in *Advances in marine biology*, Vol. 24. J. H. S. Blaxter and A. J. Southward, eds. New York: Academic Press, 1–96.

Bain, B. A. 2003. Larval types and a summary of postembryonic development in pycnogonids. *Invertebrate Reproduction and Development* 43(3): 193–222.

Bain, B. A., and F. R. Govedich. 2004. Courtship and mating behavior in the Pycnogonida (Chelicerata: Class Pycnogonida): a summary. *Invertebrate Reproduction and Development* 46(1): 63–79.

Fry, W. G., and J. H. Stock. 1978. A pycnogonid bibliography. *Zoological Journal of the Linnean Society, London* 63(1+2): 197–238.

Hedgpeth, J. W. 1947. On the evolutionary significance of the Pycnogonida. *Smithsonian Miscellaneous Collections* 106(18): 1–53.

Hedgpeth, J. W. 1982. Pycnogonida, in *Synopsis and classification of living organisms*, Vol. 2. S. P. Parker, ed. New York: McGraw-Hill.

King, P. E. 1973. *Pycnogonids*. New York: St. Martin's Press.

Stock, J. H. 1994. Indo-West Pacific Pycnogonida collected by some major oceanographic expeditions. *Beaufortia* 44(3): 17–77.

RECRUITMENT

STEVEN D. GAINES

University of California, Santa Barbara

The number of individuals of a species at any given place can increase in only two ways: new births and immigration. For most species on rocky shores, these two sources of population growth are one and the same. Since the young of most marine species are released into the sea as plankton, they are moved by ocean currents. If they survive the journey, their return to adult habitat at the shore is called recruitment. The recruits that replenish most populations on the shore are typically the offspring of adults from other sites, some from nearby, others from far away.

STATIONARY ADULTS, TRAVELING YOUNG

Most of the animals and plants found on rocky shores are either slow-moving or completely immobile. This trait makes them easy for scientists to study and easy for beachcombers to discover. Immobility, however, only characterizes the adults. As young, many are wanderers in a different habitat.

Most marine species produce young that are microscopic and cast into the sea to drift on ocean currents. Some drift for a few minutes; others drift for months. Animals release larvae; seaweeds release spores. Although they are all quite small (typically much less than 1 millimeter in length at release), they have a great diversity of shapes that often bear no resemblance whatsoever to their parents (Fig. 1).

Larvae that drift for more than a few hours typically develop through multiple stages that increase in size and may change greatly in shape. The energy needed for this growth comes most commonly from eating other plankton (in which case the larvae are called planktotrophic); but in some species the larvae receive enough nutritional stores from their mother, in the form of yolk, to complete development without eating anything else (such larvae are called lecithotrophic).

THE PERILS OF THE PLANKTON

Having the option of producing minute young who can feed on their own affords marine fish and invertebrates the opportunity to produce enormous numbers of offspring. Depending on the species, tens or even hundreds of thousands of larvae may be released in a single spawning event. The benefit of being able to make a minimal energetic investment in individual offspring is clear, but there are also inherent costs. If larvae must successfully obtain their own food to complete development, their fate depends on how much food they find and on surviving the challenges of life in the plankton long enough to complete development. These are nontrivial challenges. Larvae from intertidal mussels, for example, have been shown to vary greatly over time and space in their success at obtaining food. The consequences are acute, with large variation in the survival of larvae through development and large variation in their performance even after they recruit to adult habitat on the shore.

More daunting still are the perils of developing while drifting in the plankton. Even if food is abundant, the risk of the larvae being eaten while drifting for weeks is quite high. Potential predators range from other small zooplankton all the way up to baleen whales. Moreover, if a larva is lucky enough to make it through the planktonic gauntlet, it has to complete its development within reach of habitat where it can thrive as an adult. Imagine the challenges for

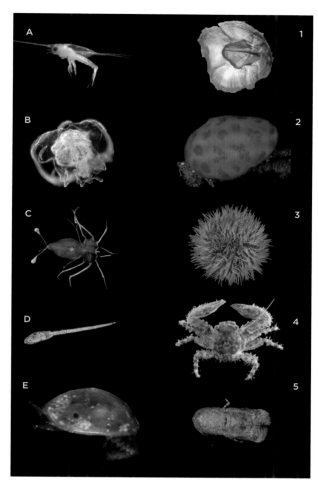

FIGURE 1 Marine ecologists commonly face the challenge of identifying the larval stages of the species they study intensively as adults. Larval forms often bear little resemblance to their ultimate adult shapes. This figure provides a small sampling of larval types and their associated adult groups. Can you match the larval type with the adult form it ultimately produces? Answers: A (porcelain crab zoea) → 4; B (Sea urchin pluteus) → 3; C (lobster phyllosoma) → 5; D (ascidian tadpole) → 2; E (barnacle cyprid) → 1. Photographs A, C, 2, 3, 4 and 5 courtesy of the Southeastern Regional Taxonomic Center/South Carolina Department of Natural Resources. Photograph B courtesy of Gerardo Amador. Other photographs courtesy D. Wendt.

LOST AT SEA

Many patterns of coastal ocean circulation are known to alter the success of larvae in ultimately reaching appropriate adult habitat. Eddies, internal waves, tides, freshwater-driven flows in estuaries, and several other circulation features have all been linked to variation in larval success through their transport of larvae either to or away from adult habitats. Perhaps the best studied of these connections, however, involves the prominent ocean process called upwelling. Coastal habitats with upwelling are some of the most productive ocean habitats on the planet. One of the key processes driving upwelling is winds. When winds blow along the coast, they can generate forces on surface waters of the ocean that are perpendicular to the coast because of the rotation of the earth and the ensuing Coriolis effect. As a result, surface waters are pushed away from the coast and replaced by deeper, colder, nutrient-rich water that is upwelled from below (Fig. 2). The infusion of nutrients into surface waters provides the fuel for blooms of phytoplankton, which subsequently support large populations of higher trophic levels. From the perspective of larvae, the blooms in phytoplankton provide a rich source of food. The problem is that while the larvae are feeding on the abundant phytoplankton, they are simultaneously getting a free ride away from shore. The offshore transport that drives the blooms also drives larvae further and further away from their ultimate destination. Variation in the strength and direction of winds along the coast can therefore play a key role in the fate of larvae from coastal species. This variability has been shown to dramatically affect the number of larvae that successfully recruit back to rocky shores in species that range from barnacles to rockfish. In cases where upwelling

larvae of species that live as adults only on rocky intertidal shores. This habitat forms the minute fringe of the sea, and larvae developing for many weeks in the plankton could easily be transported well away from the coast. If they are, and they have no pathway to return to the shore, such larvae are no better off than if they had starved or been eaten prior to completing their development. Some larvae can pause their growth once they complete development to wait for the possibility of arriving near appropriate habitat, but most species have a window of opportunity to settle before their fate is sealed.

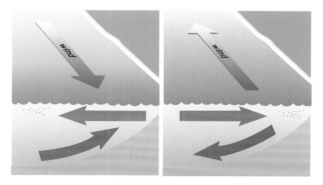

FIGURE 2 Schematic diagram of the influence of surface winds on ocean currents in upwelling ecosystems. Winds blowing parallel to the coast can create offshore or onshore transport of surface and deep waters. Larvae that reside in these waters will be moved away from or toward shore depending on their depth and the direction of the winds.

is intense and relatively persistent, recruitment of larvae to the shore can be so low that adults remain rare even if the shoreline habitat is rich and productive and risk from predators is minimal. Such populations are recruitment limited, and oceanographic variation drives their population dynamics.

ALONGSHORE TRAVEL BETWEEN POPULATIONS

Just as coastal currents can transport young away from shore, they can also move larvae and spores along the shore. This dispersal connects populations at different locations. The extent of movement varies greatly among species and locations based upon the oceanographic conditions, the length of the development period, the behavior of larvae and spores, and a variety of other factors. Studies of genetic differences among populations on different reefs suggest that some species move only a few meters, while others can move hundreds of kilometers. If the recruits at a particular site come from adults far away, the dynamics of these widely separated sites are connected. These patterns of connectivity are exceedingly difficult to measure for anything that spends more than a few hours in the plankton, but they have important consequences for a wide range of ecological processes (e.g., rates of spread of exotic species, population recovery from storms or other natural catastrophes) and human interactions with marine ecosystems (e.g., management of fisheries, protection of threatened and endangered species).

SETTLEMENT AND METAMORPHOSIS: COMPLETING THE LIFE CYCLE

The large number of biological and physical processes that affect the fate of larvae while they are in the plankton can lead to enormous variation in the number of larvae that arrive back to the vicinity of the shore. These successful larvae have more hurdles to overcome before they recruit to populations on the shore. The last steps in the recruitment of young to adult populations are settlement out of the plankton and onto the shore and, in many cases, the metamorphosis from a body plan adapted to the plankton to one adapted to life on the rocks.

For intertidal species, settlement involves transitioning from a lifestyle dominated by drifting and swimming to one secured to the rock. Since the intertidal zone is commonly awash in waves while submerged, settlement involves grasping the rock (or perhaps the shell or other

part of some other species already attached to the rock) in an extremely turbulent, energetic environment. As anyone who has been rolled by waves at the beach can attest, this is a challenging task. The ultimate success of settlement is affected by the physical characteristics of the habitat (e.g., the texture of the surface changes small-scale features of the flow, whereas the orientation of the rock to the waves affects the fluid velocities near the rock surface), the biological characteristics of the larva or spore (e.g., the strength of glues, the ability of appendages to grasp or snag things on the shore, the preferences of the species for different shoreline habitats), the biological characteristics of the shore (e.g., what space is already occupied and unavailable, what species are waiting to eat returning larvae as they tumble by) and the complex fluid dynamics of the surf zone. Each of these components can generate patterns in where and when we see different species settling into rocky intertidal habitats (Fig. 3). Many of the patterns you see combing the shore simply reflect the legacy of where settlement occurs frequently versus rarely.

Finally, once attached to the shore, most species have to undergo a radical transformation in their morphology before they can continue on the path to adulthood. This metamorphosis can be energetically quite expensive and, while in progress, generally involves increased vulnerability to predators and physical disturbance. As a result, mortality rates can be quite high and may reflect the feeding success of the larva's planktonic period. If the larva

FIGURE 3 Spatial patterns of recruitment can generate patterns of adult distributions. Gregarious settlement can clump adults together more closely than they would occur with random settlement. This may increase their competition, but could have great benefits for reproduction. Similarly, some species recruit by attaching to other species, as seen in this photograph of a seedling of the surfgrass Phyllospadix attached to the alga Gigartina. Here the alga provides a site where the surfgrass seed can attach, which ultimately may lead to the demise of the alga as the surfgrass matures. Photograph courtesy Dan Reed.

settles with limited nutritional stores, it may be unable to complete metamorphosis to a form that is capable of feeding in the adult habitat. After metamorphosis of the larva or germination of the spore, the benthic phase of the life cycle begins.

SEE ALSO THE FOLLOWING ARTICLES

Benthic-Pelagic Coupling / Dispersal / Hydrodynamic Forces / Larval Settlement, Mechanics of / Reproduction / Upwelling

FURTHER READING

Connell, J. H. 1985. The consequences of variation in initial settlement vs. post-settlement mortality in rocky intertidal communities. *Journal of Experimental Marine Biology and Ecology* 93: 11–45.

Kinlan, B., and S. D. Gaines. 2003. Propagule dispersal in marine and terrestrial environments: a community perspective. *Ecology* 84: 2007–2020.

Morgan, S. G. 2000. The larval ecology of marine communities, in *The ecology of marine communities.* M. D. Bertness, S. D. Gaines, and M. E. Hay, eds. Sunderland, MA: Sinauer Associates, 159–181.

Roughgarden, J., S. D. Gaines, and H. Possingham. 1988. Recruitment dynamics in complex life cycles. *Science* 241: 1460–1466.

Schiel, D. R. 2004. The structure and replenishment of rocky shore intertidal communities and biogeographic comparisons. *Journal of Experimental Marine Biology and Ecology* 300: 309–342.

Underwood, A. J., and M. J. Keough. 2000. Supply side ecology: the nature and consequences of variations in recruitment of intertidal organisms, The Ecology of Marine Communities. M. D. Bertness, S. D. Gaines, and M. E. Hay, eds. Sunderland, MA: Sinauer Associates, 183–200.

RED TIDES

SEE ALGAL BLOOMS

REPRODUCTION, OVERVIEW

JOHN S. PEARSE
University of California, Santa Cruz

Reproduction is a universal character of living organisms; it is the production of new bodies similar, at least genetically, to the parent bodies, and is the means by which life continues through time. How it is done varies tremendously among different life forms, almost all of which are found on wave-swept shores. When reproduction is asexual, involving various means of splitting, there is little or no genetic novelty. However, reproduction in many organisms also or exclusively involves sexual processes, which recombine genetic material to form genetically new offspring. In eukaryotic organisms this process usually involves meiosis and gametes. The formation of meiotic products, gametic fusion, and the development and dissemination of the resulting propagules vary tremendously among species but are shaped and timed by natural selection to continue the production and survival of new generations.

DIVERSITY OF INTERTIDAL LIFE AND REPRODUCTION

In terms of major kinds of organisms, the rocky intertidal is the most diverse habitat on earth; examples from the simplest to the most complex forms of life are found here in close proximity. It is to be expected, therefore, that the greatest diversity of life histories, evolutionary histories, and modes of reproduction is found here as well. Some forms reproduce with little or no sex involved at any time, while many others include sex in parts of their life cycle and replicate without sex in other parts. When sex is involved, some organisms produce specialized gametes (sperm and eggs), while others have nearly identical gametes and have seemingly only one sex; they may also have more than two sexes. When there are two sexes (male and female), individual organisms can be hermaphroditic (each producing both male and female gametes either at the same time or sequentially), or they can be gonochoric (with the different sexes found in different individuals of the same species, also known as dioecious). Usually there are approximately equal numbers of males and females in gonochoric species, but in some cases one sex greatly outnumbers the other. The two sexes can be identical in external appearance or so different that only with careful study can both sexes of a single species be recognized.

Most intertidal organisms spend most of their time closely associated with a small part of the habitat, either attached to the substrate or to each other, or living nestled within cracks and crevices. Dispersal from one site to another is usually through floating or weakly swimming propagules, larvae, or juveniles. Sometimes these dispersive phases travel many kilometers over many months and suffer high mortality before settling into new intertidal habitats. Or the adults may be dispersive, producing gametes in the open water, and the juveniles live in the intertidal, often for many years. Other organisms have almost no dispersive phase at all in their life cycles.

The enormous variety of life forms and their modes of reproduction have provided major challenges to biologists seeking to understand and generalize life processes.

The variation seen on rocky shores is not just among major groups (algae versus animals, for example) but often is seen among closely related species seemingly living nearly identical lives, or even within a single species. Why should reproductive modes vary so much along this narrow band of rock where the sea meets the land? All the organisms face the same environmental challenges, alternating between being immersed in water and exposed to air. Sunlight, water, oxygen, and nutrients are almost always not limiting, and other factors such as competition, predation, and recruitment may be more important in determining who lives where. Should there not be one optimal way to reproduce in such an environment, which all the inhabitants would follow? Apparently not. Instead, there appear to be trade-offs between optimal modes of growth, survival, and reproduction, as well as constraints reflecting diverse body plans and evolutionary histories. The following sections discuss the variety seen in the reproduction of organisms found in rocky intertidal habitats, with some of the current ideas about how it came to be.

LIFE CYCLES AND GROWTH FORMS: ASEXUAL AND SEXUAL REPRODUCTION

Unicellular Organisms

Nearly all living things consist of cells, either unicelled or multicelled, and they grow mainly by increasing cell number through cell division. In unicellular organisms, this process of clonal growth can be viewed as asexual reproduction that produces unconnected cells, which can be clumped together or widely dispersed. Clonal growth may continue indefinitely, with genetic variation occuring only through mutations. However, various forms of sexual mixing, creating increased genetic variation, also may be present, sometimes involving meiosis and gamete fusion.

In addition to the unicellular plankton that is swept through tidepools and over the surrounding intertidal rocks, many forms of unicellular organisms grow attached to the rocks and to other organisms living in this habitat.

CYANOBACTERIA

These photosynthetic prokaryotes (formerly called blue-green algae) often form a thin, dark coat on rocks of the upper reaches of the intertidal (Fig. 1). Cyanobacteria, like other bacteria that occur in the intertidal and elsewhere, grow by asexual cell division (binary fission). Often the cells stick together to form long filaments, which then fragment. Bacteria do not possess well-defined sexual processes. However, although unknown in cyanobacteria,

FIGURE 1 Black crusts of cyanobacteria covering the upper levels of intertidal rocks. Pacific Grove, California. Photograph by the author.

genetic recombination ("sex") does occur in many bacteria: genetic material is passed from one cell line to another through conjugation, transformation (picking up DNA from dead cells), or transduction (DNA transfer mediated by viruses).

SINGLE-CELLED EUKARYOTES

Unicellular eukaryotes living in tidepools and on the surrounding intertidal rocks include algae, fungi, and protozoans. These all grow by asexual cell division, forming either enormous clones of completely separate cells or colonies of cells that form discrete shapes. Sexual reproduction is rare or unknown in some forms (e.g., euglenids, choanoflagellates, naked amoebas). Others have life cycles that include sex (e.g., unicellular green algae, diatoms, foraminiferans, ciliates).

The two components of eukaryote sex, meiosis and fertilization (gamete fusion), may be separate in the life cycle of unicells, alternating between haploid cells that multiply by mitosis and eventually produces gametes, and diploid cells that multiply by mitosis and eventually undergo meiosis. For example, in foraminiferans—many of which live as sessile, calcareous-encased forms attached to rocks, shells, and other organisms in tidepools—mitotic divisions occur in the haploid phase (gamont) of the life cycle, eventually giving rise to numerous gametes. Fertilization results in a diploid phase (agamont), and after a series of mitotic divisions, meiotic divisions produce numerous haploid spores (agametes), each of which begins the haploid phase. Variations of these processes are astounding, but usually the haploid and diploid phases are morphologically very similar.

However, many unicelled algae in tidepools exist mainly in either the haploid or diploid phase. For example, diatoms are mainly diploid, and only occasionally do some cells undergo meiosis to form gametes, which fuse to form a zygote that multiplies by mitosis to form a new cell lineage of the diploid phase. In contrast, green microalgae are mainly haploid, and the zygote immediately undergoes meiosis to form haploid cells that multiply by mitosis.

Ciliates are unicellular organisms that differ in many ways from other life forms. Nearly all have a polyploid macronucleus, and division is amitotic (the micronuclei, if present, divide mitotically). Moreover, in some sessile ciliates (such as suctorians, which are common on rocks and shells in tidepools), budding pinches off small portions of the macronucleus to produce small, motile, dispersive "larval" swarmers that swim away before settling and metamorphosing into the sessile "adult" forms (Fig. 2). Sex is achieved through meiotic division of the micronuclei and the exchange of the resulting haploid micronuclei during conjugation. Fusion of the micronuclei within each parental cell forms a genetically new diploid micronucleus, which divides by mitosis. The original macronucleus disintegrates, and one or more of the new micronuclei replicates repeatedly to form a new macronucleus. In this case, sex does not result in reproduction, but rather it mixes the genes of the two parental cells, each of which become a genetically new individual that divides to begin a new lineage.

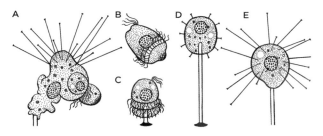

FIGURE 2 Budding in a suctorian ciliate. (A) Bud escapes from internal pouch. (B) Free-swimming bud. (C) Settlement on substrate. (D) Stalk elongates, cilia are lost, and tentacles form. (E) New mature sessile form develops. From Pearse et al. (1987).

MULTICELLULAR ORGANISMS

Multicelled organisms also grow by cell division, but their cells adhere together and differentiate into different cell types, forming integrated organisms. All types of asexual reproduction (sporulation, fragmentation, fission, budding, polyembryony, ameiotic parthenogenesis) are found among different intertidal multicellular organisms. In most cases, multicellular organisms also have sex at some stage of their life cycle, but in some species sex has never been demonstrated and probably does not occur. In addition, asexual reproduction is unknown and probably is rare or absent in some multicellular species.

Macroalgae. These organisms, commonly known as seaweeds, are multicellular, photosynthetic organisms without well-developed vascular tissues or embryos. They are the main primary producers in tidepools and on surrounding intertidal rocks. There are three main groups: green algae, brown algae, and red algae. Many have haplo-diploid life cycles and can reproduce asexually in both the haploid and diploid phases of their life cycles.

Most macroalgae regularly reproduce asexually by fragmentation, either from the holdfasts of upright-growing algae from which develop new upright portions (stipes), or from crusts growing on rocks that break apart into multiple crusts. Such lateral growth can lead to extensive areas covered by large clones (ramets) of a single genetic individual (a genet), arising either from a sporophyte (the product of a single zygote) or a gametophyte (the product of a single haploid, meiotically produced spore). In addition, in many species, both sporophytes and gametophytes can mitotically produce dispersive spores that settle and develop in the same phase as the "parent" alga. Consequently, it can be extremely difficult to determine the extent of a genetically unique individual that arose from meiosis to form a haploid gametophyte or from a zygote to form a diploid sporophyte.

In isomorphic macroalgae (many green algae, e.g., *Ulva* and *Cladophora,* and some brown algae, e.g., *Ectocarpus*) cells in the diploid phase undergo meiosis to produce haploid spores; these settle and develop by mitotic division into the haploid phase of the alga, which is morphologically identical to the diploid phase (Fig. 3). The cells in the haploid phase can become gametes, which are released and fuse with gametes from other individuals (fertilization) to form zygotes, which develop by mitotic division into the diploid phase.

In contrast, heteromorphic brown algae (e.g., kelps) have a large macroscopic diploid phase (sporophyte) with a holdfast, stipes (stems), and blades, which grows by mitotic division and produces haploid spores through meiosis (Fig. 4). These spores disperse and settle to grow by mitotic division into a microscopic haploid phase (gametophyte) that is either male or female. Male gametophytes produce by mitosis numerous biflagellated sperm, which disperse to fertilize nonmotile eggs produced and held by female gametophytes. The fertilized eggs develop into the large sporophytes.

The gametophytes are further reduced or completely absent in some green algae (e.g., *Codium*) and brown algal rockweeds (e.g, *Fucus, Pelvetiopsis,* and *Silvetia*), which are

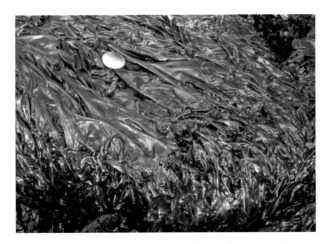

FIGURE 3 Two kinds of sheetlike algae with very different life histories. The green sheets are of a green alga *(Ulva)* that is isomorphic; the haploid gametophyte has the same appearance as the diploid sporophyte. The brown sheets are of a red alga *(Porphyra)* that is heteromorphic; the haploid gametophyte (shown) produces sperm and eggs, and the fertilized eggs develop into the diploid carposporophytes, which produce diploid spores that develop into microscopic sporophytes. The coin, included for scale, is a U.S. quarter. Fitzgerald Marine Reserve, California. Photograph by the author.

FIGURE 4 Sporophytes of heteromorphic brown algal sea palms *(Postelsia palmaeformis)*. Haploid spores are produced by meiosis and shed from the fronds to produce microscopic gametophyes below the sporophytes. Pigeon Point, California. Photograph by the author.

diploid throughout their life cycle except for sperm and eggs that are produced by meiosis; thus these species have "animal-like" life cycles.

Red algae have more complex life cycles; most have three phases, one haploid and two diploid. The zygotes remain attached to the parental female gametophyte and develop into a diploid phase, called carposporophytes. In a process analogous to polyembryony in animals (see below) these produce by mitosis diploid, genetically identical spores,

which are released to disperse, settle, and develop into free-living sporophytes. These are called tetrasporophytes, because meiosis produces clumps of four haploid spores, which are released and develop into the gametophytes. The gametophyte and free-living tetrasporophyte can be morphologically nearly identical (e.g., *Chondracanthus, Mazzaella*), or extremely different, as in the sheetlike gametophyte and microscopic sporophyte in *Porphyra* (Fig. 3) or the foliose gametophyte and encrusting sporophyte in *Mastocarpus* (Fig. 5). Some red algae have truncated life cycles, with one phase or the other lacking or inconspicuous.

FIGURE 5 Both forms of a heteromorphic red alga *(Mastocarpus papillatus)*. The upright fronds are haploid gametophytes, and fertilized eggs develop into carposporophytes, attached to the females as small bumps. These produce diploid carpospores that are released and develop into the dark, encrusting diploid sporophytes. The coin is a U.S. quarter. Pacific Grove, California. Photograph by the author.

Marine Plants. These organisms, known as seagrasses (e.g., *Phyllospadix*), are conspicuous components of rocky shores, and reproduce like other flowering plants. Rhizomes, adhering to the rocks by roots, grow out multiple ramets, which break apart to form extensive clones of one genet. Small flowers produce waterborne pollen (male gametophytes) that fertilize the ovules in the female gametophyte in the flower, which then develop into waterborne seeds.

Marine Animals. Marine animals on rocky shores, like other animals, all share similar life cycles: the diploid zygote develops as an enclosed embryo (or embryos in polyembryony) to be released as a larva or a juvenile that grows into an adult, which produces gametes (sperm or eggs) through meiosis. But there is enormous variation in the details,

FIGURE 6 Mussels and barnacles on an intertidal rock. Mussels have separate sexes and broadcast spawn; embryonic and larval development is entirely pelagic. Barnacles are hermaphrodites that copulate; embryos develop within the adults, and only the larvae are pelagic. The coin is a U.S. quarter. Pacific Grove, California. Photograph by the author.

most animals, including humans and other vertebrates; however, it is characteristic of cyclostome bryozoans, which often thrive in tidepools (e.g., *Crisia, Tubulipora*). In these cases, each zygote divides repeatedly before embryonic development begins, so that numerous genetically identical larvae are formed.

Although eggs are usually associated with sexual reproduction, they may develop parthenogenically without first undergoing meiosis or fertilization—no sex is involved. Such ameiotic (apomictic) parthenogenesis is another form of asexual reproduction. It is common in freshwater micrometazoans such as rotifers, gastrotrichs, tardigrades, and cladocerans, allowing populations to grow to very large sizes before, in some cases, undergoing sex. Ameiotic parthenogenesis almost certainly also occurs in some rocky intertidal animals, especially bdelloid rotifers, in all of which males are unknown.

even for animals living in close proximity (Fig. 6). In many marine animals the larva is a free-living, dispersive pelagic stage that settles and metamorphoses into a juvenile in the adult habitat. Asexual reproduction can occur in any of these life stages to form genetically identical ramets of the original individual genet. In some organisms the ramets remain physically and physiologically connected so that large colonies form, while in others the ramets separate into individual units called clonemates.

A common form of asexual reproduction is fragmentation, whereby a genet, usually in the form of a colony, is broken apart into independent ramets, each of which can grow into other ramets (e.g., separation of stolons connecting hydroids, bryozoans, or kamptozoans (entoprocts), or of massive colonies of sponges, corals, bryozoans, or tunicates). Fission is also common in noncolonial animals. For example, the sea anemone *Anthopleura elegantissima* of the northeastern Pacific undergoes longitudinal fission that splits an individual in two; repeated fission can result in extensive areas covered with clonemates of a single genet (Fig. 7). Fission also is common in several abundant intertidal polychaetes, such as sessile tube-builders that form large clones in calcareous reefs (e.g., *Dodecaceria, Salmacina*). Budding is a form of unequal fission, in which a small piece of an individual separates from a larger piece (e.g., the sea anemone *Metridium senile*, solitary kamptozoans).

Polyembryony is an unusual form of asexual reproduction that produces identical twins or more offspring when the zygote or early embryo divides. It is found rarely in

FIGURE 7 Genetically identical clonemates of the sea anemone *Anthopleura elegantissima* surrounding a single individual of the noncloning sea anemone *Anthopleura sola*. Both species reproduce sexually. Pebble Beach, California. Photograph by the author.

Sexual reproduction is remarkably uniform in animals compared to other organisms. Nearly all are diploid organisms that produce haploid sperm or eggs by meiosis. Exceptions are hymenopteran insects and monogonont rotifers with sex, some of which occur in the intertidal; in these animals the males are haploid, comparable to gametophytes in algae. Animal sperm and eggs are similar in morphology; sperm vary among species mainly in the presence or absence of a posterior flagellum, apical acrosome, and shape of the head; ova vary mainly in size. In most animals, the gametes are formed during gametogenesis within discrete organs, gonads. Individuals may be either male or female (gonochoric or dioecious; e.g., most cnidarians, polychaetes, molluscs, crustaceans, echinoderms, vertebrates), have both sexes simultaneously (simultaneous hermaphroditic; e.g., most sponges, flatworms, nudibranchs, barnacles, bryozoans, tunicates), or alternate between being one sex or the other (sequential hermaphroditic; e.g., oysters, owl limpets, slipper shells, wrasses) (Fig. 8).

FIGURE 8 A female owl limpet *(Lottia gigantea)*, an example of a sequential hermaphrodite that started life as a male, then switched to a female. Santa Cruz, California. Photograph by the author.

Copulation occurs in many marine animals living in tidepools (e.g., all flatworms, crustaceans, slugs, octopuses; some snails, fishes) (Fig. 9); in the case of many fish, the female lays eggs over which the male deposits sperm. In sea spiders, the female deposits eggs on the male, who then fertilizes them. In species with such intimate behavior between the sexes (or between hermaphrodite pairs), mate choice may be especially important. In many other species, the males release sperm freely into the water (spermcast spawning) and the females collect the sperm to fertilize the eggs internally, where embryonic development occurs (e.g., most sponges, bryozoans, tunicates). Still in others, both sperm and eggs are released at the same time (broadcast spawning), and fertilization occurs in the water away from both parents (most anemones, polychaetes, chitons, bivalves, echinoderms; some snails, tunicates, fishes). In broadcast-spawning species, spawning is synchronized so that both sperm and eggs are released at the same time, the eggs often emit chemicals that attract sperm, and in some cases individuals aggregate during spawning; all of these features are adaptations to maximize fertilization.

FIGURE 9 Copulating pairs of high-intertidal gonochoric periwinkles *(Littorina keenae)*. The coin is a U.S. quarter. Pacific Grove, California. Photograph by the author.

The embryos of broadcast-spawning species usually develop while drifting in the water before hatching as pelagic embryos or juveniles. In other species, the embryos are either held and protected within (in which case the animal is called viviparous or ovoviparous) or on the parent's body (brooded), sheltered and guarded by the parent (either females, e.g., octopuses, or males, e.g., some fishes, most sea spiders), or placed in protective egg cases (e.g., all nudibranchs, many snails, some flatworms, polychaetes, fishes) (Fig. 10). The embryos, whether pelagic, brooded, or in egg cases, may hatch as pelagic larvae or juveniles where they develop and drift with the currents until they settle, or if held internally,

FIGURE 10 Egg cases being deposited by a whelk *(Nucella ostrina)*. The eggs within the cases will develop directly into small snails with no intervening larval stage. The coin is a U.S. quarter. Pacific Grove, California. Photograph by the author.

brooded, or in egg cases, they may hatch as benthic juveniles from within or near the parents and have very limited dispersal.

If there is a pelagic larva in the life cycle, it can be supplied by the parent with enough nourishment to last until it settles and metamorphoses into a juvenile. These large, nonfeeding (lecithotrophic) larvae are characteristic of sponges, tunicates, limpets, abalones, most bryozoans, and many polychaetes and snails, as well as often in most other groups (Fig. 11); they typically complete pelagic larval development within days to weeks. In contrast, most bivalves, snails, barnacles, decapod crustaceans, echinoderms, and fishes produce numerous very small eggs without enough food to complete larval development (Fig. 12); these larvae need to feed in the plankton (planktotrophic) and require weeks to months before reaching a stage where they can settle and metamorphose into benthic juveniles. Even then, they usually settle at much smaller sizes than comparable species with nonfeeding larvae.

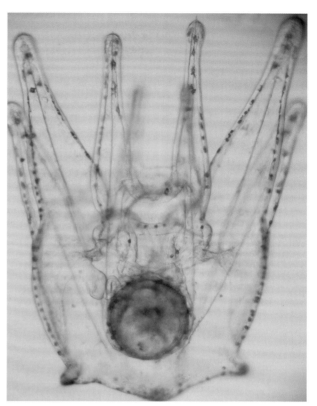

FIGURE 12 Feeding larva of a sea urchin *(Strongylocentrotus purpuratus)*. The circular object in the center is the stomach with food inside, and the long "arms" carry bands of cilia used to capture phytoplankton food. Photograph by Jason Hodin.

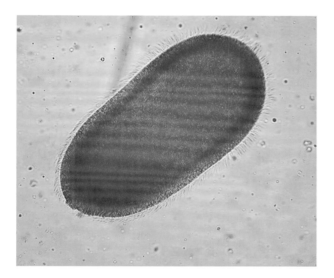

FIGURE 11 Nonfeeding larva of a sponge *(Oscarella carmela)*, 0.4 mm long, immediately after being released from the parent. Photograph by the author.

There appear to be trade-offs between producing (a) a very large number of tiny eggs that develop into small feeding larvae, needing a relatively long time in the plankton to both grow and develop before settling, (b) a smaller number of relatively large eggs that develop into nonfeeding larvae, which settle after a relatively short time in the plankton, and (c) an even smaller number of eggs that are held or brooded by the parent and avoid being in the plankton altogether. The longer in the plankton, the

higher the mortality and the higher the dispersal. Closely related species can differ between these different modes of development, or in exceptional cases (e.g., some spionid polychaetes and sea slugs) different modes of reproduction occur in the same species depending on factors not yet understood.

TIMING OF REPRODUCTION

Timing is important for reproduction of intertidal species, as for species elsewhere. Environmental conditions may favor asexual reproduction (growth) at some times and sexual reproduction at other times. During sexual reproduction, synchronized timing may be critical for fertilization or to favor the survival of new propagules (spores, zygotes, larvae, juveniles). Such synchronization of reproduction may set up natural cycles or rhythms that have a set frequency (daily, semimonthly, monthly, semiannually, annually).

The factors that determine the synchronization and frequency of rhythms can be divided into proximate and ultimate causes. Proximate causes are those factors that organisms use to synchronize activities (e.g., reaching a critical temperature or day length). Ultimate causes, on the other hand, are those that favor the establishment and survival of the resulting offspring (e.g., favorable conditions for fertilization, favorable food conditions for planktotrophic larvae, favorable growth conditions for settling juveniles). Species may integrate reproductive rhythms so that proximate and ultimate causes are combined in their reproductive strategies. For example, most individuals of a species may spawn in the early evening (daily rhythm cued by daylight) to maximize fertilization, on the new moon (monthly rhythm cued by moonlight) to minimize predation, and during the spring months (annual rhythm cued by photoperiod) to maximize survival of offspring in the favorable summer months.

Circadian Reproductive Rhythms

Many species release their spores, gametes, or larvae at particular times of the day, particularly just after sunrise or just after sunset. Synchronous release of gametes in broadcast spawning species, which enhances fertilization, is known for algae, sponges, hydroids and ascidians, and sunrise or sunset provides the timing cue. Synchronous release of larvae after sunrise is also known for hydroids, bryozoans, and ascidians, and this response to light has been used to reliably collect larvae. However, the adaptive value of synchronous larval release at a particular time during the day is unknown; perhaps it swamps potential predators.

Semilunar (Tidal) Rhythms

Intertidal organisms, of course, are exposed to tidal rhythms that regularly cover them with water and expose them to air, and many synchronize their activities to these rhythms. Reproduction synchronized to semimonthly rhythms of spring and neap tides has been demonstrated for spawning by Japanese chitons and for larval release by a variety of semiterrestrial crabs, and undoubtedly occurs in many other species as well; these rhythms may enhance fertilization by retaining the gametes during neap tides, or they may disperse propagules during spring tides. Adult emergence from pupae of intertidal midges is known to be synchronized during the lowest tides in spring and summer to allow the flying adults to escape in the air, mate, and lay eggs before the water returns.

Lunar (Monthly) Rhythms

Lunar rhythms of gamete production and spawning are known for many marine animals, including some intertidal species. A polychaete of the northeastern Pacific leaves its tubes among intertidal surfgrass beds, swim to the surface, and spawns at night on spring and summer moonless nights. Like circadian and semilunar rhythms, lunar reproductive rhythms probably maximize fertilization, but they may also decrease predation if the adults swarm on moonless nights or may simply swamp predators if they all spawn together. Variation in moonlight over the month has been shown to serve as the proximate timing cue.

Semiannual Rhythms

Semiannual rhythms of gamete production and spawning are known mainly for tropical and subtropical species and are probably a response to semiannual monsoons, during which times lowered salinities are not favorable for larvae. Some temperate species also spawn twice a year, in spring and fall, probably in response to favorable intermediate temperatures.

Annual Rhythms

Most temperate species, including those living in the intertidal, display annual rhythms of gamete production, and spawning and/or larval release. Such rhythms place larvae in the water or juveniles on the bottom during the season most favorable for their survival. Warming sea temperatures and lengthening daylight in the spring often increases production of phytoplankton, the main food of invertebrate larvae, as well as of zooplankton, the main food of fish larvae; consequently, spawning usually peaks in spring and early summer. Moreover, winter storms tend to clear rocky shores of competitors and predators, favoring the subsequent recruitment of new juveniles. The environmental cues synchronizing gamete production, which begins months before spawning, can be changes in sea temperature, phytoplankton, or day length (photoperiod). Gamete production in northeastern Pacific purple sea urchins, for example, can be stimulated in the lab to begin with short day lengths (<12 hours), corresponding to decreasing fall day lengths, and gonads full of gametes ready to spawn are present 3 months later, which in the field is in midwinter, when they normally spawn. Similar manipulations of gamete production by photoperiod have been done with intertidal polychaetes, shrimp, and a variety of sea stars. Moreover, experiments with northeastern Pacific ochre sea stars revealed that long day lengths (>12 hr) synchronized gamete production to begin in the fall, but an endogenous (internal) circannual rhythm is present, in which the annual production of gametes continues even without being synchronized each year with day length. Our understanding of how reproduction is synchronized in different organisms, including those in the intertidal, is still very incomplete.

Multiyear Rhythms

Few studies on populations of organisms have lasted more than a few years, so little is known about possible rhythms with a frequency of multiple years. However, chronobiological analyses of long-term data sets for northeastern Pacific black chitons and purple sea urchins have revealed 7-year cycles in the variation of the annual rhythm of gonadal sizes. Moreover, recruitment of juveniles is known to vary enormously from year to year, and although such variation could be caused by variation in larval mortality due to unfavorable current or food conditions among years, it could also reflect underlying multiyear reproductive rhythms.

Arrhythmic Fluctuations

Although annual rhythms occur in most temperate and polar species of intertidal organisms, sychronizing reproduction with seasonal changes, some species are reproductively active throughout the year with arrhythmic fluctuations over time. Such continuous reproduction might be expected in tropical habitats, where favorable conditions occur throughout the year. Continuous reproduction is also found in many small, short-lived species in other regions, especially those living among sand and shells. Individual life spans in such species can be very short, only a few weeks or months, but turnover would need to be continuous throughout the year for populations to be maintained.

SEE ALSO THE FOLLOWING ARTICLES

Algal Life Cycles / Fertilization, Mechanics of / Genetic Variation and Population Structure / Rhythms, Nontidal / Rhythms, Tidal / Sex Allocation and Sexual Selection

FURTHER READING

Adiyodi, K. G. and R. G. Adiyodi, eds. 1983–1994. *Reproductive biology of invertebrates*, Vols. 1–6. Chichester, UK: John Wiley & Sons (continued as a Progress series).

Giese, A. C., J. S. Pearse, and V. B. Pearse, eds. 1974–1991. *Reproduction of marine invertebrates*. New York: Academic Press (Vols. 1–5); Pacific Grove, CA: Boxwood Press (Vol. 6); Palo Alto, CA: Blackwell Scientific (Vol. 9).

Graham, L. E., and L. W. Wilcox. 2000. *Algae*. Upper Saddle River, NJ: Prentice-Hall.

Horn, M. H., K. L. M. Martin, and M. A. Chotkowski. 1999. *Intertidal fishes: life in two worlds*. San Diego, CA: Academic Press.

Lee, J. J., G. F. Leedale, and P. Bradbury, eds. 2000. *The illustrated guide to the protozoa*, 2nd ed. Lawrence, KS: Society of Protozoologists.

Pearse, V., J. Pearse, M. Buchsbaum, and R. Buchsbaum. 1987. *Living invertebrates*. Pacific Grove, CA: Boxwood Press.

Round, F. E., R. M. Crawford, and D. G. Mann. 1990. *The diatoms: biology and morphology of the genera*. Cambridge, UK: Cambridge University Press.

Young, C. M., M. A. Sewell, and M. E. Rice, eds. 2002. *Atlas of marine invertebrate larvae*. San Diego, CA: Academic Press.

RHIZOCEPHALANS

MARK E. TORCHIN

Smithsonian Tropical Research Institute, Panama

JENS T. HØEG

University of Copenhagen, Denmark

Rhizocephalans (class Cirripedia, order Rhizocephala) are highly specialized parasitic barnacles. Only the larvae are distinguishable as barnacles using morphological cirripedian characters. There are over 250 known species of these parasitic barnacles, which infect primarily decapod crustaceans. A few rhizocephalan species are known to infect isopods, cumaceans, stomatopods, caridean shrimp, and other cirripeds, but the overwhelming majority parasitize brachyuran and anomuran crabs. The parasite initially develops within its host as a network of rootlike processes that penetrate through the host's body. These roots eventually give rise to an external saclike structure (the externa) that erupts through the host's cuticle (usually under the abdomen). Rhizocephalans are parasitic castrators that block reproduction, feminize infected males, often prevent growth, and thus can profoundly influence the ecology of their hosts.

DISTRIBUTION

On any given shore, after examining the undersides of enough crabs, one is bound to find one infected with a rhizocephalan barnacle. Rhizocephalans are widely distributed, ranging from arctic to tropical regions and inhabiting coastal rocky-shore, estuarine, and deep-water environments. There are two orders of the Rhizocephala (Kentrogonida and Akentrogonida), which are distinguished by the presence of a unique life history stage, found only in rhizocephalans, called the kentrogon, discussed in the following section. All Kentrogonida are known to parasitize only decapods, while the Aketrogonida infect a wider range of hosts including cirripeds, isopods, cumaceans, and stomatopods as well as some decapods.

Although some species of rhizocephalans are found on only a single host species, others appear to have a broader host range; thus, host specificity of rhizocephalans can vary. Because adult rhizocephalans have few morphological characters, species identification is often cumbersome or impossible, and much of the existing species-level taxonomy may be unreliable. Using molecular techniques, which allow genetic differentiation, and through infection experiments it is possible to distinguish species identity and host specificity. Some species are reported from only a single species, such as *Lerneodiscus porcellanae,* parasitizing the intertidal Californian porcelain crab *Petrolisthes cabrilloi* (Fig. 1A). Others are reported from several genera of crab hosts, such as the European *Sacculina carcini* (Fig. 1B), which also occurs in the intertidal or shallow subtidal zone. Some rhizocephalan species are reported from multiple host species and occur over a wide geographic range. The hermit crab–infecting *Peltogaster paguri* (Fig. 2A) is reported from eastern Atlantic and western and eastern Pacific shores in a number of host species.

FIGURE 2 Rhizocephalans of hermit crabs, (A) *Peltogaster paguri* infecting *pagurus bernhardus* and (B) Multiple externae of *Peltogaterella sulcatus* infecting *Anapagurus chiroacanthus*. Photographs by Jens Høeg.

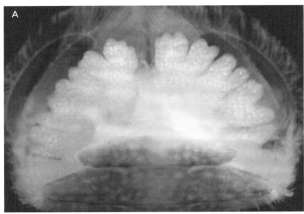

FIGURE 1 Rhizocephalan barnacles. (A) Externa of *Lerneodiscus porcellanae* under the abdomen of its host, the Californian porcelain crab, *Petrolisthes cabrilloi*. Note the developing embryos within the mantle chamber of the externa. Photograph by Jeff and Lise Goddard. (B) Externa of *Sacculina carcini* under the abdomen of the European shore crab, *Carcinus maenas*. Photograph by Todd Huspeni.

BIOLOGY AND LIFE CYCLE

To casual observers, a rhizocephalan can be mistaken for an egg mass on the underside of a crab. In fact, these parasitic castrators essentially replace the host's egg mass, most often erupting through the underside of the host's abdomen. The externa houses the parasite's reproductive

organs and serves as its brood chamber. Inside the crab's body cavity, a network of rootlike processes, called the interna, penetrate through the tissues, absorbing nutrients and supplying them to the externa. Some rhizocephalan species, such as *Peltogasterella*, develop multiple externae that connect to a common interna (Fig. 2B). The interna and externa constitute the adult female parasite (Fig. 3). Unlike most barnacles, which are hermaphroditic, rhizocephalans have separate sexes. Morphologically simplified "dwarf" males live within the female externa and provide sperm to fertilize successive broods.

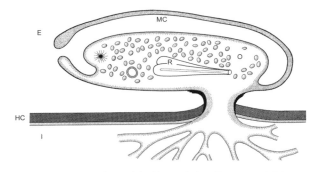

FIGURE 3 Anatomy of an adult rhizocephalan: E, externa; I, interna; HC, host cuticle; MC, mantle chamber; O, ovary; R, receptacle that holds dwarf males.

A generalized rhizocephalan life cycle is outlined in Fig. 4. Rhizocephalans begin life as nauplii and after successive molts transform into cypris (or cyprid) larvae, as do free-living barnacles. Unlike hermaphroditic free-living barnacles, however, rhizocephalan cyprids have separate sexes that are morphologically distinguishable. Female cyprids find, settle on, and cement to their crustacean host using modified antennules. The cyprids prefer

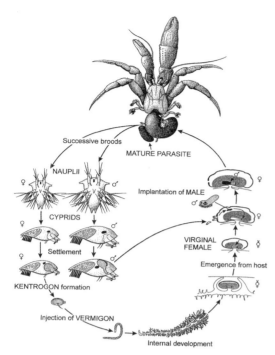

FIGURE 4 Generalized rhizocephalan life cycle.

to settle on recently molted crabs and usually attach to areas where the host's cuticle is relatively thin, such as the gills, the base of a seta, or the uncalcified cuticle at the base of a leg. The site of infection varies, depending on the species of rhizocephalan. Once attached, a unique larval stage called the kentrogon develops within the cyprid (the Akentrogonida lack this stage and the cyprid directly infects the host). Using a hollow cuticular stylet, the kentrogon penetrates the host cuticle and injects a worm-like stage called the vermigon into the host's hemocoel. The vermigon develops within the host, branching into the female interna, which upon maturation, produces a "virgin" externa that erupts through the host's cuticle upon molting. The virgin externa releases pheromones to attract male cyprids, which settle on the outer surface of the externa. Although the virgin externa may accumulate several male cyprids, which molt into a trichogon stage, a maximum of two trichogons will enter the narrow opening of the receptacle, eventually fertilizing the female, which grows into a "mature" externa within which embryos are brooded until they hatch as nauplii.

ECOLOGY

Rhizocephalans are directly transmitted parasites. Larval development is lecithotrophic (planktonic but nonfeeding) and generally of short duration, and this is especially

so in the Akentrogonida, of which all species hatch as cyprids. Dispersal ability is therefore limited, and the proportion of infected hosts in a population (prevalence) often varies over small spatial scales. Temporal variation in prevalence can also be dramatic. For some rhizocephalan species, such as *S. carcini*, externa emergence, larval production, and infection are highly seasonal. Other species, such as *Loxothylacus panopaei*, introduced in Chesapeake Bay, exhibit strong seasonal variation as well as annual variation in prevalence within mud crab populations. Because both the rhizocephalan larvae and infected mud crabs cannot survive low salinities, seasonal rains often eradicate local populations, while periods of drought may produce epidemics.

Almost all rhizocephalans castrate their hosts, and many species also interrupt their host's molt cycle, preventing growth. Although the precise mechanism remains elusive, castration is thought to be hormonally induced. Infected males generally exhibit a loss or reduction in their androgenic gland, the male sex-determining organ, causing feminization upon emergence of the externa. Behavioral modification has also been documented in infected crabs, whereby infected crabs ventilate and groom the rhizocephalan externa as if it were their own egg mass. Some rhizocephalan species, such as *S. carcini,* may induce infected crabs to exhibit seasonal migratory patterns that differ from those of healthy conspecific hosts. Rhizocephalans may control host populations. Infected hosts cannot reproduce and often do not grow; thus host abundance and population biomass may be reduced if parasite prevalence is high. The extent to which host populations are recruitment-limited and the scale at which the host and parasite larvae disperse will determine the magnitude of the impact of a rhizocephalan on its host population.

SEE ALSO THE FOLLOWING ARTICLES

Barnacles / Crabs / Diseases of Marine Animals / Dispersal / Hermit Crabs / Parasitism

FURTHER READING

Blower, S., and J. Roughgarden. 1989. Parasites detect host spatial pattern and density: a field experimental analysis. *Oecologia* 78: 138–141.

Glenner, H., and M. B. Hebsgaard. 2006. Phylogeny and evolution of life history strategies of the parasitic barnacles (Crustacea, Cirripedia, Rhizocephala). *Molecular Phylogenetics and Evolution* 41: 528–538.

Høeg, J. T. 1995. The biology and life cycle of the Rhizocephala (Cirripedia). *Journal of the Marine Biological Association of the UK* 75: 517–550.

Høeg, J. T., and J. Lützen. 1995. Life cycle and reproduction in the Cirripedia Rhizocephala. *Oceanography and Marine Biology: An Annual Review* 33: 427–485.

Høeg, J. T., H. Glenner, and J. Shields. 2005. Cirripedia Thoracica and Rhizocephala (barnacles), in *Marine Parasitology.* K. Rhode, ed. Collingwood, Victoria, Australia: 154–165.

Øksnebjerg, B. 2000. The Rhizocephala (Crustacea: Cirripedia) of the Mediterranean and Black Seas: taxonomy, biogeography, and ecology. *Israel Journal of Zoology* 46: 1–102.

Torchin, M. E., K. D. Lafferty, and A. M. Kuris. 2001. Release from parasites as natural enemies: increased performance of a globally introduced marine crab. *Biological Invasions* 3: 333–345.

Walker, G. 2001. Introduction to the Rhizocephala (Crustacea: Cirripedia). *Journal of Morphology* 249: 1–8.

RHYTHMS, NONTIDAL

KAREN L. MARTIN

Pepperdine University

The daily rhythm of the high and low tides is an obvious environmental cue to animals and plants living near ocean shores. However, many other, nontidal rhythms affect marine organisms, both in the intertidal zone and also offshore in deeper waters. These nontidal rhythms include some of the same cycles that terrestrial organisms recognize: daily cycles of sunlight and darkness, and seasonal cycles of day length, weather, and temperature changes. In addition, the daily tidal cycles themselves change dramatically with the phases of the moon and more subtly over the course of a year. The highest and lowest tides of the year occur at the full and new moons near the shortest days of winter and the longest days of summer. Chronobiology is the scientific study of biological rhythms, the predictable cyclic behavior of living things across time.

CIRCADIAN RHYTHMS AND INTERNAL CLOCKS

Many animals, plants, and even single-celled organisms such as dinoflagellates and cyanobacteria have some sort of internal clock. If they are placed in complete darkness or constant light, with no hint as to the actual time of day, they continue to maintain distinct rhythms of daily behavior over long periods of time. Without environmental cues, this cycle takes nearly but not exactly one solar day, a circadian rhythm.

When animals are in natural conditions, they "set" an internal circadian clock to synchronize normal behavior with the sun's light. Light-sensitive clocks are found in the eyestalks of crustaceans, the retina of fish, and in the brains of mammals. Animals can estimate the time of day from the amount and direction of incident light. Cycles for daily activity of feeding, moving about, or burrowing have been described in intertidal fish, shrimp, bivalves, and many types of crabs. In one dramatic case, every night the eyes of the horseshoe crab change in structure and function to become 10,000 times more sensitive than they were during the day. The eyes respond to changes in light levels and to an internal clock in the horseshoe crab's brain. Some animals possess multiple internal clocks that regulate different physiological processes. Hormones such as melatonin, growth hormone, and testosterone change over the course of a day in response to signals from the circadian clock.

BIOLOGICAL RHYTHMS AND SYZYGY TIDES

Across the 28 day lunar cycle, the tidal heights change dramatically. The greatest excursions between high and low tide occur at the new and full moon, every two weeks. These high syzygy tides reach high on shore, flooding the entire intertidal zone. A few hours later, extremely low syzygy tides leave much of the intertidal zone exposed to air. Many marine intertidal species use the syzygy tides (commonly called spring tides, but not related to the spring season) to synchronize their reproduction, including some intertidal algae. Beach-spawning fishes ride the waves of the highest tides and deposit their eggs in the upper intertidal zone so that the eggs may spend some time out of water during incubation. California grunion spawn completely out of water shortly after syzygy high tides of the full or new moon (Fig. 1), as the female emits her eggs into the wave-swept sand. The eggs incubate completely out of water buried in the damp sand, because the tidal excursion is lower on the following days. With the rising tides of the next syzygy, the eggs wash out into the surf and the hatchlings emerge, returning to aquatic life in the ocean. Numerous fish species spawn with the syzygy tides, while other species are more responsive to annual rhythms or to daily tidal cycles.

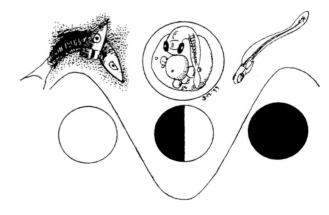

FIGURE 1 The excursions of the highest tides are affected by the moon phases, and the California grunion takes advantage of this during spawning, by placing eggs high on shore to incubate out of water between two syzygy tide series. Drawing by Gregory Martin.

INTERACTIONS AMONG DIFFERENT CYCLES AND ENVIRONMENTAL CUES

Intertidal animals and plants show many environmental rhythms of behavior; for example, horseshoe crabs forage only at night, and woolly sculpins spawn in autumn. These activity patterns may combine the circadian, annual, and tidal cycles. Two tidal cycles constitute a lunar day of 24.8 hours, slightly longer than a solar day. Thus, the time of tidal highs and lows changes every day. Careful study is necessary to understand which environmental cues are most relevant in determining behavior. Postlarval brown shrimp increase swimming activity in response to decreased sunlight but also to increases in salinity, which occur tidally near a bay. Swimming activity in May, with high tides in the late afternoon, seems to indicate that shrimp are responding to a tidal cue (Fig. 2). However in late September, high tides occur earlier in the day but swimming activity continues to peak at night, completely out of phase with the tides, indicating a strong circadian rhythm. Laboratory studies confirmed that the daily activity rhythms were based on internal cues, not simply the presence or absence of light.

Photosynthesis takes place only in sunlight, so at night oxygen levels may drop in a crowded tidepool. Carbon dioxide accumulates, causing an increase in acidity of the water. As a result, kelpfish may be found in tidepools during daytime low tides, but not at night. Probably they are less able than tidepool residents to withstand harsh conditions.

Changes in day length produce seasonal cycles of light and temperature that affect nutrient levels of the ocean and phytoplankton productivity. Animals and plants respond with cycles of reproduction, migration, and growth that are about one year long, or circannual rhythms, along with circadian and tidal rhythms.

Current areas of chronobiology research include studies of the integration of multiple environmental cycles, molecular and genetic characterization of endogenous clock function, and prediction of behavior and reproduction in different kinds of organisms from the interactions of endogenous rhythms and environmental cues.

SEE ALSO THE FOLLOWING ARTICLES

Light, Effects of / Photosynthesis / Reproduction / Tides

FURTHER READING

Battelle, B. A. 2002. Circadian efferent input to *Limulus* eyes: anatomy, circuitry, and impact. *Microscopy Research and Technology* 58: 345–55.

DeCoursey, P. J., ed. 1976. *Biological rhythms in the marine environment.* Columbia: University of South Carolina Press.

Edmunds, L. N. 1988. *Cellular and molecular bases of biological clocks.* New York: Springer-Verlag.

Martin, K. L. M. 1999. Ready and waiting: delayed hatching and extended incubation of anamniotic vertebrate terrestrial eggs. *American Zoologist* 39: 279–288.

Martin, K. L. M., R. C. Van Winkle, J. E. Drais, and H. Lakisic. 2004. Beach spawning fishes, terrestrial eggs, and air breathing. *Physiological and Biochemical Zoology* 77: 750–759.

Matthews, T. R., W. W. Schroeder, and D. E. Stearns. 1991. Endogenous rhythm, light and salinity effects on postlarval brown shrimp *Penaeus aztecus* Ives recruitment to estuaries. *Journal of Experimental Marine Biology and Ecology* 154: 177–189.

Palmer, J. D. 1995. *The biological rhythms and clocks of intertidal animals.* Oxford: Oxford University Press.

Taylor, M. H. 1999. A suite of adaptations for intertidal spawning. *American Zoologist* 39: 313–320.

Thurman, C. L. 2004. Unravelling the ecological significance of endogenous rhythms in intertidal crabs. *Biological Rhythm Research* 35: 43–67.

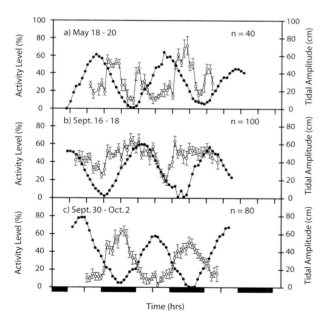

FIGURE 2 Interaction of circadian rhythms of swimming activity (open circles) with tidal cycles (filled circles) in brown shrimp postlarvae. In the first two experiments, it is difficult to distinguish whether the activity is initiated by the night time or the high tides. However in the third experiment, tidal cycles and activity cycles are completely out of phase, indicating that activity levels are primarily due to a diurnal rather than a tidal rhythm. Image courtesy of Tom Matthews.

RHYTHMS, TIDAL

STEVEN G. MORGAN

University of California, Davis

The physiology and behavior of marine organisms often cycles with the ebb and flow of tides. These rhythms are timed by one or more innate clocks that are coupled to oscillations in cues (such as hydrostatic pressure, temperature, salinity, turbulence, vibrations, or immersion) that are associated with tidal and tidal amplitude (spring–neap) cycles, and they continue, or free-run, with an approximate periodicity several times when removed from tidal conditions. Although exogenous (external) cues can stimulate activities directly, endogenous (internal) rhythms enable animals to anticipate tides and activate at the appropriate time when tides are obscured by winds.

PROTISTS AND ALGAE

Protists that reside in tidepools on rocky shores throughout their lives, including ciliates and dinoflagellates, alternate between an encysted (~18-h) and feeding swimming stage (~6-h) with an endongenous tidal rhythm that enhances remaining in tidepools during flood tides. Just before pools are flooded, swimming individuals attach to the substratum and form a cyst, which lasts for two high tides and one low tide. Just after the second high tide begins to ebb, swimming individuals emerge from cysts and remain until the next high tide reaches the pool. Swimming individuals may not be produced during the second low tide, because the extra cost imposed by two further selection events per lunar day may not be compensated by the gain in feeding during a second low tide. Temporal and spatial changes in the duration and phase of encystment occur, because tidepools are flooded at different times of the day, at different shore levels, and for different durations over the tidal amplitude cycle. Endogenous tidal rhythms are continually adjusted by natural selection on these fast-growing, short-lived individuals, as evidenced by artificial selection for a faster tidal rhythm and changes in the phase and duration of the encysted stage at different shore levels.

Gamete release by fucoid algae typically is timed to coincide with high slack tides when calm conditions enhance fertilization success. Gametes occasionally are released during low tides after several weeks of stormy, cloudy weather. A biweekly rhythm in gamete release by *Dictyota dichotma* is entrained by moonlight, and an endogenous biweekly rhythm occurs in the green alga *Halicystis ovalis*, which reproduces only during spring (maximum-amplitude) tides.

MOLLUSCS

Endogenous tidal rhythms of activity in intertidal snails are related to vertical zonation of species on the shore at temperate and tropical latitudes. Species residing in the mid-upper intertidal zone are active during high tide until the substrate dries, and they remain active longer at night when the substrate does not dry as fast *(Retina plicata, Melanerita atramentosa, Bembicium nanum, Austrococlea obtuse, Morula marginalba, Littorina planaxis, Littorina brevicula). Littorina brevicula* displays endogenous rhythms of movement for one month and spawning of fertilized egg masses for one year under constant conditions in the laboratory, peaking during the time of predicted high tide while they are immersed. *Littorina planaxis* has endogenous tidal and biweekly rhythms of spawning, which is induced by wetting. Species inhabiting the low-mid intertidal zone are active during low tide *(Amphinerita polita, Theliostyla albicilla)*, and species living at or above the upper limit or at or below the lower limit of the intertidal zone do not display tidal rhythms *(Nodilittorina pyramidalis, Melarapha unifasciata, Bembecium auratum)*. Activity patterns vary geographically for some neritid species *(R. plicata, A. polita, T. albicilla)*, occurring during daytime and nighttime high tides in Australia and during nocturnal low tides in the Indian Ocean and Red Sea.

Limpets migrate between home scars or refuges to surrounding areas to forage, and endogenous tidal rhythms of activity have been demonstrated in several species. Peak activity of *Patella vulgata* and *Helcion pectunculus* occurs during low tides, and peak activity of *Cellana grata* occurs during high tides. Spraying *C. grata* with seawater induces them to move higher on the shore during flood tide, and a tidal rhythm enables them to anticipate the time to begin and end foraging. A second ~7.2-h rhythm stops downshore movement during ebb tides, ensuring that limpets return to the same shore level. Siphonarian limpets also are active during high tides while they are awash, and they return to the same shelter during low tide, except for two species *(Siphonaria capensis, S. theristes)* that are active during low tide. Both tidal and tidal amplitude rhythms are evident for *Siphonaria theristes*. This species is active as soon as it is exposed completely by ebb tides, and it is inactive well in advance of inundation by flood tides. Further, it is active only during the lowest amplitude tide of the day throughout the year, even as the timing relative to the light–dark cycle switches. This chemically defended species is unusual in having its foot exposed by a reduced shell, living in the high intertidal zone, and being active during daytime maximum amplitude low tides in the summer. Despite increased desiccation, exceptionally low ability to grip the substrate favors activity when and where dislodgment by waves is minimized.

The behavior, morphology and defense of opisthobranch molluscs *(Onchidium, Onchidella)* have converged on that of *S. theristes*. These opisthobranchs lack shells, are chemically defended, home to refuges during high tides, and are active during daytime low tides. The parallels are particularly striking for *Onchidella binneyi*, which occurs high on the shore, is easily dislodged by waves, becomes active during ebb tides at night, and is most active during maximum-amplitude low tides. Tidal and biweekly rhythms of activity by this species are endogenous.

Chitons typically spawn gametes during high tide, and endogenous daily (24-h), tidal and biweekly rhythms of spawning occur in *Acanthopleura japonica*. Spawning peaks 30–60 min before high spring tides in the morning rather than the afternoon, regardless of the position of chitons on the shoreline, thereby synchronizing spawning during a narrow window and enhancing fertilization success. Biweekly spawning is entrained by a cycle of phase relationships between light and immersion.

Mussels open their valves, extend their siphons or foot, and filter particles from the water column when they are immersed during high tides. Tidal rhythms have been demonstrated by holding individuals collected from the field under constant conditions in the laboratory and translocating mussels between tidal regimes.

INSECTS

The beetle *Thalassotrechus barbarae* emerges at night from refuges in the supratidal zone to forage in the intertidal zone during low tide, and it exhibits endogenous tidal and daily rhythms. In a particularly well-studied group, biweekly and daily rhythms of emergence by adult midges *(Clunio)* from pupal cases that have been deposited in the intertidal zone enhance reproductive success. During the 3-h life of adult midges, males emerge from pupal cases, search for the pupae of unwinged females, assist females to emerge and mate. Females then search for and deposit eggs on appropriate exposed substrates for sedentary larvae to construct tubes and feed on diatoms and detritus. The timing of emergence among species changes geographically in response to different tidal conditions and spawning habitat availability. *Clunio tsushimensis* deposits eggs in the lower intertidal zone in Japan and Germany and emerges biweekly near maximum-amplitude low tides, whereas eggs are deposited higher in the intertidal zone and midges emerge during low tides every day in Norway. Egg deposition occurs throughout more of the intertidal zone at northern latitudes, where algae, the preferred deposition substrate, extend higher in the intertidal zone as a result of reduced physiological stress. In sheltered bays of the Baltic Sea, *C. balticus* deposits eggs at the sea surface; eggs sink to the bottom, where they always are submerged;

and emergence occurs near sunset regardless of the tides, which minimizes predation by visually feeding fishes.

A complex indirect timing system synchronizes emergence of midges with exposure of larval habitat on coasts that have strong tides. Emergence of midges during extreme low tides is timed by biweekly and daily rhythms. The biweekly rhythm ensures that midges emerge during the lowest tides of the month, when suitable larval habitat is exposed, whereas the daily rhythm ensures that midges emerge during low tides on these days. Midges are able to time emergence on low tides without a tidal rhythm because maximum-amplitude low tides occur at the same time of day every 2 weeks in semidiurnal (two tidal cycles per day) tidal regimes. However, this requires that populations adapt genetically to local tidal environments, because the time of extreme low tides varies along coastlines.

Entraining cues that synchronize biweekly emergence rhythms differ among populations (Fig. 1). A biweekly rhythm of emergence by *C. tsushimensis* in Japan, *C. mediterraneus* in Yugoslavia, and *C. marinus* in Spain and France primarily is entrained by moonlight and secondarily by vibrations of the substrate; waves breaking on the shore generate stronger vibrations during flood tides than during ebb tides. In contrast to these southern populations, biweekly emergence patterns were entrained primarily by vibrations associated with the tidal cycle and secondarily by moonlight in Germany, where moonlight is not as reliable

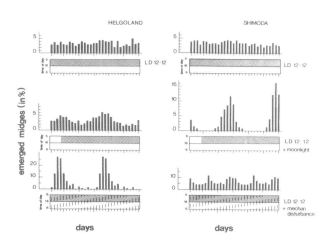

FIGURE 1 Emergence of *Clunio marinus* from Helgoland, Germany, and *C. tsushimensis* from Shimoda, Japan, under controlled laboratory conditions. Top panel, emergence in a 12:12 day–night regimen at 18 °C; middle panel, emergence under simulated moonlight (four nights of 0.3 lux illumination every 30 d in a 12:12 day–night regimen); bottom panel, emergence under a 12:12 day–night cycle and 12.4-h tidal cycle of mechanical water disturbances, resulting in equal phase relationships with the 24-h day–night cycle every 15 d. From Neumann (1985), with kind permission of Springer Science and Business Media.

during northern summers. Not only are nights brief and not particularly dark, but full moons barely rise above the horizon, so moonlight is weak, less likely to penetrate water, and more often obscured by clouds. At sheltered sites where tidal vibrations are weak, water temperature rather than wave shock entrains biweekly rhythms. Thus, several cues that are associated with the tidal amplitude cycle entrain biweekly emergence by midges from different locations.

CRABS

Endogenous tidal rhythms of activity, which peaks during high tide, were found in an anomuran hermit crab *(Pagurus geminus)* in areas that experience large tidal ranges as well as in areas that have small ones. Tidal and daily rhythms of activity also have been demonstrated for several species of brachyuran crabs that inhabit rocky shores *(Carcinus maenas, Cyclograpsus lavauxi, Hemigrapsus edwardsi)*. These crabs remain in refuges during daytime low tides and emerge during high tides, especially at night. In *C. maenas*, tidal rhythms in sensory and neuronal motor activity and the rates of gill ventilation and oxygen consumption peak just after the time of high slack tide, whereas blood sugar concentrations peak during low tide. These tidal rhythms are entrained by changes in hydrostatic pressure, temperature or salinity, and only daily rhythms are expressed in the absence of tides. Tidal rhythmicity is known to be innate, because tidal rhythms have been induced in juveniles that were raised in the absence of tides since they were embryos.

Female brachyuran crabs attach fertilized eggs beneath their abdomens, incubate embryos in refuges, and synchronize larval release endogenously relative to tidal, tidal amplitude, light–dark, and lunar cycles. Hatching is timed according to one or more of these four environmental cycles and to different phases of the cycles. Although the potential for many different patterns of larval release by crabs around the world, strong patterns occur in all of the more than 50 species studied and partially are related to the vertical zonation of adults on shorelines. Intertidal species typically remain near their refuges when releasing larvae, and they wait until high slack tides or shortly after, when larvae are swept away from shore. Species that live highest on the shore may release larvae monthly during the brief time that they are inundated. Middle-intertidal species release larvae only biweekly, when they are inundated by both maximum-amplitude tides. Low intertidal and subtidal species are immersed daily, and some of them release larvae daily, whereas others release larvae biweekly during maximum-amplitude tides. Supratidal crabs release larvae less synchronously relative to the tidal amplitude cycle than do high-intertidal species. Because supratidal crabs must leave burrows to release larvae, they are not limited to hatching on the highest tides, and peak larval release occurs biweekly, near maximum-amplitude tides, rather than monthly as in high-intertidal species. The timing of larval release relative to the light–dark cycle also is most synchronous for supratidal and high-intertidal species and least synchronous for low-intertidal species; supratidal and high-intertidal species release larvae only at night, but crabs lower on the shore sometimes release larvae during the daytime too, because light attenuates rapidly in a deeper turbid water column.

Predation by visual predators best explains the hatching patterns of crabs worldwide. Egg-bearing females, embryos, and larvae all risk predation during larval release, and the timing of larval release may reduce predation on all three life stages. Larvae are released near refuges, thereby reducing predation on embryos or females by fishes and other marine predators that forage during high tides. Females that attract the attention of predators upon venturing into the open to vigorously pump larvae into surrounding waters may quickly retreat to refuges until predators depart. Predation on females or embryos strongly affects the timing of larval release, because all intertidal crabs release larvae while their habitats are immersed regardless of phylogeny and considerable differences in body size, armor, coloration, clutch size, and speed, which affect their vulnerabilities to predators. Consequently, predation on females or embryos enforces larval release near refuges, so the vertical zonation of species partly determines the synchrony of larval release relative to the tidal amplitude cycle. Another benefit accrues from releasing larvae on maximum-amplitude high tides, when larvae are most rapidly swept away from dense assemblages of plankton-feeding fishes that aggregate along shorelines. This explains why even some low-intertidal and subtidal species with larvae that are particularly vulnerable to predation may hatch biweekly even though crabs are inundated daily, whereas other low-intertidal species, having larvae that are less conspicuously colored by yolk and chromatophores or are better defended from fishes by spines, release larvae daily. Moreover, diurnally foraging fishes select for larval release under the cover of darkness, except for those few low-mid-intertidal species that also release relatively inconspicuously pigmented larvae into a deeper murky water column during the daytime.

The timing of larval release by intertidal and shallow subtidal crabs changes across tidal regimes due to shifts in the phase relationships of tidal and tidal amplitude cycles relative to light–dark and lunar cycles. Phasing of environmental cycles also shifts throughout the year, except in semidiurnal tidal regimes, resulting in changes in the timing of larval release for species that have protracted reproductive seasons. Lastly, the timing of larval release shifts relative to the lunar cycle every 7 months as the highest-amplitude tide shifts, coinciding with new and full moons in most places in the world. Populations residing highest on the shore all make this switch, so that crabs can release larvae while immersed. However, populations occurring low on the shore may not make this switch, releasing larvae on new moons when predation by visual predators is minimized. These shifts in the timing of larval release reveal species-specific hierarchies of rhythms that are entrained by the tidal, tidal amplitude, light–dark, and lunar cycles.

Tidal rhythms initially were thought to be controlled by a single bimodal circadian (~24-h) clock with two peaks per cycle. However, research on brachyuran crabs revealed that tidal rhythms may be controlled by two lunar-day (24.8-h) clocks that are loosely coupled in antiphase (1800 apart), with each oscillator tracking one of the two tidal cycles that occurs each lunar day. Evidence for this hypothesis arose from individuals showing independence of the two daily peaks in activity related to the tides. Other investigators have found that a single, true circatidal (~12.4-h) clock may exist and operate independently from the circadian clock, because the tidal rhythm can be phase-shifted to a new time without resetting the daily rhythm. It also remains uncertain whether biweekly and monthly rhythms are controlled by separate clocks or by a combination of circatidal and circadian clocks. In the latter case, the daily amplitude would be reinforced and augmented at biweekly or monthly intervals by the coincidence of the tidal peak.

FISHES

Juvenile fishes (*Pollachius virens, Tautoglabrous adspersus, Cebidichthys violaceus*) migrate onshore during floods to forage in rocky intertidal habitats, whereas resident fishes of rocky shores, including sculpins, blennies, and gobies, remain in tidepools during low tide and migrate to surround areas to forage during high tide. Resident fishes typically stay near cover to avoid being swept away by waves during foraging forays to surrounding areas, and they often return home to their pool. Endogenous tidal rhythms occur in rocky-intertidal fishes (*Lipophrys pholis, Gobius paganellus, Coryphoblennius galerita, Chasmichthys gulosus, Oligocottus maculosus*) and can be entrained by oscillations in hydrostatic pressure or turbulence. Tidal rhythms may not be expressed in fishes living in areas with small tidal ranges. Daily spawning by the cleaner wrasse, *Labroides dimidiatus*, on a rocky shore tracks afternoon high tides, and spawning continues to occur in the afternoon at times of the tidal amplitude cycle when high tides do not occur.

SEE ALSO THE FOLLOWING ARTICLES

Fish / Predator Avoidance / Reproduction / Tides / Turbulence / Zonation

FURTHER READING

Morgan, S. G. 1995. The timing of larval release, in *Ecology of marine invertebrate larvae*. L. McEdward, ed. Boca Raton, FL: CRC Press, 157–191.

Neumann, D. 1985. Photoperiodic influences of the moon on behavioral and developmental performances of organisms. Proceedings of the 10th International Congress of Biometeorology. *International Journal of Biometeorology* 29, supp. 2: 165–177.

Palmer, J. D. 1974. *Biological clocks in marine organisms*. London: Wiley.

Palmer, J. D 1995. *The biological rhythms and clocks of intertidal animals*. New York, NY: Oxford University Press.

Skov, M. W., R. G. Hartnoll, R. K. Ruwa, J. P. Shunula, M. Vannini, and S. Cannicii. 2005. Marching to a different drummer: crabs synchronize reproduction to a 14-month lunar-tidal cycle. *Ecology* 86: 1164–1171.

RICKETTS, STEINBECK, AND INTERTIDAL ECOLOGY

SUSAN SHILLINGLAW

San Jose State University

WILLIAM GILLY

Stanford University

Marine ecology, particularly of intertidal environments, provided common ground for biologist Edward Flanders Ricketts (1897–1948) and his closest friend, writer John Steinbeck (1902–1968). Ricketts's scientific training was largely autodidactic, based on many years of careful field observations as a professional collector on the central California coast. Steinbeck, on the other hand, collaborated with Ricketts by avocation. Together they forged a unique collaborative vision that united biology, history, and philosophy, giving rise to important contributions in both science and literature.

RICKETTS AS SCIENTIFIC COLLECTOR

In 1923 Ricketts came to the Monterey Peninsula from Chicago to study intertidal life, setting up one of the first biological supply businesses on the West Coast. As a scientist, Ricketts was a superb observer of intertidal life; he was also a philosopher and a humanist who willingly explained invertebrate peculiarities to anyone interested. From these overlapping interests in collecting, cataloguing, and communicating with others—what he would call full "participation" through observation and understanding—came his idiosyncratic and highly influential book about intertidal animals, *Between Pacific Tides*, published in 1939 by Stanford University Press and still in print (Fig. 1). In a preface to the third edition (1952), Steinbeck writes: "This is a book for laymen, for beginners, and, as such, its main purpose is to stimulate curiosity, not to answer finally questions which are only temporarily answerable." Perhaps the sturdy appeal of Ricketts's intertidal gaze lies, as Steinbeck suggests, in the text's resistance to closure. *Between Pacific Tides* invites "participation." The novice learns to see "a world under a rock." In the half century following Ricketts's 1948 death, trained marine biologists have revised each new edition, updating the core text written by Ricketts with assistance from coauthor Jack Calvin.

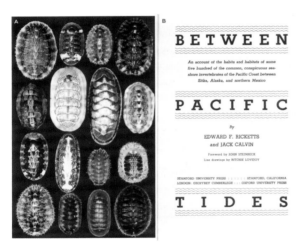

FIGURE 1 Frontispiece and title page from *Between Pacific Tides*, second edition, 1948. Photograph by Russell Cummings.

BETWEEN PACIFIC TIDES

Zonation is the organizing principle of the text: "An account of the habits and habitats of some five hundred of the common, conspicuous seashore invertebrates of the Pacific Coast between Sitka, Alaska, and northern Mexico," announces the subtitle. Different chapters study particular habitats, ranging from protected outer coast to wharf pilings. In each chapter, Ricketts asks readers to consider the principal factors that determine distribution of species and individuals within that habitat: wave shock, type of substrate, tidal exposure.

First published in an era when authoritative texts on the intertidal were organized by scientific classification of the relevant species, Ricketts's book was notable, even revolutionary, because of its focus on the environmental factors, which led to a broad, ecological picture that was readily accessible. His prose is, in turn, factual, scholarly, historical, humanistic, and practical. In the first chapter, for example, he urges readers to replace "carefully" all rocks overturned when investigating tidepools and then warns that the "un-rubber-booted" observer may get wet feet. He frequently notes contemporary field studies of the animals, having consulted more than 800 scientific articles in the course of his research. Finally, he makes note of human and regional peculiarities. "The Mexicans justly prize the owl limpet as food, for when properly prepared it is delicious, having finer meat and a more delicate flavor than abalone." Italians, he notes, relish snails "cooked in oil and served in the shell," a bit of culinary science written long before the vogue for combining text and recipes.

RICKETTS AS ECOLOGIST

An ever increasing interest in ecological relationships marked Ricketts's career as a marine biologist and led him to examine ecological implications for human societies and for the whole of life, molecular to spiritual. When Ricketts was a student at the University of Chicago from 1919 to 1922, his interest in connectivity had been nurtured by W. C. Allee, whose classic work, *Animal Aggregations: A Study in General Sociology* (1931), argues that animals behave differently in groups than as individuals and that animal aggregations could withstand stresses that individual specimens could not. This notion of cooperation profoundly influenced Ricketts's—and subsequently Steinbeck's—vision of intertidal life and of life in general. Throughout his 24 years on the Peninsula, Ricketts was a dogged and curious investigator, whose ecological studies of the dependence of various invertebrates on their environments as well as on other species, including man, had far-ranging implications that moved beyond the intertidal (Fig. 2). For example, he studied the decline of the sardine population in Monterey Bay and its relation to changing plankton levels in the eastern Pacific as well as to increasing commercial exploitation. His insights greatly preceded general scientific awareness of cycles in marine ecosystems, events that today are at the center of much marine biological research.

SEA OF CORTEZ

By far Ricketts's most famous work was *Sea of Cortez*, published in 1941 and written in collaboration with Steinbeck. Ricketts and Steinbeck had been friends for a

FIGURE 2 Edward F. Ricketts in lab, 1945. Photograph by Peter Stackpole. Copyright Peter Stackpole/Time Life/Getty Images.

decade when they decided to launch an expedition into the Gulf of California to catalogue marine life in the Gulf littoral, a relatively unexplored region in 1940. For six weeks in the spring of 1940, the crew of *The Western Flyer* collected and catalogued animals in the intertidal (Fig. 3). The text describing this expedition is, in itself, a stew of narrative techniques: half the book is a travelogue written by Steinbeck using Ricketts's field notes, and half is an annotated and referenced phyletic catalogue of the more than 500 invertebrate species collected on the voyage, carefully prepared by Ricketts. The narrative touches on history of the Gulf, the history of science, philosophical speculation, and descriptions of intertidal collecting sites and organisms. The book's narrative complexity mirrors

FIGURE 3 *The Western Flyer.* Photograph courtesy of Ed Ricketts, Jr.

the interrelationships of the littoral region being explored as well as of the replacement of an indigenous human culture with one of modern development (Fig. 4).

FIGURE 4 The only known photograph of Steinbeck (center right) and Ricketts (far right) together. Hunting expedition out of Puerto Escondido, Mexico, 1940. Photograph courtesy of Ed Ricketts Jr.

For Steinbeck and Ricketts the tidepool came to symbolize the complexity of all life. In one of the most famous passages in *Sea of Cortez*, they write, "[A] man looking at reality brings his own limitations to the world. If he has strength and energy of mind the tide pool stretches both ways, digs back to electrons and leaps space into the universe and fights out of the moment into non-conceptual time. Then ecology has a synonym which is ALL." Their sense of connectivity included invertebrates, the animals' environments, man's impact on environment, and the philosophical implications of it all—what Ricketts called the "*toto* picture." For Ricketts and Steinbeck, "breaking through" physical reality to spiritual understanding was the full measure of an ecological holism that much of their work revolves around.

STEINBECK'S CANNERY ROW

Steinbeck captured Ricketts's ecological perspective in fiction with the 1945 novel *Cannery Row*. The book's "hero," if he can be called such, is Doc, a fictional Ricketts, who is introduced into the novel collecting starfish in Monterey's Great Tide Pool and who closes the book reading a poem by Chinese poet Li Po.

Indeed, the ecological overtones of *Cannery Row* are integral to its meaning. As Steinbeck notes in his opening chapter, the tidepool is the book's central metaphor, and his art transforms Monterey's Cannery Row into a human tidepool. Specimens are the Chinese grocer Lee Chong, the madam Dora, and the bums, Mack and the boys. From the centrality of the intertidal community of Cannery Row, the book moves outward to embrace specimens of other habitats around Monterey—lonely Mary Talbot in her house full of cats and the cantankerous owner of a Carmel Valley frog pond. And in luminous moments, the text suggests what "breaking through" means: a dreamlike world connecting humans to cosmic notions of life and death. Steinbeck's tidepool world links the literal to the symbolic.

The careers of Ricketts and Steinbeck intermeshed, particularly during the 1930s. Both found in an ecological perspective the lens to study intertidal and human communities, recognizing interconnections among invertebrates, animals, humans, human histories, and environments. Their vision was fundamentally holistic—and prescient.

SEE ALSO THE FOLLOWING ARTICLES

Collection and Preservation of Intertidal Organisms / Marine Sanctuaries and Parks

FURTHER READING

Astro, R. 1973. *John Steinbeck and Edward F. Ricketts, the shaping of a novelist.* Minneapolis: University of Minnesota Press.

Beegel, S. F., S. Shillinglaw, and W. N. Tiffney, Jr., eds. 1997. *Steinbeck and the environment: interdisciplinary approaches.* Tuscaloosa: University of Alabama Press.

Ricketts, E.F Orig. undated. *The Outer Shores.* Joel W. Hedgpeth, ed. Eureka, CA: Mad River Press, 1978.

Ricketts, E. F., and J. Calvin. *Between Pacific Tides.* Stanford, CA: Stanford University Press, 1939.

Rodger, K. A., ed. 2002. *Renaissance man of Cannery Row: the life and letters of Edward F. Ricketts.* Tuscaloosa: University of Alabama Press.

Rodger, K. A., ed. 2006. *Breaking through: essays, journals, and travelogues of Edward F. Ricketts.* Berkeley: University of California Press.

Steinbeck, J. 1945. *Cannery Row.* New York: Viking Press.

Steinbeck, J., and E. F. Ricketts. 1941. *Sea of Cortez.* New York: Viking Press. Narrative portion published as *The Log from the* Sea of Cortez, 1951.

Tamm, E. E. 2004. *Beyond the outer shores: the untold odyssey of Ed Ricketts, the pioneering ecologist who inspired John Steinbeck and Joseph Campbell.* New York: Four Walls Eight Windows.

ROCK TYPES

SEE TIDEPOOLS, FORMATION AND ROCK TYPES

ROGUE WAVES

D. H. PEREGRINE

University of Bristol, United Kingdom

Rogue waves, or 'freak waves,' are those waves that are unexpectedly large or violent. Large waves are not necessarily rogue waves. If most waves are large, then another large wave is to be expected. If the sea is calm, even a relatively small wave may be unexpected and cause trouble enough to be "rogue." Offshore, in deep water, such waves are recognized as a problem and have caused substantial damage to ships and other marine structures. For deep water they are an active topic of research; although a few mechanisms for their creation are known, there is no general agreement as to how, or when, they arise. The wave environment differs at the coast.

ROGUE WAVES AT THE COAST

There are more sources of rogue waves on the coast, and yet there has been less study of them in this context. On beaches and rocks open to waves from an ocean, rogue waves are dangerous. News stories about serious injury and even death from rogue waves are regrettably common (see box). In more sheltered waters rogue waves might tend to cause inconvenient soaking rather than perilous situations. Various origins for rogue waves are given in the following paragraphs. For particular places the likelihood of each type should be assessed, so that people on the coast are aware of the danger or inconvenience that could occur.

Wind Waves

The waves on the surface of water are generally those generated by wind. For small waves this is easily seen, as they appear from individual gusts of wind. At larger scales a similar process is at work. When wind blows, waves

---

WAVE KILLS TWO IN SOUTHERN OREGON

(The Associated Press, on *The Seattle Times* web site, November 11, 2005.)

Port Orford, Ore.—A wave swept three people into the Pacific Ocean on Thursday, killing two and injuring one, the authorities said. Pamela Flynn, 72, and Thomas Flynn, 44, were pronounced dead on Agate Beach, and Brian Flynn, 42, was treated for hypothermia at a local hospital, said Lt. Dennis Dinsmore of the Curry County Sherrif's Office. They were all from Slagle, Idaho.

Deputies said the Flynns were walking on Agate Beach in Port Orford just after 2 pm. when the wave swept them into the surf. Rescue crews, responding to a 9-1-1 call, pulled Pamela and Brian Flynn from the ocean. A fishing boat in the area spotted Thomas Flynn about a half-mile offshore. A U.S. Coast Guard helicopter dropped a rescue swimmer into the ocean to retrieve him. The helicopter airlifted Flynn to the beach, but medics couldn't revive him, the Coast Guard said.

grow. The height and length of the waves increase with both wind speed and wind duration, so the most severe storms, such as hurricanes and deep depressions, produce the largest waves. This growth of waves is also affected by wave propagation. Generally, the waves travel away from the generating region, but sometimes a storm may follow the waves so that they continue to grow in size.

At the coast the water is usually shallower than offshore, and this causes waves to break and lose energy before they reach the shoreline. The longer and steeper the incident wave is, the more likely it is to break; thus, the waves that are biggest out at sea are not always the biggest at the shoreline. For waves that are being actively generated by wind, this effect tends to diminish the chance of a rogue wave at the shoreline.

It is a different matter for waves that propagate away from their region of generation. As they get further away, they become less steep, have proportionately longer crests, and are known as swell. These waves tend to focus their energy into wave groups. This is well known to surfers, who call the wave groups "sets." There may be five, seven, or more, waves in a typical set and then a long interval before the next set. This is a wave situation where a rogue wave can occur: when the number of waves in a set is just one. The long interval between such sets makes the single wave entirely unexpected to anyone who arrives at the coast when no wave is visible. Swell propagates a long way across the world's oceans: measurements have been made, in one case, swell from a storm in the southern Indian Ocean was tracked more than halfway around the world to Alaska. Thus, many ocean beaches are vulnerable to unexpected swell with a very short set. This is especially so for those waves that are first arrivals from a distant storm.

Wave Impact

Engineers who construct structures such as piers, breakwaters, and seawalls at the coast are naturally interested in the waves that may cause most damage to their work. Typically these are waves that break on or very close to the structure. Such waves are usually only a small fraction of the incident wave field and may occur only when the tide level is such as to give the right level of water at the structure. Laboratory experiments show that the most severe impacts occur only for special wave–depth combinations and then only rarely. Such potentially damaging impacts are also those that move the more substantial rocks on coasts and help shatter cliffs. On a lesser scale, examples such as the wave shown in Fig. 1 can soak an unexpectedly large region.

FIGURE 1 Wave impact on the Cornish coast on a relatively calm day in summer. Photograph by the author.

Ship Waves

In relatively sheltered waters, or anywhere on a quiet day, the largest coastal waves may be due to passing vessels. If these are large ships at very infrequent intervals, then the waves may be unexpected. Typically waves from a distant ship are well dispersed and build up to their maximum slowly. However, there is one class of ship wave that tends to be large and isolated and is known to cause trouble. Such waves are generated when vessels accelerate until their speed is equal to or greater than the maximum speed of water waves at their position. Such waves tend to form a single crest, a form that is retained as they propagate to the shore. The culprits are usually fast ferries, and nowadays operators of such ferries should be aware of the dangers of accelerating their craft in places where dangerous waves can form.

Low-Frequency Waves

When a set of waves breaks on a beach, it pushes water shorewards. On gentle beaches the whole set effectively acts together and generates a wave of long duration (i.e., low frequency), going up the beach. Such low-frequency waves also arise from the fluctuations of freshly generated wind waves. They are rarely visible except as occasional high or low water levels, but in their most extreme form can give unexpectedly large waterline fluctuations of even a meter or so in height.

The same sort of mechanism gives rise to long-period oscillations in harbors or small, almost enclosed, bays. Again the long period between wave crests arriving at one place can give unexpected increases in water levels.

Tsunamis

Tsunamis are water waves generated by ground motion, for which there are two sources: earthquakes and slides. Sometimes an earthquake stimulates a landslide or submarine slide, in which case the resulting tsunami can be much bigger than might be expected from the seismic ground motion. The following essential points are to be borne in mind when at the coast:

1. If an earthquake is felt, it is important to get to higher ground as fast as possible.
2. In areas where tsunamis from distant earthquakes occur, be aware of any warning signals.
3. If breakers are seen unexpectedly far from the shore, they could herald the arrival of a large tsunami.

SEE ALSO THE FOLLOWING ARTICLES

Hydrodynamic Forces / Ocean Waves / Storm Intensity and Wave Height / Wave Forces, Measurement of / Wind

FURTHER READING

Butt, T., P. Russell, and R. Grigg. 2004. *Surf science: an introduction to waves for surfing.* Honolulu: University of Hawai'i Press.

Grimshaw, R. J. H., ed. *ICMS workshop on rogue waves, 12–15 December 2005, Edinburgh.* Edinburgh: International Centre for Mathematical Sciences. http://icms.org.uk/archive/meetings/2005/roguewaves/sci_prog.html.

Müller, P., and D. Henderson, eds. 2005. *Rogue waves. proceedings of the 14th 'Aha Huliko'a Hawaiian Winter Workshop, 2005.* SOEST Special Publication. Manoa: University of Hawaii Press. http://www.soest.hawaii.edu/PubServices/2005pdfs/TOC2005.html.

Olagnon, M., and M. Prevosto, eds. *Rogue waves.* 2004. Brest, France: Ifremer. Proceedings of a workshop held in LeQuartz, Brest, France, October 20–22, 2004. http://www.ifremer.fr/web-com/stw2004/rw/.

Svendsen, I. A. 2006. *Introduction to nearshore hydrodynamics.* Singapore: World Scientific.

ROTIFERS

WILLEM H. DE SMET

University of Antwerp, Belgium

Rotifers (Phylum Rotifera) are tiny (40–2000 μm), multicellular, primary bilaterally symmetrical animals, having a fluid-filled cavity between the body wall and the internal organs (pseudocoelom) and a fixed number of nuclei (eutely). They are characterized by an anterior ciliary organ, the corona, and a complex masticatory apparatus called the mastax. Rotifers inhabit almost every type of aquatic and semiaquatic habitat all over the world. They often represent a consistent fraction of the biomass, and they play an important role in the energy transfer in food webs.

MORPHOLOGY AND INTERNAL ORGANIZATION

Descriptions of rotifer species always refer to the females, because the males are smaller and of much simpler organization. The body shape is very diverse, but usually comprises a head, trunk, and foot (Fig. 1). The head bears the corona, mouth, and sense organs and contains the brain and glandular retrocerebral organ. The corona is a ciliated region used for locomotion and food collection by means of its cilia. Its ground-plan consists of a ciliated area around the head and mouth opening, and two bands of cilia (trochus and cingulum). The muscular mastax is provided with sclerotized jaws (trophi), which are used to grasp, grind, or pierce and suck the food (Fig. 2). The trunk contains the digestive, excretory, and reproductive organs. The foot terminates in attachment organs (toes, adhesive disk). Pedal glands produce a secretion with

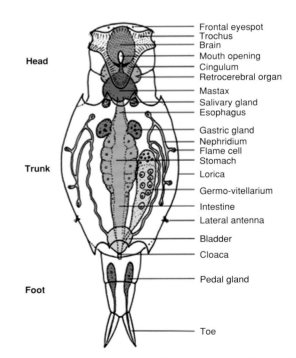

Head

Trunk

Foot

Frontal eyespot
Trochus
Brain
Mouth opening
Cingulum
Retrocerebral organ
Mastax
Salivary gland
Esophagus
Gastric gland
Nephridium
Flame cell
Stomach
Lorica
Germo-vitellarium
Intestine
Lateral antenna
Bladder
Cloaca
Pedal gland
Toe

FIGURE 1 Scheme of the anatomy of a monogonont rotifer in ventral view.

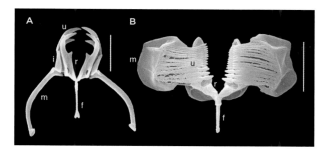

FIGURE 2 Examples of trophi types, scanning electron micrographs. (A) Forcipate type modified for grasping. (B) Malleoramate type modified for grinding. Composing sclerite elements: f, fulcrum; i, intramalleus; m, manubrium; r, ramus; u, uncus. Scale bars: 10 μm. Images by the author.

which rotifers anchor themselves to a substrate. The body is covered with an integument, containing a dense skeletal intracytoplasmic lamina, which often forms a stiff protective cover or lorica. Postembryonic tissues are syncytial. The number of nuclei is fixed during embryogenesis, constant throughout life, and species-specific (900 to 1000 in the species studied).

CLASSIFICATION, REPRODUCTION, AND OCCURRENCE

Rotifers consist of three major groups: Seisonidea (3 spp.), Bdelloidea (~375 spp.), and Monogononta (~1450 spp.). Seisonidea live epizoically on marine crustaceans; the genital apparatus is paired, and reproduction is exclusively sexual. Bdelloidea have paired ovaries and reproduce exclusively by parthenogenesis (males lacking). They are widespread inhabitants of freshwater and damp terrestrial habitats; a few are marine species. Many bdelloids can withstand unfavorable conditions through anhydrobiosis. Monogononta have a single ovary and reproduce by cyclic parthenogenesis, interrupted by sexual reproduction. The fertilized females produce resting eggs, which also can withstand harsh periods. The resting egg hatches only to a female, which can initiate a new population parthenogenetically. Monogononts are found in a wide range of both freshwater and marine habitats.

ROTIFERS FROM ROCKY SHORES AND TIDEPOOLS

About 290 taxa, mostly Monogononta, have been reported from marine environments, with some 50 of them living on wave-swept rocky shores and in tidepools, but all occurring in the marine littoral as well. Species richness and abundances are usually low on the exposed shores and greatly influenced by the degree of wave action. Only a few taxa (*Colurella* spp., *Encentrum* spp., *Proales* spp.) manage

to live in dense growths of filamentous cyanobacteria and tuft-forming seaweeds, or among the rootlike mass of holdfasts. A greater diversity and population development can often be observed in tidepools. Planktonic, periphytic, and benthic species, usually belonging to the genera *Aspelta*, *Colurella*, *Encentrum*, *Notholca*, *Proales*, *Synchaeta* and *Testudinella*, can all be present (Fig. 3). Tidepools are tremendously varied ecologically, and their rotifer community composition may vary accordingly, but few studies relate dynamics of rotifer populations to abiotic and biotic variables. The only observations show that species richness and total abundances decrease with increasing salinity; sediment loading apparently has a similar effect. Because of the significant preferences for phaeophytes or chlorophytes, occurrence of some rotifer species is also determined by the composition of the algal vegetation of the tidepools.

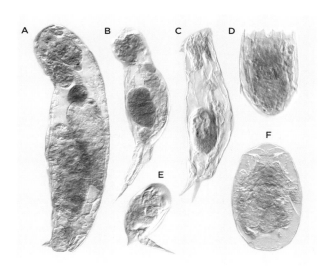

FIGURE 3 Some rotifers occurring in tidepools. (A) *Lindia gravitata*. (B) *Proales reinhardti*. (C) *Encentrum marinum*. (D) *Notholca striata*. (E) *Colurella adriatica*. (F) *Testudinella clypeata*. Images by the author.

SEE ALSO THE FOLLOWING ARTICLES

Food Webs / Salinity Stress / Wave Exposure

FURTHER READING

Fontaneto, D., W. H. De Smet, and C. Ricci. 2006. Rotifers in saltwater environments, re-evaluation of an inconspicuous taxon. *Journal of the Marine Biological Association of the UK.* 86: 623–656.

Saunders-Davies, A. 1998. Differences in rotifer populations of the littoral and sub-littoral pools of a large marine lagoon. *Hydrobiologia* 387/388: 225–230.

Wallace, R. L., T. W. Snell, C. Ricci, and T. Nogrady. 2006. *Rotifera: Volume 1: biology, ecology and systematics.* Guides to the Identification of the Microinvertebrates of the Continental Waters of the World 23. Ghent, Belgium: Kenobi Productions.

S

SALINITY, MEASUREMENT OF

STEVEN S. RUMRILL

South Slough National Estuarine Research Reserve

Salinity is a measure of the total quantity of elemental salts that are dissolved in water, expressed as a pure number derived from a dimensionless ratio that describes the electrical conductivity of seawater. Salinity values are generally elevated in marine waters, where the total amount of dissolved salt ions is high. Conversely, salinity values typically decrease in estuaries and near sources of freshwater where the total amount of dissolved salt ions is lower.

SALINITY AS A KEY OCEANOGRAPHIC PARAMETER

The salinity of seawater is a fundamental characteristic of a water mass. For example, fully marine or oceanic waters typically have salinity values in the range of 32 to 37, and differences in salinity can result from localized inputs of freshwater from a variety of sources. These inputs can include precipitation, evaporation, freezing and melting of ice, and discharges of freshwater from rivers and streams. In contrast, near-coastal waters and estuaries are frequently characterized by intermediate and lower salinity values in the range of 5 to 30 depending on site-specific differences in freshwater inputs and evaporation. The distinct salinity signatures of seawater and estuaries can be used to follow the movement, cohesion, and mixing of discrete water masses. Together, temperature and salinity combine to determine the density of seawater,

which provides a physical basis for vertical stratification of the water mass and sometimes results in formation of a distinct halocline (Fig. 1), which separates dense, high-salinity water from the more buoyant lower-salinity water. Differences in seawater salinity and density can also drive surface and subsurface currents and contribute to larger-scale processes of ocean circulation.

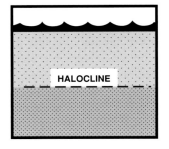

FIGURE 1 Vertical stratification of the amount of dissolved salts in seawater. A distinct halocline forms between the buoyant low-density, low-salinity water at the top of the water mass and the high-density, high-salinity water below.

ELEMENTS THAT CONTRIBUTE TO SALINITY

Seawater is composed of about 96.5% water and about 3.5% dissolved substances, which are mostly salts (Fig. 2). Chloride (Cl^-) and sodium (Na^+) ions make up the majority (about 85.6%) of the salts dissolved in seawater. The other major constituents that determine the salt content, electrical conductivity, and elemental signature of seawater are magnesium (Mg^{2+}), sulfate (SO_4^{2-}), calcium (Ca^{2+}), and potassium (K^+). Together, these six components make up 99.3% of the salt in the ocean. These six salts constitute the conservative elements of seawater, and they occur in virtually the same proportions relative to each other in seawater throughout the world's oceans. In contrast, other salts such as strontium (Sr^{2+}), bromide (Br^-), and carbonate (CO_3^{2-}) are minor constituents, and their concentrations can be highly variable between different regions of the ocean.

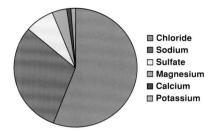

FIGURE 2 Composition of the dissolved salts in seawater. Values shown are the relative proportion of total salts for the major group of six conservative salts: chloride, 55.2%; sodium, 30.4%; sulfate, 7.7%; magnesium, 3.7%; calcium, 1.2%; potassium, 1.1%.

MEASUREMENT OF SALINITY

Salinity is technically defined in terms of the ratio of the electrical conductivity of seawater (at 15 °C and 1 atmosphere of pressure) in relation to the electrical conductivity of a standard solution of potassium chloride (KCl) that contains 32.4356 g of KCl in 1 kg of water. This worldwide standard for the repeatable measurement of salinity was established in 1978 and is known as the Practical Salinity Scale (PSS-78); it is actually a rather complicated mathematical function of this ratio. Consequently, the numerical value for the salinity of seawater is dimensionless, and standardized measurements of salinity do not have units. Because the standard measurement of salinity is a ratio, salinity values are regularly reported as a dimensionless number (e.g., "the salinity of ocean water is 33"). Salinity measurements are sometimes followed by "psu" (practical salinity units) to indicate that they were made in reference to the 1978 Practical Salinity Scale (e.g., "the salinity of ocean water is 33 psu").

Precise measurements of salinity and conductivity are made with an electrical conductivity meter. The probe of a conductivity meter measures the ability of the seawater sample to conduct an electrical charge within a given volume of water, which is directly related to the amount of salt ions dissolved in the solution. Conductivity measurements are compensated by differences in temperature and pressure to calculate salinity based on the equations of the PSS-78 Practical Salinity Scale.

More simplified measures of salinity have been expressed as the kilograms of salt dissolved in a kilogram of water, and measurements of salinity determined in this manner are reported in parts per thousand (ppt) or $^{\circ}/_{\circ\circ}$ (i.e., 0.035 kg salt dissolved in 1 kg water equals 35 ppt or 35$^{\circ}/_{\circ\circ}$). Instruments for measuring salinity on the PSS-78 scale are frequently calibrated against a reference standard known as "Copenhagen water," which is prepared from large samples of water collected in the North Atlantic, carefully diluted to 35 ppt, and distributed in sealed glass ampoules. The numerical difference between psu and ppt is typically small (about ±0.3%). Direct (but imprecise) measurements of salinity can also be made with an optical refractometer. Ions that are dissolved in seawater cause light waves to be bent (refracted) in a manner that is proportional to the total amount of salts in the water. Consequently, the refractometer measures the degree to which light is bent as it passes through a thin film of water to provide an index of salinity.

In the past, measurements of salinity were made by chemical titration of chloride ions (Cl^-) coupled with information about the relative proportions of the other elements to calculate total salinity. This method was not very precise, and computations of salinity values based on chlorine titration were reliant upon inaccurate assumptions about the composition of the various salts dissolved in the water.

SALINITY AND GLOBAL CLIMATE CHANGE

Measurements of sea surface salinity can provide important information for improved understanding of global and regional climate change in the marine environment. Changes in salinity values provide a direct indicator of changes in the global water cycle and reflect acceleration of the cycling of water through the atmosphere and ocean. Warming of the earth's climate results in a decrease in the amount of freshwater inputs (drought) and greater evaporation at low latitudes, coupled with increased rainfall and melting glaciers at high latitudes. These changes in freshwater inputs and evaporation result in differences in salinity of the ocean waters and can contribute to disruption of the circulation of ocean currents.

SEE ALSO THE FOLLOWING ARTICLES

Buoyancy / Climate Change / Evaporation and Condensation / Seawater / Water Chemistry / Water Properties, Measurement of

FURTHER READING

Curry, R., and C. Mauritzen. 2005. Dilution of the northern North Atlantic in recent decades. *Science* 308: 1772–1774.

Curry, R., R. R. Dickson, and I. Yashayaev. 2003. A change in the freshwater balance of the Atlantic over the past four decades. *Nature* 426(6968): 826–829.

Fofonoff, N. P. 1985. Physical properties of seawater: new salinity scale and equation of state for seawater. *Journal of Geophysical Research* 90: 3332–3342.

Lewis, E. L., and R. G. Perkin. 1981. The Practical Salinity Scale 1978; conversion of existing data. *Deep Sea Research* 28A: 307–328.

Mantyla, A. 1987. Standard seawater comparisons updated. *Journal of Physical Oceanography* 17: 543–548.

Schmitt, R.W., and N.L. Brown. 2005. *Long term surface salinity measurements.* Final Report NAG5-12444. Woods Hole, MA: Woods Hole Oceanographic Institution.

UNESCO. 1981. *Background papers and supporting data on the Practical Salinity Scale, 1978.* UNESCO Technical Papers in Marine Science 37. Paris: United Nations Educational, Scientific, and Cultural Organization.

SALINITY STRESS

STEPHEN R. WING AND REBECCA J. MCLEOD

University of Otago, New Zealand

Aquatic organisms are exposed to stress when there is a change in the immediate osmotic environment through variation in salinity, either in the external medium or in surface films. A cell exposed to salinity with osmolality higher than that of the intercellular fluid undergoes hyperosmotic stress and may lose water from its cellular fluid through osmosis. Conversely, a cell exposed to a medium with osmolality lower than that of the intercellular fluid undergoes hypoosmotic stress, and may suffer an influx of water and increase in hydrostatic pressure that can rupture the cell membrane. In both these cases, differences in osmolality between the medium and the cellular contents provide the potential for osmotic flux of water across the cell membrane, producing physical strain that can result in death of the cell.

PHYSIOLOGICAL AND BEHAVIORAL RESPONSES TO SALINITY STRESS

Organisms employ a variety of mechanisms to survive salinity stress by alleviating or combating osmotic pressure, the force required to halt diffusion of water across the membrane. In multicellular organisms, compartmentalization of tissues and internal fluid-filled spaces can serve to buffer cells from sharp changes in the osmotic environment. However, for those tissues in immediate contact with the external medium, such as gills or epidermis, or if exposure is extended, salinity stress is more direct. One general solution to this problem is to regulate intercellular osmotic pressure by actively extruding water or excreting salts. For example, many single celled organisms contain water expulsion vesicles or contractile vacuoles, which accumulate and extrude excess water from the cellular fluid. Many fishes and invertebrates excrete ions in the form of concentrated urine (via the kidneys), or through ion-regulating organs such

as nephridia, gills, the gut, or antennal glands. Vascular plants such as *Spartina* and mangroves that inhabit salt marshes have glands to excrete excess salt. These types of regulatory responses can be employed to respond to relatively rapid changes (on the scale of hours) in external osmotic conditions.

Another general solution to the problem is to eliminate salinity stress by modifying the internal osmolality of cells to match that of the environment. This can be achieved by incorporating ions or compatible solutes in the cellular fluid. Compatible solutes allow an organism to achieve high osmolality of cells without disrupting cellular function, as inorganic salts do, and therefore acclimate to high-salinity conditions over a matter of days or weeks. Seaweeds and salt-tolerant plants generally use low-molecular-weight osmolytes such as glycerol, sucrose, and mannitol to modulate intercellular osmolarity. Many marine invertebrates maintain high levels of some free amino acids to acclimate to external salinity conditions. For more sudden changes in salinity, organisms may buffer themselves from salinity stress by moving to a more isoosmotic environment or by employing physical or chemical barriers between vulnerable tissues and the environment. For example, bivalves can survive exposure to low-salinity events by literally clamming up until a less stressful salinity is restored. Each of these strategies requires some energy from the organism with an associated cost in terms of fitness.

The resulting tradeoff between costs and benefits of living in a harsh osmotic environment forms one of the basic evolutionary landscapes for organisms living in marginal habitats at ecotone boundaries. Diverse strategies to cope with salinity stress, and the combined effects of temperature and salinity stress, result in a wide variety of salinity tolerances and form part of the basis for distributional limits among many marine organisms. This is most apparent in habitats with strong salinity gradients or strong gradients in the variability of salinity, such as where freshwater and seawater meet in estuaries or on the intertidal shore.

ENVIRONMENTS WITH HIGH VARIABILITY IN SALINITY

In the open sea salinity varies relatively little, with a gradual decrease from the tropics to the poles of 37 to 33. However, in the coastal zone in regions where freshwater and seawater mingle, such as in estuaries, fjords, and river mouths, or in periodically isolated bodies of water such as lagoons or tidepools, organisms can experience broad fluctuations in salinity. The degree of salinity stress imposed on an organism in these environments is a function of both the rate of change in salinity and also the duration and magnitude of each exposure event. At the convergence between seawater and freshwater, there are typically large cyclic fluctuations in salinity associated with tidal flux; there are also fluctuations associated with rainfall (Fig. 1). In regions with large seasonal variation in rainfall these habitats can be ephemeral, forming only when runoff is sufficient to reach the sea. In lagoons and tidepools fluctuations in salinity may be driven by freshets that dilute the salinity, by evaporation to a more concentrated brine, and by the frequency of flushing and replenishment with seawater. Consequently, larger tidepools low on the shore offer a more stable osmotic environment than smaller pools higher on the shore.

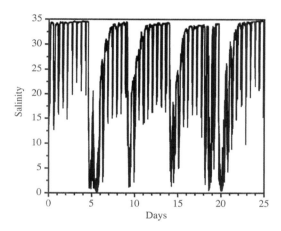

FIGURE 1 Salinity fluctuations at 6 m depth on the rock wall in Doubtful Sound, New Zealand.

The resulting spatial variability in salinity regime across these habitats influences the distribution of biological communities, with euryhaline organisms (those tolerant to a wide range of salinities) extending in their distribution into highly variable habitats, and more stenohaline organisms (those requiring a more narrow range of salinity) confined to the sea or making sojourns into these habitats only during extended penetration of seawater. For sessile organisms these range restrictions may act at the larval or settlement stage of the life cycle. Consequently, the influence of salinity stress on distribution of organisms has been a central theme in studies of benthic faunal diversity across these habitats. Generally these studies report a diminishing number of benthic species along gradients from marine to brackish water, with lowest taxonomic richness occurring in waters of

salinity 3–5 (Fig. 2). Transitions between freshwater and seawater, and isolated pools, therefore contain some habitats with exclusive access based on an organism's ability to cope with salinity stress.

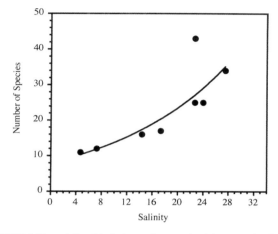

FIGURE 2 The relationship between taxonomic richness and salinity across a gradient between freshwater and seawater. Data from Rutger and Wing (2006).

ECOLOGICAL CONSEQUENCES OF SALINITY STRESS

Like many other environmental gradients, salinity variation creates a physical structure upon which ecological interactions are influenced. Differences in ability to alleviate salinity stress, in concert with other stressors, may directly restrict the distribution of a species or act indirectly to influence distribution patterns through modulation of competitive or trophic interactions among species. For example in Denmark's Limfjord, five species in the amphipod genus *Gammerus* show marked niche partitioning according to salinity tolerance. This genus, common throughout the northeast Atlantic region, has radiated into freshwater, brackish, and marine habitats across the Northern Hemisphere. Within each general salinity environment, estuarine or marine, salinity tolerance is a good predictor of niche breadth and geographic range among species; in this case the jack-of-all-trades becomes the master-of-all.

One interesting consequence of this effect of salinity tolerance on range and distribution is that areas of high salinity stress can form refuges for prey that are more euryhaline than their predators. An example of this exists in the New Zealand fjords, where up to 7 meters of annual rainfall result in a persistent low-salinity layer for the top 2–10 meters of the water column. In this surface layer, only organisms with a high tolerance to freshwater survive. As one moves deeper, a transition occurs to seawater-tolerant species. In many estuaries this transition occurs across several kilometres, but in the fjords it occurs across just a few meters because of the segregation of water by salinity. Algae and mussels living in the brackish surface waters are protected from grazing and predation by the inability of grazers such as sea urchins *(Evechinus chloroticus)* and predators such as the eleven-armed sea star *(Coscinasterias muricata)* to venture into the low-salinity water. As the freshwater layer thins during drought, sea urchins and sea stars venture upward to graze on the inhabitants (Fig. 3), but they retreat with the coming of rain and thickening of the low-salinity layer.

FIGURE 3 A feeding frenzy of eleven-armed sea stars *(Coscinasterias muricata)* on a clump of mussels *(Mytilus galoprovincialis)* exposed to seawater. Photograph by S. Wing.

The rocky intertidal zone is another environment in which salinity stress in combination with other physical stressors can have a major influence on community structure. As the tide ebbs, animals on the intertidal rocky shore become exposed to desiccation, and they may be subject to osmotic stress from changes in the salinity of surface films on exposed tissues. If there is rainfall or snow during this period, the salinity will rapidly decrease, and freshwater may accumulate in tidepools. Alternatively, a period of hot weather and low humidity on the ebb tide may result in tidepools becoming warm and hypersaline as a result of evaporation. As a consequence, intertidal organisms can be subjected to both hypo- and hyperosmotic conditions over one tidal cycle. Variability in salinity is typically highest in supralittoral tidepools, where only the most tolerant organisms can survive. These organisms may employ a variety of physiological and behavioral mechanisms to cope with the variability in salinity. At the population level

some are especially adept at persisting in the fragmented and highly variable environment of the supralittoral.

Exceptional examples of this can be found among copepods in the genus *Tigriopus*, which have a cosmopolitan distribution in supralittoral rockpools. *Tigriopus* has an amazing ability to withstand variability in salinity. For example the species *T. brevicornis* has been shown to survive salinities from 5 to 200 and is an exceptional osmoconformer over the salinity range of 10 to 100. At very low salinity, approaching 5, the species shows some osmoregulation. By incorporating compatible solutes in the form of free amino acids into the intercellular fluid, *Tigriopus* is able to alleviate salinity stress across this extreme range. This can be an expensive activity for *Tigriopus*, using approximately 10 percent of daily energy costs. However, this expense is a good investment for the copepod, because its high capacity for osmoconformity widens the range of tidepools that the species can colonize, and extends its occupancy time. This trait increases the persistence of populations across the fragmented landscape of supralittoral pools.

CONCLUSIONS

Gradients in salinity regime are an important component of the physical landscape in coastal environments. These gradients can influence distributions and ecological interactions among organisms according to differences in physiological and behavioural ability to alleviate or combat salinity stress. At the fringes, some of the most harsh environments in terms of osmotic stress provide refuges for those organisms that have paid the physiological cost of exclusive membership.

SEE ALSO THE FOLLOWING ARTICLES

Cold Stress / Desiccation Stress / Excretion / Heat Stress / Seawater / Zonation

FURTHER READING

Kirst, G. O. 1989. Salinity tolerance of eukaryotic marine algae. *Annual Review of Plant Physiology and Plant Molecular Biology* 40: 21–53.
Kowalewski, M., E. Guillermo, K. Flessa, and G. Goodfriend. 2000. Dead delta's former productivity: two trillion shells at the mouth of the Colorado River. *Geology* 28: 1059–1062.
Rutger, S. M., and S. R. Wing. 2006. Effects of freshwater input on shallow-water infaunal communities in Doubtful Sound, New Zealand. *Marine Ecology Progress Series* 314: 35–47.
Somero, G. N., and P. H. Yancey. 1997. Osmolytes and cell volume regulation: physiological and evolutionary principles, in *The Handbook of Physiology*. J. F. Hoffman and J. D. Jamieson, eds. New York: Oxford University Press, 441–484.
Witman, J. D., and K. R. Grange. 1998. Links between rain, salinity, and predation in a rocky subtidal community. *Ecology* 79: 2429–2447.

SCULPINS

CATHERINE A. PFISTER

University of Chicago

Sculpins (order Scorpaeniformes, family Cottidae) are common occupants of intertidal pools and can be the most numerically abundant fish family in tidepools. Multiple species can co-occur in tidepools, and their populations are represented by newly settled juveniles and multiyear adults. Of special interest are their abilities to home to tidepool habitat and to persist in often physiologically stressful habitats.

BIOGEOGRAPHY

Tidepool sculpins are present in cold, temperate oceans and are represented by species with relatively small body size compared with their subtidal counterparts. The species can have relatively large geographic ranges, with some ranging from Alaska to California. Although their greatest diversity can be found in the northeast Pacific, several species are also present in the north Atlantic.

LIFE CYCLE

The life cycle of sculpins is characterized by adults that are resident in a single pool or group of tidepools. They become reproductive at a body size as small as 35 mm Standard Length (a measure from the tip of the nose to the beginning of the tail fin rays). Fertilization appears to occur internally, with reproduction perhaps facilitated by modified anal fin rays in males, which may serve as claspers. Clusters of eggs are deposited in the late winter, often outside of tidepools in crevices and empty barnacle tests in the high intertidal. Eggs can show a wide range of color variation, measuring 1 to 2 millimeters in diameter. The larvae are planktonic and are thought to spend 30 to 60 days in the water column before settling back into tidepool habitat. This recruitment into rocky shore habitat occurs during the spring and summer months in the northeast Pacific (April to September), a time period that also coincides with maximal upwelling. The geographic extent of dispersal during the larval phase is not well studied, although several genetic studies suggest that dispersal has the potential to occur over scales of kilometers. However, the usual extent of dispersal and the degree to which populations are demographically connected by movement are unknown. In embayments, larval sculpins have been observed to school close to the benthos, a behavior that would reduce dispersal among populations.

INTERACTIONS WITHIN AND AMONG SPECIES

The transition from a newly settled recruit to an adult fish can be negatively affected both by the numbers of fish of the same species and by competitive interactions with other species. Although cannibalism has been observed in laboratory populations, it is unknown whether it is important in tidepools in nature. Predators, including diving birds and subtidal fishes, forage in the intertidal during high tides and are known to prey on tidepool sculpins.

Tidepool sculpins are predators on a diversity of invertebrates and even some algae. Both *Clinocottus globiceps* and *C. analis* have been shown to have some algal material in their diet. Other common prey items for sculpins include isopods, amphipods, snails, barnacles (especially the cirri), polychaetes, and terrestrial insects that alight on the tidepool surface. Both as predators and as prey, sculpins' often cryptic coloration can be an asset (Fig. 1).

FIGURE 1 The common tidepool sculpin *Oligocottus maculosus* blends into the rocky intertidal at Tatoosh Island (approximate length is 6.0 cm). Photograph by the author.

ADAPTATION TO TIDEPOOL CONDITIONS

Perhaps due to both the threat of predation and ocean waves, tidepool sculpins can be found in the same pool or group of pools for years. When removed from pools and released distances of tens of meters away, these fishes are capable of returning to the same pool, even in conditions of turbulent water motion. Some species (e.g., *Clinocottus embryum*) are capable of persisting out of the tidepool in the understory of moist algae during extensive low tides, and several species have been shown to respire out of water. Others seem capable of dealing with elevated water temperatures and low oxygen.

RESPONSES TO OCEANIC CONDITIONS

Although the planktonic larval stage of sculpins may seem to make them vulnerable and responsive to fluctuations in oceanic conditions, the evidence for this is varied. On the outer coast of Washington State, recruitment has been remarkably constant over a period of 16 years, including two El Niño/Southern Oscillation (ENSO) events, while a study in southern California has noted near-failure of recruitment during an ENSO event.

SEE ALSO THE FOLLOWING ARTICLES

Camouflage / Competition / Dispersal / Fish / Predation / Recruitment

FURTHER READING

Horn, M., and M. Chotkowski, eds. 1999. *Intertidal fishes: life in two worlds. San Diego*: Academic Press.
Pfister, C.A. 1996. Consequences of recruitment variation in an assemblage of tidepool fishes. *Ecology* 77: 1928–1941.
Pfister, C.A. 1997. Demographic consequences of within-year variation in recruitment. *Marine Ecology Progress Series* 153: 229–238.
Pfister, C.A. 2006. Concordance between short-term experiments and long-term censuses: a competition-colonization trade-off in tidepool fishes? *Ecology* 87: 2905–2914.
Yoshiyama, R.M., K.B. Gaylord, M.T. Philippart, T.R. Moore, J.R. Jordan, C.C. Coon, L.L. Schalk, C.J. Valpey, and I. Tosques. 1992. Homing behavior and site fidelity in intertidal sculpins (Pisces: Cottidae). *Journal of Experimental Marine Biology and Ecology* 160: 115–130.

SEA ANEMONES

VICKI BUCHSBAUM PEARSE

University of California, Santa Cruz

FIGURE 1 The body of a sea anemone (Cnidaria: Anthozoa: Hexacorallia: Actiniaria) is essentially a hollow tube, into which the prey is taken through the *mouth*, an opening in the *oral disk* at the tube's free end. Extending from the oral disk are *tentacles*, which bear nematocysts and capture the prey. At the other end of the tube is the *pedal disk*, by which the anemone attaches to rocks or shells. The main middle part of the body is the *column*. While all sharing this basic plan, anemones vary widely in form and color, as illustrated by the examples in this plate from Gosse (1860). This nineteenth-century naturalist painted and described the intertidal anemones of English coasts so engagingly that he popularized the keeping of aquariums in fashionable Victorian parlors. The plundering of tidepools by collectors quickly ravaged the very beauties that a chagrined Gosse had so eloquently celebrated—an early lesson in conservation.

Sea anemones (Fig. 1) are simple, soft-bodied invertebrates in the phylum Cnidaria, a group named for the microscopic cnidae, or nematocysts (thread capsules), with which all its members are armed. The sting we experience after contacting an anemone or jellyfish results from the toxins injected when nematocyst threads penetrate our skin. For humans, most anemones feel merely sticky; very few have a painful sting. For the small animals that are the anemones' prey, the sting is paralyzing or fatal. Anemones are ecologically important members of marine benthic communities from the shallows to the deep sea and nearly from pole to pole, often conspicuous and remarkably abundant, with over 1000 described species. Significant habitats include soft bottoms and coral reefs, but the focus here is on the more typical intertidal anemones of rocky shores.

SEA ANEMONES AS INTERTIDAL ANIMALS

The high diversity and great concentration of organisms in the intertidal zone testifies to its particular suitability as habitat. Competition for the limited space in this narrow zone of prime real estate is therefore stiff. Sea anemones enjoy several advantages over other organisms in this contest.

First, the large adhesive area of the bottom of a sea anemone's pedal disk confers a strong hold on the rock surface and makes the anemone extremely difficult to dislodge. Having established itself in a favorable spot, an anemone can sit tight—catching prey on its radiating tentacles during high tides and contracting down during low tides—without having to roam about in search of food, unlike, for instance, herbivorous molluscs, which must move over the rocks to graze on algae, or predatory sea stars and nemerteans, which must seek out their prey. On the other hand, unlike seaweeds and the many sessile invertebrates that cannot change locations once attached (such as sponges, hydroids, tunicates, barnacles, rock oysters, and tube polychaetes), anemones retain the option of mobility; they can creep slowly on their pedal disk to a nearby spot or even detach and float away to reattach at a new site. Thus, the basic morphology and way of life give anemones a flexibility well suited to the changes and uncertainties of the intertidal. For example, if a seaweed grew to shade an anemone that required more sunshine, the anemone might move toward the light; such behavior has been documented in *Anthopleura elegantissima* in the laboratory.

Second, the particular life history characteristics of sea anemones contribute to their ability to dominate intertidal space. In some species, the polyp that develops from the settling planula larva remains as a solitary unit but can grow to relatively large sizes; for example, individuals of *Anthopleura xanthogrammica* and *A. sola* on the west coast of North America commonly reach 15–20 cm across, and tidepools or channels may be crammed nearly full of them. In contrast, the closely related *A. elegantissima* grows large in a different mode, by cloning: each polyp divides in half, each half regenerates into a complete polyp, and hundreds of small, genetically identical polyps may eventually arise from a single founder, remaining aggregated and solidly covering a square meter or more of rock surface. Within such a clonal aggregation, the polyps display a limited division of labor, with those in the center having larger gonads and evidently allocating more energy to sexual reproduction, while those on the periphery invest more energy in specialized fighting structures (acrorhagi), used in territorial battles against adjacent, genetically different aggregations of the same species. The genus *Metridium* also includes cloning and noncloning species. Cloning is accomplished by the separation of small fragments from the edge of the pedal disk; each fragment regenerates into a small new polyp. Individuals of *Metridium* also engage in intraspecific aggression, using specialized fighting tentacles.

Such clonal replication of sea anemones and other invertebrates is not equivalent to sexual reproduction and cannot serve as an alternative, because the genetic and evolutionary consequences are entirely different. Most clonal animals regularly reproduce sexually, with the same frequency as nonclonal species do. Totally asexual offshoots of sexual lineages are rare, although they may arise and persist in a few instances (for example, a population of clonal intertidal anemones in Japan is not known to produce eggs or sperm). Cloning may aid in the establishment of an invasive species on a foreign shore. One example is a large intertidal population of the Asian anemone *Diadumene lineata* (reported as *Haliplanella luciae*) found on the Atlantic coast of North America; it appeared to consist entirely of males and was thought to represent a single clone. A study of this widespread anemone at several other sites on both coasts of the United States and in Asia suggested that populations are more commonly genetically diverse and include multiple clones, a result indicating that introductions of this invader are not uncommon events.

Generally, sea anemones that clone also engage in regular sexual reproduction, and their cloning is most usefully viewed simply as growth—like the production of polyps in the course of growth of a hydroid or coral colony. With one exception, sea anemones do not form colonies in which the polyps remain physically attached, but in many species a founding polyp generates multiple clonemates, which, whether they remain aggregated or not, provide further security that a given genetic individual will survive accidents of the challenging intertidal habitat. However, even solitary, nonclonal anemones appear to be long lived, thanks to abundant powers of healing and regeneration. Indeed, there is no evidence that sea anemones senesce in any way, and individuals kept in aquariums have lived for many decades. Short of predation or other fatal accidents, anemones are potentially immortal. The long life span of anemones may play an especially important part in their retaining control of precious intertidal space, in contrast to short-lived organisms, whose ecology is dominated by the need for regular replacement by new recruits.

Sexual reproduction in sea anemones is not known to be specifically modified in tidepool or intertidal species compared to subtidal ones. Anemones may exist as separate males and females or may be hermaphroditic. Typically, both eggs and sperm are shed into the sea, but brooding of the young by some intertidal species appears to offer an extra measure of protection against predation or environmental stresses. For example, four species of *Epiactis* found in the intertidal all brood their young. *E. prolifera* and *E. lisbethae*, which occur also in the subtidal, brood the young in an external basal groove (Fig. 2), whereas *E. ritteri* and *E. fernaldi*, which are primarily or exclusively intertidal, offer the most protection: the young are brooded internally, within the parent's coelenteron, eventually to be released through the mouth. *Actinia equina* and *A. tenebrosa* also brood young internally; the origins of these young have proved difficult to sort out, but they appear to result both from sexual outcrossing and from asexual proliferation.

FIGURE 2 Brooding sea anemone, *Epiactis prolifera*, holding its young in an external basal groove; central California coast. Photograph by James Watanabe.

CHALLENGES OF THE HABITAT

Deep subtidal habitats, although relatively constant in temperature and not subject to pounding waves, present other risks: a subtidal sea anemone is continuously available to its predators (discussed below). The rocky intertidal offers a refuge from such constant pressure: when the tide is out, intertidal anemones may enjoy a period of safety from predators that cannot survive exposure to air or are active only when submerged. At the same time, other hazards abound: exposure to strong wave action, desiccation, ultraviolet radiation, and extremes of temperature and salinity, as well as the risk of being buried for prolonged periods beneath shifting sands. Most of

these factors are treated extensively in other articles in this encyclopedia.

Although sea anemones can move slowly, their general habit is to remain in one spot for long periods, even many years, unless it proves unfavorable in some way. They cannot migrate up and down with the tides. Thus, a specimen in a tidepool lives under quite different sets of circumstances from an anemone on a rock surface that is exposed to the air for varying periods during low tides. Most species occupy both kinds of habitat, so any individual must be prepared for a great spectrum of assaults.

The tidepool habitat is in some respects an intermediate world, always more variable than well-mixed subtidal waters, but never fully exposed to air, and it might seem to be more benign than an exposed rock face. However, pools sometimes present more extreme and challenging conditions for a marine animal than simple exposure. For example, a tidepool anemone may find itself more vulnerable to any predators that happen to share its space, because the predators themselves enjoy a quiet sheltered environment and are freer to exploit their prey than they might be either in the surge of the subtidal or when themselves exposed to air. The same applies to aggression by conspecifics. During rain or transient freshwater runoff, an anemone on an exposed rock will usually sustain relatively brief and superficial exposure to low-salinity stress, whereas an anemone in a small high pool may sit for hours fully submerged in nearly fresh water while an anemone in a large low pool will experience no effects at all. Likewise, it is difficult to predict or generalize the impact of extreme temperatures. When a small pool heats up or freezes, the temperature of the anemones therein follows the same curve, whereas under the same hot-weather conditions an anemone exposed on a rock may undergo evaporative cooling, but such benefits may be offset by the stresses of ultraviolet radiation and osmotic changes.

When out of the water, a sea anemone's first defense against the environmental stresses of exposure to air and sun is a moist, semiprotected situation. Many intertidal anemones find this protection among seaweeds or in rock crevices. An anemone that has settled on the rock roof of a deep overhang may be far better protected against ultraviolet radiation and overheating than is one in a tidepool, and dense populations of anemones are often found beneath overhanging rocks. Anemones in a closely packed clonal aggregation provide a large degree of similar protection to each other. A second line of defense is behavioral: when exposed to air or other stresses, many anemones contract, shortening the sensitive tentacles; pulling the distal margin of the column over the tentacles, the mouth, and the oral surface; and holding a large reservoir of seawater inside the body. In some anemones, such as species of *Anthopleura*, the column is covered with adhesive papillae that hold a protective covering of shell fragments and small stones; such a layer reflects radiation and simultaneously slows evaporation.

Finally, when unable to mitigate extremes of desiccation or temperature, many intertidal anemones are able simply to survive them—even survive being frozen solid or baking in the sun to reach temperatures higher than the ambient air—during at least the hours that they must wait for the rising tide to cover the shore and restore more moderate conditions. Some populations of *Diadumene lineata* and *D. leucolena* have been observed to encyst in mucus and undergo a winter dormancy when temperatures fall below a given level. Another sort of enforced dormancy may occur from time to time when sand covers intertidal anemones, resulting in prolonged anoxia and starvation. Individuals of *Actinia, Anthopleura,* and *Bunodosoma* can survive this ordeal, depending on substantial stores of glycogen. For animals living so close to their own mortal limits, the relatively rapid climate change that has been documented over recent decades can bring profound impacts. Among the striking changes observed in California has been the range expansion of a southern species—the sunburst anemone, *Anthopleura sola*—over the latter half of the twentieth century, from being too minor a population to merit much notice or even a name, to becoming one of the most conspicuous intertidal animals of the central California coast (Fig. 3).

FIGURE 3 Sunburst anemone, *Anthopleura sola,* once uncommon on the central California coast, has recently expanded its range northward, probably because of rapid climate change. Photograph by John S. Pearse.

INTERACTIONS WITH OTHER ORGANISMS

Predators

As mentioned above, anemones enjoy some relief in the intertidal from predators, especially fishes that cannot pursue them at low tide and that may be inhibited by high-energy habitats even during high water. Predators on intertidal anemones thus tend to be a rather specialized lot: particular molluscs (aeolid nudibranchs and a few snails), sea stars, pycnogonids (sea spiders), and—in a few parts of the world such as Samoa—humans.

When aeolid nudibranchs feed on anemones (and other cnidarians), the molluscs obtain not only nutrition but also a source of nematocysts, which are stored—intact and capable of firing—in tentacle-like projections borne on the nudibranch's dorsal surface. The nematocysts may be used in defense by the aeolid, and *Aeolidiella alderi* was reported to use nematocysts obtained from feeding on a species of *Sagartia* to sting other anemones of the same species, before feeding on them in turn. The nudibranchs are probably also protected by their resemblance, in form and color, to their anemone prey, among which the molluscs are well camouflaged against being spotted by their own predators. Individuals of the widespread nudibranch *Aeolidia papillosa* often show strong preferences for one anemone species over another and for small prey over larger ones. Some anemones react to predatory nudibranchs by ejecting internal filaments (acontia), heavily armed with powerful nematocysts, which may successfully deter an attack. Anemones that lack acontia quickly contract, covering the tentacles and oral disk, which are the predators' preferred target. They may also begin to glide away on the pedal disk or release the pedal disk and float away. The columnar surface tissues of *Anthopleura elegantissima* contain a compound (anthopleurine) which serves as an alarm pheromone. Released into the water by a damaged anemone, or from a nudibranch that has been feeding on this anemone species, this chemical causes other anemones to contract defensively. Nearby clonemates are the first beneficiaries of such a warning, but other nearby conspecifics as well as *A. xanthogrammica* and probably *A. sola* also respond to anthopleurine, although the more distantly related *A. artemisia,* not a preferred prey of the nudibranch, does not.

The sea star *Dermasterias imbricata,* although it can reach into the lower intertidal, is largely limited to the subtidal, and this predator is thought to be the main limiting factor in the vertical distribution of *Anthopleura* species on the North American west coast. Individual sea anemones large enough to escape predation by these stars may be found subtidally.

The several micropredators that take nourishment from intertidal sea anemones exact a smaller toll. Wentletrap snails of the genus *Epitonium* snip at the tentacles, whereas those of the genus *Opalia* are seen feeding with the proboscis inserted near the base of the anemone's column. Pycnogonids (sea spiders) are also most often seen feeding on the column. Certain large ciliates appear to be specifically associated with anemones, living within the coelenteron and sometimes seen scooting around on the oral disk, but whether these are feeding on anemone tissue or on mucus and gut contents is not known. The amphipod *Orchomenella recondita,* associated with the coelenteron of sea anemones, may be a harmless commensal or may be on the road to parasitism.

Prey

The flower-like appearance of sea anemones, which their name reflects, is deceptive. All are predators, though their prey may range from microscopic plankton to small organisms encrusting ingested pieces of seaweed to large invertebrates or fishes. Anemones are not very discriminating eaters and will generally accept whatever is offered as long as the necessary chemical feeding stimuli accompany it, usually amino acids or small peptides particular to each anemone species. These chemicals trigger feeding behavior: tentacles that have captured prey fold towards the mouth, which in turn opens and protrudes toward the prey. These movements are effected by muscles in the tentacles and oral disk, while reversal of cilia in the pharynx results in ingestion of the prey.

In these general aspects, intertidal anemones do not differ from subtidal ones. The main impact of the intertidal habitat on feeding is that the time available is limited. For anemones in tidepools, the most productive feeding periods may occur while the pool is filling or emptying, because that is when moving water brings prey into contact with their tentacles; during low tide, when the anemone is sitting in the quiet water of a tidepool, it can catch only the occasional small fish or crustacean that is trapped in the pool and blunders into its tentacles. Anemones on rock surfaces outside of tidepools can feed only during high water and during that time, splashing waves and surge may strongly influence their ability to do so. Certain species take advantage of these circumstances; for example, *Anthopleura xanthogrammica* classically lives in surge channels below

mussel beds and depends on wave action to dislodge mussels and deliver them to the anemones' waiting mouths. Having successfully captured and ingested prey, anemones typically undergo a relatively long quiescent period while digestion proceeds, so the disadvantage of limited feeding bouts is not as great as it might initially appear.

Symbiosis

Among the most engaging symbioses of subtidal sea anemones are their associations with certain fish, such as clownfish, and with crabs, both of which the stinging anemone protects against predators.

Less flashy, but equally complex, is the symbiosis of many anemones with unicellular algae: zooxanthellae (dinoflagellates) and, rarely, zoochlorellae (chlorophytes). The algal symbionts live within the cells of the anemone host, in the endodermal lining of the coelenteron—in the tentacles, oral disk, and column. Under the influence of substances produced by the animal host, zooxanthellae release a portion of their photosynthetic products, and these may contribute significantly to the nutrition of the anemone, more so than with zoochlorellae. Not even all zooxanthellae are the same, however; many strains have been identified in anemones, corals, and other animal hosts. The various zooxanthellae may differ in their relationship to their hosts and have been shown to differ in their tendency to be expelled from the host during bleaching induced by increased temperatures. Such bleaching is fatal to reef corals for which the symbiosis is obligatory. In anemones, algal symbiosis appears to be facultative, and anemones that have lost their zooxanthellae, after being exposed to darkness or abnormally high temperatures, survive quite well as long as they eat enough. Under conditions of starvation, however, the nutritional boost of the symbiosis could make a critical difference.

Zooxanthellae confer a brown color on many anemones, sometimes overlain with pigments produced by the animal host, but often the dominant or only hue of an otherwise colorless anemone. Except for those individuals containing bright green zoochlorellae, the green color of symbiotic anemones is *not* due to the chlorophyll of their algae, as is often supposed, but rather to a green fluorescent protein (GFP), common also in corals and other cnidarians. The protein absorbs ultraviolet light from sunlight and emits lower-energy green light. Injected into cells or attached to other compounds as a visible tag, it has become a valuable research tool for tracing cell lineages during development or the flow of various substances in and out of cells. However, the usefulness of GFPs to their cnidarian makers is not fully understood. Because the protein is produced only when the animal is exposed to light, at least in the case of *Anthopleura elegantissima,* it may have a protective function. However, these anemones, like virtually all aquatic seaweeds and animals, also contain mycosporine-like amino acids (MAAs) that serve as sunscreens, absorbing ultraviolet wavelengths and protecting the animals against damage by ultraviolet radiation, a major hazard for intertidal species. As far as is known, animals cannot produce MAAs but acquire them from their food or symbionts, or both.

PERSPECTIVES FROM BEYOND THE ROCKY INTERTIDAL

Treating sea anemones of tidepools and rocky intertidal coasts as distinct from those of subtidal or deep-sea or soft-bottom habitats is largely arbitrary. Tidepool anemones have been studied more intensively, because they are more accessible. To a great extent, however, observations made on intertidal species apply also to many subtidal ones, and it is both difficult and important to sort out the differential factors. For instance, one can watch intertidal anemones that are contracted at low tide expand when the sea returns. But many subtidal anemones also undergo cycles of expansion and contraction, related to light and feeding. Thus, to best understand the biology of an intertidal anemone, or any organism, we must try to view it in the context of the larger taxa to which it belongs. On the other hand, all organisms living in the environment of rocky shores share certain challenges, which a comparative biological approach may help to illuminate.

SEE ALSO THE FOLLOWING ARTICLES

Adhesion / Corals / Hydroids / Octocorals / Stauromedusae / Symbiosis

FURTHER READING

Fautin, D. G. 2000. Tree of Life: Actiniaria. http://tolweb.org/Actiniaria.
Fautin, D. G., and R. W. Buddemeier. 2006. Biogeoinformatics of the Hexacorals. http://www.kgs.ku.edu/Hexacoral/.
Friese, U. E. 1972. *Sea anemones.* Hong Kong: T.F.H. Publications.
Gosse, P. H. 1860. *A history of the British sea-anemones and corals.* London: Van Voorst.
Manuel, R. L. 1988. *British Anthozoa.* Synopses of the British Fauna (New Series). D. M. Kermack and R. S. K. Barnes, eds. The Linnean Society of London. Avon: The Bath Press.
Shick, J. M. 1991. *A functional biology of sea anemones.* London: Chapman & Hall. Best general source.

SEA CUCUMBERS

GINNY L. ECKERT
University of Alaska

Sea cucumbers are cucumber-shaped invertebrate animals that make up the class Holothuroidea in the phylum Echinodermata. Echinodermata in Greek means "spiny skins"; however, most sea cucumbers differ from other echinoderms in that their spines are reduced to tiny particles, called ossicles, embedded in the skin. As a result most sea cucumbers do not appear spiny like other echinoderms such as sea urchins, sea stars, and brittle stars. Sea cucumbers maintain many of the features that are unique to echinoderms, including pentaradial symmetry, an internal skeleton made of calcite, catch connective tissue, and a water vascular system. Catch connective tissue is fascinating because it can be stiffened by nervous control in a matter of seconds and then made flexible again as quickly as it hardened. The water vascular system hydraulically powers tube feet—sucker-like projections that aid in locomotion. The most reliable methods for identification of sea cucumbers are microscopic examination of the ossicles and molecular analysis of DNA, because many species of sea cucumber are morphologically similar.

ANATOMY AND PHYSIOLOGY

Sea cucumbers are elongate (ranging from 2 cm to 5 m long) and cylindrical with the mouth at one end and the anus at the other (see Fig. 1). A ring of tentacles encir-

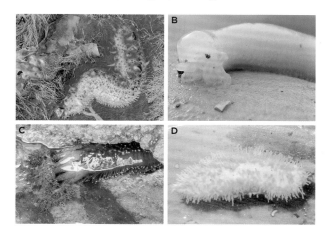

FIGURE 1 Representative sea cucumbers. Note the different arrangements of tentacles and tube feet. (A) *Parastichopus californicus,* (B) *Chirodota discolor,* (C) *Cucumaria miniata,* (D) *Eupentacta pseudoquinquesimita.* Photographs by the author.

cles the mouth and, when extended, collects food particles. The structure of the tentacles differs among species with different feeding behaviors. Suspension feeders, which collect particles from the water, have more finely branched tentacles than do deposit feeders, which either collect sediments from the surface or burrow and ingest sediment (like an earthworm). The internal anatomy of sea cucumbers includes a digestive tract, respiratory trees, nerve ring, and gonads. The body wall contains five longitudinal muscle bands and five longitudinal rows of tube feet. Sea cucumbers have several unique habits, including "breathing" through the anus, as the cloaca brings in water to the respiratory trees, and evisceration, in which they expel internal organs through the anus or through a slit in the anterior end of the body wall. Reasons for evisceration are unknown and may include predator evasion or expulsion of parasites or accumulated wastes. Some species resorb internal organs during a dormant period, presumably to reduce metabolic expenditures, and then regenerate them again at the end of dormancy. Tropical species have Cuvierian tubules, which are expelled in defense and are sticky and entangle the would-be predator.

REPRODUCTION, ECOLOGY, AND BEHAVIOR

Most species of sea cucumber have separate sexes, broadcast spawning, external fertilization, and free-swimming larval stages (Fig. 2); however other reproductive and development modes also occur. Sea cucumbers are hosts for a variety of parasitic and commensal organisms that live inside or on the sea cucumber. Some species of sea cucumber are chemically defended as a deterrent to predation; however, such chemical defense does not prevent predation by sea stars, which are the sea cucumber's principal predators. An individual deposit feeder can ingest and filter up to 45 kg of sediment a year, and as a result, large populations play an important role in recycling nutrients and cleaning the sea floor. The 1500 living species of sea cucumber are distributed worldwide and are found from the intertidal to the deepest parts of the oceans.

ECONOMIC AND MEDICINAL IMPORTANCE

Parastichopus californicus and *P. parvimensis* are commercially harvested on the west coast of the United States, dried, and sold for consumption as *trepang* or *beche-de-mer* in Asian markets. In the United States, sea cucumber is sold as a health supplement, primarily as a source of chondroitins and anti-inflammatory compounds for treating arthritis. Sea cucumber catch connective tissue is studied for medicinal purposes because of its similarity to human tendons and ligaments. Catch connective tissue collagen

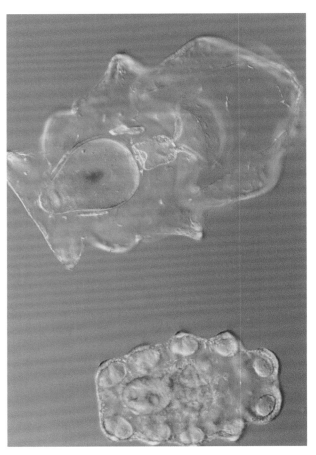

FIGURE 2 Sea cucumber (*P. parvimensis*) larvae. Larvae develop from an auricularia (top) to a doliolaria (bottom) to a pentacula (not shown) before settling as a juvenile sea cucumber.

fibers can bond and break to make the tissue rigid or fluid, therefore, sea cucumbers may hold the key to repairs for injured tendons and ligaments, such as the anterior cruciate ligament (in the human knee) and the Achilles tendon.

SEE ALSO THE FOLLOWING ARTICLES

Brittle Stars / Echinoderms / Food Uses, Modern / Materials, Biological / Sea Stars / Sea Urchins

FURTHER READING

Bergen, M. 1996. Class Holothuroidea, in *Taxonomic atlas of the benthic fauna of the Santa Maria Basin and the Western Santa Barbara Channel*. J. A. Blake, P. H. Scott, and A. Lissner, eds. Santa Barbara, CA: Santa Barbara Museum of Natural History, 195–250.

Kozloff, E. N. 1983. *Seashore life of the northern Pacific coast*. Seattle: University of Washington Press.

Lambert, P. 1997. *Sea cucumbers of British Columbia, southeast Alaska, and Puget Sound*. Vancouver: University of British Columbia Press.

Morris, R. H., D. P. Abbott, and E. C. Haderlie. 1980. *Intertidal invertebrates of California*. Stanford, CA: Stanford University Press.

Summers, A. 2003. Catch and release: sea cucumbers might put a torn Achilles tendon back together again—biomechanics. *Natural History* 112(9): 36–37.

SEAGRASSES

CAROL A. BLANCHETTE

University of California, Santa Barbara

Seagrasses are rooted vascular plants of terrestrial origin that have successfully returned to the sea with several adaptations that allow them to live submerged in the ocean. These plants occupy sandy, muddy, and rocky substrates in every sea in the world and provide habitat for entire communities of associated species. Seagrasses play an important role in the geomorphology and ecology of coastal ecosystems by stabilizing sediments, recycling nutrients, and forming the base of oceanic detrital food webs. Despite their great importance, they are currently facing many natural and anthropogenic threats from coastal development, pollution, and climate change.

SEAGRASS BIOLOGY

Although the vast majority of marine intertidal plants are algae, there is one large group of true flowering plants (angiosperms) that have evolved to live in the ocean. These are seagrasses. *Seagrass* is the common name for a large group of flowering plants that have evolved from terrestrial plants and have become specialized to live in the marine environment. Seagrasses are not true grasses and are more closely related to water lilies and terrestrial plants such as lilies. Seagrasses are commonly found on tidal mudflats in estuaries, on shallow sandy areas close to the coast, in coral reef lagoons, and in shallow rocky intertidal areas; they can grow as deep as 60 meters.

Like terrestrial plants, seagrasses harvest energy from light to carry out photosynthesis and have leaves, conducting tissues, and roots (Fig. 1). Unlike terrestrial plants, however, seagrasses do not possess the strong, supportive stems and trunks required to overcome the force of gravity on land. Rather, the natural buoyancy of water supports seagrass leaves, enabling them to remain flexible when exposed to waves and currents. Gas vacuoles keep the leaves erect and maximize light interception when submerged. Seagrasses grow like backyard grasses with roots and rhizomes (horizontal underground stems that form extensive networks below the substrate surface). The roots and rhizomes of seagrasses are often buried in sand or mud, or directly attached to rock, and they serve to anchor the grasses and absorb nutrients. Seagrasses can propagate vegetatively through growth and branching of this rhizome. Along the rhizome at intervals are erect

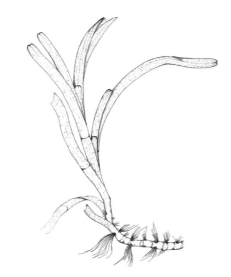

FIGURE 1 Anatomy of a seagrass plant, showing roots, rhizomes, and leaves. Illustration by Jessica Altstatt.

shoots, which bear the leaves and leaf sheaths. The leaves may vary in length from a few millimeters to several meters. Leaves, usually green, are produced on vertical branches and also absorb nutrients. The stems and leaves of seagrasses contain veins and air channels so that they can carry fluid and absorb gases. Seagrasses rely on light to convert carbon dioxide into oxygen via photosynthesis. The oxygen is then available for use by other living organisms. Seagrasses have thin, permeable cell walls to enhance external gas exchange, and gas vacuoles and aerenchyma tissue to facilitate internal gas exchange. Aerenchyma tissue forms gas conduits that run from the leaves to the roots and allows the passage of photosynthetically produced oxygen to roots and rhizomes.

Marine algae (commonly referred to as seaweeds) are plants that also colonized the sea and may be confused with seagrasses; however, they are more primitive than seagrasses. In contrast to seagrasses, seaweeds do not have a true root system (they have holdfasts) and do not have veins that carry molecules around the plant. Seaweeds have spores and do not flower or produce fruit, while seagrasses produce flowers, seeds, and fruit.

Assuming that the seagrasses gradually evolved from freshwater plants, one of their biggest hurdles to overcome was adapting to saltwater. Like celery placed in saltwater, the roots of most plants rapidly lose water if they are immersed in seawater. Salt-loving plants (halophytes), such as the seagrasses and mangroves, generally have a lower concentration of water molecules (lower water potential) in their root cells, so they can take in water. They maintain lower water potentials in their roots by having higher internal salt concentrations than seawater and by losing water at the leaf surface. Since high internal salt concentrations can be lethal to plant cells, some halophytes can excrete excess salt through their leaves and stems.

HABITAT AND DISTRIBUTION

Seagrass beds are found in the shallow coastal waters of every sea in the world. Worldwide, there are about 12 major divisions, consisting of approximately 60 species of seagrasses. They are mainly found in bays, estuaries, and coastal waters from the mid-intertidal (shallow) region down to depths of 50 or 60 meters. Seagrass distributions are extremely sensitive to variation in light and cannot live at depths lower than where photosynthetic benefits outweigh respiratory costs. Because light transmission in water is a function of water clarity, turbidity, and sediment load, seagrass distributions are strongly affected by eutrophication and other anthropogenic effects that influence water clarity. For example, seagrass beds in Chesapeake Bay historically occurred at depths of over 10 meters but are restricted today to depths of less than 1 meter because of reductions in water clarity, and therefore in light transmission, by organic fertilizer runoff and eutrophication.

Seagrasses range from the straplike blades of eelgrass (*Zostera caulescens*) in the Sea of Japan, at more than 4 meters long, to the tiny (2–3–centimeter), rounded leaves of sea vines (e.g., *Halophila decipiens*) in the deep tropical waters of Brazil. Vast underwater meadows of seagrass skirt the coasts of Australia, Alaska, southern Europe, India, eastern Africa, the islands of the Caribbean, and other places around the globe. They provide habitat for fish and shellfish and nursery areas to the larger ocean, and they perform important physical functions of filtering coastal waters, dissipating wave energy, and anchoring sediments. Seagrasses inhabit all types of substrates, from mud to rock. The most extensive seagrass beds occur on soft substrates like sand and mud. Seagrasses occupy coastal waters from tropical to temperate regions. The number of species is greater in the tropics than in the temperate zones, and only two species, *Halophila ovalis* and *Syringodium isoetifolium*, occur in both regions.

North America is home to several different species of seagrasses. *Zostera marina* (commonly known as eelgrass) is the dominant seagrass of temperate latitudes along both the Atlantic and Pacific coasts. It grows in muddy substrates along wave-protected shallow-water bays and estuaries. Along the Atlantic coast, eelgrass is replaced by two other species of seagrasses in the warmer waters from the Carolinas to the Caribbean. *Thalassia testitudinalis* (commonly known as turtlegrass) and *Halodule wrightii* (commonly known as shoalgrass) are the dominant seagrasses in this region. These species are ecologically important as nursery habitat for fishes and an important food source for endangered species such as manatees and sea turtles. Along the Pacific coast, wave-swept rocky intertidal habitats are uniquely occupied by several species of *Phyllospadix* (commonly known as surfgrass). Surfgrass roots and rhizomes can attach directly to rock, and surfgrass beds can occupy vast amounts of space in the low intertidal and shallow subtidal regions of rocky shorelines (Fig. 2).

FIGURE 2 Intertidal surfgrass bed, *Phyllospadix torreyi*. Photograph by the author.

REPRODUCTION

Seagrasses can reproduce through sexual or asexual methods. Seagrasses, which spend most, if not all, of their time submerged, have developed methods of underwater flowering and pollination for sexual reproduction. In sexual reproduction, the plants produce flowers, and water carries pollen from the male flower to the ovary of the female flower. Most, but not all, seagrass species produce flowers of a single sex on each individual, so there are separate male and female plants. The resulting fruit are often carried some distance from the parent plant before the seeds are released. In some seagrasses, such as surfgrass, the seeds (Fig. 3) have small hooks and barbs (like an anchor) and are specially designed to become entrapped in bushy algal species where they can germinate, produce roots, attach to the rock, and grow. Flowering is seasonal in most seagrasses; a few species complete their life cycle within one year and are known as annuals. These annuals produce seeds that can remain dormant in large "seed banks" for several months. Seed banks

FIGURE 3 Seed of *Phyllospadix torreyi*. Photograph by Todd Huspeni.

ensure that the species can survive until conditions return to stimulate the seeds to germinate.

Seagrasses can also grow by asexual (or vegetative) reproduction. New "plants" arise without flowering or setting seed. As mentioned previously, all seagrasses have horizontal underground stems called rhizomes. Seagrasses grow vegetatively by extending and branching their rhizomes in the same way that grass in a lawn grows. This allows significant areas of seagrass meadow to form from only a few shoots. In this way, seagrasses can recover after being damaged by grazers such as dugongs or disturbed by storms. The growth of seagrass after disturbance is critical to their survival. If all plants of a meadow are lost, a seagrass meadow needs seeds available or vegetative shoots translocated from a nearby meadow before it can recover. The ability of seagrass meadows to recover depends on the species of seagrass.

SEAGRASS ECOLOGY

Seagrass communities are highly productive and dynamic ecosystems. They provide habitats and nursery grounds for many marine animals and act to stabilize substrates and recycle nutrients. Seagrass communities may vary from a few plants or clumps of a single species to extensive single-species or multispecies meadows covering large areas of the bottom. The rhizomes and roots of the grasses bind sediments on the bottom, where nutrients are recycled by microorganisms back into the marine ecosystem. The leaves of the grasses slow water flow, allowing suspended material to settle on the bottom. This increases the amount of light reaching the seagrass bed and creates a calm habitat for many species.

Seagrasses perform a variety of functions within ecosystems and have both economic and ecological value. The high level of productivity, structural complexity, and

biodiversity in seagrass beds has led some researchers to describe seagrass communities as the marine equivalent of tropical rainforests. Although nutrient cycling and primary production in seagrasses tends to be seasonal, annual production in seagrass communities rivals or exceeds that of terrestrially cultivated areas. Within seagrass communities, a single acre of seagrass can produce over 10 tons of leaves per year. This vast biomass provides food, habitat, and nursery areas for a myriad of adult and juvenile vertebrates and invertebrates. Further, a single acre of seagrass may support as many as 40,000 fish and 50 million small invertebrates. Because seagrasses support such high biodiversity, and because of their sensitivity to changes in water quality, they have become recognized as important indicator species that reflect the overall health of coastal ecosystems.

Seagrass ecosystems provide habitats for a wide variety of marine organisms, both plant and animal; these include meiofauna and flora, benthic flora and fauna, epiphytic organisms, plankton and fish, not to mention microbial and parasitic organisms. The relatively high rate of primary production of seagrasses drives detritus-based food chains, which help to support many of these organisms. Birds, dugongs, and turtles also directly consume seagrasses. An adult green turtle eats about two kilograms of seagrass a day, whereas an adult dugong eats about 28 kilograms a day. As habitat, seagrasses offer food, shelter, and essential nursery areas to commercial and recreational fishery species and to the countless invertebrates that are produced within, or migrate to, seagrasses. The complexity of seagrass habitat is increased when several species of seagrasses grow together, their leaves concealing juvenile fish, smaller finfish, and benthic invertebrates such as crustaceans, bivalves, echinoderms, and other groups. Juvenile stages of many fish species spend their early days in the relative safety and protection of seagrasses. Seagrass meadows also help dampen the effects of strong currents, providing protection to fish and invertebrates, while also preventing the scouring of bottom areas. Additionally, seagrasses provide both habitat and protection to the infaunal organisms living within the substratum as seagrass rhizomes intermingle to form dense networks of underground runners that deter predators from digging infaunal prey from the substratum.

To appreciate the ecological importance of seagrasses, consider the sudden disappearance of eelgrass beds along the Atlantic coast during the 1930s. An epidemic infestation of the parasitic slime fungus *(Labyrinthula)*, called wasting disease, literally destroyed the rich eelgrass meadows; the results were catastrophic. Populations of cod, shellfish, scallops, and crabs were greatly diminished, and the oyster industry was ruined. There was also a serious decline in overwintering populations of Atlantic brant geese. Areas formerly covered by dense growths of eelgrass were completely devastated, and beaches, which had been protected from heavy wave action, were now exposed to storms. Without the stabilizing effects of eelgrass rhizomes, silt spread over gravel bottoms used by smelt and other fish for spawning. This resulted in a decline in waterfowl populations that fed on the fish. Without the filtering action of eelgrass beds, sewage effluent from rivers caused further water pollution, thus inhibiting the recovery of eelgrass.

THREATS TO SEAGRASS ECOSYSTEMS

As humans have encroached on the marine environment, there have been some dramatic effects on coastal ecosystems. Because of the important structural and functional role of seagrass communities, significant disturbances to seagrass beds are likely to have significant impacts on the associated ecosystem. Many recent examples exist of extensive damage done to marine and estuarine habitats through the inadvertent destruction of seagrass communities. The loss of a number of important seagrass habitats has resulted in extensive studies of some seagrass communities and an increasing awareness of the role these plants play in our coastal marine environment. Recently a number of attempts to transplant (Fig. 4) and artificially restore seagrasses have been developed, and this work continues at many research centers around the world.

Seagrasses are subject to many threats, both anthropogenic and natural. The effects of many of these stresses have on seagrasses is dependent on both the nature and severity of the particular environmental challenge. Generally, if only leaves and above-ground vegetation

are harmed, seagrasses are generally able to recover from damage within a few weeks; however, when damage is done to roots and rhizomes, the ability of the plant to produce new growth is severely impacted, and plants may never be able to recover.

Runoff of nutrients and sediments from human activities on land has major impacts in the coastal regions where seagrasses thrive; these indirect human impacts, while difficult to measure, are probably the greatest threat to seagrasses worldwide. Both nutrient and sediment loading affect water clarity; seagrasses' relatively high light requirements make them vulnerable to decreases in light penetration of coastal waters. Direct harm to seagrass beds occurs from boating, land reclamation, and other construction in the coastal zone, dredge-and-fill activities, and destructive fisheries practices. Human-induced global climate change may well impact seagrass distribution as sea level rises and severe storms occur more frequently.

Threats to seagrasses can originate long distances from the coast. Coastal agriculture in upper catchments may add to sediment and herbicide loads in runoff from the land, which has the potential to destroy large areas of seagrass. Global climate change may also threaten seagrass communities. Climate change is predicted to raise sea levels, concentrations of carbon dioxide in seawater, and seawater temperatures. Rising sea levels could increase the distribution of seagrass because more land will be covered by seawater. However, rising sea levels are likely to destabilize the marine environment and cause seagrass losses. Higher concentrations of carbon dioxide in seawater could increase the area of seagrass, because more carbon will be available for growth and seagrasses could increase their photosynthetic rates. Rising sea temperatures could cause burning or death of seagrasses in some places where they are close to their thermal limit.

SEE ALSO THE FOLLOWING ARTICLES

Algae / Ecosystem Changes, Natural vs. Anthropogenic / Habitat Restoration / Light, Effects of / Nutrients / Sea Level Change, Effects on Coastlines

FURTHER READING

Green, E. P., and F. T. Short, eds. 2003. *World atlas of seagrasses.* Berkeley: University of California Press.

Larkum, A. W. D., R. J. Orth, and C. M. Duarte, eds. 2006. *Seagrasses: biology, ecology and conservation.* Berlin: Springer Verlag.

McRoy, C. P., and C. Helfferich. 1977. *Seagrass ecosystems, a scientific perspective.* New York: Marcel Dekker.

Pettitt, J., S. Ducker, and B. Knox. 1981. Submarine pollination. *Scientific American* 244 (3): 135–143.

Phillips, R. C., and C. P. McRoy. 1980. *Handbook of seagrass biology, an ecosystem perspective.* New York: Garland STPM Press.

FIGURE 4 Divers transplanting eelgrass in a restoration project. Photograph by Carl Gwinn.

SEA LEVEL CHANGE, EFFECTS ON COASTLINES

MICHAEL GRAHAM

Moss Landing Marine Laboratories

Sea level, the recorded height of the sea surface, can vary across a broad range of spatial and temporal scales as a result of planetary motions, glacial cycles, and tectonics. A ubiquitous feature of tidal systems, sea level change affects marine organisms through the simultaneous modification of coastal geomorphology and regulation of emersion–submersion cycles.

GLOBAL VS LOCAL SEA LEVEL CHANGE

Global (eustatic) sea level changes result primarily from variation in the volume of water stored in global ice sheets, which can be driven by both natural (e.g., solar insolation) and anthropogenic factors that affect global warming and cooling. Eustatic sea level has fluctuated dramatically throughout the Earth's history, especially during the Quaternary period, when global temperatures have cycled continuously between cold (glacial) and warm (interglacial) periods (Fig. 1). Quaternary glacial periods have typically decayed rapidly into interglacial periods (~10,000–20,000 years), raising sea levels as glaciers melt; the reestablishment of glacial conditions, and subsequent decreases in sea level, have been much more noisy and gradual (~80,000–90,000 yrs).

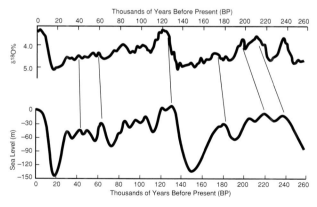

FIGURE 1 Quaternary global (eustatic) sea level curve and oxygen isotope record for the last 260,000 years. Distinct glacial periods have occurred at 18,000 and 150,000 years before present, corresponding with low sea level standstills and increased $\partial^{18}O$ levels. Sea level and oxygen isotope data are from emergent coral terraces on the Huon Peninsula, Papua New Guinea (Aharon and Chappell 1986). Reprinted with permission from Williams *et al.* (1998).

This cyclical pattern of glacial/interglacial transitions results in eustatic sea level cycles between regressing (falling) and transgressing (rising) seas, demarcated by sea level standstills corresponding to peaks in either glacial or interglacial stages. Fortunately, the oxygen isotopic ratio of seawater ($^{18}O{:}^{16}O$ or $\partial^{18}O$) is positively correlated with changes in global ice sheet volume, allowing variability in $\partial^{18}O$ from ice and lake/marine sediment cores to be used as an estimate of eustatic sea-level change (Fig. 1). During the most recent Holocene transgression (~120 m), sea level rise was rapid (1 meter per century) until 6000 years ago, when it slowed to around 10 centimeters per century; the last 3000 years have been a high-sea-level standstill.

Although the oxygen-isotope-derived eustatic sea level curve is good for estimating globally averaged sea level changes, sea level curves developed directly for specific regions can be at odds with the eustatic curve. These discrepancies are due to a variety of complex geological processes (e.g., tectonics, seismic uplift, isostatic rebound) that modify the vertical motion of reference elevations and that vary over local spatial scales and relatively short temporal scales. Regional sea level curves can be derived by various methods (e.g., radiocarbon dating of sediment deposits) and provide a more detailed view of sea level change relevant to local intertidal habitats.

SEA LEVEL CHANGE AND THE CREATION OF INTERTIDAL BENCHES AND BEACHES

Variable sea levels result in a constant state of coastal morphological change, driven by differences in the relative rates of coastal erosion and the vertical motion of coastal landmasses. Coastal erosion occurs at or below the ocean surface (as a result of waves and currents) or above the ocean surface (as a result of local precipitation and terrestrial runoff), with erosion rates dependent on the geological composition of the landmass. Wave-cut platforms are created by subsurface erosion during periods when marine transgressions slow, although most such platforms remain submerged during interglacial periods. Uplift of coastal landmasses by tectonic and seismic processes results in the permanent emergence of some wave-cut platforms, first becoming intertidal benches and, later, marine terraces (Fig. 2). Eustatic sea level changes can also synchronize the formation of intertidal benches and marine terraces at a global scale.

Another important process driving coastal geomorphic change is the interaction between erosion rates and the accumulation and deposition of sediments along coastal margins (Fig. 3). During marine regressions, terrestrial runoff travels greater vertical and horizontal distances to

FIGURE 2 Photograph of uplifted marine terraces cut by wave action during sea level standstills near San Onofre in southern California. Copyright © 2002–2006 Kenneth and Gabrielle Adelman, California Coastal Records Project, http://www.Californiacoastline.org.

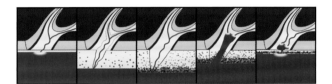

FIGURE 3 Schematic of geological transition between rocky and sandy dominated shorelines, driven by variability in sea level. Panels from left to right: (A) high-sea-level standstill; (B) regressing sea; (C) low-sea-level standstill; (D) transgressing sea; (E) beginning of high-sea-level standstill. Dark stippling represents the distribution of cobble boulders; light shading represents the distribution of sand. Reprinted with permission from Graham *et al.* (2003).

the coastline, increasing the velocity of runoff waters. The result is an increase in grain size and quantity of sediment that can be entrained in the runoff and deposited downstream. As streams travel from catchments to the sea, such erosion processes work to carve out deep coastal valleys, with coarse sediments (predominantly cobbles and boulders) transported out of the valleys and deposited along the coastline. Given a low-sea-level standstill, the cobbles begin to be distributed laterally via wave action, and the coastline takes on the general appearance of a cobbled, rocky shore. During marine transgressions, rapidly rising sea level decreases the distance between the shoreline and catchments, decreasing terrestrial runoff velocities, entraining finer sediments, and depositing them at the heads of the flooding valleys. Over time, lateral movement redistributes the sand, eventually closing off the bays, which begin to fill in with sediment and form coastal lagoons and marshes. Given an extended high-sea-level standstill, deposition of fine sediments can occur directly on the shoreline and facilitate the formation of extensive beaches. Such geomorphic processes are known to occur in temperate and subtropical regions worldwide

(e.g., California, Chile, South Australia, South Africa, and the Mediterranean), suggesting continuous cycling between the extent of sandy and rocky tidal shores at local to global scales, driven by Quaternary sea level changes.

SEE ALSO THE FOLLOWING ARTICLES

Beach Morphology / Coastal Geology / Tides

FURTHER READING

Aharon, P., and J. Chappell. 1986. Oxygen isotopes, sea level changes and the temperature history of a coral reef environment in New Guinea over the last 10^5 years. *Palaeogeography, Palaeoclimatology, Palaeoecology* 56: 337–379.

Curray, J. R. 1961. Late Quaternary sea level: a discussion. *Geological Society of America Bulletin* 72: 1707–1712.

Graham, M. H., P. K. Dayton, and J. M. Erlandson. 2003. Ice-ages and ecological transitions on temperate coasts. *Trends in Ecology and Evolution* 18: 33–40.

Inman, D. L. 1983. Application of coastal dynamics to the reconstruction of palaeocoastlines in the vicinity of La Jolla, California, in *Quaternary Coastlines and Marine Archaeology*. P. M. Masters and N. C. Fleming, eds. New York: Academic Press, 1–49.

Mastronuzzi, G., P. Sansò, C. V. Murray-Wallace, and I. Shennan. 2005. Quaternary coastal morphology and sea-level changes—an introduction. *Quaternary Science Reviews* 24: 1963–1968.

Seibold, E., and W. H. Berger. 1996. *The sea floor: an introduction to marine geology*. Berlin: Springer-Verlag.

Shackleton, N. J., and N. D. Opdyke. 1973. Oxygen isotope and palaeomagnetic stratigraphy of Equatorial Pacific Core V28-238: Oxygen isotope temperatures and ice volumes on a 10^5 year and 10^6 year scale. *Quaternary Research* 3: 39–55.

Williams, M., D. Dunkerley, P. de Decker, P. Kershaw, and J. Chappell. 1998. *Quaternary environments*. London: Arnold.

SEALS AND SEA LIONS

DANIEL COSTA

University of California, Santa Cruz

Sea lions and seals are common inhabitants of the intertidal. Although they feed at sea, they return to shore to breed, give birth, mate, rest, and molt. As mammals they belong to the order Carnivora, which is divided into two suborders, Fissipedia and Pinnipedia. The suborder Fissipedia refers to members of the order Carnivora with normal feet, such as dogs, cats, weasels, raccoons, and bears. The suborder Pinnipedia (Latin for "fin-footed") refers to the seals, sea lions, fur seals, and walrus, all with finlike feet (Fig. 1). The Pinnipedia are further separated into three families. The sea lions and fur seals, also known as the eared seals, are in the family Otariidae, the true or

FIGURE 1 Male (left) and female (right) sea lions, seals, and fur seals. Along the left are (A) male California sea lion, (B) male northern elephant seal, and (C) a male Galápagos fur seal. All males in these pictures are on territories (A, C) or defending a harem (B). Along the right are (A) Australian sea lion mother and pup, (B) a northern elephant seal female and pup, and (C) Galápagos fur seal mother and pup. Photographs by the author.

earless seals are in the family Phocidae, and the walruses are in the family Odobenidae.

SEA LIONS AND FUR SEALS

Otariids are called eared seals because of the presence of external ear flaps (pinnae) (Figs. 1, 2). Sea lions and fur seals swim with their front or pectoral flippers. They can rotate their rear flippers forward and scratch like a dog, and they can walk well on land using all four flippers and are often seen considerable distance away from the water (Fig. 2). They can also climb on top of small islands. The primary difference between fur seals and sea lions is that fur seals rely almost entirely on their fur for insulation against the cold, whereas true seals and sea lions rely on blubber. Blubber serves both as insulation and as an energy store. Fur seals have a relatively pointed snout, whereas sea lions have a broader, blunter nose (Fig. 1). Sea lions are larger than fur seals. Sea lions and fur seals

do not naturally occur along the east coast of the United States or in the North Atlantic Ocean, but they are quite common along the west coast of the United States and through much of the Pacific Ocean and oceans of the Southern Hemisphere. Sea lions are the most common pinnipeds seen in oceanariums and marine parks, because they are very food-motivated and therefore easy to train. The circus "seal" is usually a sea lion, because sea lions and fur seals are more acrobatic and agile on land than true seals are. They also make a barking sound. There are 17 species of Otariidae, including six extant sea lion species, one extinct sea lion species, and 10 fur seal species. The largest eared seal is the Steller sea lion, in which males can be up to 3.25 meters long and weigh 1120 kilograms. The smallest eared seal is the Galápagos fur seal (Fig. 1), of which males typically are 1.5 meters long and weigh 64 kilograms, while females are 1.2 meters long and weigh 27 kilograms.

FIGURE 2 A juvenile California sea lion shown scratching with its hind flipper, a behavior that only fur seals and sea lions can accomplish. Also notice the prominent external ears, a trait also unique to fur seals and sea lions. Photograph by the author.

TRUE OR EARLESS SEALS

Phocidae, also known as the true or earless seals, lack external ears and cannot rotate their hindlimbs forward, resulting in a clumsy undulating movement on land (Fig. 1). They can scratch only with their front flippers and are not able to move as well on land as sea lions or fur seals are. True seals use their hind flippers to swim, using a side-to-side motion, and are better divers than sea lions or fur seals. Phocid or true seals include 19 extant species and one extinct species. The largest seal is the southern elephant seal, which lives in waters of the Southern Ocean between Antarctica and South America, South Africa and Australia. The male may grow to be 5 meters long and may weigh up to 3600 kilograms. This seal ranks second in size only to whales among all sea mammals. The smallest seal is the ringed seal of the Arctic. It is approximately 1.4 meters long and weighs up to 91 kilograms.

ROOKERIES

Rookeries are the areas where seals and sea lions congregate to breed, molt, give birth, and haul out to rest (Fig 3). Sea lion and fur seal rookeries are only located on land, but the rookeries of true seals can be on land, on islets (rocky outcrops), or on ice. Rookeries can vary in size from a single harbor seal on a rock to the extensive beach or cobble areas of the Pribilof Islands, where over 150,000 Northern fur seals may be seen. The densest rookeries are located at higher latitudes, where it is colder. At lower latitudes, where it is warmer, seals and sea lions remain in or near the intertidal so that they can cool off in the water.

FIGURE 3 A rookery of California sea lion females on San Nicolas Island, California. Photograph by the author

BREEDING BEHAVIOR

Seals and sea lions give birth to a single highly developed, or precocial, pup. They are born with their eyes open and can vocalize and walk within minutes of birth. Incredibly, harbor seal pups can swim within 20 minutes of birth. Males do not provide any care to the pup, and twins are exceptionally rare. Phocid mothers typically remain on or near the rookery continuously from the birth of their pup until it is weaned. Body reserves stored prior to parturition fuel the female's metabolic needs and are utilized for milk production. Although some phocids—most notably harbor, ringed *(Phoca hispida)*, and Weddell seals—feed during lactation, most of the maternal investment is derived from body stores. Weaning is abrupt and occurs after a minimum of 4 days of nursing (hooded seal, *Cystophora cristata*) to a maximum

of 7 weeks in Weddell seals and Hawaiian monk seals. In some species such as northern elephant seals, the pup remains on or near the rookery, not drinking or eating for months after weaning. In other species such as harp or hooded seals, the pups undergo a prolonged migration lasting weeks to months.

In contrast, otariid mothers stay with their pups only the first week or so after parturition and then periodically go to sea to feed, returning to suckle their pup on the rookery. Feeding trips vary from 1 to 14 days, depending on the species, and shore visits to the pup, which has been fasting, last 1 to 3 days. The mother finds her pup out of the hundreds on the crowded rookery by calling for her pup. When the pup hears its mother call, it begins to call back. Each mother-and-pup pair have a unique call, and the mother responds only to her pup's call. Once together, the mother smells the pup to ensure that it is hers. The pups are weaned from a minimum of 4 months in the subpolar fur seals (Antarctic, *Arctocephalus gazella*, and northern, *Callorhinus ursinus*) to up to 3 years in the equatorial Galápagos fur seal *(A. galapagoensis)*. The remaining otariids are temperate. In these species, pups are usually weaned within a year of birth, although weaning age can vary both within and between species as a function of seasonal and site-specific variations in environmental conditions. Walruses can feed their offspring for up to 3 years, both while on shore and in the water.

LAND BREEDING

Extreme polygyny (one male breeds with many females) is characteristic of most land-breeding pinnipeds. Of the 21 species of land-breeding pinnipeds, 18 are highly polygynous and sexually dimorphic (Fig. 1). This includes all of the sea lions and fur seals; both species of elephant seal, *Mirounga*; and the gray seal, *Halichoerus grypus*. The remaining three species of land-breeding seals, two species of monk seals, *Monachus*, and harbor seals, *Phoca vitulina*, mate in the water near land and exhibit lesser degrees of polygyny. In general, land-breeding pinnipeds aggregate on beaches, rocks, or flat areas on islands. These island aggregations provide several potential advantages, including parturition sites for rearing pups, lack of terrestrial predators, and, for the lactating otariids, proximity to food resources. When females aggregate, males have an easier time controlling access to them, and therefore a single male can mate with many females. It is easier for males to control access to females on land, and in this situation large body size is favored. Large body size confers advantage in fighting as well as a greater fasting ability that allows males to hold onto their territories longer. Typically the largest, most experienced males find a mate, and in some species, such as elephant seals, the competition for mates is so fierce that only one in ten males successfully breeds. The rank of males is established early in the breeding season, when males fight to establish dominance or establish at territory. Males use their long canine teeth in these battles; although the battles are often bloody, the damage is usually minimal. In elephant seals, the male's nose signifies his age and experience, where a large nose indicates an older, more experienced male. The long nose of the elephant seal serves as both a visual and acoustic indicator of the male's status, just as the number of points on the antler of a male deer indicates his age and status. The large nose also acts as a reverberation chamber and gives their vocalizations a deep bass quality that carries well in the high background noise and crashing waves of a seal rookery.

In species that compete for females in the water, males are comparatively smaller than in those species that breed on land. This is because agility is more important than size in competing for mates underwater. It is also possible for the male to leave on short foraging trips and then return to the breeding area.

ICE BREEDING

Of the 13 species of pinnipeds that breed on ice, 11 exhibit slight polygyny or serial monogamy. Females that breed on pack ice have access to vast areas of breeding substrate, but some species clump loosely. Females tend to give birth during the short period of time when the ice is most stable and come into estrus synchronously upon weaning. This synchrony, together with the degree of spatial separation, limits the ability of males to mate with many females. The rapid growth rate of seal pups and the short period that the mother needs to stay with them until they are weaned enable them to rear their young on ice.

MOLTING

All seals and sea lions undergo an annual molt, in which they shed their hair over a period of several weeks. During the molt, seals and sea lions spend more time on shore, allowing greater blood flow to their skin. Some species, such as elephant seals, remain on shore until the molt is completed. Elephant seals undergo an "explosive" molt in which they shed the skin and hair together in large pieces (Fig. 4). In contrast, fur seals do not need to haul out to molt, because they shed a few hairs at a time, so that they maintain the insulating quality of their fur.

FIGURE 4 A molting male northern elephant seal. Notice how the skin comes off in large sheets. Also notice the lack of external ears. Photograph by the author.

SENSES

Seals have very large eyes, which enable them to feed at night or in deep water where there is little light. Large eyes function just like a camera lens; the larger the lens aperture, the greater its light-gathering ability. Most seals can see well underwater, but on land they are nearsighted. True seals have better hearing underwater than sea lions and fur seals do. The whiskers on their head are very sensitive and are probably important in finding and catching prey.

DIVING OR FORAGING BEHAVIOR

Seals and sea lions have specialized sharp-pointed teeth that allow them to grasp their slippery prey. However, they cannot chew their food, because their teeth lack flat surfaces, so they either grasp and tear their prey or swallow it whole. True or earless seals are exceptional divers and are capable of deeper, longer dives than sea lions and fur seals are. These differences reflect use of different habitats. Sea lions and fur seals feed on prey nearer the surface, while seals feed on more sedentary bottom or deeper prey. The ability of true seals to dive deeper than sea lions and fur seals is related to their ability to stay submerged longer, since the deeper an animal dives, the farther it must travel. True seals can dive longer because they can store more oxygen in their muscle and blood than sea lions or fur seals can. Seals can store three times as much oxygen in their body as can humans, whereas sea lions and fur seals can store only twice as much oxygen as humans. Northern and southern elephant seals are not only the deepest-diving pinnipeds; they are among the deepest-diving mammals. Only sperm and beaked whales can dive deeper or longer. Northern elephant seals dive continuously, day and night, for periods at sea lasting

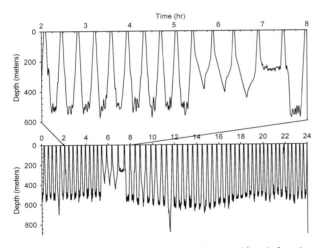

FIGURE 5 A dive record from a northern elephant seal female foraging across the North Pacific Ocean. The bottom record is a 24-hour period, with the image above a 4-hour segment of the same record. Notice the continuous diving pattern and the considerable depth to which this animal dove.

2 to 8 months (Fig. 5). They spend 90% of the time at sea submerged, average 20 minutes per dive (with maximum dive durations of over 1.5 hours) followed by less than 3 minutes at the surface between dives, and routinely diving to depths of 300 to 600 meters, with dives occasionally exceeding 1600 meters! Males travel from California to foraging areas along the continental slope from the state of Washington north to the upper reaches of the Gulf of Alaska to the eastern Aleutian Islands (Fig. 6). Female elephant seals, on the other hand, disperse widely across the northeastern Pacific, some even going as far as the international date line (180°W), in the range 44–52°N.

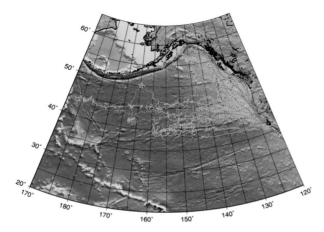

FIGURE 6 Foraging range of northern elephant seals across the North Pacific Ocean. The tracks in yellow represent female elephant seals, while the red tracks represent males. Animals were tagged at the Año Nuevo Rookery near Monterey Bay, California.

INTERACTIONS WITH HUMANS

Midden sites in coastal regions attest to a long tradition of humans hunting marine mammals for food and clothing. The commercial harvest of fur seals for their skins and seals for their oil was carried out from the late eighteenth century into the twentieth century and caused the depletion of many species. Many populations, such as elephant seals and fur seals, were harvested to the point of extinction. While not commercially harvested, the Caribbean monk seal is now extinct, and the Mediterranean and Hawaiian monk seals are highly endangered, with numbers well below 1000 individuals. The Hawaiian monk seal is listed as endangered under the Endangered Species Act, and both are protected by international agreements such as CITES (Convention in Trade of Endangered Species). In the United States, all marine mammals are protected by federal law under the Marine Mammal Protection Act, which makes it a federal crime to kill or harass (disturb) any marine mammal.

Seals, sea lions, and fishermen can compete for the same fish. Seals are often blamed for reductions in fish harvests that may in fact be due to overfishing by fishermen. It is also possible that fishermen are adversely affecting seal and sea lion populations by reducing the amount of fish available to them. For example, the Steller sea lion population in the North Pacific has been steadily declining. In 1985, 67,000 sea lions were counted in the Gulf of Alaska to the Aleutian Islands; twenty years later there were less than 25,000 sea lions in this same area. Sea lions can also have a negative effect on fish populations. For example, in the Ballard locks near Seattle, Washington, male California sea lions have learned that salmon are easy to catch when the fish get ready to go through fish ladders to go past the locks. The sea lions eat a significant proportion of the salmon, and the salmon population is in serious decline. Seals and sea lions are often accidentally caught in fishing nets used by fishermen. In most cases this does not cause significant harm to the seal or sea lion population. Sea lions and seals are also known to steal fish from fishing lines. This is especially common with salmon fishermen.

Continued increases in human population and development of coastal resources are further concerns for seal, sea lion, and fur seal populations. Pesticide residues from agricultural regions of the world and heavy metals from industrial waste have been found to build up in the tissues of many seals. In the 1970s high levels of DDT in the tissues of California sea lions caused massive premature pupping and death of these pups. Since the 1970s, when DDT production was stopped in California, the levels of

DDT in the environment have declined, as have the levels in the tissues of California sea lions. Few premature pups are seen on the rookeries today. Further concern comes from general habitat loss and degradation, as there is increasing use of the ocean for recreation, oil and gas development, and transportation. As human populations increase, there will be increased disturbance of seal and sea lion rookeries.

SEE ALSO THE FOLLOWING ARTICLES

Food Webs / Foraging Behavior / Sea Otters / Sex Allocation and Sexual Selection / Size and Scaling / Vertebrates, Terrestrial

FURTHER READING

Berta, A., J. Sumich, and K. Kovacs. 2005. *Marine mammals evolutionary biology*. 2nd ed. Burlington, MA: Elsevier.
Costa, D. P., M. J. Weise, and J. P. Y. Arnould. 2006. Potential influences of whaling on the status and trends of pinniped populations, in *Whales, whaling, and ocean ecosystems*. J. A. Estes, T. M. Williams, D. Doak, and D. DeMaster. Berkeley: University of California Press, 344–361.
Costa, D. P., and D. E. Crocker. 1999. Seals, in *Encyclopedia of reproduction*, Vol. 4. New York: Academic Press, 313–321.

FIGURE 1 Sea otter. Note the dense, water-repellent fur. Photograph by Bryant Austin.

SEA OTTERS

JAMES A. ESTES

United States Geological Survey

Sea otters are coastal living marine mammals that typically forage on a wide range of invertebrates in lower intertidal and shallow subtidal systems. They are members of the family Mustelidae and are one of several otter species that have recently radiated from freshwater into the higher-latitude coastal oceans. Sea otters occupy rocky and sand/mud habitats across the Pacific Rim from Japan to Mexico. As keystone predators, they have important direct and indirect effects on the structure and function of coastal ecosystems.

ORGANISMAL BIOLOGY

Mammals arose on the land but have invaded the sea on various occasions over the course of their evolutionary history. Most extant marine mammals have been so greatly modified for life in the sea that they bear little superficial resemblance to their terrestrial ancestors. As the most recent of these terrestrial expatriates, sea otters (Fig. 1) provide a unique view of the modifications associ-

ated with marine living in mammals. These modifications are apparent in various aspects of the sea otter's morphology, physiology, life history, and behavior.

Because mammals must maintain high core body temperatures and the thermal conductivity of water is roughly 25 times greater than that of air, elevated heat loss is among the greatest challenges faced by aquatic mammals. Increased body size is one of the most evident responses to this challenge, because heat production and heat loss increase respectively with body mass and surface area, whereas surface-to-volume ratio declines exponentially with increasing body mass. Sea otters are the largest living members of the mustelid family and the smallest fully marine living mammal, very likely thus representing the minimum possible body size for long-term existence of endothermic mammals in cold oceans. Thermoregulation in sea otters is facilitated by efficient insulation (sea otters lack blubber but have the densest fur of any extant mammal), elevated metabolic rate, and careful management of behavioral time budgets for activities that differ in the relative extent of heat loss and heat gain.

Other features of the sea otter's physiology and morphology also reflect strong selection imposed by life in the sea. For example, sea otters typically have single-young pregnancies, more closely resembling the other marine mammals (with whom they share a common environment) than the mustelids and other terrestrial mammalian carnivores (their closest living relatives). Sea otters are unique among the mustelids in having highly modified flipper-like hind limbs, an apparent adaptation for aquatic locomotion. Sea otters have an enlarged lung, modified for deep diving by cartilaginous support structures in the bronchioles that prevent compression collapse at depth. Other features of the sea otter's functional biology and body plan resemble those of their recent terrestrial ancestors more closely than those of the more highly modified cetaceans and pinnipeds. For example, sea otters do not accumulate extensive lipid energy stores and therefore have no blubber layer. This feature prevents them from fasting (among the marine mammals, sea otters are extreme "income strategists") and demands that insulation against heat loss is achieved entirely by their fur.

HISTORY AND CURRENT STATUS

The sea otter's dense, luxurious fur has motivated extensive exploitation by both aboriginal and modern humans. Sea otter remains are common in midden sites throughout the species range in the North Pacific Ocean (Fig. 2). The co-occurring abundance and size of their invertebrate prey indicate that aboriginal hunters reduced or eliminated sea otters from many areas over long periods of prehistory (see following section on Ecosystem Roles for further explanation).

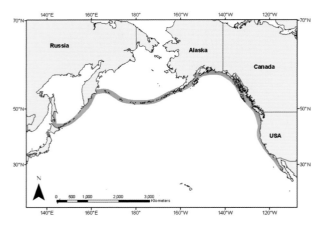

FIGURE 2 Historical range of sea otters (shown in red) in the North Pacific Ocean.

The Bering Expedition returned to Russia in 1742 with reports of bountiful sea otter and fur seal populations in the lands to the east. These reports set off the Pacific maritime fur trade, which was largely responsible for Russian colonization of the Pacific coast of North America. Sea otters had been hunted to near extinction by the early 1900s, at which time a dozen or so small and widely scattered remnant colonies survived. Further hunting was prohibited in 1913, and sea otters began to increase shortly thereafter. The scattered distribution of the surviving colonies and the sea otter's sedentary nature combined to create a recovery pattern that was highly variable in space and time. Some areas of southwest Alaska had reached environmental carrying capacity by the late 1930s or early 1940s, whereas other areas within the species' historical range remain unoccupied to this day.

Sea otters have recovered from the fur trade throughout most of their range in Russia—along the Kamchatka Peninsula, and in the Commander and Kurile Islands. By the late 1980s, the species had also recovered or was recovering throughout most of southwest Alaska (from lower Cook Inlet westward through the Aleutian archipelago). However, populations in this latter region began to decline precipitously in about 1990, the apparent consequence of increased predation by killer whales. Densities have since declined by one to two orders of magnitude, and populations are approaching extinction in some areas. Elsewhere in Alaska, numbers are stable or increasing. Relocations have been used in efforts to reestablish sea otters in southeast Alaska, British Columbia, Washington, Oregon, and at San Nicolas Island in southern California. The Oregon relocation failed, but those elsewhere are either stable (southeast Alaska) or increasing (Washington, British Columbia, and San Nicolas Island). A small remnant colony survived the fur trade in central California. This population presently ranges from about Half Moon Bay to Point Conception and contains somewhat less than 3000 individuals. Both the southwest Alaska stock and the California sea otter population are listed as Threatened under the Endangered Species Act.

BEHAVIOR AND POPULATION BIOLOGY

Sea otters feed mainly on benthic and epibenthic invertebrates in shallow rocky and soft sediment habitats. More than 150 prey species have been documented throughout the species' range. Sea otters are able to forage efficiently on well-armored shellfish because of their highly modified crushing dentition (Fig. 3) and their ability to use rocks and other objects as tools to break open their prey. Although sea otters are not adept piscivores, they eat fish in some areas, occasionally in large numbers. The sea otter's foraging range extends from the lower littoral zone at high tide to a maximum depth of about 100 meters. While dive times longer than 5 minutes have been recorded, most foraging dives are shorter than 2 minutes and occur in waters less than 30 meters deep. Longitudinal records from tagged individuals indicate extreme interindividual variation in diet. This variation appears to be more strongly driven by culture and learning than by prey availability. Individual dietary patterns and preferences are learned and transmitted across matrilines.

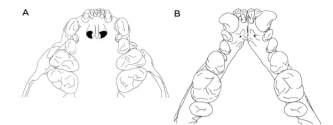

FIGURE 3 The sea otter's dentition is highly modified for crushing the exoskeletons of its invertebrate prey. (A) Roof of mouth looking up. (B) Lower jaw looking down. Note in both the large and flattened molars. Drawings by Laura Dippold.

In contrast with other otter species, which have promiscuous or serially monogamous mating systems, the sea otter is strongly polygynous. Sea otters are moderately sexually dimorphic, with body mass about 25 percent greater in males than females. Females and males attain sexual maturity at the respective ages of about 3 and 5 years, and maximum longevity in both sexes is about 20 years. Males maintain small but well-defined territories from which they aggressively exclude other males. Females enter estrus and ovulate shortly after weaning (or prematurely losing) their previous young. The average duration of pregnancy is about 6 months, and the pups are weaned approximately 6 months later. The maximum rate of population increase (r_{max}) is 17–20% yr^{-1}.

Sea otters are relatively sedentary compared with other marine birds and mammals. Female home ranges typically include less than 10 miles of coastline, and male territories are much smaller still. Longer-distance movements are occasionally undertaken. For instance, territorial males in central California commonly migrate seasonally between their territories (where females are abundant) and the ends of the range (where food resources are more abundant).

The majority of a sea otter's time (roughly 85–90%) is spent foraging and resting. The only other frequent activities are grooming and swimming. Time spent feeding varies substantially with population status and food availability, from 20% or less in food-rich environments to 50% or more in food-limited environments. Sea otters vigorously groom their fur to maintain the air layer and prevent water penetration to the skin.

ECOSYSTEM ROLES

Because of their relatively large size, high population density, high metabolic rate, and proficiency in locating and capturing food, sea otters can drastically limit their prey populations. Various studies of otherwise similar habitats with and without sea otters indicate that these prey

depletion effects usually range from one to two orders of magnitude. Since humans value and utilize many of the sea otter's prey species, conflicts with commercial and recreational shellfisheries have arisen as sea otters have reclaimed their range following the fur trade.

Various indirect effects result from the predator–prey interaction between sea otters and their invertebrate prey. For instance, sea otters help to maintain kelp forests by preying on and controlling herbivorous sea urchins (Fig. 4). This trophic cascade from sea otters to kelp forests has a broad range of effects on other species and ecosystem processes. These effects occur via three general processes—increased production, resulting from the ability of kelp to rapidly fix organic carbon through photosynthesis; changes in three-dimensional habitat structure associated with the presence or absence of kelp; and altered flow, resulting from the surface tension of kelp fronds

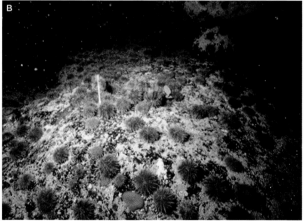

FIGURE 4 Representative landscape of kelp forest ecosystems in the Aleutian Islands with and without sea otters. Sea otters limit herbivorous sea urchins, thus permitting kelp populations to flourish (A). When sea otters are removed, sea urchins increase to the point where they overgraze kelp (B).

extending into the water column. Many other coastal species are influenced by these processes. The abundances of certain kelp forest fishes decline by roughly an order of magnitude when sea otters are lost from the system; filter-feeding invertebrates grow at elevated rates when sea otters and kelp forests are present; and the diet and foraging behavior of gulls and bald eagles in the Aleutian Islands are strongly influenced by the impacts of otters on their environments. Because of their broad ecological impacts, sea otters are known as a keystone species.

Although sea otters forage mainly in sublittoral habitats, they can influence intertidal communities in several ways. In protected habitats and during periods of calm along highly exposed coastlines, sea otters often forage in the low to mid intertidal zones, where they can strongly limit the abundance and distribution of various invertebrates including mussels, abalones, sea urchins, and clams. By enhancing the adjacent subtidal kelp forests, sea otters also indirectly influence intertidal communities by reducing wave force, increasing sedimentation, and increasing particulate organic carbon as a nutritional resource for filter-feeding invertebrates. This latter effect can produce a three- to fivefold increase in growth rate of intertidal mussels and barnacles. The diet and foraging behavior of other intertidal predators also is influenced by these processes. For example, glaucous winged gulls in the western Aleutian Islands feed mainly on intertidal invertebrates where otters are absent and on shallow-water fishes where otters are common.

The diverse and powerful influences of predation by sea otters and their recent ancestors have impacted the evolution of many co-occurring species. For instance, the chronically low intensity of herbivory in otter-dominated systems was very likely an important factor in creating a northern Pacific kelp flora that is remarkably lacking in toxic secondary metabolites. This highly productive and palatable flora may in turn have facilitated the radiation of dugongid sirenians (culminating with Steller's sea cow) and the evolution of the world's largest abalones in the North Pacific Ocean.

SEE ALSO THE FOLLOWING ARTICLES

Body Shape / Foraging Behavior / Kelps / Locomotion: Intertidal Challenges / Predation / Size and Scaling

FURTHER READING

Estes, J. A. 1989. Adaptations for aquatic living in carnivores, in *Carnivore behavior ecology and evolution.* J. L. Gittleman, ed. Ithaca, NY: Cornell University Press, 242–282.
Estes, J. A., E. M. Danner, D. F. Doak, B. Konar, A. M. Springer, P. D. Steinberg, M. T. Tinker, and T. M. Williams. 2004. Complex trophic interactions in kelp forest ecosystems. *Bulletin of Marine Science* 74: 621–638.
Kenyon, K. W. 1969. The sea otter in the eastern Pacific Ocean. *North American Fauna* 68: 1–352.
Riedman, M. L., and J. A. Estes. 1990. *The sea otter* (Enhydra lutris): *behavior, ecology, and natural history.* Biological Report 90(14). U.S. Fish and Wildlife Service.

SEA SLUGS

SEE NUDIBRANCHS AND RELATED SPECIES

SEA SPIDERS

SEE PYCNOGONIDS

SEA SQUIRT

SEE TUNICATES

SEA STARS

ERIC SANFORD

University of California, Davis

Sea stars (also known by the somewhat misleading term "starfish") are common echinoderms belonging to the taxonomic group Asteroidea. Approximately 1800 species have been described. They are distributed in all of the world's oceans, with most species living in intertidal and shallow subtidal habitats. As a result, sea stars are frequently observed in tidepools and on rocky shores. When sea stars are exposed to air during low tide, their stiff, motionless bodies show few signs of life. Thus, many first-time observers are surprised to learn that most sea stars are mobile predators or scavengers that often play important ecological roles in marine communities.

GENERAL BODY PLAN

Sea stars are invertebrates with a body plan that is radically different from those of most other animals. Their bodies consist of a central disk, typically grading into five arms, although some species have as many as 40 arms (Fig. 1). Sea stars lack a distinct head and have a mouth located on the ventral surface, in the center of the oral disk. Their bodies are radially symmetrical such that any line drawn through the central disk will divide the body into two approximately equal halves. In some species,

FIGURE 1 Diversity of sea star forms. Shown are four intertidal species common along the Pacific coast of North America. Shown are (A) the bat star *(Patiria miniata)*, (B) the ochre star *(Pisaster ochraceus)*, (C) the small six-armed star *(Leptasterias hexactis)*, and (D) the sunflower star *(Pycnopodia helianthoides)*. Photographs by Jacqueline L. Sones.

the arms are long and slender, whereas others (such as the bat star, *Patiria miniata*) have a stocky form with arms that are short relative to the central disk. Although there are exceptions, sea stars are often brightly colored, including red, orange, blue, and purple species. Some species have several color forms, such as the ochre star *(Pisaster ochraceus)*, found along the Pacific coast of North America. Size is quite variable among sea stars, with some species never exceeding a few centimeters across and other species reaching a diameter of >1 meter (such as the sunflower star, *Pycnopodia helianthoides*, Fig. 1D).

TUBE FEET AND LOCOMOTION

Sea stars crawl over surfaces using hundreds (sometimes thousands) of tiny tube feet, which are driven by hydraulic power. The tube feet are extensions of the water vascular system, a unique system of internal canals found in all echinoderms (Fig. 2). In sea stars, the tube feet are arranged in two or four rows within grooves that run along the underside of each arm. The external opening to the water vascular system is the madreporite, a conspicuous button-like plate near the center of a sea star's aboral disk. Water enters through this porous plate and passes down a short tube into a central ring canal that encircles the mouth. From here, radial canals extend along the length of each

arm. Multiple lateral canals branch from each radial canal and end in tube feet. Fluid in the lateral canals can be isolated from the rest of the water vascular system by the closing of small valves. Once water is isolated in a lateral canal, muscular bulbs called ampullae contract and force the fluid to extend the tube feet. The contraction of muscle fibers within a tube foot allows it to bend, and the synchronized movements of hundreds of tube feet allow the sea star to crawl efficiently in any direction. Most sea stars move quite slowly, although some species, such as the sunflower star *(Pycnopodia helianthoides)*, can move at the relatively brisk pace of >1 meter per minute.

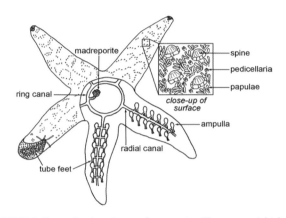

FIGURE 2 Generalized anatomy of a sea star. The upper right box shows a magnified view of the external surface. The lower half of the figure shows an internal view of the water vascular system that drives the tube feet. Drawing by Jacqueline L. Sones.

Tube feet also contribute to the success of sea stars on wave-exposed shores by providing a mechanism to grip surfaces firmly. The tip of each tube foot is a flattened disk that acts as a miniature suction cup to mechanically adhere to surfaces. In addition, the disks of tube feet secrete substances that form a chemical bond with the surface. As the sea star moves forward during locomotion, other secretions break the bond. The combined attachment strength of hundreds of tube feet allows sea stars to remain anchored even in areas of pounding surf, where other predators, such as crabs, would be easily dislodged. Sea stars that are firmly attached at low tide should be observed in place because prying them from the rock will leave behind large numbers of tube feet. In addition to locomotion and temporary attachment, the tube feet play important roles in feeding, gas exchange, and gathering sensory information. Sea stars possess specialized sensory tube feet at the tip of each arm. These tube feet lack suckers and probe ahead, exploring the environment as the sea star moves forward.

ENDOSKELETON

Sea stars have an internal skeleton made of thousands of small calcareous plates called ossicles. These plates are arranged in a lattice network, bound together and embedded within living connective tissue. The connective tissue of sea stars (and other echinoderms) has an unusual property: its rigidity is under the control of the nervous system. Thus, a sea star can rapidly stiffen or relax its skeleton by changing the fluidity of the surrounding connective tissue (also known as catch tissue). When a sea star is observed in motion (e.g., crawling across the bottom of a tidepool), its arms are noticeably flexible. However, if the sea star is lifted out of the water, it will quickly stiffen its catch tissue, locking its arms into a rigid frame. This can be an effective defense against predators. It is not unusual to see the arms of a sea star projecting from either side of a gull's mouth as the gull tries to swallow the rigid sea star whole.

EXTERNAL ANATOMY

The surface of a sea star is surprisingly complex (Fig. 2). Hard, white knobs or spines on the surface are calcareous projections of the internal skeleton. When the sea star is submerged, these skeletal structures are surrounded by soft clusters of small, finger-like projections called papulae. Gases are exchanged between the water and the sea star's body across these thin, ciliated outpockets of the body wall that emerge from the spaces between the ossicles of the endoskeleton. When the sea star is exposed to air, the papulae are generally retracted. Scattered across the surface are also specialized pincer-like structures called pedicellariae. These small structures are effective in keeping the sea star's surface free of debris and fouling organisms such as barnacle larvae. In other cases, the pedicellariae can aid in prey capture or play an aggressive role during encounters between individuals of different species.

FEEDING AND DIGESTION

Most sea stars possess a remarkable method of feeding. A sea star's mouth is located beneath its body in the center of the oral disk. The lower portion of the stomach can be everted out through the mouth to begin digesting prey outside of the body. When a sea star feeds on a clam or mussel, it crawls on top of its prey and attaches its tube feet to the two valves of the hinged shell. The sea star then stiffens its connective tissue, creating a rigid scaffolding that the tube feet can pull against. Strong forces exerted on the two valves create a narrow opening (less than 0.5 millimeter), just wide enough for the sea star to insert its stomach into the shell. The stomach releases digestive enzymes that reduce the prey's tissue into a liquid soup, which

is brought into the sea star's body along ciliated gutters in the stomach. Smaller prey such as snails are engulfed whole within the folds of the stomach. This flexible mode of feeding allows sea stars to feed on prey that span a broad range of body sizes. As a result, many sea stars are generalist predators, feeding on mussels, clams, barnacles, snails, limpets, chitons, marine worms, sea urchins, and other prey. Other sea stars are scavengers or suspension feeders.

NERVOUS SYSTEM AND SENSORY INFORMATION

Sea stars lack a centralized nervous system. Rather, their nervous system consists of an internal circular ring that surrounds the mouth, with radial nerves that branch off and extend into each arm. The radial nerves coordinate the highly synchronized movements of the tube feet. The radial nerves are also continuous with a diffuse nerve net of sensory cells that are dispersed across the sea star's entire surface. These cells are sensitive to touch and chemical signals. Many sea stars have clusters of light-sensitive cells that typically appear as a small red spot at the tip of each arm. Specialized sensory tube feet also occur at the tip of each arm. In many ways, sea stars challenge our intuitive views about the importance of a centralized nervous system. Despite lacking a brain or complex sensory structures such as image-forming eyes, sea stars are very efficient predators, able to pursue and capture fleeing predators such as snails. Their design has also proven to be quite successful; the oldest known fossil sea stars are approximately 480 million years old (Lower Ordovician).

REPRODUCTION

Sea stars reproduce sexually and almost always have separate sexes, although males and females cannot be distinguished by external appearance. Most sea stars are broadcast spawners, meaning that males and females release gametes into the water, where external fertilization occurs. Gametes (eggs or sperm) are produced by a pair of gonads located in each arm and are released into the water through pores on the aboral disk (Fig. 3A). Most species have a discrete reproductive season, often in the spring in temperate regions. Synchronized spawning events are common, with nearby individuals releasing gametes simultaneously to increase the chances of fertilization. Occasionally spawning events may be observed within tidepools; freshly released gametes may cause the water to become cloudy. More often, spawning occurs at high tide and gametes are washed away in the surf, leaving behind no hint of this hidden activity.

Although each individual may release millions of gametes, the chances of an individual gamete encountering

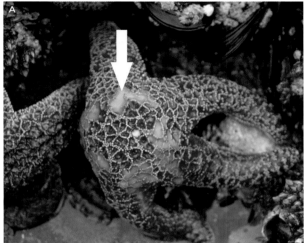

FIGURE 3 Reproductive strategies of two sea star species. (A) A female ochre star (*Pisaster ochraceus*, diameter ~25 cm) broadcast spawning unfertilized eggs (arrow) through pores on her dorsal surface. Photograph by Eric Sanford. (B) A much smaller species (the six-armed star, *Leptasterias hexactis*, diameter ~3 cm), flipped over to show that she is brooding a cluster of developing embryos (arrow) beneath her body. Photograph by Jacqueline L. Sones.

limitation, a female may have greater reproductive success if she devotes considerable care to ensuring that some fertilized embryos survive. In contrast, larger-bodied species have the capacity to release millions of gametes, and these vast numbers may outweigh the costs of low fertilization success and survival in the plankton.

In the water column, fertilized embryos develop into free-swimming larvae (Fig. 4). In some species, larvae feed on plankton, whereas other species have nonfeeding larvae that rely on stored yolk. In feeding larvae, a long band of cilia is used for both locomotion and capture of food. The alien form of these planktonic larvae bears little resemblance to an adult sea star, and the connection between these life stages was not recognized until the 1860s. Development proceeds through several larval stages and often requires several months before the settling stage returns to the shore and completes its transformation into a juvenile sea star (~2 mm diameter).

Asexual reproduction can occur in some sea star species but is not common. In these species the body splits

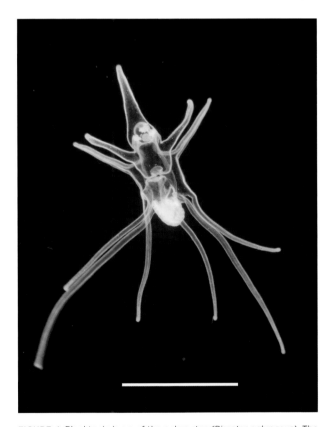

FIGURE 4 Planktonic larva of the ochre star (*Pisaster ochraceus*). The long arms of this larval stage (called a brachiolaria) are covered with cilia used for feeding and swimming. Larval arms will be reabsorbed during metamorphosis into the juvenile sea star. Scale bar = 1.0 mm. Photograph by Bruce A. Miller.

another appropriate gamete in the ocean are quite small. Brooding is an alternative strategy used by some sea stars such as the six-armed star *(Leptasterias hexactis)*. In most brooding species, females hold their spawned eggs beneath their bodies, in a space formed by arching their central disk away from the rock surface (Fig. 3B). The eggs are fertilized by sperm, which males broadcast-spawn into the water. The female continues to protect the developing embryos beneath her disk for several months until her offspring complete development and crawl away as tiny sea stars. Theory suggests that natural selection favors brooding in small-bodied species because females typically can produce only a few hundred eggs. Given this along the central disk, creating two parts, each with some of the old arms. New arms and internal organs are regenerated, resulting in two genetically identical individuals.

REGENERATION

All sea stars possess an amazing capacity for regeneration. Arms that are torn off by feeding gulls or other disturbances are slowly regrown over time (Fig. 5). It may take over a year to completely regrow a lost arm. Close inspection of sea stars on the shore will often reveal individuals that have one or more very short arms that are in the process of regenerating. Even an isolated arm may regenerate into a complete sea star if it retains some portion of the central disk. This facet of sea star biology has not always been appreciated. In the 1800s, oyster farmers in Long Island Sound, New York, removed feasting sea stars from their oyster beds with rakes. The trespassing sea stars were torn into pieces and tossed overboard, only to return in even greater numbers.

FIGURE 5 Sea star regeneration. An ochre star *(Pisaster ochraceus)* with two small arms that are being slowly regenerated. Photograph by Jacqueline L. Sones.

FORAGING AND ECOLOGICAL EFFECTS

In many intertidal and subtidal habitats, sea stars are important top predators. Sea stars may have particularly strong community effects when they prefer prey that are dominant competitors for limited space on wave-exposed rocky surfaces. For example, in the Pacific Northwest,

the ochre sea star *Pisaster ochraceus* preys on the intertidal mussel *Mytilus californianus*. Predation by the sea star keeps mussel populations from overgrowing other sessile invertebrates and algae attached to the rock and thus helps maintain a diverse assemblage of organisms in the low intertidal zone. Like most sea stars, *Pisaster* actively searches for food, and impressive aggregations of feeding sea stars may gather around patches of mussels.

In other cases, sea stars take a more passive approach to locating prey. In tidepools, sea stars are often "sit and wait" predators that rely on the action of waves to deliver prey to them. As topographic low points on the shore, tidepools often accumulate mussels or snails that have been dislodged from the rocks by pounding waves. In deep tidepools along the Pacific coast of North America, it is not unusual to see an aggregation of large *Pisaster ochraceus* feeding on hapless clumps of mussels that have been deposited there by the surf.

SEE ALSO THE FOLLOWING ARTICLES

Brittle Stars / Echinoderms / Foraging Behavior / Locomotion: Intertidal Challenges / Predation / Reproduction

FURTHER READING

Brusca, R. C., and G. J. Brusca. 2003. *Invertebrates,* 2nd edition. Sunderland, MA: Sinauer Associates.

Lambert, P. 2000. *Sea stars of British Columbia, Southeast Alaska and Puget Sound.* Vancouver: University of British Columbia Press.

Menge, B. A. 1975. Brood or broadcast? The adaptive significance of different reproductive strategies in the two intertidal sea stars *Leptasterias hexactis* and *Pisaster ochraceus. Marine Biology* 31: 87–100.

Paine, R. T., J. C. Castillo, and J. Cancino. 1985. Perturbation and recovery patterns of starfish-dominated intertidal assemblages in Chile, New Zealand, and Washington State. *American Naturalist* 125: 679–91.

Summers, A. 2003. Catch and rel ease. *Natural History* 112: 36–37.

SEA URCHINS

THOMAS A. EBERT

Oregon State University

Sea urchins are echinoderms in the class Echinoidea, which has two major taxonomic divisions. Irregular sea urchins, with an obvious front and back, include sand dollars, sea biscuits, and heart urchins. In general, these are animals of sandy or muddy environments. The other division, regular sea urchins, includes species that are more radially symmetrical and lack a clear front end. Sea urchins that live in wave-swept rocky environments and tidepools are all regular urchins and often are present in high densities. Some species are the focus of fisheries for their gonads, which are called roe or uni.

ANATOMY AND DEVELOPMENT

The skeleton of a sea urchin consists of a large number of elements including spines, the body (called a test), and the ossicles of Aristotle's lantern. The test consists of columns of plates: five double columns with pores for tube feet (ambulacral plates) and five double columns without pores (interambulacral plates). As in other echinoderms, the skeleton is internal and composed of calcite (calcium carbonate) with varying amounts of magnesium that are determined both by temperature and location in the body. Each ossicle, plate, or spine is a single calcite crystal. Growth takes place around each plate of the test as well as by the addition of new plates at the top of each column.

Aristotle's lantern is the feeding apparatus in regular sea urchins, having five teeth that can be seen at the mouth on the bottom surface. Food passes through the digestive tract, and fecal pellets are eliminated through the anus on the top of the test. As with other echinoderms, the complex water vascular system is hydraulic and extends and contracts the tube feet by a coordinated interaction of muscles in water reservoirs inside the test (ampullae) and muscles in the tube feet. Tube feet also are important for gas exchange in respiration.

Sexes are separate in sea urchins and the gonads are five-lobed. During a spawning season, eggs and sperm are shed into the sea where fertilization takes place. A swimming larva develops and in most species the larvae must feed while in the plankton. The length of the planktonic stage varies across species and may last just a few days or may require several months. The larva settles from the plankton at a size of about 0.5 millimeter and growth to reproductive size may take longer than one year. Life spans vary by species and range from just a few years to many decades.

INTERTIDAL SEA URCHINS

There is a problem with defining intertidal sea urchins because some occur at the lowest intertidal levels but are primarily subtidal. Some species occur only occasionally in the rocky intertidal and then usually only as juveniles. There also are species that are found occasionally in rocky areas, usually pools, but actually are more characteristic of sandy substrates that support seagrasses. Some authors have reported substantial numbers of such species in stud-

ies that focus on rocky sites. The important point is that sea urchins are very adaptable and so inhabit intertidal rocky pools even when, by general agreement of most biologists, they should not. Of those sea urchins that are typical of intertidal rocky habitats several genera are particularly well represented. In north temperate and arctic areas *Strongylocentrotus* is represented by *S. purpuratus*, *S. droebachiensis*, *S. intermedius*, *S. nudus*, and *S. polyacanthus* in the Pacific and *S. droebachiensis* in the Atlantic, which makes it a circumboreal genus. The genus *Echinometra* is present in all tropical seas: *E. mathaei* from the mid-Pacific through the Indian Ocean; *E. vanbrunti* from Baja California, Mexico, to Peru; *E. insularis* at Easter Island; *E. lucunter* on both sides of the tropical Atlantic; and, *E. viridis* in the Caribbean.

SPINE MORPHOLOGY

The sea urchin family with the largest number of rocky intertidal species is the Echinometridae. Several genera in this family show highly modified spine development. *Echinometra mathaei* has typical sea urchin spines (Fig. 1) that taper to sharp points. *Echinometra oblonga* has spines that are thicker and more blunt (Fig. 2). Slate-pencil sea urchins *(Heterocentrotus trigonarius)* have massive primary spines that can reach 15 centimeters in length and 1 centimeter in diameter at the base (Fig. 3). In *Colobocentrotus atratus*, the spines are highly modified and are like flat umbrellas that form a protective covering over the body (Fig. 4). These modifications are related to the environments where these species live.

FIGURE 1 *Echinometra mathaei*, Oahu, Hawaii, with spines that are typical of many sea urchin species: thickest at the base and tapering to a sharp point. Photograph by the author.

FIGURE 2 *Echinometra oblonga*, Oahu, Hawaii, with thick spines that do not have sharp points. Photograph by the author.

FIGURE 3 *Heterocentrotus trigonarius*, Ananij Island, Enewetak Atoll, showing very large and thick spines. Photograph by the author.

C. atratus lives in the highest-energy environments of any sea urchin where crashing waves are normal. Similarly, *H. trigonarius* lives at reef crests where high surf is common. *E. oblonga* lives closer to reef edges than does *E. mathaei*. Spine morphology is very conservative in sea urchins; a typical spine is widest at the base and tapers to a point, and developmental controls permit little deviation from this pattern. Having such highly modified spines suggests that one group of the Echinometridae no longer has such tight developmental controls and so spine morphology could evolve and species could colonize wave-swept environments.

EXCAVATING CAVITIES

Many sea urchin species (*Stomopneustes variolaris*, *Paracentrotus lividus*, *Echinometra* spp., *Echinostrephus* spp., *Heliocidaris* spp., *Heterocentrotus trigonarius*, and *Strongylocentrotus* spp.) excavate holes or cavities in rock

(Figs. 5, 6). How these holes are formed has fascinated both casual observers as well as some professional biologists for at least 200 years. When most people first become aware of sea urchins in cavities, they invariably ask, "How do they do it?" An additional puzzle has been based on the observation that sea urchins often appear to be larger than the opening of a cavity; they seem trapped. Suggestions for boring have included chemicals such as acid for dissolving rock, mechanical abrasion from the spines, and action of the tube-feet. Chemicals were dropped from the list of possibilities early on because holes can be formed in rocks that are not attacked by acids. By the 1850s, boring was considered to be mechanical and to involve both Aristotle's lantern and spines, and this is the current view of how cavities are formed. Even in very hard rock, teeth and spines are considered to be sufficient, a view that is supported by research that has shown that quartz sand grains and glass can be cracked by teeth that are softer than either of these materials. Abrasion also is sufficient to explain burrowing of *Strongylocentrotus purpuratus* into steel pilings in California.

FIGURE 4 *Colobocentrotus atratus*, Oahu, Hawaii, with highly modified spines that are short and flat so that the exterior of the sea urchin looks much like a limpet. Photograph by the author.

Sea urchins tend to fit their holes very well—big urchins in big holes and small individuals of the same species in small holes. The current view is that even when they appear to be larger than the opening of a

FIGURE 5 *Stomopneustes variolaris*, Negombo, Sri Lanka, living in shallow cavities. Photograph by the author.

FIGURE 6 *Strongylocentrotus purpuratus*, Sunset Bay, Oregon, in cavities. Photograph by the author.

cavity, they actually can get out if spines are flattened. Also, the current view is that small sea urchins that outgrow their cavity must move out and find a more suitable home. Some sea urchin species that excavate cavities, such as *Strongylocentrotus purpuratus*, also can be found without cavities either in pools or in gravel and shells.

FEEDING

The bottoms of cavities occupied by sea urchins are clean so some nourishment is obtained by grazing whatever grows inside the cavity. This potentially is important for species, such as *Echinometra* spp. and *Heterocentrotus trigonarius*, that form elongated cavities. The edges of these cavities often support a lush growth of algae, which indicates that the sea urchins do not venture out to graze. The most

important source of food is animal and plant fragments that are washed close to a sea urchin and captured by the tube feet and spines. Captured food is transferred to the mouth by a coordinated movement of tube feet and spines. Even when sea urchins such as *Echinostrephus* spp. fit holes very tightly, food particles can be moved between the walls of the cavity down to the oral side.

DEFENSE OF CAVITIES

Echinometra mathaei living in burrows actively defend against intruders and do so by spine waving and pushing. There are differences among species and *E. oblonga* is a better defender than *E. mathaei*. Similar aggressive behavior has been observed in the Caribbean sea urchins *Echinometra lucunter* and *E. viridis* but in addition to pushing these sea urchins also bite intruders. The defending sea urchin rotates its body so that its mouth is towards the intruder, Aristotle's lantern is extended, and the teeth are used to bite spines off the intruder. If the same size as the defender or smaller, the intruder always loses. Furthermore, there appears to be a sense of "home" because if a resident is removed and replaced by an intruder and then the resident is introduced back to the edge of the cavity, the resident will always retake its cavity if sizes were equal. *Echinometra* species are not the only sea urchins that show aggressive behavior in defense of cavities. *Strongylocentrotus purpuratus* shows aggressive territorial behavior with spine fencing and pushing.

SEE ALSO THE FOLLOWING ARTICLES

Body Shape / Competition / Echinoderms / Food Uses / Stone Borers

FURTHER READING

Ebert, T. A. 2007. Growth and survival of post-settlement sea urchins, in *Edible sea urchins: biology and ecology,* 2nd ed. J. M. Lawrence, ed. Amsterdam: Elsevier, 95–134.

Ellers, O., and M. Telford. 1991. Forces generated by the jaws of clypeasteroids (Echinodermata: Echinoidea). *Journal of Experimental Biology* 155: 585–603.

Hendler, G., J. E. Miller, D. L. Pawson, and P. M. Kier. 1995. *Sea stars, sea urchins, and allies. Echinoderms of Florida and the Caribbean.* Washington, DC: Smithsonian Institution Press.

Maier, D., and P. Roe. 1983. Preliminary investigations of burrow defense and intraspecific aggression in the sea urchin, *Strongylocentrotus purpuratus. Pacific Science* 37(2): 145–149.

Neill, J. B. 1988. Experimental analysis of burrow defense in *Echinometra mathaei* (de Blainville) on Indo-West Pacific reef flat. *Journal of Experimental Marine Biology and Ecology* 115: 127–136.

Ricketts, E. F., and J. Calvin. 1939. *Between Pacific tides.* Editions with revisions 1952, 1962 and 1968 by J. W. Hedgpeth; 5th edition 1985 revised by D. W. Phillips. Stanford, CA: Stanford University Press.

Stephenson, T. A., and A. Stephenson. 1972. *Life between tidemarks on rocky shores.* San Francisco, CA: Freeman.

SEAWATER

RICHARD T. BARBER
Duke University

Seawater is water (H$_2$O) and its natural dissolved constituents (mostly salts) from one of earth's seas or oceans. Seawater is the most abundant form of water on earth, but the intertidal zone along the edges of oceans and seas experiences a periodic absence of seawater for several hours each day. This regular variability of the presence and absence of seawater has both positive and negative impacts on the plants and animals that live in the intertidal zone.

CHARACTERISTICS AND IMPORTANCE OF SEAWATER

Seawater is distinct from freshwater, which is the water found in lakes, rivers, groundwater, and, in frozen form, glaciers and ice sheets. Both seawater and freshwater are mixtures of pure water (H$_2$O) and a number of dissolved materials, but the concentration of dissolved salts is much higher in seawater, which has an average concentration of about 3.5% salt, or, as oceanographers prefer to say, 35 parts per thousand (ppt) salinity. Freshwater is defined as water with less than 0.05 ppt salt. In seawater with a salinity of 35 ppt, the molal concentration of H$_2$O is about 54 moles per kilogram, while total molal concentration of the major ions is about 1 mole per kilogram. H$_2$O is clearly the most abundant component of seawater, and it is the properties of the pure water in seawater that are critically important for intertidal organisms. In its pure form, liquid water is a tasteless, odorless substance that is essential to all known forms of life. Because it absorbs strongly in the infrared portion of the light spectrum, water absorbs a small amount of visible red light, which gives it its slightly blue color when seen in a large body such as a lake or ocean. Water, often referred to as the universal solvent, dissolves many types of substances, especially salts, but also gases, organic compounds, and metals. Water has an unusually high specific heat capacity and heat of vaporization: it takes a lot of heat to change the temperature of water and to evaporate water. These two factors give water the ability to buffer changes in temperature.

Why is seawater important to the rocky intertidal? The rocky intertidal is one of Earth's most diverse habitats, especially with regard to the biodiversity of higher taxa. At the same time, rocky intertidal zones support dense populations of plants and animals whose biomass per unit area is among the highest known. Although it is a good place to live for many invertebrates and algae, the rocky intertidal is exposed to a set of biological stresses, such as heat and cold, desiccation, ultraviolet (UV), hypoxia, and salinity variability. Furthermore, the range of variation in these stressful processes is greater than the range experienced by organisms living full time either in the ocean or on land. How does such a rich and diverse community thrive in a habitat with such extreme stresses? The following paragraphs describe the processes by which the critical properties of seawater protect and nourish the rocky intertidal community.

AMELIORATING HEAT STRESS AND COLD STRESS

Heat stress kills by denaturing a cell's proteins; cold stress kills by freezing cellular water (removing liquid water), which dehydrates the cell and denatures its protein. Because of the huge heat capacity of water, it takes more heat to change the temperature of water than to change the temperature of air or rocky substrate. The result of this property is that seawater moderates rapid temperature changes in the intertidal environment. At low tide in the summer the sun heats rocks and organisms, but the incoming tide or an occasional big wave cools the substrate and organisms before widespread thermal mortality takes place. The overarching importance of this amelioration is illustrated by looking at the latitudinal variation in rocky-intertidal diversity and biomass from the equator to the poles. In the tropical intertidal, beneath ledges, in caves, and under rocks there are extremely diverse and high biomass communities, while the surfaces of intertidal substrate exposed to direct sunlight are nearly devoid of organisms. At low tide these exposed surfaces rapidly reach temperatures that exceed the thermal tolerance of intertidal organisms, preventing the survival of even the most hardy animals and algae on the upper surface of the rocks. Thermal death takes place before the tidal delivery of seawater can ameliorate the lethal temperatures. Tidally exposed hard substrate surfaces in polar regions are also depauperate relative to the rocky intertidal in temperate zones and to polar subtidal communities that are only a few meters offshore from the polar intertidal zone. Cold stress can make the entire intertidal in cold regions as inhospitable as heat stress makes the exposed surfaces in the tropics.

PREVENTING DESICCATION STRESS

Another lethal process to which rocky intertidal organisms are subjected when exposed to air during low tide is desiccation, the loss of pure water by evaporation. Desiccation is accelerated by the three factors that accelerate evaporation: increased temperature, increased wind, and decreased humidity. Evaporative desiccation is a threat in dry, windy, polar intertidal regions as well as in temperate and tropical intertidal zones. Seawater, with its great proportion of pure H_2O, has the capacity to reverse desiccation completely. Replacement of the pure H_2O that organisms lose to the atmosphere through evaporation is an important service of seawater that probably equals thermal buffering in survival importance. Intertidal organisms, especially certain macroalgae, can become so desiccated that they no longer look like living organisms; yet when the tide or waves deliver seawater, they rapidly rehydrate to full rigor. There is, however, a desiccation limit beyond which cell death is inevitable, so rocky-intertidal organisms have many behavioral and morphological adaptations to limit desiccation for a few hours until the tide returns.

PROTECTING AGAINST ULTRAVIOLET STRESS

Ultraviolet light is electromagnetic radiation with a wavelength shorter than that of visible light, but longer than soft X-rays. Ultraviolet light causes somatic and genetic damage in living organisms by breaking chemical bonds and making the fragments unusually reactive. This photodamage renders the changed proteins or nucleic acids incapable of carrying out their essential functions. Seawater is an extremely efficient absorber of ultraviolet radiation. Although direct exposure to solar ultraviolet radiation can damage plants and animals in less than an hour, a few inches of seawater can reduce the harmful short wavelengths (>380 nm) to levels that do little photodamage. Questions remain about the quantitative importance of ultraviolet stress for intertidal organisms. Desiccation and heat stress from solar exposure appear to be lethal well before organisms suffer significant somatic damage from ultraviolet exposure, but chronic genetic damage in the higher intertidal zone and during low tide may be a subtle but significant process for intertidal organisms.

PREVENTING HYPOXIA

Hypoxia is lack of oxygen to a degree that affects physiological processes. An important feature of seawater is that, as a universal solvent, it very efficiently dissolves oxygen. Freely circulating seawater in the intertidal is usually saturated or supersaturated with dissolved oxygen as a result of breaking waves, which provide a mechanical subsidy for the dissolution of atmospheric oxygen into seawater. Although dissolved oxygen concentrations are high in the seawater that bathes the rocky intertidal, the oxygen demand by the high biomass of respiring animals is also high. Cut off from freely circulating seawater, any isolated tidepool will rapidly undergo hypoxia because the transfer of oxygen from air to water is extremely inefficient during calm conditions without the mechanical subsidy of breaking waves. Animals in such a tidepool are stressed for a few hours by hypoxia until the tide returns with its oxygen-rich seawater. If, however, the low tide leaves organisms such as mussels at least partially exposed to air rather than submerged in isolated tidepools, these animals can absorb oxygen directly from air. The downside of using atmospheric oxygen is that exposure to air enhances desiccation.

AMELIORATING SALINITY STRESS

Salinity is the total amount of dissolved solid material in grams in one kilogram of sea water; grams per kilogram results in the units canceling, making salinity a dimensionless quantity. Salinity is a simple property in concept, but a very difficult property to measure accurately. Salinity cannot be measured by simply evaporating seawater and weighing the residual solid material, because chlorides, the most abundant ions in seawater, are volatized during drying. Conductivity is now used to measure salinity *in situ* with very high accuracy. The variability of salinity in open-ocean seawater is small, ranging for most of the ocean's seawater from 34.60 to 34.80 ppt. Marine organisms have evolved internal fluids that are in osmotic balance with seawater. When they are immersed in water that has a salinity much higher or lower than open-ocean seawater, the osmotic imbalance between the external medium and their interior fluids causes organisms to gain or lose water. When rain or terrestrial runoff substantially reduces the salinity of isolated tidepools at low tide, there can be considerable mortality of soft-bodied organisms before the incoming tide flushes out the low-salinity water. Theoretically, evaporation from isolated tidepools can also cause salinity stress, but this rarely happens because evaporation is slow, requiring a week or more to double the salinity of seawater. Tides almost always flush out the hypersaline water with normal-salinity seawater well before the hypersaline salinity stress is lethal.

SEE ALSO THE FOLLOWING ARTICLES

Evaporation and Condensation / Salinity, Measurement of / Surface Tension / Water Chemistry / Water Properties, Measurement of

FURTHER READING

Burnett, L. E. 1997. The challenges of living in hypoxic and hypercapnic aquatic environments. *American Zoologist* 37: 633–640.

De Mora, S. J., S. Demers, and M. Vernet. 2000. *The effects of UV radiation in the marine environment*. London: Cambridge University Press.

Denny, M. W. 1987. Life in the maelstrom: the biomechanics of wave-swept rocky shores. *Trends in Ecology and Evolution* 2: 61–66.

Kanwisher, J. 1957. Freezing and drying in intertidal algae. *Biological Bulletin* 113: 275–285.

Menge, B. A., and J. P. Sutherland. 1987. Community regulation: variation in disturbance, competition and predation in relation to environmental stress and recruitment. *American Naturalist* 130: 730–757.

Przeslawski, R., A. R. Davis, and K. Benkendorff. 2005. Synergistic effects associated with climate change and the development of rocky shore mollusks. *Global Change Biology* 11: 515–522.

Somero, G. N. 2002. Thermal physiology and vertical zonation of intertidal animals: optima, limits, and costs of living. *Integrative and Comparative Biology* 42: 780–789.

Tomanek, L., and B. Helmuth. 2002. Physiological ecology of rocky intertidal organisms: a synergy of concepts. *Integrative and Comparative Biology* 42: 771–775.

Witman, J. D., and K. R. Grange. 1998. Links between rain, salinity, and predation in a rocky subtidal community. *Ecology* 79: 2429–2447.

SEAWEED

SEE ALGAE

SETTLEMENT

SEE LARVAL SETTLEMENT, MECHANICS OF

SEX ALLOCATION AND SEXUAL SELECTION

STEPHEN M. SHUSTER
Northern Arizona University

The term *sex allocation* denotes the evolutionary outcome of energetic investment by individuals in sexual species when it is partitioned toward male or female function. The term *sexual selection* denotes the evolutionary process that occurs when individuals in one sex mate with disproportionate success at the expense of other individuals of the same sex.

PARENTAL INVESTMENT AND SEXUAL SELECTION

The foundation for most research on sex allocation and sexual selection is known as parental investment theory. This approach identifies a sex difference in parental investment as the source of sexual selection and as the ultimate cause for sex differences in energy allocated toward mating effort and parental care. According to this view, males and females are defined by differences in their energetic investment in gametes. In apparent confirmation of this perspective, most sexual species exhibit gamete dimorphism, or anisogamy, in which females produce few, large ova, and males produce many, tiny sperm.

An observed sex difference in initial parental investment among species is presumed by many researchers to influence sex differences in mating and parental behavior, resulting in a taxonomic bias toward parental care in females and away from parental care in males. According to parental investment theory, not only are females, with their greater initial investment in offspring, more inclined to provide care, but the small per-gamete investment in offspring by males is presumed to predispose them to pursue opportunities for additional matings rather than care for existing young. The tendency for females to provide offspring care, according to parental investment theory, makes females a limiting resource for male reproduction, a condition that appears to explain the tendency, widely noted in scientific and popular literature, for males to compete among themselves for access to females, as well as the tendency among females to be choosy when selecting mates.

A slightly different, but still complementary perspective suggests that parental care should be provided by each sex, not according to sex differences in initial parental investment, but instead according to how such care influences each individual's future fitness. According to this optimal-fitness-returns view, it is the sex difference in expected fitness available to males and females that determines which sex provides parental care and which does not. Consistent with parental investment theory, greater expected confidence of parentage coincides with female care in most species. Unfortunately, direct tests of this and related hypotheses are complicated by difficulties researchers encounter in accurately quantifying the future fitness of individuals of either sex in natural populations.

PARENTAL INVESTMENT MEASURES OF SEXUAL SELECTION

Several parameters have been suggested for quantifying the intensity of sexual selection according to the foregoing hypotheses. The operational sex ratio (OSR = $N\male/N\female$ = R_O = $1/R$) is based on the assumption that sexual selection results from competition among males for mates. The greater the number of mature males relative to the number of receptive females, the greater the value of OSR becomes, and the stronger the presumed intensity of male–male competition. The sex difference in gametic investment has been quantified as the ratio of male to female potential reproductive rates (PRR). The PRR is measured by saturating individual males and females with potential partners and calculating the ratio of maximum offspring numbers produced by these individuals. The greater this ratio is, the greater the potential difference in numbers of offspring is between a maximally successful male and a maximally successful female, although no individual of either sex may achieve this potential in nature.

Both OSR and PRR are combined in Q, the ratio of males and females "qualified" to mate. Such individuals are sexually mature, have acquired reproductive resources, and are in either the "time in" or "time out" phase of their reproductive cycle. Floaters or sneakers, that is, mature individuals not currently controlling territories or resources that attract mates, are excluded from calculations of Q, whereas mature individuals, who have reproduced at some previous time but are not currently doing so, are included in this calculation.

As explained in the following sections, when some individuals mate more than once, other individuals must be excluded from mating. For this reason, any parameter that includes only mating individuals will underestimate total selection intensity. Thus, despite Q's synthetic concept, its policy of ignoring some classes of nonbreeding males and including others can lead to inaccurate estimates of the variance in male reproductive success and of actual selection intensity.

PARENTAL INVESTMENT AND SEX ALLOCATION

How energetic investment by parents is directed toward male and female function is the basis of sex allocation theory. The three dominant themes in this research include (1) how differential energetic investment in male and female offspring leads to deviations in population sex ratio, (2) how anisogamy influences the allocation of resources toward male or female sexual function, and (3) how opportunities for multiple mating and ensuing ejaculate competition may influence gamete number and gonadal structure.

Sex allocation theory predicts that deviations in population sex ratio will occur mainly within inbreeding-tolerant species, in which parents are expected to produce only as many sons as are necessary to fertilize the ova of these same parents' more numerous daughters. Species exhibiting such "local mate competition" are rare in tidepools;

most populations have nearly equal sex ratios, or consist of simultaneous or sequential hermaphrodites, in which each individual represents both sexes. Because ova are presumed to be energetically more expensive than sperm, ovarian excess is expected, and is generally observed among intertidal hermaphrodites, including algae, cnidarians, flatworms, and annelids, as well as certain gastropods, crustaceans, and fish.

However, fine-scale adjustments in energetic allocation toward male or female function are also observed in hermaphrodites when sex differences in body size exist or when multiple mating, and thus sperm competition, may occur. Contrary to standard "optimal" allocation predictions, such adjustments vary widely among species and may favor male function, female function, or both sexual functions or may remain invariant for one or both functions. These results suggest a high degree of complexity in the genetic and evolutionary processes underlying the expression and persistence of sexual phenotypes; processes that are not specifically addressed, and are therefore difficult to explain, when evolutionary outcomes are emphasized.

SEX ALLOCATION AND SPERM COMPETITION

Game theory models, also known as evolutionary stable strategy (ESS) models, of sex allocation are consistent with parental investment and optimality predictions for simultaneous hermaphrodites with external fertilization. Certain polychaete worms and teleost fish, for example, show the predicted optimal bias toward female function just described. ESS models for hermaphrodites with multiple mating and sperm storage, as occurs in some polyclad flatworms and littoral gastropods, also predict a bias in energy allocation toward female function, particularly when male allocation depends on the ratio of sperm donor production to a sperm recipient's existing sperm stores.

Yet sex allocation models that incorporate multiple mating also suggest that resources allocated toward gonads or gametes should be optimized for group size. Larger groups are presumed to permit more multiple matings and lead to increased numbers of competing ejaculates involved in each mating episode. Such apparently competitive situations are presumed by many researchers to favor individuals who allocate their energy toward sperm numbers, as well as toward male traits that facilitate sperm transfer. However, species that digest as well as store sperm, such as nudibranchs and ascidians, consistently show more equal allocation of energy toward both sexes, a result that has been attributed, not to actual ejaculate competition, but rather to avoiding the risk of ejaculate loss.

Gonadosomatic index (GSI), the ratio of gonad size to body size, is the usual measure of energetic investment in ejaculates. A large testis mass seems necessary for a high rate of sperm production, and indeed, in laboratory experiments involving gonochorists such as fruit flies, as well as hermaphrodites such as barnacles, individuals do adjust their allocation toward sperm production in response to social conditions favoring multiple mating. Although these studies are usually interpreted in light of parental investment theory, they do not directly test the central hypothesis of this approach: that initial energetic investment in sperm or ova will determine individual tendencies to emphasize mating or parental functions. The comparatively large testes of *Pseudoceros bifurcus* flatworms, for example (Fig. 1), which may engage in hypodermic inseminations of conspecifics, seem likely to have evolved as result of the differential success sperm-transferring opportunists had in siring offspring, at the expense of less aggressive inseminators. It seems less likely, as parental investment theory implies, that facultative polygamy arose in this species so that individuals with enlarged testes could capitalize on their greater initial gamete numbers.

FIGURE 1 The hermaphroditic Indo-Pacific flatworm *Pseudoceros bifurcus* engages in hypodermic insemination of conspecifics; individuals evert their penis and attempt to transfer sperm while apparently attempting to avoid being inseminated. Image by Dave Harasti, www.daveharasti.com.

A FOCUS ON EVOLUTIONARY OUTCOME

Both parental investment and sex allocation theories emphasize evolutionary outcomes; that is, they provide detailed predictions about which traits should evolve over time. There is little doubt that these solutions have significantly advanced understanding of evolution in sexual species. However, because these theoretical approaches consider "all things equal" during the evolutionary trajectories optimal phenotypes may take, they provide few predictions about how inheritance or selection intensity may influence the evolution of sex differences or parental care. While an emphasis on optimal energetic investment and future fitness returns is amenable to theoretical analyses, it can complicate empirical analyses, particularly in hermaphroditic species, in which energetic investment in male and female function, as well as fitness returns gained through each sex, are confounded within individuals. Although parental investment and optimality theories have provided many insights into how sex differences may have evolved, this approach can, unless applied with care, lead researchers to search *for* adaptations that "should" have been favored by selection.

THE OPPORTUNITY FOR SELECTION AND SEXUAL SELECTION

Estimates of the opportunity for selection, symbolized as I, measure the variance in relative fitness, V_w, and provide an empirical estimate of selection intensity. The value of I can be calculated for any population by dividing the variance in fitness, V_W, by the squared average fitness, W^2. Fitness is easily and accurately measured as the number of offspring each individual produces. The ratio $V_W/W^2 = V_w$ describes the *opportunity* for selection because not all of the variation in parental fitness is heritable, and because by chance, an imperfect relationship exists between the actual variance in fitness, V_W, and the expected covariance between phenotype (z) and relative fitness ($w[z]$), $Cov(z,w[z])$.

Stated differently, bad things can happen to good phenotypes and vice versa, a point that addresses the possible effects of random processes on this measure of selection. The opportunity for selection places an upper boundary on the change in mean fitness itself, as well as on the standardized change in the mean value of all other phenotypic traits. In this way, it provides a dimensionless, empirical estimate of selection intensity that is useful for field and laboratory analysis.

Opportunity-for-selection theory identifies a sex difference in the variance in offspring numbers as the source of sexual selection. In contrast to parental investment theory, it identifies the magnitude of this fitness difference gained through male or female function as the ultimate cause of sex differences in energy allocated toward mating effort and parental care. The variance in offspring numbers is proportional to the strength of selection, and when some individuals mate and others do not, this variance in fitness can become large. If members of one sex have greater variance in offspring numbers than the other, a sex difference in fitness will arise. Thus, the intensity of sexual selection is determined by the magnitude of the sex difference in the variance in offspring numbers.

MEASURING THE OPPORTUNITY FOR SEXUAL SELECTION

The source of a sex difference in the variance offspring numbers is easy to see, and it provides a simple, direct measure of the intensity of sexual selection. When the sex ratio equals 1 and all males and females mate once, there can be no sex difference in fitness variance. However, if certain individuals within each sex have more than one mate, other individuals within that sex must be excluded from mating, causing the variance in offspring numbers within that sex to increase. If the fraction of individuals excluded from mating is larger in one sex than it is in the other, a sex difference in the variance in offspring numbers will arise and be recognizable as the source of sexual selection. Estimates of OSR ($= N_\delta/N_\male = R_O = 1/R$) alone can provide misleading estimates of the actual intensity of sexual selection, because they say little about the *distribution* of matings per individual except in one special case; when all individuals mate *only once*.

Note that because each individual has a mother and a father, the average number of offspring, O, as well as the average number of mates, P, must be equal for each sex. Both expressions are linked through the sex ratio, thus $O_\male = R_O O_\delta$ and $P_\male = R_O P_\delta$. This necessary limitation is not considered in the "males-are-ardent, females-are-coy" dichotomy described in parental investment theory. If average fitness and average mate numbers must be equal for both sexes, neither sex can have greater average promiscuity than the other.

The total opportunity for selection for any species can be partitioned into separate selection opportunities for each sex. These components of total selection are equal to the variance in fitness among members of each sex, V_O (= the variance in offspring numbers across all individuals within that sex), divided by the squared average in fitness among members of that sex, O^2 (= the average number of offspring per individual, squared). For males, or when considering the intensity of selection on male function, $I_\delta = V_{O\delta}/O_\delta^2$. For females, or when considering the intensity of selection on female function, $I_\male = V_{O\male}/O_\male^2$.

The relationship between I_δ and $I_\male$ is $I_\delta = R_O I_\male + I_{mates}$, where $R_O = N_\delta/N_\male$ and I_{mates} is the sex difference in the opportunity for selection that is due to differences in mate numbers among males. Note that both R_O *and* I_{mates} are part of this equation. When $R_O = 1$, subtracting $I_\male$ from both sides of the above equation yields $I_\delta - I_\male =$

I_{mates}, demonstrating that the sex difference in the opportunity for selection, that is, the opportunity for *sexual selection*, is due to a difference in the variance in mate numbers, and thus in the variance in offspring numbers between the sexes.

SEXUAL SELECTION AND SEXUAL DIFFERENCES

Strong selection within one sex leads to sexual dimorphism because traits associated with high fitness are disproportionately transmitted to the next generation. For this reason, empirical estimates of the sex difference in the opportunity for selection predict whether and to what degree the sexes will diverge in character. For example, sexual selection in the intertidal isopod *Paracerceis sculpta* (Fig. 2A, B) can be 20 times stronger in males than in females. The sex difference in fitness variance is large because each female breeds only once, making the overall variance in female offspring numbers comparatively small. Breeding females prefer to aggregate within sponges, a condition that allows males who control these aggregations to mate many times. Polygyny by some males excludes other males from mating. This, in turn, makes the variance in male offspring numbers comparatively large.

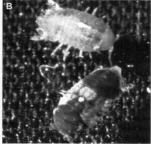

FIGURE 2 Three distinct male morphs coexist in the Gulf of California isopod *Paracerceis sculpta*: (A) α-males are largest, possess elongated uropods, and comprise 81% of aggregate male population samples; γ-males are smallest and comprise 15% of aggregate male population; β-males are smaller than α-males and comprise 4% of the male population; (B) β-males (above) also resemble females (below) in their behavior and external morphology; color patterns vary widely within and between individuals in this species. Photographs by the author.

The sex difference in fitness variance appears to be so large in this species that it allows three distinct male phenotypes to coexist. Although most males fight for aggregations, some males avoid combat altogether. They either mimic receptive females, or use their small size to enter sponges and mate. The three male morphs partition the available mates by exploiting different aspects of female tendencies to enter sponges to breed. Strong sexual selection within each male type had favored particular morphologies that are distinct from females, and are distinct from one another.

The opportunity-for-selection approach predicts weak sexual selection in hermaphrodites if each individual reproduces proportionally as a male and as a female. However, if certain individuals emphasize male function when large or when crowded, variance in the number of offspring produced through male function can become large and sexual selection can become strong, as it appears to be in *Pseudoceros bifurcus* (Fig. 1). In these simultaneously hermaphroditic flatworms, some individuals may engage in more forced matings than other individuals. A sex difference in the opportunity for selection, in turn, is expected to favor differential allocation of energy toward mating or parental functions, as it does in other hermaphrodites such as sea slugs and leeches, and as it does in gonochorists, such as sea spiders, seahorses, and pipefish. Here, the intensity of selection favoring mate acquisition or parental care appears to shape observed patterns of sex allocation and parental investment, a causal chain that is the opposite of that predicted by parental investment theory.

HARMONIC MEAN PROMISCUITY AND SPERM COMPETITION

The opportunity for selection approach suggests that when multiple mating occurs, the fitness of an individual male depends on the number of ova he fertilizes, relative to the average number of ova fertilized by other males. In externally fertilizing species, multiple ejaculates are released simultaneously near ova and sperm mixing is common. Whereas sperm precedence is known in some internally fertilizing species, the advantage gained by the first or last male to mate with a female appears to erode steadily when the number of mating males becomes larger than 2.

Because of this relationship, a male's fertilization success in many species can be quantified as the reciprocal of the arithmetic mean of the reciprocals, of the promiscuity of each of his mates. Stated differently, a male's fertilization success, H_M, equals his mates' harmonic mean promiscuity, or,

$$1/H_M = 1/N_{mates} (\Sigma 1/P_i) \qquad \text{(Eq. 1)}$$

where N_{mates} equals the number of mates a male has, and P_i is the number of mates each ith female has, that

also mates with this male. For example, if a male mates with only one female and she mates only with him, the reciprocal of her "promiscuity" is $1/P = 1$. The arithmetic mean of this value is $(1/1 \text{ mate})(1) = 1$, whose reciprocal is also 1. A male's fitness, then, equals the product of the harmonic mean promiscuity of his mates and the average number of offspring each female produces, or $O_\female/H_M$, where $O_\female$ = the average number of offspring per female.

Now, if the male must abandon his current mate to mate with another female, and if his former mate mates again, $1/H_M = 1/2(1/2+1/1) = 0.75$, i.e., his fitness drops to $0.75O_\female$ per female. If the male extends his promiscuous search for mates, as parental investment and optimality theories predict he should, his fertilization success with past mates will continue to erode. Furthermore, if the females he encounters have already mated (i.e., their $P_i > 1$), a male's fitness loss with each additional mating will soon exceed his possible fitness gain. Because the harmonic mean is more strongly influenced by small numbers than by large numbers, this approach shows why selection is likely to favor male tendencies to reduce, rather than enhance, their own promiscuity as well as that of their mates, a result supported by the ubiquity of mate guarding among gonochorists, as well as by prolonged mating associations and reduced sperm production in hermaphrodites.

The foregoing approach illustrates another principle often unrecognized in discussions of sperm competition from parental investment or optimality perspectives. While sperm numbers within multiply mated females may change the distribution of male paternity *within* broods, it need not affect the distribution of paternity *among* broods. Thus, sperm competition represents a significant evolutionary force *only* when males who mate with disproportionate success *also* have sperm that are disproportionately used to fertilize ova by each of the females with whom they mate.

If this relationship does not exist (and often it does not), multiple mating or other mechanisms that reduce paternity confidence (e.g., sperm digestion) will ameliorate rather than intensify sexual selection. This prediction is at odds with optimal expectations for increased energy allocation toward male function when multiple mating occurs. Using genetic markers to document offspring numbers, empirical values of H_M and of the sex difference in the opportunity for selection, I_{mates}, provide a means for assessing the source, as well as the actual intensity, of sexual selection on observed patterns of parental investment and sex allocation.

A FOCUS ON EVOLUTIONARY PROCESS

In contrast to parental investment and sex allocation theories, opportunity-for-selection theory focuses on *evolutionary processes*. By specifically measuring the fitness variance associated with patterns of sex allocation and parental care, this approach identifies traits on which selection may be strong or weak. Furthermore, this approach complements research on the genetic basis of sex allocation and parental care, which to date is largely unstudied, particularly among species inhabiting rocky intertidal zones. Because phenotypic change depends on the product of selection intensity and trait heritability, investigations that measure fitness variance as well as patterns of trait inheritance can identify which phenotypes are likely to respond to selection and by how much. This evolutionary genetic approach generates hypotheses about selection and inheritance that are specifically falsifiable using the data generated by analyses of fitness variance and trait expression.

By focusing on evolutionary processes rather than on presumed adaptive outcomes, opportunity-for-selection theory provides an experimental framework for investigating phenomena that are specifically avoided by parental investment and optimality approaches. In many cases, hypotheses focused on evolutionary processes are simpler, more rigorously testable, and easier to interpret than hypotheses focused on evolutionary outcome. Few studies have measured evolutionary processes in rocky intertidal habitats. There is much exciting work to be done.

SEE ALSO THE FOLLOWING ARTICLES

Fertilization, Mechanics of / Genetic Variation and Population Structure / Reproduction

FURTHER READING

Ahnesjö, I., C. Kvarnemo, and S. Merilaita. 2001. Using potential reproductive rates to predict mating competition among individuals qualified to mate. *Behavioral Ecology* 12: 397–401.

Alcock, J. 2005. *Animal behavior: an evolutionary approach*. 8th ed. Sunderland, MA: Sinauer Associates.

Charnov, E. L. 1996. Sperm competition and sex allocation in simultaneous hermaphrodites. *Evolutionary Ecology* 10: 457–462.

Cheverud, J. M., and A. J. Moore. 1994. Quantitative genetics and the role of the environment provided by relatives in the evolution of behavior, in *Quantitative genetic studies of behavioral evolution*. C. R. B. Boake, ed. Chicago: University of Chicago Press, 67–100.

Frank, S. A. 1990. Sex allocation theory for birds and mammals. *Annual Review of Ecology and Systematics* 21: 13–55.

Schärer, L., P. Sandner, and N. K. Michiels. 2005. Trade-off between male and female allocation in the simultaneously hermaphroditic flatworm *Macrostomum* sp. J. *Evolutionary Biology* 18: 396–404.

Shuster, S. M., and M. J. Wade. 2003. *Mating systems and strategies*. Princeton, NJ: Princeton University Press.

Wade, M. J. 1979. Sexual selection and variance in reproductive success. *The American Naturalist* 114: 742–764.

SHRIMPS

RAYMOND T. BAUER

University of Louisiana, Lafayette

Shrimps found on rocky intertidal shores are usually members of the infraorder Caridea (Crustacea: Decapoda). Carideans typically have two (never three) pairs of pincer-like appendages (chelipeds), the second pleura (side plates) of the tail (abdomen) overlaps adjacent pleura, and females brood their embryos below the abdomen. Shrimps can withstand little or no exposure at low tide and are thus restricted in the rocky intertidal zone to water-filled pools and channels. Carideans may be abundant in the rocky intertidal habitat and are ecologically important in trophic webs as small predators, herbivores, and detritivores as well as serving as prey of larger organisms such as fishes.

MORPHOLOGY AND TAXONOMY

The body of a shrimp (Fig. 1) is divided into an anterior cephalothorax and a posterior abdomen (tail). The first antennae bear the olfactory hairs (setae), while the second antennae have long, flexible flagella with taste and touch receptors. The three primary pairs of mouthparts

FIGURE 1 *Heptacarpus sitchensis*, a typical hippolytid caridean shrimp (a color morph with a green abdomen and striped cephalothorax). Photograph by the author.

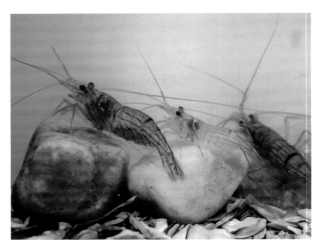

FIGURE 2 *Lysmata wurdemanni*, a hippolytid often found in pools and under ledges of rock jetties and other hard substrates in the Gulf of Mexico. Photograph by the author.

are followed by three pairs of maxillipeds and two pairs of chelipeds, used in feeding, defense, and grooming; the remaining three pairs of appendages are walking legs. The tail bears five pairs of pleopods (swimmerets) used in forward swimming, and the flexion of the muscular tail powers the paddle-like uropods of the tail fan, resulting in the rapid backward escape response ("tail flip").

Members of three caridean families (the Hippolytidae, Palaemonidae, and Alpheidae) are common inhabitants of rocky shores. "Humpbacked" shrimps such as the hippolytids *Heptacarpus sitchensis* (Fig. 1) and *Lysmata wurdemanni* (Fig. 2) and palaemonids such as *Palaemon elegans* are active epibenthic shrimps, perching on algae or boulders, often upside down under ledges. Alpheids (Figs. 3, 4) have straighter abdomens

FIGURE 3 *Alpheus heterochaelis*, an example of a "snapping" or "pistol" shrimp. Photograph by the author.

and are somewhat dorsoventally compressed. Cuticular hoods cover the eyes, perhaps protecting them from abrasion as they squeeze under rocks or within burrows made in sandy mud beneath the rocks. In the related families Hippolytidae and Alpheidae, the second pair of chelipeds is slender with small pincers, often equipped with brushes of cuticular hairs (setae) used frequent bouts of body and gill cleaning as well as feeding. The first chelipeds are heavier than the second and are the shrimp's weapons, used in fighting with conspecifics, in defense, and in prey capture. This is especially true in the Alpheidae (Figs. 3, 4), in which the very large first chelipeds are fearsome weapons (at least for a shrimp). In *Alpheus* ("snapping" or "pistol" shrimps; see Fig. 3), a highly modified first cheliped produces a popping sound, quite audible at low tide, which is a major contributor to undersea noise. In the palaemonids, it is the second chelipeds that are the major weapons, with the first serving to groom the body and gills, as well in feeding.

FIGURE 4 *Betaeus longidactylus*, with the massive first chelipeds of a male, an example of a rocky intertidal alpheid from the west coast of North America. Photograph by the author.

DISTRIBUTION

Some carideans utilize rocky tidepools as their primary habitat, while for many others the intertidal zone simply represents the very upper edge of a deeper subtidal distribution. Pools much higher than midtide level are too physiologically stressful for carideans to live in because of extreme variations in water temperature, salinity, and oxygen content.

The worldwide distribution of intertidal shrimps is influenced by the availability of suitable rocky intertidal habitat, the degree of environmental stress during low tide, and the biogeographical diversity of the taxon. For example, an abundant and somewhat diverse shrimp fauna is found on the Pacific coasts of Canada and the United States, areas with abundant tidepool habitat, relatively benign exposure conditions due to the moderating influence of the cool California Current, and a high species diversity in the family Hippolytidae. In contrast, the harsh winter conditions along the coast of Maine, bathed by the arctic Labrador Current, have resulted in a depauperate intertidal shrimp fauna. In tropical rocky intertidal habitats, harsh conditions due to intense heat and solar radiation during daytime low tides likewise create poor habitat for intertidal shrimps, in spite of the generally high diversity of caridean shrimps in adjacent subtidal habitats such as coral reefs and seagrass meadows.

LIFE CYCLE

Females mate and spawn just after a molt, incubating the fertilized eggs (embryos) until hatching. The zoeal larva, a dispersal phase, develops in the plankton in a series of molts before settling on the bottom as a postlarva to begin the benthic juvenile and then adult phases of the life cycle. Although most intertidal carideans have separate sexes, individuals of *Lysmata* spp. (see Fig. 2) first mature as males, later changing into "females" that retain male ducts and sperm, capable of reproducing as male and female (simultaneous hermaphrodites).

CAMOUFLAGE AND COLOR CHANGE

Shrimps often display striking camouflage color patterns. *Hippolyte varians* from European waters has several color variations, individuals of each resembling that of their algal perch (blending coloration). When displaced from the alga, the shrimp searches for and selects an alga of the same color. If an alga of similar color is not available, the shrimp is capable, over a period of weeks, of changing its color pattern. In contrast, the several color forms of *Heptacarpus sitchensis* are genetically fixed and composed of splotches and stripes of the colors (Figs. 1, 5) that occur within the rocky tidepool background (Fig. 6) over which they freely move. Like military camouflage, the stripes and splotches of color distract the eyes of visual predators from the telltale outline of the shrimp (disruptive coloration). Color in shrimps is due to carotenoid and other pigments within channels of chromatophores (Fig. 7). At night, the shrimps assume a bluish transparency as pigment contracts, but upon exposure to light at dawn, the pigments disperse in the chromatophore channels, restoring the color pattern. The how and why of this nocturnal color loss is not known and, like many aspects of the biology of intertidal shrimps, requires further research.

FIGURE 5 Two color morphs of *Heptacarpus sitchensis* (cf. Fig. 1) with white to pink blotches of color similar to coralline algae from the same habitat. Photograph by the author.

FIGURE 6 Rocky substrate of a rocky pool showing the color variable background (e.g., green surfgrass, pink and white encrusting and erect coralline algae, snail shells, gravel from coralline algae and shells), against which the disruptive coloration of the intertidal shrimp *Heptacarpus sitchensis* serves as a camouflage. Photograph by the author.

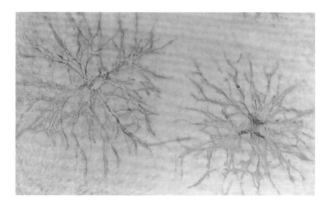

FIGURE 7 Two chromatophores with dispersed yellow pigment that, together with the Tyndall blue of scattered light in the tissues, produces the bright green color of the green and striped morph of *Heptacarpus sitchensis* (Fig. 1). Photograph by the author.

SEE ALSO THE FOLLOWING ARTICLES

Arthropods / Camouflage / Food Webs / Lobsters / Mantis Shrimps

FURTHER READING

Bauer, R. T. 2004. *Remarkable shrimps: adaptations and natural history of the Caridea.* Norman: University of Oklahoma Press.

Brusca, R. C., and G. J. Brusca. 2000. *Invertebrates,* 2nd ed. Sunderland, MA: Sinauer.

Holthuis, L. B. 1993. *The Recent genera of the caridean and stenopodidean shrimps (Crustacea, Decapoda): with an appendix on the order Amphionidacea.* Leiden: Nationaal Natuurhistorisch Museum.

Martin, J. W., and G. E. Davis. 2001. *An updated classification of the recent Crustacea.* Natural History Museum of Los Angeles County Contribution to Science 39.

Smaldon, G. 1979. *British coastal shrimps and prawns.* London: Academic Press.

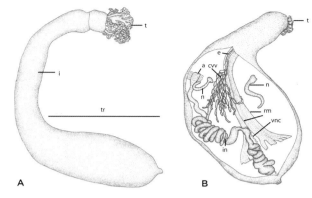

FIGURE 1 *Themiste dyscrita*, a common inhabitant of the rocky intertidal in California; up to 100 mm long. (A) External morphology, (B) internal anatomy. Abbreviations used: a, anus; cvv, contractile vessel villi; e, esophagus; i, introvert; in, intestine; n, nephridia; rm, retractor muscles; t, tentacles; tr, trunk; vnc, ventral nerve cord.

SIPUNCULANS

EDWARD B. CUTLER

Harvard University

ANJA SCHULZE

Smithsonian Marine Station, Fort Pierce, Florida

The Sipuncula (peanut worms) are a phylum of unsegmented benthic marine worms that now seem most closely related to the Annelida. The 150 species in this phylum live in oceanic habitats in sand or mud or inside protective shelters (e.g., shells, rock crevices, or burrows in coral), at all depths from intertidal to abyssal.

MORPHOLOGY

All sipunculans have a trunk and a retractable introvert (Fig. 1A). The trunks of adults range from 3 to 500 mm in length. Their shape varies from cylindrical through spindle-shaped to spherical. Simple epidermal structures (e.g., small papillae and hooks) are present in most species. The thinner introverts measure half to five times the trunk length. At the distal introvert lie the mouth and an array of tentacles of varying size and complexity. Many species have posteriorly directed introvert hooks, arranged in regular rings or randomly scattered.

INTERNAL ANATOMY

The internal anatomy of a sipunculan worm is shown in Fig. 1B. The digestive system consists of an esophagus and an intestine. Forming a double helix, the intestine first runs posteriorly, then rewinds anteriorly to the mid-dorsal anus near the trunk–introvert junction. Two saclike nephridia open ventrolaterally at the anterior end of the trunk, except in two genera that have a single organ. On the inner layer of the body wall the longitudinal muscle forms a continuous layer in half the genera, but it is divided into separate or anastomosing bundles in the remainder. Four retractor muscles control the withdrawal of the introvert. These are reduced to two or rarely one muscle in some genera. A pair of very small cerebral ganglia connects to the nonmetameric ventral nerve cord by means of circumesophageal connectives.

REPRODUCTION

Sipunculans lack sexual dimorphism. Only one species is known to be hermaphroditic, and one is known to be facultatively parthenogenetic. Eggs and sperm are produced from transient strips of gonadal tissue at the base of the ventral retractor muscles, then released into the coelom, where they mature prior to exiting through the nephridia. Following external fertilization most species produce freeswimming trochophore larvae. Transoceanic dispersal is possible in some genera in which unique, long-lived planktotrophic pelagosphera larvae develop.

ECOLOGY

The Sipuncula of California reflect the habits of the phylum in that they have been found from intertidal warm to deep, cold waters. Depending on the species, they may be living in unconsolidated sand or silt and clay sediments, shelters such as vacated mollusc shells or crevices in rocks, or under algae, mussels, or oysters, or in self-created holes bored into corals or soft rocks.

A few sipunculans are filter feeders with elaborate tentacular crowns, but most are deposit feeders. Organic detritus, bacteria, algae, and small invertebrates make up their diet. The energy is thus made available for fish, crabs, and other predators. Their boring activity can contribute to the breakdown of hard substrata, and burrowing of different species contributes to the turnover of soft, unconsolidated sediments.

SYSTEMATICS

Currently Sipuncula are grouped into two classes, four orders, six families, and seventeen genera. However, recent work based on DNA sequence analysis is raising questions about the phylogenetic accuracy of this system. A revision of several genera is likely in the near future.

SEE ALSO THE FOLLOWING ARTICLES

Dispersal / Invertebrates / Reproduction / Worms

FURTHER READING

Cutler, E. B. 1994. *The Sipuncula. their systematics, biology, and evolution.* Ithaca, NY: Cornell University Press.

Cutler, E. B., and N. J. Cutler. 1988. A revision of the genus *Themiste* (Sipuncula). *Proc. Biol. Soc. Wash.* 101: 741–766.

Rice, M. E. 1980. Sipuncula and Echiura, in *Intertidal invertebrates of California.* R. H. Morris, D. P. Abbott, and E. C. Haderlie, eds. Stanford, CA: Stanford University Press, 490–498.

Schulze, A., E. B. Cutler, and G. Giribet. 2005. Reconstructing the phylogeny of the Sipuncula. *Hydrobiologia* 535/536: 277–296.

Stephen, A. C., and S. J. Edmonds. 1972. *The phyla Sipuncula and Echiura.* London: Trustees of the British Museum (Natural History).

SIZE AND SCALING

H. ARTHUR WOODS

University of Montana

AMY L. MORAN

Clemson University

Scaling studies focus on how and why size matters to organisms. Historically, scaling has been studied by morphologists and physiologists, but its implications for other fields, especially ecology, are becoming increasingly apparent. As biologist George Bartholomew wrote in 1981, "[i]t is only a slight overestimate to say that the most important attribute of an animal, both physiologically and ecologically, is its size. Size constrains virtually every aspect of structure and function and strongly influences the nature of most inter- and intraspecific interactions." Few intertidal researchers have examined scaling as an isolated subject; yet, because of the pervasive nature of size effects on organismal biology, scaling is a fundamental component of much ecological and organismal research.

DEFINITION OF SCALING

Biological variables often vary with body size in ways that can be described by the power-law function $y = aM^b$, where y is the dependent variable, M is body mass, a is a normalization constant, and b is the scaling exponent. The constants a and b describe the height and shape of the relationship, respectively. When b equals 1, the relationship is said to be isometric. For example, in some fishes, skeleton mass increases in direct proportion to body mass—it scales isometrically. When b does not equal 1, the relationship is allometric. One of the most-studied allometric relationships concerns metabolism: in many taxonomic groups, metabolic rate scales to body mass with an exponent of ~3/4 (a relationship known as Kleiber's law, see Kleiber, 1947), although in individual groups the exponent may differ substantially from this value. That the usual value is <1 means that larger organisms have, in general, disproportionately low metabolic rates—for example, an elephant (= 5000 kg) has 300,000 times the mass of a mouse (= 0.017 kg), but its metabolic rate is only ~12,600-fold higher. Some variables do not scale with size (i.e., b = 0)—for example, in mammals capillary diameter is almost invariant with body mass.

It can be difficult to visualize and analyze relationships of the form $y = aM^b$ (Fig. 1A). Commonly, therefore, data are linearized by log-transformation (Fig. 1B). Log-transforming both sides of the power-law equation gives $\log y = \log a + b \log M$, which is the familiar equation of a line with intercept $\log a$ and slope b. Determining b for a dataset involves log-transforming body size and the dependent variable and fitting a line to the points. The slope of the line is b. Differences in a reflect differences in the height of the line in log-log space (Fig. 1).

Studies examining the scaling of size and metabolism of intertidal organisms are comparatively rare. One example, however, comes from Kenneth Sebens's (1982) work on the scaling of growth in the giant green anemone *Anthopleura xanthogrammica* (Fig. 2A). Sebens found that food intake and metabolic rate (measured as mass loss during starvation) both varied allometrically

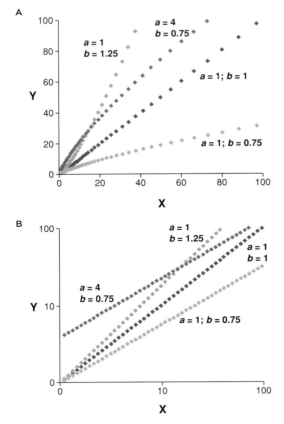

FIGURE 1 Four different scaling relationships of the form $Y = aX^b$, either untransformed (A) or $\log_{10}$-transformed (B).

with body mass, with interesting implications for possible maximal body size (Fig. 2B). Energy expenditure rose fairly steeply with body mass; the exponent b was equal to 1.08, whereas food intake rose with a substantially smaller exponent (b equaled 0.67). The consequence of these different exponents is that very large anemones increasingly cannot take in enough food to keep up with their energy expenditures, leaving smaller surpluses left over for allocation to further growth or reproduction. The point at which the two lines cross (at ~53 g in Fig. 2B) indicates the theoretical maximum body size at which food intake exactly balances metabolic costs. However, Sebens pointed out that for organisms to maximize reproductive output (a fitness parameter), they should not reach this theoretical maximum and should instead stop growing at the size that maximizes the surplus (intake minus cost).

Sebens's work found that metabolism scaled with an exponent approximately equal to 1. In general, however, metabolism scales with exponents <1, and, in particular, many studies have reported values near 3/4 (Fig. 3). Note

that this dataset does contain some marine vertebrates but no intertidal invertebrates, algae, or other plants. The broad scope of the data suggests that metabolic rates of these neglected groups would fall near or on the regression lines Sebens found—but for lack of data, this conclusion remains difficult to evaluate.

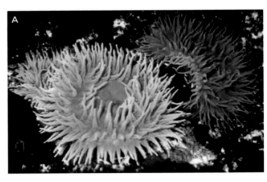

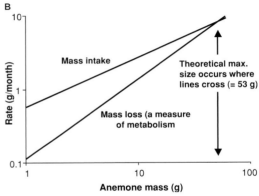

FIGURE 2 Scaling of feeding and metabolism in the giant green anemone *Anthopleura xanthogrammica*. (A) *A. xanthogrammica*. Photograph courtesy of David Secord. (B) Experimentally determined relationships between body mass and both energy intakes and expenditures, plotted in log-log space. Redrawn from Sebens (1982) by permission of The Ecological Society of America. For intake, the equation describing the scaling relationship is intake = 0.580 x body mass$^{0.67}$. For energy expenditure, the equation is expenditure = 0.114 x body mass$^{1.08}$. The lines cross when body mass is 53 g, suggesting that this mass is the maximum possible under the conditions in Sebens's study.

ORIGINS OF PHYSIOLOGICAL SCALING

Scaling studies are undergoing a renaissance from new theory, especially with respect to metabolic scaling. Understanding why metabolic rates scale as they do, often with an exponent of ~3/4, is important: metabolism is a central integrator of life processes, and it links individuals to distributions and fluxes of matter and energy in nature. What then is the origin of the value 3/4? Why not some other value?

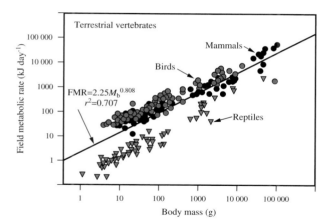

FIGURE 3 Field metabolic rates of 229 species of air-breathing vertebrates. From Nagy (2005), by permission from Company of Biologists.

Early work, before the 1930s, tried to explain metabolic scaling as functions of geometry and simple physical processes. For example, metabolism was widely believed to scale to the power 2/3—that is, as surface area scales to volume. For warm-blooded organisms (endotherms) this number made sense, as smaller organisms would need higher metabolic heat production to offset relatively higher rates of loss across their surfaces. In the 1930s, however, several summary studies established not only that mammalian metabolic rates scaled to the 3/4 power of body mass but so also did metabolic rates of various groups of cold-blooded animals (ectotherms). These unexpected deviations catalyzed scaling's empirical golden age, which found pervasive 3/4-power scaling in biology. Of phenomena not scaling to the power 3/4, many scale to other multiples of 1/4. These findings are summarized in four influential books that appeared in the early 1980s (see Further Reading).

By the mid-1980s, interest in scaling had waned, primarily for lack of fresh theoretical ideas. The doldrums lasted until the mid-1990s, when a number of new theoretical papers appeared. One of the more important was contributed by West *et al.* (1997), whose central idea was that metabolizing units (cells or mitochondria) require support from branching, hierarchical transport networks—circulatory systems in vertebrates and vascular systems in plants—and that hierarchical networks, regardless of their actual form, share certain fundamental properties.

The model was built on three main assumptions: (1) that the volume of an organism is supplied with energy and materials by a fractal-like, space-filling network; (2) that the size of the final branching unit (e.g., capillaries) is invariant across body sizes; and (3) that organisms have evolved to minimize energy costs of transporting materials through the network. Given these assumptions, West *et al.* showed that metabolism should scale to the 3/4 power of body mass for a wide range of organisms with different kinds of distribution networks. Moreover, the model predicts that many other biological rates should scale as multiples of 1/4 power. The assumptions, structure, and predictions of this model have all come under intense criticism, and it is unclear whether it will hold up against proposed alternatives.

Intertidal organisms have not played a significant role in the formulation or testing of these models. They may eventually do so, however, especially because many do not possess highly branched vascular distribution networks like those envisioned by West *et al.*, and thus they would constitute good tests of the claimed generality of the model.

USING SCALING TO TEST FUNCTIONAL HYPOTHESES

New work focused on testing biomechanical hypotheses is broadly relevant to intertidal organisms.

Algal Breaking Force and Maximum Size

In intertidal zones, many animals escape unsuitable conditions by moving. This option, however, is unavailable to sessile algae, which are important members of rocky-shore communities. Immobility forces them to take environmental insults—unusual temperatures, extreme solar insolation, crashing waves—as they come. How do such insults influence body size, form, and function?

A recent study by Kitzes and Denny (2005) of the red alga *Mastocarpus papillatus* along the central California coast provides a good example. *M. papillatus* grows in clumps attached to rocky substrates by holdfasts; individual blades are attached by stipes. Blades and stipes experience significant drag from wave-induced water motion, which may break the stipe. The authors collected stipe–blade pairs from across a wave exposure gradient and experimentally measured the force necessary to break each stipe—finding that stipes from high-drag areas had a higher breaking force (Fig. 4A). This result suggested that algae respond to their local wave environment.

This result also suggested the possibility of calculating maximum possible blade size (the main algal part that experiences drag) for a given wave exposure. Kitzes and Denny did so with a standard equation relating force of drag to water velocity, coefficient of drag, and

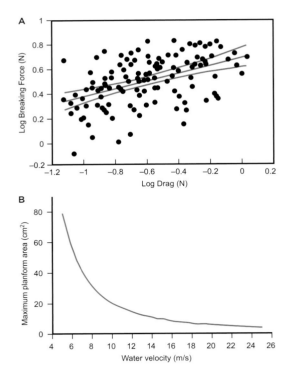

FIGURE 4 Biomechanics of a red alga, *Mastocarpus papillatus*, on wave-swept rocky shores. (A) Force required to experimentally break algal stipes, as a function of the maximum local drag forces (from waves) they experienced. (B) Calculated maximum algal blade size (as planform area, which is about half the wetted surface) as a function of water velocity. From Kitzes and Denny (2005), reprinted with permission from the Marine Biological Laboratory, Woods Hole, Massachusetts.

maximal planform area (which is about half the wetted surface area). By inserting this standard equation into their empirically determined expression for stipe breaking force, they found that the maximal planform area was equal to 0.20 divided by water velocity squared (Fig. 4B). Theoretical maximum size thus decreases steeply with water velocity. Moreover, actual observed maximum sizes fall below the theoretically determined value across the range of water velocities, suggesting that other factors besides wave exposure can further limit size—for example, rare, very high water velocities that are not reflected in the average maximum water velocities used by the authors.

Constraints on Biological Force Production

The ability to produce forces—for grasping, holding, swimming, crunching, penetrating, and so on—is important for intertidal organisms, as it is for many terrestrial organisms. Is there a limit to how much force a biological

motor of a given size can produce? In general, such limit-focused questions are difficult to answer. However, scaling studies involving a large number of independently evolving units can provide powerful means of detecting impossibilities.

For example, a recent analysis of force production by translational motors, both biological and human-made (myosin, kinesin, dynein, and RNA polymerase molecules, muscle cells, whole muscles, winches, linear actuators, and rockets), found that maximal force output scales as the 2/3 power of mass over 27 orders of magnitude—regardless of whether the motor originated by biological evolution or in an engineer's computer (Fig. 5). Surely selection favors greater force production in some biological motors, and it is certain that engineers would like to increase force per unit mass produced by human-built motors. That both origins result in motors clustering so closely around a single allometric line suggests universal constraints on motor design. It is therefore likely that no intertidal organism will ever develop a more efficient biological motor that lies to the left of the fitted line in Fig. 5.

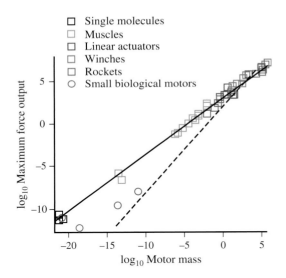

FIGURE 5 Force output by biological and mechanical motors spanning 27 orders of magnitude in mass. From Marden (2005), by permission from Company of Biologists.

SCALING OF ECOLOGICAL INTERACTIONS

A major attraction of metabolic scaling is that its principles should also apply at higher levels of organization—populations, species, communities, and ecosystems. Indeed, the models of West *et al.* (1997) and others suggest that 3/4-power scaling of metabolism can explain emergent properties of species, such as territory size, population densities, and speciation and extinction rates. Here we focus on two such properties—territory size and population density—both in intertidal systems.

Scaling and Territory Size

Many animals defend territories and rely on energy from them to fuel metabolism. For such species, home range might be expected to scale with body size as metabolism does, to approximately the 3/4 power of mass. In fact, however, home ranges of mammals scale with an exponent ≥ 1, such that home range size often increases more rapidly with body mass than metabolic rate does. Recent models suggest an explanation; as home range size increases, the ability of animals to interact with neighbors and patrol their territories declines with an exponent of −1/4. The larger an animal, the more it must share resources with neighbors; thus, to compensate, territory sizes must increase with animal size at a greater rate than expected based on metabolic demands alone.

Do these patterns hold true on wave-swept shorelines? This question is underexplored, perhaps because few intertidal organisms defend territories. Though competition for space strongly affects the ecology of rocky intertidal communities, most competitors (e.g., barnacles, anemones, mussels, and algae) are immobile as adults. Hence, a "territory" consists of the space taken up by, or within reach of, the organism itself. One exception is the owl limpet, *Lottia gigantea*, which occurs on rocky shores on the Pacific coast of North America. Individuals establish algal "garden" territories from which they actively exclude competitors, particularly other *L. gigantea*. Stimson's original description (Stimson 1970) gives the relationship between limpet territory size and length for 23 animals. The present authors calculated body volume from length (see Fig. 6 for methods) and then log-transformed both limpet volume and territory size (Fig. 6). In contrast with interspecific mammalian data, the scaling exponent was significantly lower than 1; territory size scaled to the 0.64 power of mass, with 95% confidence intervals of ±0.21 around the slope. Large limpets thus defended proportionally smaller territories than small limpets did. Perhaps limpets, unlike mammals, are better able to keep neighbors from sampling delicacies growing in their own territories.

Body Size and Population Density

In intertidal ecology, as in other environments, a central problem is to explain organismal abundance: Why are some species common and others rare? Body size

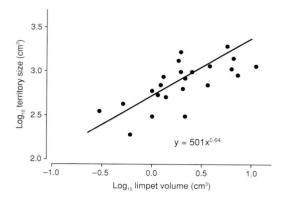

FIGURE 6 Territory size as a function of body volume in owl limpets, *Lottia gigantea*. The original body size data were given as lengths, in Stimson (1970), by permission of the Ecological Society of America. The present authors converted length data into volumes by using shape ratios from published images of *L. gigantea*.

and scaling may shed some light on this question. In many habitats, population density is negatively correlated with body size, scaling with an exponent of ~−3/4. This observation, together with Kleiber's law (that individual metabolic rates scale with body mass to the +3/4 power), has led to the "energy equivalence rule" of Damuth 1987. This rule postulates that −3/4 scaling of population density with mass mirrors macroevolutionary processes acting to keep rate of energy utilization constant among different size classes in a community. There have been numerous criticisms of the idea, including the observations that (1) many studies have failed to find a negative correlation between body mass and population density, and (2) there may be sampling bias against small, rare species.

Does organismal size in fact operate as a fundamental factor underlying population energy use in rocky shorelines? This question has been explored on Central and South American coastlines. On temperate rocky coastlines of Chile, log average density scaled to log body size with a slope of ~−3/4. The relationship appeared robust to changing community structure, because it occurred both in natural communities and in adjacent areas where keystone predators were absent. In contrast, in Panama, where overall population densities were much lower, there was no evidence of size–density scaling. The latter result was attributed to intense predation by fish, which kept numbers of all intertidal species low enough that intra- and interspecific competition for space and resources no longer structured the community. Consistent with this idea, experimental fish exclusion reestablished a negative relationship between body size and density.

SEE ALSO THE FOLLOWING ARTICLES

Body Shape / Competition / Hydrodynamic Forces / Materials:
Strength / Territoriality / Wave Forces, Measurement of

FURTHER READING

Brown, J.H., and G.B. West, eds. 2000. *Scaling in biology.* Oxford:
Oxford University Press.

Calder, W.A. 1984. *Size, function, and life history.* Cambridge, UK: Cambridge
University Press.

Damuth, J. 1987. Interspecific allometry of population density in mammals and other animals: the independence of body mass and population energy use. *Biological Journal of the Linnean Society* 31: 193–246.

Darveau, C.A., R.K. Suarez, R.D. Andrews, and P.W. Hochochka. 2002.
Allometric cascade as a unifying principle of body mass effects on
metabolism. *Nature* 417: 166—170.

Jetz, W., C. Carbone, J. Fulford, and J.H. Brown. 2004. The scaling of
animal space use. *Science* 306: 266–268.

Kelt, D.A., and D.H. Van Vuren. 2001. The ecology and macroecology of
mammalian home range area. *American Naturalist* 157: 637–645.

Kitzes, J.A., and M.W. Denny. 2005. Red algae respond to waves:
morphological and mechanical variation in *Mastocarpus papillatus*
along a gradient of force. *Biological Bulletin* 208: 114–119.

Kleiber, M. 1947. Body size and metabolic rate. *Physiological Reviews* 27:
511–541.

Marden, J.H. 2005. Scaling of maximum net force output by motors used
for locomotion. *Journal of Experimental Biology* 208: 1653–1664.

Martin, R.D., M. Genoud, and C.K. Hemelrijk. 2005. Problems of allometric scaling analysis: examples from mammalian reproductive biology.
Journal of Experimental Biology 208: 1731–1747.

McMahon, T., and J.T. Bonner. 1983. *On size and life.* New York: Scientific
American Books.

Nagy, K.A. 2005. Field metabolic rate and body size. *Journal of Experimental
Biology* 208: 1621–1625.

Peters, R.H. 1983. *The ecological implications of body size.* Cambridge, UK:
Cambridge University Press.

Schmidt-Nielsen, K. 1984. *Scaling: Why is animal size so important?*
Cambridge, UK: Cambridge University Press.

Sebens, K.P. 1982. The limits to indeterminate growth: an optimal size
model applied to passive suspension feeders. *Ecology* 63: 209–222.

Stimson, J. 1970. Territorial behavior of the owl limpet, *Lottia gigantea.*
Ecology 51: 113–118.

West, G.B., J.H. Brown, and B.J. Enquist. 1997. A general model for the
origin of allometric scaling laws in biology. *Science* 276: 122–126.

SNAILS

RON ETTER

University of Massachusetts, Boston

Snails are members of the Gastropoda, the most diverse
class in the phylum Mollusca, and are characterized by
a spiral, conical shell of calcium carbonate ($CaCO_3$), a
large muscular foot, and a rasping tongue called a radula.
Intertidal snails feed on algae or various invertebrates
(e.g., mussels, barnacles, other snails) or are scavengers.
They represent an important group of intertidal organisms that operate as both predators and prey within food
webs and play key roles in regulating the structure of
intertidal communities.

BIOLOGY

The most distinctive feature of a snail is its spiral $CaCO_3$
shell (Fig. 1), which protects the soft anatomy from abiotic and biotic stresses and can vary dramatically within
and among species in response to spatial and temporal
environmental variation. Among species, the shell can vary
from a simple conical cone as found in limpets to highly
spiraled forms found in periwinkles and whelks. The snail
is attached to the shell via the columellar muscle and cannot be removed from the shell. The shell is divided into the
body whorl—the most recent and largest whorl—and the
spire, the whorls above the body whorl. The main opening
to the shell is the aperture, where the head and foot of a
living animal extend through. When snails are disturbed,
they retract their head and foot into their shell and seal off
the aperture with the operculum (trap door).

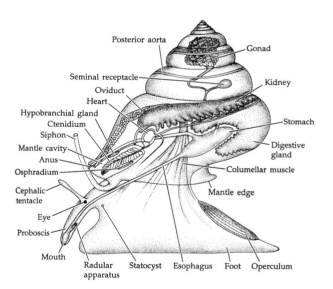

FIGURE 1 An illustration of a typical marine snail showing the major
features of the external and internal anatomy. Reprinted from Brusca
and Brusca (2003).

Most intertidal snails are highly mobile and use the
muscular foot and pedal secretions (mucus) to move over
the substrate. In addition, adhesion develops between the
substrate, pedal mucus, and foot, allowing snails to "hold
on" to the substrate and avoid dislodgement from currents
or the hydrodynamic forces that attend breaking waves.

Snails have well-developed sensory systems, including sight, smell, and touch. The eyes are located at the base of each cephalic (head) tentacle (Fig. 1) and vary in complexity from simple pits containing photoreceptors that can detect differences in light to highly developed structures with a cornea and lens that can form an image. In most snails, the eyes are used primarily to detect differences in light intensity. A pair of cephalic tentacles is used for both touch and smell. The osphradium, in the mantle cavity, provides additional chemosensory (smell) capability. The ability to detect and track smells is critical for many predatory and scavenging snails to find their prey.

Gas exchange takes place at the ctenidium (gill) located in the mantle cavity (Fig. 1). Water circulates through the mantle cavity and across the ctenidium, where gases diffuse in and out of the blood. The blood has a special copper-containing protein called hemocyanin, similar to hemoglobin, which binds with oxygen and transports it through the circulatory system. A two-chambered heart pumps blood through the ctenidium and to the other tissues.

The primary feeding structure of snails is the radula (Figs. 2, 3), essentially a tongue covered with rows of teeth that is scraped across a surface to tear off tissues for consumption. Herbivorous snails use the radula to feed on algae growing on intertidal rock surfaces, while predatory snails use it to scrape tissues from their prey and sometimes to drill through protective shells. The size, shape and morphology of the radular teeth vary within and among species and, like teeth in terrestrial mammals, are indicative of what snails eat. The radula is contained within the proboscis (Fig. 1), which is composed of the esophagus, buccal cavity, radula, and mouth. In predatory

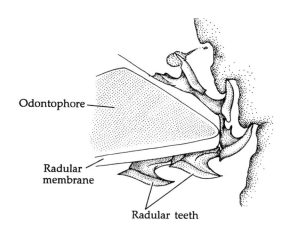

FIGURE 3 An illustration showing how the radula is scraped along the tissues of the prey. As the radula is pulled back towards the snail, the teeth dig in and tear off pieces of tissue. Reprinted from Brusca and Brusca (2003).

snails the proboscis can be highly extensible, sometimes extending three times the shell length.

AUTECOLOGY

Environmental Heterogeneity

The intertidal zone is an extremely heterogeneous environment, and snails often exhibit considerable morphological, physiological, and life history variation within and among shores. The variation develops in response to two major physical gradients. Within shores, tidal height (vertical position on the shore relative to mean low tide) determines the length of time snails are out of the water and exposed to air, which influences thermal stress, desiccation, feeding time, and predation intensity. The higher on the shore, the longer individuals are exposed to air and the more severe the physiological stresses. Because most snails feed only when immersed, those higher on the shore have greater constraints on foraging. Predation intensity is generally greater low on the shore but can be complex because snails are preyed on by both terrestrial and marine predators. When snails are immersed, lobsters, crabs, fishes, sea stars, and other snails prey on them; while emersed, predators include shorebirds, rodents, and humans.

Among shores, wave action can vary dramatically from sheltered bays and inlets with little wave energy to open coast headlands that are pounded by enormous oceanic swells. Breaking waves impart tremendous forces that directly and indirectly affect the ecology of intertidal snails. For example, the hydrodynamic forces that attend breaking waves can rip snails off the shore, depositing them in the shallow subtidal zone, where they

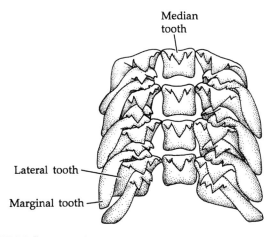

FIGURE 2 Four rows of radular teeth in a marine snail showing median, lateral, and marginal teeth. The teeth are arranged in a series of rows and attached to the odontophore. Reprinted from Brusca and Brusca (2003).

are typically consumed by predators. Indirect effects of waves are more subtle but can alter the nature and amount of food available to snails, their foraging efficiency, the efficiency of their predators, and the intensity of physiological stress.

Whether intraspecific variation develops in response to these gradients depends on the amount of gene flow relative to the spatial scale of environmental variation, the intensity of selection, and the degree of phenotypic plasticity. Gene flow represents the exchange of individuals/ genes among populations and tends to retard adaptation to local selective pressures. The mode of larval development determines, to a large extent, how much gene flow occurs. Snails that produce larvae that spend part of their development in the water column drifting with ocean currents tend to have greater gene flow among populations than do species that brood their young or deposit benthic egg capsules releasing crawl-away juveniles. Snails with greater dispersal potential generally exhibit much less phenotypic variation in response to environmental heterogeneity. For example, *Littorina littorea* is a very common periwinkle with a broad distribution encompassing the eastern and western North Atlantic. It releases larvae that drift for several weeks in the water column and shows little morphological, physiological, or life history variation among shores. In contrast, two closely related congeners with similar geographic distributions, *Littorina obtusata* and *L. saxatilis*, deposit benthic egg capsules or brood their young, respectively, and they vary considerably among shores.

Tidal Height

Because the vertical gradient within the intertidal zone is small relative to the mobility of snails, there is little opportunity for populations to diverge, and only modest intraspecific phenotypic variation emerges. For example, there is a positive correlation between tidal height and snail size for many species. This pattern may reflect higher settlement rates lower on the shore, size-specific movements, or differential mortality. Various life history characteristics (growth, reproduction, mortality) can differ between upper and lower-shore populations for some species with very limited dispersal. For example, *Littorina saxatilis* grows more quickly, attains a larger size, and produces more offspring at low levels on the shore than at higher levels.

Interspecific differences are much more apparent across the tidal gradient. Different species exploit different levels within the intertidal zone and are highly adapted to those zones. One of the key adaptive characteristics is their ability to withstand physiological stress. Snails higher in the intertidal are exposed to air longer and suffer greater thermal and desiccation stress. Thermal stress can involve higher heat loads, induced by the higher air temperatures and absorption of solar radiation, or freezing when air temperatures fall below 0 °C, both of which can severely stress or kill snails. As the tide recedes, snails begin to lose water, and this desiccation is more severe higher on the shore. Not surprisingly, the physiological tolerances of snails to thermal and desiccation stress are correlated with their height on the shore. In fact, the upper distribution limit of some species is set by their physiological tolerances.

Although physiological tolerances can set the upper limit for snails, they are not as prevalent as for sessile organisms. Instead, a number of biotic forces control the vertical distribution of intertidal snails. The reason many snails live within the intertidal zone, instead of the more benign subtidal, is because it provides a refuge from major predators that are intolerant of emersion. Predators, especially those from the subtidal, are often thought to limit the downward spread of intertidal snails. For example, the lower limit of the trochid *Tegula funebralis* in the eastern Pacific is set by predation from the sea star *Pisaster ochraceus*. The sea star is not very efficient at foraging in the intertidal, so *T. funebralis* can avoid being eaten by remaining high enough that *Pisaster* cannot feed. Interestingly, *Pisaster* also probably sets the lower limit of various whelks, both directly through feeding on them, as well as indirectly, by setting the lower limit of their prey (mussels and barnacles). A similar situation occurs for *Nucella lapillus*, a whelk in the North Atlantic. Its lower limit is set directly by predation from subtidal fishes, sea stars, and crabs, as well as indirectly by these same predators setting the lower limit of their barnacle and mussel prey.

Prey distributions can play an important role in setting the vertical distributions of both predatory and herbivorous snails. Because of their limited mobility, intertidal snails rarely live far from their prey. Thus, the biotic and abiotic forces that regulate the vertical distributions of their prey can affect indirectly the vertical distributions of snails.

Competition with other species also can delimit vertical distributions within the intertidal. For many species, their spread across the intertidal zone is constrained by other species that are competitively superior at higher or lower regions of the intertidal. For example, *Acmaea digitalis* is typically found higher on the shore than *Acmaea paradigitalis*, but when *A. digitalis* was experimentally removed, *A. paradigitalis* extended its range to higher levels on the shore. In control areas where *A. digitalis* was not removed, *A. paradigitalis*'s range did not change, suggesting that its upper boundary is set by competition with *A. digitalis*. Competition can be mediated through a variety of mechanisms, including superior feeding efficiency, greater physiological tolerances, and differential susceptibility to predation. A slightly more complex example involves *Littorina subrotundata*, which can live throughout the intertidal but is restricted to the high intertidal because it grows more slowly than *L. sitkana* and is more vulnerable to predators in the low intertidal. The faster growth rate and lower mortality rate of *L. sitkana* in the low intertidal make it competitively superior in this zone.

Because snails often are adapted to exploit particular levels within the intertidal, it is essential that they avoid moving outside their typical zone and that they can find their way back if displaced. A number of environmental cues are used to regulate their level on the shore, including light, gravity, temperature, desiccation, and chemosensory responses to predators and prey. For example, snails that are dislodged by waves and washed downshore become positively phototactic and negatively geotactic, which tends to move them back up the shore. Responses to gravity are usually stronger than those to the direction of light. In some cases, waves could displace snails upshore, depositing them in the high intertidal, where positive phototaxis and negative geotaxis would move snails in the wrong direction. However, the higher thermal and desiccation stresses experienced in the upper intertidal can switch the behavioral responses such that they become negatively phototactic and positively geotactic. In addition to light and gravity, snails use chemosensory cues from their predators and prey to help determine position in the intertidal. Both predators and prey release chemicals into the water that many snails can detect. When snails pick up these chemical cues, they tend to move towards their prey and away from their predators. Together, these behavioral patterns allow snails to maintain their level within the intertidal and, if displaced, to find suitable microhabitats.

Wave Action

MORPHOLOGY

The morphology of snails with limited dispersal varies dramatically among shores differentially exposed to waves. On sheltered shores, snails tend to be large and narrow and produce heavier shells with thick shell walls and a narrow aperture (Fig. 4). These morphological features are thought to represent adaptations to the greater intensity of predation on sheltered shores. The larger size

FIGURE 4 Examples of *Nucella lapillus* from exposed (top row) and protected (bottom row) shores. Note that the exposed shore forms are smaller with broader apertures and thinner shell walls. Photograph by the author.

and thicker shell reduces the efficacy of shell-crushing predators such as crabs, fishes, and birds. For example, when crabs are provided both thick- and thin-shelled prey, the thin-shelled morphs suffer much greater mortality because they are easier to break. Size is important because as snails increase in size, fewer predators can feed on them and eventually they may often attain a size refuge, where they are sufficiently large that most predators are unable to consume them. In addition to crushing, some crabs will hold snails by the spire with one claw and use the other to peel back the shell from the aperture to gain access to the soft tissues. The narrow aperture and thick shell wall make it more difficult for crabs to insert a claw and peel back the shell. Some snails (e.g., whelks) produce apertural teeth that reinforce the aperture wall and narrow the opening (Fig. 4). The narrow aperture also makes it more difficult for birds that prey on snails by turning them over and tearing off any tissues they can reach with their beak.

In contrast, on wave-swept shores, predation is much less, and snails are often smaller and wider with thin-shelled walls and a broad aperture (Fig. 4). These morphological features are thought to represent adaptations to the powerful hydrodynamic forces that attend breaking waves. A thick shell is unnecessary, because predation intensity is considerably reduced on wave-swept shores, and a thinner shell wall reduces the energy required to produce and transport the shell. High wave energies reduce the abundance and efficiency of the guild of mobile, durophagous predators that typically feed on snails (e.g., crabs, fishes). The pounding surf makes it difficult for even terrestrial predators to forage during low tides. Snails tend to be smaller on wave-swept shores because they grow more slowly and suffer much greater mortality (see subsequent discussion). The smaller size is favored because the intensity of hydrodynamic forces (e.g., drag) increase with increasing size, and it allows snails to exploit small cracks and crevices, where hydrodynamic forces are reduced. The wider shell and broader aperture are necessary to accommodate a larger foot, which is favored on exposed shores to reduce the likelihood of dislodgement.

Contrary to earlier views, snails use adhesion, not suction, to maintain their position on shore and resist dislodgement. Adhesion develops between two hard surfaces separated by a thin layer of fluid (e.g., a wet microscope slide cover is more difficult to remove from a slide than a dry one). In snails, the foot and substrate act as the two hard surfaces and are separated by a thin layer of pedal mucus. The adhesive force that develops is proportional to the surface area of the foot and the viscoelastic properties of the pedal mucus. The strength of attachment increases with pedal surface area, so the production of a larger foot allows snails to withstand stronger hydrodynamic forces. Pedal surface area increases with snail size but increases more quickly on exposed shores, such that snails from wave-swept shores produce a much larger foot (Fig. 5).

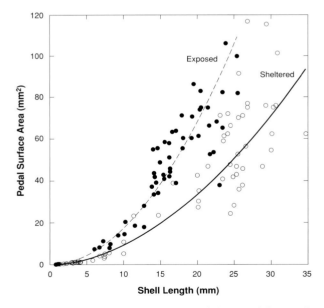

FIGURE 5 Pedal surface area as a function of shell length for *Nucella lapillus* from exposed (solid circles) and sheltered shores (open circles). Snails from exposed shores produce a much larger foot to reduce the likelihood they will be dislodged by breaking waves.

LIFE HISTORY

The intensity of waves impinging on a shoreline also has a profound influence on the life-history characteristics of snails. Life histories are a suite of characters molded by natural selection to maximize fitness and include traits such as growth rate, age and size of maturity, fecundity, and life span.

Snails on sheltered shores grow more quickly than do those on wave-exposed shores. Growth is depressed on exposed shores because the relentless pounding of breaking waves reduces foraging time and efficiency. For example, whelks that feed on barnacles and mussels forage less on exposed shores and take longer to handle a particular prey item. To avoid the powerful hydrodynamic forces that develop as waves break, snails spend most of their time in cracks and crevices, severely reducing foraging time. In addition, crawling along the shore reduces tenacity and increases the risk of being dislodged. The breaking waves continually jostle the snail, making it more difficult to drill its prey and feed, increasing prey handling time.

Despite lower predation intensity on exposed shores, snails experience much higher mortality, often because they are dislodged by waves. After dislodgement, snails are either transported high in the intertidal, where they can succumb to physiological stress, or are washed down into the subtidal, where they can be consumed by a more diverse guild of predators. Consequently, life spans tend to be much shorter on exposed shores. For instance, *Nucella lapillus,* the common predatory whelk in New England, may live for 2 to 3 years on exposed shores, while those on nearby sheltered shores live for 6 to 8 years.

The lower growth rate and higher mortality result in much smaller snails on exposed shores. The higher mortality also selects for snails to become sexually mature at a smaller size (earlier age) and channel most of their energy into reproduction. For whelks on a New England shore, 4 times more offspring were produced on exposed shores than by similar sized snails on sheltered shores. The greater reproductive effort, in part, offsets the higher mortality on wave-swept shores.

Plasticity

Many of the characters that change along these two major environmental gradients may not reflect genetic differences, but instead may be environmentally induced. For example, numerous experiments have shown that if snails are exposed to the chemical effluents of their predators, or from predators feeding on other members of the same snail species, they will produce a thicker shell. Similarly, pedal surface area is highly plastic and can be altered by changes in wave energy. Snails from a protected shore reared on an exposed shore will produce a much larger foot than will their counterparts remaining on the protected shore. Life-history characteristics will also respond to environmental changes. Snails from sheltered shores transplanted to exposed shores grow more slowly than conspecifics remaining on sheltered shores. These examples demonstrate that the morphological, physiological, and life history differences among snails from different habitats may not reflect genetic differences. The ability to produce a flexible phenotype may be an important adaptive characteristic for dealing with the pronounced spatial and temporal environmental heterogeneity of the intertidal.

COMMUNITY STRUCTURE

Because they are important components of intertidal food webs, snails play a key role in shaping the structure (abundance, composition, and diversity) of intertidal communities. As consumers, they can control the distribution and abundance of their largely sessile prey and thereby determine membership in local species assemblages. In addition to directly altering the abundances of their prey, snails can have profound indirect effects on intertidal systems via habitat modifications, trophic cascades, and positive and negative feedbacks on nontarget species. For example, by feeding on competitive dominants in local assemblages, snails can reduce competition and allow competitively inferior species to coexist (see subsequent discussion). The relative importance of herbivorous and predatory snails in controlling the composition and diversity of intertidal communities reflects a complex interplay between physical (e.g., tide levels, tide cycle, wave energy, temperature, desiccation, algal canopy) and biotic processes. For instance, snails tend to have a much greater impact in regulating communities on sheltered and moderately exposed shores because high wave energies reduce the efficacy of mobile consumers. However, even on sheltered shores, their role can be influenced by physiological stresses that change seasonally, monthly (lunar cycle), and with tidal height, or by spatial and temporal variation in the presence of their predators.

Herbivores

Herbivorous snails feed on diatoms, microalgae, algal spores, and macroalgae and can profoundly alter intertidal algal communities. For many species, the diet is fairly eclectic; the snail simply moves along the substrate scraping up whatever algae (diatoms, microalgae, and algal spores) it encounters. For others, especially those that consume macroalgae, the diet is much more restricted and clear preferences exist. For example, the fleshy unprotected tissues of ephemeral forms (e.g., *Ulva*, *Enteromorpha*, *Porphyra*) are favored over perennials (e.g., *Fucus*, *Laminaria*, *Ascophyllum*), which typically invest in structural or chemical antiherbivore defenses.

The feeding preferences of herbivorous snails can be important in regulating patterns of algal distribution and diversity. Ephemeral forms are found primarily on wave-exposed shores. On sheltered shores where herbivorous snails are abundant and efficient grazers, ephemerals are rare because they are preferred over perennials. Although ephemerals often settle on sheltered shores in winter and early spring while snails are relatively inactive, they are quickly removed once snails become active. On highly exposed shores, herbivorous snails are rare and less efficient, and ephemerals flourish.

By feeding preferentially on one particular group of algae, snails can exert a powerful control on algal diversity. A classic example of this comes from Jane Lubchenco's work on New England intertidal algal communities (Lubchenco 1978). The common periwinkle *Littorina littorea* feeds preferentially on the ephemeral algal species. She found that algal diversity within tidepools varied parabolically with the density of periwinkles, resulting in maximum diversity at intermediate snail densities (Fig. 6). In these tidepools, ephemerals were the competitive dominants. At low snail densities, the ephemerals outcompeted the perennials and eventually excluded them. As snail densities increased, herbivory on the competitive dominant

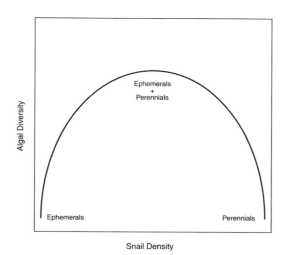

FIGURE 6 Algal diversity as a function of herbivore snail density. The major composition of the algal community is shown for various snail densities. Algal diversity peaks at intermediate levels of herbivore pressure because the snails feed on the competitive dominant algal forms preventing them from monopolizing space.

ephemeral forms prevented them from monopolizing space and allowed the competitively inferior perennial species to coexist. At high snail densities, herbivory was so intense that both the ephemerals and most of the perennials were consumed, leaving only inedible forms. Interestingly, the pattern was quite different on nearby emergent substrata (not in tidepools), where diversity simply decreased with increasing herbivore density. Although the feeding preferences were the same, the perennials were competitively superior on emergent substrata. In this case, both competition and herbivory acted to preclude ephemerals and reduce diversity. Thus, the composition and diversity of these algal communities reflect a complex interplay of wave energies, herbivore density, feeding preferences, and algal competitive abilities.

On sheltered shores, where snails can reach very high densities, grazing by limpets and periwinkles can be so intense that they effectively denude patches of the shoreline of diatoms and microalgae. If herbivorous snails are experimentally excluded, a luxuriant and dense growth of diatoms and microalgae rapidly develops. The transformation of the intertidal from a rich algal community to bare rock indicates snails can be important habitat modifiers—substantially altering the nature of the intertidal.

The impact of *L. littorea* on very sheltered southern New England cobble shores provides a stark example of how a single herbivorous snail species can lead to dramatic habitat modification. On these cobble beaches, snail densities reach 600 to 1000 individuals per square meter and impose intense consumer pressure on algal communities, essentially excluding all algae except for a couple of herbivore-resistant algal crusts. If snails are excluded, the cobble landscape develops a dense and rich algal canopy, which leads to sediment accumulation and the colonization of soft-sediment organisms. Tube-building organisms and perennial grasses further stabilize the sediment, and eventually the habitat transforms into a typical New England marsh. The presence of this snail in such high densities keeps the rock surfaces clear and prevents the typical succession from protected cobble beaches to marsh. Moreover, Mark Bertness (1984) has suggested that *L. littorea* may be responsible for shifting these habitats from soft-sediment marshes, typical of highly sheltered New England shores, to hard substrates. *Littorina littorea* was introduced to New England within the last century, so these modifications are relatively recent and ongoing.

Predators

Most intertidal predatory snails feed on shelled prey such as barnacles, mussels, oysters, and other snails. They typi-cally use the radula and acidic chemical secretions that soften the $CaCO_3$ to bore a hole through the shell of their prey. Once the shell is penetrated, they extend their proboscis (in some species three times their body length) into the prey and use the radula to rasp off tissues. In some cases (e.g., barnacles), snails can use the edge of their shell to pry open the valves and insert their proboscis to feed. Feeding times vary with the type of predator and prey but can be quite long. For instance, a whelk feeding on a mussel can take from 4 to 36 hours to drill and consume a single individual.

As with herbivores, predatory snails can exert a powerful influence on the distribution of their prey and the structure of intertidal communities. A classic experiment by Joe Connell (1961) demonstrates just how effective whelks can be in controlling the distribution of their barancle prey. He used cages to exclude the predatory whelk *Nucella lapillus* from various regions of the shore in Scotland. The results showed that the impact of *N. lapillus* was greater at lower levels of the shore and could effectively set the lower boundary of the barnacle *Semibalanus balanoides*.

Many whelks feed on mussels, the dominant competitor for space in the intertidal, especially on temperate shores in the Northern Hemisphere. On exposed shores, where mobile predators are rare and inefficient, mussels monopolize the shore, excluding most other space occupiers, including barnacles, macroalgae, tunicates, anemones, and bryozoans. On more sheltered shores, predatory snails, as well as other predators (crabs, fishes, seabirds), consume mussels, creating bare patches that allow competitively inferior species to coexist. By cropping down the competitively dominant mussels, predatory snails can regulate diversity in much the same way herbivores control algal diversity in tidepools. In fact, the predators that control the distribution of mussels often have a greater impact on the distribution of intertidal algae than do herbivores. Clearly, snails can profoundly affect the abundance, distribution, and diversity of intertidal organisms.

SEE ALSO THE FOLLOWING ARTICLES

Adhesion / Competition / Dispersal / Limpets / Predator Avoidance / Wave Exposure

FURTHER READING

Bertness, M. D. 1999. *The ecology of Atlantic shorelines*. Sunderland, MA: Sinauer.
Bertness, M. D. 1984. Habitat and community modification by an introduced herbivorous snail. *Ecology* 65: 370–381.
Brusca, R. C., and G. J. Brusca. *The invertebrates*. Sunderland, MA: Sinauer.

Connell, J. H. 1961. Effects of competition, predation by *Thais lapillus* and other factors on natural populations of the barnacle *Balanus balanoides*. *Ecological Monographs* 31: 61–104.

Denny, M. W. 1988. *Biology and the mechanics of the wave-swept environment*. Princeton, NJ: Princeton University Press.

Hughes, R. 1986. *A functional biology of marine gastropods*. Baltimore: Johns Hopkins University Press.

Koehl, M. A. R., and A. R. Wertheim. 2006. *Wave-swept shore: the rigors of life on a rocky coast*. Berkeley: University of California Press.

Lubchenco, J. 1978. Plant species diversity in a marine intertidal community: importance of herbivore food preference and algal competitive abilities. *American Naturalist* 112: 23–39.

Underwood, A. J. 1979. The ecology of intertidal gastropods. *Advances in Marine Biology* 16: 111–210.

SPONGES

SALLY P. LEYS

University of Alberta, Edmonton, Canada

WILLIAM C. AUSTIN

Khoyatan Marine Laboratory, Sidney, Canada

Sponges (phylum Porifera) are animals, often with colorful bodies that, in the intertidal, form crusts or clumps, primarily on rocky surfaces. Sponges are unique among animals in having a body perforated by canals open to the surface. This may be a significant factor in limiting species intertidal occurrence and distribution because of low tolerance of physical stresses such as desiccation, temperature, and salinity. Also, like other filter feeders, sponges feed only when fully submerged. Given that many subtidal sponges feed continuously, the mid- and upper regions of the intertidal habitat may be food-limited for sponges.

GENERAL MORPHOLOGY

Sponges are best described as encrusting, globular, massive, vase-shaped, reticulate, or branching. Of these types, the first is most common in intertidal habitats, and there the encrusting forms are usually no more than a few millimeters thick, while massive and globular sponges may be as large as 10 to 20 centimeters in diameter. Colors are often vibrant—brilliant red, violet, mustard yellow, or green. The surface of most sponges is completely clear of sediment and debris, even in muddy bays; they are soft to hard and slippery to sandpaper-like to the touch, and a few have pungent odors if scraped.

INTERNAL ANATOMY

The sponge is essentially a suction pump that draws water through minute canals, extracting food and oxygen from the fresh supply and excreting wastes into exhalent canals. There are three functional regions (Fig. 1): the apical pinacoderm, which allows water into the animal through pores (ostia); the choanosome, the bulk of the body, which contains chambers with pumping cells (choanocytes); and the osculum (a chimney-shaped extension of the canal system and apical pinacoderm), which vents water out of the animal. Each of these regions is composed of two epithelia separated by a collagenous mesohyl with some crawling cells. The outer epithelium, the exopinacoderm, is formed of flat cells called exopinacocytes, and the inner epithelium, the endopinacoderm, is formed by elongate cells called endopinacocytes.

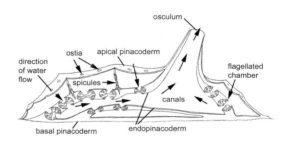

FIGURE 1 Internal anatomy of a juvenile demosponge, showing the inner and outer cell layers and the pathway that water follows as it is drawn through chambers and particles extracted for food.

INDIVIDUAL OR COLONY

Though texts often refer to sponges as colonies, the individual unit within the colony is difficult to define. One definition is that any animal enclosed by a single and continuous outer epithelium (or apical pinacoderm) can be considered an individual. Another view suggests that the individual unit within a multiosculum sponge consists of a single osculum and the chambers and canals that it drains. Typically, sponges are quantified by the percent of the substrate they cover or their volume, rather than their specific pumping and feeding capacity.

FEEDING

Choanocytes, specialized cells with a collar of microvilli surrounding a flagellum, generate the force for the feeding current. The beat of the flagellum draws water down to its base, and the combined effort of 80–100 or more choanocyte pumps in one or more chambers draws water through the sponge (Fig. 2). (While one deepsea

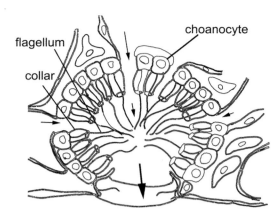

FIGURE 2 A diagrammatic view of a choanocyte chamber in a demo-sponge. The whiplike beat of the flagellum draws water around the collar of the choanocyte, allowing food to be brought past the cell body. The direction of water flow is shown by black arrows.

richer assemblage of sponges. The richness of sponge populations in caves (e.g., 50 species in one northeastern Pacific cave, 31 species in a northeastern Atlantic cave) (Fig. 3) may be related to the absence of competition with seaweeds, the lack of irradiation from harmful wavelengths of sunlight, the higher partial pressures of oxygen in the absence of seaweed respiration, lower water temperature, higher salinities in the absence of dilution from rain, lower pH, or air that is cooler and with higher humidity. Some intertidal sponges can fully or partially close the oscula and ostia when exposed to air. A few species (*Cliona* spp., *Suberites* sp., *Halichondria* sp., and *Hymeniacidon* sp.) can tolerate a modest reduction in salinity in estuaries.

FIGURE 3 Sponges in an intertidal cave at Execution Rock, Barkley Sound, British Columbia. Photograph by W.C. Austin.

Rock is not the only substrate in the intertidal. The calcareous shell of barnacles, serpulid worm tubes, and the shells of oysters and other intertidal surface-dwelling bivalves may support many sponge species. In some cases sponges may utilize seaweed holdfasts (e.g., *Halichondria panicea*; Fig. 4) or seaweed fronds (e.g., *Grantia compressa*)

group lacks chambers altogether, and feeds by carnivory, no intertidal species are known to have this habit.) The feeding activity of other intertidal and shallow subtidal benthic suspension feeders is known to substantially affect the water column properties in shallow bays. Though no experiments have examined uptake of food by intertidal sponges, studies of subtidal sponges have revealed feeding efficiencies in the range of 75–99% on plankton 0.1– 70 μm. Sponges are capable of ingesting a great variety of types and sizes of plankton. Flagellates up to 50 μm in diameter are taken up by cells in the incurrent canals, and bacteria $\leq$1 μm by the choanocytes; although most particles small enough to enter the open porocytes can be retained, studies indicate that sponges can select nutritionally favorable cells. Uptake occurs at the level of individual cells, either choanocytes or endopinacocytes, that phagocytose particles.

ECOLOGY

Habitat

Sponges, growing either on drying rocks or in tidepools, are generally restricted to the lower intertidal, while a few species extend into the mid-intertidal. The same or related genera often occur intertidally in cold temperate seas. Most orders are represented. No Hexactinellida occur in the intertidal today, although one species, *Aphrocallistes vastus*, has been found as shallow as 2 meters at one location with strong vertical water mixing in British Columbia.

Smaller tidepools exposed to sunlight may harbor fewer sponge species than the undersurfaces of adjacent flat rocks or boulders exposed to air but in the shade. On the other hand, tidepools under flat rocks or in caves often harbor a

FIGURE 4 *Halichondria panicea*, a common green demosponge species or species complex world wide. Photograph by N. Lauzon.

as substrate. In the tropics, mangrove roots are often the only substrate for sponges. Also in the tropics, a few sponge species live on mud. Some sponges can settle on other sponge species. The base of the overlying sponge may be corrugated so that the ostia and oscula of the underlying sponge are not occluded.

Form

Intertidal sponges typically have a smooth, encrusting growth form. A smooth crust is more likely to withstand wave action than an uneven surface or an erect growth form. Intertidally occurring globular or spherical-shaped sponges such as *Suberites* spp. and *Tethya* spp. have a firm rubbery texture, in contrast to the friable texture of encrusting forms such as *Halichondria* spp. and *Haliclona* spp. Some other thick intertidal sponges have a hard to stony texture (*Acarnus* sp., *Xestospongia* sp.). The arborescent spongin fibers in intertidally occurring encrusting *Aplysilla* spp. may provide elasticity in response to impacts from wave-borne material. The same may be true for some of the dictyoceratid sponges with reticulating spongin fibers, as exemplified by various "bath sponges."

In some species such as *Halichondria* spp. and *Haliclona* spp. the same sponge species may extend from a wave-protected microenvironment to a wave-exposed environment. The growth form will vary from high oscula to oscula flush with the surface with increasing exposure. Intertidal populations of *Cliona* spp. can protect themselves by boring into the shells of large barnacles, attached rock scallops, oysters, and corals. The sponge tissues ramify through the shells while the oscula extend above the surface. The oscula can contract down below the surface when a sponge is mechanically stimulated. The relatively uncommon stalked sponges in the intertidal *(Leucilla nuttingi, Grantia* sp. and *Leucosolenia eleanor, Neoesperiopsis* sp., *Haliclona occulata)* have a tough but elastic body that "goes with the flow."

Competition and Predation

Seaweeds often appear to limit the expansion of encrusting sponges over rock substrates. Other potential space competitors include bryozoans, ascidians, corals and their allies, and calcareous tube worms. Sponges may not be able to overgrow a competitor but can inhibit that competitor's overgrowth. In other cases sponges may be overgrown and smothered. Competition may also occur at the larval level. Larvae of a species of *Haliclona* appear to settle on a coral, then kill that portion of coral in the vicinity.

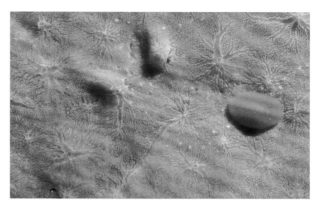

FIGURE 5 The dorid nudibranch *Rostanga pulchra* feeding on the red sponge *Ophlitaspongia pennata*. Photograph by N. Lauzon.

Predators of intertidal sponges, such as many dorid nudibranchs may be moderately species-specific. The northeastern Pacific red dorid *Rostanga pulchra* (Fig. 5) feeds on about five species of red sponges and is chemically attracted to at least some of them. Unspecific predators include species of sea urchins, sea stars, chitons, snails, segmented worms, and trigger fish. In many cases the whole sponge is not eaten, and the remaining portions may regenerate. The hard calcareous or glass skeleton, where present, may deter some potential predators. Many sponges also produce chemicals that are toxic to other organisms. These might be predators, competitors, or settlers (e.g., larvae of other species attaching to the sponge for substrate). In the tropics many sponges can irritate the skin of humans.

Growth and Regeneration

Some sponges (e.g., *Leucosolenia* sp., *Sycon* sp., *Halisarca* sp.) may grow to their final size in a month or two and, at least in cold temperate waters, die back each year. Others, notably choristids such as *Stelletta* sp. and spirophorids such as *Craniella* sp., grow slowly, and no change in size is perceptible over a 10-year period. Sponges, generally, have excellent regenerative capabilities. Some (e.g., *Halichondria panicea* and *Geodia* sp.) will periodically shed the outer skeletal layer, which serves to remove settling organisms and silt.

Locomotion

Some encrusting sponges can move short distances over time through growth in one direction. A few species (e.g., *Tethya* spp.) can send out elongate extensions at a rate of up to 5 millimeters per hour, and the whole sponge can move at rates of up to 2 millimeters per hour. Larvae of many species are motile and can control direction of movement, while some larvae can exit an osculum and slide down the parent sponge to adjacent substrate.

REPRODUCTION

Intertidal sponge species tend to be viviparous. Most species are thought to be sequential hermaphrodites: oocytes are fertilized by sperm taken in via the feeding current, and embryos develop within a follicular epithelium, either in specialized "brood" chambers or individually dispersed throughout the parent tissue. Larvae are liberated either via the canal system or, in some cases, by breaking through the adult tissue. A diversity of larval phenotypes have now been described in sponges, but most intertidal demosponge species have parenchymella larvae: ovoid propagules, 50–500 micrometers long and 30–200 micrometers wide, that are covered with short (10–20 micrometers) cilia except possibly at the anterior- and posteriormost poles. Many parenchymella larvae also have a ring of longer cilia surrounding the bare posterior pole, and in some the cells forming a band at the base of these long cilia are darkly pigmented; typically larvae are the color of the adult sponge. Halichondrid larvae are characteristically wider at the anterior than at the posterior pole, cilia are uniformly short, and pigment is absent. Although larvae have been collected using fine plankton nets in the field, natural release of larvae by intertidal sponges has not been reported. Some sponges release larvae in the morning, apparently in response to the onset of darkness the preceding day, but pieces of adult sponge carefully pried off the substrate and kept submerged in a bowl of seawater also will often release larvae over a period of several hours without additional stimuli. Larvae swim upward, rotating on their longitudinal axis or in a helical twist, and those observed in a small dish usually bounce off surfaces, readily changing direction; others keep a low profile, skating over the substrate. Most larvae settle within 12 to 24 hours, and in 4 to 6 days they form a juvenile sponge with a single osculum.

Much less is known of the larvae released by intertidal species of Calcarea such as *Clathrina* sp., which can be found at very low tides among surf grass beds on exposed shores. Larvae from a subtidal species of *Clathrina* at the Santa Catalina Islands, California, are coeloblastulae: ovoid and fully ciliated, but partly or completely hollow until settlement.

Many sponges also reproduce asexually through fragmentation or budding.

FIGURE 6 Examples of sponge spicules imaged in a scanning electron microscope. Photograph by W. C. Austin.

CLASSIFICATION

Approximately 7000 living species of sponges are known worldwide. The classification is presently in a state of flux, but there is some agreement that the phylum contains three extant classes. The Hexactinellida, with 500–600 species, are characterized by glass spicules (skeletal pieces) with six rays; Calcarea, with about 500 species, are characterized by having calcareous spicules. Demosponges comprise about 6000 species and are characterized by having glass spicules in a variety of forms (e.g., Fig. 6), but not six-rayed, spongin fibers, or, in a few species, no skeleton.

SEE ALSO THE FOLLOWING ARTICLES

Body Shape / Competition / Locomotion: Intertidal Challenges / Materials, Biological / Nudibranchs and Related Species / Stone Borers

FURTHER READING

Berguist, P. R. 1978. *Sponges*. Berkeley: University of California Press.
Boury-Esnault, N., and K. Rutzler, eds. 1997. *Thesaurus of sponge morphology*. Smithsonian Contributions to Zoology No. 596. Washington, DC: Smithsonian Institution.
Hooper, J. A., and R. W. M. van Soest. 2002. *Systema Porifera: a guide to the classification of sponges*. New York: Kluwer Academic/Plenum.
Cuénot, L. 1973. *Traité de Zoologie*. P.-P. Grassé, ed. Vol. 3. *Spongiaires*. Paris: Masson et Cie.
Mackie, G. O., *et al.* 2006. Biology of neglected groups: Porifera. *Canadian Journal of Zoology* 84(2): 143–145.

STARFISH

STARFISH

SEE SEA STARS

STAUROMEDUSAE

CLAUDIA E. MILLS
University of Washington

YAYOI M. HIRANO
Chiba University, Japan

Stauromedusae are small jellyfishes that spend their entire life attached to a substrate (usually rock or seaweed) rather than swimming freely up in the water column like most other jellyfish. Because of their attached, benthic lifestyle, they seem in some ways to have more in common with their relatives the hydroids and sea anemones than with the free-swimming, planktonic scyphozoan jellyfish to which, until recently, they have been considered more closely related. Although a few of the larger, often colorful, free-living scyphomedusae could end up stranded in tidepools (as recorded in the fictional *Adventure of the Lions Mane* by Sir Arthur Conan Doyle), most of the jellyfishes likely to be found permanently living along the rocky shore are the little attached jellyfishes known as Stauromedusae, or Staurozoa. There are also some very tiny crawling jellyfishes found along some rocky shores, which are hydromedusae.

BIOGEOGRAPHY, HABITAT, AND LOCOMOTION

Stauromedusae (Fig. 1) are primarily found in the intertidal and shallow subtidal in temperate and boreal (cold) waters, with a few species recorded from tropical or subtropical waters. These animals are found in full-salinity ocean waters along exposed to fairly protected coastlines; some are found in estuaries with slightly lowered salinities. One species of stauromedusa has recently been found living at great depth in hydrothermal vent communities, but most of the 50 or so known species of stauromedusae occur along shallow coastlines; about 40 species occur in the Northern Hemisphere, and about 8 species are known only from the Southern Hemisphere.

Stauromedusae live attached to seaweed, seagrass, gravel, or rock and may be found on exposed rocky shorelines or in tidepools. Although they appear to be static and fixed in place, stauromedusae, like sea anemones, are capable of movement so slow that the movement is not

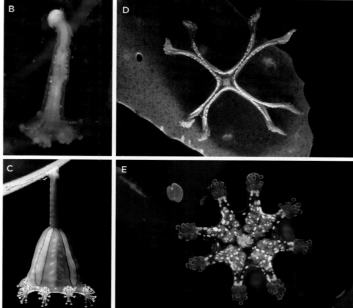

FIGURE 1 Five species of Stauromedusae. (A) Two *Haliclystus* sp. attached to eelgrass (San Juan Island, Washington), tentacle spread approximately 25 mm. (B) *Stenoscyphus inabai* attached to red alga (Misaki, Sagami Bay, Japan), height including stalk 6 mm. (C) *Manania handi*, attached to eelgrass (San Juan Island, Washington), height including stalk approximately 35 mm. (D) *Kishinouyea nagatensis* attached to kelp (Kominato, Boso Peninsula, Japan), tentacle spread approximately 30 mm. (E). Oral view of *Lucernariopsis cruxmelitensis* (Wembury, Plymouth, UK), tentacle spread 8 mm. Photographs A and C by Claudia E. Mills; B, D, E by Yayoi M. Hirano.

visible, but the acute observer will notice that individuals change their position over time. For instance, stauromedusae living on seagrass will always be found on the healthy younger parts of the blade, gradually moving down towards the basal new growth as the end of the blade becomes frayed with age. Some stauromedusae can also move rapidly by contracting or twisting the stalk and by bending the calyx or arms, with sometimes a "somersault" kind of motion. On rare occasions, stauromedusae are seen floating freely in the water, perhaps occasionally just letting go, upon which they are carried to a new location where they reattach.

MORPHOLOGY AND LIFE HISTORY

Stauromedusae (see Fig. 1) are composed of a flower-like calyx (cup), edged with clusters of tentacles in most species, on a stalk that attaches to some benthic substrate with an adhesive basal disk. There is a mouth opening on the inside center of the calyx, where food is taken in and indigestible waste is later released. Stauromedusae have four-part symmetry, usually with eight marginal arms at the top edge of the calyx. In most species, each arm is crowned with a cluster of small hollow tentacles, each with a terminal knob covered with cnidocysts—the "stinging cells" that cnidarians use to capture prey. Gonads are usually arranged radially from the base of the calyx towards the marginal arms; in some genera, gonads fill the arms between the terminal tentacles and the central mouth, in others the gonads are located only near the mouth, and in a few the gonads are restricted to the basal part of the stalk.

In general, stauromedusae are colored to match their surroundings and may be so cryptic as to be difficult to see. Some species show slight differences in color between males and females, which are otherwise of similar size and morphology and cannot be sexed without a microscope, if at all, prior to spawning.

Most, if not all, stauromedusae live less than one year. A few species may go through two generations per year, but it seems that most have an annual life cycle. Many species of stauromedusae are most abundant from summer to fall; young specimens less than 1 millimeter long emerge in the spring or early summer (although other species peak in the winter months with a shift in appearance of the young stages to the autumn) and are usually found on seaweed. Stauromedusae reach adult sizes, typically about 1 to 4 centimeters in length or calyx diameter, within several weeks. Like most other medusae, they seem to put most energy into growth until they reach near-adult size, at which point energy is allocated primarily to the production of eggs and sperm for sexual reproduction.

The sexes are separate, and, given enough to eat, most stauromedusae apparently spawn daily for a month or more, until conditions deteriorate and they die. As in many Cnidaria, at least some species of stauromedusae spawn eggs and sperm in response to a light cue after a period of darkness. Laboratory studies indicate that the eggs are sticky and probably remain on the bottom very near the parent stauromedusae. The externally-fertilized embryos develop into microscopic, wormlike, unciliated, planula larvae, which in the laboratory creep along the substrate before encysting to a resting stage that resembles a sticky flattened ball less than 1 millimeter in diameter. Field studies confirm that most species of stauromedusae disappear for several months before their young stages reappear, so encystment of the larvae seems likely also to occur in the field.

New young stauromedusae emerge the next season, which may begin, depending on the species, in spring, summer, fall, or even winter, to repeat the seasonal life cycle. We do not know exactly the environmental cues that synchronize these annual events, but there is not much variation in timing from year to year within any given species at each location. Each species operates on a clock slightly different from the others, and species of stauromedusae that co-occur on the same shore may emerge, grow up and finally disappear at substantially different times of the year.

ECOLOGY

Field studies indicate that stauromedusae eat mostly small crustaceans—primarily harpacticoid copepods, gammarid amphipods, chironomid fly larvae, and ostracods, with some species also feeding on small polychaetes (worms) and snails. Stauromedusae are able to capture prey from the plankton as well as on the bottom.

Few predators of stauromedusae are known. Like cnidarian hydroids that live along the rocky shore and are often eaten by nudibranchs, which are found on the hydroid colonies, stauromedusae have also occasionally been reported to be preyed upon by aeolid nudibranchs. Because stauromedusae in the field are often seen to be regenerating portions of their margins, such grazing-predation must be more common than has been reported. A fish has been reported to feed on stauromedusae attached to the undersides of low-intertidal boulders in the Antarctic, which were found whole in the fish's stomach.

Stauromedusa populations are occasionally very dense and monospecific, as has been seen at deep-sea hydrothermal vent sites; individuals so abundant as to be touching are also described for an intertidal species in Antarctica. More often, however, stauromedusae are relatively far apart along low-intertidal and shallow subtidal rocky shores, their cryptic coloration perhaps making them seem even rarer, and they are probably not generally among the species that drive the ecology of most rocky intertidal ecosystems.

PHYLOGENY

Within the phylum Cnidaria, stauromedusae have long been considered an order in the class Scyphozoa, but recent genetic studies (Collins and Daly 2005) point toward elevating these attached, benthic-living jellyfishes to class Staurozoa, at a rank equal to Scyphozoa, Cubozoa, Anthozoa, and Hydrozoa.

FURTHER READING

Berrill, M. 1962. The biology of three New England stauromedusae, with a description of a new species. *Canadian Journal of Zoology* 40: 1249–1262.

Collins, A.G., and M. Daly. 2005. A new deepwater species of Stauromedusae, *Lucernaria janetae* (Cnidaria, Staurozoa, Lucernariidae), and a preliminary investigation of stauromedusan phylogeny based on nuclear and mitochondrial rDNA data. *Biological Bulletin* 208: 221–230.

Dawson, M.N. *The Scyphozoan.* http://www2.eve.ucdavis.edu/mndawson/tS/tsFrontPage.html.

Doyle, A.C. 1926. *The Adventure of the Lions Mane.* http://sherlock-holmes.classic-literature.co.uk/the-adventure-of-the-lions-mane/.

Hirano, Y.M. 1986. Species of stauromedusae from Hokkaido, with notes on their metamorphosis. *Journal of the Faculty of Science, Hokkaido University, Series VI, Zoology* 24: 182–201.

Kramp, P.L. 1961. Synopsis of the medusae of the world. *Journal of the Marine Biological Association of the UK* 40: 1–469.

Mills, C.E. *Stauromedusae/Staurozoa.* http://faculty.washington.edu/cemills/Stauromedusae.html.

Mills, C.E., and R.J. Larson. 2007. Scyphozoa: Scyphomedusae, Stauromedusae, and Cubomedusae, in *Light and Smith's manual: intertidal invertebrates of the central California coast,* 4th ed. J.T. Carlton, ed. Berkeley: University of California Press 168–173.

Otto, J.J. 1978. The settlement of *Haliclystus* planulae, in *Settlement and metamorphosis of marine invertebrate larvae.* F.-S. Chia and M. Rice, eds. New York: Elsevier, 13–22.

Zagal, C.J. 2004. Population biology and habitat of the stauromedusa *Haliclystus auricula* in southern Chile. *Journal of the Marine Biological Association of the UK* 84: 331–336.

STEINBECK

SEE RICKETTS, STEINBECK, AND INTERTIDAL ECOLOGY

STONE BORERS

E. C. HADERLIE
Naval Postgraduate School

In soft sedimentary deposits such as mud or sand there are many marine animals that burrow into the sediment seeking food or shelter. There is also a great diversity of marine invertebrates that bore into solid rock along the shore and in subtidal rocky reefs. These animals derive no nutritional benefit from the material they excavate, but the burrows in which they spend their lives protect them from predators or from being dislodged by surf or surge action. Stone-boring marine animals include sponges, annelid worms, sea urchins, and a variety of bivalve molluscs. In their snug burrows, all must maintain contact with the outside seawater for respiratory purposes and to flush out wastes, and all release gametes or larvae into the seawater during reproduction. In addition, most depend on plankton in the water for food.

The rocks penetrated vary in hardness from soft mudstone and shale to sandstone, limestone, and hard, flintlike siliceous chert. Some even bore into poor-quality concrete or the shells of living and dead molluscs and barnacles. None can penetrate granite.

SPONGES

Sponges are primitive animals and are common in the sea attached to rocks and other solid substrates. One worldwide group, the genus *Cliona*, can bore into soft calcareous stone and into the shells of molluscs such as abalone, moon snails, and oysters, and the wall plates of barnacles (Figs. 1, 2). When on a suitable surface for boring, the only exposed parts of the sponge are yellow papillae protruding from small holes (1–3 mm diameter) in the substrate. Most of the sponge's body occupies tunnels below the surface. These excavations honeycomb the surface and

FIGURE 1 Boring sponge *(Cliona)* in rock oyster shell. Photograph by the author.

FIGURE 2 Red abalone *(Haliotis)* showing excavations of boring sponge. Photograph by the author.

weaken the rock or shell. Mollusc shells are often totally destroyed. The mechanism of boring is not completely understood, but when the sponge is excavating a tunnel, small chips of calcium carbonate substrate are removed after being loosened by chemical action.

WORMS

Several species of spionid worms spend their lives in tunnels they have excavated in soft rock or the shells of molluscs and barnacles. The worms are usually small (a few centimeters long) but can be very abundant. The burrows are often U-shaped and are excavated mechanically by stiff bristles and possibly by chemical action.

SEA URCHINS

In many regions in the low intertidal zone on rocky shores, on headlands subject to heavy surf action, sea urchins excavate hemispherical depressions or deep burrows, which give them protection from being dislodged by the waves. The purple sea urchin, common along the Pacific coast of North America, makes no attempt to burrow or bore in most areas of its range. But on sandstone layers of rock subject to violent surf these same urchins excavate rounded cavities 5–8 centimeters in diameter and as much as 10 centimeters deep (Figs. 3, 4). They burrow by using sharp teeth and spines. As the depression becomes deeper and the animal grows, it increases the diameter of the burrow and often becomes trapped and cannot leave. For food the urchin must then depend on capturing pieces of algae washed near its burrow opening.

BIVALVES

Of all the stone-boring marine animals, the most destructive and widely distributed are bivalve molluscs. Many bivalves, such as the clams, burrow into soft substrate using a muscular foot, which can be protruded for digging. They all have two siphons on the posterior end, which extend out of the burrow and can circulate seawater through the animal's body, bringing in oxygen and suspended food and flushing out wastes. Some clams live in fissures or cracks in the rock, and others can bore into a variety of stone from soft mudstone to hard chert (Fig. 5). The largest and most diverse group of these bivalve borers are the pholads or piddocks. The pholads, of which there are over 20 species along the Pacific coast, vary in size from small forms, with a shell a few centimeters long, to very large animals with shells over 15 centimeters long (Fig. 6). These largest pholads may have a siphon tube nearly one meter long and can form very long burrows that severely weaken the penetrated stone. All of the pholad borers have shell

FIGURES 3 Purple sea urchins *(Strongylocentrotus)* in depressions and burrows in sandstone beach rock. Photograph by the author.

FIGURES 4 Purple sea urchins *(Strongylocentrotus)* in depressions and burrows in sandstone beach rock. Photograph by the author.

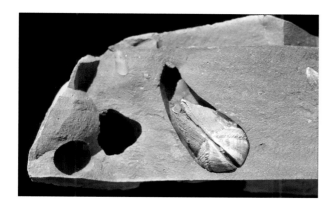

FIGURE 5 Pholad *(Penitella)* excavations in shale. Shell 5 cm long. Photograph by the author.

FIGURE 6 Mature shell of pholad *(Parapholas)*, 15 cm long. Photograph by the author.

valves with strong striations or sharp spinelike teeth on the anterior end of the valves. When the animal is young and in the boring stage, there is a broad, circular gap between the valves at the anterior end through which the borer can extend its large muscular foot, which by suction can anchor the animal to the end of the burrow (Fig. 7). When

FIGURE 7 Pholad *(Penitella)* in boring stage. 1.5 cm long overall. Muscular foot extending through gap between shell valves. Photograph by the author.

so anchored, the borer can move the valves and press them against the burrow wall, thereby eroding the rock mechanically. The sharp teeth on the edge of the valves are worn down in the process, but as the animal grows, new ones are laid down. Once the burrower has reached mature size, or runs out of room to excavate, the foot withdraws and atrophies, and a thin hemispherical dome (the callum) is formed that fills the gap between the valves. The animal can then no longer bore, but in its gonads, now mature, gametes are formed and reproduction can occur.

One species of pholad on the Pacific coast, the so-called abalone piddock, bores not only into shale rock but also into the shells of abalone or mussels. As the pholad bores deeply into an abalone shell from the outside, the abalone often secretes more thick nacreous material on the inside over the area, creating a blister pearl.

FIGURE 8 Date mussels *(Lithophaga)*, 5 cm long. Photograph by the author.

The second and smaller group of bivalve stone borers consists of the date mussels *(Lithophaga)*, so called because in size, shape, and color they resemble dates (Fig. 8). The shell is torpedo-shaped and about 5 centimeters long. The brown color is due to the presence of the thick horny layer (periostracum) over the thin shell valves. These borers are most commonly found subtidally in calcareous rock such as limestone. In the past it was postulated that date mussels bored into calcareous rock using acid secreted by their mantle, but recently it has been shown that some species have glands in the mantle tissue that can secrete a mucoprotein with calcium-binding abilities. When the tissue is protruded from between the shell valves and applied to the stone, the substrate is eroded chemically.

Many of these bivalve borers are highly destructive to the substrate and weaken the stone significantly, causing shoreline rocks to crumble and fall. They also weaken subtidal reefs. During storms and strong wave surge, large kelp plants with their holdfasts attached to reefs riddled by stone borers are often washed ashore, rafting large pieces of broken stone up on to the beach. On sandy beaches with rocky reefs immediately offshore, in shallow water one often sees many pieces of broken rock both in the surf zone and above, all severely penetrated with holes that were once the burrows of pholads or date mussels.

SEE ALSO THE FOLLOWING ARTICLES

Bivalves / Boring Fungi / Echiurans / Sea Urchins / Sponges / Worms

FURTHER READING

Haderlie, E. C. 1980. Stone boring marine bivalves from Monterey Bay, California. *The Veliger* 22: 345–354.
Turner, R. D. 1955. The family Pholadidae in the western Atlantic and eastern Pacific. *Johnsonia* 3(34): 65–160.

STORM INTENSITY AND WAVE HEIGHT

HENDRIK L. TOLMAN

NCAA/National Centers for Environmental Prediction

Water is three orders of magnitude heavier than air. This makes the surface of the ocean very stable, with a clear transition from water to air, even in the most extreme weather conditions. However, because it is an interface between fluids and not solids, disturbances of the equilibrium interface will lead to motion of the fluids under the action of gravity. This will typically result in surface gravity waves propagating over the surface of the ocean or through the entire body of water. The most visible class of waves at the ocean surface are the so-called wind waves, including the surf that is almost always present on exposed coastlines. As their name suggests, these waves are generated by winds over the oceans. To understand them, a basic understanding of winds over the oceans is required.

WINDS OVER THE OCEANS

Winds represent the motion of air in the atmosphere. These motions occur over a wide range of spatial and temporal scales, ranging from small-scale turbulence (with scales of the order of seconds and centimeters to meters), to gusts (seconds to minutes), to mesoscale weather systems (hours), to persistent large-scale circulations. In the present context, attention will be focused on three aspects of winds over the oceans: how winds are described, which are the dominant weather systems, and where and how often do these systems occur.

Air motions, or winds, cover a continuous spectrum of spatial and temporal scales and, in the boundary layer near the surface of the ocean, vary distinctly with height above the surface. It is therefore necessary to define what we mean by the wind. Meteorologists make a distinction between sustained winds and gusts. Sustained winds represent time averages of the order of 5 to 30 minutes; depending on individual definitions, gusts represent higher winds that occur at shorter time scales. The National Data Buoy Center (NDBC), which is a clearinghouse for observations from buoys on the oceans, uses an 8.5-minute average for their sustained wind speed observations and a 2-minute average as a proxy for gusts.

In the atmospheric boundary layer, the wind speed u as a function of the height z (positive upward) can generally be described as a logarithmic profile where $u(z)$ α $\ln(z/z_o)$. The roughness length z_o depends, among others, on the presence of surface waves, and on the stability of the atmospheric boundary layer. To properly compare surface wind observations, they are typically converted to a standard height of $z = 10$ m.

At higher latitudes, winds are generated by low-pressure systems (depressions), high-pressure systems, and fronts. The most extreme winds are generally associated with depressions moving in easterly directions. Traditionally the Beaufort scale is used to classify wind speeds. The Beaufort scale is represented in Fig. 1, together with high-wind warning designations used by the National Weather Service (NWS).

In the tropics and subtropics, the dominant wind systems consist of tropical cyclones (known as hurricanes, typhoons, willy-willies, etc., depending on location) and persistent large-scale circulation systems. Hurricane wind speeds correspond to Beaufort 12. Hurricanes are generally classified using their category according to the Saffir-Simpson scale, as also shown in Fig. 1. Extreme hurricanes are more intense than high-latitude storms. However, they generally have much smaller spatial scales. Persistent seasonal wind systems include the trade winds and monsoons. Weather systems with smaller spatial scales, such as thunderstorms and tornadoes, also include extreme wind speeds. However, because of their small scale, these systems have little relevance for wind waves, as will be discussed.

Beaufort scale

0	1	2	3	4	5	6	7	8	9	10	11
<1	1–3	4–6	7–10	11–16	17–21	22–27	28–33	34–40	41–47	48–53	56–63

12
>63

Saffir – Simpson scale

1	2	3	4	5
64–82	83–95	96–113	114–135	>–135

NWS non-tropical

gale	storm	hurricane force winds
34–47	48–63	>63

NWS tropical

tropical storm	hurricane (typhoon)
34–63	>63

FIGURE 1 Beaufort and Saffir-Simpson wind speed scales, and the corresponding National Weather Service (NWS) warning designations. Wind speed ranges in knots are shown in the bottom of each box. 1 knot = 0.51 m/s = 1.85 km/h.

Figure 2 illustrates where the major storm systems occur. Tropical storms are limited to selected tropical oceans, and are virtually absent from the southern Atlantic Ocean and the southeastern tropical Pacific Ocean. They typically travel in westerly directions. The strongest non-tropical storm systems occur at higher latitudes north of 40°N and south of 40°S and typically travel in easterly directions. Such storms are far more common than tropical cyclones. Typically, approximately 1250 extratropical storms and 75 tropical systems are identified worldwide each year.

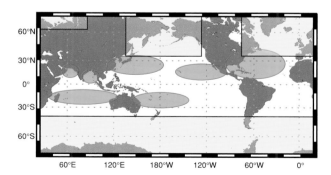

FIGURE 2 Areas dominated by high latitude storms (yellow areas) and areas with occurrence of tropical cyclones (red areas) based on data from the Goddard Institute for Space Studies (GISS/NASA) storm atlas and hurricane track data from the National Hurricane Center (NHC/NOAA) and the Joint Typhoon Warning Center (JTWC/U.S. Navy). Based on Figure 1 of Alves (2006).

WIND WAVES ON THE OCEANS: GENERATION AND PROPAGATION

A wind wave on the surface of the ocean can be described with three elementary parameters: its height H, its period T, and its length L. The wave height is defined as the vertical distance between consecutive water level maxima (crests) and minima (troughs). The length and period are defined as the distance or time between consecutive wave crests, or other periodic features of waves. The period and length are related in the dispersion relation. In deep water (depth $d > 0.5L$),

$$L = \frac{g}{2\pi}T^2 \qquad \text{(Eq. 1)}$$

where g is the acceleration of gravity, $g \approx 10$ m s^{-2}. The celerity of individual waves by definition is

$$c = L/T = \frac{g}{2\pi}T \qquad \text{(Eq. 2)}$$

Wave energy, however, travels with the so-called group velocity, which in deep water is half the wave celerity. The dispersion relation implies that waves with longer periods have longer lengths, and they travel faster.

In nature no two waves are exactly alike, and waves can be described only statistically. The wave height that is generally reported is the so-called significant wave height H_s. Traditionally, this wave height was defined as the average of the 33% highest waves. Because a trained observer generally ignores the smaller waves as being insignificant, this in fact closely corresponds to a visually observed mean wave height. The distribution of individual wave heights in a wave field can be described with the Rayleigh distribution, with the exception of the most extreme wave heights. The Rayleigh distribution implies that one in every 100 waves is expected to be as large as $1.5H_s$, and that the largest individual wave in a wave field is expected to be approximately twice the significant wave height.

In advanced observations and wave models, waves are not described individually, but with an energy spectrum. The spectrum describes the distribution of wave energy over a range of frequencies $f = T^{-1}$ and directions θ. From such a spectrum the wave energy E can be obtained by integration, and the corresponding significant wave height becomes

$$H_s = \frac{4}{\rho g}\sqrt{E} \qquad \text{(Eq. 3)}$$

Wind waves that are actively generated by the local wind are generally denoted as wind seas. The height, period, and length of wind waves are governed by three parameters: the wind speed u, the fetch x, which is distance over which the wind blows, and the duration for which the wind blows t. In shallow water, the depth d also influences wave growth. As a first approximation, deep water wave growth behavior can be approximated with empirical relations between the nondimensional wave height $\tilde{H}$, fetch $\tilde{x}$, and duration $\tilde{t}$, from which the actual wave height, fetch, and duration are computed as

$$H = \frac{\tilde{H}u^2}{g}, \ x = \frac{\tilde{x}u^2}{g}, \text{ and } t = \frac{\tilde{t}u}{g} \qquad \text{(Eq. 4)}$$

respectively. The wave height and fetch both scale with the square of the wind speed, whereas the duration scales linearly with the wind speed. This implies that a doubling of the wind speed quadruples the wave height, but also requires a fetch that is four times as long and a duration that is twice as long to reach this wave height.

Wave growth is typically addressed with the fetch-limited growth behavior of waves. Such conditions can be observed in nature for steady, obliquely offshore winds on a long, straight coastline, where x becomes the distance offshore. For short fetches observations typically indicate that

$$H \propto u\sqrt{\frac{x}{g}} \qquad \text{(Eq. 5)}$$

This implies that for a given fetch the wave height increases roughly linearly with the wind speed and with the square root of the distance to the shore. For long fetches where the wave field is fully developed and no longer varies with the fetch x,

$$H \approx 0.3\frac{u^2}{g} \qquad \text{(Eq. 6)}$$

When the winds subside, or when the waves travel out of their generation area, they travel along great circles until they encounter shallow water or a coastline. Such free-traveling waves are generally known as swell. In the generation area, waves are generated in a broad range of frequencies and directions. All these waves travel in their own direction and with their own speed. This dispersion process spreads wave energy over an increasingly large spatial domain, dramatically reducing wave heights of swell systems during propagation. Ignoring the curvature of the earth, this results in a reduction of the wave height that is inversely proportional to the distance to the generation area. However, very little of the swell energy is lost to dissipation, and swells have been observed to travel

halfway around the earth, from generation areas in the Indian Ocean to the West Coast of the United States.

WAVE HEIGHTS ON THE OCEANS

Figure 3 shows the mean significant wave height for the 2005 Northern and Southern Hemisphere winters as observed with space-borne altimeters on board of the Jason-1, GFO, and ENVISAT satellites. The largest mean wave heights occur in winter storm tracks at high latitudes identified in Fig. 2. This is a consequence of the fact that the wave height scales with the square of the wind speed and of the abundance of storm systems in these regions. Although winter conditions are similar in the Northern and Southern Hemisphere storm tracks, Northern Hemisphere summer wave conditions are much more quiescent than Southern Hemisphere summer wave conditions. Radiating from the storm track regions, swell systems cover all oceans. Even at high latitudes the wave fields are dominated by swell systems typically 75% of the time. Because swell systems predominantly travel in easterly

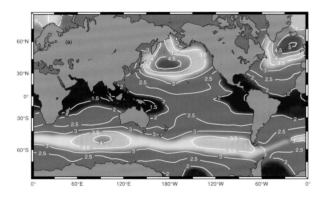

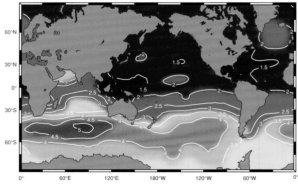

FIGURE 3 Mean significant wave heights in meters over the global oceans for December 2004 through February 2005 (top: northern hemisphere winter) and for June through August 2005 (bottom: southern hemisphere winter) as derived from the Jason-1, ENVISAT, and GFO altimeter observations. Data from the historical archive from the Naval Research Laboratory (NRL), processed at NOAA/National Centers for Environmental Prediction.

directions, eastern sides of basins are much more exposed to waves than western sides of basins.

Maximum mean significant wave heights in winter storm tracks are typically 5 m. For individual storms, however, wave heights can be much larger, with maximum observed significant wave heights of nearly 20 meters.

It should be noted that wave conditions in the deep ocean are transient in nature, severely limited by fetch and duration of winds. If winds corresponding to a category 3 hurricane could occur over an unlimited fetch with an unlimited duration, scaling laws suggest maximum significant wave height of up to 60 m, whereas in practice significant wave heights of more than 20 m are rarely if ever observed. The transient nature of the wave fields makes it difficult if not impossible to determine a physical upper limit of wave heights in the ocean.

WIND WAVES AT THE COAST

When wind seas or swell reach the coast, they encounter shallow water. This results in dissipation of wave energy in the bottom boundary layer, with decay length scales of the order of tens or hundreds of kilometers. More importantly, wave crests tend to turn parallel to the coast (refraction) and longer waves increase in height as a result of a local accumulation of wave energy related to changing propagation velocities of waves (shoaling). The refraction process tends to focus wave energy on headlands, although it tends to shelter coves and bays from wave penetration. The refraction process is similar to that of light, and it is therefore common terminology to designate such headland and bay areas as focus and shadow areas, respectively. Coastal currents, and particularly currents in tidal inlets and river mouths, can similarly result in focus and shadow areas of wave energy. The rate of variability of wave conditions along a coast depends strongly on the local bathymetry and on the local currents. This variability can be minimal on straight coastlines without clear offshore bathymetric features. An example of extreme variability of coastal wave conditions can be found on the northern shores of Oahu, Hawaii, where the local bathymetry results in the famous persistent surfing locations. This is illustrated in Fig. 4, with example model computations for this area provided by the commercial service Surfline. Offshore wave conditions to the north of Oahu are fairly uniform (map on the top for the waters around Oahu). In contrast, the near-shore wave heights are varying dramatically near the coast (map on the bottom for the North Shore). Being caused by the near-shore bathymetry, such wave height patterns are persistent, resulting

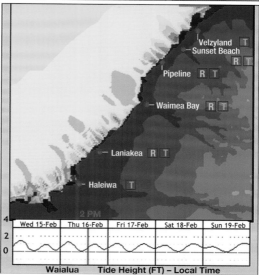

FIGURE 4 A model prediction of near-shore wave heights for the North Shore of Oahu, Hawaii, from http://www.surfline.com. Wave heights in feet.

in many famous surf beaches. Most coastlines display persistent local wave patterns or climates, depending on offshore wave conditions, the local bathymetry, and tides. Therefore, local knowledge based on observations, experience, or modeling is crucial in assessing coastal wave conditions.

SEE ALSO THE FOLLOWING ARTICLES

Currents, Coastal / Ocean Waves / Rogue Waves / Storms and Climate Change / Wave Forces, Measurement of / Wind

FURTHER READING

Alves, J.-H. G. M. 2006. Numerical modeling of ocean swell contributions to the global wind wave climate. *Ocean Modeling* 11: 98–122.
Coastal Data Information Program (CDIP). http://cdip.ucsd.edu.

Danielson, E. W., J. Levin, and E. Abrams. 2003. *Meteorology*, 2nd ed. Columbus, OH: McGraw-Hill.

Komen, G. J., L. Cavaleri, and M. Donelan. 1994. *Dynamics and modeling of ocean waves*. Cambridge, UK: Cambridge University Press.

National Centers for Environmental Prediction, Marine Modeling and Analysis Branch. http://polar.ncep.noaa.gov.

National Data Buoy Center. http://ndbc.noaa.gov.

National Hurricane Center. http://www.nhc.noaa.gov.

National Oceanic and Atmospheric Administration (NOAA). http://www.noaa.gov. Ocean Prediction Center. http://www.opc.ncep.noaa.gov.

STORMS AND CLIMATE CHANGE

PATRICK A. HARR
Naval Postgraduate School

Storms, or cyclones, in the atmosphere may be classified by their spatial scale, temporal scale, location, and physical mechanisms associated with their formation, structure, life cycle, and energetics. Most cyclones occur on the synoptic scale, which implies a space scale of thousands of kilometers and a temporal scale of days. Severe thunderstorms and tornadoes, on the other hand, occur on the mesoscale, which is a space scale of hundreds of kilometers and a time scale of hours. Synoptic-scale cyclones may be placed in two broad categories of tropical cyclones and extratropical cyclones. Climate change will affect many of the physical mechanisms associated with each cyclone classification, which may have an impact on their frequency, intensity, and geographical distribution.

STORM CHARACTERISTICS

Although the classifications of synoptic-scale cyclones imply a basic dependence on geographic location, there are many differences in the formation, structure, life cycles, and energetics associated with each synoptic-scale storm type.

Tropical Cyclones

The formation of a tropical cyclone is dependent on a complex interaction of factors that vary from large-scale pre-existing tropical atmospheric waves to transfer of heat and moisture at scales of individual cumulus clouds. Few atmospheric mechanisms are as poorly understood as those associated with tropical cyclone formation. Typically, a pre-existing disturbance that exists in a favorable atmospheric state will aid in the transfer of heat and

moisture from a warm ocean to the lower atmosphere. The moisture is transported upward in cumulus clouds, where release of latent heat contributes to a warming of the atmospheric column and reduction of surface pressure. The increased pressure gradient at low levels contributes to the increase in low-level winds. The release of latent heat and warming of the column also thickens the column, which contributes to an anticyclone at upper levels. Therefore, in the tropical cyclone low-level cyclonic inflow obtains heat and moisture from the ocean, which is transported vertically to where latent heat is released at midlevels, and the airflow exits in an anticyclonic outflow layer. Therefore, a tropical cyclone is commonly referred to as a warm-core cyclone.

Extratropical Cyclones

The formation of an extratropical cyclone is dependent on the conversion of potential energy to kinetic energy over large regions. Typically, there is an amount of potential energy that is available for conversion to kinetic energy, and this amount essentially depends on the temperature gradient between warm air at low latitudes and cold air at high latitudes. This general pole-to-equator temperature gradient is responsible for the mid-latitude westerly winds that increase in speed with height via the thermal wind relationship. The combination of the increase in westerly winds with height and the large-scale temperature gradient may lead to an instability in the flow that provides the mechanisms by which potential energy is converted to kinetic energy. In this process, warm, low-latitude air is transported poleward and upward, and cold, high-latitude air is transported equatorward and downward. The vertical displacement of each air mass contributes to the distribution of clouds and weather, and the latitudinal displacement contributes to the relaxation of the temperature gradient and westerly winds. Whereas a tropical cyclone is a warm-core system in which the column of warm air remains vertically upright over the surface low pressure center, an extratropical cyclone is a cold-core system with significant vertical tilt in which winds (cyclonic winds) increase with height.

GENERAL IMPACTS OF CLIMATE CHANGE

Due to the complexity of the earth's climate system, there will always be some uncertainty in understanding the mechanisms responsible for defining, maintaining, and altering the climate state. However, based on direct measurements of surface air temperatures and subsurface and surface ocean temperatures, it is now clear that the mean annual temperature at the earth's surface averaged over the entire globe has been increasing over the most recent 200 years. Although the increase in mean temperature is often attributed to many factors, the increases in concentrations of greenhouse gases (i.e., carbon dioxide, methane, tropospheric ozone, nitrous oxide) have been observed to closely coincide with the increase in mean surface temperature. Although there is debate over the relative roles of increased greenhouse gases and natural variability in the earth's climate system, it is generally accepted that there is a great need for increased understanding of the consequences of climate change. Different components of the climate system respond at varying rates and scales to changes in climate. These changes will have profound adverse and beneficial impacts at regional scales that will include changes in water distribution, ecosystems, and human health and well-being. However, it is often debated that rapid or accelerated changes in the climate system may be dominated by adverse impacts.

Although global climate encompasses many spatial and temporal scales, it may be argued that the climate-defined characteristics of tropical and extratropical cyclones may undergo drastic changes as the climate state shifts. Changes in cyclone frequency, longevity, and motion will impact nearly every region of the globe. Because formation, structure, energetics, and maintenance characteristics of extratropical and tropical cyclones differ, the impacts of climate change will vary for each cyclone type.

Extratropical Cyclones

Two factors have contributed to a recent increase in the understanding of climate change–induced variations in extratropical cyclone characteristics. Many studies have been conducted based on long-term integrations of global-scale numerical models. Recently, the accuracy of these models in replicating current climate conditions has been greatly increased by factors such as increased resolution over which the governing equations are integrated. Furthermore, availability of new datasets that provide global coverage of three-dimensional atmospheric conditions over multiple decades have led to studies of observed long-term changes in mid-latitude cyclone characteristics.

There has been a general agreement among numerically based simulations of changes in mid-latitude cyclone activity due to enhanced greenhouse gases. In a warming environment, the total cyclone density, which is defined as a number of cyclones per area per season, decreases over the Northern Hemisphere during December–January–February (DJF). Similarly, during June–July–August (JJA) there is a decrease in cyclone density over the Southern Hemisphere. The

change in cyclone density varies between a 5% to 10% reduction from current activity. However, when mid-latitude cyclone activity is examined with respect to intensity, there may be an increase in strong-cyclone activity over the Northern Hemisphere during summer. Also, there may be an increase in strong-cyclone activity during summer and winter over the Southern Hemisphere. The increase in strong-cyclone activity over the Northern Hemisphere is limited to the primary cyclone formation regions over the eastern coasts of Asia and North America. These are regions of locally enhanced low-level temperature gradients caused by the warm western ocean boundary currents adjacent to cold land masses. The increase in Southern Hemisphere strong-cyclone frequency occurs over the circumpolar region of Antarctica, where a strong temperature gradient occurs in association with the Antarctic continent and adjacent ocean regions. Over both hemispheres, there is a poleward shift in the location of the peak strong-cyclone activity. Recent studies of observed cyclone counts indicate that trends similar to those simulated in numerical experiments have occurred over recent decades.

The changes in cyclone activity are explained by examining the factors that are primarily responsible for cyclone formation and growth. Overall, mid-latitude cyclone activity is related to the mean baroclinicity of the atmosphere, which may be defined by the horizontal and vertical temperature distributions of the lower atmosphere. The growth of midlatitude cyclones can be expressed by a parameter defined as

$$\sigma = k|f|\frac{\partial|\vec{V}|}{\partial z}\left\{\frac{g}{\theta}\frac{d\theta}{dz}\right\}^{-1/2}$$

in which k is a constant; f is the Coriolis parameter, which accounts for latitude; V is the horizontal wind; z is the vertical height; g is the gravitational constant; and θ is potential temperature. Based on the foregoing expression, cyclone growth is dependent on the vertical change in wind (i.e., the $\partial|\vec{V}|/\partial z$ term), which is dependent on the horizontal temperature gradient as defined by the thermal wind relationship. Therefore, in a warming atmosphere the mean pole-to-equator temperature gradient decreases, which reduces the overall vertical change in wind. The equatorward temperature gradient is reduced because the high latitudes typically warm more than the tropical latitudes do. Therefore, the potential energy available for conversion to kinetic energy is generally reduced. However, some simulations have suggested that over the Southern Hemisphere the increase in air temperature over the circumpolar regions is very small, because of influences from extensive ocean coverage. This often results in an increase in the equatorward temperature gradient over the Southern Hemisphere. Therefore, over this region other contributions must contribute to the reduction in cyclone activity.

The second term in the foregoing expression (i.e., $\{(g/\theta)(d\theta/dz)\}^{-1/2}$ defines the vertical distribution of temperature, which governs the ability of air to move vertically. A warm lower atmosphere under a cold upper atmosphere will be less stable to vertical motion and contribute to enhanced vertical displacement of air masses. However, in a warming scenario, the upper troposphere warms more than the lower levels, which acts to decrease the vertical change in temperature and increase the stability with respect to vertical motion. Because there is more resistance to vertical motion, the potential for cyclone formation is reduced. As previously discussed, the small low-level temperature increases over the circumpolar region of the Southern Hemisphere act to maintain some equatorward temperature gradient. However, the maintenance of cool high-latitude air under a warming upper troposphere contributes to an increase in static stability and resistance to vertical motion. Therefore, in this region the negative impact to cyclone formation due to an increase in static stability overcomes the positive impact due to the maintenance of the equatorward temperature gradient.

Although the above reasoning explains the decrease in overall cyclone activity, the increase in strong-cyclone activity at higher latitudes appears to be more related to regionally enhanced changes in the atmospheric baroclinicity (e.g., ice/water boundaries, land/ocean boundaries that increase local temperature gradients). These results continue to be analyzed in great detail with more sophisticated numerical simulations that are required for identification of regional characteristics.

Tropical Cyclones

The understanding of the physical processes that lead to tropical cyclone formation and intensification lags far behind that of mid-latitude cyclone formation. Therefore, global-scale numerical simulations of tropical cyclone characteristics have not been very successful in that current tropical cyclone characteristics are often not accurately simulated. Recent examinations of potential changes in tropical cyclone activity due to climate change have compared changes in tropical cyclone characteristics to changes in parameters that are measured with high accuracy (i.e., sea surface temperature) and are closely related to tropical cyclones.

As discussed previously, tropical cyclone formation and maintenance are closely linked to sea surface temperatures. Recent studies have found a significant positive correlation between historical Atlantic tropical cyclone activity and sea surface temperatures over the western Atlantic Ocean. Furthermore, late-summer sea surface temperatures over the region of the North Atlantic in which the majority of tropical cyclones form are highly correlated over long time scales, with surface temperatures averaged over the entire Northern Hemisphere. Additionally, there has been substantial warming in both sea surface temperature and total Northern Hemisphere surface temperature during recent decades. Although the accuracy of the sea surface temperature data has been well established, there is much debate in the quality of historical tropical cyclone records that would allow an accurate representation of past tropical cyclone characteristics. Whereas recent increases in tropical cyclone activity have occurred over the Atlantic Ocean, there have not been corresponding increases in other ocean basins that contain tropical cyclones (i.e., western North Pacific, western South Pacific, Indian Ocean, and eastern North Pacific). Therefore, the total number of tropical cyclones globally has remained rather constant, and other important factors may therefore be governing tropical cyclone frequency. For example, there is considerable uncertainty as to the relative roles of long-term (i.e., decadal) variations and periodicities on increasing sea surface temperatures. Periodic oscillations in global-scale ocean circulations have been associated with changes in sea surface temperatures consistent with those identified with the recent increase in Atlantic tropical cyclone activity.

There are many unanswered questions with respect to potential changes in tropical cyclone intensity and tracks in a warming atmosphere. Some studies have suggested that the expected increase in sea surface temperatures due to increased greenhouse gases will act to increase average wind speeds in tropical cyclones by 5–10%. These types of changes may be evident in recent data because an increase in strong tropical cyclones over recent decades has been identified. However, there is some doubt as to the accuracy of historical data with respect to reports of tropical cyclone intensity in the presatellite era. Many observations of tropical cyclone intensity were not recorded until the storm was near shore or at landfall. Observations would be made when land influences were acting to reduce the tropical cyclone intensity. This could have caused a low bias in reported storm intensities. When satellite data became available, techniques were developed to infer intensities from cloud patterns. Therefore, the low bias in historical records together with new observations of storms at peak intensity may contribute to an artificial increase in recorded storm intensities.

Vertical wind shear is the most important atmospheric control on tropical cyclone intensity. Strong vertical wind shear prevents the vertical development of deep cumulus clouds required to provide the latent heat release for the tropical cyclone. Therefore, a decrease in vertical wind shear as discussed previously may have important impacts on tropical cyclone intensities.

Although the foregoing information summarizes the tropical cyclone activity as a response to a changing climate, tropical cyclones may also play a more active role in the climate system. Tropical cyclones make a potentially important contribution to the meridional heat transport by the oceans. Therefore, increased tropical cyclone activity could substantially increases the poleward heat transport by the ocean, which would contribute to the reduced equator-to-pole temperature gradient that was discussed in relation to extratropical cyclones. Increased poleward oceanic heat transport would warm high latitudes and moderate the warming in tropical latitudes. This reduced temperature gradient would not only contribute to the distribution of extratropical cyclones but could moderate projected increases in tropical cyclone intensity.

Whereas numerical simulations of climate change–induced variations to extratropical cyclones have provided some consensus on future scenarios, questions on observations of historical tropical cyclone activity and the remaining unknowns with respect to important mechanisms responsible for tropical cyclone formation, intensity, and intensity changes continue to fuel debate on impacts to tropical cyclones due to an increasing global temperature.

The observed increase in global mean atmospheric temperatures will have impacts on many ecosystems. In many instances, the impacts will be due to changes in atmospheric conditions over a wide variety of temporal and spatial scales. There are many unknowns with respect to the impacts of a changing climate on atmospheric features such as mid-latitude and tropical cyclones. Fortunately, new data sources and increased accuracy of numerical simulations continue to provide answers to important questions that will aid in the understanding of the impacts on storms due to climate change.

SEE ALSO THE FOLLOWING ARTICLES

Climate Change / El Niño / Heat and Temperature, Patterns of / Storm Intensity and Wave Height / Wind

FURTHER READING

AMS. 2003. Climate change research: issues for the atmospheric and related sciences. *Bulletin of the American Meteorological Society* 84: 508–515.

Bengstsson, L., K. Hodges, and E. Roeckner. 2006. Storm tracks and climate change. *Journal of Climate* 19: 3518–3543.

Emanuel, K. 2005. Increasing destructiveness of tropical cyclones over the past 30 years. *Nature* 436: 686–688.

Geng, Q., and M. Sugi. 2006. Possible change of extratropical cyclone activity due to enhanced greenhouse gases and sulfate aerosols—study with a high-resolution AGCM. *Journal of Climate* 16: 2262–2274.

Klotzbach, P. J. 2006. Trends in global tropical cyclone activity over the past twenty years (1986–2005). *Geophysical Research Letters* 33: L10805; doi:10.1029/2006GL025881.

Mann, M., and K. Emanuel. 2006. Atlantic hurricane trends linked to climate change. *Eos* 87: 233–244.

Schubert, M., J. Perlwitz, R. Blender, K. Fraedrich, and F. Lunkeit. 1998. North Atlantic cyclones in CO_2-induced warm climate simulations: frequency, intensity, and tracks. *Climate Dynamics* 14: 827–837.

Simmonds, I., and K. Keay. 2000. Variability of southern hemisphere extratropical cyclone behavior, 1958–97. *Journal of Climate* 13: 550–561.

Wang, X. L., V. Swail, and F. W. Zwiers. 2006. Climatology and changes of extratropical cyclone activity: comparison of ERA-40 with NCEP-NCAR reanalysis for 1958–2001. *Journal of Climate* 19 3145–3166.

Webster, P. J., G. J. Holland, J. A. Curry, and H. R. Chang. 2005. Changes in tropical cyclone number, duration, and intensity, in a warming environment. *Science* 309: 1844–1846.

STRESS

SEE COLD STRESS, DESSICATION STRESS, HEAT STRESS, SALINITY STRESS, ULTRAVIOLET STRESS

STROMATOLITES

PIETER T. VISSCHER

University of Connecticut

Stromatolites are layered, organosedimentary structures built by prokaryotic communities that bind and trap sediments and precipitate carbonate minerals. Stromatolites formed extensive reefs during the Precambrian but started to decline in the Paleozoic. Examples of modern marine stromatolites, which resemble the abundant fossil ones, are still found in the subtidal and intertidal zone in the Bahamas and Western Australia. Microorganisms at the surface of these stromatolites construct a microbial mat and are responsible for building the characteristic laminated sedimentary structure.

STROMATOLITE STRUCTURE

The distinct layering visible in cross section is characteristic of stromatolites (Fig. 1). This recurring layering represents

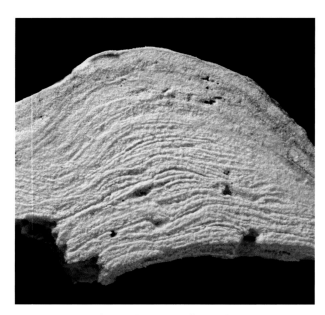

FIGURE 1 Open modern marine stromatolites (Highborne Cay, Exumas, Bahamas) in cross section. The section is approximately 6 cm across. Reprinted from Reid *et al.* (2000) by permission from Macmillan Publishers Ltd.

Western Australia, and Storrs Lake, San Salvador, Bahamas. Reports of freshwater specimens exist for a number of hard-water lakes (e.g., Pyramid Lake, Nevada, and Green Lake, New York, in the United States; Lago Sarmiento, Chile; Pavilion Lake and Kelly Lake, British Columbia, Canada), although often microbialites are misinterpreted as stromatolites.

FIGURE 2 Modern marine stromatolites viewed *in situ* underwater at Highborne Cay, Exumas, Bahamas. Dive knife (yellow, left for scale). Reprinted from *Reid et al.* (2000) by permission from Macmillan Publishers Ltd.

a record of biomineralization that distinguishes stromatolites from the other, less organized microbialites (the latter is a general designation used for lithified prokaryotic formations). The term stromatolite is a combination of *stroma*, which means "layer" (Greek) or "mat" (Latin), and *lithos*, "rock" (Greek). The mineralogical composition is predominantly calcium carbonate, although quartz and manganese oxide forms can be found as well. Ancient stromatolites, which date back approximately 3.5 billion years in the rock record, lack fossilized microorganisms or evidence of trapped and bound coarse sediments. Nevertheless, stromatolites represent the oldest macroscopic evidence of life: Ample signs for the biogenic nature of the modern analogs and the antiquity of the major stromatolite-building microbes make it likely that both have been formed by similar processes.

Modern marine stromatolites are found around Exuma Sound, Bahamas, in subtidal (Fig. 2) and intertidal environments in which they are exposed to extreme hydrodynamic conditions. These analogs of Earth's oldest biofilms are characterized by alternating hard layers, consisting of microcrystalline calcium carbonate, and soft layers, formed by bound and trapped ooids (spherical carbonate grains). Contemporary hypersaline examples of stromatolites are found in the intertidal and subtidal of lagoons in Shark Bay and Hamelin Pool,

STROMATOLITES AND MICROBIAL MATS

Contemporary, nonlithifying microbial mats resemble stromatolites, although the characteristic lamination in both has a different origin: Layering in the latter results from cycles of biomineralization, whereas in the former is the product of the community structure. Fossilized microbial mats (2.6 billion years old) are also preserved in the rock record, but are much less common than stromatolites.

Microbial mats are found in a variety of modern environments that are often viewed as extreme (as on early Earth), including hot springs and hypersaline environments. Intertidal environments are extreme because of frequent desiccation, rapid temperature and salinity fluctuations, potentially low nutrient conditions, high flow rates of the overlying water, and so on. These extreme conditions exclude many eukaryotes (grazers and macroalgae) and present ideal conditions for microbial mat formation. The lamination in mats results from the light quality and quantity and the sulfide concentration in the sediment.

Cyanobacteria (blue-green layer) are present near the surface, often covered by 0.2–0.8 millimeter of sediment and exopolymers (EPS) and use the blue and green parts of the spectrum. Occasionally, diatoms are observed near the surface as well. Underneath the cyanobacteria are anoxyphototrophs: purple and green (non)sulfur bacteria (red and green layers), which are adapted to a different part of the electromagnetic spectrum and are able to survive under higher sulfide concentrations and lower light. The deeper layers of the mat lack active photosynthesis and are often black as a result of FeS formation, which results from sulfate reduction activity.

Although most intertidal mats do not lithify, the microbial mats at the surface of the stromatolites form an exception. In modern, open marine stromatolites, the lamination is due to three different surface mat types, which form three different structural signatures: The first type is a mat that is characterized by thin filamentous cyanobacteria, which trap and bind sediments; the second type is a more developed microbial community in which heterotrophs (especially sulfate-reducing bacteria) play a key role, resulting in a thin layer of microcrystalline calcium carbonate; in the third type, additional colonization by microboring coccoid cyanobacteria results in reworking of ooids, cementing these together.

MECHANISM OF CALCIUM CARBONATE PRECIPITATION

The precipitation of calcium carbonate in stromatolites and other microbialites is accomplished through the alteration of the saturation index of calcium carbonate (SI), which is determined by the ratio of the ion activity product (IAP) and the solubility product constant (K_{sp}). If IAP > K_{sp}, precipitation may occur. Various types of microbial metabolism increase the SI (e.g., photosynthesis, sulfate reduction), enabling precipitation, whereas others decrease the SI, facilitating dissolution (e.g., aerobic respiration, sulfide oxidation). The magnitude and location of the combined metabolic processes will ultimately determine when and where dissolution or precipitation of calcium carbonate will occur. For example, photosynthesis and aerobic respiration are both coupled to light and their effects on SI cancel each other out, whereas sulfate reduction continues throughout the diel cycle and, as a result, has a major role in net carbonate precipitation.

The matrix of exopolymeric substances (EPS) is another key factor in calcium carbonate precipitation: Negatively charged carboxyl groups in newly formed EPS scavenge calcium ions, initially inhibiting precipitation. Degradation of EPS later on produces free calcium and carbonate ions, increasing the local SI and providing a nucleus for mineral precipitation. In marine and hypersaline environments, the precipitation takes place in the EPS matrix, not on the cyanobacterial sheaths. EPS has many roles in the mat: It protects microbes against stressors, including UV, and may play a role in binding and trapping of sediments.

Fully understanding how modern stromatolites form will provide a unique perspective of how life arose on earth and how it shaped our planet: Stromatolite-building cyanobacteria first oxygenated the atmosphere and changed the earth's redox conditions. This enabled diverse life to develop in the intertidal and deeper waters as well as on land.

SEE ALSO THE FOLLOWING ARTICLES

Biofilms / Fossil Tidepools / Hydrodynamic Forces / Microbes

FURTHER READING

Awramik, S. M. 2006. Respect for stromatolites. *Nature* 441: 700–701.
Dupraz, C., and P. T. Visscher. 2005. Microbial lithification in modern marine stromatolites and hypersaline mats. *Trends in Microbiology* 13: 429–438.
Grotzinger, J. P., and A. H. Knoll. 1999. Stromatolites in Precambrian carbonates: evolutionary mileposts or environmental dipsticks? *Annual Review of Earth and Planetary Sciences* 27: 313–358.
Reid, R. P., P. T. Visscher, A. W. Decho, J. F. Stolz, B. M. Bebout, C. Dupraz, I. G. Macintyre, H. W. Paerl, J. L. Pinckney, L. Prufert-Bebout, T. F. Steppe, and D. J. Des Marais. 2000. The role of microbes in accretion, lamination and early lithification of modern marine stromatolites. *Nature* 406: 989–992.
Riding, R. E., and S. M. Awramik. 2000. *Microbial sediments.* Berlin: Springer-Verlag.

SUCCESSION

STEVEN N. MURRAY

California State University, Fullerton

Succession is the sequence of species' colonizations and replacements that occur in biological communities. When initial colonization takes place on fresh, virgin substratum, the successional sequence is known as primary succession. Usually, however, colonizations occur on newly cleared substratum that has previously supported biological communities, a process known as secondary succession.

SUCCESSION ON ROCKY SHORES

In rocky intertidal habitats, new marine substratum is created by the immersion of breakwater rocks, volcanic lava flows, exposure of new surfaces by the fracturing of rocks, or the erosion of a coastal bluff that results in the deposition of rocks from long-standing terrestrial formations. Most succession on rocky shores takes place after previously occupied substratum has been cleared or overturned by physical or biological disturbance events such as the ripping away of seaweeds or mussels by storms or predation episodes that kill organisms. Succession on rocky shores also occurs after large human-generated catastrophes such as oil spills, when portions of the substratum might even be sterilized by oil removal procedures. At any rate, once freed substratum is made available on rocky shores, colonization by marine organisms will take place, and succession will proceed toward what has historically been referred to as a mature or climax community. Community development, however, can be interrupted by mortality-creating physical and biological disturbances, which occur continuously on rocky shores and create opportunities for early colonists to recruit onto successional surfaces. The rate at which succession proceeds, then, is a reflection of the frequency and scale of interrupting disturbances and the vagaries associated with the recruitment of larvae and reproductive propagules. As communities progress toward more mature states, multiple, dynamic, or what have been referred to as stable end points can develop within the context of physical limitations imposed by tidal gradients and other environmental parameters.

COLONIZATION SEQUENCES

The first studies of succession in intertidal habitats were largely descriptive reports that emphasized the nature of species replacements during colonization sequences. These reports characterized the need for microbial development on virgin substrata, such as lava, prior to colonization by macroscopic intertidal organisms. Although the types of early arrivals vary in space and time depending on larval and propagule availability, the early appearance of diatoms and short-lived, rapidly growing seaweeds, such as leafy green (ulvoids) and small red and brown algae, has been a common finding of most successional studies performed in rocky-intertidal habitats (Fig. 1). These and other species with similar biological attributes, such as rapid growth rates, short life spans, high and frequent reproductive output, and large dispersal distances, have been referred to as pioneer or opportunistic species and are predictable colonizers of newly available surfaces. As

FIGURE 1 Small, fast growing, opportunistic green seaweeds on frequently disturbed rocky-intertidal surfaces. Here the disturbance mechanism is frequent sand scour. Photograph by the author.

succession proceeds, these opportunists generally become less abundant, and larger, usually slower-growing species with longer life spans become more abundant (Fig. 2). These later successional species usually produce fewer numbers of propagules or larvae, reproduce less frequently, tend to be stronger competitors, and generally become the eventual space dominants in mature rocky-intertidal communities.

Time ⟶

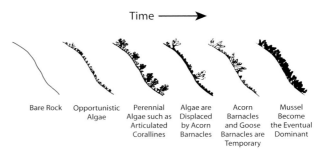

Bare Rock | Opportunistic Algae | Perennial Algae such as Articulated Corallines | Algae are Displaced by Acorn Barnacles | Acorn Barnacles and Goose Barnacles are Temporary | Mussel Become the Eventual Dominant

FIGURE 2 California mussel succession in the Pacific Northwest (after Dayton 1973). Cleared space is first colonized by small, opportunistic seaweeds, which are then replaced by larger, perennial seaweeds such as articulated coralline red algae. These perennial seaweeds are eventually replaced by larger, acorn barnacles, which overgrow and displace the algae. Gooseneck barnacles can then become established, but eventually the space becomes dominated by the California mussel (*Mytilus californianus*). Figure modified and redrawn from Carefoot (1977).

MECHANISMS OF SUCCESSION

Models

Three alternative models of species interactions during succession were described in a classic 1977 paper by J. H. Connell and R. O. Slatyer (Fig. 3). These are (1) facilitation, (2) tolerance, and (3) inhibition. The

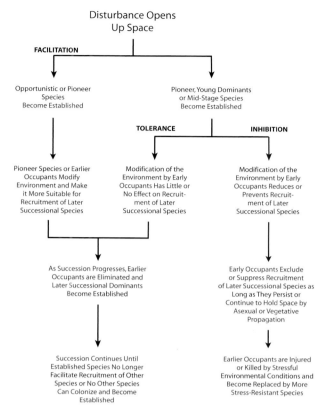

FIGURE 3 Three models of succession showing underlying mechanisms of species replacements on disturbed substratum. Figure modified from Connell and Slatyer (1977).

species that colonize substrata and are involved in successional sequences can be viewed as facilitators, tolerants, or inhibitors. In the facilitation model, early colonizing species modify the microenvironment and make it more suitable for later colonizers, which then recruit onto successional surfaces and ultimately replace them. Tolerant species are thought to be slower growing, persistent species, which are not negatively or positively affected by the initial colonizers but which eventually come to dominate a successional surface by outlasting them. In the inhibition model, early colonizers actually preempt space and resources and prevent the establishment of other species, which can colonize a successional surface only after the inhibitors become unhealthy or live out their life spans and are removed. Although Connell and Slatyer's models are thought by some to be oversimplifications because the described mechanisms can operate together in the same stage of a successional sequence (e.g., one species might facilitate another while simultaneously inhibiting a third) or can change from sequence to sequence (e.g., early colonizers might facilitate the arrival of later colonizers, which

in turn inhibit the recruitment of other species), they have proven to be useful tools for designing and interpreting successional experiments.

Application in Rocky Intertidal Habitats

Historically, most experimental successional work, including tests of successional mechanisms, has been carried out in terrestrial systems. Few experimental studies in rocky-intertidal habitats have been able to provide clear evidence for Connell and Slatyer's facilitation model, whereas several studies have supported the inhibition model and also demonstrated the importance of grazers in determining successional sequences; experimental evidence for the tolerance model is weak or lacking in rocky-intertidal communities.

Experimental tests of inhibition involve removing species and are simple to design. Species that inhibit colonization are thought to do so by preempting space and denying the ability of spores or larvae of other species to settle onto suitable substratum. For example, the presence of early colonizers in rocky-intertidal habitats, usually fast growing ephemeral algae, can inhibit the colonization of algae more characteristic of mature communities. Grazers can impact succession depending on the species they consume. If grazers preferentially consume early-colonizing, ephemeral algae, they can remove inhibition and, by allowing the arrival of later successional species, can accelerate succession. Conversely, the pace of succession can be slowed and surfaces dominated by early colonizers if grazers prefer to consume later successional species.

Facilitation of successional sequences has been demonstrated in several cases in rocky-intertidal systems. Microbial films have long been thought to facilitate succession, and the soaking of artificial surfaces in seawater to generate such films is routinely done prior to placing spore and larval collecting devices in the field. However, evidence for the role of microbial films in facilitating succession on rocky shores is based largely on observations, not experimental manipulation. In contrast, several studies have shown that certain macroorganisms can facilitate the development of other macroorganisms on rocky shores. Examples include the development of mussels, surf grasses, and algae in a barnacle–seaweed zone. Throughout the world, mussel species are known to attach preferentially to filamentous algae, byssal threads of conspecifics, other filament-like bodies, or barnacles but rarely colonize surfaces without these facilitating structures. Attachment and germination of the seeds of a surf grass on the western coast of North America are promoted by the microhabitat provided by an ephemeral, opportunistic seaweed that readily colonizes newly cleared surfaces. Barnacle cover, whether provided by artificial barnacle models or by living organisms, can reduce grazing pressure and has been shown to be necessary for algal recruitment on Oregon shores.

SCALE OF DISTURBANCE AND ALTERNATIVE PATTERNS OF COMMUNITY DEVELOPMENT

Disturbances create colonization opportunities and initiate succession in rocky-intertidal communities. Larval and spore availability and the spatial scale and timing of disturbances influence the resultant successional patterns, species interactions, and community structures.

Importance of Recruitment

Once a patch of substratum is opened, a critical step for colonization by a given species is that its larvae or reproductive propagules are available and are able to reach the cleared area. When larval settlement is low, lower adult densities usually will be realized, fewer competitive interactions will take place, and more open space will be available during early successional stages. Thus, the availability of larvae or propagules at the time of a clearance can have a significant effect on colonization patterns and subsequent successional sequences and contribute to spatial variation in intertidal communities on what appear to be physically similar shores. In recent years, increased emphasis has been placed on understanding the role of variations in recruitment, or what is referred to as supply-side ecology, in determining these patterns.

Alternative Community Outcomes

According to theory, successional patterns move toward community types representing mature or climax community states. Several studies have shown that alternative community outcomes can occur on rocky shores following patch clearance, indicating that successional patterns are time dependent and transitory. For example, in a 1981 study, Tom Suchanek showed that the size of a patch opened by disturbance affects the structure of the developing community. On western shores of North America, physical disturbances continuously open patches of substratum within California mussel *(Mytilus californianus)* beds. Smaller patches can be filled in by the immigration of adjacent California mussels, but mussel movements are too slow to fill in larger patches, which become colonized by seaweeds and another mussel species *(Mytilus trossulus)*. Suchanek argued that the tendency for *M. trossulus*, an inferior competitor to *M. californianus*, to recruit in winter was an adaptation

to patch availability in California mussel beds caused by more frequent winter storms.

On northeastern American shores, areas observed to be physically similar are covered in some cases by barnacles and mussels and in others by large growths of fucoid algae. These different community types are thought to result from year-to-year variation in algal, barnacle, and mussel recruitment and grazer densities following large-scale clearances of substratum caused by ice scour. A series of experimental clearances of different sizes, described in a 1999 study by Peter Petraitis and Steve Dudgeon, demonstrated that these different outcomes (barnacle–mussel or seaweed domination) were dependent on patch size.

Rare Events

Rare events producing large-scale disturbances can have important community consequences. In a 1999 study of recovery following an unusually strong storm in southeastern Australia, Tony Underwood found that the large-scale removal of seaweed canopies severely altered predation intensity, an event that affected resultant intertidal community structure for decades.

DIVERSITY AND SUCCESSION

Patches of substratum available for colonization and the sequence of species replacements can be interrupted by disturbances, which occur frequently in rocky-intertidal environments. Such disturbances return developing communities to earlier successional stages by causing mortality of established species (Fig. 4). When disturbances are infrequent or weak, fewer patches are available for colonization and successional processes are less interrupted. When disturbances are common or large, however, more colonization opportunities are provided, and shores often become dominated by opportunistic species (such as small, fast growing algae and barnacles) that characterize early stages of community development.

Compensatory Mortality and the Intermediate Disturbance Hypotheses

Disturbances can affect local patterns of species diversity by affecting the competitive dominance of late successional species through mechanisms referred to as compensatory mortality and intermediate disturbance. In the case of compensatory mortality, higher species diversity can be realized following a disturbance if a dominant competitor suffers greater mortality than the species it would otherwise exclude. This provides opportunities for inferior competitors to co-occur with the competitive dominant

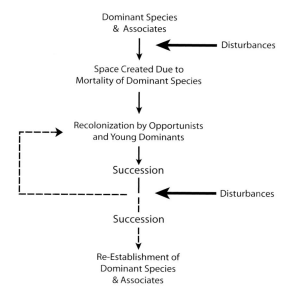

FIGURE 4 Interruption of successional sequences. When disturbances persist, succession is interrupted and instead of proceeding to stages dominated by perennial, later successional species, rocky-intertidal surfaces return to earlier stages, which usually are dominated by fast growing, rapidly colonizing, opportunistic species. Figure modified from Connell (1975).

as long as such disturbances occur with a frequency greater than the rate of exclusion. The second hypothesized mechanism by which increased species diversity can result on successional substrata results from intermediate levels of disturbance. This occurs when the disturbing agent is nonselective and intermediate in intensity, allowing species that would otherwise be outcompeted to recruit and persist together with the competitive dominants. If the disturbance is too severe, species are eliminated and diversity is reduced; if disturbances are too weak, competitive interactions will lead to the replacement of the less competitive species, which also results in lower diversity. Thus, greatest species diversity is hypothetically achieved at intermediate levels of disturbance, where habitats are occupied by mixtures of early and late successional species with different competitive abilities.

Intermediate Disturbance and Intertidal Boulder Fields

Studies performed by W. Sousa examined successional events and mechanisms and tested the applicability of the intermediate disturbance hypothesis (Fig. 5) in a Southern California intertidal boulder field. Here, the major agent of disturbance was wave energy, which rolled and overturned boulders. Larger boulders of greater mass were less frequently overturned than smaller boulders, which in turn were more frequently overturned than boulders of intermediate size.

Species living on the top surfaces of boulders are killed or injured when they are overturned, and the longer a boulder remains overturned, the greater is the space that becomes available for new colonists when the boulder is righted. Sousa found that on larger boulders, which rarely were overturned and disturbed, cover was dominated by a longer lived, turf-forming seaweed; these boulders represented late stages of succession in this habitat. For smaller boulders the interval between disturbance events was short, and the frequent rolling and turning maintained boulders at early successional stages that were dominated by only a few, mostly opportunistic species. In contrast, boulders of intermediate size and that received intermediate levels of overturning represented middle stages of succession, which were characterized by the most diverse assemblages of organisms. This is one of the best examples of how intermediate levels of disturbance can result in higher levels of species diversity in rocky-intertidal habitats through the maintenance of populations representing early, middle, and late stages of colonization.

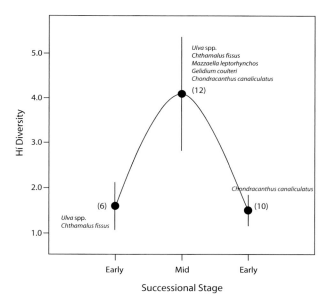

FIGURE 5 Intermediate disturbance increases diversity. Species diversity at early, middle, and late stages in an intertidal boulder field in California. Source: Sousa (1979a, b, 1980). Early successional stages were dominated by fast growing green seaweeds (*Ulva* spp.) and small, acorn barnacles *(Chthamalus fissus)*. These same species were common during middle stages of succession along with species of larger, fleshy, red algae [*Gelidium coulteri, Mazzaella leptorhynchos* (as *Gigartina leptorhynchos*), and *Chondracanthus canaliculatus* (as *Gigartina canaliculatus*)]. Early colonists were rare during late successional stages, which were dominated by a single species of fleshy, red seaweed *(Chondracanthus canaliculatus)*. Diversity (as measured by *H'*, the Shannon Weaver diversity index) was greatest at middle stages of succession. Note: Numbers in parentheses represent numbers of recorded taxa per successional stage.

SEE ALSO THE FOLLOWING ARTICLES

Facilitation / Habitat Alteration / Recruitment

FURTHER READING

Carefoot, T. 1977. *Pacific seashores: a guide to intertidal ecology.* Seattle: University of Washington.

Connell, J. H. 1975. Some mechanisms producing structure in natural communities, in *Ecology and evolution of communities.* M. L. Cody and J. M. Diamond, eds. Cambridge, MA: The Belknap Press of Harvard University, 460–490.

Connell, J. H., and R. O. Slatyer. 1977. Mechanisms of succession in natural communities and their role in community stability and organization. *American Naturalist* 111: 1119–1144.

Dayton, P. K. 1973. Dispersion, dispersal, and persistence of the annual intertidal alga, *Postelsia palmaeformis* Ruprecht. *Ecology* 54: 433–438.

Petraitis, P. S., and S. R. Dudgeon. 1999. Experimental evidence for the origin of alternative communities on rocky intertidal shores. *Oikos* 84: 39–45.

Raffaelli, D., and S. Hawkins. 1999. *Intertidal ecology,* 2nd ed. Dordrecht: Kluwer.

Sousa, W. P. 1979a. Disturbance in marine intertidal boulder fields: the non-equilibrium maintenance of species diversity. *Ecology* 60: 1225–1239.

Sousa, W. P. 1979b. Experimental investigations of disturbance and ecological succession in a rocky intertidal algal community. *Ecological Monographs* 49: 227–254.

Sousa, W. P. 1980. The responses of a community to disturbance: the importance of successional age and species life histories. *Oecologia* 45: 72–81.

Suchanek, T. H. 1981. The role of disturbance in the evolution of life histories in the intertidal mussels *Mytilus edulis* and *Mytilus californianus.* *Oecologia* 50: 143–152.

Turner, T. 1983. Facilitation as a successional mechanism in a rocky intertidal community. *American Naturalist* 121: 729–738.

Underwood, A. J. 1999. Physical disturbances and their direct and indirect effect: responses of an intertidal assemblage to a severe storm. *Journal of Experimental Marine Biology and Ecology* 232: 125–140.

SURFACE TENSION

MARK W. DENNY

Stanford University

Surface tension is a property of the interface between a liquid and a gas or between a liquid and a solid. At the edge of the ocean, the surface tension of water in contact with air provides support for insects that walk on water, and the surface tension of water in contact with rocks, plants, and animals creates capillary forces that act both as a pump and as an adhesive.

GENERAL PROPERTIES OF SURFACE TENSION

A water molecule consists of two hydrogen atoms bound to an oxygen atom: H_2O. The electrons in the molecule are attracted more to the oxygen atom than to the hydrogen atoms, and the hydrogen atoms are not both in the same line with the oxygen atom, with the result that oxygen's side of the molecule has a slight negative charge, and the hydrogens' side of the molecule has a slight positive charge. This polar nature of the water molecule in turn results in the attraction of water molecules to each other. The positive charges attract the negative charges to form hydrogen bonds. The effect of these bonds depends on the temperature. When the temperature of the water is above its boiling point, the strength of the hydrogen bonds is insufficient to hold molecules together, and water exists as water vapor, a gas. When water's temperature is below its boiling point but above its freezing point, attraction among molecules is sufficient to hold water together as a liquid. Below the freezing temperature, hydrogen bonds hold water molecules fixed to each other as ice. It is water in its liquid state that concerns us here.

At the interface between liquid water and air, water molecules are more attracted to each other than they are to the air outside. As a result, water tends to minimize its area of contact with air. Only if some energy is provided can new area of interface be created. For example, consider the apparatus shown in Fig. 1: A thin film of water is stretched across a wire frame, and one side of the frame (with length L) can slide to vary the area of water exposed to air. If we apply a force F to this sliding side and move it a distance D, we create new area of air–water interface equal to LD on each side of the frame. Because the film has both a front and a back side, the total amount of new interface is $2LD$. The energy we have expended in moving the slide is the force we have applied times the distance through which the slide has moved: FD. Thus, the energy expended per area created is $FD/(2LD) = F/(2L)$. When we conduct this experiment on pure water at room temperature, we find that 72.8 millijoules of energy must be expended to create each new square meter of air–water interface. This is the

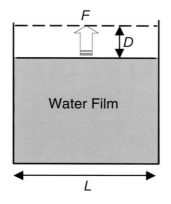

FIGURE 1 A hypothetical experiment to measure the surface energy of water. A film of water is stretched across a wire frame, and a measured force increases its area. See the text for details of the measurement.

surface energy of water in contact with air. Note, however, that this surface energy, $F/(2L)$, is equal to a force divided by a length, the same units one would use to express the tension created when a rope is stretched. In other words, surface energy is the same thing as surface tension.

The surface tension of seawater is slightly greater than that of pure water: at room temperature (20 °C) it is 73.5 millijoules per square meter (or equivalently, 73.5 millinewtons per meter). Surface tension of both pure water and seawater decreases lightly with increasing temperature. The surface tension of seawater is 76.4 millijoules per square meter at 0 °C and 70.7 millijoules per square meter at 40 °C.

Just as water has a surface energy, so do the surfaces of the solids with which it comes into contact at the ocean's edge. An indication of the magnitude of a solid's surface energy relative to that of water can be gained by placing a drop of water on the solid, and measuring the angle with which the water meets the solid (Fig. 2). Some solids (most rocks, for example) have a higher surface energy than water. In this case, the contact angle is less than 90°, and the energy of the water–rock system is reduced if the water spreads over the rock (Fig. 2A). If the surface energy of the water is greater than that of the solid (as it would be with wax, for instance), the contact angle is greater than 90°, and water is repelled by the solid (Fig. 2B).

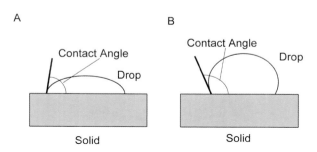

FIGURE 2 The relative surface energy of a solid in contact with water is indicated by the contact angle of a water drop. If the surface energy of the solid is larger than that of water (A), the contact angle is less than 90°. If the surface energy of the solid is less than that of water (B), the contact angle is greater than 90°.

WALKING ON WATER

Small organisms can take advantage of surface tension to walk on water. For example, the water striders *Halobates* and *Hermatobates* (members of the insect family Gerridae) are commonly found skating on the surface of tidepools throughout the tropics (Fig. 3). The physical basis for this type of locomotion is explained in Fig. 4, where the leg of a water strider is shown in cross section. Microscopic hairs on the surface of the leg trap air and give the leg a

FIGURE 3 A typical water strider. Note the dimple in the water's surface around each leg. Photograph © Andreas Just.

very low effective surface energy. As a result, the contact angle between water and leg is high (about 110°), and the surface tension of the water pulls up and out on the leg. It is this upwardly directed force (about 34% of the overall force) that resists the weight of the water strider, thereby allowing the insect to stand on the air–water interface.

Recall that units of surface tension are force per length. Therefore, by multiplying surface tension by the length of the perimeter with which the legs meet the water, we can calculate the force with which the insect is buoyed. A typical water strider has a perimeter of about 6 millimeters per foot, 36 millimeters overall. Multiplying this length by the surface tension of room-temperature seawater (73.5 millinewtons per meter), we find that the overall force acting on the strider's legs is roughly 27 millinewtons, of which 9 millinewtons is directed upward. In contrast, a typical water strider has a mass of 10 milligrams, and therefore has a weight of only 0.1 millinewton. Thus, the upward force provided by surface tension is more than sufficient to keep these small insects afloat.

However, walking on water works only for small organisms. For an animal whose shape is constant, a doubling in length results in a doubling of the perimeter of leg in contact with water but an eightfold increase in weight. Because of this problem with scaling, insects with masses greater than about 300 milligrams cannot walk on water, and those that approach this size have disproportionately long legs.

The interaction between the weight of a water strider and the water's surface tension forms a dimple in the water's surface (Fig. 4). If the strider swings its legs backward, this dimple is dragged across the water, and the force of moving the dimple propels the water strider forward. In effect, the legs of water striders row the bugs

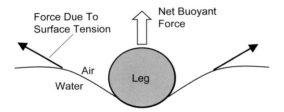

FIGURE 4 The leg of a water strider (shown here in cross section) is supported by surface tension. The interaction between the leg and the water forms a dimple in the water's surface.

across the water in the same fashion that oars (and their drag through water) propel a boat, and this type of locomotion is surprisingly efficient.

Surface tension can be a problem for small marine organisms. For example, the shells of barnacle and crab larvae are made of chitin, a material whose surface energy is less than that of water. As a result, when a larva encounters an air–water interface, it is expelled from the water into the air in the same way as the leg of a water strider is expelled. It may be difficult or impossible for the larva to overcome the surface tension and reenter the water. In this manner, surface tension can act as a trap for small animals in tidepools.

CAPILLARITY

Consider the situation shown in Fig. 5A: A small glass tube pierces the interface between water and air. Because glass has a higher surface energy than water does, water spreads over the glass and is drawn upward into the tube. This tendency for water to be drawn into a tube is known as capillarity. The smaller the diameter of the tube, the higher the water is drawn. For example, a tube with a diameter of 1 millimeter has a capillary rise of about 3 centimeters. A tube with twice the diameter has one-half the rise.

Another aspect of capillarity is explained in Fig. 5B. Again water is confined between glass walls, but in this case it is subjected to the forces of two air–water interfaces, rather than just one. Water still has the tendency to spread over the glass, but because it is now tugged equally in both directions, it cannot move. Instead, the effect of surface tension is to decrease the pressure inside the water. In the absence of capillarity, the water between the walls would be at atmospheric pressure, the same pressure that acts on the outside of the glass. Now, because of the pull of surface tension, water is at less than atmospheric pressure. The difference in pressure between water on one side of the glass and air on the other pushes the glass in. For example, when a drop of water is confined between two glass slides, atmospheric pressure forces the slides together, causing them to adhere.

Capillary rise and capillary adhesion both act at the ocean's edge. Water is drawn into the small spaces between grains of rock and among organisms, and is retained as the tide falls. In sandstone, water may be wicked by capillarity a substantial distance above the surface of a tidepool. Capillary adhesion causes the blades of intertidal algae to stick to one another at low tide, much the way the bristles of an artist's brush stick together when the brush is drawn out of water. The water thus confined between blades can help to keep the blades moist and cool at low tide.

SEE ALSO THE FOLLOWING ARTICLES

Adhesion / Seawater / Size and Scaling

FURTHER READING

Denny, M. W. 1993. *Air and water.* Princeton, NJ: Princeton University Press.
Denny, M. W. 2004. Paradox lost: answers and questions about walking on water. *Journal of Experimental Biology* 207: 601–1606.
Hu, D. L., B. Chan, and J. W. M. Bush. 2003. The hydrodynamics of water strider locomotion. *Nature* 424: 663–666.

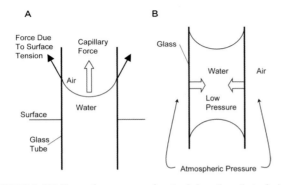

FIGURE 5 (A) The surface energy of water is less than that of glass, and the resulting capillary force pulls water upward into a small tube. (B) Capillary force reduces the pressure in water sandwiched between two glass slides, causing the slides to adhere.

SURF-ZONE CURRENTS

EDWARD B. THORNTON
Naval Postgraduate School

The interaction of ocean waves with the shore can drive currents both parallel and perpendicular to the shoreline. These currents can transport the larvae of intertidal organisms and can be dangerous to swimmers.

WAVE MOMENTUM

The surf zone is the near-shore region inside which waves break. The location of wave breaking is dependent on the water depth. Larger waves break in deeper water and smaller waves break closer to the shoreline. Wave breaking results in a loss of energy (the energy is eventually converted to heat through turbulent viscous effects). The loss of energy leads to a decrease in wave-induced momentum. A change in momentum must be balanced by a reactive force, which manifests itself in the form of near-shore currents. These currents are some of the strongest currents in the ocean, and their magnitude can vary significantly over relatively small distances (in other words, these currents can be subject to large shear). Near-shore currents are conveniently subdivided into alongshore and cross-shore flow components.

LONGSHORE CURRENTS

If breaking waves are incident to the shoreline at an angle, there is a change in alongshore momentum, which drives a current in the alongshore direction (referred to as the longshore current). This longshore current is confined to the surf zone region similarly to a river flowing with one bank at the offshore boundary where waves start to break and the other bank at the shoreline. The speed of the longshore currents increases with increasing wave angle and increasing wave height. Longshore currents do not flow like a steady river but have large horizontal vortical motions, or eddies, as the current flows along, called instabilities. These eddies extend across the surf zone and have periods of 30 seconds to 15 minutes. These instabilities are analogous to wheat blown by the wind waving unsteadily from side to side instead of simply bending over. Longshore currents can flow at speeds up to 1 to 2 meters per second, and they may be responsible for transporting the larvae of intertidal organisms along the coast and mixing them offshore.

CROSS-SHORE CURRENTS

When waves break, there is also a change in the wave momentum component in the cross-shore direction. This is balanced by a piling up of water against the shoreline, which creates an offshore-directed pressure force. The hydrostatic pressure force results in setup, an increase in the mean water level from the outer edge of the surf zone to the shoreline. The maximum setup is at the shoreline, amounting to 10% to 20% of the breaking wave height. For example, if the breaking wave height is 2 meters, then there is about a 20- to 40-centimeter increase in the mean water level at the beach. The setup then forces a return flow in the cross-shore direction, called the undertow. The undertow flows offshore in the bottom part of the water column, countering the wave-induced onshore flow at the surface. The undertow can flow offshore at speeds up to a meter per second, and provides a mechanism by which planktonic larvae of intertidal organisms can be delivered to offshore waters. By the same token, the inshore flow at the surface can transport larvae to shore across the surf zone.

RIP CURRENTS

Another, more dangerous, offshore flow is a rip current. Rip currents are seaward-directed, confined jets that flow approximately perpendicular to the shoreline extending to several surf zone widths. Rip currents originate within the surf zone with the jet usually no wider than 20 meters within the surf zone and broaden outside the breaking region. Rip currents are most often found where there are rip channels, which are indentations in the beach going offshore or cuts through a bar. The rip current flow is confined by the rip channel. Breaking waves are smaller in the deeper rip channel, so that the setup of water in back of the rip channel at the beach is less than at the adjacent beaches, just up coast and down coast where the breaking waves are larger. Water flows downhill from the regions of higher setup to lower setup converging at the rip channel, which then must flow offshore to escape the surf zone.

Rip currents often have pulsations, which can be more than twice the mean current. The pulsations are associated with long period wave motions with periods of 30 seconds to five minutes, called infragravity waves. Rip current velocity increases at lower tide as the return flow through the rip channel is more and more confined; for this reason rip currents are often erroneously referred to as rip tides. Rip currents flow offshore at speeds up to 2 meters per second, which is faster than most people can swim. These current jets account for more than 80% of lifeguard rescues and are the number one natural hazard in the state of Florida. More people fall victim to rip currents in Florida than to lightning, hurricanes, and tornadoes combined.

Rip currents are most prevalent where wave crests approach perpendicular to the shoreline. They are generally readily spotted from shore as regions where the waves are lower or not breaking within the deeper rip channel and there is a sediment plume going offshore (Fig. 1). Swimmers caught in a rip current are advised to swim parallel to shore until they are out of the jet that is carrying them offshore. As with the undertow, rip currents provide a flow that can move intertidal larvae offshore.

FIGURE 1 Snapshot of a rip current in Monterey Bay, California. The rip current is the dark patch. There is intense wave breaking on both sides of the rip currents, with little breaking within the deeper rip channel. Photograph by Ed Thornton.

SEE ALSO THE FOLLOWING ARTICLES

Currents, Coastal / Near-Shore Physical Processes, Effects of / Ocean Waves / Wave Forces, Measurement of

FURTHER READING

Bascom, W. 1980. *Waves and beaches.* New York: Anchor Press/Doubleday.

Denny, M. W. 1988. *Biology and the mechanics of the wave-swept environment.* Princeton, NJ: Princeton University Press.

Komar, P. D. 1998. *Beach processes and sedimentation,* 2nd ed. Upper Saddle River, NJ: Prentice Hall.

Svendsen, I. A. 2006. *Introduction to nearshore hydrodynamics.* Hackensack, NJ: World Scientific.

SURVEYING

LUKE J. H. HUNT

Stanford University

The discipline of surveying comprises the methods and theory needed to define a location. Surveys are used to describe an intertidal site's topography, to archive a location for future sampling, and to report habitat information such as tidal elevation and latitude. In the intertidal zone, only highly accurate surveys can describe the notable small-scale complexity that has made this habitat such a convenient study system for field ecologists. However, along coastal continental margins, where sea level references change and the earth's crustal movement may be substantial, the intertidal observer may be surprised to learn that maps and fixed locations constantly change relative to commonly used coordinate systems.

SURVEYING THEORY

Reference Systems

Surveys provide location data defined within a coordinate system. There are two types of coordinate systems used in surveying: local and global. Each type of coordinate system has intrinsic limitations, and it is important to approach surveying by first understanding the tradeoffs one makes in choosing either a local or a global coordinate system.

Local coordinates are usually distances north or south, east or west, and above or below a known point. For example, local coordinates can be used to measure limpet movement: A limpet may move 10 cm north, 3 cm east, and 2 cm up from its home scar (the site to which a limpet with homing behavior returns at low tide). Here the coordinate system is a set of Cartesian axes centered at the limpet's home scar. Although well suited for the study of a single limpet, this local coordinate system is probably not very useful outside this context. In fact, every limpet studied would likely be measured in its own coordinate system centered at its own home scar. Consequently, the observer who would like to combine these studies (for example, using a geographic information system [GIS] to compare limpet range sizes among rock types and wave climates) needs a more general coordinate system that can include all limpet study sites.

The most familiar large-scale coordinate system defines latitude, longitude, and elevation relative to a particular geodetic datum (a geodetic datum consists of a coordinate system and an elevation reference such as sea level). Recent advances in Global Positioning System (GPS) technology make it easy to measure limpet positions in global coordinates with subcentimeter accuracy. Consequently, one might conclude that one set of latitude–longitude–elevation coordinates will provide both small-scale accuracy as well as a general spatial reference. However, because the earth's crust is continually moving as a result of plate tectonics, problems arise when global coordinates are applied to local-scale problems.

Consider the following situation: The geodetic datum used for precision surveys in the United States is currently the North American Datum of 1983 (NAD83). NAD83 moves with the North American continental plate, so NAD83 coordinates on the North American plate are not changing. However, much of coastal California is actually on the Pacific plate, which moves 3–5 cm per year relative to the North American plate and therefore relative to NAD83. Because the global coordinate system moves relative to the local ground, a Southern California–limpet observer would have to cope with reference frame velocities of 3–5 cm/year when surveying within NAD83. Similar crustal movement issues exist for all global reference systems, including the most recently introduced international coordinate system (the International Terrestrial Reference Frame, ITRF). ITRF incorporates velocities into a complete statement of location—latitude, longitude, elevation, plus velocity north–south, velocity east–west, and velocity up–down. Thus, within the ITRF, positions are explicitly time dependent, whereas in older coordinate systems time-dependent errors accrue unnoticed until the coordinate system is updated. This may sound daunting to the surveyor accustomed to temporally stable local coordinates.

Fortunately, it is possible to take advantage of both the accuracy and convenience of local coordinates and the generality of global coordinates. When a global coordinate system is used, as long as at least two points with known locations are surveyed at every sampling time, the necessary rotation and translation corrections can be applied to transform the entire survey into a stationary local reference system (Fig. 1). Nationally protected reference points (such as National Geodetic control points in the United States; see Table 1 for location information) are the best choice for known reference points, because their protection and published locations ensure that future researchers will always be able to find a study's known points and recreate the survey.

TABLE 1
Sources for Data and Survey Software

United States	National Geodetic Survey (NGS)	http://www.ngs.noaa.gov
United Kingdom	Ordnance Survey	http://www.gps.gov.uk/index.asp
Canada	National Resources Canada Geodetic Survey Division	http://www.geod.nrcan.gc.ca
U.S. Tidal Data	National Ocean Service (NOS)	http://www.co-ops.nos.noaa.gov

Vertical References and Tidal Height

In the intertidal zone, elevations are usually measured from a tidal datum that defines the height of sea level. A tidal datum is computed over a period of years (as an average, a minimum value, a maximum value, etc.). Because global sea level is currently rising at about 2 mm per year and local sea levels are affected by tectonic activity, tidal datums are periodically updated (Fig. 2). When a datum is updated, all associated tide-table values are adjusted with it. In the United States, this happens approximately every 19 years. The most recent U.S. update occurred on April 1, 2003. If unnoticed, this datum adjustment could dramatically impact ongoing research. For example, take long-term intertidal studies of species' upper limits in Monterey, California. To the uninitiated observer, it would appear that every species' upper limit dropped

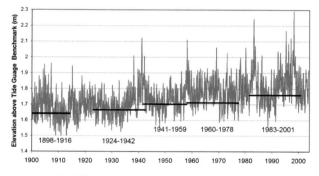

Monthly MLLW Compared to MLLW Datum, San Francisco, CA

FIGURE 2 Monthly MLLW (gray), tidal datums (horizontal lines), and tidal epochs (dates below lines) at San Francisco, California. The strong El Niño years of 1941–1942, 1957–1958, 1983, and 1997 are clearly visible as peaks in the monthly MLLW record. The most recent increase in the MLLW datum is particularly large, because the strong 1983 and 1997 El Niños both occurred during the last tidal epoch (NTDE 1983–2001) and temporarily increased the sea surface elevation during those years. The tidal datum is usually released a few years after the averaging period ends and applies until the next datum is released. For example, a measurement made in 1950 would be reported relative to the tidal datum of 1924–1942.

FIGURE 1 Two pictures can be overlaid in the same coordinate system using the coordinates of the starred points (because the photographs are the same scale). The same transformation can be made with survey data. Coordinate transformation tools are available in Geographic Information System (GIS) programs such as ArcGIS.

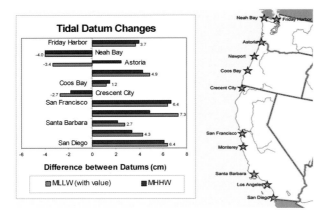

FIGURE 3 The change in tidal datum depends on the particular site and also the datum of interest. Here the difference between the most recent tidal datums (NTDE 1983–2001 and NTDE 1960–1978) is calculated. A positive difference indicates an increase in the datum elevation relative to the fixed station benchmark.

7.3 cm on April 1, 2003, when in reality, nothing remarkable happened that day: Monterey's tide-table zero was simply adjusted 7.3 cm higher to follow rising sea level.

The magnitude of the recent U.S. tidal datum adjustment varied substantially among locations because of local sea-level and tectonic changes (Fig. 3). Even at a particular location, different tidal datums (such as Mean Lower-Low Water [MLLW] and Mean Higher-High Water [MHHW]) may not change in concert (Fig. 3). As a result, the diurnal range of the tides, defined as the difference between MHHW and MLLW, is also adjusted whenever the tidal datums change. For example, in 2003, MLLW was redefined 3.4 cm lower at Astoria, Oregon, but MHHW was defined 2.4 cm higher. As a result of this most recent adjustment, the tidal range changed by 5 cm at Astoria.

Intertidal studies that involve data from different times and different places need to account for these temporal and spatial shifts in tidal datum. For this to be possible, the date of the tidal datum (in the United States this is the National Tidal Datum Epoch or NTDE) and tide-gauge location must be known. For example, a complete statement of a tidal elevation at Monterey is: one meter above MLLW (NTDE 1983–2001), Monterey, California.

SURVEY TECHNIQUE

Vertical-Only Surveys

In the intertidal zone, vertical gradients in biologically important factors may swamp horizontal variation in those same factors. In these cases, it may be acceptable to approximate horizontal positions within a few meters, but necessary to use more exact methods to determine

elevation. Fortunately, surveying elevation alone (called differential leveling in the literature) is much easier than determining horizontal position because gravity provides a conveniently measured vertical axis. (Although used as a rough horizontal reference, the earth's magnetic field has too much local variation to provide a survey-grade horizontal reference, therefore traditional horizontal surveys determine north by measuring angles to nationally maintained reference points.) For vertical-only surveys, many field observers use an inexpensive survey level, which is similar to a telescopic hunting sight mounted on a swivel. The sight is leveled atop a tripod, so as the sight is turned, the crosshairs sweep a horizontal plane. If the observer then sights on a vertical yardstick (or stadia rod), the elevation of the crosshairs above the bottom of the yardstick can be read (Fig. 4). Provided the tidal elevation of one point in the survey is known (see Table 1), the tidal elevations of every other point can be determined. For example (see Fig. 4), if a reference point is one meter above chart datum, and the crosshairs measure two meters on the stadia rod at that point, then the crosshairs sweep a plane three meters above chart datum. If a site with unknown elevation is then measured as 2.5 meters on the stadia rod (i.e., 2.5 m below the crosshairs), its elevation is 0.5 m above chart datum.

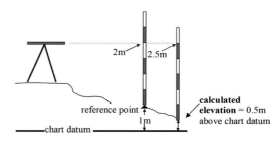

FIGURE 4 Surveying elevation with a level. Elevation relative to chart datum is calculated using a reference point with known elevation.

Traditional Methods for Horizontal Surveys

In addition to measuring horizontal and vertical angles, most modern survey instruments measure distance using a laser and reflecting prism. Such instruments are called total stations, because they record enough information (distance and angle) to calculate a relative position from a single measurement. In fact, the calculation is usually automatic and coordinate values are immediately available. The difficulty lies in finding enough reference marks to be able to set up a total station accurately, with its horizontal zero angle pointing north in the appropriate coordinate system.

To accomplish this, one must know the exact coordinates of two nearby fixed reference points and the angle between them. The instrument is then set up over one reference point and sighted on the other. The horizontal angle (or azimuth) between these two sites and north is entered into the instrument, setting the zero angle properly. In the United States, coordinates of reference points are available from the National Geodetic Survey (NGS; see references), and the program *Inverse* (downloadable from the NGS Web site) calculates the azimuth between the points.

GPS Surveys

GPS is the current standard method for surveying when horizontal accuracy is required. Along the coast, where survey reference markers may be few and far between, GPS is particularly useful. Although survey-grade GPS equipment is expensive for the casual observer, it is usually readily available within a research or university setting, and compared to traditional methods, GPS surveys are simple to conduct. The following is an introduction to the technology behind GPS surveying.

GPS relies on a constellation of orbiting satellites that broadcast a repeating code, analogous to them constantly humming a tune. The GPS receiver on the ground has this tune in memory, as well as the exact time the notes are played. By matching up the two versions of the tune and recording the time offset, the receiver determines the time difference between when the tune was broadcast and when it was received. Knowing this time difference, along with the speed of the signal, it is possible to calculate the distance to the satellite. If the distance is calculated to three different satellites (whose positions are accurately known), it is then possible to determine one's position. Thus, the accuracy of a GPS fix is really a problem of calculating transit times. Because the signal travels at the speed of light, tiny time errors represent large errors in distance. For this reason, GPS satellites contain highly accurate atomic clocks. The inaccuracies in the receiver's inexpensive clock are corrected by including a fourth satellite in the calculation of position; four satellite distance equations can calculate four unknowns: x, y, z, and the receiver's clock error.

Survey-grade GPS systems overcome two complications to achieve subcentimeter accuracies that are unattainable with inexpensive handheld units. First, because there is a period of transition between notes in the tune, it is difficult to determine precisely when one note stops and the next begins. This introduces ambiguity when aligning the notes in time and, although minimal, the distance errors amount to two to five meters. The solution is to first match the notes up as closely as possible, and then to match the peaks and valleys in the higher frequency carrier wave of the signal. The carrier wave is sinusoidal, with indistinguishable peaks. Thus, before two signals can be aligned, it is first necessary to determine which two peaks should be matched together. This is determined by the signal coded in the tune, which specifies the carrier-wave alignment within one wavelength. It is then possible for the receiver to match the carrier wave for increased precision and overcome this source of two- to five-meter error.

The second complication is due to changes in the speed of light through the atmosphere. The speed of light through air depends on the frequency of light and the density of the atmosphere along the signal's path. This effect is seen in a prism: Light is separated into a rainbow because each frequency travels at a different speed through the glass. GPS receivers capable of receiving multiple frequencies can use this same principle in a process known as multichannel correction. By calculating the difference in transit time between two frequencies, it is possible to estimate the speed correction appropriate for atmospheric conditions. Another method of correcting for atmospheric composition is to use two receivers. One receiver, set at a known location, calculates the error between its actual position and its GPS-derived position. This error can be used to correct for the atmospheric conditions above a nearby roving receiver because the signal paths to both receivers will be similar. This is called differential GPS. Carrier-phase differential corrections can be made after data collection or in real-time, if corrections are transmitted to a roving receiver via radio or cell phone. These systems are termed real-time kinematic (RTK) GPS systems. Survey-grade GPS systems differ from inexpensive handheld units largely because they employ both differential and multichannel corrections and because they process carrier wave information, often in real-time.

Survey Adjustment

A measurement is of little value unless its accuracy is known (e.g., confidence limits are required when interpreting mean values), and surveys are like any other sampling problem in that the accuracy can only be calculated by redundant measurements. However, unlike random samples, the points in a survey are geometrically related, and therefore it is possible to correct the entire network of points simultaneously. This process is termed survey adjustment. Survey adjustment is very similar to linear regression in both use and computation: Obvious errors are removed, and then residual errors are minimized across all data points. A program for automatically performing survey adjustments is available for download from the U.S. National Geodetic Survey at http://www.ngs.noaa.gov/PC_PROD/pc_prod.shtm (Table 1).

SEE ALSO THE FOLLOWING ARTICLES

Monitoring: Techniques / Sea Level Change, Effects on Coastlines

FURTHER READING

Leick, A. 2004. *GPS satellite surveying.* Hoboken, NJ: John Wiley and Sons.
Pugh, D. 2004. *Changing sea levels: effects of tides, weather and climate.* Cambridge, UK: Cambridge University Press.
Torge, W. 2001. *Geodesy.* Berlin: Walter de Gruyter.

SYMBIOSIS

JONATHAN B. GELLER

Moss Landing Marine Laboratories

Symbiosis, literally "living together," is the state of close and persistent physical association between organisms belonging to different species, without regard to the benefit or cost to participating individuals. The fate of individuals in marine communities often depends on their success in entering into, maintaining, or avoiding symbiosis.

SYMBIOSES: COMMON BUT UNDERSTUDIED ON ROCKY SHORES

Symbiosis is the state of physically intimate and non-transient association between individuals belonging to different species. Although some scholars would more narrowly reserve the term for relationships in which physiological, behavioral, or other organismal adaptations to life in partnership have been demonstrated, the more general concept enables us to consider cases in which the organismal or ecological significance of persistent physical association is currently unappreciated. Symbiotic relationships abound in rocky-intertidal communities, just as they do in all other marine habitats. Despite this, the importance of symbiosis for the function and structure of rocky-intertidal communities is largely unknown and unstudied. One reason for this neglect is a historical focus on other interspecific interactions, such as competition or predation. Another reason is that the term *symbiosis* is used popularly to mean mutualism or codependency. As a consequence, marine ecologists may have unconsciously overlooked symbiotic relationships that are not mutualistic.

Recent developments in marine ecology suggest that symbiosis will become more recognized as researchers place more attention on "positive interactions" in which individuals of one species provide access to food, refuge from enemies or harsh conditions, or living space to individuals of another species. In contrast, competition and predation are considered "negative" interactions because one species is adversely affected. Positive interactions include some symbioses. For example, associational defense is a positive interaction in which one species gains protection from predation by physical association (which could be intimate and persistent) with a better-defended species. In short, the literature is replete with examples of symbiosis, but the overall importance of symbiosis for rocky intertidal communities is poorly understood. The following examples are drawn primarily from the rocky shore of California. Similar cases exist on other shores worldwide.

GENERAL CONSIDERATIONS

Organismal and Population Outcomes of Symbiosis

The concept of symbiosis formulated by De Bary in 1879 did not specify the impact of physical association for each partner. Scientists apply additional terms to specify categories of symbiosis depending on the outcome for each partner.

A first category is parasitism or disease, in which a parasite or pathogen benefits at a cost to the host. Costs can be measured in various ways, but we are usually most interested in consequences at the level of individuals or populations. For example, when mortality rates increase or fecundity rates decrease as a result of parasitic infection, population growth is suppressed. At the level of individuals, parasitized organisms that produce fewer offspring suffer decreased genetic fitness, because they contribute fewer progeny to subsequent generations than do unparasitized individuals.

A second category is mutualism, which includes symbiotic relationships in which both partners benefit, and in which population growth increases with the prevalence of symbiosis. Individuals in a mutualistic symbiosis have higher genetic fitness than conspecific individuals living outside of symbiosis.

Third, the category of commensalism comprises situations in which partners live together and one partner benefits without obvious positive or negative consequence for the other. Typically, a commensal organism uses a host for shelter or food in a way that does not harm the host (for example, by eating shed mucus, sloughed cells, or food taken by but not usable for the host).

Spatial and Temporal Relationships of Symbiotic Partners

Symbiotic relationships all involve physical proximity, but the spatial and temporal nature of association can vary considerably. For example, one may consider the relative sizes of individuals. Symbioses are very often characterized by striking dichotomies of sizes, and this is true of most of the examples cited in this article. Usually, when sizes are very different, the larger organism is considered the host, though a "host" may facilitate, tolerate, or actively attack symbionts, depending on the biology of the situation. In a minority of symbioses, sizes are relatively similar. Decorator crabs *(Loxorhynchus crispatus)*, for example, actively attach to themselves other organisms such as sponges, bryozoans, algae, and hydrozoans, which together can quite overgrow the crabs' carapaces. The polychaete worm *Arctonoe vittata* can grow to 7 cm and lives wrapped around the foot of the keyhole limpet *Diodora aspera*, with a shell of up to 7 cm, and grows to largely fill the space where the limpet's gills lie.

When physical contact involves only external surfaces, the relationship is considered an ectosymbiosis. If the definition is not constrained by other considerations (such as specificity), hundreds of examples are known from rocky intertidal shores: some are algae, tiny tube worms, and barnacles that settle and grow on larger shelled animals such as gastropods, mussels, and chitons. Fewer cases, though not at all rare, of ectosymbiosis are more specific. Some that come to mind are the amphipod *Polycheria osborni*, which lives in concavities of the tunics of many compound sea squirts, especially in the genera *Aplidium*, *Distaplia*, and *Archidistoma*; the hydrozoan *Proboscidactyla circumsabella*, which lives on the rim of the tube of the feather duster worm *Potamilla occelata*; and the slipper snail *Crepidula perforans*, which lives inside of a shell from a snail now occupied by a hermit crab. Once an ectosymbiosis is established, the location of the partners may remain stable, as in the cases just mentioned, or partners may separate for intervals of time (for example, the goby *Clevelandia ios*, which lives in the mudflat burrow of the worm *Urechis caupo* but forays outside as well). Similarly, the pea crab *Pinnixia franciscanus* is said to move among the tubes of its worm hosts as it grows from juvenile to adult.

Symbiosis in which one partner is contained within another individual is called an endosymbiosis. In such cases, the internalized partner may be contained within a body cavity. In deep-water habitats, mutualistic chemoautotrophic bacteria live in the gills, hemocoel (blood-filled cavities), or coelom (peritoneum-lined cavity) of clams and worms. In the rocky intertidal zone, a curious example is the copepod *Diarthrodes cytsoecus*, which lives in the water-filled bulbs of the red alga *Halosaccion*. The gut is another internal habitat for symbionts; a good example is the copepod *Mytilicola orientalis* (apparently introduced to California with exotic mussels) living in the guts of the native mussels *Mytilus californianus* and *M. trossulus*. Another example is the flatworm *Syndisyrinx franciscanus* living in the intestine of sea urchins in the genus *Strongylocentrotus*. In other endosymbioses, usually involving protists or bacteria, symbionts are found in intercellular spaces or within cells. Important examples of intracellular symbiosis are the photosynthetic dinoflagellates (zooxanthellae) and unicellular green algae (zoochlorellae) that live in the cells of many cnidarians, including tidepool sea anemones in the genus *Anthopleura* (Fig. 1). In tropical regions, the symbiosis between zooxanthellae and corals is critical for the formation of reefs, which support the oceans' greatest biodiversity.

Endosymbiotic bacteria deserve special mention because they are found in and have important consequences for many rocky shore organisms. For example,

FIGURE 1 The relationship between photosynthetic protists and sea anemones (as well as with corals, bivalves, and assorted other taxa) is a paradigmatic example of symbiosis. The green or brown colors in the anemones in this assemblage of species of *Anthopleura*, come from the symbionts; brownish anemones contain zooxanthellae and green anemones contain zoochlorellae. Photograph by the author.

bacteria in the brood chamber of the common brittle star *Amphipholis "squamata"* (likely a species complex) appear to produce nutrients absorbed by brooded embryos. Sponges and bryozoans harbor bacterial symbionts that produce chemical compounds that may help defend their hosts from malignant infections or overgrowth by other organisms. Undoubtedly, there is much yet to be discovered about the relationships between bacteria and higher eukaryotes and the ecological consequences of such partnerships.

In some cases, the concepts of ectosymbiosis or endosymbiosis do not fit comfortably. For example, the bodies of parasitic rhizocephalan barnacles, which infect many species of crabs, have both internal and external portions at maturity. Some symbionts may be found on surfaces that are anatomically external (that is, on the outer epithelia or teguments of a body wall) yet are largely encased by other body surfaces. As a case in point, parasitic bopyrid isopods, such as *Argeia pugettensis,* live on the gills of shallow-water shrimp (in this example, *Crangon nigricauda);* these parasites appear to be internal because the gills are covered by the shrimp's carapace, but the parasites are not actually exposed to internal tissues. Technically, this is true of gut symbionts as well, because the lumen of the gut is anatomically exterior. However, the chemical and material exchange across the gut wall, and therefore potential to impact the host, is far more extensive than across other external surfaces, so the gut deserves special and separate consideration as a habitat for symbionts.

Necessity of Symbiosis for Partners

Symbiotic relationships may be categorized with respect to their necessity for each partner. That is, to what degree does the survival of one or both partners depend on the formation and maintenance of the symbiotic condition? Where interdependence is high, the symbiosis is considered to be obligate; where both free-living and symbiotic individuals can persist, the symbiosis is called facultative. Certainly, some species seem to be found only in a symbiotic state. Scale worms are frequent commensals, and some appear obligately so. *Hesperonoe adventor* is apparently found only within the burrows of the echiuroid worm *Urechis caupo*— a species known as the "fat innkeeper" worm because of the menagerie of commensals found in its tube, including a fish, two crabs, and a clam. *Arctonoe fragilis* is found living only in the ambulacral grooves (that is, among the tube feet) of various sea stars, whereas *A. vittata* finds home in the mantle cavities of keyhole limpets *(Diadora aspera)* and a chiton *(Cryptochiton stelleri)* as well as on sea stars. The pea crab *Fabia subquadrata* lives in the mantle

cavities of mussels and some other bivalves. True shrimps have widely exploited other organisms as habitat, and this habit is especially widespread on coral reefs. In the Californian rocky intertidal zone the shrimp *Betaeus harfordi* lives in the mantle cavity of abalone *(Haliotis), B. macginitieae* with sea urchins, and *B. ensenadensis* in the burrows of mud shrimp. In some cases, the interdependence of partners is extreme. The alga and fungus that together form the conspicuous splash-zone tar lichen *Verrucaria maura* do not thrive by themselves. For other organisms, symbiosis is optional. The limpet *Lottia digitalis* is very common on high intertidal rocks in California but is also found living on the shell plates of the stalked barnacle *Pollicipes polymerus* (Fig. 2).

FIGURE 2 The limpet *Lottia* is usually found freeliving on rock surfaces, but it also maintains significant populations on stalked barnacles *(Pollicipes polymerus).* On rock, *L. digitalis* has a brown and white pattern, but on barnacles it appears white with black lines and resembles the plates of the host barnacle. Photograph by the author.

The necessity of a symbiosis may not be symmetrical. For example, sea anemones do not require zooxanthellae to persist, whereas zooxanthellae presumably do not maintain significant populations outside of hosts. Sea star and bivalve populations always are mixtures of those with scale worms or pea crabs and those without. It should be recognized, however, that necessity of a symbiosis can be difficult to demonstrate; the failure to observe free-living partners may be a result of their rarity or occupation of unexpected habitats rather than their lack of viability outside symbiosis. Also, inability to culture organisms in a nonsymbiotic state in the laboratory may not reflect lack of viability in nature.

Perpetuation of Symbiosis

Most of the symbioses of interest to students of the rocky shore involve at least one organism that starts its life as a gamete and develops into a larva or other free-living propagule. How then do symbionts join? The answer, in most cases, is settlement of the larva of the prospective symbiont onto a potential host. Settlement of the larvae of symbiotic species is mechanistically similar to settlement of other marine organisms onto nonliving surfaces. Thus, the rich literature on larval settlement is highly relevant for studies of symbioses. Settlement, whether to biotic or abiotic surfaces, is controlled by mechanisms of sensory perception of chemical cues, and adhesion, attachment, or penetration. Chemoattraction of larvae to settlement sites has been described for a great number of marine invertebrate species, including some that are symbiotic. A classic example is the attraction of the larvae of spirorbid tubeworms to the *Fucus* algae on which adult worms live. It should be noted that evidence for a role of bacterial biofilms on living and nonliving surfaces in the settlement process is accumulating, and these may also function in the establishment of symbioses. When a symbiosis is reestablished with each generation, as by larval settlement, the pattern is referred to as horizontal transmission.

In some cases, symbioses are propagated across generations by transmission of symbionts in the gametes of the host organism. This pattern, referred to as vertical transmission, is a common feature of some bacteria–insect symbioses in terrestrial systems. This also appears to be the case for some coral–zooxanthellae relationships. However, in other coral symbioses, horizontal transmission is the rule. In those cases, larvae acquire zooxanthellae from seawater, and this appears also to be the case for rocky shore–dwelling anemones in the genus *Anthopleura*. Larvae of *A. elegantissima* have been observed to acquire zooxanthellae by ingestion, at least in laboratory studies.

Adults may also establish symbioses. For example, when worms (*Arctonoe vittata*) that live among the tube feet of the leather star, *Dermasterias imbricata*, were removed, it was found that the sea star moves toward its displaced partner. *Arctonoe* is found as a commensal on a number of other invertebrates, and the sea star may need to actively collect worms for its own benefit. Conversely, another worm, *Ophiodromus pugettensis*, is chemically attracted to its seastar host, *Patiria miniata*. *Ophiodromus* is found free-living in the lower intertidal zone, so the behavior of the worm may be essential to establish this partnership.

Specificity of Symbiotic Relationships

Specificity, in which a single symbiont species is associated with a single host species, is often a hallmark of symbiosis. For example, the limpets *"Lottia" paleacea*, *"L." depicta*, and *L. insessa* are found only on the surfgrass *Phyllospadix,* the eelgrass *Zostera,* and the kelp *Egregia,* respectively. Other symbionts, such as the slipper shell *Garnotia* (formerly *Crepidula*) *adunca*, which lives on the shell of several trochid snails in tidepools, and the aforementioned worm, *Arctonoe vittata*, have a larger, though not huge, range of hosts. Zooxanthellae (the dinoflagellates that are photosymbionts with anemones, corals, and other invertebrates) were once thought to be a single species with quite broad host range. However, molecular systematic investigations have revealed multiple evolutionary lineages among zooxanthellae with apparently broad, but not undiscriminating, use of hosts. Biological invasions, in which exotic species become established in new places, can be seen as experiments in host specificity: when a new species arrives, can it act as host for local species? Conversely, can introduced symbionts partner with local potential hosts? As marine invasions increase in frequency, these will become more important questions. Already, we remarked upon the copepod *Mytilicola orientalis*, introduced from Asia and now found in mussels native to North America.

SYMBIOSIS AND THE STRUCTURE OF ROCKY-INTERTIDAL COMMUNITIES

The role of symbiosis for the structure of rocky-intertidal communities has not been much studied. Lately, there has been some attention to the role of parasites for regulating host populations, to disease as an agent of disturbance, and to positive interactions that include some symbioses (all treated elsewhere in this volume). Nonetheless, the importance of mutualistic or commensalistic symbioses for rocky-intertidal communities is mostly unknown. Although we lack experimental evidence, we still can ask from a natural history perspective how symbiosis plausibly can influence patterns of species composition, abundance, and distribution. It appears that commensalism is the most important form of symbiosis in this regard.

A hallmark of rocky-intertidal communities is dominance of primary substrata—the rock surface—by one or few species in each intertidal zone. For example, barnacles and algal tufts often dominate the higher intertidal zones, mussels and sea anemones occupy much of the middle zones, and kelps cover lower zones. Each of these groups of dominant species provides a three-dimensional matrix that is habitat for hundreds of other species. In central California, for example, 93 species have been counted in the turf formed by the high-intertidal alga *Endocladia muricata*, and over 300 species in, on, and within beds of the mussel *Mytilus californianus*. Few, if any, of these epibiotic species are known to live only within these algal tufts or mussel beds. Yet most would not persist on the bare, exposed rock surface were the algae and mussels removed. Thus, the abundance and distribution of these species in the rocky intertidal zone are certainly enhanced by their association with structure-forming species, and the overall diversity of the rocky intertidal zone is increased.

Symbiosis could also contribute to community structure by positive or negative impacts on populations of the species that exert strong control over other populations. In the Californian rocky intertidal community, mussels, sea anemones, barnacles, and certain algae are the major space occupiers. Symbiotic relationships that modified the ability of these organisms to hold space would be quite important. Similar arguments can be made for symbioses that involve important consumers in intertidal communities. Although a case may be made that the symbiosis of zooxanthellae and sea anemones allows anemones to occupy more space than they could otherwise (though this has not been demonstrated), this would appear to be a singular case. Other recurrent symbioses involving major space occupiers or consumers seem to have no noticeable impacts for the hosts, which therefore neither expand nor contract in abundance or distribution. For example, several commensals live on or in the purple sea urchin, *Strongylocentrotus purpuratus*: the flatworm *Syndisyrinx franciscanus* lives in the urchins' guts, while the polychaete worm *Flabelliderma commensalis* and the isopod crustacean *Colidotea rostrata* live among the spines. Because sea urchins can be important grazers in the rocky intertidal zone, any symbiont that impacted their abundance could be important for community structure. Yet there is no quantitative evidence that these commensals have any impact at all on urchin populations.

One important caution is that the impact of symbionts on strong interactors in communities may be subtle and not easily detected without detailed investigation. For example, while epibiotic organisms might not directly impact a host, they might increase the host's vulnerability to predators by covering natural camouflage, by leaking attractive chemical scent, or by increasing the food value of the association to a predator. Conversely, epibionts might augment host defenses by providing camouflage or exuding defensive chemicals. Although the precise mechanism of predator deterrence is not always known, a protective effect of epibiota has been shown for several species of bivalves, including mussels and scallops.

Symbionts that positively or negatively impact populations of host species that themselves exert little influence on other species simply can have little consequence for community structure, other than increasing species richness by their presence.

HOW SYMBIOSES EVOLVE

As defined here, many symbioses involve associations that are not necessarily highly evolved. For example, the interstitial biota of mussel beds includes a recurrent assemblage of organisms that would be found in any similar protective microhabitat on the rocky shore. Yet such instances set the stage for the evolution of more sophisticated symbioses, which arise when a fitness benefit accrues to at least one member of the association. A fitness gain is more than a simple benefit: it is a benefit attributable to a genetically based trait that increases the reproductive potential of individuals relative to members of the species that lack the trait. For example, individuals in a population may differ in their tendency to enter into an association with a host because of variation in genes that are involved in locating or attaching to a host. If the association promotes greater reproductive success than the free-living condition does (for example, by providing individuals greater access to food), then individuals in symbiosis will contribute disproportionately to subsequent generations. As a consequence, the genes promoting formation of the association will increase in frequency, and the population will increasingly be composed of individuals capable of entering into symbiosis. The symbiont may evolve further sophistication in finding and exploiting hosts, eluding host defenses, or other means of increasing its reproductive success. Such means may include providing services to the host; promoting host growth or survival can benefit the symbiont by increasing its longevity or increasing its access to resources. In turn, any genetically based variation in the hosts' ability to attract or assist symbionts can spread in the host population when the hosts receive benefits by the partnership. Mutualism can then evolve in this reciprocal, stepwise fashion. Conversely, more sophisticated exploitation of hosts can lead to a decidedly parasitic relationship, and hosts can evolve more complex defense mechanisms. In some cases, host and parasite may alternate escalation of offense and defense with no net change in fitnesses of either.

These verbal descriptions of the evolution of symbiosis stress the role of natural selection for traits that increase the genetic fitness of individuals and populations. Fitness gains or losses derive from the benefits and costs of symbiosis, which in turn may depend on the physical or biotic environment. The ecology and evolution of symbiosis can therefore be dynamic: if the environment changes such that a symbiont can no longer provide a service to its host, a mutualistic interaction might become parasitic or the relationship may dissolve. Sea anemones, for example, lose zooxanthellae in dark habitats. We do not know whether anemones actively resist symbiosis in darkness or whether zooxanthellae simply cannot persist in the dark even with the assistance of a host. Yet it is clear that hosts derive no photosynthetic benefits from symbionts in darkness and would incur a cost to maintain them. Similarly, by phylogenetic analysis, we know that symbiosis has been gained and lost several times in sea anemones *(Anthopleura)*. Some extant species of *Anthopleura* harbor zooxanthellae but evolved from ancestors that did not, and conversely some extant species that never host symbionts had zooxanthellate ancestors. In the latter case, we may surmise that symbiosis was no longer an advantage even in sunlight, perhaps as a result of some change of geography, habitat, or diet. These concepts have been explored in mathematical models to help understand why the evolutionary trajectory should be toward mutualism or parasitism or indeed flip-flop between these conditions.

SEE ALSO THE FOLLOWING ARTICLES

Echiurans / Genetic Variation and Population Structure / Larval Settlement, Mechanics of / Mutualism / Parasitism / Sea Anemones

FURTHER READING

Ahmadjian, V., and S. Paracer. 1986. *Symbiosis: an introduction to biological associations.* Hanover, NH: University Press of New England.

Boucher, D. H., ed. 1985. *The biology of mutualism: ecology and evolution.* New York: Oxford University Press.

Gotto, R. V. 1969. *Marine animals: partnerships and other associations.* New York: American Elsevier.

Smith, D. C., and A. E. Douglas. *The biology of symbiosis.* Contemporary Biology. London: Edward Arnold.

Vermeij, G. J. 1983. Intimate associations and coevolution in the sea, in *Coevolution.* D. J. Futuyama and M. Slatkin, eds. Sunderland, MA: Sinauer, 311–327.

achieved by mountains, forests, rivers, oceans, and their larger inhabitants, real or imagined. Most intertidal shells are unremarkable and usually broken in the pursuit of the calories they contain. Nevertheless, shells with interesting properties that set them apart from the ordinary gained attention and value serving a common human character trait—the desire to possess objects. Out of that desire, symbolic meaning was assigned to some objects, and from that symbology evolved culture. The original intent for hanging a dead snail's shell about one's body is only a guess, but more than 100,000 years ago, littoral shells, with a hole drilled down the middle of each and strung together, were used along the eastern shore of the Mediterranean Sea. Though controversial, many archaeologists cite such behavior—personal decoration beyond practical necessity—as evidence of human culture. It is at least symbolic behavior unique to humans.

Mentioned here are just a few examples of how people have used the resources of the intertidal beyond subsistence foraging. Although beadwork predominates the appearance of shells in archaeology, this article briefly mentions some of the diverse uses of shells from the intertidal.

ORIGINS

The earliest evidence of personal adornment with marine artifacts is from Mount Carmel (Es-Skhul), Israel, and was dated earlier than 100,000 years ago and possibly as old as 135,000 years ago. Another site at Bir-el-Ater (Oued Djebbana), Algeria, may be contemporaneous with the younger layers of Es-Skhul. These sites revealed several beadlike *Nassarius gibbosulus* shells with holes through the center (Fig. 1). Fragments of three other marine species not expected to occur naturally at Es-Skhul were also present— two cockles, *Acanthocardia deshayesii* and *Laevicardium*

SYMBOLIC AND CULTURAL USES

PIETER AREND FOLKENS

Alaska Whale Foundation

The size and nature of life in the rocky intertidal have kept that environment and its fauna from attaining the kind of mythic prominence in ancient cultures that was

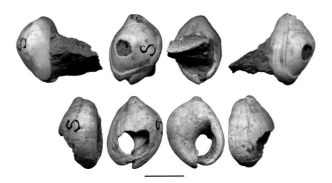

FIGURE 1 Four views of two *N. gibbosulus* shell beads from Es-Skhul. Scale bar = 1 cm. Photograph courtesy of Marian Vanhaeren and Francesco d'Errico.

crassum, and a scallop, *Pecten jaconaeus.* Es-Skhul rests about 3.5 km from the present shore of the Mediterranean Sea. Oued Djebbana is about 200 km inland. Considering the distance of the sites from the sea and the rare natural occurrence of *Nassarius* shells with holes in the center, the researchers concluded that the objects were deliberately transported inland for beadwork.

The dating of these examples encompassed a period of extreme climate change during which the sea level fluctuated more than 100 meters, from deep ice age conditions to the Late Pleistocene climatic maximum with a sea level 3 meters higher than today. Faunal remains found with the shell from Algeria (Fig. 2) indicated a savannah environment that was warmer than today, suggesting the specimens may have been deposited during brief interstadials around either 125,000 or 105,000–100,000 years ago. This "later" Middle Stone Age period in Africa included increasing innovation, more long-term occupation of habitation sites, and greater use of incised objects (symbolic artifacts).

FIGURE 2 Four views of a *N. gibbosulus* shell bead from Oued Dje-banna. Scale bar = 1 cm. Photograph courtesy of Marian Vanhaeren and Francesco d'Errico.

Forty-one *Nassarius kraussianus* shells from Blombos Cave, Still Bay, South Africa, had set the previous oldest record at around 76,000 years ago for shells with hand-made perforations and marks. The beads were apparently selected for size, have holes in similar places, and preserved wear from cordage. Some of them have traces of red ochre (a pigment material and an archaeological index of symbology) suggesting they were painted, worn by someone with body makeup, or as part of a burial. In contrast to Es-Skhul, the Still Bay culture exploited the near-shore marine environment, giving them a ready source of bead material.

Obvious evidence of shell beads is lacking between 75,000 and 42,000 years ago. This may be due to the Toba volcanic winter (~74,000 years ago) and the impact it had on humans. For example, no marine artifacts have been found in Kebara Cave, not far from Es-Skhul and Qafzeh

at Mount Carmel. Neanderthals inhabited Kebara about 60,000 years ago, but were not known for object symbolism. By 40,000 years ago shell beads appear again in the records from Africa, Australia, and Eurasia and began an essentially continuous history of intertidal artifacts in many coastal and near-coastal cultures.

Remarkable about the proliferation of early shell beads is the presence of worked marine shells far from their nearest possible natural sources. Many ancient habitation sites in Europe have revealed shell beads that had been transported hundreds of kilometers. A perforated *Olivia* shell dated more than 20,000 years ago was found in India more than 300 km from the sea. Similarly, archaeologists recovered a marine shell bead at the 11,500-year-old Lindenmeier, Colorado, site 1300 km from the nearest source. Throughout western North America, ancient habitation sites contained marine shell artifacts presumably acquired in trade. Shell beads from the eastern seaboard showed up in middens of the Dakotas. Coastal tribes, especially in the Channel Islands (California), maintained shell bead industries to create capital for trade. The use of shell beads as a medium of exchange is a hallmark of many preagrarian cultures, both coastal and inland.

BEADS AND WAMPUM

A Medium of Exchange

Trading what one has for something needed or desired has been a central human behavior since the beginning of culture, as has gifting, especially as reciprocal altruism. An agreed-upon medium of exchange has benefits over commodity trading and brings flexibility as well. Because of their convenience and portability, beads made of marine shells were commonly used for that purpose. Discoidal (disk-shaped) shell beads dominate the collected artifacts of many habitation sites, whereas tubular shell beads were favored as exchange items in others. Depending on the culture, bead sizes and shapes used as currency were often standardized.

Shell beads were much more than "Indian money" in precontact North America, and gifting them was common. When offered to European settlers, beaded belts were misinterpreted as "trade" rather than "gift" with no insight to the cultural significance. Europeans recognized the intrinsic value of beads thanks to a Dutch fur trader who demanded a ransom after taking a tribal leader hostage in 1622. The trader, Jacob Eelkes, discovered that belts of shell beads attracted substantially more pelts from the natives than European goods.

The ability of beads to procure Western goods such as iron, copper, and cloth created an explosion of available trade beads. The results were twofold. Fabricated objects using shells lost much of their cultural and spiritual significance, and a sort of bead inflation set in. Metal rasps and drills acquired from Europeans reduced the time and effort to produce beads and made smaller beads possible. Untalented individuals made their own beads, and quality fell. Merchants were challenged by this "counterfeit" currency. By the end of the seventeenth century, trade beads in New England had lost their luster, but they regained some importance during the exploration of the west coast of North America well into the nineteenth century.

Wampum

Some of the most well-known examples of a cultural use of shell beads are the objects called wampum (an Algonquin or Narragansett word meaning "white beads strung"). Wampum was essentially belts, aprons, earrings, regalia, mats, and other items of tightly strung shell beads arranged with a variety of designs and images. These meticulously fashioned mats used common shells collected during the summer from shallow mud from southern Massachusetts to northern Long Island. The columella (long central spiral) of several species of Atlantic whelk (genus *Busycon*) provided the material for white tubular beads. The more rectangular purple beads were made from the quahog, an Atlantic clam (*Mercenaria mercenaria*, family Veneridae).

Though not precisely a product of the intertidal, wampum pieces are examples of the diverse use and importance of shell beads. Originally little more than memory aids for storytellers, these shell beads strung together were assigned great symbolic significance and value by indigenous cultures of northeast America. These sacred symbols were used as gifts that fostered reciprocity, in personal decoration as status symbols, for ransom and payment of fines, as tokens of marriage proposal, as prizes, and as ceremonial objects. The utility of wampum and its importance in tribal politics were unmatched by any other object in precontact America.

BEYOND THE BEAD: ONE CULTURE'S DIVERSE USE OF THE ROCKY INTERTIDAL

Clans of the Pacific Northwest provide examples of the diverse uses of shells from the intertidal. Uses ranged from practical utensils to aids used in shamanistic practices as well as "shell money" and inlay decorations. Paired scallop half-shells, attached with small twisted strips of cedar through a hole near the hinge, served as the typical shaman's rattle from the Salish in the south (Washington State) to the Chugach in the north (south central Alaska).

Small clam shells attached to short wooden sticks worked as spoons. Larger shells served as ladles and dishes for paint, glue, and other substances. Lime extracted from burned clam shells was used as an ingredient in snuff. Knives made of blue mussel shell were used in cutting meat and fish and preparing skins. Mussel shells were also tools used in mat and basket making. The opercula of red turban shells provided material for making inlays in wooden objects from boxes to talking sticks, particularly in the artistic representation of teeth. Long, tubular scaphopod shells of the genus *Dentalia*, secured in trade from clans on the west coast of Vancouver Island, held unusually high value.

The highly iridescent interior of California abalone—especially the green abalone, *Haliotis fulgens*, and to a lesser extent the red abalone, *H. rufescens*, held great value. The iridescent part of the shell was often used for inlaying on the most valuable objects. Tightly woven basketry sometimes covered the dull exterior of abalone shells to create a bowl with an iridescent interior surface. The deep color at the center of the shell (where the muscle attaches) displays a prismatic luster and was prized as jewelry including pendants, earrings, nose rings, and headpieces.

THE INCREDIBLE LIMPET: THE DIVERSE USE OF ONE ROCKY-INTERTIDAL SPECIES

Often found in Mesolithic middens in Britain, limpets (*Patella velgata*) are usually thought of as famine food, a resource for the poorest people, and fish bait. Limpets, gathered after a bad harvest, became the focus of rioting in Stronsay, Orkney, in 1762 when it was perceived that overharvesting of kelp impacted the limpets by depriving them of shade, thus depriving the poor of food. In a cultural turnaround, limpets have become an expensive delicacy in gastronomical culture, costing $125 per gallon while feeding Hawaiian haute cuisine in the mid-1980s. Limpet dishes such as Limpet Escabèche are listed on several gourmet Internet sites.

Limpets achieved a place in superstition and have been used as charms. Water drawn from St. Govan's well with a limpet shell might be used as a cure for lameness, rheumatism, and poor eyesight. A live limpet placed at the corner of one's house in Galway Bay, Ireland, ensured successful fishing. In aboriginal Australia, it is said that a tonic made of "essence of limpet shells" helps resolve relationship problems and strengthens character. In the British Western Isles, nursing mothers drank limpet broth to increase milk production. In a more reasonable use, limpet shells were once used to regulate doses of medicine.

Similarly to other shells from the intertidal, limpets provided a source of baubles, pendants, and beads.

Definitive evidence of human manipulation of limpet shells during prehistoric times is available in Britain. More recently (early nineteenth century and before), keyhole limpet shells were fashioned into oval hair ornaments by the Nicoleño (Canaliō) people of San Nicolas Island, California, and became a characteristic object of that culture. Hats worn by Tlingits and Aleuts (Alaska) emulate the ellipsoid–conical shape of limpets.

CULTURAL USES OF MARINE MAMMALS OF THE INTERTIDAL

Marine mammals associated with the intertidal include seals, sea lions, and sea otters. Only sea otters and marine otters use rocky-intertidal invertebrates as a major food resource. Pinnipeds (largely transient in the intertidal) held only minor cultural significance, except that seal hunting had the greatest economic value (after salmon) to people of the Northwest Coast, especially farther north. Beyond food, oil, clothing, and pack material, seal and otter teeth were used for ornamentation. Sea lion whiskers adorned the tops of some carved headdresses. Seal, sea lion, and otter motifs are occasionally seen in Northwest Coast traditional graphics and carvings.

Sea otters attained an unusual prominence in the history of the eighteenth- and nineteenth-century North Pacific. Initially extirpated from the intertidal and later from the kelp forests, sea otter pelts became the most valuable commodity in trade between the West Coast and Asia, achieving upward of US$250 each. Chinese high society and royalty prized sea otter fur as features of their finest regalia. As otters became scarce in the intertidal of the lower latitudes, highly iridescent green abalone shells from the south could be traded in the north for the finest furs.

THE ROCKY INTERTIDAL AS ART GALLERY

Smooth intertidal rocks have been used as a sort of artistic canvas, gallery, and theater by coastal peoples who have left their marks in the form of petroglyphs and pictographs. In the Pacific Northwest, pictographs above intertidal habitats identified clan territory, served as memorials, and fulfilled other purposes as well. Although most indigenous rock art was created away from the intertidal zone, some were pecked into intertidal rocks and boulders above and below the tide line with an apparent connection to the water or the spirits and animals living there. In areas where tides played a part in stories and legends, petroglyphs were made below the tide line. As the tide ebbed, images were revealed.

The properties of the rock did not allow much detail in petroglyphs, so the artists often focused on general forms or a specific feature of a form. Images included people (anthropomorphs—stick figures, body forms, and heads), animals (zoomorphs—birds, whales, and fish), graphic symbols, cultural symbols (masks, revered shapes, stylized moieties), and objects. The purpose of these works is not always obvious. Dating petroglyphs is difficult as is determining the reason or reasons for making them.

MODERN CULTURAL USES

Modern Shell Collecting and Jewelry

Shell bead jewelry remains popular, without much of the traditional significance it once held. Although inexpensive glass and plastic have taken much of the shell bead market, some of the finest and most expensive modern jewelry uses marine shells. The rare, appealing, and colorful shells can command hundreds of dollars each. Adventure travel programs have developed around shell collecting in exotic places. Conchologists sponsor dozens of shell shows each year, where prizes are awarded in a variety of categories.

The Rocky Intertidal in Science, in Conservation, and as a Destination

Environmentalists frown on the crude exploitation of the intertidal for trinkets and souvenirs, especially in high-tourism areas such as Hawaii. Where obvious damage to the environment has occurred, such behavior has been outlawed. High commercial exploitation of the intertidal still exists in countries with a less-developed environmental ethic.

Scientific study of the rocky intertidal has gone beyond taxonomy, biology, and ecology. In-depth studies of particular rocky-intertidal animals are yielding information regarding such widely different disciplines as climate modeling and material engineering. Geochemical variation and sclerochronology of limpet shells (*Patella vulgata*) are being used as proxies for determining climate fluctuations during the archeological Holocene and fossilized Neogene. Biomimetic research ("science mimicking nature," the study of useful structures and material properties in nature with a view toward reproducing them in technology) is looking at the calcium carbonate "tile" laminate and protein adhesive the red abalone uses to build a shell structure able to withstand considerable blunt force. The work may lead to the development of new materials for lightweight and effective body armor from the microscopic level.

A few intertidal locales have been preserved in part for their cultural significance. The area encompassing the James V. Fitzgerald Marine Reserve (San Mateo County,

California) is a prime example of an intertidal site with cultural significance in the modern era. Marine reserves with rocky-intertidal habitats include Bodega Bay (California) and Kermadec Islands/Goat Island Marine Reserve (also known as Leigh Marine Reserve and Cape Rodney–Okakari Point, New Zealand) and Poor Knights Island (New Zealand).

SEE ALSO THE FOLLOWING ARTICLES

Abalones / Food Uses, Ancestral / Limpets

FURTHER READING

Gastronomical culture from the intertidal. http://theworldwidegourmet. com/fish/shell/

Henshilwood, C.S., ed. The Blombos Cave Project. http://uib.no/sfu/ blombos, January, 2004.

Moratto, M.J. 1984. *California archaeology*. Orlando, FL: Academic Press.

Ogden, A. 1941. *The California sea otter trade 1784–1848*. Berkeley, CA: University of California Press.

Vanhaeren, M., F. d'Errico, C. Stringer, S.L. James, J.A. Todd, and H.K. Mienis. 2006. Middle Paleolithic shell beads in Israel and Algeria. *Science* 312: 1785–1789.

T

TEACHING RESOURCES

SEE EDUCATION AND OUTREACH

TEMPERATURE CHANGE

LAUREN SZATHMARY AND BRIAN HELMUTH
University of South Carolina

Global climate change encompasses many indirect and direct impacts on the physiology and ecology of plants and animals. Changes in precipitation, sea level rise, nutrient availability, and rates of erosion are all predicted to occur in the coming decades over a range of spatial and temporal scales. One of the most direct and damaging effects of climate change, however, is the impact of global warming on the body temperatures of organisms, and the subsequent impacts of changes in organism temperature on survival, physiological performance, and patterns of species distribution. Understanding how climate drives patterns of body temperature and how changes in global climate are likely to alter spatial and temporal patterns of organism temperature, is therefore key for forecasting the likely impact of climate change on species distribution patterns.

A MODEL SYSTEM TO STUDY EFFECTS OF TEMPERATURE CHANGE

Patterns of both environmental parameters and organism body temperatures in the rocky intertidal are quite complex and vary considerably spatially and temporally; thus, examining how changing climate will affect this ecosystem is a complex undertaking. Additionally, different ectothermic species achieve different body temperatures under identical sets of environmental conditions because of the species' distinct physical properties. The complex interaction of climate with the characteristics of an organism and its habitat in driving patterns of body temperature therefore calls for a systematic, thorough approach to determining where and when to look for the direct impacts of climate change in the rocky intertidal.

The rocky intertidal zone is a model system for exploring how climate change will affect species distributions in nature. Organisms in rocky intertidal zones are thought to live very close to their thermal tolerance limits, resulting in regular patterns of zonation. Several studies have demonstrated strong correlations between temperature and the distribution of intertidal species and populations over a range of spatial and temporal scales. Therefore, changes in the body temperatures of these organisms are expected to have large effects on the patterns of zonation and geographic distribution of species in this ecosystem. Specifically, if rocky intertidal species are already living at the limits of their physiological tolerances, then any significant change in the physical environment should be reflected by shifts in upper zonation limits. Namely, in the face of warming we should see a downward shift in the upper zonation limits of species wherever this limit is set by some aspect of climatic stress. Second, if some aspect of physiological stress related to body temperature sets species range boundaries, then we should similarly observe a geographic shift in the distribution of warm- and cold-acclimated species. Such shifts have been observed at geographic range boundaries. Recent evidence, however, suggests that we should be looking at the centers of species distributions in addition, because environmental conditions in the middle of range boundaries can in some cases be as extreme as they are at range edges.

DRIVERS OF INTERTIDAL
TEMPERATURE CHANGE

Intertidal ecosystems exist at the interface of the marine and terrestrial environments (Fig. 1). Because organisms in these ecosystems are alternately submerged during high tide and aerially exposed during low tide, they must contend with stresses imposed by both environments. These species are therefore affected by changes in both water temperature and terrestrial climatic parameters (e.g., air temperature or solar radiation). Thus, both aquatic and terrestrial climatic influences are ecologically important and will result in complex patterns of ecosystem change. Only by examining the effects of both changing water temperature and changing aerial conditions will we be able to gain a clear understanding of how temperature change as a whole will affect the rocky intertidal.

FIGURE 1 Rocky intertidal at Strawberry Hill, Oregon. Photograph by Brian Helmuth.

Water Temperature

Water temperature has a large effect on the physiology of species and species interactions in marine ecosystems. Animals that inhabit rocky intertidal zones are ectotherms, so while underwater, their body temperatures are generally assumed to approximate the temperature of the surrounding water. In some cases water temperature extremes can lead to mortality of adult organisms during high tide, and several studies have shown that elevated or decreased water temperatures can lead to spawning failures or larval mortality. Body temperature extremes during submersion also have significant effects on physiological rate functions, such as changes in metabolism and feeding rates.

Water temperature further has an indirect influence on aerial body temperature by setting the initial temperature of organisms prior to aerial exposure and by serving as a heat source or sink to the substratum. In fact, some recent models have suggested that in some parts of the intertidal, water temperature can contribute as much to the heat budgets of some intertidal organisms as air temperature.

Aerial Temperatures

The body temperatures of intertidal organisms are generally much higher during aerial exposure than they are during immersion. While exposed to aerial conditions, body temperatures are the result of many interacting environmental variables, such as solar radiation, wind speed, air temperature, relative humidity, and cloud cover. Air temperature is often used as a proxy for intertidal organisms' body temperature, but there is often a very poor correlation between air temperature and body temperatures on any given day. For example, an animal's body temperature can be as much as 20 °C higher than the surrounding air temperature. Exposure to aerial conditions can therefore cause body temperature extremes, which can lead to mortality events from heat stress as well as more chronic forms of thermal damage.

FORECASTING IMPACTS OF
TEMPERATURE CHANGE

While the complexity of thermal regimes presented by the intertidal environment poses a challenge to intertidal ecologists, it also presents an unparalleled opportunity for testing hypotheses regarding the effects of climate and climate change on species distribution patterns. By identifying areas where body temperatures are most likely to be physiologically damaging or lethal, we can anticipate where climate change is most likely to have ecological impacts. Importantly, recent research has shown that the complexity of quantifying patterns of body temperature and thermal stress is not as straightforward as previously assumed.

As the tide recedes and the organism is aerially exposed, its temperature is driven by the interaction of environmental factors such as solar radiation, wind speed, air and ground temperature, and relative humidity. On a sunny summer day, the body temperature of an intertidal ectotherm can easily exceed 30–35 °C; whereas it can fall to below freezing during exposure on clear, cold nights. Intertidal body temperatures can strongly depend on the amount of incident solar radiation, and even small differences in the angle of the substratum relative to the sun can have large impacts on body temperature. Animals and plants on horizontal, upward-facing substrata can experience temperatures that are over 10 °C higher than those experienced by animals on adjacent vertical, pole-facing slopes located only a few centimeters away. Moreover, these differences can exceed those observed over the scales of hundreds to thousands of kilometers of coastline.

While the body temperature of intertidal organisms during aerial exposure at low tide is driven by terrestrial climate conditions, the timing and duration of that exposure are driven by the dynamics of the tidal cycle and modified by the intensity of wave splash. Therefore, geographic patterns of intertidal body temperature are complex because of differences in the timing of low tide. For example, in Washington and Oregon, low tide in summer often occurs at midday, when climatic conditions are hottest. In southern California, low tides in summer seldom occur near midday. As a result, patterns of thermal stress do not necessarily become hotter as ones moves south along this coast; instead, a thermal mosaic of "hot spots" emerges. At these sites, where low-wave-splash regimes coincide with periods of hot climatic conditions and low tides in summer, conditions may be extreme. Critically, these sites may occur in the middle of species ranges as well as at range edges and thus are likely locations to observe impacts of climate change.

Importantly, the rate of heat transfer between an organism and its physical environment is to some extent driven by its size and morphology and can be strongly affected by characteristics such as color and material properties. Exposure to aerial climatic conditions at low tide therefore affects species differently, and the body temperatures reached by these organisms are determined in part by the physical properties of the animals themselves. Thus, species experience different body temperatures under identical sets of environmental conditions. For example, the sea star *Pisaster ochraceus* has the ability to evaporatively cool; whereas the mussel *Mytilus californianus* does not. These body temperature differences across species could have important impacts on species interactions. In the *Pisaster–Mytilus* example, if *Mytilus* is more thermally stressed than its predator, the foraging success of *Pisaster* may improve. Conversely, in a case where a predator is more thermally stressed than its prey, climate change may lead to increases in prey populations. In this example, because *Pisaster* is a keystone species, changes to foraging rates could affect the ecosystem as a whole.

The complexity of thermal regimes in the intertidal is driven by differential body temperatures reached by individual species, how these differences in body temperature affect ecological interactions, and tidal cycle and wave splash regimes. To understand fully how climate change will impact the intertidal, we must take all of these factors into account and keep in mind that evidence of change will likely be seen both at range limits and throughout the ranges.

SEE ALSO THE FOLLOWING ARTICLES

Climate Change / Heat and Temperature, Patterns of / Heat Stress / Rhythms, Tidal / Temperature, Measurement of / Tides

FURTHER READING

Denny, M. 1993. *Air and water: the biology and physics of life's media*. Princeton, NJ: Princeton University Press.
Gates, D. 1980. *Biophysical ecology*. New York: Springer-Verlag.
Kondratyev, K. Ya., and A. P. Cracknell. 1998. *Observing global climate change*. Bristol, PA: Taylor and Francis.

TEMPERATURE, MEASUREMENT OF

JENNIFER JOST AND BRIAN HELMUTH
University of South Carolina

Temperature plays a crucial role in the function of proteins and enzymes and, therefore, in cellular processes such as metabolism. Recent innovations at the molecular and biochemical levels have precipitated major advances in our understanding of how temperature drives organismal biology and ecology. One major gap in our knowledge is a quantitative understanding of what the body temperatures of most organisms are under natural field conditions and how geographic patterns in body temperature may translate into species distribution patterns. Importantly, as is true for the terrestrial environment, the temperature of intertidal ectotherms is often very different from the temperature of their surrounding habitat. Accurately measuring the temperatures of organisms and their environment at appropriate temporal and spatial scales is therefore important for understanding the impacts of climate change on species distribution patterns in nature. To some extent, the current lack of accurate field body temperature data is the result of difficulties inherent in deploying instrumentation in the physically harsh intertidal environment. The recent commercial availability of rugged, low-cost instruments capable of recording temperatures at the scale of centimeters and at high rates has made the measurement of temperature in intertidal zones commonplace. However, each type of instrument carries with it a different set of potential pitfalls that must be confronted when measuring the temperatures of intertidal organisms.

DETERMINANTS OF TEMPERATURE IN THE INTERTIDAL ZONE

Environmental Determinants

Multiple climatic factors drive the flux of heat into and out of organisms and their surrounding environment, including solar radiation, wind speed, air and ground temperature, cloud cover, and relative humidity. The size, color, mass, material properties and morphology of an organism can all affect the temperature that it experiences, so that two organisms exposed to the same microclimates can experience very different temperatures. Importantly, all of these factors also affect the temperature recorded by instrumentation as well and must be considered when recording field temperatures.

Organismal Determinants

An important distinction is the difference between heat (energy) and temperature (a measure of average kinetic energy). Characteristics of an organism such as surface area, color, and morphology interact with environmental parameters such as solar radiation, wind speed, and air temperature to determine the rate of heat flux into and out of the plant or animal. The amount of heat energy needed to raise the temperature of the organism in turn depends on the mass and material properties of which the plant or animal is composed. All materials, regardless of shape or size, can be defined on the basis of their specific heat capacity, a description of the amount of heat energy (in joules, J), needed to raise one kilogram of material 1 K (kelvin; a change of 1 K = a change of 1 °C). Therefore, to convert from the heat stored in an organism's body to the temperature (T_b) of its body, it is necessary not only to keep track of the fluxes of heat into and out of the body but also of the organism's mass (m) and specific heat (c_p):

$$\text{Heat stored} = m\, c_p\, T_b = \text{Heat gained} - \text{Heat lost} \quad (1)$$

Thus, the flux of heat (energy) into an object increases the temperature of the body at a rate proportional to the product of the object's mass and specific heat (mc_p). A larger, more massive organism, or one composed of a material with a high specific heat, requires more heat energy to raise its temperature to the same degree as a smaller organism. The ability to dampen the response in body temperature to fluctuations in the environment

is termed "thermal inertia," a ratio of factors that resist changes in temperature (mass and specific heat) to those that enhance them (such as areas of exposure and the coefficient of heat transfer). Thus, when environmental conditions fluctuate at a rate faster than the thermal response time of an organism, the thermal inertia of the body will tend to decrease the temperature response of the body to those fluctuations, and the organism will be less likely to experience rapid changes in body temperature. However, when conditions change at a rate that is slower than the thermal response time of the organism, body temperatures will tend to track environmental conditions more closely.

MEASURING BODY TEMPERATURE DURING EMERSION

Data Loggers: Types and Uses

These same principles apply to the design of data-logging instruments (Fig. 1). Depending on the placement of the sensor, the response time of most commercially available data loggers (which is now usually reported by the manufacturer) ranges from less than a second (thermocouples) to several minutes in water, and the response time is usually many minutes in air. Due to the high thermal inertia of water, temperature changes are often slower than the thermal response time of the instrument, and thus the size of the logger can be ignored. The type of logger used, however, is of paramount importance in air, where the size, color, and morphology of the instrument can significantly alter the temperature it records.

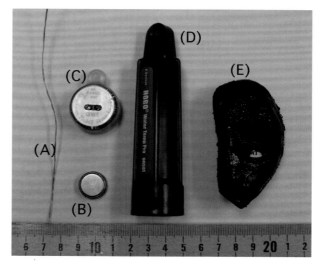

FIGURE 1 Data-logging instruments: thermocouple (A), iButton (B), Tidbit (C), Water Temp Pro (D), and a Robo Mussel (E). Scale shown in centimeters. Photograph by Jennifer Jost.

During aerial exposure at low tide, intertidal invertebrates and algae are exposed to fluctuating climatic conditions and may experience a large range of body temperatures in a short period of time. During high tide a water layer buffers intertidal invertebrates from changes in air temperature, wind speed, and solar radiation. However, during low tide the body temperature of an intertidal invertebrate is driven by multiple factors (e.g., mass, shape, specific heat capacity, solar radiation, air and surface temperatures) and is often considerably warmer or colder than the temperature of the surrounding air or substrate. Therefore, simply measuring air temperature does not always provide an accurate estimate of body temperature during low tide.

Importantly, the rate of heat transfer between an organism and its physical environment is to a large degree determined by its size and morphology and can be strongly affected by characteristics such as color and material properties. Subsequently, organisms exposed to identical environmental conditions can experience quite different body temperatures (Fig. 2).

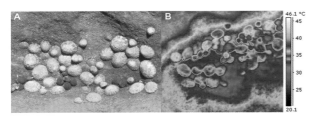

FIGURE 2 Limpets attached to rock substrate (A), and the same field of vision as seen through an infrared camera (B). Note body temperature differences between dark- and light-shelled limpets. Images by Brian Helmuth.

Because of these issues, whenever possible, temperature loggers must be thermally matched to organisms in order to accurately predict body temperature in the field. For organisms with temperatures closely linked to the surrounding sediment or rock substratum (e.g., barnacles), small data loggers that closely mimic surface temperature may be used; in all cases this assumption should be verified in the field.

Experimental studies have shown that failure to account for these effects can lead to large errors in body temperature estimation. A recent comparison of body temperatures of live mussels (*Mytilus californianus*) to unmodified Tidbit loggers showed that the average difference in temperature was approximately 14 °C. When the same instruments were encased in epoxy so that the thermal inertia, color, and morphology of the logger were made to closely approximate that of a living mussel, the average difference dropped to approximately 2.5 °C. Similar approaches have

been used to develop biomimetic models for intertidal sea stars, saltmarsh mussels, and snails, by using different materials (such as foam and silicone-filled shells). In all cases, the relationship between the temperature of the data-logging instrument and the temperature of the organism must be verified in the field. This is generally accomplished by using a thermocouple-based thermometer, which has a probe that has a high frequency response and is small enough to be placed even in small organisms. These instruments are generally too fragile to be deployed over long time spans in the field; otherwise they would avoid many of the problems noted.

An important implication of this discussion is that a single instrument design cannot provide accurate temperature estimates for multiple species of organisms. Moreover, there is no such thing as a single "habitat" temperature for any particular site. Instead, it is necessary to either create physical models with similar thermal properties (e.g., specific heat capacity, reflectivity, porosity, color, size) for each species of interest, or else to develop mathematical models for these species in order to relate climatic data to body temperatures.

Infrared Thermography

All objects at temperatures above absolute zero (0 K, -273 °C) emit radiation. The wavelength at which this radiation is emitted is related to the temperature of the object. Furthermore, the hotter an object is, the more energy it emits in a particular wavelength band. For objects in the biologically relevant range of temperatures (0–100 °C), energy is emitted in the infrared spectrum (wavelength >2000 nm). By measuring the energy in this spectrum, and by understanding some aspects of the object's emissivity, we can measure the temperature of an object via noncontact methods. Although expensive, such instrumentation opens up the door for an exploration of temperature variability over a wide range of temporal and spatial scales in nature.

New methodologies in data-logging technology have created a vast array of opportunities for the measurement of temperature in the physically harsh intertidal environment. However, the application of these techniques must be approached with caution and with an understanding of how organisms and instruments interact with their physical environment. For a more full understanding, see some of the works in the Further Reading list.

FURTHER READING

Arya, P. 2001. *Introduction to micrometeorology,* 2nd ed. International Geophysics Series. San Diego, CA: Academic Press.
Denny, M. 1993. *Air and water: the biology and physics of life's media.* Princeton, NJ: Princeton University Press.
Gates, D. 1980. *Biophysical ecology.* New York: Springer-Verlag.
Hochachka, P., and G. Somero. 2002. *Biochemical adaptation: mechanism and process in physiological evolution.* New York: Oxford University Press.
Unwin, D., and S. Corbet. 1991. *Insects, plants, and microclimate.* Slough, UK: Richmond Publishing Company.

TEMPERATURE PATTERNS

SEE HEAT AND TEMPERATURE, PATTERNS OF

TERRITORIALITY

JOSÉ M. ROJAS AND F. PATRICIO OJEDA

Pontificia Universidad Católica, Santiago, Chile

Territoriality is a behavior related to the active defense of a delimited area by an individual. The motivation behind the establishment of this aggressive behavior is usually tied to survival, through the exclusive use of food or refuges in the area, or to reproduction, through the access to mates and protection of offspring from an intruder. An underlying fact of these behaviors is that the reaction is not rigidly determined by location alone (e.g., it may be tied to a particular point in the mating period or to the competitive ability of the defender relative to the adversary). In this manner, different degrees of aggressiveness or behavioral displays can be described in a defender as a function of the threat level that an intruder represents, given the reproductive states, state of ontogenetic development, or position in the social hierarchy. Based on this, it is assumed that territoriality is governed by a decision-making process in which the individuals are capable of recognizing their invaders and consequently evaluating the costs associated with defending the area in terms of the possibility of physical damage in the case of a confrontation.

GENERALITIES IN STUDIES OF TERRITORIALITY

Basically there are two approaches used in studies of territoriality. On the one hand are those approaches focused on behavioral strategies used by individuals for maintaining an area of exclusivity, with a special emphasis on the aspects tied to the result of an antagonistic interaction between individuals. The other approach is directed toward the community or population consequences of the interaction result of the establishment of the exclusive use territories. This latter approach involves aspects such as the implications in structuring social groups through hierarchies of dominance in the access to resources or patterns of distribution of the population in the area. An interesting element of this last approach is the importance of social elements to the definition of territoriality, which implies a third axis in the already-established definition of territoriality based on the availability of the resource in space and the motivation of the defender at a given point in time.

TERRITORIALITY IN THE ROCKY INTERTIDAL

The literature on behavioral patterns of intertidal animals is limited, and studies that relate behavior to the spatial structure of the populations are even more critically limited. It is in this context that the examples of territoriality in the rocky intertidal described herein focus on the mechanisms of territorial defense and the motivations for abandoning it.

One of the common interests in territoriality studies is establishing the basis in which territory size is defined. In general terms the results show a positive relationship between territory size and body size of the defender, regulated by an increase in the energetic requirements of the individual with body size and the costs of defense associated with the increase of the territorial perimeter. One of the pioneer examples comes from studies by Stimson (1969) with the limpet *Lottia gigantea*, one of the most important grazers on Californian rocky shores. Stimson describes the occurrence of individual feeding areas on the rock that were actively defended in the face of an intruder, with an intensity that was proportional to the size of the defender. The observations show a large variability in the defensive mechanisms used by *Lottia* in response to intruders. The mechanisms used depend on whether the intruder is a predator or a competitor and, in the latter case, the resource for which they are competing. Mytilid mussels, barnacles, and anemones are *Lottia's* primary competitors for space, and when confronted with one of these species, the *Lottia* bulldozes over individuals that recruit into its territory or overgrow the boundary into its territory. On the other hand, when faced with other herbivores, such as *Acmaea*, *Tegula*, or other limpets, *Lottia* responds by destabilizing the intruder, pushing and hitting it in the foot to increase the probability of the intruder getting washed off by a wave (Fig. 1). When confronted with a predator, such as *Thais marginata* or *Acanthina spirata*, *Lottia* responds by a behavior known

FIGURE 1 Two adult limpets (*Scuttellastra argenvillei*; 9.0 cm in length) violently confront one another in battle for the right to a prime feeding territory on the wave-exposed west coast of South Africa. With raised shells, the two rivals are poised in wrestling position, thrusting and pushing their opponent until the weaker abandons the area. Such contests can be fatal if they occur between challengers of contrasting size; the larger lifts its shell and slams it down to crush the smaller. Location is everything for these limpets, which must trap live kelp fronds that sway in the surf. Photograph by Evie A. Wieters.

as mushrooming, which consists of raising its shell up in a menacing strategy, and later bringing the anterior end of the shell down on the foot of the predator so that the latter retracts its foot. In consequence, *Lottia* manages to weaken the attachment force of the predator, making it more susceptible to dislodgement by waves. This territorial behavior pattern has been widely documented in a wide range of intertidal herbivorous gastropods, particularly those that present low tidal migrations. These animals, as adults, establish territories around scars in the rock and react particularly aggressively toward conspecifics. An interesting aspect of this pattern corresponds to the decrease in the level of physical contact that the defender establishes when faced with an intrusion in relation to the magnitude of the threat that it may receive at the beginning of the interaction.

Similarly, a common intertidal limpet in the Chilean coast, *Scurria araucana*, displays a variety of behavioral patterns related to food acquisition, although the most elaborated responses involve only a fraction of the animals at each population. In such cases, animals develop a scar and possess an individual foraging area (garden) proportional to their body size, where they graze on small or juvenile algae (Fig. 2). Inside the garden, continuous grazing maintains an early successional algal assemblage acting as a self-renewing food source, actively defended against conspecific or interspecific intrusions independently of the size of the intruder. Nonetheless, experiments demonstrated that all limpets can develop a scar and display agonistic behavior regardless their size, sex, or the possession of a garden, and especially in particular

microhabitats such as high tidepools. Thus, all *S. araucana* individuals have the potential to acquire and defend a territory, although the expression of such behavior seems to depend on food availability and environmental conditions.

FIGURE 2 An intertidal limpet (*Scurria araucana*; 2.5 cm in length) patrols its garden, a triangle-shaped territory established among chthamalid barnacles at the rocky shores of Lirquén, southern Chile. The limpet forms a home scar (the discolored mark to the left), which it leaves only to expel intruders and to graze on the small algae colonizing the rock surface inside the garden. Photograph by Patricio A. Camus.

Ontogenetic Factors in Territoriality

The display of territoriality behaviors can change through ontogeny, as exemplified by small *Lottia*, in which no adverse reactions are observed when faced with competitors for space, apparently because of their low energetic requirements, which are fulfilled by algae that even settle on the shells of the competitors. A less extreme case of ontogenetic behavioral alterations has been observed in porcelain crabs such as *Allpetrolisthes spinifrons*. This species is a permanent habitant of two species of anemones (*Phymactis clematis* and *Phymacthea pluvia*) in the mid to low zones of the Chilean rocky intertidal. The anemone provides the crabs with food, as a consequence of the way in which the anemones feed, and refuge from predators. The number of crab residents on the anemone depend on their body size and the size of the host. When the number of smaller crabs increases too much or large-sized crabs are present, antagonistic displays develop among individuals. Following a period during which invader and defender mutually examine one another with their antennae or periopods, defenders use their chelae to push the invaders out. The final result of the interaction between the crabs is usually determined by the sizes of the individuals; this can erroneously lead one to believe that the development of territorial behaviors in these porcelain crabs is exclusively in the large-sized individuals.

Intertidal Territoriality in Fish

Gibson (1982) reviewed the literature on the biology of intertidal fish, which included a thorough analysis on the knowledge of territorial behaviors up to the time, and concluded that the display of territoriality in intertidal fish is highly variable. For example, in the goby *Gobiosoma chiquita*, males are responsible for defending territory, whereas in the Hawaiian rockskipper, *Istiblennius zebra*, the females also have to defend the territory, but the males do so much more aggressively. In the case of the Caribbean blenny *Ophioblennius atlanticus*, both sexes are consistently aggressive when faced with any intruder. The motivation behind the development of competitive behaviors principally include the defense of feeding and nesting sites, refuges from predation, and, as in the porcelain crabs, the results of the interactions are usually determined by the body sizes of the individuals. The amplitude of the territory, however, is not dependent exclusively on the individual energetic costs and requirements for defending the area but also on the costs and requirements for maintaining it.

Studies on the activity patterns of fish have shown that, because of the greater damage that would be incurred if the fish get washed away, turbulence is a key factor in defining the daily activity rhythms. Comparisons made between intertidal and subtidal species of blennies show both a decrease in the size of the territory as well as the frequency of policing in the intertidal. One element that stands out in these results is relevance of physical conditions as a component affecting the cost of maintaining an area of exclusion, from which the availability of refuges, and therefore the protection from predators or wave action, can be considered as limited resources. The hierarchical population structures have been described for the gobies *Lepidogabious lepidus*, *Gobiosoma chiquita*, and *Chasmodes bosquianus*, in which only the dominant individuals have territories in which they have priority access to refuges and food, and subordinate individuals lack them. Territorial behaviors linked to social hierarchies have also been described in a reproductive context. In the three-finned blenny, *Tripterygion delaisi*, the nest sites are protected by males, and the females select their mates based on the ability of the males to defend the resources for the offspring and defend the offspring from predators. The males select and actively defend cavities or cracks in the rock as possible nesting sites, and then make direct attacks against any intruder, either conspecific or not. This type of behavior is only developed during reproductive periods by dominant males, identified by visible coloration patterns that apparently signal to other conspecifics their position in the social hierarchy. This use of coloration also carries costs, as it makes them more conspicuous to predators; but the benefits, in terms of minimizing the possibility of damage from physical confrontation between antagonistic individuals and in terms of final reproductive success, are apparently sufficient to maintain this feature.

CONSEQUENCES OF TERRITORIALITY ON THE DISTRIBUTION OF ANIMALS

The defense and consequent exclusive use of resources by an individual carries direct repercussions on the partitioning of resources into a population or community based on the defense capabilities of the individuals. Among the consequences, the most evident and quantifiable is the spatial distribution of individuals, where it is expected that less able defenders or weaker competitors should occupy areas of lower-quality resources. In this context, the high physical variability of the intertidal environment means that a large part of the evidence on the occurrence of territorial behaviors and their consequences are extrapolations from laboratory studies, and in a relatively small fraction of the animals that inhabit the intertidal. Moreover, the available information indicates that the majority of territorial behaviors are actually occurring during high tides, when it can also be expected that antagonistic interactions are occurring. Therefore, the high degree of uncertainty regarding the scenario of natural behavioral of intertidal species is evident, where aspects such as the specific conditions under which they occur and their consequences for the communities and populations in the system are yet to be fully understood.

SEE ALSO THE FOLLOWING ARTICLES

Competition / Fish / Predator Avoidance / Recruitment / Reproduction / Size and Scaling

FURTHER READING

Branch, G. M. 1975. Mechanisms reducing intraspecific competition in *Patella* spp.: migration, differentiation and territorial behaviour. *Journal of Animal Ecology* 44: 575–600.

Gibson, R. N. 1982. Recent studies on the biology of intertidal fishes. *Oceanography and Marine Biology Annual Review* 20: 363–414.

Maher, C. R., and D. F. Lott. 1995. Definition of territory used in the study of variation in vertebrate spacing systems. *Animal Behavior* 49: 1581–1597.

Stimson, J. 1969. Territorial behavior of the owl limpet, *Lottia gigantea*. *Ecology* 51: 113–118.

Verner, J. 1977. On the adaptive significance of territoriality. *American Naturalist* 111: 769–775.

THERMAL STRESS

SEE HEAT STRESS

TIDEPOOLS, FORMATION AND ROCK TYPES

GARY GRIGGS

University of California, Santa Cruz

Tidepools form where resistant bedrock is exposed in the intertidal zone. The type of rock exposed, the tidal range, as well as wave impact, abrasion, chemical weathering, and grazing and boring by intertidal organisms are all important processes affecting the formation of tidepools.

GEOLOGIC SETTING FOR TIDEPOOLS

Several geologic factors are necessary for the formation of tidepools, and these conditions do not occur along all coastlines. Tidepools require the existence of resistant rock outcrops in the intertidal zone as well as some minimal tide range. There are many low-relief, usually geologically older coasts, such as the Atlantic and Gulf Coasts of the United States from New Jersey to Texas, where the landscape has been lowered by millions of years of erosion, and sediment deposition dominates the coastal zone and the landforms. Typically the coastline of these areas consists of a very wide, low-relief coastal plain, back bays, barrier islands, and dunes with sandy beaches forming the shorelines. There is no bedrock exposed, and therefore there are no rocky intertidal zones or tidepools to study or enjoy.

Along the rocky coast of England and much of the northern Mediterranean, Nova Scotia, Maine, or along geologically active coastlines such as the west coast of North America, the intertidal zones are typically rocky, and tidepools are common features. There are still many areas of the California coast (the sandy beaches of much of the Los Angeles County coast, for example) where tidepools do not exist, but there are many more areas where geologic conditions are ideal and tidepools, with their diversity of intertidal life, have formed.

FORMATION OF TIDEPOOLS

The best-developed tidepools form where hard or resistant rock occurs in the intertidal zone and where wave action and other erosional processes have sculpted or otherwise worn away portions of the exposed rock, leaving holes or depressions where seawater can collect at high tide (Fig. 1). Tidepools can form in crystalline rock such as granite, volcanic rock such as basalt, or, very commonly, in a wide variety of sedimentary rocks such as sandstones, shales, or conglomerates.

FIGURE 1 Holes or depressions worn in resistant bedrock along weaker zones or material form ideal tidepools. Photograph by the author.

What is necessary for tidepools to form is that the rock exposed to wave action is hard enough to survive the regular attack of the waves, but has sufficient weaknesses, such as joints, fractures, stratigraphic or bedding differences, that the exposed rock surface is worn away or eroded unevenly. Weaker layers, beds, or surfaces will be removed more rapidly than the adjacent more resistant rock, so depressions, hollows, or cracks can develop and increase in size over time, allowing seawater to collect and organisms to colonize and occupy these areas (Fig. 2).

Erosion of intertidal rock outcrops and the formation of tidepools can occur through several processes including

FIGURE 2 Tidepools have formed due to differential erosion along bedding planes in sedimentary rock. Photograph by the author.

wave impact, abrasion, solution or chemical weathering, and bioerosion. The hydraulic forces of breaking waves themselves can be very large and are sufficient to dislodge weaker or less coherent rock fragments and initiate the formation of irregular surfaces in the intertidal zone.

Abrasion can also be a powerful agent for grinding or wearing away solid rock. Abrasion depends upon what "tools," such as sand, gravel, cobbles, or shells, are available for the breaking waves and surf to wash, drag, or roll across an intertidal bedrock surface. Some tidepools are quite deep and are often called potholes. They have been created by harder cobbles that were either transported into the tidepool by wave action, or they may be worn fragments of the adjacent intertidal rock that have been washed back and forth and around the bottom of the hole or pool, much like the action of a mortar and pestle. With time, the scouring action of these tools can enlarge and deepen the tidepool.

Solution or chemical weathering results from the action of salt water, solar heating, and the atmosphere. Salt water is very corrosive, and when intertidal rock is submerged in cold seawater at high tide and then is exposed to sunlight and heats up during low tides, it will gradually break down or decompose the weaker minerals in the rock. The constant wetting and drying, heating and cooling, expansion and contraction, as well as the chemical action of the salt water, are all effective in chemically and physically breaking down the rocks exposed in the intertidal zone, making them weaker and even more susceptible to wave impact and abrasion.

Bioerosion is not often thought of as an important process, yet over time, limpets, snails, chitons, and sea urchins all graze and scrape the rock surfaces as they feed, removing algae, fungi, and lichens, which are pioneer colonizers in the intertidal zone. Other bioerosion is accomplished by barnacles and worms, and other boring organisms that not only directly remove rock material but also, by enlarging crevices and creating weaknesses in rock structure, render the residual rock more susceptible to wave action and weathering. Kelp and other large algae anchor themselves to the bedrock, and where the bedrock is weak or partially detached, wave action on the algae can break loose fragments of bedrock, thereby degrading or lowering the surface of the bedrock in the intertidal zone.

Wave and tidal action are very important factors in the formation of tidepools (Fig. 3). Waves are the driving force behind the hydraulic impact and the abrasion that serve to grind, scrape, or otherwise erode or degrade the bedrock surface, such that depressions or hollows can form in the softer or weaker material and fill with seawater. A significant tidal range is also important in the

FIGURE 3 Shore platforms elevated above the low-tide terrace are good geological locations for tidepools. Photograph by the author.

development of a large intertidal zone, because the width of the rock shelf or intertidal zone exposed is dependent upon the differences in elevation between high and low tides. Although the tidal range changes monthly in response to the pull of gravity resulting from the differing positions of the sun and moon, a maximum spring tidal range of at least 2 meters will produce a more extensive intertidal area and increase the potential for tidepool formation than a tidal range of a meter or less.

Many areas of well-developed tidepools form where a rocky low-tide terrace or a shore platform exists. Extensive low-tide terraces typically occur where the bedrock is layered sedimentary rock, where the bedding is nearly horizontal, and where there is little sand in the intertidal zone to cover the bedrock. These conditions allow wave erosion to create a wide, nearly horizontal, rocky terrace, exposed at low tide, with irregularities or weak areas having been eroded more extensively to create pools or hollows. Shore platforms typically form about 2 meters above sea level and are therefore not submerged at all high tides, but they are splashed at most high tides and can be completely washed over by large storm waves at high tides (see Fig. 2). They are believed to be a result of a process called water table weathering, whereby the rock that remains wet or saturated is hard and quite resistant to wave attack. The rocks at the base of the sea cliff, in contrast, which are periodically wetted and dried, heated and cooled, undergo physical and chemical breakdown and progressive failure. Elevated shore platforms are common worldwide, where nearly horizontally bedded sedimentary rocks are exposed along the coastline, and are ideal areas for tidepool formation. The greater the tidal range and the more wave energy, the wider the platform or low-tide terrace will be, which will provide more area for tidepool formation.

SEE ALSO THE FOLLOWING ARTICLES

Coastal Geology / Fossil Tidepools / Salinity Stress / Stone Borers /
Wave Forces, Measurement of

FURTHER READING

Griggs, G. B., and A. S. Trenhaile. 1994. Coastal cliffs and platforms, in
   Coastal evolution—late Quaternary shoreline morphodynamics. R. W. G.
   Carter and C. D. Woodroffe, eds. Cambridge, UK: Cambridge
   University Press, 425–450.
Trenhaile, A. S. 1987. *The geomorphology of rocky coasts.* Oxford: Clarendon/
   Oxford University Press.
Woodroffe, C. D. 2003. *Coasts—form, process and evolution.* Cambridge,
   UK: Cambridge University Press.

TIDES

MARK W. DENNY
Stanford University

Tides are the "heartbeat of the ocean," and nowhere is that
pulse more evident than in the intertidal zone of rocky
shores. Under the competing pull of centrifugal force
and gravitational attraction, sea level rises and falls periodically. Typically, there are two high tides each day, one higher
than the other, and two low tides per day, one lower than
the next. The range of the tides (the difference in height
between the higher high tide and the lower low tide) fluctuates with the phase of the moon and the time of year,
and varies drastically from one location to another—from
less than 0.3 meter in places like the Mediterranean Sea to
more than 15 meters in the Bay of Fundy. The future pattern
of the tides can be predicted from celestial mechanics, but
changes in atmospheric pressure and the effects of storms
can cause measured tides to differ from those predicted.

CENTRIFUGAL AND GRAVITATIONAL FORCES

The mechanics of tides were first explained by Sir Isaac
Newton in his *Principia Mathematica* (1687) and serve as
a classic application of two basic laws of physics:

1. Acceleration of an object—a change in its speed or
   direction—occurs only under the application of a net
   force.
2. If object *A* places a force on object *B*, object *B* places
   an equal force on object *A*, but this reaction force acts
   in the opposite direction.

Consider an object traveling in a circle—a bucket of
water swung on the end of a rope, for instance (Fig. 1).

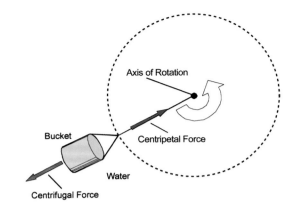

FIGURE 1 A bucket of water swinging on a rope serves as an example
of circular motion and the forces involved.

The bucket and water travel with constant speed, but
their direction continuously changes as they move
along their circular path. That is, the bucket and
water are continuously accelerated, and from law 1, we
know that a force must be applied. In this case, the
force is a tension in the rope, which pulls the bucket
towards the center of the circle. This is a "center-
directed" (centripetal) force, and wherever there is circular motion, a centripetal force must be present. The
bucket, in turn, imposes a centripetal force on the water
contained within it.

From law 2 we know that, because the bucket imposes
an inward force on the water, the water places an outward
(= centrifugal) force on the bucket. If one were to punch
a hole in the bottom of the bucket, water would flow out,
away from the center of circular motion. (Actually, the
water attempts to travel in a straight line while the bucket
tries to force it to move in a circle. But the concept of
centrifugal force is an alternative statement of the physics and is perhaps more intuitive.) It is centrifugal force
that provides much of the thrill when one rides a Ferris
wheel. At the top of the wheel, the upwardly directed centrifugal force opposes the downward force of gravity, and
one feels nearly weightless. At the bottom of the wheel,
centrifugal force adds to gravity, and one is pushed down
into one's seat.

A Ferris wheel provides useful insight into another
aspect of centrifugal force. Consider the Ferris wheel car
shown in Fig. 2A. As the car travels, its axle moves in
a circle around the center of the wheel, and centripetal
forces—always directed toward the center of the wheel—
are applied as shown by the dashed arrows,. Each of these
centripetal forces is accompanied by an outwardly directed
centrifugal force of equal magnitude (the solid arrows), as
described above. Focus now on the bottom of the car. It

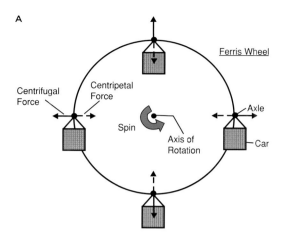

A

Ferris Wheel

Centrifugal Force

Centripetal Force

Axle

Spin

Axis of Rotation

Car

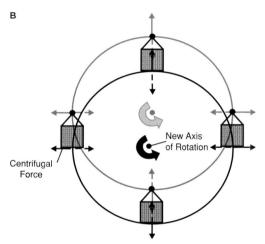

B

Centrifugal Force

New Axis of Rotation

FIGURE 2 The forces involved in the circular motion of a Ferris wheel car.

too travels in a circle, but in this case the circle's center is one car length below the center of the wheel (Fig. 2B). The centripetal force acting on the bottom of the car is directed toward this different center, producing centrifugal reaction forces as shown. Notice that the direction of the centrifugal forces acting on the bottom of the car are parallel to those acting on the car's axle. In other words, as the car moves in its circular orbit, centrifugal forces acting on any part of it are the same in both magnitude and direction. As will be seen, the earth and its oceans move in a circular path similar to that of the Ferris wheel car and are subjected to the same sort of centrifugal forces.

Let us digress for a moment to explore the concept of center of gravity. The center of gravity of an object (or system of objects) is the point at which one could locate all the mass of the system without changing its dynamics. Take, for instance, the system shown in Fig. 3A—two masses (m_1 and m_2) balanced on a massless beam. Under

the acceleration of gravity, m_1 pulls down on the left side of the beam, but given correct placement of the fulcrum, this action is counteracted by the weight of m_2 pulling down on the right side of the beam. Now, if we were to move both masses directly above the fulcrum, they would remain balanced (Fig. 3B). Thus, the position of the balance point locates the center of gravity.

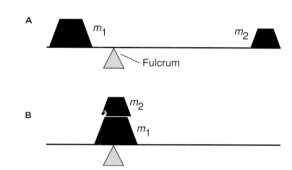

A

m_1

m_2

Fulcrum

B

m_2

m_1

FIGURE 3 The fulcrum of a balanced system of weights locates the center of gravity.

Now consider the motion of the earth and moon. The earth is 85 times more massive than the moon, and as a result, the center of gravity of the earth–moon system is actually inside the earth, about three-fourths of the way from earth's center to its surface (Fig. 4A). One is likely to think of the moon as traveling in an orbit around the earth, but in reality the earth and moon each travel in circles around their common center of gravity (Fig. 4B).

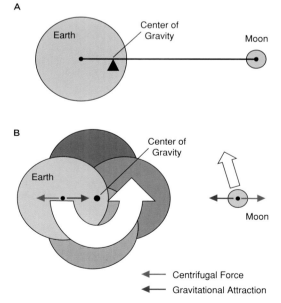

A

Earth

Center of Gravity

Moon

B

Earth

Center of Gravity

Moon

Centrifugal Force

Gravitational Attraction

FIGURE 4 The earth and moon move in orbits around their common center of gravity.

(The orbits are actually elliptical, a fact that will be taken into account later, but the deviation from a circular orbit is small enough that it need not concern us at this point.)

The fact that the earth and moon travel in circles is evidence that they are subjected to centripetal forces like those on a Ferris wheel car. Before, when we considered the circular motion of a Ferris wheel car, centripetal force was supplied by the wheel. Obviously, there is no wheel attaching the earth to the moon; instead, there is gravitational attraction. As Newton discovered, the force of gravity between two masses (M_1 and M_2) is directly proportional to the product of their masses and inversely proportional to the square of the distance between their centers (R):

$$\text{Attractive Force} \propto \frac{M_1 M_2}{R^2} \qquad \text{(Eq. 1)}$$

The distance separating the centers of the earth and moon as they move in their orbits is such that the gravitational force pulling these celestial objects together is exactly equal to the centripetal force required for circular motion. From law 2, we know that a centrifugal force—equal in magnitude (but opposite in direction) to the centripetal force—acts on both the earth and moon (Fig. 4B).

This equality of forces (centrifugal = gravitational) is true only at certain points on the earth or moon, however. For example, in Fig. 5, the gravitational force acting on a bit of ocean water at Point 1 is greater than the force acting on an equal mass of water at Point 2 because Point 1 is closer to the moon. In contrast, the centrifugal force acting on both masses of water is the same, because, in this type of circular motion, centrifugal force is equal everywhere (recall the discussion of the Ferris wheel car). At Point 1, gravitational force exceeds centrifugal force, and the resulting net force causes water to move toward

the moon. At Point 2, centrifugal force outweighs gravitational force, and water moves away from the moon. The overall effect of the interplay between gravitational and centrifugal force is the creation of two equal "bulges" of water on earth's surface (Fig. 6). One bulge has its peak under the moon, and the second has its peak on the side of earth opposite the moon. As ocean water moves into these bulges, "valleys" are left behind. These bulges and valleys are the tides.

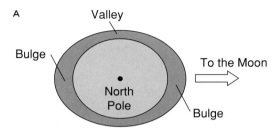

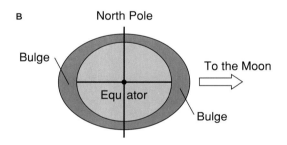

FIGURE 6 Two perspectives on tidal bulges: (A) a view from above the North Pole; (B) a view from above the equator.

THE PERIOD OF THE TIDES

To this point, we have treated the earth as moving in a fashion analogous to that of a Ferris wheel car—it travels in a circle—but we have ignored the fact that, once every twenty-four hours, the earth rotates around an axis more or less perpendicular (actually 18.3° to 28.6° from perpendicular) to the plane of the moon's orbit (Fig. 7). Consequently, as the earth rotates, a shoreline fixed to its surface is carried through the tidal bulges and valleys. When the shore is in a bulge (for instance, when the moon is high overhead), the tide is high, and when the shore is in a valley (when the moon is near the horizon), the tide is low.

From this explanation, one might expect that the period of the tides (the time from one high tide to the next) would be precisely half a day—12 hours. But the earth is not the only object in the system that moves. It takes the moon approximately 29 days to complete an orbit around

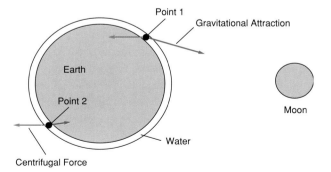

FIGURE 5 Water at different points on earth is subjected to different combinations of forces. At Point 1, gravitational attraction outweighs centrifugal force. At Point 2, centrifugal force outweighs gravitational attraction.

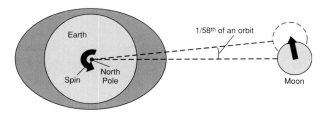

FIGURE 7 In the course of a day, the earth completes one rotation about its axis, and the moon moves through 1/58 of its orbit.

the earth (full moon to full moon), so in the half-day it takes the earth to rotate from one tidal bulge toward the next, the moon has advanced 1/2 × 1/29 = 1/58 of a circle. Because the moon travels in the same direction as the earth rotates, the earth must rotate that extra 1/58 of a circle to catch up to the bulge. 1/58 of a day is approximately 25 minutes. Therefore, the period from one high tide to the next is approximately 12 hours and 25 minutes, and one lunar tidal day (the period from one high tide under the moon to the next high tide under the moon) is approximately 24 hours and 50 minutes. Because there are two high tides in approximately a day, the tidal pattern just described is termed semidiurnal.

MIXED TIDES

This simple explanation would lead one to believe that one high tide is just like the next (Fig. 8A), whereas in fact, alternate high tides are typically of unequal height (a pattern known as *mixed tides*, Fig. 8B). Mixed tides occur as a consequence of the tilt of the moon's orbit relative to the plane of earth's equator.

Before we can productively discuss this tilt, we need a method to describe it. Imagine a line drawn from the

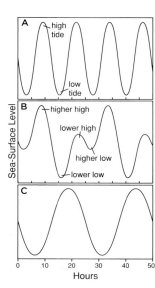

FIGURE 8 Different types of tidal patterns: (A) semidiurnal tides; (B) mixed semidiurnal tides; (C) diurnal tides.

center of the earth to the center of the moon (Fig. 9). At any given time, this line intersects earth's surface at a particular latitude. This latitude is known as the moon's *declination*, and it can vary through time.

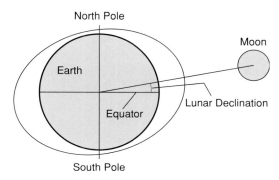

FIGURE 9 A schematic diagram defining lunar declination.

If the plane of the moon's orbit were in the plane of earth's equator, the moon's declination would always be 0°. However, if the moon's orbital plane is tilted relative to the plane of the equator, the moon's declination varies as the moon goes through its orbit. The maximum declination during this cycle is equal to the angle between the plane of the moon's orbit and the plane of the equator. The moon's maximum declination, as previously mentioned, varies from approximately 18.3° to 28.6°.

One other important fact is evident from Fig. 9. The line from earth to moon—which we have used to define the declination—also passes through the peak of the tidal bulge. Thus, the latitude and longitude where this line intersects earth's surface mark the location on the globe subjected to the highest tide (determining the precise location of the highest tide is a bit more complex, but this simplification is sufficient for the moment). Points on earth with the same longitude (position east–west) but different latitude (position north–south) from that of the lunar declination are subjected to high tides that are lower by an amount proportional to their angular distance from the bulge's peak.

Consider the situation shown in Fig. 10. In this example, the moon's maximum declination is 23.5° (an average maximum value), and we explore the pattern of tides that occur at San Francisco, California, at 36°N. When the earth is rotated such that San Francisco is on the side toward the moon (point A), the city is only 36° − 23.5° = 12.5° of latitude from the tidal bulge, and consequently the tide is quite high. This is the *higher high tide* of that tidal day. Twelve hours and 25 minutes later, San Francisco is again at the same longitude as a tidal bulge (point B in Fig. 10), but it encounters the bulge 36° + 23.5° = 59.5° of

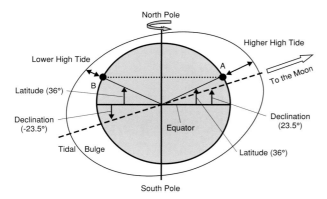

FIGURE 10 Because of the tilt of earth's axis relative to the plane of the moon's orbit, the tide is higher at Point A than at Point B.

latitude from its peak. Thus, although the tide is high at this point, the high is lower than one would expect from the simple explanation given in Fig. 7. This is the *lower high tide*. In summary, because of the tilt of the moon's orbit, the pattern of high tides typically follows that shown in Fig. 8B. On many shores, successive low tides are also of unequal height (as shown in Fig. 8B), but this inequality is less easily explained; we will revisit it shortly.

The precise pattern of the tides depends on the shape of the local ocean basin, and varies substantially from place to place. Along some shores—the mid-Atlantic coast of North America, for instance—successive high tides are approximately the same height, as are successive lows (Fig. 8A). In a few places (e.g., polar seas and some shores in Vietnam), the higher low tide is very nearly the same height as the lower high tide, so there appears to be only a single high tide and a single low tide each day (Fig. 8C). These are diurnal tides.

THE ROLE OF THE SUN

In addition to the moon, the sun exerts a gravitational pull on Earth's oceans and creates tidal bulges and valleys. The sun is 27 million times more massive than the moon, but it is also 389 times farther away from earth, with a net result that the bulges and valleys of the solar tides are only 46% as high or low as those due to the moon.

Overall tides on earth are due to the combined effects of solar and lunar tidal bulges. For example, when the moon is new (situated between the earth and the sun), the lunar and solar tides coincide (Fig. 11A). The combined tidal bulge is higher than it otherwise would be, and the combined tidal valley is lower. The same effect occurs when the moon is full (on the opposite side of earth from the sun). These high-amplitude tides are known as spring tides, and they occur approximately twice per month.

We noted above that low tides tend to occur when the moon is near the horizon, and when new or full, the moon is on the horizon at sunrise and sunset. Thus, the spring lower low tides prized by intertidal biologists typically tend to occur near sunrise and sunset.

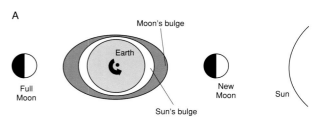

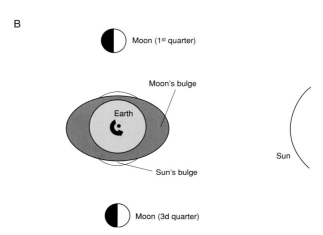

FIGURE 11 (A) At new and full moon, the lunar and solar tidal bulges reinforce. (B) At the quarter moons, the lunar and solar bulges counteract.

When the moon is one-quarter or three-quarters full, lunar tidal bulges coincide with solar tidal valleys, and the two bulges tend to cancel each other out (Fig. 11B). Because the solar valley is smaller than the lunar bulge, there is still a high tide, but it is not as high as it otherwise would be. By the same token, the solar bulge coincides with the lunar valley, and the low tides are not as low as they otherwise would be. These small-amplitude tides are known as neap tides, and they too occur approximately twice per month, alternating with the spring tides.

We have seen that the motion of the moon around the earth makes the lunar tidal day longer than 24 hours. The same effect applies to earth's motion around the sun. However, this effect is already taken into account in the measurement of the length of a day, and the average solar tidal day is exactly 24 hours. The difference in length between the lunar and solar tidal days (24 hours versus 24 hours 50 minutes) causes the lunar and solar tides to combine differently from one day to the next. The result is a complex pattern of tides that repeats itself (more or less) once per month.

KELVIN WAVES AND AMPHIDROMIC POINTS

The physics described in the preceding sections—collectively known as the equilibrium model—presents a severely simplified picture of the tides, and there are several important phenomena it cannot predict. For example, the tidal range in Hawaii is very small (<0.3 meter), whereas the tidal range at the same latitude on the coast of Mexico is quite large (>3.0 meter). In contrast, the equilibrium model (e.g., Fig. 10) predicts that these tidal ranges should be the same. Along the west coast of North America (which runs nearly north–south), the time of low tide gets later as one travels north, but the equilibrium model predicts that the time of the tide should be the same. And as we have noted, the heights of successive low tides are commonly unequal, whereas the equilibrium model predicts that they should be equal.

These and similar disparities can be explained by an alternative model of the tides, first proposed by Simon Laplace in the late 1700s. Laplace suggested that the tides could be treated as long-wavelength ocean waves, each tidal bulge being the crest of a wave and each tidal valley being a trough. These waves travel across the ocean under the gravitational urging of the moon and sun, but they are forced to interact with the shores of the ocean basin in which they travel. In effect, the waves of the tides reflect back and forth across an ocean the way waves slosh in a bathtub. Furthermore, the tidal waves (not to be confused with tsunamis, waves caused by earthquakes and volcanic eruptions) are subjected to gyroscopic centrifugal forces as the earth rotates about its axis. The combination of reflection and centrifugal forces causes the waves to travel as shown in Fig. 12, a pattern known as a Kelvin wave. In the Northern Hemisphere, the crest of the Kelvin wave (the high tide) travels counterclockwise around the ocean basin, as does the trough of the wave (the low tide). At the center of the Kelvin wave system is a point (the amphidromic point) where the water level never changes. This pattern of flow explains the disparities mentioned at the beginning of this section. Hawaii is near an amphidromic point in the North Pacific Ocean, and therefore has very low tidal amplitude. In contrast, the coast of Mexico is at the

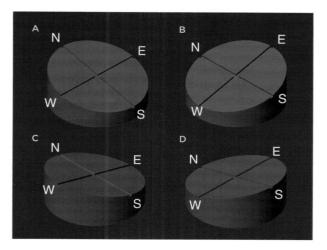

FIGURE 12 A schematic diagram of a Kelvin wave in an ocean basin. The tilted, flat surface with the cross etched into it represents the ocean's surface. For example, in A there is a high tide at the northern edge of the ocean and a low tide at its southern edge. The tide is at midlevel at the eastern and western edges. The amphidromic point is at the center of the cross. In the Northern Hemisphere, the Kelvin wave rotates counterclockwise (A → C → D → B → A), and the high tide moves from the north to the west to the south to the east and back again to the north. In the Southern Hemisphere, the tidal Kelvin wave moves clockwise (A → B → D → C → A).

ocean's border, where the amplitude of the Kelvin wave is highest. Because the wave travels counterclockwise, tides on the west coast of North America occur later at higher latitudes.

In the Southern Hemisphere, tidal Kelvin waves travel clockwise. Thus, the tides on the coast of Chile occur later the farther south one travels.

A single system of Kelvin waves (as shown in Fig. 12) seldom occupies an entire ocean basin. Consider for example, the pattern of Kelvin waves off the shores of northern Europe. There is an amphidromic point in the North Sea between Scotland and Norway, and another between England and Denmark. A third is found in the English Channel. As a result, the timing of tides in northern Europe is complex.

The wave nature of tides can also explain the inequality of successive low tides, although the explanation is beyond the purview of this article.

Note that the pattern of Kelvin waves can shift the time of high and low water from what one might expect from the equilibrium model. For example, summertime low tides occur early in the morning in California (which one would expect), but occur near midday in Oregon (which one might not). In the tropics, because of the wave nature of tides, low tides can occur when the moon is directly overhead.

ADDITIONAL PATTERNS

There are several additional effects that inject complexity into the tidal rhythm.

The Moon's Orbit

In reality, the moon travels in an elliptical orbit around the earth, moving closer and farther away. When the moon is closest to earth (perigee, which happens once every 27.5 days), its gravitational pull on the oceans is maximal, and the tidal range is about 20% larger than average. When the moon is farthest away (apogee), the tidal range is 20% less than average.

Earth's Orbit

Similarly, earth moves in an elliptical orbit around the sun. It is closest to the sun in early January, and farthest away in early July. Thus, the solar tidal range is maximal in January, and the spring tides near that time are the most extreme of the year, especially when those spring tides coincide with lunar perigee.

Lunar Declination

As noted previously, the maximum declination of the moon varies from 18.3° to 28.6°. The reason is that the moon's orbital plane is tilted approximately 5.15° relative to the plane of the earth's orbit around the sun (called the plane of the ecliptic), and the earth's equator is tilted approximately 23.45° to the plane of the ecliptic. As the alignment of the moon's orbital plane relative to the earth's axis varies, the angle between the earth's equator and the moon's orbit, which is the maximum declination of the moon, varies from 23.45° − 5.15° = 18.3° to 23.45° + 5.15° = 28.6°. This fluctuation repeats itself every 18.6 years as the moon's orbital plane precesses, or wobbles, around the earth's orbital plane as the sun pulls on the moon. When maximum lunar declination is large, the disparity in height between higher high and lower high tides is large (see Fig. 10). This long-term pattern may have important effects on intertidal communities. At some heights in the intertidal zone, organisms may spend nearly 50% less time submerged at certain points in the cycle of lunar declination than at others. Lunar declination was at its maximum in 1988 and will be maximal again in 2007.

MEASURING TIDES

Because sea level changes with the tides, there is an intrinsic problem in measuring the location of the water's surface. With respect to what level should this measurement be made? Sailors (who have a personal stake in the tides)

tend to care most about low tides—if the tide is too low, their boats run aground. So, as a practical matter, an index of low tide is typically taken as the reference level (known as the chart datum) for tidal measurements. In the United States, chart datum is the average level of the lower low tides measured across a specific 19-year period (known as a tidal epoch, representing a complete cycle of lunar declination variation). In other words, the zero point for measuring tides at a particular site on a U.S. shore is the long-term average of the lower low tides at that site—a value known as mean lower low water (MLLW). A "plus tide" is above MLLW. A "minus tide" is below MLLW and therefore exposes organisms that seldom experience terrestrial conditions. The definition of chart datum varies from one country to another. In Canada, for instance, chart datum is the average of the 19 yearly lowest low tides of an epoch, a slightly lower tidal level than MLLW.

It is important to note that tidal datum values can shift through time. Sea level is currently rising worldwide, and as it does, so does the height of lower low water. Thus, the 19-year mean lower low water for the tidal epoch 1960–1978 was lower than the average taken for epoch 1983–2001, and consequently, the U.S. chart datum moved upward. This latest U.S. tidal epoch included several severe El Niño/Southern Oscillation events, which temporarily raised sea level on the west coast of North America. As a result, at many sites the datum for this epoch is substantially higher than the previous datum.

PREDICTING TIDES

Tides lend themselves to prediction. The period at which tidal bulges move across the ocean is set by celestial mechanics (e.g., the orbital periods of the moon and earth, the period of oscillation in maximum lunar declination). Knowing these periods, mathematicians can apply a method of analyzing cyclical variations, known as Fourier analysis or harmonic analysis, to characterize the historical pattern of the tides at a given site on the shore. The numerical coefficients gained from this analysis (harmonic coefficients) can then be used to predict the height of the tide at any time in the future.

This method for predicting tides was first refined and applied in the late 1800s by William Thomson (Lord Kelvin, for whom the Kelvin temperature scale and Kelvin waves are named) and Sir George Darwin (son of Charles Darwin). Based on their calculations, analog computers were built to calculate predicted tides, and these predictions were sold to mariners. Today, digital computers make predictions, and software for predicting tides is readily available. Alternatively, official governmental predictions are available online (e.g., for the United States, at http://www.tidesandcurrents.noaa.gov).

These methods predict the astronomical tides, which are the tides that would occur in the absence of wind, storms, changes in atmospheric pressure, and El Niño/Southern Oscillation events. However, meteorological and oceanographic effects can substantially change the actual height of the tide (deviations of ±0.2 meter are common), and these environmental effects cannot be predicted from celestial mechanics and site history. As a result, predicted values for tides should be viewed as rough approximations of reality.

TIDAL EFFECTS ON DAY LENGTH

As the waves of the tides surge back and forth across an ocean, energy is lost to friction, primarily in shallow seas and estuaries. The effect of this frictional "drag" is to gradually slow the rotation of the Earth. In other words, under the influence of the tides, the length of a day slowly increases. Evidence for the change in day length is found in sediments and fossil organisms (stromatolites and corals, for instance), which show both daily and yearly patterns of growth. The length of a year has remained nearly constant, so by counting the number of daily patterns laid down in a year, the length of an ancient day can be determined. Five hundred million years ago, the day was only 21 hours long, and the tidal rhythm then was substantially different from that seen today.

COMPLICATIONS

The brief overview presented here barely scratches the surface of the complex subject of tidal variation. On any particular shore, one or more of the generalities derived from simple tidal theory is likely to be violated. To understand the reasons for these deviations, one should consult the texts cited in the Further Reading.

SEE ALSO THE FOLLOWING ARTICLES

El Niño / Ocean Waves / Rhythms, Tidal / Sea Level Change, Effects on Coastlines / Surveying

FURTHER READING

Cartwright, D. E. 1999. *Tides: a scientific history.* Cambridge, UK: Cambridge University Press.
Denny, M. W., and R. T. Paine. 1998. Celestial mechanics, sea-level changes, and intertidal ecology. *The Biological Bulletin* 194: 108–115.
Macmillan, D. H. 1966. *Tides.* New York: American Elsevier.
Open University. 1989. *Waves, tides, and shallow-water processes.* Oxford: Pergamon Press.
Schureman, P. 1941. *A manual of the harmonic analysis and prediction of tides.* U.S. Department of Commerce Special Publication 98. Washington, DC: U.S. Government Printing Office.

TUNICATES

TODD NEWBERRY

University of California, Santa Cruz

RICK GROSBERG

University of California, Davis

The tunicates (also known as the Urochordata) are one of three major subphyla in the phylum Chordata (the other two being the cephalochordates, commonly called lancelets or amphioxus, and the vertebrates). At some point in their life cycles, all chordates possess gill slits that perforate the pharynx; a longitudinal skeletal rod, the notochord, flanked by blocks of muscle; and a hollow neural tube, lying dorsal to the notochord, swollen anteriorly into a brain of some sort. In most tunicates, only the pharyngeal gill slits conspicuously remain through the adult phase of the life cycle, the notochord and neural tube being either lost or greatly reduced as juveniles develop into adulthood. The thousands of species in the three classes of the Urochordata all live in the sea. Two of these classes are wholly pelagic: the big thaliaceans (salps, doliolids, and pyrosomes), which occasionally litter the beach with their flabby tunics; and the little appendicularians (or larvaceans), with their persistent notochords and neural tubes, rarely noticed but abundant in the plankton. A third class, the scarcely known sorberaceans, lives exclusively on deep-sea bottoms.

DISTRIBUTION AND DIVERSITY

On rocky shores we encounter only ascidians, the so-called sea squirts. By far the majority of the roughly 2000 described ascidian species are subtidal. Most have limited geographical ranges, but many occur widely in similar habitats, especially disturbed ones such as harbors and estuaries. Boats spread species that foul hulls and ballast tanks, exotic oyster fisheries and other shellfish culture probably introduce invasive ascidians, and still other kinds doubtless ride to new places on flotsam.

Ascidians of some sort live in most benthic habitats from the abyss to the edge of the sea. Most extensive rocky coasts at temperate latitudes will boast dozens of species, even if only in those lowest rocky channels and pools where tidal isolation is very brief. There, pools, channels, and shaded, fairly alga-free crevices and overhangs protected from the full surf may harbor little "tunicate heavens" of intermingled species, especially ones that form encrusting colonies. In the tropics harsh solar radiation restricts even low-intertidal ascidians to the most shaded crannies; in fact, even the subtidal diversity of tropical waters may be limited down to several dozen meters' depth, below which the assortment of species often strikingly increases. At any latitude, however, those seashore ascidians hardy enough to endure tidal exposure may occur in huge numbers and biomasses. Thus, millions of the large solitary ascidian *Pyura praeputialis* (aka *P. stolonifera*), locally dubbed "cunjevoi," dominate the mid-intertidal of some subtropical Australian shores (Fig. 1), and now, introduced into the Chilean Bay of Antofogasta, have displaced a formerly dominant native mussel there.

FIGURE 1 Dense aggregations of the large solitary ascidian called "cunjevoi" (species of the stolidobranch genus *Pyura*) growing on rocky outcrops along an Australian shore. The white arrow in the lower left points to an individual, most of which are several cm tall. Photograph courtesy of Andy Davis.

BODY PLANS, FEEDING, GROWTH, AND REPRODUCTION

When first encountered, a solitary ascidian, even a gaudily colored one, can be puzzling enough: an often contorted ovoid mass encased in a leathery or gelatinous tunic and pierced by two siphons that, disturbed, may contract to mere wrinkles and punctures (Fig. 2A, B). Most colonial ascidians are even more perplexing. The colony's surface, with its many oral and atrial pores, may suggest a sponge (Figs. 3A, 4A). Rubbing a discreetly sampled fragment of the colony between one's fingers will often reveal the phylum. As a rule of thumb, an abraded bit of a sponge crumbles, while one of a colonial ascidian feels slippery and stays intact. Of course, ascidians that incorporate calcareous spicules into their tunic or sponges that are rich in fibers fail this casual test. As it does so often in the intertidal, a well-used hand lens then settles the matter: the close look at the fragment's freshly cut surfaces reveals the zooids' elaborate little bodies, especially their gonads

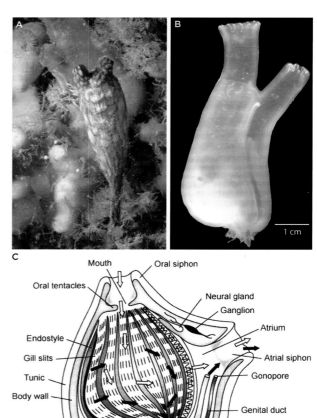

FIGURE 2 Solitary ascidians. (A) The stolidobranch *Styela clava*, invasive around the world; this individual is about 4 cm tall. The compact body, topped by its siphons, has a stalk of tunic. Photograph courtesy of Marco Faasse. (B) The phlebobranch *Ciona savignyi*, in another ascidian genus that is widespread in calm waters. Below the large, slightly frilled oral siphon, the pharynx and some of the gut loop are apparent through the animal's transparent tunic, and the white sperm duct is visible near the base of the atrial siphon. Photograph courtesy of California Academy of Sciences. (C) Schematic body-plan of a phlebobranch (e.g., *Ascidia*, of somewhat different proportions than *Ciona*). The stolidobranch body-plan differs in some structural ways: pleated pharyngeal walls in solitary forms, and multiple gonads in the body wall rather than a single one in the loop of the gut. White arrows show water flow through the oral siphon, pharynx, and atrium; black arrows show movement of pharyngeal mucus and trace food and feces through the gut. Modified from Brusca and Brusca (2003).

and food-laden gut loops. The hand lens will reveal bryozoan and cnidarian imposters for what they are, too.

All tunicate bodies, even larval ones, lie within more or less substantial tunics of protein and carbohydrate, including the animal oddity, cellulose. Adult ascidians have a large pharynx, the branchial sac, the walls of which are pierced by ranks and rows of gill slits. Cilia around the gill slits passes through these mucus sheets and so is stripped of microscopic food. Cilia along the dorsal sill roll the food-laden mucus into a cord and trundle it to a complex, U-shaped gut that empties into the peribranchial atrium. There, fecal wastes join the exhalant current.

The atrium may serve another purpose, as well. All ascidians are hermaphrodites, even if sperm and eggs ripen at different times. Some release both eggs and sperm via the atrial siphon to the sea. But others, such as most colonial species, retain their internally fertilized eggs even as they release their sperm. These ascidians use their atriums or derivative pouches as well-irrigated brood chambers from

FIGURE 3 A colonial aplousobranch, *Aplidium elegans*; all aplousobranch ascidians are colonial. (A) Surface view of an entire colony, about 3 cm across, growing on a hidden substrate. Note the few wide-open common atrial siphons, each serving a complex system of many zooids, and the many small oral siphons of individual zooids. Photograph courtesy of Dirk Shories. (B) Schematic side view body plan of an *Aplidium* zooid. Some aplousobranch genera, lacking a post abdomen, carry the gonads in the gut loop, with the heart close by. Note the larva brooded in the atrium. Modified from Brusca and Brusca (2003).

these gill slits draw water through an oral siphon into the pharynx, drive it through the gill slits into a peribranchial atrium, and expel an exhalant current through a dorsal atrial siphon (Fig. 2C, white arrows). Other cilia lining the pharynx carry mucus sheets that have been secreted from a midventral glandular groove and ride up across the internal walls of the pharynx to a mid-dorsal sill (Fig. 2C, black arrows). The water expressed through

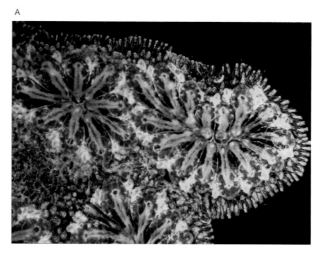

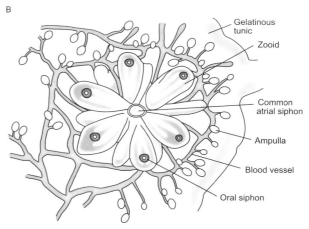

FIGURE 4 A colonial stolidobranch ascidian, *Botryllus*. (A) Systems of *Botryllus schlosseri*. The zooids in each system assume radiating patterns with oral siphons peripheral and common atrial siphons central. Numerous tiny saccular ampullae of the tunic's intrazooidal blood-circulatory complex fringe the colony. (B) Schematic surface-view plan of a botryllid system, showing zooids and the tunic's circulatory elements. After Milkman (1967).

which structurally advanced, robustly swimming, and very short-lived tadpole larvae eventually emerge (Fig. 3B).

Unfortunately, striking species aside, many ascidians—unlike, say, snails and crustaceans—reveal their identities more by internal traits than by field marks. So sometimes only a dissection will suffice, or at least a cut that (alas, lethally) removes the body from its tunic. Even so, a hand lens again may come to the rescue: deft handling and close examination often permits even the tidepooler to see through a small ascidian's clear tunic to its pharynx, gut, and gonads—the three principal organs used for identification.

The relationship of gut to gonads decides the ascidian order: Enterogona (gonads in the loop of the gut) or Pleurogona (gonads in the atrial body wall). But the complexity of the sievelike pharynx provides a more convenient way to arrange the class into three suborders. Ascidians with the simplest pharynx, the enterogonid aplousobranchs, have an elongate body whose abdomen accommodates the pendulous loop of the gut and the gonads alongside it. In some aplousobranch families the body even extends a saccular postabdomen, where the descendant gonads mature (Fig. 3B).

A much more compact body typifies the two other ascidian suborders. The enterogonid phlebobranchs, with their more numerous and elaborate pharyngeal gill slits and blood vessels, and the pleurogonid stolidobranchs (Fig. 2A, C), whose pharyngeal walls are thrown into longitudinal folds that greatly increase their filtering surface, both have a relatively immense pharynx and atrium that dominate their anatomy. The tightly looped gut and the gonads of a phlebobranch ascidian lie together in a stubby abdomen (as in the famous *Ciona*; Fig. 2B) or in the body wall beside the atrium (as in the genus *Ascidia*). The gut loop of a stolidobranch (such as *Styela* or *Pyura*) lies in the outer wall of the atrium; the gonads mature either in the gut loop or as one or many isolated organs elsewhere in the atrial wall.

Some familiar ascidians, such as the stolidobranchs *Styela* and *Pyura* and the phlebobranch *Ciona*, are solitary all their lives; the fertilized egg develops into a single body (Fig. 2A, B). These grow sometimes into an adult a mere centimeter across, but some solitary stolidobranchs get big: the spheroid *Pyura praeputialis* (Fig. 1) can reach the size of a football. And some solitary species grow stalks beneath their bodies. A colleague has reminded us that the aptly named *Pyura pachydermatina* can appear to be balancing on a little elephant's trunk, altogether a meter long.

But all aplousobranchs and many ascidians in the other two suborders (such as the stolidobranch *Botryllus*; Fig. 4A, B) clone in diverse ways (e.g., from buds protruded out of the body wall or from chambers that bulge from blood vessels in the tunic or by strobilation of an elongate zooid) to form colonies. The zooids of colonial ascidians are small (at most a few centimeters, as in some species of *Clavelina*; Fig. 5A, B) or even tiny (barely a millimeter, as in many species of the family *Didemnidae*). However, the colonies they generate range from little plaques to broad sheets (Fig. 4A), elaborate lobes (Fig. 3A), or encrustations of berry-like zooids. A colony's zooids show little or no polymorphism, but, immersed together in the colony's common tunic, they may arrange themselves in

FIGURE 5 Another aplousobranch, *Clavelina lepadiformis*. (A) One zooid, about 1 cm tall. The tunic and most of the body are glassily transparent. Lines of brightly reflective pigment mark the lip of the oral (at top) and atrial (on zooid's dorsal side) siphons, the front rim and dorsal midline of the pharynx, the endostyle, and a few other structures. Fecal pellets reveal the intestine. This ascidian has a relatively huge thorax, a tiny abdomen, and no post-abdomen. Photograph courtesy of Duncan Vliet. (B) A *Clavelina* colony growing on rock. A thin tunic covers all the zooids but does not deeply embed them (in contrast to the thickly embedded *Aplidium*). Zooids bud by extending short vessels from their bases to form a delicate clonal bouquet. Photograph courtesy of Duncan Vliet.

linear or radiating patterns so that groups of them pool their individual exhalant water currents into shared cloacal chambers in the tunic (Figs. 3A, 4A). The combined exhalant currents propel wastes and sperm and eggs or released larvae far from the colony surface and so beyond the water from which the zooids draw their food.

ENIGMAS AND ODDITIES

Ascidians have a remarkable array of traits that challenge or confound our assumptions about how animals work. Some of these traits can be seen during any scrutiny; but others—such as tunicates' famous notochords and hollow dorsal nerve cords—are ones that come and go over the course of the animal's development from egg to adult; or, like reproductive and developmental "strategies," they are not so much structures as events in the life cycle. As such, ascidians exemplify John Tyler Bonner's dictum that the organism is the whole life cycle, and Gavin de Beer's that natural selection inspects and shapes all of development: that what evolves are entire life cycles.

Reversing Heart

The heart (Figs. 2B, 3B) comprises an outer epithelial tube surrounding a heavily muscled inner tube that spirally wrings itself out every few seconds and then, to the observer's astonishment, reverses the direction of its contractions and thereby that of the blood flow throughout the body: arterial sinuses and spaces become venous ones and vice versa.

Blood Cells and Metals

The blood contains a great and enigmatic variety of cells. Apparently multipurpose amoebocytic lymphocytes and purine-laced nephrocytes are two such sorts, with perhaps misleading names. Bladder cells of intriguing composition occur, too, in many ascidians. In some colonial species, during the growth of each zooid, what appear to be ordinary amoebocytes clump amidst the sinusoidal body wall's loose meshwork of fibers at predictable places to form the somatic rudiments of the gonads—ovaries here and testes there—and evidently even can turn into gametocytes.

Additionally, in several ascidian families the animals concentrate rare metals in the blood: most famously vanadium, but also niobium, titanium, and chromium.

Endostyle

An elaborate mid-ventral pharyngeal glandular groove, the endostyle (Figs. 2C, 3B), exudes the food-trapping mucus that rides up the pharynx's walls. Marginal cells along the endostyle bind iodine into a compound resembling vertebrate thyroxine. In both its detailed structure and its function, the endostyle closely resembles those of lamprey eels and hagfish, whose endostyles in turn apppear to be rudimentary vertebrate thyroid glands.

Spicules

In a few families, ascidians produce calcareous spicules that rival in abundance and diversity those of sponges. Stellate spicules in the tunic of didemnid ascidians may make it brittle or, along with symbiotic algae, turn colonies remarkable colors. Discoid spicules may encapsulate the abdomen in the polycitorid genus *Cystodytes*. In some pyurids, spicules line the siphons and migrate to the tunic surface, forming bristles, perhaps as antipredator armor.

Larval Notochord, Adult and Larval Dorsal Nerve System

The ascidian's tiny larva, rarely longer than a few millimeters, is tadpole-shaped, with a stout trunk and a muscular, swimming tail stiffened by a dorsal notochord derived (as in vertebrates) from the roof of the embryonic gut. This notochord is lost at larval settlement, when, as the tadpole settles on a surface, a catastrophic metamorphosis transforms the little body into a sac. A hollow nerve cord rides above the notochord and swells anteriorly into a hollow brain. The dorsal nerve cord is also lost at metamorphosis, and the anterior larval "brain" becomes the adult's neural complex: a prominent antero-dorsal ganglion and a neurosecretory gland with a ciliated opening to the roof of the animal's mouth. This complex is often likened to the vertebrate posterior hypophysis and pituitary gland.

Genetic vs. Morphological Identity

Many ascidians, including most colonial forms, brood and release advanced larvae that settle almost at once, sometimes within minutes and thus within the immediate vicinity of their mother. It follows that many colonies in a tidepool or on a pier must be closely related and that succeeding sexual generations there will be all the more highly inbred. In *Botryllus*, at least, such colonies are known to use precise kin recognition systems to "choose" or "refuse" to fuse with one another into genetically compatible, chimeric colonies, one perhaps even parasitizing the other.

The Significance of Colonial Habits Themselves

When "the body" is a morphologically reiterated module in a larger complex, an intricate question poses itself: In what ways do the interacting levels of developmental organization—reiterated module, unified or dispersed "ramet" colony, genetically chimeric colony, whole "genet" life cycle—differ as targets of natural selection? Ascidian coloniality appears to have arisen time and again. Species that form tiny colonies—if not in complex ascidians, then certainly in simpler designs such as bryozoans and hydroids—seem to be as successful in evolution as ones that form big ones. What advantages accrue, then, to coloniality, and at what level of developmental organization, and under what ecological circumstances, is it advantageous? And if advantages do accrue, then why do solitary ascidians, too, so diversely prosper side by side with colonial forms?

In one internal, ecological, and evolutionary aspect after another, ascidians present puzzles that today are only partly solved: organs that continue to mystify, substances that baffle in their formation and role, developmental strategies of uncommon variety and consequence, and traits that remind us of our own invertebrate ancestry. Yet many of the ascidians we encounter along the shore are just enigmatic slabs of tunic—inviting a closer look, yes, but gelatinously defying it—until, with that closer look, their strange ways reward the patient inquirer.

SEE ALSO THE FOLLOWING ARTICLES

Body Shape / Circulation / Introduced Species / Materials, Biological / Sponges

FURTHER READING

Abbott, D. P., and A. T. Newberry. 1980. Urochordata: the tunicates. in *Intertidal invertebrates of California.* R. H. Morris, D. P. Abbott, and E. C. Haderlie, eds. Stanford, CA: Stanford University Press, 177–226.
Berrill, N. J. 1950. *The Tunicata, with an account of the British species.* London: The Ray Society.
Brusca, R. C., and G. J. Brusca. 2003. *Invertebrates.* Sunderland, MA: Sinauer.
Cloney, R. A. 1990. Urochordata–Ascidiacea, in *Reproductive biology of invertebrates.* K. G. Adiyodi and G. Adiyodi, eds. New Delhi: Oxford and IBH, 361–451.
Kott, P. 1989. Form and function in the Ascidiacea. *Bulletin of Marine Sciences* 45: 253–276.
Lambert, C. C., and G. L. Lambert, eds. 1982. The developmental biology of the ascidians. *American Zoologist* 22: 751–849.
Millar, R. H. 1970. *British ascidians.* London: Academic Press.
Millar, R. H. 1971. The biology of ascidians. *Advances in Marine Biology* 9: 1–100.
Newberry, A. T. 1999. Tunicata (Urochordata), in *Encyclopedia of reproduction,* Vol. 4. E. Knobil and J. D. Neill, eds. New York: Academic Press, 872–879.
Satoh, N. 1994. *Developmental biology of ascidians.* Cambridge, UK: Cambridge University Press.
Sawada, H., H. Yokosawa, and C. Lambert, eds. 2001. *The biology of ascidians.* Tokyo: Springer Verlag.
Zeng, L., and B. J. Swalla. 2005. Molecular phylogeny of the protochordates: chordate evolution. *Canadian Journal of Zoology* 83: 24–33.

TURBULENCE

STEPHEN MONISMITH
Stanford University

Most flows in nature are turbulent. By this we mean that they exhibit significant stochastic, that is, random or chaotic, temporal and spatial variability in velocities, pressures, temperatures, and so forth. Despite that the fundamental equations of motions for viscous fluids such as seawater have been known for more than 160 years, and despite the recent development of incredible computing power, this randomness means that the instantaneous details of any turbulent flow are inherently unpredictable. Current supercomputers allow direct, brute-force computation of turbulent flows in a domain perhaps the size of a shoebox—not in something as large as a tidepool! The results of this randomness are important to the life in and on the shores of coastal oceans, enhancing the dissipation of energy contained in surface waves as well as dramatically increasing rates of exchange of mass between solid substrates such as those of the rocky seashore (or coral reefs) and the overlying flowing water and rates of mixing and dispersal of materials contained in that water, such as the gametes of intertidal and subtidal organisms.

ORIGINS OF TURBULENCE

In careful laboratory experiments turbulent flows can be seen to develop from smooth, laminar flows as the result of a sequence of instabilities. For example, the interface between two fluid layers flowing at different speeds evolves through what is known as the Kelvin–Helmholtz instability from a series of waves to a periodic array of spirals that closely resembles a popular Celtic decorative motif (Fig. 1). These initially concentrated vortices eventually twist and become fully three-dimensional, losing almost all semblance of order—that is, the flow becomes turbulent. What makes prediction of these flows difficult is that (as mathematicians would say) turbulent flows depend sensitively on their initial conditions. Imagine that we do two otherwise identical experiments in which we somehow start up a two-layered flow of this kind and observe how it evolves. This sensitivity means that slight changes in the timing or strength of the small disturbances that grow into Kelvin–Helmholtz billows between the two experiments will lead to different realizations of the flow between the two experiments.

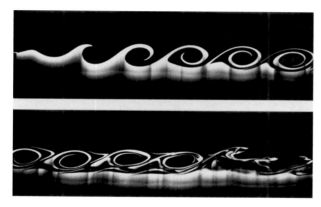

FIGURE 1 Formation and breakdown of Kelvin–Helmholtz instability in a laboratory experiment. Photograph courtesy of G. A. Lawrence. In this experiment, the lighter, upper layer is flowing faster than the denser lower layer. Both are flowing to the right. Fluorescent dye dissolved in the lower layer and illuminated with a sheet of white light is used to visualize the development of the instability.

This transition to turbulence generally is found to depend on a dimensionless number that measures the relative importance of whatever acts to stabilize turbulence. For example, flow through a circular pipe becomes turbulent when the Reynolds number Re = UD/V, where U is the mean velocity in the pipe, D is the pipe diameter, and V is the kinematic viscosity (stickiness of the fluid), is greater than about 2000, meaning that flow of syrup through a small capillary tube might be laminar, whereas the flow of water in the pressure main that connects to the author's house is very likely turbulent.

For flows relevant to tidepools and the near-shore ocean, variations in density caused by variations in salinity or temperature also are important. This leads to a second dimensionless number, the Rayleigh number, which comes into play when the water surface is cooled. Surface cooling produces water that is heavier (denser) at the surface than below and so tends to fall through the fluid below as narrow plumes. This flow is easily observed (albeit upside down) by heating water in a pot or by putting ice cubes in a glass of water.

THE SYMPTOMS OF TURBULENCE

Any attempt at measuring fluid properties (for example, salinity, temperature, or oxygen concentration) or hydrodynamic variables such as velocity or pressure in the ocean or on its shores will reveal temporal variability with frequencies ranging continuously from tens of cycles per second to one cycle per entire record length. In the late 1880s Osborne Reynolds sought to provide a practical approach for dealing with the temporal randomness of turbulence by noticing that while flows (for example)

in pipes fluctuated instantaneously, one might define an average velocity that was sensibly constant, at least if the same conditions were maintained for periods longer than some time scale characterizing the temporal variation of the turbulent fluctuations, or eddies. This notion of averaging was useful in light of the fact that in most cases there is little interest in the details, and rather more interest in predicting the net outcome, for example, how much water will flow through a pipe of a given size for a given applied pressure difference.

In addition to temporal variability, turbulent flows are also characterized by a wide range of spatial scales. If one looks at a turbulent flow (Fig. 2), one can see very small eddies, very large eddies, and a continuous range of sizes in between. For example, in turbulent flows in the kelp forests of Monterey, the smallest eddies might be of the order of 1 millimeter in spatial extent, whereas the largest eddies are comparable to the water depth, that is, of the order of 10 meters in spatial extent. It should be noted that the human eye is remarkably good at discerning this kind of spatial organization. Mathematically, we speak of correlation scales as being defined by the variation with spatial separation of the correlation of the fluctuations at two points. The motions at two points would be well correlated if, whenever the velocity changed at one point by some amount, it changes by an equal and opposite amount nearby (imagine two sides of a bicycle wheel). Thus, the velocity at given point is perfectly correlated with itself (zero separation), whereas the velocity at a point in Monterey Bay very likely has no relation whatsoever to the velocity at some point in Florida Bay (very

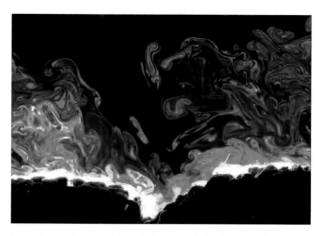

FIGURE 2 Turbulent mixing of fluorescent dye in wavy flow over a laboratory coral reef. In this case, the overlying flow goes back and forth horizontally, while slowly drifting to the left. The fluorescent dye is injected uniformly under the coral heads and is mixed vertically by turbulence produced by flow over the coral heads. Photograph courtesy of M. A. Reidenbach.

large separation). In fact, turbulent motions in the near-shore coastal ocean might be correlated with each other only on scales of a few meters.

It has long been recognized that this range of spatial scales is made possible by what is referred to as the energy cascade, the transfer of energy from the largest scales to the smallest scales. If an eddy of scale L is characterized by velocity U, it has a kinetic energy proportional to U^2. In general, it tends to lose energy at a rate such that in one turnover = rotation period, $T \sim L/U$, so that the rate of transfer of energy from the larger scales to smaller scales, at least at scale L, is found to vary as U^3/L. This rate of energy transfer is called the rate of dissipation of turbulent kinetic energy and is usually denoted by the Greek symbol ε. It is measured in units of watts per kilogram (W/kg). This transfer continues until the spatial scale is so small that viscosity smears out any variations in velocity that the turbulence can create, effectively converting mechanical energy into heat at a rate v. The Russian mathematician Kolmogorov showed in the early 1940s that in terms of v and the kinematic viscosity v, this smallest scale, now called the Kolmogorov scale, is equal to $(\varepsilon^3/\varepsilon)^{1/4}$. To be fair, the nature of the cascade was recognized well before Kolmogorov. For example, the British meteorologist L. F. Richardson described the cascade in 1922 by the short poem:

> Big whorls have little whorls,
> Which feed on their velocity;
> And little whorls have lesser whorls,
> And so on to viscosity
> (in the molecular sense).

EFFECTS OF TURBULENCE

Perhaps the most striking physical effect of turbulence is the fact that it dramatically enhances the rate at which mass, heat, and momentum are dispersed or mixed in a given flow. Literally, this is largely the result of stirring, the stretching and folding of fluid elements caused by spatial variations in turbulent velocity. If one placed a small blob of dye in a glass and stirred the water in the glass with a spoon, one would see that the blob would become distorted such that surface area of the blob would increase by many times its initial area and at the same time it would grow thinner by about the same ratio so as to conserve volume. Thus, by Fick's law, which states that the irreversible flux of a given quantity is proportional to the surface area through which that flux might take place times the gradient of concentration normal to that surface, the molecular mixing of the dye would increase by orders of magnitude over what might be accomplished by molecular diffusion alone.

The ecological consequences of turbulence are substantial. For example, increased mixing of particulate matter in water by turbulence can be important to sessile organisms such as filter-feeding bivalves (e.g., oysters). When these animals graze, they clear the water near them of the organic particles they feed on. In the absence of turbulence, they would quickly exhaust their local food supply; however, if the flow over them is turbulent, mixing quickly replaces this food-depleted water with relatively food-rich water from above, enabling them to feed at what are sometimes striking rates. In a similar fashion, turbulent flows over coral reef communities can facilitate nutrient transfer that promotes relatively productive ecosystems.

Turbulent mixing of momentum significantly retards many flows in the near-shore and intertidal region. In the case of momentum, mixing can be thought of as greatly increasing the fluid viscosity, that is, producing an "eddy viscosity" that is 10^2 to 10^6 times larger than the kinematic viscosity. This increased "viscosity" means that any velocity variations that might be created, for example by winds blowing on the water, tend to be quickly smeared out, although never completely. For the relatively generic case of turbulent flow over a rough bottom, turbulence leads to a Reynolds-averaged velocity (also known as the mean velocity) that varies logarithmically with height. This log-layer is found not only in places such as tidal channels but wherever there is turbulent flow over a solid surface, for example, on the wing of an airliner. Flows in the near shore can be complicated by the presence of aquatic vegetation, most notably kelp, which can reduce turbulence by increasing friction and damping currents. However, kelp also increases turbulence generation because flow around the stipe and along the underside of the canopy can produce turbulent wakes. As a consequence of this spatial change in sources of turbulent production, flows in kelp forests should not be expected to show the log law.

TURBULENCE AND TIDEPOOLS

Tidepools and the near-shore ocean have one source of turbulence that is relatively unique: wave breaking. As waves propagate inshore, they steepen and eventually break, converting large amounts of energy stored in the wave into turbulent motions (Fig. 3). In the absence of reflection, the energy flux of the waves must balance the total rate of dissipation by turbulence integrated over the surf. From a fluid mechanics perspective, this form of turbulence generation is especially challenging, because

FIGURE 3 Wave breaking and near-shore turbulence near Hopkins Marine Station. Photograph courtesy of Luke Miller.

disruption of the free surface that separates air and water is involved, as are the large volumes of air entrained by the break. Similar processes take place offshore in deep water, where wind-generated waves can also locally steepen and break, although never with the simple periodicity of waves on the shorelines.

Another source of turbulence particularly relevant to tidepools and the ocean nearby is the breaking of internal waves: waves that propagate on the interface between the warm surface waters of the ocean and cooler waters below. Internal waves seem to be a ubiquitous feature of the coastal ocean. When they propagate into shallow water, similar to their surface relatives, they too break, resulting in turbulence generation and local mixing of the vertical thermal structure.

In a like fashion, variations in density with height, such as occur in many flows in the coastal ocean, can dramatically reduce vertical mixing, decreasing connectivity between different parts of the water column and increasing velocity shear. This effect is parameterized by the Richardson number (Ri), the ratio of the stabilizing effect of stratification to the destabilizing (turbulence-producing) effects of shear. When Ri is greater than 1/4, turbulence is suppressed and dies. This can have a dramatic effect in the ocean, where rates of growth of phytoplankton, and hence the supply of food to higher trophic levels, can be strongly coupled to physical forcing that changes vertical density stratification.

Finally, what about turbulence in tidepools? Certainly, when the tidepools are fully submerged, they are presumably subjected to the full effects of any waves. On the other hand, when a tidepool is not submerged, it seems likely that the most important source of turbulence would be turbulence produced by turbulent bores, the remnants of breaking waves that might flow up to the tidepool before receding. However, since there have been no studies of flows in tidepools, the true nature and behavior of turbulence awaits elucidation.

SEE ALSO THE FOLLOWING ARTICLES

Buoyancy / Diffusion / Hydrodynamic Forces / Tides / Wave Forces, Measurement of / Waves, Internal

FURTHER READING

Denny, M. W. 1988. *Biology and the mechanics of the wave-swept environment.* Princeton, NJ: Princeton University Press.
Gargett, A. E. 1989. Ocean turbulence. *Annual Review in Fluid Mechanics* 21: 419–451.
Kundu, P. K., and I. Cohen. 2004. *Fluid mechanics*, 3rd ed. New York: Elsevier.
Monismith, S. G. 2007. Hydrodynamics of coral reefs. *Annual Review in Fluid Mechanics* 39: 37–55.
Peregrine, D. H. 1983. Breaking waves on beaches. *Annual Review in Fluid Mechanics* 15: 149–78.
Pope, S. B. 2000. *Turbulent flows.* Cambridge, UK: Cambridge University Press.
Richardson, L. F. 1992. *Weather prediction by numerical processes.* Cambridge, UK: Cambridge University Press.
Van Dyke, M. 1982. *Album of fluid motion.* Stanford, CA: Parabolic Press.
Vogel, S. 1996. *Life in Moving Fluids*, 2nd ed. Princeton, NJ: Princeton University Press.

ULTRAVIOLET STRESS

J. MALCOLM SHICK

University of Maine

Solar ultraviolet (UV) radiation, because of its high energy, damages biological molecules such as DNA, proteins, and lipids. UV radiation penetrates coastal seawater, where it may kill organisms outright, adversely affect diverse physiological processes (embryological development, growth, photosynthesis, immune response), and evoke avoidance behaviors. Marine organisms have evolved biochemical defenses against the direct and indirect effects of UV radiation. The metabolic costs of maintaining these defenses represent a stress, particularly when the defenses must cope with enhanced UV irradiance.

SOLAR UV RADIATION AT THE EARTH'S SURFACE AND UNDER WATER

Because of atmospheric absorption and scattering, solar radiation reaching the earth is lowered in intensity and its spectral distribution is truncated (Fig. 1A). The shortest ultraviolet wavelengths (UVC, <280 nanometers) are absorbed by ozone (O_3) and molecular oxygen (O_2) and do not reach the earth's surface. UVB (the band of wavelengths from 280 to 320 nm) is also greatly attenuated by stratospheric O_3, whereas UVA (320–400 nm) and photosynthetically available radiation (PAR, 400–700 nm) are not.

Owing to the net degradation of stratospheric O_3 by reactions with polluting halocarbons (molecules containing halogen atoms such as chlorine, fluorine, and

bromine), most notably chlorofluorocarbons (manufactured as refrigerants and aerosol propellants), levels of UVB reaching the earth's surface have increased since the 1970s—on average by 50% and 15% in the Antarctic and Arctic, respectively, during seasonal ozone depletion, and seasonally by 4–7% in midlatitudes of the northern hemisphere. UVB, already high in the tropics, has not changed significantly there.

Infrared radiation (>700 nm), which has a low energy content per photon (Fig. 1B) and is manifested as heat, constitutes just under half of the total solar energy incident on the earth (Fig. 1A). PAR (also involved in vision) accounts for roughly half. Combined UVA and UVB constitute only about 5%. Despite this low incidence, because of their high energy content per photon (Fig. 1B), UV wavelengths have large biological effects.

The variable penetration of UV radiation into seawater is caused primarily by regional and temporal differences in its absorption by biologically derived dissolved organic matter and particulate organic matter (including detritus and living plankton). The attenuation of UV radiation is inversely related to wavelength, so that UVA penetrates deeper than does UVB. In clear seawater that is low in productivity, UV radiation penetrates to several tens of meters, whereas in productive coastal waters it reaches maximally to about 10–20 meters, with much variability among waters having different optical properties. In the coastal Gulf of Maine (United States), UVA is 1% of its surface value at about 5 m; the 1% depth there for UVB is 2–5 m, but off Southern California (United States) UVB is detectable to 22 m. Organisms living intertidally and in tidepools potentially are exposed to levels of UVR ranging from a few percent up to 100% of the local surface intensity.

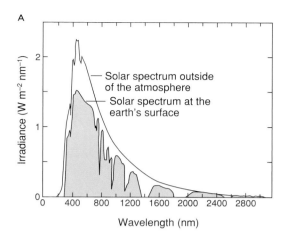

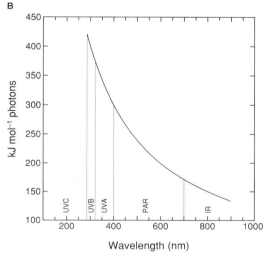

FIGURE 1 (A) The solar spectral irradiance (energy flux at various wavelengths) outside the atmosphere and incident on the earth. Modified from M. Blumthaler and A. R. Webb, UV radiation climatology, in Helbling and Zagarese (2003). Reproduced by permission of The Royal Society of Chemistry. (B) The energy content of a photon of light is inversely related to its wavelength (λ). UVC = ultraviolet C; UVB = ultraviolet B; UVA = ultraviolet A; PAR = photosynthetically available radiation; IR = infrared radiation.

BIOLOGICAL EFFECTS OF UV RADIATION

Because their cyclic molecular structures contain conjugated double bonds (i.e., those that alternate with single bonds: C–C=C–C=C–C), in which the electrons are loosely bound, nucleotide bases (especially) and aromatic amino acids containing cyclic side groups (to a smaller extent) readily absorb UVB. The same is true of polymers containing these building blocks: the nucleic acids DNA and RNA, and proteins. These molecules are structurally altered when the loosely bound electrons are raised to higher energy levels on absorbing these energetic wavelengths. In DNA, the most common UVB photoproducts are pyrimidine

dimers, in which two adjacent molecules of the pyrimidine nucleotide thymine are covalently linked in a cyclic structure that blocks transcription of the genetic information into RNA. UVB radiation causes mutations in DNA and can produce melanomas in mammals and fishes.

Deleterious effects on proteins (e.g., loss of function, particularly in enzymes involved in photosynthesis) can be caused by both UVB and UVA, in the latter case primarily via the action of intracellular photosensitizing molecules that transfer the absorbed radiant energy to O_2, leading to the production of reactive oxygen species such as singlet oxygen, hydrogen peroxide (H_2O_2), and superoxide ($O_2^{\cdot-}$) and hydroxyl (OH·) free radicals. Free radicals are chemicals having at least one unpaired electron and in consequence are highly reactive. Oxygen free radicals and other reactive oxygen species in turn oxidatively degrade proteins, DNA, photosynthetic pigments, membrane lipids, and other cellular constituents, with widespread physiological effects. Reactive oxygen species normally are held in check by natural antioxidants such as carotenoids, ascorbic acid (vitamin C), and enzymes such as superoxide dismutase (SOD) and catalase. When there is an imbalance between the production of reactive oxygen species and the defenses against them, oxidative stress is the result.

Reactive oxygen species are also produced in seawater through the interactions of dissolved organic matter with UV radiation and trace metals, with unknown effects on organisms. Any effects might be pronounced in tidepools, whose isolation at low tide and large biomass could lead to higher concentrations of precursors (dissolved organic matter) and products (reactive oxygen species). At the same time, high concentrations of dissolved organic matter (including algal exudates known to absorb UV radiation) in tidepools might help to protect organisms there from the direct effects of UV radiation.

In rare cases it has been possible to infer the proximal cause of the detrimental biological effects of UV radiation by comparing the absorption spectra of biological molecules with the action spectra—the relative effectiveness of different wavelengths of monochromatic radiation—causing a specified biological effect. However, few empirically determined action spectra are available for marine processes. More commonly, biological weighting functions (polychromatic action spectra generated by successively filtering out shorter wavelengths) have been used to assess the effects of ultraviolet and interacting longer wavelengths on marine processes. The biological weighting functions for mortality in embryos of the copepod *Calanus finmarchicus*, the codfish *Gadus morhua*, and the anchovy *Engraulis mordax* closely match that for damage to isolated DNA (Fig. 2A), indicating that mortality is related to genetic damage. UVB wavelengths <312 nm are the most damaging, whereas UVA has no effect. Conversely, inhibition of photosynthesis in phytoplankton and corals also involves UVA (probably mediated by reactive oxygen species), an inhibition that is offset by intracellular sunscreens that absorb in this waveband (Fig. 2B and see the section "Mechanisms of Repair and Defense").

Most commonly, mortality, photosynthesis, growth, or another biological process is measured in the presence and absence of broadband UV radiation, or UVB in particular (UV wavelengths are removed from solar or artificial light sources using cutoff filters), or under artificially enhanced UVB to simulate the effect of ozone depletion. Such studies indicate that shallow-water marine organisms may be killed by levels of UV radiation they likely encounter in nature. These include viruses (whose infectivity is also reduced by UV radiation), bacteria, eukaryotic phytoplankton, macroalgae (seaweeds), diverse invertebrates (planktonic crustaceans and larvae of echinoderms have been most studied), and the larvae of fishes. Some generalizations are possible. UVB is more consistently lethal than is UVA (Fig. 2A), and early developmental stages are more susceptible than later ones.

The intertidal copepod *Tigriopus californicus*, an inhabitant of supralittoral pools in the highest reaches of the intertidal zone and normally exposed to high fluxes of solar UV radiation, is more resistant to UVB than is the planktonic copepod *Acartia clausii*. In the latter species, individuals that survive exposure to UVB produce fewer offspring than do unirradiated individuals. It is unknown whether chronic exposure to UVB in intertidal organisms affects their long-term fecundity. Hatching size in surviving codfish larvae is smaller when they have been exposed to UVB during embryonic development.

Broadband UV radiation reduces growth in brown, red, and green seaweeds, as well as skeletal growth in corals having unicellular algal endosymbionts living within the host animals' cells, consistent with the many demonstrations that UV radiation also diminishes photosynthesis in such organisms. Impaired growth is not simply an effect of a diminished supply of photosynthetically fixed carbon, however, and as in the cases of stunted cod larvae and of nonsymbiotic sea anemones *(Actinia equina* and *Metridium senile)* that grow more slowly under UV radiation, it may reflect a reallocation of resources from growth to resistance against and repair of UV-induced damage. UV radiation reduces the grazing on small phytoplankton

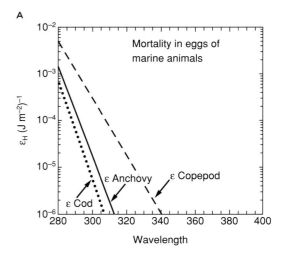

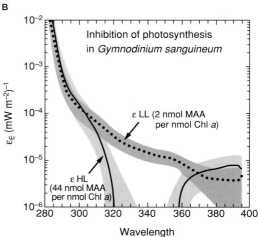

FIGURE 2 (A) Biological weighting functions (εH), which show the relative effect of different UV wavelengths, indicate that mortality in the eggs of copepods and fishes is related to damage to their DNA by UVB radiation. For comparison, the spectrum for anchovy egg mortality is the action spectrum for damage to naked DNA normalized to the exposure causing 50% egg mortality. Units of εH are the reciprocal of UV dose (J m^{-2}) because mortality is best predicted from cumulative exposure. Here, the eggs of a copepod are more sensitive than those of fishes. (B) Biological weighting functions (εE; units are the reciprocal of dosage rate, mW m^{-2} = mJ s^{-1} m^{-2}, used in short-term exposures) indicate that the inhibition of photosynthesis by UVA ($\approx$320–360 nm) in the dinoflagellate *Gymnodinium sanguineum* grown under low light (LL) is offset by the 22-fold higher concentration of UV-absorbing mycosporine-like amino acids (MAAs; see section "Mechanisms of Repair and Defense") per nanomole of chlorophyll a occurring in cells previously grown in bright light (HL). UVB also exerts strong negative effects in both groups at wavelengths shorter than those absorbed by MAAs. Shaded areas represent the 95% confidence belts for the biological weighting functions. A and B modified from P. J. Neale and D. J. Kieber, Assessing biological and chemical effects of UV in the marine environment: Spectral weighting functions, in Hester and Harrison (2000). Reproduced by permission of The Royal Society of Chemistry.

by heterotrophic nanoflagellates, which should also affect the consumers' growth.

In the case of inhibition of photosynthesis, specific effects of UV radiation include the degradation of the structural D1 protein in photosystem II (a complex of light-trapping molecules, a photochemical reaction center, and electron transfer molecules), and of Rubisco, the primary CO_2-fixing enzyme. Intracellular reactive oxygen species are particularly implicated in photooxidative damage during the exposure of photosynthetic organisms to UV radiation. This includes not only free-living algae but also corals and sea anemones that harbor unicellular algal endosymbionts. Cytological abnormalities, damage to DNA, and oxidative stress in the host or the symbiotic algae have been demonstrated in the anemones *Aiptasia* spp., *Anthopleura elegantissima*, *Cereus pedunculatus*, and in the corals *Montipora faveolata*, *Porites porites*, and *Stylophora pistillata* that have been exposed to UV radiation.

Subtidal seaweeds are generally more susceptible to inhibition of photosynthesis and growth by UV radiation than intertidal macroalgae are. Unlike phytoplankton in a deep, vertically mixed water column that affords a periodic reprieve from UV radiation, algae living intertidally and in tidepools are continuously subjected to UV inhibition of photosynthesis in daylight, although they recover more quickly than subtidal algae when the stress is removed. Differential susceptibility to UV radiation among species is one factor affecting the vertical zonation of seaweeds.

UV radiation inhibits the uptake of inorganic nutrients such as ammonium, nitrate, and phosphate by phytoplankton and macroalgae by an unknown mechanism, perhaps damage to ion transporter proteins or to membrane lipids (important targets of reactive oxygen species). This may be important because the algae normally are nitrogen limited. The pools of algal storage products, including carbohydrates, lipids, and amino acids, are also affected.

Damage to DNA by UVB has been studied in marine bacteria, phytoplankton, and zooplankton, including larvae of invertebrates and fishes, as well as benthic invertebrates such as corals. The relationship between the extent of DNA damage (pyrimidine dimers) and mortality has rarely been assessed, but the two are correlated in larvae of sea urchins. Damage to DNA, and perhaps to actin microfilaments and components of cell signaling pathways, delays the first cell division in the eggs of sea urchins and tunicates, and results in abnormal embryonic and larval development in sea urchins. Delays in the cycle of cell division allow for repair of damaged DNA, which may implicate the UVB-dependent activation of genes controlling the cycle. Exposure of sea urchin and cod fish embryos to UVB leads to enhanced transcription of the tumor suppressor gene *p53*, whose product delays cell division while DNA is repaired. Irreversibly damaged cells may be removed by apoptosis (programmed cell death), which increases in sea urchin embryos exposed to UVB. The UV inhibition of germination and cellular proliferation in spores of intertidal brown seaweeds is sensitive to temperature, an important consideration in this thermally variable environment.

First demonstrated in mammals, suppression by UV radiation of the immune response to invading substances is also seen in fishes. This might lead to a greater susceptibility to infections.

MECHANISMS OF REPAIR AND DEFENSE

Organisms have been exposed to solar UV radiation since early in the history of life and have evolved mechanisms to deal with its damaging effects. Mechanisms of repairing UV-induced damage to DNA are found from viruses to eukaryotes. Photoreactivation involves the cleavage of covalently linked thymine dimers by a photolyase enzyme that is activated by UVA and blue wavelengths of 370–450 nm in a dose-dependent manner. Therefore, a concern about stratospheric ozone depletion is that it increases the amount of UVB reaching organisms and damaging their DNA without a corresponding increase in the longer photoreactivating wavelengths, which are unaffected by ozone. It also follows that a realistic assessment of the net biological effects of UVB should include allowance for photoreactivation.

Downward movement in the water column (which attenuates damaging UVB more rapidly than it does photoreactivating UVA or blue wavelengths) through vertical mixing or active migration of plankton tips the balance in favor of repair relative to damage, but this defense is less available to sessile intertidal organisms or those confined to tidepools. Photoreactivation has been demonstrated in diverse marine organisms by varying the availability of photoreactivating wavelengths after a UVB insult. A single study of three genera of opisthobranch molluscs (sea slugs), however, did not reveal a correlation between their capacities for photoreactivation by photolyase and their normal environmental UVB regimes.

Unlike photoreactivation, nucleotide excision (dark repair) does not require light and is more complex, involving a suite of enzymes. Dark repair can correct not only pyrimidine dimers but also other structural damage to DNA. It occurs in all taxa examined but has not been systematically studied in marine organisms.

Rather than risking damage that may have further effects before it is repaired, it seems safer to avoid damage in the first place, although the relative metabolic costs of these alternatives have not been compared. Behavioral avoidance of UV radiation is widespread and involves, for example, migrating vertically in the water column (a limited option in the intertidal), seeking shade under the seaweed canopy, under rocks, or in empty shells (employed by tidepool fishes), retracting vulnerable body parts (done by sea anemones, as shown in Fig. 3), and covering the body with reflective shells and other debris (seen in sea anemones—Fig. 3—and sea urchins).

FIGURE 3 Symbiotic sea anemones such as *Anthopleura xanthogrammica* must expose their tentacles and oral disk to some sunlight to allow their algal endosymbionts in those tissues to photosynthesize, as seen in the specimen on the far right. However, sea anemones also reduce the damaging effects of the brightest sunlight by retracting their tentacles and shielding them and the oral disk by contracting the sphincter muscle at the top of the body column like a drawstring (as done by the sea anemone at left). Even the shaded individual on the right has begun to contract its margin where full sunlight has just reached. This avoidance is mediated by PAR and UV radiation. These sea anemones also have attached reflective gravel to the verrucae (adhesive structures) of their columns. Attachment of gravel and shell debris by immersed individuals such as these in tidepools is evoked by UV radiation, so the material serves as a UV sunshade or parasol, particularly when the coverage is extensive (inset). When exposed to air at low tide, the related species *A. elegantissima* responds to low humidity by attaching more debris, which provides a boundary layer that restricts the evaporation of water and reduces desiccation, as well as being a shield against direct sunlight.

Sessile organisms must rely on mechanisms to block or screen out UV radiation. These include shells, tests, egg capsules, and other coverings, as well as biochemical sunscreens that absorb UV radiation and harmlessly dissipate its energy. The latter include melanins and mycosporine-like amino acids (MAAs), and less demonstrably, carotenoids and macroalgal phlorotannins (polyphenolic compounds consisting of multiple six-carbon rings containing three conjugated double bonds and attached –OH groups). Carotenoids seem to protect more by quenching reactive oxygen species (especially singlet oxygen) and photo-excited chlorophylls than by dissipating UV radiation directly. The phlorotannin content of the brown seaweed *Ascophyllum nodosum* increases under UV exposure, but its sunscreen effectiveness (it absorbs in the UVB) has not been evaluated.

Melanins (complex pigments formed by the oxidation and polymerization of the aromatic amino acid tyrosine) broadly absorb UVA and UVB and occur in the epidermis of marine invertebrates, including sea anemones, flatworms, polychaetes, echinoderms, and arthropods (crustaceans and insects), as well as in the integument of fishes. Increased deposition of melanin following exposure to UV radiation (the familiar tanning response in littoral humans) occurs in bony fishes and sharks, as well as in the sea anemone *Metridium senile*, where it presumably enhances photoprotection.

Mycosporine-like amino acids are, taxonomically, the most widespread sunscreens in marine organisms. They are small cyclic molecules in which variation in a side chain changes their absorption characteristics (Fig. 4). The 20-odd identified MAAs absorb UVA and UVB radiation in the range 309–360 nm, providing a broadband UV filter. They absorb UV radiation efficiently (far more so than DNA and proteins in this waveband), and thus intercept UV radiation before it damages other biological molecules, dissipating the energy as heat and without forming potentially damaging active intermediates. MAAs are colorless and transparent to PAR, important

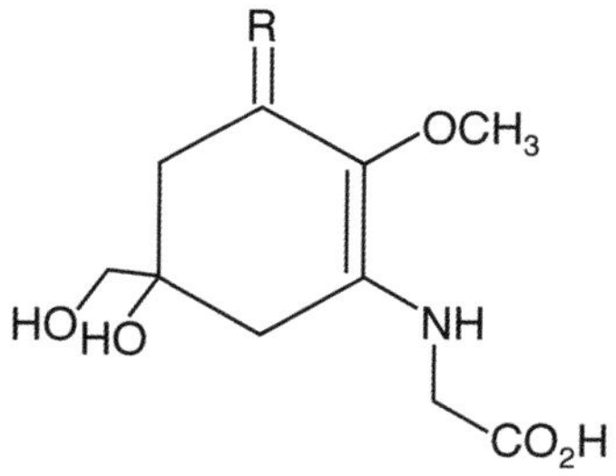

Mycosporine-glycine: R = O; λ_{max} = 310 nm
Palythine: R = NH; λ_{max} = 320 nm
Diverse MAAs: R = N of amino acid or other amine;
λ_{max} = 330–360 nm

FIGURE 4 General structure of mycosporine-like amino acids (MAAs), natural sunscreens that are widespread among marine organisms. Chemical substitutions (R) on the central cyclic structure modify its wavelength of maximum absorption (λ_{max}), whereby organisms that contain multiple MAAs have a broadband UV sunscreen.

because algae and symbiotic invertebrates such as sea anemones must transmit these solar wavelengths in order to photosynthesize.

MAAs are natural products of metabolism in algae and cyanobacteria, but are also accumulated by marine consumers from their food and subsequently concentrated in epidermal tissues and in eggs, where they afford a concentration-dependent protection of the embryos against UV radiation. Tidepool sculpins (fishes in the family Cottidae) contain UV-absorbing compounds in their skin's mucous layer, the absorption spectra of which are indicative of MAAs that occur also in the ocular lenses of fishes. Overall, the degree of UV absorption by sculpin mucus is correlated with differences in incident UV radiation associated with latitude and intertidal height. Corals and seaweeds in shallow water accumulate higher concentrations of MAAs than do deep-dwelling representatives of the same species, and the biosynthesis of MAAs is stimulated primarily by UV radiation and also by intense PAR. Biological weighting functions indicate that MAAs protect photosynthesis from UVA in free-living dinoflagellates (Fig. 2B) and corals. MAAs released from algae are an important fraction of the UV-absorbing dissolved organic matter in seawater.

The activities of antioxidant enzymes (notably superoxide dismutase) also increase with exposure to UV radiation in algae, invertebrates (including nonsymbiotic but especially symbiotic species), and larval fishes. The upregulation of SOD indicates that it is a defense against the secondary effects of UV radiation acting via the intracellular generation of superoxide radicals and other reactive oxygen species derived from them.

UV RADIATION AND THE INTERTIDAL ENVIRONMENT

Although the intertidal zone has not been extensively studied with particular reference to UV radiation, littoral organisms are exposed to the highest local solar irradiances in the marine environment. Accordingly, they have evolved adaptations to avoid or repair UV-induced damage. The available evidence suggests that, although the increased UVB irradiance since the 1970s has demonstrable short-term effects on shallow-water organisms, it has not caused major changes in littoral communities, probably because the organisms' defensive mechanisms thus far have been able to cope with this increase.

Nevertheless, maintaining elevated defensive and repair capacities imposes a cost, and the manifestations of long-term UV stress on community productivity are uncertain. Subtidal bottom-dwelling organisms, and those living deeper in the water column, have proved more sensitive

to UV radiation and seem more at risk from its effects under ozone depletion. Confined tidepools may provide natural reaction vessels for studying the photochemistry of dissolved organic matter and its dual attributes of attenuating UV radiation and generating reactive oxygen species, as well as natural laboratories for evaluating the potential biological effects of these chemicals.

SEE ALSO THE FOLLOWING ARTICLES

Light, Effects of / Metamorphosis and Larval History / Photosynthesis / Reproduction / Symbiosis / Water Chemistry

FURTHER READING

Cockell, C. S., and J. Knowland. 1999. Ultraviolet radiation screening compounds. *Biological Reviews* 74: 311–345.

De Mora, S., S. Demers, and M. Vernet, eds. 2000. *The effects of UV radiation in the marine environment.* Cambridge, UK: Cambridge University Press.

Franklin, L. A., and R. M. Forster. 1997. The changing irradiance environment: consequences for marine macrophyte physiology, productivity and ecology. *European Journal of Phycology* 32: 207–232.

Helbling, E. W., and H. Zagarese, eds. 2003. *UV effects in aquatic organisms and ecosystems.* Cambridge, UK: The Royal Society of Chemistry.

Hester, R. E., and R. M. Harrison, eds. 2000. *Causes and environmental implications of increased UV-B radiation.* Issues in Environmental Science and Technology 14. London: Royal Society of Chemistry.

Lesser, M. P. 2006. Oxidative stress in marine environments: biochemistry and physiological ecology. *Annual Review of Physiology* 68: 253–278.

Shick, J. M. 1991. *A functional biology of sea anemones.* London, UK: Chapman & Hall.

Shick, J. M., and W. C. Dunlap. 2002. Mycosporine-like amino acids and related gadusols: biosynthesis, accumulation, and UV-protective functions in aquatic organisms. *Annual Review of Physiology* 64: 223–262.

UPWELLING

JOHN A. BARTH

Oregon State University

Upwelling is the vertical motion of seawater from depth toward the sea surface. Water moves upward to supply regions where wind causes surface waters to diverge. Upwelling fuels the incredibly productive food web found in the coastal ocean, and the circulation it generates is key to the transport of organisms across and along the continental margin.

WIND FORCING OF UPWELLING

Coastal upwelling is driven by the wind, more specifically by the divergence of surface currents driven by the wind. As wind blows over the ocean surface, it pushes a thin surface layer downwind. The wind stress is communicated down into the upper ocean by the viscosity of water, that is, the friction between water layers, with lower layers moving slower than those above. Because the earth is rotating, there is a Coriolis force to the right (to the left in the Southern Hemisphere) of the wind-driven currents when facing downstream. The following discussion is valid for the Northern Hemisphere. Eventually a balance is achieved between the wind stress and the Coriolis force, and the resulting upper-ocean currents are to the right of the wind. Water velocities are strongest at the surface and decrease with depth while turning in a clockwise direction. The currents near the sea surface are 45° to the right of the wind. Looking down from above, a line connecting the tips of the velocity vectors is a spiral, the so-called Ekman spiral, named after V. W. Ekman, who detailed this motion in 1905 during his Ph.D. research, motivated by the observation by F. Nansen and others that icebergs in the Arctic Ocean moved to the right of the wind rather than directly downwind.

The Ekman layer is a surface boundary layer, and the depth to which the wind-driven currents penetrate, typically 20–50 m, is called the Ekman depth. When the currents in the entire Ekman layer are summed up, the net transport of water, referred to as the Ekman transport, is exactly 90° to the right of the wind. This remarkable result is the driving force behind coastal upwelling. Alongshore winds, blowing with the coast to the left when facing downwind, drive surface currents offshore. To compensate for the loss of water at the surface near the coast, water is upwelled from depth (Fig. 1).

Upwelling is forced primarily by alongshore winds according to the argument just stated, but there are at least two other ways that wind can drive upwelling. Additional upwelling can be driven by cross-stream variations in the strength of the wind. Where alongshore winds are stronger, say offshore or near coastal headlands, the resultant Ekman layer transport is greater than in adjacent regions with weaker winds. To conserve mass, water is upwelled from below. In some regions of coastal upwelling, this "curl-driven" upwelling can account for one-third or more of the total upwelling. A second source of upwelling originates from winds blowing in the offshore direction. These winds can drive a thin (a few meters thick) surface layer directly offshore in a surface "log layer," requiring a supply of water from below to conserve mass near the coast. This effect is small compared with the efficiency of alongshore winds in driving upwelling.

Coastal upwelling occurs anywhere where winds blow parallel to the coast, but it is especially prevalent along the eastern boundaries of the ocean basins at mid to high latitudes. Alongshore winds blowing toward the equator are created by air moving around high-pressure systems (clockwise in the Northern Hemisphere) over the eastern ocean basins, for example around the North Pacific High. These winds are intensified by the presence of low-pressure systems over the adjacent warm continents. There are four major eastern boundary current regions dominated by upwelling in the world's oceans: the Benguela Current region off southwest Africa, the Canary Current region off northwest Africa and the Iberian Peninsula, the Humboldt Current region off Peru and Chile, and the California Current system off the west coast of North America (Fig. 2). The remaining eastern boundary, along the west coast of Australia in the Indian Ocean, is dominated by the Leuwin Current, which is not primarily wind-driven.

QUANTIFYING UPWELLING VELOCITY AND TRANSPORT

Upwelling velocities are measured in tenths of millimeters per second but, summed over a day, result in water parcels moving upward 10–30 m, a significant fraction of the total water depth over the continental shelf. Such small vertical velocities are virtually impossible to measure directly with a mechanical or acoustic current meter, but the effect of upwelling is easy to recognize by the presence of cold water at the surface near the coast (Fig. 2).

The offshore surface Ekman transport ($M = \tau/\rho f$, in $m^2\ s^{-1}$) is estimated by dividing the wind stress (τ, in newtons m^{-2}) by the density of water ($\rho \approx 1024\ kg\ m^{-3}$) and the Coriolis parameter (f), equal to twice the rotation rate of the earth times the sine of the latitude (about

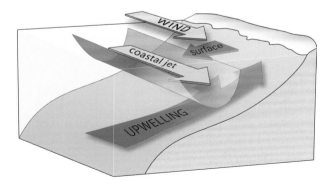

FIGURE 1 Schematic wind-driven upwelling and ocean currents. Surface water is driven offshore by an alongshore wind and the Coriolis force. Cold, salty, nutrient-rich water is upwelled to resupply the offshore surface flow. The resulting cross-shelf temperature (density) gradient creates a strong, surface-intensified coastal upwelling jet flowing in the direction of the wind. Drawing by J. Barth and D. Reinert.

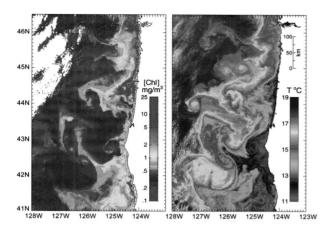

FIGURE 2 Sea-surface temperature (right) and chlorophyll (left) measured by satellite along the U.S. West Coast on September 26, 1998. Image by J. Barth and D. Reinert. Data courtesy of T. Strub.

0.0001 s⁻¹ at mid-latitudes). Wind stress is equal to the wind speed squared, multiplied by the density of air (1.3 kg m⁻³) and a drag coefficient, which is dependent on wind speed and details of the atmospheric boundary layer structure but is about equal to 0.0014. As an example, a 20-knot wind yields an offshore Ekman transport of 1.9 m² s⁻¹. If this transport is distributed over a 30-m-deep layer, the offshore velocity is 0.06 m s⁻¹, or about 5 km per day. This formula for offshore surface Ekman transport is used to compute the "Bakun Upwelling Index," a value commonly used in studies of wind-driven coastal upwelling and ecosystem response.

Since the offshore surface Ekman transport must be balanced by upwelling from below, an estimate of the cross-shelf width (L) over which the upwelling occurs may be used together with a conservation-of-mass argument to calculate the vertical upwelling velocity. In the example above, using L = 20 km yields an upwelling velocity of one tenth of a millimeter per second (10⁻⁴ m s⁻¹) or about 10 meters per day.

COASTAL CIRCULATION IN AN UPWELLING REGION

Upwelling dramatically influences seawater properties near the coast. Because sea water is colder, saltier, and higher in nutrients at depth, water upwelled near the coast is considerably different from surface waters farther offshore. The boundary between cold, salty, nutrient-rich waters inshore and warm, fresh, nutrient-poor waters offshore is called the upwelling front. Upwelling fronts can be strong and sharp with temperature differences of a few to 10 °C over a half to 10 km, easily detected,

for example, in satellite sea-surface temperature images (Fig. 2). Note that because salinity is "conserved" (that is, it is not altered by surface warming, as is temperature, nor consumed or produced through biological and chemical processes, as are nutrients), it is the most faithful tracer of upwelling.

Another physical consequence of upwelling is that sea level goes down by tens of centimeters as surface waters are forced offshore, an effect measurable with a tide gauge. In addition to the few to ten cm s⁻¹ offshore surface Ekman velocities described in the preceding section, an even weaker compensatory onshore flow at depth—weaker because the required onshore transport is distributed over a layer thicker than the surface Ekman layer—is established to supply the upwelled water (Fig. 1). This compensatory or return flow can occur at mid-depth, in the bottom boundary layer, or both, depending on details of the vertical stratification of the water column and the bottom slope. The proportion of upwelled water supplied from mid-depth or from near the bottom has important consequences for the water properties supplied to the nearshore and the rocky intertidal.

The presence of cold water next to the coast creates a density gradient across the shelf (the upwelling front) because it is denser than warm water found offshore. This density gradient adds to the difference in sea level across the shelf to form an onshore pressure force, which is balanced by an offshore Coriolis force associated with a strong coastal upwelling jet flowing in the direction of the wind (Fig. 1). The horizontal scale over which this balance is achieved is called the Rossby radius of deformation and is a fundamental scale in geophysical fluid dynamics. For coastal upwelling in a stratified fluid, it is the first internal Rossby radius of deformation (L_i), determined by the strength of the stratification (N in s⁻¹), the local water depth (H), and the Coriolis parameter ($L_i = NH/f$), that is most important. Over the continental shelf and slope, L_i varies from 5 to 20 km. From this follows a typical cross-shelf scale of the upwelling front and the width of the alongshore coastal upwelling jet.

The coastal upwelling jet is surface intensified and strongest in a core, associated with the upwelling front, often located over the mid-shelf (about 80–100 m bottom depth, for example off the U.S. West Coast). The upwelling jet can reach speeds of 0.5–1.0 m s⁻¹ and carry about 0.25–0.5 million meters cubed per second of water downwind, equal to 15–30 times the transport in the Mississippi River. Although the surface and deep compensatory currents can move material across the shelf at about 5 cm s⁻¹, the wind-driven alongshore currents are ten or more times faster, so that water parcel trajectories are not two-dimensional but are, in fact, helical motions along the coast.

ECOSYSTEM RESPONSE TO COASTAL UPWELLING

Upwelling has profound consequences for coastal pelagic and benthic ecosystems. Nutrients are supplied to nearshore and intertidal habitats. Wind-driven currents transport the larvae of rocky-intertidal organisms to and from the coast and, potentially, over great distances alongshore. Phytoplankton blooms occur in continental shelf waters as nutrient-rich water encounters the euphotic zone. Upwelled nutrients include nitrate, phosphate, and silicate, necessary for phytoplankton growth, and trace metals such as iron, which influence the rate of photosynthesis in phytoplankton. These phytoplankton form the base of the food web, supporting zooplankton populations that are, in turn, preyed upon by fish and other large consumers. This efficient upwelling-driven food web supports over 25% of the world's fish catch.

The cold, salty, nutrient-rich upwelled water is the source of water to the rocky intertidal zone. This is demonstrated by the strong relationship between temperatures measured right on a rocky intertidal outcrop adjacent to tide pools off Oregon and temperatures measured directly offshore on the open, wind-driven continental shelf (Fig. 3). Cold, upwelled waters near the coast also help produce fog, which can be a hazard to navigation or supply life-giving moisture to arid regions inshore of upwelling regions, for example the Namib Desert. Cold ocean temperatures from upwelling also increase the strength of the daily sea breeze, intensifying the land–ocean temperature contrast, which forces air landward during the late afternoon.

Besides supplying nutrients to the euphotic zone, the wind-driven surface velocities transport the larvae of rocky-intertidal organisms offshore. The onshore, compensatory

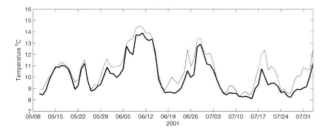

FIGURE 3 Time series of temperature measured on a rocky intertidal outcrop (thick curve) adjacent to tidepools off Oregon, showing high correlation with temperatures measured directly offshore at the surface (thin curve) in 30 m of water. Drawing by J. Barth, A. Kirincich, and F. Chan. Data courtesy of Partnership for Interdisciplinary Studies of Coastal Oceans (PISCO).

flow feeding the upwelling transports water and the material it contains (eggs, larvae) shoreward. These cross-shelf motions are critical in removing and supplying the eggs and larvae of rocky-intertidal organisms to and from the shore. The swift, alongshore coastal upwelling jet can similarly transport eggs and larvae hundreds of kilometers alongshore. This alongshore transport and its variability are important for the connectivity of rocky-intertidal populations through dispersal of their larvae.

TIME- AND SPACE VARIABILITY OF COASTAL UPWELLING

The description above assumes steady, time-independent wind forcing. In reality, coastal winds vary over a wide range of periods. The dominant variability occurs with 2- to 10-day periods, the "weather band," as atmospheric weather systems transit coastal regions. Weather-band variability can break up periods of strong, upwelling-favorable winds with intervals of low winds (relaxation) or downwelling-favorable winds. When upwelling-favorable winds relax, alongshore flow on the shelf, especially near the bottom and nearest the coast, can reverse and flow opposite to the previously established downwind currents. During wind relaxation, the across-shelf surface Ekman transport ceases, and during downwelling-favorable winds, surface transport is shoreward. Thus, relaxation and downwelling events are important for onshore recruitment of invertebrate larvae (crabs, barnacles, bivalves, and molluscs).

Depending on latitude, coastal winds can also vary strongly on a seasonal cycle, with sometimes rapid transitions between summertime upwelling-favorable winds and fall–winter downwelling-favorable winds. Lastly, year-to-year or "interannual" variability is introduced to coastal upwelling from events, such as El Niño/La Niña cycles or enhanced transport of high-latitude, subarctic water into coastal upwelling regions. This interannual climate variability influences the properties of upwelling source waters (warmer, nutrient-poor water during El Niño; colder, nutrient-rich water during enhanced subarctic transport events) and the strength, timing, and duration of coastal winds.

Further complexity to the circulation and ecosystem consequences described in the previous paragraphs is introduced by three-dimensional processes such as flow instability, alongshore variations in wind speed and flow–topography interaction. The coastal upwelling jet and front are unstable and can break down into meanders and eddies, which increase the amount of frontal habitat and expedite across-shelf transport of nutrients, larvae, phytoplankton, and other water properties (Fig. 2).

This instability process is analogous to how the atmospheric jet stream meanders to create high- and low-pressure weather systems (eddies). When the upwelling jet interacts with coastal topographic features (capes, bays, submarine canyons and banks), strong three-dimensionality is introduced into the system, including the formation of upwelling hot spots with increased primary production; the separation of the jet from the coast, thus injecting water and the material it contains into the deep ocean; and the creation of retention or low-flow regions where organisms can be preferentially held in place near the coast. Each of these processes contributes to the variability in ocean "weather" and, hence, the influence of the coastal ocean on rocky intertidal habitats.

SEE ALSO THE FOLLOWING ARTICLES

Currents, Coastal / El Niño / Fog / Nutrients / Seawater / Wind

FURTHER READING

Bakun upwelling index. http://www.pfeg.noaa.gov/products/PFEL/modeled/indices/upwelling/upwelling.html.

Huyer, A. 1983. Coastal upwelling in the California Current system. *Progress in Oceanography* 12: 259–284.

Huyer, A., R. L. Smith, and J. Fleischbein. 2002. The coastal ocean off Oregon and northern California during the 1997–8 El Niño. *Progress in Oceanography* 54: 311–341.

Lentz, S. J. 1992. The surface boundary layer in coastal upwelling systems. *Journal of Physical Oceanography* 22:1517–1539.

Mooers, C. N. K., and A. R. Robinson. 1984. Turbulent jets and eddies in the California Current and inferred cross-shore transports. *Science* 223: 51–53.

Open University. 2001. *Ocean circulation,* 2nd ed. Oxford: Butterworth-Heinemann.

Roughgarden, J., S. Gaines, and H. Possingham. 1988. Recruitment dynamics in complex life cycles. *Science* 241: 1460–1466.

Small, L. F., and D. W. Menzies. 1981. Patterns of primary productivity and biomass in a coastal upwelling region. *Deep-Sea Research* 28A: 123–149.

Tomczak, M., and J. S. Godfrey. 2003. *Regional oceanography,* 2nd ed. Delhi: Daya Publishing House. http://www.es.flinders.edu.au/~mattom/regoc/pdfversion.html.

V

VARIABILITY, MULTIDECADAL

FRANCISCO CHAVEZ

Monterey Bay Aquarium Research Institute

Changes in climate, ocean circulation, and ocean ecosystems with periods of about 50 years have been recently recognized and referred to as Pacific multidecadal variability. These climate or regime shifts have particularly large impacts on small pelagic fish such as anchovies and sardines. These populations shift synchronously in Japan, California, and Peru, with sardines dominating for about 25 years and anchovies for the following 25 years. Impact on other ocean ecosystems, such as those on rocky shores, are unknown but likely.

NATURAL CLIMATE CHANGE

As farmers and fishermen have known for centuries, climate fluctuations are both important and normal. Climate varies on daily, weather-system, and seasonal time scales, and we are now learning about longer term fluctuations that occur within the bounds of natural variability. These cycles strongly affect humans, their economies, and the ecosystems on which they depend. But humans are also affecting climate by increasing atmospheric CO_2, and the biological consequences of the resulting global warming are not at present predictable and are potentially catastrophic. Understanding human-induced climate change will require characterization of natural climate variability and the use of natural cycles as models of climate change.

Prior to this growing awareness of natural climate variations, ecologists viewed the physical environment as a stable background against which biotic interactions drive population change and structure communities. Over the past two decades, strong El Niños, the ozone hole, and the looming specter of global warming forced the uncomfortable realization that the physical environment is changing, even on the relatively short time scales of ecological study, and that human activities may affect climate in unforeseen ways. Climate and the physical environment have thus reemerged as major themes in ecological science. In the oceans, it is clear that natural climate variability can have large impacts on ecosystem structure and biological productivity. The correlation between climate variability and the productivity and structure of ocean ecosystems has been well established. Climate-driven changes in ocean circulation, ocean mixing, dust deposition, or both, can regulate the overall productivity of an ocean ecosystem by changing the supply of a limiting nutrient. Changes in primary productivity then cascade through every trophic level. Climate can also influence animal populations directly through effects on recruitment, competitive advantages, or predation. Of particular interest are relationships between abiotic (bottom-up climate impacts on overall ecosystem productivity) and biotic (top-down climate impacts on competition and predation) effects.

CLIMATE IS NEVER AVERAGE

In general dictionaries, climate is defined as the average course or condition of the weather at a place, usually over a period of years as exhibited by temperature, wind velocity, and precipitation. However, the average itself changes, depending on the period used. For example, sea surface temperature (SST) for the coasts of Peru and California for the years 1993–1996 was warmer than during 1999–2003, and during the 1997–1998 El Niño, SST was extremely warm. These changes in temperature were accompanied by fluctuations in ocean productivity, with warmer years being less productive than cooler years. What causes climate to change from year to year and decade to decade? What are the consequences of this climate variability? Complete answers to these questions are still forthcoming, although significant progress has been made over the past several decades, particularly in understanding the consequences of climate variability. For example, the warm SST observed during 1997–1998 was a result of a strong El Niño that was well documented in the equatorial Pacific and along the west coast of the Americas. El Niño is a prime example of the insight gained on the consequences of climatic variability. Although El Niño had been recognized off Peru since the ancient civilizations of the Incas, it was the large 1957–1958 El Niño that brought international attention to the phenomenon. Following the 1982–1983 El Niño it became clear that the oceanic perturbations in the tropical Pacific had global effects on climate. Oceanic effects were originally thought to be restricted to the tropical and eastern Pacific, but careful studies in the center of the Pacific Ocean close to Hawaii uncovered El Niño effects there as well. The past several decades have seen growing awareness of La Niña, the counterpart or opposite condition of El Niño.

THE STORY OF EL VIEJO

Recently, focus has shifted to longer decade-scale changes that show remarkable basin-wide coherence and, again, strong impacts on oceans ecosystems. These multidecadal changes help explain the differences in SST observed before and after the 1997–1998 El Niño. The period prior to 1997–1998 was associated with a warm quarter century; the period after may be a cool one. This particular cycle, which is often referred to as the Pacific Decadal Oscillation (PDO), has a period of approximately 50 years (Fig. 1). Because of the similarity to El Niño and La Niña, the name El Viejo (the old man) for the warm eastern Pacific regime, and La Vieja (the old woman) for its counterpart, has been suggested. When the eastern Pacific is warmer than average, during El Viejo, El Niños may be more frequent and of greater intensity. Similarly, during La Vieja, La Niñas may be frequent or stronger. Longer time series, both historical and present, will enlighten us further on El Niño frequency and intensity and the potential effects of anthropogenic perturbations.

In a simplified conceptual view of the Pacific, the trade winds set up a basinwide slope in sea level, thermal structure, and—importantly for biology—nutrient structure. The shallow thermocline in the eastern tropical Pacific leads to enhanced nutrient supply and productivity. The El Viejo/La Vieja fluctuations have basinwide effects on SST and thermocline slope that are similar to El Niño and

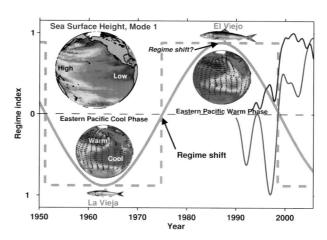

FIGURE 1 Hypothetical oscillation of a regime index with a period of 50 years. From the early 1950s to about 1975, the Pacific was cooler than average and anchovies dominated. The cool phase is referred to as La Vieja. From about 1975 to the late 1990s, the Pacific was warmer and the sardines dominated. The warm phase is referred to as El Viejo. The large-scale spatial pattern of sea surface temperature (SST) and atmospheric circulation anomalies are shown for each regime. The spatial pattern shows that warming or cooling is not uniform and the eastern Pacific is out of phase with the central North and South Pacific. Some indices suggest that the shifts are rapid (dashed) whereas others suggest a more gradual shift (solid). Regime shifts are commonly associated with a change in index sign, but populations may also exhibit changes in abundance when the index stops increasing or decreasing. The first empirical orthogonal function (EOF) of global TOPography EXperiment (TOPEX) for Ocean Circulation sea surface height (SSH) is shown above the cool, La Vieja regime. Low SSH implies a shallow thermocline/nutricline when the coefficient (blue line) is positive. The coefficient is shown in blue together with surface chlorophyll anomalies (mg m⁻³) for the eastern margin of the California Current system from 1989 to 2005. Note the high chlorophyll after 1997–1998, consistent with the shallow thermocline of the eastern Pacific. Modified from Chavez et al. (2003).

La Niña but on longer time scales. During the cool eastern boundary La Vieja regime, the basin-scale sea level slope is accentuated (sea level is lower in the eastern Pacific, higher in the western Pacific). Lower sea level is associated with a shallower thermocline and increased nutrient supply and productivity in the eastern Pacific; the inverse occurs in the western Pacific. In addition to thermocline and sea surface temperature, there are regime shift changes in the transport of boundary currents, equatorial currents, and of the major atmospheric pressure systems. Changes in the abundance of anchovies and sardines are only one of many biological perturbations associated with regime shifts, and these are reflected around the entire Pacific.

The northeast Pacific may be the most studied area in terms of regime shifts. During El Viejo from the late 1970s to the early 1990s, zooplankton and salmon declined off Oregon and Washington but increased off Alaska. The flip-flop between ecosystems in the Gulf of Alaska and the California Current is one of the conundrums associated with these longer-term cycles. During El Niño the California Current and the Gulf of Alaska seem to be in phase in that productivity decreases in both locations. The El Niño changes are associated with thermocline depth, which decreases in both cases lowering nutrient supply and primary production. Spring mixed-layer depths, on the other hand, decrease in the Gulf of Alaska while they increase in the California Current during the positive or El Viejo phase of the PDO. It is surmised that mixed-layer decreases may favor primary production in the North Pacific by reducing light limitation. In the California Current the deeper mixed layers are associated with a deeper thermocline and lower primary production. A possible explanation for the different responses of the California Current and the Gulf of Alaska to El Niño and El Viejo might be that (1) during strong El Niños thermocline displacements dominate in both systems, reducing productivity at both locations, and (2) weaker thermocline anomalies during El Viejo reduce productivity in the California Current, but these effects are counteracted in the Gulf of Alaska by changes in mixed-layer depth resulting in increases in productivity at this location. This suggests fundamental differences between low-latitude stratified oceans, where thermocline displacements and upwelling regulate, and high-latitude environments, where mixing is the dominant process.

Seabird populations decrease off California and Peru during El Viejo. The California Current weakens and moves shoreward during the warm phases, and the subarctic gyre intensifies. Warmer temperature and lower salinity near the California coast support the weakening of the California Current. A stronger and broader California Current, brought about during the cool La Vieja regime, is associated with a shallower coastal thermocline from California to British Columbia, leading to enhanced primary production. It should be noted that the multidecadal changes in the circulation of the California Current system are not fully resolved and some argue that regime-shift changes in the position rather than the intensity of the currents are the primary mechanism for some of the observed patterns. What is very clear is that the boundaries of warm-water species move poleward during the warm regimes, and the boundaries of cold-water species move equatorward during cool regimes. In the southeastern Pacific biological variability is similar to that observed in the northeast Pacific albeit much less well documented.

In the northwest Pacific off Japan, the depth of the thermocline, nutricline, sea surface temperatures, and the winter mixed layer have shown changes on multidecadal time scales. During the El Viejo regime, sea surface temperatures cooled, sea level dropped, the thermocline and nutricline shoaled, and mixed layers deepened. Transport by the Kuroshio Current weakened. Primary production increased and sardine populations expanded from coastal waters eastward across the North Pacific to beyond the International Date Line. It remains unclear why sardines increase off Japan when local waters cool and become more productive, whereas they increase off California and Peru when those regions warm and become less productive.

In the warm pools of the western and northeastern tropical Pacific, physical variability has been harder to elucidate, partly because temperatures are warm and homogenous there. However, there is evidence of lower recruitment of yellow fin tuna during the cool La Vieja regime. The northeastern tropical Pacific is surrounded by regions with strong multidecadal fluctuations (California Current, Peru Current, equatorial Pacific, subtropical gyre). Tuna in the warm waters of the western Pacific seem to be similarly affected. Populations of yellow fin tuna in the western Pacific may have increased during the cool regimes. Highly mobile organisms such as the blue fin tuna migrate on basin scales, spending significant periods in areas altered by these large-scale climate and ocean changes. These organisms must respond in complex ways to regime shifts.

Episode-to-episode differences for El Viejo and La Vieja are just beginning to emerge. This should come as no surprise. After several iterations of El Niño it became clear that no two El Niños were alike. In 1982 the development of a canonical El Niño in the tropical Pacific was described for data from the 1950s to the 1970s. Following this description (and the 1976 regime shift) no El Niño resembled the development of the canonical El Niño. It will be interesting if after the regime shift in the mid- to late 1990s, El Niño development once again follows the canonical El Niño; the first one, the moderate 2002–2003 event, has apparently not.

SEE ALSO THE FOLLOWING ARTICLES

Climate Change / El Niño / Heat and Temperature, Patterns of / Rhythms, Tidal

FURTHER READING

Chavez, F. P. 2005. Biological consequences of interannual to multidecadal variability, in *The sea*. Vol. 13. A. Robinson and K. Brink, eds. Cambridge, MA: Harvard University Press, 643–679.
Chavez, F. P., J. P. Ryan, S. Lluch-Cota, and M. Ñiquen [Carranza]. 2003. From anchovies to sardines and back—multidecadal change in the Pacific Ocean. *Science* 299: 217–221.
Hare, S. R., S. Minobe, and W. S. Wooster, eds. 2000. The nature and impacts of North Pacific climate regime shifts. *Progress in Oceanography* 47: 99–408.

Klyashtorin, L. B. 2001. *Climate change and long-term fluctuations of commercial catches: the possibility of forecasting.* FAO Fisheries Technical Paper 410. Rome: Food and Agriculture Organization of the United Nations.

Mantua, N. J., and S. R. Hare, 2002. The Pacific decadal oscillation. *Journal of Oceanography* 58: 35–44.

Mantua, N. J., S. R. Hare, Y. Zhang, J. M. Wallace, and R. C. Francis. 1997. A Pacific interdecadal climate oscillation with impacts on salmon production. *Bulletin of the American Meteorological Society* 78: 1069–1079.

McKinnell, S. M., R. D. Brodeur, K. Hanawa, A. B. Hollowed, J. J. Polovina, and C.-I. Zhang, eds. 2001. Pacific climate variability and marine ecosystem impacts. *Progress in Oceanography* 49: 1–6.

McPhaden, M. J., and D. X. Zhang, 2002. Slowdown of the meridional overturning circulation in the upper Pacific Ocean. *Nature* 415: 603–608.

Miller, A. J., D. R. Cayan, T. P. Barnett, N. E. Graham, and J. M. Oberhuber. 1994. The 1976–77 climate shift of the Pacific Ocean. *Oceanography* 7: 21–26.

Minobe, S. 1997. A 50–70 year climatic oscillation over the North Pacific and North America. *Geophysical Research Letters* 24: 683–686.

VERTEBRATES, TERRESTRIAL

THOMAS P. PESCHAK

University of Cape Town, South Africa

A surprisingly diverse community of terrestrial vertebrates, which includes baboons, porcupines, rats and even bears, forage on wave-swept rocky shores, consuming marine invertebrates and fish on all continents except Antarctica.

DIVERSITY OF TERRESTRIAL VERTEBRATES ON ROCKY SHORES

The phenomenon of terrestrial vertebrates as consumers and agents of energy transfer on wave-swept rocky shores is underreported and poorly understood. This lack of understanding is primarily due to the fact that the subject matter falls in the gap between marine and terrestrial science, resulting in the majority of information being presented in anecdotal form. To date, in excess of 35 species of mammal, reptile, and amphibian have been recorded using wave–swept rocky shores as feeding grounds, and as scientific scrutiny of this phenomenon increases, this number is likely to increase.

Primates

In South Africa, Chacma baboons *(Papio ursinus)* consume limpets, mussels, crabs, and shark egg cases, whereas on Koshima Island off Japan, macaques *(Macaca fuscata)* prize barnacles and limpets from rocks and enter shallow water to capture small octopuses.

Rodentia and Lagomorpha

Porcupines *(Hystrix africaeaustralis)* along Namibia's coast and European rabbits *(Oryctolagus cuniculus)* on South African sea-bird islands graze on seaweeds. In Chile Norway rats *(Rattus norvegicus)* consume more than 40 species of intertidal organisms, with keyhole limpets being the most frequently consumed marine prey.

Carnivora

In North America, raccoons *(Procyon lotor)* feed on crabs, sea urchins, gastropods and small fishes. Arctic foxes *(Alopex lagopus)* from Iceland, Greenland, and Alaska also consume a wide variety of intertidal species ranging from polychaetes and mussels to starfish and kelp. Coyotes *(Canis latrans,* Fig. 1) in Baja California, Mexico, feed on

FIGURE 1 The coyote *(Canis latrans)* is ordinarily regarded as a creature of the desert rather than the seacoast but feeds on rocky-shore invertebrates in Baja California. Photograph by Will Funk, Alpine Aperture.

gastropods and crabs, and the American mink *(Mustela vison)* feeds on a wide variety of intertidal prey ranging from crabs to sea urchins. On the coasts of Alaska and Canada, American black bears *(Ursus americanus)* scrape barnacles off rocks and feed on mussels, and Grizzly bears *(Ursus arctos)* flip boulders in search of crabs. Eurasian otters *(Lutra lutra)* off the west coast of Scotland feed on crabs and intertidal fish, South Africa's Cape clawless otter *(Aonyx capensis)* consume rock lobsters, crabs, and clinids, and along South America's west coast the diet of the marine otter *(Lutra felina)* is fish, crustaceans, and molluscs.

Artiodactyla

A wide variety of seaweeds are consumed by red deer *(Cervus elaphus)* on the Isle of Rum off the U.K. coast, by mule deer *(Odocoileus hemionus)* on islands off the west coast of North America, by reindeer *(Rangifer tarandus)* on the sub-Antarctic island of South Georgia, and by mountain goats *(Oreamnos americanus)* on Alaska's coast.

Reptilia and Amphibia

The intertidal lizard *Uta tumidarostra* feeds on intertidal crustaceans on islands off the coast of Baja California, whereas the Galápagos marine iguana *(Amblyrhynchus cristatus)* grazes seaweed. A species of yet undescribed frog consumes marine invertebrates in the intertidal off Brazil's Atlantic coast.

FORAGING BY TERRESTRIAL VERTEBRATES

There are two possible scenarios as to why terrestrial vertebrates would repeatedly use rocky-shore resources: (1) They inhabit perennially impoverished coastal ecosystems, and the ocean provides year-round trophic subsidies. Porcupines, for example, would be unlikely to survive the marginal conditions of Namibia's Skeleton Coast were it not for the inclusion of seaweeds in their diets. (2) They inhabit seasonally impoverished coastal ecosystems, and the ocean provides only seasonal trophic subsidy, especially at times when the availability of terrestrial foods is limited. Mule deer and Artic foxes, for example, forage more frequently on rocky shores during winter, but Chacma baboons and rabbits increasingly rely on marine foods during the summer dry season.

IMPACT ON ROCKY SHORES

Few terrestrial vertebrates feed exclusively on marine resources, and in most species intertidal foraging is of short duration and intermittent in nature. This is largely because the most productive zones of wave-swept shores are accessible only for a few hours during bimonthly spring tides. The impact that terrestrial vertebrates have on wave-swept rocky shores is therefore limited; however, there are some exceptions. Raccoons appear to have a significant impact on sea urchin populations on the Washington State coast, and rats remove nearly 20% of small keyhole limpets from some Chilean rocky shores. In South Africa large baboon troops leave behind clearly visible patches of lower-density mussel cover and bare rock after feeding on mussel beds.

THE FUTURE OF TERRESTRIAL VERTEBRATES ON ROCKY SHORES

Development and urbanization processes are intense along the world's coastlines and there is mounting evidence that terrestrial vertebrates have been and are being extirpated from the seashore. Compared to previous centuries, the numbers of terrestrial vertebrates and the frequency with which they prey on intertidal marine life appear to be decreasing. In South Africa, for example, the number of baboon troops that forage on rocky shores has decreased dramatically since the beginning of the twentieth century. In addition, human activities have also made some prey species rare or completely unavailable on rocky shores. The seventeenth-century observation of bears feeding on lobsters in the intertidal zone off Maine is unlikely to be repeated today, not only because bears no longer range to the coast, but also because large lobsters are unlikely to remain in any quantities on those same rocky shores.

SEE ALSO THE FOLLOWING ARTICLES

Foraging Behavior / Iguanas, Marine / Predation / Seals and Sea Lions / Sea Otters

FURTHER READING

Carlton, J.T., and H. Janet. 2003. Maritime mammals: terrestrial mammals as consumers in marine intertidal communities. *Marine Ecology Progress Series* 256: 271–286.

Conradt, L. 2000. Use of a seaweed habitat by red deer *(Cervus elaphus)*. *Journal of Zoology (London)* 250: 541–549.

Moore, P.G. 2002. Mammals in intertidal and maritime ecosystem: interactions, impacts and implications. *Oceanography and Marine Biology: An Annual Review* 40: 491–608.

Navarette, S.A., and J.C. Castilla. 1993. Predation by Norway rats in the intertidal zone of central Chile. *Marine Ecology Progress Series* 92: 187–199.

Peschak, T. P. 2004. Sea monkeys: Chacma baboons as intertidal predators. *BBC Wildlife* 22 (8): 50–55.

Peschak, T. P. 2005. *Currents of contrast: life in Southern Africa's two oceans.* Cape Town, South Africa: Struik.

VISION

TODD H. OAKLEY

University of California, Santa Barbara

Most intertidal animals possess vision, photoreception, or both, which are involved in critical life events such as settling, feeding, escaping predators, and mating. Vision is distinguished from simple photoreception by the ability to form images, as opposed to simply detecting the direction or intensity of light. Understanding vision involves knowledge from physics, genetics, physiology, neurobiology, and behavior.

FUNDAMENTALS: LIGHT, PHOTORECEPTOR CELLS, AND OPSIN

All photoreception—including vision—begins with light. As electromagnetic radiation, light exhibits physical properties of both waves and particles. Different wavelengths of light, ranging from about 300 nanometers to about 700 nanometers, are interpreted by many animals as different colors, often allowing color vision. The particles of light, photons, cause the first step in photoreception and vision: When a photon strikes a light-reactive molecule called a chromophore, the molecule changes physical shape, causing a shape change in a protein, opsin, that is bound to the chromophore. The opsin shape change starts a chain of signaling events (phototransduction) inside photoreceptor cells, leading to a nervous impulse, which is ultimately transmitted to the central nervous system for processing (Fig. 1).

Photoreceptor cells are neurons specialized for light sensitivity by expressing the proteins involved in phototransduction, including opsin. Animal photoreceptor cells are divided into two distinct classes, ciliary and rhabdomeric, which differ physiologically and morphologically. Physiologically, the cell types differ by the presence of opposite electrical polarities. Morphologically, ciliary photoreceptors use cilia to increase the surface area of the cell that is exposed to light, whereas rhabdomeric

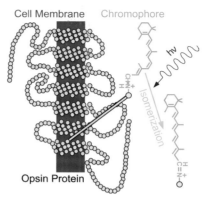

FIGURE 1 Opsin is a protein involved in the first step of photoreception. The proteins are embedded in photoreceptor cell membranes and are coupled to a chromophore. The chromophore (11-*cis*-retinal) is a light-reactive chemical that is a vitamin A derivative. When the chromophore is struck by a photon, 11-*cis*-retinal undergoes a change in shape to all-*trans*-retinene. The all-*trans* form of the chemical is reactive with the next protein in the phototransduction cascade, a chain of events that results in a nervous impulse generated by the photoreceptor cell. After Rowe (2000).

photoreceptors use membranous folds called microvilli to increase surface area. Many organisms possess both ciliary and rhabdomeric photoreceptor cells. In vertebrates, ciliary photocells dominate the visual system and include dim light–specialized rods and bright light–specialized cones. In many invertebrates, including arthropods, rhabdomeric photocells dominate the visual system.

IMAGE FORMATION

Many intertidal animals possess simple photoreceptors, often called ocelli, which do not form an image and therefore do not mediate true vision. Forming an image requires the input of multiple separate photoreceptors that sample light reflected from objects in the environment from different directions. Often, different photoreceptor cells use different opsin proteins, specialized for responding to different wavelengths of light and thus allowing for color discrimination. The integration of multiple photoreceptors is accomplished differently in different animals. For example, vertebrates, such as intertidal fishes, possess a hemispherical field of photoreceptor cells, called the retina, at the back of the eye. A lens at the front of the eye focuses light on the retina (Fig. 2). These spherical eyes equipped with lenses are termed camera-type eyes, and similar eyes are also found in cephalopod molluscs (squid and octopus).

Unlike the continuous field of photocells in a camera-type eye, compound eyes, such as those of crustaceans or fan worm annelids, are composed of discrete units called ommatidia. The ommatidia sample light from different directions to form an image. Each ommatidium contains several photoreceptor cells. The individual ommatidia of compound eyes are separated and integrated in many different ways in different organisms. Two major types of compound eye are called apposition and superposition. Apposition eyes are typical of diurnal arthropods, including crabs. In these eyes, each ommatidium contributes to one group of photoreceptor cells; the ommatidia are shielded from each other by dark pigments through which light does not pass. In superposition compound eyes, multiple ommatidia contribute light to one group of photoreceptor cells, sometimes using lenses and sometimes reflectors.

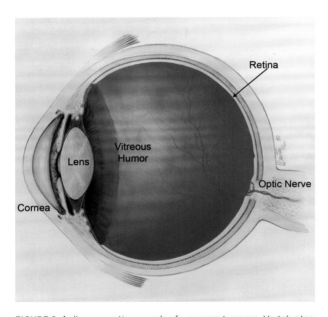

FIGURE 2 A diagrammatic example of a camera-type eye. Vertebrates, including humans, use eyes with a lens that focuses light onto a hemispherical field of photoreceptor cells at the back of the eye called the retina. Eyes of cephalopod molluscs are similar in structure.

In some organisms, vision is not mediated by camera or compound eyes. For example, some intertidal animals, such as chitons and echinoderms, possess many photoreceptors spread across the surface of the body. In some cases, these have been shown to form images, because the animal is able to integrate signals from multiple receptors that gather light from different

FIGURE 3 Some animals possess a form of vision mediated by diffuse photoreceptors scattered around the body. The pictured sea urchin has not specifically been tested for vision, but another urchin responded in experiments as if it has diffuse vision. Image ©Photographer: Asther Lau choon siew | Agency: Dreamstime.com

directions. The sea urchin *Echinometra lucunter* may use this form of vision to leave and return to dark shelters (Fig. 3).

USES AND IMPORTANCE OF VISION

Many behaviors are mediated by vision, including feeding, mating, and finding habitat. Predators have an obvious use for vision, using highly sensitive and acute vision to locate prey. Many animals, such as crabs, perform mating rituals similar to dances that are viewed by potential mates. Finding appropriate habitat is also important for intertidal organisms and may be mediated by vision or photoreception. Many animals seek shady crevices, which may provide some shelter from predators. In addition, numerous sessile intertidal organisms begin life as planktonic larvae, whose settlement may in part be guided by photoreception or vision.

SEE ALSO THE FOLLOWING ARTICLES

Larval Settlement, Mechanics of / Light, Effects of / Octopuses / Predation / Sea Urchins

FURTHER READING

Land, M. F., and D. E. Nilsson. 2002. *Animal eyes.* Oxford, UK: Oxford University Press.

Rowe, M. P. 2000. Inferring the retinal anatomy and visual capacities of extinct vertebrates. *Palaeontologia Electronica.* http:// palaeo-electronica.org/ 2000_1/ retinal/issue1_00.html.

WebVision. http://webvision.med.utah.edu/.

WATER

SEE SEAWATER

WATER CHEMISTRY

PATRICIA M. SCHULTE

University of British Columbia, Vancouver, Canada

Water chemistry is the collective term used to describe the types and amounts of all of the substances present in a particular sample of water, including dissolved salts, minerals, gases, nutrients, and other chemicals. Along with temperature, desiccation, and wave action, water chemistry is one of the critical abiotic factors that affect the type and number of organisms living in the rocky intertidal. Four aspects of water chemistry have substantial effects on these organisms: (1) the salinity of the water, (2) the concentration of dissolved gases in the water, (3) the pH of the water, and (4) the amount of dissolved nutrients available to photosynthetic microorganisms, plants and algae—the primary producers that form the base of the food web in the rocky intertidal.

WATER CHEMISTRY IN TIDEPOOLS

At high tide the water chemistry of the rocky intertidal is similar to that of the surrounding seawater, but as the tide recedes, the water that is trapped in the rocky depressions that form tidepools can undergo very large changes in water chemistry, and these changes can have profound effects on tidepool life. The extent of change in water chemistry during the tidal cycle varies among tidepools, depending on their location within the intertidal zone and other physical characteristics such as their size, shape, and orientation with respect to the sun. As a result, even adjacent tidepools can have somewhat different water chemistries at low tide, varying in salinity, dissolved gases, pH, and nutrient availability.

SALINITY

As shown in Table 1, seawater contains a mixture of various dissolved and ionized salts. Six of these (chloride, sodium, sulfate, magnesium, calcium, and potassium) make up more than 99% of the dissolved salts in seawater, but there are actually more than 70 different elements or compounds in a typical sample of seawater. Many of these trace elements are very important for biological reactions. For example, cobalt is a component in vitamin B12, and selenium is needed for the synthesis of thyroid hormones. The relative proportions of the various dissolved substances in seawater are fairly constant among oceans, but the total concentration of salts may vary.

The term *salinity* is used to describe the total concentration of salts in a particular sample of water, and

TABLE 1
Composition of Seawater

Constituent	Concentration (mg/L)	Percent of total solids
chloride (Cl^-)	19,000	55%
sodium (Na^+)	10,500	30%
sulfate ($SO4^{2-}$)	2,700	8%
magnesium (Mg^{2+})	1,350	4%
calcium (Ca^{2+})	400	1%
potassium (K^+)	380	1%
bicarbonate (HCO_3^-)	142	0.4%
bromide (Br^-)	65	0.2%
other solids	34	0.1%
total dissolved solids	34,500	
water (balance)	965,517	

is equivalent to the total number of grams of salts per kilogram (or liter) of water. Seawater in offshore areas typically has a salinity of 35 grams of salts per liter of water, whereas in near-shore areas with substantial freshwater input, salinity is often much lower. Biologists usually express salinity in units of parts per thousand (abbreviated ppt or ‰). Thus, offshore seawater is said to have a salinity of approximately 35 ‰.

Although generally similar to the salinity of the adjacent seawater, the salinity of a tidepool can vary with the tidal cycle, depending on the influence of three main factors: (1) the height of the pool on the shore, (2) the amount of sunshine during the low-tide period, and (3) the amount of freshwater input due to precipitation.

The height of a pool on the shore, and its shape and orientation with respect to the tides, influence how long the pool is submerged each day. As a result, intertidal habitats are divided into various zones, including the subtidal, low intertidal, mid-intertidal, high intertidal, and splash zone, depending on how much of the tidal cycle they spend under water.

Figure 1 shows how salinity varies during the day in tidepools at different heights in the intertidal on a rainy day. Pools in the low intertidal zone are underwater for much of the tidal cycle, and salinity varies only slightly during the day. Pools in the high intertidal are underwater only during high tide, and on rainy days rain falling in the pool can lower the salinity during low tide. On sunny days, however, evaporation of water due to wind or sunshine can increase the salinity of the tidepool. Pools in the splash zone receive seawater input only through wave action, and on rainy days much of the water in these pools is freshwater from the rain. Thus the salinity of pools in the splash zone can be much lower than that of the surrounding seawater. Alternatively, on very dry days, water evaporates from these pools, increasing their salinity

to levels much higher than that of seawater, sometimes causing a crust of salt to form around the edges of the tidepool.

The size, shape, and surroundings of a tidepool can also affect the degree to which salinity fluctuates during the tidal cycle. For example, large pools typically have more stable salinity than do small pools. But the salinity fluctuations of even two very similarly sized pools can differ during the tidal cycle. Consider two similarly sized tidepools at the same height in the intertidal—one located at the top of a rock outcropping, and one located in a depression surrounded by a large area of rock that slopes down toward the tidepool. The second tidepool will undergo much larger fluctuations in salinity during a rainstorm because rainwater falling on the surrounding rocks will drain into this tidepool, whereas the other tidepool will receive freshwater only from the rain falling directly into it. In fact, tidepool salinity can drop from above 30‰ to below 5‰ in less than an hour during heavy rain in tidepools that get large amounts of water input as a result of rain draining from the surrounding rocks.

Salinity is an important abiotic factor that affects animals and plants because it influences the movement of water into and out of cells. Maintaining control of water movement is critical for both animals and plants because water movement causes cells to swell or shrink. When an animal cell is exposed to high salinity, water moves out of the cell, causing the cell to shrink. This shrinkage results in intracellular crowding, which can cause the death of the cell. In contrast, exposing an animal cell into freshwater causes water to move into the cell, causing the cell to swell and burst. Similar phenomena occur in plants and algae, but because of the presence of a rigid cell wall surrounding the flexible internal components of the cell, the overall size of the cell does not always change appreciably. Instead, the flexible components of the cell within the cell wall shrink or swell. As these components shrink, they pull away from the cell wall (a process termed plasmolysis). As these components swell, they push against the cell wall. This process of shrinking or pushing changes the turgor pressure, the pressure exerted by the components of the cell on the cell wall. Turgor pressure is an important factor in allowing plants and algae to maintain their shape. When turgor pressure declines, plants wilt.

Water moves across cellular membranes via a process termed osmosis. This osmotic movement of water is influenced by the difference in osmolarity between the inside and outside of the cell. Osmolarity is a measure of the total number of dissolved molecules or ions in a solution

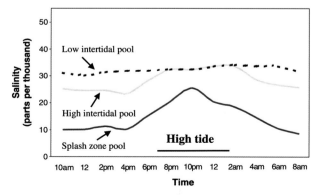

FIGURE 1 Changes in salinity in tidepools at different heights in the rocky intertidal on a rainy day.

(including salts, organic molecules, and so on). In seawater, dissolved salts make the primary contribution to the osmolarity of the solution, but intracellular osmolarity is influenced by the concentrations of many substances. Offshore seawater has a salinity of approximately 35‰ and an osmolarity of approximately 1,000 milliosmoles per liter (mosm/L), whereas the brackish water found in estuaries often has a salinity of around 18‰ and an osmolarity of around 500 mosm/L.

How organisms respond to changes in the salinity (and thus the osmolarity) of seawater varies among species. Osmoconformers alter the osmolarity inside their cells to match that of the environment, thus preventing net water movement. However, these organisms do not alter the intracellular concentrations of salts such as sodium, chloride, or calcium, because changes in ion concentration have negative effects on cellular functioning. Because the osmolarity of a solution is the result of the total concentration of all dissolved substances, regardless of their type, it is possible to match the changes in external osmolarity that are caused by changes in ion concentrations by altering the intracellular concentrations of other substances. Thus, osmocoformers vary the level substances termed compatible solutes, which do not interfere with biological reactions. Compatible solutes include substances such as glucose, trehalose, amino acids, and other small organic molecules. Most unicellular organisms and many multicellular organisms are osmoconformers. Marine plants, algae, invertebrates, and elasmobranchs (sharks, skates, and rays) are examples of multicellular osmoconformers that make use of compatible solutes to prevent loss or gain of water. The elasmobranchs use a particularly interesting approach. They have high levels of two chemicals (urea and trimethylamine oxide) in their tissues to match the osmolarity of seawater. Alone, each of these chemicals has negative effects on protein functioning, but together their effects cancel each other out, and so the pair can be considered a compatible solute mixture.

Most vertebrate and many invertebrate animals use an alternative strategy, termed osmoregulation, to cope with changes in water salinity. These organisms maintain a relatively constant osmolarity in their extracellular fluids despite changes in the osmolarity of the external environment. As a result, the cells within these organisms do not experience changes in extracellular osmolarity as environmental salinity changes, because the osmolarity of the blood remains constant. However, this constancy of the internal environment means that these animals must cope with constant movement of water across their external surfaces. In order to counteract this passive movement of water, osmoregulators have specialized tissues that are responsible for maintaining the osmolarity of the blood and extracellular fluids. For example, bony fish osmoregulate using their gills, kidneys, and gut to regulate water balance and to maintain the osmolarity of their blood relatively constant. When a fish is in full-strength seawater (with high salinity and osmolarity), it tends to gain ions across the gill and gut and lose water across these same tissues. To compensate, the fish drinks seawater to replace the lost water, and actively pumps ions out of the body via the gills and kidney. When a fish is in freshwater, however, it tends to lose ions and gain water. Under these conditions it must actively pump ions into its body. Only a relatively small fraction of fish species can survive in tidepools in the high intertidal regions, where salinity varies greatly, because they must be able to quickly alter their strategies for maintaining ion and water balance, pumping ions either into or out of their body as the salinity of the water changes.

DISSOLVED GASES

Seawater contains a diverse mixture of dissolved gases, including nitrogen, oxygen, carbon dioxide, and trace gases such as argon. Gases can enter seawater by dissolving directly from the atmosphere. The dissolution of atmospheric gases in water is governed by Henry's law (one of the ideal gas laws), which states that the concentration of gas in water is proportional to the partial pressure of the gas in the atmosphere multiplied by the solubility of the gas in water. We can write Henry's law as follows:

$$[G] = p_{gas} \times s_{gas} \qquad \text{(Eq. 1)}$$

where $[G]$ is the concentration of gas dissolved in a solution, p_{gas} is the partial pressure of the gas in the atmosphere above the solution, and s_{gas} is the solubility of the gas in that solution. The solubilities of gases in water vary greatly. For example, carbon dioxide (CO_2) is almost 30 times more soluble in water than is oxygen (O_2). Thus, at equivalent partial pressures, water contains 30 times more carbon dioxide than oxygen. In addition, the solubility of a gas varies inversely with the temperature and salinity of the water. For example, the solubility of oxygen in water decreases by almost 50% between 0 and 40 °C. Thus, water in cold Arctic and Antarctic habitats will typically have higher concentrations of dissolved gases than water in warm tropical habitats. If we factor in the effects of both salinity and temperature, seawater at 20 °C has almost the same oxygen content as freshwater at 30 °C.

Nitrogen gas enters seawater from the atmosphere or from the air bubbles created by breaking waves, and thus the levels of dissolved gaseous nitrogen do not vary greatly during the tidal cycle. In contrast, both oxygen and carbon dioxide can enter seawater via exchange with the atmosphere and as a result of biological processes such as photosynthesis and respiration. Photosynthesis is a process by which marine plants, algae, and many microorganisms use the energy from light to convert carbon dioxide into carbohydrates, producing oxygen as an end product. The process of respiration (which organisms use to obtain energy from carbohydrates and other organic molecules) results in the consumption of oxygen and the production of carbon dioxide. Thus, the activities of living organisms can have a profound effect on the concentrations of dissolved oxygen and carbon dioxide in seawater.

When a tidepool is submerged, the levels of dissolved oxygen and carbon dioxide are similar to those of the surrounding seawater because wave action causes the water to be well mixed, but during low tide periods the levels of these dissolved gases can vary widely. During the day, plants and algae photosynthesize and produce oxygen, causing dissolved oxygen to reach very high levels—a condition termed hyperoxia. In fact, the water that is close to photosynthesizing algae can become so supersaturated that bubbles of oxygen may form during the day. In contrast, at night, when both plants and animals respire and consume oxygen, dissolved oxygen levels drop—a condition termed hypoxia. The levels of carbon dioxide follow the opposite trajectory to that of oxygen, with high levels at night and lower levels during the day because plants are net producers of carbon dioxide at night and net consumers of carbon dioxide during the day. The amount and nature of the animals and vegetation within a pool will influence the extent of these daily fluctuations, and thus even adjacent pools can exhibit great variation in these cycles if the numbers and types of organisms vary between them. Similarly, there is substantial seasonal variation in these parameters as a result of changes in temperature and light levels, because temperature affects the rates of both photosynthesis and respiration, and light influences photosynthetic rates.

Animals living in tidepools display a variety of interesting adaptations to cope with these fluctuating oxygen levels. Most tidepool organisms are quite tolerant of hypoxia compared to their open-water relatives, but these animals may also use a variety of mechanisms to avoid hypoxia. Many fishes avoid the intertidal zone during low tides, instead remaining in subtidal habitats during most of the tidal cycle and entering the intertidal areas only when they are fully submerged. Species that remain in tidepools throughout low tide often utilize a strategy termed aquatic surface respiration, in which they swim up to the surface of the water and ventilate their gills with the thin layer of well-oxygenated water at the tidepool's surface. In addition, a number of intertidal fish species can leave the water for brief periods and breathe air. In most tidepool fishes, this aerial respiration is accomplished using the gills rather than a lunglike structure. The gills in these species tend to be somewhat thick and stiffened to prevent the fine gas exchange structures of the gills from collapsing in air, allowing continued gas exchange.

PH

The pH of a solution is inversely proportional to the activity of hydrogen ions (H^+) in that solution. Thus pH is a measure of the relative acidity or alkalinity of a solution. The logarithmic pH scale ranges from 0 (very acidic) to 14 (very basic), with each unit increase representing a tenfold decrease in the activity of H^+ ions. The pH of seawater typically ranges between about 7.5 and 8.5, and thus seawater is slightly basic (as a result of the buffering effects of the presence of carbonate salts).

The pH of seawater is greatly affected by the amount of carbon dioxide (CO_2) dissolved in the water, because when CO_2 dissolves in water, it sequentially forms carbonic acid (H_2CO_3), bicarbonate (HCO_3^-), and carbonate (CO_3^{2-}) according to the following equations:

$$CO_2 + H_2O \rightleftharpoons H_2CO_3 \rightleftharpoons HCO_3^- + H^+ \rightleftharpoons CO_3^{2-} + H^+ \qquad \text{(Eq. 2)}$$

Changes in CO_2 levels influence the amounts of all of the chemical participants in these equilibria, including the H^+ ions. Thus, changes in CO_2 can change the pH of a solution. For example, increases in dissolved CO_2 result in increases in the concentration of hydrogen ions in seawater, thus decreasing the pH of the water.

Ocean pH is thought to have dropped by almost 0.1 pH units since the beginning of the Industrial Revolution as atmospheric carbon dioxide has increased. By the year 2100 some scientists predict that ocean pH will drop by as much as 0.4 pH units as a result of continuing anthropogenic CO_2 production. A variety of studies have demonstrated that changes in pH can have deleterious effects on organisms. For example, snails and sea urchins have been shown to exhibit reduced growth as a result of exposure to pHs only 0.03 units lower than normal levels. These deleterious effects of pH may be the result of both direct and indirect mechanisms. The direct effects of pH changes are the result of the effects of pH on protein function. Proteins typically have a relatively narrow pH range within which they function optimally, and thus organisms must closely regulate their intracellular pH as external pH changes. Changes in external pH thus are likely to impose an energetic cost to fuel this regulation. Indirect negative effects of pH change can occur because toxicity of metals such as copper and nickel varies with pH. A final indirect effect of changes in pH relates to the effect of pH on the availability of salts such as calcium carbonate. The carbonate (CO_3^{2-}) formed as a result of CO_2 dissolution reacts with various ions to form carbonate salts such as calcium carbonate ($CaCO_3$). Calcium carbonate is a critical mineral needed for the formation of the hard skeletons of animals such as corals, coralline algae, and sea urchins. A decrease in pH can reduce the level of carbonate salts in the water and thus may alter the ability of organisms to form calcified hard structures such as those found in coralline algae or the exoskeletons and shells of many invertebrates.

In the confined environment of a tidepool, respiration by plants and animals can produce substantial CO_2, particularly at night, while during the day photosynthesis consumes CO_2. As a result, the pH of tidepools fluctuates on a daily basis, with pH being lowest at night and highest during the day. For example, high intertidal pools may have a pH as low as 6.9 at night and 10.3 during the day. Tidepool organisms must have mechanisms for tolerating these short-term pH changes, although these mechanisms are not well understood.

NUTRIENTS

A nutrient is an element or compound that is necessary for the growth of organisms. Primary producers such as plants, algae, and a variety of planktonic photosynthetic organisms rely on dissolved nutrients for growth and metabolism, whereas consumers such as animals obtain nutrients from their food. Thus, the amounts of dissolved nutrients in seawater can have a direct effect on primary producers such as algae, and changes in the type or abundance of these primary producers can affect the type and abundance of the organisms that consume them.

Nutrients are usually subdivided into two categories: major nutrients and minor nutrients. Major nutrients are usually required in large quantities and have the potential to limit growth. Minor nutrients are required in much smaller quantities and, although required to support growth, are not usually limiting. Nitrogen (N), phosphorus (P), and silica (Si) are considered to be the major nutrients dissolved in seawater. Nitrogen is a component of both amino acids (the building blocks of proteins) and of nucleic acids (the building blocks of DNA and RNA). Phosphorus is a component of nucleic acids, phospholipids, ATP, and other energy storage molecules and is a major component of the bones and teeth of vertebrates

and the shells of some invertebrates. Silicon is considered to be a major nutrient for diatoms, a group of phytoplankton with a hard external shell composed of silica (silicon dioxide). Diatoms are at the base of many marine food webs, and thus silicon can be a major factor in structuring marine communities.

Marine organisms generally use nitrogen in the form of nitrate. A variety of bacteria and cyanobacteria are able to convert nitrogen into nitrate. This process typically occurs in two steps: (1) nitrogen-fixing bacteria convert nitrogen into ammonium ion, and (2) nitrifying bacteria convert this ammonium into nitrites and nitrates. These processes are thought to occur primarily in terrestrial habitats, and the resulting nitrate leaches into the ocean, although more recent estimates suggest that marine cyanobacteria may provide a substantial amount of the available nitrate, at least in open-ocean habitats. Nitrates are also produced as fecal waste and from decaying plant and animal matter. Organic waste materials, including feces and dead plants and animals, tend to sink to the bottom of the ocean, carrying their nutrients with them, and thus surface waters are often somewhat nitrogen limited.

The biologically significant form of inorganic phosphorus is phosphate (PO_4^{3-}). The major source of this phosphate is the weathering of rocks. Primary producers such as photosynthetic microorganisms then incorporate this phosphate into organic molecules, at which point it is available for use by other organisms.

Nitrate and phosphate levels in tidepools vary in both space and time. One major source of spatial variation in these nutrients is the presence of ocean conditions termed upwellings. Upwellings are wind-driven water movements that carry deep, nutrient-rich water up to the surface. Upwellings result in local areas of very high primary productivity because plants and algae are provided with additional nitrogen, which acts as a fertilizer and promotes growth. Thus the rocky intertidal areas that experience periodic upwellings have different communities than those in other areas.

Another major source of variation in the nutrients available in tidepools is the presence of seabird nesting areas or marine mammal rookeries. The organic wastes from these animals provide a rich fertilizer for the water of the tidepools. Concentrations of nitrogen and phosphorus in tidepools can be increased by up to several hundredfold when rainwater washes these nutrients into a tidepool. Anthropogenic inputs of chemical fertilizers can also alter the levels of nitrates and phosphates in runoff water, which can increase primary productivity and alter the community composition of tidepools.

Nutrient levels can also vary on a daily cycle. During daytime low tides, nutrient levels in a tidepool can drop precipitously as the nutrients are consumed by plants and algae, only to be replenished again at high tide.

INTERACTION OF FACTORS

All of the abiotic factors discussed in the preceding section interact to cause the complex pattern of changes in water chemistry that are observed in tidepools during the tidal cycle. These abiotic factors in turn influence the biota of the tidepool, which themselves can influence the tidepool's water chemistry. Dense mats of algae can shade the tidepool, reducing evaporation on sunny days and blunting changes in salinity. But at the same time, submerged algae consume carbon dioxide and produce oxygen during the day (and produce carbon dioxide at night), resulting in changes to the dissolved gases and nutrients within the tidepool. As a result, tidepools are incredibly dynamic habitats that pose special challenges to the plants and animals that live within them.

SEE ALSO THE FOLLOWING ARTICLES

Evaporation and Condensation / Nutrients / Photosynthesis / Salinity Stress / Seawater / Zonation

FURTHER READING

Garrison, T. S. 2004. *Oceanography: an introduction to marine science.* Toronto: Brooks-Cole.

Horn, M. H., K. L. Martin, and M. A. Chotkowski, eds. 1999. *Intertidal fishes: life in two worlds.* San Diego: Academic Press.

Morris, S., and A. C. Taylor. 1983. Diurnal and seasonal variation in physico-chemical conditions within intertidal rock pools. *Estuarine, Coastal and Shelf Science* 17: 339–355.

Snoeyink, V. L., and D. Jenkins. 1980. *Water chemistry.* Baltimore: John Wiley and Sons.

WATER PROPERTIES, MEASUREMENT OF

BRIAN A. GRANTHAM

Washington State Department of Ecology

Water property measurements are critical for understanding the dynamics of coastal ecosystems. Most relevant to rocky-shore studies are physical processes, including temperature, currents, tides, upwelling, and internal waves; chemical properties, such as salinity and nutrient availability; and biological properties such as phytoplankton abundance. The water properties of interest and environmental conditions at the study site dictate the choice of sampling design and instrumentation.

SAMPLING DESIGN

Physical, biological, and chemical processes exhibit widely varying periodicities, making the choice of sampling rate, sampling period, and measurement technique critically important. Measurements must be taken often enough and for long enough to adequately resolve the parameters of interest, but these needs must be balanced against the constraints imposed by data storage capacity, battery power, and the practicality of deploying and maintaining suitable instruments.

SAMPLING RATE AND TIME SERIES LENGTH

Sampling rates must be based on the characteristics of the property being measured (Fig. 1). For a periodic process, the highest resolvable frequency with a sampling interval ΔT is known as the Nyquist frequency, defined as $fN = 1/(2\Delta T)$. More intuitively, this means that the shortest period that can be resolved is equal to twice the length of the sampling interval. In practice, however, the noise and sampling errors in real data necessitate the use of sampling rates three to four times higher than the highest frequency of interest. For example, although hourly tidepool temperatures could theoretically be resolved with samples every 30 minutes, sampling at least every 15–20 minutes is recommended. It is important to note that if the

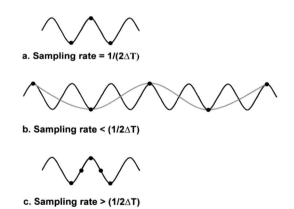

a. Sampling rate = 1/(2ΔT)

b. Sampling rate < (1/2ΔT)

c. Sampling rate > (1/2ΔT)

FIGURE 1 Effects of sampling rate on resolution of a periodic process. (a) The Nyquist frequency (1/2ΔT) is the lowest resolvable frequency for a given sampling rate ρt. This means that a minimum of two time periods (three samples) are required to sample a single oscillation in the process. (b) An aliased signal (red line) resulting from sampling at a lower frequency than the process being measured (black line). (c) Example of sampling at a rate sufficiently high to avoid aliasing and resolve the process, despite noise and error in the data. (See text for further discussion.)

sampling rate is too low, high-frequency fluctuations that are not measured can generate an artificial signal in the data that cannot be distinguished from the true signal—a phenomenon known as aliasing.

Just as sampling rate determines the highest frequency that can be resolved, the length of a time series determines the longest. To conduct meaningful statistical analyses, multiple cycles of the process must be sampled, so the time series must be at least twice as long as the longest period of interest—preferably longer. For example, if daily cycles are of interest, then the time series must be at least two days long.

MEASUREMENT METHODS

Bottle Samples

Laboratory analysis of discrete (bottle) samples is the primary method for measuring many water properties, including nitrate, nitrite, phosphorus, ammonium, silicate, chlorophyll, salinity, dissolved oxygen, and others. Properly implemented, analytical methods provide the most accurate measurements, so discrete samples are routinely collected to verify the functioning of instruments used to make *in situ* measurements. Near-surface bottle samples can be collected by hand or using a pole with a bottle clamp, and samples at greater depths can be obtained using a Niskin bottle on a hand line. Samplers that can be programmed to collect and preserve multiple water samples for later analysis are also an option in some locations.

Handheld Instruments

Handheld instruments range from simple thermometers and refractometers to sophisticated, multiparameter field instruments measuring temperature, conductivity, salinity, fluorescence, pH, and dissolved oxygen. They can be used to sample surface waters directly or, for some properties, to test discrete-depth samples collected with Niskin bottles. The advantages of handheld instruments are portability and an immediate readout of measurements, but as in bottle sampling, collecting high-frequency time series is difficult.

Stand-Alone Instruments

Instruments capable of taking high-frequency measurements and storing them in memory have revolutionized marine research by making it possible to obtain long time series or collect spatially explicit data rapidly.

The oceanographer's workhorse for measuring water properties is the CTD—which stands for conductivity, temperature, and depth—an instrument that takes multiple measurements per second as it is lowered through

FIGURE 2 Hand deployment of a medium sized CTD measuring conductivity, temperature, depth, chlorophyll a fluorescence and dissolved oxygen.

the water column (Fig. 2). CTDs can be outfitted with a range of sensors, as well as Niskin bottles that can be triggered at specific depths. When used with an electromechanical cable, a CTD can display real-time data and allow operator-directed triggering of Niskin bottles. Compact CTDs that can be deployed by hand are routinely used for near-shore research.

MODULAR INSTRUMENTS/SENSORS

Temperature. Small, durable, inexpensive loggers that can be attached to rocks, fixed structures, or even embedded in model organisms can be used to measure temperature in almost any coastal environment. In many loggers, low power requirements and abundant memory allow the collection of high-frequency time series over extended periods of time.

CONDUCTIVITY/SALINITY

Most instruments for measuring conductivity must be continuously submerged and have fragile conductivity cells, precluding their use in intertidal locations subject to strong wave forces or sediment scour. However, many high-quality instruments are available for use in more protected waters.

DISSOLVED OXYGEN

Dissolved oxygen (DO) sensors are usually incorporated into CTDs or other instrument packages. Two sensor types are available: Membrane-based sensors are an older, but proven technology; solid state optical sensors are newer but have shown promising results. The latter do not require membrane replacements and may suffer less from fouling, but for extended deployments, it is recommended that all sensors have some type of antifouling treatment.

CHLOROPHYLL FLUORESCENCE

Fluorometers provide estimates of phytoplankton standing stock by measuring the fluorescence generated when chlorophyll *a* is excited by blue (455 nm) light. Many do not require pumps and can be used as stand-alone sensors or integrated into larger instrument packages. The amount of chlorophyll in a phytoplankton cell and its excitation response vary with factors such as species composition, time of day, and self-shading, so inferences based on the relationship between fluorescence and phytoplankton must be made with caution. For this reason, bottle samples are often taken to provide an analytical check on instrument accuracy.

LIGHT TRANSMISSION

Phytoplankton, suspended particles, bacterial, and dissolved organic matter all affect water clarity by scattering and absorbing light. For decades Secchi disks have been used to visually estimate near-surface water clarity, but differences between users and factors such as angle of the sun can strongly influence the results. Nonetheless, many long-time series are available and the methodology is inexpensive and simple, so Secchi data are still routinely collected. Transmissometers, which measure the amount of light transmitted through the water between a light source and a detector located a short distance away, allow accurate data to be collected throughout the water column.

OPTICAL BACKSCATTER/TURBIDITY

Backscatter sensors provide an estimate of the concentration of suspended solids in the water by measuring the quantity of light reflected back to a sensor.

PHOTOSYNTHETICALLY ACTIVE RADIATION (PAR)

Photosynthetically active radiation data indicate the amount of light available for the growth of phytoplankton and marine vegetation, as well as the depth of the photic zone—the maximum depth at which photosynthesis can occur. Sensors are available for measuring PAR in air and water.

CHEMICAL AND OTHER SENSORS

Two technologies are available for obtaining time series of chemical water properties. Systems that use wet-chemistry methods to analyze samples *in situ* are available for nitrate, nitrite, ammonium, silicate, and phosphorus. However, these systems are bulky and therefore best suited for use on moorings or fixed structures. The newest generation instruments are solid-state nitrate sensors that use spectrophotometric methods to measure the analyte-specific absorption of light. These relatively small sensors can be used in instrument packages or stand-alone configurations, making it possible to deploy them in a wide range of environments. Similar instruments for measuring other analytes are under development.

CURRENT MEASUREMENTS

Measurements of currents are needed to understand the transport of nutrients, larvae, food particles, sediments, and other compounds in coastal environments. Instruments based on mechanical, electromagnetic, or acoustic technologies are available, giving researchers various options.

Acoustic Doppler Current Profilers (ADCPs). ADCPs are often the first choice for measuring currents because they can make rapid measurements throughout the water column. ADCPs work by emitting short sound bursts or pings, which bounce back off particles in the water and are received by the instrument. Doppler shifts in the frequency of the returning sound waves are used to calculate how fast the water is moving and the travel time for the waves allows velocities to be assigned to depth bins throughout the water column (Fig. 3).

Fixed-Current Meters. Fixed-current meters usually cost less than ADCPs but measure current velocities at only one depth. Several types are available. Mechanical current meters have been in use for many years, but fouling can be a problem on long deployments and the units are difficult to maintain and calibrate. Electromagnetic current meters are robust and more resistant to fouling, so they are well suited for use in extreme environments. Fixed-depth acoustic current meters based on the same principles as ADCPs are the newest technology available.

Other Methods. Two alternative methods for studying water movements are the use of free-floating drifters and dye releases. Drifters consist of a surface float and a

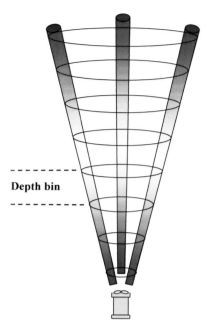

FIGURE 3 Acoustic Doppler current profilers (ADCPs) use the Doppler shift in sound waves bouncing off particles in the water to determine current velocities in depth bins throughout the water column.

Depth bin

tethered drogue or sea anchor and can be outfitted with various sensors, as well as GPS and satellite transmitters for tracking. Because most drifters are relatively large, their use is typically restricted to deeper waters. Fluorescent dyes—most commonly rhodamine and fluorescein—can also be used to examine short-term water movements. Dye near the surface can be seen for short distances, but tracking is usually done using a specially designed fluorometer. Fluorescein is much cheaper and more easily detectable than rhodamine, but it decomposes in light, limiting its suitability for shallow-water use.

SAMPLING APPROACHES FOR DIFFERENT ENVIRONMENTS

Fixed instruments allow collection of high-frequency data at times and in places that cannot be sampled by hand. However, the research environment limits the choice of instrument and deployment method, particularly on exposed shorelines where wave forces, scour, and the potential for damage by waterborne debris are high. Human traffic can also be a factor, because instruments used in high-use areas may be subject to vandalism or theft.

Intertidal and Tidepool Measurements

The options for sampling in exposed intertidal areas are limited by wave forces and wet–dry cycles for which most instruments are not designed. Sampling by hand is always an option, but some automated instruments can be used if they can be adequately secured and protected. Temperature loggers, for example, can be bolted to rock surfaces, but may require protection by wire cages to prevent damage from floating debris. Similarly, compact, self-contained fluorometers encased in PVC and bolted to rock surfaces have been used in the intertidal (Fig. 4). It may also be possible to mount conductivity sensors and solid-state nutrient analyzers in protected tidepools that retain water throughout the tidal cycle. Collecting data in high-energy environments is always a challenge and the success of a project often hinges on the researcher's ingenuity. When considering a study, researchers should check with suppliers to see if instruments that meet their needs are available.

FIGURE 4 An example of a fluorometer protected in a PVC case and bolted to a rock surface in the intertidal zone. The sensor window is visible inside the open end of the case.

Near-Shore Waters (<25–30 m)

In near-shore waters, although sea conditions can be rough, sampling from a vessel gives the researcher great flexibility, with the time and spatial extent of coverage limited only by available ship time and endurance of the scientific staff. Most commonly, CTDs equipped with multiple sensors are used to make water column profiles at selected locations. Water samples can be taken by hand, with Niskin bottles, or by triggering bottles attached to a CTD. Samples can then be analyzed on board using handheld instruments or laboratory methods, or preserved for future analyses. Ships can also be equipped with flow-through systems that pump water from below the hull to onboard instruments. Finally, currents can be measured using ship-mounted, downward-looking ADCPs.

For long-time series in near-shore waters, instruments can be securely attached to piers and pilings, but these may not be located in areas of interest. Moorings and bottom mounts can be placed where needed, but require a substantial investment in design, construction, and servicing time. However, moorings can carry multiple

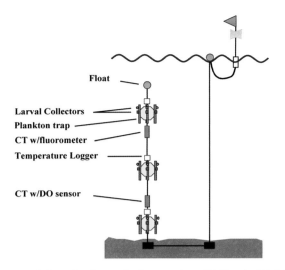

FIGURE 5 Schematic of a mooring for near-shore (15–25 m depth) data collection. Instrumentation includes two conductivity and temperature units (one with a fluorometer and one with a dissolved oxygen sensor), several temperature data loggers, and three racks carrying plankton traps and devices for measuring larval recruitment. If required, fixed-depth current meters could also be added to the array.

Float

Larval Collectors

Plankton trap

CT w/fluorometer

Temperature Logger

CT w/DO sensor

Pumped systems are also an option for sampling in near-shore areas where pipes can be run from the water to instruments in a secure location onshore. Such sites are limited, but where available they can be excellent for long-term monitoring.

SEE ALSO THE FOLLOWING ARTICLES

Currents, Coastal / Nutrients / Phytoplankton / Salinity, Measurement of / Temperature, Measurement of / Water Chemistry

FURTHER READING

Emery, W. J., and R. E. Thompson. 1998. *Data analysis methods in physical oceanography.* New York: Elsevier.

Grasshoff, K., K. Kremling, and M. Ehrhardt. 1999. *Methods of seawater analysis.* Weinheim: Wiley-VCH.

Legendre, P., and L. Legendre. 1998. *Numerical ecology.* Amsterdam: Elsevier Science.

U.S. Environmental Protection Agency. *EPA methods for analysis of marine and estuarine water samples.* http://www.epa.gov/nerlcwww/methmans.html#Marine%20&%20Estuarine.

Valiela, I. 1995. *Marine ecological processes.* New York: Springer-Verlag.

sensors and other sampling equipment (Fig. 5). Acoustic Doppler current profilers can be mounted on moorings as well, but bottom-mounted systems are more stable, less likely to be damaged by passing vessels, and can be used in shallower waters than moorings (Fig. 6). During long deployments biofouling can affect data quality seriously. To minimize these impacts, all sensors should be protected with antifouling treatments and instruments should be serviced frequently.

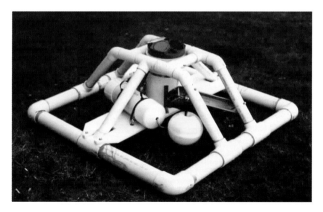

FIGURE 6 An acoustic Doppler current profiler (ADCP) in a PVC bottom mount. The orange float, which is tethered to the mount, is part of a recovery system that includes an acoustic release mechanism. An acoustic signal transmitted through the water triggers the mechanism to release the tethered float, allowing the instruments to be retrieved from a vessel.

WAVE EXPOSURE

MATS LINDEGARTH
Göteborg University, Sweden

Exposure of shores to the effects of ocean waves can have wide-ranging effects on intertidal plants and animals, and the concept of wave exposure has played an important role in our understanding of nearshore ecology. For example, many studies have documented differences in species and their interactions between "exposed" and "protected" sites, and ecological models incorporate wave exposure as a variable when predicting how intertidal communities function. However, despite its utility as a guide to the variation of communities on the shore, wave exposure has seldom been quantified. The concept of wave exposure is most useful when it is carefully defined and measured.

BIOLOGICAL EFFECTS OF WAVE EXPOSURE

The effects of ocean waves on intertidal assemblages are often obvious to the casual observer. Shores exposed to large waves are covered by different plants and animals than those found nearby on rocks protected from the waves (Fig. 1). The most intuitive mechanism underlying this effect is direct physical disturbance causing death or

FIGURE 1 Waves crash into a rocky shore in South Australia, shaping the morphology and behavior of individual organisms and affecting the structure and composition of animals and plants in these habitats. Photograph courtesy of Prof. Jon Havenhand.

damage to individual animals and plants. Often, this disturbance is caused by hydrodynamic forces (i.e., drag and lift) related to the velocity of the water in breaking waves. Alternatively, disturbance can be caused by the impact of objects (such as logs and rocks) carried by waves. Many species of intertidal animals and plants have evolved morphologies and behaviors specifically adapted to cope with these forces. Those species better adapted to wave-imposed forces can live on shores of greater exposure than other species. Despite these adaptations, hydrodynamic and impact forces can at times cause massive damage that fundamentally alters the structure and function of exposed rocky habitats, creating changes that persist for several years. The magnitude of physical disturbance is less on protected shores, and as a result, the structure of protected communities is different from that of exposed assemblages.

In addition to such conspicuous physical effects, exposure to waves may also act though more subtle mechanisms. For example, wave exposure may act indirectly by releasing prey species from a voracious predator that is sensitive to physical forces. Thus, if the prey can itself resist wave forces, it can thrive in the presence of waves but will be eaten in their absence. Waves also affect the level of turbulence in the water bathing the shore and thereby can affect the fluxes of nutrients and propagules. As a result, the level of wave exposure on intertidal shores can modify fundamental processes such as primary production and recruitment.

Another important mechanism of wave exposure is wave runup and splash, which moistens the upper intertidal zone and prevents desiccation of organisms living there. The effects of splash help to determine the height to which plants and animals can grow on the shore, and thereby they affect the characteristic vertical zonation observed on many rocky shores. By affecting the turnover of water and therefore the temperature and chemical environment, the effects of runup are also of particular importance for the dynamics of assemblages in pools in the upper intertidal zone. Through these mechanisms, the biological effects of wave exposure are widespread, profound, and highly relevant to our understanding of a variety of patterns and processes in rocky intertidal environments.

CONCEPTUAL DIFFICULTIES

Development of general ecological theories based on wave exposure is difficult for a number of reasons. First, wave exposure is a complex physical concept. For example, waves approaching a shore commonly vary in their height, period, and direction, and each of these factors can alter how waves break and subsequently interact with intertidal organism. However, we rarely know which of these quantities is most important and whether it is their extreme values or their averages that matter.

Second, as discussed in the preceding section, waves affect biota through a diverse set of mechanisms; and these mechanisms often interact. For example, the large hydrodynamic forces on a wave-exposed shore are often accompanied by increased splash. In this case, an increase in physical disturbance might be offset by a decrease in desiccation stress, and it is often difficult to predict which effect will be more important. Furthermore, the effects of wave exposure may be obscured by responses to environmental gradients other than waves. For example, biological interactions such as predation and competition may operate independent of wave exposure, causing spatial and temporal variability in abundance and community structure. Because of these complications, it is difficult to define and measure wave exposure.

DEFINITIONS OF WAVE EXPOSURE

If you arrive at a rocky shore on a windy day, it is often easy to categorize some parts of the shore as "protected" while others are perceived as more "exposed." This classification may be based, for instance, on the fact that there is more splash on some parts of the shore than on others. If you are simply trying to decide where it is safe to walk without getting wet, such coarse judgments are usually sufficient, but if you are a marine biologist trying to figure out whether and how waves affect animals and plants on this particular seashore and possibly on seashores in general, more stringent definitions of wave exposure are needed.

Problems arise for two reasons. First, casual definitions of wave exposure (such as that based on wave splash) are relative rather than quantitative: one part of the shore has more splash than the other, but this ranking does not place wave splash on an absolute scale. Second, wave exposure is a continuous variable, but it is often forced into a small number of classes, typically "exposed" and "protected." Imagine the problems you would encounter in describing an object's color (a continuous variable) if the only choices available were the ends of the spectrum: blue and red.

The need for objective measurement of at least some aspect of wave exposure is fundamental. Without such measures we cannot even be certain whether there exists any consistent difference in environmental conditions between locations perceived as either "protected" or "exposed." Consider two studies examining the relationship between wave exposure and the abundance of a species of animal at "protected" and "exposed" locations (Fig. 2). If wave exposure is defined only in relative terms in each of the studies, it may be that the average "exposed" locality in the first study (E1) is in fact more protected than the average "protected" locality in the second study (P2). If the abundance of organisms on the two shores varies in proportion to wave exposure (something that cannot be taken for granted in real life), the two studies will show qualitatively similar patterns, but the magnitude of effects will be very different between them. Furthermore, if wave exposure is not defined quantitatively, it is also possible that the range of environmental conditions will be very different among studies. In this particular example, the definition of what constitutes an "exposed" locality in

the second study (E2) is much broader than in the first. Therefore, the responses to wave exposure of organisms at the second site will appear more variable and unpredictable than those at the first. Thus, though conclusions about ecological patterns in relation to relative measures of wave exposure may be correct, their value for comparative studies and general syntheses is limited. A quantitative measure of wave exposure is needed.

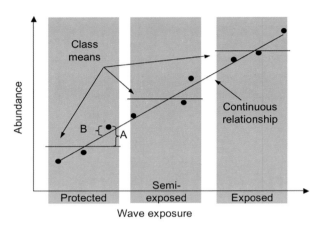

FIGURE 3 Consequences of classification vs. continuous definitions of wave exposure for the explanatory power of statistical models. In a model based on classes of wave exposure, the success is measured by the deviations (A) between each individual measurement and their associated class means (horizontal lines). In the continuous model, deviations (B) are measured from each measurement to the estimated relationship symbolized by the diagonal line. The A deviations are on average substantially larger than the B deviations

Even quantitative measures can be mishandled. In a common example, wave exposure is measured quantitatively, but localities are subsequently classified into broader categories, such as "protected," "semi-exposed," or "exposed" (Fig. 3). The quantitative measurement alleviates the problem with lack of comparability among studies; but the subsequent classification unnecessarily introduces a kind of measurement error that may mask any existing, true relationship between wave exposure and the biological variable of interest. A measured relationship between wave exposure and, for instance, the abundance of some organism is bound to be less precise if levels of exposure are classified into a small number of classes than if wave exposure is considered as a continuous variable (Fig. 3). This is because the success of a scientific model, say a model developed to describe the relationship between the abundance of intertidal organisms and wave exposure, can be expressed in terms of how precisely the abundance can be predicted from information about

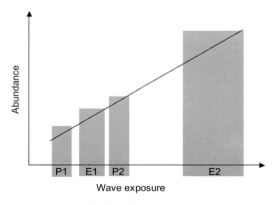

FIGURE 2 Hypothetical relationship between the abundance of an intertidal species and some measurable quantity of wave exposure (solid line). In each of two independent studies examining the effect of wave exposure, "protected" and "exposed" locations are defined in relative terms (Study 1: P1 and E1; Study 2: P2 and E2). The classifications are inconsistent among studies.

the level of wave exposure. If levels of wave exposure are classified as "protected," "semi-exposed," or "exposed," the predicted abundances correspond to the estimated mean abundances for the different classes, and the precision of the model can be measured as deviations from class means (the distance labeled A in Fig. 3). If wave exposure is instead treated as a continuous variable in an ecological model, the abundance at any level of exposure can be predicted from the estimated continuous relationship, and its success is measured as deviations from this relationship (the distance labeled B in Fig. 3). The A deviations are on average substantially larger than the B deviations.

In a study on the effects of wave exposure on intertidal assemblages on rocky shores in two areas on the Swedish coast, both of these difficulties were evident. Physical measurements of wave exposure revealed that sites originally defined as "exposed" in one area were in fact more protected than those defined as "protected" in the other area. Furthermore, simultaneous statistical analyses using categorical ("protected" vs. "exposed") and continuous (mean wave height) definitions of wave exposure showed that the latter can account for a larger proportion of the variability not only in theory but also in practice. The most striking example was that the continuous model explained up to 80% of the variability in species richness, while only 23% was explained by the categorical analysis.

Thus both logical reasoning and empirical evidence suggest that quantitative definitions are important components for the development of a general understanding of the effects of wave exposure in intertidal habitats. Having said this, one clearly has several options as to how a quantitative index of wave exposure can be constructed. Existing approaches include measurement of (1) aspects of coastal morphology (the shape of the shore), (2) maximum wave height, (3) maximum force acting on some object, and (4) dissolution rate of plaster attached to the rock. While each of these may have its particular benefits, indices that can readily be translated into water velocity (e.g., wave height and maximum force) appear to be most fruitful. This is because of the tight mechanistic links between water velocity and the physical forces acting on animals and plants in the intertidal zone. Which index will be more useful and efficient in practice can be determined only if quantitative definitions of wave exposure become more frequently applied.

SEE ALSO THE FOLLOWING ARTICLES

Hydrodynamic Forces / Ocean Waves / Storm Intensity and Wave Height / Surf-Zone Currents / Turbulence / Zonation

FURTHER READING

Dayton, P. 1971. Competition, disturbance, and community organisation: the provision and subsequent utilization of space in a rocky intertidal community. *Ecological Monographs* 41: 351–389.

Denny, M. W. 1988. *Biology and the mechanics of the wave-swept environment.* Princeton, NJ: Princeton University Press.

Denny, M. W. 1995. Predicting physical disturbance: mechanistic approaches to the study of survivorship on wave-swept shores. *Ecological Monographs* 65: 371–418.

Dethier, M. N. 1984. Disturbance and recovery in intertidal pools: maintenance of mosaic patterns. *Ecological Monographs* 54: 99–118.

Lindegarth, M., and L. Gamfeldt. 2005. Comparing categorical and continuous ecological analyses: effects of "wave exposure" on rocky shores. *Ecology* 86: 1346–1357.

Underwood, A. J. 1999. Physical disturbances and their direct effect on an indirect effect: responses of an intertidal assemblage to a severe storm. *Journal of Experimental Marine Biology and Ecology* 232: 125–140.

WAVE FORCES, MEASUREMENT OF

MICHAEL L. BOLLER

Stanford University

Because of the high variability and large magnitude of wave-induced hydrodynamic forces, their measurement requires specialized equipment. These forces have been measured in the rocky intertidal zone by both mechanical and electronic devices. Generally, mechanical devices have been used to record the maximum force over the deployment period, and electronic sensors have been used to collect more detailed measurements of the variation in force over time. Spring-scale dynamometers are the most popular and reliable mechanical devices, and a variety of electronic systems have been built by researchers for use in the intertidal zone.

SPRING-SCALE DYNAMOMETERS

Spring-scale dynamometers (Fig. 1A, B) are the most common mechanical devices used to quantify the maximum wave force or wave exposure of a site. The main advantages of spring-scale dynamometers are their low cost and relatively easy construction. However, the collection of data is labor intensive, because each dynamometer produces one datum (maximum force) per deployment, requiring repeated deployments to characterize the wave forces of a site. Also, the timing and direction of force are not measurable with these devices. Additionally, the reaction time of spring-scale dynamometers may be too slow to record short-duration forces such as the impingement force.

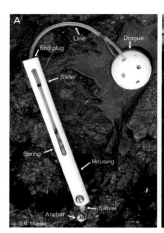

FIGURE 1 Spring-scale dynamometers. (A) Surface anchored dynamometer deployed in upper intertidal zone of Friday Harbor Laboratories, Washington. Photograph by G. M. Moeser. (B) Countersunk dynamometers at Hopkins Marine Station, California. The left dynamometer is what is seen when it is removed from a mount, while the right dynamometer is installed into a hole drilled into the rock face. Photograph by P. Martone.

Spring-Scale Dynamometer Construction

The spring-scale dynamometer consists of a simple, home-made spring-scale that records the maximum force applied to its drogue (Fig. 1A, B). The plastic housing is constructed using 1/2-inch CPVC water supply pipe. A slot is cut along half of the length of the pipe, and a hole is drilled through the opposite end. A heavy-duty swivel and one end of the spring are fastened at the hole with a bolt. A line is tied to the opposite end of the spring and fed through a rubber slider and end plug. The line is then tied to a drogue (usually a practice golf ball or a roughened plastic sphere). Before each deployment, the rubber slider is pushed to the end plug. Force acting on the drogue pulls the string and slider against the end plug. When the force is removed, the spring returns to is resting position, and the displacement of the slider away from the end plug is measured through the slot using calipers. The maximum force that a spring-scale dynamometer can measure can be adjusted by varying the stiffness of the spring. Further, the magnitude of forces imposed by waves is proportional to the size of the drogue, so a smaller drogue can be used at exposed sites. The force can then be scaled up to what would be seen by a larger drogue.

Calibration

Calibration of spring-scale dynamometers is necessary to know the absolute magnitude of force imposed by the waves. The instruments are easily calibrated by hanging weights from the drogue and recording deflection of the slider. The force applied by the weight is equal to the mass × g (the acceleration of gravity = 9.8 m s^{-2}). A force-versus-slider displacement calibration curve can then be made.

Deployment

Spring-scale dynamometers can be built either to mount on the surface of the substrate (Fig. 1A) with a plastic dry-wall anchor, large fishing swivel, and stainless steel screw, or to be countersunk flush with the substrate so that only the drogue extends into the water (Fig. 1B). Surface mounting can be much easier (only a small hole is required for the anchor), but it exposes the housing of the device to the waves and causes the device to reorient with the flow, extending the reaction time and possibly causing peak forces to be missed. Countersinking of the dynamometer is more difficult (a larger hole in the rock is needed) but results in a better response time and prevents the devices from being lost through failure of the anchor. Countersunk dynamometers have a modified construction (Fig. 1B).

ELECTRONIC DEVICES

Electronic measurement provides much greater detailed data of the forces generated by breaking waves. A high-frequency recording can be used to determine the magnitude, direction, frequency, and duration of wave forces. Additionally, the distribution of forces over time can be determined, because each wave can be characterized. No out-of-the-box systems for measuring the large forces of breaking waves are commercially available, but commercial components have been modified and laboratory-made systems have been used to measure the force on drogues and organisms (Figs. 2–4). Generally these devices require (1) a force transducer to convert the mechanical force imposed by the wave into an electronic signal, (2) an amplification of that signal to a level that can be easily measured, and (3) a system for recording the data (a chart recorder, computer with data-logging capability, or data logger). Further, these devices need to be waterproofed to protect the electronics.

Force Transducer

A force transducer is an analog electronic device that converts force applied to the transducer into a voltage signal. It can be configured to measure one, two, or more axes; multiaxis measurements can be used to determine both magnitude and direction of the wave forces. Data are collected from each axis, and the resulting force vector can be calculated.

Laboratory-built force transducers have been used to measure forces on both drogues and organisms (Fig. 2).

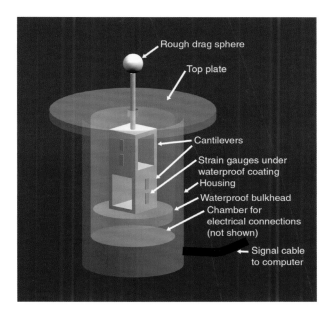

FIGURE 2 Multiaxis drag sphere flow probe schematic (redrawn from Gaylord 1999, with permission from Elsevier). The housing has been drawn as transparent so that the force transducer can be seen.

With a little practice, good single and multiaxis force transducers can be built in the laboratory using foil strain gauges. Strain gauges are basically long, zigzagged traces of fine wire that change resistance when stretched. They are attached to a stiff beam so that force applied to the beam causes the strain gauge to stretch. When the gauges are arranged in a circuit known as a Wheatstone bridge, the small change in resistance is converted into a change in voltage. The flexibility of a force transducer influences its response time, so that stiffer transducers can be used to measure shorter duration forces (such as the impingement force).

Commercial force transducers can include expensive multiaxis load cells that can measure tensile stress as well as shear and torsion. Single-axis load cells have been used to measure the force transmitted through the stipe of large kelp (Fig. 3).

Miniature force transducers originally designed as human input devices for computers and other electronics

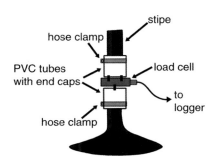

FIGURE 3 Schematic of a stipe mounted load cell used to measure wave forces (redrawn from Stevens *et al.* 2002, with permission from Elsevier).

(e.g. the "mouse button" that comes built into the keyboard of some computers) have also been used to measure wave forces on organisms (Fig. 4). These small sensors have the advantages of being relatively inexpensive and working on the scales of smaller organisms seen in the intertidal zone. However, implementing these devices requires the design and assembly of circuits on printed circuit boards, a process much more difficult than the use of a commercially purchased force transducer and amplifier.

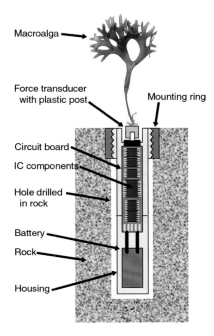

FIGURE 4 Schematic of an autonomous wave force sensor. Image redrawn from Boller and Carrington 2006, with permission from Elsevier.

Amplifier

Amplification of the force signal is necessary because the changes in voltage that are output from a force transducer are very small and difficult to measure directly. For a force transducer circuit, the amplifier is set up to magnify the voltage between two inputs: the voltage of the Wheatstone bridge and a reference voltage. The reference voltage sets the zero level of the force transducer. When the force transducer is deflected, the voltage of the bridge is then amplified to a level that can be recorded. The overall amplification needed (called the "gain") depends on the range of forces that are to be measured, the sensitivity of the force transducer, and the sensitivity of the datalogger which records the voltage.

Commercial amplifiers are available for corresponding force transducers. Very inexpensive, but less versatile amplifiers can be built with integrated circuit (IC) instrumentation amplifiers. With a single chip and a few other components, a force transducer amplifier can be built and integrated into the force transducer circuit on a single printed circuit board.

Data Recording

Once the force signal is amplified, it needs to be measured and recorded to become useful data. Because the force signal is converted into a voltage by the force transducer, there are a wide variety of methods that can be used to make a record. Most simply, an analog chart recorder can be used to draw a time-series trace from which the force can be calculated. An analog-to-digital (A/D) converter and computer can directly convert a force signal into a digital format. The smallest voltage increment that an A/D converter can detect is determined by the resolution in bits (binary digits, e.g., 16-bit) and the range of voltages (e.g., ± 10 V). The converter divides the full range into 2^n divisions, where n = the number of bits of the converter. More divisions equal a smaller voltage step between divisions, thus higher resolution. For example, a 12-bit, ±5 V converter has 4096 increments (2^{12}) across the 10 V range, resulting in the smallest detectable change in voltage 0.002 V. When using an A/D converter, it is important to understand the relationship between the resolution of the converter, the range of voltages that are being sensed, and the range of voltage output by the amplifier.

Dedicated data loggers can also be purchased to record force data. With this type of system, the data can be recorded at the field site and then transferred back to the lab for analysis. Additionally, miniature data loggers can be built and integrated into the force transducer circuit. This type of data logger will have a microcontroller unit (MCU) as the brain of the device and nonvolatile memory chips (ICs that do not lose the data if the power fails) to store data. The MCU can have an on-board A/D converter and real-time clock such that it can control the rate of data collection and sense the change in voltage without additional components.

The rate of data capture is an important consideration when using a digital system. Aliasing can occur when the sample rate is too low. To correctly measure the frequency of the force signal from waves, the data logger should sample at least twice the frequency of that signal. Further, to measure the peak of transient forces, the sensor should record fast enough for several measurements to be taken during the shortest-duration peaks (Fig. 5).

Tethered versus Autonomous

The most common electronic systems for measuring wave forces consist of a force transducer tethered to a land-based computer via a cable (Figs. 2, 3). The cable

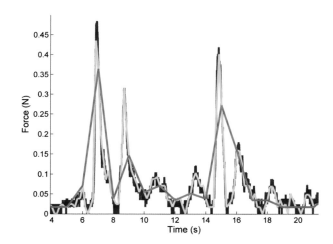

FIGURE 5 120-Hz wave-force data with resampling at lower frequencies. At low frequencies peak forces are underestimated and peaks are confounded.

provides the power for the force transducer and carries the signal back to the amplifier, A/D converter, and computer. With this system, long records of wave forces can be collected, limited only by the storage available on the computer. However, the use of a tethered system has some draw backs. First, the cable is subjected to forces as the waves break and must be secured to the substrate and probably protected from abrasion. Second, the need for a computer nearby limits the feasibility of this system to locations where a computer can be housed, or to times when the researcher can be in the field.

An autonomous system, where the force transducer, amplifier, data logger, and power supply are housed together in a waterproof enclosure (Fig. 4), can allow for more flexibility in field sites. With a compact design and no cables to get in the way, these sensors can be placed in a habitat with little disturbance to the surrounding organisms. This setup does have drawbacks; the deployment time is limited by both the memory and battery capacities.

Calibration

Calibration of electronic force sensors is important to know the absolute magnitude of the force imposed by the waves. A calibration curve can be made by measuring the voltage output from the amplifier when holding the relevant axis of the force transducer perpendicular to the ground and hanging weights from it. The force applied to the transducer (= mass × g) is then plotted versus voltage. A separate calibration should be performed on each axis of the transducer to ensure that forces are calculated properly.

Electronic force transducer calibration is influenced by other factors. First, changing the length of the transducer by attaching a drogue may change the output; thus calibration should be done with the drogue attached. Further, the weight should be consistently hung from the center of the drogue, because attaching the weight to different points may change the effective length of the transducer, changing the calibration. Second, force transducers can be influenced by temperature drift, in which the voltage output of the circuit changes with temperature. Thus, calibration of the force transducer is best done at field temperature.

SEE ALSO THE FOLLOWING ARTICLES

Hydrodynamic Forces / Salinity, Measurement of / Temperature, Measurement of / Wave Exposure

FURTHER READING

Bell, E.C., and M.W. Denny. 1994. Quantifying "wave exposure": a simple device for recording maximum velocity and results of its use at several field sites. *Journal of Experimental Marine Biology and Ecology* 181: 9–29.

Boller, M.L. 2005. Hydrodynamics of marine macroalgae: biotic and physical determinants of drag. Ph.D. dissertation, University of Rhode Island.

Boller, M.L., and E. Carrington. 2006. *In situ* measurements of hydrodynamic for us imposed on *Chondrus crispus* Stackhouse. *Journal of Experimental Marine Biology and Ecology* 337: 159–170.

Denny, M.W. 1988. *Biology and the mechanics of the wave-swept environment*. Princeton, NJ: Princeton University Press.

Denny, M.W., and D.S. Wethey. 2001. Physical processes that generate patterns in marine communities. In *Marine community ecology*. M. D. Bertness, S. D. Gaines, and M. E. Hay, eds. Sunderland, MA: Sinauer, 3–37.

Fraden, J. 2004. *Handbook of modern sensors*. New York: Springer-Verlag.

Gaylord, B. 1999. Detailing agents of physical disturbance: wave-induced velocities and accelerations on a rocky shore. *Journal of Experimental Marine Biology and Ecology* 239: 85–124.

Scherz, P. 2000. *Practical electronics for inventors*. New York: McGraw-Hill.

Stevens, C.L., C.L. Hurd, and M.J. Smith. 2002. Field measurement of the dynamics of the bull kelp *Durvillaea antarctica* (Chamisso) Heriot. *Journal of Experimental Marine Biology and Ecology* 269: 147–171.

WAVES, INTERNAL

JAMES J. LEICHTER

Scripps Institution of Oceanography

Internal waves can be broadly defined as gravity waves occurring within density-stratified fluids. They are common in both the coastal and open oceans, potentially occurring anywhere that the local water column shows stable stratification with respect to density and where there are generating mechanisms such as, but not limited to, tidal currents flowing over abrupt topographic features. The interactions of internal waves with sea floor topography can have important physical and biological consequences, especially in near-shore environments. These consequences include enhanced near-bottom turbulence and mixing, redistribution of dissolved nutrients, and vertical and horizontal transport of sediments and biological particles such as plankton.

GENERAL PROPERTIES OF INTERNAL WAVES

A useful conceptual model of ocean internal waves and their propagation can be motivated by considering discrete water parcels in a water column that is stratified with respect to density (see Fig. 1). Under conditions of stable density stratification—in which density increases either continuously or discretely with depth—any vertical displacement of a given water parcel either upward or downward will be opposed by a restoring force, due to gravity, that is proportional to the difference in density between the water parcel and its new surroundings. The restoring force will cause the parcel to move back toward its original position. The water parcel's momentum will tend to make it overshoot this original position, again setting up an opposing restoring force, with the result that a

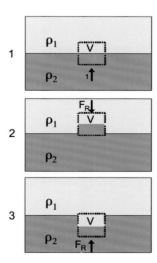

FIGURE 1 Schematic view of the generation of a propagating internal wave at the interface between two fluid layers of densities, ρ_1 and ρ_2, where $\rho_1 < \rho_2$ such that the vertical stratification is stable. In (1) an upward force, F_1, acts on water parcel V originally located at the density interface. The work of F_1 acting on the water parcel moves V to a new location in (2). A gravitational restoring force, F_R, acts to move V downward to a new location in (3). In the absence of energy loses to friction and mixing V will continue to oscillate vertically above and below the density interface. The frequency of the oscillation will be proportional to the magnitude of the difference between ρ_1 and ρ_2.

vertical oscillation is established that can propagate along lines of constant density (isopycnals) in the case of discrete stratification or at angles to the horizontal in cases of continuous stratification. In fact, the more familiar surface gravitational waves can be conceptualized as internal waves at the interface between two fluids (air and water) with a very large difference in densities. Because density gradients within the water column are very much smaller than that between the water and air, the wave periods of internal waves are typically much longer than those of surface waves, varying from minutes to multiple hours.

The minimum wave period (highest frequency) of internal waves in a given setting is controlled by the degree of density stratification in the water column, and can be predicted by the Brunt–Väisälä buoyancy frequency:

$$N = (-g/\rho \; d\rho/dz)^{1/2} \qquad \text{(Eq. 1)}$$

where N is the frequency (in radians), g is the acceleration due to gravity, ρ is density, and z is depth. Typical values for the minimum Brunt–Väisälä period (1/BV frequency) in the ocean range from 2 to 20 minutes.

The maximum wave period (lowest frequency) for internal wave propagation is the local inertial period:

$$T = 2\pi/f \qquad \text{(Eq. 2)}$$

where f, the Coriolis parameter, is a function of latitude:

$$f = 2\,\Omega \sin \phi \qquad \text{(Eq. 3)}$$

with ϕ being the earth's angular velocity about the local vertical and ϕ being latitude. The inertial period describes the time a body traveling in a rotating reference frame, with apparent deflection due to the Coriolis effect, takes to travel in a complete circle. Conceptually, a wave form with period longer than the local inertial period would be unable to propagate away from the location of generation. Maximum inertial period occurs at low latitudes, and minimum inertial period occurs near the poles. For example, at 10°N, T = 69 hours; at 30°N, T = 24 hours; and at 45°N, T = 16.9 hours. Thus, under typical ocean stratification at 30°N, internal waves can exist with periods varying between the Brunt–Väisälä periods (typically on the order of several minutes) up to the inertial period of 24 hours.

PROPAGATION AND GENERATION

The speed at which the waveform of internal waves travels is also much smaller for internal waves than for surface waves, typically on the order of 10 to 30 centimeters per second. However, the vertical displacements, which are greatest when the vertical density gradient is strongest

(the pycnocline), are typically quite large, on the order of 10 meters or more. Wave heights for internal waves at tidal periods can be very much larger, with reports as large as 200 meters. Whereas surface waves tend to occur during multiday bouts of intense activity (for example, associated with offshore storms) interspersed with relative calm, internal waves appear to be far more temporally persistent and are almost always present except for periods when stratification breaks down; for example, during strong or persistent wind mixing in winter. The most common mechanisms of internal wave generation in the coastal oceans and near the shelf break is the displacement of isopycnals associated with tidal currents flowing over abrupt topographic features. Tidal reversals of the currents lead to the generation of internal waves at the tidal frequency, termed internal tides.

The leading edge of an internal tide is often accompanied by packets of higher frequency, nonlinear, internal waves, and horizontal current velocities associated with internal tides can be comparable to or larger than those associated with surface tides. Propagating internal waves can also take on the form of half waves of depression or elevation only, sometimes termed solitary waves or solitons. The dynamics of solitons are highly nonlinear (meaning they cannot be accurately described by the equations of linear wave theory), and net horizontal water velocities can be large, especially in shallow water. Internal waves generated at or near the shelf break tend to travel inshore, and as they progress into shallow water over a gradually sloping bottom, the waves tend to steepen and shoal. Shear at the leading edge of the traveling wave can lead to instability and breaking, with the resulting turbulent bores of water from below the pycnocline continuing to travel inshore if the bottom angle is less than a critical angle for wave reflection. The resulting mixing across isopycnal surfaces may be an important source of nutrients to the euphotic zone. The shear associated with currents generated by internal tides is thought to be a major source of vertical mixing within the interior of the oceans.

BIOLOGICAL IMPORTANCE

Although ocean internal waves are a subsurface phenomenon, manifestations at the surface such as slicks, where water is converging over traveling internal waves, are observed frequently by remote sensing from airplanes and satellites (see Fig. 2). Surface slicks are caused by damping of surface gravity and capillary waves in the convergence zone over traveling internal waves. Understanding of the biological and ecological effects of internal waves for nearshore communities is limited, but they likely extend to both the subtidal and intertidal communities, including

FIGURE 2 Surface slicks formed over internal waves propagating into shallow water near shore in Southern California. Slicks result from dampening of surface gravity and capillary waves in the convergence zone over traveling internal waves. Linear slicks formed offshore become deformed as the underlying internal waves refract around bottom features and break in shallow water. Photograph by J. Leichter.

tidepool communities along rocky shores. These effects likely include both direct consequences of rapid thermal variability and transport of materials such as nutrients and zooplankton, and indirect effects associated with mixing and enhanced biological production. Vertical oscillations and mixing within the water column can concentrate and redistribute phytoplankton and zooplankton. It has been suggested that onshore transport of zooplankton at the surface can occur in slicks over nonlinear internal waves. Onshore transport of fronts associated with surface warm bores associated with internal waves very near shore has been shown for the coast of Southern California and may contribute to pulses of barnacle larvae arriving on shore. Oscillations of the pycnocline have been described as a mechanism of nutrient transport to kelp forests and coral reefs. The downward transport of subsurface layers of high chlorophyll concentration may be important in rocky-subtidal habitats. The temporal variability associated with internal waves in near-shore communities can also be accompanied by strong spatial variability associated with the interaction of waves with rough topographic features such as the spur and groove formations on coral reefs. Internal tides can be a major source of sediment resuspension near the bottom, and their interaction with the continental shelves is thought to play a major role in determining the angle of the shelf slopes.

OPEN RESEARCH QUESTIONS

There are a number of open and significant research questions associated with the mechanisms and effects of internal waves. The dissipation of energy in the subtidal

and consequences of mixing are likely to have an impact on the distribution of nutrients and zooplankton in a variety of habitats, but these processes are poorly understood. Modulation of the amplitude of internal waves by mechanisms, such as alongshore currents, that affect the mean depth of thermocline may produce synchronized, episodic variability at multiple sites alongshore at scales of tens of kilometers or more. Understanding the sources of spatial variability in the impact of internal waves offers a context for experimental work to understand the biological consequences of internal waves at kilometer to regional scales.

SEE ALSO THE FOLLOWING ARTICLES

Near-Shore Physical Processes, Effects of / Nutrients / Ocean Waves / Tides / Turbulence

FURTHER READING

Cacchione, D. A., L. F. Pratson, and A. S. Ogston. 2002. The shaping of continental slopes by internal tides. *Science* 296: 724–727.

Garrett, C. 2001. Internal waves, in *Encyclopedia of ocean sciences.* J. H. Steel, S. A. Thorpe, and K. K. Turekian, eds. San Diego: Academic Press.

Haury, L. R., M. G. Briscoe, and M. H. Orr. 1979. Tidally generated internal wave packets in Massachusetts Bay. *Nature* 278: 312–317.

Leichter, J. J., H. L. Stewart, and S. L. Miller. 2003. Episodic nutrient transport to Florida coral reefs. *Limnology and Oceanography* 48: 1394–1407.

Pineda, J. 1999. Circulation and larval distribution in internal tidal bore warm fronts. *Limnology and Oceanography* 44: 1400–1414.

Pond, S., and G. L. Pickard. 1983. *Introductory dynamical oceanography*, 2nd ed. Oxford: Pergamon Press.

Shanks, A. L. 1995. Mechanisms of cross-shelf dispersal of larvale invertebrates and fish, in *Ecology of marine invertebrate larvae.* L. McEdward, ed. Boca Raton, FL: CRC Press, 323–367,

Wolanski, E. 1994. *Physical oceanographic processes of the Great Barrier Reef.* Boca Raton, FL: CRC Press.

WIND

WENDELL A. NUSS
Naval Postgraduate School

Wind is the movement of air that is created by differences in pressure or the distribution of mass in the atmosphere. Wind speed and direction are determined by the horizontal distribution of mass, which produces a gradient of pressure. Pressure gradients arise in a variety of ways and strongly influence the evolution of the wind at all levels in the atmosphere. The winds near the surface play a significant role in shaping the environment on rocky shores.

BASIC CHARACTERISTICS OF WIND

Wind is the movement of air and is characterized by a speed and direction of motion. The wind direction is the direction from which the wind is coming; for example, a west wind is wind blowing from the west toward the east. The wind speed is the distance the air moves over a unit of time and can be expressed as meters per second, miles per hour, or, most commonly around water, as nautical miles per hour or knots. The wind direction and speed vary horizontally, vertically, and over time. These basic characteristics of the wind depend upon the atmospheric pressure gradient, which results from the spatial distribution of atmospheric mass.

Atmospheric pressure represents the weight of air above any given point, which varies both vertically and horizontally for a variety of reasons. Although the largest variation in pressure is in the vertical, the vertical pressure gradient is nearly balanced by the earth's gravity, and much smaller horizontal pressure variations are most important in driving the wind. The mass of the air above a point on earth's surface is determined primarily by its temperature, so that warm air produces lower pressure and cold, dense air produces higher pressure. Because the air is moving, this relationship between temperature and pressure is more complex, but, to a first approximation, low-pressure regions are warm and high-pressure regions are cold. Changes in the distribution of atmospheric pressure result from changes in the temperature of the air and by the redistribution of mass that occurs through organized circulations in the atmosphere that represent weather systems over a broad range of scales. Figure 1 shows an example of the horizontal distribution

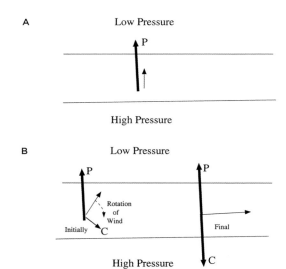

FIGURE 2 Distribution of atmospheric pressure (thin lines) going from higher to lower pressure. Pressure gradient force is shown as heavy solid arrow labeled P. (A) Thin arrow shows the winds that result from the pressure gradient force acting on the air. (B) Thin arrow shows the winds initially (left) and then finally (right) as the Coriolis force due to the earth's rotation causes them to turn. Coriolis force is marked by medium thick arrow labeled C.

of pressure near the earth's surface. Regions of high and low pressure are identified.

The basic influence of the pressure gradient is to produce a force on the air to cause it to move or accelerate toward low pressure (down the pressure gradient). This is illustrated in Fig. 2A, which shows a parcel of air moving in the direction of the arrow toward lower pressure. If no other forces were to act on this moving air parcel, it would produce a wind in the direction toward the low pressure and its speed would continue to increase over time. However, the motion of the air is influenced by the earth's rotation, which produces an additional force (the Coriolis force) and causes air to turn to the right (in the Northern Hemisphere) so that over time the wind tends to blow parallel to the pressure contours rather than across them. Figure 2B illustrates this situation in which the pressure gradient force is balanced by the Coriolis force due to the earth's rotation, producing a wind direction parallel to the pressure contours (perpendicular to the pressure gradient). The speed of the wind in this balanced state, known as geostrophic balance, is proportional to the magnitude of the pressure gradient. This adjustment to a balanced state (geostrophy) takes time to occur, and so changes in pressure gradient first cause the wind to accelerate toward low pressure and then slowly turn to be parallel to the pressure contours.

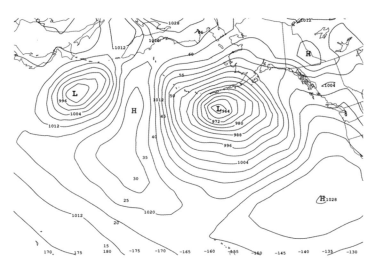

FIGURE 1 Distribution of sea-level pressure over the North Pacific Ocean on a typical winter day. The contours are of constant pressure in millibars or hectopascals. Centers of low (L) and high (H) pressure are identified.

The winds near the earth's surface are strongly influenced by the surface itself, to modify them compared to that which would be expected for the same pressure gradient aloft, away from the surface. The effect of drag on the land or water surface produces a frictional force that slows the movement of air and reduces the wind speed. Because of this slowing and the introduction of this additional force, the wind direction also changes to always have a component toward lower pressure. The effect of friction decreases away from the surface to essentially vanish at a height of 1000–1500 m above the surface. The magnitude of the frictional force depends on the character of the underlying surface as well as the structure of the overlying atmosphere. Because of the ability of water to move, the amount of surface drag over the water is reduced and the effect of friction tends to be less over the ocean compared to the adjacent land areas. Consequently, winds over the ocean may be 20–40% higher than those over the land for the same pressure gradient.

WIND DUE TO WEATHER SYSTEMS

Organized large-scale weather systems in the form of high- and low-surface-pressure centers produce the first order changes in the winds for a given location. Figure 1 illustrates a typical distribution of high and low pressure across the North Pacific on a winter day. Low-pressure centers are associated with storms, and the wind tends to blow counterclockwise around areas of low pressure in the Northern Hemisphere (clockwise in the Southern Hemisphere). The pressure gradient can get very strong in low-pressure regions, as seen near the low that is east of Japan in Fig. 1. Consequently the strongest winds, sometimes in excess of 60 knots, most often occur in association with storms. In contrast, high-pressure regions are usually associated with fair weather and lighter winds. The wind blows clockwise around high-pressure centers in the Northern Hemisphere. The pressure gradient around highs tends to be weaker than around lows, and so light winds are common with high centers.

Migratory weather systems tend to move from west to east and produce changes from low to high pressure as they pass. This movement is a consequence of a dynamic process in the atmosphere that results in pressure changes ahead and behind high- and low-pressure centers to force them to move. This movement produces an effect on the winds, causing them to change and adjust as the pressure gradient changes. Because the movement of weather systems tends to be slow, the distribution of winds around a high- or low-pressure center tends to be balanced with the pressure gradient and not intensify or weaken rapidly over time. Consequently,

the wind direction and speed changes at a specific location primarily represent the movement of the weather system and the observer's different relative position in the weather system. Figure 3 illustrates this effect, where an observer along the Central California coast would see south winds at the first time ahead of the storm, followed by west winds at the middle time as the storm passes, and then northwest winds after the low passes to the north at the later time. Important changes do occur as weather systems develop and decay as well to cause wind speed increases and decreases over time in addition to those due to the movement of the storm.

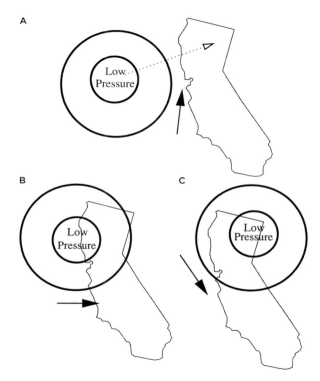

FIGURE 3 Sequence of three times showing a low pressure center (storm) crossing the coast of California. Direction of movement of the storm is indicated by dashed arrow. (A) Low is offshore, and south winds along the coast, for an observer at the location marked by the asterisk, occur ahead of the storm (solid arrow). (B) Storm is just onshore, producing west winds along the coast at the same location south of the storm (solid arrow). (C) Storm is inland, producing northwest winds along the coast at the point identified by the asterisk to the southwest of the storm (solid arrow).

COASTAL INFLUENCES ON THE WIND

An important aspect of the wind on rocky shores is the influence of the coast on the character and distribution of winds. The land-versus-water surface characteristics across the coast produce heating differences that influence the winds in the coastal region. In addition, the coast may have significant topography that also influences the wind along the coastline. The impact of heating on the coastal winds results from the tendency of the land to heat more readily than the adjacent ocean. This results in the lowering of pressure over the land relative to the water and acceleration of the wind across the coast. This effect produces the well-known sea breeze circulation where the wind blows across the shore toward the warmer land. This onshore wind is weak or nonexistent in the morning and then increases in the afternoon as the heating reaches a peak. Because the heating occurs over a rather short time period (less than 12 hours), the wind cannot adjust to the decreasing pressure over the land, and the wind blows across the coast toward lower pressure and not parallel to the pressure contours. At night as the land cools, the sea breeze wind decays and the winds drop. If the land cools sufficiently, the land may end up cooler than the adjacent ocean, resulting in higher pressure over land and an offshore-directed wind, the land breeze.

Coastal mountains produce a wealth of effects on the winds in the coastal environment. Critical to many of these effects is the presence or absence of a low-level temperature inversion below the top of the coastal mountains. A temperature inversion occurs when warm air is located above colder air. The presence of a low-level temperature inversion produces a very stable layer that limits the ability of air to move vertically, because cold air cannot rise into the overlying warm air. When next to coastal mountains this stable layer can effectively prevent air from flowing over the topography and force it to flow around it. The effect is referred to as flow blocking, and flow blocking is most prevalent when strong static stability (temperature inversion) and weak cross-mountain winds combine. The important impact of flow blocking on the winds is that the wind direction will be constrained to be mostly parallel to the mountain barrier (coast). The wind speed in these situations tends to be enhanced as well, because the winds fail to reach a balance with the Coriolis force and therefore accelerate toward lower pressure along the barrier. The impact is to produce a near-surface jet of high winds along the barrier, known as a barrier jet or coastal jet, as illustrated in Fig. 4.

Coastal jets or barrier jets are common features along mountainous coastlines, particularly the U.S. West Coast. During the summer months, warm air occurs above a cold ocean to form a shallow layer of marine air that is cooler and moister than the air above. This layer is capped by a rather strong inversion below mountaintop level to

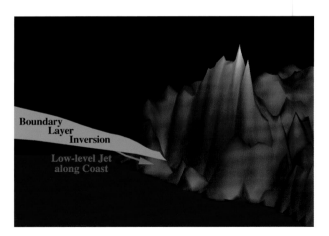

FIGURE 4 Schematic showing coastal mountains with a low-level temperature inversion (labeled inversion). Below the inversion along the coastal mountains, strong winds occur to produce a coastal jet illustrated by the arrow below the inversion (labeled coastal jet).

effectively block any cross-coast flow. A rather strong, 50-to-60-knot, coast-parallel jet occurs 100–500 meters above the ocean surface along the coast. Friction reduces the surface speeds to around 30–35 knots. When this coastal jet adjusts to points and capes along the coast, winds tend to be enhanced to produce localized regions of higher winds. These regions become important for coastal upwelling. The wind direction is generally coast-parallel, and the strongest winds tend to occur just offshore from coastal promontories. However, as diurnal heating occurs, these strong winds tend to get turned toward the land, producing rather strong afternoon sea breezes. Similar effects occur along other mountainous coastlines as well, particularly along the west coasts of continents.

Barrier jets are not strictly limited to summertime, because similar features are observed in advance of approaching winter storms. In this case, the jet winds tend to be poleward instead of equatorward along the coast, and the duration is limited by the time period over which the storm interacts with the coast. This type of topographic interaction in winter storms accounts for some of the strongest coastal winds (outside of hurricanes), and speeds in excess of 100 knots have been observed.

Another important effect of coastal mountains on the local winds is the production of enhanced flow through gaps in the mountains. These gap flows occur through relatively narrow coastal valleys, channels, and between islands. The basic setup to produce these types of winds is the occurrence of high pressure on one side of a narrow gap and low pressure on the other. This pressure difference acts to accelerate the flow through the gap, producing strong winds on the low-pressure side of the gap. The winds blow parallel to the gap, typically cross coast. Figure 5 illustrates gap winds near the San Francisco Bay, where the winds blow through the break in the coastal mountains at the mouth of the bay to produce strong localized winds just offshore.

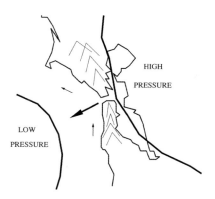

FIGURE 5 Schematic showing the San Francisco Bay with low pressure offshore and higher pressure inland. This results in gap winds that exit through the mouth of the bay to produce stronger winds (heavy solid arrow) just out from the bay compared to weaker winds (lighter solid arrows) north or south of this break in the mountains.

IMPACTS ON ROCKY SHORES

The surface wind speed and direction in the coastal zone very much influence the local microclimate of rocky shores. Persistent stronger surface winds tend to promote greater evaporation as well as increased near-shore wind wave generation. As larger scale weather systems move across the coast, the time scale of the wind variations due to these systems tends to be over periods of days to weeks but may account for the strongest events over a season. Coastal jets and consistent diurnal sea breeze forcing tend to contribute most to day-to-day wind forcing, particularly along the U.S. West Coast. Gap winds, particularly if directed offshore, can produce significant drying and possibly even warming in the coastal region. Although these events tend not to be persistent, they may contribute to important shorter-term changes in the tidepool environment.

SEE ALSO THE FOLLOWING ARTICLES

Near-Shore Physical Processes, Effects of / Storms and Climate Change

FURTHER READING

Ahrens, C. D. 1999: *Meteorology today: an introduction to weather, climate, and the environment*, 6th ed. Pacific Grove, CA: Brooks/Cole Publishing.

WORMS

ANDREAS SCHMIDT-RHAESA

Zoological Museum, Hamburg, Germany

"Worm" is a general, and unsystematic, term for any animal that is distinctly longer than broad. This feature is probably very ancient and characteristic for the ancestor of all bilaterian animals. The wormlike body organization was conserved in many taxa: Platyhelminthes, Nematoda, Nematomorpha, Priapulida, Acanthocephala, Nemertini, Sipunculida, Echiurida, Annelida, Phoronida, and Enteropneusta. In many other groups, some members also express a wormlike habitus, such as the "worm-molluscs" (Aplacophora), wormlike gastropods, and bivalves among molluscs. On rocky shores and in tidepools, one can expect to find species from the taxa Platyhelminthes, Nematoda, Nemertini, Sipunculida, Echiurida, Annelida, and Enteropneusta, and these taxa are presented in more detail.

PLATYHELMINTHES, THE FLATWORMS

Flatworms are either free-living in the marine or freshwater environment (only exceptionally on land) or are parasitic. Most of the free-living species are very small (less than 1 mm) and live in the interstitial system between sand grains. Such species are abundant on sandflats but can be found on rocky shore only where sandy sediments accumulate. Larger, epibenthic species are found in the taxon Polycladida, and members may be found on rocky shores. Many flatworms are predators, although some feed on algae. They are themselves food to many larger organisms. Some larger polyclad species resemble opisthobranch molluscs (mimicry), which might prevent them from being regarded as food.

Polyclads are, like most flatworms, dorsoventrally flattened. They measure usually a few centimeters in length. The mouth opening is more or less central on the ventral surface, it is the only opening of the intestinal system and therefore also serves as the anus. The intestine is strongly branched and distributes nutrients in the entire body. Most species are carnivorous; a few species live commensally on echinoderms or crustaceans. Most polyclads crawl, but some species are able to swim with undulating movements of the margins of their body. Species from the taxon Cotylea have a sucker with which they can attach to different substrates. The anterior end usually contains eyes and, in many species, sensory appendages. Polyclads are, like almost all flatworms, hermaphrodites, but male gametes mature before female gametes and are transferred in a copulation. Uniquely among free-living flatworms, polyclads have a microscopic, planktonic larva. Many polyclad species are brightly colored and very distinctive.

Polyclads are distributed mainly in the sublittoral but can also be found in the eulittoral. On rocky shores, polyclads can be found under stones, on macroalgae, in mussel (*Mytilus*) beds, or between barnacles.

NEMATODA, THE ROUNDWORMS

Roundworms have an anterior mouth opening, a muscular sucking pharynx, and a straight intestine leading to an anus on the posterior end. The body is covered by a cuticle, which is molted four times during development. Nematodes occur abundantly in marine, freshwater, and terrestrial habitats, and a number of species are parasites. With few exceptions, the free-living forms are very small (below 1 mm) and live interstitially between sand grains. Therefore, they can be expected on rocky shores where sediment accumulates, but their investigation requires extraction from the sediment and a microscopic investigation. In most sediments, nematodes have high abundances and belong to the numerically dominating organisms.

NEMERTINI, THE RIBBON WORMS

Nemertini, or Nemertea, are small (a few millimeters) to large (several centimeters) worms capable of extreme changes in body size and form. They have well-developed circular and longitudinal musculature surrounding the body and, uniquely among invertebrates, an epithelially lined blood vascular system. The most conspicuous feature of nemerteans is the proboscis, which can be everted to great lengths and totally retracted into the body. The proboscis lies within a coelomic cavity; the rhynchocoel, and pressure upon the coelomic fluid by the body wall muscles drives the proboscis out. Retraction is performed by a retractor muscle, which attaches at the anterior end of the proboscis and the caudal end of the rhynchocoel. In the nemertean taxon Anopla, the proboscis is located at the anterior end of the body separately from the mouth opening, while in the taxon Enopla, mouth and proboscis have the same opening. In hoplonemerteans (which belong to the Enopla), the proboscis includes a stylet. Most nemerteans are carnivorous. Large species capture their prey with the proboscis and swallow them as a whole, while species with a stylet can puncture and anesthetize their prey before sucking them out. Therefore, nemerteans are important predators, especially at night.

There are approximately 900 species described, the majority of which live in or on marine sediments. A considerable number of species occurs in the intertidal, either burrowing in sandy sediments or living under stones, in crevices, in algal holdfasts, or between algae. On rocky shores, several species from the genera *Amphiporus*, *Cerebratulus*, *Lineus* (Fig. 1), *Micrura*, and *Paranemertes* can be expected. *Tubulanus* species forms delicate tubes. There are some species that live commensally and can be expected in the rocky intertidal, especially *Carcinonemertes* spp., which live on the gills of decapod crustaceans *(Cancer, Hemigrapsus, Pachygrapsus)* and feed on their eggs.

FIGURE 1 A nemertean (*Lineus bilineata*; dark animal) and a stationary polychaete (*Terebella lapidarius*; red animal) from an opened narrow crevice close to the low-water line on the French Atlantic coast. Photograph by the author.

SIPUNCULIDA, THE PEANUT WORMS

Most sipunculids are large worms, measuring several centimeters (Fig. 2). About 160 exclusively marine species are known. They have a muscular trunk and an introvert that can be completely withdrawn into the trunk. At the anterior end of the introvert are short tentacles, which are used for food uptake and respiration. The intestine is U-shaped, and the anus is in the anterior region of the trunk. Sipunculids possess a spacious coelom in the trunk and a smaller coelom in the tentacles.

Several sipunculid species burrow and therefore occur in sandy sediments, but some species occur under rocks or in crevices. For example, *Phascolosoma agassizii*, *Nephasoma*, and *Themiste*-species can be found in muddy crevices between rocks. Species such as *Phascolion strombus* and

FIGURE 2 Three sipunculids (*Golfingia elongata*; above right) and one echiuran (*Thalassema thalassemum*; center) taken from rock crevices on the French Atlantic coast. Photograph by the author.

Aspidosiphon muelleri live in empty shells of gastropods or in serpulid tubes.

ECHIURIDA, THE SPOONWORMS

Echiurids are large, exclusively marine worms. About 150 species are known, living within sandy sediments or between rocks. The body is divided into a trunk and a proboscis, which can elongate and contract but cannot be completely withdrawn. Most echiurids remain in "safe" crevices or in the sediment and use their proboscis to move over the sediment and collect organic material. Such material is enclosed in mucus and transported by ciliary action to the mouth opening at the base of the proboscis. An exception is *Urechis caupo*, which burrows in sandy sediments and pumps water through a mucus net, which is ingested from time to time together with all organic particles that stick to it. The anus is at the posterior end of the body. Echiurids possess a spacious coelom. They are closely related to annelids and may even be a taxon within annelids.

Bonellia viridis, occurring in the Mediterranean, the Red Sea, and the northwestern Atlantic, is particular in that the female is large (trunk up to 15 centimeters) and has an extremely long proboscis, but the male is only few millimeters long and lives inside the nephridia of a female to fertilize her eggs as they are released. This is the most extreme case of sex dimorphism in animals. Echiurids have pelagic trochophore larvae, and in *Bonellia viridis* the fate of each larva depends on whether it comes into contact with an adult female or not. If coming into contact, the larva develops into a male, remains small, and migrates into the female's nephridia. If not, it settles and grows to become a female.

Most echiurid species can be found in sandy sediments, but *Bonellia* occurs between rocks, from where it

sticks out its proboscis. *Thalassema* species can be found in mud-filled crevices between rocks (Fig. 2).

ANNELIDA, THE SEGMENTED WORMS

The segmented worms form a large group, with about 20,000 species described to date. Three subtaxa can be distinguished, of which Polychaeta are almost exclusively marine, while Oligochaeta and Hirudinea (leeches) live predominantly in freshwater or terrestrial soil. Intertidal species from rocky shores are mainly polychaetes, although some oligochaetes can also be found burrowing in sediment or under stones.

As the name suggests, segmented worms have segments—that is, they have repeated body regions, well separated from each other by external rings and internal septa. Each such segment contains (ideally) a pair of coelomic cavities, segmentally arranged musculature, a pair of segmental swellings (ganglia) of the nervous system, and a pair of excretory organs. In most species, segments differ somewhat between different body regions. Some polychaetes and oligochaetes ingest sediment unspecifically and digest their organic components. Others collect organic matter from the sediment surface with the aid of ciliated tentacles. Several species are predators and have a large muscular pharynx or even jaws.

Reproduction among polychaetes is very diverse. Several species are able to reproduce asexually, often by forming buds that develop into juvenile worms. This can result in chains of individuals with the "mother animal" followed by a chain of juveniles of increasing size and age. Sexes are different, and in sexually reproducing species gametes are generally released into the water. Few polychaetes have a direct transfer of spermatozoa to females. Some polychaetes exhibit morphological changes toward maturity or a special behavior such as a synchronized spawning of gametes. In most species, the fertilized egg develops into a so-called trochophore larva that is pelagic at first but later settles to the sediment. Segments are formed in a terminal growth zone within the larva. Oligochaetes are hermaphrodites but transfer their spermatozoa to a partner. Their eggs are deposited in cocoons, and juveniles develop directly, without a larval stage.

Polychaetes are either mobile animals, burrowing or crawling over the sediment surface or they are (more or less) stationary and often live in tubes of some kind. The mobile polychaetes depend on sediment and are therefore abundant in sand flats and are found on rocky shores where sediment accumulates: under stones, in crevices, or in holdfasts of macroalgae. The stationary polychaetes either live in protected regions such as crevices and expand their tentacles from these

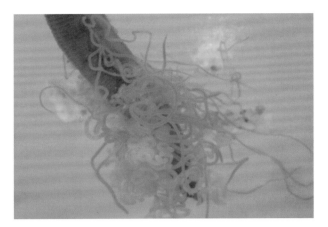

FIGURE 3 Anterior end of a stationary polychaete (Annelida) with tentacles used for food collection. Photograph by the author.

hiding places (Fig. 3), or form quite stable, sometimes calcified, tubes, which are present on more exposed hard substrates. Examples are the spiral tube worm *(Spirorbis)* and *Pomatoceros triqueter* (Fig. 4) in Europe. A few species, such as *Polydora ciliata*, are boring and live in mollusc shells. Mobile polychaetes are often predators, whereas sedentary forms are fine-particle feeders. Such particles are collected by tentacles from the sediment surface in the vicinity of the worm, trapped in mucus, and transported to the mouth by the action of cilia along the tentacle.

FIGURE 4 Rock covered by numerous calcified tubes of *Pomatoceros triqueter* (Polychaeta, Annelida). Photograph by the author.

PHORONIDA, THE HORSESHOE WORMS

Phoronids are a small group (about 10–15 species) of marine worms. They measure from 6 millimeters up to 25 centimeters. The adult worms live in chitinous tubes, from which they can extend their tentacles, which are arranged in an oval, horseshoe-shaped, or spiral pattern. These tentacles are used to filter fine particles from the water current. The tentacles surround the mouth opening, the gut is U-shaped, and the anus opens close to the tentacles. Phoronids possess a conspicuous planktonic larva, which was originally considered as a separate species and named *Actinotrocha branchiata*. Although now known to be the phoronid larva, it is still called the actinotrocha. It possesses larval tentacles covered by a hood and a trunk. The adult body forms during a radical metamorphosis from a special region of the trunk when the actinotrocha settles to the sediment and most of the larval tissues are absorbed. Phoronids are related to brachiopods and probably to bryozoans (ectoprocts).

Phoronid tubes of most species stick in sandy or muddy sediments, but some species attach to hard substrate and may therefore be found on rocky shores. The Australian *Phoronis australis* lives in the tube walls of *Cerianthus* (Cnidaria); *Phoronis ovalis* from the North Sea and the Baltic Sea lives on mollusc shells or in shell pores created by other organisms.

ENTEROPNEUSTA, THE ACORN WORMS

About 70 species of enteropneusts are known, all of which live in marine sediments. Species measure from a few centimeters up to a maximum of two meters. The body is divided into three regions: a proboscis, a collar, and a trunk. The proboscis is used for locomotion and feeding. Size increase by swelling anchors it in the sediment, and the remaining body can be pushed forward. Enteropneusts are fine-particle feeders; the particles are collected by cilia on the proboscis and directed toward the ventral mouth opening at the border between proboscis and collar. Enteropneusts have gill slits in the anterior part of the intestine; these additionally filter fine particles from the water and may also aid in respiration. Sexes are separate (gonochoric), and fertilization takes place in the water. Many species possess a larva called tornaria.

Most species create U-shaped burrows in sandy or muddy sediments. On rocky shores, entropneusts can be expected only exceptionally, for example *Ptychodera* spp., which live under stones and shells.

SEE ALSO THE FOLLOWING ARTICLES

Echiurans / Micrometazoans / Phoronids / Sipunculans

FURTHER READING

Hayward, P. J., and J. S. Ryland. 1995. *Handbook of the marine fauna of North-West Europe.* Oxford: Oxford University Press.
Ricketts, E. F., J. Calvin, and J. W. Hedgpeth. 1985. *Between Pacific tides*, 5th ed. Stanford, CA: Stanford University Press.
Riedl, R. 1983. *Fauna und Flora des Mittelmeeres.* Hamburg: Verlag Paul Parey.
Ruppert, E. E., and R. S. Fox. 1988. *Seashore animals of the Southeast.* Columbia: University of South Carolina Press.

Z

ZONATION

CHRISTOPHER HARLEY

University of British Columbia, Vancouver, Canada

Biological zonation occurs when certain species occupy discrete bands or zones along an environmental gradient. Examples of zonation may be found from mountain tops, where plant species are zoned according to elevation, to deep-sea hydrothermal vents, where microbes are zoned along thermal and chemical gradients. Perhaps nowhere is zonation more striking, however, than at the interface between land and sea. On rocky coasts, conditions shift from fully terrestrial to fully marine over the space of just a few meters. As a result, intertidal species occur in discrete zones according to their ability to cope with physical stress and interactions with other species (Fig. 1).

THE INTERTIDAL GRADIENT

Tides and Emersion Patterns

The rise and fall of the tides creates a vertical gradient in the amount of time spent in or out of the water (Fig. 2). Depending on the tidal dynamics, waves, wind, and atmospheric pressure, certain positions on the shore may be exposed for several days or longer, while others are never exposed for more than a few minutes or hours. For descriptive purposes, approximate emersion characteristics can be used to divide the region between the highest occurrence of marine life and the level of the lowest low tide into four zones: the *supralittoral zone*, the *high intertidal zone*, the *mid-intertidal zone*, and the *low intertidal zone*. Note that these distinctions are made

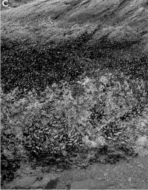

FIGURE 1 Examples of intertidal zonation. (A) Tatoosh Island, Washington. (B) Bodega Bay, California. (C) Beagle Channel, Tierra del Fuego, Argentina. Despite differences in species composition, climate, and tidal dynamics, zonation patterns in California and Tierra del Fuego are remarkably similar. All photographs by the author.

for the sake of convenience; they do not have any exact relationship to a measurable quantity such as emersion time, and they vary somewhat in usage among intertidal biologists.

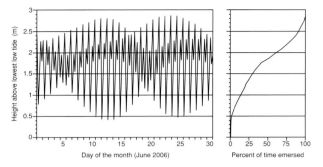

FIGURE 2 Example of tidal patterns and emersion time. Left panel: predicted tidal patterns for San Francisco, California, in June 2006. Right panel: predicted percentage of time during that month spent emersed (out of the water) at all shore levels. Actual emersion patterns will differ as predicted tidal patterns are modified by wind, waves, and atmospheric pressure. Tide predictions courtesy of the National Oceanographic and Atmospheric Administration.

The supralittoral zone, also known as the splash zone, is reached only occasionally by extreme high tides. In the absence of regular tidal immersion, it is wetted only intermittently by large waves and salt spray. The high intertidal zone is emersed periodically by the tides and thus receives predictable (if not necessarily frequent) immersion. Importantly, this zone is submerged for long enough to receive large numbers of waterborne larvae and algal propagules, allowing sessile (attached) marine species to colonize this shore level. The mid-intertidal zone is exposed to air and water in about equal proportions, and, depending on the tidal regime, is more or less guaranteed to be both immersed and emersed at least once per day. The low intertidal zone is rarely emersed for more than a few hours at a time. This limited exposure to air allows for the persistence of many species that cannot survive at higher shore levels.

Physical Stress

Differences in emersion time result in extreme changes in physical conditions over a few vertical meters. At the seaward boundary of the intertidal zone, conditions are fairly stable and generally quite moderate. Ocean temperature, salinity, and pH are relatively constant, most harmful ultraviolet radiation is absorbed near the water's surface, and the basic resources for living things (light, nutrients, and suspended food) are generally available. Because most intertidal species are marine in origin, exposure to terrestrial conditions at low tide presents a number of challenges. During emersion, intertidal organisms are vulnerable to desiccation, extremes in temperature, potentially dangerous levels of ultraviolet radiation, and osmotic stress related to high or low salinities (due to evaporation and precipitation, respectively). The higher

on the shore an organism lives, the longer it spends out of the water, and the more severe these stresses become. (Fig. 3). For example, in the lower intertidal zone, an organism's body temperature rarely strays more than a few degrees from the temperature of the ocean. In the high intertidal zone, body temperatures may vary by more than 25 °C over the course of a single day, and thermal extremes can range from below freezing during the winter to over 40 °C in the summer.

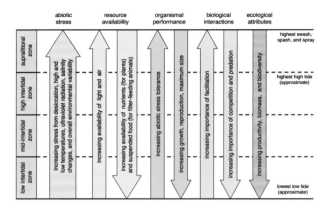

FIGURE 3 Factors that vary with vertical position in the intertidal. Overall trends are indicated with arrows; however, variables such as abiotic stress and biological interactions may be important for specific species at any given shore level. Oxygen and CO_2 are more easily aquired from air than from water, which is important for some plants and air-breathing animals. However, because most marine species require a moist gill membrane or body surface for effective gas exchange, desiccation usually overrides oxygen or carbon dioxide availability in terms of where in the intertidal it is easiest for an organism to breathe or photosynthesize.

Resource Availability

The time available for the acquisition of certain resources is also dependent on shore level. Rocky intertidal algae lack roots and must absorb nutrients from seawater. Algae living higher on the shore are emersed for longer and have less time to acquire water-borne resources. Similarly, many intertidal animals filter suspended food from seawater at high tide, and those higher on the shore have less time to feed. The only resources that increase with height on the shore are light and air. Some light-limited algae do grow faster at higher shore levels, but since photosynthesis requires hydrated tissues, increased light availability is generally offset by increasing desiccation stress. Most intertidal organisms can respire underwater and, in fact, must be wet to respire effectively; however, some intertidal insects, arachnids, and pulmonate snails require access to air.

General Ecological Patterns

Species that live higher on the shore tend to be much more tolerant of physical stress than are species that live lower down. The higher stress and lower rates of resource acquisition result in lower growth rates and (often) smaller maximum sizes at higher shore levels. These unfavorable conditions also limit the total number of species that can persist at higher shore levels. In general, diversity, growth rates, and total productivity increase from the splash zone down to the low intertidal zone.

CHARACTERISTIC FLORA AND FAUNA BY ZONE

Certain types of organisms dominate different zones on rocky shores (Table 1), and these broad patterns are surprisingly geographically consistent. Although the patterns described in the following paragraphs hold fairly well for temperate shores, important differences exist in tropical and polar regions. In the tropics, extreme thermal and desiccation stress may greatly reduce the abundance of certain types of organisms (e.g., algae and nonshelled invertebrates). In addition, herbivorous and carnivorous fish are more important in the tropics. In polar regions, freezing temperatures and ice scour are major sources of mortality, and the biomass and diversity of intertidal organisms is low relative to temperate shores.

Supralittoral Zone

Primary producers include lichens and microscopic algae, although ephemeral macroalgae may become abundant during the winter, when temperatures are cooler and wave splash and spray are more regular. Depending on the frequency with which the rock is wetted, herbivorous limpets and snails may be abundant at this shore level.

High Intertidal Zone

Unless climatic conditions are overly hot and dry, algal biomass is dominated by macroalgae. Ephemeral algae may be seasonally abundant, and desiccation-resistant perennials (e.g., fucoids and red algal turfs) persist in this zone year-round. The most obvious invertebrate inhabitants at this shore level are often acorn barnacles, which provide important habitat and refuge space for countless small plants and animals that would otherwise suffer severe desiccation stress. Herbivorous littorinid snails and limpets are often extremely abundant in the high intertidal zone. Invertebrate predators include whelks, shore crabs, and nemertean worms.

Mid-Intertidal Zone

The mid-shore is often dominated by large fucoid algae, particularly on wave-protected shores. The mid-intertidal algal canopy provides an important refuge for many additional species. More exposed shores are usually dominated by filter-feeding invertebrates, particularly mussels. Mussels and other abundant filter feeders (such as tunicates in Australia and Chile) create prime habitat for hundreds of species of algae, small invertebrates, and even fish. The most common mid-intertidal (non-filter-feeding) herbivores are molluscs (limpets, snails, and chitons) and crustaceans (amphipods, isopods, and crabs). Predators include sea stars and drilling whelks, which are permanent

TABLE 1
Types of Organisms That Occupy Emergent Surfaces in Different Zones on Rocky Shores

Zone	Primary producers	Sessile filter feeders	Mobile herbivores	Carnivores
supralittoral	lichens, microalgae, some winter ephemeral algae	none	terrestrial arthropods, occasional (often seasonal or nocturnal) limpets, snails, isopods, crabs	terrestrial arthropods
high intertidal zone	microalgae, ephemeral algae, desiccation-resistant algae (e.g. fucoids and turfs)	barnacles	limpets, snails, amphipods, isopods, insects, some crabs	whelks, nemertean worms, insects
mid-intertidal zone	turf-forming algae, branching algae, encrusting algae	mussels, barnacles, tunicates	limpets, snails, chitons, isopods, amphipods, crabs	sea Stars, various gastropods, nemertean worms, fish
low intertidal zone	large kelps, articulated and crustose coralline algae; filamentous, turf-forming, and branching algae; seagrasses	sponges, anemones, hydrocorals, bryozoans, tunicates, polychaete worms, barnacles, mussels	limpets, snails, chitons, abalone, sea urchins, isopods, amphipods, crabs	sea stars, various gastropods, crabs, lobsters, octopuses, fish

NOTE: Many predatory birds and a few mammals can forage at any shore level. In hot or unusually dry climates, foliose algae are uncommon except in the low intertidal zone. Herbivorous and carnivorous fish may forage extensively in the low and mid-intertidal zones in the tropics.

residents of this zone, and crustaceans such as crabs and lobsters, which may migrate into the mid-intertidal zone at high tide.

Low Intertidal Zone

The low shore is often typified by large kelps, which provide an extensive canopy and substantially modify conditions for organisms living in the subcanopy or on the rock itself. Coralline algae, which are prone to desiccation stress and thus less common at higher shore levels, often form extensive stands. A diverse group of noncalcified red algae may also be found at this shore level. Reduced emersion stress and prolonged feeding times favor sessile filter-feeding animals. The low intertidal zone is often a colorful mix of sponges, tunicates, bryozoans, and hydrocorals. Large limpets, chitons, abalone, and sea urchins are the predominant grazers. Ecologically important predators include sea stars, crabs, lobsters, gastropods, nudibranchs, and octopuses.

CAUSES OF ZONATION

Understanding the causes of intertidal zonation requires going beyond observation-based correlations between species distributions and environmental conditions. Distinguishing among competing causal hypotheses requires experimental tests. Although some early experiments were carried out in the first half of the twentieth century, the 1960s and 1970s witnessed a revolution in intertidal experimental ecology. Since then, our understanding of the precise mechanisms that control the upper and lower limits of intertidal species has improved greatly.

Factors That Set Upper Limits

The position of a species' upper limit may be influenced by a wide variety of physical and biological factors. It is often impossible (or at least impractical) to determine what specific aspect of the environment determines a species' upper limit, particularly as these environmental factors may interact with one another (for example, desiccation occurs more rapidly at higher temperatures). Furthermore, the upper limit of a species may be set by different factors at different places and times. Despite these complexities, physical stress consistently emerges as a major determinant of upper limits in the intertidal zone. Experiments that transplant organisms above their natural zone usually find that those individuals eventually perish from physical stress. Physical stress is especially important for young plants and animals, which are disproportionately vulnerable to heat, desiccation, and ultraviolet radiation. Stress does not have to be lethal, however, to determine

an upper limit. Physiological mechanisms for combating sublethal stress are energetically expensive, and mobile organisms may avoid high shore levels to avoid paying these physiological costs.

Although physical stress is thought to set a great number of upper limits, additional factors are also important. The upper limits of certain species may be set in the absence of physical stress if individuals cannot attain enough food or nutrients to grow and reproduce at higher shore levels. Larval habitat selection is also key, particularly for sessile species. A larva may "taste" the rock in several different places before deciding where to attach, using cues such as the presence of conspecifics and chemical signals from microscopic organisms. In theory, a species may fail to occur at a perfectly suitable shore level simply because larval rejection of the habitat precludes the establishment of an adult population there.

Finally, interactions among species may be involved in determining the position of upper limits. In Australia, ecologists have found that seaweeds are capable of living higher on the shore than they naturally occur. They are prevented from living at those shore levels by the grazing activities of herbivorous gastropods. At lower shore levels, the growth of the algae outpace rates of consumption, allowing algae to persist. Conversely, facilitation (e.g., by the presence of an algal canopy) allows some species to persist at shore levels above where they would live in the absence of the facilitator. These types of positive and negative interspecific interactions may be important in modifying the maximum vertical position of many species, including those whose upper limits are more noticeably controlled by abiotic factors such as desiccation and temperature.

Factors That Set Lower Limits

As with upper limits in the intertidal, lower limits may be set by a variety of different factors. Some types of algae and air-breathing invertebrates may fail to occur in low intertidal habitats because of limited light or limited oxygen availability. The lack of larval settlement may also be important in some cases. However, the majority of lower intertidal limits are determined in large part by interactions among species. The first demonstration of this now widely accepted generality came from pioneering research on barnacles in Scotland. There, the barnacle *Chthamalus* occurs in a distinct intertidal zone above the barnacle *Semibalanus*. *Chthamalus* was not prevented from living at lower shore levels by adverse physical conditions. Rather, any *Chthamalus* that settled lower on the shore were smothered or crushed by the faster-growing *Semibalanus*. When the competitively dominant barnacles were removed, the lower limit of *Chthamalus* moved down the shore.

In addition to competition, trophic interactions (herbivory and predation) define the lower limits of many species. In the Northeast Pacific, the mid-intertidal zone is dominated by beds of the mussel *Mytilus* (see Fig. 1). The sea star *Pisaster* is a voracious consumer of mussels and can often be found eating mussels along the lower limit of the mussel bed. When sea stars are experimentally removed, the lower limit of the mussel bed moves slowly but steadily down the shore through the growth and movement of adult mussels and the recruitment of new individuals into the existing bed. The location of some of the experimental work concerning *Pisaster*'s effect on the lower limit of mussels is still known as "the Glacier," so named for the inexorable downhill march of the mussel bed in the area from which sea stars had been removed.

Sharp versus Fuzzy Zonation Patterns

By definition, intertidal zonation is the occurrence of organisms in discrete zones along the vertical gradient of emersion time. The "discreteness" of the distribution of individuals of a particular species may be obvious or quite subtle. This depends in part on the natural history of the species. Many of the species with sharp vertical limits exhibit positive density dependence near the edges of their vertical distribution. More simply put, these organisms enhance the survival of their own species by reducing physical stress or vulnerability to consumers. For example, barnacles and mussels both provide cool, moist, and relatively safe nooks and crannies into which vulnerable conspecific larvae can settle and grow. This positive feedback means that mussel recruitment and survival are high in the mussel bed, and low both above and below it, where juvenile mussels are vulnerable to physical stress and predators, respectively. Many algal canopies operate in a similar way; dense canopies and turfs facilitate the survival of individual algal fronds, which dry out and die if their neighbors are removed.

Although sharp zonation patterns immediately catch an observer's eye, many species do not exhibit sharp upper or lower limits. A species' vertical limits may be fuzzy for several reasons. Most obviously, the sharpness of zonation patterns depends on species abundance. When a species is rare, it may be difficult to define its biological zone simply because there are not enough individuals to produce a sharp upper or lower limit. The facilitative positive feedback loops just described also require sufficient population densities to operate. For example, the high-intertidal barnacle *Balanus glandula* is very abundant in Washington

and Oregon, and the upper limit of their distribution is quite distinct. In California, where barnacle recruitment is limited as a result of upwelling and offshore larval transport, barnacle population densities are low, and it can be very difficult to define a discrete upper limit for *Balanus*. Of course, even abundant species, particularly when they are mobile, may have blurry zonation limits with abundance decreasing gradually to a few scattered outlying individuals.

Vertical limits may also be blurred when certain individuals are "missed" by the physical and biological forces that define zonation patterns. The existence of patchy favorable microhabitats, such as cracks and pits in the rock, can allow scattered individuals to persist well above the position of the upper limit on uniform bedrock. Similar fuzziness exists at lower species limits—predation intensity, for example, can vary even over small spatial scales, and predators may miss scattered individuals found below the lower limit of the majority of the prey population. At small spatial scales, lower limits are often more variable and less distinct than upper limits because their root causes (biological interactions) are also more variable at these scales.

ZONATION PATTERNS AMONG TIDEPOOLS

The patterns and processes described in the foregoing sections apply to emergent surfaces (i.e., those that are exposed to air during low tide). Organisms that live in tidepools do not suffer stresses related to emersion. However, tidepools at the highest shore levels can be isolated from the ocean for several days during neap tides, whereas low-shore pools are seldom isolated for more than a few hours. This difference in isolation time creates physical and chemical differences among tidepools along the vertical intertidal gradient. For example, during long periods of isolation, tidepools high on the shore can reach temperatures that are much higher, or lower, than the temperature of the ocean. Tidepools lower on the shore are not isolated from the ocean for long enough for temperature changes to become severe. High-shore pools are also subject to more dramatic swings in salinity as a result of evaporation, precipitation, and runoff. Additional differences are related to tidepool chemistry. Tidepool animals (and, during the night, seaweeds as well) take up oxygen from the water, and oxygen levels in tidepools can become uncomfortably low. In fact, some fish will even crawl out of high-shore tide pools during nighttime low tides in order to breathe. Tidepool residents also alter tidepool chemistry when they excrete their wastes. The ammonium and nitrate excreted by tidepool animals are excellent nutrients for algae but can alter tidepool pH and become dangerous at high concentrations. As with physical fluctuations such as temperature, these chemical fluctuations are more pronounced in tidepools that are isolated from the ocean for longer periods of time. These patterns in the physical and chemical environment, along with biological interactions, lead to unique sets of tidepool species at different shore levels.

VARIATION IN ZONATION PATTERNS

Variation in Space

Intertidal zonation patterns can vary at scales ranging from thousands of kilometers down to a few meters or less. At the largest spatial scales, zonation patterns vary with the biogeographic distribution of species and with regional climatic conditions. At smaller scales (meters to kilometers), the most common cause of spatial variability in biological zonation is variation in wave exposure, which refers to the force of waves breaking against the shore and therefore the degree of wave run-up (swash), splash, and spray. The mechanism by which waves elevate intertidal zones is straightforward. Upper limits are generally related to some aspect of being out of water, particularly extremes of temperature and desiccation stress. Waves act to reduce the amount of time that an organism sits "stranded" at low tide, waiting for the incoming water to cool it off and rehydrate it. Thus, at a site with a consistent 50 cm of swash, a point 200 cm above the low-tide line would be similar in many ways to a point 150 cm above the low-tide line in an area with no swash. In cases of extreme wave exposure, the upper limits of some intertidal species may be several meters higher than the level of the highest tides.

Several other variables can cause spatial variability in zonation patterns. Substrate orientation is one such variable, and upper limits are often higher in the shade or on north-facing surfaces (in the Northern Hemisphere) because these microhabitats are cooler. Light intensity is another important variable, and the zonation of the endosymbiotic algae that live within green sea anemones in the Northeast Pacific changes from sunlit areas into dim sea caves, and from relatively bright, low-latitude sites in California to less bright, higher-latitude sites such as Washington State. The timing of low tides is also critical. In California, where summertime low tides occur around dawn, low intertidal thermal and desiccation stress are much lower than on the Atlantic coasts of New England or Argentina, where summertime low tides occur in the middle of the day. Other variables that are known to influence zonation patterns include salinity, rock type, tidal currents, sand inundation, and ice scour.

With regard to "horizontal" variation in patterns of vertical zonation, it is important to remember that a species' upper and lower limits are set by different factors, and these factors may not shift in unison through space or time. Thus, zones can be wider or narrower from place to place. In some locations, a species' upper limit may meet its lower limit; in effect, the vertical range is zero and the species cannot persist. This "squeeze effect" is thought to prevent certain species from occupying either warmer or cooler stretches of shoreline, depending on their thermal tolerance relative to that of their consumers and competitors.

Shifts through Time

Just as zonation patterns are not constant in space, they are also variable in time. In Hong Kong, the limpet *Cellana* migrates up and down the shore with the tide so as to always remain at the water's edge. Except when stranded by a rapidly retreating tide, the *Cellana* zone is always on the move. More commonly, temporal zonation shifts are observed in strong seasonal patterns in the vertical position of mobile species and those sessile species with rapid generation times (such as ephemeral seaweeds). Along the coast of West Africa, intertidal shores feature a belt of barnacles above a belt of competitively dominant seaweed. During the dry season, the upper limit of the algae is low on the shore, and the width of the barnacle zone is wide. During the rainy season, the reduction in physiological stress allows the upper limit of the algae to shift upward, with a corresponding decrease in the width of the barnacle zone. At the end of the rainy season, the upper limit of the seaweed shifts back downshore, and the barnacles recover their former position in the space vacated by the algae.

With the increased attention now being given to the impacts of regional climate cycles (e.g., the El Niño/Southern Oscillation and the North Atlantic Oscillation) and global warming, long-term changes in marine ecosystems are attracting great interest. At present, the long-term datasets needed to demonstrate climate impacts on zonation are rare or absent. However, two general predictions regarding the effects of climate change can be made. First, due to ice melt and thermal expansion of the oceans, global sea level is rising by approximately 2 mm per year. Intertidal zones are predicted to shift upwards in response to rising sea level. (Earthquakes in Chile and Alaska, which raised or lowered coastlines by several meters, have provided dramatic examples of the effects of changes in relative sea level.) Second, rising temperatures should act to compress the vertical zones of many species. As temperatures rise, heat-sensitive intertidal organisms will not be able to spend as much time out of the water and will thus be forced to live lower on the shore (relative to the position of the low-tide line). Because upper and lower limits can change independently, certain species may greatly expand their vertical ranges while others may go locally extinct if their vertical zone is reduced to zero.

SEE ALSO THE FOLLOWING ARTICLES

Competition / Desiccation Stress / Predation / Recruitment / Ultraviolet Stress / Wave Exposure

FURTHER READING

Knox, G. A. 2001. *The ecology of seashores.* Boca Raton, FL: CRC Press.

Kozloff, E. N. 1996. *Seashore life of the northern Pacific coast.* Seattle: University of Washington Press.

Lewis, J. R. 1964. *The ecology of rocky shores.* London: English Universities Press Ltd.

Menge, B. A., and G. M. Branch. 2001. Rocky intertidal communities, in *Marine community ecology.* M. D. Bertness, S. D. Gaines, and M. E. Hay, eds. Sinauer Associates, Inc., Sunderland, Massachusetts.

Raffaelli, D., and S. Hawkins. 1996. *Intertidal ecology.* London: Chapman & Hall.

Ricketts, E. F. 1985. *Between Pacific tides.* Stanford, CA: Stanford University Press.

Stephenson, T. A., and A. Stephenson. 1972. *Life between tidemarks on rocky shores.* San Francisco: WH Freeman.

The glossary that follows defines over 900 specialized terms that appear in the text of this encyclopedia. Included are a number of terms that may be familiar to the lay reader in their common sense but that have a distinctive meaning within this field of study. Definitions have been provided by the encyclopedia authors so that all of these terms can be understood in the context of the articles in which they occur.

abalone A large edible marine gastropod of the genus *Haliotis,* having an ear-shaped shell with a row of holes along the outer edge and a large muscular foot used to hold the animal tightly onto a rock.

abiotic Due to physical processes; not due to biological processes. An ABIOTIC FACTOR is any nonbiological environmental condition, such as temperature, salinity, pH, or desiccation, as opposed to biotic (biological) factors such as competition or predation.

aboral surface The top (dorsal) side of a sea star; the side opposite the sea star's mouth.

absorptance For algae, the fraction of solar irradiance absorbed.

acceleration The rate at which velocity changes through time, typically expressed as meters per seconds squared (m/s^2). The change can be in either the direction or the magnitude of velocity.

acceleration reaction A force, proportional to volume, exerted on an object along the axis of flow, caused by the relative acceleration of the body and fluid. Also, ACCELERATIVE FORCE.

acclimation The ability of an organism to modify its environmental (e.g., thermal) optimum or tolerance limits when exposed to a change in its environment; as measured in controlled studies in the laboratory, where only a single factor such as temperature is varied.

When observed in the field (e.g., in response to seasonal changes in the environment), this ability is termed ACCLIMATIZATION.

acidity A term describing the hydrogen ion concentration of a solution when the pH is below 7 (the pH of neutrality).

acoelomate Lacking a fluid-filled body cavity between the outer body wall and the gut.

acoustic Doppler current profiler (ADCP) An instrument that uses the Doppler shift in sound waves to measure current velocities throughout the water column.

acrosome reaction Release from the sperm cell of a secretory granule containing sperm proteins that consequently form long, narrow extensions, which can bind to sperm receptors on an egg.

actinula The specialized planktonic larva of some Hydrozoa.

advection The horizontal movement of an object (e.g., seawater or larvae).

aeolid A nudibranch belonging to the suborder Aeolidoidea, having cerata on the back and lacking a rhinophoral sheath.

algae Any of various chiefly aquatic, eukaryotic, photosynthetic organisms, ranging in size from single-celled forms to the giant kelp.

aliasing The erroneous appearance of a low-frequency signal, caused by too low a sampling rate.

alkalinity A term describing the hydrogen ion concentration of a solution when the pH is above 7 (the pH of neutrality). Oceanographers define seawater alkalinity as the number of milliequivalents of H^+ ions required to titrate one kilogram of seawater to the bicarbonate equivalence point ($\sim$pH 4.5).

allele A version of a gene that codes for a particular trait; a specific DNA sequence occupying a given locus (position) on a chromosome.

allelopathy The use of chemical compounds to defend space against interspecific competitors (e.g., by inhibiting their growth or larval settlement).

alternative states The ability of more than one type of community to develop and be maintained in the same place at different times or in adjacent, physically similar habitats at the same time. Also, MULTIPLE COMMUNITY OUTCOMES.

altricial Of young animals, born or hatched in a state that is not highly developed. In contrast to marine mammals, most terrestrial carnivores are born in an altricial state.

ambient light The light existing in a subject area, prior to any added by a photographer.

ambulacral groove A groove on the underside of each limb of an echinoderm, through which the tube feet of the water vascular system extend.

amino acids Organic compounds containing amino (NH_2) and carboxyl (COOH) groups. Amino acids are linked together by peptide bonds to form proteins.

amoebocyte A morphologically and functionally diverse group of blood cells that typically exhibit amoeboid locomotion. In tunicates, some amoebocytes carry vanadium granules, while others may be used for storage, phagocytosis, and defense against pathogens.

amplitude Of a sinusoidal wave, half the vertical distance between the crest and trough. Wave height is twice the wave amplitude.

ampullae (*singular* **ampulla**) Internal, muscular bulbs that contract to elongate the tube feet of sea stars.

anabolic metabolism The processes that consume energy in order to generate new molecules (proteins, lipids, nucleic acids, etc.) necessary for growth of an organism.

anaerobic metabolism Cellular processes of an organism carried out in the absence of molecular oxygen.

analog Expressing information in the form of a continuously variable voltage signal.

analyte A substance or chemical constituent being measured in an analytical procedure.

anchor ice Sea ice that forms on the sea floor; can remove organisms and material when its buoyancy exceeds its attachment strength.

angiosperms A major group of terrestrially derived flowering plants in which the ovules, or young seeds, are enclosed within the ovary to produce a true fruit.

angle of incidence The angle between wave crests and the shoreline. Waves approaching shore with their crests parallel to shore have 0 angle of incidence.

anhydrobiosis Literally, life without water; a cryptobiotic (i.e., ametabolic) state promoted by extreme dehydration of an organism.

Annelida A phylum of bilaterally symmetrical, segmented worms that inhabit marine, freshwater, and terrestrial habitats.

antagonistic interaction The inhibition of one species by the action of another.

antenna pigments Photosynthetic pigment molecules (chlorophylls, carotenoids, and phycobilins) that absorb and transfer light energy to photosynthetic reaction centers.

antennules The first pair of antennae on a crustacean.

anthropogenic Induced or altered by the presence and activities of humans, rather than occurring in the natural environment.

anticyclonic flow Flow having a rotation about the local vertical axis that is in the opposite sense as that of the earth's rotation; when viewed from above, clockwise in the Northern Hemisphere and counterclockwise in the Southern Hemisphere.

antifouling treatment A mechanical, physical, or chemical treatment to reduce biofouling of instruments or other equipment during long periods of submersion.

apical On or toward the apex (tip) of a branch.

apomixis A process of reproduction without gamete production or fertilization.

aposematic Signaling a warning or alert that the emitter of the signal is unprofitable or dangerous to the receiving aggressor.

apparent competition Patterns of resource use (including spatial distributions) that are not caused by competition but that mimic its effects.

aquarist A person who maintains aquariums.

aragonite A mineral form of calcium carbonate with a distinctive crystal structure that is deposited by many corals and coralline algae.

Aristotle's lantern A structure of calcite (calcium carbonate) parts at the mouth of a sea urchin, including the ever-growing teeth as well as the muscles associated with orienting and moving the teeth.

artisanal commercial fishers A subgroup of small-scale fishers operating in shallow inshore and close offshore waters, using small- and medium-sized boats, on expeditions normally lasting more than one day, with more mechanized fishing gear than used by artisanal fishers. Their primary objective is sale of resources.

artisanal fishers A subgroup of small-scale fishers operating mainly in inshore waters, usually less than 2 miles from shoreline, using small boats (wooden or fiberglass) equipped with oars, and generally fishing only during the daytime and at short distances from the base port. The objective of these locally oriented fisheries in developing countries is sale of resources, with part of the catch used for the fishers' own consumption or shared among kin.

Aschelminthes A likely polyphyletic group consisting of Nematoda, Nematomorpha, Gastrotricha, Loricifera, Kinorhyncha, and Priapulida, the members of which have a body surface covered by a cuticle, internal organs that lie in a body cavity (pseudocoel), a digestive tract with terminal mouth and ventral anus, and which are dioecious, with no asexual reproduction.

asexual Of reproductive processes, not involving sex.

associational defense A type of species interaction that occurs when organisms are protected from their enemies when living in association with other organisms.

atrium The chamber, encased by the body wall and tunic, that surrounds the phanryngeal basket of a tunicate; filtered water, wastes, gametes, and larvae exit the atrial space through an exhalent (or atrial) siphon.

autotomy The separation of a body part from the rest of a biological specimen, or the breakup of a specimen into two or more portions; an unfortunate outcome when a living specimen is placed rapidly and directly in fixative or preservative without relaxation first.

autotroph An organism capable of autotrophy.

autotrophy The ability of an organism to nourish itself by making organic material from simple inorganic compounds.

backwash The return flow of water down the beach face after a broken wave has washed up the beach.

bag limit The maximum limit, in number or amount, of a given group of bird, mammal, fish, amphibian, invertebrate, or plant species that may be lawfully taken by any one person during a specified period of time.

baroclinicity The state of stratification in a fluid in which surfaces of constant pressure intersect surfaces of constant density.

barrier island A long narrow, wave-built island consisting of sand that is separated from the mainland by a bay or lagoon.

bauplan Pattern of general morphology of a taxon. (From a German word for *blueprint*.)

benthic 1. Referring to, associated with, or attached to the sea floor. 2. Living on or in the sea floor, rather than floating in water.

benthic microphytes Microscopic primary producers (autotrophs), often comprising benthic diatom taxa that adhere to both the substratum and other benthic organisms.

benthos The bottom of a sea, lake, or deep river and the animals and plants that live there.

berm The high, wide part of a summer sandy beach, which provides recreational area for beachgoers.

bifurcated Divided into two branches.

biodiversity The number of species in a defined area; sometimes used interchangeably with *diversity* and *species richness.*

bioerosion The removal of rock material by various biological processes, such as the grinding or scraping of the rock surface during feeding by marine invertebrates.

biofilm The thin film of bacteria, cyanobacteria, microalgae, and microscopic settling stages of macroalgae and invertebrates that covers the rock surface.

biogenous Of beach material, derived from organisms, such as beach sand consisting of broken-up coral or shell material.

bioirrigation The mixing of oxygenated waters into sediments via echiuran burrow irrigation.

biological interaction Any process by which animals and/or plants affect each other; e.g., competition or predation.

biomass The mass of living organisms in a population or area.

biomimetic Replicating an organism's characteristics (e.g., thermal characteristics).

bioremediation The use of organisms, usually microorganisms but also plants, to break down pollutants in an area.

biotic Due to biological processes; not due to physical processes.

bioturbation The reworking of sediments during echiuran burrow construction and feeding.

blastocoel In a developing embryo, the fluid-filled space that lies between the outer cell layer and the embryonic gut.

blastopore In a developing embryo, the primary opening of the embryonic gut cavity.

blood The circulatory fluid in closed circulatory systems, consisting of an inert plasma matrix (primarily water) with dissolved organic and inorganic substances and specialized living cells, including those for oxygen transport.

blooms Large accumulations of phytoplankton biomass caused by their rapid growth.

bottleneck An evolutionary event in which a significant percentage of a population or species dies or is otherwise prevented from reproducing; can result in a reduction in genetic diversity in the subsequent population, since all individuals share a small group of common ancestors (those present at the bottleneck).

Bouin's solution A highly corrosive solution, bright yellow in color (made from 15 parts saturated aqueous picric acid, 5 parts 100% formalin, and 1 part glacial acetic acid), that tends to harden soft tissue and dissolve hard parts of calcium carbonate; used as an initial fixative for histological work or for some dissections.

branchial plume A secondary gill, usually located posteriorly on dorid nudibranchs.

breakers or breaking waves Water waves that have steepened so much that their crests have fallen forward and water falls down in front of them.

brightness histogram In photography, a graph of all the brightness levels in a scene, ranging from black to white; found on digital cameras and in editing programs.

broadcast spawning Release of eggs and sperm from the body cavity to the external environment.

buffered solution A solution whose acidity has been neutralized by the addition of a basic substance (a chemical buffer) so as to make it resistant to pH changes caused by the additions of small amounts of acid or base.

buoyancy 1. The net vertical force on an object surrounded by fluid; positive (upward) if the density of the fluid is greater than the density of the object, negative (downward) if the density of the fluid is less than that of the object, and neutral (no force) if the density of the fluid equals that of the object. **2.** The aggregate of all mechanisms used to determine an organism's location within the water column.

byssal threads Fibers produced by mussels that allow them to anchor to the substratum. Also, BYSSUS.

byssate Of bivalve molluscs, attaching themselves to the substrate by a byssus, proteineous threads secreted by the foot.

calcareous ooze Marine sediments with >30% calcium carbonate.

calcareous stone Sedimentary rock composed mainly of calcium carbonate.

calcified Incorporating or coated with calcium carbonate (limestone).

calyx The cuplike open structure, usually bearing tentacles and with a central mouth, attached to the narrower and often elongate stalk (peduncle) of a stauromedusa.

camouflage Concealment by disguise, coloration, odor, or tactile properties.

capital strategist An organism that can store food energy, enabling it to survive and reproduce without continual feeding.

carapace The calcified shell covering the cephalothorax of higher crustacea.

carbonate saturation state A measure of the potential for carbonate to precipitate into solid salts, primarily by complexation with calcium ions. At the sea surface, the concentrations of calcium and carbonate are generally supersaturated, which helps promote the precipitation of calcium carbonate.

carcinogens Cancer-causing agents.

cardiovascular system The circulatory system, consisting of a heart, vessels and fluid. Its primary role is for transport of substances to and from the tissues.

carotenoid A fat-soluble yellow, orange, or red accessory pigment that absorbs light energy and transfers it to the reaction center of chlorophyll *a* for use in photosynthesis.

carpogonium The gamete produced by red algal female gametophytes.

carposporophyte A third, reduced diploid phase that lives on the female haploid blade of most red algae and produces carpospores when reproductively mature.

catch connective tissue Mechanically active collagenous tissue in echinoderms that stiffens or softens under nervous control.

catchment A depression above sea level in which rainwater accumulates and from which it flows to the sea.

cellular respiration The series of metabolic processes by which living cells produce energy through the oxidation of organic substances.

cement gland A gland found pycnogonids (males only) that produces a sticky substance used to glue the eggs to the ovigers; usually located on the femur or femur and tibia 1.

central disk The middle portion of a sea star, from which the arms extend.

centrifugal force The outwardly directed reaction force that accompanies centripetal (center-directed) force.

centripetal force The force directed toward the center of rotation when a mass moves in a circle.

Cephalopoda The class within the phylum Mollusca to which octopuses and squid belong.

cephalothorax The anterior half of the body of a decapod crustacean, composed of the fused head and thorax.

cerata (*singular* **ceras**) Fingerlike respiratory organs containing branches of the digestive gland and located on the dorsal surface of certain aeolid nudibranchs; may contain sequestered pigments or chemicals sequestered from prey.

chart datum The tidal datum that is chosen as the tide-table zero elevation and used for nautical charting (e.g., mean lower low water); varies from country to country depending on the tidal datum used.

chelicerae (*singular* **chelicera**) The first pair of anterior appendages in arthropods; in pycnogonids they are often chelate, but they can be variously modified or missing altogether; in older literature, often called CHELIFORES.

chelipeds Appendages in crustaceans that terminate in chelae (pincers).

chemical gradient A vector quantity expressing the spatial rate that a concentration changes and the direction of maximum change. It has units of mass length^{-4}.

chemoautotrophic Using energy derived from oxidation of inorganic carbon sources.

chemotaxis Directed movement with respect to an increasing chemical concentration.

chemotroph An organism that derives its energy from breaking chemical bonds.

chert Very hard sedimentary rock composed mainly of crushed siliceous diatoms.

chimera An organism that comprises two or more genotypes. In colonial ascidians, chimeras form when two genetically distinct colonies fuse by their tunics and circulatory systems.

chlorophyll A fat-soluble green pigment. Various chlorophylls exist (chlorophyll *a*, *b*, *c*, etc.), but all seaweeds and true plants contain chlorophyll *a*.

Chlorophyta A division (phylum) of algae, the "green algae."

choanocyte A cell with a collar and flagellum, whose beat generates the feeding current through a sponge.

chromatophore A color-changing cell in the skin of an octopus or squid.

chromophore A light-reactive chemical whose conformational change after photon strike is the first step of vision.

chronobiology The scientific study of biological rhythms and cyclic behavior.

circadian Of a biological rhythm, having a period that approximates one solar day (~24 h) and persists in the absence of environmental cues.

circannual Of a biological rhythm, having a period of approximately one year; usually associated with seasonal migration, reproduction, or growth.

circatidal Of a biological rhythm, about one tidal cycle (~12.4 h) long.

cirral net The fan-shaped feeding structure of a barnacle.

cirri (*singular* **cirrus**) **1.** Small flexible appendages used by barnacles to collect food from the water column. **2.** Fleshy, antenna-like projections often present on the foreheads of blennies.

climate The average condition of the weather at a place over a period of years, measured by temperature, wind velocity, and precipitation.

clonal Referring to the division of one individual into two or more physiologically separated individuals

(**clonemates**), with nearly identical genetic composition, which as a group form a clone.

clone 1. To produce physically separate, genetically identical ramets, primarily by fission, budding, and fragmentation. **2.** The collection of all genetically identical ramets descended from a sexually produced zygote.

clutch size The maximum number of eggs present in an individual nest during the breeding season.

Cnidaria A phylum of radially symmetrical metazoans with two principal cell layers (ectoderm and endoderm) separated by an acellular or partly cellular mesenchyme; includes the sea anemones, corals, jellyfish, hydroids, and sea fans.

cnidocyst A microscopic cell organelle present in great numbers and often in distinct patterns on the tentacles, mouth, and body surface of a jellyfish, hydroid, or sea anemone, comprising a capsule and a single-use everting thread; used in feeding (to capture prey) or for defense. Also, NEMATOCYST.

cnidocyte In cnidarians, a cell in which a nematocyst, spirocyst, or ptychocyst is produced and housed.

cnidosac A small sac at the distal end of a ceras in aeolid nudibranchs that stores unexploded nematocysts recovered from cnidarian prey.

coding region The portion of DNA that is transcribed to RNA and translated into proteins. There may be multiple coding regions within a gene, separated by sections of DNA that is not transcribed to RNA (known as introns).

coelenteron The internal body cavity of a cnidarian, connected to the outside through the mouth.

coelom A secondary body cavity that arises wholly within the mesodermal layer; a fluid-filled space between the outer body wall and gut that is completely lined with mesodermal epithelium.

coenenchyme Tissue that surrounds the polyps of octocorals and usually contains sclerites and a network of gastrovascular canals.

coenosarc Living tissue inside the diploblastic cnidarian, joining individual hydranths or polyps together and integrating these individual units into a colony.

coldspots Areas exhibiting low species richness.

colonial Referring to the division of one individual into two or more physiologically connected (e.g., via a common gastrovascular system) individuals that live within a colony. Individuals within the colony may be called many names (e.g., zooids, polyps) depending on the taxonomic group.

co-management A management system in which there is sharing of responsibilities regarding management of resources by government and users (stakeholders). Forms vary in degree of user authority from consultative co-management to delegated co-management.

commensal 1. Describing or referring to a symbiotic relationship in which one species benefits while the other is unaffected. 2. An organism living on or within another, but not causing injury to the host.

commensalism A relationship between organisms in which one lives on the other to its benefit (often in the form of living space or food) without inflicting harm or providing benefits to the other species.

common cloaca The chamber formed by the coalescence of the exhalent siphons or atria of multiple zooids in colonial tunicates.

community structure A term referring to the distribution and abundance of co-occurring species.

compatible solutes Organic osmolytes (e.g., free amino acids, polyols, and methylamines) that are accumulated within cells during hyperosmotic stress. These compounds increase cellular osmolality without impeding enzymatic function.

compensatory mortality A mechanism hypothesized to maintain local diversity whereby disturbance-induced mortality of a late successional dominant precludes exclusion of competitively inferior species that are more resistant to the disturbance.

competent Of a larva, physiologically capable of undergoing metamorphosis to the juvenile stage.

competition A relationship between two individuals or species striving for the same limited resource (e.g., food or space).

competitive exclusion The inability of two species to coexist when their use of shared, limited resources overlaps sufficiently, causing one species to exclude the other from the habitat.

conchiolin The protein material secreted by the mantle that bonds the crystals of calcium carbonate or nacre in shells of gastropods and other molluscs; usually brownish in color and laid down in response to an irritant or damage to the shell.

conduction The process by which heat energy is transferred directly through a stationary material.

conductivity The capacity of seawater to conduct an electrical charge within a given volume of water.

conjugation The exchange of genetic material through temporary connections between two cells.

connectivity A description of the ability of organisms from one geographic region to move to, and interact with other organisms from, different geographic regions.

conspecific Of individuals, belonging to the same species.

consultative co-management A form of co-management in which the state creates mechanisms of dialogue and informs users of management decisions.

contraction Shrinkage or reduction of size of a specimen of an organism without introversion.

contramensalism An interaction between two organisms in which the outcome of the encounter results in a positive effect on the fitness of the first organism but a negative effect on the fitness of the second.

convection The process by which heat energy is transferred between a surface or organism and a gas or liquid by the circulation of fluid (e.g., air) from one region to another.

coralline An alga in the order Corallinales whose cell walls contain calcium carbonate, making the alga hard; generally, pink in color.

Coriolis effect An apparent deflection of a moving object as seen from a rotating frame of reference; because of this effect, ocean currents deflect.

Coriolis force The force exerted on a moving body due to the rotation of the earth, which is responsible for turning the wind to the right in the Northern Hemisphere and to the left in the Southern Hemisphere.

Coriolis parameter Twice the vertical component of the earth's angular velocity about the local vertical at a given latitude.

coronate larva A nonshelled, nonfeeding larva produced by bryozoans. The coronate larvae of gymnolaemates are different from the coronate larvae produced by stenolaemates.

correlation The extent to which two variables are related; i.e., variation in one is accompanied by corresponding variation in the other.

cosmopolitan species A species found throughout the world, often as the result of human agency.

Couette cell A device for observing the effects of turbulence in fluids, consisting of two concentric cylinders. One or both cylinders rotate, exposing fluid (and suspended particles such as gametes) between the cylinders to predictable velocity gradients.

crest The topmost point of a water wave.

crosslinks Chemical or physical linkages joining two polymers together.

crust An alga, cyanobacterium, or lichen whose overall growth form is flat, with one whole side adherent to the substratum, and that grows as a laterally expanding disk.

Crustacea A subphylum of arthropods (animals with jointed chitinous exoskeletons) in which individuals bear two pairs of antennae and the exoskeleton (cuticle) is often calcified.

crustose Growing as a crust, prostrate along the substratum.

cryptic Of animals, hidden from potential predators by form or coloring.

crystalline rock Rock such as granite or marble, that is formed by crystallization or growth of the individual constituent minerals from a fluid.

CTD An instrument used to measure and record water column profiles of *c*onductivity, *t*emperature, and *d*epth.

ctenidium (*plural* **ctenidia**) The type of gill found in most molluscs, consisting of a central axis from which individual plates or filaments (lamellae) arise; in bivalves, usually bipectinate (having plates on opposing sides of the axis) and used for respiration and feeding; in some primitive bivalves, used only for respiration; in other molluscan classes may be similar or modified.

cultivated Of macroalgae, farmed in open-ocean or land-based operations.

curl Variations of velocity in the cross-stream direction such that a water paddle placed in the flow would spin.

Cuverian tubules Tubules that are attached to the cloaca of some tropical sea cucumbers and become sticky when extruded, presumably as a defense against predation.

cyanobacteria A group of photosynthetic bacteria, formerly referred to as "blue-green algae." They are mostly unicellular, although they can grow in chains or colonies, making them large enough to see with the unaided eye.

cyclonic flow Flow having a rotation about the local vertical axis that is in the same sense as that of the earth's rotation, when viewed from above, counterclockwise in the Northern Hemisphere and clockwise in the Southern Hemisphere.

cyphonautes larva A shelled, planktotrophic larva produced by a few gymnolaemate bryozoans, including members of the genera *Membranipora* and *Electra*.

cyprid The final (nonfeeding) planktonic larval stage of barnacles, which undergoes the process of settlement.

cystid The nonprotrusible portion of the body wall of a bryozoan, consisting of both cellular and noncellular components.

cytoplasm The living matter of a cell, excluding the nucleus.

Decapoda A crustacean group that includes shrimps, crayfishes, lobsters, and crabs and that is defined by covered gills and the division of thoracic limbs into three pairs of food-handling limbs (maxillipeds) and five other limb pairs. (Literally, having ten feet.)

defoliation The removal of leaves from a plant or, more broadly, the removal of vegetation.

deglaciate To remove glacier ice covering an area of land.

delegated co-management A form of co-management in which the government transfers management decisions to user groups, who then inform the government of decisions made at the local level.

demersal Living on or near the sea floor.

demography The study of characteristics of population size, including reproductive dynamics such as birth, death, migration, and overall geographic distribution, that may cause population size to increase or decline.

denaturation A change in the folded structure of proteins and other biomolecules caused by heat, acids, bases, or detergents. Denaturation of enzymes causes them to lose their catalytic activity.

denitrification The process by which nitrate is converted to gaseous NO, N_2O, or N_2 end products; this process represents a loss of nutrient nitrogen from an ecosystem.

density 1. The mass per unit volume of a fluid or solid. 2. The number of individuals per unit area.

density-driven circulation Water movement that is based on differing densities of saltwater.

desiccation Loss of water from the surfaces of marine organisms caused by exposure to air. Desiccation results in physiological stress and is made more severe by factors that may increase the rate of water loss, such as temperature and air movement.

determinate cleavage A type of embryonic cleavage in which the fates of cells are determined at a very early stage of cell division; typical of many protostome phyla.

detritivore An organism that feeds on detritus. Also, **deposit feeder.**

detritus Organic waste material from decomposing dead plants or animals.

dew point The temperature at which water vapor condenses; the dew point increases with increasing vapor concentration in the air.

diffusion 1. The net spontaneous transport of molecules along a concentration gradient within a fluid such as air or water. 2. Mixing caused by eddies that vary in size from the smallest scales (Kolmogorov scale) to subtropical gyres.

diffusion coefficient A constant used to describe the rate of diffusion, expressed in units of length^2 time^{-1}. Also, DIFFUSIVITY.

digital Using a numerical representation of a signal that is repeatedly sampled at a specific interval.

dimer 1. A molecule composed of two similar subunits (monomers) linked together. 2. In genetic analysis, two bases in a strand of DNA that occur together in a repeating sequence (e.g., CACACACA comprises a CA dimer repeated four times).

dioecious Possessing either male or female sex organs, but not both.

diogenid A crab of the family Diogenidae.

Diogenidae A large family of left-handed hermit crabs.

diploblastic Describing or referring to a body plan that has only two well-defined tissue layers (ectoderm and endoderm); a middle layer of cells and connective tissue may be present but is not sufficiently well-organized to be considered a true tissue.

diploid Having two paired (homologous) sets of chromosomes in each cell.

dipole A molecule with a positive and a negative end, the net electrical charges being separated by some distance.

disease A set of symptoms associated with pathology to an organ or system of an individual.

dissimilarity A measure of how much difference there is between two or more replicates in samples of a set of measurements of different variables (e.g., numbers of many species).

dissipation The conversion of mechanical energy into heat by viscous effects associated with very small-scale motions.

disturbance The injury, displacement, or mortality of organisms caused by an external physical or biological agent (other than predation).

diurnal Having a recurrence period of about one day.

divergence Variation in velocity in the alongstream direction so as to create a deficit or excess of water at a point.

diversity An index that combines the number of species in a community (richness) and the equitability of their abundances (evenness). Also, SPECIES DIVERSITY.

division One of the major groups used by botanists to classify organisms; the equivalent of phyla used to classify animals.

doldrums A belt of low atmospheric pressure and light and variable winds near the equator.

dominance hierarchy A social order of dominance sustained by aggressive or other behavior patterns.

domoic acid A neurotoxin produced by several species of marine diatoms of the genus *Pseudo-nitzschia*.

dorid A nudibranch belonging to the suborder Doridoidea, which have secondary branchial gills.

dorsum The back or upper surface of an animal.

drag A force (proportional to exposed area) that resists the velocity of a body relative to water or other fluid.

drag coefficient A dimensionless measure of the resistance of a body to velocity relative to a fluid medium.

drift algae Pieces of seaweed that have become detached and drift with ocean currents.

drogue An object used as a standard shape on which drag can act.

durophagous Feeding on hard-shelled prey such as clams, snails, and sea urchins.

dynamic viscosity SEE VISCOSITY.

ecological model A mathematical or conceptual construct that can explain an observation in nature.

ecosystem An ecological system comprising an assemblage of organisms living together with their environment.

ecosystem functioning The transformation and flux of energy and materials in ecosystems. Because organisms play important roles in mediating biogeochemical processes, there is growing realization that diversity influences ecosystem functioning.

ecotone A boundary between distinct ecological communities.

ectoderm The outermost tissue layer; gives rise to the epidermis, outer cuticle (if present), nerve tissue, and some types of excretory organs.

ectosymbiosis A symbiotic relationship in which only external surfaces are in contact.

ectotherm An organism that regulates its body temperature largely by exchanging heat with its surroundings.

eddy Water flowing contrary or in circular motions to the main current.

eddy diffusion The dispersal of material in a turbulent environment.

effective population size A simplifying theoretical descriptor of a population, representing the number of individuals in a population that are contributing heritable information to the next generation; the relationship between effective population size (symbolized N_e) and the actual census size of a population may depend on gender bias, population age structure, overall variance among individuals in reproductive success, and the stability of the population.

egg jelly An extracellular matrix containing complex carbohydrates and glycoproteins. In the egg jelly of sea urchins, species-specific factors activate sperm.

einstein A unit of measure of irradiance equal to one mole of photons, regardless of their frequency, designated as einsteins per square meter per second.

Ekman layer 1. The layer near the sea surface that is directly forced by winds. 2. The benthic layer directly forced by turbulence generated at the seafloor by deep currents.

electromagnetic radiation Self-propagating waves in space with electric and magnetic components that oscillate at right angles to each other and to the direction of propagation and that carry energy and momentum that can be transferred to matter.

electronegative Tending to attract electrons to form a chemical bond.

El Niño A weakening in the strength of global ocean current patterns that usually causes lower water temperatures on the western sides of ocean basins and warmer water temperatures on the eastern sides of ocean basins. (Originally named in the late 1890s by Peruvian fishermen, who observed the changes in currents and weather associated with a small, warm coastal current that intensified during Christmas time and so named it "The Little Boy" for the Christ Child.)

El Niño/Southern Oscillation (ENSO) A large-scale oceanic-atmospheric perturbation that has global consequences on climate; during an ENSO event, the easterly trade winds weaken because of a reversal in the pressure system between the western Indonesia high and the Eastern Pacific low system, and as a result, warmer water flows from the western to the eastern Pacific, deepening the thermocline and reducing upwelling.

emergent Out of the water and exposed to air; not submerged.

emersion The condition of being out of water (e.g., at low tide).

emissivity The proportion of potential infrared energy actually emitted by an object.

endemics Animals or plants restricted or peculiar to a region, locality, or place.

endobiont A symbiont living within a host's body.

endogenous clock An internal free-running clock that initiates species to perform functions at a certain time, in a rhythmic fashion.

endopinacocytes Elongate, squamous (flat) cells that line the internal canals of sponges. Also, ENDOPINACODERM.

endosymbiosis A symbiotic relationship in which external surfaces of a symbiont contact internal surfaces (within tissues or cells) of a host.

energy budget The balance of absorptance, transmittance, and reflectance of irradiance as a function of wavelength for an alga; these values sum to 1.0.

ENSO El Niño/Southern Oscillation.

entraining cue A stimulus associated with an environmental cycle that initiates or rephases a biological rhythm.

ephemeral 1. Existing or continuing for a short time only. **2.** Of algae, short-lived (often less than a year).

ephyra (*plural* **ephyrae**) The juvenile medusa of Scyphozoa, produced through strobilation of a scyphistoma.

epibenthic Living attached to the sea floor.

epibiont An organism that grows directly on the surface of another organism (the host).

epibiontic Living on or attached to another organism, but without benefit or detriment to the host organism.

epibiotic Living on living organisms.

epifluorescent microscopy The use of special fluorescent dyes to tag molecules of interest (e.g., DNA) in a sample, which is then illuminated with ultraviolet light to make the dyes fluoresce brightly, revealing the target molecules' existence and identity under the microscope.

epiphytic Growing on plants.

EPS Exopolymeric substances, a gel-like matrix of sugars, amino acids, and other organic compounds that forms a structured polymer in the microbial mat community and that is produced by cyanobacteria and heterotrophic bacteria and mediates calcium carbonate precipitation.

esthetes or **aesthetes** Small sensory organs that occur in large numbers in the upper (tegmentum) shell layer of all chiton valves, known to have photosensory capabilities in a broad range of chiton genera but also potentially performing mechanosensory, chemosensory, or secretory roles.

estivation A physiological state of dormancy induced by exposure to desiccating conditions (xeric conditions, high temperature, or both); sometimes considered a state of torpor in summer.

ethanol A colorless and flammable liquid (C_2H_5OH), used to preserve once-living tissue; used for long-term storage, most material for dissections, and genetic/molecular work. Also, ETHYL ALCOHOL, GRAIN ALCOHOL.

eukaryote Any organism with cells containing membrane-bound vesicles including nuclei and mitochondria. All members of the protist, fungi, plant, and animal kingdoms are eukaryotes.

eukaryotic Having cells containing membrane-bound nuclei and organelles.

eulamellibranch A type of ctenidium in which individual lamellae are elongated and folded, with this position fixed by tissue junctions within and between adjacent lamellae.

eulittoral [zone] That part of a shore that is between the low- (at low tide) and high-water line (at high tide), and the water covering of which therefore changes with tides; the intertidal zone.

euphotic zone The surface ocean layer where light penetration is sufficient to support photosynthesis.

euryhaline Able to tolerate a wide range of salinity.

eustatic Referring to worldwide changes in sea level.

eutrophication The addition of nutrients such as nitrogen and phosphorus to water bodies (naturally or by pollution), making them nutrient-rich and altering their ecological state.

evenness A measure of the equitability of the abundances of species in a community.

evolution Changes in the characters of a population or species over successive generations determined by shifts in the frequency of their genes.

exaptation A trait that evolved for some selective or historical reason other than that of its current function.

excretion The process of discharging metabolic wastes (primarily byproducts of nitrogen metabolism) from the body.

exobiont A symbiont living on the outside of a host's body.

exopinacocytes Flat or platelike (pavement) cells that form the outer epithelium of sponges. Also, EXOPINACODERM.

exoskeleton The calcified cuticle (or shell) of a crab (or other arthropod), composed of layers of polysaccharide (chitin), protein, lipid, and mineral (calcium carbonate).

exploitation competition Negative effects of organisms on one another caused by the consumption of shared, limited resources when organisms generally do not interact directly.

exponential phase The period of growth during which the population grows at an exponential rate. Also, LOGARITHMIC GROWTH PHASE.

extensibility The maximum strain (relative extension) at which a material fails.

externa The external reproductive saclike structure of a rhizocephalan.

external fertilization Fertilization outside the body, in which eggs and sperm interact in the water column after being released by parents.

extirpation The elimination of a species from a particular area, but not from its entire range.

extracellular Of molecules, outside the cell membrane.

extratropical storms Storms that develop at latitudes outside the tropics by the interactions of cold and warm air masses and that are thereby associated with the paths of the jet streams.

exuvium (*plural* **exuviae**) The shed chitinous outer covering of crustaceans and other arthropods.

eyespot A small mass of pigmented cells associated with nerve cells and serving as a light receptor of sorts. Also, **ocellus.**

facilitation A relationship between two individuals or species in which one benefits from the presence or activities of another.

facultative 1. Optional rather than obligatory; may or may not be present. 2. In the context of symbiosis, capable of, but not restricted, to living in symbiosis.

facultative mutualism A mutualism in which each species can live without its partner in some locations but grows faster or occupies a wider range of habitats when the partner is present.

female mate choice Mate preference exerted by females for certain males over other males.

fertilization The genetic fusion of two gametes, usually a sperm from one individual with an egg from another (called SELF-FERTILIZATION if the sperm and egg are from the same hermaphroditic individual).

fetch The length of water over which storm winds blow, a factor important in the generation of ocean waves.

fidelity In homing behavior, a return to the same site more often than would be expected by chance.

filibranch A type of ctenidium in which individual lamellae are elongated and folded, with this position fixed by tissue junctions within lamellae and ciliary junctions between adjacent lamellae.

fission A process of asexual reproduction in which an individual pulls itself apart into two new, independent individuals.

fitness The evolutionary success of an individual, measured as the probability that its alleles are passed on to the next generation.

fixation A procedure in preparation of a museum specimen, intended to terminate life processes quickly and with a minimum of distortion to cytological and tissue detail.

fluid A substance that is capable of flowing and does not have a defined shape; can be a liquid (such as seawater) or a gas (such as air).

flume A steady-flow channel.

flux The rate at which mass, heat, momentum, or any other quantity flows through a defined area in a unit interval of time.

food preference The choice exercised by consumers in which possible prey are ranked by relative suitability before attack and consumption.

food web A network of feeding relationships among species in a community.

formaldehyde A colorless and irritating gas (CH_2O); used in aqueous solution to make the preservative formalin.

formalin A toxic liquid preservative made from the gas formaldehyde in aqueous solution; sometimes used as an initial fixative or for long-term storage for organisms with delicate tissues; often used as a 10% solution.

fouling The growth of epibionts on the surface of a host.

foundation species Common and abundant species in a community that create and maintain habitats on which other community members depend to live and persist in the habitat.

Fourier analysis A mathematical method that combines sine and cosine waves to approximate a given time series of data.

fractal A shape or form that is self-similar at all scales; i.e., it looks the same regardless of magnification.

free radical An atom or molecule with an unpaired electron, making it highly reactive with other molecules or atoms.

frequency 1. The number of repetitions over time of a sampling event. **2.** For waves, the number of wavelengths that pass a fixed point in one unit of time.

friction A force that resists motion of two bodies in contact.

frond An erect, often leaflike portion of an alga.

front A boundary between two water masses with distinct properties.

frustule The siliceous cell wall of diatoms.

F_{ST} Fixation index, a measure of the fraction of the total genetic variation that is distributed among populations, ranging from 0 to 1. A higher value reflects greater genetic differentiation between populations, suggesting that most of the total genetic variation is due to differences among populations.

fucoid A large brown alga of the order Fucales, having a simple life history featuring production of eggs that, after fertilization, grow into sporophytes.

functional-form group A grouping of species according to similarities in morphology and role in a community.

functional group A collection of species that share similar functional traits, such as prey, resource needs, or ecosystem function.

fungi (*singular* **fungus**) Saprophytic and parasitic organisms that lack chlorophyll and include molds, rusts, mildews, smuts, mushrooms, and yeasts.

fusiform Shaped like a spindle, tapering at both ends.

gametes Haploid cells that fuse during fertilization to form zygotes (e.g., eggs and sperm); the direct product of meiosis in most animals but usually formed by mitosis in gametophytes of plants and algae.

gametophyte The haploid portion of a multicellular algal life cycle that produces gametes (sperm and eggs); the gametes fuse and develop into diploid sporophytes.

gastrovascular cavity In organisms belonging to the Cnidaria and the Platyhelminthes, a body cavity that functions in both digestion and the transport of nutrients to all parts of the body.

gastrulation The embryonic event in which the gut first forms; typically occurs by invagination (inward growth of tissue from a single area) that forms a saclike gut, but it may occur by several other processes besides.

gel A dilute network of polymers that has solid characteristics despite consisting mostly of water.

gene A DNA sequence that codes for a specific protein, differentiated from the rest of the genome because it is transcribed into RNA, which is translated into a protein; may contain coding regions and "nonsense" DNA.

gene flow The movement of alleles between populations as a result of dispersal.

generalist A consuming species feeding on a taxonomically broad spectrum of prey, often belonging to different phyla.

genet A genetically distinct lineage descending from a zygote, or in the case of many unicellular organisms, algae, and some land plants, from a meiotically produced haploid spore.

genetic Referring to the genes, which encode information that can be passed from one generation to the next.

genetic drift The change in gene frequency within a population purely as a result of chance. This process is more pronounced in small populations and may result in loss of genetic diversity.

geniculum (*plural* **genicula**) A flexible joint that separates calcified segments in an alga.

geodetic datum A coordinate system used for precise surveying that is composed of two parts: a Cartesian coordinate system (usually right-handed, with the origin at the earth's center of mass, the z-axis through the north pole, and the x-axis through the Greenwich meridian), and a mathematical surface (an ellipsoid) that is used for elevation reference.

geomorphology The scientific study of the landforms on the earth's surface and of the processes that have fashioned them.

georeferenced Having knowledge of the spatial relationship of different objects to each other in a particular environment, so that a location may be determined; requires recognition of landmarks.

geostrophic balance A flow state in which the pressure gradient and Coriolis forces acting on a current are equal and opposite; an ocean current in geostrophic balance flows perpendicular to horizontal pressure gradients.

geotactic Moving in response to gravity.

germ layer Embryonic tissues that arise during early development: the (outer) ectoderm, (inner) endoderm, and mesoderm in between.

girdle Muscular mantle tissue surrounding and supporting the valves of a chiton, typically ornamented with various elements such as scales, setae, or spines.

glacial/interglacial cycle A cycle of colder and warmer temperatures and larger and smaller ice cover, on the order of tens of thousands of years long.

glacial period Part of a glacial/interglacial cycle characterized by cold temperatures, ice sheet growth, and lowered sea level.

glaciation The advance of ice sheets from polar regions toward the equator.

glycoprotein A macromolecule consisting of a protein chain to which carbohydrate groups are covalently attached.

gonochoric Having separate male and female sexes.

gonochorist An organism that has the ability to reproduce either as a male or as a female, but not both.

gonophore The entire reproductive complement of a hydroid.

gonopore An opening on an arthropod through which gametes move.

G protein–coupled cAMP transduction pathway A cascade of biochemical events induced by the binding of chemical signal molecules to receptor proteins, ultimately leading to an influx of sodium and calcium ions and action potential generation, thus converting an environmental chemical signal into a language ultimately mediating animal, plant, or cell behavior.

grazer An animal that feeds on photosynthetic organisms such as plants, algae, and phytoplankton. Also, HERBIVORE.

grazing halo A bare area around the home scar of an herbivore, created by the herbivore's grazing of the seaweed around the scar.

greenhouse gases Gases that absorb thermal radiation in the earth's atmosphere. Naturally occurring water vapor is the most important greenhouse gas, but carbon dioxide and methane are strong greenhouse gases that are increasingly released into the atmosphere by humans through the burning of fossil fuels, deforestation, and agriculture (especially livestock and paddy rice farming).

gross photosynthesis The total amount of carbon fixed, or oxygen evolved, by photosynthesis per unit time. Gross photosynthesis equals net photosynthesis plus respiration.

group A number of successive water waves that are all somewhat larger than usual. Also, SET.

habitat alteration Changes in the condition of the physical or biological characteristics of a habitat; can be caused by acute changes, such as an oil spill, or gradual changes over a long period of time, such as changes in salinity due to increased runoff into an intertidal area.

habitat-ameliorating positive interaction A type of species interaction that occurs when the presence of an organism reduces potentially limiting physical stresses, such as heat, desiccation, and wave forces, on another organism.

halocline The region of the water column where there is a strong vertical gradient in salinity.

haploid Having one set of chromosomes in each cell.

harmonic mean promiscuity The reciprocal of the arithmetic mean of the reciprocals of the promiscuities of an individual's mates; for a male, this quantity, $1/H_M$, multiplied by the fecundity of his mates equals his fertilization success.

Hartford loop An arrangement of piping between an aquarium exhibit and the drain through which water flows out of the exhibit. The vertical height of this piping relative to the exhibit controls the water level in the exhibit.

harvested Of macroalgae, collected from natural populations.

heart A muscular pump with valves that causes the blood or hemolymph to circulate.

heat The kinetic energy associated with the disordered motion of molecules.

heat shock proteins (HSP) Proteins that help repair other proteins that have been damaged by heat stress or other stressors such as toxins, starvation, or oxygen deprivation. Heat shock proteins are found in a similar form in most plants and animals, suggesting that they play a vital evolutionary role in protecting organisms from environmental stress.

helical klinotaxis A type of chemotaxis whereby organisms swim in helices and repeatedly sample the chemical environment over time.

hemocoel A pseudocoelom that serves as the major blood-circulatory space as well as the main body cavity and occurs predominantly in arthropods and molluscs.

hemocyanin A circulating respiratory protein that has two copper atoms at its center involved in oxygen binding. Molluscan hemocyanin has a characteristic molecular structure, conserved pattern of amino acid sequence, and chromosomal arrangement not found in non-molluscs.

hemolymph 1. The circulatory fluid of open systems that performs the functions of both the blood and the lymph of closed systems. **2.** The extracellular body fluid of marine invertebrates.

hepatopancreas The multifunctional digestive organ in a crustacean that secretes digestive enzymes and absorbs nutrients from the digestive tract and also synthesizes oxygen transport and molting proteins into the circulating hemolymph.

herbarium A collection of dried, pressed, or preserved plant specimens with associated relevant data.

herbivore SEE GRAZER.

hermaphrodite An organism that has the ability to reproduce both as a male and as a female, either simultaneously (at the same time) or sequentially (changing from one sex to the other over time).

hermaphroditic Having both male and female reproductive organs present in a single individual. Male and female organs may be present at the same time, or male features may precede female features (protandrous) or vice versa (protogynous).

heteromorphic Of an alga or plant, having a life history in which one of the phases (gametophyte or sporophyte) is different in size and structure from the other phase.

heteromyarian Describing or referring to a condition in bivalves in which the two adductor (shell closing) muscles are unequal in size.

heterotroph An organism that derives its carbon from organic material. Thus, HETEROTROPHY.

heterozooid A nonfeeding individual of a bryozoan colony, specialized for sexual reproduction, attachment of the colony to a substrate, or colony defense.

higher taxa Taxonomic groups above the genus and species level.

holdfast A specialized basal structure that anchors the body (thallus) of a macroalga to the rock surface.

home scar A depression in the rock or seaweed created by an herbivore that returns to the depression at low tide and leaves at high tide to forage; for some limpets, an exact fit to the shell of a particular individual.

home site A location on the substratum to which an organism returns. In contrast to a home scar, the fit is not necessarily precise.

hominid Any of several species within the genus *Homo*. Often refers to ancestors of *Homo sapiens*, including the geographically widespread *Homo erectus*.

horizontal transmission A life history in which the progeny of symbionts have a free-living stage prior to entering into a symbiotic association.

host An organism that provides habitat for another organism either in its tissues, on its surface (epibiont), or within the interstitial spaces formed by its branches.

hotspots Areas exhibiting high species richness.

HSP Heat shock protein.

hull In chitons, an extracellular egg covering that is partly produced by material from the developing egg (oocyte) and shaped by the overlying follicle cells on the surface of the egg; hulls are smooth and thick or else thin but elaborately shaped as spines, cones, or cups.

hybrid A cross-breed between individuals of different species of plants or animals.

hydrocolloid A substance extracted from red and brown macroalgae that has gel-like properties when mixed with water.

hydrodynamic lift Force, perpendicular to the direction of flow, created when water moves over the an object.

hydrophilic Of substances, water-loving; capable of fully dissolving in water.

hydrophobic Of substances, tending to repel water; incapable of being completely dissolved in water.

hydrophobic interaction A type of bonding in which nonpolar materials interact to minimize their contact area with water.

hydrostatic lift Upthrust created by inclusion of low-density materials in the body.

hydrostatic pressure Pressure caused by water overlying an organism.

hyperosmotic **1.** Having greater osmotic pressure than the reference solution. **2.** Of fluid outside an animal's cells, having a higher concentration of solutes than the cells, causing water to be drawn out of the cells. Also, HYPERTONIC.

hyperstability In fishery science, a phenomenon in which an observed index of stock abundance (e.g., catch per unit of effort, or CPUE) remains stable although the abundance of the stock in question is actually declining.

hyphae The tubular filaments composing the body of a fungus.

hypoosmotic **1.** Having lesser osmotic pressure than the reference solution. **2.** Of fluid outside an animal's cells, having less solute (more water) than the cells, causing water to flow into the cells. Also, HYPOTONIC.

hypothesis SEE PREDICTION.

hypoxia A deficiency of dissolved oxygen in the water surrounding an animal.

ice age **1.** A period of generally lower temperatures and ice cover over large areas of the earth, on the order of tens to hundreds of millions of years long, often with cycles of glacial and interglacial periods. **2.** (*Capped.*) Specifically, the latest glacial epoch (Pleistocene).

ice foot The fringe of accumulated ice attached to the shoreline.

immersion The condition of being completely underwater (e.g., at high tide).

immunoreaction The recognition and binding of an antigen by its specific antibody.

impingement force A hydrodynamic force generated by the initial "slap" of the face of a wave on an emerged object.

income strategist An organism that must have continual access to food in order to survive and reproduce.

indeterminate cleavage A type of embryonic cleavage in which the fates of cells are not determined until later in development; typical of many deuterostome phyla.

indirect effect The effect that one species has on another species that is not exerted through direct physical, chemical, or visual contact, but through the alteration of the density or behavior of a third species. Also, INDIRECT ECOLOGICAL EFFECT.

indirect life cycles Life cycles in which the larval stage and adult stage differ greatly in form and function and where the transition between the two stages usually occurs at an abrupt metamorphosis that also often coincides with transition from planktonic to benthic habitat.

induced defense Phenotypic changes causing altered prey behavior or morphology, thus leading to diminished success of subsequent attacks.

inertia The tendency for an organism (or other object) to resist a change to its state of motion; the reason it takes a force both to accelerate a mass from rest and to decelerate a mass from an initial velocity.

infaunal Living within soft sediments of the sea floor.

infectious Of a pathogen, capable of moving from its host to another host.

infragravity waves Water surface oscillations that have periods in the range 30 seconds to five minutes, at a shoreline propagating either parallel to shore, as EDGE WAVES, or as STANDING WAVES perpendicular to shore.

instability The tendency of a flow to change from one state to another.

intensity In a study of parasitism, the number of parasites on a host deriving from separate infections.

interference competition Negative effects of organisms on one another caused by direct interactions in which individuals harm one another in the process of seeking a shared, limited resource.

interglacial period Part of a glacial/interglacial cycle characterized by warmer temperatures, retreat of ice sheets, and higher sea levels. The earth is currently in the latter portion of an interglacial period that peaked approximately 10,000 years ago.

intermediate disturbance Disturbance on a scale high enough to prevent a few late successional species from predominating and too low to allow only a few disturbance-tolerant or opportunistic species to prevail, hypothesized to yield maximal species diversity by allowing opportunistic as well as late successional, competitively dominant species to co-occur in mixed stands.

interna The internal trophic root system of a rhizocephalan.

internal wave 1. A wave that propagates through the interior of a fluid in which the density varies with height and that is a direct consequence of buoyancy. **2.** A wave that propagates below the sea surface along the pycnocline. An INTERNAL TIDE is a type of internal wave that is generated by tidal forces.

interspecific Between individuals of different species.

intertidal invertebrates Invertebrate species living in habitats such as sand and mud flats and rocky shores in the intertidal zone.

intertidal zonation SEE VERTICAL ZONATION.

intertidal [zone] That part of the coast that is covered by water at high tide but exposed to air at low tide.

intracellular Of molecules, within the cell membrane.

intraspecific Between individuals of the same species.

introduced species A species (including infectious agents) that has expanded its historical range through the intentional or inadvertent activities of humans. Also, **nonnative species.**

introvert A body region that can be completely withdrawn into the trunk.

invertebrates Multicellular animals that do not have backbones or verterbral columns.

in vivo **absorbance** The capacity of living tissue to absorb specific wavelengths of radiation.

ion An atom, or a group of atoms, that has a net electric charge; negatively charged ions, which have more electrons than protons, are called ANIONS, whereas positively charged ions, which have more protons than electrons, are called CATIONS.

iridophore A color cell in the skin of an octopus that reflects a blue-green color.

irradiance A measure of light intensity, defined as the radiant power per unit area incident upon a surface.

isoosmotic At osmotic equilibrium relative to the reference medium.

isopycnal A surface of constant density.

ISO speed A rating adopted by the International Organization for Standardization (ISO) to represent the sensitivity with which photographic film or digital sensors respond to light; doubling the ISO speed halves the necessary exposure.

isostatic rebound Uplift of landforms caused by a decrease in mass of an overlying object, such as a melting or retreating glacier.

isotonic Of a solution, consisting of mixed substances that have equal osmotic pressures and being of uniform concentration throughout.

IUCN SEE WORLD CONSERVATION UNION.

joints Cracks or fractures in rocks that are usually oriented in similar directions as a result of stress on the rocks over time.

joule The energy required to exert a force of 1 newton through a distance of 1 meter.

juvenile The free-living stage in the life cycle of an alga that develops from a zygote or spore, and of an animal that is born from a parent, hatched from an egg, or metamorphosed from a larva; the juvenile is morphologically similar to the adult but is sexually immature.

kelp Large brown algae of the order Laminariales having an alteration of generations between a haploid, microscopic gametophyte and a diploid, macroscopic sporophyte.

Kelvin waves Gravity-driven waves generated following the relaxation of the trade winds in the Pacific, first traveling equatorially from Indonesia to South America and then coastally along the coasts of South and Central/North America with a speed of about 250 km a day.

kentrogon A larval life history stage that is unique to the Rhizocephala; it uses a stylet to inject the parasite (vermigon) into the host.

keystone species A consumer species whose effect on the biological composition and function of a community is disproportionate to its abundance. The interaction usually involves preference by the consumer for a competitively superior prey. Also, KEYSTONE PREDATOR.

kinetic energy The energy of a mass in motion, equal (at velocities much less than that of light) to half the product of an object's mass and the square of its velocity.

kleptoparasite An individual that steals food caught by another individual before that individual can digest it.

knot One nautical mile (one minute of latitude, or 1.852 km) per hour, equal to approximately 0.5 meter per second.

Kolmogorov scale The smallest scale of velocity variability present in a turbulent flow; determined by the energy dissipation rate and by the fluid viscosity.

lamellae (*singular* **lamella**) Plates or filaments in the ctenidia of molluscs.

laminar-shear flow Velocity gradient within flow that moves as a stack of sheets (laminae) (no turbulent mixing).

larva (*plural* **larvae**) The free-living stage in the life cycle of many animals, which is born from a parent or hatched from an egg and eventually metamorphoses into a juvenile; many marine larvae spend this life stage suspended in the water column and metamorphose into a site-attached adult form.

larval export The net export of larvae from the larger, more abundant adults inside a marine reserve to unprotected areas beyond the reserve boundary.

latent heat The amount of energy required for a material to undergo a change of phase. LATENT HEAT OF VAPORIZATION is potential energy that can be converted to heat energy when a vapor condenses.

latitude The angular distance north or south of the equator.

LCD monitor Liquid crystal display monitor, an electronic-powered information-viewing screen on the back of digital cameras that allows images to be composed or reviewed.

lecithotrophic Of larvae, depending on yolk reserves rather than feeding to complete development between hatching and metamorphosis.

lek A small symbolic territory defended by a male (such as a marine iguana) for mating purposes; the lek is devoid of material resources other than the space and the male himself.

leucophore A color cell in the skin of an octopus that reflects white light.

LGM Last Glacial Maximum, *ca.* 18,000 to 20,000 years ago, when the previous glacial period was at maximum development and ice sheets were most extensively developed worldwide.

life history events The timing and rates of key events in the lives of organisms that affect demography and population growth rates. Examples include timing of reproduction, average longevity, growth rates, time to reach maturity, size of newly born offspring or eggs, and probability of death as a function of age or size.

lift A force on an object moving with respect to a fluid, exerted at a right angle to the flow; not necessarily upward with respect to gravity.

light-harvesting pigments Pigments that absorb light energy from the ambient environment and funnel that energy to the reaction centers of photosynthesis.

ligula An organ at the end of a male octopus's third right arm that is used in mating.

limited resource A resource (e.g., food or space) used by a species that, when in short supply, limits the population growth of a species.

limiting nutrient The element or compound (e.g., nitrogen) whose availability limits the growth or distribution of a species. In most temperate coastal ecosystems, growth of primary producers is limited by nitrogen.

linear wave A water wave that transmits energy with no net transport of mass; as a linear wave passes through the ocean, water travels in a circular orbit and returns to its starting position.

lithification The process by which a soft, unconsolidated sediment turns into a hard, rocklike structure that can be preserved in the geological record.

littoral 1. Intertidal. 2. Describing or referring to organisms living on or adjacent to the seashore.

littoral drift The transport of sand or other beach material along the shoreline by wave-induced longshore currents.

locus The position of any given genetic sequence on the chromosome. (Latin for *place.*)

logarithmic growth phase SEE EXPONENTIAL PHASE.

log layer A fluid layer in which velocities increase logarithmically with distance from a solid surface.

longitude The angular distance east or west of the prime meridian that passes through Greenwich, England.

lophophore The set of ciliated tentacles that surround the mouths of bryozoans, phoronids, and brachiopods; used for feeding and respiration.

Lophotrochozoa A higher taxonomic category proposed through molecular phylogenetic studies; includes several traditional animal phyla such as Annelida, Bryozoa, Mollusca, and Phoronida.

luciferase The enzyme (an oxygenase) in a luminescent organism that catalyzes the light-emitting reaction with luciferin.

luciferin The molecule in a luminescent organism that becomes oxidized, producing light.

lymph In organisms with closed circulatory systems, the transparent fluid (containing lymphocytes but not oxygen transport cells) that is derived from tissues, collected from all parts of the body, and returned to the blood via lymphatic vessels.

lymphocyte In tunicates, small (usually a few microns or less), morphologically undifferentiated blood cells with a large nucleus and little cytoplasm; may be multipotent, giving rise to other types of blood cells and playing an important role in budding and regeneration; may also play a role in self/nonself recognition.

mabé A pearl formed by layers of iridescent nacre overgrowing a dome-shaped insert placed between the mantle and inner shell of abalone. Often referred to as a BLISTER PEARL.

macroalgae Relatively large algae, usually easy to **see** with the unaided eye.

macroevolution Evolution over long time scales or involving large changes in trait values or evolution of new traits.

macroparasite An individual parasite that feeds on an individual host, on which it has a small impact; pathology suffered by the host is intensity-dependent. Also, **typical parasite.**

macrophytes Macroscopic primary producers, which include macroalgae and seagrasses.

macroscopic Visible to the eye without using a microscope or other magnification device.

madreporite A porous plate that connects the water vascular system of a sea star to exterior seawater; sometimes referred to as the SIEVE PLATE because of its filtering function.

maerl A sedimentary deposit of small strongly branched nongeniculate corallines.

magnetite An extremely hard magnetic ferrous oxide mineral (Fe_3O_4), which is the most mature state of mineralization on the cusps of the major (or second) lateral pair of teeth of the chiton radula.

male combat Competition among males for access to females or to resources that females require for reproduction. Also, MALE–MALE COMPETITION.

mantle The tissue in many molluscs, including all bivalves, that underlies the shell and secretes new shell material to the shell's edge.

mariculture The commercial intensive farming of marine organisms for food, usually in manmade facilities (e.g., ponds) outside of the natural environment, but also including the growth of marine species on rafts and ropes in suitable natural surroundings, such as sheltered embayments and estuaries.

marine regression The retreat of the sea to uncover land areas, caused by falling sea level.

marine terrace A flat or nearly horizontal wave-cut bench that has been exposed above sea level as a result of uplift of the coastline.

marine transgression The advance of the sea to cover land areas, caused by rising sea level.

maritime 1. Referring to the sea or involving seagoing ships. 2. In archaeology, entailing intensive use of a variety of marine resources, particularly those that must be acquired using watercraft and technology such as hook and line, nets, or harpoons.

medusa (*plural* **medusae**) A jellyfish. (From the Greek *Medousa*, a female monster with snaky hair, capable of turning those who beheld her into stone.)

meiofauna Microscopic metazoa passing through a sieve of 1-mm mesh size, in particular those that reach adulthood at this size (PERMANENT MEIOFAUNA) as opposed to macrofauna juveniles (TEMPORARY MEIOFAUNA).

meiosis Division of eukaryotic diploid cells (or nuclei) that results in haploid cells that are gametes in animals and some algae, haploid micronuclei in ciliates, or spores that develop into gametophytes in most algae and plants.

meridional In a direction parallel to a line of longitude or meridian; i.e., in a northerly or southerly direction.

meristem The rapidly growing, embryonic, and undifferentiated portion of the plant from which new cells are formed. The APICAL MERISTEM is located at the tip of the growing thallus, INTERCALARY MERISTEMS are located between the base and the tip of the thallus, and BASAL MERISTEMS are located at the base of the thallus.

meristematic Of fungi, having a pattern of growth in which a thallus is developed by isodiametric enlargement and division of cells; occurs in combination with thick, melanized cell walls.

mesoderm The middle tissue layer of triploblastic animals; the final tissue to appear during embryonic development, giving rise to muscles, reproductive organs, and other important structures.

mesograzer A small animal that lives on a host seaweed and feeds on either the host or fouling organisms on the host, or on a combination of the two.

mesopelagic zone That portion of the oceanic water column from about 200 to 1000 meters depth, where light is very dim.

mesopsammon Protozoa and micrometazoa (meiofauna) that inhabit the interstitial space between sand grains and move there without burrowing.

mesoscale eddies Circular movements of water on a scale approximately 2 to 200 km in horizontal extent.

metagenesis Life history that includes both an asexually reproducing polyp and a sexually reproducing medusa.

metameric Having a body consisting of a series of segments.

metamorphosis A relatively abrupt change in physical form associated with a transition from one life stage to the next; it often involves a change in habitat (e.g., from plankton to rocky shore).

Metazoa Multicellular animals with cells organized into tissues and organs.

microatoll A coral colony with a dead and often eroded summit caused by exposure to the atmosphere and tissue death.

microbialite General term for a lithified microbial mat, the mineral precipitate of which can be preserved in the rock record, but that does not necessarily have a regular laminated sedimentary structure.

microbial loop The flow of carbon and energy through microbial communities.

microbial mat A self-sustaining, organosedimentary structure formed by cyanobacteria, aerobic heterotrophs, fermenters, sulfate reducers, and sulfide oxidizers, which require only the input of light. By excreting copious amounts of exopolymeric substances, microbial mats form sedimentary biofilms.

microbiota Microscopic organisms.

microclimate The climate of a small, specific place within an area as contrasted with the climate of the entire area.

microhabitat A very small, specialized habitat, such as a clump of grass or a space between rocks.

micron One-millionth of a meter.

micropredator A consumer that moves from host to host like a predator but, like a parasite, consumes only a small part of the host in a feeding; the impact of micropredators increases with their intensity.

microwatt per square centimeter (μW cm^{-2}) The unit of measure for light intensity, as from luminescent organisms; one millionth of a watt per square centimeter.

midden Deposits significantly altered through the deposition of organic refuse related to prehistoric human habitation. In coastal settings, middens often contain a high concentration of shells mixed with bones of marine fauna.

millijoule A measure of energy equivalent to one thousandth of a joule.

millinewton The force required to accelerate 1 gram at 1 meter per second per second; a measure equivalent to one thousandth of a newton.

milliosmole A unit of measure of osmolarity, approximately equal to one millimole of dissolved particles.

mitochondria Double-membraned structures or organelles found within eukaryotic cells, which they serve as respiratory centers, providing energy for the cell.

mitochondrial DNA DNA found in the mitochondria of eukaryotic cells. It is generally inherited exclusively from the mother, and it changes more rapidly over generations than does nuclear DNA. Mitochondrial genes are used to track the history of populations because the mutation rate is easily measured and predictable.

mitosis Division of eukaryotic cells (or nuclei) that results in cells (or nuclei) that are genetically identical (except for mutations) to the parent cells (or nuclei); a major mechanism of organismal growth.

mixed layer The relatively warm, upper-ocean layer, which is often fully mixed by its interaction with wind.

mixing Changes in fluid property produced by diffusion alone or by diffusion acting in concert with stirring.

mixing ratio The mass of water vapor in grams contained in a kilogram of air. Also, SPECIFIC HUMIDITY.

mixotrophic Exhibiting the characteristics of both autotrophy and heterotrophy.

mobile organism An animal that is able to move across the substrate during all its life history stages.

modular organism An organism whose body plan consists of repeating, semi-independent, multicellular units. Such organisms commonly grow as colonies or highly branched individuals, and asexual reproduction is predominant.

module A potentially self-sustaining multicellular entity that is the basic iterated structural unit in modular, colonial organisms; often used interchangeably with *zooid* (in colonial bryozoans, entoprocts, cnidarians, and pterobranchs) or *polyp* (primarily in cnidarians).

modulus of elasticity SEE STIFFNESS.

mole 6.02×10^{23} particles, the mass numerically equal to the molecular mass of a substance. For a substance with molecular weight x, 1 mole of molecules weighs x grams; e.g., 18 grams of water contains 1 mole of water molecules. Also, GRAM MOLECULE, GRAM MOLECULAR WEIGHT.

Mollusca The large group of soft-bodied invertebrates, including octopuses, snails, clams, and limpets.

moment The turning effect exerted at a pivot point; the magnitude (in newtons) of force applied multiplied by the perpendicular distance (in meters) to the pivot point.

momentum The product of mass and velocity.

monitoring The process of collecting information in a planned program of sampling, to answer questions (i.e., to test predictions) about the ecology of rocky shores.

monogamy A breeding system in which one male breeds with one female.

monomyarian Describing or referring to a condition in bivalves in which one large and central adductor (shell closing) muscle is present.

monophyletic Of organisms, descended from a single common ancestor.

multidimensional scaling A method of illustrating dissimilarity in samples of complex sets of data.

multiple community outcomes SEE ALTERNATIVE STATES.

multivariate data Information about many variables (e.g., numbers of many species) analyzed together.

mutualism A type of species interaction that occurs when two species benefit each other; a reciprocal facilitation that can be either facultative or obligatory and highly evolved.

myoglobin An oxygen-carrying pigment found in high concentrations in the muscles of diving mammals and birds.

nacre Iridescent calcium carbonate in its aragonitic crystalline form; found in the inner shell layer of many gastropods and bivalves. Also, MOTHER OF PEARL.

nanometer (nm) A unit of length or distance equal to 10^{-9} meters; commonly used to express the wavelengths of light.

natal Associated with one's birth or place of birth.

Natural Resources Damage Assessment (NRDA) The process of documenting, assessing, and rehabilitating damages for injury to, destruction of, loss of, or loss of use of natural resources, including the reasonable costs of assessing the damage; conducted formally in the United States under the Comprehensive Environmental Response, Compensation and Liability Act (CERCLA).

natural selection The process by which individual organisms with favorable traits are more likely to survive and reproduce, thus causing a change in gene frequency toward genes that are beneficial in a given environment.

nauplius The feeding, planktonic larval stage of barnacles such as rhizocephalans. Thus, NAUPLIAR.

Neanderthal people Archaic *Homo sapiens* occupying Europe and the circum-Mediterranean area prior to the expansion of anatomically modern humans *(Homo sapiens sapiens)* about 40,000 years ago.

neap tides Tides of the smallest amplitude (minimum difference between high and low tide), occurring twice each month, around the first and last quarter-moon phases.

nematocyst SEE CNIDOCYST.

Neogene The last 23 million years of earth history, including the Miocene, Pliocene, Pleistocene, and Holocene (Recent) epochs.

nephridia (*singular* **nephridium**) Osmoregulatory or excretory organs; tubules through which ultrafiltrate (blood without cells or proteins) passes from the coelom to the exterior of an animal.

net ecosystem production The difference between net primary production (NPP) and respiration by heterotrophs within an ecosystem; net ecosystem production is zero when NPP and heterotrophic respiration are in complete balance.

net photosynthesis The amount of carbon fixed for photosynthesis that is available for growth, after accounting for respiration. Net photosynthesis equals gross photosynthesis minus respiration.

neutral theory The theory of genetic variation that describes allelic variation patterns when there are no fitness advantages (i.e., in terms of viability or fecundity) conferred to any particular genotype at a locus.

new production Primary production associated with nitrogen inputs from outside an ecosystem. These inputs are usually in the form of upwelled nitrate from deep ocean waters or atmospheric N_2 gas, which is fixed (converted to NH_4^+) by marine prokaryotes.

newton The force required to accelerate 1 kilogram at 1 meter per second per second.

niche The environmental space (biological and physical) within which a species can survive and reproduce.

Niskin bottle A device for collecting water samples at discrete depths, having a spring-loaded valve that is closed by sending a messenger device down a cable to the bottle.

nitrogen fixation The process by which nitrogen gas is reduced into biologically available inorganic nitrogen compounds (e.g., ammonia, NH_3).

nonlinear wave A wave that transports mass in addition to energy.

nonnative species SEE INTRODUCED SPECIES.

nuclear DNA DNA found in the nuclei of eukaryotic cells.

nutrients Substances that provide nourishment for living organisms; e.g., the minerals taken up from the ocean by phytoplankton.

obligate In the context of symbiosis, restricted to living in symbiosis during some part of a life cycle.

obligate mutualism A mutualism in which both species require each other's presence in order to live, grow, and complete their life cycle.

ocelli (*singular* **ocellus**) 1. SEE EYESPOT. 2. Shell eyes present in radiating rows in the valves of some specific chiton genera, not large enough to form images but with an important role in alerting the chiton to potential predators.

offshore Of water movement, away from the shoreline.

olfaction The chemical sense of "smell" in which potentially stimulating compounds (e.g., from food or a mate) that are dissolved in the surrounding medium (water or air) are perceived at a distance by specialized receptors of an organism.

oligotrophic Nutrient-poor.

onshore Of water movement, toward the shoreline.

ontogeny The course of growth and development of an individual to maturity.

oogonium (*plural* **oogonia**) The gamete produced by brown algal female gametophytes.

operational sex ratio (OSR) The ratio of sexually mature males to sexually receptive females; the greater the number of mature males relative to the number of receptive females, the greater the value of the OSR becomes, and the stronger the presumed intensity of competition among males for mates.

operculum (*plural* **opercula**) The opening in a barnacle's test through which it extends its cirral net.

opportunistic species Faster-growing species, generally characterized by shorter life spans, high reproductive outputs, and strong dispersal abilities that frequently recruit onto freshly disturbed or cleared substrata.

opportunity for selection The variance in relative fitness, symbolized as I; provides an empirical estimate of selection intensity. The value of I can be calculated for any population by dividing the variance in fitness, V_W, by the squared average fitness, W^2.

opportunity for sexual selection A sex difference in the opportunity for selection, indicating that selection is operating more strongly in one sex than the other.

opsin A protein that couples with a light-reactive chemical (chromophore), which becomes activated and interacts with the next protein in the phototransduction cascade when the chromophore is struck by a photon of light.

optimal foraging theory A theory that attempts to predict when a consumer will feed and what it will choose to feed on, based on the potential energy it may gain and the energy losses it will incur in doing so.

oral surface The underneath (ventral) side of a sea star, where the mouth is located.

oral veil In nudibranchs, a membranous, hoodlike extension of the head.

organic carbon Originally, a molecule that came from, or comes from, living organisms. All organic molecules consist of a basic skeleton made up of carbon atoms, with atoms of hydrogen, oxygen, and nitrogen (in proteins), arranged around the carbon skeleton.

oscillatory Of flow, moving back and forth.

osculum (*plural* **oscula**) The opening through which water leaves a sponge.

osmoconformer An organism that matches the osmotic pressure of the surrounding water by allowing flow of water into or out of its tissues.

osmolyte A solute molecule that serves the special purpose of raising the osmotic pressure of a body fluid.

osmoregulation The process of maintaining the concentration of substances dissolved in body fluids within a narrower range or at a different level than the range found within the environment.

osmoregulator An organism whose body fluids are maintained at an osmotic concentration independent of the surrounding fluid.

osmotic balance The water balance in organisms.

osmotic pressure The pressure required to halt diffusion of water across a membrane.

ossicles Microscopic skeletal elements made of calcite that are embedded in the body wall of echinoderms.

ostium (*plural* **ostia**) Any opening through which water enters a sponge.

otolith A calcified ear stone in the internal ear of vertebrates.

ovicell A specialized chamber or modified individual in a gymnolaemate bryozoan colony, used for brooding embryos.

ovigerous legs or **ovigers** The third pair of anterior appendages in pycnogonids.

oxidation–reduction reaction A chemical reaction in which energy is transferred from one molecule to another via the exchange of electrons. The molecule losing the electron is said to be oxidized, and the molecule receiving the electron is said to be reduced. Also, REDOX REACTION.

oxygen debt Extra oxygen consumption necessary to oxidize the products of anaerobic metabolism that accumulate as a result of intense exercise or exposure to oxygen limitation.

oxygenic photosynthesis Photosynthetic reactions involving the production of free oxygen, in addition to organic carbon. Except for some purple bacteria, all photosynthetic organisms generate oxygen.

pagurid A crab of the family Paguridae.

Paguridae A large family of right-handed hermit crabs.

palps The second pair of anterior appendages in arthropods; in pycnogonids they consist of 2 to 10 segments (depending on family and genus) and are absent in females from one subgroup of callipallenids and absent in both sexes in some other groups (e.g., Phoxichilidiidae) Also, PEDIPALPS.

PAM fluorescence Pulse-amplitude-modulation fluorescence, measured by PAM fluorometry, a technique that can be used to quantify photosynthetic primary productivity.

panmictic Of a population, having a state of uniformly high gene flow with no statistical difference in allelic composition among regional samples.

papillae (*singular* **papilla**) Small projections from the main body of an animal.

papulae (*singular* **papula**) External, finger-like outpockets of the body wall of a sea star, used in respiration.

paracoelomate An animal at the pseudocoelomate grade of organization (triploblastic; body cavity, if present, derived from blastocoel and not fully lined with mesodermal epithelium); many small-bodied paracoelomates have secondarily lost their body cavity and are functionally acoelomate.

paralarva A newly hatched octopus that swims in the plankton.

parapodial lobes Flattened thin upward extensions of the foot in anaspid opisthobranchs, often used for swimming.

parasite Broadly, an organism that feeds off another organism but, in contrast with a predator, attacks one host per life stage. Also in contrast to predators, typical parasites (unlike parasitoids) do not need to kill their host as a part of their transmission.

parasitic castration The trophic "strategy" of a parasite in which the host's reproductive resources are shunted into parasite growth and reproduction.

parasitoid An individual consumer that attacks a single host and inevitably causes the death of the host, after which the parasitoid usually emerges as an adult.

parenchymella The oblong ciliated larva of many intertidal sponge species.

parental investment Energetic investment by parents in gametes or offspring.

parthenogenesis A mode of reproduction in which the egg develops into a new individual without fertilization. Thus, PARTHENOGENETIC.

partial pressure The pressure exerted by one of the components in a gas mixture; according to Dalton's law of partial pressures, the total pressure of a gas mixture is the sum of the partial pressures of all of the gases in that mixture.

pathogen A type of parasite (typically a bacterium, a virus, or a protozoan) that multiplies within a single host and may cause disease.

pascal (Pa) A unit of pressure or force per area exerted on a surface; one pascal is equivalent to a force of one newton on one square meter.

pascal second (Pa s) A unit expressing (force × time)/area, equivalent to N s/m^2 or kg/m s; the unit of dynamic viscosity (the tendency of a fluid to resist shear).

Peclet number A dimensionless number used to describe the relative contributions of water motion and diffusion to the transport of mass.

pedicellariae Small, jawlike appendages of sea stars; used in defense and feeding.

peduncle 1. The bulbous lower section (stalk) of a sea pen, lacking polyps and usually buried in sediment. **2.** SEE STALK.

peel failure Failure as a result of the force concentration resulting from reduction of a two-dimensional area of application nearly to a one-dimensional line of application.

pelagic 1. Referring to the mid- and upper water layers of the oceans. 2. Living in the open ocean.

pelagic larva The developmental stage of many invertebrate and fish species that lives freely in the water column of the ocean before metamorphosing into an adult.

pelagophyte A class of minute unicellular phytoplankton that are ball-shaped.

pelagosphera A unique larval form in sipunculan worms, having a pronounced postoral ciliary band, a distinct lower lip, a dorsal anus, and usually a terminal organ for attachment.

pentaradial symmetry A type of radial symmetry in which five parts are arranged roughly evenly around a central axis, as in the body plan of many echinoderms. Also, PENTAMERISM.

perennial Of algae, long-lived (several years).

pereopods The external appendages of crabs, commonly referred to as legs and claws, consisting of the following segments: coxa, basis-ischium, merus, carpus, propodus, dactyl (in order of moving from the body to the most distal tip of each appendage).

pericardium The coelomic sac in the Mollusca that contains the heart.

periplasmic membrane In bacteria, a thin structure separating the cytoplasm from the cell wall.

permeability The rate at which water or a fluid can flow through a material (e.g., sand); depends upon the size of the spaces or voids between the grains.

perturbation Any event that disrupts ecosystem, community, or population structure and changes resources, substrate availability, or the physical environment.

Phaeophyta A division (phylum) of algae, the "brown algae."

Phanerozoic The geological period of time beginning approximately 545 million years ago and extending to the present time.

pharmaceuticals Drugs or medications.

pharynx A muscular region in the anterior digestive tract, used for respiration and feeding. In some worms, the pharynx can be everted. In tunicates, it is perforated by numerous gill slits.

phase A portion of a cycle.

phase shift In ecology, a major and often sudden change in community structure caused by a large physical or biological disturbance or both.

phenoloxidase A copper enzyme that catalyzes the hardening of an arthropod's newly formed exoskeleton and functions in the immune response through the formation of melanin.

phenotypic plasticity Flexible changes in the traits of an organism in response to environmental cues.

pheromone A substance or chemical messenger produced by one organism that evokes a sexual or other behavioral response in another of the same species.

photic zone The depth of water through which light penetrates.

photoautotroph An organism that obtains energy and carbon from photosynthesis.

photon A discrete unit of light carrying a fixed amount of energy (quantum) that depends inversely on wavelength.

photoreceptor A nerve cell specialized for sensing light; the site of phototransduction.

photosynthesis The process by which carbohydrates are synthesized from carbon dioxide and water using light as an energy source; most forms release oxygen as a byproduct.

photosynthetically active radiation (PAR) Light having wavelength between 400 and 700 nm, used by plants in photosynthesis.

photosynthetic carbon fixation The use of light energy captured by photosynthesis to convert carbon dioxide into organic carbon, initially in the form of carbohydrates.

photosynthetic reaction center That part of the photosynthetic apparatus responsible for converting light energy absorbed by the pigments into chemical energy in the form of electrons capable of participating in redox reactions.

phototactic Moving in response to light.

phototransduction A cascade of protein interactions beginning with the activation of opsin by a light-induced chromophore shift and ending with a change in ion concentration in a photoreceptor cell.

phototroph An organism that derives its energy from sunlight.

phycobilins A group of water-soluble accessory pigments found in the red algae and the cyanobacteria.

phylogeny The evolutionary relationships between an ancestor species and its known descendants. Thus, **phylogenetic.**

phylum (*plural* **phyla**) One of the major groups used by biologists to classify organisms (especially animals); equivalent to the division used to classify plants.

physicochemical Relating to both physical and chemical properties.

phytoplankton Minute, free-floating aquatic plants.

picophytoplankton Phytoplankton that are less than 2 μm in diameter.

piscivorous Feeding on fish.

pituitary In vertebrate chordates, the master endocrine gland located at the base of the brain, beneath the hypothalamus, controlling many major metabolic functions; the development, position, and patterns of gene expression of the tunicate neural complex (cerebral ganglion, neural gland, and ciliated funnel) suggest a shared ancestry with the vertebrate pituitary gland.

plankton Small-bodied animals (zooplankton) and plants (phytoplankton) that live suspended in the water column and are at the mercy of ocean currents for transport (i.e., they cannot swim strongly). Thus, PLANKTONIC.

planktotrophic Of larvae, obtaining the bulk of their energy for growth and development by feeding on other plankton.

planula The ciliated, typically benthic larva characteristic of most Cnidaria.

plate tectonics A widely accepted and unifying theory describing the evolution of the major surface features of the earth (mountain ranges, volcanoes, faults, and trenches, for example), which are a result of the movement of a number of semirigid plates that make up the earth's crust. This movement has resulted in changes in the geographic positions of continents and size and shape of ocean basins.

Pleistocene epoch A geological period extending from 1.65 million years ago to 11,000 years ago. During this time, 19 to 20 glacial/interglacial cycles occurred.

pneumatocyst A gas-filled bladder found in some algae. The gas provides buoyancy to maintain an upright posture.

polar Referring to the two regions of earth poleward of the polar circles (66.5° latitude).

polarity The presence of an unequal distribution of charge in a molecule.

pollutogen A living agent that may cause disease but cannot move from one host to another in the environment into which it is placed.

polychaetes Segmented tube worms belonging to a class of the phylum Annelida. Each segment has lateral appendages (parapodia) and bundles of setae on each side.

polyembryony A form of asexual reproduction in which the zygote divides once or repeatedly and the resulting cells develop into genetically identical, diploid offspring.

polygyny A mating system in which one male mates with many females. Thus, POLYGYNOUS.

polymer A molecule consisting of repeating groups of atoms, often forming a long chain. Stretchy polymers are called ELASTOMERS.

polymorphism In colonial organisms, the co-occurrence of morphologically and functionally specialized zooids or modules in a ramet.

polyp 1. The sedentary or sessile stage of any cnidarian, typically attached to rock by its basal end and with a mouth and tentacles at its free end. **2.** SEE MODULE.

polyphenolic compounds A group of chemical compounds found in plants, including chemicals such as tannins (bitter tasting chemicals) and flavonoids (secondary metabolites with antioxidant activity).

polyphyletic group A group whose members share certain characteristics but derive from separate evolutionary origins outside the group.

polypide Portions of a bryozoan individual including the lophophore, U-shaped digestive tract and its musculature, cerebral ganglion, and tentacle sheath.

polyploid Of nuclei, having more than one (haploid) or two (diploid) sets of chromosomes.

polysaccharide A polymer of simple sugar molecules having the general formula $(CH_2O)n$.

polyspermy The fertilization of an egg by more than one sperm; leads to abnormal development. The fast block to polyspermy involves depolarization of the egg membrane. The slow block to polyspermy involves the vitelline envelope, a physical barrier to other sperm.

population A group of individuals of the same species living in a given area.

population structure A measure of genetic differences among populations within a species.

potential energy In meteorology, the energy that a parcel of air has by virtue of its position; potential energy is transformed to kinetic energy as the parcel moves.

potential reproductive rate (PRR) The ratio of maximum offspring numbers produced by a maximally successful male and a maximally successful female in a given species.

potential temperature The temperature that an unsaturated parcel of dry air would have if brought from an initial state to a standard pressure, usually taken as 1000 hPa (hectopascals), without interacting with its environment.

Practical Salinity Scale (PSS) A worldwide standard for measurement of salinity that is based on the electrical conductivity of seawater.

Practical Salinity Unit (PSU) A "unit" of measurement given with salinities computed according to the Practical Salinity Scale.

precocial Of young animals, born or hatched in a highly developed state. In mammals, for example, a precocial young animal's eyes are open and it can vocalize and can be quite independent.

predation A relationship between two individuals or species in which one consumes the other. Thus, PREDATOR.

prediction A logically structured statement derived from an explanation (or model) about ecological patterns or processes. Also, HYPOTHESIS.

present atmospheric level (PAL) The present-day mass fraction of diatomic oxygen in the atmosphere (20.9%), which does not change with elevation.

preservation A process or protocol in the preparation of museum specimens that saves organic substances from decay. Common preservatives include 5%–10% formalin, 70% ethanol, or isopropyl alcohol.

pressure Force applied over an area, usually expressed as newtons per meter squared (N/m^2) or pascals ($1\ Pa = 1\ N/m^2$).

pressure gradient The rate of change of pressure over distance, expressed as pascals per meter or similar ratio of pressure to distance. The pressure gradient is directed toward higher pressure.

prevalence The proportion of a host population that is infected with a given parasite species.

primary consumer An animal that eats a primary producer; an herbivore.

primary producer Any organism that converts inorganic carbon compounds such as CO_2 into sugar and more complex molecules. In marine ecosystems, primary producers include phytoplankton (unicellular algae and cyanobacteria), seaweeds (macroalgae), and seagrasses (marine angiosperms). They occupy the bottom of the food web, and most are photosynthetic.

primary production The production of biological organic compounds from inorganic materials through photosynthesis.

primary productivity The rate of carbon fixation (e.g., by photosynthesis) in a given area over a given time period.

proboscis (*plural* **proboscides**) An elongated or extensible tubular feeding structure extending from the head of an animal.

prokaryote A bacterial or cyanobacterial organism having cells that lack a nucleus and other organelles. Thus, PROKARYOTIC.

promiscuity The number of mates, P_i, each ith individual has.

pronucleus The nucleus of an egg or sperm prior to fertilization.

propagules 1. A dispersing early life history stage; in a marine context, typically gametes, animal larvae, or plant spores. 2. The dispersal units of seaweeds and seagrasses, including single cells produced by sexual and nonsexual events and multicellular vegetative fragments.

protists Organisms in a somewhat artificial taxonomic grouping that contains a wide array of unicellular and multicellular eukaryotes.

Protostomia A taxon of animals in which the mouth develops from the blastopore; includes the molluscs, annelids, and arthropods.

proxy study A study that approximates conditions of interest when it is not feasible to manipulate those conditions; e.g., studying the warm conditions generated during El Niño years as a proxy for expected future global warming caused by human activities.

pseudocoelom A fluid-filled body cavity between the outer body wall and gut that is not fully lined with mesodermal epithelium; the idealized pseudocoelom is contiguous developmentally with the embryonic blastocoel.

ptychocyst A type of cnidocyst (nematocyst) consisting of a single-walled capsule that contains a spineless, closed tubule that is folded rather than coiled; found only in Cerianthia.

pycnocline A region of rapid change in density with increasing depth.

quadrat A sampling unit, often consisting of a square or rectangular frame of appropriate size for quantifying abundances of one or more species.

Quaternary The geological period that comprises the Pleistocene (~1.8 million years ago until ~10,000 years ago) and Holocene (last 10,000 years; also called Recent) epochs; a sub-era of the Cenozoic.

rachis The upper, polyp-bearing section of a sea pen, which projects into the water column.

radiance The radiant power per unit source area per unit solid angle (the amount of electromagnetic radiation leaving or arriving at a point on a surface), usually expressed as watts per meter squared per steradian ($watts/m^2/steradian$).

radiation In biology, diversification in which various different lines of organisms (i.e., species) arise from a single ancestral line of organisms.

radula (*plural* **radulae**) The flexible coiled ribbon of chitinous teeth, often with lateral protrusions, in gastropods and most molluscs; used by grazers to scrape algal food and by predatory snails to bore into shells.

ramet An individual body that is produced asexually from the parental body; large numbers of genetically identical ramets can form clones or colonies of a genet.

raptorial Adapted to reach out and seize prey.

Rayleigh number The dimensionless number that determines when convection starts and what form it takes, in general representing the relative importance of the destabilizing influence of density versus the stabilizing influences of viscosity and molecular diffusion.

recruitment The input of new individuals into a population; in marine benthic communities, follows a period of settlement of larvae or propagules.

recruitment limitation A situation in which the size of a population is primarily determined by the number of new recruits arriving per unit time.

Red List A list maintained by the World Conservation Union that classifies species in different categories according to their risk of extinction.

redox reaction SEE OXIDATION–REDUCTION REACTION.

reef flat A shallow horizontal section of a coral reef that is subject to tidal fluctuations and constant wave action.

reflectance 1. The ratio of radiant energy reflected by a body to the energy incident upon it. **2.** Specifically, the fraction of irradiance reflected by an alga.

refractometer A handheld instrument for measuring seawater salinity through its effect on the water's index of refraction.

regenerated production Primary production that is fueled by local-scale nitrogen recycling within an ecosystem, usually excreted ammonium or organic nitrogen-containing compounds.

regeneration Regrowth of an appendage (such as an octopus's arm) after it has been bitten off by a predator.

regression SEE MARINE REGRESSION.

relative humidity The ratio of the observed mixing ratio for a volume of air to the maximum mixing ratio, given the temperature of the air (saturation mixing ratio); expressed as a percentage.

remineralization The decomposition of organic compounds into inorganic constituents.

replicate The units measured as part of a sample; e.g., a single tidepool in a set being monitored.

replicate-copy growth A more precise term for asexual reproduction; growth of a genet by producing ramets.

reproduction The production of new bodies from existing bodies.

reproductive success The number of offspring of an individual surviving at a given time.

resilience In ecology, the speed at which an ecological system returns to its former state after a perturbation.

restoration The process of assisting the recovery of an ecosystem that has been degraded, damaged, or destroyed. Ecological restoration initiates or accelerates the recovery of an ecosystem with respect to its health, integrity, and sustainability.

retraction The state of introversion, inversion, or withdrawal of a part of an organism.

Reynolds number A dimensionless number used to describe the relative forces imposed by water's inertia and viscosity.

rhinophore A sensory tentacle located on the head or anterior end of an opisthobranch mollusc.

rhodolith A free-living, usually spherical, accretion of predominantly nongeniculate corallines.

Rhodophyta A division (phylum) of algae, the "red algae."

rhythm A physiological or behavioral cycle that is timed endogenously by one or more clocks that approximate periodicities of environmental cycles.

Richardson number The dimensionless number that represents the relative importance of the stabilizing influence of density and the destabilizing influence of velocity shear.

richness SEE SPECIES RICHNESS.

rockpool British English term for tidepool.

rogue wave An extremely high but short-lived ocean wave that is formed by the chance summation of several waves such that their heights combine, but that quickly disappears as the waves then separate.

rugosity A measure of the variability in vertical relief or roughness of a surface.

salt A chemical compound consisting of cations (positively charged ions) and anions (negatively charged ions).

saltwater wedge A wedge of dense saltwater that sinks below less dense freshwater.

sample A set of replicated units of study chosen to represent a larger group of interest that cannot all be measured.

sampling design The logical plan to collect appropriate samples to fit with requirements of statistical analyses so that interpretations are logical and reliable.

sampling point A sampling unit consisting of a single point under which the type(s) of organism present is noted.

sarcodines An eclectic group of protists that possess fingerlike, filamentous, or netlike cytoplasmic extensions called pseudopods. Known informally as the **amoebae.**

saturated colors Chromatically pure colors, not diluted with white, gray or black; reflectance concentrated over a narrow bandwidth as opposed to equal reflectance across the entire spectrum.

saturation mixing ratio The maximum mass of water vapor in grams contained in a kilogram of air for a particular temperature and pressure.

Scleractinia Cnidarians with a delicate to massive calcareous (aragonite) exoskeleton having platelike skeletal extensions or septa that occur in multiples of six.

sclerite A minute calcareous spicule embedded in a tissue matrix, as in the tissue of octocorals.

SCUBA Acronym for *self-contained underwater breathing apparatus.*

scyphistoma The primary polyp of Scyphozoa; undergoes strobilation to produce ephyrae.

sea 1. An ocean or a division of an ocean. 2. The wave conditions within the area of a storm, being highly irregular because of the presence of waves having wide ranges of heights and periods.

sea anemones Plantlike animals of the phylum Cnidaria.

sea ice Any ice formed from the freezing of sea water.

Secchi disk A disk, usually with black and white quarters, used to assess water clarity by measuring the depth at which the disk is no longer visible from the surface.

sedimentary rock A rock (such as sandstone or shale) formed by the accumulation of mineral grains transported by wind, water, or ice to the site of deposition, or by chemical precipitation at the depositional site.

seismic activity Movements of the earth's crust, usually associated with earthquakes. Such movements can cause shorelines to be displaced upward or downward, resulting in rapid and permanent shifts in environmental conditions. Also, SEISMIC GROUND MOTION.

selection events Bouts of natural selection acting on organisms, such as greater predation on larvae by visually feeding fishes during the daytime than at night.

semi-altricial Of animal young, dependent on the parents for survival for an extended period after birth even though having some mature characteristics; for example, penguin chicks are down-covered, but they are otherwise blind, are unable to thermoregulate at hatching, and have limited physical co-ordination.

semidiurnal Having a recurrence period of about one-half day.

senescence The period near the end of an organism's life cycle. In an octopus, a period marked by behavior changes similar to those associated with Alzheimer's disease in humans.

sensible heat Heat energy that can be measured, or sensed, using a thermometer.

sessile organism A plant or animal that, after it arrives on the shore and settles on a substrate, does not move during its remaining life history stages.

setae (*singular* **seta**) Hairlike projections from the surface of an animal.

settlement The life phase characterized by the transition from a suspended larva to a site-attached adult.

setup Superelevation of the water level at the shoreline where the increased hydrostatic pressure balances the decrease in wave momentum as the waves break across the surf zone.

sex Any process that involves mixing or redistributing genetic material to produce genetically novel individuals (genets).

sex allocation The partition of energetic investment by individuals in sexual species toward male or female function.

sex ratio The ratio of the number of females to the number of males in a population. This ratio also estimates the average number of mates per male.

sexual selection An evolutionary process that occurs when individuals in one sex mate with disproportionate success at the expense of other individuals of the same sex.

shape coefficient A characteristic value used to account for the amount of surface exposed to friction. The value is directly proportional to the force applied (i.e., drag or lift) and inversely related to the exposed or cross-sectional area of an object.

shear Spatial variations in fluid velocity. In the ocean, the most important shears involve vertical gradients of horizontal velocities.

shear stress The force exerted as one layer of fluid moves past another; causes the fluid to deform.

shear velocity An index of turbulent mixing near the seabed.

shepherd's crook SEE STRIGILIS.

Sherwood number A dimensionless number used to describe the total transport of material to an object relative to the transport expected if only molecular diffusion is operating.

shock pressure A transient increase in pressure that occurs when water in a breaking wave is obstructed or stopped for an instant before the impinging water can escape.

shorebirds A group of birds, known as waders in Britain, belonging to the order Charadriiformes and generally associated with coastal or wetland environments.

shoreline Stretches of shore formed by rock, sand, gravel, other physical components, usually experiencing tides and harboring different ecosystems (rocky shores, sandy beaches, estuaries, mangroves, corals, and others).

sieve elements Cells with porous end walls, through which products of photosynthesis (sugars) and other organic molecules are transported.

significant wave height 1. The average height of the highest one-third of waves passing a particular point. 2. The average of the highest one-third of the ocean waves that have been measured or predicted as having been generated by a storm.

siliceous ooze Marine sediments that contain >30 % biogenic silica.

sinuses Cavities or pouches in the body, which, in organisms with open or partially closed circulatory systems, enable circulatory fluids to come into direct contact with organs and tissues.

slack tide A period with little or no tidal movement occurring at the transition between incoming and outgoing (or outgoing and incoming) tides.

small-scale fisheries Fisheries based on intertidal or inshore waters, usually operating within a few miles from the shoreline, in waters less than approximately 100–200 meters in depth; includes traditional or subsistence, artisanal, and artisanal commercial fisheries.

soliton A self-reinforcing solitary wave, often taking the form of a half wave of depression or elevation only, with highly nonlinear wave dynamics.

solute A substance that is dissolved in a fluid, thus forming a solution.

somatic Referring to the functioning body of an organism, as distinct from its genetic material.

specialist A consuming species preying on a single species or on a highly restricted range of often closely related prey.

species A population or group of populations that in a particular time and place are in reproductive contact but are reproductively isolated from all other populations.

species diversity SEE DIVERSITY.

species interaction strength The effect of one species on another, usually measured by comparing results of predator, grazer, or competitor exclusion experiments to controls.

species richness A measure of diversity; the number of species in a specified unit area or volume.

specific gravity A measure of relative density, calculated as the density of an object divided by the density of water at the same temperature.

specific heat The amount of heat, measured in joules, required to raise the temperature of 1 kg of a substance by 1 °C. Also, SPECIFIC HEAT CAPACITY.

specific humidity SEE MIXING RATIO.

spectral irradiance The radiant flux incident on a unit area of a surface per wavelength interval (in nanometers, nm), usually expressed as watts per meter squared per nanometer (watts /m²/nm).

specular highlights Brilliant glints of light reflected from surfaces so smooth that they act as tiny mirrors.

spermatium The gamete produced by red algal male gametophytes.

spermatophore A sperm packet passed into a female octopus by the male during mating.

spermatozoid The gamete produced by brown algal male gametophytes.

sperm competition A condition in which a limited number of ova are available for fertilization by sperm within the ejaculates of more than one male.

Spheniscidae The bird family comprising the penguins.

spicules Small calcareous or siliceous bodies that, together, form all or part of the skeleton of a sponge.

spillover The movement of adult fish and invertebrates from the protection within marine reserves to unprotected areas beyond the reserve border.

spine A calcite part of the skeleton of a sea urchin that is attached by a socket in the spine base with muscles and connective tissues to a ball on the body. Spines are covered with tissue and can repair and regenerate if broken or abraded.

spiral cleavage A characteristic pattern of early embryonic cell cleavage, in which some cleavage spindles become oriented perpendicular to other cleavage spindles within the same embryo, most often contrasted with radial cleavage; spiral cleavage is typical of molluscs and other related phyla in the superphylum Lophotrochozoa, such as Annelida, Sipuncula, Echiura, Nemertea, and Platyhelminthes, but not lophophorate phyla such as Brachiopoda and Phoronida.

spirocyst A type of cnidocyst (nematocyst) consisting of a single-walled capsule that contains a coiled tubule with sticky microfilaments; found in Hexacorallia.

spore A cell produced by division or budding from a parental cell; can be a resistant stage in bacteria and some unicellular eukaryotes, the beginning of a new multicellular body in algae that is genetically the same as the parent, or if produced meiotically, the beginning of a genetically novel haploid lineage in some protozoans and algae (developing into gametophytes in multicellular algae).

sporophyte The diploid portion of a multicellular algal life cycle that produces haploid spores by meiosis; the haploid spores are released, settle, and develop into haploid gametophytes.

spring tides Extreme high and low tides that occur about twice monthly, coinciding with the full and new moons.

stability The ability to withstand perturbations without large changes in community properties such as biomass, cover, or productivity.

standard deviation A measure of variability among measurements (replicates) in a sample; the square root of the variance.

standard error A measure of the variability of average values among repeated samples of the same measurements.

stalk The narrow, sometimes quite elongated portion of a stauromedusa below the calyx, terminating in an adhesive basal disk, with which it attaches to the substrate. Also, **peduncle**.

state waters The zone at least sometimes covered by seawater and subject to state regulation.

static stability The ability of a fluid at rest to become turbulent (i.e., having reduced or negative stability) or laminar (i.e., having increased or positive stability) as a result of the effects of buoyancy.

stationary phase A plateau of the growth curve after log growth, during which cell number remains constant; the production of new cells at the same rate at which older cells die.

statolith A calcified ear stone in the statocyst or auditory organ of many invertebrates.

stenohaline Able to tolerate only a narrow range of salinity.

stiffness A measure of a material's rigidity (the inverse of flexibility), calculated as the slope of the stress–strain curve. Also, MODULUS OF ELASTICITY, YOUNG'S MODULUS.

stipe A stalklike structure that connects the sheetlike or branched blade of a macroalga to its attachment structure, the holdfast.

stirring The distortion, stretching, and thinning of fluid elements by velocity gradients, which leads to enhanced molecular diffusion and thus increases mixing.

stolon 1. In octocorals, a runner-like extension of coenenchyme from which polyps arise. 2. In boring fungi, any of the thin hyphae passing through the rock and forming satellite colonies at their ends.

strain A normalized measure of material deformation, defined as the change in an object's length divided by its original length.

strain energy The amount of energy that a material absorbs, measured as the area under the stress–strain curve.

stratified Having distinct layers.

strength The stress at which a material fails.

stress 1. In mechanics, the force on an object, whether tensile, compressive, or shearing, divided by the area over which the force is applied. 2. In ecology, any factor that results in a reduction in an organism's performance (growth, survival, etc.). Abiotic (or physical) stress refers specifically to stress related to nonbiological factors such as temperature, moisture, and oxygen availability.

strigilis Segments 7 through 10 of the oviger in pycnogonids, which in many species resembles a shepherd's crook; hence, the alternate name SHEPHERD'S CROOK.

strobilation Serial transverse fission of a scyphozoan polyp to produce juvenile medusae.

sublittoral [zone] The region on a shore that is below the low-water line; the uppermost region of a shore that is constantly covered by water or falls free only during extremely low tides.

subsistence fishers A subgroup of small-scale fishers consisting of individuals, families, or tribes using non-mechanized gear; the harvested marine resources are primarily consumed by the fishers themselves or shared among kin.

subsurface chlorophyll maximum (SCM) layer A layer of concentrated phytoplankton in the water column, usually associated with the thermocline and pycnocline.

succession Transitions that take place in a community as it undergoes a series of species colonizations and replacements over time.

sulfide system The biotic community of deeper anoxic layers in subtidal and sheltered intertidal sediments where sulfide produced by sulfate-reducing bacteria is present; the sulfide system is separated from superficial oxic layers by the redox potential discontinuity (RPD) layer.

supply-side ecology A term used to describe the consequences of variations in larval or propagule supply to the development of community structure and community dynamics.

supralittoral [zone] The area of shore immediately above the highest mean tidal level. Also, SPLASH ZONE, SPRAY ZONE.

surge channel A narrow gully open at one or both ends to water and wave action.

suspension feeder An organism that feeds on particles of organic material suspended in the water column.

swell 1. Regular waves having fairly well-defined crests and troughs, characterized by one dominant period and uniform wave heights. 2. Water waves that have been generated by winds at an appreciable distance from the point of observation.

symbiont An organism that lives in close association with another.

symbiosis A durable and intimate interspecific interaction, literally "living together"; the interaction may be positive for both species (mutualism), negative for one species (parasitism), or neutral for one species and positive for the other (commensalism).

syncytium (*plural* **syncytia**) A multinucleate mass of protoplasm without cell membranes separating individual cells.

syzygy The time of the lunar cycle at which the tidal excursions are greatest: the phases of the full and new moon.

taxon (*plural* **taxa**) The named classification unit (e.g., *Homo sapiens*, Hominidae, or Mammalia) to which individuals, or sets of species, are assigned.

tegmentum The upper shell layer of a chiton valve, having aragonitic calcium carbonate combined with other minerals as well as living tissue (especially sensory organs known as esthetes), sometimes also with larger photosensory ocelli.

temperate Describing or referring to the two regions of the earth that are between the polar circles (66.5° latitude) and the tropics (23.5° latitude).

temperature A measure of the average kinetic energy of the particles in a sample of matter, expressed in terms of units or degrees designated on a standard scale.

tenacity A measure of how much force is required to dislodge an animal from the substratum.

tentacles In sea cucumbers, feeding structures with sticky mucus that collects particles.

terrigenous Of beach material, derived from terrestrial or continental rock or mineral sources.

test In barnacles, the calcareous outer shell.

tetrapyrrole An organic molecule composed of four molecules of pyrrole (a heterocyclic aromatic organic compound organized into a five-membered ring, with the chemical formula C_4H_5N). Tetrapyrroles are important components of chlorophylls, phycobilins, hemoglobins, and bilirubin pigments.

thallus (*plural* **thalli**) The body of an alga that is not differentiated into specialized cell types.

theca (*plural* **thecae**) A chitinous exoskeleton found in some hydroid species that extends to cover otherwise naked hydranths and gonophores.

thermal conductivity The ability of heat to move along a temperature gradient within a material such as air or water.

thermal inertia The ability to damp the response in body temperature to fluctuations in the environment; depends on a ratio of factors, such as mass and specific heat, that resist changes in temperature.

thermal wind The wind difference between two pressure levels p_1 and p_2 such that the pressure at level p_1 is less than the pressure at level p_2, which implies a difference in the horizontal distribution of temperature.

thermocline A region of rapid vertical change in ocean temperature that separates the warm, nutrient-poor upper layer from the cool and nutrient-rich deeper layer.

thin layer A concentration of plankton in a layer that has a small vertical thickness but covers a large horizontal area.

thrust The force generated against the environment for propulsion.

tidal amplitude The height difference between high slack and low slack tides.

tidal datum A level surface with its zero elevation defined relative to the measurements of a local tide gauge (thus it is local in scope); measurements or predictions are made over a period that includes the most important tidal cycles (often 19 years), and a summary statistic is chosen. Common tidal datum values are mean-lower-low water, mean-sea level, and lowest astronomical tide.

tissue A group of similar cells arranged into a functional unit. A tissue may be arranged in sheets attached to an extracellular basement membrane, as a matrix of connective tissue, in bands or sheets of muscles, or in units of nervous systems.

titration A laboratory method used to determine the concentration of dissolved substances in a solution.

torsion 1. The twisting of an object, or the strain produced by twisting. **2.** The counterclockwise rotation of the gastropod visceral mass 180° relative to the antero-posterior axis of the head-foot complex. This developmental event in the larva rotates the posterior mantle cavity anteriorly, where it resides over the head of the adult snail.

total allowable catch (TAC) The quantity of a species that may be set as a quota to restrict the annual exploitation of a fishery; usually refers to the commercial catch.

trade winds Tropical winds driven by the differential heating of the lower atmosphere near the equator; air from higher latitudes moves toward the equator to replace equatorial air that has been heated and lifted out of the lower atmosphere, but the incoming air is deflected westward by the Coriolis effect.

transect A line or band along which sampling units (quadrats or sampling points) are laid to obtain estimates of abundance of species in that location.

transgression SEE MARINE TRANSGRESSION.

transmittance The fraction of irradiance transmitted through and emerging from a translucent body such as an alga.

transporter A protein that shuttles biologically important molecules across cell or nuclear membranes or both.

Each different type of transporter protein is specific in the molecule(s) it recognizes and to which it binds.

Triassic Period The geologic period 245–202 million years ago, beginning the Mesozoic Era, in which sea urchins and ammonites predominate in the marine fossil record.

tridacnid bivalve A giant clam in the family Tridacnidae. Anchored to hard substrates, these clams expose their fleshy mantle to sunlight through a large shell gape and derive nutrition from the photosynthetic activity of symbiotic zooxanthellae contained in the mantle.

tripeptide A molecule consisting of three amino acids in a chain; it may, as in the case of certain luciferins, form a ring structure.

triploblastic Describing or referring to a body plan that has three well-defined tissue layers (ectoderm, endoderm, and mesoderm in between).

trochophore A larval form found in annelids, molluscs, and sipunculids. The body is spherical and has a horizontal ciliary ring and ciliary bundles at the top and bottom. Larvae of some other animal groups have a similar form, indicating an evolutionary relationship to trochophores.

trophic Describing or referring to feeding relationships between organisms (food chain).

trophically transmitted parasites Parasites transmitted from their host to a predator on that host; thus, death of the first host is required for completion of the parasite's life cycle; pathology in the predator is intensity-dependent.

trophic cascade A chain reaction of effects extending downward through food webs from apex predators to plants.

trophic level The position of a species in a food web as determined by how many consumption links are below it.

tropical Describing or referring to the region of the earth that straddles the equator, with poleward boundaries at the Tropic of Cancer (23.5°N) and the Tropic of Capricorn (23.5°S).

tropical storms Storms that develop close to the equator during the summer to fall, their energy being derived from the warm surface water; the most severe of these storms are known as hurricanes, typhoons, or cyclones, depending on where they occur.

troposphere The lowest 10–20 km of the atmosphere, bounded above by the tropopause, and characterized by decreasing temperature with height, appreciable vertical wind motion, appreciable water vapor, and weather.

trumpet hyphae In some kelps, colorless, elongate cells in the medulla that are expanded (like the bell of a trumpet) at the cross-walls and are involved in the transport of sugars and other organic compounds.

tsunami A water wave, or waves, generated by ground motion such as an earthquake, landslide, or submarine slide.

tube feet In echinoderms, cylindrical external structures having suction cups at their ends; used in locomotion and feeding.

turbulence Random, irregular motion in fluid flow, in which vortices break up into smaller eddies, causing rapid mixing. Also, TURBULENT MIXING.

typical parasite SEE MACROPARASITE.

ultraplankton Unicellular planktonic organisms <2 microns in size.

unconformable Of geological strata, not lying in a parallel position; not sequential.

undertow Offshore flowing water beneath the surface that balances the onshore flow of water carried between the crest and trough of waves approaching the beach.

unitary organisms Organisms that are noncolonial individuals. They are not constructed of repeated multicellular units and exhibit little or no branching.

universal solvent A liquid that can dissolve polar and nonpolar solutes.

unsaturated colors Colors that are diluted with large amounts of white, gray, or black, reflecting light across all wavelengths in a spectrum.

uprush The water carried up the beach face as a wave breaks and washes shoreward.

upthrust An upward force on an object in water.

upwelling The movement of deep, cold, nutrient-rich waters to the surface through either the action of winds or the effect of bottom topography.

urchin barren An area where all seaweeds except crustose coralline algae have been removed by sea urchin grazing.

uropods The laterally placed sixth (terminal) pair of abdominal appendages in a decapod crustacean.

valve Any of the (normally) eight shell plates of a chiton.

van der Waals attractions Weak attractive forces between molecules caused by transient dipole interactions.

variance A measure of variability among measurements (replicates) in a sample, the square root of which is called the standard deviation.

vector In ecology, the physical means by which a species is transported, such as ballast water, ships' hulls, or the movement of oysters.

veliger The free-swimming larval stage of gastropods and other molluscs, in which the shell, foot, and velum appear between the initial trochophore stage and post-settlement stages.

velum A pair of ciliated lobes in the veliger larval stage of gastropods and other molluscs, used for swimming and feeding, and lost immediately after settlement.

velocity gradient A difference in velocity between adjacent layers of fluid.

vermigon An unsegmented, limbless, wormlike rhizocephalan stage that is injected into the host, migrates through the hemocoelic system, and grows into the adult parasite.

vertebrates Animals with a backbone; includes mammals, birds, reptiles, amphibians, and fishes.

vertical transmission A life history in which the progeny of symbionts are passed to the progeny of hosts (in gametes, embryos, or juveniles) without any free-living stage.

vertical zonation The distribution of organisms along the vertical marine-to-terrestrial gradient that is characteristic of the intertidal zone. Also, INTERTIDAL ZONATION.

visceral mass The body cavity containing digestive, renal, circulatory, and reproductive organs of the mollusc.

viscosity The measure of a fluid's resistance to flow, or more precisely, its resistance to deformation by shear stress; a result of the internal friction of molecules in the fluid. Also, DYNAMIC VISCOSITY.

viscous extrusion medium A thick substance containing gametes in some species.

vitelline envelope A glycoprotein layer (not a membrane) formed by the contents released during cortical vesicle fusion with the egg plasma membrane during the slow block to polyspermy. Sperm receptors on the vitelline envelope bind to bindin on the sperm head.

volant Able to fly.

vulnerable A designation by the World Conservation Union for species that are facing high risk of extinction in the wild in the medium-term future.

wake The relatively slow-moving region found behind an object in a flowing fluid.

water column The water between the bottom and the surface of the ocean.

water mass A parcel of water having uniform characteristics such as temperature, salinity, and nutrient content.

water vascular system In echinoderms, a series of water-filled canals that function in locomotion.

wave celerity The speed of advance of individual ocean waves.

wave climate The wave conditions experienced at a specific location in the sea over the years, including the range of wave heights, periods, and directions of travel.

wave dispersion The dependence of the speed of advance of ocean waves on their periods and water depth, with the longer-period waves traveling faster than the shorter-period waves; dispersion causes waves to sort themselves out after they leave the area of the storm and cross the ocean, becoming more regular waves (swell).

wave energy The kinetic energy contained within an ocean wave as a result of its orbital motions of the water, plus the potential energy associated with the difference in the form of the wave compared with a flat water surface.

wave exposure 1. A general description of the size and force of waves as they reach the shore. Wave exposure is related to swash, splash, and spray, all of which can reduce physical stress. 2. The effect of topography on propagation of wave energy to the shore; the shallower sea floor in front of points and headlands slows the wave at the base, causing the wave front to refract toward the obstruction; in addition, areas with low wave exposure may be protected by reefs or headlands that intercept approaching waves.

wave height The vertical distance from the trough of an ocean wave to its crest; twice the wave amplitude.

wavelength The horizontal distance between successive crests in an ocean wave.

wave period The interval of time between successive waves in a series.

wave power The rate at which the energy of a wave is carried across the ocean. In shallow water, the wave's energy multiplied by its celerity. In deep water, the wave's energy multiplied by half its celerity.

wave refraction The bending of the crests of ocean waves and changes in the direction of their advance, due to variations in depth and orientation relative to the coastline, as the waves cross the shallow waters of the continental shelf.

wave runup A surge of water that flows up the shore after a wave has hit the shore.

webover A behavior of an octopus in which it expands the webbing between its arm over a rock to catch prey underneath.

weight The force exerted on a mass by the acceleration of gravity (9.8 meters per second per second); for example, a mass of 1 kilogram has a weight of 9.8 newtons.

wind direction The direction from which the air is moving.

wind shear The local horizontal or vertical variation of the wind.

wind speed A measure of how fast the air is moving, expressed as nautical miles per hour (knots), meters per second (m/s), or miles per hour (mph).

wind stress The force per unit area due to action of the wind on the sea surface.

World Conservation Union An international organization committed to conserving the integrity and diversity of nature and to ensuring equitable and sustainable use of natural resources. Also, INTERNATIONAL UNION FOR THE CONSERVATION OF NATURE AND NATURAL RESOURCES (IUCN).

Young's modulus SEE STIFFNESS.

zeitgebers Factors such as day length, tidal cycle, and the like, used by species as a cue to maintain and set their internal clocks.

zonation SEE VERTICAL ZONATION.

zooid SEE MODULE.

zooplanktivorous Feeding on zooplankton (small animals that live suspended in the water column). Thus, **zooplanktivory.**

zoospore A spore that is mobile (has flagella).

zooxanthellae Symbiotic unicellular algae (dinoflagellates) that live in the inner gastrodermal cells of corals and some other tropical invertebrates.

zooxanthellate corals Stony corals that harbor photosynthetic algae (dinoflagellates) in their gastrodermal tissues.

zygote A diploid cell resulting from the fusion of two gametes.

algal color, 33–37
    evolutionary diversity, 33–34
    functional anatomy, 35
    interpreting, 34–36
    major groupings, 33–34
    reasons for, 33, 36–37
algal crusts, 37–41
    defenses against herbivores, 40
    desiccation, 38
    disturbance, 38–39
    favorable environments for, 38–40
    growth, 39
    lichens, 40, 175
    life histories, 40–41
    recognizing, 37–38
    reproduction, 40–41
    variation, 39–40
algal economics, 41–45
    algal uses and products, 42
    cultivation and harvesting, 41–42
    cultural value and indigenous uses, 42–43
    ecosystem services, 44
    tourism, aesthetic and economic value,
        43–44
algal extracts, 42
algal life cycles, 45–47
    ecological implications, 47
    reproduction, 47
    types of, 45–47
algal ridges, 22, 163
algal turfs, 47–50, 204
    coral reefs, 49
    defined, 47
    herbivory and, 48–49
    intertidal, 48
    subtidal, trophic importance of, 48
    temperate, 48
    tropical, 48–49
alginates, 12
Allee, W. C., 474
ameiotic parthenogenesis, 461
American black bears, 618
American mink, 618
amino acid 3,4-dihydroxyphenylalanine
        (DOPA), 7
ammonium, 216
amnesic shellfish poisoning (ASP), 33
amphidromic point, 593
amphipods, 50–53
    environmental physiology, 51
    food habits, 50–51
    life cycles, 52
    reproduction, 60
    taxonomy and morphology, 50
amplification, defined, 635
amplifiers, 635–636
Anderson School of Natural History, 353
anemones. See sea anemones
Animal Aggregation: A Study in General
        Sociology, 474
anoxia, 32, 33
anisogamy, 516, 517
anthropogenic disturbance, 192
anthropogenic ecosystem changes, 202–204.
        See also human impacts

apomixis, 47
apparent competition, 156–157
applied ecology, 381
aquaria, public, 53–56
    artificial substrates, 54
    biological environment, recreating, 54–55
    cost of admission to, 44
    exhibit lighting, 54
    life support systems, 54
    live touch experiences, 54–55
    physical environment, recreating, 54
    reproductive programs, 55
    rocky-shore exhibits, creating, 53
    role of in minimizing human impacts, 55
    temperature, 54
    visitor experience, 55–56
    water motion, 54
aqueous solubility, 124
Archaea, 367
Archimedes' principle, 118
Arctic foxes, 617, 618
areas of high disturbance, defined, 38
Aristotle's lantern, 196, 362, 511, 513
Arrhenius, Svante, 137
arthropodins, 125
arthropods, 56–60
    Chelicerata, 56
    classification of, 56
    Crustacea, 56
    Hexapoda, 56
    key characteristics of, 57
    molting, 57–59
    Myriapoda, 56
    Pycnogonids, 56
    reproduction, 59–60
    sclerotization, 58
    tidepool crustaceans, 56–57
articulated coralline algae, 21, 22–23, 48
artificial reefs, 260
artificial tagging, 184–185
ascidians. See tunicates
asexual reproduction, 461, 509–510
    budding, 461, 540
    fragmentation, 15, 47, 461, 540
    vegetative reproduction, 495
associational defenses, 218, 220
astropods, 378
atmospheric composition, historical variation
        in, 9–10
atmospheric pressure, 640
avascular circulatory system, 135
average flow speed, 111

backscatter sensors, 628
Bacteria, 367
Bakan Upwelling Index, 611
Baker, Katherine Drew, 42
barnacles, 56, 61–64, 245
    adhesion by, 4, 5, 6
    defenses against, 63
    growth, 63–64
    interactions, 62–64
    life cycle, 61–62
    limpets, presence of, 63
    parasitic (rhizocephalans), 465–468

reproduction, 59, 61
research, new directions in, 64
settlement cues, 62
Bartholomew, George, 525
bathymetric upwelling, 32
beaches, 65–67
    cleanup efforts, 44
    defined, 65
    erosion of, 65
    human impacts on, 65, 67
    nourishment, 67
    sand
        beach shape and movement of, 66–67
        formation and sources of, 65–66
        slope differences, 65–66
Beaufort wind speed scale, 546, 547
bêche-de-mer, 197
benthic-pelagic coupling (BPC), 68–71
    defined, 68
    ecological importance of, 68
    internal waves, 70–71
    Langmuir circulation, 71
    sedimentation, 68
    upwelling, 68–70
Bernoulli's principle, 278
Bertness, Mark, 536
Between Pacific Tides, 56, 151, 474
biodiversity
    as biological insurance, 82, 84
    complementarity, 81
    consumers and, 84
    defined, 81–82
    ecosystems, importance for, 84
    effects of on ecosystem structure and
        function, 82
    exotic species, invasion of, 83–84
    facilitation, 79, 82
    global patterns of, 72–73
    intertidal cultivation and harvesting, effects
        of, 233–234
    latitudinal gradients, primary causes of, 73–75
        Cenozoic radiation hypothesis, 75
        evolutionary speed hypothesis of, 75
        niche–assembly hypothesis, 74–75
        species–energy hypothesis, 74
        temperature and diversity, 74–75
    maintenance of, 76–81
        competition for limited resources,
            species approach to, 78–79
        disturbance and, 80
        facilitation and, 79
        postsettlement factors, 77–80
        predation in, 79–80
        scale of observation, increasing, 80–81
        species diversity within dispersal distance,
            77
        supply of individuals, 77
    measurement of, 71–72, 75–76
    redundancy, 81, 84
    sampling effect, 81
    scale dependent patterns of, 73
    seaweeds and seagrass, 19–20, 82–83
    sessile invertebrates, 83–84
    significance of, 81–85
    sustainable food uses and, 233–234

impact of on predation by spiny lobsters, 334–335
impingement force, 280
lift, 119, 278, 282
locomotion and, 337–338
organismal response to, 361–362
patch formation in space-limited habitats, 282
scaling considerations, 283
shoreline fluid forces, generalities of, 277
tissue properties and force, 281
virtual buoyancy, 279
hydroids, 284–285
  defined, 284
  distribution of, 284
  diversity of, 284
  morphology and reproduction, 284–285
  species interactions, 285
hydromedusae, 286–288
  ecology, 287
  form and structure of, 286–287
  habitat and distribution, 286
  reproduction, 287
  species, list of, 286
  taxonomy and phylogeny, 287
hydrostatic lift, 118–119
hydrostatic skeleton, 307
hypoxia, preventing with seawater, 515

ice, 256–257
  free-floating, 289
  marginal ice zone (MIZ), 32
icebergs, 289
ice edge algal blooms, 32
ice foot, 289, 290
ice scour, 80, 289–291
  defined, 289
  effect of on communities, 290
  general attributes, 289–290
  and global warming, 290
  temporal and spatial variation, 290
iguanas, marine, 291–294, 618
  algae grazing by, 292
  coping with climate variation and El Niño, 208, 291, 293
  ecological role of, 292
  evolution, 291
  hybridization, 292
  mortality, reductions in, 293
  physiological adaptations, 292
  reproductive traits, 291–292
  shrinking to survive, 293
impact pressure, 104–105
impingement force, 280
indicator species, 393
inducible defenses, 441
infrared thermography, 582
inhibition model of succession, 557
innkeeper worms. See spoonworms
inner-shelf flow, 322–323
  insects
    beetles, 471
    flies, 227–228
    midges, 471
    tidal rhythms and, 471–472
intensive sampling, 393

interference competition, 154
Intergovernmental Panel on Climate Change (IPCC), United Nations, 137
Intermediate Disturbance Hypothesis, 80, 163, 191, 558–559
internal waves, 637–639
  benthic-pelagic coupling (BPC), 70–71
  biological importance, 638–639
  defined, 637
  general properties of, 637–638
  open research questions, 639
  propagation and generation, 638
  and transport of plankton into the intertidal, 400–401
International Terrestrial Reference Frame (ITRF), 565
interspecific competition, 154–155
intertidal lizard, 618
intracellular crowding, 622
intraspecific competition, 154–155, 157
introduced species, 204, 294–297, 571
  in British Columbia and Washington State, 296
  in the Canadian Maritimes and New England, 294–295
  dispersal and, 183
  invaded shores, ecology of, 294–296
  limpets, 332
  mariculture and, 233–234, 384
  seaweeds, 20
  in South Africa, 101, 295–296
  in South America (Chile), 296
  transport vectors, 294
  undetected invasions, history and scale of, 296
invertebrates, 297–314
  arthropods, 56–60
  basic characteristics, summary of, 303–305
  biofilms, settlement effects of, 86
  body cavities, 307–309
  body size, 306–307
  body symmetry, 306
  brachiopods, 112–113, 298
  circulatory system in, 133–134
  classification, 299
  coelom, 307, 308
  commensalism, 312
  developmental features, 309–310
  dispersal, 313, 314
  distribution of, 297
  diversity of, 297
  echinoderms, 194–197, 306
  ecology, 310–314
  ectoderm/endoderm/mesoderm, 300, 302, 306
  egg cleavage, 309
  endosymbiosis, 312–313
  evolution of, 298
  excretion of ammonium, and seaweed growth, 216–217
  extinction rates, 298, 314
  feeding, 310–311
  habitat, 311–312
  hermaphroditism, 313
  hydrostatic skeleton, 307

larval types, 310
life histories, 314
mutualism, 312
octopuses, 415–418
parasitism, 312
phyla, 297–298, 299–310
  with representatives on rocky shores, list of, 301–302
  present and past, 297–299
preserving, 152
pseudocoelom, 307–308
reproduction, 313–314
sessile, diversity in, 83–84
sponges, 300
symbiosis, 312
isopods, 50–53
  camouflage, 122
  environmental physiology, 51
  food habits, 50–51
  life cycles, 52
  reproduction, 60
  taxonomy and morphology, 50

jaw worms (gnathostomulida), 371
jellyfishes, 359
  hydromedusae, 286–288
  stauromedusae, 541–543
jewel box shells, 99
jingle shells, 99

Keeling, Charles, 137
kelps, 14, 55, 315–320
  community and ecosystem effects, 319–320
  ecological significance of, 315–316
  El Niños and, 319, 520
  forests, 12, 139, 209, 260, 317, 505
  geographic distributions, 317–318
  life history and morphology, 316
  local distributions, 318–319
  predator diversity and, 84
  recruitment, 317
  reproduction, 44, 316–316, 459
  sea urchins, deforestation by, 319, 442
Kelvin waves, 172, 207, 593
keystone predation, 139, 197, 202, 205, 333, 439–440, 503
kinetic energy, 211, 581
king crabs, 166–167
kingfishers, 92
kinorhyncha, 371
Kleiber's law, 525, 529
Kolmogorov scale, 602
Kuroshio Current, 616

laminar flow, 111
landslides, 260, 351
Langmuir circulation, 71
La Niña, 207–208, 293
LaPlace, Simon, 593
larval settlement, 321–325
  conceptual framework, 321–322
  dispersal in inner-shelf waters, 322–323
  larval supply to the intertidal, 323
  mechanics of, 321, 323–325
  symbiosis and, 571

measurement of, 482, 515, 627–628
micrometozoans and, 373
stress, 482–484
  ameliorating with seawater, 515
  ecological consequences of, 483–484
  physiological and behavioral responses
    to, 482
  salinity variability and, 482–483
  solutions to, 482
sampling effect, 81
Sanctuary Integrated Monitoring Network,
  352
sand, 18, 257
sand cycle. *See* beaches
sand dollars, 194
sandpipers, 92, 93
saturation mixing ratio, 228
saturation vapor concentration, 214
saxotoxins, 32
scaling
  algal breaking force and maximum size,
    527–528
  body size and population density, 529
  constraints on biological force production,
    528
  defined, 525–526
  of ecological interactions, 528–529
  physiological, origins of, 526–527
  size and, 525–530
  to test functional hypotheses, 527–528
  and territory size, 529
scallops, 99
scientific method, intertidal opportunities for
  using, 205–206
sclerotization, 58
scoters, 92
scramble competition, 154
sculpins, 223, 224, 485–486
  adaptation to tidepool conditions, 485–486
  air-breathing abilities, 226
  biogeography, 485
  interactions within and among species, 485
  life cycle, 485
  recruitment, 485, 486
  reproduction, 485
  responses to oceanic conditions, 486
sea anemones, 486–491
  body shape, 105, 486
  cloning by, 489
  defenses, 488–489
  desiccation, 489
  exposure to air, effects of, 9
  food gathering by, 282
  habitats, 486–487, 488–489
  interactions with other organisms, 489–491
  as intertidal animals, 487–488
  perspectives from beyond the rocky
    intertidal, 491
  as predators/prey, 489–490
  reproduction, 487–488
  symbiosis, 490–491
  water storage by, 174
The Sea Around Us, 151
sea breeze circulation, 642

sea chubs, 223
sea cucumbers, 114, 194, 196, 197, 491–492
  anatomy and physiology, 491–492
  ecology and behavior, 492
  economic and medicinal importance, 492
  locomotion, 491
  reproduction, 460, 492
seagrasses, 493–496
  biodiversity of, 19–20, 82–83
  biology, 493–494
  defined, 493
  ecological value, 495–496
  habitat and distribution, 494
  reproduction, 460, 494–495
  restoration of, 260
  and seaweeds compared, 493
  threats to, 496
seals and sea lions, 498–503
  breeding behavior, 500
  classification, 498–499
  diving or foraging behavior, 501–502
  eared seals, 499
  earless (true) seals, 499–500
  fur seals, 499
  ice breeding, 501
  interactions with humans, 502–503
  land breeding, 500–501
  molting, 501
  rookeries, 500
  sea lions, 499
  senses, 501
sea level change, 497–498
  coastal erosion and, 497–498
  and coral reefs, 163–164
  global versus local, 497
  global warming and, 138
Sea of Cortez, 474–475
sea otters, 503–506, 618
  as abalone predators, 3
  behavior and population biology, 504–505
  distribution of, 503, 504
  ecosystem roles, 84, 505–506
  feeding, 504
  foraging, 505–506
  fur trade and, 504, 575
  grooming, necessity of, 505
  history and current status, 504
  mating and reproduction, 505
  organismal biology, 503–504
  physiology, 503–504
  prey depletion by, 505
  relocation efforts, 504
  trophic cascade role, 505–506
sea palm, 13, 17, 350
sea slugs. *See* nudibranchs and related species
sea squirt. *See* tunicates
sea stars, 8, 113, 114, 196, 506–510
  camouflage, 123
  distribution, 506
  endoskeleton, 508
  external anatomy, 508
  feeding and digestion, 508
  foraging and ecological effects, 510
  general body plan, 506–507

locomotion, tube feet and, 507
nervous system and sensory information,
  508
pedicellariae, 508
predation by, 202
regeneration, 510
reproduction, 508–510
sea surface temperature (SST), 207, 265,
  614–615
sea urchins, 23, 113, 114, 194, 196, 510–513
  adhesion by, 5
  anatomy and development, 195, 511
  Aristotle's lantern, 196, 362, 511, 513
  defense of cavities, 513
  excavating cavities, 512–513, 544–546
  feeding, 511, 513
  intertidal, 511
  irregular vs. regular, 510
  kelp deforestation by, 319, 442
  reproduction, 511
  spine morphology, 511–512
seawater, 514–516
  and air compared, 8–9, 546
  characteristics and importance of, 514
  composition of, 480, 621
  density, determinants of, 480
  desiccation stress, preventing, 514–515
  dissolved gases in, 623–625
  and freshwater compared, 514
  heat and cold stress, ameliorating, 514
  hypoxia, preventing, 515
  pH, 624–625
  relative density, 8
  salinity, measurement of, 480–482
  salinity stress, ameliorating, 515
  surface tension, 561
  ultraviolet stress, protecting against, 515
seaweeds. *See also* algae
  architecture, growth, and functional forms,
    12–15
  biodiversity of, 19–20, 82–83
  boulder fields, 18
  carotenoids in, 12
  chemical defenses, 19
  chlorophyll in, 11–12
  Chondrus genus, 28
  cobbles, 18
  color groups, 11–12
  conservation issues, 20–21
  consumption, strategies to avoid, 19
  controlled seeding, 42
  crashing waves, influence of, 17–18
  cultivation and harvesting, 20, 41–42, 233
  cultural value and indigenous uses of,
    42–43
  defenses against herbivory, 273
  desiccation, response to, 175
  dispersal, 182, 183
  distribution, general patterns of, 26–27
  El Niño, effects of, 20–21
  Fucus genus (rockweeds), 28
  global demand for, 41
  global warming and, 20
  health risks associated with, 44

seaweeds (*continued*)
  human impacts on, 20, 29
  interacting with other seashore denizens, 18–19
  intertidal zonation of, 16–17
  as invasive species, 20
  invertebrate-excreted ammonium as a nitrogen source for, 216–217
  life cycles, 15–16
  reproduction, 15, 459–460
  sand and, 18
  and seagrasses compared, 493
  as shape-shifters, 19
  storing energy and cellular walls, 12
  sunlight, influence of, 17
  tidal waves and environmental gradients, 17–18
  tourism and, 43–44
  trampling, vulnerability to, 259
  understory species, 19
  uses and products derived from, 42
Sebens, Kenneth, 525
secondary succession, 555
sedimentation, 68, 192, 257, 320, 497–498
segmented worms (annelida), 645–646
"sense of smell," 125
serpent stars. *See* brittle stars
sessile invertebrates, diversity in, 83–84
settlement, larval, 321–325, 456–457
sex allocation, 516–521
  defined, 516
  evolutionary outcome, 518
  game theory models of, 517
  parental investment and, 517
  and sperm competition, 517–518
sexual reproduction, 461–463
sexual selection, 516–521
  defined, 516
  evolutionary outcomes, 517
  evolutionary processes, 517
  harmonic mean promiscuity and sperm competition, 520–521
  measuring the opportunity for, 519
  opportunity-for-selection theory, 518–519, 520, 521
  optimal-fitness-returns theory, 516
  parental investment measures of, 516–517
  parental investment theory, 516
  and sexual differences, 519–520
shape coefficient, 364
shear, 109
shear stress, 109–110, 222, 323, 324, 360
shell habitat webs, 273
ship groundings, 260, 351
ship waves, 477
shore crabs, 167–168
shrimps, 56, 521–524
  camouflage and color change, 523
  distribution, 522–523
  life cycle, 523
  mantis, 341–344
  morphology and taxonomy, 521–522
sieve elements, 12
sipunculans, 524–525
  classes and orders, 525

ecology, 524–525
form and structure, 524
internal anatomy, 524
reproduction, 524
"sit-and-wait" predators, 121, 510
size and scaling. *See* scaling
size-selective harvesting, 203
skimming flow, 111
slicks, 400, 638
slimes, 356–358
small-scale fisheries, 199–201
snailfishes, 223
snails, 530–537
  adhesion, 7, 534
  biology, 530–531
  community structure, 535–536
  environmental heterogeneity, 531–532
  growth, 534
  herbivorous, 442, 531, 535–536
  plasticity, 534–535
  predators, 532, 536
  radula, 530, 531, 536
  shell, 530, 533
    as homes for hermit crabs, 274
  tidal height, 532–533
  water loss, response to, 174–175
  wave action, 531, 533–534
Sousa, W. P., 156, 559
Southern Oscillation. *See* El Niño/Southern Oscillation (ENSO) events
spatial foraging, 241
spatial variation
  competitive intensity reduced by, 158
  in foraging behavior, 241
  impact of ice, 290
  in regime of disturbance, 187
specialization, cnidaria, 143
species distribution, British Isles and Europe as baseline for, 383–384
species–energy hypothesis of biodiversity, 74
species name, 299
specificity, 571
sperm dilution, 221
sperm–egg interactions, 117, 126
sperm–egg recognition, 220
spider crabs, 168
spiders. *See* pycnogonids
spiny lobsters
  as keystone predators, 333–335
  predation manipulation, 440
sponges, 300, 537–540
  classification, 540
  colors, 537
  competition and predation, 539
  feeding, 537–538
  form, 539
  general morphology, 537–538
  growth and regeneration, 539
  habitat, 538–539
  individual or colony, 537
  internal anatomy, 537
  locomotion, 539–540
  reproduction, 540
  stone boring, 543–544
spoonworms (echiurans), 197, 645

spring-scale dynamometers, 633–634
spring tides, 468, 592
stable density stratification, 637
starfish. *See* sea stars
stauromedusae, 541–543
  biogeography, 541–542
  ecology, 542
  form and structure, 542
  habitat, 541–542
  life cycle, 542
  locomotion, 541–542
  phylogeny, 543
  reproduction, 542
Steinbeck, John, 150, 349, 350, 473–476
steppingstone model of population exchange, 247–248
stereom, 194
stiffness, defined, 356
still-water environments, locomotion in, 336
stolon production, 47
stone borers, 543–546
  sea urchins, 544–546
  sponges, 543–544
  worms, 544
stone crabs, 166–167
storm intensity
  wave heights, 546–550
  wind waves
    at the coast, 549
    generation and propagation, 547–548
    winds over the oceans, 546–547
storms
  characteristics, 550
  and climate change, 550–553
  effects of on shore habitats, 256
  extratropical cyclones, 550, 551–552
  tropical cyclones, 410, 550, 552–553
strain, defined, 356
strain energy, 356
strength (materials), 360–366
  adhesive, 7
  causes of, 362, 364
  ceramics, 362
  composites, 362
  defined, 356, 360
  elastomers, 362
  gels, 362
  measurement of, 364–365
  mechanical properties, intertidal materials, 360–364
  polymers, 362
  substratum vs. organism, 364
stress, 356, 360–362
  cold, 148–150, 514
  defined, 106
  desiccation, 173–177, 514–515
  as disturbance, 186
  environmental, and larval survival, 366–367
  and facilitation, limits on, 219
  heat, 266–270, 514
  lethal, 148–149
  material, 356, 360–362
  salinity, 482–484, 515
  shear, 109–110
  sublethal, 148, 149–150

MARK W. DENNY is the DeNault Professor of Marine Sciences at Stanford University's Hopkins Marine Station in Pacific Grove, California.

His research explores the application of physics and engineering principles to the study of intertidal ecology, a model system for the emerging discipline of "ecomechanics." He has published more than 80 articles on subjects ranging from locomotion in invertebrates to hydrodynamic forces on surf-zone seaweeds and animals to the prediction of thermal death in limpets.

He is the author of four books, one of which (*Biology and the Mechanics of the Wave-Swept Environment*) deals with life on rocky shores and another of which (*Air and Water: The Biology and Physics of Life's Media*) was awarded an Association of American Publishers award in 1993 for the Best New Book in Astronomy and Physics. Dr. Denny has received several awards for excellence in teaching and is a past president of the Western Society of Naturalists.

STEVEN D. GAINES is Director of the Marine Science Institute and Professor of Ecology, Evolution, and Marine Biology at the University of California, Santa Barbara.

He is a marine ecologist with ongoing research on marine conservation, the design of marine reserves, sustainable fisheries, and the impact of climate change on marine habitats. His research projects include several large collaborative efforts including *PISCO* (Partnership for Interdisciplinary Studies of Coastal Oceans) and the new Sustainable Fisheries Group, which is developing market-based solutions to promote sustainable fisheries.

He has published more than 110 articles and four books on a diverse array of subjects. His prior books explore *Chance in Biology*, *Marine Community Ecology*, and *Species Invasions: Insights into Ecology, Evolution, and Biogeography*. Dr. Gaines has received several research awards including being named a Pew fellow in Marine Conversation.